Microbial Technology

SECOND EDITION/VOLUME II

Second Edition/Volume II

Microbial Technology

Fermentation Technology

Edited by

H. J. PEPPLER

Universal Foods Corporation
Milwaukee, Wisconsin

D. PERLMAN

School of Pharmacy
University of Wisconsin
Madison, Wisconsin

ELSEVIER

AMSTERDAM • BOSTON • HEIDELBERG • LONDON
NEW DELHI • NEW YORK • OXFORD • PARIS • SAN DIEGO
SAN FRANCISCO • SINGAPORE • SYDNEY • TOKYO
Academic Press is an imprint of Elsevier

AP

Microbial Technology: Microbal Processes/2ed Vol. II
Peppler and Perlman

ACADEMIC PRESS
An Imprint of Elsevier
111 Fifth Avenue, New York 10003

Original ISBN: 0-12-551502-2

First Printed in India 2004
Reprinted 2006
Reprinted 2008
Reprinted 2009
Reprinted 2012

Indian Reprint ISBN: 978-81-8147-496-4

Published by Elsevier, a division of Reed Elsevier India Private Limited.

Registered Office: 305, Rohit House, 3 Tolstoy Marg, New Delhi-110 001.
Corporate Office: 14th Floor, Building No. 10B, DLF Cyber City, Phase-II, Gurgaon-122002, Haryana, India.

Printed and bound at Rajkamal Electric Press, Kundli, Haryana.

Contents

List of Contributors

Numbers in parentheses indicate the pages on which the authors' contributions begin.

Maynard A. Amerine (131), Department of Viticulture and Enology, University of California, Davis, California 95616

William B. Armiger (375), BioChem Technology, Inc., Great Valley Corporate Center, Malvern, Pennsylvania 19355

William H. Bartholomew (463), Research Department, Stauffer Chemical Company, Westport, Connecticut 06880

Paul A. Belter (403), Fermentation Research & Development, Fine Chemical Division, The Upjohn Company, Kalamazoo, Michigan 49001

W. T. Blevins (303), Auburn University, Botany and Microbiology Department, Auburn, Alabama 36830

L. T. Chang (243), Bristol-Myers Company, Industrial Division, Syracuse, New York 13201

Ichiro Chibata (433), Research Laboratory of Applied Biochemistry, Tanabe Seiyaku Co., Ltd., 16-89 Kashima-3-chome, Yodogawa-ku, Osaka, Japan

N. D. Davis (303), Auburn University, Botany and Microbiology Department, Auburn, Alabama 36830

Richard P. Elander (243), Bristol-Myers Company, Industrial Division, Syracuse, New York 13201

Stanley M. Finger (179), Department of Chemical Engineering, University of Maryland, College Park, Maryland 20742

Larry L. Gasner (211), Department of Chemical Engineering, University of Maryland, College Park, Maryland 20742

Randolph T. Hatch (179), Department of Chemical Engineering, University of Maryland, College Park, Maryland 20742

C. W. Hesseltine (95), Northern Regional Research Center, Federal Research, Science and Education Administration, U.S. Department of Agriculture, Peoria, Illinois 61604

N. J. Huige* (1), Jos. Schlitz Brewing Company, Milwaukee, Wisconsin 53212

Arthur E. Humphrey (375), School of Engineering, University of Pennsylvania, Philadelphia, Pennsylvania 19104

W. F. Maisch (79), Research Department, Hiram Walker & Sons, Inc., Peoria, Illinois 61601

Irving Marcus (497), Oblon, Fisher, Spivak, McClelland & Maier, Arlington, Virginia 22202

Gerard J. Moskowitz (201), Dairyland Food Laboratories, Inc., Waukesha, Wisconsin 53187

G. B. Nickol (155), Research Division, U.S. Industrial Chemicals Co., Cincinnati, Ohio 45237

L. K. Nyiri† (331), Department of Chemical Engineering, Lehigh University, Bethlehem, Pennsylvania 18015

Norman F. Olson (39), Department of Food Science, University of Wisconsin, Madison, Wisconsin 53706

D. Perlman (173), School of Pharmacy, University of Wisconsin, Madison, Wisconsin 53706

A. J. Petricola (79), Technical Division, Hiram Walker & Sons Limited, Walkerville, Ontario N8Y 4S5, Canada

Herman J. Phaff (131), Department of Food Science and Technology, University of California, Davis, California 95616

Harold B. Reisman (463), Stauffer Chemical Company, Westport, Connecticut 06880

Tadashi Sato (433), Research Laboratory of Applied Biochemistry, Tanabe Seiyaku Co., Ltd., 16-89 Kashima-3-chome, Yodogawa-ku Osaka, Japan

M. Sobolov (79), Technical Division, Hiram Walker & Sons Limited, Walkerville, Ontario N8Y 4S5, Canada

L. P. Tannen** (331), New Brunswick Scientific Co., Edison, New Jersey

* Present address: Miller Brewing Company, Milwaukee, Wisconsin 53208.

† Deceased.

** Present address: The Great Atlantic and Pacific Tea Co., Inc., Manufacturing Group, Horseheads, New York, 14845.

Tetsuya Tosa (433), Research Laboratory of Applied Biochemistry, Tanabe Seiyaku Co., Ltd., 16-89 Kashima-3-chome, Yodogawa-ku, Osaka, Japan

George T. Tsao (223), School of Chemical Engineering, Purdue University, West Lafayette, Indiana 47907

Hwa L. Wang (95), Northern Regional Research Center, Federal Research, Science and Education Administration, U.S. Department of Agriculture, Peoria, Illinois 61604

D. H. Westermann* (1), Jos. Schlitz Brewing Company, Milwaukee, Wisconsin 53212

* Present address: Jayto, New Berlin, Wisconsin 53151.

Preface

In the decade since the first edition of "Microbial Technology" appeared, applied microbiology has changed, expanded, and diversified. As new products were introduced in this period, and greater demand for some of the old ones developed, the total fermentation capacity increased at about the same rate noted in the previous 25 years. The number of fermentation products, it is estimated, has quadrupled, while the volume of products manufactured has increased tenfold. This growth has prompted publication of a second edition, completely revised and enlarged.

To accomplish a worldwide survey of industrial microbiology and to describe its contributions to agriculture, industry, medicine, and environmental control, the editors are indebted to 57 willing and expert contributors. Their comprehensive reviews of traditional fermentations and propagations, as well as newly developed microbe-dependent processes and products, are presented in a two-volume set.

Volume I, subtitled "Microbial Processes," describes the production and uses of economic bacteria, yeast, molds, and viruses, and reviews the technologies associated with products of microbial metabolism.

Volume II, subtitled "Fermentation Technology," deals principally with fermentations and modifications of plant and animal products for foods, beverages, and feeds, while reviewing salient aspects of microbial technology: general principles, culture selection, laboratory methods, instrumentation, computer control, product isolation, immobilized cell usage, economics, and microbial patents.

H. J. Peppler
D. Perlman

Preface

In the decade since the first edition of "Microbial Technology" appeared, applied microbiology has changed, expanded, and diversified. As new products were introduced in this period and greater demand for some of the old ones developed, the total fermentation capacity increased at about the same rate noted in the previous 20 years. The number of fermentation products, it is estimated, has quadrupled, while the volume of products manufactured has increased tenfold. This growth has prompted publication of a second edition, completely revised and enlarged.

To accomplish a worldwide survey of industrial microbiology and to describe its contributions to agriculture, industry, medicine, and environmental control, the editors are indebted to 57 willing and expert contributors. Their comprehensive reviews of traditional fermentations and propagations, as well as newly developed microbe-dependent processes and products, are presented in a two-volume set.

Volume I, subtitled "Microbial Processes," describes the production and uses of economic bacteria, yeast, molds, and viruses, and reviews the techniques for assay, and [illegible] waste products of microbial fermentations.

Volume II, subtitled "Fermentation Technology," deals principally with [illegible] of food and animal products, foods, beverages, and feeds, while reviewing allied aspects of microbial technology: general principles, culture selection, laboratory methods, instrumentation, computer control, product isolation, immobilized cell usage, economics, and microbial patents.

H. J. Peppler
D. Perlman

Contents of Volume I

MICROBIAL PROCESSES

Chapter 1

Beer Brewing

D. H. WESTERMANN
N. J. HUIGE

MICROBIAL TECHNOLOGY, 2nd ed., VOL.II
 ISBN 0-12-551502-2

I. INTRODUCTION

A. Historical Background

History indicates each society has developed an alcoholic beverage of some type based on the indigenous sources of starch or sugar. Wheat, corn, rye, millet, rice, oats, barley, potatoes, and other vegetables and fruits have been converted to potable alcoholic beverages. It is generally conceded that beer originated in the twin cradles of civilization, Mesopotamia and Egypt, about 6000 years ago.

According to Weeks (1949), the etymology of the word beer, as we know it today, indicates it originated from the Latin verb *bibere,* to drink. Similarly, the Spanish word for beer, *cerveza,* apparently originated from *cerevisia,* which combines the Latin *ceres,* goddess of grain, and *vis,* vigor.

It is not clear from historical records whether this vigor, which can be interpreted as the physiological effect of alcohol, or its preservative contribution to grains was the principal stimulus for long-term growing acceptance by societies. However, it is known that alcoholic beverages became closely associated with religious ritual during the infancy of early civilized societies.

The knowledge that a concoction of grains and water could ferment naturally to produce an alcoholic beverage followed the advance of agriculture across Europe with the Celts. The Teutons, who eventually settled in the Rhine area and became the Germanic Tribe, had followed the Celts in their westward migration. Some of the Teutons eventually settled in England by the fifth century and were followed by the Saxons. At least the possible sequence by which the brewing art could have spread across the continent and England appears historically logical.

The influence of the church on the art and science of brewing grew rapidly over a 1000-year period, reaching a climax in the sixteenth and seventeenth centuries. The influence of religious and civil law on brewing in Europe and England is a fascinating story, well documented by Arnold (1911).

Major brewing centers were eventually established in Pilsen, Czechoslovakia; Munich and Dortmund, Germany; Burton-on-Trent, England; and Dublin, Ireland. The local water supplies of each area gave rise to unique product characteristics associated only with that area.

Although the American Indians were already making a brew from maize, English-type beer was brought to America on the *Mayflower*, according to Weeks (1949). A journal of the *Mayflower* contained the following record: "We could not now take time for further search . . . . our victuals being much spente, especially our beer."

English ale brewing traditions accompanied the colonists, and it was not until the 1840s that German lager beer became more widely accepted in the United States (Singruen, 1938).

English ales were fermented at ambient temperatures (21°–27°C) with top-fermenting yeast, while the German lager beers were fermented with bottom-fermenting yeast at the lowest temperatures which could be sustained year-round (10°–15°C). The lager beer was stored in cool caves for several months, resulting in improved physical stability (Baron, 1962). The superior keeping quality and preferred taste of lager beer rapidly led to its domination of the American market.

B. Current Practice

Beer retains its unique taste and character because of historical acceptance coupled with legal restraints of composition. In the United States Federal Regulations, beer is defined as a malt beverage resulting from an alcoholic fermentation of the aqueous extract of malted barley with hops. It may include other sources of carbohydrates called adjuncts. Legally, malt liquor, ale, porter, stout, and sake are also malt beverages.

Current American beers are made from malted barley, with corn grits or syrup or rice grits as the adjunct carbohydrate source, and hops. Purified water, to which brewing salts are added, and a cultured yeast complete the five principal ingredients of brewing.

Barley which has been malted comprises the basic raw material for brewing. In the malting process, the barley kernel, separated from the stalk and chaff, is germinated under controlled temperature and humidity to generate enzyme systems and partially degrade endosperm starch and protein. Growth of the germinated kernel is stopped by drying of kilning in hot air, during which time some flavor and color components are also formed. The dried malt is milled before it is mashed with brewing water.

In the mashing process, milled malt and adjunct, such as corn grits or rice, are each made into a mash with brewing water, subjected to individual controlled time–temperature cycles, and combined for further enzymatic digestion and subsequent extraction of sugars and proteins.

The extract produced in the combined mashing process is separated from the nonsoluble husk and grain portions in a lauter tub or filter press. This extract is called wort.

Wort is boiled in the brew kettle, along with hops and syrup, if used as an adjunct. The dried hop cones used for brewing contain lupulin glands which are the source of bitter resins and essential oils that contribute some of the characteristic bouquet to beer.

The boiled wort is cooled, inoculated with a pure yeast culture, and fermented. Each brewery maintains a yeast strain considered to be unique to its own process and product character. The fermentable sugars are converted to ethyl alcohol and trace quantities of flavor ingredients, while the nonfermentable carbohydrates remain in the beer. Carbon dioxide from the fermentation is recovered, purified, and returned to the finished product.

The yeast generated during fermentation is separated from the beer by centrifugation or natural settling. After storage and aging for several weeks, recarbonation, and two or more filtrations, the finished product is packaged.

II. RAW MATERIALS

A. Barley

For centuries barley was the "staff of life" in Europe, as evidenced by barley kernels found in ancient ruins and tombs. Until the fifteenth century it was the principal ingredient of bread. However, because barley flour is not readily leavened, wheat, which makes a lighter bread, became the favorite of bakers.

Botanically, barley is a grass or member of the Graminae family. However, it has unique properties which make it ideally suited for the brewing of beer. The barleys used for brewing possess a very tough husk which is firmly cemented to the kernel. This husk provides protection for the kernel during handling and subsequent germination. Later in the brewing process the husks form a filter bed for the separation of extractable carbohydrates and protein from the mash and contribute a characteristic flavor. Barley can be readily malted, a process in which the events of plant reproduction are simulated and later stopped by the removal of moisture. During the malting process barley produces large amounts of amylases, proteases, and other enzymes which partially degrade starch, proteins, and some types of cellulose. The biochemistry of the malting process is described in Section III.

Two distinct barley types and many varieties of each type have been grown for brewing. The two-row and six-row designations of type describe the number of rows which form around the axis of the kernel head. In two-row barley, one kernel develops at each of two nodes, while three kernels develop at each node in six-row barley. Each type of barley has its own brewing characteristics. Historically, two-row barley produced a more mellow beer and contained less protein and enzymatic potential than six-row barley. This resulted largely from the greater plumpness of the two-row kernel (i.e., less surface to volume ratio). Agronomic cross-breeding during the last decade has tended to minimize this difference.

To produce typical light American beers, barleys must be used which provide ample enzymatic activity to hydrolyze adjuncts. The six-row barley varieties, Larker, Dickson, and Conquest, are typical of those grown in North and South Dakota and Minnesota. These varieties provide about 80% of the barley for the American brewing industry. They have high enzymatic potential and protein contents of up to 13.5%. While higher enzymatic potential and protein content can be generated, brewers usually avoid such barley because of product instability. A western six-row barley with less desirable brewing characteristics is also grown primarily in California.

The two-row barleys, known for their contribution of mellowness and flavor, include varieties such as Piroline, Betzes, Klages, Firlbecks III, Vanguard, Hannchen, and Shabet. They are grown in the western states of Washington, Oregon, Idaho, and Montana.

The proximate composition of the three major barley types, described by Winton and Winton (1932), is shown in Table I, and the analysis of malts from these barleys in Table II.

B. Adjuncts

Approximately 70% of the malted barley kernel is potentially available as soluble carbohydrate. It is this carbohydrate, principally starch, which is converted to fermentable sugar and nonfermentable dextrins. The fermentable sugars are ultimately converted primarily to ethyl alcohol. The maximum alcohol and remaining solids content of an all-malt beer is therefore established by the milled malt solids concentration which can conveniently be handled in an all-malt mash. The ratio of protein to carbohydrate and amount of alcohol in the finished beer is primarily a fundamental characteristic of the malted barley.

All-malt beers historically were more satiating and had poor physical

TABLE I. Proximate Composition of Three Barley Types

Analysis	Midwestern six-row	California six-row	Western two-row
Kernel Weight (mg)	36	44	40
Husk (%)	12	14	10
Protein (%)	12	11	10
Fat (%)	2	2	2
Starch (%)	58	58	60
Fiber (%)	5.7	6.6	5.2
Ash (%)	2.7	3.0	2.5
Enzyme potential after malting	High	Low	Medium

TABLE II. Typical Analyses of Malts from Three Barley Types

Malt analysis[a]	Midwestern six-row	California six-row	Western two-row
Kernel weight, dry basis (mg)	30.0	39.0	37.0
Growth; length of acrospire			
0–¼ (%)	0	1	1
¼–½ (%)	1	5	1
½–¾ (%)	5	9	6
¾–1 (%)	92	85	91
Overgrown (%)	2	0	1
Assortment			
On 7/64 screen (%)	32	68	85
On 6/64 screen (%)	53	26	10
On 5/64 screen (%)	14	6	1
Through screen (%)	1	6	1
Moisture (%)	4.2	4.4	4.2
Extract, fine grind, dry basis (%)	76.8	77.0	80.5
Extract, coarse grind, dry basis (%)	74.8	75.0	79.0
Difference (%)	2.0	2.0	1.5
Wort color, ½ inch cell, °Lovibond	1.6	1.4	1.2
Protein (N × 6.25), dry basis (%)	12.3	11.0	10.5
Soluble protein as percentage of total	40.0	35.0	38.0
Diastatic power, dry basis, °L	135	65	90
α-Amylase, dry basis, 20° units	38	25	30

[a] From American Society of Brewing Chemists (1976).

stability due to their high concentration of soluble protein. American malts had three properties which were different from European malts. Protein content and enzymatic activity were substantially higher, and the husk was thicker. Early American brewers recognized that these properties could be exploited to use other sources of starch as a supplemental brewing material. The higher enzymatic activity would allow hydrolysis of gelatinized starch from other cereal grains, and the heavy husk would provide an efficient filter bed for extraction of solubles from mashes containing adjuncts. The limited residual proteins from these supplemental materials, such as corn and rice, were not readily hydrolyzed by malt proteases. Therefore, beers produced with these supplemental adjuncts had lower protein content for a given alcohol content, resulting in better physical stability with a lighter character.

With the exception of specialty products, all American brewers use an adjunct for brewing, as described by Matz (1970). While wheat, milo, and unmalted barley have been used, corn grits, corn syrup, and rice grits are dominant. Of the 155,000,000 barrels of beer currently produced, two-thirds are brewed with corn grits or corn syrup adjunct.

When corn grits, refined grits, flakes, or rice are used by the brewer, they are pregelatinized by boiling in a "cooker" before adding to the malt mash. Cooker mashes generally contain about 10% malt to provide the enzymes which partially hydrolyze the gelatinized starch to reduce viscosity.

Corn syrups produced for the brewing industry by the corn wet milling industry are made from acid and enzymatic hydrolysis of starch slurries. Dependent upon the method of hydrolysis, it is possible to produce syrups with 0–100% fermentability. Typical analyses of brewing adjuncts are shown in Table III.

Brewing technologists generally acknowledge that to produce presently accepted beers, yeast requires 120–140 mg/liter of free amino nitrogen for proper nutrition. Since the nitrogen is available only from malt, this limits the malt to adjunct ratio theoretically to approximately 50/50 for lager beers. American industry beers vary from 100% malt for the specialty products to about 60/40, malt/adjuncts, for lager beers.

C. Hops

Since medieval times herbs have been used to flavor beers, but only hops are used commercially today. The characteristic aroma and bitterness imparted to beer by the oils and resins of hops make beer and similar malt beverages quite unique.

According to Arnold (1911), records exist indicating hops were grown in the seventh or eighth century, but the first recorded use in brewing was made by a nun, St. Hildegard (1098–1197), in a convent at Rupertsberg, Bingen-on-Rhine. The record states: "The hop is of a heating and drying nature, but does contain some little moisture; is, however, of slight benefit to man, since it promotes melancholy and creates in man a sad mood, and also it affects his bowels unpleasantly by reason of its heating properties. Its bitterness, though, when added to beverages, prevents in the latter putrification, and gives to them a longer durability." (Sic)

While the preservative value of hop resins is limited, it is known to contribute some microbiological stability during beer processing. Whether this attribute or its uniquely acceptable flavor contribution stimulated its use may never be put in proper perspective. Both attributes were important.

About 20% of the world's hops are grown in the United States, in the states of Washington, Oregon, California, and Idaho. The Old World famous centers of hop growing still prevail in Kent, England; Saaz, Bohemia; and Hallertau, Bavaria.

The natural family of Cannabinaceae consists of two genera, *Humulus* and *Cannabis*. *Cannabis* is represented by *C. sativa,* which includes Indian Hemp, marihuana, and hashish, while *Humulus* consists of *H. lupulus* and *H. japonicus*. The *H. lupulus* provides resins for brewing,

TABLE III. Typical Analyses of Brewing Adjuncts

Composition	Amount
Cereal Adjuncts	
Corn Grits	
Moisture (%)	10.9
Extract, dry basis (%)	91.4
Oil, dry basis (%)	0.76
Corn Flakes	
Moisture (%)	9.0
Extract, dry basis (%)	92.4
Oil, dry basis (%)	0.50
Refined Grits	
Moisture (%)	9.6
Extract, dry basis (%)	103.3
Oil, dry basis (%)	0.03
Rice	
Moisture (%)	12.0
Extract, dry basis (%)	93.0
Oil, dry basis (%)	0.86
Liquid Adjuncts	
Corn Syrup	
Extract, as is (%)	82.0
Extract, dry basis (%)	100.0
Fermentable extract, as is (%)	60.2
Reducing sugars (as dextrose), as is (%)	50.5
Ash, as is (%)	0.19
pH (10% solution)	4.95

whereas the *H. japonicus* is an ornamental plant devoid of brewing value. It is the dried fruit or flower of the female hop plant which is used for brewing.

The hop plant possesses an extensive root system, which each year sends out a large number of shoots which grow very rapidly to a length of 20–25 ft. These climbing vines are supported on a wire network, and in late August or early September the flowers or cones of these female plants are picked, dried to 9–10% moisture, and baled. Extensive effort is taken to eliminate or minimize the growth of male plants to prevent pollination and seed production in the female flowers. The typical composition of a dried hop cone is given in Table IV, from Hough *et al.* (1971).

The structure of the hop cone is shown in Fig. 1. The cone consists of bracts and seed-bearing bracteoles which are attached to a supporting structure or strig. Lupulin glands, small resinous beads which contain the materials of brewing value, develop along with seeds at the base of the bracteole as the hop cone ripens. The lupulin glands contain the

TABLE IV. Proximate Composition of Hops[a]

Component	Percent of total weight
Water	10.0
Total resins	15.0
Essential oils	0.5
Tannins	4.0
Monosaccharides	2.0
Pectin	2.0
Amino acids	0.1
Proteins (N × 6.25)	15.0
Lipids and Wax	3.0
Ash	8.0
Cellulose, lignin, etc.	40.4
Total	100.0

[a] From Hough *et al.* (1971).

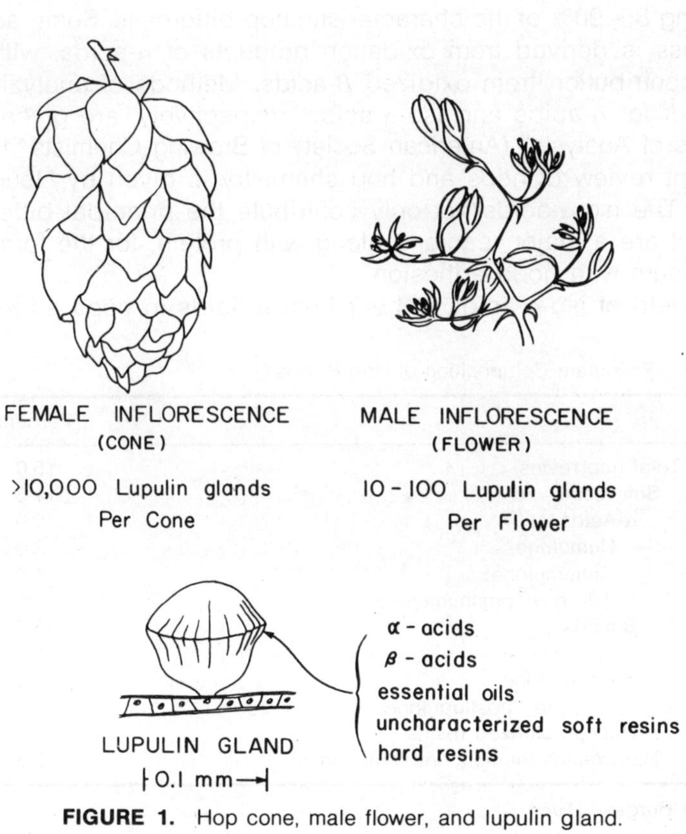

FIGURE 1. Hop cone, male flower, and lupulin gland.

resins and essential oils which characterize the hop variety and determine its value in brewing.

The hop resins are broadly classified by their solubility, as indicated by Findlay (1971), and the specific compounds which comprise the approximate 15% total resins are shown by Burgess (1964) in Table V. The major constituents of hop oil are the hydrocarbons myrcene, farnesene, humulene, and caryophyllene, which account for 50–80% of the oil and the 80 or more oxygenated compounds which are currently identified.

The lupulin of hop varieties differs largely in its volatile oil composition and percentage of cohumulone in the α-acids. The European hop varieties Hallertau and Saaz contain about 20% cohumulone in α-acids, while the American varieties Bullion and Brewer Gold contain about 40%. These hops also contain more humulene and less myrcene than American hops.

When the hop cones are added to the brew kettle the water-soluble resins and oils are dissolved in the wort. The α-acids are isomerized, providing 85–90% of the characteristic hop bitterness. Some additional bitterness is derived from oxidation products of α-acids, with a very minor contribution from oxidized β-acids. Methods of analyzing hops and beer for α-acids and iso-α-acids, respectively, are given in "The Methods of Analysis" (American Society of Brewing Chemists, 1976). An excellent review of hops and hop chemistry is given by Hough *et al.* (1971). The iso-α-acids not only contribute the principal bitterness to beer but are a major reactant, along with protein, for the formation of stable foam with good adhesion.

The yield of iso-α-acids in beer from α-acids in hops added to the

TABLE V. Proximate Composition of Hop Resins[a]

	Percent
Total hop resins	15.0
Soft resins	11.5
α-Acids	5.0
Humulone	3.0
Cohumulone	1.5
Ad-, pre-, posthumulone	0.5
β-acids	3.0
Lupulone	1.5
Colupulone	1.2
Ad-, pre-, postlupulone	0.3
Uncharacterized resins	3.5
Hard resins (humulinone, hulupones)	3.5

[a] From Burgess (1964).

brew kettle is about 25–30%, so the financial incentive for improvement is substantial. One simple method of increasing yield is to grind the hop cones, thus increasing surface exposure of fractured lupulin glands to wort in the brew kettle. This increases yield to 35–40%. Solvent extracts of hops are also made with hexane or dichloromethane. These extracts added to the brew kettle give yields of about 40–45%.

Some commercially available extracts contain isomerized α-acids which have been prepared in aqueous systems with sodium or potassium hydroxide, calcium carbonate, or magnesium carbonate. Isomerization rates are very rapid at higher pH compared to the pH of 5.0–5.4 which prevails in the brew kettle.

The yield of purified potassium salts of α-acids properly added to beer after fermentation and filtration, as described by Westermann (1976), reaches 95+%. These salt extracts are void of the essential oils and β-resins normally contained in whole hops or whole hop extracts. The residual portion of extract from which the purified salts are made is added to the kettle to retain the contribution of remaining hop ingredients to beer character.

D. Water

The chemical composition and concentration of salts in brewing water has a profound effect on the brewing process and resulting beer characteristics. Two ions, calcium and bicarbonate, are critical because of their impact on pH during the brew house operations. The pH of the water and resulting malt extract governs the activity of numerous enzyme systems of the malt, the malt yield, the solution of tannins and other coloring ingredients, the precipitation of proteins in the brew kettle, the rate of browning reactions, isomerization rate of humulone, and the physical and foam stability of the beer. Indeed, brewing water composition is of vital importance. Siebel (1950) has described the composition of water as suitable for brewing lager beer as shown in the tabulation below.

Composition	Amount
Sodium carbonate	None
Calcium and magnesium carbonate	Up to 50 ppm
Sodium and magnesium sulfate	Up to 50 ppm
Calcium sulfate	240 to 400 ppm
Sodium chloride	Up to 100 ppm
Calcium and magnesium chloride	Not over 75 ppm
Iron, as Fe^{2+}	Less than 0.2 ppm
Silica, as SiO_2	Less than 10 ppm

The composition of brewing waters of the world famous brewing centers has been compiled by Scholefield (1956), as shown in Table VI.

The high permanent hardness of water used for Burton bitter ales should be noted. The sweeter, darker beers of Dublin, London, and Munich contain less calcium sulfate and more calcium carbonate. Pilsen pale lager is brewed with very soft water.

An ideal water for American lager beers would contain a hardness level midway between Dortmund and Pilsen water, i.e., about 100–125 ppm of Ca^{2+} with generally less than 20 ppm of bicarbonate hardness. At these levels the mash would have a pH of less than 5.6, minimizing the extraction of silicates, tannins, and other harshly flavored materials from the malt. Similarly, the extract in the brew kettle (wort) would ultimately have a pH of 5.1–5.2, producing a beer with potentially good foam and physical stability.

Since waters theoretically ideal for brewing lager beer are not naturally available, two corrective actions can be taken when appropriate. When temporary hardness (bicarbonate) is too high, a mineral acid is added to the water, followed by heating and gas stripping to remove CO_2 from the decomposition of carbonic acid. In addition, gypsum ($CaSO_4$) is added to grains sparge water and to the brew kettle to raise the Ca^{2+} content of the resultant wort.

The desired pH in both mashing and the brew kettle result from the following reactions. The carbonate ion, bicarbonate ion, and CO_2, or carbonic acid, exist in equilibrium, depending upon pH of the water. At high pH the carbonate ion prevails, at neutral pH the bicarbonate, and at low pH, dissolved CO_2 exists. When heating water which contains the bicarbonate ion, an undesirable rise in pH occurs as follows:

$$Ca(HCO_3)_2 \rightleftharpoons Ca^{2+} + 2\,HCO_3^-$$

$$HCO_3^- + H_2O \rightleftharpoons H_2CO_3 + OH^-$$

$$H_2CO_3 \rightleftharpoons H_2O + CO_2 \uparrow$$

Therefore, as previously mentioned, bicarbonate ion concentration is usually reduced before water is used for brewing.

The calcium ion from gypsum ($CaSO_4$) is used to lower the pH in mashing and the brew kettle due to its reaction with secondary phosphates, phytate, and some proteins, peptides, and nucleic acids.

$$3\,Ca^{2+} + 2\,HPO_4^{2-} \overset{\Delta}{\rightleftharpoons} 2\,H^+ + Ca_3(PO_4)_2 \downarrow$$

A similar reaction occurs with phytic acid. The calcium salts tend to precipitate on boiling, exceeding the concentration allowed by virtue of their solubility product constant. If an increase in sweetness or palate

TABLE VI. Ionic Concentrations of Typical Brewing Liquors[a,b]

Ions	Burton-on-Trent	Dortmund	Edinburgh (sandstone)	London (off chalk) domestic	Munich (Dublin-type)	London (deep-well)	Pilsen
Na^+	1.3	3.0	4.0	1.1	0.1	4.3	0.1
Mg^{2+}	5.2	1.9	3.0	0.3	1.6	1.6	0.1
Ca^{2+}	13.4	13.0	7.0	4.5	4.0	2.6	0.4
NO_3^-	0.5	—	0.5	0.1	0.1	—	—
Cl^-	1.0	3.0	1.7	0.5	0.1	1.7	0.1
SO_4^{2-}	13.7	5.9	4.8	1.2	0.1	1.6	0.1
CO_3^{2-}	4.7	9.0	7.0	4.1	5.5	5.2	0.3
$(Ca^{2+} + Mg^{2+})/CO_3^{2-}$	100/25	100/60	100/70	100/86	100/77	100/24	100/71
Total salts (ppm)	1226	1011	800	319	273	463	30.8

[a] Figures rounded to one decimal place. Concentration of ions in millivals (N/1000).
[b] From Scholefield (1956).

fullness is desired, $CaCl_2$ can be used to obtain the apparent results from the chloride ion. An interesting review of ion impact on product properties and processes is given by Comrie (1967).

E. Yeasts

Yeasts are unicellular fungi which are classified by their spore-forming characteristics into one of three out of four fungi groups: Ascomycetes, Basidiomycetes, or Fungi Imperfecti. Of the approximately 350 species of 39 genera of yeasts in these three groups, only two species of the genus *Saccharomyces* are important in brewing. These are *S. cerevisiae* and *S. carlsbergensis*. "There is no clear indication from yeast ecology how *S. carlsbergensis* and *S. cerevisiae* came to be universally employed for beer production" (Hough *et al.*, 1971). *Saccharomyces cerevisiae* is traditionally used in baking, distilling, brewing of sake and ale, production of industrial alcohol, and for animal feed.

According to Lodder and Kreger-Van Rij (1952), Stelling-Dekker established in 1931 that top-fermenting yeasts used for the production of ales and the bottom-fermenting yeasts used for the production of lager beer were actually different. She classified those ale yeasts which rose to the top of the fermenter at the completion of fermentation as *S. cerevisiae*. Those yeasts, traditionally used for lager, which agglomerated and settled to the bottom at the completion of fermentation were classified as *S. carlsbergensis* (now *S. uvarum*). The continued application of these species to their respective brewing process conditions may have enhanced divergence of their physiological and morphological characteristics.

In order to assure continued use of the desired yeast species and strain, it is necessary to have clear criteria for identification. This is accomplished through a wide variety of measurements and procedures, none of which is solely adequate. Cell shape, growth on selected solid media, spore formation, and colony geometry are the principal morphological criteria. The cell's growth kinetics, carbohydrate and nitrogen assimilation characteristics, and nature of metabolite formation aid in characterizing a given strain. Identification of cell wall components and immunological characteristics are also used. These scientific assessments, along with yeast performance in the brewery, are used to evaluate the yeast strain.

Yeast is generally added to the fermentor at a level of 10–15 million cells per milliliter of wort, or a range of 1.3–2 lb of yeast slurry per barrel of wort. During fermentation, cells multiply about four- to fivefold, depending upon initial wort aeration and wort composition. The size and density of the yeast cell varies according to the strain and the process time at which measurement is made. During the initial portion of a fermentation cycle, when glycogen reserves are being built up after budding, the cell has a greater mass and the ratio of carbohydrate to protein is high. At the end of the fermentation cycle, when glycogen stores are depleted and excretion occurs, the mass is decreased. The cell mass fluctuates through a sine curve cycle during the course of fermentation, but protein mass stays fairly constant. The percentage of protein (Kjeldahl N $\times$ 6.25) varies from 35 to 50%, whereas cells per gram of yeast solids ranges from 7 to 16 billion. Yeast cells range from 4 to 14 μm in length and from 3 to 10 μm in width.

At the end of fermentation, yeast cells flocculate or form aggregate clumps. This physical phenomenon of producing clumps lighter than or heavier than the beer is a principal characteristic which distinguishes the ale or top-fermenting yeast from the lager or bottom-fermenting yeast. The flocculation tendency can be measured by a standard test (Burns test), but quantitative data are highly speculative. The lager yeasts which settle are classed as flocculating (rapid settling) or powdery (slow settling). The rate of flocculation is important since it drastically affects metabolism of some important secondary yeast by-products and the efficiency of filtration following fermentation due to the blinding effect of yeast on filters.

The solids content of the yeast slurry recovered from the fermentor is totally dependent upon the mechanism of recovery. Well-settled yeast slurry will contain 15–17% solids, whereas centrifuged yeast contains 18–26% solids, depending upon machine type and process conditions. Yeast pressed at 40–50 psi will contain about 25–28% solids.

The yeast harvested from the fermentor can be added at once to another batch of wort, but it is often washed prior to being recycled. Washing frees the yeast of mechanical impurities, such as coagulated protein, hop resins, and some dead yeast cells. Treatment with ammonium persulfate adjusted to pH 2.2 with phosphoric acid has been shown by Bruch *et al.* (1964) to eliminate most bacterial contamination in the harvested yeast.

Laufer and Schwarz (1936) have described apparatus for growing a pure culture yeast in the plant under sterile conditions to obtain sufficient pitching yeast for a full-size brew or a portion of it. It consists of a series of two or three vessels, each about ten times larger than the preceding vessel. After 1–2 days of growth, the culture is transferred to the next largest vessel containing cold, sterilized wort. Rather than start a new laboratory culture each time, a small portion of the contents of the second vessel is sometimes transferred back to the starter vessel.

The amount of fresh culture yeast grown is dependent upon the viability of the yeast after multiple fermentations, the amount of contamination of the recovered yeast, and the method of recovery. In commercial brewing practice, new yeast tends to decrease in viability or becomes contaminated with continued recycle; therefore, recycled yeast is replaced with a new culture after from 5 to as many as 100 fermentation cycles.

III. MALTING PROCESS

Malting is the process in which grains are intentionally germinated and subsequently dried by the application of heat. Although the term "malt" is commonly used for barley malt only, wheat and rye are also commercially malted in small quantities in the United States. Reviews of the malting process and design of malting plants are given by Macey (1963), Kellett (1965), and Bradee and Westermann (1975).

During the malting process the enzyme systems necessary for reproduction of the barley plant are fully developed and proteins and carbohydrates of the barley kernel are modified or made soluble for subsequent hydrolysis in the mash. The resulting carbohydrates furnish the energy source for the production of alcohol, while the amino acids and other trace materials provide the nutrition for yeast growth. During the kilning process, color and flavor ingredients are developed, and the moisture content is lowered to allow storage of the malted barley.

Before malting, barley is stored several weeks to overcome dormancy. During storage it is kept at 10–12% moisture and is frequently transferred to maintain low temperatures and break the dormancy. The moisture content must be kept low to protect the barley from mold growth.

The malting process can be divided into five basic steps listed below.

1. Barley cleaning and grading.
2. Steeping in water to increase moisture content.
3. Germination of moist grain under controlled conditions.
4. Kilning to arrest growth, develop color and flavor, and reduce moisture to a suitable storable level.
5. Rootlet removal.

Figure 2 is a schematic flow diagram of the malting process, including a material balance.

A. Cleaning and Grading

Barley received by the malster is cleaned to remove weed seeds, other grains, broken kernels, chaff, sand, dust, and foreign objects such as iron. Iron particles are removed by an electromagnet. Small particles,

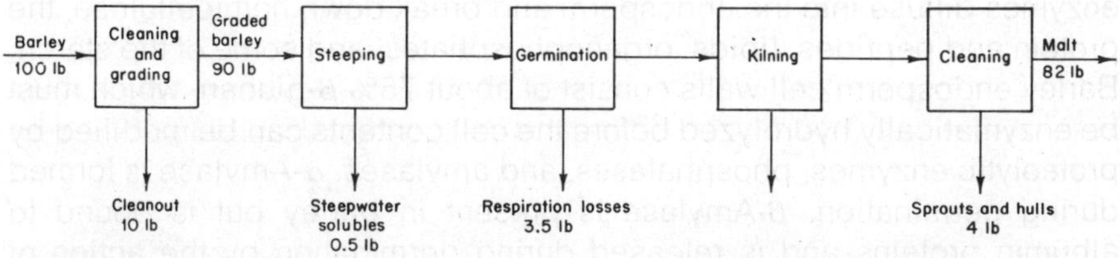

FIGURE 2. Schematic flow diagram of the malting process; material balance on a bone-dry basis.

such as sand and seeds, are usually removed through a fine vibrating screen. Broken kernels and other types of grains are often removed in inclined rotating cylinders with inner periphery pockets that will carry the smaller grains upward, subsequently emptying into a trough for removal.

Size separation according to kernel width is accomplished by passing the barley over three width-separation graders, producing A, B, and C grades. Undersize kernels passing through the C grader are collected and sold as feed. The small C-grade barley is used in the production of distillers malt. Brewers malt is generally produced from the larger A and B grades.

B. Steeping

Graded barley is conveyed to cold water-filled (10°–15°C) steep tanks by pneumatic, bucket, belt, or screw conveyors. The first stage of the steeping process is used to wash the barley and to remove dirt, loose hulls, and other foreign material by overflowing the tank with water. Air under pressure is injected to mix the barley and water and initiate respiration.

During steeping the grain absorbs moisture and dormant embryos become activated. As soon as the kernel absorbs moisture, the respiration rate of the embryo increases rapidly. Aeration may be intermittent or continuous.

Steep water is changed every 10–15 hours. Steeping is continued until the moisture content of the grain reaches 44–45%, which may take from 48 to 60 hours. A total of 1.5–2 lb of water per pound of barley is generally used.

C. Germination

During the germination period the endosperm serves as a nutrient source for the embryo growth. The embryo synthesizes gibberellic acid, which diffuses into the aleurone layer as described by MacLeod (1976). Various hydrolytic enzymes are synthesized by the aleurone cells. These enzymes diffuse into the endosperm and break down hemicellulose, the protein and peptides, lipids, organophosphates, and some of the starch. Barley endosperm cell walls consist of about 75% β-glucan, which must be enzymatically hydrolyzed before the cell contents can be modified by proteolytic enzymes, phosphatases, and amylases. α-Amylase is formed during germination. β-Amylase is present in barley but is bound to albumin proteins and is released during germination by the action of proteolytic enzymes. During malting, only 5–10% of the starch is hydrolyzed to various degrees.

Modification of the endosperm starts near the embryo and progressively extends toward the distal end of the kernel. Daily samples are taken to test the progress of the modification, which is usually complete after about 5 days.

Relatively low temperatures, from 15° to 21°C, depending on the type of malt, must be maintained during the entire germination time to reduce respiration and rootlet losses and still bring about the desired physical and chemical changes. Fresh air is necessary to supply oxygen for respiration, prevent the accumulation of carbon dioxide, and control temperature in the grain bed by removing heat formed during respiration. The moisture content during germination is kept constant by humidifying the fresh incoming air or by water sprayers.

Germination is currently carried out in either drum or compartment systems. Drum systems have a capacity of 600–700 bushel of barley per drum. Drums are equipped with a device for distributing conditioned air and revolve slowly to prevent matting of rootlets.

In compartment malting, the steeped barley is put in long, rectangular, concrete open-top boxes equipped with perforated floor plates to permit flow of conditioned air. The grain bed, about 18–30 inches deep, is slowly turned over by mechanical screws that move back and forth through the compartment.

Drum systems require a higher initial investment and have higher operating and maintenance costs than the compartment system.

D. Kilning

After the desired degree of modification has been attained, the so-called "green" malt is transported to the kiln. The kilning process is carried out in two phases: (1) drying and (2) heating. During the drying phase, the moisture content is reduced from 45 to about 6% to terminate respiration, arrest biological activity, prevent bacterial spoilage, and preserve the enzymes and substrates which will be required in the subsequent mashing process. Additionally, during the first part of the drying phase, some of the characteristic flavor and aroma components are synthesized in a series of complex biochemical and chemical reactions. Amino acids are formed during further breakdown of proteins and react with sugars to form the colored and flavored melanoidins.

The enzymatic activity and flavor characteristics of the malt are markedly affected by the moisture content of the malt and the kiln air temperature applied at a given moisture content. The moisture content of regular brewers malt is reduced to about 6% with lower temperature air before the final kilning temperature is raised to about 82°C, reducing moisture to about 4.5%. Conversely, a crystal or caramel malt, from which very dark beer is made, is rewetted after light kilning and held at 66°–72°C for several hours without ventilation. It is kilned off up to 121°C and contains about 3% moisture. It is apparent that a wide variety of moisture contents and temperatures can be used to produce specialty malts.

Some kilns consist of an oven below two or three perforated floors, through which hot air is passed. Green malt is first deposited on the top floor for the initial drying process. After 12–24 or more hours, depending on the number of decks and type of malt, the malt is dropped to a lower floor and dried at the desired temperature to final moisture content. Each floor is equipped with turning machines.

Some malsters introduce sulfur dioxide into the air stream, which passes through the malt for a short period each day. This sulfur dioxide serves as a bactericide and fungicide for surface organisms, bleaching the malt, and lowering mash pH. This procedure has been minimized for ecological reasons.

The rootlets formed on the malt become dry and brittle during kilning and are removed in a scalper provided with a dust removal system. The rootlets are accumulated and sold as a valuable food and feed product.

The finished malt is transferred to storage, where it is held until scheduled for brewing.

IV. BREWING PROCESS

After storage, malt is elevated to the mill house, where cleaning and milling take place. In the commonly used batch brewing process (see Fig. 3), a batch of milled malt is weighed and mixed with water in the mash tub. A portion of the malt is mixed with adjunct and water in the cooker. After liquefaction the cooker mash is added to the main mash. After the enzymatic conversion of starch and proteins, wort is separated and extracted from the spent grains. The strained wort is then boiled with hops or hop extract. During boiling a precipitate is formed. It is

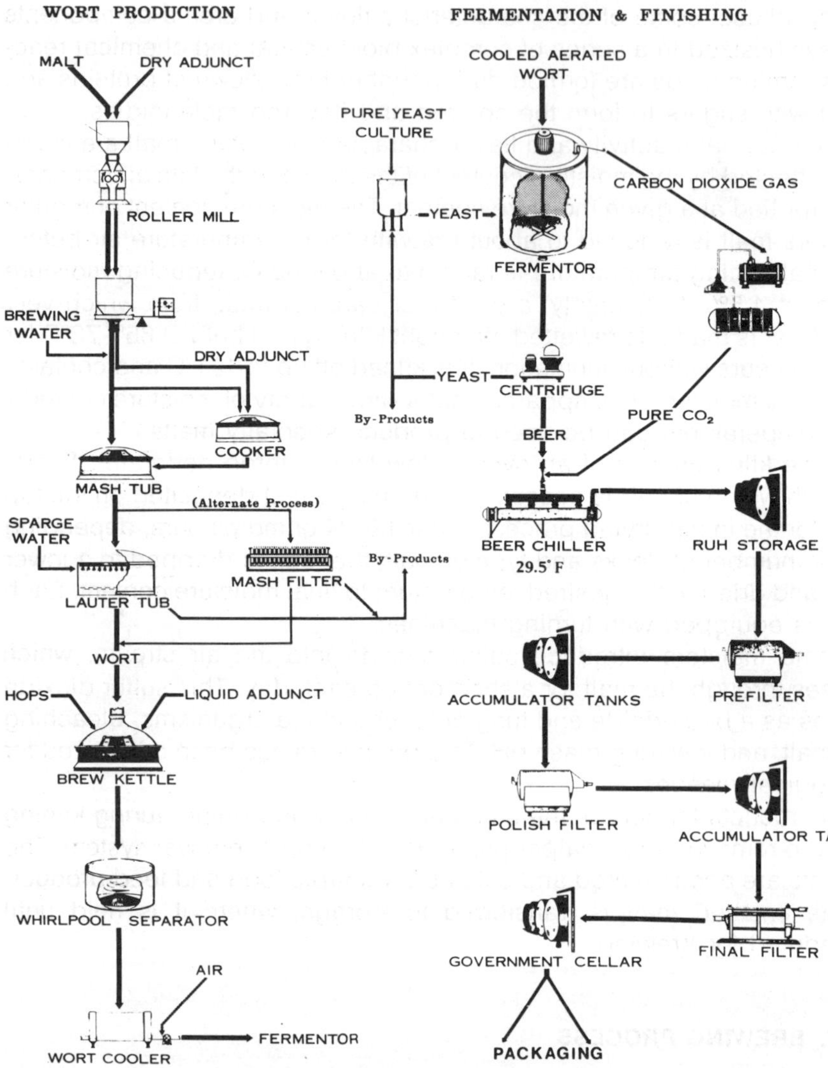

FIGURE 3. Flow chart of the brewing process.

removed after the wort is transferred to the hot wort settling tank. During subsequent transfer to the fermentor, wort is cooled and aerated. Each of these process steps is described in detail below.

A. Malt Preparation, Milling, and Weighing

Malt blended to proper specification is stored and kept at 4–5% moisture in silos of capacities up to 800,000 lb of malt. Elevation of malt to the mill house is done pneumatically or by bucket elevator. A considerable amount of abrasion may take place during pneumatic conveying. Typically, the amount of loose husks and fines after pneumatic elevation is 3–5%, while after bucket elevation, it is 1–2%.

Prior to milling, the malt is passed over a magnetic separator and sometimes through an aspirator. In the aspirator, loose husks and dust are collected to bypass the mills.

The purpose of milling is to provide ground malt with a particle size distribution that will give the brewer the optimum extract yield after mashing and wort separation within an allotted time. Generally, the degree of conversion and extraction increases with decreasing particle size, whereas the rate of wort separation decreases with decreasing particle size, as reported by Huige and Westermann (1975). Every brewer establishes his preferred optimum balance between yield and rate.

In most breweries, malt is milled in roller mills consisting of three pairs of rollers which operate using both direct pressure and shear. Double screens are located between the pairs of rollers to separate the various particles when they have attained proper size. Large mills have capacities of up to 13,000 lb of malt per hour. The three sets of rollers are generally adjusted to progressively decreasing gaps of about 0.040–0.030 inches.

The endosperm is typically crushed to produce fine grits rather than flour, though some flour production is inevitable. Schöffel (1972) indicated that the optimum particle size not only depends on the gap between the rollers, but also on the type, quality, and moisture content of the malt and the type of wort separation equipment that is used. Generally, lauter tubs and Strainmasters require a smaller fraction of flour than mash filters or continuous lautering devices, such as centrifuges.

To produce less flour and a faster wort separation, it is advantageous to minimize husk fracture. This can be accomplished by soaking the malt in cold or warm water or treating with steam for a period of up to 30 minutes. The moisture content of the malt increases up to 35%. Special mills are used for this so-called wet milling process, as described by Schauss (1964) and Narziss (1974, 1976, 1977).

B. Adjunct Cooker

Since cereal adjuncts have not gone through a malting process, their cell walls have not been modified. They contain starch and proteins that are difficult to solubilize because of their high molecular weight. Before these adjuncts are added to the main mash, therefore, they are milled, mixed with water, and brought to a boil in the steam-heated adjunct cooker to bring about gelatinization of starch. A small percentage of malt (approximately 10%) is added to aid liquefaction and decrease viscosity. The ratio of water to adjunct is about 2.5–3.5 lb/lb, and the total time to heat and boil 1–1½ hours. The cooker is agitated to increase heat transfer rates and to prevent burn-on.

C. Mashing

The purpose of mashing is to extract from the grains the available soluble substances (about 17% of malt) and to convert additional insoluble solids through controlled enzymatic reactions (additional 58% of malt). Yields for adjuncts such as corn grits may be as high as 87%. Because temperature and pH affect enzymatic activity, they must be optimized and controlled during mashing.

The pH of the mash is controlled at 5.2–5.5 by the addition of lactic acid and/or mineral acids and calcium sulfate. Calcium sulfate reacts with phosphates in the malt to provide buffer action, as described in the section on brewing water.

Many different time–temperature cycles may be used during mashing, depending on the raw materials and the type of wort desired. Temperatures are increased by heating with steam or by adding mash from other vessels, such as the cooker. Between temperature rises the mash is held for certain rest periods at temperatures that are optimum for certain enzymatic reactions. During mashing-in (mixing with water) and during heating periods, the mash is slowly agitated.

In a typical American mash schedule, known as the "double mash method," ground malt is mashed with water at a temperature of 35°C. During the so-called "protein rest" that follows, the temperature is kept constant for about 1 hour. The completed cooker mash is then added to raise the temperature of the combined mashes to 67°–68°C and begin starch hydrolysis. A rapid temperature rise (10 minutes) will increase the wort dextrin content, while a slow rise (30 minutes) will increase wort fermentable sugars. The mash temperature is kept constant for 20–30 minutes to allow completion of starch hydrolysis. After the absence of starch is confirmed by an iodine test, the mash temperature is increased to the "mashing-off" temperature of 75°–80°C, completing the mashing process.

The biochemical reactions occurring during mashing are an extension of those started during malting. Most of the soluble nitrogenous compounds of wort are formed during malting. During the protein rest when the temperature is close to the optimum for most proteolytic enzymes, a further breakdown of proteinaceous material takes place. Amino acids, peptides, peptones, and ammonia are formed. However, proteolysis is far from complete, and soluble higher molecular weight proteins are also extracted into the wort. Some of these high molecular weight proteins are subsequently precipitated during kettle boil. The final molecular weight distribution of proteins determines palate fullness, physical stability, and foam properties of the beer. Excessive hydrolysis of proteins may result in a beer with little foam stability, whereas too little hydrolysis might give rise to a cloudy, unstable beer.

The most important enzymatic reactions involve the degradation of starch to the fermentable sugars glucose, maltose, and maltotriose, and to nonfermentable higher molecular weight dextrins. The ratio of fermentable sugars to dextrins can be controlled by proper adjustment of the temperature–time cycle.

Malt starch consists of 20–25% amylose, a linear polymer composed of 20–25 glucose units joined by 1—4 chemical bonds. Amylose is hydrolyzed by α-amylase to linear dextrins containing about six glucose units, which are hydrolyzed further to maltose by the action of β-amylase. The other 75–80% of starch is amylopectin, a branched polymer composed of several thousand glucose units joined by both 1—4 and 1—6 chemical bonds. α-Amylase catalyzes the liquefaction of amylopectin by attacking the 1—4 bonds and bypassing the 1—6 bonds, yielding branched dextrins of lower molecular weight. This results in a rapid viscosity decrease. β-Amylase partially hydrolyzes the branched dextrins by attacking the 1—4 bonds and removing maltose from the linear portion of the chain.

The enzyme phytase is active during the "protein rest" period. This enzyme hydrolyzes phytin to inorganic phosphates, which act as buffers during mashing and fermentation, and inositol, an essential growth factor for yeast nutrition.

D. Wort Separation

After the mash cycle, malt husks and other insoluble material must be separated from the dissolved solids solution, wort, before it flows into the brew kettle. The process consists of two steps: (1) the principal separation of wort from insoluble solids by filtration, and (2) washing of the remaining insoluble solids bed to displace residual wort between the particles and to extract dissolved solids from within the particles. Two types of commercial wort separators include: (a) equipment in which the particles themselves form the sole filter medium. The lauter tub, in which wort is withdrawn from a slotted false bottom vessel, and the Nooter Strainmaster, in which wort is withdrawn from a plurality of horizontal strainer pipes at different heights of the bed, are widely used. (b) Equipment in which a filter screen or filter cloth is used as the main separating medium. Examples of this type are the plate-and-frame filter press and continuous lautering devices, such as a screening centrifuge.

A lauter tub is a cylindrical tank with a height to diameter ratio of about 1:5 and a maximum diameter of approximately 32 ft. The false bottom consists of plates with 0.020-inch-wide slots which provide 8–10% of open area and form a support for the filter bed of undissolved grain particles. The lauter tub is provided with vertical knives mounted on an arm that rotates around a central shaft. The knives, which can be lowered to cut into the bed while rotating, are used to prevent the bed from getting packed too tightly. After the mash has been pumped to the lauter tub, the particles are allowed to settle for 10–30 minutes. The first wort that is withdrawn is cloudy and is recycled until the desired clarity is obtained. After most of the wort has run off to the brew kettle, sparging is started with water of about 76°C, adjusted to the desired pH. Sparging is usually continued until the dissolved solids content of the withdrawn liquid drops below 1%. A typical analysis of wort brewed with corn grits is shown in Table VII.

The rate of wort withdrawal is generally about 0.25 gal/minute/ft^2 of cross-sectional area and may be as high as 0.04 gal/minute/ft^2. At a high rate of wort withdrawal, care has to be taken that the pressure differential across the bed is kept below 10 inches of water to prevent the grain bed from becoming too compressed. Also, upon increasing the runoff rate, the contact time between sparge water and grains to be extracted is reduced, thereby decreasing the degree of extraction of dissolved solids. After lautering is completed, the so-called spent grains are removed from the lauter tub by the action of the knives which are positioned to sweep out the bed. Spent grains are sold as cattle feed, either wet or dry. Before drying, the moisture of spent grains is usually reduced from about 80% to 60–72% in grain presses or centrifuges. The liquid containing 2–5% dissolved solids is sometimes returned to the cooker to be used as part of the cooker water, as reported by Coors and Jangaard (1975).

TABLE VII. Typical Wort Analysis

Wort color (°Lov. 52–1 inch cell)	6.0
Specific gravity (20°C/20°C)	1.04545
Extract (°Plato)	11.3
Limit attenuation (°Plato)	2.5
Fermentable extract (%)	63.0
Wort protein (N × 6.25) (%)	0.41
Soluble nitrogen (gm/100 gm)	0.065
Wort pH	5.35
Titr. acidity; ml N/10 alkali/100 gm	9.0
Calcium (ppm)	72
Iodine reaction	Negative

E. Brew Kettle Boil

After clarification the wort is heated in the brew kettle and boiled for 1–1½ hours, during which 5–8% of the water evaporates. During the brew kettle boil, hops or hop extracts are added all at once or in increments. When liquid adjunct, i.e., corn syrup, is used, it is generally added at the start of boil to maximize brewing reactions.

Historically, brew kettles have been constructed of copper because of material availability and ease of construction of the peculiar shape of the dome and draft stack. Some brewers believe that copper reacts with some sulfur-containing compounds that could adversely affect beer quality. All modern brew kettles are made with stainless steel. Kettles are steam heated by a bottom jacket and/or a central heat transfer surface that acts as a percolator. Heating can also be accomplished in outside heat exchangers, a technique more frequently practiced by European brewers. The objectives of the brew kettle boil are discussed below.

1. Wort Sterilization

Since wort is an ideal growth medium for microorganisms, sterilization is required before wort enters the fermentors. Hops that are added to the kettle also aid in preservation.

2. Enzyme Inactivation

Boiling will destroy any enzyme system that has not been completely inactivated at the end of the mash cycle. Inactivation is necessary to fix the ratio of fermentable sugars to dextrin to control the final alcohol content.

3. Protein Precipitation

High molecular weight proteins are precipitated during kettle boil, forming large flocs called "hot break" or trub. This precipitation is essential to enhance the colloidal stability of the beer. Precipitation occurs by (a) heat denaturation of those proteins which have an isoelectric point of the wort (pH 5.1–5.2), and (b) complexing with tannins and their oxidation products from malt husk or hops.

The pH of wort, which drops by 0.3–0.5 units during kettle boil, must be controlled to provide sufficient precipitation of "undesirable proteins" for a colloidally stable beer but no precipitation of "desirable proteins" which would result in a beer with poor foam quality.

4. Hops Extraction and Isomerization

During kettle boil, water-soluble resins, oils, tannins, amino acids, and monosaccharides are extracted from hops. Tannins aid in the precipitation of proteins.

Some of the about 100 hop oil constituents were originally thought to directly contribute to beer hop aroma, but since none of those components can be identified in finished beer, it is now believed that hop oil degradation products impart the hoppy aroma, according to Likens and Nickerson (1964).

The soft resins that contribute to beer bitterness in varying degrees are α- and β-acids and their isomerization, oxidation, and degradation products. The main contributors to bitterness are isomerized α-acids. The rate of isomerization increases with increasing pH. At a kettle pH of about 5.2, it takes about $1\frac{1}{2}$ hours for 90% of the important α-acid humulone to isomerize to isohumulone. Some isomerization may take place in the hot wort settling tank. Only about 25–30% of α-acids appear as isomerized products in beer due to incomplete extraction, incomplete isomerization, absorption to colloidal particles, and further degradation. For better control and utilization, preisomerized hop extracts are sometimes added after kettle boil, as noted by Westermann (1976) in a United States patent.

5. *Distillation of Volatiles*

During kettle boil, about 90% of the essential hop oils are removed by steam distillation, along with some fatty acids and low molecular weight degradation products.

6. *Color Development*

About two-thirds of the final wort color is produced during kettle boil. One-third is developed during malting, as reported by Nakayama (1962). Color and flavor development take place by the following types of reactions: (a) oxidation of phenolic compounds; (b) Maillard reaction between reducing sugars and amino acids; (c) caramelization of polyhydroxycarbonyl compounds.

F. Trub Separation, Wort Cooling, and Aeration

The protein precipitate formed during kettle boil and the spent hops material have to be separated from wort prior to fermentation. This is usually done in a Whirlpool separator or a combination of settling tank and filter or centrifuge. If whole hops are used, a hop strainer is often used to remove most of the hop solids prior to trub separation.

Trub consists of 50–60% protein, 16–20% hop resins, 20–30% polyphenols, and 2–3% ash by weight (Hough *et al.*, 1971). The particle size is 30–80 μm. To aid separation, some brewers add materials to the kettle that enhance floc formation, such as Irish moss or bentonite.

During the last 10 years the Whirlpool separator has become popular.

In this sedimentation tank, which has a height to diameter ratio of about 1:4, wort is fed tangentially, thereby creating a whirlpool pattern. Due to the centripetal force, the wort moves down the sides and up the middle. Particles with a settling velocity greater than the upward liquid velocity will settle out in the middle, forming a trub cone. Wort is drawn off from the outer periphery of the bottom. The trub solids are flushed to the sewer, are recovered as by-products, or are recycled back into the process.

After separation of the trub, wort is cooled in a plate heat exchanger to the starting temperature required for fermentation. As the temperature reaches about 48°C, an additional protein precipitate starts forming (particles of <5 μm) known as the "cold break." This cold break is removed by some brewers, who feel that it alters yeast metabolism. Separation is done by cold wort diatomaceous earth filtration or in a flotation tank.

Wort flowing to the fermentor is aerated to bring the oxygen concentration up to 8–10 ppm. Oxygen is necessary in the initial growth phase of fermentation.

G. Fermentation

1. Fermentation Process

After cooling and aeration, wort is inoculated with yeast, usually obtained from a previous fermentation. about 1–2 lb of yeast slurry per barrel of wort is used, equivalent to 7–15 million yeast cells per milliliter of wort. At higher yeast concentrations the start of active fermentation is faster and the fermentation rate increased. Wort is pumped directly to the fermentors or to starting tanks where fermentation begins and dead yeast cells, hop resins, and part of the cold break are separated.

Changes taking place during a typical lager fermentation are graphically depicted on Fig. 4. A lag period of about 12 hours is necessary for yeast to acclimatize to its new environmental conditions. After this time, yeast multiplication starts. The dissolved oxygen originally present in wort is sufficient to give a four- to fivefold increase in yeast cell population in 3–4 days. The fermentable carbohydrates furnish energy and carbon to produce yeast metabolites of ethanol, carbon dioxide, and other minor by-products, resulting in a decrease in specific gravity. Since the reaction is exothermic, 24 kcal are liberated per mole of glucose, the temperature rises. When a desired maximum temperature (in the example of Fig. 4, 13°C) is reached, it is kept constant by cooling the wort.

Within 12–26 hours after pitching, the wort becomes saturated with

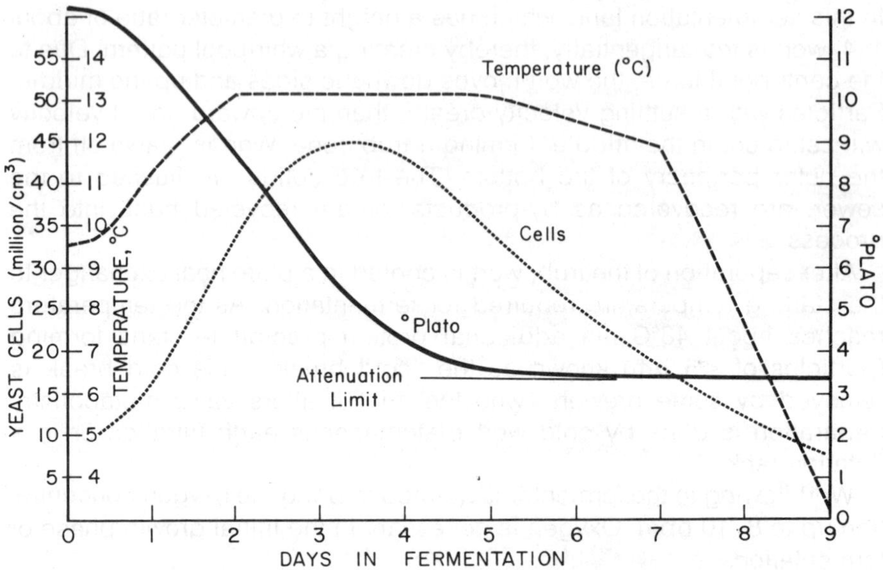

FIGURE 4. Brewery batch fermentation.

carbon dioxide and small bubbles appear on the surface, forming a creamy head called kraeusen. As the bubbles rise to the surface, they expand, thereby agitating the wort and causing the yeast to be suspended. Agitation and suspension of wort affect the rate of various reactions that take place during fermentation. Agitation is sometimes aided by providing mechanical agitators. Also, the use of large vertical fermentors promotes agitation.

As fermentation proceeds the foam head becomes thicker and yellowish due to precipitation of malt, hops substances, and dead yeast cells. The foam also contains brown spots from oxidized hop resins.

After 4–5 days the rate of CO_2 production decreases, the foam head begins to collapse, and some of the yeast cells start settling out. In the traditional process, cooling is further intensified.

The so-called primary fermentation is complete after 4–5 days, but virtually complete utilization of fermentables may take as long as 12 days. At this point all fermentable sugars, except for a small portion of the maltotriose, are fermented and the yeast has settled.

The flocculation tendency of the yeast is an important parameter. If a yeast flocculates too quickly, fermentation may not have proceeded far enough; whereas a yeast that flocculates with difficulty and stays in suspension too long may partly autolyze, imparting an undesirable yeasty flavor to the beer. The degree of flocculation of a yeast strain, although a genetic characteristic, can vary widely due to mutation within that strain and the concentration of divalent ions, such as calcium, outside the cell. Stewart (1975) concluded, "The contribution of calcium to flocculation is not the absolute amount of this ion absorbed by the yeast cell wall, but rather the stereo-specific manner by which it is bound. . . ."

After fermentation, yeast is removed by decantation, slurry centrifugation, or a combination of the two.

2. Fermentation Biochemistry

Yeast, specifically *S. uvarum,* requires fermentable carbohydrates for adequate growth and maintenance and also growth factors, amino acids, vitamins, and essential minerals for proper nutrition. In a typical lager wort, the proportion of solids in the substrate is given in the tabulation below.

Monosaccharides (principally glucose)	8–10%
Disaccharides (principally maltose)	50–55%
Trisaccharides (maltotriose)	10–12%
Higher dextrins	20–25%
Proteins (peptones, peptides, polypeptides, amino acids)	4–6%
Minerals (calcium, potassium, zinc, magnesium, etc.)	2%
Growth factors (B complex vitamins)	Traces

The addition of adjuncts, particularly syrup, can change the amounts of glucose, maltose, and maltotriose in the wort.

Yeast is facultative, i.e., it can grow aerobically as well as anaerobically. If sufficient oxygen is present, all fermentable carbohydrates are oxidized to water and carbon dioxide, and about ten times the amount of heat is liberated as during anaerobic respiration. The energy not dissipated as heat is used for cell growth and multiplication. During fermentation of wort, anaerobic conditions prevail, although some oxgen is initially required for cell mass production. Fermentable carbohydrates are transported into the cell, converted to glucose, and metabolized to produce principally ethanol and carbon dioxide.

The fermentation proceeds primarily via the glycolytic or Embden–Meyerhof-Parnas metabolic pathway. In the first step in this pathway, glucose is phosphorylated and isomerized to fructose phosphate, which is phosphorylated again and cleaved into two phosphorylated compounds. Several oxidation–dephosphorylation–reduction reactions take place to yield ethanol and carbon dioxide.

The initial amount of glucose in the wort (*Ig*) is an important factor in the carbohydrate assimilation kinetics, as shown by Westermann (1976). For glucose assimilation, which occurs first, *Ig* is the rate-limiting factor at 5–6 gm/100 gm of wort. Above this level the yeast concentration is limiting. Sucrose is hydrolyzed early by invertase, which is exogenous to the yeast cell. Maltose is first transported into the cell by a permease system and is subsequently hydrolyzed to glucose by maltase. Since glucose inhibits the activation of the permease system, a higher initial glucose concentration increases the lag time for initial maltose uptake and decreases the maltose assimilation rate. The effect of glucose on maltotriose is quite similar, although maltotriose assimilation occurs at a much slower rate. In most American beers about 10–20% of the maltotriose remains unfermented. Maltotetraose and dextrins are not fermented.

Amino acids, necessary for yeast nutrition, are assimilated in the cell at various rates, as described by Jones and Pierce (1969). Purines and pyrimidines necessary for nucleus formation can be synthesized by the yeast cell but are usually present in wort.

Many of the desirable and undesirable aroma and flavor compounds of beer are products of amino acid assimilation. Yeast strain, amino acid profile and concentration in wort, and pH are all important factors.

Higher alcohols, such as propanol and isobutanol, are generated from amino acids, but also from keto- acids derived from carbohydrate metabolism. The concentrations of individual higher alcohols in finished beer range from several to 60 ppm.

Aldehydes are precursors of alcohols. The concentration of acetaldehyde, the precursor of ethanol, can be as high as 10 ppm. Organic acids, such as acetic, lactic, pyruvic, citric and malic acids, are all derived from the tricarboxylic acid cycle. Fatty acids (C_3–C_{22}) are extracted from grains or are synthesized by yeast. The unsaturated fatty acids from wort may partially supplant oxygen required for yeast viability, as shown by Thompson and Ralph (1967).

Ester formation occurs largely during the earlier stages of active yeast growth from the condensation of alcohols and an enzyme intermediate. Little ester formation occurs due to condensation of alcohols and organic acids during aging.

Hydrogen sulfide is formed during the enzymatic breakdown of cysteine or methionine by yeast. Most of the hydrogen sulfide is purged out with CO_2.

Diacetyl (2,3-butanedione) and 2,3-pentanedione, collectively called vicinal diketones, impart a buttery flavor which is generally considered undesirable in American beers. Since the threshold level for detection of diacetyl by the consumer is about 0.1 ppm, substantial effort is made to control its concentration below this level. α-Acetolactic acid, the precursor of diacetyl, is formed during yeast cell metabolism from pyruvic acid. Diacetyl is formed by the oxidative decarboxylation of α-acetolactic acid. The decarboxylation rate increases with increasing temperature and decreasing pH and is independent of yeast cell concentration. Once diacetyl is formed, it can be assimilated by yeast to produce acetoin or 2,3-butanediol, which are not readily detected in beer. The assimilation of diacetyl occurs principally during secondary fermentation or aging. The assimilation rate of diacetyl is strongly yeast cell concentration dependent, and continued agitation during secondary fermentation to keep yeast suspended has been shown to be beneficial by Westermann (1976).

H. Lagering, Aging, and Finishing

The primary objectives of these final steps of the brewing process are: (1) flavor maturation; (2) physical stabilization; (3) clarification; and (4) adjustment of carbon dioxide level.

Ruh storage, or lagering, and aging are terms referring to different time periods for maturing beer. In the traditional ruh storage or lagering process, which is accomplished in closed vessels, maturation of the beer involves a secondary fermentation to assimilate the small amount of residual maltotriose and to complete diacetyl assimilation. This is accomplished by the action of 6–10 million cells per milliliter or residual yeast transferred from the primary fermentor. During the 2- to 6-week process the temperature may start at 3°–6°C and gradually be decreased to −1°C.

In the more recent aging process, completely fermented beer is matured at temperatures of about −1°C in the presence of 1 million or less yeast cells per milliliter.

1. Saturation with CO_2

Beer can be saturated with CO_2 by one or more of the following methods:

1. Natural buildup of CO_2 during secondary fermentation or kraeusening by applying counter pressure. In the kraeusening process, carbon dioxide is produced in an after-fermentation. Usually 10–20% of wort in the active, or so-called kraeusen, stage of fermentation is added to the partially fermented beer to bring about secondary fermentation.
2. Sparging with CO_2 in storage tanks.
3. Sparging with CO_2 in line. The carbon dioxide that is used during sparging is usually CO_2 collected during primary fermentation and

purified by distillation or by water scrubbing, permanganate scrubbing, and carbon absorption.

2. Flavor Maturation

Diacetyl, hydrogen sulfide, mercaptans, and dimethyl sulfide produced during fermentation are decreased during secondary fermentation and aging. The removal of diacetyl is yeast cell dependent, and higher temperatures and high cell concentration are favored. Hydrogen sulfide and undesirable acetaldehyde can be removed by purging with CO_2.

Purging with CO_2 is also important to minimize oxygen concentration. Low oxygen levels will minimize staling reactions in packaged beer and reduce the formation of haze precursors.

The concentrations of some esters, such as ethyl acetate and isoamyl acetate, that are responsible for the fruity background flavor in beer increase slightly during lagering, as Hashimoto and Kuroiwa (1972) have reported.

3. Beer and Foam Stabilization

Three types of stability are required for a good beer: (1) microbiological; (2) physical, and (3) foam.

Microbiological stability is obtained by pasteurization (to be discussed later) or by sterile filtration of beer through very fine sheet or membrane filters. The filters which are used for beer have a pore size of 0.45–0.8 μm, become clogged very easily, and require adequate prefiltration. The technique was popular for the production of "draft beer" in packages in the 1960s. It has only very limited use today.

Wainwright (1974) describes two types of haze. Chill haze (0.1–2 μm) is defined as a haze that forms at 0°C and redissolves at 20°C, and permanent haze (1–10 μm) is defined as that remaining above 20°C. Haze is composed of polypeptides, polyphenols, and carbohydrates in various ratios, with a small percentage of ash. Stable linkages between polypeptides and polyphenols can be formed after oxidation of the latter to quinones, according to Bishop (1975). Methods used to prevent or delay haze formation are: (1) hydrolysis of polypeptides by papain to low molecular weight components that cannot form hazes, (2) absorption of polypeptides by silica gel or bentonite, (3) precipitation of polypeptides by tannic acid, (4) absorption of polyphenols by PVPP or Nylon 66.

Foam stability is improved with increased viscosity and increased surface tension and elasticity. Foam-promoting substances are proteinaceous-type materials, surface-active hop components, and gums that increase beer viscosity (Runkel, 1976). Glycoproteins increase foam stability considerably. Foam adhesion or cling is promoted by iso-α-acids from hops.

4. Clarification

After lagering or aging, beer still contains yeast cells, proteinaceous-type precipitates, microorganisms, and colloidal matter. A first crude clarification which sometimes takes place prior to aging is usually carried out by centrifugal separation. Batch or continuous centrifuges are used with capacities of up to 250 barrels of beer per hour.

Clarification by filtration is carried out in sheet filters, diatomaceous earth filters, or pulp filters.

Pulp filters use a medium which contains cotton or wood pulp fiber made into a mat several inches thick. The filter mat is repulped, washed, sterilized, and reformed frequently. Because of their small capacity and high labor cost, pulp filters are rarely used in the United States.

Sheet filters consist of typical plate and frame filters containing filter sheets of cellulose, with or without a small percentage of asbestos. The sheets clog easily if beer has not been prefiltered. Depending on the density of the filter sheets, a high degree of beer clarity and sterility can be achieved.

Diatomaceous earth leaf filters are the main type of filter used for prefiltration. Diatomaceous earth, the calcined skeletons of the single-cell diatoms, is an excellent filter aid because of its high porosity. Diatomaceous earth is available in different grades. The grade that is selected depends on desired clarity of the filtered beer and the size and nature of the particles to be removed from the beer. At the start of the filtration cycle, the filter support, a mesh screen or a filter pad, is precoated with a thin layer (several millimeters) of filter aid. After 10–15 minutes of recycling, the main filtration starts. Filter aid is injected in beer that is going to the filters to build up a porous filter bed. After the pressure drop exceeds approximately 50 psi, the cycle is ended and the filter cake is removed manually or automatically. Generally about 25–40 lb of diatomaceous earth per 100 barrels of beer is used. Flow rates are in the order of 0.4–0.5 gal/minute/ft^2 of filter area. Some of the larger filters of approximately 1000 ft^2 can handle 900 barrels of beer per hour.

V. PACKAGING

The brewing industry has historically been the pacesetter in high-speed, high-efficiency packaging. Modern 12-oz can and bottle lines produce 1500 cans per minute and 900 bottles per minute. Today, 87.6% of all beer sold in the United States is packaged in bottles or cans. The

26.0 billion steel and aluminum cans used for beer account for over 25% of the can industry's total production. The 11 billion returnable and nonreturnable bottles account for 39% of packaged sales. Cans account for over 60%. The ratio of nonreturnable to returnable bottles shipped is 18.8 : 1.

In 1934, 2 years after the repeal of prohibition, 75% of the beer sold was draught. By 1937, 44% was sold in bottles, all returnable. In 1935 the first cone top can was introduced and found acceptance because it could be filled and crowned on bottling equipment. Its relative bulk and higher cost contributed to its rapid replacement by the flat top steel can.

Can competition forced the glass industry to develop the "one-way" ($7\frac{1}{2}$-oz glass) and the throwaway bottles ($6\frac{1}{2}$-oz glass) during World War II. In 1959 the "glass can" was developed. This short-necked stubby bottle concept prevails today for disposable bottles.

The can industry introduced the aluminum "soft-top can" for easy opening, and in 1962 introduced the "pop-top" lid. The bottle industry countered with the "twist-off" crown in 1963 to retain market share.

By 1964 aluminum cans became established, and the competitive battle for container market share continued. "Tin-free steel," welded side seams, cemented side seams, and drawn and iron aluminum and steel cans reflected technological improvements fostered by intense competition.

Imprudent disposal of the tabs from "pop-top" lids fostered the introduction of various ecology lids on which the opening device is an integral part of the lid.

Plastic beer bottles, although evaluated by some major brewers, are not yet a reality.

Returnable bottles are washed in a compartmented bottle washer in which the bottle, carried in a multipocket conveyor, is subjected to caustic solutions and freshwater rinses. Bottles and cans are filled on rotary continuous operating fillers containing up to 140 valves with maximum speeds of 1500–1600 containers per minute.

Both bottles and cans are pasteurized while being conveyed through a multicompartment pasteurizer in which the package is subjected to sprayed water at temperatures up to 71°C. The three principal organisms to be inhibited are wild yeast (*Saccharomyces diastaticus*), pediococci (*Pediococcus cerevisiae*), and long rods (*Lactobacillus brevis*). The pasteurization effect required is comparable to about 6–8 minutes at 60°C. Following pasteurization, bottles are labeled, and both bottles and cans are automatically placed in a proliferation of secondary packages. Unpasteurized draught beer, principally in half and quarter barrels, accounts for about 12% of the beer sales.

TABLE VIII. Beer Production and Per Capita Consumption

Country	Production in barrels	Per capita consumption in gallons
West Germany	79,628,000	39.7
England	55,056,000	30.4
Russia	51,131,000	6.2
Japan	33,482,000	9.6
Czechoslovakia	19,066,000	40.3
France	19,017,000	11.2
Canada	18,117,000	24.8
East Germany	17,214,000	31.0
Mexico	16,509,000	8.7
Australia	16,481,000	37.8
Brazil	14,775,000	4.3
Spain	14,162,000	12.4
Belgium	11,902,000	37.8
Poland	10,959,000	10.2
Netherlands	10,596,000	24.2
Denmark	7,567,000	45.9
Yugoslavia	7,224,000	10.5
Colombia	6,580,000	8.4
Austria	6,478,000	26.7
Romania	6,348,000	9.3

VI. AMERICAN INDUSTRY STATISTICS*

In 1975, the 102 breweries in the United States produced a total of 160,575,908 barrels (31 gal/barrel) and sold 148,632,852 barrels, resulting in a per capita consumption of 21.6 gal. Although many other countries produce less, their per capita consumption is substantially higher, as shown in Table VIII.

The Unites States brewing industry produces retail product sales in excess of $15 billion per year and employs approximately 51,000 persons. Over $850 million worth of agricultural products are used, including 4.2 billion lb of malt, made from 125 million bu of barley, valued at $610 million; other grains, chiefly corn and rice, valued at $188 million; and $49 million worth of hops.

The direct salaries, excluding fringe benefits, amounted to $860 million in 1975.

Close to $2.5 billion is spent annually on packaging and containers.

* The statistics provided in this section have been taken from the "Brewers Almanac."

The industry annually spends $80 million on fuel power and water, $360 million on transportation, $600 million on equipment and improvements, and $230 million on advertising.

The present Federal Excise Tax on beer is $9 per barrel, set in 1951. In 1975, the United States Government received $1,308,673,000 in excise taxes on malt beverages.

With the state taxes levied in addition to the $9 per barrel Federal tax, each barrel of beer sold earns an average of $13 in taxes. The state taxes range from $0.62 per barrel in Wyoming to $23.81 per barrel in South Carolina. The 1976 "Brewers Almanac" states, "By contrast, the brewing industry's average profit per barrel after taxes has been estimated at about a dollar and one-half."

REFERENCES

American Society of Brewing Chemists (1976). "Methods of Analysis," 7th ed., Am. Soc. Brew. Chem., St. Paul, Minnesota.

Arnold, J. P. (1911). "Origin and History of Brewing." Wahl-Henius Inst. Technol., Chicago, Illinois.

Baron, S. (1962). "Brewed in America—A History of Beer and Ale in the United States." Little, Brown, Boston, Massachusetts.

Bishop, L. R. (1975). *J. Inst. Brew.* **81,** 444–449.

Bradee, L. H., and Westermann, D. H. (1975). Internal communication, p. 37. Jos. Schlitz Brewing Co., Milwaukee, Wisconsin.

Brewers Almanac (1976). United States Brewers Association, Washington, D.C.

Bruch, C. W., Hoffman, A., Gosine, R. M., and Brenner, M. W. (1964). *J. Inst. Brew.,* **70,** 242.

Burgess, A. H. (1964). "Hops—Botany, Cultivation and Utilization." Wiley (Interscience), New York.

Comrie, A. A. D. (1967) *J. Inst. Brew.* **73,** 335–341.

Coors, J. H., and Jangaard, N. O. (1975). *Proc. Eur. Brew. Conv.,* p. 311. Elsevier Scientific Pub Co., Amsterdam.

Findlay, W. P. K., ed. (1971). "Modern Brewing Technology." Macmillan, New York.

Hashimoto, N., and Kuroiwa, Y. (1972). *Brew. Dig.* **47,** 64–71.

Hough, J. S., Briggs, D. E., and Stevens, R. (1971). "Malting and Brewing Science." Chapman & Hall, London.

Huige, N. J., and Westermann, D. H. (1975). *MBAA Tech. Q.* **12,** 31–40.

Jones, M., and Pierce, J. S. (1969). *Eur. Brew. Conv. Proc. Congr.* p. 315.

Kellett, O. S. (1965). *MBAA Tech. Q.* **2,** 69.

Laufer, S., and Schwarz, R. (1936). "Yeast Fermentation and Pure Culture Systems." Schwarz Lab. Int., Inc., Mt. Vernon, New York.

Likens, S. T., and Nickerson, G. B. (1964). *Annu. Proc. Am. Soc. Brew. Chem.* **4,** 5.

Lodder, J., and Kreger-Van Rij, N. J. W. (1952). "The Yeasts—A Taxonomic Study." Wiley (Interscience), New York.

Macey, A. (1963). *Brew. Guardian* **92,** 41.

MacLeod, A. (1976). *MBAA Tech. Q.* **13,** 193–198.

Matz, S. A. (1970). "Cereal Technology." Avi Publ. Co., Westport, Connecticut.

Nakayama, T, O. M. (1962). *Annu. Proc. Am. Soc. Brew. Chem.* 137–139.

Narziss, L. (1974). *Brauwelt* **114,** 851–892.
Narziss, L. (1976). *Brauwelt* **116,** 1736–1740.
Narziss, L. (1977). *Brauwelt* **117,** 50–53.
Runkel, U. D. (1976). *Monatschr. Brau.* **29,** 248–260.
Schauss, O. O. (1964). *MBAA Tech. Q.* **1,** 112–119.
Schoffel, F. (1972). *Brauwissenschaft* **25,** 301–312.
Scholefield, A. J. B. (1956). "The Treatment of Brewing Water." Published privately, Rockcliff, Liverpool.
Siebel, F. P., Jr. (1950). *Brew. Dig.* **25,** 45.
Singruen, E. (1938). *Mod. Brew.* **19,** 40.
Stewart, G. G. (1975). *Brew. Dig.* **50,** 42–56.
Thompson, C. C., and Ralph, D. J. (1967). *Eur. Brew. Conv., Proc. Congr.* p. 177, Elsevier Scientific Pub. Co., Amsterdam.
Wainwright, T. (1974). *Brew. Dig.* **49,** 38–48.
Weeks, M., Jr. (1949). "Beer and Brewing in America." U.S. Brew. Found., New York (Designed and produced by Appleton, Parsons, and Co., New York).
Westermann, D. H. (1976). *Abstr. Pap., 171st Centen. Meet., Am. Chem. Soc., 1976.*
Westermann, D. H., Chicoye, E., and Hoffman, D. R. (1976). U.S. Patent 3,965,188.
Winton, A. L., and Winton, K. B. (1932). "The Structure and Composition of Foods," Vol. I. Wiley, New York.

Chapter 2

Cheese

NORMAN F. OLSON

MICROBIAL TECHNOLOGY, 2nd ed., VOL.II
 ISBN 0-12-551502-2

I. INTRODUCTION

A. History

The exact origin of cheese is difficult to determine since it undoubtedly preceded use of written records. Cheese of some sort would have been made when milk was allowed to stand undisturbed until it coagulated. Use of animal stomachs to carry milk may have provided humans with the first cheese curd which was formed by the action of enzymes in the stomach walls and by metabolism of microbial contaminants. With a little ingenuity, humans learned that stomach linings and various plant extracts could curdle milk if they were soaked in or added to milk.

Early records indicate that the art of cheese making spread from the Middle East to the northern Mediterranean area and northern Europe. Romans introduced cheese making into England. The art of making many of the important present-day European types of cheese was preserved and at times jealously guarded in monasteries during the Middle Ages.

The taste for and techniques of making cheese were carried by English colonists to the United States, Australia, and New Zealand. Cheese making was a farm industry in England and America until the middle of the nineteenth century. Housewives converted surplus milk, from their farm and possibly neighboring farms, into cheese for family consumption. Centralization of cheese operations into manufacturing plants was inevitable so that this system eventually replaced most of the farm-based manufacturing.

B. Classification of Cheese

The long history of cheese, the accidental modification of the cheese-making processes, its development in diverse areas during an era of limited communications, and global production has resulted in an enormous number of cheese varieties. Over 400 varieties have been described and more than 800 indexed by Walter and Hargroves (1972). Attempts have been made to categorize these varieties. Classification by countries in which varieties originated is satisfactory from a merchandising standpoint but is unsatisfactory since the method does not characterize the variety. In addition, varieties may be essentially the same and differ only in their national name. Cheese varieties can be grouped into the following eighteen types based on differences in manufacturing: brick, Camembert, Cheddar, cottage, cream, Edam, Gouda, hand, Limburger, Neufchatel, Parmesan, Provolone, Romano, Roquefort, Sapsago, Swiss, Trappist, and whey cheeses (Walter and Hargrove, 1972). This system is logical and informative but is incomplete since two varieties may be made by the same process but differ markedly from the mode of curing. A third system of classifying differentiates on the basis of techniques of curing and firmness of the cheese varieties (Walter and Hargrove, 1972).

II. FUNDAMENTAL PROCESSES IN CHEESE MANUFACTURE

The manufacture of cheese can be divided into five general phases: (1) control of properties of milk, (2) coagulation of milk, (3) separation of the solid curd from fluid whey, (4) forming curd into its final physical form, and (5) controlled storage of cheese to maintain the flavor of curd or to develop desired changes in flavor and rheological characteristics of cheese before consumption. Variations in each of these phrases result in cheese with different characteristics and explain the diversity in types of cheese.

A. Control of Milk Properties

Some of the most important factors in controlling the characteristics and quality of cheese are the biological, chemical, and physical properties of milk. The type of animal used to produce milk affects all three properties and is the basis for differentiating certain cheese varieties. Cheese usually is made from bovine milk but some varieties are made from milk of sheep, goats, water buffalo, mares, and camels.

1. Biological Properties

a. Adventitious Microorganisms. The biological properties of milk can determine the type of cheese made from that milk but can have a greater influence on quality of any given type. Milk with relatively low numbers of microorganisms is preferred for cheese making, even though certain contaminants can contribute to flavor development in cheese (Mocquot, 1971). Problems with excessive gas production, development of undesirable flavors, and potential contamination with pathogenic microorganisms dictate against use of milk with a high microbial population.

Numbers and types of contaminants in milk depend upon health of the cows and sanitary practices during the production and processing of milk (Foster *et al.*, 1957). Numbers of psychrotrophic bacteria have assumed greater importance because of their ability to grow in refrigerated milk, especially at temperatures exceeding 7°C, and their lipolytic and proteolytic activities. Growth and activities of these species reduce yield of cottage cheese and cause spoilage during storage (Cousin and Marth, 1977a). Psychrotrophs can hydrolyze casein in milk, resulting in greater losses of nitrogenous materials in whey, and decrease yield of other cheese varieties (Cousin and Marth, 1977b; O'Leary *et al.*, 1977; Nelson and Marshall, 1977).

b. Enzymes in Milk. In addition to microorganisms, milk contains enzymes that are elaborated by the mammary tissue or by microorganisms contaminating milk. The principal, naturally occurring enzymes in milk are catalase, peroxidase, xanthine oxidase, alkaline and acid phosphatases, amylases, proteases, lipases, and aldolase (Jenness and Patton, 1959). Several of these enzymes have direct or indirect effect on cheese making. Catalase has no direct effect but is an indicator of mastitic milk, which retards normal acid production and may affect rheological properties of cheese curd. Lactoperoxidase is inhibitory to certain lactic acid bacteria used in cheese making (Reiter, 1973). Alkaline phosphatase serves as an index of adequate pasteurization since it is destroyed during normal pasteurization of milk. Lipases and proteases hydrolyze casein and milk fat, affecting flavor and rheological properties of cheese. It is doubtful that hydrolytic enzymes naturally occurring in milk contribute greatly to ripening of cheese varieties such as Cheddar since only a small fraction of the total activity has an optimum in the pH range of cheese while it is curing (McGugan, 1975). Proteolytic and lipolytic activities of added and adventitious microorganisms contribute significantly to the characteristics of many varieties of cheese, especially the mold-ripened and surface-ripened cheeses (Mocquot, 1971). Esterases isolated from the oral glands of calves, kids, and lambs produce the desired lipolysis and flavor in Italian varieties of cheese such as Romano and Provolone (Nelson *et al.*, 1977).

c. Control of Biological Properties. The cheese maker can exert some control over biological properties of milk by certain treatments and additives. The most common is heat treatment, including pasteurization (Wilster, 1969). To be labeled as pasteurized-milk cheese, cheese must not contain levels of alkaline phosphatase greater than those specified in federal and state regulations (U.S. Department of Health, Education and Welfare, 1974). Certain varieties of cheese, such as cottage and cream, must be made from pasteurized milk. Other varieties need not be pasteurized but must be stored for at least 60 days at temperatures of 2°C or higher if they are made from nonpasteurized milk. Pasteurization treatments of 72°C for 15 seconds and 63°C for 30 minutes or equivalent treatments have been selected to ensure destruction of pathogenic microorganisms which might contaminate milk. These treatments are used also to reduce numbers of microorganisms in milk and to give greater control over fermentations during manufacturing and curing. However, such treatments reduce the rate of flavor development in cheese since microorganisms involved in ripening are destroyed. Subpasteurization treatments are used as a compromise to attain faster flavor development but still control undesirable fermentations. Hydrogen peroxide treatment is used also to reduce numbers of microorganisms in milk (Kosikowski, 1977). It is not a substitute for pasteurization; the resulting cheese must be stored as is that made from unpasteurized milk.

2. Physical Properties

Physical properties of milk may be modified by homogenization, filtration and clarification, and vacuum treatments (Bradley, 1969; Harper and Hall, 1976). Milk fat globules are reduced in size (diameter) from about 4 μm to 1 μm by homogenization by forcing milk through a small orifice or specially designed wire mesh. The treatment results in a weaker curd formed by milk-clotting enzymes and decreased whey syneresis from curd but reduced fat losses during cheese making and less fat leakage from cheese (Ernstrom and Wong, 1974). Homogenization is used in the manufacture of cream cheese, blue cheese, certain soft types, and process cheese spreads. In the manufacture of cream cheese, cream is homogenized prior to cheese making and the curd may be homogenized if the "hot-pack" method is used. Use of this treatment on the finished cream cheese or other soft types and cheese spreads increases smoothness of the product. Decreasing the size of fat globules, with concomitant increase in surface area, accelerates lipolysis and flavor development in blue cheese. Even though fat losses are reduced by homogenization, problems of reduced curd strength and whey syneresis and development of hydrolytic rancidity have prevented application of the treatment to most varieties of cheese. Mild homogenizing treatments of milk have been used experimentally in making Cheddar cheese and in an accelerated direct-acidification procedure for cheese making (Peters, 1964; Quarne *et al.*, 1968).

Filtration and clarification affect physical properties of milk by removing extraneous material from milk. Clarification can be more efficient than filtration, removing materials such as leukocytes in addition to particulate material. Milk is clarified by flowing through a revolving bowl similar in construction to a milk separator (Wilster, 1969; Harper and Hall, 1976). All substances which are more dense than normal milk are sedimented out by the centrifugal force. This process is used widely in making Swiss cheese; more uniform eye formation results from removal of extraneous matter which can act as foci for developing holes or eyes.

Vacuum pasteurization of milk removes gases and volatile odors from milk (Wilster, 1969). This can be beneficial in areas plagued with weed or similar flavors in milk. The removal of some water vapor from milk reduces milk volume and can have economic advantages by increasing milk throughput in a large plant.

3. Chemical Properties

a. Treatments of Milk. Altering relative amounts of the principal milk constituents is the most universal method of controlling chemical properties or composition of milk. Removal of fat from milk by centrifugation using a "separator" is a practical method of adjusting composition (Bradley, 1969). Cream cheese is made from cream having a fat content higher than normal milk; cottage cheese is made from skim milk.

Chemical properties of milk can be affected by chemical additives permitted under regulations promulgated by the U.S. Food and Drug Administration (1974). Food-grade colors are used to enhance color of certain cheese varieties. They do not affect flavor or rheological properties of cheese but their use is dictated by tradition and personal or regional preferences of consumers. Several types of colors are certified under federal regulations but annatto, a carotenoid pigment extracted from the seeds of the tree *Bixa orellana,* is used most widely. Conversely, intensity of natural milk pigments is reduced in certain cheeses (blue, Swiss, and Italian varieties). This is done by masking the color with harmless artificial blue or green coloring or by bleaching milk with benzoyl peroxide. That portion of vitamin A potency of milk destroyed by bleaching must be restored.

Calcium chloride may be added at levels not exceeding 0.02% of milk weight to aid in rennet coagulation, especially milk heated to higher than normal pasteurization temperatures. Food-grade acids can be used to aid in curd formation in certain types of cheese, such as Ricotta, and as acidulants in process cheese products. Accelerated, continuous procedures have been suggested using direct acidification of milk (Olson, 1975). Hydrogen peroxide may be used, at levels not to exceed 0.05%, to treat milk prior to the manufacture of Cheddar, Colby, and Swiss cheeses. Residual hydrogen peroxide must be destroyed by catalase before cheese making.

b. Yield of Cheese. When milk is converted into cheese, milk constituents are partitioned between curd and whey as illustrated in Table I (see Section II, C). The principal milk constituents contributing to yield of cheese are milk fat and casein. Based upon a large number of analyses, a formula was developed which related yield of Cheddar cheese to casein and fat contents of milk (Van Slyke and Price, 1952).

$$\text{Yield} = \frac{(0.93F + C - 0.1)\,(1.09)}{1.00 - W}$$

where yield = kilograms of cheese per 100 kg of milk
F = kilograms of fat per 100 kg of milk
C = kilograms of casein per 100 kg of milk
W = kilograms of water per kilogram of cheese

It assumes that 7% of fat and 0.1 kg of casein are lost in whey, pressings, curd fines, etc. Higher losses with some manufacturing methods necessitate changes in both factors for fat and casein. The factor of 1.09 corrects for solids in cheese other than fat and casein.

D. B. Emmons (personal communication, 1977) adapted the above formula to cheeses of varying fat contents. In the formula below, it is assumed the casein constitutes 78% of total milk protein and that 3.89% of casein is lost in whey.

$$\text{Yield} = 0.93F + \frac{0.907P}{1.00 - MNFP}$$

$$= 0.93F + \frac{1.163C}{1.00 - MNFP}$$

where yield = kilograms of fat-reduced cheese per 100 kg of fat-reduced milk
F = kilograms of fat per 100 kg of fat-reduced milk
P = kilograms of protein per 100 kg of fat-reduced milk
C = kilograms of casein per 100 kg of fat-reduced milk
$MNFP$ = kilograms of water per kilogram of nonfat portion of cheese

In addition to milk composition, yield is affected by manufacturing practices which affect retention of constituents (Olson, 1977).

TABLE I. Partition of Milk Solids between Cheese and Whey during Manufacture of Cheddar Cheese Containing 37% Moisture[a]

Constituent	In cheese[b]	In whey[b]
Fat	3.7	0.3
Casein	2.4	0.1
Lactose	0.2	4.9
Whey proteins and salts	0.4	1.0
Water	3.9	83.1
Total	10.6	89.4

[a] Adapted from Van Slyke and Price (1952).
[b] Kilograms of milk constituents from 100 kg of milk containing 4.0% fat.

B. Coagulation of Milk

In the manufacture of almost all varieties of cheese, milk proteins are coagulated to form a continuous solid curd in which milk fat globules, water, and water-soluble materials are entrapped. The casein fraction of milk normally forms this curd but in a few varieties of cheese, whey proteins are coprecipitated with the casein fraction. Coagulation may be induced (1) by adjusting the pH of milk to the isoelectric point of casein, (2) by action of milk-clotting enzymes, (3) by application of heat, or (4) by addition of salts. The first two systems are commonly used alone or in combination to form cheese curd. Application of heat is used for only a small number of cheese varieties and salt addition is rarely used because of effects on flavor.

1. Lactic Starter Cultures

Acidification of milk is usually accomplished by fermentation of lactose to lactic acid by selected cultures of bacteria. *Streptococcus cremoris* and *Streptococcus lactis* are used in cheese varieties heated to 40°C or less during the acid fermentation. High-temperature homolactic bacteria, such as *Streptococcus thermophilus, Lactobacillus bulgaricus,* and *Lactobacillus helveticus,* are used in manufacture of cheese heated to higher temperatures (Ernstrom and Wong, 1974).

Lactic acid-producing bacteria are usually selected, propagated, and distributed in the United States by industrial firms. Research on physiology, propagation methods, and preservation of lactic cultures and improvements in technology of preparation, distribution, and use of these cultures have significantly improved their activity and reliability.

Commercial mesophilic cultures used for such cheese varieties as Cheddar and cottage may be single strains of *S. cremoris* or *S. lactis* or combinations of strains of these two species, or combinations of these two species with low numbers of *Leuconostoc* species or *Streptococcus diacetilactis.* Strains are selected for acid-producing activity, bacteriophage sensitivity, growth compatibility with other strains, lack of gas production, and effect on flavor of cheese. Lactic cultures are supplied to cheese plants in the lyophilized or frozen state, with the latter being more prevalent in North America. Freezing, preservation, and distribution is accomplished by use of liquid nitrogen or dry ice. Traditional lyophilized and frozen cultures are propagated several times to attain the volume of culture to inoculate milk for cheese making. Concentrates of lactic cultures have been prepared by centrifuging cells out of a growth medium and resuspending in menstrua to give a culture "paste" with approximately 100 times more cells than a normal culture (Gilliland and Speck, 1974; Reinbold, 1973). The concentrated cultures have been inoculated directly into bulk cultures for cottage and Cheddar cheese manufacture and more recently inoculated directly into milk in the cheese vat (Gilliland, 1977; Volume I, Chapter 3).

One of the principal causes of lack of acid production during cheese making is destruction of lactic cultures by the lytic action of bacteriophage (Sandine, 1977). Effects of bacteriophage can be limited by using properly designed equipment and facilities, asepsis in culture handling, sanitation, proper handling of whey, use of bacteriophage inhibitory media, and a culture rotation program (Collins, 1977: Lawrence *et al.*, 1976). The latter technique is effective if the cultures used in sequence are not attacked by the same bacteriophage. Collins (1952) felt that attack by bacteriophage would not be noticed during cheese manufacture if at least 50% of the strains were resistant to the infecting phage strains.

Lactic acid-producing cultures have significant effects on the curd-making and cheese-curing phases of cheese manufacture (Lawrence *et al.*, 1976; Ernstrom and Wong, 1974). Acid production effects coagulation of milk at the isoelectric point of casein (pH 4.6) in the manufacture of such varieties as cottage cheese and accelerates enzymatic coagulation of milk in other cheese varieties. Further acid production and reduction in pH of rennet curd during cheese making aids in whey syneresis from curd, controls undesirable microorganisms, and increases the plasticity of curd (Kosikowski, 1977; Ernstrom and Wong, 1974). Acids (lactic, acetic, and propionic), ketones and aldehydes or their precursors, and esters produced in total or in part by lactic cultures contribute directly to the flavor of cured cheese (McGugan, 1975; Ernstrom and Wong, 1974; Lawrence *et al.*, 1976). Indirect effects on flavor and appearance of cured cheese result from control of the growth and metabolism of adventitious bacteria by reduction of pH of curd. Certain strains of lactic streptococci are undesirable for cheese making because of production of excess gas, antibiotics against other strains of lactic bacteria, bitterness, and fruity flavors (Lawrence *et al.*, 1976).

2. *Milk-Clotting Enzymes*

Enzymes used to clot milk are classified as acid proteases (Ernstrom and Wong, 1974). Requirements considered important for milk-clotting enzymes are: clotting and proteolytic activities, proteolytic specificity, properties of curd, presence of contaminating enzymes, and heterogeneity of proteinases (Green, 1977). Milk-clotting enzymes traditionally have been selected on the basis of their high ratio of milk-clotting to proteolytic activities. Low proteolytic activity is essential to avoid excess protein degradation in cheese or in curd during manufacturing which results in reduced yields of cheese. Proteolytic specificity refers to relative rates of activity on α_s- and β-caseins with the resulting release of bitter peptides. Commercial enzyme preparations are tested for this tendency before wide-scale commercial use. Rates of increase in curd firmness vary with different enzymes but cutting the curd at the same firmness apparently negates these differences and any effect on properties of curd during subsequent manufacturing (Ramet *et al.*, 1969; Richardson *et al.*, 1971). Some preparations of microbial milk-clotting enzymes reportedly contained contaminating lipases and proteases (Green, 1977). Purification procedures were instituted to eliminate these contaminants in commercial enzymes. Commercial milk-clotting enzyme preparations may contain several proteases. Differing responses of these enzymes to pH, temperature, and salts must be taken into consideration by the cheese manufacturer. Adjustments in the manufacturing schedule, such as the time of cutting the curd, may have to be made.

A number of different enzyme preparations are used commercially to clot milk: rennet extract, porcine pepsin, and proteases from selected microorganisms. The most widely used microbial proteases are derived from *Mucor miehei, Mucor pusillus,* and *Endothia parasitica* (Ernstrom and Wong, 1974; Green, 1977). Rennet is the partially purified enzyme extract from the fourth stomach of calves. The principal milk-clotting enzyme in the extract is chymosin (rennin). A second enzyme, bovine pepsin, contributes a small portion of the total activity of extracts from young, suckling calves. The relative concentration of pepsin increases as the calf ages. A mixture of rennet and porcine pepsin called 50:50 is used to manufacture cheese which is not ripened extensively.

The action of milk-clotting enzymes on milk can be divided into three phases: the primary or enzymatic, secondary or clotting, and tertiary or proteolytic phase (Ernstrom and Wong, 1974; Green, 1977). The total effect of the primary and secondary phases is the conversion of one of the milk protein fractions, casein, from a colloidal suspension to a fibrous network. The casein fraction in milk is heterogeneous; its main components are designated as α_s-, β-, γ-, and κ-caseins. These components interact in milk to form a stable suspension of spherical particles called micelles; stability of the suspension in the presence of calcium depends on κ-casein. Milk-clotting enzymes attack κ-casein causing a release of macropeptides from this fraction; loss of the peptides destroys the stabilizing effect of κ-casein and sensitizes micelles to calcium. Coagulation of casein in the presence of calcium constitutes the second phase of enzyme action. The first two phases of milk clotting are accelerated by lowering the pH and by increasing temperatures up to 45°C. The rate of the secondary phase is more temperature-dependent than the primary phase so that coagulation of milk can be delayed at low temperatures for a considerable time after the primary phase is completed.

Instantaneous coagulation occurs upon warming the milk. This characteristic is used in some continuous cheese-making systems in which milk is cold-renneted and then coagulated continuously in special devices (Kosikowski, 1975).

Proteolytic action of milk proteases, lactic cultures, and adventitious bacteria have complicated investigation of the tertiary phase of chymosin action in cheese. Studies on proteolytic action of chymosin on purified casein suggest that chymosin hydrolyzes protein in cheese to polypeptides (Green, 1977; Lawrence *et al.*, 1976; O'Keefe *et al.*, 1976). Lactic cultures and adventitious microorganisms further degrade polypeptides to lower molecular weight peptides and amino acids.

C. Separation of Curd and Whey

Subsequent to coagulation, partition of the solid curd and whey (water plus water-soluble compounds) occurs. This is accelerated in the cheese-making process by heating, a decrease in pH, and physical disruption and manipulation of the curd. In the procedures for most varieties of cheese, the milk is coagulated into a contiguous curd mass in large vats (Fig. 1). Special curd knives, steel frames strung with wire, are used to cut the curd into particles of the desired size. These may be small irregularly shaped particles resulting from extensive cutting or discrete cubes from more precise cutting procedures. Cutting curd greatly accelerates initial syneresis of whey by increasing the surface to volume ratio of curd. Hence, rates of syneresis are inversely related to curd size.

Production of acid by lactic starter cultures lowers the pH and water-holding capacity of curd. Temperature and pH probably are the most important factors in controlling whey syneresis or moisture expulsion from curd during manufacture and in establishing moisture levels for a given variety of cheese (Patel *et al.*, 1972). Stirring curd in whey also accelerates whey syneresis, presumably from physical stresses on curd (Lawrence, 1959). The curd is stirred in whey until the pH has decreased sufficiently and an optimum amount of whey expulsion has occurred. Then the curd is physically separated from whey by various means which differ between cheese varieties. Whey can be removed from curd by settling the denser curd and allowing whey to flow from the curd or by placing curd and whey in perforated forms (hoops) in which curd is trapped and whey escapes. The first method is used in manufacture of Cheddar and cottage cheeses; the second in brick and blue cheeses. Curd can be removed from whey by gathering it in a cloth and lifting the cloth and curd out of the whey and into a drainage form as was done in the traditional method for Swiss cheese.

FIGURE 1. Horizontal vats used in cheese making. Revolving agitators shown in the center of vats traverse the length of vats and stir the curd and whey.

The separation of curd and whey results in a partition of milk solids in the two phases as shown in Table I. Approximately 50% of the milk solids are lost in the whey in the manufacture of Cheddar cheese. This percentage loss would be higher for cheese made from milk with a lower fat content since most (90%) of the milk fat is recovered in curd. This value varies between cheese varieties and depends upon treatment of curd during manufacturing. Extensive cutting of curd, exposure to high temperatures, and excessive physical stress on curd increases fat losses. Loss of casein in whey usually is low but increases with use of inferior quality milk and poor manufacturing techniques. Whey proteins, as indicated in Table I, partition primarily into whey because of their high solubility under conditions of normal cheese manufacture. Levels retained in curd depend on amounts of whey retained in the curd; more are retained in high-moisture cheese. Exposing milk to high temperatures before or during coagulation causes an interaction of whey proteins with casein and their precipitation with curd (Ernstrom and Wong, 1974). Such treatments are used in making Ricotta and certain Norwegian and South American varieties of cheese. It has been suggested for cottage cheese but not adopted widely because of problems with curd formation and firmness (Emmons and Tuckey, 1967).

Water-soluble salts will partition in the same manner as whey proteins. A portion of the calcium is bound to casein or is associated with casein as colloidal calcium phosphate. Retention of calcium with curd varies inversely with the pH of curd at the time of whey drainage; cottage cheese has low levels, whereas Cheddar and Swiss have higher levels. The milk sugar, lactose, is the greatest contributor to solids loss in cheese making and constitutes 70% of the whey solids. A few cheese varieties made from whey and whey–milk mixtures contain high levels of lactose but are not normally cured after manufacture.

Most (95%) bacteria present in milk are trapped in the curd (Ernstrom and Wong, 1974). Consequently, lactic acid fermentation occurs primarily within curd particles and the lactic acid diffuses out into whey. The environment for bacteria within cheese curd during cheese making has not been completely characterized but is dynamic. Acid is being produced within the curd, causing a decrease in pH of the serum in curd and an increase in soluble calcium salts in the serum and whey. Serum or whey is being expelled, carrying with it lactose, the substrate for lactic acid fermentation. However, diffusion is not unidirectional since Czulak *et al.* (1969) found that lactose diffused back into Cheddar cheese curd from whey at about the same rate as it was fermented in curd. The equilibration was not rapid enough in cheese in which lactic acid fermentation is abnormally fast. Consequently, composition of curd at the end of whey expulsion depended upon the lactic fermentation and the rate and time for diffusion of constituents in and out of curd.

The low molecular weight of milk-clotting enzymes precludes their entrapment in curd. Retention of rennet in Cheddar cheese is low, with only 6% of the original activity evident in curd after pressing (Holmes *et al.*, 1977). Approximately 35% of the activity is destroyed during the manufacturing process up to the point of whey drainage; the remainder is partitioned in whey. No activity of proteases from *M. miehei* and *M. pusillus* was lost during cheese making with 2–3% remaining in the cheese and the rest in the whey.

After physical separation of curd and whey, the curd is subjected to a series of treatments to allow acid production and attainment of optimum pH levels, to induce sufficient moisture loss from curd, to apply salt to the curd, and to fuse the curd into its final shape. The techniques used in these steps vary between varieties and types of cheese and influence characteristics of the varieties. Certain cheese varieties are distinguished by shape alone. The Italian varieties, Provolone and salami, may be made from the same curd but molded into different shapes.

D. Cheese Ripening

The transformation of the elastic or chalky cheese curd with an acid flavor to a ductile, full-flavored cheese is accomplished by controlled storage called ripening. Agents responsible for these changes are milk enzymes, adventitious microorganisms, and added microbial cultures (Ernstrom and Wong, 1974). The changes which occur are extremely complex in almost all cheese varieties. They involve hydrolysis of a portion of the fat and protein with subsequent decarboxylation, deamination, and dehydrogenation of some of the hydrolysis products. Cheese flavor has been attributed to carbonyls, nitrogenous compounds, fatty acids, sulfur compounds, and miscellaneous minor components.

In certain cheese varieties, flavor originates predominately from a specific group of microorganisms and enzymes. Propionibacteria contribute significantly to flavor and eye formation in Swiss cheese; pregastric esterases are vital for the "piccante" flavor of Romano and Provolone cheeses; *Penicillium roqueforti* and *Penicillium camemberti* produce the typical flavors and appearance of blue and Camembert cheeses. Progressive growth of yeast, micrococci, and *Brevibacterium linens* impart characteristic flavors to smear-ripened cheese such as Limburger. Other varieties, such as Cheddar and Gouda, do not depend on a specific class of compounds for their typical flavor (Ernstrom and Wong, 1974). A balance of various components is essential for the overall flavor of these varieties. In such a complex system, addition of single bacterial species has not consistently accelerated the development of typical Cheddar flavor.

Fermentation patterns in most varieties of cheese have not been defined sufficiently to allow direct control of cheese ripening. Indirect control of ripening is attained by establishing proper conditions in cheese and in the external environment. Low numbers of microorganisms in milk make such control of cheese quality more effective. Fermentations are regulated by pH, moisture, salt, osmotic pressure, oxidation–reduction potential, and depletion of reducing sugars in cheese. Temperature affects biological changes occurring throughout cheese; relative humidity influences the type and extent of microbial growth on the cheese surface.

Improvements in packaging and in ripening rooms have contributed significantly to standardization of cheese quality. Development of mechanical refrigeration for temperature control obviated many quality defects associated with high ripening temperatures. Even though ripening temperatures are fairly low there is considerable variation between varieties of cheese; Cheddar cheese is ripened at 2°–10°C but certain firm-bodied Italian varieties may be ripened at higher temperatures.

Swiss cheese is held initially at 21°C to induce eye formation and then ripened at lower temperatures. Surface-ripened cheeses (brick, Limburger, and Bel Paese) are held at 13°–15°C and under high relative humidities to allow surface microbial growth before final storage at lower temperatures.

E. Cheese Packaging

Cheese is normally covered with a protective coating at some stage of the ripening process. Coatings vary from vegetable oils to special plastic films. The shape and characteristics of cheese and changes to be induced in cheese dictate the type of packaging material. Cheese which is to be slightly dehydrated, such as Parmesan and Romano, may be rubbed with a vegetable oil to retard surface microbial growth but allow loss of moisture from the cheese. Waxes have been formulated for cheese packaging which cling to cheese to minimize surface growth and have low water-transmission rates. It is necessary to dry the surface of cheese slightly before waxing; this causes some loss of moisture and yield and necessitates some special handling. Waxes are especially useful for cheese varieties with unusual shapes where application of molten wax is a simple method of packaging. Waxes are also used in combination with plastic films.

The most significant change in packaging of cheese prior to ripening was the development of the rindless block style of cheese. Most Cheddar and Swiss cheese manufactured in the United States is made in this form. The cheese is pressed in rectangular hoops or forms and wrapped in a plastic film. The films are usually laminants of different plastic films and waxes with specific properties (National Research Council, 1958). This system of packaging has several advantages. It is not necessary to dry the cheese surface as thoroughly as is done in waxing of cheese, which eliminates yield loss. Ripening of film-wrapped cheese can be done at lower humidity levels without excessive moisture losses from the cheese.

Cheddar and Cheddar-type cheeses are pressed into larger units such as large cylinders (barrel cheese) and blocks. The barrel style of Cheddar cheese is made by placing 215 kg of curd, after sufficient whey drainage and salting, in a cylindrical barrel. A uniform fusion of the curd, without external pressure, is attained by inverting the barrel at appropriate intervals; or pressure may be applied. Vacuum treatments can be used also to improve fusion of this style and "block" Cheddar cheese (Price and Olson, 1965; Robertson, 1966). The trend to pressing cheese curd into large forms with the aid of vacuum undoubtedly will continue.

III. CHEESE VARIETIES IN WHICH MILK IS CLOTTED BY ACID

A. Cottage Cheese

1. Curd Formation

Cottage cheese curd is formed by acid coagulation of skim milk (Emmons and Tuckey, 1967). Traditionally, acid production results from fermentation by added lactic starter cultures containing *S. cremoris*, *S. lactis*, and small amounts of *Leuconostoc cremoris*. Lactic fermentation in skim milk may be carried out at a lower temperature, 22°C, for a longer time (16 hours), using a low inoculum (0.5–1.0%) of a normal lactic culture or at a higher temperature of 32°C for 4–5 hours using 4–5% culture. Intermediate conditions may be used to fit manufacturing schedules. Reduction of pH causes aggregation of one fraction of milk proteins, casein, to form a continuous gel in the cheese vat. When the pH has decreased to 4.6–4.7 and the proper firmness of the gel has been attained, the gel is cut into cubes with curd knives. The size of cubes may be 0.63 cm to make "small-curd" or 1.3–1.9 cm to make "large-curd" cottage cheese.

2. Cooking the Curd

Whey expulsion from curd begins immediately after cutting and is accelerated by stirring and heating the curd–whey slurry. Heating or "cooking" is done slowly to avoid case-hardening, formation of an excessively impervious skin on the curd, which retards further whey syneresis. The final cooking temperature will approximate 55°C. Growth and metabolism of the lactic culture diminishes and ceases when the temperature is raised beyond 40.6°–43°C (Emmons and Tuckey, 1967). The combination of low pH and high cooking temperature destroys virtually all contaminating psychrotrophic bacteria and lactic bacteria in the curd and whey (Collins, 1961). Consequently, microbial spoilage of packaged cottage cheese results from postcooking contamination.

3. Acidity Effects

The pH attained at cutting and the minimum pH reached when cooking temperature exceeds 43°C have significant effects on characteristics of the finished cheese. If the pH drops below the isoelectric point (pH 4.6) of casein, the curd will retain more water and the cheese may have a soft, pasty body. Excess acidity may also produce a mealy body if the curd is firm. A high pH may produce curd that is too firm and rubbery and may become translucent. Susceptibility to spoilage is greater in a low-acid curd with spoilage manifesting itself as a slimy, gelatinous surface growth, discoloration, and fermented, fruity, and unclean flavors.

4. Whey Draining and Curd Washing

The curd and whey are stirred, after attaining the desired cooking temperature, until the curd has become adequately firm. The curd particles are allowed to settle but not fuse while most of the whey is drained. The curd is then subjected to a series of washings with progressively colder water. This cools the curd, reduces acid flavor, and raises the pH of curd slightly.

Microbial contamination of cheese occurs most frequently from contact with water. This can be minimized by treating water to adjust available chlorine levels to 5–10 ppm. Acidification of water prior to chlorination increases effectiveness of the sanitizer and prevents curd defects resulting from excessively high pH.

5. Creaming the Curd

A cream mixture is added to the washed and drained curd to yield a finished creamed cottage cheese. The minimum fat content of creamed cottage cheese in the United States is 4% but "low-fat" cottage cheese is also produced. When cream is mixed with curd, a portion of the serum phase of cream is absorbed by curd, leaving a desirable, viscous coating of cream on the cheese. Insufficient absorption, excess cream, or low viscosity of cream results in "free cream" which flows from the curd.

Flavor intensity of creamed cottage cheese has been increased through citrate fermentation by heterofermentative lactic streptococci. A widely used method involves growth of *L. cremoris* in skim milk, acidfying to pH 4.3 with sterile citric acid, and further incubation to develop flavor (Mather and Babel, 1959). The cultured skim milk containing desired concentrations of diacetyl and acetoin is added to the cream dressing. A special strain of *S. diacetilactis* has been used also to increase flavor intensity (Lundstedt and Fogg, 1962).

6. Mechanization

Alternative and mechanized systems for cottage cheese manufacturing have been used experimentally or on a commercial basis (Ernstrom and Kale, 1975; Kosikowski, 1977). Replacement of biological production of acid with direct acidification methods has been done on laboratory, pilot-plant, and commercial scales. A process of partial acidification of skim milk with food-grade acids and final acidification with D-glucono-δ-lactone is used commercially (Ernstrom and Kale, 1975). This system is feasible because hydrolysis of lactone to gluconic acid is slow enough to allow mixing with warm milk and clotting of the milk in traditional vats. A continuous system of heating acidified skim milk in a tubular heat exchanger to cause clotting and equipment to process the stream of curd has been evaluated on a pilot-plant basis (Ernstrom and Kale, 1975).

An enclosed system of washing, cooling, draining, creaming, and packaging cottage cheese is used commercially (Ernstrom and Kale, 1975). Curd and a portion of the whey, at the end of heating (cooking) is pumped to closed, cylindrical tower-vats where it is drained and washed before flowing through the rest of the system.

B. Cream and Neufchatel Cheeses

Cream cheese is a soft, fresh variety made from cream which is standardized to yield a cheese containing at least 33% fat and not more than 55% moisture (U.S. Food and Drug Administration, 1974). Cream cheese curd is formed by acid coagulation, as is cottage cheese, but is thoroughly disrupted by agitation instead of being cut into cubes (Emmons and Tuckey, 1967; Kosikowski, 1977). The same species of lactic bacteria are used for starter as those used for cottage cheese.

Curd and whey are separated by differential centrifugation in special separators or, to a limited extent, by drainage in cloth bags and filter presses. The curd can be packaged and marketed after whey removal or it can be processed by the "hot-pack" method. This involves heating, homogenizing, and packaging hot curd. Cheese made by this process has better storage stability than cheese packaged immediately after whey drainage.

Neufchatel cheese is similar to cream cheese but differs in composition; the minimum milk fat content according to United States standards is 20% and the maximum moisture content is 65% (Emmons and Tuckey, 1967; Kosikowski, 1977). Manfacturing methods are similar to cream cheese with the "hot-pack" method being most prevalent.

C. Baker's Cheese

Skim milk is clotted by lactic fermentation using *S. cremoris* type culture and small amounts of rennet (4 ml/454 kg of skim milk) and procedures similar to those for cottage cheese (Emmons and Tuckey, 1967; Kosikowski, 1977). The curd is disrupted in a manner similar to that used for cream cheese. Traditionally, the curd was drained in filter bags under pressure but centrifugal separation has been used in some operations.

IV. CHEESE VARIETIES IN WHICH MILK IS CLOTTED BY PROTEASES

A. Cheddar Cheese

Cheddar cheese is named after the village of Cheddar in Somersetshire, England. It has evolved into the most important single variety of cheese in the world. In the United States, Cheddar plus its modifications, Colby, granular Cheddar, washed curd, Jack, and Monterey are referred to as American cheese. The manufacturing techniques for Cheddar cheese have undergone and are still undergoing modifications (Robertson, 1966; Olson, 1971, 1975). Cheddar cheese was first made by a "stirred curd" procedure, described in Section IV, B, but evolved into the typical Cheddar procedure outlined in Table II.

1. Manufacturing Procedures

This variety is made from raw, pasteurized, or partially pasteurized whole milk (Kosikowski, 1977; Price *et al.*, 1971; Wilson and Reinbold, 1965; Wilster, 1969). In the United States, the fat content of milk can be standardized so that the finished product contains a minimum of 50% fat in the dry matter. Lactic acid-producing cultures normally used contain *S. lactis* and *S. cremoris,* the latter species being preferred. The rate of growth and acid production by the culture must be controlled and correlated with rate of whey syneresis and time schedule of manufacturing. This is accomplished in modern plants by using cultures of known activity, accurately measuring the desired amount of culture into cheese milk, and standardizing manufacturing times and temperatures.

Milk is clotted with suitable proteases and it is cut into 0.63-cm cubes with curd knives. Syneresis of whey occurs rapidly after cutting and is accelerated by stirring, heating, and acid development. Maximum cooking temperatures, 40°–41°C, are fixed by effects on metabolism of lactic streptococci. Collins (1977) found that holding strains of *S. cremoris* at 38.9°C, to simulate conditions of cooking Cheddar cheese, arrested growth of most strains but harmed only two out of nine test strains if the holding time was 45 minutes. A third strain was adversely affected with a 90-minute holding period. Growth rates differed between strains at temperature simulating those of Cheddar cheese curd after cooking and after whey drainage. The above differences in temperature responses of strains could affect rate of acid production in later stages of cheese making.

After the curd has reached proper pH and firmness, it is allowed to settle in whey to form a cohesive block at the bottom of the vat. This aids in producing the smooth, closely knit appearance of Cheddar cheese.

TABLE II. Procedure of Manufacturing Cheddar Cheese Containing 37–38% Moisture from 1000 kg of Milk[a]

Steps in making	Time of step	Temperature (°C)	Acidity measurements %	Acidity measurements pH	Comments
Add starter	8:15	31	0.16	6.56	7 kg
Add color	8:45	31	0.16	—	65 ml
Add rennet	9:00	31	0.17	6.50	195 ml
Coagulation	9:12	31	—	—	
Cut curd	9:30	31	0.10	—	0.64-cm knives
Steam on	9:45	31	0.10	—	
Steam off	10:15	39	0.11	6.4	Stir slowly
Drain portion of whey	10:30–10:45	—	—	—	
Drain whey	11:00	39	0.13	6.2	
End dipping	11:30		0.15	6.0	45-cm trench
Pack	11:45		0.17	5.9	Blocks 17-cm wide
Pile 2 high	12:30		0.25	5.7	Cut blocks in half
Pile 3 high	1:00		0.32	5.5	
Mill	1:30		0.45	5.4–5.5	
Salt	1:50		—	—	2.7 kg
Hoop	2:30		—	—	All salt dissolved
Prepress	2:50		—	—	Full pressure in 15 min
Dress	3:20		—	—	
Press	3:30	Press for 5–20 hours at full continuous pressure			

[a] Adapted from Price *et al.* (1971).

Stirring curd after whey removal maintains the granular form, thereby allowing more openings in the finished cheese. After whey drainage, the large mass of curd is cut into smaller blocks which are turned and piled in a process called "cheddaring." The blocks of curd exhibit plastic flow and deform under the influence of increased acidity and physical stress. Individual curd particles in the blocks change from a granular form to contiguous flat platelets, producing a fibrous network of protein which imparts some of the typical elasticity and plasticity to Cheddar cheese. Cheddaring is continued until optimum acidity and firmness levels are reached. Mechanized systems of cheddaring cheese have been developed; one example is illustrated in Fig. 2. After cheddaring, the blocks are cut or "milled" into small pieces with special cutting devices called "curd mills." Subdividing the blocks cools the curd to avoid excessive fat losses during subsequent pressing and facilitates mixing and uniform salting. The salted curd is put into forms (hoops) and pressed into the desired shape by mechanical pressure.

FIGURE 2. Mechanized cheddaring system. Whey is drained and curd is fused and flows to obtain the "cheddared" structure.

2. *Ripening*

Cheese is usually packaged in various types of plastic films which are sealed to act as oxygen and moisture barriers. The cheese is placed in storage at selected temperatures ranging between 2° and 7°C to undergo ripening. This is a series of complicated and interrelated reactions involving metabolism of carbohydrates, protein, and fat. Simple sugars in cheese are fermented by the lactic bacteria to reduce the pH to 5.0–5.2 during the first few days after manufacturing (Brown and Price, 1934). This minimum pH value depends upon the pH attained during manufacturing, the amount of moisture or serum, and consequently the concentration of sugars in curd, buffering capacity of curd, and any inhibition of culture (Breene *et al.*, 1964). These interrelationships are taken into consideration when Cheddar cheese is made with varying moisture contents. A lower pH will have to be attained at milling when "low-moisture" Cheddar is manufactured since subsequent decrease in pH will be less because of low amounts of sugar to be fermented and the slightly higher buffering capacity of cheese. High-moisture cheese will be milled at a higher pH value since this cheese will have a higher concentration of fermentable sugar. Consequently, the drop in pH will be greater during the first few days of ripening.

Initial proteolysis of Cheddar cheese probably results from action of milk-clotting enzyme (Ernstrom and Wong, 1974; Lawrence *et al.*, 1976; McGugan, 1975). The rubbery, elastic structure of fresh curd is converted to a smooth consistency by hydrolysis of casein in the cheese. This change occurs within the first 1–2 months of ripening. Proteolysis also contributes to flavor by releasing amino acids which impart a background flavor and are precursors for other flavor compounds (McGugan, 1975).

The flavor defect, bitterness, results from proteolysis, but there is disagreement as to whether milk-clotting enzymes or certain lactic streptococcal strains are causative agents (Lawrence *et al.*, 1976). It has been demonstrated that cheese made by chemical acidification without lactic bacteria can become bitter, presumably from the action of milk-clotting enzymes (Micketts and Olson, 1974; Perry and McGillivray, 1964). However, conditions of manufacture and composition of this cheese did not duplicate that of Cheddar cheese. A direct role of lactic bacteria in producing bitterness was suggested by use of bacteriophage to restrict growth of *S. cremoris* HP but not acid production near the end of manufacture (Lowrie, 1977). Cheese made under these conditions was very slightly bitter after 6 months of aging, whereas control cheese made with the same strain was definitely bitter.

Conditions of manufacture and cheese composition may also affect development of bitterness. Fox and Walley (1971) found that 5–10% NaCl significantly reduced hydrolysis of one casein fraction, β-casein, by chymosin but had little effect on hydrolysis of α_s-casein. It was suggested that NaCl in cheese might control bitterness since β-casein hydrolysates are bitter whereas α_s-casein hydrolysates are not.

3. Microbiology of Ripening

The relative importance of various groups of bacteria in development of flavor of Cheddar cheese has been investigated but has not been resolved. Reiter (1973) summarized studies conducted in England and Canada which indicated a marked increase in flavor intensity of cheese made with lactic starter culture compared to that made by chemical acidification using glucono-δ-lactone. However, the greatest flavor intensity was obtained when a mixed bacterial flora, isolated from fresh cheese curd, was used with the lactic starter. The importance of nonlactic flora was questioned in other studies. Lawrence *et al.* (1976) pointed out that choice of appropriate strains of lactic bacteria is important in attaining flavor intensity. Results of several studies suggested that added nonstarter bacteria produced off-flavors or had no effect on flavor (Lawrence *et al.*, 1976).

4. Cheddar Flavor

The complex flavor profile of Cheddar cheese has complicated identification of important flavor components and mechanisms of forming them. Inability to find a compound which imparts cheese flavor prompted the "component balance theory" of cheese flavor (Mulder, 1952; Kosikowski and Mocquot, 1958). As the name indicates, flavor depends upon a proper blend of components, none of which completely possesses the unique cheese flavor.

Factors affecting flavor development and compounds contributing to Cheddar cheese flavor were discussed in a review by McGugan (1975). The role of milk fat was thought to be direct as a source of free fatty acids or indirect as a solvent for flavor components. Sulfur-bearing amino acids provide sulfur moieties which are converted to compounds implicated in cheese flavor.

Several compounds appear to be especially important in Cheddar cheese flavor: methanethiol, free fatty acids, acetate, diacetyl and methyl ketones. Many other compounds identified in Cheddar cheese may contribute to flavor. These include aldehydes, H_2S, dimethyl sulfide, lactones, keto acids, phenolics, and pyrazines. However, the importance of various components has been disputed. The proper ratio of free fatty acids to acetic acid was felt to be essential by some workers but other research suggested that free fatty acids did not play an important role. McGugan (1975) suggested that part of the dilemma may result from differences in perception and preferences of typical Cheddar cheese. Diversity between countries and effects of different ages of cheese probably complicate a concise definition of Cheddar cheese flavor.

B. Colby and Stirred-Curd (Granular) Cheeses

Colby cheese is the most popular modification of Cheddar in the United States and is made like Cheddar cheese up to whey drainage (Kosikowski, 1977; Wilson and Reinbold, 1965; Wilster, 1969). Instead of fusing the curd under whey, a portion of the whey is drained and cold water is added to reduce the temperature of curd and whey to approximately 31°C. The curd and whey–water mixture are stirred and drained after 10–20 minutes. The granular curd is salted, hooped, and pressed by the methods used for Cheddar.

The water treatment reduces curd temperature, resulting in more openings in the finished cheese, and leaches out a portion of the acids, salts, and lactose to give a mild flavored cheese with a shorter shelf-life than Cheddar. Lower water temperatures produce higher moisture levels in cheese, the legal maximum being 40%. Water treatment of curd is used also in the manufacture of Monterey and washed-curd cheese. Curd-washing treatments are essential for these high-moisture cheeses to reduce the lactose content and thereby control the pH of the finished cheese (Price and Buyens, 1967).

The granular or stirred-curd modification of Cheddar is made like Colby but the curd is not washed with cold water; cooking temperatures are similar to those used for Cheddar cheese (Price *et al.*, 1971; Wilson and Reinbold, 1965; Wilster, 1969). Its composition is similar to Cheddar but its appearance or openness resembles Colby. The mechanical openings can be eliminated with vacuum treatments (Price and Olson, 1965).

C. Surface-Ripened Cheeses

1. General Description

These are soft and semisoft varieties which are ripened under conditions to induce a progression of microbial growth on the cheese surface (Kosikowski, 1977; Olson, 1969). Examples are brick from the United States, Limburger from Belgium, Port du Salut from France, and Bel Paese from Italy. All use the same basic procedure of coagulation of milk with milk-clotting enzymes, drainage of curd in perforated hoops, brine-salting or application of dry salt to cheese, and initial ripening at 12°–18°C under high relative humidities. Differences in varieties result from the shape of cheese and the amount and, to a slight extent, type of surface growth. Microorganisms on the surface usually are not added as pure cultures but develop under ripening conditions and from carryover contamination in the ripening room.

2. Brick Cheese

Two methods have been used to manufacture brick cheese (Olson, 1969). In one method, *S. cremoris*-type lactic culture and washing of curd with water are unique features (Price and Buyens, 1969). A "whey" culture is used and the curd is not washed in the second method (Olson, 1969). The culture for the latter method is made by heat-treating whey from previous day and incubating it to allow growth of lactic acid-producing bacteria surviving the heat treatment. The "washed-curd" method is the most widely used procedure in the United States.

The basic manufacturing procedure for "washed-curd" brick cheese to the point of whey drainage is similar to that of Cheddar cheese with the exception that smaller amounts of lactic starter are used, ripening or incubation of milk before addition of milk-clotting enzymes is eliminated, very little acid is produced prior to whey drainage, and lower cooking temperatures are generally used. When the desired cooking temperature is reached and the curd has attained the proper firmness, it is allowed to settle but not fuse and the whey is drained to within 2.5 cm of the curd surface. Water at curd temperature and equivalent to 50% of the original milk volume is added and the curd slurry is stirred for 15–20 minutes. The liquid is again drained down to 2.5 cm above the curd level. The curd and whey slurry is pumped over to perforated draining hoops, where the curd fuses into a cohesive mass. The fused cheese mass contains a number of mechanical openings, presumably from the high pH of curd, which reduces its ductility, and because of the rapid drainage of whey before the curd mass completely fuses. The blocks of curd are turned routinely to obtain smooth surfaces and are allowed to drain for approximately 6 hours before brine-salting. The pH of cheese decreases from 6.4 to about 5.6 during this period. Salting is accomplished by immersion of cheese blocks in 23% NaCl brine for 18–24 hours.

The salted cheese is held at 16°C and at a high relative humidity (ca. 95%) to allow microbial growth on its surface. The growth is a progression of yeasts, micrococci, and *Brevibacterium linens* (Langhus *et al.*, 1945; Lubert and Frazier, 1955). Like many food fermentations, the type and extent of growth is not controlled by inoculation but by conditions such as high NaCl concentration on the cheese surface, low pH of 5.0–5.2, reduced water activity of curd, and high relative humidity in the room. Yeasts metabolize lactic acid in the surface layer of curd, raise the pH, and synthesize vitamins to stimulate subsequent growth of the micrococci and *B. linens* (Olson, 1969).

The microbial flora produces flavor components on the cheese surface which diffuse into the cheese. Hydrogen sulfide and methyl mercaptan released from sulfur-bearing amino acids are important flavor components in surface-ripened cheese and presumably are important in brick cheese. Slight lipolysis is necessary for acceptable flavor development in surface-ripened varieties (Olson, 1969).

The extent of ripening to develop surface growth depends upon the desired flavor intensity. A mellow to moderately intense flavor develops in curing periods of 4–10 days. It is necessary to maintain moist cheese surfaces to favor yeast and bacterial growth and limit mold growth. After sufficient surface growth has developed, it is washed off and the cheese is covered with a wax coating or packaged in plastic film. The cheese then is ripened at 2°–8°C for 1–3 months. During this time, the initial elastic structure of cheese is converted by proteolysis to a smooth consistency. This ripening period also allows flavor components to diffuse in from the surface.

3. Other Surface-Ripened Varieties

Varieties such as Limburger, Port du Salut, Bel Paese, and Tilsit are manufactured and ripened in a manner that is essentially the same as surface-ripened brick cheese (Kosikowski, 1977; Olson, 1969). Acidity control of cheese is accomplished by washing curd and limiting acid production before whey drainage. Whey syneresis is accelerated in manufacture of Port du Salut by cutting curd into smaller cubes and using higher cooking temperatures as compared to brick cheese. High-temperature lactic cultures composed of *S. thermophilus* and *Lactobacillus* species, used for some surface-ripened varieties, produce slightly different flavor profiles as compared to *S. cremoris* cultures. *Streptococcus diacetilactis* has been isolated from St. Paulin cheese and presumably is responsible for eye formation and diacetyl which would contribute to flavor.

Higher than normal clotting temperatures (40°–45°C) are partially responsible for the smooth, elastic body of Bel Paese, an Italian soft to semisoft cheese. Lower temperatures during development of surface growth on this variety results in greater dominance of yeast. Temperatures and relative humidities recommended for ripening Port du Salut vary widely. Lower temperatures favor growth of yeasts over *B. linens;* lower humidities allow growth of mold such as *Geotrichium* species. Consequently, flavor characteristics of this and similar varieties can be modified significantly by ripening conditions.

D. Other Semisoft Cheeses

A great number of varieties such as Edam, Gouda, Monterey, Muenster (in the United States), and Danish-type cheeses are included in this class. The type of lactic culture and methods of milk clotting, cutting, and cooking curd are similar to the Cheddar schedule shown in Table II (Kosikowski, 1977; Olson, 1969). Higher moisture contents in these varieties necessitate lower cooking temperatures, minimal acid production prior to whey drainage, and washing of curd with water to control pH of cheese. Washing treatments remove lactose and thereby limit the amount of lactic acid produced after whey drainage. It may not be necessary to wash curd of varieties such as Edam and Gouda if sufficient whey (lactose) is removed from curd by syneresis and if the pH of curd is high enough at whey drainage. With the proper balance of lactose retained and desired pH of curd, the drop of pH will be optimal (see Section IV, A, 2).

Depending upon the variety, curd handling and pressing can be done by procedures similar to those for Cheddar or brick cheese. These processes have been mechanized, especially for those varieties in which individual pieces of cheese are small (Olson, 1975). The size and shape of cheese varies widely between and within each variety (Olson, 1969).

Most varieties in this class are salted by immersion in NaCl brine. Cheese is waxed, covered with a "plastic" solution, or wrapped in heat-shrinkable plastic film with application of a vacuum treatment. Ripening of cheese is done at 2°–10°C for a few months to develop desired flavor intensity. The "component balance theory" can be applied to flavor development in most of these varieties. Metabolism of *S. diacetilactis* may produce "eyes" in Edam and Gouda cheeses and some Danish varieties.

E. Swiss Cheese

1. Manufacturing Methods

The basic steps in manufacturing (milk clotting, cutting curd, cooking curd and whey, and separation of curd and whey) are similar to those for Cheddar cheese. A cylindrical, copper kettle was the traditional vat but the horizontal or vertical tower-vats are used in many operations (Reinbold, 1972). The fat content of milk is usually standardized to a slightly lower level as compared to that for Cheddar cheese.

High-temperature lactic cultures, *S. thermophilus* and *Lactobacillus* species, are necessary for Swiss cheese because of the high cooking temperatures (50°–53°C) attained during manufacturing. Rennet extract was the traditional milk-clotting enzyme but a protease preparation from *E. parasitica* was found to be very suitable for manufacture of Swiss cheese because of its low heat stability (Ernstrom and Wong, 1974). High cooking temperatures and high pH values of curd during cooking result in very low residual enzyme levels in cheese. This minimizes proteolytic attack on the cheese structure which is beneficial for good "eye" formation.

The curd formed by milk-clotting enzymes is cut into small particles with wire knives (harps) by hand or more recently by mechanically operated knives (Reinbold, 1972). The small curd particles result in rapid whey syneresis which compensates for decreased whey expulsion because of low acid development during initial stages of manufacture. The small curd particles also form a tightly knit structure in the finished cheese which produces more uniform "eye" formation.

Cooking temperatures of 50°–53°C are commonly used to attain the proper moisture level, firmness, and plasticity of curd. In the traditional method, the warm curd was dipped from the kettle with a cloth and transferred to a wooden circular hoop if the curd was to be pressed into a traditional Swiss wheel (Kosikowski, 1977). This form has been replaced with a rectangular hoop to yield a rindless block style (Reinbold, 1972). More highly mechanized manufacturing systems consist of the horizontal or tower-vats in which curd is formed, cut, and cooked. After the curd has attained optimum firmness, it flows by gravity or is pumped with some whey to special draining forms. The curd settles and fuses, under applied pressure, into dense blocks while the whey drains from the perforated forms. The method of pressing and condition of curd are critical for proper eye formation in the finished cheese.

The pressed blocks or wheels are salted by immersion in a NaCl brine and the surfaces of blocks are dried before packaging in plastic film. The packaged cheese is ripened through the series of stages described in Section IV, D, 3.

2. Lactic Bacteria

Lactic acid bacteria in Swiss cheese serve the typical functions of production of lactic acid, control of undesirable fermentations, and contributing to flavor development. Presence of both *S. thermophilus* and high-temperature lactobacilli appears to be essential to obtain a satisfactory cheese (Langsrud and Reinbold, 1973a). Only one-half as much acid was produced in cheese made with *S. thermophilus*, whereas acidity levels were higher in cheese made with *L. helveticus,* but the maximum level was reached fastest in cheese containing both species. The complexity of symbiosis between high-temperature lactic bacteria necessitates careful selection of compatible strains to obtain predictable growth rates and acid production during cheese manufacturing.

The two genera of lactic bacteria exhibit different growth patterns during cheese manufacturing. Limited growth of *S. thermophilus* occurs during initial stages of manufacture up to the stage when the cheese is being pressed (Frazier *et al.*, 1934, 1935). Rapid growth of this species occurs during the first 4–8 hours after whey drainage and then numbers decline slowly. Numbers of lactobacilli and propionibacteria may decline during initial stages of manufacturing but begin to grow midway through pressing. The high metabolic activity, with concomitant acid production by lactic bacteria and high temperature of curd, suppresses growth of undesirable microorganisms during pressing.

Limited production of acid occurs during the early stages of manufacture; the pH of curd at the time of draining whey (dipping) is 6.3–6.5. The pH of curd should not be below 5.3 at the end of pressing since excessive acidity inhibits growth of propionibacteria and eye formation (Langsrud and Reinbold, 1973a). Also, cheese structure may lack ductility resulting in abnormal eye formation or cracks in cheese. High pH values can cause soft cheese structure, excessive eye formation, and possibly abnormal fermentations during ripening.

3. Ripening

Ripening of Swiss occurs in three phases. After salting and packaging, the cheese may be stored at 7°–13°C for up to 10 days to allow some diffusion of NaCl into the cheese block. The blocks of cheese are held at 21°–25°C for 2–7 weeks to induce growth and gas (carbon dioxide) production by *Propionibacterium* species to form the characteristic eyes (Reinbold, 1972). Since these bacteria are dispersed throughout the curd, gas must diffuse through the curd matrix and collect at points of weakness. Sufficient gas pressure develops to stretch the protein structure surrounding weak focal points and form smooth, round eyes. Thus the number, size, and uniformity of eyes depends on rate of gas production, presence of contaminating gas-producing bacteria, number of weak points in curd, and plasticity of curd (Langler and Reinbold, 1974; Reinbold, 1972). Lactic bacteria continue to function during ripening in symbiosis with other bacteria. They ferment most of the simple sugars in curd to lactic acid via the homofermentative Embden–Meyerhof–Parnas (EMP) pathway but use a heterofermentative pathway to a slight extent since small amounts of acetic acid and CO_2 are produced (Langsrud and Reinbold, 1973b). Growth of propionibacteria is stimulated by CO_2 and by the reduced oxidation–reduction potential resulting from metabolism by the lactic starter.

Numbers of lactic bacteria decrease during the "warm-room treatment" phases of ripening presumably because of lack of fermentable sugars as energy sources (Langsrud and Reinbold, 1973b). Numbers of propionibacteria increase 3- to 980-fold over levels in curd at pressing. The population attained is correlated directly with pH values of cheese. Although propionic and acetic acids may be produced by metabolism of propionibacteria on several substrates, lactate is the principal substrate in cheese (Hettinga and Reinbold, 1972).

The component balance theory of cheese flavor is considered applicable to Swiss cheese (Langsrud and Reinbold, 1973b). However, high-quality cheese contains relatively large amounts of acetic and propionic acids, very little or no butyric acid, and small amounts of higher fatty acids. The level of acetaldehyde in Swiss cheese exceeds its flavor threshold and is thought to be an important flavor component. Amino acids capable of imparting a sweet taste are considered important for this flavor note in Swiss cheese.

F. Italian-Type Cheeses

Commercially important varieties of Italian-type cheeses can be classified as hard (grating) and "pasta filata" varieties. Parmesan and Romano cheese are examples of grating cheeses. "Pasta filata" means filamentous curd; Provolone and Mozzarella are examples of this type.

1. Parmesan Cheese

Parmesan is typical of several very hard Italian cheese varieties which are commonly grated for use in foods. The fat content of milk for this variety usually is reduced since U.S. Standards of Identity require only 32% of the cheese solids must be milk fat. Manufacturing methods are similar to Swiss up to the point of whey drainage and the same species of lactic bacteria are used but no propionibacteria are added (Kosikowski, 1977; Reinbold, 1963). The ratio of *Lactobacillus* species to *S. thermophilus* in the lactic starter is about 1 : 1, which is higher than that used in starters for Swiss cheese. Very small amounts (about 5 gm/454 kg milk) of lipases (pregastric esterases) may be added to milk to produce the desired flavor intensity in the ripened cheese as described in Section IV, F, 5. Calf-derived enzymes are usually used for Parmesan cheese since the desired flavor is more subtle.

Deviation from procedures for Swiss occurs at whey drainage since the curd is not fused under whey into a dense block. It is either stirred to prevent fusion of curd particles or may be matted into a porous mass. After the whey has drained, the curd is placed in cylindrical forms and pressed. The wheels of cheese are brine-salted for approximately 14 days depending upon the size of cheese. The long salting time is required because of the low moisture content of the cheese which retards salt diffusion. The cheese is allowed to dry slowly during initial stages of ripening so that the final moisture content is 32% or less. The minimum curing time for this variety in the United States is 10 months; curing for 18–24 months produces a more intense sweet, nutty flavor.

2. Romano Cheese

Romano cheese is made in essentially the same way as Parmesan cheese but has higher moisture and fat contents (Kosikowski, 1977; Reinbold, 1963). Lipases derived from kid goats or mixtures of kid and lamb enzymes are commonly used to produce fairly intense flavors (see Section IV, F, 5). Enzymes are added in greater amounts to cheese milk as compared to Parmesan, resulting in higher levels of free fatty acids in cheese. After whey drainage, the curd is pressed into cylindrical forms, brine-salted for 1–2 days, and salted for several weeks by application of salt to the cheese surface. This variety usually is not ripened as long as Parmesan and can be used as a table cheese after 5–8 months or for grating purposes with longer ripening.

3. Other Hard Italian Cheeses

There are many other Italian-type cheeses that are made by a process similar to Parmesan but differ in composition, flavor, regional name, and the breed of animal used for milk source. Asiago may be less firm, may possess a more delicate nutty flavor, and is used as table cheese. Inclusion of the term Pecorino in a varietal name indicates that it was made from goat's milk. The terms Parmigiano, Sardo, and Reggiano indicate regions of Italy in which these particular types are made or originated. Fontina is a "table cheese" which can be cut or sliced but is used as a grating cheese when aged. It has eyes similar to Gouda cheese and may have a surface flora of *B. linens* (Kosikowski, 1977).

4. Provolone Cheese

This variety is one of several pasta filata types of cheese. The name "pasta filata" can be translated as filamentous dough or curd, which describes the structure of the cheese produced by stretching the curd during manufacture. Manufacturing procedures for Provolone are similar to Cheddar up to the milling stage except for use of high-temperature lactic cultures and higher cooking temperatures (Kosikowski, 1977; Reinbold, 1963). The ratio of streptococci to lactobacilli is about 1:1; lactobacilli are important to attain the low pH at later stages of manufacturing.

When the "cheddared" curd reaches a pH of 5.2–5.3, it is warmed to 50°–55°C. This usually is done in a mechanical system using hot water or steam as the heating medium. The warm curd is mixed mechanically until its texture resembles taffy candy. The filamentous curd is then manually or mechanically molded into its final shape (Kosikowski, 1977). Smaller, cylindrical pieces are made by extruding warm curd into forms; larger pieces are formed manually. The final shape is used as a basis of naming many pasta filata cheeses, including the cylindrical salami, pear-shaped Provolone, large round Mandarini Provolone, and small Provolone styles, Provoletti and Bocchini.

Provolone styles are salted by immersion in salt brine; the duration of salting varies indirectly with the surface to volume ratio of cheese pieces. A smoke flavor can be imparted to cheese by exposure to smoke or by applying a liquid concentrate of wood smoke.

5. *Flavor Development*

Lipolysis, which is important in flavor development in Italian-type cheeses, results to a great extent from action of lipolytic enzymes in rennet paste or by pregastric esterases (Neelakantan *et al.*, 1971; Nelson *et al.*, 1977). Pregastric esterase preparations are liquid extracts of selected oral tissue of young calves, kid goats, or lambs. Rennet pastes are prepared from the abomasum of the above species which are slaughtered immediately after suckling. The engorged stomachs are processed into a product comprised of stomach tissue and milk-curd contents. Lipolytic activity of the paste arises from lipases (pregastric esterases) which were released into milk by the oral tissues. Esterases from *Mucor miehei* were used experimentally for Fontina and Romano cheeses (Peppler *et al.*, 1976). Fungal esterases produced equivalent flavor intensities as pregastric esterases in Fontina cheese when both preparations were standardized to equivalent lipolytic activities. Five times more fungal esterases as compared to pregastric esterases were required to attain desired flavor intensities in Romano cheese.

6. *Mozzarella Cheese*

This variety is a soft, usually unripened, pasta filata-type cheese (Kosikowski, 1977; Reinbold, 1963). It is manufactured by the same basic procedures and with the same type of lactic cultures but cooking temperatures may be lower since moisture content is generally higher than in Provolone cheese. Standards of Identity in the United States specify four types of Mozzarella cheese which differ in moisture and fat contents: Mozzarella, part-skim Mozzarella, low-moisture Mozzarella, and part-skim, low-moisture Mozzarella (U.S. Food and Drug Administration, 1974).

Mechanical systems for stretching and molding curd are universally used (Olson, 1975). Size of the finished cheese varies but a 2.3-kg rectangular loaf is dominant. Brine-salting is accomplished by floating cheese loaves in NaCl brine or by immersing mechanized trolleys holding cheese loaves. Cheese is packaged in heat-shrinkable bags or in pouches; air is removed by vacuum prior to sealing. Storage of cheese is sufficient to obtain adequate proteolysis of the protein structure to attain desired physical characteristics; this may require 7–30 days. Most Mozzarella is used as a food ingredient which makes sliceability, meltability, cohesiveness, and optimum firmness important physical characteristics.

G. Mold-Ripened Cheeses

1. Blue-Veined Cheeses

These types of cheese have been characterized as being unique because an almost pure culture of *Penicillium roqueforti* grows within the cheese (Morris, 1969). Growth of mold in cheese is facilitated by control of curd structure, pH, NaCl concentration, and the O_2–CO_2 interchange in cheese. It is critical to handle curd properly during manufacturing to obtain an open texture within the cheese so that mold can grow throughout the openings and produce the "blue veining" which is typical of these varieties.

Several varieties of cheese fall within the blue-veined class, of which blue (Bleu), Roquefort, Gorgonzola, and Stilton are examples. Manufacture of blue cheese will be described in this section; principles of manufacturing for this variety apply to other blue-veined cheeses.

a. Milk Treatment and Clotting. Milk is prepared by blending pasteurized or heat-treated skim milk with homogenized cream containing about 12% fat. Homogenization pressures of 35 kg/cm² for the first stage and 70 kg/cm² for the second have been suggested (Kosikowski, 1977). The cream may be bleached with benzoyl peroxide, or chlorophyll may be added to produce the whiteness desired for this variety.

Lactic culture (*S. cremoris* and *S. lactis*) is added and the milk is ripened at 30°–31°C for up to 1 hour. *Penicillium roqueforti* mold powder is inoculated into the milk to obtain a uniform distribution in curd. The powder may be sprinkled on curd at hooping but this is less convenient.

Milk-clotting enzyme is added at the rate of 165–200 ml per 1000 kg of milk. The firmness of curd at cutting is equal to or greater than that of Cheddar cheese. This aids in maintaining the cube shape of the curd and increases openness of the finished cheese. Cutting curd in large 1.5-cm (⅝-inch) cubes also increases openness.

b. Curd Handling. Stirring of curd and whey commences after sufficient whey is expelled to act as a lubricant between cubes of curd. Care must be exercised in stirring to maintain the large, cube shape which is important in obtaining open texture in cheese. The temperature of curd and whey may be raised slightly to adjust the moisture content of the finished cheese.

When the desired whey expulsion and curd firmness is attained, the curd is settled and a portion of the whey is drained off. The curd and remaining whey are transferred manually to draining forms (hoops) with perforated scoops or by special curd pumps; a higher proportion of curd to whey is required for pumping. Distortion (flattening) and breakage of curd during whey drainage and curd transfer must be avoided to obtain open texture in the finished cheese. The curd settles and fuses while whey drains through the cylindrical, perforated hoops. Turning hoops during initial stages of draining produces smooth outside surfaces. Drainage in hoops continues overnight.

The wheels of cheese are salted by daily application of crystalline NaCl over a 5-day period or by immersing wheels in a 23% NaCl brine for 24–48 hours and then three daily applications of crystalline salt. The concentration of NaCl in the final cheese is about 4% or 8–9% in the serum phase. The salt content is much higher initially on the cheese surface which controls the extent and type of microbial growth.

One day after salting, the wheels of cheese are pierced with a gang of needles. This allows subsequent escape of CO_2 and entrance of O_2 which is essential for mold growth and metabolism. The cheese is stored at 10°–13°C for 2–3 months to induce mold growth; subsequent ripening is carried out at lower temperatures of 2°–5°C.

Lactic fermentation continues during draining so that the minimum pH of 4.5–4.7 is reached 24 hours after manufacturing (Coulter *et al.*, 1938; Foster *et al.*, 1957). Metabolism of *P. roqueforti* and other microorganisms causes the pH of cheese to increase to 6.0–6.3 during ripening.

c. Ripening. Proteolysis and lipolysis of blue cheese results, to a great extent, from enzymes elaborated by *P. roqueforti* (Ernstrom and Wong, 1974). The water-soluble lipase system from this species hydrolyzes milk fat to yield fatty acids; caproic, caprylic, and capric acids and their salts impart a sharp, peppery flavor to blue cheese (Currie, 1914). Methyl ketones, formed by oxidation of fatty acids to β-keto acids and subsequent decarboxylation, are thought to be the principal components contributing to aroma of blue cheese (Girolami and Knight, 1955). A mixture of acetone, pentanone-2, heptanone-2, and nonanone-2 possessed an aroma similar to blue cheese (Patton, 1950). Heptanone-2 is the predominant ketone in this variety and Roquefort and Camembert cheeses (Anderson and Day, 1965; Schwartz and Parks, 1963; Schwartz *et al.*, 1963).

2. Cheese with Surface Mold

A wide variety of cheeses especially of French origin are made with surface mold growth. Typical varieties are Camembert, Coulommiers, and Brie.

a. Camembert. Initial stages of manufacture of this variety are technologically uncomplicated but strict sanitation is essential (Kosikowski, 1975;; Mocquot, 1955). Pasteurized, whole milk is inoculated with lactic bacteria (*S. cremoris* type) and *Penicillium camemberti* or *Penicillium caseicolum*. Curd is formed by milk-clotting enzymes and may be cut, when fairly firm, into large cubes. The temperature of curd and whey may be raised slightly to obtain desired moisture levels in cheese. Approximately 1.5–2 hours after the lactic culture is added, the curd and a portion of whey are dipped into draining forms. Special vats are used in some operations which allow dispensing of curd mass into draining forms. The diameter of forms is the same as that of the finished cheese so each form of curd yields one wheel of cheese.

After drainage in hoops, the wheels are salted by application of dry salt or immersion in a salt brine. A suspension of mold spores may be sprayed on the wheels prior to salting as an alternative or supplement to inoculation of milk. The cheese is placed on wire racks to allow maximum exposure of surface to facilitate mold growth. Ripening may be a sequence of storage at 14°C and relative humidity (RH) of 75% for 1–2 days to dry the surface, followed by storage at 14°C at an RH of 95% for 12 days. A cottony, white mat of mold develops on the cheese; the cheese is turned on the racks to obtain uniform growth. Initial storage at low RH is eliminated in some operations.

The mold-covered wheels of cheese are wrapped in foil and ripened at about 5°C. Ripening is rapid; the cheese is ready for consumption within 2–5 weeks depending upon consumer preference. Physical characteristics of cheese change dramatically during ripening. Initially, the cheese is brittle and has a white, grainy appearance which is transformed into a smooth consistency and translucent appearance. The change occurs from the surface inward, presumably from action of mold proteases.

b. Coulommiers and Brie. Coulommiers is very similar to Camembert in methods of manufacturing and microbial population but is formed into larger wheels. Brie is also made in larger wheels and is a soft cheese but *B. linens* grows on the cheese surface with *P. camemberti*. This bacterial species, which also is important in ripening of Limburger cheese, produces distinctive flavor notes.

V. PROCESS CHEESE

This type of cheese is made by melting and mixing desired blends of cheese, with the aid of heat, to give a smooth homogeneous product (Kosikowski, 1977). There are three general types of processed cheese defined under the Food, Drug and Cosmetic Act: pasteurized process cheese, pasteurized process cheese food, and pasteurized process cheese spread (U.S. Food and Drug Administration, 1974). These differ in moisture and fat contents and in number and kinds of food ingredients allowed.

A. Manufacture

Process cheese can be made from most varieties of cheese, but Cheddar is used most commonly. A blend of different lots of cheese is selected to obtain desired composition and physical and flavor characteristics. The cheese is comminuted and heated in a steam-jacketed kettle or by infusion of culinary steam directly into the cheese mass. Two types of cookers have been used, namely, jacketed kettles with paddle agitators or horizontal, cylindrical cookers equipped with screw mixers. Processing is done on a batch basis through the cooking operation but a sufficient number of cookers are used to supply a continuous flow of cheese to subsequent packaging operations.

Emulsifying salts are mixed with cheese at cooking to increase smoothness and aid in dispersion of fat in the finished cheese. Citrates and phosphates of varying chain length are used depending upon the characteristics desired in process cheese. Softer cheese is produced when disodium phosphate is used rather than sodium citrate or certain polyphosphates.

B. Packaging and Mechanization

Process cheese, cheese food, and firmer cheese spreads are commonly sold in 0.5–2.3 kg loaves, stacks of slices, or individually wrapped slices. Loaves or large stacks of slices are sold in institutional markets. Slices are formed by extruding molten cheese on a cooled belt and allowing the cheese to solidify before the slices are mechanically cut and stacked (Kosikowski, 1977). Softer process cheese spreads are packaged in jars, plastic tubes, and aerosol cans. The moisture level in such products must be sufficiently high for easy removal from the package.

C. Cold-Pack Cheese

In addition to processed cheese, natural cheese is converted to cold-pack cheese and cold-pack cheese food by grinding or comminuting without the application of heat (U.S. Food and Drug Administration, 1974). Equipment with knives revolving at high speeds, adapted from the sausage industry, reduce the structure of natural cheese to a smooth spreadable paste. Additives such as dehydrated milk products, acidifying and sweetening agents, and gums can be added to modify the flavor and rheological characteristics of cold-pack cheese food.

REFERENCES

Anderson, D. F., and Day, E. A. (1965). *J. Dairy Sci.* **48,** 248–249.

Bradley, R. L. (1969). "Milk," Encycl. Sci. Technol., pp. 499–510. McGraw-Hill, New York.

Breene, W. M., Price, W. V., and Ernstrom, C. A. (1964). *J. Dairy Sci.* **47,** 840–848.

Brown, L. W., and Price, W. V. (1934). *J. Dairy Sci.* **17,** 33–45.

Collins, E. B. (1952). *J. Dairy Sci.* **35,** 381–387.

Collins, E. B. (1961). *J. Dairy Sci.* **44,** 1989–1996.

Collins, E. B. (1977). *J. Dairy Sci.* **60,** 799–804.

Coulter, S. T., Combs, W. B., and George, S. T. (1938). *J. Dairy Sci.* **21,** 239–245.

Cousin, M. A., and Marth, E. H. (1977a). *Cult. Dairy Prod. J.* **12,** 15–18 and 30.

Cousin, M. A., and Marth, E. H. (1977b). *J. Dairy Sci.* **60,** 1048–1056.

Currie, J. N. (1914). *J. Agric. Res.* **2,** 429–434.

Czulak, J., Conochie, J., Sutherland, B. J., and Van Leeuwen, H. J. M. (1969). *J. Dairy Res.* **36,** 93–101.

Emmons, D. B., and Tuckey, S. L. (1967). "Cottage Cheese and Other Cultured Milk Products." Chas. Pfizer & Co., Inc., New York.

Ernstrom, C. A., and Kale, C. G. (1975). *J. Dairy Sci.* **58,** 1008–1014.

Ernstrom, C. A., and Wong, N. P. (1974). *In* "Fundamentals of Dairy Chemistry" (B. H. Webb and A. H. Johnson, eds.), 2nd ed., pp. 662–769. Avi Publ. Co., Westport, Connecticut.

Foster, E. M., Nelson, F. E., Speck, M. L., Doetsch, R. N., and Olson, J. C., Jr. (1957). "Dairy Microbiology." Prentice-Hall, Englewood Cliffs, New Jersey.

Fox, P. F., and Walley, B. F. (1971). *J. Dairy Res.* **38,** 165–170.

Frazier, W. C., Sanders, G. P., Boyer, A. J., and Long, H. F. (1934). *J. Bacteriol.* **27,** 539–549.

Frazier, W. C., Burkey, L. A., Boyer, A. J., Sanders, G. P., and Matheson, K. J. (1935). *J. Dairy Sci.* **18,** 373–387.

Gilliland, S. E. (1977). *J. Dairy Sci.* **60,** 805–809.

Gilliland, S. E., and Speck, M. L. (1974). *J. Milk Food Technol.* **37,** 107–111.

Girolami, R. L., and Knight, S. G. (1955). *Appl. Microbiol.* **3,** 264–267.

Green, M. L. (1977). *J. Dairy Res.* **44,** 159–188.

Harper, W. J., and Hall, C. W. (1976). "Dairy Technology and Engineering." Avi Publ. Co., Westport, Connecticut.

Hettinga, D. H., and Reinbold, G. W. (1972). *J. Milk Food Technol.* **35,** 358–372.

Holmes, D. G., Duersch, J. W., and Ernstrom, C. A. (1977). *J. Dairy Sci.* **60,** 862–869.

Jenness, R., and Patton, S. (1959). "Principles of Dairy Chemistry." Wiley, New York.

Kosikowski, F. V. (1975). *J. Dairy Sci.* **58,** 994–1000.

Kosikowski, F. V. (1977). "Cheese and Fermented Milk Foods," 2nd ed. Edwards, Ann Arbor, Michigan.

Kosikowski, F. V., and Mocquot, G. (1958). *FAO Agric. Stud.* **38,** 141–144.

Langhus, W. L., Price, W. V., Sommer, H. H., and Frazier, W. C. (1945). *J. Dairy Sci.* **28,** 827–838.

Langler, T., and Reinbold, G. W. (1974). *J. Milk Food Technol.* **37,** 26–41.

Langsrud, T., and Reinbold, G. W. (1973a). *J. Milk Food Technol.* **11,** 531–542.

Langsrud, T., and Reinbold, G. W. (1973b). *J. Milk Food Technol.* **36,** 593–609.

Lawrence, A. J. (1959). *Aust. J. Dairy Technol.* **14,** 169–172.

Lawrence, R. C., Thomas, T. D., and Terzaghi, B. E. (1976). *J. Dairy Res.* **43,** 141–193.

Lowrie, R. J. (1977). *J. Dairy Sci.* **60,** 810–814.

Lubert, D. J., and Frazier, W. C. (1955). *J. Dairy Sci.* **38,** 981–990.

Lundstedt, E., and Fogg, W. B. (1962). *J. Dairy Sci.* **45,** 149–150 (abstr.).

McGugan, W. A. (1975). *J. Agric. Food Chem.* **23,** 1047–1050.

Mather, D. W., and Babel, F. J. (1959). *J. Dairy Sci.* **42,** 1045–1056.

Micketts, R., and Olson, N. F. (1974). *J. Dairy Sci.* **57,** 273–279.

Mocquot, G. (1955). *J. Soc. Dairy Technol.* **8,** 17–24.

Mocquot, G. (1971). *Proc. Int. Symp. Conversion Manuf. Foodstuffs Microor., 1971.* pp. 191–197. SaiRon Publ. Co., Ltd., Tokyo, Japan.

Morris, H. A. (1969). Paper presented at *6th Annu. Marschall Invitational Ital. Cheese Semin.* April 30, 1969, Madison, Wisconsin.

Mulder, H. (1952). *Ned. Melk- Zuiveltijdschr.* **6,** 157–168.

National Research Council (1958). N.A.S.—N.R.C., *Publ.* **645,** 1–48.

Neelakantan, S., Shahani, K. M., and Arnold, R. G. (1971). *Food Prod. Dev.* **5,** 52–58.

Nelson, J. H., Jensen, R. G., and Pitas, R. E. (1977). *J. Dairy Sci.* **60,** 327–362.

Nelson, P. J., and Marshall, R. T. (1977). *Proc. 72nd Annu. Meet. Am. Dairy Sci. Assoc.* p. 35.

O'Keefe, R. B. *et al.* (1976). *J. Dairy Res.* **43,** 97–107.

O'Leary, J. O., Hicks, C. L., and Bucy, J. (1977). *Proc. 72nd Annu. Meet. Am. Dairy Sci. Assoc.* p. 55.

Olson, N. F. (1969). "Ripened Semisoft Cheeses." Chas. Pfizer & Co., Inc., New York.

Olson, N. F. (1971). *Int. Dairy Congr.* [Proc.], *Congr. Rep., 18th, 1970* Vol. 2E; pp. 334–347.

Olson, N. F. (1975). *J. Dairy Sci.* **58,** 1015–1021.

Olson, N. F. (1977). *Dairy Ind. Int.* **42,** 14–19.

Patel, M. C., Lund, D. B., and Olson, N. F. (1972). *J. Dairy Sci.* **55,** 913–918.

Patton, S. (1950). *J. Dairy Sci.* **33,** 680–684.

Peppler, H. J., Dooley, J. G., and Huang, H. T. (1976). *J. Dairy Sci.* **59,** 859–862.

Perry, K. D., and McGillivray, W. A. (1964). *J. Dairy Res.* **31,** 155–165.

Peters, I. I. (1964). *Dairy Sci. Abstr.* **26,** 457–461.

Price, W. V., and Buyens, H. J. (1967). *J. Dairy Sci.* **50,** 12–19.

Price, W. V., and Olson, N. F. (1965). *Wis., Agric. Exp. Stn., Spec. Bull.* **10,** 1–14.

Price, W. V., Calbert, H. E., and Olson, N. F. (1971). *Wis., Agric. Exp. Stn., Res. Bull.* **464.**

Quarne, E. L., Larson, W. A., and Olson, N. F. (1968). *J. Dairy Sci.* **51,** 527–530.

Ramet, J. P., Alais, C., and Weber, F. (1969). *Lait* **49,** 40–52.

Reinbold, G. W. (1963). "Italian Cheese Varieties." Chas. Pfizer & Co., Inc., New York.

Reinbold, G. W. (1972). "Swiss Cheese Varieties." Chas. Pfizer & Co., Inc., New York.

Reinbold, G. W. (1973). *Milk Ind.* **73,** 13–16.

Reiter, B. (1973). *J. Dairy Res.* **26,** 3–15.

Richardson, G. H., Gandhi, N. R., Divatia, M. A., and Ernstrom, C. A. (1971). *J. Dairy Sci.* **54,** 182–186.

Robertson, P. S. (1966). *J. Dairy Res.* **33,** 343–369.

Sandine, W. E. (1977). *J. Dairy Sci.* **60,** 822–828.

Schwartz, D. P., and Parks, O. W. (1963). *J. Dairy Sci.* **46,** 989–990.

Schwartz, D. P., Parks, O. W., and Boyd, E. N. (1963). *J. Dairy Sci.* **46,** 1422–1423.

United States Food and Drug Administration (1974). "Cheese, Processed Cheese, Cheese Foods and Related Foods," Federal Food, Drug and Cosmetic Act, Part 19, Title 21. Code of Fed. Regul., U.S. Dept. of Health, Education and Welfare, Washington, D.C.

Van Slyke, L. L., and Price, W. V. (1952). "Cheese." Orange Judd Publ. Co., Inc., New York.
Walter, H. E., and Hargrove, R. C. (1972). "Cheeses of the World." Dover, New York.
Wilson, H. L., and Reinbold, G. W. (1965). "American Cheese Varieties." Chas. Pfizer & Co., Inc., New York.
Wilster, G. H. (1969). "Practical Cheesemaking." 11th ed. Oregon State Univ. Bookstores, Inc., Corvallis.

Chapter 3

Distilled Beverages

W. F. MAISCH
M. SOBOLOV
A. J. PETRICOLA

I. INTRODUCTION

"Bourbon Whisky". . . is whisky produced at not exceeding 160° proof from a fermented mash of not less than 51 percent corn . . . and stored at not more than 125° proof in charred new oak containers. [Code of Federal Regulations, Title 27, Sub Part C 5.22(b) Class 2(1)(i).]

The above citation is representative of the large body of regulations which both protect and control the beverage alcohol producers of the world. It is important to understand this point when studying the technical data of this old and distinguished industry, for virtually all of the

MICROBIAL TECHNOLOGY, 2nd ed., VOL.II
 ISBN 0-12-551502-2

distilled products consumed in the United States are manufactured according to legally defined production methods.

The history of the individual production region is also an important consideration. Scotch whisky is unique in taste and aroma because those distillers began with the malt, the peat, and the water that were uniquely available in Scotland. Law now protects that uniqueness, while technology has been assigned the role of controlling the quality, assuring the reproducibility, and collecting the taxes. These pursuits ultimately have produced a body of observations which has explained the processes and has recently permitted meaningful improvements.

One must be equally cautious when interpreting economic data of the distilling industry. New patterns are developing in consumer purchasing of beverage alcohol. This can produce real or imaginary surpluses or shortages among the various categories and skew raw material usage figures.

However, over the past 20 years, grain and grain products usage at all United States' distilled spirits plants has been roughly 40,000,000 bu (56 lb/bu) per year. Corn represents about 68% of the total; milo, 13%; and rye and malt approximately 8% each (Annual Statistical Review, 1973).

Worldwide, the total output of fermentation-based ethanol in 1976 is estimated at 85 million gal (Anonymous, 1977). The United States' total of public taxes and fees generated by beverage alcohol in 1976 was in excess of $10 billion (DISCUS Facts Book, 1976).

II. PROCESSING

The production of beverage alcohol involves the interrelated operations of mashing, saccharifying, fermenting, and distilling. Each step contributes significantly to the success of the next.

A. Mashing

The typical process begins with cleaned, sound grains (corn), which are milled to a predetermined particle size to increase the surface area for contact with water during cooking. Raw starch is quite resistant to the enzymatic action of most amylases. The mashing step causes hydration and swelling of the starch granules in the grain, resulting in a gel. Coarser grinds are more difficult to gelatinize, as are older, drier grains. The interrelatedness of the steps begins to be seen here. Too fine a grind can affect subsequent feed recovery, while too coarse a grind necessitates higher temperature in order to gelatinize the remaining 2–5% of the starch.

Liquefying amylases are necessary to reduce the gel strength in the cookers and prepare the starch for subsequent saccharification. For this, the enzyme (premalt) is added with the grain. After cooking, the mash is cooled and additional enzyme is added to the mash to achieve further liquefaction, ultimately cleaving 15–20% of the starch linkage in the case of malt enzymes. Approximately 95% of the starch linkages are α1—4; the remaining 5% represent 1—6 linkage.

Mashing procedures are varied to meet conditions. For example, cooking times and temperatures can both be reduced if the saccharifying system is optimized. The source of the starch (type of grain) and the product type are also important variables.

The batch process for bourbon whiskey calls for separate cooking and conversion (saccharifying) of the corn and rye mash. For flavor considerations, this is generally low-temperature (atmospheric) cooking. Each batch of each grain receives its own malt enzyme slurry, sufficient for conversion. Nevertheless, the rye starch is not totally saccharified. Thus, flavor considerations can influence yields.

In the manufacturing of grain neutral spirits, one seeks to optimize yields and efficiency. Pressure cooking is usually chosen for this process because gelatinization is improved and more grain can be added to the mash. By coupling "continuous" cooking with high temperature, one can develop a shorter heating cycle than that of the batch process. A shorter conversion stage is also possible, for one needs only that drop in viscosity which facilitates the pumping of the mash. Saccharification must be continued in the fermentor in order to obtain optimum yields.

Scotch malt whisky is made using a wort rather than a whole-grain process. Here, the grains are cooked, converted, screened, and pressed to produce a liquid filtrate for fermentation.

Canadian whisky distillers employ generally the same practices as their counterparts in the United States.

B. Saccharifying

1. Malt Enzymes

The experience of the distilling industry has clearly established the correlation between α-amylase in malt and alcohol yield from grain. Thus, in addition to its initial thinning of the starch viscosity, it is the primary saccharifying enzyme in malt. β-Amylase is rapidly destroyed at temperatures used for most distillery saccharification. The major products of the malt amylases on starch are maltose, isomaltose, and isomaltotriose. Fermentation filtrates often contain appreciable residues of di-, tri-, and oligosaccharides, thereby suggesting there is still more to be learned.

Gibberellin treatment of malt has increased the α-amylase content threefold over that in the original malt varieties. Activity in excess of 100 dextrinizing units (DU)/gm (dry basis) is now routine.

The bacterial load of malts is also very important to the distiller. Bacterial counts of 1×10^6/gm can be expected. Depending upon the heating/converting conditions selected, significant numbers of these organisms can survive and enter into the fermentation. Lactic acid organisms can compete there with the yeast, resulting in both flavor defects and yield reductions. This is especially serious in continuous conversion with shorter cooking times.

The pH of the mash is important at this stage. First, the optimum pH value for α-amylase activity is approximately 5.0, with the minimum near 4.2. Lower values will stop α-amylase conversion in the fermentor. This underlines the importance of controlling the lactic acid organism population.

2. Microbial Enzymes

Federal regulations permit the use of both malt and microbial amylase for the starch conversion. (Malt, however, must predominate for whiskey production.) Until the late 1950s there was little practical interest in using these enzymes.

Bacterial α-amylase from *Bacillus* sp. is commercially available and has the advantage of more heat stability than malt α-amylase. These enzymes permit the use of higher starch concentrations in the mash, also. However, there are disadvantages to their use for saccharifying from a distiller's standpoint. For example, they are more sensitive to the low pH conditions which prevail in mashes.

During the 1960s, distillers began in earnest to adapt "aerobic fermentation" techniques, and many now produce glucoamylase (amyloglucosidase) from a variety of fungal strains (*Aspergillus, Rhizopus*). The organisms are cultivated in submerged culture, much as in antibiotic production. Glucoamylase ferments containing 10–14 glucoamylase units (GU)/ml can be produced and stored under aseptic conditions on the distillery premises.

For spirits production, the addition of 3000 GU/bu of grain can reduce the malt usage to as little as 2% of the weight of the grain or one-half of 1% for gibberellin-treated malt (Van Lanen and Smith, 1968). A concomitant increase in yield is also experienced with this substitution. Enzymes of this group attack the starch by separating single glucose units and can break both α1—4 and 1—6 links. Since the distillers yeasts utilize glucose more readily than maltose, fermentation can go to completion more rapidly.

A factor of interest is the stability of these enzymes under distillery fermentation conditions. In fermentors converted with glucoamylase, there is essentially no loss of enzyme activity through 96 hours, even under extreme pH conditions (less than 3.5). Malt enzymes become completely inactivated under ever less stringent conditions. Lower maltose, isomaltose, and oligosaccharide residues are the end result and likely account for the increased yield. Experience suggests that glucoamylase saccharifies the starch more rapidly and more completely than malt.

For flavor considerations and by law, distillers must use malt and rye (small grains) in the bourbon mash bills. Experimentation has demonstrated that glucoamylase is not detrimental to whiskey production, but today we know of no distillers who are producing bourbon in this way.

Though previously unreported, a new combination enzyme system is currently in use in the distilling industry. This improvement involved the cooperative efforts of Midwest Solvents Corporation and Hiram Walker & Sons, Inc.* In this process, a high α-amylase and a high glucoamylase titer are achieved within the same *Aspergillus* fermentation broth. By measured feeding of gaseous ammonia into corn or milo mash, via the aeration system, one can increase the yield of glucoamylase to levels in excess of 20 GU/ml. The addition is balanced to maintain the pH of the fermentation above 4.5, and the yield increase is associated with the availability of nitrogen. Ammonium salts are also stimulative.

The pH control results in an accumulation of α-amylase as well, and yields in excess of 100 DU/ml are easily maintained.

A finished ferment can be held under aseptic conditions in excess of a week at ambient temperatures and used for both thinning of starch slurry and saccharifying of mash, totally eliminating the need for malt in spirits production.

C. Yeast

Most distillers add their yeast inoculum in the form of a grain mash. This mash is the result of yet another process. The operation begins in the laboratory where the yeast cultures are maintained and where they are generally checked daily for purity before transfer (Coulter, 1964). Trial fermentation may also be run to assure strain integrity. Malt syrup and wort agar are traditional yeast laboratory media.

Yeast propagation is an anaerobic operation in the distillery, and rye and barley malt make up the traditional medium. Other grains do not provide proper nitrogen for rapid yeast growth, which is the intent of this operation. There is a staged buildup in inoculum quantity throughout the

* Patent applied for.

process. A typical increase would be from 25 ml in the laboratory to 9000 gal in the plant in six steps. The time intervals are generally 12–24 hours for each step, depending upon incubation temperature.

With a population in excess of 1×10^8/ml of bacteria-free yeast, the mash is then ready to inoculate the distillery fermentation mash for ethanol production. A rate of 3% by volume of such a mash would be suitable for bourbon inoculation. One-half percent would be adequate for a glucoamylase converted corn mash for spirits production.

Distillers sour their yeast mash during its preparation to facilitate final sterilization and to aid thinning; rye mash can become quite viscous. The pH of this mash is lowered by culturing a lactic acid-producing organism in the mash for a few hours. These organisms are subsequently destroyed by the heat and acid during sterilization. The mash is then cooled and is ready for yeast propagation. Direct addition of acid can be substituted.

The introduction of glucoamylase production into the distillery has changed yeast production also. The glucoamylase ferment of *Aspergillus* can upgrade an inexpensive corn mash into an excellent yeast propagation medium (Van Lanen *et al.*, 1975). One increases the glucoamylase usage to 6000 GU/bu and adds the ferment to the corn mash along with the normal lactic culture. The mash (20° Balling) thus treated is then held at 50°C (122°F) for 4 hours to develop souring and to permit hydrolysis. During this time, most of the starch in the mash is converted to sugar by glucoamylase activity. Protease activity present in the same ferment modifies the corn protein, increasing the α-amino nitrogen content. Combined, these actions free virtually all of the oils in the corn. At this point, the medium is sterilized and may be stored sterile as needed.

To prepare the yeast inoculum, this sterile mash is inoculated with yeast at 27°C (80°F) and is maintained at that temperature for a 4°–5° drop in Balling, usually 6–8 hours. At this point, the mash is cooled to 15°C (60°F) and held there for use. Yeast populations of $3\text{–}5 \times 10^8$/ml are obtained. This permits reduction of the volume of inoculum required for each fermentor. Also, this inoculum remains unusually stable for a yeast mash. Through several days of usage, these yeasts retain nearly 100% viability. The precise mechanism remains to be defined. Nevertheless, this represents a distinct advantage over traditional rye mash inoculum.

D. Fermentation

The distillers mash must be cooled to fermentation temperature before the addition of yeast. Mash coolers, by the nature of their function, provide areas of optimum temperature for lactic acid bacteria development. Thus, the lighter the bacteria load entering the cooler, the cleaner the mash emerging. Cleaner malt-free mash offers a decided benefit for spirits production.

It is at this stage that spent stillage (alcohol-stripped, fermented mash) is added to the mash stream either at the cooler or at the fermentor. This dilutes the sugar mixture to the consistency the distiller has determined to be most suited for his particular process. Yeast strain, equipment, and final product all influence his decision. The pH of the grain mash is also lowered by this addition into the range for best yeast growth. Trace nutrients and buffering capacity from earlier fermentations thus benefit the current ones. However, all water added to the process must ultimately be removed again in the final drying of the stillage into distillers feed grains.

The initial temperature of the grain mash in the fermentor must also be set for the prevailing conditions. Highly converted corn mashes ferment rapidly with sudden CO_2 production, and they are strongly exothermic. Bourbons generally begin slowly. Unless the fermentor is open-topped and/or is provided with adequate cooling, a maximal temperature of some 32°C (90°F) will likely be exceeded, resulting in a loss of yield and/or abnormal flavor development.

Few whiskey distilleries are equipped to hold a constant optimum fermentation temperature. Some actively choose not to. The usual alternative is to set at a lower temperature (21°–24°C, 70°–75°F).

Very little *in vivo* work has been published by distillers on yeast enzyme patterns, etc. (Oura, 1977). Ethanol production can be detected almost from the onset, and a pH drop is characteristic of the fermentation. The number of yeast cells increases and ultimately reaches the same level regardless of inoculum size. There is a 30-billion cell increase associated with the production of each ounce of whiskey (De-Becze, 1964).

The sugar concentration of the mash is usually depleted within 48 hours, but to optimize spirit yields or to develop characteristic batch method whiskey flavors, most beverage alcohol undergoes a 3-day fermentation.

Distillery scheduling usually calls for some short-term (60–70 hours) and some long-term (approximately 90 hours) fermentation in order to fit distillation scheduling into a 5-day week. Fermentors scheduled for long term would be set at lower temperatures and held over the weekend, perhaps with cooling.

The mash in the fermentor is never bacteria-free, and rapid pH drops are possible with contaminated fermentors. A normal flora may contribute significantly to the final aroma profile of a distillate (Suomalainen and Lehtonen, 1976; Reazin *et al.*, 1970).

The reaction products of yeast fermentations and the flavor development which accompanies these fermentations are beyond the scope of this chapter. However, many of these compounds have been researched (Suomalainen and Nurminen, 1975; Harrison and Graham, 1970). Troublesome off-odors are frequently associated with abnormal fermentations. Acrolein-producing bacteria can develop during fermentation if malt quality is not properly maintained (Serjak *et al.*, 1954). Diacetyl, a vicinyl diketone, and related compounds are a problem which distillers share with the brewing industry (Rice and Helbert, 1975; Branen and Keenan, 1971). Normal control procedures include rate and duration of fermentation and good sanitation. Improper yeast strain selection is also a contributing factor. Excessive aldehydes in the fermentation beer can be rectified by distillation techniques. Phenolic compounds are best controlled by adjusting the cooking temperatures of the grain.

III. PLANT

Local grain surplus likely forced many of the early millers to attempt distilling. Present-day distillery operators are on a bit more solid footing (Prescott and Dunn, 1959). Technology is now available with which he can, and must, control his process. The water and the grains still are the essential raw material, energy and transportation still may determine profitability, but proper design will produce a quality product. The superior product requires skill and experience.

A. Milling

Both product type and mode of operation are very important to design and ultimately determine the choice of process equipment. For example, hammer mills equipped with screens for particle sizing offer consistent operation over long periods with less operator attention, and they mesh nicely with continuous cooking equipment. A 12-mesh screen is standard for control, and approximately 5% of the grain particles should remain on the screen when milling corn for atmospheric cooking (100°C for approximately 40 minutes). Pressure cooking of corn (above 121°C for 5–15 minutes) permits a larger percentage, 12–25%, on the 12-mesh screen. Particle retention values for small grains average about one-half that of the corn standards.

Roller mills can cause striation of barley malt hulls, which tend to plug spiral plate mash coolers. However, they offer good control. Minimizing flour aids dried grain production later on. Roller mills find application primarily in batch cooking operations and where traditional mashing practices are used to preserve distillate flavors.

B. Cooking

Batch cooking vessel design can be horizontal or vertical and includes mixers to suspend the grain particles. Typically, steam is injected into these vessels for heating, and cooling is provided via jacketing or flash evaporation under vacuum. Either barometric or surface condensers would be provided to condense the generated vapors. Operating capacities are approximately 10,000 gal (four hundred 56-lb bushels of grain per charge). The entire cycle may take from 2 to 3 hours.

Continuous cooking equipment draws a uniform, lower steam load than batch equipment. Space and time requirements are also less for the same throughput. The entire process of grain from the bin to mash into the fermentor can be 10 minutes when pressure cooking is used. Higher capacity distilleries, which mash over 5000 bu/day, are well-suited to this operation. In this equipment, a prewarmed grain slurry (15–18 gal/bu) enters the pipeline vessel and is instantly heated to over 149°C (300°F) in a steam jet. After 4–5 minutes of pipeline flow, the mash begins to cool. Careful design can develop considerable heat conservation at this point. A partial recycle of thin stillage (residues from the still bottoms which are screened or centrifuged to remove suspended particles) as liquid for slurrying the grain for the cooker will furnish prewarming and reduce the water load to the driers. Quality considerations or operating schedules may prevent these adaptations, however.

Recycling blowdown vapors from the continuous cooking holding tanks, as preheat for grain slurries, affords significant energy savings also.

C. Saccharifying

After cooking, the gelatinized starch is enzymatically converted into sugars. Batch cooked mash is generally flashed to 63°C (145°F) at the end of the cycles while still in the cooker. Enzyme is added and the mixture is transferred into a receiving vessel (drop tub) for the conversion step. This stage lasts 30 minutes for malt enzyme.

Mash leaves a continuous cooker at a higher temperature and usually passes through a flashing vessel. An enzyme slurry stream is metered in at the cool end discharge, and conversion takes place as the mixture flows through a second pipeline to the final coolers.

Glucoamylase offers the distiller a tailored enzyme system for saccharifying grain mashes. The enzyme ferment is more tolerant of distillery variations in pH and heat, and it saccharifies more starch into a more utilizable form.

The production techniques are new to most distillers. The typical 5000-gal tub of mash would have set solids of about 18%, made up from approximately 10,000 lb of ground corn, 50 lb of malt, and 4000 gal of water or grain stillage. Such a mash would be cooked to 121°C (250°F) for 5 minutes, cooled to receive 100 lb of liquefying malt, held at 65.5°C (150°F) for 10 minutes, and finally sterilized at 121°C (250°F) for 60 minutes. The final pH runs to 4.5–5.0.

These culture vessels are generally pressurized with height two times the width and are equipped with cooling, air sparge, and a rotary blade agitator. After inoculation with an *Aspergillus* culture, the fermentor is incubated at 35°C (94°F) with aeration (0.25–1.0 v/v per minute) with sterile air. Within 4–8 days, the unitage of glucoamylase is sufficient for use and the vessel is cooled. Unless contamination occurs, the ferment is quite stable.

As in the case with yeast, commercial sources of the enzyme are also available. Standard enzyme usage rates are 3000 GU/56-lb bushel, regardless of the form. Addition may be made into the cooling mash or directly into the fermentor.

A new microbial technology for yeast propagation and the production of high α-glucoamylase is outlined in Section II, B.

After conversion, mash must be further cooled to the fermentation temperature. Mash cooler designs are variable, and spiral plate exchangers, flash coolers, and double pipe exchangers are all employed. The mash has higher viscosity at this point and is vulnerable to bacteriological contamination, and both must be considered in design. Plate coolers require less heat transfer surface and less space than most other types.

D. Yeasting

Microbial technology has progressed to the point that a distiller today need not design a yeast process at all. Active dry distillers yeasts are available commercially in several convenient package forms. These are stable and of respectable quality. However, considerations of distillate flavor and "real" costs lead most distillers to cultivate their own supply.

The pure culture yeast process begins in the laboratory, where the initial step is the selection of the type of yeast that is desired. Routine enrichment, isolation, and trial fermentation techniques produce the master culture(s) upon which the distiller depends. These cultures can be lyophilized or, more likely, maintained on wort agar medium slants and stored at reduced temperature until needed. Cultures are transferred at 6- to 12-month intervals. A standard regime follows below.

A fresh master culture is streak-plated on wort agar and individual isolated colonies are picked to produce the working slants.

Each day a working slant is inoculated into sterile, nondiastatic malt syrup (25 ml at 20° Balling), which is incubated for 24 hours at 24°–27°C (75°–80°F). The contents of this culture are, in turn, used to inoculate 900 ml of malt syrup, which is again incubated at 24°C (75°F). These cultures would typically contain 150–200 × 10^6 viable cells per milliliter at 24 hours. In this time, the sugar content would be reduced to about 9%. These flasks are carefully screened for bacterial contamination, since once the culture leaves the laboratory and is cultivated in grain mash, detection is delayed and so is a very difficult process.

The plant medium traditionally is rye mash with not less than 10% barley malt. Distillery yeast propagation, in contrast to bakers' yeast production, does not work with sugar-to-nitrogen ratios; nor is it concerned with absolute yields. If the yield of yeast is below 75% of the weight of the raw material, the "loss" is recovered in subsequent alcohol production and/or feed recovery.

The requirements for available carbon, energy, and nitrogen sources are accounted for in the grain selection; phosphates and vitamins seldom become limiting, because cultivation until the Balling drops to one-half that in the original mash is the rule of thumb.

Plant propagation of yeast is also staged. The initial tank would contain 50 gal of mash; after 12–24 hours this would furnish inoculum for a 500-gal tank and, similarly, on to a 10,000-gal tank. Incubation temperatures are adjusted to yield an active inoculum of yeast at 1 × 10^8 cells/ml in schedule with the distillery's fermentor sets. Inoculation rates vary as to product: 0.5% for spirit and 3.0% for whiskey would be average. A 20° Balling mash would be soured with lactic acid by permitting lactic acid-producing bacteria to incubate in the medium at about 49°C (120°F) until titratable acidity reached 1.2% (pH 3.7–4.0). Nine thousand gallons of such a mash can be sterilized in the propagation vessel by heating to 88°C (190°F) for 2 hours.

Obviously, for a pure culture system, all yeasting equipment, such as connecting lines and valves, should be of sanitary construction, avoiding pockets, dead-end lines, and pumps. Yeast culture delivery from vessel to vessel would be by gravity or air pressure (filtered), with transfer lines maintained under steam when not in use. Cleaning and sterilization programs vary, but 1 hour of hot water wash, followed by 2 hours at 100°C (212°F), is a common practice (Coulter, 1964).

E. Fermenting

The typical set fermentor data for bourbon whiskey production averages: (1) temperature, 24°C (75°F); (2) Balling, 13–14; (3) pH, 4.5–4.9; and contents of approximately 25% back set stillage. Final data would show a (1) temperature of about 32°C (90°F); (2) Balling, 0.4–0.5; (3) pH, 4.0; (4) lactic acid, 0.3–0.5%; and (5) an alcohol proof over 15.5°.

Table I presents the data which are required to determine production efficiencies. The chemistry of the reactions dictates the maximum amount of product which can be formed (A and B). The input quantities forecast what should be formed (C). Recovered product permits the calculations (D).

TABLE I. Alcohol Yields

A. Stoichiometry

Starch + water → glucose

$C_6H_{10}O_5$ (162) + H_2O (18) → $C_6H_{12}O_6$ (180)

Glucose → 2 alcohol + 2 carbon dioxide

$C_6H_{12}O_6$ (180) → 2 C_2H_5OH (92) + 2 CO_2 (88)

B. Theoretical yields—dry basis (70% starch for calculation)

100 kg dry corn → 70 kg dry starch + 30 kg dry feed

70 kg starch → 77.7777 kg glucose (7.777 kg hydrolysis gain)

77.7777 kg glucose → 39.7531 kg alcohol + 38.0246 kg CO_2

Summary:

100 kg corn + 7.7777 kg hydrolysis gain → 107.7777 kg

39.7531 kg alcohol + 38.0246 kg CO_2 + 30.0 kg feed

C. Typical analysis of distillery grain receipts

		Dry basis		
	% Moisture	*Starch*	*Protein*	*Fat*
Barley malt	5.4– 6.0	57.0–61.5	12.2–13.8	1.6–2.1
Rye	11.6–12.9	64.2–67.7	13.3–14.6	1.3–2.0
Corn	13.5–16.1	71.2–73.0	9.4–10.3	3.9–4.9
Milo	13.1–14.6	72.4–75.9	9.7–11.9	2.2–3.4

D. Yield calculations

1. *Moisture*

Grain into fermentor × moisture in grain = moisture in mash

Corn 99%	×	14.08	=	13.94
Malt 1%	×	6.00	=	0.06
				14.00%

2. *Proof gallon (PG)*

Volume of mash × proof by sample distillation ÷ 100 = proof gallon

98,115 gal × 16.43 = 16,120

3. *Yield*

Proof gallons ÷ bushels in fermentor = PG/bu (56 lb)

16,120 ÷ 3060 = 5.27

4. *Dry basis*

Proof gallons ÷ dry matter

5.27 ÷ 0.86 (100 − 14) = 6.13 dry basis yield

5. *Alcohol/bushel (56 lb) (theoretical)*

56 lb Corn → 39.2 lb starch → 22.262 lb alcohol

22.262 ÷ 3.305 (conversion factor) = 6.76 PG

There is no typical vessel. Open-top, closed-top, wooden, and steel vessels all exist, often side by side. Successful operation may mean optimizing yields or repetitive production of critical flavor distillate. Most newer installations resemble Fig. 1; they are designed to facilitate temperature control and cleaning. Mechanical agitators, coupled with cooling coils or circulating mash coolers located alongside the tanks, are the two most common installations. Pressure washers with rotating heads and steaming aid in fermentor cleanup.

F. Distilling

The pot still is considered the earliest form of distilling equipment. It is of the simplest design, consisting of a kettle to hold the mash, a steam coil, and a condenser. This was used for single-stage batch distillation. While there have been improvements, refluxing is still limited; rather large heads and tails cuts are generally removed to obtain the desired quality.

Today, most distillers employ continuous stills. In this design the fermented beer enters the upper end of the column and flows downward via an overflow weir and downpipes, criss-crossing a series of perforated trays. Steam is introduced from the bottom and rises with sufficient velocity to strip the beer of its volatile components. Vapors recondense higher up in the column, depending upon their volatilities. Thus, liquids can be drawn off or recycled at the various plates, as need be. If higher efficiency is desired, rectifying stills can be added which will concentrate volatile impurities within a narrow region of the still.

Heat exchangers are generally coupled with continuous stills to conserve energy by preheating the incoming beer with the outgoing stillage.

FIGURE 1. Fermentors used in distilled beverage production. (Photo courtesy of Hiram Walker & Sons, Ltd.)

G. Drying

"Stillage" is the term applied to dealcoholized grain residues that remain after distillation. As this material emerges from the bottom of a still, it is typically sent by pipeline to the drying equipment, where the insoluble solids are separated from the soluble stream.

This liquid portion can be dried as distillers solubles, or, as is most common, all fractions are individually dewatered and then recombined and dried.

The initial step involves screening or centrifugation. The "throughs" can then be processed through a triple effect evaporator, where the moisture is reduced under vacuum from 95–97% down to 75%. A finishing pan completes the task and the resulting syrup, 35–40% solids, is then mixed back into the grain residues. Rotary drum or flash driers complete the drying. The resulting feed concentrate, distillers dried grains with solubles, is a valuable feedstuff, with annual production averaging 400,000 tons and a 1976 price of approximately $140/ton. Table II shows a typical analysis.

For total nutrient composition, the reader is referred to the Distillers Feed Research Council (Distillers Feed Research, 1970).

IV. PRODUCT

The culmination of properly controlled process and well-designed plant is a quality product. The product type determines which process and which design are chosen.

Vodka is a neutral, unaged spirit distillate. The fermentation medium may contain virtually any starchy substance, but it is almost always corn or milo. Both the yeast and the fermentation conditions are chosen to maximize yield. Being a spirit, it is distilled at or above 190° proof, and the distillate may be treated with charcoal to render it more neutral in taste and odor.

TABLE II. Composition of Distillers Feed Dried Grains with Solubles (Corn)

Composition	Percentage
Moisture	9.0
Protein	27.0
Fat	8.0
Fiber	8.5
Ash	4.5

Whiskies are generally distilled at no more than 160° proof, and the distillates contain numerous flavor constituents (congeners). Depending upon which grain makes up the major portion of the mash bill, one produces rye, corn, or malt whiskey. At this time, no North American producer is utilizing microbial enzymes to produce whiskey, as far as is known. Thus, malt would make up a small but significant portion of all whiskey mash bills. *Bourbon whiskey* is further defined and must contain at least 51% corn, with rye and malt making up the remainder of the grain, generally 15–30% and 5–15%, respectively.

An aging process is necessary to fully develop the flavoring of whiskey. Only charred white oak, *Quercus alba,* will produce the desired mellowing effect. This is accomplished by prolonged storage in barrels or casks. The changes which occur during aging can be monitored chemically, and data show that the wood char removes some compounds and adds others. Still other flavor elements develop from the oxidation–reduction reactions which take place within the barrel as it breathes during aging. A significant loss of ethanol occurs during aging due to barrel soakage and some evaporation. Ultimately, barreling proof and solubility of the congeners together determine their final concentration. Scotch, Canadian, and American light whiskey may claim age for storage in reused cooperage, while bourbon whiskey must, by law, be aged in new barrels.

Tequila is a distinctive product of the Jalisco area of Mexico and is produced from agava cactus. *Rum* is the distillate from the fermented sugar cane products, distilled between 80° and 190° proof. *Brandy* is also distilled at less than 190° proof from fermented fruit or fruit juices.

Extensive chemical analyses have been made on aged as well as unaged distillates. Table III presents quantitative data for the different classes of compounds found in bourbon whiskey.

TABLE III. An Example of Whiskey Analysis for 5-Year Bourbon

Specific gravity 20°C/20°C	0.93
Proof	100.00
pH	4.1
Extract[a]	136.0
Tannins[a]	42.0
Furfural[a]	1.3
Aldehydes[a]	5.2
Esters[a]	38.7
Other congeners[a]	218.0
Total acids as acetic[a]	54.0

[a] All values given are in grams per 100 liters.

REFERENCES

Annual Statistical Review (1973). Distilled Spirits Inst., Washington, D.C.
Anonymous (1977). *Chem. Week.* **120,** 26–28.
Branen, A. L., and Keenan, T. W. (1971). *Appl. Microbiol.* **22,** 517–521.
Coulter, C. J. (1964). *Proc. Am. Soc. Brew. Chem.* pp. 64–69.
DeBecze, G. I. (1964). *Biotechnol. Bioeng.* **6,** 191–221.
DISCUS Facts Book (1976). Distilled Spirits Council U.S., Washington, D.C.
Distillers Feeds—A Progress Report (1970). Distillers Feed Res. Counc., Cincinnati, Ohio.
Harrison, J. S., and Graham, J. C. J. (1970). *In* "The Yeasts" (A. H. Rose and J. S. Harrison, eds.), Vol. 3, pp. 283–348. Academic Press, New York.
Oura, E. (1977). *Process Biochem.* **12,** 19.
Prescott, S. C., and Dunn, C. G. (1959). "Industrial Microbiology," 3rd ed. McGraw-Hill, New York.
Reazin, G., Scales, H., and Andreason, A. (1970). *J. Agric. Food Chem.* **18,** 585–589.
Rice, J. F., and Helbert, J. R. (1973). *Proc. Am Soc. Brew. Chem.* pp. 11–17.
Serjak, W. C., Day, W. H., Van Lanen, J. M., and Boruff, C. S. (1954). *Appl. Microbiol.* **2,** 14–20.
Suomalainen, H., and Lehtonen, M. (1976). *Kem.–Kemi* **3,** 69–77.
Suomalainen, H., and Nurminen, T. (1975). Some Aspects of the Structure and Function of the Yeast Plasma Membrane. Aikon Keskuslaboratorio Report 8101, Aiko, Helsinki, Finland.
Van Lanen, J. M., and Smith, M. B. (1968). U.S. Patent 3,418,211.
Van Lanen, J. M., Smith, M. B., and Maisch, W. F. (1975). U.S. Patent 3,868,307.

GENERAL REFERENCES

Aiba, S., Humphrey, A. E., and Millis, N. F. (1965). "Biochemical Engineering," 1st ed. Academic Press, New York.
Gutcho, M. H. (1976). "Alcoholic Beverage Processes." Noyes Data Corp., Park Ridge, New Jersey.
Haas, G. J. (1976). *In* "Industrial Microbiology" (B. M. Miller and W. Litsky, eds.), pp. 165–191. McGraw-Hill, New York.
Miller, D. L. (1975). *Biotechnol. Bioeng.* Symp. **5,** 345–352.
Packowski, G. W. (1963). *Kirk-Othmer Encycl. Chem. Technol., 2nd ed.* **1,** 1.
Solomons, G. L. (1969). "Materials and Methods in Fermentation." Academic Press, New York.

Chapter 4

Mold-Modified Foods

HWA L. WANG
C. W. HESSELTINE

MICROBIAL TECHNOLOGY, 2nd ed., VOL.II
 ISBN 0-12-551502-2

I. INTRODUCTION

In earliest times, man was plagued with either feast or famine, so any means he could discover to conserve food when it was bountiful was a great step forward in his survival and his conquest of the earth. Since man was, of necessity, a wanderer and a hunter, he learned about the drying and smoking of meat. Certainly these methods not only conserved his supplies, but they also reduced weight, enabling him to carry more food with him. At the time of the discovery of North America by the Europeans, most of the Indians were in this stage of development. The discovery of these two methods—drying and smoking, like the invention of the wheel, was perhaps by serendipity. Early man might not know why foods spoiled, he only knew they did. Later we can speculate that he discovered the use of salt in combination with drying and smoking.

Man's next discovery for preserving food was the fermentation of foods; although he had no idea what happened when microbial growth occurred, he learned that meat and plant material could be kept for long periods of time. It was also essential that he knew how to use salt, a necessary agent in the successful fermentation of materials, because it inhibits the development of toxins from microorganisms. Undoubtedly, many an early ancestor died from botulism or was made ill by *Staphylococcus aureus*. When salt was added and natural fermentation occurred, he knew definite changes in odor, appearance, and taste occurred that helped the product remain wholesome. Probably the first fermentations were discovered accidentally when salt was incorporated with the food material, and the salt selected certain harmless microorganisms that fermented the product to give a nutritious and acceptable food. If we speculate along these lines, we might expect the first fermented food to have been fish. With the advent of certain religions in which meat was excluded from the diet, the use of salt and fermentation was adapted to certain plant products. For instance, Bush (1959) states that Buddhism was well established in China and Korea by the fourth century and was introduced into Japan between A.D. 500 and 600. It may very well be that the cultivation of soybeans and their use in food, including fermented foods, were then introduced in Japan. In fact, today, food fermentations in the Orient are spoken of as the "traditional fermentations," which include the very important soy sauce and miso (soybean paste). From these traditional food industries have developed the modern industrial enzyme industry and the flavoring agents such as monosodium glutamate and nucleotides. We would like to suggest that one reason Japan is now the leader in the industrial fermentation field is in large part due to the food fermentation base from which it launched its industrialization of microorganisms. This background in food fermentations supplied Japan with the knowhow and the technicians familiar with the handling of microorganisms.

Fish and soybean fermentations continue to be the most important fermented food industries in the Orient. Fish fermentations remain indigenous, and the products are strictly traditional and limited to local consumption. The soybean fermentations, on the other hand, are changing rapidly through modern microbiology and technology. Although fermentations carried out in the traditional way at home still exist, with the art passed on from one generation to the next, some of the most modern and sophisticated fermentations in the world are found in the Oriental fermented soybean foods industry, especially in Japan.

Some soybean fermentations are neither strictly fermentation of soybeans alone nor carried out by only one kind of microorganism; they may involve a substrate consisting of both cereals and soybeans and an inoculum consisting of bacteria, yeasts, and molds. In this chapter, we shall discuss an area of industrial microbiology that concerns the fermentation of raw agricultural products into foods mainly by the action of diverse species of fungi.

II. SOY SAUCE

A. General Description

Soy sauce is a dark-brown liquid with a salty taste and distinct pleasant aroma, which is made by fermenting soybeans, wheat, and salt with a mixture of mold, yeast, and bacteria. The fermentation is essentially a process of enzymatic hydrolysis of proteins, carbohydrates, and other constituents of soybeans and wheat to peptides, amino acids, sugars, alcohols, acids, and other low-molecular compounds by the enzymes of the microorganisms. In addition to the fermentation technique, a chemical method in which acid hydrolyzes the protein and the carbohydrates has been used in the United States. However, the flavor of the chemically hydrolyzed product is quite different from that of the product brought about by microbial enzymes.

The brewing of soy sauce originated in China many centuries ago and later was introduced to Japan and other Oriental countries. Consumed in almost every Oriental country, soy sauce is used as a seasoning agent in preparation of food and as a table condiment. The product is known as chiang-yu in China, shoyu in Japan, kecap in Indonesia, kanjang in Korea, toyo in the Philippines, and see-iew in Thailand. In the Western World, the product is often referred to as soy sauce.

Japan leads the soy sauce industry in the world. Japan not only has the largest fermentation plant but also employs the most advanced technology, developed through comprehensive research on all aspects of soy sauce fermentation. According to Ebine (1976), 14,000 metric tons of whole soybeans and 176,000 metric tons of defatted soybean products were used in Japan in 1974 to produce more than 1 million kl of soy sauce. Annual per capita consumption in Japan is estimated at about 10.2 liters. Although there are more than 4000 soy sauce producers in Japan, the largest four or five companies supply about 50% of the total produced. In 1972, Kikkoman, one of the largest producers in Japan, built a plant in Walworth, Wisconsin, and has since been in full operation to supply soy sauce for the American market.

B. Method of Preparation

Many improvements have been made since the early development of the fermentation, but the basic method of manufacture is almost unchanged (Yokotsuka, 1960; Hesseltine and Wang, 1972; Yong and Wood, 1974) and is illustrated in Fig. 1.

1. Raw Materials and Their Treatment

a. Soybeans. Traditionally, soy sauce is made from whole soybeans; however, in recent years, defatted soybean meals and flakes have taken their place. Today, more than 90% of soy sauce production in Japan is from defatted soybean products.

In preparation for fermentation, selected beans are washed, soaked overnight with several changes of water to prevent bacterial growth, drained, and steamed at 10 lb/in.^{2} for several hours. Cooking conditions have been known to greatly affect enzymatic digestion of soybean protein which, in turn, affects nitrogen recovery. Tateno and Umeda (1955) and the Noda Soy Sauce Company, Ltd. (1955) claimed that soybean protein was best utilized by the enzymes when the beans were soaked in water for 10–12 hours at room temperature and then steamed at 10–13 lb for about 1 hour. So the current trend has been to cook the soaked beans at a higher temperature for a shorter time (Yokotsuka, 1971). For this purpose, a batch-type cooker was developed so that the temperature could be raised rapidly to 120°C and lowered quickly immediately after cooking.

When defatted soybean meals or flakes are used, they are first moistened by spraying with water amounting to about 130% of soybean weight and then are steamed at 13 lb/in.^{2} for 45 minutes.

b. Wheat. Whole wheat is generally preferred. It is first roasted and then coarsely crushed. The roasting process adds flavor and color to the

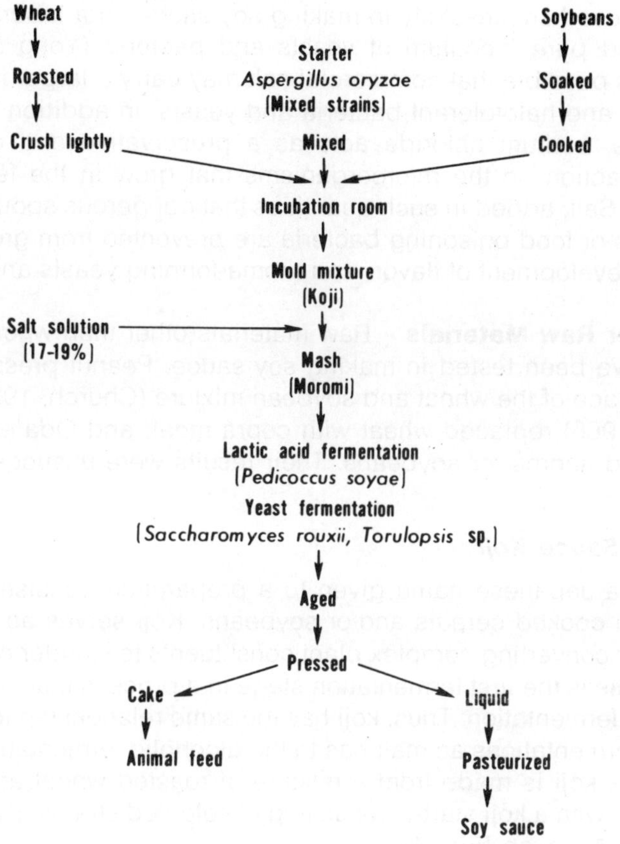

FIGURE 1. Flow sheet for the manufacture of soy sauce.

resulting soy sauce and in addition destroys surface microorganisms and facilitates enzymatic hydrolysis. According to Yokotsuka (1964), the addition of wheat to the fermentation mixture serves several purposes. First, the mold grows better and produces more enzymes on a mixture of wheat and soybeans than on wheat or soybeans alone. Second, the addition of roasted, crushed wheat to the cooked soybeans would minimize the growth of undesirable bacteria. Moisture of cooked soybeans is 60%, ideal for bacterial growth, whereas the moisture of the soybean–wheat mixture (1 : 1) is about 45%, adequate for mold growth but not for bacteria. Third, wheat serves as a precursor of sugars, alcohols, organic acids, and flavor compounds. Last, wheat is rich in glutamic acid.

c. Salt. Although commercial (not chemically pure) sodium chloride is generally used in making soy sauce, chemically pure sodium chloride has been used successfully in making soy sauce on a laboratory scale with added pure inoculum of yeasts and bacteria (Yong and Wood, 1974). It is possible that commercial salt may carry a larger inoculum of halophilic and halotolerant bacteria and yeasts. In addition to giving a salty taste, sodium chloride acts as a preservative and also has a selective action on the microorganisms that grow in the fermentation substrate. Salt, added in such quantities that dangerous spoilage microorganisms or food poisoning bacteria are prevented from growing, permits the development of flavor- and aroma-forming yeasts and bacteria.

d. Other Raw Materials. Raw materials other than wheat and soybeans have been tested in making soy sauce. Peanut press cake was used in place of the wheat and soybean mixture (Church, 1923). Baens-Arcega (1966) replaced wheat with copra meal, and Oda *et al.* (1949) substituted acorns for soybeans. Their results were unsuccessful commercially.

2. *Soy Sauce Koji*

Koji is a Japanese name given to a preparation consisting of mold growth on cooked cereals and/or soybeans. Koji serves as an enzyme source for converting complex plant constituents to simpler compounds. Usually this is the first fermentation stage in a process that also involves a second fermentation. Thus, koji has the same relationship to the Oriental food fermentations as malt has to the alcoholic fermentation of grain. Soy sauce koji is made from a mixture of roasted wheat and steamed soybeans, with a koji starter consisting of selected strains of *Aspergillus oryzae* or *A. soyae* (Fig. 1).

a. Koji Starter. Koji starter is known as tane koji in Japan. Different types of tane koji are available for commercial use in making soy sauce koji, miso koji, and others. Strains of *A. oryzae* and *A. soyae* have varying morphological and physiological characteristics. Each strain is selected for its special abilities by natural selection or by induced mutation to give a desirable starter for a particular fermentation. Although the strains of molds used for each type of tane koji may be different, the method for preparing tane koji is the same; it has been outlined as follows by Hesseltine and Wang (1972).

Polished rice is soaked in water overnight, drained, steamed for 1 hour, and mixed thoroughly with about 2% of wood ash as a source of trace elements. Addition of ash of certain trees is important because the RNA/DNA ratio of the spores varies; a high RNA/DNA ratio gives spores with greater viability and vigor. The rice is then inoculated with spores of a selected strain of *A. oryzae,* spread out in trays in a layer approximately 1.5-cm deep, and covered with damp cloths to favor the growth of mold mycelium. After incubation at 30°C for 5 days, the rice is well covered with mycelium and with green to yellowish green spores of *A. oryzae*. The spores are harvested, dried at 50°C, and stored at 15°C. A koji starter is usually composed of spores from several different strains of *A. oryzae* blended together in a definite proportion, so that various enzymes may be produced in the proper amounts during the preparation of koji. In Japan, each gram of commercial tane koji has about 25×10^7 viable spores.

The literature on preparation of mold spores for food fermentation has been reviewed by Hesseltine *et al.* (1976).

b. Preparation of Koji. Raw materials, soybeans, and wheat are treated as described above. The proportion of soybeans to wheat used in fermentation mixtures may vary from one manufacturer to another, but according to Yokotsuka (1960) the best soy sauce is generally made from a soybean-to-wheat ratio of 50 : 50 by weight, or 52 : 48 by volume based on that of raw materials. An appropriate amount of tane koji, which is about 0.1–0.2% of soybean–wheat mixture, is first mixed thoroughly with roasted wheat and then added to cooked soybeans. The inoculated mixture is distributed in shallow wooden boxes, covered with cloth, and then incubated at 30°C. After 24 hours of incubation, the mixture is covered with a thin white growth of mold. As mold growth continues, the temperature of the mixture could rise above room temperature to 40°C or higher. Therefore, the mixture should be turned or stirred periodically to maintain uniform temperature, moisture, and aeration. The mixture should also be free of lumps to minimize bacteria propagation. As the incubation time increases, the molds continue to grow and their growth turns yellow and dark green. Also, the moisture of the mixture gradually decreases. After about 72 hours of incubation, the soy sauce koji is matured. In recent years, automatic koji-making processes have been developed to replace the traditional way of making koji. The new equipment includes automatic inoculator, automatic mixer, large perforated shallow vats in closed chambers equipped with forced filtered-air devices, temperature controls, and mechanical devices for turning the substrates during incubation.

Soy sauce koji of superior quality has a dark-green color, pleasant aroma, sweet but bitter taste and, most important of all, high protease and amylase activity. The relationship between enzyme yield and culturing conditions varies with the organisms. *Aspergillus oryzae* (Maxwell, 1952) and *A. soyae* (Yamamoto, 1957) produce greater amounts of proteases at temperatures lower than their optimum growth temperature. Similar findings with other fungi were reported by Wang *et al.* (1974). The effect of temperature on enzyme production observed in these studies emphasizes the importance of a common practice in koji making, i.e., frequent turning of the growth mass or use of thin layers of solid substrates. Otherwise, the heating which results from active growth will increase incubation temperature and affect enzyme production. Harada (1951) found that it is necessary to keep the moisture content of koji at 27–37% in order to obtain high enzyme activity. Thus, moisture and temperature are the two most important factors in making koji of superior quality. Current industry practice is to increase moisture content of the soybean–wheat mixture and to lower incubation temperature.

3. Brine Fermentation

Soy sauce koji is transferred to a deep vessel to which an equal volume of salt solution (20° Be) is added to make a mash, or moromi as it is called in Japan. However, it was found (Yokotsuka, 1960) that moromi made with a ratio greater than 1:1 of salt solution to koji utilized total nitrogen of the raw material better. Also, the lower the concentration of the salt solution, the better the utilization of nitrogen. Therefore, the volume of salt solution has been increased to 1.1–1.2 times that of the koji and the concentration of salt solution has been reduced accordingly. The sodium chloride content of the mash is about 17–19%; below 16% putrefaction may occur.

Moromi is fermented in large concrete tanks or wooden vats for 8–12 months. Frequent agitation and aeration of moromi are necessary for normal fermentation to proceed and to prevent the growth of undesirable anaerobic microorganisms, to maintain uniform temperature, and to facilitate the removal of carbon dioxide. However, too much stirring hinders the fermentation. The change of temperature is said to be important for normal progress of fermentation. Therefore, shoyu fermentation in Japan usually starts in April and takes a year to complete. In general, low-temperature fermentation gives better results; because the rate of enzyme inactivation is slow, the enzymes remain active longer (Komatsu, 1968). Watanabe (1969) indicated that good-quality shoyu can be obtained by 6-month fermentation when the temperature of moromi is controlled as follows: start at 15°C for 1 month, followed by 28°C for 4 months, and finish at 15°C for 1 month.

Since koji is not prepared under aseptic conditions, one would expect the presence of yeasts and bacteria in the moromi. However, pure cultures of yeasts and bacteria are sometimes added to the mash to accelerate the fermentation and to improve the flavor of the final product. Strains of *Saccharomyces rouxii*, *Torulopsis* yeasts, and *Pediococcus soyae* were found to be important flavor producers. Representative strains maintained in the AR Culture Collection have been designated as *S. rouxii* NRRL Y-6681, *T. etchellsii* NRRL Y-7583, *T. versatilis* NRRL Y-7584, and *P. halophilus* NRRL B-4243 and NRRL B-4244. The initial pH of the mash (6.5–7.0) gradually decreases as the lactic acid fermentation advances, and at pH of around 5.5 yeast fermentation takes place.

4. Pressing

The matured moromi is separated by pressing into a liquid part, known as raw soy sauce, and a solid cake. When whole soybeans are used as raw material, soy sauce oil, consisting chiefly of ethyl esters of higher fatty acids, is produced during fermentation and appears at the upper layer of the raw soy sauce. This layer of oil must be removed and has no potential use. This is one of the reasons that defatted soybean products have replaced whole soybeans as raw material in soy sauce fermentation.

5. Pasteurization and Bottling

The raw soy sauce is pasteurized at 70°–80°C, filtered to remove precipitates, and bottled for market. In Japan either benzoic acid or butyl ester of *p*-hydroxybenzoic acid is added as a preservative.

C. Composition

According to Watanabe (1969), about 5000 liters (specific gravity, 1.18) of soy sauce is obtained from 1 ton of defatted soybean meal, 1 ton of wheat, and 1 ton of salt. The composition of soy sauce varies with the raw materials used. Table I shows the average composition of soy sauce made from whole soybeans and from defatted soybean meal by 11 factories in Japan (Umeda, 1963; Umeda *et al.*, 1969). A good soy sauce has a salt content of about 18%. Its pH is between 4.6 and 4.8; below that the product is considered too acid, suggesting that acid has been produced by undesirable bacteria. It is also generally recognized in Japan that the quality and price of shoyu are determined by nitrogen yield, total soluble nitrogen, and the ratio of amino nitrogen to total soluble nitrogen. The nitrogen yield is the percentage of nitrogen of raw material converted to soluble nitrogen, which shows the efficiency of enzymatic conversion. The total soluble nitrogen is a measure of the concentration of nitrogenous material in the shoyu and indicates a standard of quality. A ratio of greater than 50% of amino nitrogen to total soluble nitrogen is also evidence of quality. These results can be affected by many factors such as raw materials, steaming conditions, tane koji, koji making, and brine fermentation. Technology to improve these values is constantly being sought. During the last 20–30 years, a shorter cooking time at higher temperatures has been adopted. Koji-making

TABLE I. Average Composition of Soy Sauce Made from Whole Soybeans and Defatted Soybean Meal[a]

Conditions	Raw material	
	Whole	Defatted meal
Baume (°)	22.7	23.4
NaCl (%)	18.5	18.0
Total nitrogen (%)	1.6	1.5
Amino nitrogen (%)	0.7	0.9
Reduced sugar (%)	1.9	4.4
Alcohol (%)	2.1	1.5
Acidity I (ml)[b]	10.1	14.0
Acidity II (ml)[c]	9.8	13.6
pH	4.8	4.6
Glutamic acid (%)	1.3	1.2
Nitrogen yield (%)	75.7	73.7

[a] From Umeda (1963); and Umeda *et al.* (1969).
[b] Milliliters of 0.1 *N* NaOH required to neutralize 10 ml of soy sauce to pH 7.0.
[c] Milliliters of 0.1 *N* NaOH required to bring the pH of 10 ml soy sauce from 7.0 to 8.3.

technology has been improved to increase enzymatic activity. New strains of koji molds, yeasts, and bacteria have been developed by induced mutations or by diploid formation. Temperatures for moromi fermentation have been controlled. As a result of these improvements, nitrogen recovery has increased from 60% in 1945 to about 90% in 1975 (Ebine, 1976), fermentation period has been reduced significantly, and flavor has been greatly improved. More recently, an enzymatic process for making soy sauce has been attempted. The raw materials were subjected to enzymatic treatment but did not change to koji (Nakadai *et al.*, 1975).

The chemical changes in the production of soy sauce and in its flavor are complicated. Yokotsuka (1960) has written a complete review on these changes. He states that of the total nitrogen about 40–50% is amino acids, 40–50% peptides and peptones, 10–15% ammonia, and less than 1% protein. Seventeen common amino acids are present, with glutamic acid and its salts being the principal flavoring constituents. The organic bases, believed to be hydrolyzed products of nucleic acids, are adenine, hypoxanthine, xanthine, guanine, cytosine, and uracil. Sugars present are glucose, arabinose, xylose, maltose, and galactose, as well as two sugar alcohols, glycerol and mannitol. Organic acids reported in soy sauce are lactic, succinic, acetic, and pyroglutamic. The color of soy sauce is generally recognized to be the result of a nonenzymatic browning reaction. More than a hundred compounds have been reported as flavor components of soy sauce; however, the compounds that decisively characterize soy sauce remain unknown. The guaiacyl compounds seem to have an important effect upon the overall flavor of Japanese soy sauce.

III. MISO

A. General Description

Miso, the most important and popular fermented soybean food in Japan, is also made and consumed in other parts of the Orient. This product is generally known as bean paste. Like soy sauce, bean paste is made from cereal, soybeans, and salt by the action of molds, yeasts, and bacteria. With respect to microorganisms and fermentation principles, soy sauce and bean paste fermentations are also similar.

Bean paste has the consistency of peanut butter, some smooth and some chunky, and its color varies from light-yellow to reddish-brown. It has a distinctive pleasant aroma resembling that of soy sauce, and it is typically salty, although the degree of saltiness may vary; it sometimes even has a sweet taste. Like soy sauce, bean paste is used as a flavoring agent in cooking as well as a table condiment. These products blend well with varieties of foods, including seafoods, meats, and vegetables. They can be used in place of soy sauce and salt.

There are as many types of bean paste as there are different varieties of cheese in the United States. They are made by varying the cereal used, the ratios of beans to cereal, salt content, length of fermentation, and addition of other ingredients, such as hot pepper which is very popular in China and Korea. According to Ebine (1971), Japanese miso is categorized into three major types based on the raw materials used: rice miso made from rice, soybeans, and salt; barley miso from barley, soybeans, and salt; and soybean miso from soybeans and salt. These three types are further classified on the basis of salt content into sweet, medium salty, and salty groups; each of these groups is again divided by color into white, light yellow, and red depending upon the length of fermentation time. About 590,000 metric tons of commercial miso and 150,000 metric tons of homemade miso are produced in Japan annually. Of the total industrial production, approximately 80% is rice miso. The per capita yearly consumption in Japan, which probably is the highest in the world, is estimated at 6.7 kg.

B. Method of Preparation

Although the manufacturing method may differ from country to country, or from variety to variety, the principle is believed to be the same. This account will be general in nature and follows the procedures for making

Japanese rice miso. The production process, as shown in Fig. 2, generally consists of cooking soybeans, preparing koji, mixing cooked soybeans with rice koji and salt, fermenting and ripening in a tank, blending, pasteurizing, and packing. Laboratory methods of making miso have been investigated in this country by Hesseltine and his co-workers (Shibasaki and Hesseltine, 1962; Hesseltine, 1965). The characteristics on three types of rice miso in relation to the fermentation conditions are presented in Table II (Ebine, 1967).

1. Treatment of Soybeans

Whole soybeans are generally used for making miso. Dehulled soybeans or full-fat soybean grits are sometimes used for making white or light-yellow rice miso. A patent by Smith *et al.* (1961) covers the use of grits. Unlike soy sauce fermentation, of which over 90% produced in Japan is made from defatted soybean products, defatted soybean products are not suitable for making miso of good quality.

In general, the miso industry prefers to use soybeans of a large size (more than 170 gm/1000 seeds) because the ratio of hull to cotyledon is lower in the large beans. The soybean should have high water-absorbing capacity and, when cooked under described conditions, the beans should be homogeneously soft with fine texture and bright color. The Japanese miso makers found the domestic soybeans to be the most suitable, followed by Chinese and then United States soybeans. The major difference between Japanese and United States soybeans is in the amounts of oil and carbohydrates (total sugar); the Japanese varieties have higher carbohydrate and lower oil content than United States varieties. Tests have shown no correlation between capacity to absorb water and protein or oil content, but there is a high correlation between carbohydrate content and water-absorbing capacity. Among the United States varieties tested, Kanrich, Mandarin, and Comet are the most promising ones (Ebine, 1967).

To prepare whole soybeans for fermentation (Fig. 2), they are washed, soaked in water for about 20 hours at 16°C, and drained. The soaked beans are then cooked in water (white miso) or steamed at a temperature of 115°C for about 20 minutes in a closed cooker and slightly mashed. The new batch-type cooker used in soy sauce fermentation, in which the temperature can be raised rapidly to 120°C and lowered quickly immediately after cooking, has also been introduced to the miso industry.

2. Preparation of Rice Miso Koji

Polished rice is soaked in water (15°C) overnight or until the moisture content is about 35%. Excess water is drained off, and the soaked rice is steamed at atmospheric pressure for 40 minutes. A continuous cooker

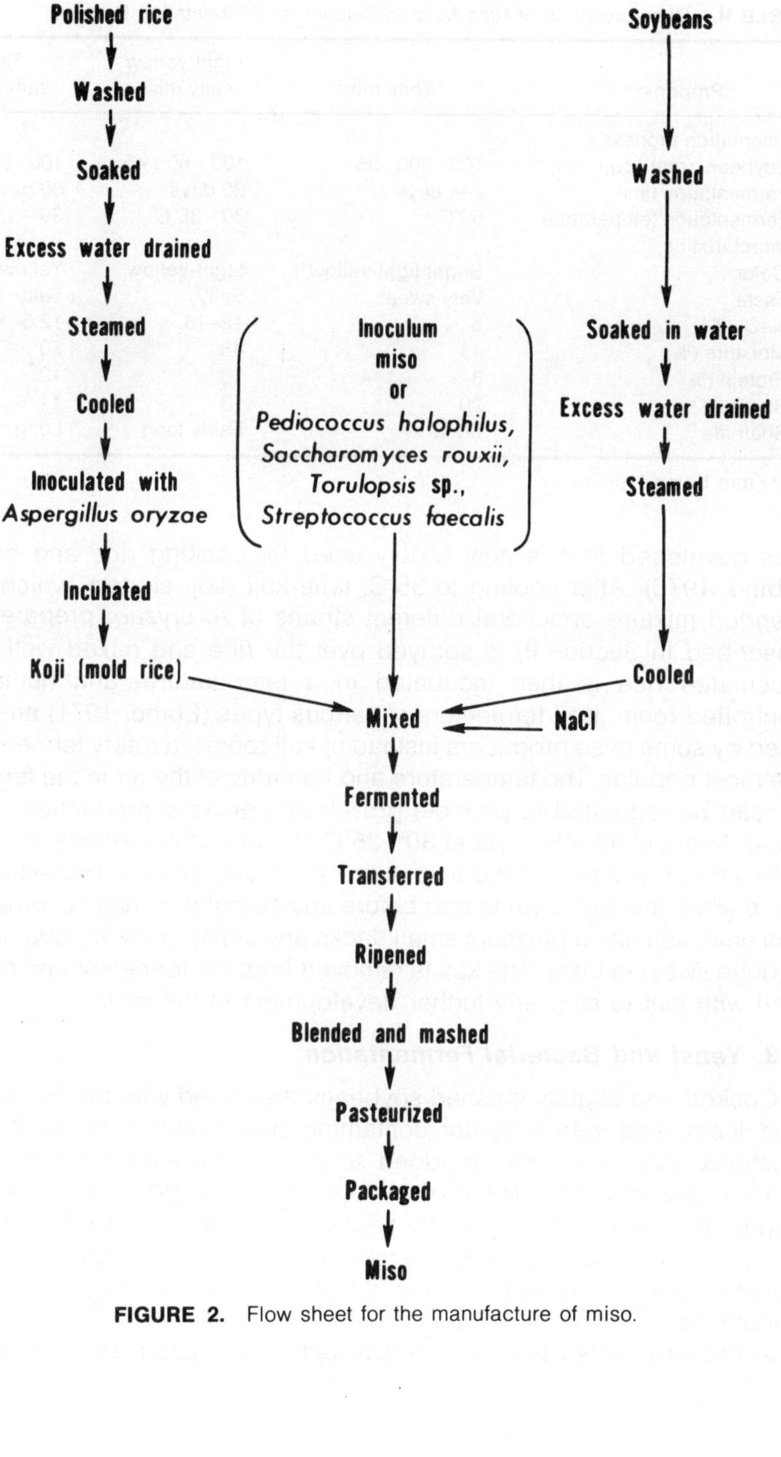

FIGURE 2. Flow sheet for the manufacture of miso.

TABLE II. Characteristics of Rice Miso in Relation to Fermentation Process[a]

Properties	White miso	Light-yellow salty miso	Red salty miso
Fermentation process			
Soybean : rice : salt	100 : 200 : 35	100 : 60 : 45	100 : 50 : 48
Fermentation time	2–4 days	30 days	60 days
Fermentation temperature	50°C	30°–35°C	30°–35°C
Characteristics			
Color	Bright light-yellow	Light-yellow	Yellow–red
Taste	Very sweet	Salty	Salty
NaCl (%)	5	12–13	12.5–13.5
Moisture (%)	43	48	50
Protein (%)	8	10	12
Sugar (%)	20	13	11
Shelf-life	Short	Fairly long	Long

[a] From Ebine (1967).

was developed and is now widely used for cooking rice and barley (Ebine, 1976). After cooling to 35°C, tane koji (koji starter), which is a blended mixture of several different strains of *A. oryzae*, prepared as described in Section II, is sprayed over the rice and mixed well. The inoculated rice is then incubated in a temperature- and humidity-controlled room. Koji fermentors of various types (Ebine, 1971) are now used by some miso producers instead of koji rooms; a rotary fermentor is the most popular. The temperature and humidity of the air in the fermentor can be regulated to promote growth and enzyme production of the mold. In about 40–48 hours at 30°–35°C, the rice is completely covered with white mycelium of the inoculated *A. oryzae* strains. Harvesting is done while the koji is white and before any sporulation has occurred. At this time, koji has a pleasant smell, lacks any musty or moldy odors, and is quite sweet in taste. The koji is removed from the fermentor and mixed well with salt to stop any further development of the mold.

3. *Yeast and Bacterial Fermentation*

Cooked and slightly mashed soybeans are mixed with the salted koji and inoculated with a starter containing pure cultures of yeasts and bacteria. Sufficient water is added so that the moisture content of the mixture is about 48%. The mixture, now known as green miso, is thoroughly blended and then tightly packed into a vat or tank for fermentation at 25°–30°C. During the fermentation period, the green miso is transferred from one vat to another at least twice to improve fermentation conditions.

In the past, miso from a previous batch was used as inoculum of bacteria and yeasts; however, the present pure culture starters speed up the fermentation from 6–12 months to 2–3 months and reduce the influence of weed yeasts and bacteria. Strains of *Saccharomyces rouxii, Torulopsis, Pediococcus halophilus,* and *Streptococcus faecalis* are the most important yeasts and bacteria in the miso fermentation.

4. *Aging and Packaging*

At the end of fermentation, the fermented mass is kept at room temperature for about 2 weeks to ripen. The aged product is then blended, mashed, pasteurized, and packaged. Generally, 3300 kg of light-yellow salty miso with a moisture content of 48% is made from 1000 kg of soybeans, 600 kg or rice, and 430 kg of salt.

Traditionally, miso is sold in wooden kegs of various sizes. Presently, miso is also sold in sealed polyethylene bags or tubes. For packaging in plastic bags, miso must first be pasteurized at 60°–70°C for 30 minutes to prevent swelling. Sorbic acid or its potassium salt is also added at a level of less than 1 gm/kg of miso (Watanabe, 1969).

Dehydrated miso powder has become increasingly popular. The dehydration, which is carried out by freeze-drying, does not affect the characteristic flavor of miso. Therefore, the dehydrated product has a great potential as an ingredient for instant mix products. The traditional miso can also be used as an ingredient for dips, salad dressings, and sauces in Western cookery.

C. Composition

Since different types of miso are made by varying the ratios of the raw materials, the composition of miso varies with the type. Table III presents the standard chemical composition of several types of miso as reported by Ebine (1971). Miso has a fairly narrow moisture range of 44–50%. Protein content ranges from 8 to 19% and fat from 2 to 10%, reflecting the

TABLE III. Composition of Various Types of Miso[a]

Variety	Moisture (%)	Protein (%)	Reducing sugar (%)	Fat (%)	Sodium chloride (%)
White miso	44	8	33	2	5
Edo sweet miso	46	10	20	4	6
Salty light-yellow miso	49	11	13	5	12
Salty red miso	50	12	14	6	13
Salty barley miso	48	12	11	5	12
Soybean miso	47	19	2	10	10

[a] From Ebine (1971).

increase of soybean used in the fermenting mixture. Except for the white and sweet types, miso contains more than 10% salt. Because of the high salt concentration, miso may be kept for considerable periods without refrigeration; but, for the same reason, the use of miso as a protein food is limited.

The Japanese soybean fermentation industry as well as government research institutes are actively engaged in research on all aspects of soy sauce and miso fermentation. There are literally hundreds of references on these subjects; therefore, no attempt is made to review the literature. Much of the literature being published has appeared in Japanese reports and journals, such as *Fermentation Technology, Agricultural and Biological Chemistry,* and annual reports from government research institutes.

IV. HAMANATTO

Hamanatto is the Japanese name for a salty product made by fermenting whole soybeans with strains of *A. oryzae.* A similar product, however, is widely produced and consumed in other Oriental countries: tou-shih in China, tao-si in the Philippines, and tao-tjo in the East Indies. In the United States, the fermented beans are often referred to as black beans because of their color. The fermented beans have a salty taste and a pleasant flavor resembling soy sauce. They are used as a side dish to be consumed with bland foods such as rice gruel, or they can be cooked with meats, seafoods, and vegetables as a flavoring agent.

The methods of preparing soybeans for fermentation and the composition of the brine vary from country to country, but the essential features are similar. A typical process for making hamanatto in Japan based on information furnished by Dr. T. Kaneko (Hesseltine and Wang, 1972) of Nagoya University, Japan, is as follows. Soybeans are washed, soaked, steamed until soft, drained, cooled, and then mixed with parched wheat flour (soybeans : wheat flour, 2 : 1). The wheat flour-coated beans are inoculated with a short or medium stalked strain of *A. oryzae.* After incubation for 1–2 days, the molded beans (koji) are packed in a container with a desirable amount of salt, water, and other spices, e.g., 2.5 kg koji, 650 gm salt, and 3.6 liters water. The container is tightly covered and the beans are aged from 6 to 12 months. After that, the beans are dried. The composition of the brine solution may vary with each manufacturer; thus, the final product differs somewhat in taste and appearance. In general, Japanese hamanatto is rather soft, having a high moisture content. Chinese tou-shih has a much lower moisture content than hamanatto and therefore is not so soft. Tao-tjo tends to have a sweet taste because sugar is often added to the brine.

Investigations on the hamanatto fermentation are scanty; thus, methods of producing the product have not been modernized. Recently, Kon and Ito (1975) found that the main microorganisms considered responsible for the hamanatto fermentation are strains of *A. oryzae, Streptococcus,* and *Pediococcus.* The *A. oryzae* strain is dark olive-green and produces strong proteolytic, but not amylolytic, enzymes. Kiuchi *et al.* (1976) found that the gas chromatographic pattern of fatty acid composition of hamanatto was similar to that of soybeans; however, 78% of total lipids in hamanatto is free fatty acids.

V. SUFU

A. General Description

Sufu, a traditional Chinese food, is a soft cream cheese-type product made from cubes of soybean curd (tofu) by the action of a mold. In the Western world, sufu has been referred to either as Chinese cheese or as bean cake. Because of the numerous dialects used in China, the product is also known as fu-ju, tou-fu-ju, and others (Wang and Hesseltine, 1970a). Sufu has a salty taste and its own characteristic flavor and aroma that develops during the brining and aging process. It is consumed directly as a relish or is cooked with vegetables or meats. Either way, sufu adds zest to the bland taste of a rice–vegetable diet. Because sufu has the texture of cream cheese, it would be suitable to use in Western countries as a cracker spread or as an ingredient for dips and dressings.

B. Method of Preparation

The process of making sufu had been considered as a natural phenomenon until 1929, when a microorganism believed to be responsible for sufu fermentation was isolated and described by Wai (1929). Almost 40 years later, Wai (1968) reinvestigated the sufu fermentation and developed a pure culture fermentation. The following information on sufu was obtained mainly from his study.

Three steps are involved in sufu fermentation: preparing soybean curd (tofu), molding, and brining and aging (Fig. 3).

1. Preparing Soybean Curd

Soybeans are washed, soaked, and ground with water; a water-to-dry bean ratio of 10 : 1 is commonly used. After the ground mass is boiled or

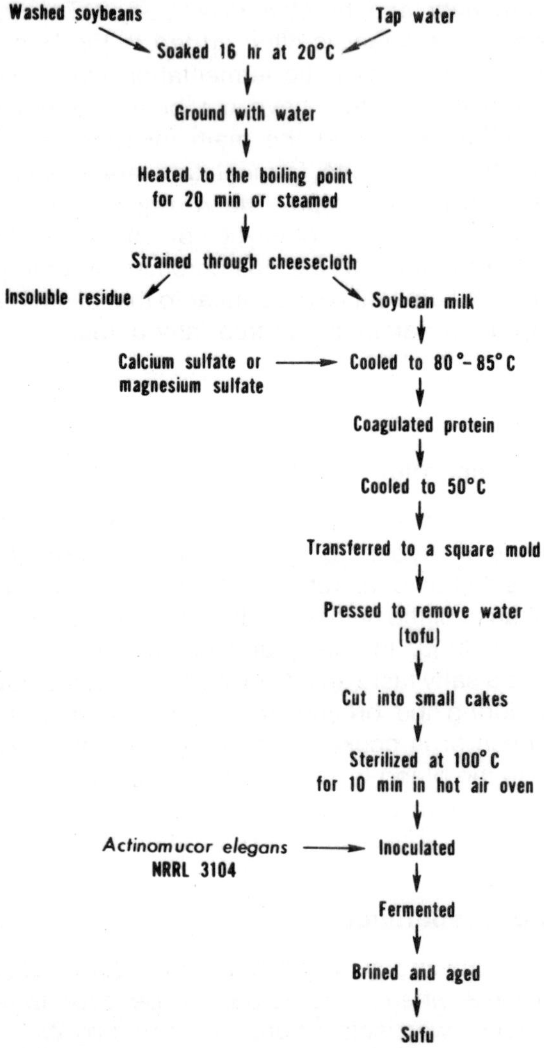

FIGURE 3. Flow sheet for the preparation of sufu.

steamed for about 20 minutes to facilitate filtering, to inactivate growth inhibitors, and to reduce the beany flavor, the hot mass is strained through a cloth bag of double-layered cheesecloth or a fine metal screen to separate the soluble fraction (soybean milk) from the insoluble residue. A coagulant, such as calcium sulfate, magnesium sulfate, or acetic acid, is then added to the milk to coagulate the proteins. The curd is transferred into a cloth-lined wooden box and pressed with weight on top to remove whey. A soft, but firm, cakelike curd (tofu) forms. Typically, tofu used for sufu fermentation has a water content of 83%; protein, 10%; and lipids, 4%.

2. Molding

The fungus responsible for sufu fermentation initially originated on rice straw. In the traditional method of preparing sufu, rice straw was used to cover the tofu cubes that were placed in a large bamboo tray. After 3–4 days at 10°–15°C, tofu cubes were overgrown by the mold mycelia.

To prevent the growth of contaminating bacteria, Wai (1968) suggested soaking the tofu cubes (2.5 × 3 × 3 cm) in a solution containing 6% NaCl and 2.5% citric acid for 1 hour, followed by hot air treatment at 100°C for 15 minutes. The treated tofu cubes are mounted on sticks, separated from one another, and placed in a tray with pinholes in the bottom and top to aid air circulation, because mycelia must develop on all sides of the cubes. After cooling, the cubes are inoculated over their surface by rubbing with pure culture of an appropriate fungus grown on filter paper impregnated with a culture solution, and then incubated at 20°C or lower for 3–7 days depending on the culture used. At this time, the cubes are covered with a luxurious growth of white mycelium and have no disagreeable odor. The molded cube has a water content of 74%; protein, 12.2%; and lipid, 4.3%.

The mold used in sufu fermentation has to have certain qualities. It must elaborate enzyme systems having high proteolytic and lipolytic activity to act upon the rich protein and lipid substrate. The organism probably utilizes the carbon in lipids as an energy source, because little carbohydrate is readily available in bean curd. The mold must have a white or light yellowish white mycelium to ensure an attractive final product. The texture of mycelial mat should be dense and thick to form a firm film over the surface of the fermented tofu cubes to prevent any distortion in their shape. Of course, it is also important that the mold growth does not develop any disagreeable odor, astringent taste, or mycotoxin. Wai (1968) confirmed that *Actinomucor elegans, Mucor hiemalis, M. silvaticus,* and *M. subtilissimus* possess all these characteristics and can be used to make sufu of good quality. Among these, *Actinomucor elegans* is the best species and the one always used commercially.

3. Brining and Aging

Conventionally, the molded tofu cubes are first salted by putting them into a tub in alternate layers with salt for a day. The salted cubes are rinsed with water and packed into glass or earthenware containers; these containers are then filled with a brine solution containing varying amounts of distilled liquor, salt, and flavoring agents, depending upon the type of sufu. After aging for 1–3 months or longer, the product is ready for market. The exact composition of the brine and the length of aging vary from one manufacturer to another, and from one type of sufu to another. According to Wai (1968), a basic and most common brine suitable for Chinese taste contains 12% NaCl and about 10% alcohol. Other additives, such as red rice, fermented rice mash, large amounts of wine, hot pepper, and rose essence are frequently incorporated into the brine either to give color or flavor. Therefore, the characteristically mild taste and aroma of sufu can be easily enhanced or modified by the ingredients of the brine solution.

C. Changes Occurring during Fermentation

During the brining and aging process, the enzymes elaborated by the mold act upon their respective substrates and yield various hydrolytic products. After 30 days of aging at room temperature, total soluble nitrogen increased from 1.00 to 2.74% and total insoluble nitrogen decreased from 7.89 to 6.05%, whereas total nitrogen changed only slightly. Lipids in the molded cubes were also partially hydrolyzed through the aging period. Free fatty acids increased from 12.8 to 37.1% and total lipids remained unchanged. Wai (1968) also demonstrated that enzyme digestion occurred mostly during the first 10 days of aging.

Since sufu fermentation has a rather simple substrate—55% protein and 30% lipids on a dry basis—it is likely that the hydrolytic products of protein and lipids provide the principal constituents of the mild characteristic flavor of sufu. The added alcohol reacts with the fatty acids chemically or enzymatically to form esters that provide the pleasant odor of the product. Ethanol also prevents growth of some microorganisms.

The added salt imparts a salty taste to the product, as well as retarding mold growth and growth of contaminating microorganisms. Most important, the salt solution releases the mycelia-bound proteases. In sufu fermentation, the mold growth is limited to the surface of the cubes, and the mycelium does not penetrate into the tofu cubes. The enzymes produced by the mold, on the other hand, are not extracellular but are loosely bound to the mycelium, possibly by ionic linkage (Wang, 1967b), and can be easily eluted by NaCl or other ionic salt solutions. Therefore, the NaCl brine solution also serves to elute the enzyme from mycelia which, in turn, penetrate into the molded cubes and act upon the substrate protein. The minimal salt concentration to efficiently release the mycelia-bound proteases is about 3%. Thus, a less salty product than the traditional ones, which are too salty to suit Westerners' taste, may be obtained by reducing the salt content of brine solution from 12% to not less than 3%.

VI. TEMPEH

A. General Description

Tempeh, or tempe kedelee, originating in Indonesia, is made by fermenting dehulled and briefly cooked soybeans with a mold, *Rhizopus;* the mycelia bind the soybean cotyledons together in a firm cake. The raw tempeh has a clean, fresh, and yeasty odor. When sliced and deep-fat fried, it has a nutty flavor, pleasant aroma, and texture that are familiar and highly acceptable to almost all the people around the world. Tempeh does not possess the beany flavor some people find unpleasant in soybeans, and it is easy to cook and digest. Unlike most of the other fermented soybean foods, which are usually used as flavoring agents or relishes, tempeh is used as a main dish in Indonesia.

B. Method of Preparation

1. Traditional Method

Tempeh fermentation is characterized by its simplicity and rapidity. Making tempeh in Indonesia is a household art. The procedure may vary slightly from one household to another, but the principal steps are similar. Soybeans are soaked in tap water overnight until the hulls can be easily removed by rubbing between the hands. Some prefer to boil the soybeans for a few minutes to loosen the hulls and then to soak the beans overnight. After dehulling, the beans are boiled with excess water, drained, and surface-dried. Small pieces of tempeh from a previous fermentation, wrappings of the previous tempeh, or ragi tempeh (commercial starter) are mixed with the soybeans, which are then wrapped in wilted banana or other large leaves and incubated in a warm place for 24–48 hours until the beans are covered with white mycelium and bound together as a firm cake. The cake is either sliced into thin strips, dipped in a salt solution, and deep-fat fried in coconut oil or cut into pieces and used in soups much as we use chunks of meat.

2. Pure Culture Method

In the late 1950s scientists at the New York Agricultural Experiment Station, Geneva, New York, and at the Northern Regional Research Center, Peoria, Illinois, began to study this centuries-old fermentation. As a result, a pure-culture fermentation method for laboratory or industrial scale was developed.

Earlier reports (Stahel, 1946; Van Veen and Schaefer, 1950; Steinkraus *et al.*, 1960) indicated that the mold responsible for tempeh fermentation was *Rhizopus oryzae.* We have received cultures isolated from different lots of tempeh in Indonesia, and we found that only *Rhizopus* could make tempeh in pure-culture fermentation. Of the 40 strains of *Rhizopus* received, 25 of them are *R. oligosporus* Saito; others are *R. stolonifer* (Ehren) Vuill, *R. arrhizus* Fischer, *R. oryzae* Went and Geerligs, *R. formosaensis* Nakazawa, and *R. achlamydosporus* Takeda. Apparently, *R. oligosporus* is the principal species used in Indonesia for tempeh fermentation; a strain identified as *R. oligosporus* Saito NRRL 2710 is one of the better producers of a good product (Hesseltine *et al.*, 1963b). This strain is characterized by sporangiospores showing no striations and very irregular in shape under any condition of growth. The sporangiophores are short and unbranched and arise opposite rhizoids that are very reduced in length and branching. All isolates show large members of chlamydospores (Hesseltine, 1965).

In making tempeh, Hesseltine *et al.* (1963b) used a spore suspension of *R. oligosporus* NRRL 2710 grown on potato–dextrose–agar slant at 28°C for 5–7 days. Steinkraus and his co-workers (1965) used a powdered lyophilized inoculum made by growing pure culture on sterilized, hydrated soybeans; Rusmin and Ko (1974) made an inoculum by growing pure culture on steamed rice, but the inoculum also contained a considerable amount of bacterial spores (10^4–10^5 per gm). Wang *et al.* (1975a) developed a tempeh inoculum having a high viable spore count that would maintain its viability for a long time with minimal attention. The spores of *R. oligosporus* Saito NRRL 2710 were made by fermenting rice at 40% moisture level for 4–5 days at 32°C. The fermenting mass was made into a slurry by blending with sterilized water and then was freeze-dried. On a dry basis, the viable spore count per gram of preparation was about 1×10^9 before freeze-drying and 1×10^8 after freeze-drying. When the freeze-dried preparation was kept in a closed plastic bag at 4°C for up to 6 months, the spore counts showed typical experimental variations and were comparable to their original counts. At room temperature, a significant decrease in viability was noted after 2 months (from 2×10^7 to 1×10^6); thereafter, no further decrease was observed. The bacterial count of the preparation was minimal; therefore, bacterial contamination was not found to be a problem either during the process of fermentation or in storage. Since the beginning of 1977, the Northern Regional Research Center has distributed nearly 30,000 samples of this inoculum and the starter is now available commercially.

The preparation of soybeans for pure-culture fermentation is similar to that for the traditional method: soaking, dehulling, cooking, and surface-drying. The soybeans are soaked in water overnight at room temperature, dehulled by hand or by a simple roller mill, and then boiled in water for 30 minutes (Hesseltine *et al.*, 1963b; Rusmin and Ko, 1974). In 1964, Martinelli and Hesseltine introduced full-fat soybean grits (soybean cotyledons that have been mechanically cracked into four to five pieces) for tempeh fermentation (Fig. 4). Since soybean grits absorb water easily, the soaking time can be reduced from more than 20 hours to 30 minutes. Furthermore, since the hulls are removed mechanically in producing grits, much labor can be saved. Steinkraus and his co-workers (1960) suggested soaking the beans in 0.85% lactic or acetic acid solution. The soaked beans are dehulled with an abrasive vegetable peeler and then boiled in the acid solution for 90 min. With this treatment, the cooked beans have an acidity around 4–5 which, according to Steinkraus *et al.* (1960), is necessary to prevent the development of the spoilage bacteria during the mold fermentation. However, other investigators have not encountered bacterial growth in their processes. Because *R. oligosporus* produces an antibacterial agent (Wang *et al.*, 1969), and because this organism also has the unique characteristic of growing fast, there is probably little chance for bacteria to gain ground before the tempeh fermentation is complete.

The cooked beans are drained, spread to cool and to surface-dry, and thoroughly mixed with an inoculum of pure culture. A recommendation was made to use 10^6 spores per 100 gm of cooked beans (Wang *et al.*, 1975a). The inoculated beans are tightly packed into an appropriate container for incubation to obtain a final product in which a white mycelium has developed abundantly but black spores are minimal.

Like many other molds *R. oligosporus* requires air to grow, but it does

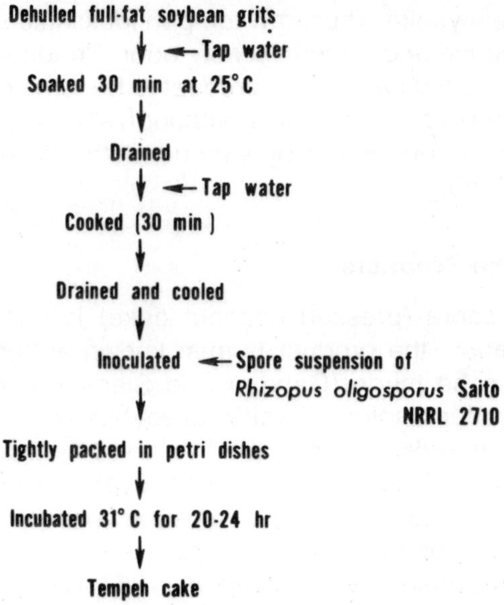

FIGURE 4. Flow sheet of tempeh fermentation.

not require as much aeration as many other molds; in fact, too much aeration will cause spore formation and may also dry up the beans, resulting in poor growth. Therefore, it is important to properly pack the beans for fermentation.

Hesseltine *et al.* (1963b) have carried out pure-culture fermentation in petri dishes, a testing procedure which proved to be very satisfactory. Shallow aluminum foil or metal trays, with perforated or woven mesh bottoms and covers or perforated plastic film covers, or perforated plastic bags and tubings have also been successfully used in tempeh fermentation (Steinkraus *et al.*, 1960, 1965; Martinelli and Hesseltine, 1964). Thickness of the package and the size and the distance of the perforations are important. Containers of 2.5–3.0 cm in depth with holes having a diameter of about 0.6 mm and a distance apart of about 5 mm are most desirable.

Tempeh fermentation can be carried out at temperatures ranging from 25° to 37°C; however, the time required for fermentation decreases as temperature increases, 80 hours at 25°C and 15–18 hours at 35°–38°C (Martinelli and Hesseltine, 1964; Steinkraus *et al.*, 1965). At higher temperatures the soybeans tend to dry out; consequently, the mold growth is suppressed and there is a greater chance for bacteria to grow. We found (Wang *et al.*, 1975a) that at the inoculum level of 1×10^6 spores per 100 gm of cooked substrate tempeh fermentation can be best carried out at 32°C for 20–22 hours.

When fermentation is complete, the beans are covered with and bound together by white mycelia. Thus, raw tempeh looks like a firm white cake and has an attractive and slightly yeasty odor. Prolonged fermentation, on the other hand, often causes the product to become obnoxious due to the enzymatic breakdown of proteins. Although shelf-life of the product is also short, it can be prolonged by various methods, such as freezing, drying, and canning.

C. Tempeh-like Products

In Indonesia, copra (pressed coconut cake) is sometimes used in tempeh fermentation; the product is then known as tempeh bongkrek. Tempeh can also be made (Gandjar and Slamet, 1972; Wang *et al.*, 1975a) from the water-insoluble fraction of soybeans which is the residue of making soybean milk and tofu, the two main food products derived from water extraction of soybeans (Wang, 1967a). However, the moisture content of this water-insoluble fraction must be reduced to less than 80% so that the texture appears crumbly before it is suitable for fermentation. We have also developed (Hesseltine *et al.*, 1967) new tempeh-like products by fermenting cereal grains such as wheat, oats, barley, rice, or mixtures of cereals and soybeans with *Rhizopus*. However, it is essential to slightly modify the surface of the grain or to crack the grain for good growth of the mold. Fermentation of starch tubes such as cassava with *Rhizopus* can also result in a food product with significant increases in protein content (Stanton and Wallbridge, 1969).

Another traditional Indonesian food closely related to tempeh is known as ontjom or oncom. Ontjom is made from peanut press cake by strains of *Neurospora* or *Rhizopus; Neurospora* fermentation results in a pink or orange cake, and *Rhizopus* in a cake of ash-grey color (Van Veen and Graham, 1968). The preparation and the use of ontjom are quite similar to that of tempeh.

D. Biochemistry and Physiology of *Rhizopus oligosporus*

The utilization of various carbon and nitrogen compounds by *R. oligosporus* Saito NRRL 2710 was investigated by Sorenson and Hesseltine (1966). They found that the principal carbohydrates of soybeans, i.e., stachyose, raffinose, and sucrose, are not utilized as sole sources of carbon, whereas common sugars such as glucose, fructose, galactose, and maltose supported excellent growth, as does xylose. Various vegetable oils can be substituted for sugars as sources of carbon with excellent growth. Since the soybean sugars are not utilized by *R. oligosporus*, and since strong lipase activity has been reported for *Rhizopus* cultures used in tempeh fermentation (Wagenknecht *et al.*, 1961; Wang and Hesseltine, 1966), it is likely that lipid materials, and particularly fatty acids, are the primary sources of energy for the tempeh fermentation. Ammonium salts and amino acids such as proline, glycine, aspartic acid, and leucine are excellent sources of nitrogen. Other amino acids are less suitable, and tryptophan supports no growth at all. However, the fungus does not depend upon the presence of any specific amino acid in the medium for growth.

Hesseltine *et al.* (1963a) found a water-soluble, heat-stable mold inhibitor in soybeans which inhibits the growth of *R. oligosporus*. Later, Wang and Hesseltine (1965) found that the water-soluble and heat-stable fraction of soybeans also inhibited the formation of proteolytic enzymes by *R. oligosporus*. Therefore, soaking and cooking of soybeans in excess water, which is later discarded, are essential in making tempeh.

Rhizopus oligosporus is highly proteolytic, which is important in tempeh fermentation because of the high protein content of the substrate. The ability of *Rhizopus* to produce proteolytic enzymes varies greatly between different strains of the same species as well as between species (Wang and Hesseltine, 1965). The proteolytic enzyme systems have optimal pH at 3.0 and 5.5, with the pH 3.0 type predominating in submerged cultivation and pH 5.5 type predominating in tempeh fermentation. Both enzyme systems have maximum activities at 50°–55°C and are fairly stable at pH 3.0–6.0 but are rapidly denatured at pH below 2 or above 7. The proteolytic enzyme having optimum pH at 3.0 has been purified and separated into five active fractions; crystalline enzymes were obtained from the two major fractions (Wang and Hesseltine, 1970b).

In addition to high protease activity, the mold possesses strong lipase activity (Wagenknecht *et al.*, 1961; Wang and Hesseltine, 1966) but low amylase activity and no detectable pectinase (Hesseltine *et al.*, 1963b). Since starch is seldom found in mature soybeans, it is not particularly important that this species produces amylase during tempeh fermentation. Lipase is produced by the mold to hydrolyze soybean lipids.

Although antibiotics have frequently been isolated from fungal cultures, only a few have been reported from cultures of the class Phycomycetes. We found that strains of *Rhizopus* and *Actinomucor* do produce antibacterial agents (Wang *et al.*, 1972a; Ellis *et al.*, 1974). The compound produced by *R. oligosporus* (Wang *et al.*, 1969) is especially active against some gram-positive bacteria, including both microaerophilic and anaerobic bacteria, e.g., *Streptococcus cremoris, Bacillus subtilis, Staphylococcus aureus, Clostridium perfringens,* and *C. sporogenes*. The compound contains polypeptides having high carbohydrate contents. Its activity is not affected by pepsin or *R. oligosporus* proteases, is slightly decreased by trypsin and peptidase, but is rapidly inactivated by pronase. Calloway *et al.* (1971) found that tempeh did not increase gas production over baseline values of healthy young men and caused a significant delay in the time of gas forming, which suggests temporary suppression of intestinal bacteria. The delay could well be due to the presence of the antibiotic substance produced by the mold, *Rhizopus*.

E. Changes Occurring during Fermentation

The effects of *R. oligosporus* on soybeans have been studied by several investigators and reviewed by Hesseltine and Wang (1972) and Iljas *et al.* (1973). Steinkraus *et al.* (1960) found that the temperature of fermenting beans rises to above that of the incubators as fermentation progresses, but that it falls as the growth of mold subsides. The pH increases steadily, presumably because of the protein breakdown. After 69 hours of incubation, soluble solids rise from 13 to 28%; soluble nitrogen also increases from 0.5 to 2.0%, whereas total nitrogen remains fairly constant and reducing substances slightly decrease, probably due to utilization by the mold. Similar changes were observed when wheat was fermented by *R. oligosporus* (Wang and Hesseltine, 1966). A decrease in ether-extractable substances of soybeans after fermentation was reported by Murata *et al.* (1967) and Wang *et al.* (1968), indicating that the mold uses the soybean oil as its energy source (Sorensen and Hesseltine, 1966). Wagenknecht *et al.* (1961) reported that one-third of the total ether-extractable soybean lipid is hydrolyzed by the mold after 69 hours of incubation, and among all the fatty acids, 40% of the linolenic acid is utilized by the mold. The phytic acid content of soybeans is reduced by about one-third as a result of this fermentation (Sudarmadji and Markakis, 1977).

Although total nitrogen remains fairly constant during fermentation, the free amino acid in tempeh increases. The amino acid composition of soybeans, on the other hand, is not significantly changed by fermentation (Smith *et al.*, 1964; Stillings and Hackler, 1965). Perhaps the amount of mycelial protein present in tempeh is not high enough to alter greatly the amino acid composition of the soybeans, nor does the mold depend upon any specific amino acid for growth as suggested by Sorenson and Hesseltine (1966).

Niacin, riboflavin, pantothenic acid, and vitamin B_6 contents of soybeans increase after fermentation, whereas thiamin has no significant change (Roelofsen and Talens, 1964; Murata *et al.*, 1967). Wang and Hesseltine (1966) also noticed in fermenting wheat with *R. oligosporus* that the amount of niacin and riboflavin of the wheat tempeh greatly exceeds that of unfermented wheat, while thiamin appears to be less. Apparently, *R. oligosporus* has a great synthetic capacity for niacin, riboflavin, pantothenic acid, and vitamin B_6, but not for thiamin.

Lipids in tempeh were found to be more resistant to autoxidation than those in control soybeans. The peroxide value of tempeh was 1.1, whereas that of control soybeans was 18.3–201.9 (Iljas, 1969). Ikehata *et al.* (1968) found that the peroxide value of lyophilized tempeh stored at 37°C for 5 months increased from 6 to 12 compared with the increase of from 6 to 426 in unfermented soybeans. The antioxidant activity of tempeh was further substantiated by Packett *et al.* (1971), who reported that corn oil containing 50% tempeh showed higher antioxidant potential than oils containing 25% tempeh, 0.01% α-tocopherol, or 0.03% α-tocopherol. In 1964, György *et al.* isolated a new isoflavone from tempeh designated as "Factor 2," which was then identified as 6,7,4′-trihydroxyisoflavone. 6,7,4′-Trihydroxyisoflavone was later chemically synthesized and proved to be a potent antioxidant for vitamin A and for linoleate in aqueous solution at pH 7.4 (Ikehata *et al.*, 1968). However, when the isoflavone was mixed with soybean powder or soybean oil, it did not prevent their autoxidation. These authors speculated that the insolubility of the isoflavone in the oil and the difficulty of dispersing into soybean powder may be some of the reasons for its failure to prevent autoxidation.

Although *R. oligosporus* does not synthesize substances that inhibit the activity of trypsin, the trypsin-inhibitory activity of the soybeans was found to increase as the tempeh fermentation progressed (Wang *et al.*, 1972b). The compounds responsible for this increased trypsin-inhibitory activity have been identified as free fatty acids, primarily oleic, linoleic, and linolenic acids (Wang *et al.*, 1975b). The free fatty acids are liberated from oil in the soybeans by fungal lipase, and they differ from other reported soybean trypsin inhibitors that are protein in nature. Free fatty acids have been previously reported to inhibit various enzymes, such as glycolytic, glyconeogenic, lipogenic, and also proteolytic enzymes. Their effect appears to be a nonspecific type of inhibition.

In the production of tempeh, soybeans are only partially cooked and they remain nearly as firm as the soaked beans. After fermentation the beans are soft and similar in texture to completely cooked soybeans. An earlier cytological study (Steinkraus *et al.*, 1960) showed only slight penetration of the mycelia into the underlying tissue of the bean, suggesting that the digestion was mainly enzymatic. However, a recent study (Jurus and Sundberg, 1976) revealed hyphae infiltration to a depth of 742 μm or about 25% of the average width of a soybean cotyledon. These authors speculated that the extreme depth of mycelial infiltration partially explains the rapid physical and chemical changes occurring during tempeh fermentation. The hyphae may mechanically push the bean cells apart prior to, or in conjunction with, enzymatic digestion; thus, the beans become soft. Likewise, the penetration of enzymatic activity could also be enhanced, since the distance over which diffusion of enzyme must occur is greatly reduced.

F. Nutritional Value

Indonesians consider tempeh to be a nourishing and easily digestible food, and it is used as a main dish. Van Veen and Schaefer (1950) observed beneficial effects of tempeh on patients with dysentery in the prison camps of World War II, and they suggested that tempeh was much easier to digest than soybeans. However, animal feeding experiments have not substantiated this conclusion (Hackler *et al.*, 1964; Smith *et al.*, 1964; Murata *et al.*, 1967), even though more than half of the soybean protein, fat, and N-free extract could be solubilized by 72-hour fermentation (Van Buren *et al.*, 1972). The increase in vitamin content, as previously mentioned, must have a great nutritional significance to the people of those countries whose food industries do not enrich their foods with vitamins as do those in the United States.

The protein quality of tempeh as measured with the protein efficiency ratio (PER) is not significantly different from that of unfermented beans (Hackler *et al.*, 1964; Smith *et al.*, 1964; Wang *et al.*, 1968; Murata *et al.*, 1971). However, the quality of tempeh protein can be improved by making tempeh from mixtures of cereals and soybeans. For example, the PER value of wheat–soybean (1 : 1) tempeh was comparable to that of casein (Wang *et al.*, 1968). Not only does a mixture of wheat and soybean (1 : 1) have a balanced pattern of amino acid, but the authors also found that the PER of wheat was greatly increased by *Rhizopus* fermentation.

The superior nutritive value of tempeh over unfermented soybeans has been noted by György (1961) on animals fed low-protein diets. His results resemble those obtained with animals fed antibiotics added to their protein source. Therefore, the finding of antibacterial agents produced by *R. oligosporus* may offer a better understanding of the value of tempeh in the diet of Indonesians, and, perhaps, of fermented foods in the diets of all Orientals.

The cooking procedures also affect the nutritional value of tempeh, and the PER value of tempeh significantly declined after more than 3 minutes of frying in oil; on the other hand, steaming for up to 2 hours had no effect (Hackler *et al.*, 1964).

Because of the high acceptability of taste and texture, lack of beany flavor, nutritional advantages, and its simple, low-cost processing techniques, tempeh appears to be a good candidate for any country searching for a low-cost, high-protein food. With the recently increasing interest of vegetarians in foods of vegetable origin, tempeh consumption has been on an upsurge in this country. In addition to tempeh making as a home project, several commercial tempeh producers have been established. Tempeh may soon be a regular item in the United States market.

VII. ANG-KAK

A. General Description

Unlike other mold-modified products, which are usually used as a flavoring agent or as a protein source, ang-kak or red rice is a color agent. It was originated in China and is a product made by fermenting rice with strains of *Monascus purpureus.* Because of its red color, ang-kak is used for coloring various foods and as a color additive for manufacturing fermented foods such as sufu, red wine, fish sauce, and fish paste in the Orient.

B. Cultures

Only those strains (Church, 1920) that produce a dark-red growth throughout the rice kernels, at low enough moisture levels to allow the individual grains to remain separate from one another, are suitable for the fermentation. *Monascus purpureus* NRRL 2897 maintained in the AR Culture Collection was isolated from an ang-kak sample bought in the Philippines market and has demonstrated the ability to carry out the fermentation successfully. Palo *et al.* (1960) studied various conditions of the fermentation and found that the optimum temperature for pigment formation is about 27°C. Growth will occur at as low as 20°C and as high as 37°C, but at these extremes poor pigmentation results. The mold will produce the pigment over a wide range of pH values (3–7.5).

Lin (1973) isolated a strain of *Monascus* sp. F-2 from kaoliang koji and found that it produced large amounts of pigment by submerged culture with rice as a sole carbon source. *N*-Methyl-*N'*-nitro-*N*-nitrosoguanidine treatment and successive isolation (Lin and Suen, 1973) greatly improved the yield of the pigment production by *Monascus* sp. F-2. Two hyperpigment-producing strains, R-1 and R-2, were isolated which produced five times more red pigment than the original parent strain F-2 did.

C. Method of Preparation

All varieties of rice are suitable except the glutinous ones, which are unsatisfactory because the rice becomes gluey and the grains stick together. The procedure for preparing ang-kak on a laboratory scale as developed by Palo *et al.* (1960) is to first wash the rice, soak it in water for 24 hours, and then drain it thoroughly. The rice is placed in a beaker or other suitable container that is large enough to have considerable air space above the rice. It is then covered with a milk filter disk, autoclaved for 30 minutes at 121°C, cooled, and inoculated with a sterile water suspension of ascospores removed from 25-day-old cultures of *M. purpureus* grown on Sabouraud's agar. After the suspension is thoroughly mixed with the rice, it should be incubated at 25°–32°C. At the time of inoculation the rice will seem rather dry; in fact, the rice should never be wet or mushy. This is the most important state for the good production of ang-kak. In 3 days, the rice should begin to redden. At this time, the material should be stirred and shaken to redistribute the rice in the center and bottom of the fermentor and to retain an even moisture at the surface. We have found that an occasional vigorous shaking of the fermentation jar is quite adequate. Addition of some sterile water may be required from time to time to replace lost moisture. Care must be taken only to moisten and not to soak the rice. In 3 weeks, the rice should be deep purplish red, with each rice kernel unattached to its neighbors. The material is then dried in an oven at 40°C. When the kernels are broken, the pigment should be completely through the rice. Each grain should crumble easily between one's fingers.

D. Pigments

The ang-kak studied by Nishikawa (1932) consisted of two pigments, principally monascorubrin, a red coloring material with a formula of $C_{22}H_{24}O_5$, and a small amount of monascoflavin, a yellow pigment ($C_{17}H_{22}O_4$). The pigments tend to accumulate in the microorganisms because of their poor solubility. Recently Yamaguchi *et al.* (1973) reported that the pigment can be solubilized by reacting it with a water-soluble protein, a water-soluble peptide, an amino acid, or a mixture of these. The reaction is carried out at a pH from 5 to 8.5 and requires from several minutes to several hours. As the reaction progresses, the color of the pigment changes from a brownish red to a deep scarlet and the pigment becomes water soluble. The solubility of the pigment in water varies with the type of protein, peptide, or amino acid used. The same authors also indicated that the water-soluble pigment can be produced directly by the mold grown in a medium containing soluble protein, peptide, or amino acid, at a pH of 7–9 under aerobic conditions, and at a temperature of about 27°C for 48–72 hours. The mechanism of this reaction is not clear; Yamaguchi *et al.* (1973) speculated that the water-soluble pigment is a complex resulting from binding between the pigment and the chosen amino group. The water-soluble pigment is suitable for coloring foodstuffs, particularly sour-type drinks, candy, milk products, and meat products.

VIII. ABSENCE OF MYCOTOXINS IN FERMENTED FOODS

The mycotoxin in which most interest has developed is the carcinogen, aflatoxin, produced by *Aspergillus flavus* and *A. parasiticus.* These species are in the same group as *A. oryzae,* a species widely used in Oriental food fermentations. Therefore, the Japanese scientists have looked extensively at the aflatoxin problem in fermented products.

Yokotsuka *et al.* (1967a,b) investigated 73 industrial strains of *Asper-*

gillus, including *A. oryzae, A. sojae, A. niger,* and *A. usami,* for production of fluorescent compounds after cultivation on a Czapek–Dox medium containing zinc. No aflatoxin was found, but about 30% of these strains produced compounds that were very similar to those of aflatoxin in fluorescence spectra and R_f values in thin-layer chromatograms. However, the fluorescent compounds proved to be nontoxic (Manabe *et al.*, 1968, 1972; Manabe and Matsuura, 1972a,b). Some strains produced aspergillic acid in modified Mayer's medium at 30°C by surface culture after more than 10 days, and on solid soy sauce koji fermentation after 4–10 days (Yokotsuka, 1971). However, the koji is usually harvested within 2–3 days in manufacturing soy sauce and miso. The same author also reported that a small amount of β-nitropropionic acid was produced by one of the tested strains in the beginning of the culturing, but it rapidly disappeared. Kojic, oxalic, or formic acids were formed by some molds during koji making, but only in trace amounts. Considering the low toxicity of these compounds, they create no problems. Similar results were also obtained by other investigators (Murakami *et al.*, 1967, 1968a,b; Matsuura *et al.*, 1970).

Earlier, our laboratory (Hesseltine *et al.*, 1966) examined samples of soy sauce, miso, and tou-shih and of soybean, rice, and wheat tempehs, but we found no aflatoxin. Rice koji, miso, and soy sauce samples collected from various producers in Japan were examined and found negative for aflatoxin (Matsuura, 1970).

Maing *et al.* (1973) prepared soy sauce with a strain of *A. oryzae* NRRL 1988 and with an aflatoxin-producing strain of *A. parasiticus* NRRL 2999. No aflatoxin was produced in the soy sauce prepared with *A. oryzae,* but large amounts of aflatoxin were formed by *A. parasiticus* when it was grown with *A. oryzae.* After brining for 6 weeks, large amounts of aflatoxin remained in the fermentation in which *A. parasiticus* was grown. Obviously, if an aflatoxin-producing strain of *A. flavus-* or *A. parasiticus*-contaminated culture was used to make koji for soy sauce, the soy sauce would contain toxin.

IX. CONCLUSIONS

Among many mold-modified foods consumed around the world, we have been concerned only with those that have been scientifically studied. Information on the nature of those products, the methods of manufacture, the correctly identified cultures, and the properties of the microorganisms that make them unique in their particular fermentation have been discussed. Although future studies are needed to develop uniform high-quality products and well-defined economical processes to manufacture them, pure culture methods for preparing those foods have been developed which would facilitate further studies.

Obviously, the preparation of foods by fermentation has certain advantages that can be summarized as follows: Fermentation (1) produces desirable enzymes, (2) destroys or masks undesirable flavor and odors, (3) adds flavor and odor, (4) preserves, (5) synthesizes desirable constituents such as vitamins and antibiotics, (6) increases digestibility, (7) changes physical states, (8) produces color, and (9) reduces cooking time.

If properly fermented, these foods are not hazardous to health. The microorganisms responsible for these processes are not toxin producers.

REFERENCES

Baens-Arcega, L. (1966). Philippines Patent 2,553.

Bush, L. (1959). "Land of the Dragonfly," pp. 28–51. Robert Hale, Ltd., London.

Calloway, D. H., Hickey, C. A., and Murphy, E. L. (1971). *J. Food Sci.* **36,** 251–255.

Church, M. B. (1920). *J. Ind. Eng. Chem.* **12,** 45–46.

Church, M. B. (1923). *U.S., Dep. Agric., Bull.* **1152,** 1–26.

Ebine, H. (1967). U.S. Dep. Agric. Final Tech. Rep., Public Law 480. Project UR-A11-(40)-2. Natl. Agric. Libr., Beltsville, Maryland.

Ebine, H. (1971). *In* "Conversion and Manufacture of Foodstuffs by Microorganisms—A Symposium," pp. 127–132. Saikon Publishing Co., Ltd., Kyoto, Japan.

Ebine, H. (1976). *In* "Expanding the Use of Soybeans for Asia and Oceania—A Conference," (R. M. Goodman, ed.), INTSOY Ser. No. 10, pp. 126–129. University of Illinois, Urbana.

Ellis, J. J., Wang, H. L., and Hesseltine, C. W. (1974). *Mycologia* **66,** 593–599.

Gandjar, I., and Slamet, D. S. (1972). *Penelitian Gizi dan Makanan Jilia* **2,** 70–79.

György, P. (1961). *N. A. S.–N. R. C., Publ.* **843,** 281–289.

György, P., Murata, K., and Ikehata, H. (1964). *Nature (London)* **203,** 870–872.

Hackler, L. R., Steinkraus, K. H., Van Buren, J. P., and Hand, D. B. (1964). *J. Nutr.* **82,** 452–456.

Harada, Y. (1951). *Rep. Tatsuno Inst. Soy Sauce* **2,** 51–55.

Hesseltine, C. W. (1965). *Mycologia* **57,** 149–197.

Hesseltine, C. W., and Wang, H. L. (1972). *In* "Soybeans: Chemistry and Technology" (A. K. Smith and S. J. Circle, eds.), Vol. 1, pp. 389–419. Avi Publ. Co., Westport, Connecticut.

Hesseltine, C. W., Decamargo, R., and Rackis, J. J. (1963a). *Nature (London)* **200,** 1226–1227.

Hesseltine, C. W., Smith, M., Bradle, B., and Djien, K. S. (1963b). *Dev. Ind. Microbiol.* **4,** 275–287.

Hesseltine, C. W., Shotwell, O. L., Ellis, J. J., and Stubblefield, R. D. (1966). *Bacteriol. Rev.* **30,** 795–805.

Hesseltine, C. W., Smith, M., and Wang, H. L. (1967). *Dev. Ind. Microbiol.* **8,** 179–186.

Hesseltine, C. W., Swain, E. W., and Wang, H. L. (1976). *Dev. Ind. Microbiol.* **17,** 101–115.

Ikehata, H., Wakaizumi, M., and Murata, K. (1968). *Agric. Biol. Chem.* **32,** 740–746.

Iljas, N. (1969). M.S. thesis, Ohio State University, Columbus.

Iljas, N., Peng, A. C., and Gould, W. A. (1973). *Ohio Agric. Res. Dev. Cen., Hortic. Ser.* **394.**

Jurus, A. M., and Sundberg, W. J. (1976). *Appl. Environ. Microbiol.* **32,** 284–287.
Kiuchi, K., Ohta, T., Itoh, H., Takabayashi, T., and Ebine, H. (1976). *J. Agric. Food Chem.* **24,** 404–407.
Komatsu, Y. (1968). *Seas. Sci.* **15,** No. 2, 10–20 (in Japanese).
Kon, M., and Ito, H. (1975). *Rep. Natl. Food Res. Inst. (Tokyo)* **30,** 232–237.
Lin, C. F. (1973). *J. Ferment. Technol.* **51,** 407–414.
Lin, C. F., and Suen, S. J.-T. (1973). *J. Ferment. Technol.* **51,** 757–759.
Maing, I. V., Ayers, A. C., and Koehler, P. F. (1973). *Appl. Microbiol.* **25,** 1015–1017.
Manabe, M., and Matsuura, J. (1972a). *J. Food Sci. Technol. (Tokyo)* **19,** 268–274.
Manabe, M., and Matsuura, J. (1972b). *J. Food Sci. Technol. (Tokyo)* **19,** 275–279.
Manabe, M., Matsuura, S., and Nakano, M. (1968). *J. Food Sci, Technol. (Tokyō)* **15,** 341–346.
Manabe, M., Ohnuma, S., and Matsuura, J. (1972). *J. Food Sci. Technol. (Tokyo)* **19,** 76–80.
Martinelli, A. F., and Hesseltine, C. W. (1964). *Food Technol.* **18,** 167–171.
Matsuura, S. (1970). *JARQ* **5,** 46–51.
Matsuura, S., Manabe, M., and Sato, T. (1970). *Proc. U.S.-Jpn. Conf. Toxic Micro-Org.: 1968,* pp. 48–55.
Maxwell, M. E. (1952). *Aust. J. Sci. Res., Ser. B* **5,** 43–55.
Murakami, H., Takase, S., and Ishii, I. (1967). *J. Gen. Appl. Microbiol.* **13,** 323–334.
Murakami, H., Takase, S., and Kuwabara, K. (1968a). *J. Gen. Appl. Microbiol.* **14,** 97–110.
Murakami, H., Sagawa, H., and Takase, S. (1968b). *J. Gen. Appl. Microbiol.* **14,** 251–262.
Murata, K., Ikehata, H., and Miyamoto, T. (1967). *J. Food Sci.* **32,** 580–588.
Murata, K., Ikehata, H., Edani, Y., and Yojanagi, K. (1971). *Agric. Biol. Chem.* **35,** 233–241.
Nakadai, T., Nasuno, S., and Iguchi, N. (1975). U.S. Patent 3,914,436.
Nishikawa, H. (1932). *J. Agric. Chem. Soc. Jpn.* **8,** 1007–1015.
Noda Soy Sauce Company, Ltd. (1955). "The Treatment of Soy Sauce Fermentation." Noda-shi, Chiba-ken, Japan.
Oda, M., Ikeda, K., and Tanimoto, M. (1949). *J. Ferment. Technol.* **27,** 16–23.
Packett, L. V., Chen, L. H., and Liu, J. T. (1971). *J. Food Sci.* **36,** 798–799.
Palo, M. A., Vidal-Adeva, L., and Maceda, L. M. (1960). *Philipp. J. Sci.* **89,** 1–22.
Roelofsen, P. A., and Talens, A. (1964). *J. Food Sci.* **29,** 224–226.
Rusmin, S., and Ko, S. D. (1974). *Appl. Microbiol.* **28,** 347–350.
Shibasaki, K., and Hesseltine, C. W. (1962). *Econ. Bot.* **16,** 180–195.
Smith, A. K., Hesseltine, C. W., and Shibasaki, K. (1961). U.S. Patent 2,967,108.
Smith, A. K., Rackis, J. J., Hesseltine, C. W., Smith, M., Robbins, D. J., and Booth, A. N. (1964). *Cereal Chem.* **41,** 173–181.
Sorenson, W. G., and Hesseltine, C. W. (1966). *Mycologia,* **58,** 681–689.
Stahel, G. (1946). *J. N. Y. Bot. Gard.* **47,** 261–267.
Stanton, W. R., and Wallbridge, A. (1969). *Process Biochem.* **4,** No. 4, 45–51.
Steinkraus, K. H., Hwa, Y. B., Van Buren, J. P., Provvidenti, M. I., and Hand, D. B. (1960). *Food Res.* **25,** 777–788.
Steinkraus, K. H., Van Buren, J. P., Hackler, L. R., and Hand. D. B. (1965). *Food Technol.* **19,** 63–98.
Stillings, B. R., and Hackler, L. R. (1965). *J. Food Sci.* **30,** 1043–1049.
Sudarmadji, S., and Markakis, P. (1977). *J. Sci. Food Agric.* **28,** 381–383.
Tateno, M., and Umeda, I. (1955). Japanese Patent 204858.
Umeda, I. (1963). U.S. Dep. Agric. Final Tech. Rep., Public Law 480. Project UR-A11-(40)-(C). Natl. Agric. Libr., Beltsville, Maryland.
Umeda, I., Nakamura, K., Yamato, M., and Nakamura, Y. (1969). U.S. Dep. Agric. Final Tech. Rep., Public Law 480 Project UR-A11-(40)-21. Natl. Agric. Libr., Beltsville, Maryland.

Van Buren, J. P., Hackler, L. R., and Steinkraus, K. H. (1972). *Cereal Chem.* **49,** 208–211.
Van Veen, A. G., and Graham, D. C. W. (1968). *Cereal Sci. Today* **13,** 96–99.
Van Veen, A. G., and Schaefer, G. (1950). *Trop. Geogr. Med.* **2,** 270–281.
Wagenknecht, A. C., Mattick, L. R., Lewin, L. M., Hand, D. B., and Steinkraus, K. H. (1961). *J. Food Sci.* **26,** 373–376.
Wai, N. (1929). *Science* **70,** 307–308.
Wai, N. (1968). U.S. Dep. Agric. Final Tech. Rep., Public Law 480. Project UR-A6-(40)-1. Natl. Agric. Libr., Beltsville, Maryland.
Wang, H. L. (1967a). *Food Technol.* **21,** 115–116.
Wang, H. L. (1967b). *J. Bacteriol.* **93,** 1794–1799.
Wang, H. L., and Hesseltine, C. W. (1965). *Can. J. Microbiol.* **11,** 727–732.
Wang, H. L., and Hesseltine, C. W. (1966). *Cereal Chem.* **43,** 563–570.
Wang, H. L., and Hesseltine, C. W. (1970a). *J. Agric. Food Chem.* **18,** 572–575.
Wang, H. L., and Hesseltine, C. W. (1970b). *Arch. Biochem. Biophys.* **140,** 459–463.
Wang, H. L., Ruttle, D. I., and Hesseltine, C. W. (1968). *J. Nutr.* **96,** 109–114.
Wang, H. L., Ruttle, D. I., and Hesseltine, C. W. (1969). *Proc. Soc. Exp. Biol. Med.* **131,** 579–583.
Wang, H. L., Ellis, J. J., and Hesseltine, C. W. (1972a). *Mycologia* **64,** 218–221.
Wang, H. L., Vespa, J. B., and Hesseltine, C. W. (1972b). *J. Nutr.* **102,** 1495–1499.
Wang, H. L., Vespa, J. B., and Hesseltine, C. W. (1974). *Appl. Microbiol.* **27,** 906–911.
Wang, H. L., Swain, E. W., and Hesseltine, C. W. (1975a). *J. Food Sci.* **40,** 168–170.
Wang, H. L., Swain, E. W., Wallen, L. L., and Hesseltine, C. W. (1975b). *J. Nutr.* **105,** 1351–1355.
Watanabe, T. (1969). "UNIDO Expert Group Meeting on Soybean Processing and Use." USDA, Peoria, Illinois.
Yamaguchi, Y., Ito, H., Watanabe, S., Yoshida, T., and Komatsu, A. (1973). U.S. Patent 3,765,905.
Yamamoto, K. (1957). *Bull. Agric. Chem. Soc. Jpn.* **21,** 319–324.
Yokotsuka, T. (1960). *Adv. Food Res.* **10,** 75–134.
Yokotsuka, T. (1964). "Oilseed Protein Foods—a Symposium," pp. 31–48. Mt. Fuji, Japan.
Yokotsuka, T. (1971). "Conversion and Manufacture of Foodstuffs by Microorganisms—a Symposium," pp. 117–125. Saikon Publishing Co., Ltd., Kyoto, Japan.
Yokotsuka, T., Sasaki, M., Kikuchi, T., Asao, Y., and Nobuhara, A. (1967a). *In* "Biochemistry of Some Foodborne Microbial Toxins" (R. I. Mateles and G. N. Wogan, eds.), pp. 131–152. MIT Press, Cambridge, Massachusetts.
Yokotsuka, T., Sasaki, M., Kikuchi, T., Asao, Y., and Nobuhara, A. (1967b). *J. Agric. Chem. Soc. Jpn.* **41,** 32–38.
Yong, F. M., and Wood, B. J. B. (1974). *Adv. Appl. Microbiol.* **17,** 157–194.

Chapter 5

Wine

HERMAN J. PHAFF
MAYNARD A. AMERINE

MICROBIAL TECHNOLOGY, 2nd ed., VOL.II
 ISBN 0-12-551502-2

I. INTRODUCTION

A. Definition

Wine is, by definition, "produced by normal alcoholic fermentation of the juice of sound ripe grapes (including restored or unrestored pure condensed grape must), with or without added fortifying grape spirits or alcohol, but without other addition or without the addition or abstraction except as may occur in cellar treatment: Provided, that the product may be ameliorated before, during or after fermentation." (U.S. Internal Revenue Service, 1974). The regulations also provide for maximum limits for volatile acidity, amount of sulfur dioxide, and ethanol limits for various types of wines. Amelioration (addition of sugar and/or water) is permitted in most states, but not in California.

Fruit wines are also provided for in the regulations. In this country, few have been produced. Fruit and citrus wine must also be "produced by the normal alcoholic fermentation of the juice of sound, ripe fruit (including restored or unrestored pure condensed fruit must), with or without the addition, after fermentation, of pure condensed fruit must, and with or without added fruit brandy or alcohol, but without other addition or abstraction except as may occur in cellar treatment: Provided, that . . . the product may be ameliorated." (U.S. Internal Revenue Service, 1974).

Federal and state regulations (see, for example, California, 1970) in this country also control many winery operations, wine labeling, tax levels, etc. In other wine-producing countries, regulations of varying degrees of complexity apply (Amerine *et al.*, 1973).

B. Size of the Industry

While the wine industry includes fruits other than grapes, the volume of nongrape wine is very small—probably less than 1%. World wine production (from grapes) has been increasing since World War II. This is due to greater yields per acre as well as to large new acreages (as in Australia, California, and the Soviet Union). Currently about 9 billion gal (or 335 million hectoliters) of wine are produced annually. Nearly 80% of this is produced in Europe, with 14% from North and South America, over 4% from Africa, and only 1% each from Asia and Oceania. To supply the raw material, more than 22 million acres of grapes (10 million ha) are planted. Some of this acreage, perhaps 15%, is used for table grapes and raisin production.

The economic value of the world wine industry has not been calculated. It would include grape producers, wineries, auxiliary industries (producing bottles, corks, etc.), transportation to national and international markets, wholesalers, retailers, restaurant wine services, and local, state, and national taxes and law enforcement. The economic value of the small California wine industry (FOB the winery) has been estimated at nearly a billion dollars annually. The retail value is, of course, much greater.

C. Historical Background

Fermentation predates recorded history. Honey and grains were probably fermented as household beverages before grapes. The early wine industry of the Fertile Crescent, about 3500 B.C., spread west (around the Mediterranean) and north (to Hungary, Germany, and France), and in the post-Columbus period to the Americas, South Africa, and Oceania. The Romans advanced the art of wine making, but it was an industry of large risks due to spoilage until the mid-nineteenth century. The research of Louis Pasteur revolutionized the wine industry. From a high-risk labor-intensive industry, it has been converted into a low-risk industry with all the accoutrements of a large-scale modern food production. While more and more technology has been introduced into the industry, in variety selection, location of vineyards, crop level, process control, aging, and blending, considerable art remains in quality evaluation.

The technology of the industry has been summarized in books by Amerine *et al.* (1973), Amerine and Joslyn (1970), Joslyn and Amerine (1964), Ribéreau-Gayon and Peynaud (1972–1976), Troost (1971), and others. A useful list of bibliographies on wines and related subjects has been published by Amerine and Singleton (1971).

II. MICROBIOLOGICAL ASPECTS

A. Nature of the Microbial Process

Wine fermentation is basically the transformation of the various sugars of grapes by yeast under anaerobic conditions into ethanol, carbon dioxide, and small amounts of by-products. D-Glucose and D-fructose, the two principal sugars of grape juice, yield essentially equimolar proportions of ethanol and carbon dioxide. Desirable strains for this process are representatives of the species *Saccharomyces cerevisiae* Hansen or several closely related species, such as *Saccharomyces oviformis* Osterwalder. The nomenclature of yeast species mentioned in this chapter is based on the usage in "The Yeasts—A Taxonomic Study" (Lodder, 1970). Yeast species and strains vary considerably in their alcohol tolerance, rate of fermentation, completeness of sugar to ethanol conversion, and nature and proportion of desirable by-products.

Natural fermentations, in which a pure-culture inoculum of a wine yeast is not used and where the operator relies on the natural yeast flora of the crushed grapes and fermentation equipment, are now less commonly used. In such fermentations a succession of yeast and bacterial populations is found in the fermenting grape must. Strains of *Kloeckera apiculata* (and its sporogenous counterparts *Hanseniaspora guilliermondii* and *H. uvarum*) and *Candida pulcherrima* (sporogenous form *Metschnikowia pulcherrima*) are common during the early stages of fermentation. Because of their low alcohol tolerance (5–7%) they are overgrown later on by more alcohol-tolerant species such as *Saccharomyces rosei* and *S. cerevisiae.* Numerous species of yeasts have been isolated from grapes in various parts of the world (Amerine and Joslyn, 1970). As pointed out by these authors, under uncontrolled conditions a high-quality wine is produced only occasionally, i.e., when these different microorganisms occur in proper proportion and sequence and undesirable spoilage organisms are absent or present in very low numbers. Because of the difficulty of controlling the natural microbial population and the danger of a high relative proportion of undesirable organisms, most wines, especially in new viticultural areas, are fermented with the aid of essentially pure culture inocula using strains of wine yeast of proven quality. Such strains are usually used in conjunction with proper concentrations of sulfur dioxide (see Section II, E) to which the wine yeast is resistant and wild yeasts and bacteria are sensitive. Under these conditions the fermentation starts rapidly and goes to completion in a relatively short time. The yeast starter may be taken from a slant culture in a suitable container and the growth is introduced into a gallon of pasteurized must. After active fermentation ensues the fermenting liquid is transferred to progressively larger containers and finally into large tanks of must. In recent years some yeast companies have introduced compressed yeast or active dried yeast of suitable wine yeast strains. The latter product, with about 7.5% moisture content, is stable and can be stored for many weeks. After rehydration at the proper temperature (usually 43°C) it can be added in high concentration to the must to be fermented; fermentation then starts rapidly. Since a heavy inoculum is used, yeast growth at the expense of sugar from the grape is reduced.

B. Fermentation and Its By-Products

Alcoholic fermentation takes place in the yeast cell by a series of reactions, usually referred to as the Embden–Meyerhof–Parnas pathway (Sols *et al.*, 1971). The overall reaction $C_6H_{12}O_6 \rightarrow 2\ C_2H_5OH + 2\ CO_2$ theoretically yields 51.1% ethanol by weight and 48.9% CO_2. In practice the ethanol yield is somewhat lower, approximately 48%. The formation of several by-products and conversion of about 1% of the sugar into yeast cell mass account for the lower yield. In addition some alcohol is lost by entrainment in the carbon dioxide and by evaporation.

Compounds normally recognized as by-products of alcoholic fermentation of grape must are the following (percentages are based on fermentable sugar transformed): glycerol (2.5–3.0%), acetic acid (0.05–0.65%), acetaldehyde (0.01–0.04%), 2,3-butanediol (0.06–0.10%), succinic acid (0.02–0.05%), and higher alcohols (highly variable but ranging from approximately 0.01 to 0.04%). The level of all of these by-products is strongly influenced by yeast strain and environmental conditions, especially temperature.

Glycerol, lactic acid, acetaldehyde, 2,3-butanediol, and succinic and acetic acids are constant by-products of alcoholic fermentation. Glycerol accumulates mostly during the initial stages of fermentation and is formed by the reduction of dihydroxyacetone phosphate to glycerol phosphate followed by enzymatic hydrolysis to free glycerol and phosphate. The enzymes involved are an NADH-dependent dehydrogenase and α-glycerophosphatase. Lactic acid is thought to arise by reduction of pyruvate during the course of fermentation. Acetaldehyde represents primarily a small fraction of that intermediate not reduced by NADH and alcohol dehydrogenase to ethanol. Some acetaldehyde is assumed to undergo condensation with another molecule to form acetoin ($CH_3 \cdot CO \cdot CHOH \cdot CH_3$) and the latter is reduced to 2,3-butanediol.

Succinic acid is thought to be formed by the action of phosphoenolpyruvate carboxykinase, which enzyme with the aid of high-energy phosphate catalyzes the condensation of carbon dioxide and pyruvate to form oxaloacetic acid, which is then converted to succinic acid via malic acid. Under anaerobic conditions acetic acid is formed only during the initial stages of fermentation by dismutation of acetaldehyde. During this reaction one molecule of acetaldehyde is oxidized to acetate at the expense of a second molecule that is reduced to ethanol.

Higher alcohols, formerly believed to be formed solely in wine by deamination, decarboxylation and reduction of preformed amino acids, are now known to arise also from keto-acid precursors of amino acids. A small portion of these keto-acids undergoes decarboxylation followed by reduction to the corresponding alcohol. The alcohols most prevalent in wine are 1-propanol, 2-methyl-1-propanol (isobutanol), 3-methyl-1-butanol (isoamyl alcohol), and 2-methyl-1-butanol (optically active amyl alcohol).

C. Other Compounds Formed by Yeast Fermentation

Hydrogen sulfide, methyl mercaptan, and ethyl mercaptan are undesirable odoriferous compounds found in some wines. Their levels vary greatly with the use of different strains of wine yeast. Hydrogen sulfide is formed primarily by reduction during fermentation of elemental sulfur sprayed on grapes as a fungicide. Mercaptoethanol is believed to be formed by a reaction between hydrogen sulfide and acetaldehyde. These sulfur compounds impart an objectionable flavor to wines in very low concentration (1–5 ppm). Low levels of hydrogen sulfide can often be removed by aeration of the wine. Higher concentrations can be reduced by addition of sufficient sulfur dioxide to provide free sulfur dioxide which is capable of oxidizing hydrogen sulfide to sulfur.

D. Effect of Temperature on Fermentation

Alcohol yield and the rate of fermentation, as well as the concentrations and proportions of the fermentation by-products, are affected by temperature. Temperature is also among the most important factors in flavor formation. Most enologists agree that more bouquet is formed in a wine by a long, slow fermentation at a low temperature, than by a short rapid fermentation at a temperature nearer to the maximum temperature for fermentation. Fermentation at relatively low temperatures produces more aromatic compounds than at higher temperatures. There is no single optimum temperature for this process since it varies with the strain of yeast used. It has been suggested (Amerine and Joslyn, 1970) that a desirable temperature schedule for white wines is a starting temperature between 7° and 16°C and a finishing temperature of about 19°–20°C.

Alcoholic fermentation generates not only 2 moles of ATP per mole of glucose (used for cell mass biosynthesis) but also a large amount of heat. The generation of 2 moles of ATP represents only 14 kcal out of a total of 56 kcal free energy loss per mole of glucose fermented. Thus it has been estimated that 75% of the energy of anaerobic metabolism is lost in the form of heat (Sols *et al.*, 1971). Large fermentation tanks, in particular, need cooling, since without it the temperature of the fermenting must rises to a point where the fermentation sticks (i.e., ceases because of damage to the yeast cells). Control of fermentation temperature can be achieved by cooling coils in the fermentation vessel through which cold water or a refrigerant is circulated. Other procedures use a refrigerant in double-walled tanks, or the fermenting must may be pumped through a heat exchanger.

E. Use of Sulfur Dioxide

This chemical has been used in wine making since antiquity to disinfect containers, control contaminating microorganisms, and protect wine against excessive oxidation during storage and aging. Its use is worldwide and if properly employed is beneficial to the quality of even the finest wines, especially white table wines. In wine making sterilization by sulfiting is done prior to the addition of a pure wine yeast starter.

Sulfur dioxide, besides binding to microorganisms and causing their death or inhibiting their growth, also combines readily with carbonyl compounds (e.g., acetaldehyde), unsaturated aliphatic compounds, and proteins of the must. In the combined form it is much less toxic to microorganisms and less effective as an antioxidant. Analytical procedures (Amerine and Ough, 1974) are available which distinguish between free and bound sulfur dioxide in a wine.

In large wineries sulfur dioxide is added to must or wine from pressure cylinders where the gas is stored in liquefied form. For smaller quantities of wine potassium bisulfite is often used. The concentration of free SO_2 in finished wines is generally 10–40 mg/liter or less. The effectiveness of sulfur dioxide increases with lowering of the pH, and wines with high total acidity are therefore given the lowest level of sulfur dioxide. Freshly pressed musts are supplied with approximately 100–200 mg/liter of sulfur dioxide. The concentrations normally used in must have little if any effect on the activity of the cultured strains of wine yeast. This resistance has been acquired by adaptation of these strains to certain levels of free sulfur dioxide. The inhibitory action of sulfur dioxide on spoilage bacteria and wild yeasts is attributed to its inhibition of metabolic enzymes containing S—H groups and to its inhibition of the action of the important coenzyme nicotinic acid adenine dinucleotide (NAD).

F. Yeast Nutrition

Yeast multiplication during complete fermentation of grape must depends to some extent on the amount of inoculum used. For growth under anaerobic conditions, yeast requires a fermentable source of carbon and energy, a source of nitrogen, minerals (both macronutrients and trace elements), and several accessory factors, such as vitamins, traces of fatty acids, and sterols (Phaff *et al.*, 1978). Nutrients such as potassium salts, phosphates, and other minerals, as well as accessory factors, are rarely limiting in grape must, but nitrogenous compounds occasionally are. Small amounts of ammonium salts [e.g., $(NH_4)_2HPO_4$] are therefore added sometimes to stimulate sluggish yeast growth. Ammonium ions are rapidly assimilated by yeasts and in addition are known to stimulate fermentation.

Wine yeasts normally ferment hexose sugars, sucrose, and maltose although at different rates. It is also known that most strains ferment glucose faster than fructose but some do the reverse and prefer fructose. This appears to be related to the rate of uptake of the different sugars or the affinity they have for the common carrier responsible for their facilitated diffusion into the cell. Wine yeasts as well as other yeasts cannot ferment pentose sugars.

G. Yeast Autolysis

Autolysis is defined as self-digestion of yeast cells. It starts usually with the breakdown of the cell envelope resulting in cell death and this is followed by hydrolysis of the cytoplasmic proteins and nucleic acids by proteases A, B, and C and nucleases, respectively, contained in the vacuoles of the cell. The released amino acids and nucleotides impart an undesirable yeasty flavor to wine. Autolytic tendencies are promoted by high fermentation temperatures and autolysis also varies greatly with the strain of yeast used. An added problem of autolysis is that the released amino acids, nucleotides, and vitamins from yeast can serve as nutrients for fastidious spoilage bacteria, in particular, lactic acid bacteria. For this reason wine should be separated from the sedimented yeast as early as possible upon completion of fermentation. A possible exception to early racking may be in cases of high-acid wines, where autolysis of yeasts is desired in order to release amino acids to stimulate the growth of the bacteria involved in the malolactic fermentation (see Section IV, C).

H. Deacidification of Wines

Organic acids affect the sensory properties of wines, particularly tartness. Tartaric and L-malic acids are the major acids in grapes, the former being quantitatively the most important. However, grapes grown in cool regions, such as the eastern United States, sometimes contain high levels of L-malic acid, leading to an excessive titratable acidity in the wine produced.

Two methods are known to reduce by fermentation excessive tartness due to L-malic acid. The first involves the conversion of malic acid (a dicarboxylic acid) to lactic acid (a monocarboxylic acid) and carbon dioxide during the so-called bacterial malolactic fermentation. The second method depends on the ability of certain yeasts to convert L-malate to ethanol and carbon dioxide in a mixed fermentation together with a strain of wine yeast.

The malolactic fermentation has been recognized since 1890 and is used in many countries (Kunkee, 1967). It is not recommended in regions where wines are low in total acidity, but this is not always avoidable, since the fermentation often occurs spontaneously by bacteria present in the winery or on grapes. The taxonomy of the lactic acid bacteria associated with the malolactic fermentation is quite confusing and probably more than a single species has the capacity of decomposing malic acid. Species have been placed in *Lactobacillus, Pediococcus, Leuconostoc,* and various other genera. Some of the complex growth requirements for the malolactic bacteria are supplied by yeast cells carrying out the alcoholic fermentation. Some of these are probably the result of yeast autolysis (see Section II, G). The malolactic fermentation most commonly occurs during the middle to late stages of alcoholic fermentation. The conversion of malic to lactic acid takes place in two steps as shown in Eqs. (1) and (2).

$$\text{L-Malic acid} + \text{NAD} \underset{\substack{\text{malic}\\ \text{enzyme}}}{\overset{Mn^{2+}}{\rightleftarrows}} \text{pyruvic acid} + CO_2 + \text{NADH} + H^+ \quad (1)$$

$$\text{Pyruvic acid} + \text{NADH} + H^+ \underset{\substack{\text{lactic}\\ \text{dehydrogenase}}}{\rightleftarrows} \text{L-lactic acid} + \text{NAD} \quad (2)$$

Others have presented evidence that malolactic fermentation is a simple decarboxylation process without intermediates (Peynaud *et al.*, 1968). In either of the two mechanisms there is no production of energy (ATP) which can support the growth of the malolactic bacteria. It is therefore the view of most investigators that these bacteria obtain their energy for growth from the metabolism of small amounts of residual sugars or citric acid present in the fermenting must or wine (Amerine and Kunkee, 1968).

Yeast fermentation of malic acid constitutes another procedure for lowering the total acidity of wine. Several species of the genus *Schizosaccharomyces* (the fission yeasts), e.g., *Sch. pombe* and *Sch. malidevorans,* have the interesting potential to convert L-malic acid under fermentative conditions to 1 mol of ethanol and 2 moles of carbon dioxide. It appears that some strains of *Saccharomyces* can also carry out this reaction but with low efficiency. Little is known about the biochemistry of this conversion. This principle has been applied in recent years to fermentation of wines of excessive titratable acidity with mixed cultures of a normal wine yeast and a strain of *Schizosaccharomyces.* According to Gallander (1977) the best procedure was to inoculate the must initially with *Sch. pombe* (which can also carry out alcoholic sugar fermentation) and to reinoculate the fermenting must with a strain of *Saccharomyces cerevisiae.* According to several authorities, however, the effect of such mixed fermentations on the quality of the wine needs further investigation.

Chemical reduction of acidity using calcium carbonate is also used. This treatment primarily reduces the level of tartaric acid by precipitation as the calcium salt.

I. The Fermentation of Flor Sherries

When newly fermented wine containing about 15% ethanol is exposed to the air, a thick surface film of yeast may develop with the capacity of slowly oxidizing a small portion of ethanol to acetaldehyde. This compound and its reaction product with ethanol, acetal, are the most important flavor compounds in Spanish flor sherry and similar products made in other countries. The levels of free aldehyde may rise to 200 or 300 mg/liter and total aldehyde (including bound) may reach 400 mg/liter. The procedure of making flor sherries consists of two fermentation steps, (1) fermentation of the must from desirable grape varieties (e.g., Palomino grapes) to an ethanol level of close to 15% and (2) development of a film yeast to develop the required aldehyde level.

If the sugar content of the grapes is insufficient to reach 15% ethanol, fortification with distilled wine spirits is necessary to avoid the growth of spoilage bacteria (e.g., *Acetobacter*) or wild yeasts which would raise the acetic acid level of the wine. The two basic types of Spanish sherry are the *fino* type, whose flavor is developed largely by a secondary film (flor) fermentation, and the *oloroso* type, which is not subjected to this secondary fermentation. The secondary flor fermentation used for the *fino*-type of Jerez sherries of Spain requires from one to several years for completion. The secondary fermentation is done in partially filled (about three-quarters full) oak barrels where the wine is covered by a thick layer (1- to 1.5-cm thick) of film yeast. The barrels are arranged so that progressive fractional blending during maturation can be done (the so-called solera system) with as little disturbance of the film yeast layer as possible. The wine undergoes a continuous aging and improvement in quality during its prolonged storage in the oak casks. The details of the solera process are described by Joslyn and Amerine (1964).

In recent years a number of procedures have been worked out to reduce the long oxidative secondary fermentation. Considerable success has been achieved by using a submerged fermentation process in which a strain of flor yeast is mixed with wine of 15% alcohol in pressure tanks where air or oxygen is introduced in finely dispersed form to promote the required aldehyde formation. This process can be made semicontinuous by removing 10% of the volume each day and replacing it by fresh wine. The submerged process has the advantage of saving a considerable amount of time, less alcohol is lost by evaporation, and the long-term investment in the product is reduced. However, an acceptable, good-quality sherry still requires aging in full oak casks although not nearly as long as in the *fino* solera process. For further details of the process see Ough and Amerine (1960).

The taxonomy of the yeasts found in the flor is quite confusing. Numerous investigators in various countries have attached names to these film-forming yeasts. These include *Saccharomyces beticus* (various races), *S. cheresiensis*, *S. montuliensis*, and even strains of the better known species *S. oviformis*, *S. fermentati*, *S. delbrueckii*, and even *S. cerevisiae*. It seems evident that flor yeasts are made up of a number of different species of yeast either in pure culture or as a balanced population of different strains. For further information on this interesting microbiological process and commercial practice the reader is referred to Joslyn and Amerine (1964).

J. Wine Spoilage Organisms

Bacteria as well as yeast may be responsible for the spoilage of wines. Important factors influencing microbial spoilage are the composition of the wine (sugar, alcohol, and sulfur dioxide content primarily), storage conditions (e.g., temperature and amount of air space in the container), and, perhaps most important, the extent of initial contamination during storage in vats or bottles.

Spoilage by yeasts is often more of an aesthetic problem than a marked deterioration of the flavor and aroma of a wine. Occasionally a yeast may be desirable or undesirable depending on the conditions. For example, *Saccharomyces oviformis*, which is very effective in producing a very dry wine, becomes an undesirable contaminant when it is found in wines where a residual sugar is desired. *Saccharomyces bayanus* also has been found in bottled dry table wines with 0.5–1.0% residual sugar where it causes turbidity or a sediment. In young table wines *Pichia membranaefaciens* (an aerobic yeast) is sometimes found when the bottle contains sufficient oxygen to permit limited development of this species (or its nonsporogenous counterpart *Candida valida*). The most serious spoilage yeasts are species of *Brettanomyces* (sporogenous forms are placed in the genus *Dekkera*). These are slow-growing yeasts which can develop in both white and red table wines causing turbidity and slight off-flavors. If oxygen is present they can oxidize ethanol to acetic acid. They are more prevalent than had been suspected for many years since plating of a turbid wine may not produce visible colonies for 10–12 days on agar media. A suitable differential medium is malt agar with 100 mg/liter of cycloheximide (Actidione). Wine yeasts are very sensitive to this antibiotic while *Brettanomyces* species are resistant. The most common species of that genus in wines is *B. intermedius*.

Bacterial spoilage of wine (Vaughn, 1955) is less of a problem today than it was in the past. This is due to observance of good sanitary practices in most wineries and the proper use of sulfur dioxide.

Aerobic spoilage may be caused by acetic acid bacteria which are common on grapes. They oxidize ethanol to acetic acid and some ethyl acetate may be formed as well. Of greater importance is anaerobic spoilage by various lactic acid bacteria. These organisms are capable of producing metabolic products other than lactic acid which can profoundly affect the quality of a wine. For further details see the excellent review on this subject by Vaughn (1955).

III. KEY DEVELOPMENTS

From the microbiological point of view, the most significant recent developments have been large-scale use of dried pressed yeasts, industrialization of the flor sherry process, and greater knowledge of and control of the malolactic fermentation. Most recently great interest has been shown in yeasts producing lower sulfide and greater flavor. Should the wine industry have to cut back on use of sulfur dioxide, the need for a new antimicrobial agent would become acute. Since sulfur dioxide has a variety of useful properties other than its antiseptic and preservative value (antioxidative and enzyme inhibition; clarification; bleaching and discoloration; promotion of solubilization of the coloring matter of grape skins; increasing the fixed acidity, the extract, and the alkalinity of the ash because of its solvent effect on potassium acid tartrate), a single agent may not be available and agents for specific needs may be needed. In any case grapes will have to have less microbial infection than at present and fermentation and aging must be conducted under increasingly stringent sanitary conditions.

Since the advent of dried pressed yeast, few wineries have used yeast culture tanks. Those that employ such vessels use stainless steel equipment with built-in sterilizing and cooling facilities. Most fermentors built in recent years have been of plastic, stainless steel, or epoxy-lined iron. Except for small fermentors, they have built-in or auxiliary cooling equipment. Many are operated manually, but a few have automatic controllers or even attached computers. Many wineries now routinely centrifuge musts for white wine production (Fig. 1). Many wines are now germ-proof filtered into the bottles (Fig. 2). Pasteurization of high-quality table wines is not an acceptable practice in the wine industry because of its unfavorable effect on wine quality. To prevent microbial spoilage of any origin all winery equipment, including cooperage, must be kept scrupulously clean. The wine must be inspected periodically for contamination by direct microscopic inspection and especially by filtration of measured volumes of wine through membrane filters followed by plating of the filters on a suitable agar medium.

Semidry table wines (with 1–3% reducing sugar) have become in-

FIGURE 1. Large centrifuge (rear left) in a California winery, used for clarifying musts and wines. (Courtesy Paul Masson Vineyards, Saratoga, California.)

FIGURE 2. Large tubular membrane (Millipore) filter in California winery used for germ-proof filtration of wine. (Courtesy Guild Wineries and Distilleries, San Francisco, California.)

creasingly popular. Such wines are particularly subject to recurrence of yeast fermentation.

The wine industry is making increasing use of filtration through large membrane filters to remove microbial contamination before bottling (Fig. 2). In addition, a number of chemical preservatives (not including sulfur dioxide) have been advocated in recent years for the preservation of wine. The latter include diethyl pyrocarbonate (DEPC) and sorbic acid. The former compound, claimed to decompose into ethanol and carbon dioxide when added to wine, is a potent inhibitor of yeast growth. Although still used in some countries, it has been prohibited in the United States since 1972 because of some evidence for the formation of harmful complexes between DEPC and organic components of wine.

Sorbic acid is an effective inhibitor of yeast growth but it does not affect bacterial development. Since the ban of DEPC, sorbic acid or its potassium salt has become the only practicable fungistatic agent legally authorized for wine use in the United States. The usual concentration of sorbate in wine is 150–200 ppm. A disadvantage of sorbate is the possibility of its conversion by the action of lactic acid bacteria to an ethyl ether (2-ethoxyhexa-3,5-diene) which has an intense odor of crushed geranium leaves (Crowell and Guymon, 1975). Such an off-odor associated with bacterial spoilage of sorbate-containing wines is easily detected by sensory analysis. For further details on chemical preservatives, their use, and appropriate literature references see Amerine and Kunkee (1968).

IV. PROCESS TODAY

A. Yeast Selection

Pure yeast cultures are available from private and public laboratories in wine-producing countries. Some wineries maintain stock cultures and propagate them as pure cultures. Others purchase dried yeast strains from commercial producers. A variety of yeast strains of *S. cerevisiae* is being used: Burgundy, Montrachet, Chablis, Champagne, Steinberg, etc. Mixed cultures of different species have been tried but are not widely used.

B. Process Flow

Almost all wine is produced by a batch process. The main reason for use of a batch rather than a continuous process is the short period of availability of the raw product (usually only 6–10 weeks). The expensive equipment and controls needed for a continuous process would thus be unused for most of the year—a heavy overhead. Furthermore, the raw material is of variable suitability for producing different types of wines. However, there are several cases where continuous systems have been used. Where the wine is refermented for a new product, continuous systems may be considered. Sparkling wine production in a series of tanks is used in the Soviet Union, but the process has not found favor in the West. Submerged culture of flor yeasts in a continuous fashion has been successful in California, the Soviet Union, and elsewhere. For ordinary quality wines, continuous fermentors have been used in Argentina and Italy. Distilling material has also been successfully produced in continuous systems. Producers have avoided such systems for quality wines because of the problems of controlling the types of grapes entering the system and because (unless strict controls are maintained) of the possibility of spreading bacterial and yeast contamination.

Another possibility for continuous fermentation would be the conversion of grape juice to grape concentrate by vacuum concentration and using diluted concentrate for wine production throughout the year. This might be particularly useful for wine production in areas where grapes are not grown. It would be most appropriate for white wines and sherry stock.

The process of white wine production involves crushing the grapes in crusher–stemmers and separation of the seeds and skins from the juice. The most common method of separation is pressing, usually in screw-presses, although a number of hydraulic and other types are employed. Where grapes are cheap and there is a large demand for wines for distilling, only the juice which drains off from the crushed grapes—the free-run—is normally used for wine. The rest is fermented and distilled for wine spirits.

Red wine production differs in that the grapes must be fermented on the skins for 3–10 days or longer. This is necessary for the alcohol formed to extract the red anthocyanin pigments from the skins. Unfortunately, the skins tend to float on the surface of the juice and reduce contact of the skins with the fermenting juice. The traditional method of securing contact has been to push the floating skins down into the fermenting liquid. In large containers this is very difficult and at present the liquid is pumped from the bottom of the container over the skins. When grapes contain mold, a number of polyphenoloxidase enzymes may be present. These cause an oxidation of the red anthocyanin pigments to compounds with a brown color. To prevent this there is an increasing practice of heating red grapes to inactivate the enzymes before fermentation. Heating also helps to release the color from the skins.

Pink (or rosé) wines are produced in three ways: from pink-colored grapes, by separating the skins of dark grapes from the juice after 12–36 hours of fermentation on the skins, or by blending white and red wines. Some are finished without residual sugar; others may be sweet.

In the production of white or of red dessert wines, the process is the same except that wine spirits are added after pressing, during fermentation. The time of addition of the spirits is determined by the level of residual sugar which is desired in the finished wine. For "dry" sherry production, this may be less than 1%. For muscatels, it may be 12–14% sugar. The level of fortification also varies: for dry sherry intended for flor-yeast fermentation, 15% ethanol is the maximum. For sherries that are to be baked (i.e., heated at 55°C for 90 days), 17–18% is the usual level. For muscatels and ports, 18–19%, or more, is common.

Following the primary fermentation, table wines which are not protected from air may develop four types of surface growth (i.e., a film): of *Acetobacter,* of *Pichia* sp. (usually *P. membranaefaciens*), *Candida vini,* or of *Saccharomyces* sp.

Acetobacter is sensitive to alcohol and develops slowly on wines with more than 12% ethanol. These bacteria produce acetic acid and other acids, ethyl acetate, and minor products. The various strains of *Pichia* and *Candida* are likewise sensitive to ethanol. They do not carry out alcoholic fermentation, but they will grow (flower is the colloquial term) aerobically on the surface of wines of 10–13% ethanol. They produce acetaldehyde and traces of organic acids.

As a beneficial process we are only interested in the aerobic film stage of *Saccharomyces* sp. These yeasts can grow on the surface of wines of up to 15 and 16% ethanol. This is of great industrial importance because, by keeping the ethanol content at 15%, growth of *Acetobacter* sp., *Pichia* sp., and other undesirable film-forming yeasts is inhibited. As discussed in Section II in more detail, the primary oxidative metabolic product of *Saccharomyces* species is acetaldehyde.

The domestication of the film stage of *Saccharomyces* as the producer of a special type of wine dates from the nineteenth century in southern Spain. Doubtless it existed earlier and elsewhere (Armenia, for example), but it was first used for producing large amounts of dry fino-type sherries in Spain.

Besides the specialty product of flor sherry (by the film or submerged culture process), sparkling wines require special processing. The excess carbon dioxide needed in sparkling wines can come from direct carbonation or by a secondary fermentation of added sugar. Very few carbonated wines are produced. The fermentation-produced sparkling wines are made by two basic procedures: fermentation in large closed tanks (500–25,000 gallons; 19–950 hl) or fermentation in bottles of less than 2 liters capacity. Fermentation in tanks is by far the cheaper process since following fermentation the wine is simply filtered into special bottles, labeled, and sold.

The bottle fermentation process may take two forms. Following fermentation the wine can be transferred from the bottle into a tank and then filtered into another bottle. As with the tank process, this filtration is necessary to remove the yeast which was added to secure the secondary fermentation. The producer is able to label such wine as "bottle-fermented." In the more traditional bottle process, the wine is left in the bottle for 1–3 years. The yeast cells autolyze during this aging period and this is believed responsible for the development of the additional bouquet in the wine. To remove the yeast cells it is necessary to get the yeast deposit onto the cork (done by turning the bottles on end and by a complicated process taking 1–6 weeks of getting the yeast deposit to slide down through the neck of the bottle onto the cork). The neck of the bottle, including the deposit, is then frozen and the cork removed. The ice plug is then disgorged. The small amount of wine lost by disgorging is then replaced and a new cork inserted. About 1–2 atmospheres of pressure are lost. Nevertheless, commercial sparkling wines usually retain 2–3 atmospheres of pressure. In this case the producer can label the wine "fermented in this bottle." For further details see Amerine and Joslyn (1970).

C. Processing

Following fermentation the treatment of table and dessert wines may take a variety of forms. The new wines are usually separated from the yeast sediment (racked) soon after fermentation. In the case of red wines where a malolactic fermentation is desired, the initial racking may be delayed. This is believed due to the desire to secure release of amino acids by autolysis of the yeast. The additional amino acids favor the growth of the bacteria that are responsible for the malolactic fermentation.

Large-scale rack-and-cloth filters are used for the initial clarification. Recalcitrant wines (i.e., wines that do not clarify well) may be centrifuged (Fig. 1). Following the rough filtration, the wine is "fined." This involves addition of some agent which physically and by adsorption helps clarify the wine of suspended material. A slurry of clay material, bentonite, is now the main agent used. Other fining agents that have been used in special cases are PVPP (polyvinylpolypyrrolidone), gelatin, casein, isinglass, and others.

The wines are again filtered (Fig. 3) following fining. They are aged in wood, plastic, or stainless steel or epoxy-lined metal tanks. These vary in

FIGURE 3. Large plate filter for filtration of finished wines in a California winery. (Courtesy E. & J. Gallo Winery, Modesto, California.)

size from about 50 gal (2 hl) to 250,000 gal (9500 hl), or more. The time of aging may vary from a few weeks to two or more years, depending on the type of wine.

Most new wines are supersaturated with potassium acid tartrate (cream of tartar). With time and especially when the wine is stored at low temperatures, the excess potassium acid tartrate crystallizes and settles. However, since many wines are sold when very young, there may not be sufficient time for their precipitation. In this case, prior to bottling, the wines may be cooled to a temperature just above freezing for one to several weeks to hasten the precipitation of the excess potassium acid tartrate.

V. PACKAGING AND DISTRIBUTION

In many parts of the world, a large amount of wine is distributed directly in bulk (e.g., wooden barrels) to the consumer. In Austria, this may be as much as 50%. Shipment to bottlers may be in wooden barrels or even in glass- or plastic-lined tank cars, trucks, or, for long distances, in tankers.

However, increasingly, wines are sold in glass bottles of about 750-ml capacity. Mechanically operated bottling machines are commonly employed. The bottles may be closed with metal screw caps or with corks. The labels and capsules (to cover the cork) are usually applied at the same time as the wine is bottled. In fact, in the larger wineries, the wines may be bottled, labeled, packaged, and placed on shipping pallets in a continuous mechanically operated line. Shipment of the cases is by truck or rail. During shipment, the cased wines should not be exposed to extremes of cold or warm temperature.

VI. EVALUATION OF WINES

Examination of wines in the laboratory is essential to assure a sound product of high quality. The procedures include tasting and other sensory observations, microbiological examination, and determination of such constituents as ethanol, total acidity, volatile acidity, sugar, extract, tannin, sulfur dioxide, and depth of color. Such analyses also assist the wine maker in correcting certain deficiencies by blending of different lots of wine. The details of the sensory evaluation and various analytical procedures are described by Amerine and Ough (1974), Amerine and Roessler (1976), and Amerine *et al.* (1973). Below we present briefly the highlights of the various procedures.

A. Sensory Examination

Sensory examination is an important aspect of cellar operation. A good winery operation requires at least one individual who is responsible for collecting and evaluating the records of the sensory evaluation of its wines. Persons on the evaluation panel should have a keen palate and discriminating judgment, should know wine types, and should be able with reasonable accuracy to classify the cellar's wines. Wines are judged by the following criteria: (1) Appearance (e.g., clarity and freedom from sediment) and color (appropriateness for a particular type of wine). (2) Odor, aroma, and bouquet. Particular attention is paid to the bouquet (which may reveal the extent of aging) and to the presence of odors of sulfur dioxide, hydrogen sulfide, acetic acid, diacetyl, and other volatiles resulting from bacterial action. (3) Taste. Here the main variables are acidity (sourness), sugar (sweetness), bitterness, and astringency (caused by the tannin content of wine). (4) Flavor. Flavor expresses the overall impression of the wine and is based on a combination of taste and bouquet plus aroma. It includes the aftertaste, the sensory notes that linger on the palate after the wine is tasted. Sensory evaluation is of great importance in blending and in deciding when a wine is ready for bottling or shipment.

B. Microbiological Examination

This includes both microscopical inspection and plating. Its main purpose is to detect excessive numbers of spoilage bacteria (mainly acetic and lactic acid bacteria) and wild yeasts. Their presence can be detected amid a large preponderance of wine yeasts by supplementing suitable agar media with 100 mg/liter of cycloheximide. Bacteria and many wild yeasts (e.g., *Kloeckera* and *Brettanomyces*) are resistant to this antibiotic, whereas wine yeasts are totally inhibited. Low degrees of contamination in wines can be detected by filtering a liter or so of wine through a Millipore or similar filter and placing the filter on a suitable agar medium.

C. Chemical and Physical Analyses

Detailed procedures have been published (Amerine and Ough, 1974) to determine various important constituents of both must and wines. Space does not permit describing the various procedures in detail. We will merely present a list of the most important determinations normally performed in wineries having good laboratory facilities.

1. *Soluble Solids*

Most of the soluble solids in grape must are sugars. A determination of this property indicates ripeness of the grapes and the maximum alcohol level that can be expected after fermentation. The reducing sugar content of wines is a measure of their dryness or sweetness.

2. *Acidity and Individual Acids*

Methods differentiate between volatile and nonvolatile acids; together they represent the total (titratable) acidity. Estimation of changes in malic and lactic acids provides information on the course and extent of the malolactic fermentation. The pH is also determined.

3. *Alcohols.*

It is obvious that the determination of ethanol (Fig. 4) is important in connection with required maximum and minimum levels in table wines, dessert wines, and appetizer wines. Less routine is the determination of glycerol, 2,3-butanediol, methanol, and higher alcohols.

4. *Carbonyl Compounds*

The principal carbonyl compound is acetaldehyde, which is particularly important in flor sherries. Minor compounds include acetoin, diacetyl, and hydroxymethylfurfural.

5. *Phenolic Compounds*

Among the compounds in this group that are important in winery operations, the following may be mentioned: grape pigments (including anthocyanins), flavonoids and nonflavonoid phenols, hydrolyzable tannin phenols, catechins, and others.

FIGURE 4. An automated gas chromatography system constitutes one method for the analysis of ethanol. Samples of wine are precisely diluted with a diluent containing an internal standard. The diluted samples are loaded in the tray of the automatic sample injector. The peak areas for ethanol and the internal standard are electronically integrated. Corrections are made for response factors and the ethanol content is calculated and printed out for the sample by the integrator. The integrator is not shown in the photograph. (Courtesy E. & J. Gallo Winery, Modesto, California.)

6. *Nitrogen Compounds*

These include total nitrogen, ammonium ion, amino acids, α- amino nitrogen, protein, and nitrate. The presence of excessive levels of amino acids, protein, and nucleotides indicates that yeast autolysis may have occurred during late stages of fermentation.

7. *Mineral Constituents and Gases*

Analyses may be needed for cations (potassium, sodium, calcium, magnesium, iron, and copper), anions (bromide, chlorine, phosphate, and sulfate), and dissolved gases (oxygen, carbon dioxide, hydrogen sulfide).

8. *Color Determination*

Color is determined most accurately by using tristimulus values. However, for white wines a simple measurement of light absorbance in a spectrophotometer is adequate to follow development of a brown color due to oxidation and storage at high temperature.

9. *Chemical Additives*

Such analyses include the determination of free and total sulfur dioxide. If other chemical preservatives are used the chemist may have to test at least for benzoic and sorbic acids.

VII. USES

Probably no other food product has as many cultural patterns of use as wine. These arise from historical precedents and the nature of wine itself. The Mediterranean concept of wines is for their consumption as table beverages, that is, as accompaniments of food. These are wines of moderate ethanol and with little sugar. Since many white table wines are tart (have an appreciable acid taste), they are generally considered to be a good accompaniment of fish and moderately flavored meats (such as chicken). Red wines, with their stronger flavor, are considered best with the stronger flavored meats and cheeses. However, personal predilections vary, and who is to say that a dry white table wine is better with shrimp, chicken, chops, or roasts?

Later sweet wines of low or high alcohol content were produced. The latter are made either by harvesting of specially ripened grapes or by addition of distilled spirits to wines during fermentation. These were more properly considered as dessert wines—that is, wines to be consumed with desserts or after meals. Traditionally wines have been used in the kitchen, as for *coq au vin* or *boeuf bourguignonne*. Small amounts of sherry are added to clear soups. Sparkling wines or dessert wines are sometimes added to fruits.

REFERENCES

Amerine, M. A., and Joslyn, M. A. (1970). "Table Wines: The Technology of Their Production," 2nd ed. Univ. of California Press, Berkeley and Los Angeles.

Amerine, M. A., and Kunkee, R. E. (1968). *Annu. Rev. Microbiol.* **22,** 323–358.

Amerine, M. A., and Ough, C. S. (1974). "Wine and Must Analysis." Wiley, New York.

Amerine, M. A., and Roessler, E. B. (1976). "Wines: Their Sensory Evaluation." Freeman, San Francisco, California.

Amerine, M. A., and Singleton, V. L. (1971). "A List of Bibliographies and a Selected List of Publications That Contain Bibliographies on Grapes, Wines, and Related Subjects," pp. 1–39. University of California, Agricultural Publications, Berkeley.

Amerine, M. A., Berg, H. W., and Cruess, W. V. (1973). "The Technology of Wine Making," 3rd ed. Avi Publ. Co., Westport, Connecticut.

California (1970). "Regulations Establishing Standards of Identity, Quality, Purity, Sanitation, Labeling, and Advertising of Wine." California Administrative Code, Title 17, Chapter 5, Article 14, Sections 17,000–17,105. California State Printing Office, Sacramento.

Crowell, E. A., and Guymon, J. F. (1975). *Am. J. Enol. Vitic.* **26,** 97–102.

Gallander, J. F. (1977). *Am. J. Enol. Vitic.* **28,** 65–68.

Joslyn, M. A., and Amerine, M. A. (1964). "Dessert, Appetizer and Related Flavored Wines—The Technology of Their Production." University of California, Division of Agricultural Sciences, Berkeley.

Kunkee, R. E. (1967). *Adv. Appl. Microbiol.* **9,** 235–279.

Lodder, J., ed. (1970). "The Yeasts—A Taxonomic Study." North-Holland Publ., Amsterdam.

Ough, C. S., and Amerine, M. A. (1960). *Food Technol.* **14,** 155–159.
Peynaud, E., Lafon-Lafourcade, S., and Guimberteau, G. (1968). *Mitt. hoeheren Bundes-lehr- Versuchsanst. Wein- Obstbau, Klosterneuburg* **18,** 343–348.
Phaff, H. J., Miller, M. W., and Mrak, E. M. (1978). "The Life of Yeasts," 2nd ed. Harvard Univ. Press, Cambridge, Massachusetts.
Ribéreau-Gayon, J., and Peynaud, E. (1972–1976). "Sciences et techniques du vin," Vols. 1, 2, and 3. Dunod, Paris.
Sols, A., Gancedo, C., and DelaFuente, G. (1971). *In* "The Yeasts: Physiology and Biochemistry of Yeasts" (A. H. Rose and J. S. Harrison, eds.), Vol. 2, pp. 271–307. Academic Press, New York.
Troost, G. (1971). "Die Technologie des Weines," 3rd ed. Ulmer, Stuttgart.
United States Internal Revenue Service (1974). "Wine," Part 240, Title 26, Code, Federal Regulations, U.S. Govt. Printing Office, Washington, D.C. (ATF Publ. 5120.2, 12/74).
Vaughn, R. H. (1955). *Adv. Food Res.* **6,** 67–108.

Chapter 6

Vinegar

G. B. NICKOL

I. HISTORY AND DEVELOPMENT

A. Derivation of Vinegar

It is very likely that the first vinegar used by ancient peoples was simply spoiled wine. In fact, the word *vinegar* comes from the French words *vin* ("wine") and *aigre* ("sour"). Wine making dates back at least 10,000 years, and it can be assumed that vinegar making goes back just as far. Evidence exists indicating that most ancient peoples who used

MICROBIAL TECHNOLOGY, 2nd ed., VOL.II
 ISBN 0-12-551502-2

vinegar derived it from wines made from the numerous fermentable materials available in the areas where they lived (Hildebrandt *et al.*, 1967; Allgeier *et al.*, 1974a, b). Both the Old and New Testaments contain references to vinegar. The best known of these tells of vinegar offered as a drink to Jesus on the cross.

Early records indicate that vinegar had many uses other than as a food or food preservative. Medicinal uses were numerous and varied from the use of vinegar as an antibiotic in wet compresses to its use in treatment of plague. In early times most households made their own vinegar, but eventually its production was taken over by wineries and breweries. Vinegar produced by breweries was known as alegar.

B. Advances in Vinegar Making

Initially vinegar was commercially produced by the *let-alone process,* which was simply a matter of holding wine in open, partially filled containers (Allgeier *et al.*, 1960). An advance was made when it was discovered that some finished vinegar added to partially filled casks of wine substantially speeded up the wine's transformation into vinegar. This procedure was known as *field process,* since the barrels were held in open fields. The *Orleans process,* developed in 1670, was a mechanical adaptation of the field process that made it a continuous one. The cask was provided with a spigot for withdrawing finished vinegar, and a tube was inserted through the bung to the bottom of the container permitting the addition of wine. When the contents of the cask had changed to vinegar, only part of it was drawn off, leaving some vinegar behind to mix with the new wine added. This *slow process,* as it is also called, is still used in some parts of the world.

The idea of making vinegar by trickling the liquid through packing materials such as branches and stems dates to the early eighteenth century and is credited to both H. Boerhaave and Kastner. It was improved by J. S. Schüzenbach in 1823 to make it the *quick process.* His improvement was nothing more than the provision of ventilation holes near the bottom of the generator. Sufficient warmth is maintained in the generator to cause a continuous draft up through the packing, thus supplying the oxygen required for the chemical reactions that produce vinegar (see Section II). Schüzenbach's version of the quick process predominated in commercial vinegar production for about 100 years.

The next major improvements of the quick process were made in 1929. At that time, forced aeration and temperature controls were introduced, and the trickling generator widely used today resulted. The most recent process to be devised and developed for commercial use was first publicized in 1949. This *submerged culture process,* as it is called, involves the aeration of the liquid itself, and its success hinges greatly on the efficiency of that aeration. The improved quick process and the submerged culture process are both used extensively today and will be described in detail in Section IV.

II. MECHANISM OF ACETIC ACID FERMENTATION

Vinegar is a solution of which acetic acid is the essential ingredient, the others being water and various impurities. By FDA standards, vinegar must contain at least 4% acetic acid (see Section V). In all the processes described in Section I, the acetic acid is produced from ethanol by microorganisms. The conversion of ethanol to acetic acid is relatively simple compared to other biological processes. It is a two-step reaction. Ethanol is converted to acetaldehyde and concurrently hydrated, as shown in Eqs. (1) and (2) below. The resulting hydrated acetaldehyde is then dehydrogenated to acetic acid, as shown in Eq. (3). The conversion of ethanol to acetic acid is exothermic. One gram of ethanol theoretically produces 1.304 gm of acetic acid and 0.391 gm of water.

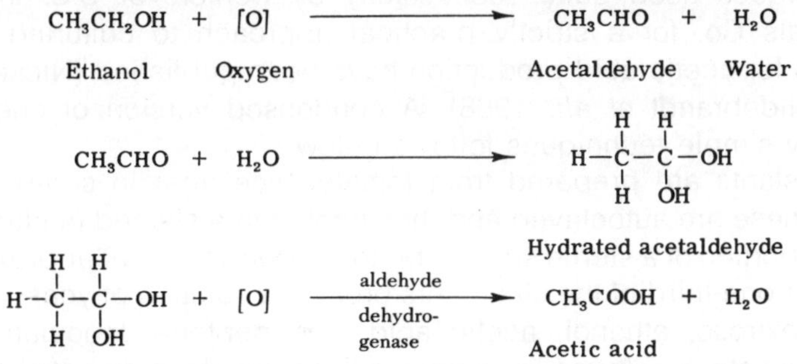

III. ACETIC ACID ORGANISMS

A. Classification of *Acetobacter*

For thousands of years, people who produced vinegar were making use of acetic acid-producing organisms, although the microbiological nature of the vinegar process was not suspected until F. T. Kützing theorized in 1837 that microorganisms converted ethanol to acetic acid. His theory was confirmed by Pasteur in 1868.

Although the acetic acid-producing organisms, or *Acetobacter*, were among the first bacteria to be studied, vinegar production has essentially been accomplished with chance contaminants. The reason for this has been the relative difficulty of isolating pure cultures from active vinegar processes. Also, the nature of the industry up until very recently has been conducive to operating profitably in a nontechnical manner.

Over 100 species, subspecies, and varieties of the genus *Acetobacter* have been classified. Many of these species probably would be acceptable in a vinegar generator with the exception of *A. xylinum,* a heavy slime former, and any species that oxidizes acetic acid to carbon dioxide (overoxidation). The production of vinegar by pure-culture techniques is, strictly speaking, not done. However, in the submerged culture process some operations have been started with pure cultures, although aseptic conditions have not been maintained to prevent contamination. One of the first discussions of pure-culture vinegar production was published by Shimwell (1954). The most extensive treatise on the biochemical activities and taxonomy of acetic acid bacteria is that of Asai (1968).

B. Laboratory Techniques for Culturing *Acetobacter*

1. Slant Cultures

Techniques used quite successfully by workers at U.S. Industrial Chemicals Co. for a strictly practical approach to culturing the organisms for acetic acid production have been published (Nickol *et al.*, 1964; Hildebrandt *et al.*, 1968). A condensed version of one of the relatively simple techniques follows below.

Agar slants are prepared from tomato juice agar in screw-capped tubes. These are autoclaved and then cooled in a slanted position, after which a portion of a sterile special broth is added to cover approximately the lower one-third of the slant. This broth is a solution of yeast nitrogen base, dextrose, ethanol, acetic acid, and peptone. Inoculations are made from stock cultures or vinegar using a wire loop and streaking the culture through the liquid up over the agar slant. The slants are then incubated at 30°C until good growth is attained, after which they are stored at 5°–10°C. Cultures must be transferred monthly to retain viability.

2. Culture Purification

To purify a culture, tomato juice agar petri dishes are numbered from 1 to 6. One inoculating loop of culture is placed on plate 1. Cultures are spread over plate 1 with a sterile bent glass rod using a circular motion about five times so as to cover the entire plate. Without flaming the bent rod, plate 2 is smeared uniformly with the culture in a similar manner. Still without flaming the rod, plates 3 to 6 are smeared. The effect is to diminish the inoculum with each succeeding plate. Plates are inverted, placed in polyethylene bags, closed with a rubber band, and incubated at about 28°C for approximately 48 hours or until colonies appear. If growth is confluent and single colonies cannot be isolated, the procedure is repeated using fresh plates with fewer bacteria in the initial inoculum.

A typical well-isolated colony is fished to a slant, as above, and incubated 48 hours or until satisfactory growth appears. Once a culture is purified and shown to be efficient in vinegar production, it need not be repurified unless it is inadvertently contaminated. Care is exercised in purifying cultures so as to avoid genetic variation. This may be somewhat overcome by picking several typical colonies and recombining them in a purification process to get rid of an obvious contaminant.

All cultures, purified or nonpurified, are gram stained and checked under the microscope. Most good vinegar cultures are gram-negative short rods.

3. Maintenance of Cultures

Although vinegar bacteria survive under natural conditions in orchards and vineyards, under pure-culture conditions they tend to die rather quickly. Even under refrigeration at 5°–10°C, cultures normally should be transferred each month. The development of the lyophilization process enables cultures to be maintained for extended periods without transfer (Hildebrandt *et al.*, 1968). Such cultures remain unaltered both qualitatively and quantitatively for all practical purposes. Contamination is precluded, once the ampules have been sealed. Lyophilized cultures can be shipped without concern as to viability. Therefore, subculturing is not necessary. Further, there is no need for special storage upon receipt.

Two to five years is a reasonable time to expect lyophilized cultures to survive, although one culture of *Acetobacter oxydans* was viable after 20 years. Care, however, must be exercised in ascertaining that the lyophilization process is carried out properly.

4. Shake-Flask Cultures

The acid-producing characteristics of cultures can be evaluated in shake flasks. A convenient medium is sweet cider (no preservative added) with sufficient alcohol added to give an initial alcohol concentration of about 4% by volume. Additional alcohol must be added as acetification proceeds to prevent complete depletion of alcohol. If alcohol is unavailable for the organisms, the culture will die. However, the 4% alcohol concentration should not be exceeded. A good acid producer will reach a 10% acidity within 3 days. *Acetobacter* cultures that attain a 10% acidity in shake flasks will produce 12% or higher acidities in trickling or submerged culture generators. Laboratory-scale generators can be readily started with active shake-flask cultures.

IV. COMMERCIAL VINEGAR PRODUCTION

A. Trickling Generators

1. Description of Operation of Frings-Type Generator

The most widely used generator is of the Frings type (Fig. 1). It is a tank constructed of cypress or redwood and is packed with about 2000 ft^3 of curled beechwood shavings. A false bottom supports the shavings above the lower one-fifth of the tank, which is a storage reservoir with about a 3500-gal capacity. A pump circulates an ethanol–water–acetic acid mixture from this storage space up through a cooler to a distributing sparger arm in the top of the tank. The flow of the liquid through the sparger propels it and keeps it turning continuously, resulting in uniform distribution of the in-process vinegar over the packing. The liquid trick-

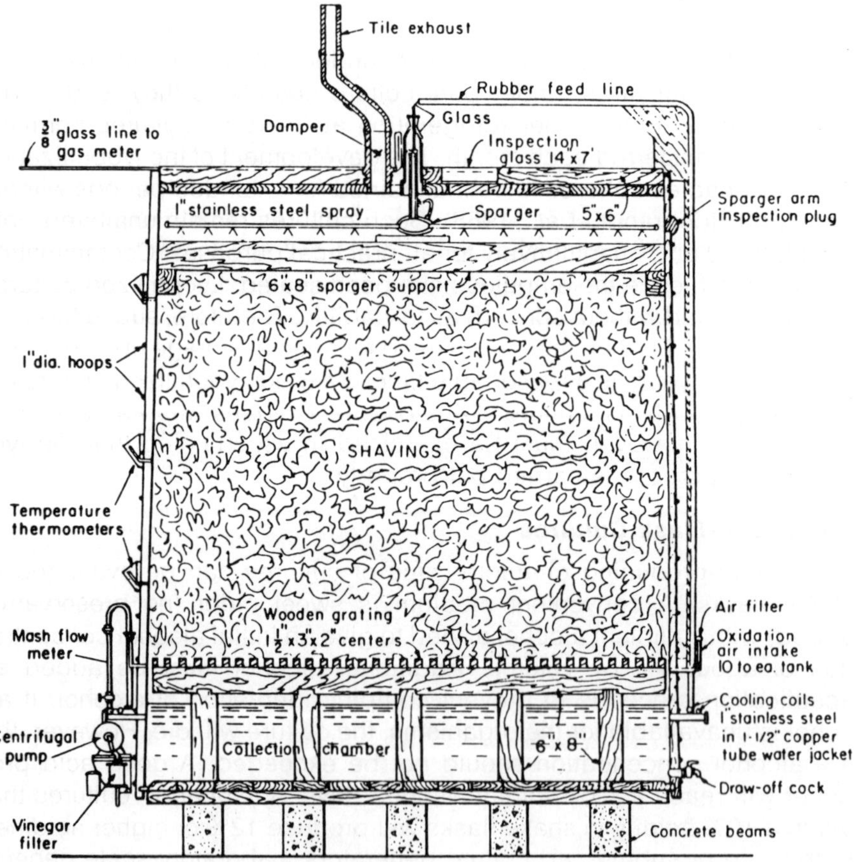

FIGURE 1. Trickling generator.

les down through the packing and returns to the storage reservoir. The temperature of the generator is regulated by flow rate and cooled to be between 29°C at the top of the generator to a maximum of around 35°C at the bottom of the packing. A blower is used to force air into ports in the false bottom and up through the packing. The air rate is valve-controlled and monitored by means of a rotameter. The top of the generator is covered but vented to the outside air. Excessive air rates waste alcohol and vinegar through the exhaust. An air-flow rate of about 80–90 ft^3/hour/100 ft^3 of packing is considered adequate. The oxygen content of the exhaust can be measured with an oxygen meter. If this is done, the air rate should be regulated to maintain an oxygen concentration in the exhaust slightly above 12%.

The packing used in most generators consists of shavings made of air-dried beechwood planed to form coils about $1\frac{1}{4}$ inches in diameter and about 2 inches long. Although these coils tend to soften somewhat with age, the degree of settling is relatively minor compared to that obtained with cheaper packing material, such as corn cobs or other forms of wood shavings. Extractives from the fresh beechwood also are far more acceptable than those from other woods. While some supplemental packing may be required to compensate for settling, the life of a properly operated packed generator is well over 20 years.

Ethanol concentration is another critical factor in the generator operation. The amount of alcohol charged must be sufficient to yield the desired acetic acid concentration plus an amount that allows a residual of over 0.2% unconverted alcohol. If all alcohol is depleted from the charge, *Acetobacter* die and the generator is inactivated. The amount of finished vinegar withdrawn from the reservoir should be limited to an amount that, when replaced with the ethanol-contained charge (mash), will have a combined alcohol concentration of less than 5% by volume. This is especially important when the finished acidity is 12% or higher.

2. *Efficient Operation of Generators*

A common way of expressing acid production rates for this type generator is in terms of grain gallons per 100 ft^3 of packing per day. The term grain gallon used in the industry in the United States is defined as 1 gal of 100 grain (10% acetic acid) vinegar. Percentage acidity is expressed as grams of acetic acid per 100 ml of vinegar. Production rates vary greatly in commercial generators; however, when producing 120 grain vinegar, a rate of 30 grain gal/100 ft^3 of packing per day would be very good. These production rates may be calculated as follows:

$$P = \frac{24GV}{PH}$$

where G = grain of draw-off
V = draw-off volume in gallons
P = packing volume in 100 ft^3
H = cycle time—feed to draw-off in hours
24 = factor converting hours to days

Calculations are applicable only when all conditions such as volumes and draw-off acidities are the same for extended periods. In addition to good operating practices, the final acidity or grain strength of the vinegar greatly affects the attainable acid production rate. With low acidities, high rates are readily produced. When acidities exceed 12%, however, rates tend to drop drastically. Generators are at times maintained on standby by operating at 13–14% acidities.

In trickling generators, wine and cider vinegar can be produced without adding supplementary nutrients, since sufficient nutrients already exist naturally in wine and cider. However, when a pure ethanol–water–acetic acid mixture is used, supplementary nutrients are necessary. A number of good prepared nutrients are commercially available. These very often contain yeast as a source of vitamins and other growth factors (Hildebrandt *et al.*, 1961). The formula for a simple, easily prepared, and completely soluble nutrient developed by USI researchers is as listed in the following tabulation.

Ingredient	Weight/1000 gallon charge
Corn sugar	1 lb–13 oz
Ammonium phosphate, dibasic	13 oz
Magnesium sulfate	$3\frac{1}{4}$ oz
Potassium citrate	$3\frac{1}{4}$ oz
Calcium pantothenate (*d* form)	1 gm

The efficiency of the conversion of ethanol to acetic acid is another important factor for consideration in the operation of the trickling generator. From the reactions shown in Section II, it can be seen that 1 mole of ethanol is oxidized to 1 mole of acetic acid and 1 mole of water. From this it can be calculated that 1 gal of 12% ethanol by volume theoretically yields 1.035 gal of 11.98% acetic acid, which is equivalent to 1 gal of 12.4% acetic acid.

In actual practice some residual alcohol must always be retained. Thus, in making efficiency calculations this must be considered by either deducting it from the ethanol in the charge or adding its acid equivalent to the final product. A good efficiency in commercial trickling generators is considered to be in the range of 88–90% of the theoretical. Losses by evaporation and some conversion of alcohol to compounds other than acetic acid cannot be avoided.

3. Common Disorders and Problems

The vinegar eel (*Anguillula aceti*), a nematode, is a frequent source of trouble during vinegar manufacture, occurring more commonly in trickling generators than submerged culture operations. Eels are undesirable, for in large numbers, putrefaction of the dead organisms can cause off-odors. They feed on the zoogleal mat and organic matter in the generator, gaining access via dirt and insects. At one time they were so common in generators that they were thought to be essential for good acid production. Good filtration and pasteurization eliminates them from bottled vinegar. Only sterilization of the generators with live steam will eliminate them from infected tanks.

The troublesome vinegar mites may also appear in large numbers in trickling generators (Schierback, 1951). Species commonly found are *Tyroglyphas longior* and *T. siro*. These, too, can only be eliminated by steaming.

Infection with *Acetobacter xylinum* can result in coating the packing in generators with slime which can eventually lead to complete blockage. This is much more likely to occur when the generators are operated at acidities under 10%.

Excessive amounts of phosphates added with the nutrients can result in turbidity of the finished vinegar if hard water is used for dilution.

B. Submerged Culture Generators

1. Description of Frings Acetator

The most widely used submerged culture vinegar generator is the Frings Acetator (Ebner, 1976) produced by the Heinrich Frings Company of Bonn, Germany (see Fig. 2). It is a skillfully engineered stainless steel fermentor that is equipped with a high-speed bottom-entering agitator, temperature control, and a foam breaker. The unique feature of this equipment is its highly efficient method of supplying air. This is accomplished by means of a high-velocity self-aspirating rotor on the bottom of the tank that pulls air in through a pipe that is connected to a rotameter and control valve on the outside. The rotor is hollow and star-shaped, with slots cut on the trailing side of the star points, and on turning distributes finely divided bubbles of the aspirated air. Stator vanes

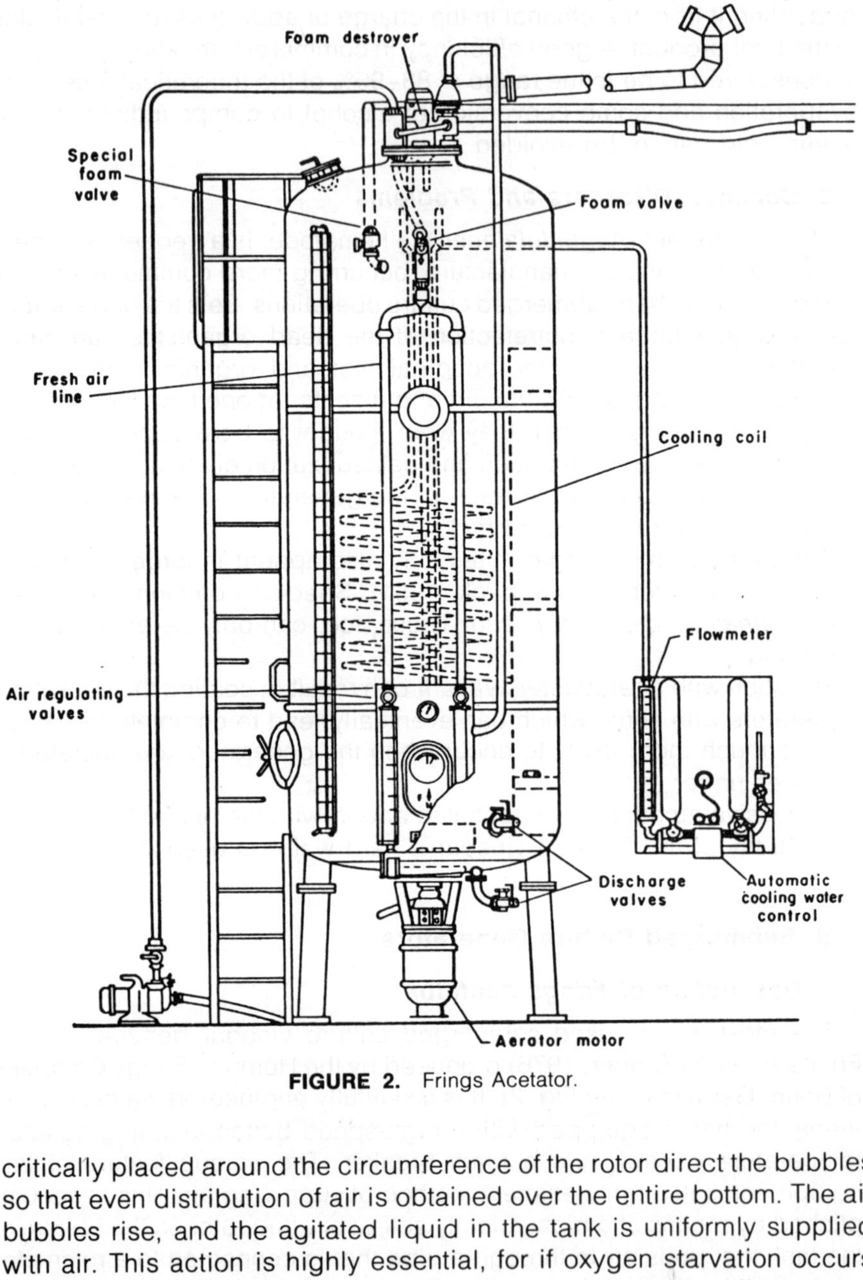

FIGURE 2. Frings Acetator.

critically placed around the circumference of the rotor direct the bubbles so that even distribution of air is obtained over the entire bottom. The air bubbles rise, and the agitated liquid in the tank is uniformly supplied with air. This action is highly essential, for if oxygen starvation occurs at any time, the *Acetobacter* organisms are adversely affected immediately.

The temperature of this type of generator is controlled with cooling water circulated through internal coils. With the rapid production rates obtained in these generators, an adequate means of dissipation of the heat of oxidation is necessary. Temperature of operation is generally about 30°C, although the optimum temperature appears to vary inversely with the total concentration of alcohol and acid in the mash. Some strains of *Acetobacter* are more tolerant than others of higher operating temperatures, but most strains will die above 38°C.

Foam formation is a common problem in submerged fermentation generators, and foam-breaking mechanisms are absolutely essential. In the Acetator a centrifugal foam breaker is used that is automated by an electrode that senses both type and volume of foam.

The Acetator is operated batchwise and can be regulated by means of an Alkograph, an instrument that constantly monitors the alcohol concentration of the mash and is used to signal the completion of the batch. The automated system can be set to be triggered when the residual alcohol drops to 0.2% by volume. About 35% of the finished product is then pumped out. Fresh mash replaces this vinegar, and the cycle starts again. Cycle times for 12% vinegar are usually about 35 hours. The rate of production can be as much as ten times as great as that obtained in trickling generators of equivalent size.

Vinegar produced by the submerged process is extremely turbid, since it contains the bacteria that produced it. For filtration, large-capacity filters, filter aids, fining agents, and settling tanks are required. In contrast, when vinegar is produced in trickling generators, the vinegar organisms are retained on the shavings, and relatively small filter capacities are required.

2. Operation of the Acetator

To start a submerged culture generator, "warm" vinegar from a similar operation can be used as seed. This is freshly prepared vinegar that will contain some live organisms, which are activated under ideal conditions of aeration, alcohol concentration, and nutrients. This startup could take as much as 7–21 days. The Frings Company manufactures small laboratory (1.7 liters) and pilot-plant (5 liters) fermentors, which can be used for the cultivation of bacteria starting from lyophilized cultures of proven acid producers. The pilot units are constructed to be battery operated when normal current supply is interrupted. This assures the availability of a supply of organisms if the commercial units are shut down because of electrical failures.

The formula for a soluble nutrient that was developed in the USI vinegar laboratory for submerged culture follows in the tabulation below.

Ingredient	Submerged culture weight/1000 gallon charge
Corn sugar	9 lb
Ammonium phosphate, dibasic	4 lb
Magnesium sulfate	1 lb
Potassium citrate	1 lb
Calcium pantothenate (*d* form)	0.16 oz (5 gm)

In this type of operation, where the bacteria are drawn off with the finished vinegar, a considerably greater concentration of food is required. In the formula above, the same ingredients are used as were given in Section III for trickling generators, but five times the amounts are required.

3. Comparison of Trickling and Submerged Operations

The desirable features and advantages of the Acetator over the trickling generator are numerous: (1) about one-sixth the amount of space is required to produce an equal quantity of vinegar; (2) invested capital as compared with productive capacity is much lower than for packed generator operations; (3) the efficiency is greater, and yields are 5–8% higher; (4) it is easy to switch from one type of vinegar to another and to start up to full production; and (5) generators can be highly automated.

C. Other Vinegar Generators

1. Yeomans Cavitator

The Yeomans Cavitator (Beaman, 1967) is a submerged culture-type system somewhat similar to the Acetator but differing in the manner in which air is supplied. In this generator, which is no longer being manufactured, the rotor–agitator is top driven and withdraws liquid and air from a centrally located draft tube. The continuous liquid overflow into the draft tube is mixed with the finely divided air in an upward current and given uniform distribution of air through the tank. Some cavitators are still in use in the United States and Japan.

2. Bourgeois Process

The Bourgeois process is a submerged culture system used in Spain and Italy in which compressed air is distributed to the tanks through a rotor. It is a two-generator system wherein one is used to supply inoculum for the other. Batches are finished when all alcohol has been converted to acetic acid. Then the entire contents of the tank are withdrawn. Since vinegar bouquet and flavor are attributed to esters, some alcohol is subsequently added to the finished product to enhance its quality through esterification. The Bourgeois process is satisfactory only where low acidities and slow production rates can be tolerated.

3. Fardon Process

Fardon process generators are also of the submerged culture type. Liquid is pumped from the tank through a venturi nozzle for aeration and returned. Like the Bourgeois process, it is slow and inefficient and used mainly in Africa for making malt vinegar.

4. The Tower Fermentor

The Tower Fermentor is a relatively recent development in aeration systems patented in England (Greenshields and Smith, 1974) and used in the United Kingdom. It is a continuous trickling-type generator about 2 ft in diameter and 20-ft tall. Plastic perforated plates covering the cross-section are used for the distribution of air. It has been reported that this is satisfactory for producing all types of vinegar.

V. FINISHED VINEGARS

A. Federal Regulations

There are numerous types of vinegar on the market, including many with added spices and flavors. Basically, the major types are those listed in Food and Drug Administration Compliance Policy Guide 7109.04, Chapter 9—Condiments.

Policy:

FDA considers the following to be satisfactory guidelines for the labeling of vinegars:

Natural vinegars as they come from the generators normally contain in excess of 4 grams of acetic acid per 100 ml. When vinegar is diluted with water, the label must bear a statement such as "diluted with water to ____ percent acid strength," with the blank filled with the actual percent of acetic acid—in no case should it be less than 4 percent. Each of the varieties of vinegar listed below should contain 4 grams of acetic acid per 100 ml (20°C). Maximum residual alcohol concentration permitted is 0.5% by volume of the total product as sold.

Vinegars:

1. VINEGAR, CIDER VINEGAR, APPLE VINEGAR. The product made by the alcoholic and subsequent acetous fermentations of the juice of apples.
2. WINE VINEGAR, GRAPE VINEGAR. The product made by the alcoholic and subsequent acetous fermentations of the juice of grapes.
3. MALT VINEGAR. The product made by the alcoholic and subsequent acetous

fermentations, without distillation, of an infusion of barley malt or cereals whose starch has been converted by malt.
4. SUGAR VINEGAR. The product made by the alcoholic and subsequent acetous fermentations of sugar sirup, molasses or refiners sirup.
5. GLUCOSE VINEGAR. The product made by the alcoholic and subsequent acetous fermentations of a solution of glucose. It is dextrorotatory.
6. SPIRIT VINEGAR, DISTILLED VINEGAR, GRAIN VINEGAR. The product made by the acetous fermentation of dilute distilled alcohol.
7. VINEGAR, made from a MIXTURE OF SPIRIT VINEGAR AND CIDER VINEGAR. The product made by preparing a mixture of distilled vinegar and cider vinegar. This product should be labeled as a blend of the products with the product names in order of predominance. This labeling is applicable to a similiar product made by acetous fermentation of a mixture of alcohol and cider stock.
8. VINEGAR MADE FROM DRIED APPLES, APPLE CORES OR APPLE PEELS. Vinegar made from dried apples, apple cores or apple peels should be labeled as "vinegar made from ________," where the blank is filled in with the name of the apple product(s) used as the source of fermentable material.

B. Popular Types of Vinegar

1. White Distilled Vinegar

For white distilled vinegar it is the ethanol used as the raw material that has been distilled. In the United States, ethanol purchased for vinegar making is denatured (U.S. Industrial Chemicals Company, 1969). The specially denatured alcohol (SDA 29) used by most manufacturers contains 1 gal of ethyl acetate in 100 gal of 190° proof ethanol. Other SDA formulas permitted but seldom used for vinegar production are SDA 18 and SDA 35-A. In 1975, 19.6 million gal of SD alcohol were sold for vinegar processing. The presence of the ethyl acetate denaturant in SDA 29 has no deleterious effect on acetic acid production. In fact, it is considered as an essential flavoring in vinegar. It is always present whether added or not, its concentration varying directly with residual alcohol content.

The alcohol can be either synthetic ethanol, derived from natural gas, or fermentation alcohol, usually from corn or molasses. Price normally determines the type of ethanol that is used, since both synthetic and fermentation alcohol can be of excellent quality. Vinegar produced from the two types can be differentiated by their ^{14}C contents using a liquid scintillation counter (Hildebrandt *et al.*, 1969). Relatively high concentrations of ^{14}C are found in vinegar made from alcohol originating from recently photosynthesized carbohydrates, while low concentrations are found in vinegar from synthetic alcohol from fossilized materials. This method cannot distinguish between vinegar produced from fermentation ethanol and acetic acid (synthetic) produced by the destructive distillation of wood.

The flavor of distilled vinegar produced from high-quality ethanol, whether from the trickling or the submerged culture processes, is generally unaromatic in comparison with fruit or grain vinegars. Unaromatic vinegar is desirable in products that contain vinegar solely for its acid content. Its flavor and aroma can be even more subdued by holding the residual alcohol at a bare minimum so as to limit the ester content of the vinegar. On the other hand, if a highly aromatic distilled vinegar is desired, it can be obtained by adding to the 190° proof ethanol about 0.25–0.5% grain fusel oil. The higher alcohols that make up the fusel oil are partially esterified and add a fruity aroma to the vinegar.

Few studies have been made on the composition of distilled vinegar. Kahn *et al.* (1972), using gas chromatography–mass spectrometry, identified 27 compounds in samples of submerged and trickling vinegars. Other than the acetic acid these were mainly esters, alcohols, halogenated compounds, hydrocarbons, carbonyls, ethers, and acetals, with no significant differences attributed to the production method.

2. Cider Vinegar

Cider vinegar is made from the fermented juice of apples or apple-juice concentrate. Apple juice containing 11% soluble solids yields vinegar of approximately 55 grain. Most cider vinegar is bottled for retail trade and must be sparkling clear. Thus, special care is required in its filtration. Even after careful filtration, cider vinegar has a tendency to form a certain amount of colloidal turbidity, which in time will form a sediment. This is sometimes overcome by storage of the vinegar for a considerable period of time prior to filtration and bottling. Tannins from the apple that are present in cider vinegar make it extremely sensitive to blackening from iron contamination. As little as 1 ppm iron can result in a pronounced color change (Conner and Allgeier, 1976).

As would be expected, gas–liquid chromatography (GLC) analysis of cider vinegars shows the presence of many compounds not found in distilled vinegar. Unfortunately, none of these is in sufficient and uniform concentration to make possible the detection of adulteration of cider vinegar with the cheaper distilled vinegar. Complaints that some adulterated products are on the market are occasionally directed at vinegar producers.

3. Wine Vinegar

Lower quality grape wines are commonly used for acetous fermentation to wine vinegar. Both white and red wines are used. The comments made above regarding cider vinegar are also applicable to wine vinegar. Treatments that are specifically permitted for wine vinegar and not for others are the addition of sulfur dioxide (140 ppm) and the addition of wine (to give a maximum of 3.0% residual alcohol) to the finished 40-grain vinegar. The sulfur dioxide is added for preservation of color and for stability, and the wine is added for bouquet.

4. Malt Vinegar

Malt vinegar is made by the alcoholic and acetous fermentation of a malt mash or mash containing corn and barley in addition to the malt. Enzymes in the malt convert the grain starches to sugar, and the sugar is fermented by yeast to ethanol. This semiclarified fermented beer is then subjected to the usual acetous fermentation.

VI. PROCESSING OF VINEGAR

A. Filtration and Clarification

Distilled vinegar from trickling generators is relatively free of insoluble material, so filters of very small capacity suffice. However, other vinegars from the trickling process and all vinegars from the submerged process require the use of filter aids and fining agents to accomplish the desired clarity. Large-capacity filters are required when these additives are used.

B. Bottling

In order to prevent bacterial growth subsequent to bottling, vinegar must be pasteurized. The filled bottles are tightly sealed, heated to 60°–65°C, and held at that temperature for 30 minutes prior to cooling. A somewhat less desirable procedure is sometimes used that consists of heating the vinegar to 65°–70°C and filling the bottles with this hot vinegar and sealing immediately.

C. Concentration of Vinegar

Vinegar can be concentrated by a freezing process. The vinegar is frozen to a slush, which is centrifuged to separate the concentrated vinegar from the low-acid-content liquid. The water from this melted ice contains about 1–2% acid and can be used for the dilution of 190° ethanol in making up mash for the generators. This process was developed and commercialized by Girdler Process Equipment Division, Chemtron Corporation, Louisville, Kentucky. Vinegars of 200-grain strength are readily obtainable from 120-grain raw vinegar. Although 300-grain vinegar can be made, the production cost is relatively high.

There are a number of advantages to 200-grain vinegar. Transportation costs are substantially reduced by the reduction in volume; also, it is of special value to pickle processors, since it enables them to build up the acid strengths and reuse brine solutions that otherwise would have to be discarded because of dilution by the juice of the pickles.

VII. ANNUAL PRODUCTION AND USES

The total sales of all types of vinegar in 1976 in the United States amounted to about 148 million gal. Broken down as to types, the values are given in the tabulation below.

Type	Millions of gal (approximate)
Distilled	110 (100 grain)
Cider	26 (50 grain)
Wine	6 (50 grain)
Others	6 (50 grain)

Approximately 68% of all vinegar produced is used in products commercially prepared (in order of the volume sold to each) by the dressing and sauce industry, the pickle industry, tomato products manufacturers, and the mustard industry.

VIII. APPENDIX

The Vinegar Institute* is a trade organization of vinegar manufacturers. It is made up of approximately 25 active members, who are the principal manufacturers and bottlers of vinegar. The organization has worked closely with the FDA to establish high standards and exacting specifications for all types of vinegar. Environmental Protection Agency standards and labeling regulations applicable to vinegar manufacture are closely followed and explained to members.

The principal vinegar producers in the United States are: California-Omega Foods, Emeryville, California; Heinz U.S.A., Pittsburgh, Pennsylvania; Hunt Foods, Inc., Fullerton, California; Indian Summer, Inc., Evansville, Indiana; National Vinegar Company, St. Louis, Missouri; A. M. Richter Sons Company, Manitowoc, Wisconsin; Speas Company, Kansas City, Missouri; and Standard Brands, Inc., Baltimore, Maryland.

* 64 Perimeter Center East, Atlanta, Georgia 30346.

REFERENCES

Allgeier, R. J., and Hildebrandt, F. M. (1960). *Adv. Appl. Microbiol.* **11,** 163–181.
Allgeier, R. J., Nickol, G. B., and Conner, H. A. (1974a). *Food Prod. Dev.* **8**(5), 69–71.
Allgeier, R. N., Nickol, G. B., and Conner, H. A. (1974b). *Food Prod. Dev.* **8**(6), 50–53 and 56.
Asai, T. (1968). "Acetic Acid Bacteria." Univ. Park Press, Baltimore, Maryland.
Beaman, R. G. (1967). *In* "Microbial Technology" (H. J. Peppler, ed.), pp. 344–376. Van Nostrand-Reinhold, Princeton, New Jersey.
Chemetron Corporation (undated). "Crystallization and Concentration. Subject: Vinegar," Tech. Bull. Votator Div., Chemetron Corp., Louisville, Kentucky.
Conner, H. A., and Allgeier, R. J. (1976). *Adv. Appl. Microbiol.* **20,** 82–133.
Ebner, H. (1976). *Ullmans Encyklo. Tech. Chemi.* **11,** 41–55.
Greenshields, R. N., and Smith, E. L. (1974). *Process Biochem.* **9,** 11–13, 15, 17, and 28.
Hildebrandt, F. M., Nickol, G. B., Dukowicz, M., and Conner, H. A. (1961). "Vinegar Newsletter," No. 33. U.S. Ind. Chem. Co., New York.
Hildebrandt, F. M., Nickol, G. B., and Conner, H. A. (1967). "Vinegar Newsletter," No. 50. U.S. Ind. Chem. Co., New York.
Hildebrandt, F. M., Weber, G. R., and Nickol, G. B. (1968). "Vinegar Newsletter," No. 54. U.S. Ind. Chem. Co., New York.
Hildebrandt, F. M., Nickol, G. B., and Kahn, J. H. (1969). "Vinegar Newsletter," No. 56. U.S. Ind. Chem. Co., New York.
Kahn, J. H., Nickol, G. B., and Conner, H. A. (1972). *Agric. Food Chem.* **20,** 214–218.
Nickol, G. B., Conner, H. A., Dukowicz, M., and Hildebrandt, F. M. (1964). "Vinegar Newsletter," No. 43. U.S. Ind. Chem. Co., New York.
Schierback, J. (1951). "The Manufacture of Vinegar," Tech. Inf. Serv., Rep. No. 17. Natl. Res. Counc., Ottawa, Canada.
Shimwell, J. L. (1954). *J. Inst. Brew.* **60,** 136–141.
U.S. Industrial Chemicals Company (1969). "Ethyl Alcohol Handbook." U.S. Ind. Chem. Co., New York.

Chapter 7

Ketogenic Fermentation Processes

D. PERLMAN

I. INTRODUCTION

The ketogenic fermentation processes to be reviewed in this chapter are those in which polyhydric alcohols are converted into ketoses. The most important processes are the oxidation of D-sorbitol into L-sorbose, and of glycerol into dihydroxyacetone, both operations being carried out on an industrial scale using strains of *Acetobacter suboxydans*.

The type reaction for this bacterial oxidative ketogenic process is:

$$-\overset{|}{\underset{|}{C}}-OH \longrightarrow -\overset{|}{\underset{|}{C}}=O$$

The product of the reaction is governed by the choice of substrate and by the specificity of the enzyme system of *Acetobacter suboxydans*. The major requirements are high oxygen levels in the solution and an adequate nutrient supply.

II. THE SORBOSE FERMENTATION

L-Sorbose is produced from the polyhydric alcohol D-sorbitol by the action of several species of bacteria of the genus *Acetobacter*.

MICROBIAL TECHNOLOGY, 2nd ed., VOL.II
 ISBN 0-12-551502-2

Acetobacter suboxydans appears to be the species most widely used industrially, although many other species, including *A. xylinum, A. gluconicum, A. xylinoides, A. pasteurianum, A. kutzingianum, A. orleanse,* and *A. hoshigaki,* also produce sorbose readily.

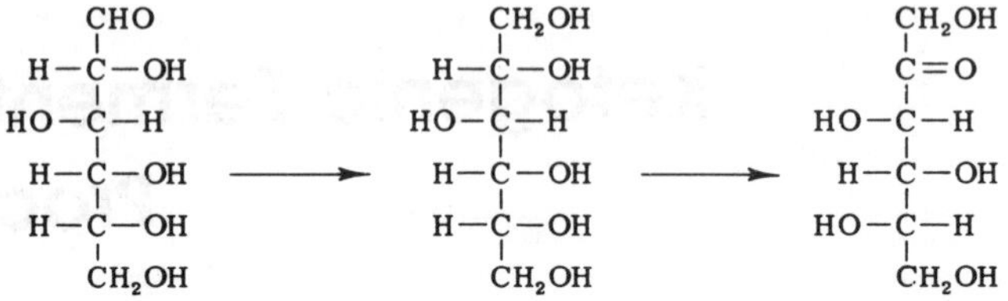

D-Glucose D-Sorbitol L-Sorbose

The starting material in the sorbose fermentation is the sorbitol obtained when the aldehydic terminal group of the D-glucose molecule is reduced to a primary alcohol by either electrolytic or catalytic methods. When a nickel catalyst is used for the hydrogenation, care must be used to remove it from the purified product since *A. suboxydans* is very sensitive to nickel ion.

A procedure for the production of L-sorbose from D-sorbitol by *A. suboxydans* described by Wells *et al.* (1937, 1939) uses sorbitol solutions of 20–30% concentration, supplemented with 0.5% yeast extract (or the cheaper corn steep liquor) and sufficient $CaCO_3$ to neutralize the acidity of the medium (mainly from the corn steep liquor and a small amount of gluconic acid formed by oxidation of the traces of glucose remaining in the sorbitol syrup). Inoculum was a cell suspension of *A. suboxydans* grown in a medium containing sorbitol, glucose, yeast extract, and $CaCO_3$.

The fermentation conditions found optimal for this oxidative process included vigorous aeration, agitation, operation of the vessel at 30 psig air pressure, and incubation at 30°C. Wells *et al.* (1937, 1939) used small amounts of octadecanol to control foaming. They noted almost quantitative conversions of 20, 25, and 30% solutions of sorbitol in fermentation periods of 24, 30.5, and 45 hours, respectively.

At the completion of the incubation period, the sorbose-containing solution was mixed with decolorizing charcoal and filter aid and the suspension was filtered through a press. This was followed by concentration of the filtrate *in vacuo* to a syrup, which crystallized on cooling to 15°C. The crystals were separated by centrifugation, washed with ice water, and dried. A second crop of sorbose was obtained from the mother liquor and washings. The total recovered yield was about 70% based on the L-sorbose in the fermented medium.

More recently, attention has been focused on shortening the incubation time and increasing the process efficiency. Elsworth *et al.* (1959) have pointed out the potential of a continuous fermentation process as being more economical than a batch operation, and this has been explored by Muller (1966). Bull and Young (1977) reported enhanced efficiency in continuous fermentations on a laboratory scale with microbial cell recycle, obtaining both higher yields and reduced incubation intervals. Aeration using oxygen-enriched air was required to obtain maximal efficiency of the fermentation process.

Nearly all of the 70,000,000 lb of L-sorbose produced annually by this process is used as an intermediate in the chemical synthesis of L-ascorbic acid. The major manufacturer is Hoffmann-LaRoche Inc. (with operations in the United States and West Germany), and significant quantities are produced by E. Merck (West Germany), Pfizer, Inc. (United States), and Takeda Chemical Industries Ltd. (Japan).

III. THE DIHYDROXYACETONE FERMENTATION

Dihydroxyacetone is produced by the microbial oxidation of glycerol with selected strains of *Acetobacter suboxydans*.

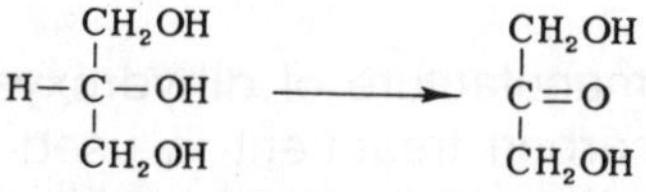

This conversion was noted as early as the end of the 19th Century (Bertrand, 1898) and some realization of its use as a pharmaceutical intermediate followed (Cathcart and Markowitz, 1927; Lambie, 1927; Underkofler and Fulmer, 1937). The most recent interest (Andreadis and Miklean, 1960; Anonymous, 1960) is derived from its utility as a suntanning agent.

Initial studies (Visser t'Hooft, 1925) showed that aeration of the fermentation medium for 6 days increased the rate of conversion of glycerol to dihydroxyacetone as well as the efficiency of conversion—90% with aeration and 57% under static conditions. Virtanen and Barlund (1926) also noted highest conversions when the cultures were aerated and the pH was maintained at about pH 5.

Although a series of studies showed that increasing the initial concentration of glycerol from about 6 to 12% resulted in decreased efficiency of conversion (Underkofler and Fulmer, 1937; Low and Krema, 1928), this may now be explained on the basis of limited oxygen levels in the medium and also limited nutritional requirements (Flickinger and Perlman, 1977). Underkofler and Fulmer (1937) have also noted that conversions of 91% of the glycerol (6% solution) could be obtained in 3–7 days using submerged aeration when the basal medium contained 0.5% yeast extract and 0.25% KH_2PO_4. On the other hand, Bernhauer and Schon (1928) noted that a 30-day incubation period was needed to convert a 12% glycerol solution with only a 43% conversion efficiency.

The current process requirements for commercial production of dihydroxyacetone include a conversion of a 10% glycerol solution efficiently, e.g., 95% or better, to dihydroxyacetone in a short fermentation cycle. Green (1960; Green *et al.*, 1961) has described a process in which the mash contains 0.5% KH_2PO_4, 0.5% brewers' yeast, 2% $CaCO_3$, and 10% glycerol. If the initial pH was set at 5.5, about 60% of the fermented glycerol was converted in the 40-hour incubation period to dihydroxyacetone. He also commented that pilot-plant trials provided assurance that dihydroxyacetone yields could reach 93% at a plant level.

Flickinger and Perlman (1977) reported that *Gluconobacter melanogenus* IFO 3293 converted glycerol to dihydroxyacetone in a 25-liter batch fermentor when grown in a yeast extract–phosphate medium. The efficiency of conversion was increased over threefold when oxygen tension was controlled by increasing the partial pressure of oxygen in the aeration. Under optimal conditions, 0.05 atm dissolved oxygen, quantitative conversion of 10% glycerol solutions were obtained.

Current commercial manufacture of dihydroxyacetone involves a recovery system where carbon treatment is used to decolorize the fermented medium, the cells are removed by filtration, and the clarified liquor is concentrated *in vacuo.* Suitable solvents, e.g., acetone, isopropyl alcohol, are added to induce crystallization of the desired product and care is taken to prevent polymerization. The white crystalline dihydroxyacetone produced commercially has an ash content of not over 0.2% and a purity of over 98%.

Companies currently in commercial operations include G. B. Fermentation Industries, Inc. (United States), E. Merck (West Germany), S. B. Penick Company (United States), Roussel-UCLAF (France), Schering Corporation (United States), Tanabe Seiyaku Company, Ltd. (Japan), and Wallerstein Laboratories (United States).

REFERENCES

Andreadis, J. T., and Miklean, S. (1960). U.S. Patent 2,949,403.
Anonymous (1960). *Chem. Eng. News.* Feb. 22; Aug. 15.
Bernhauer, K., and Schon, K. (1928). *Hoppe-Sayler's Z. Physiol. Chem.* **177,** 107.
Bertrand, G. (1898). *C. R. Hebd. Seances Acad. Sci.* **126,** 842.

Bull, D. N., and Young, M. D. (1977). *Abstr. Pap. Meet., 174th, Am. Chem. Soc.* p. 01.
Cathcart, E. P., and Markowitz, J. (1927). *Biochem. J.* **21,** 1419.
Elsworth, R., Telling, F. C., and East, D. N. (1959). *J. Appl. Bacteriol.* **22,** 138.
Flickinger, M. C., and Perlman, D. (1977). *Appl. Environ. Microbiol.* **33,** 706.
Green, S. R. (1960). U.S. Patent 2,948,658.
Green, S. R., Whalen, E. A., and Molokie, E. (1961). *J. Biochem. Microbiol. Technol. Eng.* **3,** 351.
Lambie, C. G. (1927). *J. Soc. Chem. Ind.* **46,** 300.
Low, A., and Krema, A. (1928). *Klin. Wochenschr.* **7,** 2432.
Muller, J. (1966). *Zentralbl. Bakteriol., Parasitenkd., Infektionskr. Hyg., Abt. 2,* **120,** 349.
Underkofler, L. A., and Fulmer, E. I. (1937). *J. Am. Chem. Soc.* **59,** 301.
Virtanen, A. I., and Barlund, B. (1926). *Biochem. Z.* **169,** 169.
Visser t'Hooft, F. (1925). Dissertation. Technical University, Delft, Holland.
Wells, P. A., Stubbs, J. J., Lockwood, L. B., and Roe, E. T. (1937). *Ind. Eng. Chem.* **29,** 138.
Wells, P. A., Lockwood, L. B., Stubbs, J. J., Roe, E. T., Porges, N., and Gastrock, E. A. (1939). *Ind. Eng. Chem.* **31,** 1518.

Chapter 8

Mushroom Fermentation

RANDOLPH T. HATCH
STANLEY M. FINGER

I. INTRODUCTION

The worldwide demand for edible mushrooms has increased significantly in the last decade to the point that there are now in excess of 1.5 billion lb produced annually. The demand for fresh mushrooms in the United States has been particularly strong. The per capita consumption has increased from 1.6 to 2.2 lb over the last decade. This is reflected in the nationwide production of mushrooms which has grown from 108 million to 399 million lb since 1959 as shown in Table I. The demand for imported mushrooms (primarily canned) has grown even faster to the point that imports now account for 24% of the United States market. This has led to a recent shift in the nature of the industry from that of the small

MICROBIAL TECHNOLOGY, 2nd ed., VOL.II
 ISBN 0-12-551502-2

TABLE I. United States Mushroom Production and Imports[a]

Crop year	U.S. Production	Canned imports: Production	Canned imports: Percentage of U.S. Market
1959–1960	108	3	2.7
1960–1961	115	5	4.2
1961–1962	127	15	10.6
1962–1963	132	16	10.8
1963–1964	131	21	13.8
1964–1965	140	16	10.3
1965–1966	156	21	11.9
1966–1967	165	26	13.6
1967–1968	181	34	15.8
1968–1969	189	31	14.1
1969–1970	194	42	17.8
1970–1971	207	43	17.2
1971–1972	231	62	21.2
1972–1973	254	74	22.6
1973–1974	279.5	70	20.0
1974–1975	299.1	77	20.5
1975–1976	309.8	88	22.1
1976–1977	347.1	107	23.6
1977–1978	399	—	—

[a] Million pounds, fresh weight.

independent grower toward an industry dominated by the large producer. Although there are more than 500 independent mushroom growers, the Butler County Mushroom Farm, Ralston Purina, Castle & Cook, and Campbell Soup account for more than 50% of the total production. The geographic distribution of the producers as shown in Fig. 1 is nationwide, with the center of the largest growing area at Kennett Square, Pennsylvania.

A number of different mushrooms are cultivated worldwide. The most common of western countries are *Agaricus bisporus* and *Agaricus bitorques*. The variety Golden White of *A. bisporus* is grown by the Butler County Mushroom Farm due to the ease of production and quality of mushroom produced in its underground caves. The optimum fruiting temperature lies in the range of 14°–18°C. The *Agaricus bitorques*, which fruits at a higher temperature, is grown in Holland. The *A. bitorques* has the commercial advantages of freedom from viruses and ease of hybridization. Because of its higher fruiting temperature, *A. bitorques* is ideally suited to above-ground production in climatically controlled rooms.

The mushroom traditional to Japan is *Lentinus edodes* (Hayes, 1974), otherwise known as Shii-take. This mushroom is unique in the method of

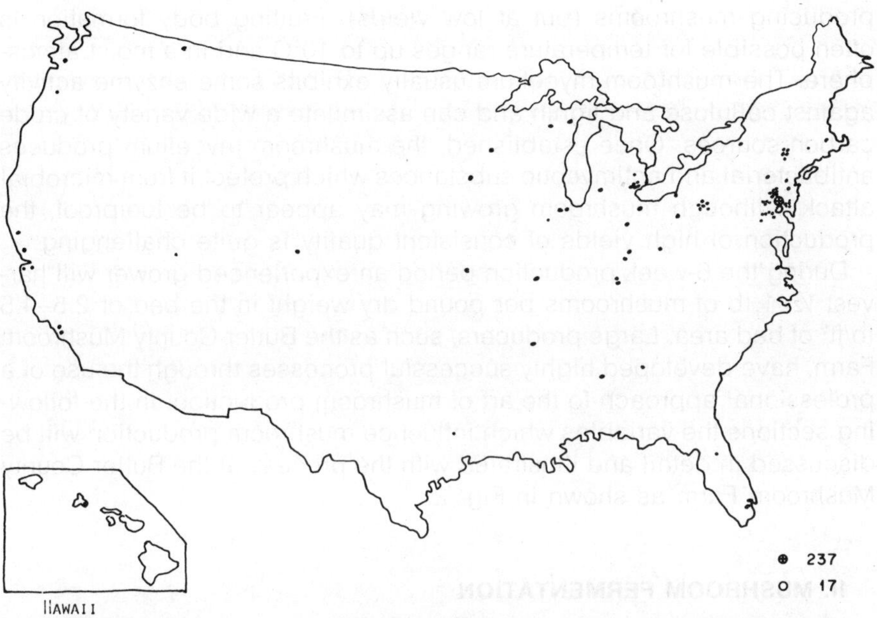

FIGURE 1. Geographical distribution of United States mushroom growers.

its culture. Hardwood logs are inoculated through plugs drilled into the logs. After 8 months the logs are moved to a site which is optimum for fruiting body formation. The mushroom caps can then be harvested for up to 6 years. This is a low-cost process which takes advantage of natural environmental conditions. The fruiting bodies form in the spring and fall when the ambient temperatures are in the range of 12°–20°C.

Other traditional mushrooms of the orient are the oyster (*Pleurotus oestreatus*) and padi-straw (*Volvaria volvaceae*) mushrooms (Hayes, 1974). The oyster mushroom can be grown on a variety of composted plant residues and industrial wastes. Its production differs from that of *Agaricus* in that fruiting body formation requires low light intensity and no casing layer. The name is derived from the oyster shape of the fruiting bodies. The padi-straw species is a tropical mushroom which is well suited to the high temperatures of the tropical climates. The traditional method of production utilizes rice straw as the substrate with minimal composting before inoculation with *V. volvaceae* mycelium. The modern methods of cultivation used in *A. bisporus* production are now being adopted for *V. volvaceae* due to its ability to achieve much higher yields.

A great deal of controversy exists concerning the way mushrooms are, and should be, grown. This is due to the differing requirements of the wide variety of mushrooms produced around the world and the ease in producing mushrooms (but at low yields). Fruiting body formation is often possible for temperature ranges up to 10°C and in a moist atmosphere. The mushroom mycelium usually exhibits some enzyme activity against cellulose and lignin and can assimilate a wide variety of crude carbon sources. Once established, the mushroom mycelium produces antibacterial and antimycotic substances which protect it from microbial attack. Although mushroom growing may appear to be foolproof, the production of high yields of consistent quality is quite challenging.

During the 6-week production period an experienced grower will harvest ½–1 lb of mushrooms per pound dry weight in the bed or 2.5–4.5 lb/ft² of bed area. Large producers, such as the Butler County Mushroom Farm, have developed highly successful processes through the use of a professional approach to the art of mushroom production. In the following sections the variables which influence mushroom production will be discussed in detail and illustrated with the process at the Butler County Mushroom Farm as shown in Fig. 2.

II. MUSHROOM FERMENTATION

It is possible to grow mushroom mycelium on a wide variety of substrates in submerged culture (Litchfield, 1967) and on semisolid mate-

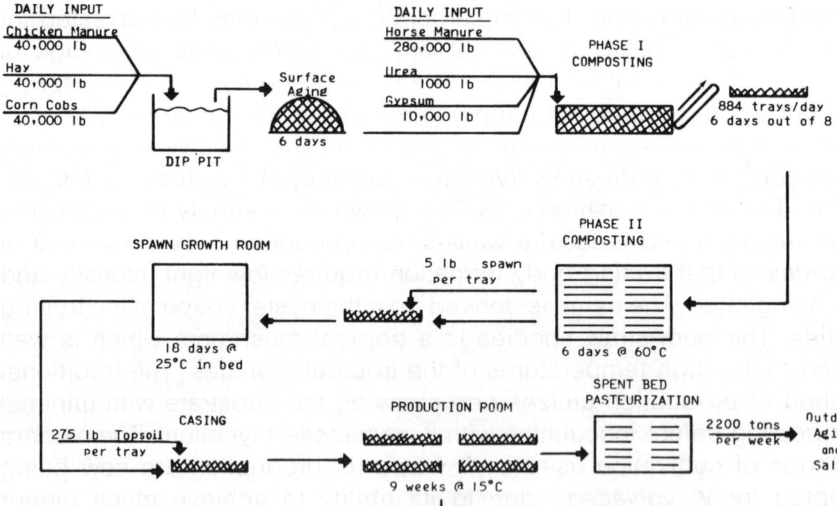

FIGURE 2. Mushroom process flowsheet, Butler County Mushroom Farm, Butler County, Pennsylvania.

rials. Since the mycelial growth is quite slow, the most practical and most common method of growth is on a composted bed of semisolid materials. Although the nutritional requirements are not completely understood, it has been established that the mushroom can metabolize a wide variety of carbon sources, including cellulose and lignin (Rancheva, 1971).

The production of mushrooms consists of essentially three steps: (1) Substrate preparation, (2) inoculation, and (3) induction of fruiting body formation. Each of these steps will be examined in detail.

A. Substrate Preparation

Mushroom growing on a commercial scale always starts with the production of a suitable compost material. Proper composting is essential for successful mushroom growing since the resulting yield of mushrooms is strongly affected by the availability of nutrients and the presence of toxic substances, competing weeds, and disease microorganisms. In a general sense the composting process can be said to convert the starting materials to a form suitable for mushroom growth and unsuitable for the growth of contaminants. The substrate materials usually consist of manure, corn cobs, straw, and other agricultural residues, although municipal garbage has also been used (Franz, 1972). The one notable exception is the use of cow manure which has been found to result in poor mushroom yields. (This is probably due to the efficiency of the cow's digestive system.) The composting is accomplished in two steps: (1) Phase I, usually conducted outdoors for the initial breakdown of the raw materials; and (2) phase II, conducted indoors at controlled elevated temperatures for the conversion of free ammonia to protein and the growth of thermophilic microorganisms (Regan and Jeris, 1970; Shilesky and Maniotis, 1969; Siu, 1951; Tribe, 1962) (possibly with an amino acid profile closer to the optimum for mushroom growth).

During the initial stages of a compost pile, decomposition is quite rapid, causing temperatures to rise to the thermophilic region and stay there for an extended period. During this time materials that are easily converted are preferentially reacted. Easily transformed materials include such carbohydrates as sugars and starches (Keller, 1961; Gray, 1967; Braun, 1962), hemicelluloses, polysaccharides, and protein (Fuller and Bosma, 1965). As much as 80–95% of the easily decomposed portion of the substrate in the pile is converted (Keller, 1961). As the easily decomposed material is being used up, the microorganisms begin to attack the more resistant parts of the substrate, such as cellulose and lignins (Gotaas, 1956; Keller, 1961; Gray, 1967; Fuller and Bosma, 1965; Braun, 1962; Fergus, 1969).

B. Carbon-to-Nitrogen Ratio (C/N)

Carbon and nitrogen are both needed for cellular growth. Living organisms utilize about 30 parts of carbon for every part of nitrogen (Mercer *et al.*, 1962; Waksman, 1938). An excess of one or the other of these elements during aerobic microbial growth causes undesirable results. If carbon is in excess, several generations of organisms are needed to consume the additional carbon (Gotaas, 1956; Fuller and Bosma, 1965), thereby extending the time necessary for decomposition. If nitrogen is in excess, all the carbon is consumed and the extra nitrogen is released as ammonia (Gotaas, 1956; Fuller and Bosma, 1965). Several groups have studied the question of what is the optimum value of C/N for the most rapid and effective composting. It is not surprising that most of the research shows an optimum of about C/N = 30, considering this is approximately the stoichiometric proportion at which these elements are utilized by living organisms (Mercer *et al.*, 1962; Waksman, 1938). It is generally agreed that composting time "increases considerably with increases in C/N ratios above 40." (Fuller and Bosma, 1965). On the low side, a C/N of 10 results in satisfactory composting, while values of 14–20 are common (Gotaas, 1956). It should be pointed out that the carbon and nitrogen may exist in unbiodegradable states and the unavailable portion does not belong in a calculation of C/N. For example, carbon in the form of fuel remains (coal, coke) and nitrogen in plastics, rubber, etc., are very resistant to aerobic decomposition (Gotaas, 1956; Keller, 1961).

C. Water

Microorganisms require sufficient water in order for metabolism to proceed. However, if the moisture content of a compost pile is too high, the water will displace the air in the voids, causing anaerobic conditions (Gotaas, 1956). Thus, there is an optimal range of water content for most effective composting. This range has been reported to be 40–60% (Gotaas, 1956), 60% (Shulze, 1961), 52–58% (Anon., 1965), 55–69% (Wiley and Peace, 1955), and 30–75% (California, 1953). In general, it is desirable to have as much water as possible adsorbed into the straw and other materials being composted, but free water in the voids of the pile should be kept to a minimum.

Anyone who has seen an active compost pile cannot help but notice the prodigious amounts of water vapor generated. Water loss measurements have shown that the water content in a compost pile decreases quite significantly during composting unless additional water is added (Mercer *et al.*, 1962; Kochitzky *et al.*, 1969; Rose *et al.*, 1965).

D. Acidity

As with many of the other physiological variables, the pH of the compost changes during the composting process. Initially the pile becomes more acidic as long as the substrate is not highly buffered (Keller, 1961). This is due to the fact that the decomposition products of the easily attacked components of the substrate are simple acids (Gray, 1967). As the composting progresses, the pile becomes alkaline and ammonia is produced. If the initial pH of the substrate is low, however, as in the case with fruit wastes (pH of 4.5–5.0), there is an initial lag in the development of microbial flora and hence a lag in the rise of the temperature to the thermophilic region (Mercer *et al.*, 1962; Rose *et al.*, 1965).

E. Phase I Composting

The first phase is usually carried out in windrows, which are simply large piles of the substrate material. A windrow is made up each day of the following mixture:

Component	Quantity (lb)
Race horse manure	280,000
Corn cobs	40,000
Hay	40,000
Chicken manure	40,000
Urea	1,000
Gypsum	10,000

The hay, corn cobs, and chicken manure are first dipped in water and allowed to age 6 days before being mixed with the other ingredients. The compost is mixed and formed into rectangular piles 6.5 ft high × 7.5 ft wide × 200 ft long, as shown in Fig. 3. The compost piles are turned four times over a period of 8–9 days, 50 hours between turnings. The turning or mixing of the compost can be done with agricultural equipment or, in large-scale composting, with equipment especially designed to mix and reform the pile in one continuous operation is often used. After forming the pile, the temperature increases rapidly to the thermophilic region (>45°C) and will reach temperatures of about 60°–70°C (Gotaas, 1956). During this time the pile shrinks and the color and physical appearance of the substrate changes. The thermophilic stage is also marked by a rise in the carbon dioxide concentration and a drop in the oxygen concentration often down to 1% or less (Keller, 1961).

FIGURE 3. Compost windrows. (Courtesy of Butler County Mushroom Farm, Butler County, Pennsylvania.)

F. Temperature

Aerobic microbial growth during composting generates a good deal of heat. Since composting materials are relatively good insulators, a temperature distribution occurs within the pile with the highest temperatures being in the interior. Temperature distributions in actual compost piles have been measured as illustrated in Fig. 4 (Lambert, 1934). It is interesting to note that measured temperature distributions are quite irregular. This is due to the nonhomogeneity of the pile from one point to another in terms of composition, density, water content, etc. Recently, a mathematical model of the composting process was developed (Finger, 1975; Finger *et al.*, 1976) for the prediction of conditions and reaction rates

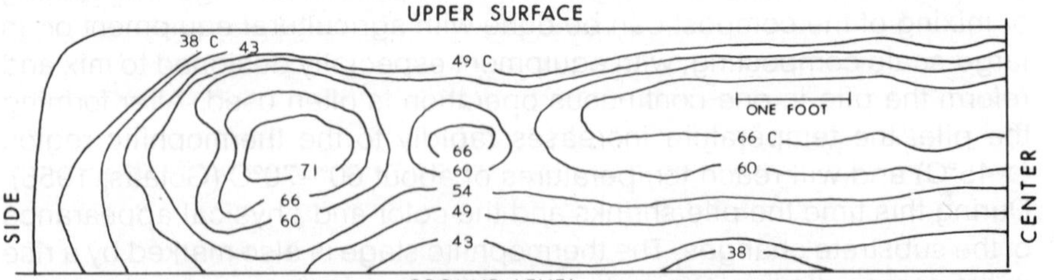

FIGURE 4. Measured temperature distribution of a compost pile.

over the cross-section of the pile. It was assumed in the development of the model that the rate of substrate decomposition is limited by the transfer rates of heat out of and oxygen into the pile. Extensive data taken from compost piles were then used to establish the rate constants in the model. Computer predictions of the temperature distributions within the pile at various times after turning are shown in Fig. 5a–d.

The temperatures reached in the pile depend on the ambient temperature, water content, degree of aeration, substrate materials, and pile size. The maintenance of aerobic conditions is very important in achieving thermophilic conditions as anaerobic decomposition processes generate much less heat than aerobic processes. The common way to maintain aerobic conditions is by periodic turning of the pile. After the turning, the temperature in the pile drops somewhat but returns to its original high value within a few hours (Gotaas, 1956).

The effect that pile size has on temperature is quite significant; a moderate increase in pile size will increase the temperatures measured within the pile. In one experiment the height of a compost pile was changed from 5 ft to 2 ft and back again. When the pile was 5-ft high the maximum interior temperature was 70°C (ambient temperature was 15°–20°C). When the height was reduced to 2 ft the maximum temperature dropped to 65°C within 3 hours. When the pile was reconstructed to its original 5-ft height, the temperature returned to 70°C (Gotaas, 1956). Thus, the interior temperatures in the pile can be controlled somewhat by varying pile size and, indeed, this technique is used commercially to obtain suitable interior temperatures. The effect of size on compost temperatures is directly related to the fact that compost substrates are good thermal insulators. Thus, the larger the pile, the more insulation there is to keep in the heat generated during the decomposition reactions. Convective heat loss at the surface should also be minimized by using a tightly packed outer shell. This leads to more uniformly high temperatures over the cross-section.

G. Oxygen

Aerobic microbial growth requires oxygen to be present for the enzymatic reactions to proceed (Aiba *et al.*, 1965). This oxygen comes from the surrounding atmosphere and diffuses into the compost pile. Although it might seem that the pile could be made sufficiently porous, analyses of the air in compost piles "showed that even with high porosity the gas exchange was retarded" (Tietjen, 1964); that is, concentration gradients were observed. Since the biochemical reactions occur in an aqueous environment, the oxygen must diffuse into the aqueous layer around the substrate and perhaps even diffuse into the pores of the substrate before

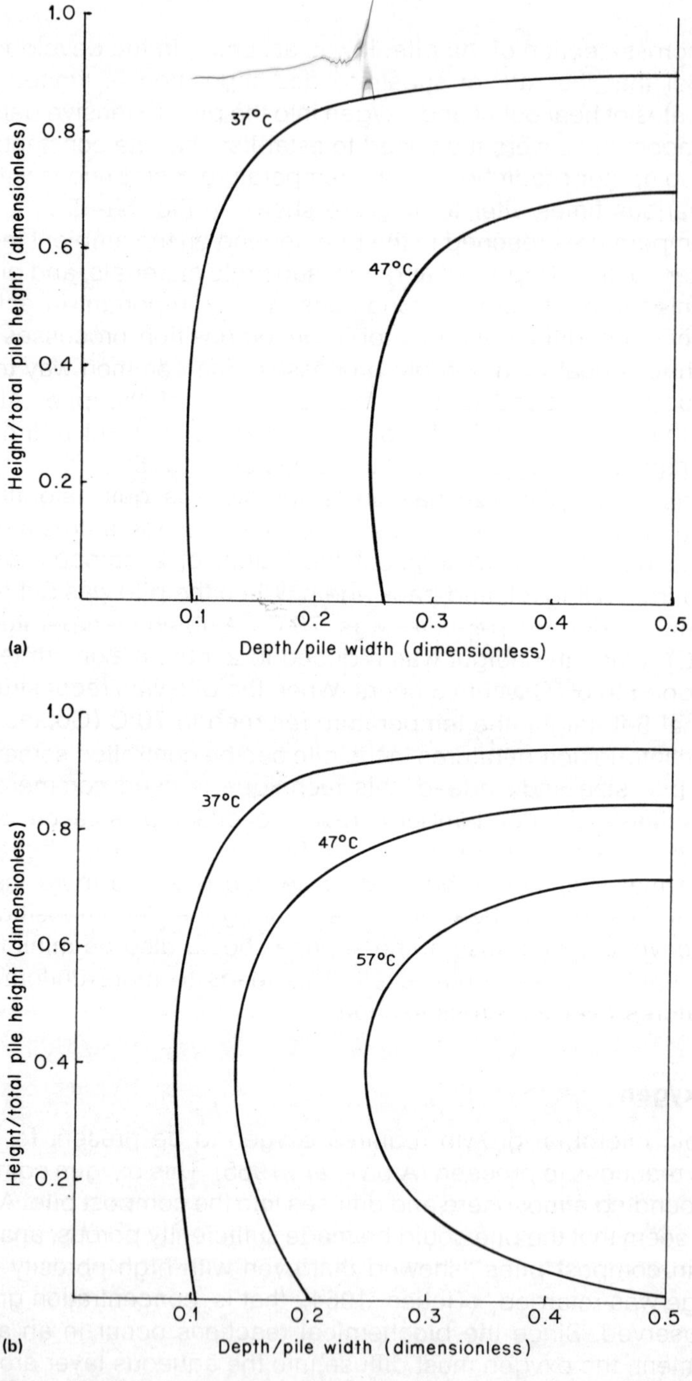

FIGURE 5. Predicted temperature distribution (a) 12 hours after turning; (b) 24 hours after turning.

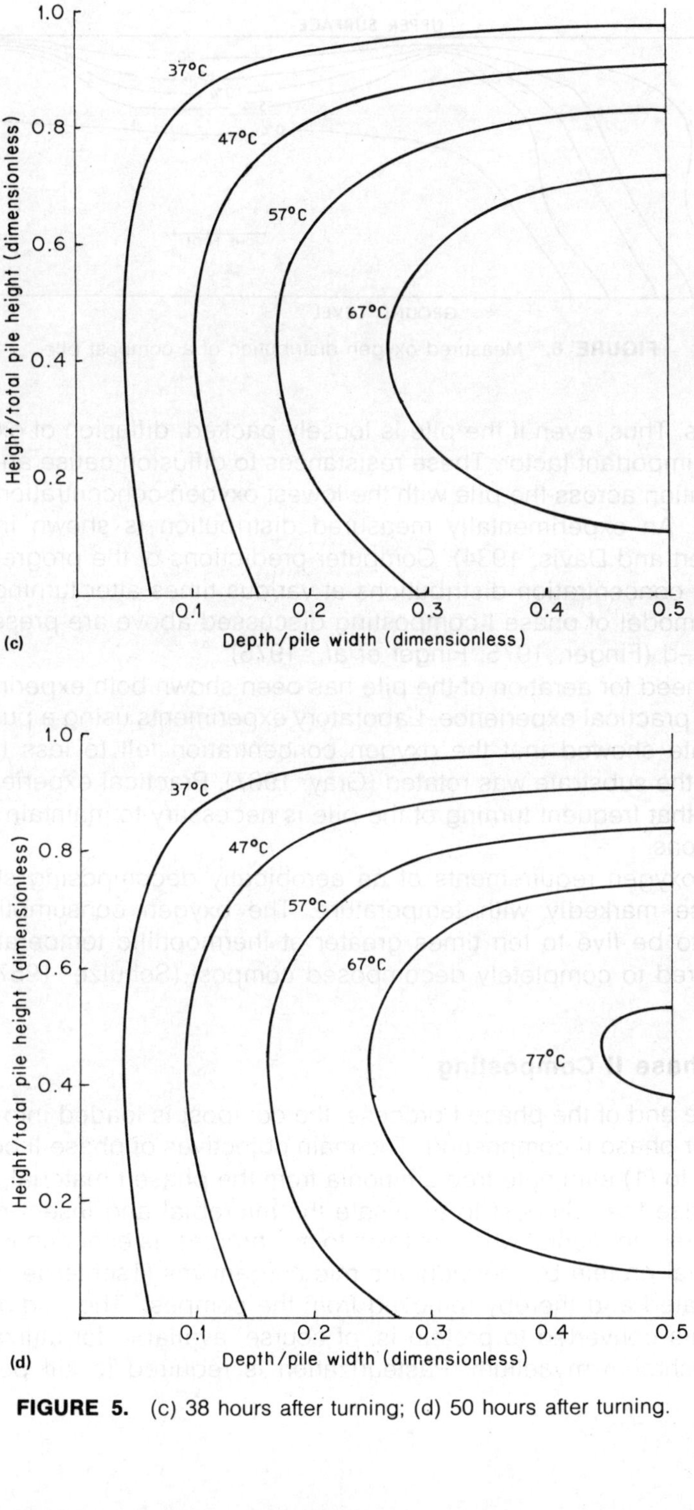

FIGURE 5. (c) 38 hours after turning; (d) 50 hours after turning.

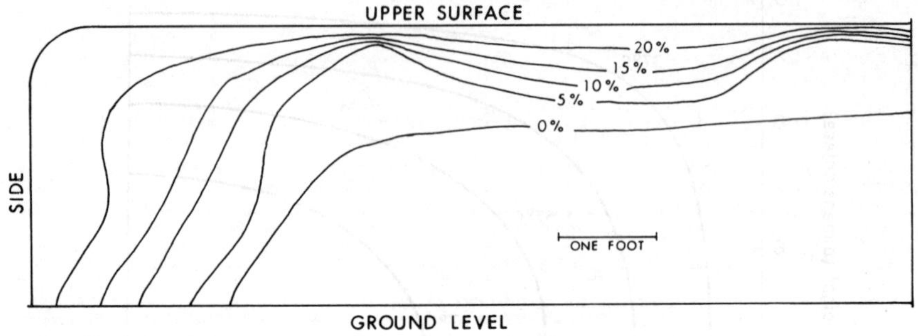

FIGURE 6. Measured oxygen distribution of a compost pile.

it reacts. Thus, even if the pile is loosely packed, diffusion of oxygen is still an important factor. These resistances to diffusion cause an oxygen distribution across the pile with the lowest oxygen concentrations in the interior. An experimentally measured distribution is shown in Fig. 6 (Lambert and Davis, 1934). Computer predictions of the progression of oxygen concentration distributions at various times after turning based on the model of phase I composting discussed above are presented in Fig. 7a–d (Finger, 1975; Finger *et al*., 1976).

The need for aeration of the pile has been shown both experimentally and by practical experience. Laboratory experiments using a pulverized substrate showed that the oxygen concentration fell to less than 1% unless the substrate was rotated (Gray, 1967). Practical experience has shown that frequent turning of the pile is necessary to maintain aerobic conditions.

The oxygen requirements of an aerobically decomposing substrate increase markedly with temperature. The oxygen consumption was found to be five to ten times greater at thermophilic temperatures as compared to completely decomposed compost (Schulze, 1957).

H. Phase II Composting

At the end of the phase I process, the compost is loaded into ricks or trays for phase II composting. The main objectives of phase II composting are to (1) eliminate free ammonia from the phase I material, and (2) pasteurize the compost to eliminate the microbial and insect pests.

The free ammonia, which is toxic to mushrooms, is either converted to microbial protein by thermophilic microorganisms (Tschierpe, 1973) or evaporated and thereby removed from the compost. The portion of the ammonia converted to protein is, of course, available for utilization by the mushroom mycelium. Pasteurization is required to kill pests and

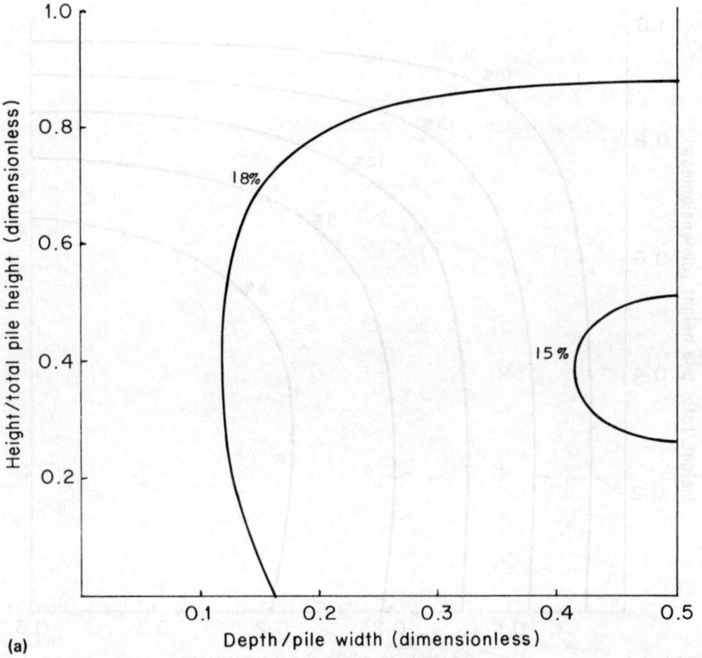

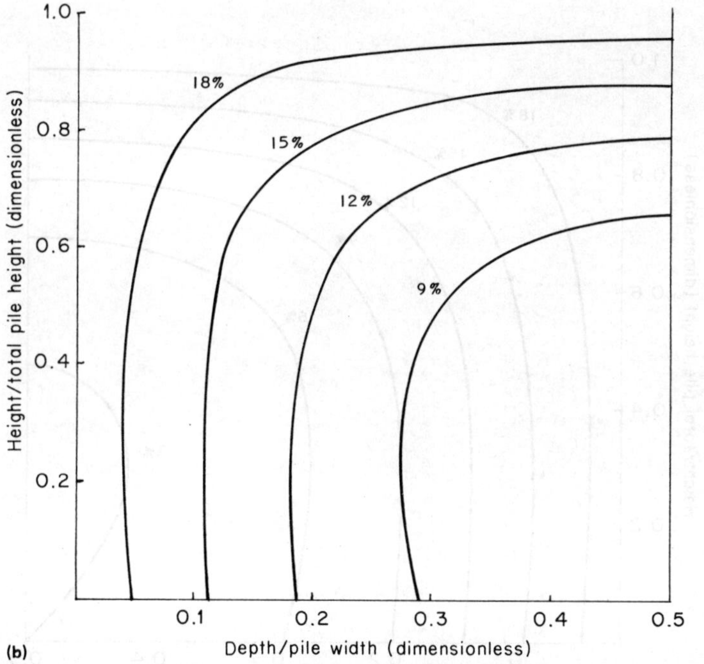

FIGURE 7. Predicted oxygen distribution (a) 12 hours after turning; (b) 24 hours after turning.

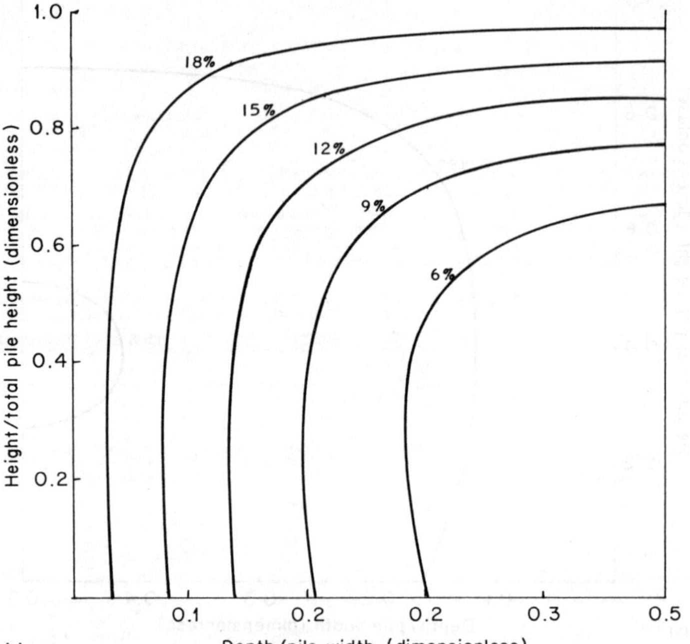

(c)

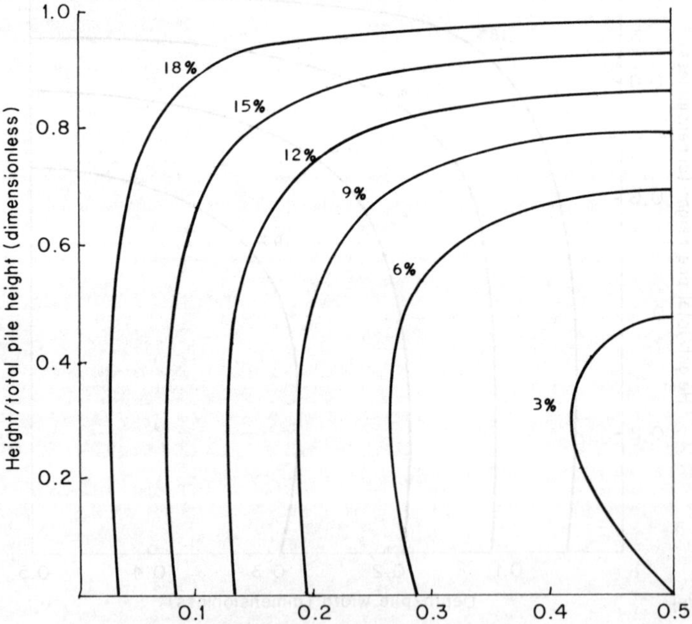

(d)

FIGURE 7. (c) 38 hours after turning; (d) 50 hours after turning.

weeds which could compete with mushroom growth or injure the mushroom.

The phase II process is much more controlled than the first phase of composting. Phase II is generally performed in an enclosed chamber or room. Environmental conditions, such as temperature, humidity, and gas content of the air, are controlled. Room temperatures up to 71°–77°C have been reported (Franz, 1972). The humidity is high (close to 100% relative humidity) and is controlled by blowing air through the room. This also controls the oxygen and carbon dioxide levels and removes any vaporized ammonia. Ricks or trays are used for the phase II process so that relatively uniform compost conditions can be achieved. The temperatures reached by the compost during phase II are sufficient to promote the growth of thermophilic microorganisms. It is important, therefore, that sufficient carbohydrate remain after phase I composting to promote strong thermophilic growth during phase II.

The compost is loaded into deep wooden trays (10 in × 10 ft × 4.5 ft). The first day live steam is used to bring the room temperature up to 60°C. Thermophilic growth is then usually sufficient to maintain the tray temperature for the next 6 days. The room temperature is then lowered and after 48 hours the temperature in the bed (tray of compost) is 21°–29°C. Close control of the phase II process is very important in maximizing mushroom yield. Once the ammonia has been eliminated and pasteurization has taken place, the temperature should be dropped to the spawning temperature. Prolonged phase II composting will adversely affect nutrients in the compost, resulting in lower mushroom yields. The nitrogen content of the bed prior to spawning is approximately 2–2.5% of dry weight.

III. SPAWN PRODUCTION

Once the phase II composting is complete, the compost bed is ready for inoculation. The inoculum is prepared from preserved mushroom spores selected from a variety according to the following criteria: (1) Type of mushroom growth, (2) mushroom yield, (3) growth rate, (4) mushroom size, and (5) mushroom sensitivity to environmental conditions.

Mushroom spores are collected on a card placed under the mushroom cap. Once collected, the spore viability can be preserved indefinitely under refrigeration. [*Agaricus bisporus* spores have been reported by J. Sinden (Personal communication, 1977) to remain viable after storage at

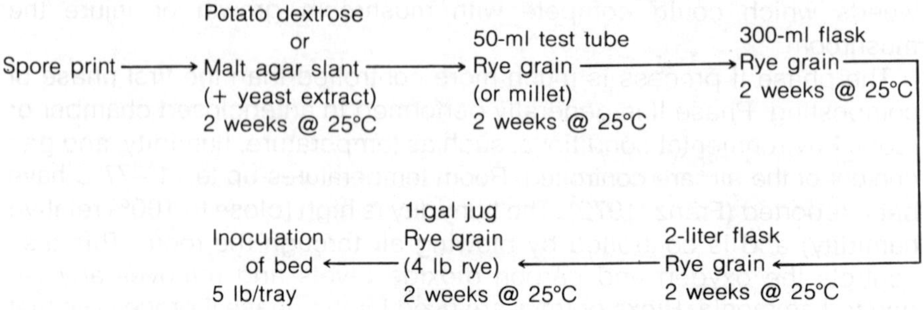

FIGURE 8. Spawn propagation from mushroom spores.

room temperature since 1937.] The mushroom spores are then propagated as illustrated in Fig. 8.

The rye grain medium is quite simple and composed of the following: Rye, 49%; water, 50%; and calcium carbonate, 1% (precipitated chalk). The calcium carbonate is necessary to remove organic acids such as oxalic acid in order to prevent autolysis. The rye acts as a support for mushroom growth as much as a nutrient source since very little rye is actually consumed by the mushroom mycelium.

The spawn is used to inoculate the phase II compost bed once the bed temperature has dropped to approximately 27°C. The spawn is added at the ratio of 5 lb spawn to each tray (1500 lb total weight per tray, 360 lb dry weight). The trays are then placed in an incubation room with a controlled air temperature of 17°C and a bed temperature of 21°–29°C.

After inoculation of the bed, the mycelial growth goes through a lag phase of several days. By the end of the first week, however, there is sufficient growth to produce large amounts of carbon dioxide and heat. In a typical growth room 36,000 ft^3/m of saturated 13°C air are required to maintain the room air temperature at 17°C (100% humidity) for 450 trays. This level of air flow is required for the last 11 days of the 18-day spawn growth phase.

IV. MUSHROOM FORMATION

Once the spawn is fully developed in the tray (usually after 18 days), the tray is ready to support the healthy production of mushrooms. The tray is removed from the spawning room and cased. This involves the addition of a 2-inch (~6 lb/ft^2) layer of top soil to the top of the bed. The casing is moisturized to about 50% saturation to allow the even spreading over the surface of the tray. (At the Butler County Mushroom Farm the casing is made from the spent mushroom trays which are aged outdoors for 2 years.)

After casing, the trays are moved into production rooms where the air temperature is maintained at 14°–17°C. The first step in the formation of a mushroom cap is mycelial cell differentiation to the growth of fruiting body cells. The fruiting body starts as a pinhead-sized mass of mushroom tissue at or near the surface of the bed which is the start of the mushroom.

Although the various factors which affect the development and growth of fruiting bodies are only partially understood, the effect of carbon dioxide has been found to be quite important as shown in Table II. Usually the carbon dioxide level in the production room is raised to 0.5% in order to suppress fruiting for the first week until the mycelium completely penetrates the casing.

Other factors which influence the fruiting include the air temperature and the nature of the casing. At a temperature above 22°C, *Agaricus bisporus* stops fruiting. Some varieties of *Agaricus bitorques* will, however, form fruiting bodies at higher temperatures. The function of the casing in the formation of fruiting bodies has been determined to be a combination of effects (Hayes, 1973), including (1) starvation of mycelium; (2) removal of mycelial metabolites; (3) diffusion barrier for mycelial metabolites; (4) carbon dioxide gradient from bed to casing; and (5) microbial activity which produces substances required for fruiting, such as reduction of Fe^{3+} to Fe^{2+} by *Pseudomonas putida,* and/or removal of mycelial by-products. Although sterile casing does not promote fruiting, the use of activated carbon has been found to work as well or better than topsoil.

After 3 weeks in the production room, the mushrooms are ready to be picked from the trays as shown in Fig. 9. Mushrooms are all hand picked and trimmed. It is necessary to remove the butt of the stem (~20% of the total mushroom) due to the casing particles which are intertwined with the mushroom tissue. Approximately 180–190 lb of salable mushrooms are harvested per tray over a 6-week period. During this time the surface of the tray is moistened with a fine water spray several times a week to

TABLE II. Effect of Carbon Dioxide on Fruiting Body Formation

Percentage CO_2 content of air	Effect
0.5	Fruiting inhibited
0.2	Fruiting disturbed
0.1–0.15	Optimum for good fruiting
0.1	Fruiting is slower

FIGURE 9. Mushroom picking in production room. (Courtesy of Butler County Mushroom Farm, Butler County, Pennsylvania.)

maintain the moisture content of the casing. Approximately half of the water content of the mushroom comes out of the casing and the remainder from the original bed.

At the end of the 9-week period in the production room, the trays are removed and steamed to kill mushroom viruses, any insects, nematodes, and weed molds which may have entered the bed. Although the spent bed has a nitrogen content of 2.5% (dry weight) the residual nutrient content is not generally exploited. The primary uses are (1) top dressing for nurseries, (2) topsoil, and (3) casing soil for mushroom production. Approximately one-half of the spent bed material is aged outdoors for 2 years before recycle as casing soil. The remainder is sold immediately as top dressing or aged and sold as topsoil.

V. PROCESSING PROBLEMS

A wide variety of difficulties can affect the production of mushrooms. The process must be carried out in a consistent manner to assure reasonable yields and high quality. The most difficult variable to control in this regard is the compost. The raw materials vary in their nutrient content, water-holding ability, and microbial activity. The phase I composting step is undertaken year-round outside and is affected to some extent by the weather. The raw materials are also stored unprotected outside. The specific problems with phase I composting tend to be (1) too much water, (2) too large an anaerobic region, (3) too loose a pile, (4) too large a pile, (5) too low or too high a nitrogen content, and (6) overcomposting. The first four problems affect the uniformity of the compost. The nitrogen content of the starting materials is usually adjusted to 1.5% (by dry weight). This will rise to 2.0–2.4% nitrogen by the end of the composting step due to the weight loss from carbon dioxide evolution and drainage of the pile. If the starting materials are overdosed with nitrogen, the free ammonia content will be too high (>0.05%) or the level of free amines will be too high. If the phase I composting is carried out too long, the remaining nutrient level will be too low for sufficient thermophilic microbial growth during phase II composting.

After spawning, a number of contaminants can infect the mycelial bed and the mushrooms. The most common problems are caused by (1) viruses, (2) bacterial diseases, (3) fungal diseases, (4) weed molds, and (5) insects (small flies, mites, etc.) and (6) nematodes. There are three different viruses which have been found to infect the mushroom and cause the following effects: (1) complete inhibition of mycelial growth; (2) formation of long, spindly mushrooms with small caps; (3) reduction of shelf-life; (4) reduction in mushroom yield; and (5) death of mushroom cap followed by softness and rotting. The viral particles tend to concentrate in the mushroom cap and enter the mushroom spores. The spores released by the infected cap will then act as the vector for transmission of the viruses.

The bacterial diseases are most commonly caused by *Pseudomonas* species. Two of the more common diseases are known as brown blotch and mummy disease. Brown blotch is a mahogany colored surface infection caused by *Pseudomonas tolaasi*. The mummy disease (also caused by a *Pseudomonas* species) can stop all cap formation and result in dried-out mushrooms and curved stems. The mummy disease is not transmitted through spores but has been shown to be carried by water drops from ceilings of spawning and pasteurizing rooms.

VI. SUMMARY

Mushrooms have become an important, high-protein food in the United States; and annual production exceeded 349 million pounds in 1977–1978. The annual growth in sales has averaged 8.5% over the last 5 years based upon the pounds produced domestically and 9.2% when imports are included. The price has also risen significantly with a doubling of the fresh sales price since 1968. This sales growth has resulted in an increased competition for the raw materials. The annual consumption of horse manure now exceeds 350,000 tons and is rapidly approaching the supply limit. This is reflected in a doubling of the horse manure price over the last 2 years. The mushroom industry is countering the raw material problems by the use of a more scientific approach to the entire process in order to increase mushroom yields. Over the next decade, the composting step will undergo significant change. Out of necessity, less horse manure will be used and new raw materials will be substituted.

ACKNOWLEDGMENT

The authors express their appreciation to John Yoder, president of the Butler County Mushroom Farm, and Dr. James Sinden for their significant contributions to the background information of this process.

REFERENCES

Aiba, S., Humphrey, A. E., and Millis, N. F. (1965). "Biochemical Engineering," 1st ed. Academic Press, New York.

Anonymous, W. (1965). "Preliminary Report on a Study of the Composting of Garbage and Other Solid Wastes," Civ. Sanit. Eng. Rep. Michigan State University, Ann Arbor.

Braun, R. (1962). *Compost Sci.* **3,** 36.

Bulboaca, M. (1970). *Lucr. Stiint., Inst. Agron.* **13,** 53.

California (1953). "Composting for Disposal of Organic Refuse," Sanit. Eng. Res. Proj., Tech. Bull. No. 9. University of California, Berkeley.

Fergus, C. L. (1969). *Mycologia* **61,** 120–129.

Finger, S. M. (1975). Ph.D. thesis, University of Maryland, College Park.

Finger, S. M., Regan, T. M., and Hatch, R. T. (1976). *Biotechnol. Bioeng.* **18,** 1193.

Franz, M. (1972). *Compost Sci.* **13,** 6.

Fuller, W. H., and Bosma, S. (1965). *Compost Sci.* **6,** 26.

Gotaas, H. B. (1956). "Composting." World Health Organ., Geneva.

Gray, K. R. (1967). *Compost Sci.* **7,** 29.

Hayes, W. A. (1973). *Mushroom News.*

Hayes, W. A. (1974). *Process Biochem.* **9,** 21–28.

Keller, P. (1961). *Compost Sci.* **2,** 20.

Kochitzky, O. W., Seaman, W. K., and Wiley, J. S. (1969). *Compost Sci.* **9,** 5.

Lambert, E. B. (1962). *Mushroom Dig.* **47.**

Lambert, E. B., and Davis, A. C. (1934). *J. Agric. Res.* **48,** 587.

Litchfield, J. H. (1967). *In* "Microbial Technology" (H. J. Peppler, ed.), p. 107. Van Nostrand-Reinhold, Princeton, New Jersey.

Mercer, W. A., Chapman, J. E., Rose, W. W., Katsuyama, A., and Dwinnel, F., Jr. (1962). *Compost Sci.* **3,** 9.

Rancheva, T. (1971). *Gradinar. Lozar. Nauka* **8,** 135–140.

Regan, R. W., and Jeris, J. S. (1970). *Compost Sci.* **11,** 17.
Rose, W. W., Chapman, J. E., Roseid, S., Katsuyama, A., Porter, V., and Mercer, W. A. (1965). *Compost Sci.* **6,** 13.
Schulze, K. L. (1957). "Aerobic Decomposition of Organic Waste Material," Interim Reports No. 1, No. 2, and No. 3. College of Engineering, Michigan State University, East Lansing.
Schulze, K. L. (1961). *Compost Sci.* **2,** 32.
Shilesky, D. M., and Maniotis, J. (1969). *Compost Sci.* **9,** 20.
Siu, R. (1951). "Microbial Decomposition of Cellulose." Van Nostrand-Reinhold, Princeton, New Jersey.
Tietjen, C. (1964). *Compost Sci.* **5,** 8.
Tribe, H. T. (1962). *Compost Sci.* **2,** 38.
Tschierpe, H. J. (1973). *Mushroom J.* No. 1, p. 30.
Waksman, S. A. (1938). "Humus," 2nd ed., Williams & Wilkenson Co., Baltimore, Maryland.
Wiley, J. S., and Peace, G. W. (1955). *Proc. Am. Soc. Civ. Eng.* **81,** Paper No. 846.

Chapter 9

Inocula for Blue-Veined Cheeses and Blue Cheese Flavor

GERARD J. MOSKOWITZ

I. INTRODUCTION

Blue-veined cheeses, such as Roquefort, Gorgonzola, and Stilton, have been known for centuries. Their discoveries have been attributed to fortunate accidents where shepherds, having left sheep or goats' milk curd in a moist and cool cave, returned sometime later to find that a greenish blue mold had developed throughout the cheese. The mold was responsible for imparting the desirable flavor of the cheese. References to blue-veined cheeses were reported as early as A.D. 79 while

MICROBIAL TECHNOLOGY, 2nd ed., VOL.II
 ISBN 0-12-551502-2

records of the Monastery of Conques mention Roquefort as early as the year 1070. Stilton cheese was of more recent vintage, the first records dating back to 1720 in England (Davis, 1965).

In the United States, blue cheese production has been primarily concentrated in Minnesota, Iowa, Illinois, and Wisconsin. In 1976, the quantity of blue variety cheese produced in the United States was 33,875,000 lb. This showed an increase of 19% over the 1975 figure of 28,506,000 lb and was comparable with an overall cheese production increase of 19%. The quantity of blue cheese produced in the United States has remained at about 1% of the total United States cheese production for a number of years (Dairy Products Annual Summary, United States Department of Agriculture, 1976).

The characteristic flavor of blue cheese has been attributed to the metabolic products of blue molds such as *Penicillium roqueforti*. Recent research on strain selection has resulted in the isolation of several *P. roqueforti* strains that have desirable properties for cheese manufacture. These include the ability of the organism to produce an adequate quantity of low molecular weight ketones and alcohols, which are responsible for the characteristic flavor. This, however, must be balanced with an appropriate amount of proteolytic activity, which has been shown to be important for the texture and background flavor of the cheese.

II. PRODUCTION OF BLUE CHEESE

A. Manufacture

In the United States, blue-veined cheeses are produced primarily from cow's milk, although goat milk is occasionally used. Milk is generally pasteurized and homogenized. Pasteurization destroys most of the microorganisms native to the milk, thereby allowing the cheese maker to more closely control the microbial flora of his cheese. Homogenization uniformly distributes the milk fat and makes the fat more susceptible to enzymatic hydrolysis. Milk fat is the substrate that provides the fatty acids from which the characteristic blue cheese flavor develops. This is described in greater detail in Section IV.

Lactic acid bacteria are first added to the milk. At the cheese maker's option, mold spores may also be added. When acid development has reached the desired stage, rennet is added to form the curd. After 60–90 minutes, the coagulum is cut, the whey drained, and the curd transferred to perforated hoops. Some cheese makers split the mold spore inoculum, adding one portion to the milk and the remainder to the drained curd with 1–2% salt.

The hoops are drained at room temperature, with frequent turning, usually overnight. The wheels of cheese are rubbed with salt daily, up to 5 days, or bathed in saturated brine for 24–48 hours. At 1 or 2 days of age the loaves are pierced, top and bottom, with a gang of stainless steel needles. The penetration of air is essential for mold growth and flavor development. Blue-veined cheeses are cured for at least 60–90 days at about 8°C in 95% relative humidity. For additional details see this volume, Chapter 2, and Davis (1976).

B. Standards of Identity

The 1977 Code of Federal Regulations, Section 133.106, defines blue cheese as a cheese produced from cows' milk by methods similar to those described above and that contains a bluish green mold throughout (Fig. 1). Moisture should not exceed 46% and the cheese solids must contain not less than 50% milk fat. The mold prescribed for use is *Penicillium roqueforti*. In addition, a harmless preparation of enzymes may be added to help develop the flavor during curing (less than 0.1% of the weight of the milk used). The standards also allow for the use of $CaCl_2$, at not more than 0.02%, to aid curd formation and for the use of benzoyl peroxide to bleach the milk.

III. PRODUCTION OF *Penicillium roqueforti* SPORES FOR BLUE-VEINED CHEESES

A. Strain Selection

In 1972, Dairyland Food Laboratories purchased the Midwest Blue Mold Company of Stillwater, Minnesota. Midwest produced a selected spore inoculum for blue cheese. Through years of experimentation, two strains were chosen because they possessed superior cheese-making properties. The Minnesota strain was first isolated at the University of Minnesota and the Iowa strain was first isolated at Iowa State University,

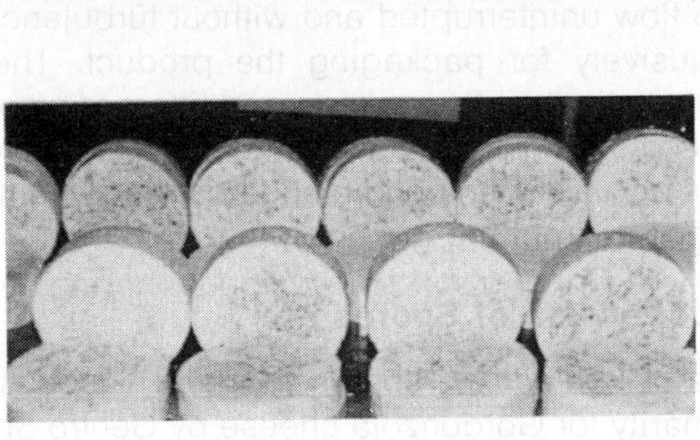

FIGURE 1. Blue cheese.

Ames. Both were sold by Midwest Blue Mold and constitute part of the current product line.

B. Spore Inoculum Production

1. *Procedures*

The techniques used at Dairyland Food Laboratories to produce blue mold start with the propagation of pure cultures on agar slants. Spores are washed from the slants and used to inoculate the substrate. These substrates include sterilized bread (Hussong and Hammer, 1935) or sterilized bran. Growth occurs under controlled conditions of time, temperature, and humidity. After the mold has covered the substrate and the spore count is optimal (usually 16–22 days), the material is harvested, dried, and milled to produce a fine powder. The product is then blended and standardized to produce a uniform product and packaged at a concentration of not less than 5.0×10^8 spores/gm.

Bread was the original substrate used by Hussong and Hammer (1935) for mold growth. Work by Nelson and Zanzig at Dairyland Food Laboratories indicated that bran would also adequately serve as a substrate for mold growth and that the mold produced on bran had similar properties to bread-grown mold (J. H. Nelson, unpublished observations).

2. *Production Facility Design*

Conditions for blue mold production must be carefully controlled to produce a uniform and consistent product. Special precautions must also be taken to prevent contamination of both the production facility and the external environment. A specially designed facility has been built to accomplish this (Zanzig, 1975). The 2750-ft² area contains both laboratories and a sophisticated production area. The inoculation room contains a vertical laminar air-flow system that prevents outside air penetration by positive pressure. Specially designed perforated tables allow the air to flow uninterrupted and without turbulence. One room is set aside exclusively for packaging the product. The laboratory is equipped with a vertical air flow system and filters to prevent contamination. All air is incinerated prior to exhausting to the outside. In addition, this area has its own air conditioning and heating systems which isolate it almost completely from the general plant.

C. Liquid Suspension of Spores

Concentrates of *Penicillium* spores in liquid suspension have been developed primarily for Gorgonzola cheese by Centro Sperimentale del

Latte, Milan, Italy. Four strains of *Penicillium,* differing in ripening characteristics, are available in bottles containing 25 ml, sufficient for the inoculation of 1000 gal of milk. Shelf-life of these preparations is reported to be 2 months when stored at 2°–5°C.

IV. BLUE CHEESE FLAVOR

A. Chemistry

The flavor of blue cheese is a chemically complex system. A great deal of effort has gone into the elucidation of the flavor components of blue cheese.

Hammer and Bryant (1937) identified the methyl ketone 2-heptanone in blue cheese. A number of workers, including Patton (1950a,b), Morgan and Anderson (1956), Coffman *et al.* (1960), and Scarpellino and Kosikowski (1961), have described the major role methyl ketones play in the flavor of blue cheese. The characteristic flavor of blue cheese is primarily due to the presence of various methyl ketones and secondary alcohols containing from 5–11 carbon atoms (Nelson and Richardson, 1967; Arnold *et al.*, 1975). The most abundant methyl ketones identified are 2-heptanone, followed by 2-nonanone, 2-pentanone, and 2-undecanone. Higher molecular weight ketones may also be present (Schwartz and Parks, 1963). Morris *et al.* (1963) and Anderson and Day (1965) identified the free fatty acids, acetic through stearic, along with several unsaturated C_{18} fatty acids in blue cheese.

Jackson and Hussong (1958) have identified the secondary alcohols 2-pentanol, 2-heptanol, and 2-nonanol in blue cheese. These are probably produced by reduction of the above-mentioned methyl ketones. The products of proteolysis are important in the background flavor. However, the role of these peptides, amino acids, and other metabolites in flavor development has not been well characterized.

Table I (Nelson and Richardson, 1967) lists a number of compounds identified in blue cheese. It is quite likely that all of these are responsible for the characteristic flavor although, as indicated previously, the methyl ketones are probably the major contributors.

B. Mold Metabolism

Penicillium roqueforti produces both a protease and a lipase. The protease is, in large part, responsible for the smooth, full flavor and texture. The amino acids produced by proteolysis act as buffers for the cheese at pH 6.5 and are undoubtedly used by the mold for growth.

TABLE I. Summary of Compounds Positively or Tentatively Identified in Blue Cheese[a,b]

Acetic acid*	Ethyl formate
Butanoic acid*	Ethyl acetate
Hexanoic acid*	Ethyl butanoate*
Octanoic acid*	Ethyl hexanoate
Acetone*	Ethyl octanoate
2-Pentanone*	Ethyl decanoate
2-Hexanone	Ethyl 2-methylnonanoate
2-Heptanone*	Isopropyl hexanoate
2-Octanone	Isopropyl decanoate
2-Nonanone*	Pentyl hexanoate
2-Decanone	3-Methylbutyl butanoate
2-Undecanone*	Acetaldehyde
2-Tridecanone	2-Methyl propanal
2-Propanol	3-Methylbutanal
2-Pentanol*	Furfural
2-Heptanol*	Hydrogen sulfide
2-Octanol	Methyl mercaptan
2-Nonanol*	Diacetyl
Methanol	Diethyl ether
Ethanol	Benzene
2-Methylbutanol	Toluene
3-Methylbutanol	Dimethylcyclohexane
1-Pentanol	Cresyl methyl ether
2-Phenylethanol*	δ-Octalactone
Methyl acetate	δ-Decalactone
Methyl butanoate	
Methyl hexanoate*	
Methyl octanoate*	
Methyl decanoate	
Methyl dodecanoate	

[a] From Nelson and Richardson (1967).

[b] Compounds marked with an asterisk indicate blends of compounds which contribute to the flavor of blue cheese (Anderson, 1966).

Excessive proteolysis results in a soft, bitter cheese while insufficient proteolysis produces tough, crumbly cheese (Kinsella and Hwang, 1976).

The lipase hydrolyzes the triglycerides of milk fat to produce the free fatty acid precursors of the low molecular weight methyl ketones and secondary alcohols. Both the milk lipase, if it has not been inactivated by pasteurization, and the *P. roqueforti* lipase are involved. *Penicillium roqueforti* lipase has been isolated and studied.

The properties of the enzyme vary depending on the substrate used, the conditions of assay, and the source of the enzyme (Morris and Jezeski, 1953). Thus, Shipe (1951) found that the enzyme had a temperature optimum of 30°–35°C and a pH optimum of 5.0. Morris and Jezeski (1953) found that the enzyme had a pH optimum on butterfat of 6.5–6.8 and a temperature optimum of 30°–32°C. Further variations were found depending on whether the enzyme was isolated from the medium (soluble) or was released by disrupting the mycelia. This observation, that soluble lipase seemed to differ from mycelial lipase, indicated the possible existence of more than one lipase. Imamura and Kataoka (1967) have confirmed this through the isolation of both an intracellular and an extracellular lipase from *P. roqueforti*.

The soluble enzyme has a greater specificity for low molecular weight fatty acids than high molecular weight fatty acids (Shipe, 1951; Morris and Jezeski, 1953; Arnold *et al.*, 1975). These fatty acids serve both as flavor components and as precursors for methyl ketones, which in turn can be reduced to secondary alcohols as shown in Fig. 2 (Kinsella and Hwang, 1976).

The fatty acid is first activated with coenzyme A to form an acyl-CoA which is then oxidized via the fatty acid β-oxidative pathway to β-ketoacyl-CoA. The β-ketoacyl-CoA is deacylated and decarboxylated to form a methyl ketone with one less carbon atom than the fatty acid precursor.

C. Production of Blue Cheese Flavor by Submerged Fermentation

1. Theoretical Aspects

Because of the differences in the physical form of the substrate and the need to develop the flavor in a matter of hours, it was not possible to

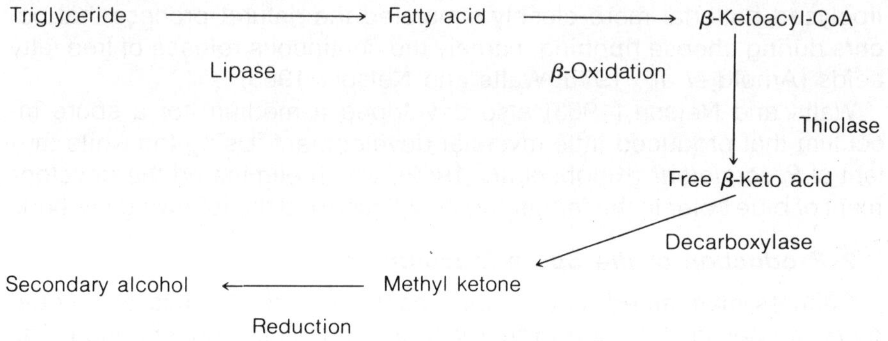

FIGURE 2. Production of methyl ketones and secondary alcohols from free fatty acids by *P. roqueforti*.

directly translate the conditions found in blue cheese to a submerged fermentation.

The flavor in blue cheese is due, in part, to the metabolism of the mycelia or vegetative cells. The use of mycelia in submerged fermentations would result in a heavy fungal growth that would be difficult to remove from the product. Heavy mycelial growth is inhibited and spore development enhanced in both the spore inoculum and submerged fermentation by the addition of salt (Watts and Nelson, 1963). Knight (1963) found that in the presence of free fatty acids, *P. roqueforti* spores were more capable of converting fatty acids to methyl ketones than were vegetative cells. However, spores themselves did not produce lipase enzyme and therefore could not be efficiently used with nonlipolyzed milk fat.

Dairyland Food Laboratories, under license from the Wisconsin Alumni Research Foundation, has developed a commercial process for blue cheese flavor production based on Knight's discovery (Watts and Nelson, 1963). The process closely parallels conditions found in the manufacture of blue cheese, namely the use of similar dairy substrates, salt levels, aeration conditions, *Penicillium* mold, and lactic acid.

Free fatty acids are provided either by activation of the milk lipase, or preferably by pretreatment of the milk with pregastric esterase (Italase), to produce low molecular weight free fatty acids (Watts and Nelson, 1963). The use of the pregastric esterase more closely parallels the mold lipase activity pattern as both are more specific for low molecular weight fatty acids than for higher molecular weight fatty acids. Milk lipase is not specific for low molecular weight fatty acids (Nelson, 1972). Watts and Nelson (1963) further refined this procedure by using pregastric esterase-treated butterfat (LBO-50, Farnham, 1950, 1957) in programmed additions to provide concentrated substrate. High levels of free fatty acids inhibit spore metabolism. This programmed addition of lipolyzed butterfat more closely simulated the natural process that occurs during cheese ripening, namely the continuous release of free fatty acids (Arnold *et al.*, 1975; Watts and Nelson, 1963).

Watts and Nelson (1963) also developed a medium for a spore inoculum that produced little mycelial development. Using the white mutant of *P. roqueforti* (Knight *et al.*, 1950), which eliminated the development of blue color in the fermentor, they described the following method.

2. Production of the Spore Inoculum

Cultures maintained in V-8* juice containing agar slants are propagated further in V-8 juice to increase the inoculum concentration. The

*V-8 is the Trademark of the Campbell Soup Company, Camden, New Jersey.

suspension is transferred to a salt-containing milk substrate, its pH is adjusted to 4.5–5.5 with lactic acid, and it is incubated on a rotating shaker for up to 4 days to further develop the culture and condition it to a dairy substrate. After 4 days, additional salt is added and the incubation is continued for another 24 hours.

3. Fermentation of the Blue Cheese Flavor

The fermentation medium consists of nonfat dry milk or other milk products, such as skim milk, whey, or whole milk, salt, and lipolyzed milk fat (LBO-50). The medium is blended to produce a homogeneous mass and then sterilized. After cooling to 24°–26°C, the fermentor is inoculated with the spore preparation and fermented at 78°F for up to 66 hours. Programmed additions of LBO-50 are made periodically. The product has 7–12 times the ketone content of blue cheese and four times the flavor level (Arnold *et al.*, 1975). The product is subjected to 130°C for 4 seconds to kill the mold and inactivate the enzymes. The final product contains 4% fat, 5% protein, 8% carbohydrates, 5% salt, and 78% water.

Dwivedi and Kinsella (1974a,b) described a procedure in which *P. roqueforti* mycelia, acting on a lipolyzed milk fat substrate, produced methyl ketones in a continuous fermentation. They further found that the inhibition due to high concentrations of free fatty acids could eventually be overcome by the mold. In addition, the concentration of free fatty acids present affected the relative ratio of methyl ketones produced.

Jolly and Kosikowski (1975) recently described a whey-based submerged fermentation supplemented with coconut oil. A microbial lipase plus *P. roqueforti* spores were used.

4. Chemical Composition of Dairyland Food Laboratories' Blue Cheese Flavor

Dairyland Food Laboratories' blue cheese flavor contains 2-pentanone, 2-heptanone, 2-octanone, 2-nonanone, and 2-undecanone, much the same aş that found in blue cheese itself.

5. Fortified Blue Cheese Flavor

A fortified blue cheese flavor is prepared by the addition of compounds normally found in blue cheese. This product can have between 6 and 20 times the flavor level of cheese itself but must be considered an artificial flavor. This addition of flavor compounds to a fermented base provides a superior product as compared to simple chemical compounding due to the presence of the suitable background components.

Blue cheese flavor can be used in salad dressing, dips, appetizers and blue cheese powder. The flavor can be used to increase the potency of blue cheese solids and to reduce cheese solids' content in formulations.

REFERENCES

Anderson, D. F. (1966). Ph.D. thesis, Oregon State University, Corvallis.
Anderson, D. F., and Day, E. A. (1965). *J. Dairy Sci.* **48,** 248–249.
Arnold, R. G., Shahani, K. M., and Dwivedi, B. K. (1975). *J. Dairy Sci.* **58,** 1127–1143.
Coffman, J. R., Smith, D. E., and Andrews, J. S. (1960). *J. Food Res.* **25,** 663–669.
Davis, J. G. (1965). "Cheese," Vol. I, pp. 3–15. Am. Elsevier, New York.
Davis, J. G. (1976). "Cheese," Vol. III, pp. 621–622. Am. Elsevier, New York.
Dwivedi, B. K., and Kinsella, J. E. (1974a). *J. Food Sci.* **39,** 83.
Dwivedi, B. K., and Kinsella, J. E. (1974b). *J. Food Sci.* **39,** 620.
Farnham, M. G. (1950). U.S. Patent 2,531,329.
Farnham, M. G. (1957). U.S. Patent 2,794,743.
Hammer, B. W., and Bryant, H. W. (1937). *Iowa State Coll. J. Sci.* **11,** 281.
Hussong, R. V., and Hammer, B. W. (1935). *J. Dairy Sci.* **18,** 599.
Imamura, T., and Kataoka, K. (1967). *Dairy Sci. Abstr.* **29,** 215.
Jackson, H. W., and Hussong, R. V. (1958). *J. Dairy Sci.* **41,** 920.
Jolly, R., and Kosikowski, F. V. (1975). *J. Food Sci.* **40,** 285–287.
Kinsella, J. E., and Hwang, D. (1976). *Biotechnol. Bioeng.* **18,** 927–938.
Knight, S. G. (1963). U.S. Patent 3,100,153.
Knight, S. G., Mohr, W. H., and Frazier, W. C. (1950). *J. Dairy Sci.* **33,** 929.
Morgan, M. E., and Anderson, E. O. (1956). *J. Dairy Sci.* **39,** 253.
Morris, H. A., and Jezeski, J. J. (1953). *J. Dairy Sci.* **36,** 1285–1298.
Morris, H. A., Jezeski, J. J., Combs, W. B., and Kuramoto, S. (1963). *J. Dairy Sci.* **46,** 1.
Nelson, J. H. (1972). *J. Am. Oil Chem. Soc.* **49,** 559–562.
Nelson, J. H., and Richardson, G. H. (1967). *In* "Microbial Technology" (H. J. Peppler, ed.), pp. 82–106. Van Nostrand-Reinhold, Princeton, New Jersey.
Patton, S. (1950a). *J. Dairy Sci.* **33,** 400.
Patton, S. (1950b). *J. Dairy Sci.* **33,** 680.
Scarpellino, R., and Kosikowski, F. V. (1961). *J. Dairy Sci.* **44,** 10.
Schwartz, D. P., and Parks, O. W. (1963). *J. Dairy Sci.* **46,** 989.
Shipe, W. F. (1951). *Arch. Biochem.* **30,** 165–179.
United States Department of Agriculture (1976). *Dairy Prod. Annu. Summ.* p. 21.
Watts, J. C., and Nelson, J. H. (1963). U.S. Patent 3,072,488.
Zanzig, C. E. (1975). *Food Process* **36**(12), 98–99.

Chapter 10

Microorganisms for Waste Treatment

LARRY L. GASNER

I. INTRODUCTION

The increased sophistication of our society in the past several decades has led to dramatic increases in the volume and complexity of treatment of waste effluent streams and spills. This is only now beginning to be matched by increased sophistication of waste stream and spill biodegradation microbiology. Inoculum cultures for this use are marketed by about 20 companies with a 1978 total market value of about \$2–4 million. One estimate of the total potential domestic market is \$200 million (Anonymous, 1976, 1977a,b).

Although some role of microorganisms was recognized since the

MICROBIAL TECHNOLOGY, 2nd ed., VOL.II
 ISBN 0-12-551502-2

activated sludge process was first developed (Andern and Lockett, 1914), the process is still not well enough understood to be able to define the respective roles of the various components of the microbial population (Jones, 1976). Indeed, a form of "spontaneous generation" is frequently relied upon to develop the appropriate complex population mix. The development of the art of inoculation of sewage systems is a recent phenomenon that is rapidly increasing in activity and dollar volume. This experience is also leading to increased efficacy of some of the microbial cultures sold. In like manner, it was the huge oil spill of the *Torrey Canyon* near the coast of England in 1967 that led to extensive efforts in the United States to develop enriched "seed" cultures for hydrocarbons (Anonymous, 1970), although septic tank additives and drain cleaner mixtures containing microorganisms had been marketed considerably earlier.

A. Scope

This chapter is limited to the technology of isolation, production, and application of those microbial cultures which are commercially available or imminent as inocula for the treatment of wastes. This does not include description of processes which do not involve the marketing of microbial cultures. For example, *Candida utilis* production from sulfite waste liquors, ethanol production from whey wastes or waste cellulose, or single-cell protein production from hydrocarbon, food, or other wastes are not included. Also not included are novel waste-processing schemes or fermentor designs that have been proposed for waste treatment.

B. Products

1. Characteristics

Products are available as freeze-dried or air-dried solids or as stabilized microbial suspension in liquid form. Solids are available in various size shipments and containers with the most common container size being 25 lb. Costs vary between about $7.00 and $22.00 per lb, FOB, which frequently includes technical service. Stabilized liquid formulations typically contain of the order of 5×10^{10} microbes per liter and sell in the above price range per gallon in plastic gallon containers. Products can be stored at room temperature for years and retain useful potency. Dry products are activated by mixing at least 2 gal of water with each dry pound. Both liquid and activated dry products can be sprayed onto the surface or otherwise well mixed into the system. Drip feeders are available for continuous feeding of microbial suspensions.

2. *Types of Cultures*

Formulations including live microorganisms are available with specific activity for one specific component such as phenol, aromatics or glycol, or one specific enzyme activity such as amylase activity, protease activity, lipase activity, or cellulase activity. On special order specific biodegradation activities can be obtained against DDT, polychorinated biphenyls, and other more exotic compounds. The above are generally nonclassified isolates; however, pure identified microbial cultures are also available, for example, *Bacillus subtilis* KJ, which produces high levels of amylolytic and proteolytic enzymes; *Bacillus thuringiensis,* which produces high levels of lipolytic enzymes and low levels of amylolytic enzymes; and *Nocardia corallina,* which degrades phenol and phenolic compounds. Other unidentified products have broader ranges of activity, such as the degradation of cutting oils and crude oil, and the treatment of ironworks wastewater.

Most preparations consist of mixtures of microbes. These mixtures are of three types as follows: (1) natural mixtures, for example rumen bacteria with high cellulase activity, (2) selected mixtures which have been propagated together in a single fermentor, or (3) mixtures which have been propagated individually and combined in the final formulation. As many as 20 microorganism types can be found in a general purpose-type preparation. Also available are products of unspecified contents and unspecified preparation methods.

Aerobic, anaerobic, or facultative organisms are available as are low (4.5) or high (9.5) pH-tolerant preparations.

C. Application Guidelines

Actual applications must be guided by several considerations. Mixing with the waste stream is necessary. A wetting agent in a few cases is incorporated in the product. For application to oil spills a natural dispersant, such as a lignosulfonate, may be supplied as a component of the product or should be applied separately. For oil spills on land it may be necessary to build a dike and flood the area. Topical preparations can be applied directly to surfaces, such as in dumpsters for odor control.

Necessary nutrients are required for any microbial growth. In certain cases supplementary nitrogen and phosphorus may be provided in the formulation or may be added separately.

The idea of providing an inoculation requires adding an initial high dose followed by a continuous maintenance low dose or several maintenance periodic doses. Actual recommended dose size varies widely. For activated sludge systems cost estimates for a 1 million gal/day plant range between $20 and $100 per million gallons. Because periodic fluctuations upset small waste treatment facilities more than larger waste treatment plants, the recommended maintenance dosage level may be severalfold larger for the smaller vessel.

A major problem area is that most waste material is not well defined, nor is it easily classified. The mechanical, nutritional, or chemical environment may not be identical from site to site. Also, interactions between species within the inoculum and those native to the site are influenced by subtle, little understood differences in the environment. Therefore, it cannot be reliably predicted that any given culture mixture will reproducibly perform well under superficially identical conditions. Before any product should be used in significant amounts, therefore, it is necessary to have verification tests performed and to view as speculation any advance claims as to efficacy on an untested waste. Lack of confidence on the part of the purchaser due to prior overzealous claims on the part of some suppliers is the biggest weakness in this entire industry. This lack of confidence is a combination of disappointment with the microbial enzyme products which had promised similar performances and overpromotion by those firms which are not soundly based in the area.

There should be caution in dealing with firms that have limited documented similar case histories and cannot refer prospective customers to several other satisfied clients with recent success in similiar systems. Gimmicks in the promotion of products should also lead to caution. Advertising the storage of stock cultures in a bank vault, emphasizing the remoteness of the earth from which certain cultures were isolated, or enriching cultures the normal way but adding a trace of mutagenic agent so the resulting cultures are called "mutants" are gimmicks to imply unproved uniqueness and undemonstrated value.

II. PRODUCTION

A. Origin of Cultures

Sources of soil and water samples vary according to the type of waste degradation expected. For example, soil samples were taken which were contaminated by a natural oil seep in Cape Simpson, Alaska (Cundell and Traxler, 1974). Fifteen hydrocarbon-degrading bacteria were isolated from natural saline waters of low temperature (0°–25°C). The facultative psychrophilic bacteria represented a wide range of genera and grew on aliphatic, aromatic, and naphthenic materials. In fact, the coastal waters of the world contain a large fraction of microorganisms, up to 50%, that are capable of oxidizing hydrocarbons (Anonymous, 1973). A good medium for marine hydrocarbon microbial isolation and growth has been demonstrated to be enriched seawater medium with a small volume of seawater added (Katos *et al.*, 1971).

Soil organisms have also been found to readily degrade oily wastes under prevailing conditions. A rate of 1 lb/month-ft^3 of soil was reported for soil at a Houston refinery when excess nitrogen and phosphorous were added (Franche and Clark, 1974). It appears that microorganisms are widely distributed in oil field soils and anywhere that is particularly subject to frequent petroleum pollution. Once a population becomes established in an oil-enriched soil, it will remain, either dormant or utilizing other substrates, and it will readily respond to the reapplication of oil.

Because organisms generally degrade hydrocarbons according to type, paraffinic first, naphthenic second, and aromatic third, it is generally believed that mixed cultures of organisms of various selectivities, such as are found in nature, are the best to use for rapid and complete degradation. On the other hand, the possibility of genetically introducing multiple selectivities into a single strain, producing a "supermicrobe," has resulted in some exciting new prospects, which will be discussed in Section IV. Mixed cultures of hydrocarbon- and petroleum-oxidizing microbes are used in petroleum spills on water and on land, treating industrial and refinery wastes, and cleaning ship tanks.

Additional isolates with lipolytic activity and proteolytic activity can be obtained from grease traps of meat packing plants, sewage treatment plants, septic tanks, and food-processing waste treatment, in addition to common sources of samples. Cultures obtained are aerobic, facultative aerobic, or anaerobic. Selection of facultative aerobic microorganisms has advantages in that they can be effective both in anaerobic sludges, grease trap lagoons, lift stations, and tanks as well as maintain activity when the lagoon or grease trap becomes aerobic. Facultative aerobes can be selected to not produce malodorous H_2S when anaerobic. These sources of microbes combined with the hydrocarbon-utilizing microbes are used in food-processing waste streams, grease traps, waste lines, municipal sewer lines, and places where deposits of food fats and oils occur. They are also effective with newly refined petroleum oils.

Microbial isolates with activity against specific chemical or industrial wastes come from soil, water, and sediment samples where repeated or continuous exposure to similar chemistries is suspected. Waste treatment systems where a particular chemical, such as a phenol, a cyclic or aromatic compound, or mixture are being successfully treated is a particularly good source of such isolates. Cultures can be isolated that will degrade phenol, pesticides, and other chemicals, including complex industrial wastes characteristic of an industry. For example, pulp and paper mill effluent treatment can be enhanced by microbial isolates that attack the various forms of lignin (Stern and Gasner, 1973, 1974).

B. Isolation and Enrichment

In general, microbial cultures found in these products are isolates from natural soils and waters where normal enrichment techniques have been used. A normal enrichment method acclimates the organisms to a synthetic or actual waste environment, selecting those to grow which can best accommodate to the chemistry and the nutrients and which reproduce most rapidly. Temperature, ionic composition, pH, and dissolved O_2 are among the variables. Selection is generally for those cultures where the desired degradative enzyme system is constitutive and not subjected to substrate inhibition, repression, or toxicity. Isolation of pure cultures is usually on selective media streak plates. Stock cultures are maintained on slants or as lyophilized powder according to standard techniques.

Specific procedures and tests include acclimatization of cultures and treatability tests of specific wastes. This may use the enrichment cultures mentioned earlier, municipal sewage inocula, or natural soil samples, water samples, or other natural sources of microorganisms. These same techniques can be used to obtain microbial cultures for any specific waste.

The first step in a "treatability" test is the development of a "synthetic sewage" medium. A typical formula consists of a stock salt solution containing (per liter) 18.6 gm $MgCl_2 \cdot 6H_2O$, 36.4 gm $CaCl_2 \cdot 2H_2O$ and 0.2 gm $FeCl_3 \cdot 6H_2O$; and a nutrient stock solution containing (per liter) 20 gm glucose, 20 gm nutrient broth, 30 gm KH_2PO_4, and 10 gm sodium benzoate. One milliliter of the stock salt solution and 20 ml of the nutrient stock solution are added to each liter final volume. The pH is adjusted to within 6.5–7.5 with 1.0 *N* H_2SO_4 or 1.0 *N* NaOH. Distilled water is used throughout.

Toxicity is commonly assessed in a series of parallel shake flasks with the suspected toxic component or the whole raw chemical waste added in geometrically increasing concentrations to the above "synthetic sewage" medium. Inoculation of each flask, including the control flasks, is uniform. Microbial growth is measured daily, often by turbidity. The concentration at which the toxic component begins to inhibit growth is noted and is later used as a warning of an upper concentration limit. A certain amount of experience is helpful in interpreting these experiments and repeat experiments are frequently carried out to verify the preliminary results. In some cases the initial shock to the inoculum may dominate the response or a long acclimatization lag may delay the growth response.

Treatability tests are commonly carried out in a laboratory-sized continuous flow treatment apparatus. Batch test information is much less useful. A 3- to 5-liter chamber with a sloping bottom is used, fitted with an air-stone sparger. About three-fourths of the chamber is aerated. This is separated from the overflow sludge settling zone by an underflow vertical partition.

Synthetic sewage plus a nontoxic level of wastes are added and inoculated with up to 2000 mg/liter suspended solids inoculum. An initial influent flow rate of about 0.3–0.5 v/v vessel is used. This flow rate is increased stepwise by two- to fourfold during the test. The toxic component or raw waste stream content is increased slightly each day. The pH and salt content are allowed to change in the direction of the natural tendency. The variables are gradually changed individually in order to approach a natural optimum. Additives, such as nutrients, dispersants, or flocculants, may be used. Influent and effluent samples are taken daily for tests. Tests run include suspended solids (dry weight), volatile solids (dry weight minus ash), microscopic examination (cell growth observation), BOD (biochemical oxygen demand), COD (chemical oxygen demand), TOC (total organic carbon), and dissolved oxygen. Suspended solids are particularly susceptible to toxicity. Total organic carbon reductions are also generally accompanied by BOD and COD reductions. Therefore, all of the above tests may not be necessary.

Treatability tests such as these take at least 1 month to complete. Because of the variability of many industrial wastes and the cyclic chemical nature of the influent, it may be necessary to operate the laboratory unit for up to 2 years. This type of test is also used by industrial wastewater design engineers to simulate an actual wastewater treatment plant and aid in the design of new facilities.

C. Cultivation and Processing

It should be noted that many suppliers do not produce their own cultures. Rather, the fermentation is done by contract fermentation companies. In a few cases, commercially available bulk-produced pure-culture dried biomass is simply purchased, as is, or is marketed directly by the producer as a sideline business.

For those suppliers who carry out their own fermentations, procedures vary, but the basic slant, loop, inoculum preparation, inoculation sequence is generally unremarkable. Fermentors up to 2000 gal are used. Aseptic conditions are maintained even where mixed cultures are simultaneously propagated, largely as a guard against *Salmonella, Staphylococcus,* and *Streptococcus* contamination. Some care is taken to select fermentation conditions, temperature, and substrates that will produce microbes derepressed, induced, and conditioned to their ultimate environment. Separation methods are by centrifuge or filter. Air-drying is less expensive than freeze-drying; it is used for those cultures whose viability is not severely affected. Spore formers, such as *Bacillus* species, can be air-dried, whereas non-spore formers usually are freeze-dried.

Details of fermentation and preservation and application of a live mixed culture liquid product are provided in a patent (Horsfall and Gilbert, 1976). Seven bacteria with complementary activities were isolated from soil and water samples. These have been identified as *Aerobacter aerogenes, Bacillus subtilis, Nitrobacter winogradskii, Cellulomonas biazotea, Rhodopseudomonas polustris, Pseudomonas stutzeri,* and *Pseudomonas* species. Of these only the *Bacillus* is a spore former. The *Pseudomonas* species contribute to denitrification. Individual inocula are prepared and inoculated into a sterile 400-gal steam-jacketed fermentor. The medium contains 0.1% K_2HPO_4, 0.1% NH_4Cl, 0.05% $MgSO_4 \cdot 7H_2O$, 0.2% $CH_3COONa \cdot 2H_2O$, and 0.1% yeast. This is a minimal medium with acetate as the carbon source designed specifically to derepress the organisms to a maximum number of substrates. Aerobic aseptic growth continues for about 24 hours at 20°–25°C.

Following this, the addition of 0.1% sodium sulfide, $Na_2S \cdot 9H_2O$, causes the cessation of aerobic metabolism and growth. The broth is then transferred to translucent bottles and anaerobic growth under warm light is continued for 24–72 hours. Protosynthetic growth of the *Rhodopseudomonas* during this time is indicated by the development of its characteristic red color. The product contains 5×10^7 viable microorganisms per milliliter, which has been found to be maintained for 3 years when stored in plastic gallon containers at about room temperature. The viability maintained is such that upon dilution in a nutrient medium growth begins immediately without a lag phase. In one variation it is formulated with a trace of polymeric flocculant.

After final preparation of the cultures they may be blended with additives before packaging. Wetting agents, emulsifiers, and dispersants can be added to aid in dispersing the microorganisms throughout a hydrocarbon phase and to aid in dispersing the hydrocarbon into droplets such that oxygen transfer is facilitated. The addition of flocculants is occasionally practiced. Most frequently nitrogen and phosphorus are added to supply nutrients.

The final stage in production is to take a sample for microbiological testing. The U.S. Department of Agriculture requires a certification that any product used in or about a food-processing plant to be free of *Salmonella*. Most products carry this certification.

III. APPLICATIONS

A. Municipal Sewage Treatment

Modern sewer systems deal with large amounts of food wastes, laundry wastes, and petroleum wastes. The greasy materials build up a coating on the inner surface of the sewer lines and become embedded with nonsoluble materials, reducing the design capacity of the line. Particularly at locations where the flow velocity is low this leads to foul odors, overflow, and backup. Mechanical removal is difficult and hazardous.

Microbial cultures can be drip fed into manholes at the extremities of the sewer system. Populations of microbes increase rapidly, consuming the grease. When the grease has been significantly removed, the population decreases. A maintenance level of microorganism feed is maintained. A 1974 "rough and conservative" estimate for the Washington Suburban Sanitary Commission for 3000 miles of pipe with a 150 million gal/day flow was that mechanical cleaning would cost approximately $4 million, whereas the use of bacteria would cost $1.3 million (Hoover, 1974). Extrapolation of these estimates from a service area of 3 million population to the entire nation gives an idea of the total magnitude of this market.

These cultures have been used in a number of cities with good results. Arvada, Colorado which uses a digester for production of methane gas, reported that the production of gas tripled within 60 days after cleanup. The Perry, Florida system which consists of primary treatment, normally removed 28–30% of the BOD (biochemical oxygen demand) and suspended solids. This increased to 50–70% within 1 year of the use of bacterial inoculation. In Erie, Pennsylvania the energy requirements for operation of the pumping stations was reduced by a factor of 3 and all offensive odors were eliminated. Stockton, California also reported a tripling of methane production from 6.5 to 18.8 ft^3/lb volatile solids while the BOD reduction of the primary stage of the treatment plant increased from 18.9 to 26.4%.

Numerous reports of improvements in operation of a normal activated sludge plant can be provided by the various suppliers of inoculation cultures. These claim digestion of grease and oil, elimination of odors, improved BOD removal rates, improved settling, greater treatment capacity in the lagoon, reduction of sludge volume, increased removal of complex substrates, energy savings through reduced aeration, improved predictability of results, reduced corrosion, and improved safety. Some references are Baig and Grenning (1976), Bower (1972), Chambers (1977), and Kirkup and Nelson (1977).

B. Edible Oils and Greases

Food processors, meat processors, restaurants, army bases, schools, homes, and other places where food is processed may have grease buildups in their drain lines, downpipes, grease traps, or septic tanks leading to backups and odors. This problem can also be treated with microbial cultures. In addition refined oils often are not consumed by those microorganisms that attack crude oil and tars. Therefore, special cultures have been developed which degrade these materials.

The major problem to be confronted with respect to small installations is that the water does not stay in the piping system long enough to establish a biodegradation flora. In addition, these pipes may be frequently flushed with hot water containing toxic bleaches and detergents which flush out the cultures which have been established. Therefore, it may be necessary to add an inoculum at the low-flow time of day, i.e., at closing time. It also may require adding nitrogen (5–20 ppm) and phosphorus (1–4 ppm) to the water to supply nutrients. Regular retreatment (three to seven times/week) is usually specified. The microorganisms in the liquid products may revive faster than dried microorganisms, which is an advantage.

C. Heavy and Crude Oils

Oily wastewaters aboard ships cause problems because they must be contained and treated before release. The Queen Mary contained 800,000 gal of oily wastewater (20% was lubricating oil and waste fuel) when it was docked at Long Beach in 1967 (Queen Mary Report, 1971). One hundred fifty pounds of dried microbial cultures were mixed with water (1 gal/lb) and poured into the bilge compartments without agitation. Within 6 weeks the water was pure enough to discharge into the harbor. Since then, CHT (collection, holding, and transfer) tank degreasers have been routinely used by the U.S. Navy (McPhee and Geyer, 1977). For 10,000-gal tanks, 3.5 lb of dried microbial culture are added each day for 5 days followed by a 1.0 lb/day maintenance dose.

Refinery wastes can be handled in lagoons or flow-through systems. The process requires inoculation with microorganisms followed by emulsification of the oil to increase interfacial area. Mixing is usually required. Full decomposition can be achieved in several months (Caswell, 1971). Oil-contaminated soil can be treated by damming and flooding; inoculation and continuous mixing speed the process. Inoculation by spraying directly on the soil surface can be used to seed soils that receive occasional spills with hydrocarbon-utilizing microbial populations. Fertilization with nitrogen and phosphorus is desirable.

D. Chemical and Industrial

The wastes of the chemical industry usually can be handled by dilution and pH adjustment along with the inoculation of appropriate acclimatized cultures in an activated sludge system. Recalcitrant wastes, such as pesticides and halogenated aromatics, often are concentrated enough to support microbial growth and to be biodegradeable, whereas they are not in the dilute form found in nature. In certain cases the inoculum supplier will provide some microbial isolation and enrichment services.

Other industrial wastes consist of mixtures of materials that are characteristic to that particular industry. Cultures are available for many industries although the recommended method of application varies widely. Certain fungi, for example, were found to be efficient and practical converters of organic materials in commercial corn and soy food-processing waste streams (Church and Nash, 1970).

IV. FUTURE PROSPECTS

As the technology grows in sophistication and integrity, the market penetration will also undoubtedly increase. The application of selected microorganisms to just a few large municipal systems will have a large impact on the total market size.

The possibility of genetic engineering of cultures to achieve specific activities and multiple activities is an exciting concept which could lead to improved culture applicability to new systems. In March, 1978 the Washington D.C. Court of Customs and Patent Appeals held that a corporation may patent a new form of life that has been bred in its laboratories (Anonymous, 1978). This case represents a landmark in that it now is possible to patent a microbial species if it can be shown that it has been genetically manipulated to give a novel organism which could not be found in nature. The case involves a General Electric Co. patent of a *Pseudomonas putida*-type bacteria into which was incorporated multiple compatible plasmids from other strains, giving it unique multisubstrate capability (Chakradarty, 1974). The specific plasmids were for octane, xylene, metaxylene, camphor and salicylate degradation. The plasmids were specifically bound to the genome by ultraviolet (uv) radiation (see this volume, Chapter 19).

It can be speculated that many systems now requiring multiple organisms for attack or those having complex interrelationships can be manipulated genetically to give more efficient, patentable microorganisms. The patentability also would lend exclusivity to the culture marketer and may make this a more attractive market in which to compete.

REFERENCES

Andern, E., and Lockett, W. T. (1914). *J. Soc. Chem. Ind.* **33,** 1122–24.

Anonymous (1970). *Chem. & Eng. News.* Sept. 7, pp. 48–49.

Anonymous (1973). *Oceanology (Engl. Transl.)* **13,** 725.

Anonymous (1976). *Bus. Week,* July 5, p. 28D.

Anonymous (1977a). *Chem. Week* **121,** 47.

Anonymous (1977b). *Food Eng.* **49,** 138.

Anonymous (1978). Newsweek, March 13, p. 71.

Baig, N., and Grenning, E. M. (1976). *In* "Biological Control of Water Pollution" (J. Tourbier and R. W. Piersow, eds.), pp. 245–252. Univ. of Pennsylvania Press, Philadelphia.

Bower, G. C. (1972). *Proc. Ann. Meeting Chesapeake Water Poll. Control Assoc.*

Caswell, C. A. (1971). *Purdue Ind. Wastewater Conf., 6th, 1971* pp. 1020–1022.

Chakradarty, A. (1974). U.S. Patent 3,813,316.

Chambers, J. (1977). In *5th Annu. Ind. Pollut. Conf., Water Wastewater Equip. Manuf. Assoc.*

Church, B. D., and Nash, H. A. (1970). "Use of Fungi Imperfecti in Waste Control," Final Rep., Grant No. 12060 EHT. Federal Water Quality Administration, U.S. Dept. of the Interior; US Printing Office, Washington, D.C.

Cundell, A. M., and Traxler, R. W. (1974). *Dev. Ind. Microbiol.* **15,** 250–255.

Franche, H., and Clark, F. (1974). *Nucl. Sci. Abstr.* **30/31,** 6590.

Hoover, J. R. (1974). Montgomery County Planning Board, News release dated July 12, 1974; Silver Spring, Maryland.

Horsfall, F., and Gilbert, B. (1976). U.S. Patent 3,963,576.

Jones, G. L. (1976). *Process Biochem.* **11,** 3–5 and 24.

Katos, H., Oppenheima, C., and Miget, R. (1971). *Biol. Conserv.* **14,** 12.

Kirkup, R. A., and Nelson, L. R. (1977). *Public Works* **10,** 74–75.

McPhee, A. D., and Geyer, A. T. (1977). "Aeration, Bottom Turbulence and Bacteriological Studies of Naval Ship Sewage Collection, Holding and Transfer Tanks," Rep. 77-0043. Commander, Nav. Ship Eng. Cent., Washington, D.C.

"Queen Mary Report" (1971). Bower Industries, Orange, California.

Stern, A. M., and Gasner, L. L. (1973). U.S. Patent 3,737,374.

Stern, A. M., and Gasner, L. L. (1974). *Biotechnol. Bioeng.* **16,** 789–805.

Chapter 11

Elementary Principles of Microbial Reaction Engineering

GEORGE T. TSAO

I. INTRODUCTION

Efficiency and reproducibility are major concerns of reaction engineering. Applications of chemical engineering principle and practice in microbiological processes constitute the main activities of fermentation engineers. In the development of an industrial process, good engineering could mean the following tasks undertaken in a logical manner:

1. Identification of main products and substrates and, if possible, also main intermediates. In the case of a fermentation process, such information is usually provided by microbiologists and biochemists working with laboratory microbial cultures.

MICROBIAL TECHNOLOGY, 2nd ed., VOL.II

ISBN 0-12-551502-2

2. Stoichiometry of the process. For a fermentation, an exact material balance of the process is not always possible. Nevertheless, information should be gathered in as much detail as possible regarding the ratios among products, substrates (e.g., carbon source, nitrogen source, oxygen supply), and known intermediates and regarding the variations of these ratios responding to environmental changes. In some fermentation processes such as those for producing single cell protein (SCP) from hydrocarbon, energy balance is also of great importance.

3. Kinetics and process rate. Often, problems of (1) and (2) cannot be fully answered without the consideration of a time scale. In batch processes, the accumulated changes in products, intermediates, and substrates are of great concern. However, the time span and the manner by which such changes take place, and thus the kinetics and rate of the involved processes, are also necessary information. In the case of continuous fermentation which is still gradually gaining in its popularity, the design and analysis of the reactors are usually based upon the rate of change of these quantities and dilution rate.

4. Reactor design. Information on (1), (2), and (3) is a prerequisite for the ultimate objective of microbial reaction engineering which deals with the design and analysis of fermentors. Even though stirred tanks still represent the most popular geometry of fermentors, an increasing number of other physically shaped vessels, such as tower fermentors, do appear in the industry. Even with the same stirred tank, considerations regarding a proper choice from different modes of operation, including batchwise, fed-batchwise, continuous, and others, are also a part of the reaction engineering activity.

II. IDEALIZED REACTOR DESIGN I: PERFECTLY MIXED BATCH FERMENTOR

An industrial fermentation carries out microbiological changes in a finite physical container. The design of a fermentor to carry out a fermentation in an efficient and predetermined manner is the most important objective and function of fermentation engineering. In chemical engineering curricula the teaching of reactor design is often started by analyzing three idealized situations: perfectly mixed batch reactor, perfectly mixed continuous flow reactor, and plug flow reactor. By examining how reactors are analyzed, one can identify the types of information that are needed in order to have good engineering of the fermentation system.

A material balance of a selected chemical species (e.g., a product, a substrate, or an intermediate) is usually the starting point of reactor analysis.

$$\text{Accumulation} = (\text{input} + \text{production}) - (\text{output} + \text{consumption}) \quad (1)$$

In the case of a perfectly mixed batch fermentor, where input = 0 and output = 0, Eq. (1) can be rewritten as

$$V\,\Delta C = VR_p\,\Delta t - VR_c\,\Delta t \quad (2)$$

where V = volume of the fermentation broth; R_p and R_c = rates of production and consumption of the chemical species under consideration, respectively; ΔC = the change of concentration; and Δt = time span over which the change takes place. If the chemical species under consideration is the carbon source, say, glucose, and R_p = 0, Eq. (2) becomes, with G to be glucose concentration and R_G to be the rate of consumption of glucose,

$$\Delta G = -R_G\,\Delta t \quad (3)$$

or in a differentially small time span, dt,

$$dG = -R_G\,dt \quad (4)$$

In theory, if Eq. (4) can be integrated from $t = 0$ to any t, we will have the knowledge of how glucose concentration varies.

$$\int_0^t dt = t = \int_{G_0}^{G_t} \frac{dG}{-R_G} \quad (5)$$

In order to integrate Eq. (5) we need to know the functional form of R_G, i.e., how R_G varies with other parameters of the fermentation system. Knowledge of R_G in a batch fermentation is, however, often incomplete, and thus integration of Eq. (5) is not always an easy matter. Similarly, if the chemical "species" is the total microbial biomass in the fermentor, Eq. (2) will yield, with X to be biomass per unit volume of fermentation broth,

$$dX = R_x\,dt \quad (6)$$

where R_x = rate of production of biomass. In writing Eq. (6) microbial cell lysis and other processes that destroy the biomass are assumed to be absent; i.e., the rate of consumption of biomass is zero. Again, integration of Eq. (6) from 0 to t will, in theory, predict the biomass production during the time span.

$$\int_0^t dt = t = \int_{X_0}^{X_t} \frac{dX}{R_x} \quad (7)$$

Equations (5) and (7) thus identify knowledge of R_G and R_x as important prerequisites to good fermentor design.

In practice, how G and X vary with t are determined in experimental batches in laboratory or pilot-plant fermentors and expressed as "growth curves." From the slopes of growth curves, R_G and R_x can be extracted. When the results do not yield some simple mathematical expressions for R_G and R_x, as often is the case, one then uses graphical or numerical techniques to integrate Eqs. (5) and (7) to predict G and X at various t. How accurately one can use R_x and R_G from laboratory or pilot-plant fermentors to predict "growth curves" (i.e. G and X at various t) in a different fermentor is the problem of "scaleup" (or "scaledown"). Since R_G and R_x depend upon many parameters of the fermentation system in addition to G and X, our incomplete knowledge of these parameters is the source of difficulty in translating laboratory culture results to industrial operations. It does not appear that we can overcome this difficulty, in the near future, before we develop the necessary capability of defining and monitoring many more parameters of a fermentation system than we can presently.

III. MICROBIAL CELL GROWTH

The Monod equation, even though it resembles those of Michaelis–Menten in enzyme kinetics and Langmuir in surface adsorption, is basically an empirical equation which correlates surprisingly well with many sets of published data of microbial cell growth. The Monod equation states, when the growth is limited by one single substrate [Eq. (8)],

$$\mu = \frac{\mu_{\max} S}{K_s + S} \tag{8}$$

where $\mu = R_x/X$ = specific rate of biomass production, $\mu_{\max}$ = maximum value of μ when S is very large, S = substrate concentration, and K_s = a constant. Similar to the case of Michaelis–Menten equation, by plotting $1/\mu$ against $1/S$, the values of $\mu_{\max}$ and K_s can be determined from experimental data. When S is much larger than K_s, μ approaches $\mu_{\max}$ and becomes independent of substrate concentration; i.e., when

$$S >> K_s \quad \text{and} \quad \mu \rightarrow \mu_{\max} \tag{9}$$

$$R_x = dX/dt = \mu_{\max} X$$

Equation (9) leads to

$$1\text{n}\, X = (\mu_{\max})t + 1\text{n}\, X_0 \tag{10}$$

where $X_0 = X$ at $t = 0$. Equation (10) predicts exponential growth in microbial batch cultures when substrate is in good supply.

Equation (8) has been expanded to Eq. (11) to account for microbial cell growth that shows multiple exponential growth phases.

$$\mu = \mu_{max}\left(\frac{k_1 S_1}{K_1 + S_1} + \frac{k_2 S_2}{K_2 + S_2} + \cdots + \frac{k_i S_i}{K_i + S_i}\right)\left(\frac{1}{\sum_{j=1}^{i} k_j}\right) \quad (11)$$

When substrates S_1 through S_i are all in excess; i.e., $S_1 >> K_1$, $S_2 >> K_2$, . . . , $S_i >> K_i$, $\mu = \mu_{max}$ and the growth will show one exponential phase. When all but one substrate are in excess and when the one (for instance, glucose) is exhausted (i.e., $S_1 = 0$)

$$\mu^1 = \left(\sum_{j=2}^{i} k_j \quad \sum_{j=1}^{i} k_j\right)\mu_{max} \quad (12)$$

cell growth will show a second exponential phase with a different specific growth rate, μ^1.

In some practical cases, such as methanol fermentation for SCP production, the specific growth rate is subjected to substrate inhibition. A commonly written equation for this situation is

$$\mu = \mu_{max}\frac{S}{(K + S + S^2/K_i)} \quad (13)$$

where K_i is known as the inhibition constant.

When microbial growth is inhibited by a metabolic product, the following Eq. has been written:

$$\mu = \mu_{max}\frac{S}{K + S}\frac{K_p}{K_p + p} \quad (14)$$

where p = product concentration. Another equation for product inhibition is given below:

$$\mu = [S/(K + S)]\,(\mu_{max} - K_a A) \quad (15)$$

where A = product concentration and K_a = the toxicity index of the product.

All equations listed above are empirical in nature. They are useful in quantifying the effects of substrate and product concentration on cell growth rate. From these empirical equations, expressions of R_x can be derived and substituted in, for instance, Eq. (7) in calculating biomass production.

Modeling of cell growth is an area of active research from which one gains additional insight into the complicated cellular metabolic activities which are the basis of all industrial fermentation processes. All equations given here treat biomass as an undifferentiated whole. An additional number of growth models do exist in the literature in which biomass is differentiated according to biological activities. In some recent cases, enzyme repression and induction are also considered in modeling cell growth. As one can predict, such sophisticated models will become increasingly useful in future engineering design of fermentation systems.

IV. YIELD, METABOLIC QUOTIENT, AND PRODUCTION RATE

Growth yield is defined by the quotient

$$\Delta X/\Delta S = Y_{X/S} \tag{16}$$

where ΔX is the increase in biomass and ΔS is the consumption of substrate. As ΔS approaches zero,

$$Y_{X/S} = \frac{dx}{ds} = \frac{dx/dt}{ds/dt} = \frac{R_x}{R_s} \tag{17}$$

where R_s is the rate of substrate consumption.

The yield concept can be easily broadened to specific products other than the total biomass. If P is a specific product,

$$Y_{P/S} = \frac{dP}{dS} = \frac{R_p}{R_s} \tag{18}$$

where R_p is the rate of P formation. The Y factors are useful parameters in quantifying growth and product formation. It would be a mistake to read much more meaning into a Y factor than what is defined above. For instance, even if we define a $Y_{A/G}$ to be the "yield" of an antibiotic compound when glucose is consumed by the microbes, this definition does not necessarily imply a direct metabolic linkage between glucose and the compound. However, the manner by which $Y_{A/G}$ responds to environmental changes, including the change in glucose concentration in a fermentor, may help reveal some metabolic linkage between A and G.

The specific consumption rate of a substrate is also known as the metabolic quotient of the substrate:

$$q = (1/X)\,(ds/dt) \tag{19}$$

Well-known examples are q_{O_2} and q_G for oxygen and glucose consumption, respectively. Metabolic quotient, like the yield factor, is again a useful tool in quantifying fermentation phenomena.

Depending upon the type of fermentation, the rate of product formation may relate to biomass in more than one way. The most commonly quoted relationship is the following:

$$dP/dt = k_1\,(dx/dt) + k_2 x \tag{20}$$

If k_1 is a constant and k_2 is equal to zero, the product formation is said to be directly linked to cell growth. Again, however, a word of caution is in order. Until justified by other evidence, one should not read too much physiological meaning into the relationship of product formation, Y factor, and q factor. Even though they can be helpful in quantifying fermentation phenomena, they are basically empirical parameters.

V. IDEALIZED REACTOR DESIGN II: PERFECTLY MIXED CONTINUOUS FLOW REACTORS

In addition to the perfectly mixed batch reactor discussed previously, two idealized continuous flow reactors will be analyzed. This section deals with the perfectly mixed continuous flow reactor.

The phrase "perfectly mixed" requires that concentration and temperature are uniform throughout the liquid fermentation broth in the reactor. As in the case of the idealized batch fermentor, material balances can be conveniently done over the whole broth in a perfectly mixed continuous flow reactor.

Applying Eq. (1) and assuming steady state,

$$\text{Input} + \text{production} = \text{output} + \text{consumption} \tag{21}$$

For biomass X (see Fig. 1),

$$FX_0 + VR_x = FX \tag{22}$$

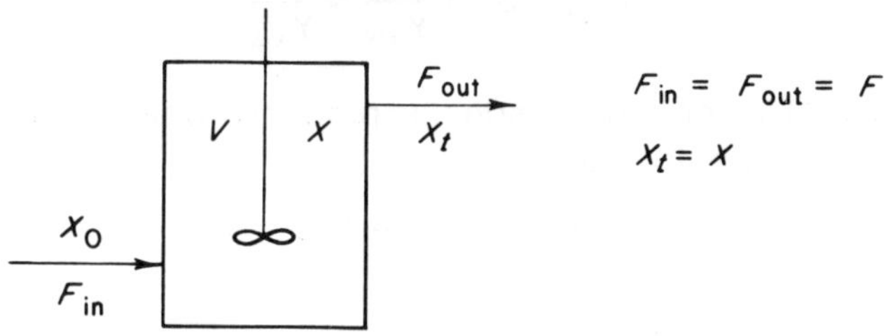

FIGURE 1. Perfectly mixed continuous flow reactor.

with the consumption term set to be zero, and F being the liquid volumetric flow rate in and out of the fermentor, and X_0 the microbial cell concentration in the feed. The condition of "perfectly mixed" requires that the cell concentration in the outlet, X_f, is identical to the cell concentration X in and throughout the fermentation broth.

Equation (22) can be changed to

$$D(X - X_0) = R_x = X\mu \tag{23}$$

where $D = F/V$ = dilution rate. With $\tau = 1/D$ = average residence time in the reactor,

$$1/D = \tau = (X - X_0)/R_x \tag{24}$$

If $X_0 = 0$, i.e., no cells in the feed, Eq. (23) reduces to

$$D = \mu \tag{25}$$

Equation (25) states that a steady state, perfectly mixed continuous flow fermentation can be maintained as long as the specific growth rate is equal to the dilution rate. According to Eq. (25), however, the dilution rate alone does not determine the cell concentration X in a perfectly mixed continuous flow reactor.

When Eq. (21) is applied to a specific substrate S which is consumed but not produced in the fermentor,

$$FS_0 = FS + VR_s \tag{26}$$

or

$$D(S_0 - S) = R_s \tag{27}$$

or

$$\frac{1}{D} = \tau = \frac{S_0 - S}{R_s} \tag{28}$$

Combining Eqs. (17) and (27),

$$D(S_0 - S) = \frac{R_x}{Y_{X/S}} = \frac{\mu X}{Y_{X/S}} \tag{29}$$

When Eq. (21) is applied to a specific product p which is produced but not consumed in the fermentor,

$$FP_0 + VR_p = FP \tag{30}$$

or

$$D(P - P_0) = R_p \tag{31}$$

or

$$1/D = \tau = (P - P_0)/R_p \tag{32}$$

Combining Eqs. (18) and (31),

$$D(P - P_0) = Y_{P/S}R_S \tag{33}$$

Combining with Eq. (17),

$$D(P - P_0) = Y_{P/S}\frac{\text{RX}}{Y_{X/S}} = Y_{P/X}R_X \tag{34}$$

where

$$\frac{Y_{P/S}}{Y_{X/S}} = Y_{P/S}Y_{S/X} = Y_{P/X} \tag{35}$$

Equation (34) can be rewritten as

$$D(P - P_0) = Y_{P/X}\mu X \tag{36}$$

Equations (25), (29), and (36) describe the steady state concentrations of biomass, substrate, and product in a perfectly mixed continuous flow reactor. For microbial cell growth limited by one substrate obeying the Monod equation, a unique set of steady state conditions is determined by the dilution rate and the substrate concentration in the feed. By substituting Eq. (8) into Eq. (25),

$$D = (\mu_{\max}S)/(K_S + S) \tag{37}$$

or

$$S = S \text{ at steady state} = \frac{DK_S}{\mu_{\max} - D} \tag{38}$$

Combining Eqs. (25) and (29),

$$X = Y_{X/S}(S_0 - S) \tag{39}$$

Combining Eqs. (25), (36), and (39) and letting $P_0 = 0$,

$$P = Y_{P/X}X = Y_{P/X}Y_{X/S}(S_0 - S)$$

or

$$P = Y_{P/S}(S_0 - S) \tag{40}$$

When the dilution rate D is increased, S will increase and approach S_0 ($S \leq S_0$) according to Eq. (38). When S approaches S_0, according to Eq. (39) X will approach zero, which is the condition of "washout."

The biomass productivity of the reactor can be measured by $FX/V = DX$, i.e., the amount of biomass produced per unit time per volume of fermentation broth. According to Eq. (23), with $X_0 = 0$,

$$DX = \mu X - (\mu_{max}SX)/K_S + S) \tag{41}$$

Substituting Eq. (39) into Eq. (41), one will have

$$DX = \frac{\mu_{max}S}{K_S + S} Y_{X/S}(S_0 - S) \tag{42}$$

To optimize biomass productivity, Eq. (42) is differentiated with respect to S and the first derivative set to zero:

$$\frac{d(DX)}{dS} = 0 = (\mu_{max}Y_{X/S})\left[\frac{S_0 - S}{K_S + S} - \frac{S}{K_S + S} - \frac{S(S_0 - S)}{(K_S + S)^2}\right]$$

The condition of maximum productivity is, therefore,

$$S_{max} = (K_S^2 + S_0K_s) - K_s \tag{43}$$

Substituting Eq. (43) into Eq. (38), $D_{max} = D$, at which maximum DX occurs $= \mu_{max}\,[1 - (K_s/K_s + S_0)]$. $D = \mu_{max}$ is the condition of washout. There seems to be a misconception often stated in textbooks; i.e., the maximum biomass productivity of a perfectly mixed continuous flow reactor always exists "near" washout. According to Eq. (44), this is not necessarily true because how close the maximum productivity condition is to that of washout depends upon the ratio of K_S/S_0 of which, at least, the value of S_0 can be arbitrarily set.

The biomass yield of the fermentor can be measured by $FX/FS_0 = X/S_0$, i.e., the amount of biomass produced by unit of substrate fed into the reactor. According to Eq. (39), the maximum X and thus the maximum X/S_0 occurs at the trivial condition of $S = 0$, the condition of batch growth (i.e., $D = 0$).

$$\max\frac{X_f}{S_0} = \lim_{S\to 0} Y_{X/S}(S_0 - S)\,\frac{1}{S_0} = Y_{X/S} \tag{44}$$

If the microbial cell is subjected to substrate inhibition (for instance, growth on methanol) according to Eq. (13), then at steady state,

$$D = \mu = \mu_{max}\,\frac{S}{K + S + S^2/K_i} \tag{45}$$

$$S = S\text{ (at steady state)} = \frac{K_i}{2}\,\frac{\mu_{max} - D}{D} \pm \frac{1}{2}\left[k_i^2\left(\frac{\mu_{max} - D}{D}\right)^2 - 4KK_i\right]^{\frac{1}{2}} \tag{46}$$

Therefore, there are two possible steady state values of S at the same dilution rate.

A further analysis of these two potential steady states will reveal another interesting problem of reactor design. This deals with the problem of stability, which can best be explained with the diagram in Fig. 2. The curve in the diagram is a graphical representation of Eq. (45). At a particular specific growth rate μ_1 (= D), the two possible steady states are represented by P_1 and P'_1 with S_1 and S'_1, respectively. First, we will consider steady state P_1. Suppose a small disturbance in the reactor conditions takes place resulting in a new μ_2 which is slightly smaller than μ_1 which is equal to D that remains unchanged. The rate of substrate consumption is smaller at μ_2 than that at μ_1. Consequently, substrate concentration will increase in the fermentor. Because of substrate inhibition, at a higher S value a smaller μ will result which will further decrease the substrate consumption rate and lead to an even higher S value. Therefore, P_1 represents an unstable steady state at which a small decrease in μ will enlarge itself and lead to cessation of growth completely at $S = S_2$.

Suppose a small disturbance of P_1 results in a new μ_3 which is slightly larger than μ_1 which equals D that remains unchanged (see Fig. 2). The rate of substrate consumption is larger at μ_3 than that at μ_1. Consequently, substrate concentration will decrease, leading to an even faster specific growth rate, which will result in an even smaller S. Therefore, with a small increase in μ the reactor condition deviates away from that of P_1 and, in fact, will eventually change to P'_1 which represents a stable steady state.

Point P'_1, being stable, can be analyzed in an exercise identical to that described above for analyzing P_1. In the case of P'_1, a small disturbance will decay itself and the reactor will return automatically to P'_1.

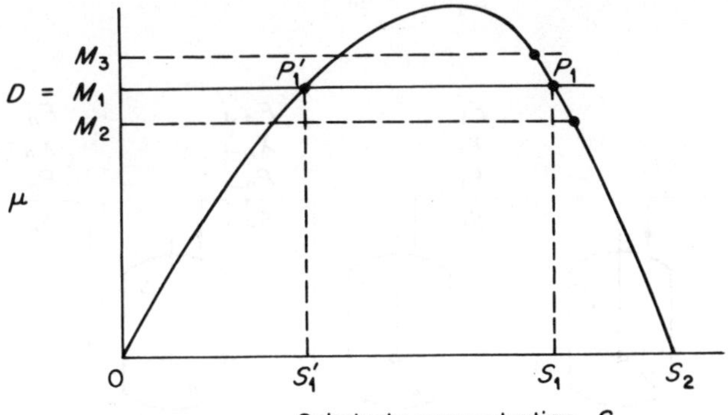

FIGURE 2. Diagram for stability analysis.

When a number of perfectly mixed continuous flow reactors are operated in series, the overall performance can be analyzed by referring to the diagram in Fig. 3. At steady state.

$$F = F_1 = F_2 = \cdots = F_n \tag{47}$$

Let

$$V_1 = V_2 = \cdots = V_n = V/n \tag{48}$$

Making a biomass balance of the nth reactor,

$$FX_{n-1} + V_n R_{x,n} = FX_n \tag{49}$$

or

$$D_n(S_n - X_{n-1}) = R_{x,n} = X\mu \tag{50}$$

or

$$\frac{X_n - X_{n-1}}{R_{x,n}} = \frac{1}{D_n} = \tau_n \tag{51}$$

where $R_{x,n}$ is rate of biomass production which is a function of the conditions in the nth reactor. Equation (51) can be written for each of n reactors. Summing all the n equations will yield

$$\sum_{i=1}^{n} \frac{X_i - X_{i-1}}{R_{x,i}} = \sum_{i=1}^{n} \tau_i = \tau = \frac{1}{D} \tag{52}$$

where $D = F/V = F/V_i n$. Similarly, one can write:

$$\sum_{i=1}^{n} \frac{S_i - S_{i-1}}{-R_{s,i}} = \tau \tag{53}$$

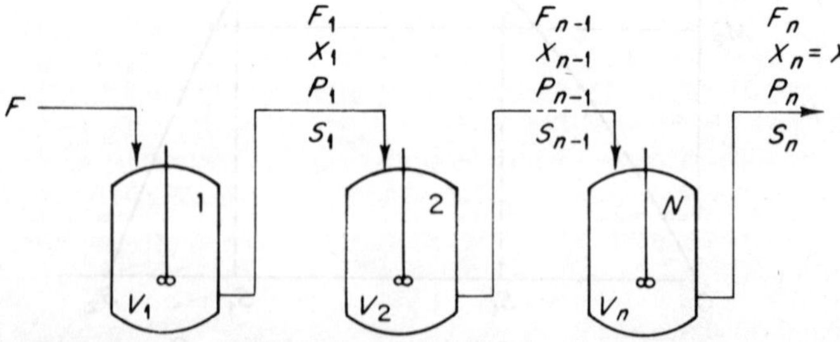

FIGURE 3. Multiple perfectly mixed continuous flow reactors in series.

and

$$\sum_{i=1}^{n} \frac{P_i - P_{i-1}}{R_{p,i}} = \tau \tag{54}$$

VI. IDEALIZED REACTOR DESIGN III: PLUG FLOW REACTOR

Idealization of a plug flow reactor is made in that no backmixing takes place. Consequently, every infinitesimal quantity of feed that enters into the reactor will remain a separate entity, reside in the reactor for a fixed length of time, and then exit from the reactor. Numerous such infinitesimal entities move through the reactor, with no interchange of material or energy among different entities, each of which behaves like a separate perfectly mixed batch reactor. In the real physical world, idealized plug flow conditions are approached by those in packed bed reactors and reactors in the shape of long tubes. Therefore, plug flow reactors are also called idealized tubular reactors. A plug flow reactor may be visualized as a column of soldiers marching with a uniform pace down an avenue through a city block; each soldier remains a separate entity and spends an equal length of time passing through the particular city block.

A plug flow reactor is diagrammed in Fig. 4, with a length L from the entrance to the exit. The volumetric liquid feed rate is designated by F and the total liquid volume of the reactor by V. Assuming that the tubular reactor has a uniform liquid cross-section area of A, the liquid feed that has moved from the entrance and reached the differential reactor volume marked dV in Fig. 4 will have resided in the reactor for a time of AL/F; and by the time it exits, its total residence time in the reactor will be $\tau = V/F = AL/F$. Therefore, at any instance, the liquid content in the reactor is of nonuniform residential age, with the oldest about to exit, the youngest just at the entrance, and the inbetween residential ages increasing from

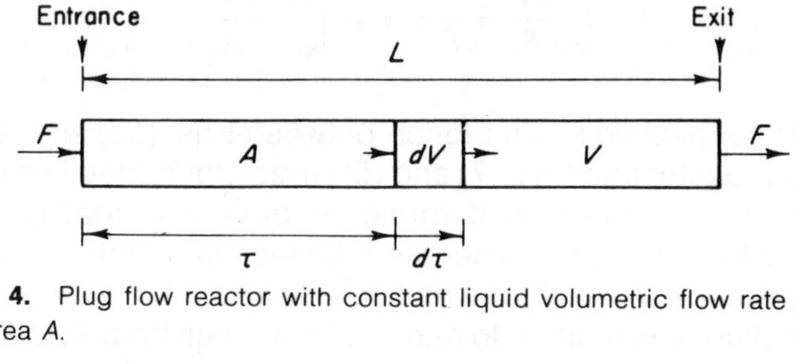

FIGURE 4. Plug flow reactor with constant liquid volumetric flow rate F and cross-sectional area A.

the entrance end to the exit end. Liquid feed that has resided for different lengths of time in the reactor naturally has chemical and/or biological reactions carried out to different extents. It should be obvious that the concentrations of substrates, intermediates, and products in a plug flow reactor are thus nonuniform and vary from the entrance to the exit of the reactor. When material balance equations are written in analyzing a plug flow reactor, they are usually first written not for the entire reactor but only for a differential section (say dV in Fig. 4) of the reactor. Applying the steady state Eq. (21) to dV of Fig. 4, one can write the following biomass balance equation:

$$FX + (dV)R_x = F(X + dX) \tag{55}$$

In writing Eq. (55), X is the biomass concentration in the inlet to dV and ($x + dx$) is the biomass concentration in the outlet from dV. Equation (55) simplifies to

$$dV/F = dX/R_x \tag{56}$$

Equation (56) is then integrated to describe the variation of biomass concentration through the whole plug flow reactor:

$$\int_0^v \frac{DV}{F} = \frac{V}{F} = \tau = \int_{x_0}^{x_f} \frac{dx}{R_x} \tag{57}$$

where X_0 = original x at inlet to the whole reactor and x_f = final x at exit of the reactor. Similarly, for a substrate, from Eq. (21),

$$FS = (dV)R_s + F(S + dS) \tag{58}$$

or

$$dV/F = dS/(-R_s) \tag{59}$$

or

$$\int_0^V \frac{dV}{F} = \frac{V}{F} = \tau = \int_{S_0}^{S_f} \frac{dS}{-R_s} \tag{60}$$

The close respective resemblance between Eqs. (57) and (60) written for plug flow reactor and Eqs. (7) and (5) written for perfectly mixed batch reactor should come as no surprise. A plug flow reactor has been analyzed as the totality of numerous infinitesimal batch reactors with a total liquid volume V, each of which is allowed a time τ, the residence time, in the plug flow reactor, to carry out reactions from S_0 to S_f and from X_0 to X_f, while a perfectly mixed batch reactor of a liquid volume V carries out reactions over a time t to change from S_0 (or G_0) to S_f (or G_f) and from X_0 to X_f. All changes in *S, X,* and *P* that take place in a perfectly mixed batch reactor at different times *t* can be found in a plug flow reactor at different *L* from the entrance, and thus at different residence times in the reactor. In theory, therefore, conditions in a perfectly mixed batch reactor along a time axis can be exactly reproduced in an idealized plug flow reactor along an axis representing the distance from reactor entrance. As discussed previously in Section V (Eq. 44), the maximum biomass yield, $\max(X_f/S_0)$, is achieved at zero dilution rate to a perfectly mixed continuous flow reactor, i.e., when it is operated as a batch reactor. The same maximum biomass yield can be achieved, therefore, in a plug flow reactor, which a perfectly mixed continuous flow reactor is, in theory, unable to achieve (unless $D = 0$).

Even though a plug flow reactor can in theory outperform a perfectly mixed continuous flow reactor as we have just described, in reality, a truly plug flow fermentor does not exist. The idealization of "plug flow" requires absolutely no backmixing, while in a real fermentation, requirements of aeration and/or agitation destroy this condition. In order to approach the performance of a plug flow reactor, multiple perfectly mixed continuous flow reactors in series provide a practical compromise. The performance of multiple-stage, perfectly mixed, continuous flow reactors of a total liquid volume *V* will exactly equal to that of a plug flow reactor having a liquid volume *V* when the number of stages approaches infinity. This behavior can be seen from Eq. (52) by letting $\Delta X = X_i - X_{i-1}$, and $n \to \infty$, and comparing with Eq. (57).

$$\tau = \lim_{n\to\infty} \sum_{i=1}^{n} \frac{X_i - X_{i-1}}{R_{x,i}} = \lim_{n\to\infty} \sum_{i=1}^{n} \frac{\Delta x}{R_{x,i}} = \int_{x_0}^{x_f} \frac{dx}{R_x} \tag{61}$$

In reality, of course, only a finite number of stages of perfectly mixed continuous flow reactor can be used, which should have a performance in X_f/S_0 intermediate to those of a plug flow reactor and a one-stage perfectly mixed continuous flow reactor.

When a different performance index, the specific volumetric biomass output (biomass productivity DX_f) is used in comparing the two idealized flow reactors, the relative superiority of the two can be different. This can be seen by the following analysis.

Growth is a special case of autocatalytic reactions. The growth rate R_x behaves in general according to the curve in Fig. 5 with a minimum at an intermediate extent of substrate conversion. R_x equals $X\mu$ and is a function of both *X* and *S*. When the extent of conversion is low, *X* is low but *S* is high, and when the extent of conversion is high, *X* is high but *S* is low. Therefore, generally speaking, at an intermediate extent of conver-

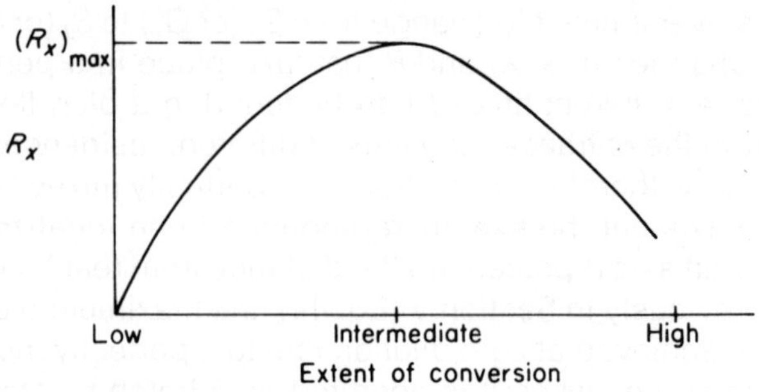

FIGURE 5. Growth rate at different conversion.

sion, a maximum R_x exists. The extent of conversion can be measured and expressed in many ways. One convenient way is to express conversion in the concentration of substrate left. This could be the value of S in a batch reactor after a reaction time t or in a plug flow reactor after a residence time τ. This could also be the value of S in the exit stream of a perfectly mixed continuous flow reactor or in the exit stream of several stages of such reactors in series.

Taking the curve in Fig. 5, and expressing the extent of conversion in S, one can plot l/R_x versus S in Fig. 6. From Eqs. (17) and (60), assuming $Y_{X,S}$ to be a constant and $X_f >> X_0$,

$$\frac{V}{F} = \int_{S_0}^{S_f} \frac{dS}{-R_s} = Y_{X/S} \int_{S_0}^{S_f} \frac{dS}{-R_x} \tag{62}$$

and

$$Y_{X/S}(S_f - S_0) = X_f - X_0 = X_f \tag{63}$$

The biomass productivity of a plug flow reactor is, therefore,

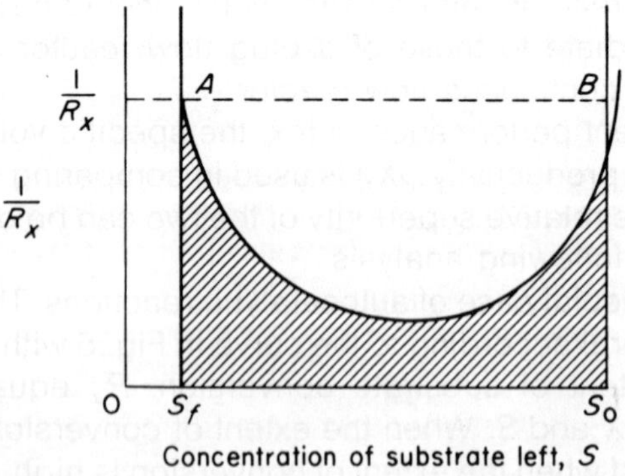

FIGURE 6. Inverse growth rate at different conversion.

$$DX_f = \frac{FX_f}{V} = X_f / \left[Y_{X/S} \int_{S_0}^{S_f} dS/(-R_x) \right] \tag{64}$$

The shaded area in Fig. 6, according to the rules of graphical integration, is directly proportional to the expression in the denominator on the right-hand side of Eq. (64) and is inversely proportional to biomass productivity DX_f.

Similarly, the biomass productivity of a perfectly mixed continuous flow reactor can be shown as follows, by combining Eqs. (17) and (27), to be inversely proportional to the area of the rectangle ABS_0S_f [area $ABS_0S_f = (S_0 - S_f)X\ 1/R_x$]:

$$DX_f = \frac{FX_f}{V} = \frac{X_f}{Y_{X/S}[(S_f - S_0)/(-R_x)]} \tag{65}$$

knowing that for a perfectly mixed continuous flow reactor

$$X_f = X \text{ and } S_f = S$$

In the case shown in Fig. 6, the area of rectangle ABS_0S_f is obviously much larger than the shaded area. This means a plug flow reactor has a better performance in biomass productivity, DX_f, than a perfectly mixed continuous flow reactor operating under the same S_0, S_f, and $Y_{X/X}$. However, being autocatalytic and having a maximum R_x at an intermediate extent of conversion, a perfectly mixed continuous flow reactor can give a better biomass productivity than a plug flow reactor. This is shown by a comparison between the shaded area and the area of rectangle $A°B°S_0S_b$ in Fig. 7, if the desired fermentation is to run from S_0 and stop at S_b.

In practical cases, we usually would like to finish a fermentation at a high extent of conversion, i.e., a low concentration of substrate left, S_f (S_f

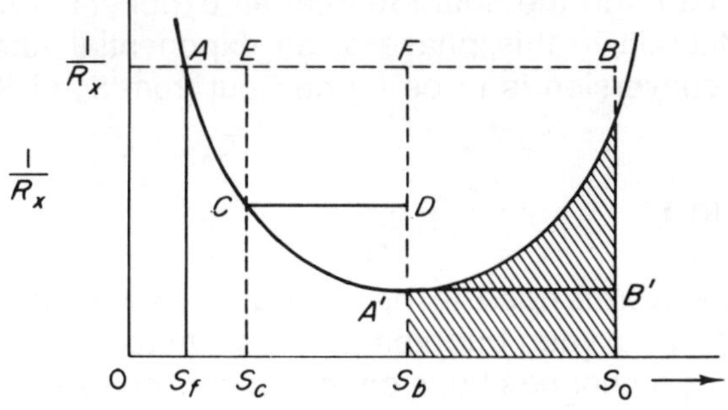

FIGURE 7. Comparison of reactor performance.

$< S_b$). The situation as shown in Fig. 7 suggests the use of a perfectly mixed continuous flow reactor to bring S_0 to S_b and finish the conversion from S_b to S_f in a plug flow reactor with its biomass productivity inversely proportional to the area bounded by $AA°S_bS_f$.

As we described previously, it is very difficult to realize truly plug flow conditions in a real-world fermentor which is often aerated and agitated. A compromise solution could be the use of a two- or multistage perfectly mixed continuous flow reactors in series to bring conversion from S_b to S_f. In Fig. 7, an overall three-stage perfectly mixed continuous flow reactor series is shown with S_b, S_c, and S_f being the concentrations of substrate left after the first, second, and third stage, respectively. The total area of three rectangles $A'B'S_0S_b$, CDS_bS_c, and AES_cS_f is inversely proportional to the overall biomass productivity DX_f of the reactors in series.

VII. FED-BATCH REACTOR

In situations shown in Figs. 6 and 7, a multiple-stage, perfectly mixed, continuous flow reactor has been offered as a compromise solution for obtaining high biomass productivity. Another solution could be the use of a "fed-batch reactor." A fermentor operated in a fed-batch mode is first charged partially full with a broth volume of V_0. After a controlled length of time liquid feed is added continuously at a controlled rate until the fermentor is full at V_f. The fermentor is then left to run as a batch until S_f, after which the fermentor is discharged to V_0 and the cycle is repeated. The fermentor is cyclic in fed and batch phases with a short discharge phase in between cycles. A fully detailed analysis of a fed-batch fermentor is very intriguing and is beyond the scope of this chapter. By controlling the liquid feed, the fed phase may be made to behave like a perfectly mixed continuous flow reactor with a nearly constant S_b. For instance, one can add the liquid feed at an exponential rate to sustain biomass production in this phase at an exponential rate. During the batch phase, conversion is to be carried out from S_b to S_f.

VIII. AERATION

Finally, a few words need to be said about aeration which, due to dispersion of air or oxygen bubbles, creates a heterogeneous mixture in a fermentor. This author has been careful, in this chapter, in distinguishing the liquid volume from the total fermentor volume. The latter includes those of the head space and gas bubbles. All material balances that we have described apply to the fermentation liquid volume which is assumed to be homogeneous and ignoring the fact that microbial cells actually form yet a third material phase. Aeration is, in a way, a special problem. No batch fermentation is truly "batch" because oxygen is always continuously added and carbon dioxide often continuously removed. Similar complications also exist in liquid hydrocarbon fermentation, in which an additional liquid phase exists. The analysis of this chapter applies to the volume of aqueous broth. In analyzing the overall fermentation system involving several material phases, one can write material balance equations for each of all the phases. These equations are coupled by mass transfer in between phases. These more complicated problems of modeling and reactor design and analysis are not included here.

GENERAL REFERENCES

Levenspiel, O. (1972). "Chemical Reaction Engineering," 2nd ed. Wiley, New York.
Lim, H. C., Chen, B. J., and Creagan, C. C. (1977). *Biotechnol. Bioeng.* **19,** 425.
Perlmutter, D. D. (1972). "Stability of Chemical Reactors," Prentice-Hall, Englewood Cliffs, New Jersey.
Pirt, S. J. (1975). "Principles of Microbe and Cell Cultivation." Halsted Press, New York.
Smith, J. M. (1970). "Chemical Engineering Kinetics," 2nd ed. McGraw-Hill, New York.

Chapter 12

Microbial Culture Selection

RICHARD P. ELANDER
L. T. CHANG

MICROBIAL TECHNOLOGY, 2nd ed., VOL.II
 ISBN 0-12-551502-2

I. INTRODUCTION

Microbial culture selection is involved in the initial isolation of microorganisms from nature and later in the purification of strains of interest due to the synthesis of desirable metabolic products. In most large-scale screening programs, rare, unusual cultures are selected by taxonomists since the probability of new interesting activities is maximized in sensitive specific assays using rare, fastidious microorganisms (Conover, 1971). The purification of unusual organisms is difficult and involves considerable technique in culture selection. Often interesting cultures must be constantly repurified, because of slowly-growing commensal organisms found associated with the organism of interest.

Pure cultures of microorganisms are inherently variable in their growth characteristics and metabolic activities. Therefore, the initial activity of the microbiologist is to minimize the genetic variability of the microorganisms by selecting out stable and genetically uniform isolates which produce a minimum number of unwanted metabolites and copious amounts of the desired component. The second important activity is to develop high-producing strains. At first, a simple genetic approach is used such as the selection of strains from untreated spore populations, or the exposure of spores or cells to mutagens followed by the random selection of isolates via monospore isolation and their testing in limited fermentation test systems. In order to find the few strains capable of higher productivity, a large number of cultures must be evaluated. This approach probably has resulted in the discard of interesting strains, but the methodology is simple and requires relatively little genetic experience.

Although natural selection and mutation selection are generally successful in improving titers of described metabolites, culture stability, growth and sporulation characteristics, etc., often represent additional critical problems to be resolved. These characteristics are also inherent genetic problems and must be resolved in a culture improvement program. Loss in titer (termed "culture rundown" or "strain degeneration") often occurs with improved strains and such variants are frequently encountered during growth following vegetative transfer, storage, or long-term preservation. Another associated problem is the biosynthesis of undesirable metabolites in a multicomponent complex of related metabolites. The elimination or decrease in titers of unwanted components can be resolved by culture selection or by changes in the chemical and physical environment of the fermentation.

A high-producing stable strain is essential both to fundamental studies carried out to establish pathways of biosynthesis and to fermentation optimization through control of the chemical and physical environments. The use of selection to develop a stable productive culture and the acquisition of knowledge of the strain–environment relationships are important to allow the fermentation specialist to control and direct the microbe to more efficient productivity of the desired metabolite.

The purpose of this chapter is to review genetic and selection methodology which is currently utilized in the selection, purification, and stabilization of high-productivity mutants as exemplified by currently important fermentation microorganisms. We also review the present status of genetic recombination with regard to the generation of improved fermentation microorganisms. We also highlight the recent progress attained in cloning foreign DNA through plasmid, phage, or transformation systems, to generate new "hybrid" organisms. This new technology of genetic engineering further expands man's control of microorganisms and allows further exploitation of the microbial world to better serve man's needs.

II. MUTAGENESIS AND IMPROVED PRODUCT YIELD

Mutagenesis followed by the subsequent selection and purification of superior strains represents the most important initial activity in improving the yield of a fermentation product. Mutation programs are vital to fermentation companies in the higher productivities are essential in reducing or maintaining product cost in periods of inflation and increasing labor and raw material costs. Therefore, mutation and genetic programs are being expanded in both industrial and applied microbiological laboratories. It is clear that mutation and selection to enhanced strain productivity represents the major factor involved in the manyfold improvement of fermentation products.

A. Natural Selection

As soon as a metabolite shows clinical potential, programs are initiated in fermentation development areas for strain improvement and the subsequent long-term preservation of improved variants. It is well established that the most effective method initially for improving the yield of a fermentation product is the application of mutation followed by selection.

General reviews of this subject are available and recent reports by Fantini (1976) and Elander and Espenshade (1976) may be useful.

Most microorganisms exhibit a great capacity for natural variation. Therefore, the initial approach in selection work is to survey this background variability for improved cultures. Colonies are generally examined in laboratory fermentations which originate from untreated spore suspensions (also termed monospore isolates). Cultures isolated from nature are generally heterokaryotic and the selection of single spores with uninucleate conidia generally leads to a variety of colony types (termed "population pattern"), many of which may be superior in their capacity to synthesize a fermentation product. A classic paper dealing with penicillin and initial strains of *Penicillium chrysogenum* is by Backus and Stauffer (1955).

B. Mutation Selection

The exposure of spores to mutagenic agents and the subsequent selection of colony populations is known as "mutation selection;" it has two aspects when applied to industrial microorganisms. One aspect, *major mutation,* involves the selection of mutants which manifest a profound change in a biochemical character of practical interest. These strains are usually isolated from spores surviving prolonged exposures to mutagens and generally exhibit a distinct change in some morphological or biochemical character. In contrast, *minor mutations* in strains represent only subtle changes in a particular character and are superficially indistinguishable from the parent phenotype. Usually the biochemical changes are so slight that only the trained expert can distinguish these variants. Minor variants are commonly exploited in programs devoted exclusively to yield improvement.

1. The Role of Major Mutation in Culture Selection

Examples of major mutation in strain development are numerous. Alikhanian (1962) cites a role for major mutation in the streptomycin-producing organism *Streptomyces griseus.* The initial strain synthesized large amounts of a substance with low activity in addition to small quantities of streptomycin. The low-activity substance is mannosidostreptomycin (streptomycin B), which competes with streptomycin for biosynthetic intermediates and also interferes with the isolation of the antibiotic. A variant was isolated which produced negligible amounts of the undesired substance, thus allowing greater synthesis and recovery of the desired moiety. The development of nonpigmented strains of *P. chrysogenum* is another example of major mutation in an industrially important fungus. Strains blocked in the synthesis of the yellow pigment chrysogenin were isolated which allowed for a more efficient recovery of penicillin (Backus and Stauffer, 1955).

A careful study of variants exhibiting impaired antibiotic productivity may elucidate biosynthetic pathways and contribute to the identification of precursors. Studies on cobalamin by Barchielli *et al.* (1960), and on tetracycline by Miller *et al.* (1965), and by Hošťálek and Vaněk (1973), reveal interesting data on precursor molecules involved in reaction steps prior to terminal ring closure.

The use of major mutation has acquired particular significance and, in certain instances, has led to new and more efficacious products. The tetracycline-producing organisms appear to be particularly amenable to this approach. McCormick *et al.* (1957) described an interesting modification of tetracycline synthesized by a mutant strain of *Streptomyces aureofaciens*. The antibiotic was shown to be changed at the C-5a position and was almost devoid of antibiotic activity. However, they also reported that *S. aureofaciens* (strain S-604) synthesized 6-demethyltetracycline, a new antibiotic material not elaborated by the Duggar strain A-377. The new molecule has several advantages over the methylated form and today is one of the leading commercial forms of tetracycline.

The employment of mutants for the synthesis of modified antibiotic molecules appears to be a fertile area for major mutation in strain development. Mutants have been mentioned already that may accumulate presursor molecules which aid in the elucidation of pathways. A minor modification of an antibiotic molecule could lead to new biological and therapeutic properties for a known antibiotic. Some years ago, Kelner (1949) published a paper which has great interest in this connection. He examined a series of streptomycete cultures which failed to inhibit certain test bacteria. The negative strains were then exposed to heavy doses of ultraviolet (uv) and x-ray radiation. After examining several thousand irradiated strains, Kelner found mutant lines which exhibited antimicrobial activity. In fact, certain weak antibiotic producers then showed a change in spectrum. The difference in antimicrobial spectrum indicated that in certain cases this change might have resulted in the qualitative modification of an antibiotic. Structural modifications with a resulting change in spectrum have been demonstrated for oxytetracycline and chlortetracycline.

2. The Role of Minor Mutation in Culture Selection

Minor mutation usually plays the dominant role in yield improvement programs. These mutations generally affect only quantitatively the amount of product synthesized. Such mutations are subtle, and variants exhibiting such features are generally similar in characteristics, other than yield, to the parent form. Quantitative definition of what constitutes a significant yield increase is somewhat relative and depends on the reliability and accuracy of the assay procedure and the productivity level of the metabolite under investigation. Usually, a 10–20% increase is implied. The variants are usually selected from spore suspensions exposed to small or moderate doses of a mutagen. If one repeatedly selects "minor" (positive gain) variants and uses each succeeding improved strain for further mutation and selection work, after many stages of "stepwise selection," a large yield increase may be obtained (see Section VII, A).

Minor variants exhibiting a slight change in a quantitative feature may vary also in other features due to pleiotropic effects, i.e., pigmentation, conidiation, and radial growth rate. However, since the variation in yield is slight, great dependence must be placed on efficient and accurate selection techniques. The population to be tested must be large and the assay must be highly accurate and specific for the desired product. Problems of this sort are statistical and are discussed in reviews by Davies (1964), Brown and Elander (1966), and Fantini (1975).

Examples of gradual stepwise improvement in antibiotic production are numerous and are reviewed in detail for penicillin and cephalosporin later in this chapter (see Section VII, B). Vezina *et al.* (1976) have recently summarized the development of a series of improved antimycin A-producing strains at the Ayerst Research Laboratories. Their study emphasized the power of selection procedures. First, much attention was focused on developing a rapid, accurate, and convenient assay system. Such assay procedures are now usually chemical in nature and are often automated spectrophotometric or high-pressure liquid chromatographic systems capable of handling large numbers of samples per day with computerized data handling systems. Second, at each round of mutagenesis (usually a cyclical use of different mutagens having differing modes of action), several improved strains were compared not only for productivity but also for early dense conidiation, absence of pigment, and growth rate; only one strain, which represented a distinct advantage over the parent, was chosen for further selection. Third, since the fermentation environment plays an important role in the expression of genetic determinants, the fermentation medium and conditions were periodically revised to best suit the organism for further enhanced titer improvement. Fourth, strains showing productivity in laboratory flasks were later evaluated in small pilot-scale aerated–agitated fermentors. Throughout their antimycin improvement program, a rigid maintenance, propagation, and preservation scheme was adhered to. The results of this industrial strain improvement program resulted in a major yield improvement from 0.75 to 9.75 gm/liter, an overall 125-fold augmentation. This program is summarized in Fig. 1.

Many refinements in the techniques of mutation and selection have been introduced during the past two decades. Today, there are scores of mutagens available, and for many of them a mode of action is known. Excellent reviews on the chemical basis of mutation are available (Orgel, 1965; Drake and Baltz, 1976). Unfortunately, the phenomenon of mutation is random and not directable, so that mutation as applied to strain

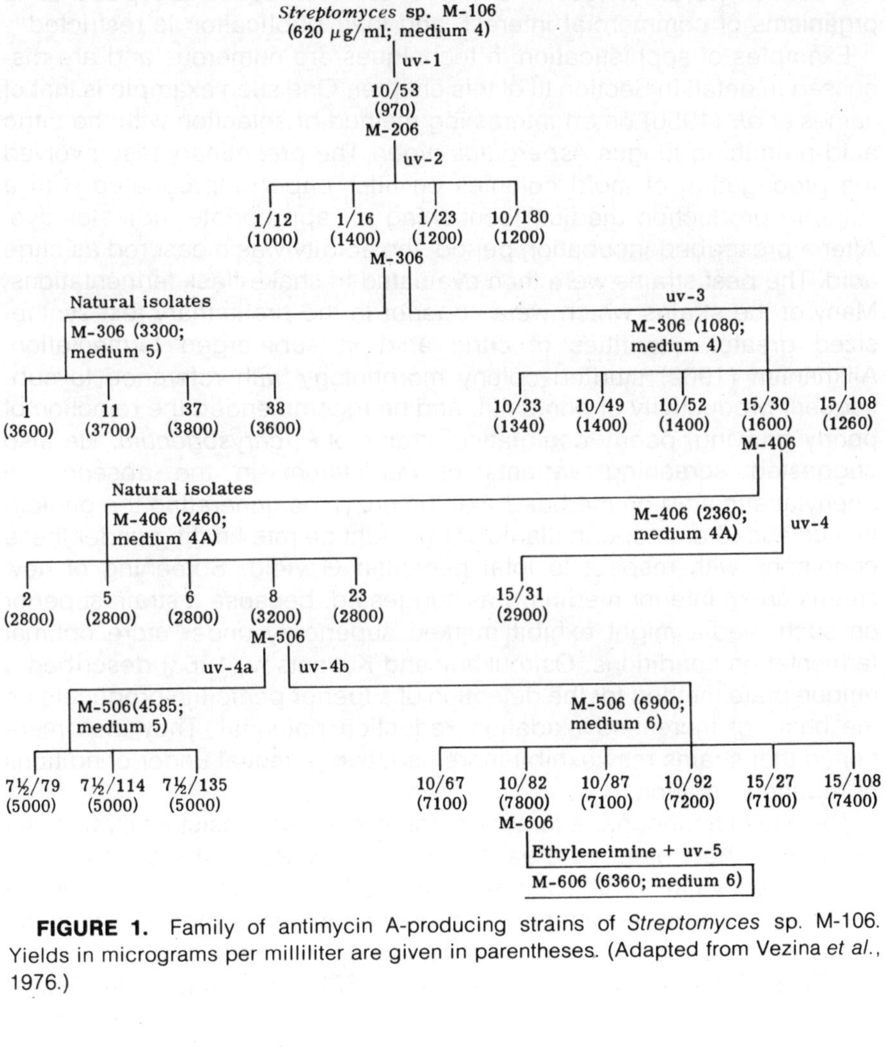

FIGURE 1. Family of antimycin A-producing strains of *Streptomyces* sp. M-106. Yields in micrograms per milliliter are given in parentheses. (Adapted from Vezina *et al.*, 1976.)

development is largely empirical. Little is known of the kind of mutants desired. The limited knowledge of antibiotic biosynthesis is shown by the fact that a biosynthetic pathway is not fully elucidated for a single antibiotic.

In contrast, considerable refinement in mutation and selection techniques has been demonstrated over the past two decades. Refinements in evaluation have been numerous, and these probably represent the key to a successful strain development program. The importance of evaluation is illustrated by the problem of selecting relatively few enhanced variants from a population of hundreds of thousands of individuals showing control yields or less. Many of these techniques are peculiar to organisms of commercial interest, and their publication is restricted.

Examples of sophistication in techniques are numerous and are discussed in detail in Section III of this chapter. One such example is that of James *et al.* (1956) on an interesting method of selection with the citric acid-producing fungus *Aspergillus niger.* The preliminary test involved the propagation of mold colonies on filter paper impregnated with a suitable production medium containing an appropriate indicator dye. After a prescribed incubation period, the acidity was measured as citric acid. The best strains were then evaluated in shake-flask fermentations. Many of the strains which were superior in the preliminary test synthesized greater quantities of citric acid in submerged fermentation. Alikhanian (1962) studied colony morphology with reference to subsequent productivity of penicillin, and he recommended the rejection of poorly growing, poorly conidiating strains of *P. chrysogenum.* He also suggested screening variants of *Penicillium* in the absence of phenylacetic acid on the basis that the enzymes generating the penicillin nucleus (6-aminopenicillanic acid) might be rate limiting under these conditions with respect to total penicillin G yield. Screening of new strains on an inferior medium was suggested, because a strain superior on such media might exhibit marked superiority under more optimal fermentation conditions. Ostroukhov and Kuznetsov (1963) described a unique plate method for the detection of superior penicillin producers on the basis of increased oxidation–reduction potential. They also mentioned that strains may exhibit more negative potential under conditions of reduced aeration.

The use of actinophage for the selection of phage-resistant strains first arose with the violent outbreak of phage infection in the streptomycin industry (Carvajal, 1952). Carvajal observed that an occasional resistant colony which survived phagolysis produced more streptomycin than the sensitive production strain. A pronounced mutagenic effect of actinophage was observed by Alikhanian (1962), who claimed that 99% of the *Actinomyces olivaceus* colonies cultivated from a phagolyzate were similar morphologically to uv mutants and certain of the mutant types were not found in usual uv-survivor populations.

Sophistication at the selection stage is essential for effective strain improvement; the following generalities can be made:

1. Strains selected as obvious variants following exposure to mutagens usually are inferior in their capacity to elaborate metabolites. Those strains with enhanced capacity for accumulation of products are extremely few in number, and selection and evaluation play extremely important roles in their detection.
2. Mutagen dose is important in strain selection methodology. The rate of mutation is a function of dose. Strains experiencing major mutation are best isolated from populations surviving prolonged doses of mutagens. Variants to be employed for increased productivity are generally isolated from populations surviving intermediate dose levels.
3. Strains with enhanced capacity for synthesis generally exhibit wildtype morphololgy and growth habit. Strains with altered morphology, etc., may be inherently better producers but may require considerable fermentation development. Since positive variants are extremely few in number, it is better to screen a large number of variants under a few fermentation conditions rather than to screen a small number of variants under a wide variety of environmental conditions. This is especially true when one seeks only quantitative change in the variants.
4. The philosophy of stepwise selection accepts small increments in increased productivity. The range of increased productivity at each step is generally 10–15% after the initial strains are obtained by natural selection. As productivity increases, the probability of finding superior strains decreases; hence accurate and more sophisticated evaluation procedures play increasingly important roles. The development program is only as effective as the mutation, selection, and evaluation procedures coupled to it.
5. Variant strains often require special propagation and preservation procedures. Actual production gains depend largely on stability and reliability of performance. Maintenance through continued selection and purification plays an important role in production laboratories.
6. Although a strain may meet the numerous necessary criteria of superiority in the laboratory, there is no guarantee that enhanced productivity will occur in production fermentors. Aeration–agitation patterns, nutrient availability, etc., are often unbalanced for variant cultures. Scaleup experimentation in pilot-plant studies is often necessary before any enhanced strain potential may be realized in actual production.

III. ISOLATION OF MUTANT CLASSES AND THEIR USE IN MICROBIAL PROCESSES

Although the application of a suitable mutagenic treatment procedure is essential to effective mutation methodology, the development of rapid, selective isolation procedures and the tests employed to identify the desired class of mutants is of far greater importance. Procedures used for the isolation of mutant classes can be considered as variations on the proverbial problem of "discovering the needle in the haystack." The simplest procedure, and probably the most tedious, is to examine a large number of clones in liquid fermentations in the hopes of finding one mutant which shows some improvement in a particular biochemical character over the parent strain. This approach has often been utilized with industrial microorganisms. However, it has become increasingly important to consider new methods which allow for the selective isolation of particular classes of mutants, including overproducers of desired products. The following discussion will be concerned with the isolation of stable mutants exhibiting robust growth which can be used on an industrial scale.

The chance of discovering the desired mutant may be improved by increasing the frequency of mutation by a cyclical application of mutagens, either singly or in combination. It is also possible to increase the number of potentially interesting mutants by utilizing selective enrichment procedures which allow the mutants to grow more rapidly than the parent culture. These procedures include alternate subculture in selective and nonselective media either in stationary or in continuous culture. Selection procedures can employ color reactions to identify mutant colonies and are very useful. The use of replica-plating procedures allows for the efficient transfer of hundreds of colonies in a single transfer on a variety of media which may be selective in themselves or reveal the presence of higher concentrations of desirable products.

Positive selection procedures utilizing nutrient media which restrict the growth of undesirable clones, but which allow for the expression of particular mutant classes, are highly desirable. Selection for growth on novel substrates often restricts the growth of the parent and allows the mutants to grow, albeit at reduced rates. Often, leaky auxotrophs (bradytrophs) may be more sensitive to inhibition by metabolic analogs and this represents a way to discover derepressed mutants for the production of biosynthetic intermediates. The use of synchronized cells grown in chemostat culture is an excellent procedure for obtaining mutants resistant to high concentrations of toxic inhibitors, metabolic analogs, antibiotics, etc. The theoretical basis for mutant isolation is that mutant cells capable of producing larger quantities of enzymes or possessing greater permeability to substrates will have a distinct growth advantage and eventually outgrow and replace the less efficient parental strain.

We now review the procedures for the enrichment and isolation of distinct classes of industrially important mutants and describe how these mutant classes have been utilized in microbial fermentation processes. For background reading, we recommend important reviews by Hopwood (1970), Abe (1972), Clarke (1976), and Hopwood and Merrick (1977).

A. Auxotrophic Mutants

Auxotrophic mutants are recognized by their growth on a complex multicomponented, "complete" medium and their lack of growth on a chemically defined "minimal" medium. The mutants are then classified into categories by propagation on minimal medium supplemented with amino acids (amino acid-requiring), vitamins (vitamin-requiring), or nucleic acid components (purine- or pyrimidine-requiring). Holliday (1956) developed a single-step characterization procedure which includes the replication of a masterplate of mutant strains to a number of petri dishes, each containing a pool of substances, arranged in such a manner that each substance is present in two different pools arranged in such a way that only 12 plates are needed to test 36 different compounds.

Auxotrophic mutation generally leads to lower productivity in antibiotic-producing microorganisms. This is especially true when one selects mutants which are deficient in growth factors which are also precursor molecules for antibiotic synthesis. Antibiotic productivity is usually low for auxotrophic mutants, even when tested in nutritionally complex media containing copious amounts of the required growth factor. Auxotrophy often affects antibiotic synthesis in the negative direction, even when the required growth factor is not a precursor of the secondary metabolite. Often such auxotrophs grow poorly and, therefore, exhibit decreased antibiotic formation. Other auxotrophic strains appear to have multiple genetic lesions affecting growth and antibiotic synthesis. Despite all these problems, high-producing auxotrophic strains have been obtained for producers of tetracycline, erythromycin, tylosin, actinomycin, and streptomycin. Multiple-mutant cultures can be readily selected from populations following long exposure to potent mutagenic agents, i.e., diazomethane and nitrosoguanidine.

Conidial color variants and biochemical mutants are routinely encountered in survivor populations of *P. chrysogenum.* In one detailed study, albino and yellow conidial mutants were reported to produce 30% less penicillin in flask fermentations. Auxotrophic mutants exhibited reduced vegetative development, with yields varying from 60 to 90% of the productivity of the prototrophic parental culture (Elander, 1967). A large number of morphological and biochemical variants of the cephalosporin C-producing fungus, *Cephalosporium acremonium* (also classified as *Acremonium chrysogenum),* showed a decrease in relative potency with increasing mutagenesis (Elander, 1975).

Auxotrophic mutants of the glutamic acid bacterium, *Corynebacterium glutamicum,* have been exploited in Japan for the commercial production of a number of amino acids listed in Table I.

The lysine-producing strain is a homoserine (or threonine plus methionine) auxotroph and is blocked in that part of the branched pathway leading to methionine, threonine, and isoleucine. The total world market for L-lysine is in excess of 15,000 tons/year, the amino acid being used primarily as a feed supplement. Fermentation potencies of 44 gm/liter of L-lysine·HCl utilizing cane molasses have been reported accompanied by yields of 30–40% in relation to the initial sugar concentration (Nakayama, 1976). The blocking of homoserine synthesis via mutation in the homoserine dehydrogenase gene results in the release of concerted feedback inhibition by threonine plus lysine on aspartokinase as long as threonine is restricted in the medium. In this case, aspartic semialdehyde is converted to lysine and the amino acid is overproduced.

Phe, tyr, and *phe tyr* mutants of *C. glutamicum* have been utilized for the commercial production of L-tyrosine, L-phenylalanine, and L-tryptophan, respectively. Figure 2 shows the remarkable progress in productivity of mutants producing high yields of these amino acids when subsequent analog-resistance mutations are imposed upon the auxotrophs.

B. Analog-Resistant Mutants

The first step in isolating mutants resistant to structural analogs of metabolites is to determine, by plating on a graded series of inhibitor levels, the minimal concentration that prevents growth of the wildtype strain. Resistant mutants are then selected by plating out large numbers of mutagenized cells on plates containing a series of increasing concentrations of the inhibitor. Gradient-plating techniques represent useful alternative procedures for obtaining resistant mutants (Hopwood, 1970).

Mutants resistant to amino acid analogs have been shown to owe their resistance to mutation (a) in a regulatory or operator gene, which has resulted in derepression of biosynthetic enzymes, thereby overproducing the antagonized amino acid and thus overcoming competitive inhibition by the analog; (b) in the structural gene of a permease for the antagonized amino acid, which has resulted in reduced capability to incorporate the analog; (c) in the structural gene of an amino acyl-tRNA

TABLE I. Amino Acid Accumulation by Auxotrophic Strains of *Corynebacterium glutamicum*[a]

Auxotrophic markers	Amino acid accumulated
arg/cit	Ornithine
hom; thr; leu; ile; met + thr; ile + leu	Lysine
ile; leu; thr; arg	Valine
thr	Homoserine
tyr	Phenylalanine
lys	Diaminopimelic acid
leu	Isoleucine
phe	Tyrosine
arg	Citrulline

[a] From Abe (1972).

synthetase, which has resulted in discrimination between the antagonized amino acid and the analog during protein synthesis; and (d) in the structural gene of a biosynthetic enzyme subject to end-product inhibition, which has resulted in loss of feedback control.

Selection of mutants resistant to analogs of primary metabolites often overproduce end product since the mutants are no longer subject to feedback repression. Such a concept has been applied by Elander *et al.* (1971) to the pyrrolnitrin fermentation, an antifungal phenylpyrrole derived from tryptophan in strains of *Pseudomonas*. D-Tryptophan is a direct precursor of this antibiotic but was impractical to use in large-scale fermentation because of its high cost. An effort was made to select fluoro- and methyltryptophan-resistant mutants no longer subject to feedback inhibition by tryptophan. Resistant mutants were ultimately selected which produced nearly threefold more antibiotic than the sensitive parent culture and, in addition, D-tryptophan was no longer necessary for pyrrolnitrin formation (Fig. 3).

Nüesch *et al.* (1976) utilized analogs of methionine in studies on the relationship between methionine and cephalosporin C synthesis in strains of *C. acremonium*. They isolated a series of mutants both sensitive and resistant to selenomethionine and found reduction in methionine uptake which correlated with reduction in cephalosporin C synthesis and excretion of 2-hydroxy-4-methiolbutyric acid, a deamination product of methionine.

The use of auxotrophic mutants which require one or more amino acids in the synthesis of a desired amino acid and the additional induction of resistance to an analog of the desired metabolite often results in increased overproduction levels of amino acids. As shown in Fig. 2 for the production of L-tyrosine by mutants of *C. glutamicum*, first-stage screening was initiated by isolating phenylalanine-requiring mutants (*Phe*$^-$).

The *Phe*⁻ mutant was then further mutagenized and selected for resistance to a series of analogs, including 3-aminotyrosine (3AT), *p*-aminophenylalanine (PAP), *p*-fluorophenylalanine (PFP), and tyrosinehydroxamate (TyrHx). A mutant strain was eventually isolated (Pr-20) which produced 17.6 gm/liter of L-tyrosine, a sixfold increase when compared to the KY10233 (Phe^-) strain (Arima, 1977). The stepwise increases in productivity of L-phenylalanine and L-tryptophan are also summarized in Fig. 2.

C. Reversion Mutants

1. Reversion of Nonproducing Strains

To increase the efficiency of selection methods utilizing tetracycline-producing strains of *Streptomyces rimosus* and *Streptomyces viridifaciens,* procedures were developed based on the selection of high-producing variants from various types of mutant populations. The greatest number of these variants were obtained from revertants of non-producing mutants, i.e., mutants in which a previous mutagenic treatment had induced lesions in loci controlling tetracycline biosynthesis. Reversion mutants which cannot produce tetracycline can be detected easily by plating the nonproducing culture after mutagen treatment, followed by an overlay assay using a sensitive assay organism. Revertants will produce clear zones, whereas nonproducers will show no zones of inhibition. Utilizing this technique, Dulaney and Dulaney (1967) obtained many revertants of a nonproducer parent which produced more than twice as much chlortetracycline as the original grandparent culture. One revertant strain showed more than a sixfold increase over its grandparent strain.

2. Reversion of Auxotrophic Mutants

Overproduction of precursor molecules is another potential means to increase antibiotic formation. This overproduction of the primary metabolite precursor can be achieved by relieving feedback regulation of its biosynthetic pathway. Modification of the structure of a feedback-sensitive enzyme through auxotrophic mutation followed by enzyme replacement via a reversion mutation is a common method for relieving such regulation. Dulaney and Dulaney (1967) used this technique to increase chlortetracycline production in a low-producing strain of *S. viridifaciens*. Reverse mutation of a homocysteine auxotroph resulted in 88% of the revertants producing threefold more tetracycline than the original prototrophic culture (Fig. 4).

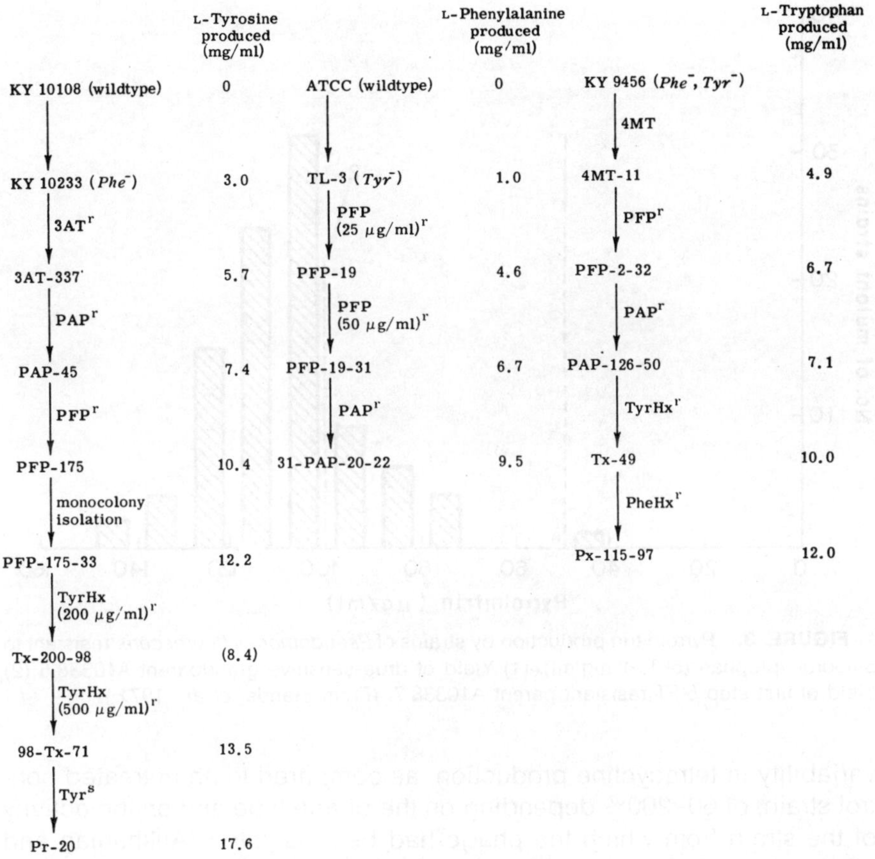

FIGURE 2. Overproduction of L-tyrosine, L-phenylalanine, and L-tryptophan by mutants of *Corynebacterium glutamicum*. 3AT = 3-aminotyrosine, PAP = *p*-aminophenylalanine, PFP = *p*-fluorophenylalanine, PyrHx = tyrosinehydroxamate, 4MT = 4-methyltryptophan, PheHx = phenylalaninehydroxamate. (After Arima, 1977.)

D. Phage-Resistant Mutants

The use of phage as a mutagenic agent has often been claimed, although evidence for discrete mutagenesis based on a well-designed model has not been reported. However, microbial populations surviving phagolysis show extreme variability in both morphology and productivity. Viruses are known to induce chromosomal aberrations and breakage in mammalian cells and to induce point mutations in *Drosophila*. Mutagenic effects of actinophage have been demonstrated in production strains of *S. aureofaciens*. Phage-resistant variants showed increased

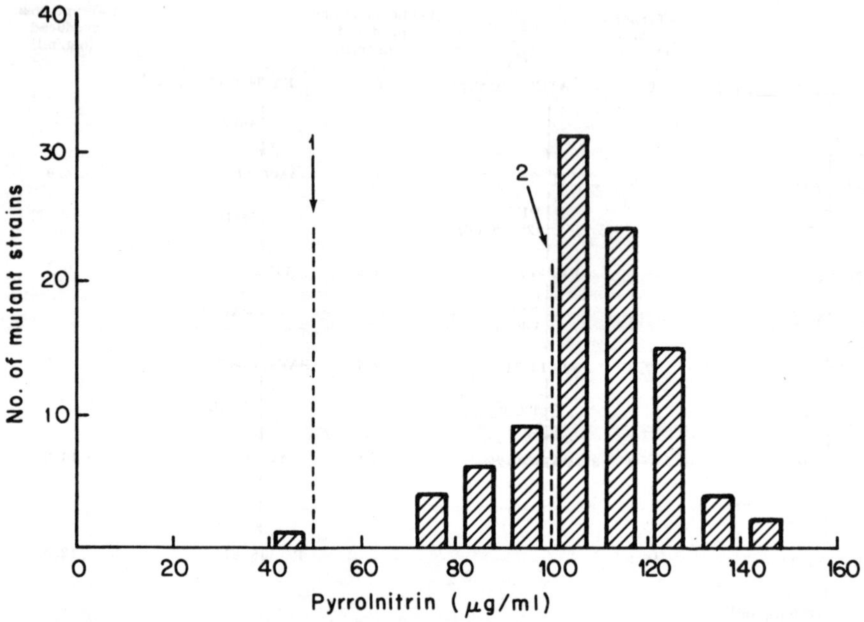

FIGURE 3. Pyrrolnitrin production by strains of *Pseudomonas fluorescens* resistant to 5-fluorotryptophan (5FT, 1 mg/ml). (1) Yield of drug-sensitive grandparent A10338.5 (2) Yield of first step 5-FT resistant parent A10338.7. (From Elander *et al.*, 1971.)

variability in tetracycline production, as compared to an untreated control strain, of 60–200% depending on the phage type and on the activity of the strain from which the phage had been isolated (Alikhanian and Iljina, 1958). Perlman and Ogawa (1976) reported that streptomycin-negative strains of *S. griseus* resistant to phage isolated from producing strains synthesized streptomycin, which can be interpreted as an example of transduction of antibiotic-producing capacity. Increase in antibiotic titer could well be a pleiotropic effect of resistance mutations selected by phage action. The selection of phage-resistant populations in screening for gain variants has proved to be of utility in a number of actinomycete-based antibiotic programs.

It should be reiterated that phage-resistant strains are inevitably subject to the possibility of infection by new contaminating phages and by host range mutants. Even though the use of resistant strains may contribute to stabilization of the fermentation process, it is important to eradicate all sources of contaminating phages from the fermentation area. At present, the most effective method aside from rigorous tank sterility procedures is to keep in the fermentation culture repository a variety of resistant strains with differing sensitivities to each of the phage types

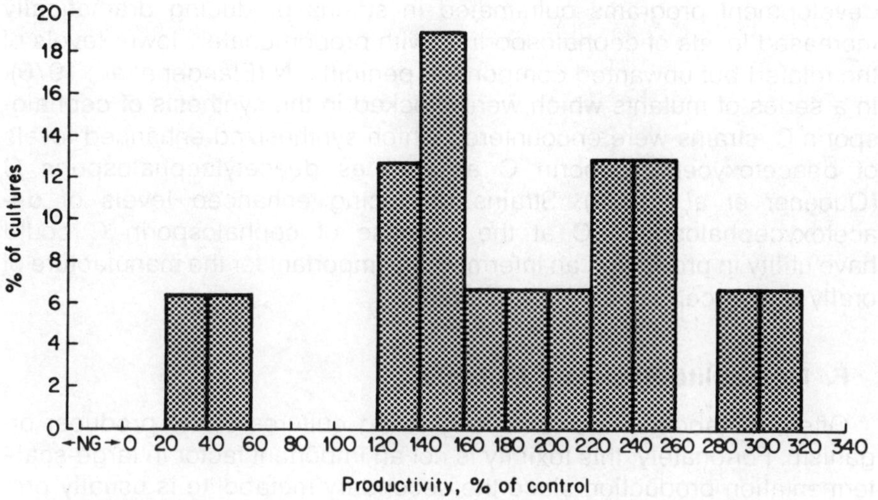

FIGURE 4. Spread in tetracycline productivity of prototrophs recovered from a homocysteine-requiring mutant of *Streptomyces viridifaciens*. (From Dulaney and Dulaney, 1967.)

detected in the fermentor. The strains can then be utilized alternately during periods of repeated contamination.

E. Mutation to Nonproduction of Undesirable Metabolites and the Use of Blocked Mutants

Biosynthesis of antibiotic mixtures containing a variety of related or unrelated antibiotics is commonly encountered with producer strains. The problem is to produce the desired antibiotic component exclusively or to select variants which selectively produce more of the desired component and less of the others. Mutation and medium manipulation with the nebramycin-producing organism *Streptomyces tenebrarius*, a culture producing a number of nebramycin components, yielded a variety of strains, some of which elaborated a single component in fermentations (Stark *et al.*, 1971). A methionine auxotroph of *Streptomyces erytheus*, the producing culture of erythromycin, when fed varying concentrations of the required amino acid, synthesized increased levels of erythromycin C relative to the erythromycin A component (Ostrowska-Krysiak, 1971). *Streptomyces noursei* produces cycloheximide and the unrelated metabolite, nystatin. Mutants were selected which no longer synthesized cycloheximide and which concomitantly produced markedly increased levels of nystatin (Spizek *et al.*, 1965). Strains of *C. acremonium* produce a variety of different antibiotic moieties. Intensive development programs culminated in strains producing dramatically increased levels of cephalosporin C with proportionately lower levels of the related but unwanted component, penicillin N (Elander *et al.*, 1976). In a series of mutants which were blocked in the synthesis of cephalosporin C, strains were encountered which synthesized enhanced levels of deacetoxycephalosporin C as well as deacetylcephalosporin C (Queener *et al.*, 1974). Strains producing enhanced levels of deacetoxycephalosporin C at the expense of cephalosporin C could have utility in producing an intermediate important for the manufacture of orally active cephalosporin antibiotics.

F. Metabolite-Resistant Mutants

Often, metabolites are toxic to growing cultures of the producer organism. Fortunately, this toxicity is not an important factor in large-scale fermentation production since the secondary metabolite is usually produced in idiophase after the growth phase of fermentation is essentially completed. Antibiotics which inhibit the growth of the producer organism include tetracycline, novobiocin, actinomycin, streptomycin, and nystatin. Differences in biosynthetic activity may be affected by genetically determined resistance to the producing culture's own antibiotic. Therefore, increasing resistance either by adaptation to gradually increasing concentrations of antibiotics or by subjecting mutagen-treated populations to increasing gradients of antibiotic is a useful procedure for improving the biosynthetic capacity of production strains. Katagiri (1954) obtained a fourfold increase in chlortetracycline (CTC) productivity in *S. aureofaciens* by repeated transfers in media containing 200–400 μg CTC/ml. The sensitivity of the surviving strains to increasing concentrations of chlortetracycline was in direct proportion to their capacity to synthesize antibiotic. Whereas low concentrations of tetracycline led to increased frequency of low-producing variants, exposure to high concentrations resulted in an increased number of active mutants. These results can be explained by the lethal effect of the antibiotic which selectively deprived the natural population of low-producing sensitive clones. This method is most effective during the early phase of strain improvement or in cases when a population of a high-producing strain must be repurified.

G. Mutations Affecting Permeability

The production of large amounts of glutamic acid in bacteria can be attained by selecting mutants having an altered cell membrane with increased outward permeability of the amino acid. The increased permeability can be attained by biotin deficiency or addition of penicillin or fatty acid derivatives to wildtype strains of *C. glutamicum,* or by oleic acid limitation of oleic acid-requiring auxotrophs or by glycerol limitation of glycerol-requiring mutants (Nakayama, 1976). A change in membrane chemistry correlates with increased permeability of cells and with excretion of high levels of glutamic acid (Arima, 1977).

Permeability may also be important in the production of L-lysine. Mutants of *Escherichia coli* resistant to *S*-(β-aminoethyl)cysteine excrete lysine and are defective in the lysine transport system (Halsall, 1975). The overproduction of lysine was not due to mutations affecting either of two regulated enzymes, aspartokinase and dihydrodipicolinic acid synthetase, but rather to active transport systems which enable an efflux of lysine against a fivefold higher concentration of extracellular lysine compared to the intracellular concentration (Ibuki and Takinami, 1975).

H. Miscellaneous Classes of Mutants

The metabolic poison monofluoracetate (MFA) is known to inhibit aconitate hydratase, an enzyme which converts citrate to isocitrate in strains of *Candida lipolytica.* Mutants sensitive to MFA produced less aconitate hydratase and, when used to produce citric acid on *n*-paraffin, synthesized citric acid and isocitric acid at a ratio of 97 : 3 compared to the usual ratio of 60 : 40 (Akiyama *et al.*, 1972).

Selenomethionine-sensitive strains of *C. acremonium* have increased capacity to synthesize cephalosporin C due to enhanced methionine uptake (Nüesch *et al.*, 1976). Okanishi and Gregory (1970) using mutant strains of *Candida tropicalis* obtained cells whose protein content had an increased methionine content by the ingenious selection of small colonies on a sulfur-deficient medium. They postulated that methionine-rich mutants would have a greater sulfate requirement than parental cells and would, therefore, form smaller colonies on a medium in which sulfate was limiting.

The production of glucose isomerase (GI) in strains of *Streptomyces,* used commercially for the conversion of glucose to fructose, was repressed by glucose and required the presence of D-xylose as an inducer. Mutants resistant to D-lyxose, an analog of glucose, were found to produce GI constitutively without xylose but were still repressed by glucose. Additional selection for resistance to a glucose analog, 3-*o*-methylglucose, resulted in the isolation of a unique mutant which produced GI in the presence of glucose, thereby being insensitive to catabolite repression (Sanchez and Quinto, 1975).

Manipulation of the parasexual cycle in filamentous fungi can be used to select for high-yielding diploid or polyploid strains. Heterozygous diploid strains of *A. niger* have been reported by Ikeda *et al.* (1957) to be superior producers of citric acid and kojic acid, and Ilczuk (1971) reported that other higher ploidy strains of *A. oryzae* and *A. sojae* were superior amylase and protease producers. Pathak and Elander (1971) reported that diploid strains of *P. chrysogenum* were heterotic with respect to β-galactosidase, glucose oxidase, and alkaline protease.

I. Blocked Mutants and Product Cosynthesis

Cosynthesis depends upon the use of two blocked mutants which, when propagated together, are able to synthesize the end product. Intermediates which accumulate in the blocked mutants diffuse out of the cell and are used by mutants blocked earlier in the pathway. The phenomenon has been reported for *S. aureofaciens* (Delic *et al.*, 1969) and *S. rimosus* (Alikhanian *et al.*, 1961), which synthesize tetracyclines. Kahler and Noack (1974) have recently reported cosynthesis of turimycin in strains of *S. hygroscopicus*. Delic *et al.* (1969) demonstrated cosynthesis between *S. rimosus* and *S. aureofaciens* and arranged the mutants into complementation classes. McCormick (1967) and Mitscher (1968) utilized cosynthesis in the elucidation of the biosynthetic pathway of tetracyclines.

Cosynthesis has the potential of detecting new antibiotics, as demonstrated in strains producing tetracyclines. Growing a mutant of *S. aureofaciens* blocked in the synthesis of demethyltetracycline together with a mutant of *S. aureofaciens* blocked in the synthesis of chlortetracycline resulted in the synthesis of an antibiotic not produced by *S. aureofaciens:* tetracycline (Mitscher, 1968). Čižmek (1976) cultivated two blocked mutants of *S. psammoticus* together and detected a compound active against a tetracycline-resistant strain of *Bacillus subtilis*. The uncharacterized compound was reported to be an unknown product of tetracycline biosynthesis. Cosynthesis between colonies of two mutants separated by a distance of 5 mm on agar eliminates genetic exchange as a mechanism for the phenomenon. A unique technique of sandwiching filter membranes between two layers of agar, each inoculated with one of two blocked mutants, indicates the diffusability of the compound.

J. Localized Mutagenesis and Comutation

Localized mutagenesis in small, selected regions of the chromosome offers a promising new approach for the industrial microbial geneticist. Mutation programs can be directed for maximizing mutations in any marked area on the chromosome, especially areas known to affect the formation of end products. *N*-Methyl-*N'*-nitro-*N*-nitrosoguanidine (NG) induces multiple mutations on the bacterial chromosome which tend to occur in clusters near the replication points. Localized mutagenesis, achieved by treating synchronized bacterial cultures with NG, was accomplished in *E. coli* (Hohfeld and Vielmutter, 1973) and extended to the genus *Streptomyces* by Randazzo *et al.* (1973), Godfrey (1974), and Matselyukh and Mukvic (1973).

The comutation procedure, discoveıed by Guerola *et al.* (1971) in *E. coli,* depends upon NG-induced mutations occurring in clusters regardless of culture synchronization. A specific mutation is selected, such as reversion of an auxotroph, and the revertants are scored for a second mutation. A high frequency of nonselected mutants is usually found. The nonselected mutations are referred to as *comutations* and regularly occur in genes closely linked to the selected mutation.

The comutation technique has been effectively applied to the isolation of temperature-sensitive mutants by Oeschger and Berlyn (1974). Randazzo *et al.* (1973) used comutation to select a large number of nutritional mutants within the histidine operon of *Streptomyces coelicolor.* In a study of four mutations in *S. coelicolor* (*cys* A1, *met* A2, *his* A1, and *arg* A1), Randazzo *et al.* (1976) found all comutations were due to mutations in genes linked to the revertant locus; therefore a comutation region exists and mutation outside the comutation region is restricted. Mutations induced by NG are localized in a short segment without effecting the residual genome. Isolation of comutants in unknown loci linked to the revertant site can be done by a heterokaryon method or by the use of temperature-sensitive mutants. Comutation also appears promising as a relatively simple tool for gene linkage and mapping studies.

K. Mutational Biosynthesis and New Metabolite Formation

Mutation of microorganisms producing secondary metabolites has resulted in the selection of strains capable of producing new modified metabolites either directly or in response to some added precursor analog. The modified metabolites usually possess structural features of the parent compound but often lack certain functional groups. Other compounds contain structurally modified functional groups which convey differing biological activities. This biogenetic approach to generate new secondary metabolites has been especially useful in the generation of new aminoglycoside antibiotics.

The phenomenon was first reported by Shier *et al.* (1969) utilizing a mutant strain of *Streptomyces fradiae,* the organism that synthesizes neomycin. Wildtype strains of *S. fradiae* normally synthesize neomycins A, B, and C, incorporating the biosynthesized diaminocyclitol subunit deoxystreptamine into all the neomycin components. They obtained a mutant strain derived from an NG-treated population which was unable to synthesize deoxystreptamine and, therefore, was dependent on an outside source of diaminocyclitols for antibiotic synthesis. When the mutant was cultured with added streptamine instead of deoxystreptamine, two new antibiotic substances were produced which were called hybrimycin A_1 and A_2. Epistreptamine, the C-2 epimer of streptamine, gave two more antibiotics, hybrimycin B_1 and B_2 (Fig. 5).

New biosynthetic analogs of butirosin, another aminoglycoside antibiotic, have been reported by Claridge *et al.* (1974), using mutant strains of *Bacillus circulans*. Other examples of new biosynthetic products produced by mutant strains are listed in Table II.

IV. GENETIC SYSTEMS IN ECONOMICALLY IMPORTANT MICROORGANISMS

A. Bacteria and Actinomycetes

Both eubacteria and actinomycetes are prokaryotic organisms; i.e., the nucleus is a single, naked chromosome devoid of a nuclear membrane. Bacteria have been important tools for research in molecular biology which has provided significant fundamental information for the

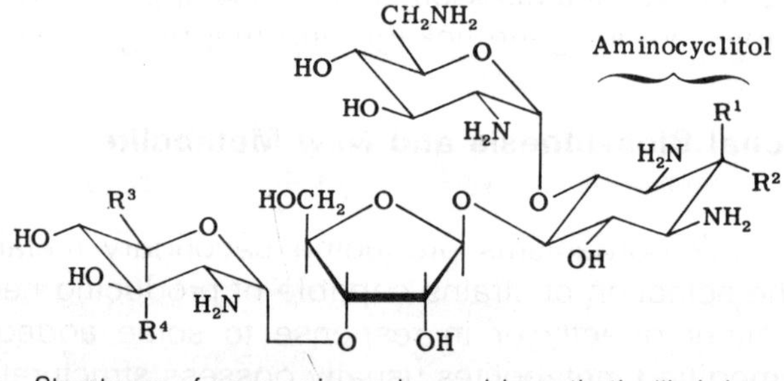

FIGURE 5. Structures of neomycin and new biosynthetic "hybrimycin" antibiotics. (From Shier *et al.*, 1969.)

Antibiotic	Aminocyclitol	R^1	R^2	R^3	$R4^4$
Neomycin B	Deoxystreptamine	H	H	H	CH_2NH_2
Neomycin C	Deoxystreptamine	H	H	CH_2NH_2	H
Hybrimycin A1	Streptamine	H	OH	H	CH_2NH_2
Hybrimycin A2	Streptamine	H	OH	CH_2NH_2	H
Hybrimycin B1	Epistreptamine	OH	H	H	CH_2NH_2
Hybrimycin B2	Epistreptamine	OH	H	CH_2NH_2	H

TABLE II. Mutational Biosynthesis and New Product (Idiolite) Formation

Organism	Original product	New product	Reference
Pseudomonas aureofaciens	Pyrrolnitrin	4′-Fluoropyrrolnitrin	Gorman *et al.* (1968)
Streptomyces peucetius	Daunomycin	Adriamycin	Arcamone *et al.* (1969)
Nocardia mediterranei	Rifamycin B	Rifamycin W	White *et al.* (1974)
Micromonospora inyoensis	Sisomicin	Mutamicins	Testa *et al.* (1974)
Streptomyces griseus	Streptomycin	Streptomutin	Nagaoka and Demain (1975)
Cephalosporium acremonium (ATTC 20389)	Penicillin N	6-(D)-[(2-amino-2-carboxy)-ethylthio]-acetamido-penicillanic acid	Troonen *et al.* (1976)
Micromonospora purpurea	Gentamicin	Hydroxy- and deoxy gentamicins	Rosi *et al.* (1977)

fermentation industry during the past two decades. Among bacteria of economic importance are strains of the genus *Bacillus* (producers of antibiotics, insecticides, and enzymes), *Corynebacterium* (amino acid production), and *Pseudomonas* (hydrocarbon degradation and antibiotic production). Bacteria generally do not exhibit an alternation of haploid and diploid phases in their life cycles and were considered to be asexual organisms until Lederberg and Tatum (1946) demonstrated the existence of genetic recombination. Other mechanisms for gene transfer have since been discovered and are discussed briefly in the following sections. For a more detailed account of bacterial genetic systems, the reader is referred to the excellent text by Hayes (1968).

1. Conjugation

Lederberg and Tatum (1946) showed that when two strains of *E. coli* K-12 carrying different genetic markers were mixed together, some recombinants were formed. It was later found that not all *E. coli* strains could mate to give recombinants. On the basis of ability to transfer genetic material, *E. coli* K-12 can be classified into two sexes, male (♂ or F^+) and (♀ or F^-) (see Fig. 6). The character of "maleness" is conferred by a sex factor called F (for fertility) which is composed of DNA and determines the capability to form a conjugal union with and to transfer genetic material to ♀ bacteria.

The F factor exists in ♂ bacteria in one of two mutually exclusive states. In one state (F^+), the sex factor is unassociated with the bacterial chromosome and replicates independently of it. In the other state (Hfr), the sex factor is inserted into the chromosome to form a single unit of

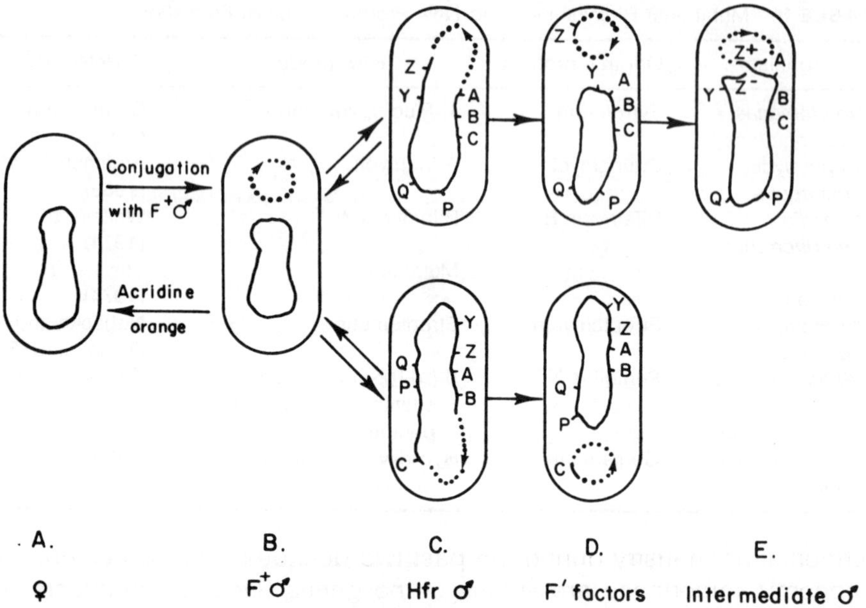

FIGURE 6. Properties and interrelationships of F^+, *Hfr* and F′ donors. The interrupted lines represent the bacterial chromosome, and the dotted lines the sex factor; the arrow head indicates the polarity of transfer. The letters A–Z show the location of various bacterial genes. (From Clowes and Hayes, 1968.)

replication. When F^+ bacteria come in contact with F^- cells, the sex factor is transferred with high efficiency to females, converting them to the F^+ state. However, chromosome transfer is insignificant. Thus, F factor is clearly an extrachromosomal element (plasmid). On the other hand, the Hfr strain can transfer chromosomal genes at high frequency, depending on their location on the chromosome. During conjugation, the circular bacterial chromosome opens up at a specific point (origin) and enters the F^- cells at a constant rate in the same sequence as the arrangement on the chromosome. In this fashion, a gene can be mapped according to the time required to transfer into F^- bacteria during conjugation.

Some Hfr cells tend to revert to the F^+ state under normal growth conditions, presumably due to reversal of the act of recombination which led to insertion of F factor into the chromosome. When this occurs a piece of the bacterial chromosome is often incorporated into the sex factor. Such modified sex factors are termed F-prime (F′) factors. A widely used F′ factor is one which carried a dominant allele of a gene, such as *lac*⁺ (F′ *lac*⁺). When the F′ factor is introduced into a bacterial host which carried the recessive allele of the gene, a complementing heterogenote (F′ lac^+/lac^-) is formed. The F′ factor is very useful for monitoring the presence of F factor in bacterial cells and for studying the dominant and recessive relationships of genes. Figure 6 illustrates the various types of sex in bacteria and their origin. As shown, the F factor, like many other plasmids, can be cured by acridine orange, ethidium bromide, etc.

The F factor from *E. coli* K-12 can also be transmitted to other bacteria including *Shigella, Salmonella,* and *Serratia. Salmonella* strains which acquire F factor from *E. coli* can now transfer chromosomal markers to both *Salmonella* and *E. coli* recipients. Genetic recombination via conjugation has also been reported in *Pseudomonas aeruginosa, Serratia marcescens,* and *Vibrio cholerae*.

2. Transduction: Phage-Mediated Gene Transfer

The phages which are capable of transduction are temperate, meaning they can establish themselves in a host bacterium without killing the cell. A bacterium harboring a temperate phage is called lysogenic. The phage in a lysogenic strain can be induced to multiply and lyse (kill) cells by a variety of treatments of which the most convenient is uv irradiation. Bacteria infected with a temperate phage may either lyse at once or become lysogenic depending on conditions. Figure 7 illustrates lysis and lysogenization in enteric bacteria.

In generalized (or unrestricted) transduction, a phage-sensitive bacterial strain is infected with transducing phage under conditions favoring lysis. The phage lysate is then freed from the bacterial cells and used to infect a recipient strain under conditions favoring lysogenization. The transductants are detected by plating on a selective medium.

In specialized transduction, a temperate phage called lambda (λ) has the special property of transducing the *gal* gene exclusively. The reason λ has this special transducing ability is because it actually integrates with the bacterial chromosome at a site close to the *gal* locus. Upon lysogenic induction, the λ is excised from the bacterial chromosome along with adjacent bacterial genes.

3. Transformation: DNA-Mediated Gene Transfer

Transformation is the process by which the genetic traits of bacterial strain A (donor) are transferred to strain B (recipient) by incubating the recipient cells with DNA from a donor strain (Fig. 8). The physiological characteristic of cells which can take up DNA and be transformed is called competence. Although transformation is not a common phenomenon, the demonstration of transformation in most bacterial species is probably a matter of discovering the proper conditions for competency. Genera in which species have been reported to undergo transformation

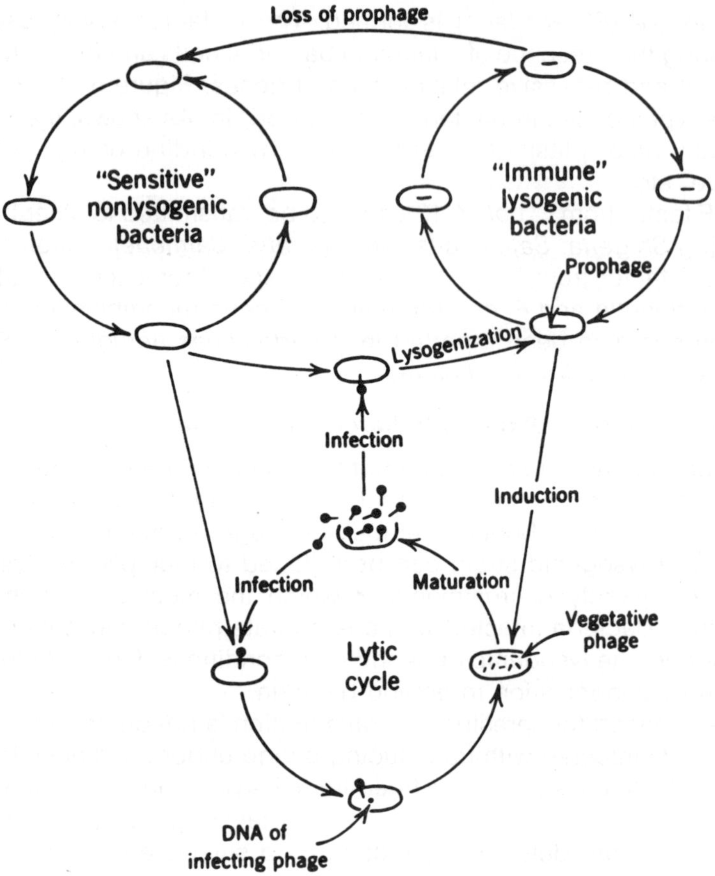

FIGURE 7. Lysis and lysogenization in enteric bacteria. (From Wagner and Mitchell, 1964.)

include *Bacillus, Hemophilus, Pneumococcus, Neisseria,* and *Rhizobium.* In *E. coli* the failure to demonstrate genetic transformation until recently appears to be due to the presence of hydrolytic nucleases which rapidly destroy incoming DNA. It has now been shown that when suitable nuclease-deficient mutant strains of *E. coli* are used as recipients, transformation can be readily demonstrated. Also, recently it has been possible to transform antibiotic-sensitive *E. coli* strains into antibiotic-resistant clones by incubating the recipient cells with antibiotic-resistant plasmid DNA under special conditions. (See section VIII on molecular cloning.)

Streptomycetes are distinguished from eubacteria by the formation of filamentous cells and aerial mycelia. Genetically, this implies that a longer time is required in streptomycetes to establish genetic homogeneity following gene transfer. As in *E. coli, Streptomyces*

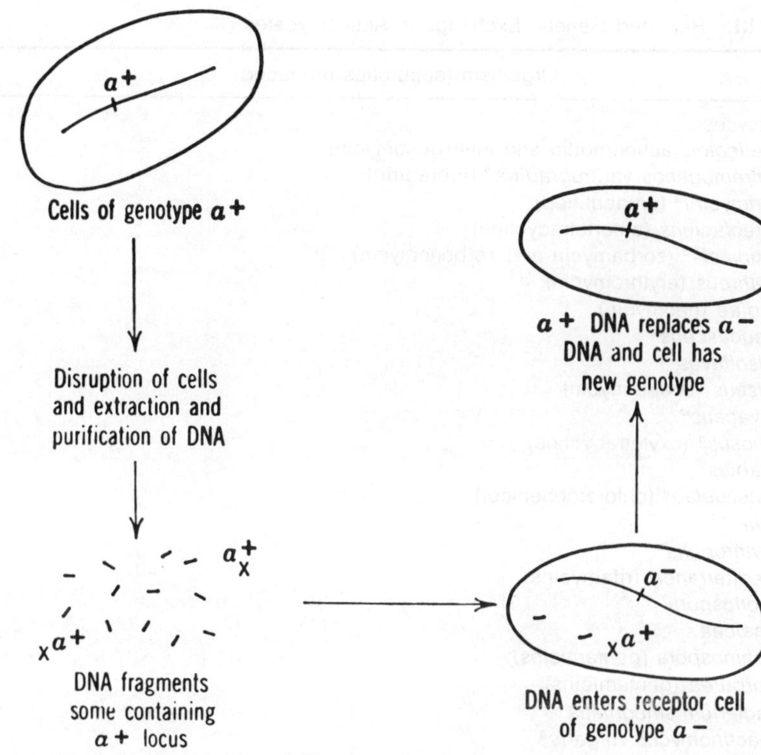

FIGURE 8. Generalized process of transformation in bacteria. (From Wagner and Mitchell, 1964.)

achieves its gene transfer mainly via conjugation. Transformation has been claimed in some streptomycetes but it has not been confirmed. It does occur, however, in a related species, *Thermoactinomycetes vulgaris* (Hopwood and Wright, 1972).

Hopwood (1973) suggested that the nontransformability of *Streptomyces* may be due to the excessive quantity of extracellular DNAase which readily degrades the donor DNA. Phages infecting *Streptomyces* (called actinophages) and transduction in *Streptomyces* have been reported. However, a useful transduction system in *Streptomyces* has not been demonstrated unequivocably.

Table III shows a list of actinomycetes (including *Streptomyces*) in which gene exchange via conjugation has been demonstrated. A detailed review on the genetics of *Streptomyces* has been published by Hopwood *et al.* (1973). In *S. coelicolor,* which is the most thoroughly studied *Streptomyces* species, fertility is mediated by a plasmid *scp*-1 [equivalent to F factor in *E. coli* K-12 (Hopwood *et al.*, 1973)]. As shown in Fig. 9, *scp-1*$^+$ strains (a) bearing the autonomous plasmid scp 1 can

TABLE III. Reported Genetic Exchange in Actinomycetes

Organism (antibiotics produced)
Streptomyces
S. coelicolor (actinorhodin and methylenomycin)
S. achromogenes var. *rubradiris*[a] (rubradirin)
S. acrimycini[a] (unidentified)
S. aureofaciens (chlortetracycline)
S. bikiniesis[a] (zorbamycin and zorbonomycin)
S. erythreus (erythromycin)
S. fradiae (neomycin)
S. glaucescens[a]
S. grisoflavus
S. griseus (streptomycin)
S. olivaceus[a]
S. rimosus[a] (oxytetracycline)
S. scabies
S. venezuelae[a] (chloramphenicol)
Nocardia
N. erythropolis[a]
N. mediterranei[a] (rifamycins)
Micromonospora
M. chalcea
M. echinospora (gentamicins)
M. purpurea (gentamicins)
Mycobacterium smegmatis
Thermoactinomyces vulgaris[b]

[a] A circular map with close resemblance to *S. coelicolor*.
[b] *Thermoactinomyces vulgaris* has a transformation system; in other cases a "conjugation" mechanism of recombination is implicated. (After Hopwood, 1976.)

transfer random chromosomal markers to scp 1^- (b) at low frequency. Strains in which scp 1 is integrated into the chromosome (c) can transfer the markers adjacent to the site of plasmid integration at very high frequency when mated with scp 1^- (equivalent to Hfr in *E. coli*). Scp 1′ strains are those in which a piece of chromosome has been inserted into the autonomously replicating plasmid (d) (equivalent of F′ in *E. coli*). Such strains can transfer the chromosomal markers as well as scp 1 plasmid to scp 1^- strains at high frequency.

Linkage maps for several species of *Streptomyces* have been shown to be circular (Hopwood *et al.*, 1973). All the maps of *Streptomyces* show an approximate equivalent location for most loci, as exemplified by the genetic map of *S. coelicolor* and *S. rimosus* (Alacevic, 1976; see Fig. 10).

An important aspect of recombination in *Streptomyces* is interspecific or interstrain recombination, particularly between two strains producing

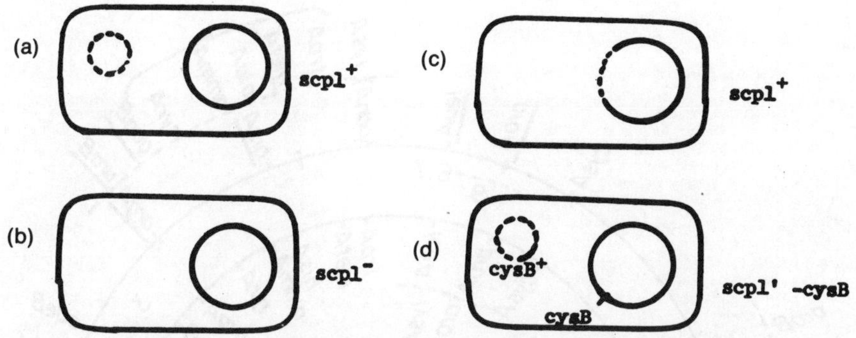

FIGURE 9. Various fertility types in *Streptomyces coelicolor*.

different antibiotics. Not only can interspecific crossing be used as a test for taxonomic relatedness between strains but it also could have great potential for breeding new strains for the synthesis of structurally modified antibiotics (Alacevic, 1963). Hopwood *et al.* (1977) have recently reported on genetic recombination involving protoplast fusion in streptomycetes. Their technique involves the use of polyethylene glycol (PEG)-induced protoplast fusion and regeneration with recombination frequencies so high that selectable markers were no longer necessary, thereby greatly facilitating the routine use of recombination in strain improvement programs.

B. Filamentous Fungi and Yeasts

In contrast to eubacteria and actinomycetes, the yeasts and molds possess true nuclei and, in some species, regular sexual cycles comparable to those existing in higher plants and animals. The major difference between the life cycles of higher organisms and those of fungi is the prolonged and generally predominant haploid phase in fungi. *Neurospora crassa* has been the fungal organism which has been thoroughly exploited by geneticists. The life cycle of *N. crassa* is shown in Fig. 11. The vegetative mycelium of *N. crassa* is haploid and can be propagated almost indefinitely by serial transfer of hyphal fragments or asexual spores (conidia). Two strains of opposite mating types (*A* or *a*) are required to initiate the sexual cycle. Fertilization is effected by mixing the conidia of *a* mating type with mycelia of *A* mating type growing on appropriate media. After a period of nuclear division and nuclear migration, fusion between *A* and *a* nuclei takes place. Each fused nucleus (diploid) immediately undergoes meiosis to form four haploid nuclei which, in turn, undergo two successive mitotic divisions to form eight ascospores. The eight ascospores are contained linearly in

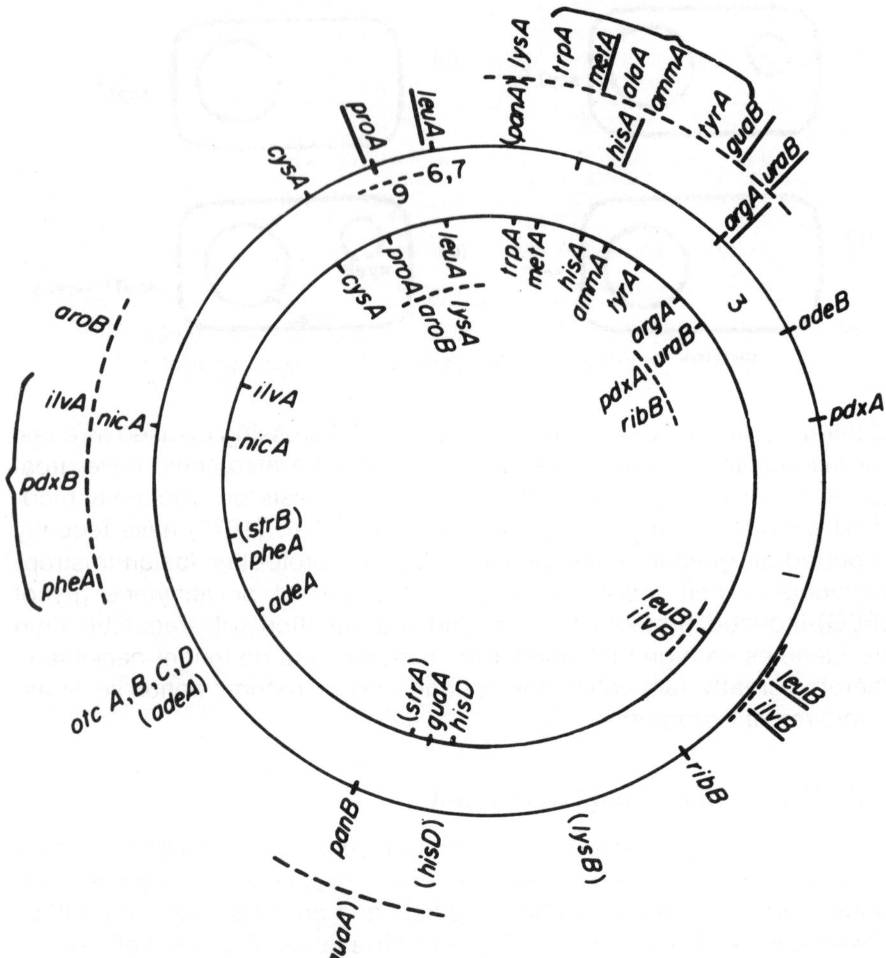

FIGURE 10. The map of *Streptomyces rimosus* (outer circle) compared to that of *S. coelicolor* with empty regions omitted (inner circle). (After Alacevic, 1976.)

a sac (ascus). Analysis of the linear segregation of each genetic marker as well as the recombination between markers in each ascus provide a unique way of mapping genes in *N. crassa*.

Few of the industrially important fungi, however, are endowed with such conventional sexual cycles as found in *N. crassa*. Fortunately, the filamentous fungi often form heterokaryons in which rare diploid nuclei result from the fusion of two haploid nuclei. This process is called parasexuality and is illustrated in Fig. 12. Although recombination (mitotic recombination) is much less frequent in the parasexual cycle compared

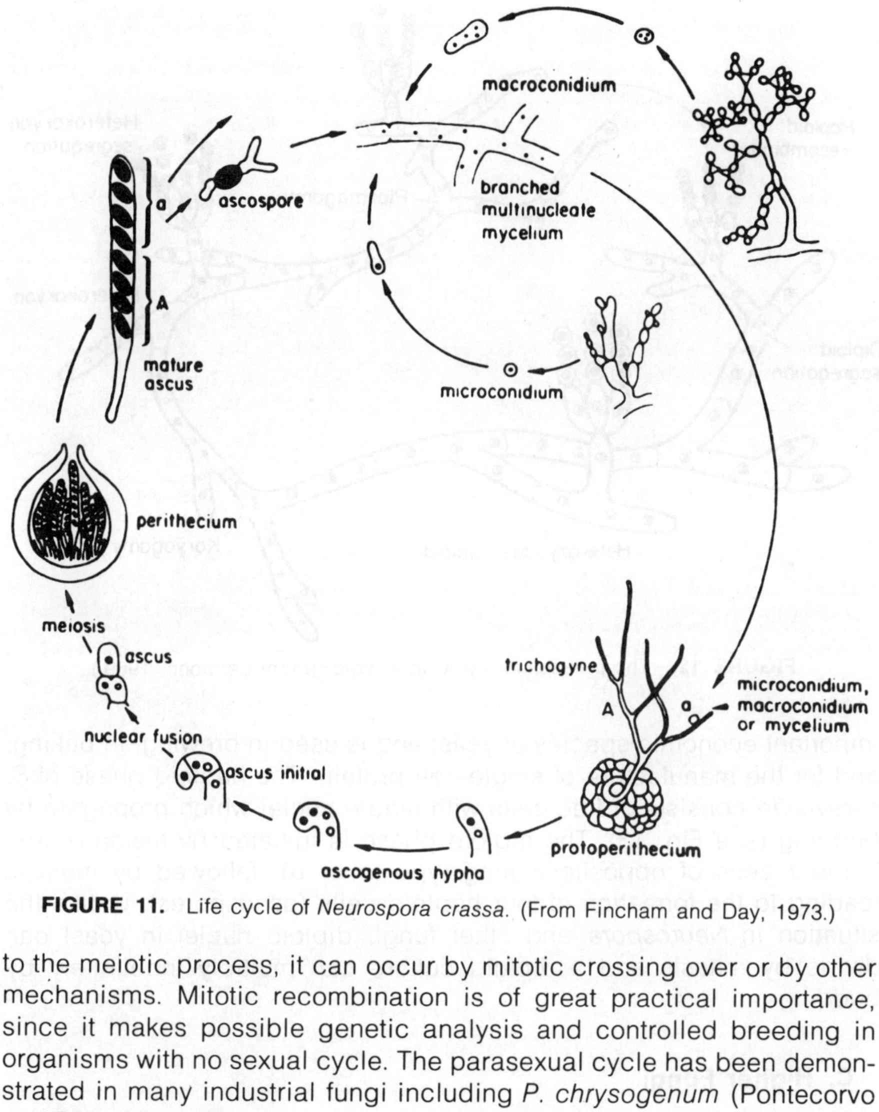

FIGURE 11. Life cycle of *Neurospora crassa*. (From Fincham and Day, 1973.)

to the meiotic process, it can occur by mitotic crossing over or by other mechanisms. Mitotic recombination is of great practical importance, since it makes possible genetic analysis and controlled breeding in organisms with no sexual cycle. The parasexual cycle has been demonstrated in many industrial fungi including *P. chrysogenum* (Pontecorvo and Sermonti, 1953), *C. acremonium* (Nüesch *et al.*, 1973), and *Aspergillus oryzae* (Ikeda *et al.*, 1957). A successful attempt to apply parasexual genetics to strain improvement in *P. chrysogenum* has been reported by Elander (1967). In this study, a homozygous diploid representing genome duplication of the haploid parent was an efficient producer of penicillin V.

Yeasts possess a similar life cycle to that of *N. crassa* except that they are morphologically more simple. *Saccharomyces cerevisiae* is the most

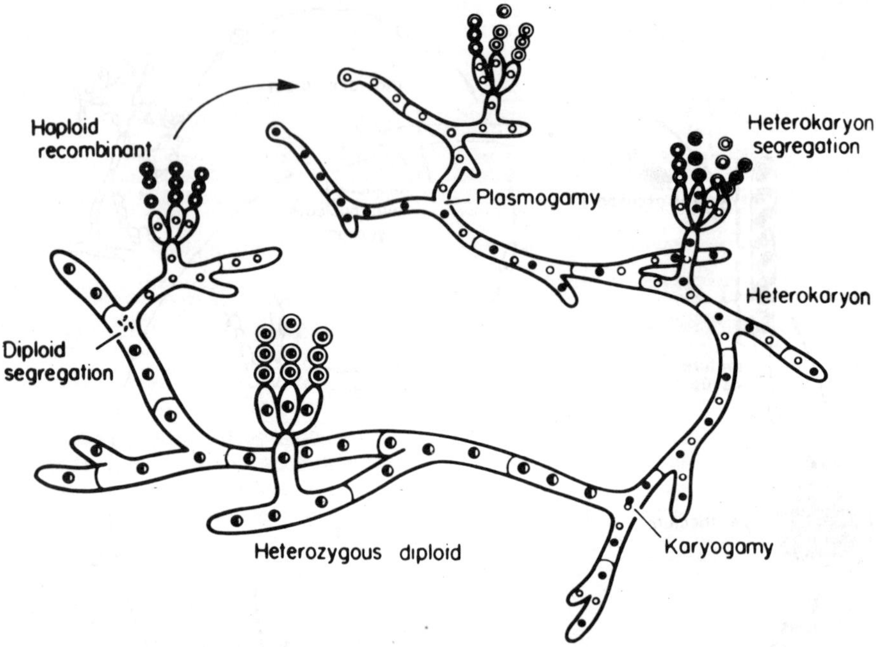

FIGURE 12. The parasexual cycle in a mold. (From Sermonti, 1969.)

important economic species of yeast and is used in brewing, in baking, and for the manufacture of single-cell protein. The haploid phase of *S. cerevisiae* consists of free cells with single nuclei which propagate by budding (see Fig. 13). The diploid phase is initiated by fusion of free haploid cells of opposite mating type (*a* or α), followed by meiosis leading to the formation of four haploid cells (ascospores). Unlike the situation in *Neurospora* and other fungi, diploid nuclei in yeast can divide by mitosis and a diploid culture can multiply indefinitely by budding.

C. Higher Fungi

The cultivated mushroom, *Agaricus bisporus,* is a member of the group of higher fungi called basidiomycetes. Many basidiomycetes have two distinct phases in their life cycle, a haploid homokaryotic phase and a dikaryotic phase in which each cell contains two nuclei of different mating types. During the dikaryotic phase, special cells called basidia are formed on which the products of meiosis (basidiospores) are born externally. The number of basidiospores on each basidium can be either two or four in *A. bisporus*. The life cycle of a representative

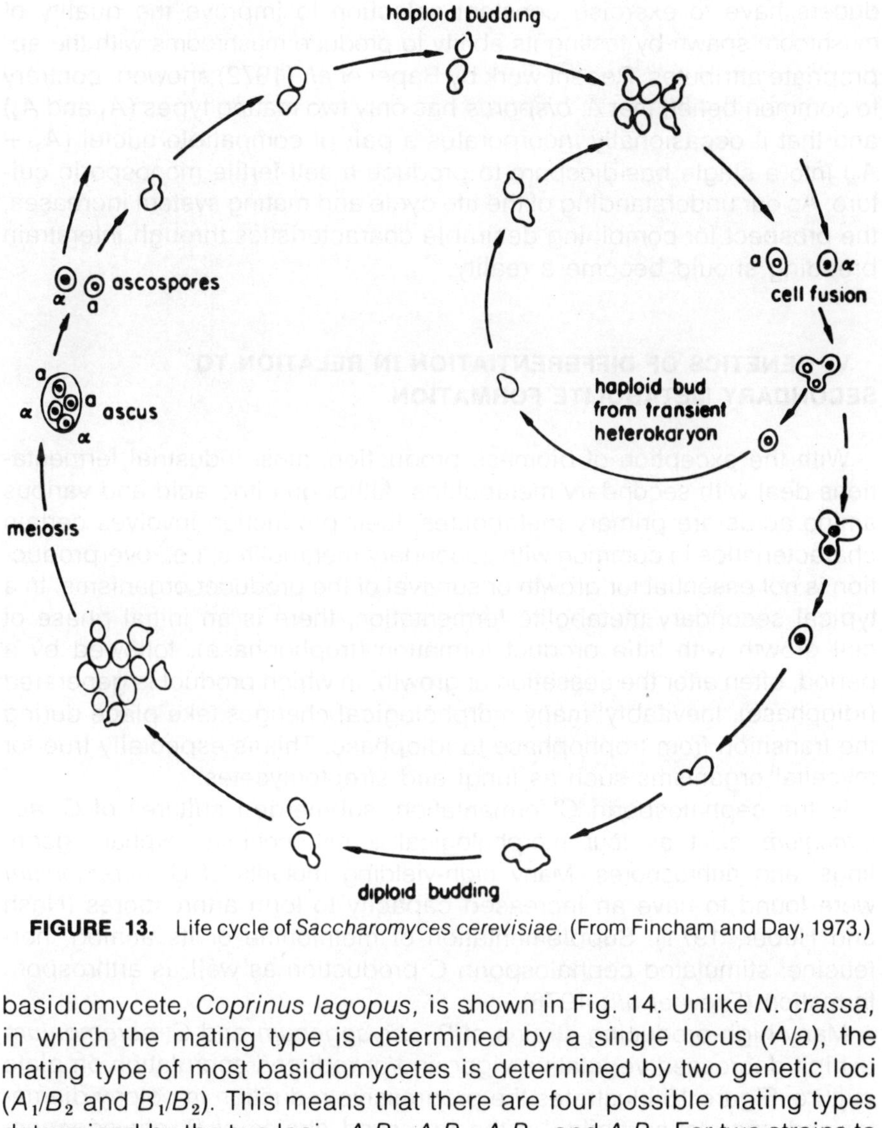

FIGURE 13. Life cycle of *Saccharomyces cerevisiae*. (From Fincham and Day, 1973.)

basidiomycete, *Coprinus lagopus*, is shown in Fig. 14. Unlike *N. crassa*, in which the mating type is determined by a single locus (*A*/*a*), the mating type of most basidiomycetes is determined by two genetic loci (A_1/B_2 and B_1/B_2). This means that there are four possible mating types determined by the two loci—A_1B_1, A_1B_2, A_2B_1, and A_2B_2. For two strains to be genetically compatible (to form dikaryons and subsequent fruiting), each of the two nuclei should possess different alleles on each locus, i.e., A_1B_1 versus A_2B_2, A_1B_2 versus A_2B_1.

In mushroom cultivation, the most serious problem has been the occurrence of spontaneous mutations leading to strain variability and instability. Thus, heterokaryotic cells resulting from spontaneous mutations appear to be a common phenomenon. Accordingly spawn producers have to exercise constant selection to improve the quality of mushroom spawn by testing its ability to produce mushrooms with the appropriate attributes. Recent work by Raper *et al.* (1972) showed, contrary to common belief, that *A. bisporus* has only two mating types (A_1 and A_2) and that it occasionally incorporates a pair of compatible nuclei (A_1 + A_2) into a single basidiospore to produce a self-fertile monosporic culture. As our understanding of the life cycle and mating system increases, the prospect for combining desirable characteristics through interstrain breeding should become a reality.

V. GENETICS OF DIFFERENTIATION IN RELATION TO SECONDARY METABOLITE FORMATION

With the exception of biomass production, most industrial fermentations deal with secondary metabolites. Although citric acid and various amino acids are primary metabolites, their production involves certain characteristics in common with secondary metabolites; i.e., overproduction is not essential for growth or survival of the producer organisms. In a typical secondary metabolite fermentation, there is an initial phase of cell growth with little product formation (trophophase), followed by a period, often after the cessation of growth, in which product is generated (idiophase). Inevitably, many morphological changes take place during the transition from trophophase to idiophase. This is especially true for mycelial organisms such as fungi and streptomycetes.

In the cephalosporin C fermentation, submerged cultures of *C. acremonium* exist as four morphological forms: conidia, hyphae, germlings, and arthrospores. Many high-yielding mutants of *C. acremonium* were found to have an increased capacity to form arthrospores (Nash and Huber, 1971). Supplementation of methionine or its analog, norleucine, stimulated cephalosporin C production as well as arthrospore formation (Drew *et al.*, 1976).

Many high-producing strains of *P. chrysogenum* and *C. acremonium* exhibit decreased vegetative vigor and/or reduced sporulation on plate culture. Figure 15 illustrates the progressive reduction in colony diameter and sporulation ability in the improved strains of *P. chrysogenum* within the Wisconsin family of strains (Sermonti, 1969). However, it is possible to derive a well-sporulating isolate from a poorly sporulating parent without sacrificing the productivity of the parent. For example, strain Wis 51-20 was a good producer of penicillin but showed poor growth and sporulating ability. After repeated mutations and reisolation, strains with superior sporulation capacity and high productivity were derived from strain Wis 51-20 (Backus and Stauffer, 1955). Similar re-

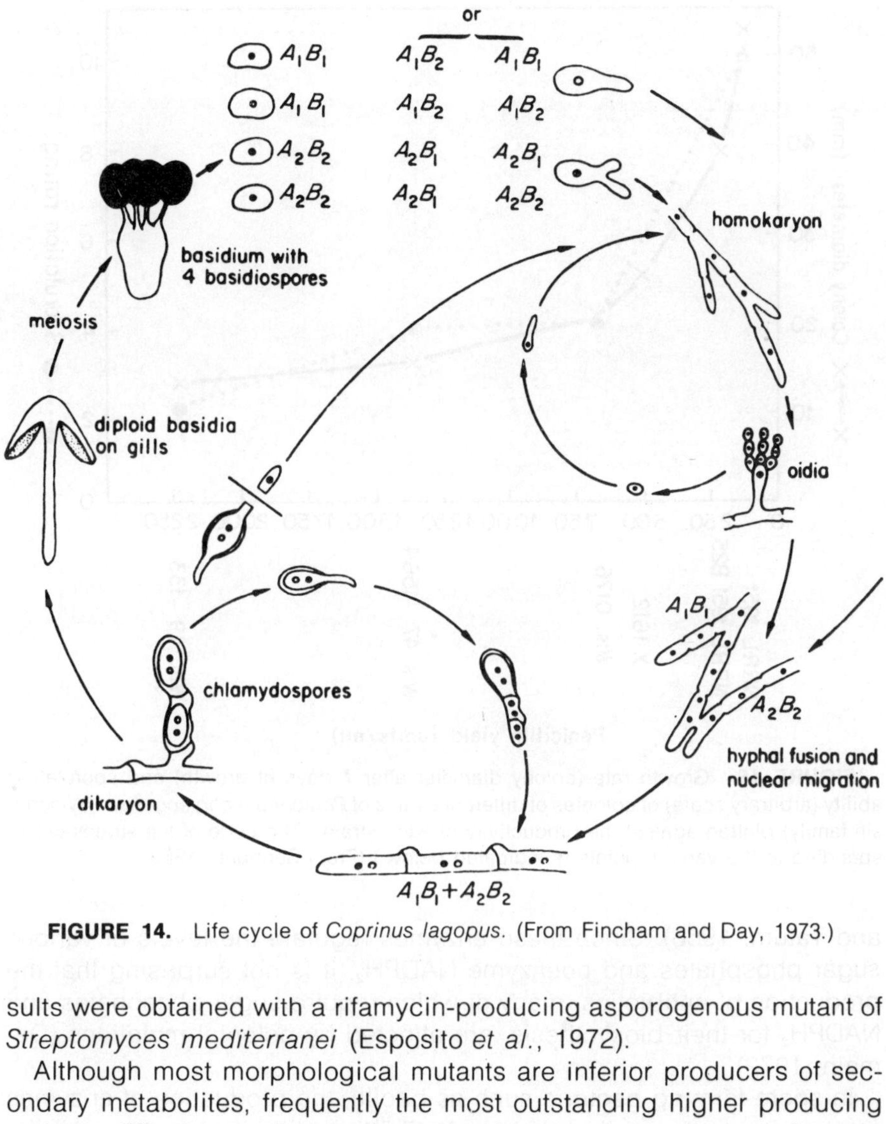

FIGURE 14. Life cycle of *Coprinus lagopus*. (From Fincham and Day, 1973.)

sults were obtained with a rifamycin-producing asporogenous mutant of *Streptomyces mediterranei* (Esposito *et al.*, 1972).

Although most morphological mutants are inferior producers of secondary metabolites, frequently the most outstanding higher producing mutants differ in morphology from their less productive parent. As seen from a penicillin strain improvement program (Backus and Stauffer, 1955), the types of morphological mutants which exhibited high productivity were those which gave restricted growth (colonial mutants) on plates (Fig. 10). In *Neurospora,* the biochemical lesion(s) for colonial mutations has been traced to altered enzymes in the pentose phosphate pathway, such as glucose-6-phosphate dehydrogenase, 6-phosphogluconate dehydrogenase, and phosphoglucomutase (Brody

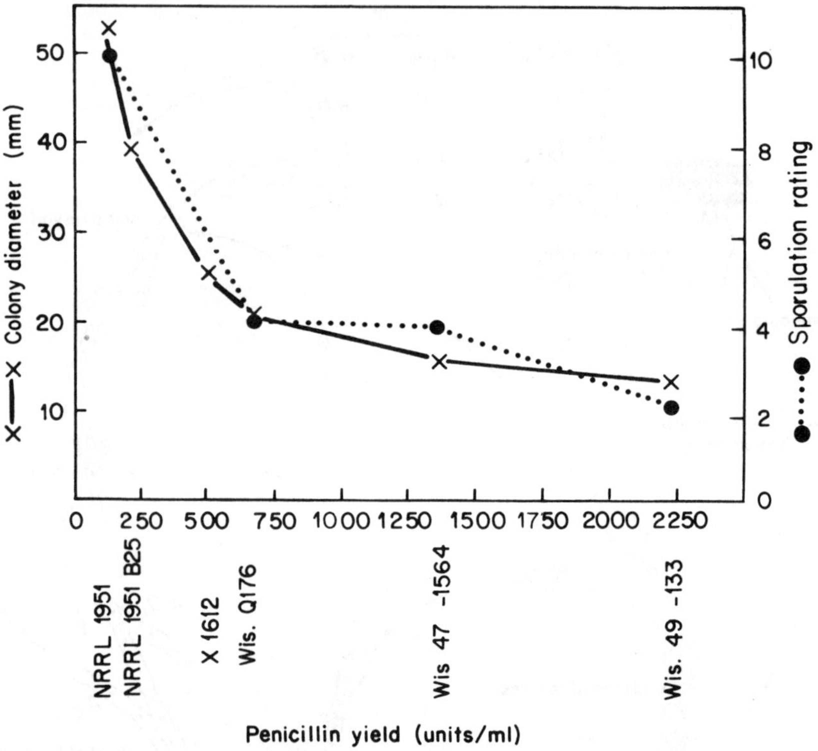

FIGURE 15. Growth rate (colony diameter after 7 days of growth) and sporulating ability (arbitrary scale) of colonies of different strains of *Penicillium chrysogenum* (Wisconsin family) plotted against the productivity of each strain. (The code of the strains corresponding to the various points is indicated below.) (From Sermonti, 1969.)

and Tatum, 1966). Since these enzymes regulate the levels of various sugar phosphates and coenzyme $NADPH_2$, it is not surprising that the production of antibiotics, many of which require sugar phosphates and $NADPH_2$ for their biosynthesis, are affected by colonial mutations (Demain, 1972).

In spore-forming bacteria such as bacilli, the production of enzymes (proteases and amylases) and antibiotics (bacitracin, gramicidin, polymyxin, circulin) are often associated with onset of sporulation (Schaeffer, 1969). In *Penicillium griseofulvum,* the conditions determining maximum sporulation are quite similar to those promoting griseofulvin production (Morton, 1961). Although mutations resulting in simultaneous loss in sporulation and ability to synthesize enzymes or antibiotics (pleiotropic mutations) are fairly common, there appears to be no causal relationship between these two processes. In bacilli, many asporogenous mutants (*sp*⁻) still produced copious amounts of amylase, protease, or antibiotics and, conversely, many mutants selected as nonproducers of amylase, protease, or antibiotics (bacitracin or polymyxin) sporulated normally (Schaeffer, 1969).

The formation of aerial hyphae and antibiotics, and other characteristics, in many strains of *Streptomyces,* appear to be regulated by plasmid genes (Okanishi *et al.*, 1970; Redshaw *et al.*, 1976; Pogell *et al.*, 1976; Wright and Hopwood, 1976). Such strains can be cured of ability to form aerial mycelium as well as their ability to produce antibiotics by treatment with heat, acridine orange, or ethidium bromide. A chromosomal mutation called *bld* which eliminated aerial mycelium also had the pleiotropic effect of eliminating methylenomycin A production, which is plasmid determined (Pogell *et al.*, 1976).

In submerged fungal fermentations, mold mycelia can grow in either filamentous or pellet forms depending on physiological conditions, such as medium, aeration, and inoculum size (Whitaker and Long, 1973). The filamentous form is more difficult to aerate efficiently because of its higher viscosity. Thus, it is not surprising that pellet forms with diameters of 0.1–0.2 mm are more desirable for citric acid production than filamentous forms in *A. niger*. In the case of the penicillin fermentation, there appears to be no consensus as to which of the two forms is more desirable. In an experiment employing two morphologically distinct strains of *P. chrysogenum* fermented with excess oxygenation under identical conditions, the pellet variant produced twice the penicillin titer of the filamentous strain (Carilli *et al.*, 1961).

VI. THE ROLE OF PLASMIDS IN MORPHOLOGICAL DIFFERENTIATION AND SECONDARY METABOLITE FORMATION

Plasmids are extrachromosomal genes which replicate and segregate independently of chromosomal (nuclear) genes. Besides the well-known plasmids which confer drug resistance (R factors), fertility (F and *scp*-1 factors), and toxin production (colicin E_1, E_2) in eubacteria, there are many other characteristics in prokaryotes and eukaryotes which are determined by plasmids. In the eubacteria, covalently closed circular (ccc) DNA can be isolated from strains suspected of plasmid gene involvement. Detailed accounts of bacterial plasmids, which are beyond the scope of this report, can be found in recent reviews on the subject (Helinski, 1973; Falkow, 1975). In *Streptomyces,* plasmid genes have been reported to be involved in the formation of aerial hyphae, fertility, enzymes, and antibiotics. Table IV lists various characteristics and

TABLE IV. Characters and Products of *Streptomyces* Determined by Plasmid Genes

Character or product	Species	Reference
Fertility	*S. coelicolor*	Hopwood *et al.* (1973)
Aerial hyphae	*S. alboniger*	Redshaw *et al.* (1976)
	S. venezuelae	Okanishi *et al.* (1970)
Resistance to antibiotics	*S. rimosus*	Boronin and Sadovnikova (1976)
	S. coelicolor	Kirby *et al.* (1975)
	S. bikiniensis	Shaw and Piwowarski (1977)
Melanin excretion	*S. glaucescens*	Baumann and Kocher (1976)
Tyrosinase (melanin formation)	*S. scabies*	Gregory and Huang (1964a,b)
	S. venezuelae	Okanishi *et al.* (1970
	S. kasugaensis	Okanishi *et al.* (1970)
β-Lactamase	*Streptomyces* spp.	Oganwara and Nozaki (1977)
Kasugamycin	*S. kasugaensis*	Okanishi *et al.* (1970)
Aureothricin	*S. kasugaensis*	Okanishi *et al.* (1970)
Chloramphenicol	*S. venezuelae*	Akagawa *et al.* (1975)
Methylenomycin A	*S. coelicolor*	Kirby *et al.* (1975)
Turimycin	*S. hygroscopicus*	Kahler and Noack (1974)
Streptomycin	*S. bikiniensis*	Shaw and Piwowarski (1977)

products in species of *Streptomyces* which have been reported to be determined by plasmid genes.

In *S. coelicolor,* plasmid scp 1 not only confers fertility to strains that harbor this plasmid but also determines the production of and resistance to methylenomycin A (Kirby *et al.*, 1975) (see Section IV for fertility determined by scp 1). Attempts to isolate scp 1 plasmid DNA have not been successful to date. However, a 10–20 × 10^6 dalton ccc DNA has been isolated from both scp 1^+ and scp 1^- strains (Schrempf *et al.*, 1975). This ccc DNA appears to correspond to a new sex factor, scp 2, in *S. coelicolor.*

Most of the evidence for plasmid involvement in genetic control has been based on curing experiments, i.e., demonstration of increased frequency of nonproducing strains (as compared to mutation frequency) after treatment with agents such as acridine orange, ethidium bromide, or high temperature (Okanishi *et al.*, 1970; Redshaw *et al.*, 1976). In the case of *S. scabies* (tyrosinase), *Streptomyces venezuelae* (chloramphenicol), and *S. coelicolor* (scp 1), crosses provided additional evidence that extrachromosomal genes were involved. In *S. glaucescens,* genetic analysis showed that the tyrosinase gene was located on the chromosome, whereas the excretion character of melanin was determined by an extrachromosomal gene (Baumann and Kocher, 1976).

Among other plasmids of potential economic importance are the plasmids involved in the biodegradation of organic compounds in

Pseudomonas (Chakrabarty, 1976). Many of these plasmids are transmissible and most are compatible with one another. Thus, it is possible to maintain several hydrocarbon-degrading plasmids in a single bacterial strain. Whether such a strain will have an enhanced capability to utilize crude oil remains to be seen. Also noteworthy is the recent discovery that the ability of *Agrobacterium tumefaciens* to induce crown gall tumors in plants can be attributed to the presence of a large plasmid (1.2×10^8 daltons). The plasmid has also been detected in crown gall tumors. The tumor-inducing ability of the bacterium can be eliminated irreversibly by propagation at high temperature and can be transferred from a virulent to an avirulent strain by means of plasmid transfer (Watson *et al.*, 1975).

Plasmids or plasmidlike genes have also been reported in eukaryotic organisms. Some strains of *S. cerevisiae* contain copies of 2-μm ccc DNA (Livingston and Klein, 1977). A single-step spontaneous mutant of *S. cerevisiae,* resistant simultaneously to oligomycin, venturicidin, chloramphenicol, cycloheximide, and triethyltin, was also found to contain 2-μm DNA (Guérineau *et al.*, 1974). Oligomycin resistance appears to be controlled by a chromosomal gene, even though its expression was correlated with the presence of 2-μm DNA within the cells. The resistance to other drugs can be eliminated by ethidium bromide, suggesting their extrachromosomal nature (although unrelated to 2-μm DNA). A plasmidlike behavior has been recently demonstrated in transformed strains of *N. crassa* (Mishra, 1976). Two classes of inositol-independent (*inl*$^+$) transformants were observed; one class showed the Mendelian transmission of *inl*$^+$ character in meiotic crosses, while the other class showed non-Mendelian transmission. The latter class can be cured of the *inl*$^+$ character by treatment with ethidium bromide.

VII. INDUSTRIAL STRAIN IMPROVEMENT PROGRAMS IN PENICILLIN- AND CEPHALOSPORIN-PRODUCING FUNGI

A. The Penicillin Fermentation

Large-scale programs concerned with the induction, selection, and utilization of superior penicillin-producing variants of *P. chrysogenum* have been carried out in government, university, and industrial laboratories for over 30 years. From the cumulative screening of hundreds of thousands of strains, a series of very superior penicillin producers has been developed, certain of which are utilized today for the commercial manufacture of penicillin (Elander, 1967; Elander *et al.*, 1973). Over the years, "mutation breeding" has been undertaken with strains of *P. chrysogenum* more than with any other industrially important microorganism. The genealogy of the Wisconsin family of strains of *P. chrysogenum* and two important modern industrial lineages is presented in Fig. 16. Although the initial screening was based on natural (nonmutated) colony populations, subsequent natural populations failed to lead to major yield increases. For example, at Stanford University some 60,000 cultures were screened for improved productivity with no success (Sermonti, 1969). Key strains in the early ancestry of the Wisconsin series were the famous Q-176 culture, which had significantly improved antibiotic titers, and strain BL3-D10, which did not produce the characteristic and troublesome chrysogenin pigment (Backus and Stauffer, 1955). All further mutant selections were derived from the Q-176 culture over the next decade.

Three distinct selection lines were established from the pigmentless mutant. One line was based on the selection of natural variants, another based on uv-radiation survivors, and a third highly successful line based on treatment with a nitrogen mustard mutagen. Industrial lines were later derived from the 51-20 strain utilizing a variety of different mutagens, including NG, diepoxybutane, and nitrous acid (Elander *et al.*, 1973).

Throughout the Wisconsin, Lilly, and Wyeth programs, it was noted that the highest morphological mutation rate with uv radiation did not necessarily coincide with the highest rate of kill. The Wisconsin group reported that lower doses, i.e., treatments yielding 25–30% survivors, were often effective in inducing higher productivity mutants than a dose which would effect a maximum kill (Backus and Stauffer, 1955).

During the period of intensive selection work, attention was focused on searching for strain characteristics which appeared to correlate with high yield. In general, the Wisconsin group demonstrated a correlation between increased productivity and reduced sporulation and growth, characteristics which also appear to correlate with improved cephalosporin C variants of *C. acremonium* (Elander, 1975; Queener *et al.*, 1975). In the Wisconsin series, the greatest change was observed in the early ancestry. Between the NRRL-1951 and Q-176 strains, growth and sporulation were reduced by 60%, with a concomitant yield increase of nearly sixfold. Later, an additional threefold increase in antibiotic titer was associated with only an additional 10% reduction in mycelial vigor. These changes may represent a correlated response due to linkage of loci determining growth with those influencing penicillin titer. Selection can also be utilized to isolate strains with improved growth and sporulation characteristics, the latter being most important for long-term preservation and providing for vigorous vegetative seed development for large-tank fermentations (Elander *et al.*, 1973). The correlation between strain productivity, decreased sporulation, and weak vegetative de-

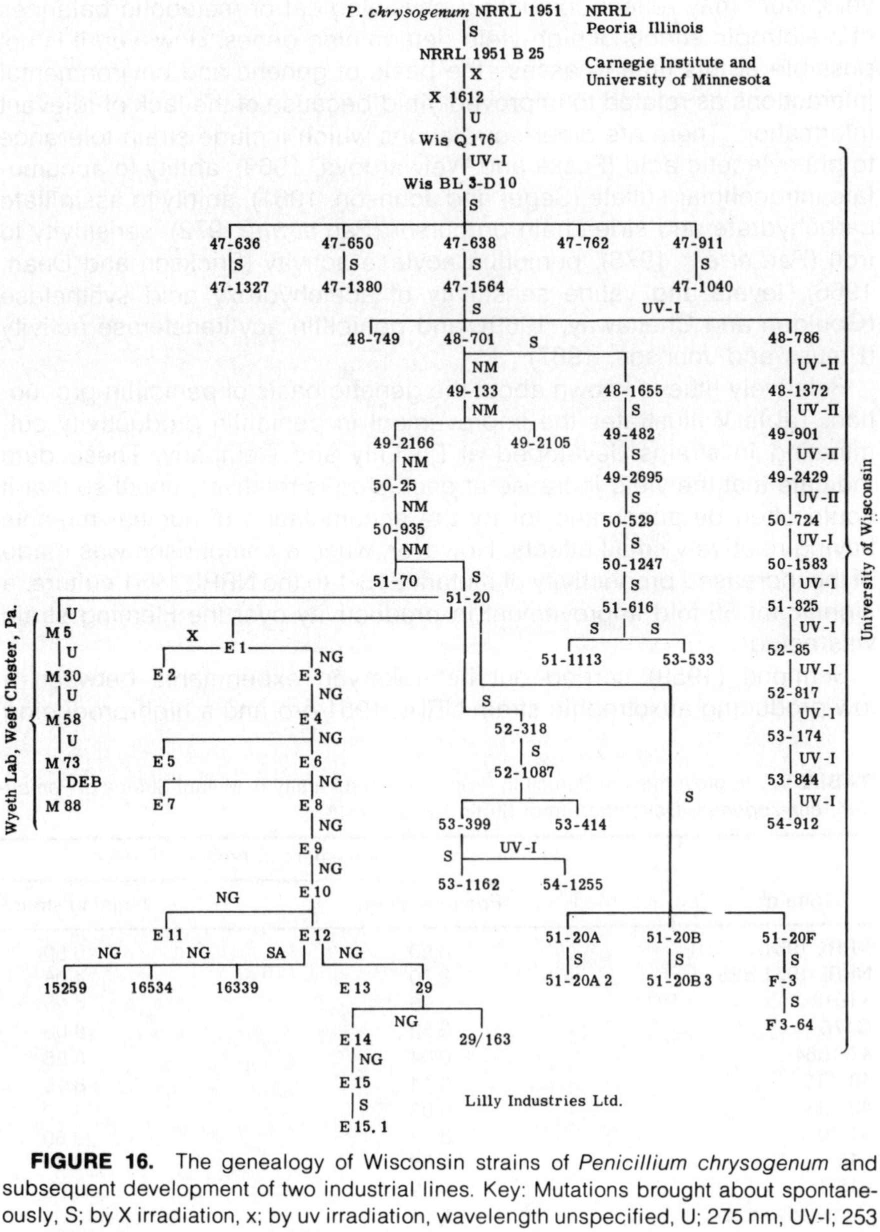

FIGURE 16. The genealogy of Wisconsin strains of *Penicillium chrysogenum* and subsequent development of two industrial lines. Key: Mutations brought about spontaneously, S; by X irradiation, x; by uv irradiation, wavelength unspecified, U; 275 nm, UV-I; 253 nm, UV-II; nitrogen mustard, NM; nitrosoguanidine, NG; diepoxybutane, DEB. (Based on Stauffer, 1961; and Elander, 1967.)

velopment may reflect conflicting physiological or metabolic balances of pleiotropic effects of high-yield determining genes. However, it is not possible at this time to assess the basis of genetic and environmental interactions as related to improved yield because of the lack of relevant information. There are other correlations which include strain tolerance to phenylacetic acid (Fuska and Welwardova, 1969), ability to accumulate intracellular sulfate (Segel and Johnson, 1961), ability to assimilate carbohydrate and side-chain precursor (Pan *et al.*, 1972), sensitivity to iron (Pan *et al.*, 1975), penicillin acylase activity (Erickson and Dean, 1966), levels and valine sensitivity of acetohydroxy acid synthetase (Goulden and Chattaway, 1969), and penicillin acyltransferase activity (Preuss and Johnson, 1967).

Relatively little is known about the genetic basis of penicillin production. Table V illustrates the improvement in penicillin productivity culminating in strains developed at Eli Lilly and Company. These data indicate that the yield increase at each step is relatively small so that it could often be accounted for by the accumulation of nuclear mutants having relatively small effects. However, when a comparison was made of the increased productivity of mutant E15-1 to the NRRL 1951 culture, a significant 55-fold improvement in productivity over the Fleming strain was noted!

Sermonti (1959) carried out heterokaryon experiments between a low-producing auxotrophic strain NRRL 1951 *pro* and a high-producing

TABLE V. Improvement in Penicillin Production in a Family of Mutant Strains of *Penicillium chrysogenum* Developed from Strain NRRL 1951[a]

Strain[b]	Improvement in production over: Previous strain	Fleming strain[c]
NRRL 1951	0.50	0.50
NRRL 1951 B25	2.70	4.05
X1612	0.06	4.90
Q176	0.52	9.00
47-1564	−0.04	8.55
48-701	−0.14	8.15
49-133	0.08	12.70
51-20	0.73	22.60
E15	1.05	50.64
E15-1	0.09	55.00
Cumulative total	5.55	176.09

[a] From Elander (1967).

[b] The strains concerned can be identified in the family lineage in Fig. 16.

[c] The penicillin titers are from shake-flask fermentations using modern fermentation conditions.

strain Wis 49-133 *nic*. Homokaryotic segregants from the heterokaryon showed a clear association between the nuclear marker and yield; i.e., *pro* segregants were low-producers and *nic* segregants were high-yielding. Thus, penicillin yield is determined by nuclear genes and there is evidence from both heterokaryons and parasexual recombinants that several different genes are involved. Both the *y met* and *w ade* heterokaryons of Elander (1967) (also see Fig. 17) and the "New Hybrid" strain of Alikhanian and Borisova (1956) show significantly higher yields than their parent homokaryons, suggesting complementation between nonallelic genes.

Caglioti and Sermonti (1956) attempted to map *pen*-1, a determinant

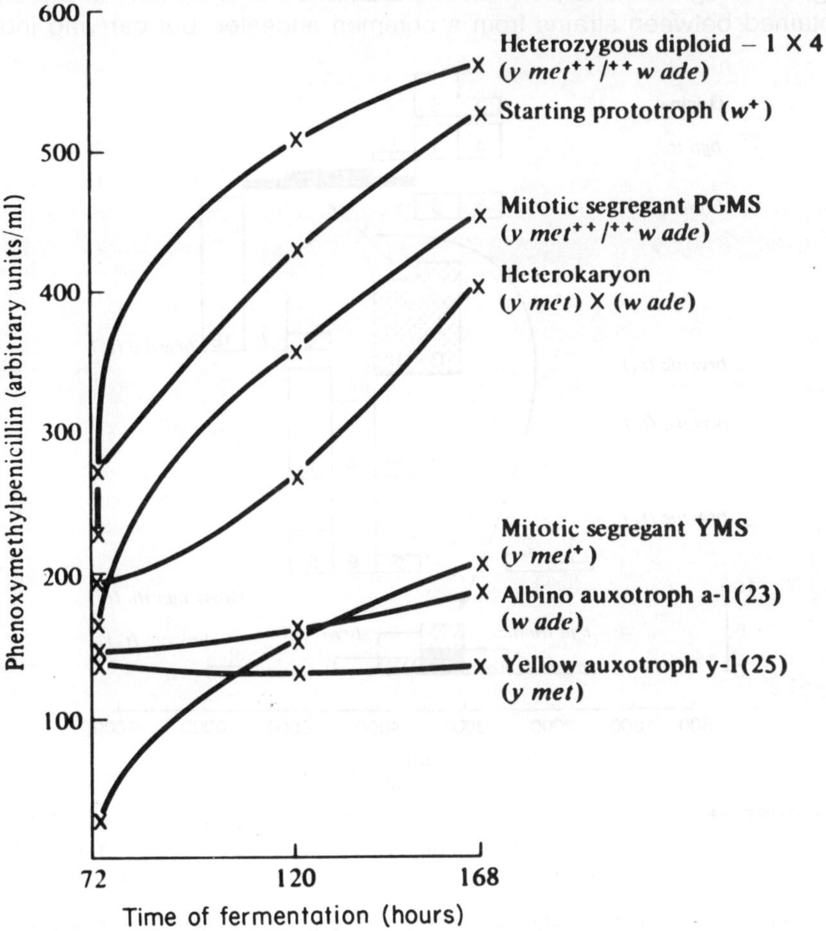

FIGURE 17. Penicillin production by recombinant strains of *Penicillium chrysogenum* E15. (From Elander, 1967.)

for increased yield by mitotic recombination. They showed its locus was distal to *pro*-1 and *met*-1, two auxotrophic markers in one of three linkage groups they tentatively identified. More recently Ball (1971, 1973a,b) confirmed by haploidization analysis the existence of three linkage groups in a Glaxo strain. Ball induced increased yield in five separate strains, each of which carried conidial color or auxotrophic markers (Fig. 18).

Two phenomena have been described which suggest that genes determining increased yield of penicillin are recessive and that independently induced mutations could be allelic. In most cases when a diploid strain is derived via the parasexual cycle between a low-yielding and a high-yielding strain, its yield is comparable to that of the former. Diploids obtained between strains from a common ancestor, but carrying inde-

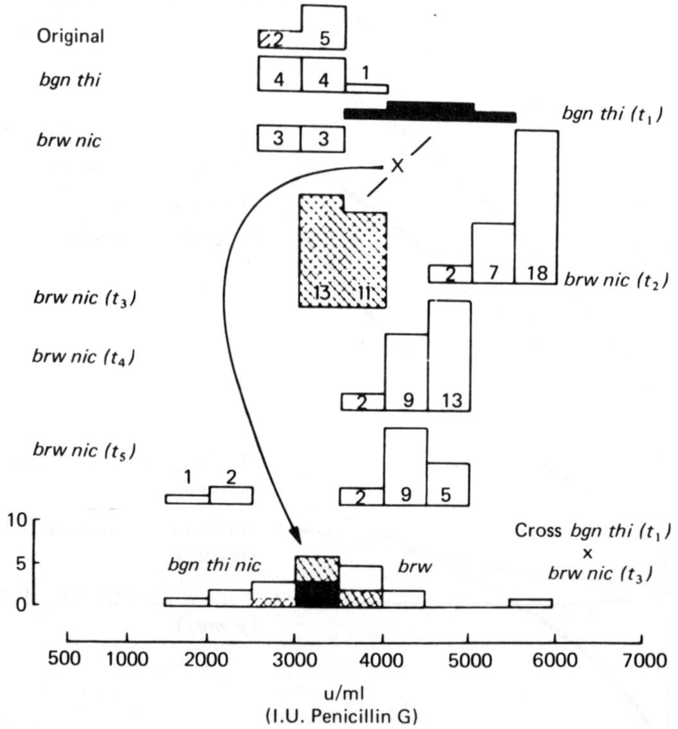

FIGURE 18. Penicillin production of some genetically marked strains of *Penicillium chrysogenum* and the yield of a cross between two of them (*bgn thi* t_1 × *brw nic* t_3). Key: Number in histograms indicate number of strains. Symbols: *bgn*, bright green and *brw*, brown colonies; *nic thi*, auxotrophic for nicotinic acid and thiamin, respectively; t_1–t_5, determinants of penicillin yield. Solid histogram, *bgn thi* t_1 phenotype; crosshatched histogram, *brw nic* t_3 phenotype, others indicated. (Based on data of Ball, 1973a, as outlined by Burnett, 1975.)

pendently induced mutations for higher yield, gave yields equivalent to or not much lower than their parental strain, but notably higher than the yield of the common ancestor (Macdonald *et al.*, 1964; Elander, 1967; Elander *et al.*, 1973). An explanation for these observations is that the strains carried a number of allelic mutant sites in common, even though the sites had been independently mutated. It appears there are a number of genetic determinants, recessive in their expression and located on more than one chromosome, which determine increased penicillin yield in *P. chrysogenum*. Whether these determinants are oligogenes, polygenes, or a combination of the two has not been ascertained. Their expression can be modified by genetic background, for most conidial color and auxotrophin markers reduced yield drastically.

To date, the technique of mutation and selection is the most reliable procedure for improving penicillin titers. Breeding involves the parasexual cycle and, therefore, the peculiarities of the cycle in *P. chrysogenum* present difficulties in achieving recombination and subsequent unimpaired segregation. Strain variability and instability cause frequent problems. It has proved difficult to initiate the parasexual cycle because heterokaryons are not produced easily with the Wisconsin strains of *P. chrysogenum*. Even when heterokaryons are produced, numerous difficulties have been encountered with the instability of diploids and their segregants. Elander (1967) described a highly stable diploid strain at Eli Lilly derived from the haploid production strain E-15 (Fig. 17). Spontaneous variation in the diploid and haploid was 9.2 and 31.9%, respectively, as assessed by the color types which segregated. The difference was even more striking after exposure to uv irradiation, where the diploid : haploid derived variants were 11.6 and 41.1%, respectively. In this case, the diploid strain was far more stable than its haploid progenitor.

In contrast, Ball (1971, 1973b) reported that diploids show greater instability than their haploid progenitors but within the same range. The diploid showed twice as many poor sporulating types as the haploid, while densely sporulating types were increased tenfold. Ball (1973a) and Roper (1973) have compared this phenomenon with that of "mitotic nonconformity" described by Nga and Roper (1969) in *A. nidulans*. Mitotic nonconformity stems from the existence of duplicate segments of small fractions of the genome which have probably originated through translocations. If the difficulties described in producing diploids can be overcome, they offer many attractions for the fungal geneticist. First, there is the possibility of a heterotic effect on yield (Sermonti, 1959; see Fig. 17). Even if this does not occur, the possibility exists of selecting productive segregants from diploid strains. If the latter approach is to be used, it is recommended to start initially with strains lacking chromosomal rearrangements or to induce variants in increased yield from an existing strain without inducing chromosomal aberrations. Since spontaneous dediploidization can always occur, systems have been proposed to reduce such effects. Macdonald (1964) has suggested the use of parental haploids which grow poorly in complex fermentation media so that the parental segregants arising during fermentation would be selected against. Azevedo and Roper (1967) suggested the induction of recessive lethals in the diploid so that haploid segregants would be eliminated as they arose. It has been suggested that the Lilly diploid (Fig. 17) may be a balanced lethal diploid in that it was extremely stable but, even after treatment with uv irradiation, the proportion of viable segregants it produced was hardly increased. Provided that diploid stability and unrestricted parasexual recombination can be achieved, it should be possible to make further selection progress. Elander (1967) described selection No. 2, a spontaneous segregant from the Lilly diploid, which produced nearly 25% more antibiotic than its parent, which itself yielded better than the production haploid E-15. Ball (1973a,b) has also described recombinant segregants with improved yield (see Fig. 18).

The first 30 years of mutation selection and scientific breeding have demonstrated the potentials of mutation breeding and the dangers present in utilizing mutagens which induce chromosomal aberrations (Ball, 1973b). The next 25 years should exhibit more extensive application of the parasexual cycle to breeding (Merrick, 1976; Hopwood, 1977).

B. The Cephalosporin Fermentation

An intensive strain improvement program was initiated in England in 1957 to improve, through mutagenesis and fermentation development, the poor production of cephalosporin C in Brotzu's Sardinian strain of *C. acremonium* (Elander, 1975). Mutagenesis of the Brotzu isolate resulted in the eventual selection of the improved mutant, M-8650, which served as the progenitor strain for large-scale industrial programs (Fig. 19). The synthesis of cephalosporin C in laboratory fermentations using the early British fermentation conditions for a series of improved uv variants is shown in Fig. 20.

Strain CW-19 produced threefold more antibiotic than the Brotzu culture and, when fermented under more optimal fermentation conditions, the culture was capable of synthesizing nearly 15-fold more antibiotic than the progenitor strain. The CW-19 variant also had a significantly improved cephalosporin C/penicillin N ratio (Elander *et al.*, 1976).

Biometric considerations of the overall data concerning "normal" and "mutant" (morphological and biochemical) populations of uv-survivor strains of *C. acremonium* and their potential to synthesize cephalosporin

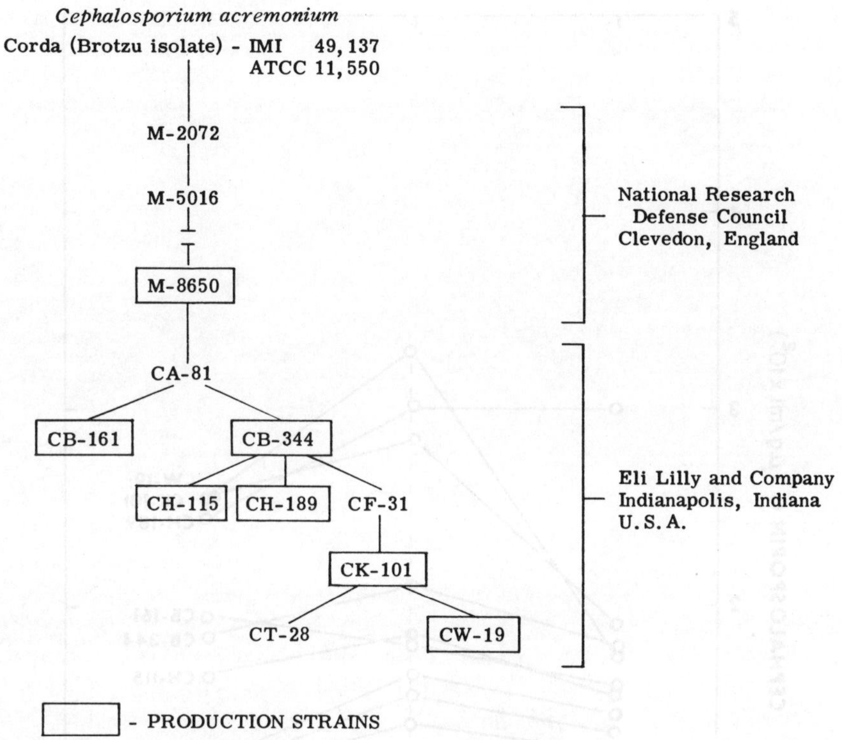

FIGURE 19. Lineage of early Lilly uv (275 nm) variants of *Cephalosporium acremonium*. (After Elander, 1975.)

C showed an approximate 11 to 1 advantage for "normal" clones versus "mutant" clones in searching for mutants possessing the capacity to produce 20% more antibiotic. The above probability statements were calculated for many survivor populations where suitably large numbers of "normal" and "mutant" populations were evaluated on a statistical basis (Brown and Elander, 1966).

The improved uv variants differed markedly from the progenitor strains in cultural characteristics and biochemical properties. Untreated colony populations of the improved variants showed a progressive reduction in colony diameter and decreased vegetative development. The improved mutants were also characterized by decreased sporulation vigor, features which were also characteristic of the improved penicillin variants (Elander *et al.*, 1976). Other mutagenic agents have also been reported to elicit improved antibiotic-producing variants.

Recent patent literature from Japan has reported that strains of *C. acremonium* or *C. polyaleurum* resistant to the polyene antibiotics nystatin, kabicidin, or trichomycin produced greater than 10 gm/liter of

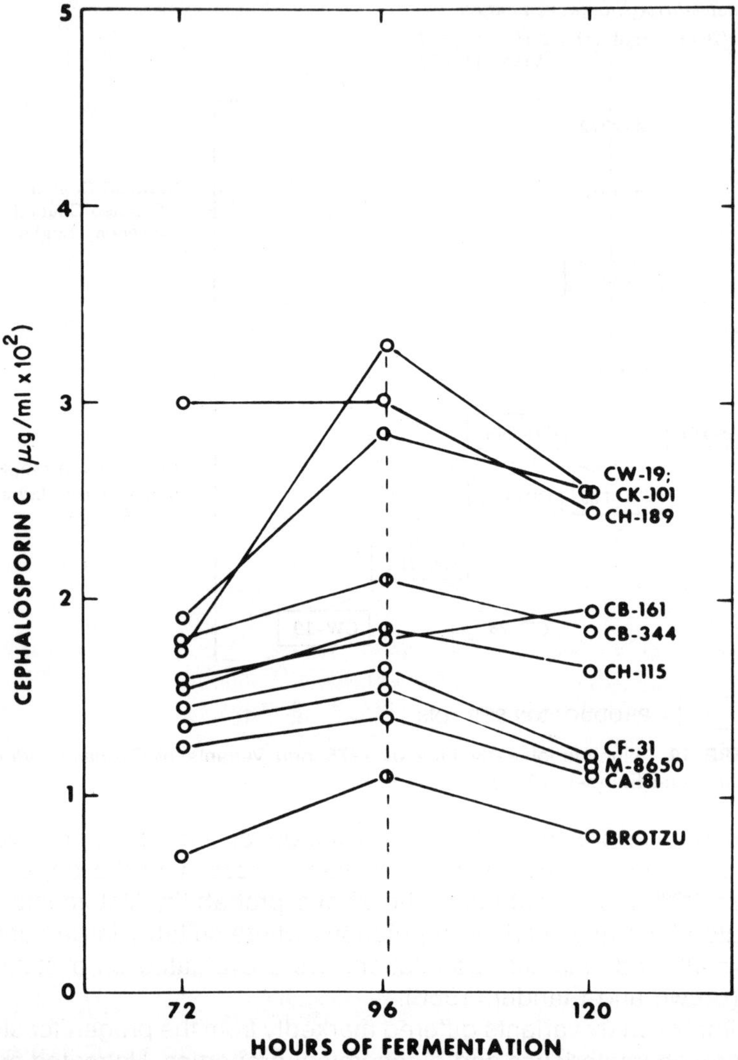

FIGURE 20. Cephalosporin production by Brotzu strain and early improved Lilly variants of *Cephalosporium acremonium*. (After Elander, 1975.)

cephalosporin C (Japanese Patent JA-110723, issued 1/16/75 to Takeda Chem. Ind. Ltd.). In another recent patent issued to the same company, certain polyploid forms of *C. acremonium* or *C. polyaleurum* were reported to be potent cephalosporin C producers. The higher ploidy clones were induced following exposure to camphor and acenaphthene and the subsequent selection of giant cells (Japanese Patent JS-109680, issued 1/16/75 to Takeda Chem. Ind. Ltd.).

TABLE VI Comparison of Random Selection versus Directed Selection Procedure in Strains of *Penicillium chrysogenum* and *Acremonium chrysogenum*[a]

Type	Treated	Organisms Examined	Selection procedure	No. Tested	No. Retained	% Retained[b]
Random selection	uv, NG	*P. chrysogenum* and *C. acremonium*	None	860	7	0.81
Directed selection	uv, NG X-ray	*P. chrysogenum* and *C. acremonium*	1. Colony-plate	438	6	1.36
			2. Auxotrophs	35	2	5.71
			3. Haploidization-inducing agents	503	8	1.59
			4. Mitotic inhibitors	225	3	1.33
			5. Mercury	162	2	1.23
			6. Amino acid analogs	567	8	1.41
			7. Sulfur analogs	452	22	3.98
		C. acremonium (only)	1. Methionine analogs	605	22	3.64
			2. Increased sensitivity to methionine			
			a. Growth	247	4	1.62
			b. β-lactam synthesis	52	2	3.85

[a] Adapted from Chang and Elander (1979).
[b] Superior on both primary and secondary screening tests.

Nash and Huber (1971) reported that submerged cultures of *C. acremonium* differentiate into small swollen fragments termed arthrospores and the differentiation coincides generally with the maximal rate of cephalosporin C synthesis. In their improved variants, arthrospore formation was proportional to increased antibiotic formation. Methionine supplementation enhanced the onset of differentiation and the requirement for methionine was increased for higher yielding mutants. Methionine and sulfur metabolism are very important for cephalosporin C synthesis and the metabolism of methionine, its analog norleucine (Drew *et al.*, 1976), and sulfate may trigger antibiotic synthesis through cellular differentiation. Dennen and Carver (1969) have reported that strains with increased capacity to synthesize cephalosporin C have decreased capacity to degrade methionine for cellular requirements and their arylsulfatase derepression is increased proportionately.

The probable pathway for sulfur metabolism in *C. acremonium* has been described by Drew and Demain (1975). Cysteine appears to be the immediate donor of sulfur to cephalosporin C, but the amino acid is not highly stimulatory for cephalosporin C formation in media containing sulfate (Demain, 1973, 1974). Methionine shows excellent stimulation of antibiotic synthesis and its stimulation is not due to sulfur donation but to some unresolved role in antibiotic regulation. Nüesch *et al.* (1973) and Drew and Demain (1975) have obtained mutants with genetic blocks between sulfate and cysteine and cystathionine and homocysteine. The mutants (designated 274-1 and 8650 *slp*) grow on cysteine, cystathionine, and homocysteine (but not on sulfate) but these compounds do not replace methionine with respect to cephalosporin C stimulation. Nüesch *et al.* (1973) reported on a Ciba *slp* mutant blocked in the sulfate reduction pathway prior to sulfide formation. The mutant was able to assimilate more exogenous methionine and had potential to synthesize nearly fourfold more antibiotic than its sulfide-proficient parent. Revertant strains of the mutant assimilated less methionine and synthesized only low levels of cephalosporin C. Drew *et al.* (1976) showed a similar result for the 274-1 mutant. It produced over 1100 μg/ml of cephalosporin C, whereas its prototrophic parent strain CW-19 formed only 400 μg/ml under identical fermentation conditions. Nüesch and co-workers have proposed that higher levels of cephalosporin C obtained with nonsulfate-utilizing mutants are due to their inability to synthesize cysteine, an amino acid which acts as a repressor and inhibitor of methionine permease.

In a study on a series of improved cephalosporin C mutants, Queener *et al.* (1975) reported that the specific activity of glutamate dehydrogenase (GDH) was derepressed upon entry into the stationary growth phase, whereas two mutants in a low-yielding series had repressed levels of

GDH at the same growth phase. The altered regulation pattern for GDH may have removed a nitrogen limitation for cephalosporin C synthesis. In this study and in a previous report by Elander *et al.* (1976), there appears to be an inverse relationship between vegetative development and enhanced cephalosporin C synthesis.

Komatsu *et al.* (1975) isolated an improved mutant of *C. acremonium* which synthesized copious amounts of cephalosporin C in the presence of sulfate. The mutant was also reported to be sensitive to methionine in that it displayed potency levels more than twofold that of the parent in the presence of sulfate, but its productivity was severely inhibited by more than 0.5% of methionine which gave high cephalosporin C production with the parent strain. Recent studies by Komatsu and Kodaira (1977) showed that the improved mutant had enhanced L-serine sulfhydrylase activity compared to the parent. They hypothesize that the mutant is superior to its parent in cephalosporin synthesis in that it has an improved ability to maintain a high level of the cysteine pool. Although there is no evidence to show that L-serine sulfhydrylase actually functions in the biosynthesis of cysteine in *C. acremonium,* they maintain that a second enzyme in the cysteine pathway, cystathionine β-synthase, may facilitate the conversion of methionine to cysteine, thereby enabling the mutant to maintain a high pool of cysteine, which appears to be essential for high cephalosporin synthesis in the presence of sulfate.

Rational selection procedures have been recently reported to be more efficient in selecting superior penicillin V- and cephalosporin C-producing mutants in large-scale selection programs at the Bristol-Myers Company (Elander, 1978; Chang and Elander, 1979). The rational selection procedures are based on known or probable biochemical mechanisms, thereby removing much of the empiricism commonly associated with labor-intensive random screening and selection. Table VI shows a comparison of rational *vs* random selection for the selection of superior fermentation variations of *P. chrysogenum* and *C. acremonium* (also classified as *Acremonium chrysogenum*).

VIII. POTENTIAL ROLES FOR RECOMBINANT DNA, CLONING, AND GENE AMPLIFICATION IN MICROBIAL SELECTION TECHNOLOGY

A series of recent major technological breakthroughs in molecular biology has made it possible now to introduce foreign DNA into *E. coli* or other bacteria using plasmids or phages as cloning vehicles (Cohen, 1975). Two key developments have contributed more than anything else to this new technology. First, the observation that treatment of *E. coli* cells with $CaCl_2$ renders them competent for transformation by plasmid DNA (Cohen and Chang, 1975), and second, the discovery of a battery of restriction endonucleases, which cleave DNA at specific sites (Roulland-Dussoix *et al.*, 1975). A new field of genetic engineering aiming at introducing foreign DNA into bacteria (also termed "gene splicing," "molecular cloning," or "recombinant DNA") has evolved over the past 4 years.

A typical recombinant DNA experiment is diagrammed in Fig. 21. Briefly, the experiment consists of the following steps: (1) treatment of the plasmid DNA (vector) and foreign DNA, respectively, with endonuclease to generate fragments with complementary, overlapping single-strand ends ("cohesive ends"); (2) enzymatic ligation of fragments of plasmid DNA's with those of foreign DNA to form recombinant DNA's; and (3) tranformation of *E. coli* cells with recombinant DNA's. The use of drug-resistant plasmid provides a convenient means of selecting for clones containing recombinant DNA (Cohen *et al.*, 1973).

Since many plasmids, such as Col E_1 (determines colicin E_1 production and resistance), are present as multiple copies in cells (25–30 copies for Col E_1 plasmid), a gene can be greatly amplified in cells after being cloned in *E. coli* cells. Such gene amplification has been observed with *E. coli* tryptophan operon genes cloned into Col E_1 plasmid (Hershfield *et al.*, 1974). When this plasmid was introduced into an *E.*

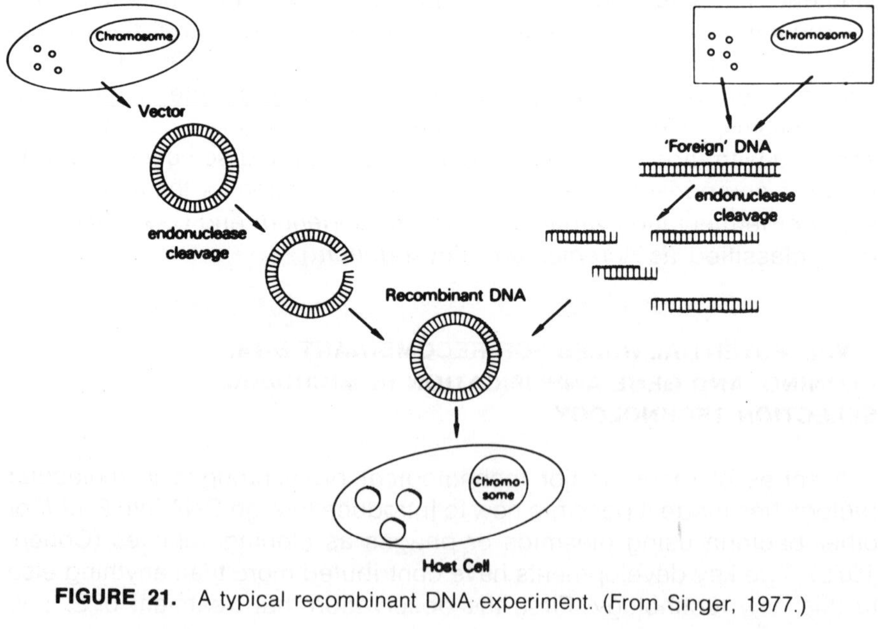

FIGURE 21. A typical recombinant DNA experiment. (From Singer, 1977.)

coli strain and the tryptophan operon induced by addition of inducer, the specific activity of the tryptophan operon enzymes increased more than 10-fold. Additional gene amplification through increase in plasmid copy number was achieved when plasmid-containing bacteria were grown in the presence of chloramphenicol. Since chloramphenicol inhibits protein synthesis, no amplification of gene products (proteins) was obtained. Future research may provide a means for amplifying plasmid gene products (proteins) as well as gene copies.

To date, many plasmid cloning vehicles as well as restriction endonucleases are readily available for cloning experiments (Helinski *et al.*, 1975). Genes from totally different biological classes, such as *Xenopus* and *Drosophila,* have been cloned in *E. coli* (Cohen *et al.*, 1973). In many instances, the foreign genes are able to replicate and are even transcribed, but not translated. Efforts are also being made to introduce into *E. coli* genes specifying such synthetic functions as nitrogen fixation, antibiotic and enzyme production, and antibody and hormone synthesis (Wade, 1977). Hormones such as insulin, somatostatin, and human growth hormone are particularly amenable to cloning since they are all relatively small peptides whose sequences are known.

Recombinant bacterial plasmids have been constructed that contain various regions of the rat insulin gene sequence (Ullrich *et al.*, 1977). In this instance the foreign DNA (rat insulin genes) was prepared enzymatically using reverse transcriptase (RNA-dependent DNA polymerase) with rat insulin messenger RNA (mRNA) as template. It remains to be seen whether the *E. coli* clones containing insulin gene segments are capable of producing insulin or insulin mRNA. Another recent development in recombinant DNA research is the successful transfer of nitrogen fixing gene (*nif*) from *Klebsiella pneumoniae* to *E. coli* by means of a recombinant plasmid (Dixon and Postgate, 1972). Even though the new strain of *E. coli* was able to synthesize nitrogenase, it could not fix nitrogen unless it was protected from oxygen. It appears that the technology for creating a higher plant capable of efficient nitrogen fixation is still in the distant future.

As can be seen from the above examples, industrial microbiologists are now faced with enormous problems in their attempts to apply recombinant DNA techniques to "tailor make" microorganisms. Some of the more obvious problems are as follows:

1. Even if one discounts the evolutionary barriers between species, the "spliced" genes in the new host cell have to overcome various regulatory limitations in order to be expressed.
2. The clones containing foreign DNA may not be genetically stable; the foreign DNA may be eliminated by segregation unless certain selec-

tion pressure is applied. An example of this situation is the mouse–human hybrid cells which tend to lose human chromosomes during propagation (Weiss and Green, 1967).

3. To clone a whole array of antibiotic pathway genes (e.g., more than ten genes are required for cephalosporin C biosynthesis) onto a plasmid could be an awesome task unless the genes are clustered. However, it should be less difficult to clone the structural genes for hormones or enzymes (amylase or protease) or genes with regulatory functions.

4. Proper cloning vehicles have not been developed for industrial microorganisms (mostly actinomycetes and fungi). Although plasmids have been found in streptomycetes, none has been reported to be useful as a cloning vehicle. A recent report that *Staphylococcus aureus* plasmids coding for tetracycline and chloramphenicol resistance can be introduced by transformation into *B. subtilis* (Ehrlich, 1977) could mean that some bacterial plasmids can be transferred to *Streptomyces* and be used as cloning vehicles. Recently, the *E. coli* gene coding for leu-2 has been introduced and expressed in *Saccharomyces cerevisiae* (Hinnen *et al.*, 1978).

The commercial manufacture of somatostatin and of human insulin using recombinant DNA technology may soon be possible as scientists at the University of California at Los Angeles and Genentech Inc. have inserted chemically synthesized genes into a bacterial host. The human gene coding for somatostatin is a small protein consisting of 14 amino acids. The artificial gene is then attached to a gene segment (operon) coding for β-galactosidase which is transferred into the *E. coli* cell via a plasmid known as pBR322 (Itakura *et al.*, 1977).

The chemical synthesis of the gene sequence coding for human insulin posed a more formidable task in that the gene coding for the A chain alone consists of 21 amino acids. The B chain gene sequence consists of some 30 amino acids. The synthetic A and B gene sequences were then linked to the β-galactosidase operon and inserted separately into different *E. coli* host cells (see Fig. 22). The *lac* operon was then "switched on" by the addition of an appropriate inducer substrate and the message was translated into the appropriate human insulin protein fragment. The A and B insulin chains were next purified and rejoined chemically to generate active human insulin (Anon., 1978a). Other areas currently under development using recombinant DNA technology include the eventual commercial synthesis of L-asparaginase, human interferon and other hormones, vaccines, and antiviral products (Anon., 1978b).

Finally, although the potential benefits provided by applying recombinant DNA techniques may be substantial, the potential biohazards

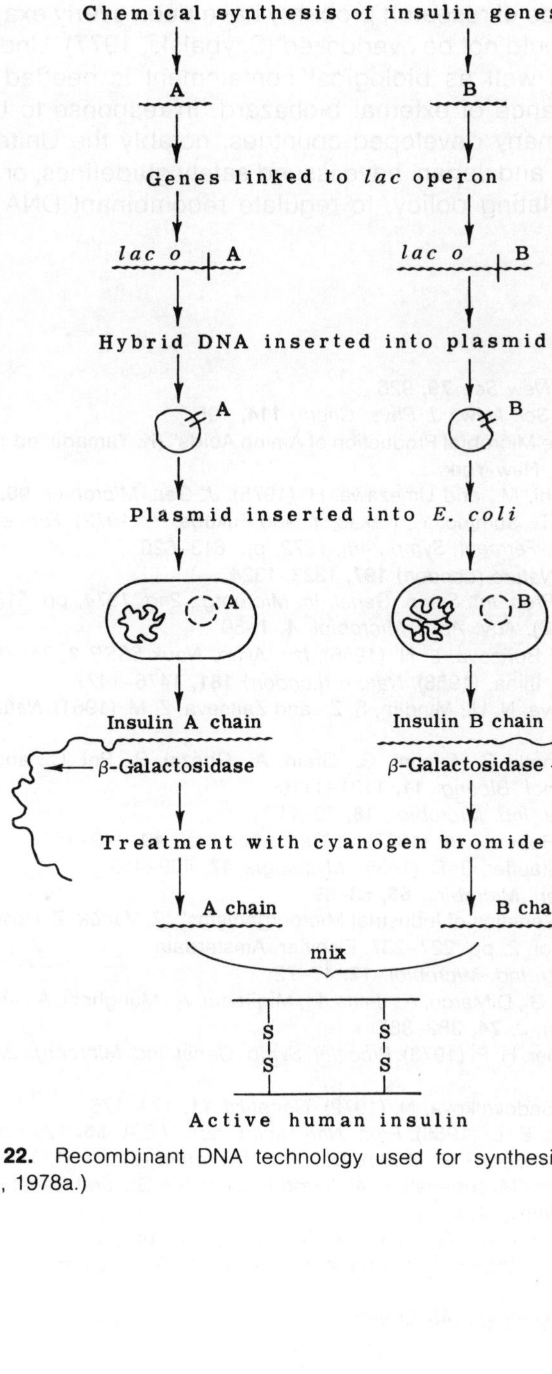

FIGURE 22. Recombinant DNA technology used for synthesizing human insulin. (From Anon., 1978a.)

inherent in this type of research probably have been overly exaggerated; however, they should not be overlooked (Szybalski, 1977). Undoubtedly, both physical as well as biological containment is needed so as to minimize any chance of external biohazard. In response to this need, governments in many developed countries, notably the United States, Germany, Britain, and Japan, have issued safety guidelines, or are in the process of formulating policy, to regulate recombinant DNA research (Singer, 1977).

REFERENCES

Anonymous (1978a). *New Sci.* **79,** 926.

Anonymous (1978b). *Sci. News J. Phys. Chem.* **114,** 195.

Abe, S. (1972). *In* "The Microbial Production of Amino Acids" (K. Yamada, ed.), pp. 39–66. Wiley (Halsted), New York.

Akagawa, H., Okanishi, M., and Umezawa, H. (1975). *J. Gen. Microbiol.* **90,** 336–346.

Akiyama, S., Suzuki, T., Sumino, Y., Nakao, Y. and Fukuda, Y. (1972). *Ferment. Technol. Today, Proc. Int. Ferment. Symp., 4th, 1972,* pp. 613–620.

Alacevic, M. (1963). *Nature (London)* **197,** 1323–1324.

Alacevic, M. (1976). *Proc. Int. Symp. Genet. In. Microorg., 2nd, 1974,* pp. 513–519.

Alikhanian, S. I. (1962). *Adv. Appl. Microbiol.* **4,** 1–50.

Alikhanian, S. I., and Borisova, L. N. (1956). *Izv. Akad. Nauk SSSR* **2,** 74–79.

Alikhanian, S. I., and Iljina, (1958). *Nature (London)* **181,** 1476–1477.

Alikhanian, S. I., Orlova, N. U., Mindlin, S. Z., and Zaitseva, Z. M. (1961). *Nature (London)* **189,** 939–940.

Arcamone, F., Cassinelli, G., Fantini, G., Grein, A., Onezzi, P., Pol, C., and Spalla, C. (1969). *Biotechnol. Bioeng.* **11,** 1101–1110.

Arima, K. (1977). *Dev. Ind. Microbiol.* **18,** 79–117.

Azevedo, J. L., and Roper, J. A. (1967). *J. Gen. Microbiol.* **49,** 149–155.

Backus, M. P., and Stauffer, J. F. (1955). *Mycologia* **47,** 429–463.

Ball, C. (1971). *J. Gen. Microbiol.* **66,** 63–69.

Ball, C. (1973a). *In* "Genetics of Industrial Microorganisms" (Z. Vaněk, Z. Hošťálek, and J. Cudlín, eds.), Vol. 2, pp. 227–237. Elsevier, Amsterdam.

Ball, C. (1973b). *Prog. Ind. Microbiol.* **12,** 47–72.

Barchielli, R., Baretti, G., DiMarco, A., Julita, P., Migliacci, A., Munghetti, A., and Spalla, C. (1960). *Biochem. J.* **74,** 382–387.

Bauman, R. and Kocher, H. P. (1976). *Proc. Int. Symp. Genet. Ind. Microorg., 2nd, 1974* pp. 535–552.

Boronin, A. M., and Sadovnikova, N. (1972). *Genetika* **11,** 174–176.

Brody, S., and Tatum, E. L. (1966). *Proc. Natl. Acad. Sci. U.S.A.* **56,** 1290–1297.

Brown, W. F., and Elander, R. P. (1966). *Dev. Ind. Microbiol.* **7,** 114–123.

Burnett, J. H. (1975). *In* "Mycogenetics: An Introduction to the General Genetics of Fungi," pp. 235–258. Wiley, New York.

Caglioti, M. T., and Sermonti, G. (1956). *J. Gen. Microbiol.* **14,** 38–46.

Carilli, A., Chain, E. B., Ghulandi, G., and Morisi, G. (1961). *Sci. Rep. Ist. Super. Sanita* **1,** 177–189.

Carvajal, F. (1952). *Mycologia* **45,** 209–234.

Chakrabarty, A. M. (1976). *Annu. Rev. Genet.* **10,** 7–30.
Chang, L. T., and Elander, R. P. (1979). *Dev. Ind. Microbiol.* **20** (*in press*).
Čižmek, S. (1976). *Abstr. Int. Ferment. Symp., 5th, 1976,* p. 216.
Claridge, C. A., Bush, J. A., DeFuria, M. D., and Price, K. E. (1974). *Dev. Ind. Microbiol.* **15,** 101–113.
Clarke, P. H. (1976). Proc.—*Int. Symp. Genet. In. Microorg., 2nd, 1974* pp. 15–28.
Clowes, R. C., and Hayes, W. (1968). "Experiments in Microbial Genetics." Wiley, New York.
Cohen, S. N. (1975). *Sci. Am.* **233,** 24–33.
Cohen, S. N., and Chang, A. C. Y. (1975). *In* "Microbiology—1974" (D. Schlessinger, ed.), pp. 187–198. Am. Soc. Microbiol., Washington, D.C.
Cohen, S. N., Chang, A. C. Y., Boyer, H. W., and Hellwig, R. B. (1973). *Proc. Natl. Acad. Sci. U.S.A.* **70,** 3240–3244.
Conover, L. H. (1971). *Adv. Chem. Ser.* **108,** 33–80.
Davies, O. L. (1964). *Biometrics* **20,** 576–591.
Delic, V., Pigac, J., and Sermonti, G. (1969). *J. Gen. Microbiol.* **55,** 103–110.
Demain, A. L. (1972). *Ferment. Technol. Today, Proc. Int. Ferment. Symp., 4th, 1972* pp. 239–245.
Demain, A. L. (1973). *Adv. Appl. Microbiol.* **16,** 177–202.
Demain, A. L. (1974). *Lloydia* **37,** 147–163.
Dennen, D. W., and Carver, D. D. (1969). *Can. J. Microbiol.* **15,** 175–181.
Dixon, R. A., and Postgate, J. R. (1972). *Nature (London)* **237,** 102–103.
Drake, J. W., and Baltz, R. H. (1976). *Annu. Rev. Biochem.* **45,** 11–37.
Drew, S. W., and Demain, A. L. (1975). *Antimicrob. Agents Chemother.* **8,** 5–10.
Drew, S. W., Winstanley, D. J., and Demain, A. L. (1976). *Appl. Microbiol.* **31,** 143–145.
Dulaney, E. L., and Dulaney, D. D. (1967). *Trans. N. Y. Acad. Sci.* [2] **29,** 782–799.
Ehrlich, S. D. (1977). *Proc. Natl. Acad. Sci. U.S.A.* **74,** 1680–1682.
Elander, R. P. (1967). *In* "Induced Mutations and their Utilization" (H. Stubbe, ed.), pp. 403–423. Akademie-Verlag, Berlin.
Elander, R. P. (1975). *Dev. Ind. Microbiol.* **16,** 356–374.
Elander, R. P. (1979). *In* "Genetics of Industrial Microorganisms." (O. K. Sebek, and A. I. Laskin, eds.), pp. 21–35. American Soc. Microbiol., Wash., D.C.
Elander, R. P., and Espenshade, M. A. (1976). *In* "Industrial Microbiology" (B. M. Miller and W. Litsky, eds.), pp. 192–256. McGraw-Hill, New York.
Elander, R. P., Mabe, J. A., Hamill, R. L., and Gorman, M. (1971). *Folia Microbiol. (Prague)* **16,** 156–165.
Elander, R. P., Espenshade, M. A., Pathak, S. G., and Pan, C. H. (1973). *In* "Genetics of Industrial Microorganisms" (Z. Vaněk, Z. Hošťálek, and J. Cudlín, eds.). Vol. 2, pp. 239–253. Elsevier, Amsterdam.
Elander, R. P., Corum, C. J., DeValeria, H., and Wilgus, R. M. (1976). *Proc. Int. Symp. Genet. Ind. Microorg., 2nd, 1974,* pp. 253–271. Academic Press, New York.
Erickson, R. C., and Dean, L. D. (1966). *Appl. Microbiol.* **14,** 1407–1448.
Esposito, A., Licciardello, G., Murthy, Y. K. S., Sacerdoti, S. A., and Sparapani, P. (1972). *Ferment. Technol. Today, Proc. Int. Ferment. Symp., 4th, 1972* Abstract, p. 217.
Falkow, S. (1975). "Infectious Multiple Drug Resistance." Pion Ltd., London.
Fantini, A. A. (1976). *In* "Methods in Enzymology" (J. H. Hash, ed.), Vol. 43, pp. 24–41. Academic Press, New York.
Fincham, J. R. S., and Day, P. R. (1973). "Fungal Genetics," 3rd ed. Blackwell, Oxford.
Fuska, J., and Welwardova, F. (1969). *Biologia (Bratislava)* **24,** 691–698.
Godfrey, O. (1974). *Can. J. Microbiol.* **20,** 1479–1486.

Gorman, M., Hamill, R. L., Elander, R. P., and Mabe, J. (1968). *Biochem. Biophys. Res. Commun.* **31,** 294–297.
Goulden, S. A., and Chattaway, F. W. (1969). *J. Gen. Microbiol.* **59,** 111–118.
Gregory, F. K., and Huang, J. C. C. (1964a). *J. Bacteriol.* **87,** 1281–1286.
Gregory, F. K., and Huang, J. C. C. (1964b). *J. Bacteriol.* **87,** 1287–1294.
Guérineau, M., Slominski, P. P., and Auner, P. R. (1974). *Biochem. Biophys. Res. Commun.* **61,** 462–469.
Guerola, N., Ingraham, J. C., and Cerda-Olmedo, E. (1971). *Nature (London), New Biol.* **230,** 122–124.
Halsall, D. M. (1975). *Biochem. Genet.* **3,** 109–116.
Hayes, W. (1968). "The Genetics of Bacteria and their Viruses," 2nd ed. Wiley, New York.
Helinski, D. R. (1973). *Annu. Rev. Microbiol.* **27,** 437–470.
Helinski, D. R., Lovett, M. A., Williams, P. H., Katz, L., Kupersztock-Portnay, Y. M., Guiney, D. G., and Blair, D. G. (1975). *In* "Microbiology—1974" (D. Schlessinger, ed.), pp. 104–114. Am. Soc. Microbiol., Washington D.C.
Hershfield, V., Boyer, H. W., Yanovsky, C., Lovett, M. A., and Helinski, D. R. (1974). *Proc. Natl. Acad. Sci. U.S.A.* **71,** 3455–3459.
Hinnen, A., Hicks, J. B., and Fink, G. R. (1978). *Proc. Natl. Acad. Sci., U.S.A.* **75,** 1929–1933.
Hohfeld, R., and Vielmutter, W. (1973). *Nature (London), New Biol.* **242,** 130–133.
Holliday, R. (1956). *Nature (London)* **178,** 987–990.
Hopwood, D. A. (1970). *Methods Microbiol.* **3A,** 363–433.
Hopwood, D. A. (1973). *In* "Actinomycetales" (G. Sykes and F. A. Skinner, eds.), pp. 131–153. Academic Press, New York.
Hopwood, D. A. (1976). *In* "Microbiology—1976" (D. Schlessinger, ed.), pp. 558–562. American Society for Microbiology, Washington, D.C.
Hopwood, D. A. (1977). *Dev. Ind. Microbiol.* **18,** 9–21.
Hopwood, D. A., and Merrick, M. J. (1977). *Bacteriol. Rev.* **41,** 595–635.
Hopwood, D. A., and Wright, A. M. (1972). *J. Gen. Microbiol.* **71,** 383–398.
Hopwood, D. A., Chater, K. F., Dowding, J. E., and Vivian, A. (1973). *Bacteriol. Rev.* **37,** 371–405.
Hopwood, D. A., Wright, H. M., Bibb, M. J., and Cohen, S. N. (1977). *Nature (London)* **268,** 171–174.
Hošťálek, Z., and Vaněk, Z. (1973). *In* "Genetics of Industrial Microorganisms" (Z. Vaněk, Z. Hošťálek, and J. Cudlín, eds.), Vol. 2, pp. 353–371. Elsevier, Amsterdam.
Ibuki, A., and Takinami, K. (1975). *Abstr. Meet., Agricl. Soc. Jpn., 1975,* p. 341.
Ikeda, Y., Nakumura, K., Uchida, K., and Ishitani, C. (1957). *J. Gen. Appl. Microbiol.* **3,** 93–126.
Ilczuk, Z. (1971). *Nahrung* **15,** 381–386.
Itakura, K., Hirose, T., Crea, R., Riggs, A., Heyneker, H., Bolivar, F., and Boyer, H. W. (1977). *Science* **198,** 1056–1063.
James, L. V., Rubbo, S. D., and Gardener, J. F. (1956). *J. Gen. Microbiol.* **14,** 223–277.
Kahler, R., and Noack, D. (1974). *Z. Allg. Mikrobiol.* **14,** 529–535.
Katagiri, I. (1954). *J. Antibiot.* **7,** 45–52.
Kelner, A. (1949). *J. Bacteriol.* **57,** 73–92.
Kirby, R., Wright, L. F., and Hopwood, D. A. (1975). *Nature (London)* **254,** 265–267.
Komatsu, K., and Kodaira, R. (1977). *J. Antibiot.* **30,** 226–233.
Komatsu, K., Mizumo, M., and Kodaira, R. (1975). *J. Antibiot.* **28,** 881–888.
Lederberg, J., and Tatum, E. L. (1946). *Nature (London)* **158,** 558–560.
Livingston, D. M., and Klein, D. M. (1977). *J. Bacteriol.* **129,** 472–481.
McCormick, J. R. D. (1967). *In* "Antibiotics" (D. Gottlieb and P. D. Shaw, eds.), Vol. 2, pp. 113–121. Springer-Verlag, Berlin and New York.

McCormick, J. R. D., Sjölander, N. L., Hirsch, U., Jensen, E., and Doershuk, A. P. (1957). *J. Amer. Chem. Soc.* **79,** 4561–4563.
Macdonald, K. D. (1964). *Nature (London)* **204,** 404–405.
Macdonald, K. D., Hutchinson, J. M., and Gillett, W. A. (1964). *Antonie van Leeuwenhoek* **30,** 209–224.
Matselyukh, D. P., and Mukvic, N. J. (1973). *Mikrobiol. Zh.* **35,** 411–417.
Merrick, M. J. (1976). *Proc.—Int. Symp. Genet. Ind. Microorg., 2nd, 1974* pp. 229–242.
Miller, P. A., Hash, J. H., Lincks, M., and Bohonos, N. (1965). *Biochem. Biophys. Res. Commun.* **3,** 575–577.
Mishra, N. C. (1976). *Nature (London)* **264,** 251–253.
Mitscher, L. A. (1968). *J. Pharm. Sci.* **57,** 1633–1638.
Morton, A. G. (1961). *Proc. R. Soc. London, Ser. B* **153,** 548–569.
Nagaoka, K., and Demain, A. L. (1975). *J. Antibiot.* **28,** 627–635.
Nakayama, K. (1976). *Process Biochem.* **11,** 4–9.
Nash, C. H., and Huber, F. M. (1971). *Appl. Microbiol.* **22,** 6–10.
Nga, B. H., and Roper, J. A. (1969). *Genet. Res.* **14,** 63–70.
Nüesch, J., Treichler, H. J., and Liersch, M. (1973). *In* "Genetics of Industrial Microorganisms" (Z. Vaněk, Z. Hošťálek, and J. Cudlín, eds.), Vol. 2, pp. 309–334. Elsevier, Amsterdam.
Nüesch, J., Hinnen, A., Liersch, M., and Treichler, H. J. (1976). *Proc.—Int. Symp. Genet. Microorg., 2nd, 1974* pp. 451–472.
Oeschger, M. P., and Berlyn, M. K. B. (1974). *Mol. Gen. Genet.* **134,** 77–83.
Ogawara, H., and Nozaki, S. (1977). *J. Antibiot.* **30,** 337–339.
Okanishi, M., and Gregory, K. F. (1970). *Can. J. Microbiol.* **16,** 1139–1143.
Okanishi, M., Ohta, T., and Umezawa, H. (1970). *J. Antibiot.* **23,** 45–47.
Orgel, L. E. (1965). *Adv. Enzymol.* **27,** 289–346.
Ostroukov, A. A., and Kuznetsov, V. P. (1963). *Antibiotiki* **8,** 33–35.
Ostrowska-Krysiak, B. (1971). *Acta Microbiol. Pol., Ser. B* **3,** 23–29.
Pan, C. H., Hepler, L., and Elander, R. P. (1972). *Dev. Ind. Microbiol.* **13,** 103–112.
Pan, C. H., Hepler, L., and Elander, R. P. (1975). *J. Ferment. Technol.* **53,** 854–861.
Pathak, S. G., and Elander, R. P. (1971). *Appl. Microbiol.* **30,** 750–758.
Perlman, D., and Ogawa, Y. (1976). *J. Antibiot.* **29,** 1112–1113.
Pogell, B., Sankaran, L., Redshaw, P. A., and McCann, P. A. (1976). *In* "Microbiology—1976" (D. Schlessinger, ed.), pp. 543–547. Am. Soc. Microbiol., Washington, D.C.
Pontecorvo, G., and Sermonti, G. (1953). *Nature (London)* **172,** 126–127.
Preuss, D. L., and Johnson, M. J. (1967). *J. Bacteriol.* **94,** 1502–1508.
Queener, S. W., Capone, J. J., Radue, A. B., and Nagarajan, R. (1974). *Antimicrob. Agents & Chemother.* **6,** 334–337.
Queener, S. W., McDermott, J., and Radue, A. B. (1975). *Antimicrob. Agents & Chemother.* **7,** 646–651.
Randazzo, R., Sermont, G., Carere, A., and Bignami, M. (1973). *J. Bacteriol.* **113,** 500–509.
Randazzo, R., Sciandrello, G., Carere, A., Bignami, M., Velcich, A., and Sermonti, G. (1976). *Mutat. Res.* **36,** 291–297.
Raper, J. R., Raper, C. A., and Miller, R. E. (1972). *Mycologia* **64,** 1088–1117.
Redshaw, P. A., McCann, P. A., Sankaran, L., and Pogell, B. M. (1976). *J. Bacteriol* **125,** 698–705.
Roper, J. A. (1973). *In* "Genetics of Industrial Microorganisms" (Z. Vaněk, Z. Hošťálek, and J. Cudlín, eds.), Vol. 2, pp. 81–88. Elsevier, Amsterdam.
Rosi, D., Goss, W. A., and Daum, S. J. (1977). *J. Antibiot.* **30,** 88–97.
Roulland-Dussoix, D., Yoshimori, R., Green, P., Betlack, M., Goodman, H. M., and Boyer, H. W. (1975). *In* "Microbiology—1974" (D. Schlessinger, ed.), pp. 187–198. Am. Soc. Microbiol., Washington, D.C.

Sanchez, S., and Quinto, C. M. (1975). *Appl. Microbiol.* **30,** 750–758.
Schaeffer, P. (1969). *Bacteriol. Rev.* **33,** 48–71.
Schrempf, H., Bujard, H., Hopwood, D. A., and Goebel, W. (1975). *J. Bacteriol.* **121,** 416–421.
Segal, I. H., and Johnson, M. J. (1961). *J. Bacteriol.* **81,** 81–98.
Sermonti, G. (1959). *Ann. N.Y. Acad. Sci.* **81,** 950–956.
Sermonti, G. (1969). "Genetics of Antibiotic-Producing Microorganisms." Wiley, New York.
Shaw, P. D., and Piwowarski, J. (1977). *J. Antibiot.* **30,** 404–408.
Shier, W. T., Rinehart, K. L., Jr., and Gottlieb, D. (1969). *Proc. Natl. Acad. Sci. U.S.A.* **63,** 198–204.
Singer, M. (1977). *Gene* **1,** 123–139.
Spizek, J., Malek, I., Dalezilova, L., Vondracek, M., and Vaněk, Z. (1965). *Foli Microbiol.* **10,** 259–262.
Stark, W. M., Knox, N. G., and Wilgus, R. M. (1971). *Folia Microbiol. (Prague)* **16,** 205–217.
Stauffer, J. F. (1961). *Sci. Reps. Ist. Super. Sanita* **1,** 441–447.
Szybalski, W. (1977). *Gene* **1,** 181–183.
Testa, R. T., Wagman, G. H., Daniel, P. J. L., and Weinstein, M. (1974). *J. Antibiot.* **27,** 917–921.
Troonen, H., Roelants, P., and Boon, B. (1976). *J. Antibiot.* **29,** 1258–1267.
Ullrich, A., Shine, J., Chirgwin, J., Pictet, R., Tischer, E., Rutter, W. J., and Goodman, H. M. (1977). *Science* **196,** 1313–1319.
Vezina, C., Bolduc, C., Kudelski, A., and Sehgal, S. N. (1976). *J. Antibiot.* **29,** 248–264.
Wade, N. (1977). *Science* **195,** 558–560.
Wagner, R. P., and Mitchell, H. K. (1964). "Genetics and Metabolism," 2nd ed. Wiley, New York.
Watson, B., Currier, T. C., Gordon, M. P., Chilton, M., and Nester, E. W. (1975). *J. Bacteriol.* **123,** 255–264.
Weiss, M., and Green, H. (1967). *Proc. Natl. Acad. Sci. U.S.A.* **58,** 1104–1111.
Whitaker, A., and Long, P. A. (1973). *Process Biochem.* **8,** 27–31.
White, R. J., Martinelli, E., and Lancini, G. (1974). *Proc. Natl. Acad. Sci. U.S.A.* **71,** 3260–3264.
Wright, L. F., and Hopwood, D. A. (1976). *J. Gen. Microbiol.* **95,** 96–106.

Chapter 13

Methods for Laboratory Fermentations

N. D. DAVIS
W. T. BLEVINS

I. INTRODUCTION

Fermentation processes applicable to industrial use are first investigated in the laboratory prior to scale-up and development of production

MICROBIAL TECHNOLOGY, 2nd ed., VOL.II
 ISBN 0-12-551502-2

processes. Although methods used in the fermentation laboratory for these small- to intermediate-scale fermentations are not necessarily unique to this specialty of microbiology, certain equipment and philosophies are routinely applied. These constitute the subject matter of this chapter.

II. EXPERIMENTAL DESIGN

A. The Hypothesis

The fermentation technologist is an experimental scientist who must design each experiment with great attention to detail. Carelessly designed experiments are a waste of resources. An experiment tests a hypothesis, and the hypothesis should be precisely written and its assumptions and implications thoroughly examined and understood in planning the experiment.

B. Fermentation Experiments

An experiment should be designed in such a manner that it unequivocally demonstrates that the hypothesis is either true or false; such an experiment is quite rare. In practice it is usually a matter of to what relative degree the data support or contradict the hypothesis.

In designing an experiment, one must decide in advance which measurements will be made, how they will be made, and what data format will be required. The hypothesis dictates the general form the experiment must take. The nature of the data to be accumulated is the dominant factor in making such decisions as the number of replications, method of randomization, and number of repetitions. Inherent biological variation and precision of measurements and analyses are always factors to be considered. The experimenter should be acquainted with range, variance, standard deviation, significant differences among measurements, and confidence limits for interpreting data. Thus, the fermentation scientist needs a thorough grasp of the principles of biostatistics (Campbell, 1974) and should consult with a statistician during the design of an experiment.

C. Record Keeping and Data Processing

Complete, accurate, and detailed record keeping is essential. Recording data directly into bound record books, dating and initialing entries, and recording all data, observations, and conclusions should be routine.

Experiments should be numbered and data organized in a manner that allows easy access to original data, analyses, hypotheses, and conclusions years later. Information control, record keeping, and methods of good laboratory management are detailed by Hockenhull (1977).

Computer facilities for data processing are available in many laboratories and are considered a necessity rather than a luxury. Also, the age of "smart" instruments coupled to mini- and midicomputers is here and growing rapidly; computer analysis and control of fermentations is becoming widespread. A basic knowledge of computer technology is becoming increasingly important to the fermentation technologist. Computer control of fermentations is discussed in this volume, Chapter 15. Computer technology applicable to laboratory fermentations has recently been discussed by Humphrey (1976) and Isenberg and MacLowry (1976).

III. SCREENING TECHNIQUES

A. Primary Screening

1. General Considerations

A mixed population of microorganisms may contain a number of species which have potential for industrial applications. Primary screening programs involve highly selective procedures for detecting and isolating individual species from a mixed population. The most frequently used screening procedure is to isolate microorganisms from soil dilutions. Soil is useful since it is a habitat where organisms are involved in decomposition and resynthesis of complex organic materials and in various transformations of inorganic materials. The spread-plate or crowded-plate technique commonly is used; these methods involve use of various differential and selective media. General principles for microbial culture selection are reviewed in this volume, Chapter 12. Hopwood (1970) has discussed the isolation of mutants.

2. Biologically Active Substances

a. General Techniques. During growth of a mixed population of microorganisms, a chemically defined liquid culture medium may be utilized so that a given species will grow at a faster rate than the other organisms. Successive transfers to this specialized medium permit the desired organisms to become the predominant species. Selection may also develop if unwanted organisms are inhibited, while organisms of interest are permitted to grow. One can, therefore, "enrich" a medium in a manner that favors the desired species or physiological type, and these species can be isolated to pure culture by plating on solid medium of the same composition (Stanier *et al.*, 1976). For example, an organism capable of using atmospheric nitrogen (N_2) as the sole source of nitrogen may be isolated by using a medium devoid of fixed nitrogen. If N_2 is available, only those organisms that can fix atmospheric nitrogen can grow. Enrichment culture experiments utilize such factors as the carbon source, energy supply, temperature, oxygen level, and hydrogen ion concentration. Microorganisms capable of specific biotransformations may be isolated by this method, as may degraders of cellulose or other polymers, hydrocarbons, and other compounds of interest.

Selective pressures can also be exerted by using osmotic pressure, an antibiotic, various inhibitors, or other factors as desired. For direct isolation, a mixed inoculum containing several species is spread directly on the surface of a selective solid medium; all microorganisms which can grow will produce colonies. The maximum possible number of types of microbes that can grow in a particular environment can be isolated by this method.

Once an organism capable of producing some amino acid, vitamin, or enzyme is isolated and purified, strain improvement techniques are usually employed to improve the growth of the organism and/or the production of the desired fermentation product. These methods are discussed by Fantini (1976).

b. Antibiotics and Other Growth-Inhibiting Agents. The discovery of penicillin by Alexander Fleming in 1929 and the demands for antibiotics during World War II stimulated technological advances for the detection and large-scale production of antibiotics and other antagonistic compounds. Intensive screening programs were developed to find microorganisms capable of producing chemotherapeutically useful antibiotics (Spooner and Sykes, 1972).

Antibiotics generally are defined as those organic compounds produced by microorganisms that are antagonistic in low concentrations to the growth or survival of microorganisms (Casida, 1968). Other metabolic products, such as ketones, acids, and amines, may be inhibitory to growth but are not considered to be antibiotics. Antibiotics are produced primarily by bacteria belonging to the genera *Streptomyces, Nocardia,* and *Micromonospora* (the actinomycetes); by *Bacillus* (a eubacterium); and by fungi, notably *Penicillium.* Of over 3000 known antibiotics, approximately 70% are produced by actinomycetes, 20% by fungi, and 10% by eubacteria.

A detailed survey of laboratory techniques for isolation of antibiotic producers and the detection and identification of antibiotics has recently been published (Hash, 1976). In a screening program, several general items are of importance. Although there are a few thermophilic and psychrophilic antibiotic producers, most prokaryotic organisms are mesophilic, growing best over a temperature range of 23°–37°C. Antibiotic-producing fungi grow best at 20°–22°C; actinomycetes grow optimally around 28°C. Fungi can grow at a lower pH than most of the bacteria, so acidity can be used to inhibit bacterial growth during the screening process. Furthermore, the pH for optimal growth may be different from the pH favoring optimal production of the antibiotic, and the pH of assay media can drastically affect the activity of the antibiotic. Also, most of the antibiotic producers are highly aerobic, so appropriate aeration techniques are employed.

The nutritional conditions used for growth of antibiotic producers are also important. The isolation media for actinomycetes are usually enriched with carbon sources, such as glycerol, chitin, or starch, with organic nitrogen sources such as casein, arginine, and asparagine. Nitrate or ammonia also may serve as nitrogen sources.

In screening soil or mud, samples may be pretreated in several ways to increase the proportion of actinomycetes. Air-drying the sample will reduce the number of vegetative bacterial cells while allowing actinomycete spores to survive. Mixing soil with $CaCO_3$ for several days also may be effective. Overgrowth by fungi can be suppressed by using antifungal agents in the growth medium. For isolation, appropriate dilutions of the sample are plated onto enriched agar by the spread-plate method. Agar plates should be relatively dry on the surface to prevent the spread of motile bacteria. Following incubation for 5–7 days at 25°–28°C, actinomycetes appear as tough, chalky, white to grey colonies. Those that appear to be antagonistic to surrounding bacteria should be removed for further testing. Antagonism is best observed using a crowded-plate technique, in which plates having 400 or more colonies are examined. Antibiotic-producing colonies are subcultured to a similar medium and purified by streaking.

Nonantibiotic compounds that are inhibitory to growth of test bacteria may be produced by various organisms. Organic acids or amines, for example, may antagonize other organisms growing around colonies producing these compounds. These organisms initially might be presumed to produce antibiotics, so further testing is always necessary. These basic or acidic compounds can be detected by incorporating various pH indicators into poorly buffered agar media. Colonies producing these compounds may cause a color change in the surrounding agar due to changes in acidity or alkalinity.

c. Vitamins and Other Growth Factors. Primary screening for organisms that produce vitamins or other growth factors requires a bioassay organism that has an absolute requirement for the particular growth factor in question. If an isolate produces the growth factor extracellularly, its metabolic activities will permit growth of the bioassay organism. The detection of organisms that produce growth factors needed in human nutrition relies on microbiological assays using auxotrophs that have a specific requirement for a particular growth factor in order to grow. This can encompass requirements for vitamins, amino acids, purines, pyrimidines, etc.

Vitamins serve as precursors for coenzymes that function as catalysts in various enzyme systems; thus, deficiencies can lead to metabolic blocks reversible only when the vitamin is present in the nutrient supply (Hanson, 1967; Skeggs, 1976). These vitamin-requiring organisms may occur naturally (for example, lactic acid bacteria) or may be auxotrophic mutants isolated in the laboratory (Hopwood, 1970). Various factors must be considered in choosing an assay system for specific growth factors. The assay organism may be able to grow if only a precursor of the growth factor is present. For example, the thiazole or pyrimidine moiety of thiamine can be used by *Saccharomyces cerevisiae* during growth; thus, this presumptive thiamine-requiring organism can grow without the vitamin actually being supplied. Furthermore, end products of enzymatic reactions catalyzed by the growth factor might support growth if they are present in the medium. This occurs with some assay organisms for vitamin B_{12}, wherein deoxyribosides may support growth (Skeggs, 1976).

The medium for assaying for a particular growth factor should contain all nutrients necessary for optimal growth except for the one factor being measured. The medium should ideally support no growth in the absence of the growth factor, so that no turbidity develops in culture tubes lacking the factor. Most vitamin assays can be done using commercially prepared dehydrated media.

A screening approach similar to that used for antibiotics can be employed to isolate microorganisms that can synthesize extracellular vitamins, amino acids, or other growth factors from natural sources. For example, glutamic acid-producing cultures were first isolated in Japan by screening feces of various birds and mammals. Cross-feeding of the growth factor-requiring auxotroph by the growth factor producer will occur on a medium deficient in the particular required growth factor. Large numbers of isolates can be screened using the appropriate microbiological assay organism, and zones of growth stimulation by streaks or colonies of the isolate is indicative of growth factor synthesis.

Plates of a medium favoring production of a particular growth factor (if this is known) are allowed to incubate for 2–3 days at 28°C following inoculation of an appropriate dilution of some natural source. These plates may be replicated to a medium of the same composition using the replica-plate method of Lederberg and Lederberg (1952); colonies on newly inoculated plates will arise at the same site as on the original plate. After 2–3 days growth, the microorganisms on the replicated plates can be killed by exposure to lethal doses of ultraviolet (uv) light. The plates are then overlaid with a melted medium containing the specific assay organism. After incubation for 24 hours, the assay organism will grow in the immediate areas where any isolated colonies had produced the desired growth factor. By comparing the locations of the turbid areas with those on the original plate, the viable organisms which were responsible for production of the growth factor can be isolated (Miller and Litsky, 1976). These strains can be purified by the streak-plate technique and grown in a liquid medium; subsequent isolation and identification of the growth factor can be accomplished.

Amino acid producers have come primarily from species of bacteria; thus, fungal inhibitors can be added to the medium to prevent overgrowth by these eukaryotes (Dulaney, 1967). On the other hand, vitamins have been produced mostly by eukaryotes such as yeasts, fungi, and algae, especially for large-scale productions.

d. Enzymes. Enzyme-catalyzed reactions using intact growing cells present problems that have stimulated studies involving the use of cell-free enzyme preparations in fermentations. Enzymes are widely used in pharmaceutical, brewing, textile, cheese, leather, meat, corn syrup, wine, baking, and paper industries. Plants, animals, and microorganisms may serve as sources of these enzymes, but the convenience and economy of using microorganisms warrants special emphasis on this aspect of enzyme production (Underkofler, 1976).

Species of bacteria, molds, and yeasts have been useful in large-scale production of enzymes. As outlined by Beckhorn (1967), the organism must be nonpathogenic, produce no toxic products, and be genetically stable with respect to the production of the enzyme in question. Cultures may be screened from stock collections or from natural sources. A useful technique is the enrichment culture method discussed previously; another technique is one in which the potential substrate is incorporated into the growth medium. For instance, production of amylase and protease by various isolates can be detected by incorporating soluble starch or casein, respectively, into the agar medium. Clear zones of substrate degradation are observed around the isolates that are synthesizing the extracellular enzymes. Lack of a detection system on agar requires enzyme analysis of the supernatant fraction following growth in liquid media. Isolates that show potential can be mutated to try to improve production (Hopwood, 1970).

3. Biomass Production

Aspects of biomass production procedures have been treated extensively in earlier chapters and also in the 1975 International Symposium on Single Cell Protein (Tannenbaum and Wang, 1975). Filamentous fungi, yeasts, bacteria, and algae all have been used for production of single-cell protein (SCP). Bacteria and yeast have the briefest generation times, thus producing more cells faster. Bacteria have a comparatively high nucleic acid content that contributes to toxicity of bacterial SCP intended for human consumption. However, slower growth rates can limit nucleic acid content, thus minimizing this disadvantage. Yeasts and algae are rich sources of B vitamins, while fungal mycelia generally contain lesser amounts. Also, bacteria are higher in certain vitamins and in sulfur-containing amino acids than are yeasts. The ideal organism for biomass production should be thermoduric and have a high protein content, relatively low nucleic acid content, a rapid growth rate, and high survivability or robustness. Toxic and pathogenic properties of the organism must be determined by feeding studies when evaluating its potential as a food supplement.

4. Fermentation Products

The products of a fermentation process may be tremendously diverse, requiring separation, identification, and quantitation of many compounds. A complete analysis of fermentation products can be accomplished by methods discussed by Dawes *et al.* (1971). Gaseous products are best analyzed by gas chromatography–mass spectrometry (GC–MS). Headspace vapors of fermentation vessels or fermentation tubes may be sampled and analyzed directly for methane, CO_2, etc. Separation of volatile products from cleared fermentation media can be achieved by distillation after adjustment to the desired pH (7.0–7.5 for neutral volatiles). Volatiles that may be present are methanol, ethanol, isopropanol, butanol, acetoin, diacetyl, acetone, acetaldehyde, or volatile fatty acids (steam distilled at pH 2.8–3.0) such as formic, acetic, propionic, and butyric acids. Also, various volatile organic compounds can be trapped on a chromatography column (or freezing trap) connected in tandem to a closed fermentation train. These compounds can subsequently be eluted and analyzed by GC–MS.

Ether extraction of a neutral, cleared solution will leave sugars, sugar alcohols, and organic acids in the aqueous phase; these can be acidified and subjected to continuous ether extraction. Organic acids will be removed in the ether phase with sugars and sugar alcohols remaining in the residual aqueous phase. The nonvolatile components may be separated and quantitated by various techniques (Dawes *et al.*, 1971).

5. Biological Techniques

A biological assay may be necessary for detection or assay of some fermentation products (Roberts and Boyce, 1972). A sensitive test organism may be observed for effects of the compound on growth, or the concentration of the compound may be determined by enzymatic analysis. Use of auxotrophic test organisms that require specific compounds for growth has been discussed earlier. The amount of growth observed is directly proportional to the concentration of the required compound, as long as that compound is the limiting growth factor. The amount of growth can be measured turbidimetrically. Bacteria, yeast, and fungi have been used as test organisms in the analysis of amino acids, antibiotics, vitamins, etc.

Employing a solid medium inoculated with a test organism, diffusion assays are carried out in which zones of growth or inhibition will occur as a particular compound diffuses from a cup or disk into the surrounding agar. The diameter of the zone can be compared with a standard concentration (Cooper, 1963).

A turbidimetric end-point determination can be made on inhibitory substances by observing the inhibitory effect on growth of twofold serial dilutions of the agent. The minimal dilution at which growth of the test organism is still inhibited can be used to determine the relative amount of agent in the undiluted sample (Kavanaugh, 1963).

Metabolic responses to the presence of a fermentation product may be diverse, and various detection methods may be used. Among these are titration of acid produced, measurement of oxygen absorbed or CO_2 evolved, or measurement of a variety of enzyme activities. These techniques may require use of respirometers, radioisotopes, spectrophotometers, chromatography or other methods.

B. Secondary Screening

1. General Considerations

Secondary screening qualitatively and quantitatively expands information gained from the primary screening program. Secondary screening represents a major commitment of time and resources to determine whether a given microorganism possesses the potential for use in a new or improved industrial fermentation. As such, secondary screening must serve to eliminate worthless microorganisms and fermentation procedures from further consideration as quickly and as inexpensively as possible.

Secondary screening involves many of the same techniques used in primary screening. Usually small flasks are utilized during early stages, small fermentors during intermediate stages, and pilot-plant fermentors for scaleup during later stages. While pH indicators or calcium carbonate may have been used in primary screening to reveal microorganisms that have produced acids, secondary screening may be used to determine precisely which acids are produced and the quantity of each acid accumulated. Effects of various environmental parameters on total and individual acids produced may be studied in order to develop a procedure that maximizes the production of a specific acid. Thus, much more elaborate experimentation is required for secondary than for primary screening.

2. Product Recovery

Standard techniques for biomass recovery include centrifugation and filtration. These and other methods for harvesting biomass are detailed by Thomson and Foster (1970). Recovery of a fermentation product may be a simple procedure, such as adjusting the acidity of the medium to bring about flocculation or precipitation; the desired product then can be recovered by centrifugation or filtration. In some cases the product can be precipitated as water-insoluble calcium or barium salts. Precipitation of proteins from saturated salt solutions is also a useful technique. Concentration and precipitation in an organic solvent, in which the product of interest is insoluble, is best for certain metabolites, i.e., addition of a concentrated chloroform solution of aflatoxin into hexane, whereupon the aflatoxin is precipitated.

Recovery techniques also may be based on vacuum distillation, steam distillation, and fractional distillations that are much used in the well-equipped fermentation laboratory.

Solvent extraction following pH adjustment is another fundamental technique. Adjustment of pH can suppress ionization of acids or bases in aqueous solutions. Many nonionized organic molecules may then be extracted into nonmiscible organic solvents. Evaporation of the solvent then concentrates the metabolites. Product recovery methods are ideally designed not only to harvest the desired product from the fermentation medium, but also to concentrate and purify it if possible.

3. Product Detection

A product detection method is designed to take advantage of known chemical, physical, or biological properties of the fermentation product of interest. Product accumulation can be monitored indirectly by taking advantage of any correlation that exists between metabolite formation and some other variable, such as medium pH, optical density, color, and titratable acidity. Growth can be monitored or measured by a large variety of methods that are discussed in detail by Humphrey (1976), Isenberg and MacLowry (1976), and Kubitschek (1969). Biomass estimations may involve use of centrifugation to determine packed cell volume or weight, nephelometry to determine cell numbers via standard curves, cell counting chambers of several types, and electronic instruments. Other methods are based on dry weight measurements, nitrogen determinations, radionuclide uptake, and on-stream monitoring of optical density (OD) using a micro-OD unit coupled to a digital display device or a recorder via fiber optics. Evaluation of growth by physical and chemical means is discussed by Mallette (1969). The pH and other environmental parameters can be measured on-stream by special electrodes connected to appropriate meters or other devices. Oxygen electrodes are discussed by Beechey (1972). Electrode measurements of carbon dioxide are reviewed by Nicholls (1972). Samples may be withdrawn periodically and the pH or other variables measured. Titratable acidity of samples is more informative than simple pH measurements and also can form the basis of a method of quantitative analysis. Volatile titratable acidity, due to acetic acid, can be determined separately from nonvolatile acidity due to other acids following steam distillation.

Visible and uv spectrophotometric monitoring methods are very sensitive and convenient where applicable. Compounds that absorb visible or uv radiation at specific energy levels can be monitored using appropriately equipped spectrophotometers. Fluorometry is a sensitive method that can be used to follow the accumulation of fluorescing metabolites such as riboflavin. Biologically active substances can be monitored using various biological assays, as previously discussed.

Spot testing of culture media is sometimes convenient for monitoring fermentation products (Feigl, 1977), although test sensitivity and small concentrations of the product of interest often prevent development of a useful procedure. For example, kojic acid forms a deep red color in the presence of alcoholic ferric chloride. This test is very sensitive and the intensity of the color produced can be used as a quantitative analysis for kojate. Obviously, sensitivity of such tests can be increased by extraction and concentration of the product. Most end products can be monitored by various combinations of chromatography and qualitative chemical analyses. Instrumental methods are useful for monitoring as well as for analyses (Gouw, 1972; Weber, 1973). It is a challenge to the fermentation technologist to select from among the many techniques available and to develop a monitoring system for a specific microbial end product.

4. General Analytical Methods

Qualitative analyses of fermentation products generally consist of isolation, concentration, purification, and identification or characterization of the product. Dawes *et al.* (1971) comprehensively discuss methods of analysis of fermentation products; Herbert *et al.* (1971) consider chemical analyses of microbial cells. Concentration of metabolites is often essential and can be accomplished by standard separation methods (Miller, 1975), such as precipitation and filtration or centrifugation, as previously discussed. Other techniques include extraction (Sutherland and Wilkinson, 1971) or removal of water through evaporation, or by specialized techniques such as freeze drying, dialysis, hollow fibers, absorption onto silica, charcoal, cellulose, etc. Liquid–liquid or soxhlet extractions are also common techniques for concentration and partial purification of various classes of compounds. After extraction, organic solvents are easily evaporated , and the residue containing the fermentation product(s) is further processed.

Identification of components in concentrated extracts is frequently accomplished following some form of chromatography (Perry *et al.*, 1972). These include gas, paper, thin-layer (TLC), and column chromatography procedures, basic types of which are adsorption, ion-exchange, partition, and exclusion chromatography. High-performance liquid chromatography (Brown, 1973) is also a standard qualitative and quantitative tool in many analytical laboratories.

Thin-layer chromatography is a particularly useful and versatile tool for the fermentation specialist who must isolate and identify small quantities of individual products from a mixture of microbial metabolites (Stahl, 1969). Preparatory-scale TLC generally can quickly provide sufficient quantities of purified materials to achieve identification. Also, spot testing directly on TLC plates may reveal the class of organic compounds to which a particular metabolite belongs, thus facilitating other qualitative analyses (Krebs *et al.*, 1969; Feigl, 1977). Many products can be easily and quickly isolated by preparatory-scale TLC and identified by comparing uv or visible absorption spectra with published spectra or with spectra of authentic samples of suspected metabolites. Infrared (ir) spectrophotometry also is convenient on quantities of material isolated from preparatory TLC plates. Published spectra and computer or manual searches of published ir indices are convenient for identifying a great many organic compounds. Mass spectrometry (MS) is another powerful tool for identification of organic compounds. Low- and high-resolution MS analyses may be purchased from commercial or university laboratories. Molecular weight and elemental analyses coupled with uv, visible, and ir spectra are frequently sufficient for positive identification of a fermentation product. Additional data, such as melting points, TLC R_f values in different solvent systems, spot chemical tests, and fluorescence analyses usually provide more than sufficient data for identification except for previously unidentified metabolites.

IV. MEDIA SELECTION

A. General Considerations

The screening and selection of an organism for fermentation processes require selective and differential media for detection and isolation of the desired microorganism. The chemical composition of the medium to be used for the fermentation and the environmental conditions to be employed are very important. The medium must contain nutrients that supply carbon and energy, necessary growth factors such as vitamins and amino acids, inorganic salts, and any precursors essential to product formation. The choice of a good medium is of interest from the standpoint of utility and economy. Booth (1971) summarized information pertaining to culture media for fungi. Lapage *et al.* (1970) have compiled information on media for growth of microorganisms in general. Calam (1969), Bridson and Brecker (1970), and Pirt (1975) can be consulted for detailed information on most aspects of the culture of microorganisms in liquid media. Handling and storage of media and equipment are discussed by Elliott and Georgala (1969).

B. General Purpose Media

1. Solid or Semisolid Media

a. Solidifying Agents. Primary screening for potentially useful microorganisms requires the use of solid media in many instances (Codner, 1969). Some agent is added to a liquid medium, heated to the appropriate temperature for solubilizing the agent, mixed thoroughly, and allowed to solidify. The ideal solidifying agent is one which is not degraded by the microorganism(s) to be studied, so that growth is totally at the expense of added nutrients.

The most commonly used solidifying agent is agar, a polysaccharide not degraded by most bacteria. Normally a 1.5–2.0% concentration of the dehydrated agent is added to a liquid, heated to boiling, then allowed to solidify at approximately 42°C. For semisolid media, a concentration of 0.7% or less can be employed.

Other semisolid culture media contain a high concentration of bran or some cereal adjusted to the desired moisture content. The microorganisms are grown either in stationary culture on the surface of the medium in trays or while being agitated in rotating drums.

b. Synthetic, Semisynthetic, and Complex Media. Fermentation media can be classified as synthetic, semisynthetic, or complex. A synthetic medium is one in which the exact composition and quantity of ingredients is known; a semisynthetic medium is one which has very few unknown ingredients. A synthetic or semisynthetic medium is very useful in determining the specific capabilities of an organism. Such media can be manipulated by varying levels of known compounds to observe the effects of such alteration on growth and/or product formation. A medium solidified with agar technically is a semisynthetic medium, although the exact chemical nature of other ingredients is known, since agar is not a pure polymer as normally employed.

A complex medium is one in which specific components are unknown because of the addition of crude ingredients such as molasses, corn syrup, wheat bran, corn steep liquor, yeast extract, or protein digests. In later stages of screening, these crude ingredients may be valuable in promoting growth and/or product formation and would thus be important in large-scale operations. These complex media may stimulate growth and product formation because of unknown growth factors. Conversely, inhibitory effects may be observed when certain crude substrates are used due to feedback control mechanisms operating in the microorganism. Another disadvantage is that addition of proteins or other crude ingredients may increase foaming in liquid culture; this is not a problem when solid or semisolid media are used. Usually, addition of specific inorganic nutrients is unnecessary when complex media are used, since such media contain sufficient quantities of anions and cations to sustain growth. If a synthetic medium is to be used, these ions must be added. If toxic ions are present in some crude ingredient, chelating agents to bind these ions may be added to decrease inhibitory effects.

The utility of solid or semisolid media in primary screening is obvious; however, in large-scale operations submerged liquid culture is often preferred because it is more flexible and easier to control.

2. Liquid Media

a. Complex and Semisynthetic Media. During screening, the capabilities of an organism initially may be determined using a synthetic medium, the composition of which can be easily altered. In intermediate scale-up studies, submerged cultures with liquid media must be employed, since this probably will be the method of choice in a large-scale operation. Use of complex media will probably be cheaper. Other advantages and disadvantages have already been discussed. The complex medium must support sufficient growth and also promote enzyme or product formation. Crude sources of carbohydrates (starch wastes, cellulosic wastes, molasses, corn syrup, sulfite waste liquor), nitrogenous compounds (casein, gelatin, protein digests), and growth factors (yeast extract, corn steep liquor, etc.) may be added and the effects on growth and/or product formation observed. On a laboratory scale, liquid cultures can be incubated in shake flasks or small fermentors following sterilization of the media and inoculation with the organism. Solubility of the crude additives is not a great problem since extensive agitation probably will be employed. However, shallow layers of a liquid medium may be necessary for growing aerobic organisms in stationary culture.

b. Synthetic Media. As previously discussed, media for which the exact composition and quantities of added compounds are known are called synthetic media. Such media often are required for studies on growth, metabolism, and product formation. Liquid synthetic media are employed for substrate specificity and growth factor studies. Also, the effect of precursors of some product are best observed in a synthetic medium in which other interfering compounds have been eliminated. Results of any study are more often reproducible when a synthetic medium is used. A synthetic medium is more convenient when detecting intermediates or end products, since extracted samples will be cleaner than if a complex medium were used. Also, results from using radioisotopes in metabolic studies are more meaningful if synthetic media are employed.

3. *Miscellaneous Problems*

a. pH and Buffers. During fermentation processes, microorganisms synthesize metabolic products which may drastically change the pH of the media. Excessive production of acid end products, ammonia, CO_2, or amines may cause pH changes. Altered pH may be unfavorable for optimal growth and/or product formation; thus, it is often critical to maintain the medium at a certain pH. Agents that possess buffering capacity may be added to retard drastic pH changes. The addition to media of $CaCO_3$ to minimize effects of acid production has already been mentioned. Complex media with high levels of proteins may possess good buffering capacity near neutral pH. Various inorganic compounds, such as mono- and dihydrogen potassium phosphates, are commonly used for buffering, as are organic acids or amino acids.

b. Oxidation–Reduction Potentials. The oxygen tension required for a fermentation process determines the importance of adding components that help maintain a very low oxygen tension as in anaerobic fermentations. In aerobic fermentations, the oxidation–reduction potential presents little problem. However, large-scale operations often experience problems with insufficient aeration. In larger fermentors, adequate aeration can be achieved by forced air or mechanical agitation. Methods for supplying oxygen demands are detailed by Weinshank and Garver (1967).

c. Precipitates and Other Problems. Combinations of certain salts sometimes cause precipitations to occur which reduce the availability of an essential ion in growth media. This often can be alleviated by separately adding pure solutions of the presterilized salts following autoclaving of the salt-free medium. The pH of the medium may also be important in preventing formation of precipitates. In high-protein media, low pH may cause denaturation and precipitation of some proteins. Precipitation may not be a significant problem if the fermentation requires extensive agitation or aeration.

V. EQUIPMENT

A. Static Cultures

1. Culture Vessels

The choice of containers for growth of cultures, sterilization and storage of media, inoculum preparation, etc., varies with the type of organism and fermentation (aerobic or anaerobic). In primary screening, hundreds of samples may be screened using small volumes of a medium in tubes, bottles, or flasks. Significant cell mass in static culture is obtained with larger volumes incubated in 1- to 3-liter flasks, 5-gal carboys, or in other large vessels with shallow layers of medium to provide adequate oxygen exchange. Flat bottles incubated on their sides also are commonly used to provide a large surface area for growth. Special side arm flasks are used for gaseous substrates.

Fungal fermentations may be conducted by growing cultures as surface mats, although submerged fermentations that are agitated or aerated are more common. Intermediate-scale, static cultures generally are incubated in trays or shallow pans that have been inoculated with a spore suspension. Calam (1969) has reviewed the culture of microorganisms in liquid medium, including special techniques, i.e., anaerobic and hydrocarbon fermentations.

2. *Incubators and Environmental Cabinets*

Investigations may require static cultures using solid or semisolid substrates and precise control of atmospheric conditions. Where temperature is the important parameter, refrigerator-style incubators may be used. However, many studies dealing with fermented foods, tropical deterioration, mycotoxin production, etc., require maximal aeration coupled with maximal nutrient concentration and optimal moisture. Control of relative humidity, temperature, oxygen, carbon dioxide, etc., can best be obtained in environmental cabinets, which are capable of controlling temperatures from 5° to 90°C and relative humidity from 40 to 99%. Some environmental cabinets may provide a means for monitoring and regulating gaseous composition, i.e., O_2, N_2, and CO_2 content. Grain on trays in closed cabinets can be pasteurized or sterilized in place, or grain in flasks with cotton plugs can be sterilized separately in an autoclave. Recorders provide a permanent record of environmental parameters.

B. Agitated Cultures

1. *Shakers and Incubator Shakers*

Bench-top and floor-model variable-speed gyratory or reciprocating incubator shakers with interchangeable platforms are available commercially for aerobic culture. Low-temperature incubators, water bath shakers, and incubators with light sources for photosynthesis investigations also can be purchased. Wrist action or twist action shakers may be more desirable in some studies. For incubation of large numbers of flasks, multitier shakers with changeable holders or platforms for various sizes of flasks also are available.

2. *Rotating Drums*

A rotating-drum method serves to minimize limiting effects of various diffusion gradients. Substrate moisture, nutrient concentration, and substrate–organism contact all can be maximized. Also, effects of aeration and agitation can be independently adjusted and investigated, as can the effects of temperature. This equipment consists of a horizontally mounted drum which rotates constantly during fermentation. Baffles on the inner wall facilitate agitation. Sterile air of predetermined relative humidity and temperature may be introduced into the system and effluent gases may be scrubbed of toxic particles. Also, back pressure may be regulated if desired. Various sizes of bottles and carboys also may be continuously rolled at varied speeds by roller platforms or cabinets to provide highly aerated thin films of media suitable for culturing many microbial and eukaryotic cells. This equipment may be relatively simple, open systems or may be integrated into controlled environmental cabinets.

The rotating-drum method has been used successfully for such processes as gluconic acid production, mold amylase production on wheat bran, and production of bacterial or eukaryotic cells. A chief limitation of the method is the difficulty of scale-up to pilot-plant or larger sized industrial fermentations.

C. Submerged Cultures

1. *Flasks and Carboys*

If a fermentor is not available for growth of submerged cultures, large flasks or carboys can be used. Aeration is facilitated if flasks have baffles extending from the bottom or sides which increase aeration during shaking or stirring. Aeration is best achieved using spargers and filtered sterilized air delivered under pressure to produce fine bubbles. Brown (1970) discusses details of aeration in submerged culture. The size of holes in the sparger may be varied to avoid clogging by mycelia or bacterial clumps. Aeration via magnetic stirring or mechanical agitation of carboys also can be accomplished with proper equipment.

2. *Fermentors*

Fermentors up to about 16 liters typically are used during final stages of secondary screening and for subsequent investigations to determine whether satisfactory scale-up of fermentations is possible. This is essential because microorganisms do not always perform in submerged culture as they do in stationary culture. Yields of desired metabolites may be drastically reduced and in many cases the entire spectrum of metabolites is changed. Small to intermediate fermentors (Figs. 1 and 3) also are required to determine the effects of aeration, temperature, medium composition, pH, time, etc., on the progress and rate of a submerged culture fermentation and on yields of particular products. Precise effects of various factors influencing a fermentation are best determined using a chemostat (Fig. 2) as discussed below. Design and use of fermentors are discussed by Wang and Fewkes (1976). One-liter fermentors are mostly useful for preparation and adaptation of inoculum, whereas 4- to 28-liter fermentors (Fig. 1) or chemostats are convenient for fundamental inves-

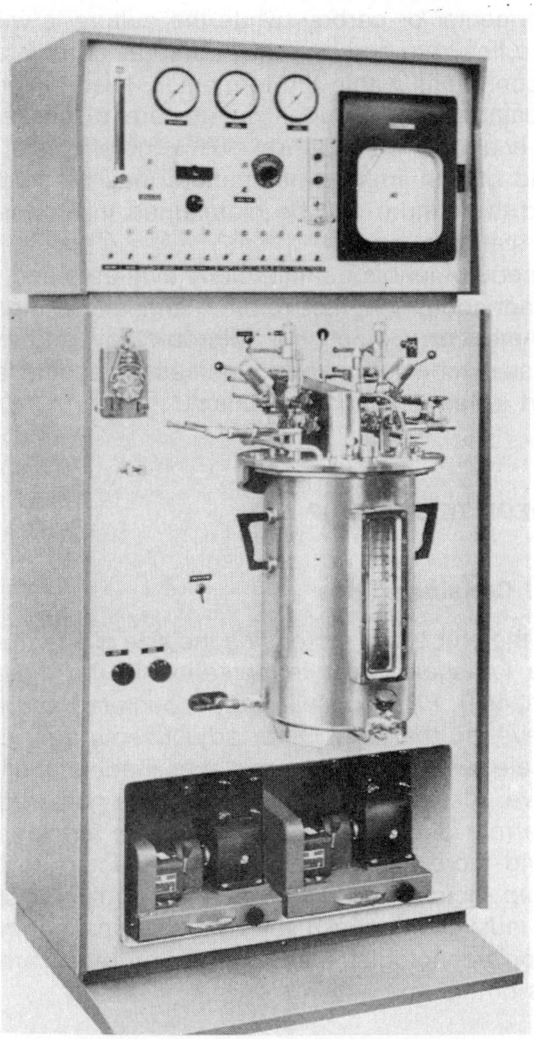

FIGURE 1. A steam sterilizable in-place fermentor. (Courtesy New Brunswick Scientific Co. Inc., Edison, New Jersey.)

tigations. Intermediate scale-up utilizes larger fermentors generally of from 30 to 150 liters or more (Figs. 3 and 4). These intermediate and pilot-plant fermentors bridge the gap between the laboratory and the production-size fermentors.

3. Continuous Culture

Precise control of fermentations can be achieved in a chemostat (Fig. 2) using continuous culture techniques. Fresh medium is continuously added to a fermentor or carboy, while the culture is withdrawn at the same rate. The flow rate can be adjusted so as to maintain the desired steady state conditions within the fermentor. Theory, construction, and use of chemostats for continuous culture are discussed by Tempest (1970), Evans *et al.* (1970), and Pirt (1976). The effects of environmental conditions and growth-limiting parameters may be studied using this technique and the culture can be maintained in a constant or steady physiological state.

Most of this equipment is commercially available and, except for the larger pilot-plant equipment, is exhibited annually at the national meetings of the American Society for Microbiology and other scientific societies. These exhibits afford opportunities to become familiar with the state of the art in fermentation equipment.

VI. INTERMEDIATE SCALE-UP

A. General Considerations

Scale-up is the process of increasing the size of a fermentation system to arrive at a full-scale plant fermentation in the most efficient and economical manner. Problems will be encountered and resolved at the small-scale level so that only minor adjustments are necessary at the production-scale level. Obviously, mistakes made at the larger level are more expensive. Also, additional information is gathered during scale-up so that the role of each factor affecting the fermentation is clearly understood and more amenable to control. Factors to be considered during scale-up include those of the small fermentor stage plus new problems that must often be resolved by a chemical engineer rather than the fermentation technologist. Many of these problems are considered in other sections of this text.

B. Media

Prior to scale-up of a fermentation one must determine the exact composition of the nutrient solution. This is usually determined in small-scale fermentors. Major changes should not be necessary during scale-up, although minor changes undoubtedly will be required. One or more alternatives in the basic medium composition should be developed in the event that ingredient shortages or other unforeseen economic factors might force changes in composition.

Convenience of medium preparation and prevention of contamination

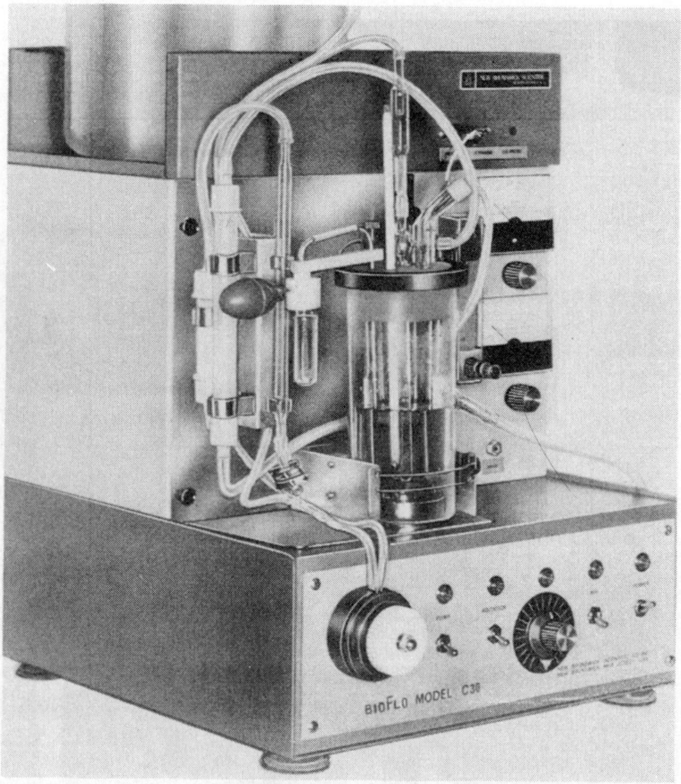

FIGURE 2. A continuous culture bench-top chemostat with a 350-ml working capacity. (Courtesy New Brunswick Scientific Co. Inc., Edison, New Jersey.)

are important economic factors to be considered during or before scale-up. Medium composition should be simple and should require the fewest manipulations and the least possible amount of preparation time. Steam sterilization is costly and troublesome and is not always necessary. Inoculation techniques are discussed by Meyrath and Suchanek (1972). The use of mass inoculum solves many problems during scale-up. A 1–5% inoculum of a fast growing organism is added insurance against contamination by unadapted wild strain microorganisms. In some cases, it is possible to recycle biomass one or more times reducing the time and expense of inoculum buildup and the total time and cost of the fermentation. Shortest fermentation lag times result when the inoculum is grown in and adapted to the same medium as that used for production. However, some fermentations require different media for growth and production, and it is not possible to use an adapted inoculum.

FIGURE 3. A small to intermediate-sized pilot-plant fermentor with control units (Courtesy Fermentation Design, Inc., Edison, New Jersey.)

C. Equipment

Scale-up of a fermentation is usually accomplished via submerged culture, pilot-plant fermentors of from 30 to 150 liters capacity (Figs. 3 and 4). Pilot-plant fermentation units generally provide monitoring equipment and controls that allow sufficient control of fermentation parameters to maximize production under controlled and reproducible

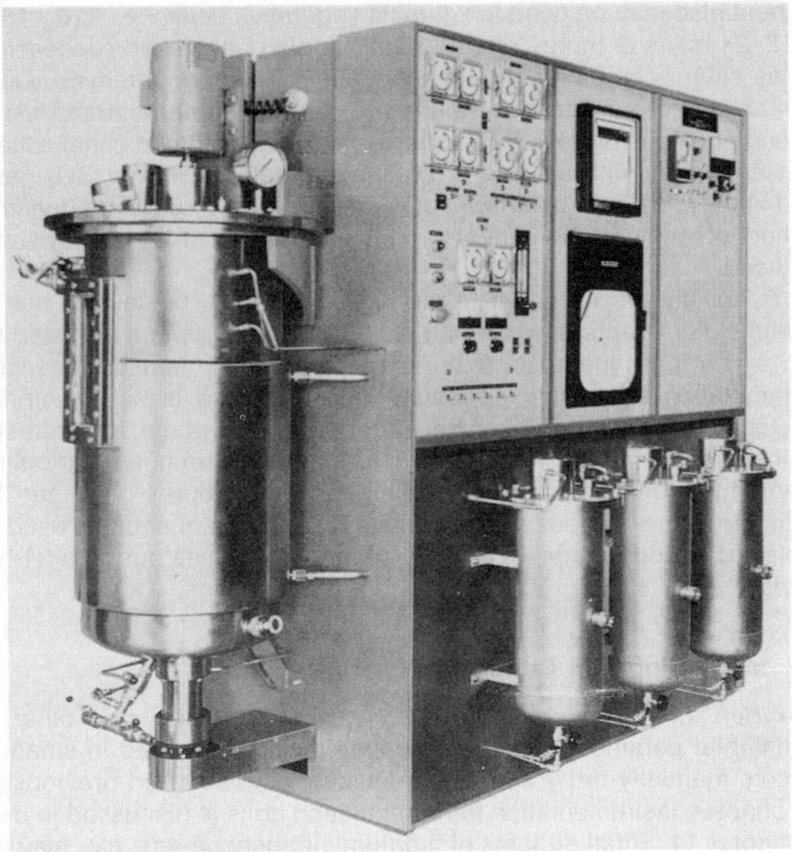

FIGURE 4. A 150-liter capacity pilot-plant fermentor with control unit. (Courtesy Fermentation Design, Inc., Edison, New Jersey.)

conditions. Various aspects of the fundamentals of fermentor design are discussed by Blakebrough (1969) and in this volume, Chapters 11 and 14.

D. Sterilization

Laboratory fermentation media and equipment are generally sterilized using steam under pressure to attain a temperature of 121°C for 15 minutes or more; other temperatures and times also may be used. Principles and practices for sterilization of laboratory apparatus and media are discussed in detail by Stumbo (1976). Nonpressurized steam (100°C and atmospheric pressure) may be used for from 2 minutes to several hours to sterilize equipment depending on the organisms to be killed.

Dry heat also may be used to kill most vegetative cells, i.e., 120°–180°C for 12–24 hours or more. Infrared heat has also been used successfully, as has steam combined with chemicals such as formaldehyde. Gases such as ethylene oxide can be used to sterilize equipment and media, although noxious or toxic fumes may be hazardous to personnel. Chemical sterilization with sodium or potassium metabisulfite or SO_2 gas is used in the wine and other industries. Sterilization by ultrafiltration is the method of choice where heat-sensitive nutrients must be separately sterilized.

It is not always essential to sterilize or even to pasteurize nutrient solutions. For example, where the pH of a medium can be adjusted to 1 or 2, and a mass inoculum of an adapted organism is used, chance of contamination is greatly reduced. Where expense is not prohibitive, various antibiotics also may be added to the medium for protection against contamination. Complex natural media such as agricultural commodities or foods may require higher temperatures or longer than customary time periods for sterilization. A technique often used for semisolid or solid media is to sterilize them two times during a 24-hour period.

E. Monitoring and Control

Oxygen, pH, temperature, electromotive force, CO_2, and other environmental parameters may be monitored and controlled in small fermentors manually or by automatic devices, as discussed previously in this chapter. Instrumentation for fermentation units is discussed in detail in Chapter 14. Small aliquots of medium, headspace gas, etc. may also be withdrawn manually and analyzed, and adjustments may be made as desired by the operator. Although expensive monitors and controllers generally are necessary for chemostats, pilot-plant fermentors, and production-size fermentors, manual monitoring and control is often sufficient for small units. Foaming is a common problem in most fermentors, especially where the medium contains ingredients such as proteins, casein, peptone, beef extract, corn steep liquor, whey, and yeast extract. Foam control is discussed in detail by Bryant (1970). Foaming problems are best controlled automatically by use of appropriate probes, detectors, pumps, and tubing. They are often only required during the first 24–48 hours of a fermentation. A variety of silicone-based control agents is available commercially. Also, various combinations of natural vegetable oils or long-chain fatty acids can be used as foam control agents. Occasionally foam control agents are toxic to a microorganism, especially if used too liberally.

F. Product Recovery

Problems of harvesting the biomass and clarifying medium generally become more numerous and difficult to deal with on scale-up except where special equipment is available. Biomass is commonly harvested by centrifugation, by spraying onto heated drums, or by filtering with use of filter aids such as Hyflo Supercel, kieselguhr, or various forms of silica, diatomaceous earth, asbestos, or other materials. Filter presses using cloth or asbestos–paper mixtures are widely used. Newer technology also makes use of stacks of membrane filters or cartridges. Large-scale continuous and noncontinuous centrifuges are generally used for clarifying cultures of pathogenic bacteria. A disadvantage of some filter systems is that the biomass cannot subsequently be separated from filter aid materials. The centrifuge does not have this disadvantage. Various techniques for biomass recovery are discussed in detail by Thomson and Foster (1970), Mulvany (1969), and Belter (this volume, Chapter 16).

Scale-up of product recovery methods is generally a matter of selection of the most suitable alternative from among those investigated during development of the laboratory fermentation. Plant engineers must be consulted at this stage of development. Important ancillary factors to be considered are those pertaining to economics, efficiency, safety, and convenience. Also, methods for safe disposal of waste materials and recovery of valuable by-products should be considered. Again, attention should be paid to the general principle that at each stage of scale-up, only minor adjustments should be necessary. All aspects of a fermentation should be planned in advance with the final plant process in mind in order to minimize the problems that invariably arise on scale-up.

REFERENCES

Beckhorn, E. J. (1967). *In* "Microbial Technology" (H. J. Peppler, ed.), pp. 366–380. Van Nostrand-Reinhold, Princeton, New Jersey.

Beechey, R. B. (1972). *Methods Microbiol.* **6B,** 25–53.

Blakebrough, N. (1969). *Methods Microbiol.* **1,** 473–504.

Booth, C. (1971). *Methods Microbiol.* **4,** 1–47.

Bridson, E. Y., and Brecker, A. (1970). *Methods Microbiol.* **3A,** 229–295.

Brown, D. E. (1970). *Methods Microbiol.* **2,** 125–174.

Brown, P. R. (1973). "High Pressure Liquid Chromatography: Biochemical and Biomedical Applications." Academic Press, New York.

Bryant, J. (1970). *Methods Microbiol.* **2,** 188–212.

Calam, C. T. (1969). *Methods Microbiol.* **1,** 255–326.

Campbell, R. C. (1974). "Statistics for Biologists." Cambridge Univ. Press, London and New York.

Casida, L. E., Jr. (1968). "Industrial Microbiology." Wiley, New York.

Codner, R. C. (1969). *Methods Microbiol.* **1,** 427–454.
Cooper, K. E. (1963). *In* "Analytical Microbiology" (F. Kavanaugh, ed.), pp. 1–86. Academic Press, New York.
Dawes, E. A., McGill, D. J., and Midgley, M. (1971). *Methods Microbiol.* **6A,** 53–215.
Dulaney, E. L. (1967). *In* "Microbial Technology" (H. J. Peppler, ed.) pp. 308–343. Von Nostrand-Reinhold, Princeton, New Jersey.
Elliott, E. C., and Georgala, D. L. (1969). *Methods Microbiol.* **1,** 1–20.
Evans, C. G. T., Herbert, D., and Tempest, D. W. (1970). *Methods Microbiol.* **2,** 277–327.
Fantini, A. A. (1976). *In* "Methods in Enzymology" (J. H. Hash, ed.), Vol. 43, pp. 24–41. Academic Press, New York.
Feigl, F. (1977). "Spot Tests in Organic Analysis." Elsevier, Amsterdam.
Gouw, T. H. (1972). "Guide to Modern Methods of Instrumental Analysis." Wiley, New York.
Hanson, A. M. (1967). *In* "Microbial Technology" (H. J. Peppler, ed.), pp. 222–250. Von Nostrand-Reinhold, Princeton, New Jersey.
Hash, J. H., ed. (1976). "Methods in Enzymology," Vol. 43. Academic Press, New York.
Herbert, D., Phipps, P. J., and Strange, R. E. (1971). *Methods Microbiol.* **5B,** 210–344.
Hockenhull, D. J. D. (1977). *Adv. Appl. Microbiol.* **21,** 125–159.
Hopwood, D. A. (1970). *Methods Microbiol.* **3A,** 363–433.
Humphrey, A. E. (1976). *Dev. Ind. Microbiol.* **18,** 58–70.
Isenberg, H. D., and MacLowry, J. D. (1976). *Annu. Rev. Microbiol.* **30,** 483–505.
Kavanaugh, F. ed. (1963). "Analytical Microbiology." Academic Press, New York.
Krebs, K. G., Heusser, D., and Wimmer, H. (1969). *In* "Thin-Layer Chromatography" (E. Stahl, ed.), 2nd ed., pp. 854–909. Springer-Verlag, Berlin and New York.
Kubitschek, H. E. (1969). *Methods Microbiol.* **1,** 594–610.
Lapage, S. P., Shelton, J. E., and Mitchell, T. G. (1970). *Methods Microbiol.* **3A,** 1–33.
Lederberg, J., and Lederberg, E. M. (1952). *J. Bacteriol.* **63,** 399–406.
Mallette, M. F. (1969). *Methods Microbiol.* **1,** 522–566.
Meyrath, J., and Suchanek, G. (1972). *Methods Microbiol.* **7B,** 159–209.
Miller, B. M., and Litsky, W. (1976). "Industrial Microbiology." McGraw-Hill, New York.
Miller, J. M. (1975). "Separation Methods in Chemical Analysis." Wiley, New York.
Mulvany, J. G. (1969). *Methods Microbiol.* **1,** 205–253.
Nicholls, D. G. (1972). *Methods Microbiol.* **6B,** 55–63.
Perry, S. G., Amos, R., and Brewer, P. I. (1972). "Practical Liquid Chromatography." Plenum, New York.
Pirt, S. J. (1975). "Principles of Microbe and Cell Cultivation." Wiley, New York.
Pirt, S. J. (1976). *Dev. Ind. Microbiol.* **17,** 1–10.
Roberts, M., and Boyce, C. B. C. (1972). *Methods Microbiol.* **7A,** 153–190.
Skeggs, H. R. (1976). *In* "Industrial Microbiology" (B. M. Miller and W. Litsky, eds.), pp. 47–59. McGraw-Hill, New York.
Spooner, D. F., and Sykes, G. (1972). *Methods Microbiol.* **7B,** 211–276.
Stahl, E., ed. (1969). "Thin-Layer Chromatography," 2nd ed. Springer-Verlag, Berlin and New York.
Stanier, R. Y., Adelberg, E. A., and Ingraham, J. (1976). "The Microbial World." Prentice-Hall, Englewood Cliffs, New Jersey.
Stumbo, C. R. (1976). *In* "Industrial Microbiology" (B. M. Miller and W. Litsky, eds.), pp. 412–450. McGraw-Hill, New York.
Sutherland, I. W., and Wilkinson, J. F. (1971). *Methods Microbiol.* **5B,** 345–383.
Tannenbaum, S. R., and Wang, D. I. C. (1975). "Single Cell Protein II." MIT Press, Cambridge, Massachusetts.
Tempest, D. W. (1970). *Methods Microbiol.* **2,** 260–276.

Thomson, R. O., and Foster, W. H. (1970). *Methods Microbiol.* **2,** 377–405.
Underkofler, L. A. (1976). *In* "Industrial Microbiology" (B. M. Miller and W. Litsky, eds.), pp. 128–164. McGraw-Hill, New York.
Wang, D. I. C., and Fewkes, R. C. J. (1976). *Dev. Ind. Microbiol.* **18,** 39–56.
Weber, D. J. (1973). "Principles and Application of Instrumentation in The Biological Sciences." Brigham Young Univ. Printing Service, Provo, Utah.
Weinshank, D. J., and Garver, J. D. (1967). *In* "Microbial Technology" (H. J. Peppler, ed.), pp. 417–449. Van Nostrand-Reinhold, Princeton, New Jersey.

Chapter 14

Instrumentation of Fermentation Systems

L. P. TANNEN
L. K. NYIRI

I. CONCEPTS OF BIOLOGICAL PROCESS ANALYSIS AND CONTROL

A. Introduction

The objectives of applying instrumentation to fermentors and other types of biochemical reactors are: (1) to analyze the process status and

MICROBIAL TECHNOLOGY, 2nd ed., VOL.II
 ISBN 0-12-551502-2

(2) to establish an optimum set of environmental conditions in order to maximize specific cellular or enzymatic activities. For the most part the information which follows directs its attention to pilot- and semiworks-scale equipment requirements but the data can be equally applied to large commerical production equipment. In many instances the problems noted for small-scale equipment would not be a significant factor at the larger scale. Our rationale in concentrating on the smaller scale in this chapter is due to the fact that we are still, unfortunately, in an embryonic stage of instrumentation development in fermentation and it is on the pilot scale that much of the systems work must be tested and proven before commercial application can be seriously considered.

From an electronic systems engineering point of view, fermentation does not differ significantly from typical chemical processes (Considine, 1957). The difference between chemical and biological process analysis and control is *what* and *how* to measure in order to obtain an adequate picture of the biological process and its dynamics on the basis of which a process-control scheme can be implemented. In this review, biological system is defined as any *in vitro* culture of microbial, plant, or animal cells. Because of the characteristics of biological systems, we are faced with problems where very complex processes are analyzed and controlled by electronic means which were developed for relatively simple (e.g., chemical) processes. This general discrepancy, despite developments over the past decade, has not yet been satisfactorily resolved for the fermentation industry.

B. Terminology

The relationship between process status and controlled variables makes the definition of certain terms necessary. There have been attempts to adapt standard terminologies generally used in control engineering to biochemical engineering. Constantinides *et al.* (1970a) used the expression *state variable* by which the state of the system is described and the variable is not under direct control. Further, *control variables* are those variables which can be directly controlled by the experimenter and which influence the way the system changes from its initial state to that at any later time, e.g., temperature, DO (dissolved oxygen) concentration. Nyiri *et al.* (1975a) explored this concept further by defining state variables for fermentation processes as *process status indicators* incorporating all off-line or on-line obtainable signals which directly or indirectly describe the conditions of the biological process. Process status indicators are then sugar concentration, cell mass, oxygen uptake rate, respiratory quotient, etc., values of which either can be directly measured in the broth (e.g., cell mass with an on-line turbidometer), derived from combination of available on-line signals (e.g., combining off-gas O_2 and CO_2 measurements to obtain respiratory quotient), or determined by wet chemical analysis of samples. The expression for control (or controlled) variable remains unchanged with the understanding that these control variables individually, or in combination, create the environmental conditions in which the cells express a certain metabolic activity. Therefore, in this context, the terms *control variable* and *environmental variable* will be synonymous.

II. TYPES OF INSTRUMENTATION USED IN BIOLOGICAL PROCESS ANALYSIS

Analysis of conditions in a biological process is performed by two general methods; namely, off-line operating instruments and on-line operating instruments.

A. Off-Line Operating Instruments

Off-line operating instruments are not connected directly to the fermentor or biochemical processor; consequently, sampling, sample preparation or treatment, and sample analysis are performed and the analysis results provide information on the process status. However, the elapsed time between the sampling and the assay results can be significant. Information obtained by this method is for the most part useful for postexperimental evaluation of the process but cannot normally be used for process-control purposes. Current examples of off-line operating instruments include spectrophotometers, mass spectrometers, gas chromatographs, enzyme activity analyzers, and electron microscopes, as well as specialized wet chemical and microbiological procedures for product analyses (e.g., reducing sugar or dry weight measurements). Recently, there has been a trend toward connecting certain of these instruments to fermentor units by means of automatic samplers and in-line sample treatment devices, using electronic conversion of the analyzer signals for data logging. A typical example of such an arrangement is the application of a Technicon Auto Analyzer for measuring in-process sugar levels where the elapsed time between sampling and results (approximately 20 minutes) may allow the data to be used for process-control action in certain biological processes.

B. On-Line Operating Instruments

On-line operating instruments are directly connected to the fermentor or biochemical processor unit and have the capability to sense a specific variable which directly or indirectly reflects a process condition and/or the performance characteristics of a controller. It is noted here that there are relatively few on-line direct sensing devices which give indication of the process status at the cellular level (e.g., NAD–$NADH_2$ measurement belongs to this category). The predominant types of on-line operating instruments indirectly reflect a process condition in the whole broth culture; e.g., a dissolved oxygen sensor connected to an analyzer indicates the dissolved oxygen level in the broth, which is a consequence of the metabolic activity of the culture and the oxygen transfer capacity of the fermentor system. Most of these instruments therefore are reflecting the performance characteristics of a controller which interacts with a process variable. A dissolved oxygen sensor connected to a dissolved oxygen controller reflects the performance characteristics of the method by which the transfer of oxygen from the gas phase into the liquid phase is performed while the culture consumes the available oxygen.

III. MEASUREMENT AND CONTROL OF ENVIRONMENTAL VARIABLES

Biological processes are essentially under the control of two major factors: genetics and environmental conditions. These two factors do interact, however, in the everyday practice of fermentation technology during *in vitro* culture of a given mutant where only the environmental conditions are controlled. The environmental variables include temperature, hydrogen ion concentration (pH), partial pressure of dissolved oxygen (DO), and concentration and proportion of certain nutrients (carbon and nitrogen sources as well as salts). The composite condition of these individual variables creates an environment which influences the cells' metabolic activity. A given value of an environmental variable is usually maintained by conventional controllers such as are used in typical chemical engineering practice. An example of such a control scheme is shown in Fig. 1, which describes the control of temperature in a reactor. This is considered the most elementary system with which essentially every fermentor is equipped. The example shows that, in fact, the fermentor's temperature is controlled by measuring the temperature (which is the result of metabolic heat production and removal) via an electronic signal-producing thermometer, by comparing the deviation between the actual and desired temperature, and by manipulating a control valve, the position of which controls the flow rate of cooling water to a cooling surface. In a complex situation more than one manipulated variable performs the control of a process variable. This is exemplified in Fig. 2, where the schematics of control of dissolved oxygen are demon-

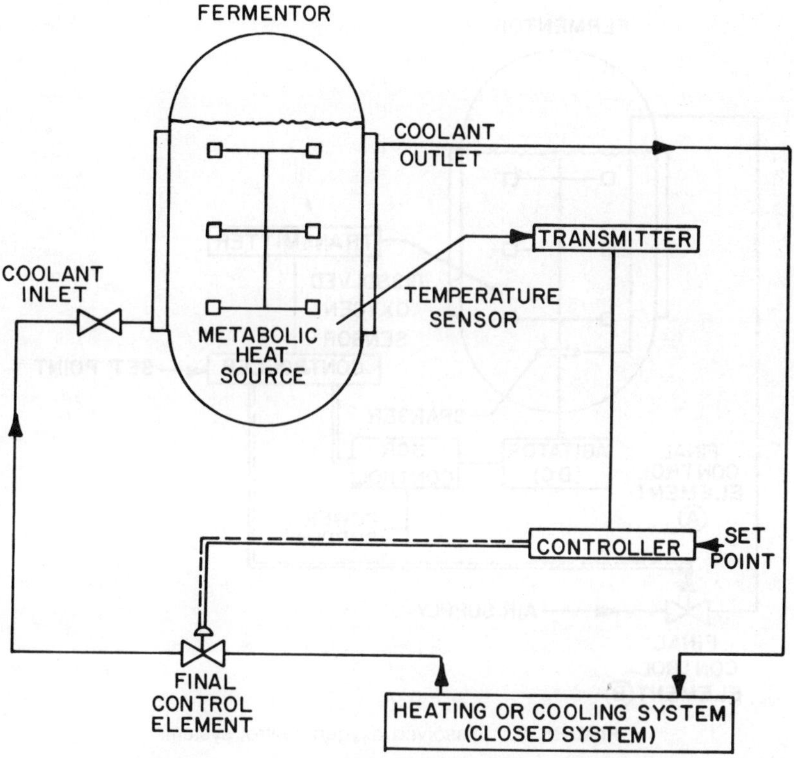

FIGURE 1. Basic fermentor temperature-control system.

strated. In this case the controller, based on a DO level measurement and comparison with a set point, governs both the inlet air (O_2) flow rate control valve and the impeller motor speed controller in an interactive way to maintain a dissolved oxygen level under dynamic conditions where the oxygen consumption and oxygen input rates must be in balance. A close examination of the process, however, reveals that, in fact, the air (O_2 gas) flow rate and the agitation speed interactively influence the oxygen mass transfer coefficient (K_La), which results in the dissolution of gaseous O_2 into the culture broth (Aiba *et al.*, 1973).

In view of these facts, fermentation or biochemical process control can be described as adaptation of environmental conditions to genetically controlled physiological conditions to influence cellular or enzymatic activities. It is a series of events which ultimately depends on the proper on-line analysis of the existing process status and the manipulation of final control elements (valves, meters, pumps).

During the culture of cells there are certain process conditions which

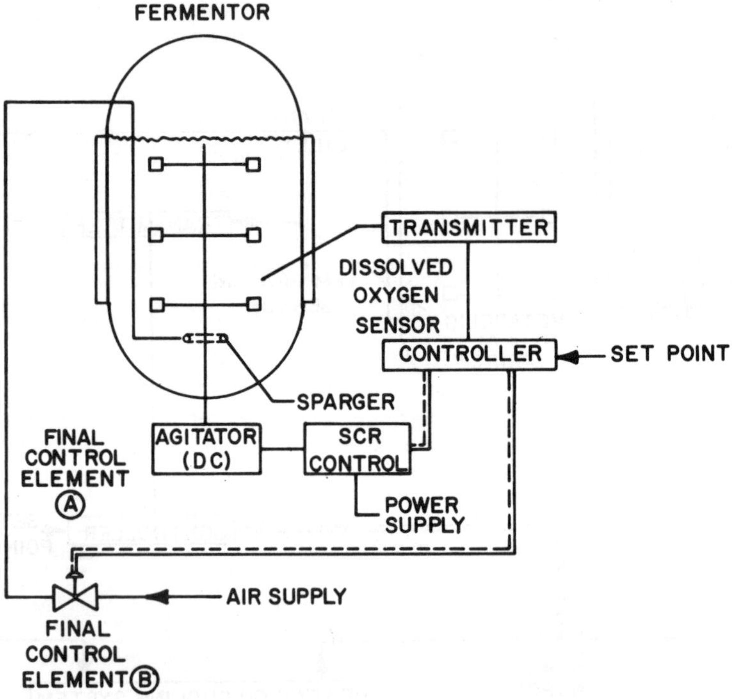

FIGURE 2. Dissolved oxygen-control system.

undergo change. These conditons can be grouped into four major categories: (1) physical conditions, (2) physicochemical conditions, (3) cellular physiology conditions, and (4) cell biochemistry conditions. All of them require analysis and control in order to achieve optimum productivity of a cell culture. Examples of some typical systems for measurement and control of functions within these categories will now be presented.

A. Physical Conditions

1. Volume Measurement

The first important physical condition of a liquid culture is its volume or working volume. Information on liquid volume (or level) is essential in liquid flow control (e.g., filling up vessels, continuous culture feeds, or introduction of liquid nutrients) which ultimately affect metabolic activity. There are two systems which are generally used for on-line volume determinations: (1) liquid level sensors, and (2) weight or mass measurement devices.

For sensing of liquid level experience indicates fairly reliable performance can be obtained in small-scale equipment with capacitance probes. As an example, the principle of operation of the Robertshaw capacitance level probe is described.

The sensor uses an electronic circuit which compares the rapid cyclical rate of charge of two capacitor systems, one reference capacitance internal to the electronic system, and the other, the external sensor capacitance probe itself, located in the fermentor. When the liquid level in the fermentor changes, the sensor probes capacitance and hence rate of charge changes. This rate difference is measured and provides an output signal proportional to the change in liquid level. Experience has shown that capacitance probe systems will provide level measurement with an accuracy of ±2–4% (full scale) depending on the liquid mixing surface turbulence and foaming conditions of the fermentor.

For larger scale systems (5000 liters and above) the ΔP measurement method is usually used, where a flush mounted diaphragm pressure cell is located in the base of the vessel for liquid height, and hence volume, measurement. Among suppliers of volume measurement systems are Ranco Controls, Drexelbrook, and Robertshaw Controls.

2. Weight/Mass Measurement

Weight or mass of the liquid can be determined by scales of various types. In this method the vessel is suspended on a scale and the combined weight of the vessel and liquid is electronically measured, usually by strain-type gauges. Several methods for electronically or mechanically eliminating the tare weight of the system can be used. Advantages of this method are that (a) the system is external to the vessel and need not be steam sterilizable, (b) its calibration is relatively simple, and (c) its accuracy is within ±1–2% (based on field experience). Among the disadvantages are the following: (a) Special engineering is needed for suspension of the vessel on the scale; consequently, this approach cannot be readily applied to existing installations. (b) There is a lower volume limit (approximately 250 liters) below which measurement accuracy falls off rapidly as the dynamic load becomes less than the tare weight. (c) Certain systems are sensitive to vibration and ambient temperature changes and compensation must be allowed for this in design. (d) The system is more costly than the simpler level probe method by a factor of four to five times. Suppliers of strain gauge systems are BLH Electronics, Transducers, Inc., and Micro-Strain, Inc.

3. Liquid Metering

Liquid metering is an important physical process variable since it serves as the basis for material balance analysis when fresh nutrient feed is added during the culture. There are several devices from industrial and biological waste treatment applications which have been adapted to fermentation processes. Solomons (1969) treated this subject previously and unfortunately there have been no major developments reported during the past 10 years.

Difficulties in measurement and control of liquid flows to fermentation units stem from the following factors: (1) There is significant difference in sensing flow values on laboratory, (small) pilot, and semicommercial scales. Consequently, accuracy of flow measurement is scale-dependent. (2) There are significant differences between handling (pumping) solutes of low or high viscosity and those with or without suspended solid particles.

Neither the problem of scale nor the rheological dependency of the flowmeter measurement and subsequent control has been systematically studied; consequently, no conclusions can be drawn. In Table I some representative examples of available liquid flow measurement devices are shown with their claimed accuracy. It should be cautioned that attempts to use diaphragm-type steam sterilizable metering pumps for very low liquid flow rates (milliliters to several liters per hour) can be highly inaccurate. Care should be exercised when such pumps are used and sufficient performance testing carried out with the fluid to be pumped to develop a measure of true accuracy.

4. Gas Flow Measurement

Introduction of air into aerobic biological processes serves essentially two purposes: (1) It supplies the necessary oxygen (Aiba *et al.*, 1973); and (2) It assures the removal of carbon dioxide (ventilation) (Lengyel and Nyiri, 1966) and other volatile gaseous products of metabolism.

In certain cases introduction of pure oxygen (generally up to 40% of total gas volume) or CO_2 (generally up to 5% of total gas volume) (Kruse and Patterson, 1973), is necessary, which requires specially calibrated measuring devices.

On-line monitoring of gas flow can be performed by means of a simple *variable area meter*. This is usually a glass tube with a tapered bore resulting in a variable cross-sectional area and containing a float (a small sphere or a hollow thimble). This is the simplest system and is generally used for manual airflow control. The flowmeter device is equipped with a throttling valve to adjust the gasflow rate. Field experience indicate ±2% (full-scale) accuracy, depending on the inlet gas pressure. Such gas flowmeters are available in unit ranges from as low as 10 cm³/min and with turndown ratios of 10 : 1. These simpler units do not provide any outputs for use with data logging or electronic computa-

TABLE I. Liquid Metering Systems

Manufacturer	Model	Type	Range	Accuracy[a]
A. Flow monitors (meters)				
Brooks	Series 22	Propeller	0–17,610 gal/min	± 2%
Fisher & Porter	10A Series	Rotameter	0.1 cm^3/min–4400 gal/min	± 2%
Foxboro	Series 81 and 82	Turbine	Variable	± 0.2%
Foxboro	Series 2000	Electro-magnetic	0.07–115,000 gal/min	± 1%
B. Metering pumps (steam sterilizable in-place)				
Interpace Corp.	Lapp	Diaphragm	0.5–5500 liter/hour	± 1% of set point
Chemcon, Inc.	Series A52	Diaphragm	0–570 liter/hour	± 1.5% of set point
	Series 1100	Diaphragm	0–1700 liter/hour	± 0.5% of set point
Milton Roy Co.	Milroyal Diaphragm simplex	Diaphragm	0.7–900 liter/hour	± 1% of set point
	Milroyal Diaphragm duplex	Diaphragm	13.5–4500 liter/hour	± 1% of set point

[a] Manufacturer's published claim over 10–100% scale range.

tion systems. A large number of companies supply such devices, including Fisher-Porter, Brooks, and Foxboro.

Gas flowmeters with electrical transducers which can provide output signals for use in data logging and control are available. These include devices such as (a) magnetic flow rate sensing elements, (b) thermal mass flow rate sensing elements, and (c) laminar flow rate sensing devices. A typical thermal mass gas flowmeter is shown in Fig. 3.

The thermal mass flowmeter has a ±1% full-scale (FS) accuracy and is based on the principal of detecting a temperature differential across a heater device placed in the path of the gas flow. Two temperature sensors, one upstream and one downstream of the heat source are made part of a bridge circuit which becomes unbalanced when there is flow past the heat source. This type of device is most suited to flow rates of under 500 liters/minute. The output of the transducer is in the range of 0–5 V.

Another type of gas flow metering device which can provide similar signal outputs to the thermal mass type are laminar gas flow elements (Fig. 3). Laminar flow measurement devices operate on capillary flow principles. The sensors flow versus differential pressure curves approach linearity and this makes them applicable over wide ranges of flow measurement. Mechanically, a matrix device inserted in the line breaks the flow into multicapillary paths. As the gas flows through these "capillaries" the friction created causes a pressure drop which is measured across the entire matrix device. Design of most devices is based on a 4- to 8-inch permanent water differential pressure loss. Units are available to measure flow over the range 0.000092 to 2250 Standard cubic feet per minute (SCFM) and are accurate to within ±0.5% error, FS. Sources of thermal mass and laminar flow measuring devices are Brooks Instrument, Fisher-Porter, and Meriam Instruments.

5. Temperature Measurement

Heat energy is considered to be the only environmental variable which freely penetrates the cell wall and directly influences intracellular metabolism. Therefore, temperature measurement (and control) is considered to be the most important direct process variable. Because of the hundreds of available variations, proper temperature sensor selection for biological processes depends on the specific use of the signal.

Typical sensor devices which can be used for fermentation application are described below.

a. Platinum Resistance Sensor (RTD). This device has a range capability of −150°C to 600°C with a linearity error of approximately 1% over the range 0°–300°C. Because current is passed through this type of

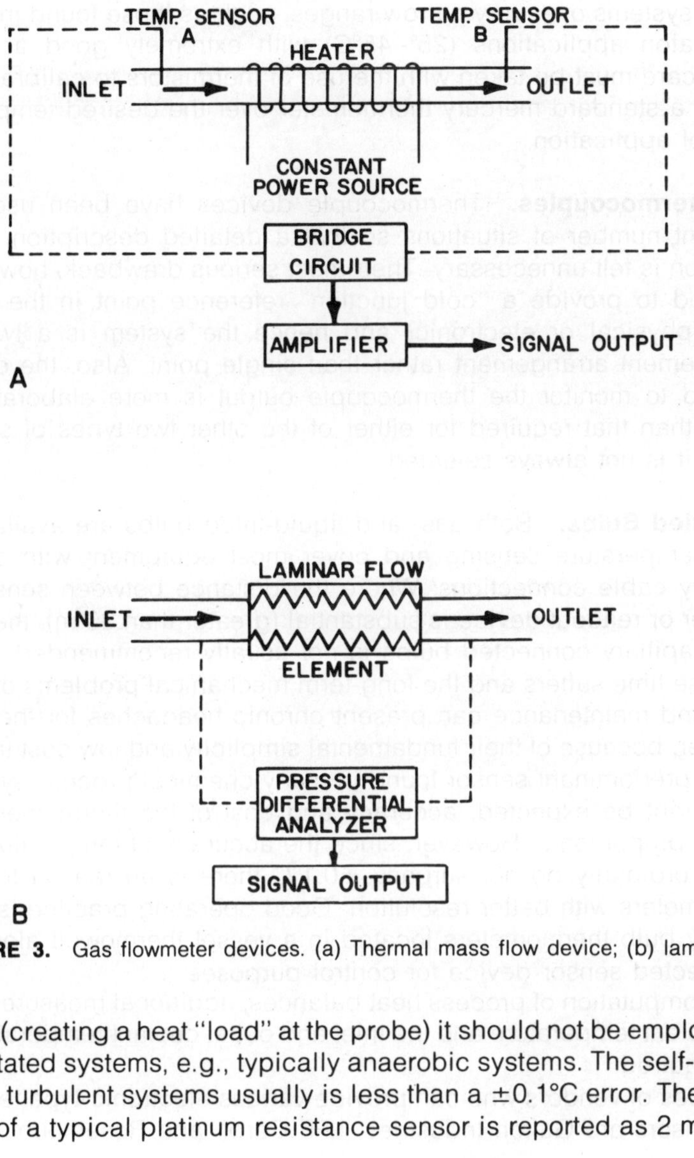

FIGURE 3. Gas flowmeter devices. (a) Thermal mass flow device; (b) laminar flow device.

device (creating a heat "load" at the probe) it should not be employed for nonagitated systems, e.g., typically anaerobic systems. The self-heating error in turbulent systems usually is less than a ±0.1°C error. The output signal of a typical platinum resistance sensor is reported as 2 mV/°C at 200°C.

b. Thermistors. These semiconductor devices are relatively inexpensive and quite sensitive to temperature changes. However, their output (resistance–temperature relationship) is highly nonlinear over large temperature ranges and numerous calibration points are required. Because of their large outputs they can be employed for temperature control systems over very narrow ranges, such as those found in typical fermentaton applications (25°–45°C) with extremely good accuracy. Again, care must be taken with the use of thermistors to calibrate them against a standard mercury thermometer over the desired temperature range of application.

c. Thermocouples. Thermocouple devices have been used in a sufficient number of situations so that a detailed description of their operation is felt unnecessary. Their most serious drawback, however, is the need to provide a "cold junction" reference point in the system (either physical or electronic) and hence the system is a two-point measurement arrangement rather than single point. Also, the circuitry required to monitor the thermocouple output is more elaborate (and costly) than that required for either of the other two types of sensors; hence, it is not always selected.

d. Filled Bulbs. Both gas- and liquid-filled bulbs are available for direct temperature sensing and cover most equipment with suitable capillary cable connections. Where the distance between sensor and analyzer or readout device is substantial (greater than 30 m), the use of these capillary connected bulbs is not usually recommended as their response time suffers and the long-term mechanical problems of installation and maintenance can present chronic headaches for their user. However, because of their fundamental simplicity and low cost they still are the predominant sensor found in many chemical process systems.

As might be expected, accuracy and cost of the thermometers are directly proportional. However, since the accuracy of temperature controllers ordinarily do not surpass ±0.1°C there is no reason to select thermometers with better resolution. Good operating practice is to use mercury bulb thermometers located in a vessel thermowell along with the selected sensor device for control purposes.

For computation of process heat balances, additional measurement of the ambient temperature and the inlet and outlet coolant temperature are also required.

Sources of vendors who can provide various electronic-type temperature sensors are Gulton Industries, LFE Corp.; and Rosemount, Inc.

6. *Pressure Measurement*

Static head pressure inside the bioreactor influences the partial pressure conditions and, hence, solubility of gases (O_2, CO_2, in particular) and volatile compounds. A positive head pressure (about 1.2 atm absolute) can also contribute to the maintenance of sterility inside bioreactors operating under aseptic conditions. In addition it is important to measure internal vessel pressure conditions during sterilization as well as the pressure condition of the main supply services (steam, water, and air). Depending on the mode of application, there are two pressure measuring systems generally accepted for biochemical reactors: (1) pressure devices without electrical signal outputs, and (2) pressure transducers.

In their applications there are several points to note.

1. At nonsterile points, generally Bourdon gauges are used. For steam pressure measurement a gauge protector is usually employed.
2. For determination of vessel pressure under aseptic conditions a diaphragm is placed between the sterile environment and the gauge.
3. For fermentor head pressure measurement, which also provides an electronic signal useful for on-line analysis as well as for automatic control of a given head pressure when gas flow rates are changed, pressure transducers are required.

An example of a typical pressure transducer with an electronic output is shown in Fig. 4. Pressure entering the port moves the diaphragm, causing a frictionless floating cone to be repositioned within the magnetic field of a linear variable differential transformer (LVDT). Movement of the transformer cone alters the inductive coupling between the primary and the secondary coils, resulting in a voltage change proportional to the displacement caused by variations in the input pressure. Typical specification for such a device are (a) a range of 3–15 to 0–3000 psi, (b) a linearity of ±0.25%, (c) a stability of 0.01%/°F (from +10°F to 140°F), and (d) a response time of 100 msec. One source of pressure transducer devices is Schaevitz Engineering.

7. Agitation Speed Measurement

The measurement of agitation speed can be by either direct or indirect monitoring devices. Indirect monitoring devices include both magnetic and optical methods. Magnetic sensors are noncontact transducers whose output voltage reflects the rate of change that is taking place in their magnetic field. The sensors are easily mountable adjacent to the external drive shaft (in a nonferrous housing) and opposite special gear teeth devices mounted directly to the drive shaft. These devices come in both analog and digital outputs. A typical supplier of such magnetic sensor systems is Electro Corporation, Inc.

Direct gear or belt-driven tachometers are also available for monitoring of agitator speed but are only advisable for use at low rpm (<250). For higher speed monitoring, indirect magnetic devices are preferred. Note that output meters generally read in revolution per minute (rpm). Caution must be exercised in comparing agitation speed literature data from different experimenters in view of the fact that the diameter of the

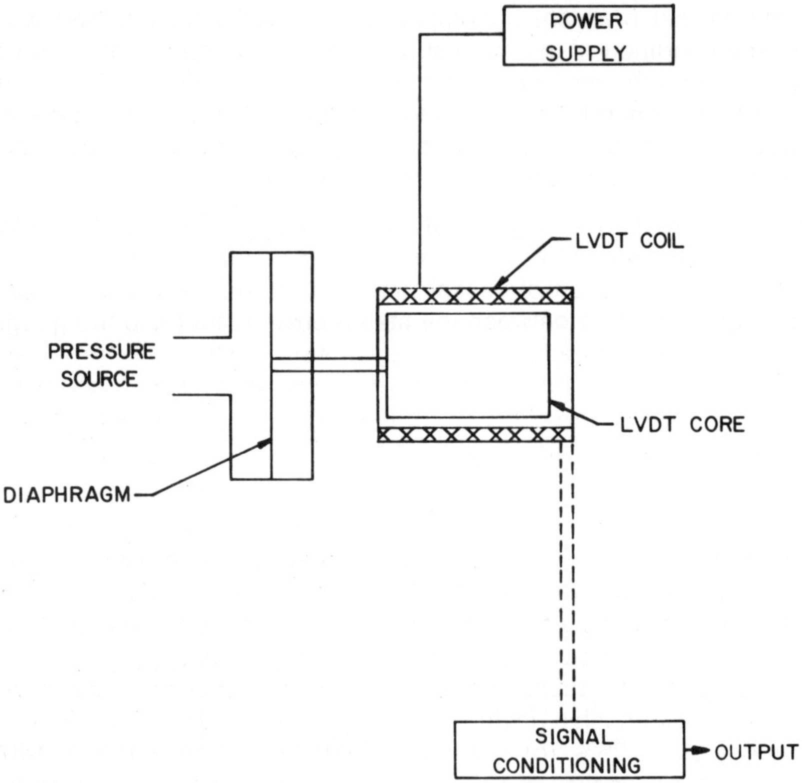

FIGURE 4. Schematic of a diaphragm pressure transducer. LVDT, linear variable differential transformer.

impeller will determine the absolute rotation speed of the tip of the impeller, which in turn defines the liquid shear rate and the liquid turnover rate in the culture vessel (Metzner and Taylor, 1962). In order to develop scale-up correlations between agitation and energy uptake by this method, it is necessary to compare tipspeed (*TS*) values derived from Eq. (1):

$$TS = ND_i \quad (1)$$

where N = rotation speed of impeller (per minute)
D_i = diameter of impeller (linear units: ft, cm)

Accuracy of agitation measurement devices ranges between +2% (FS).

8. Motor Power Uptake

Total energy consumed in operation of the agitator is generally measured at the armature of the drive motor in units of watts consumed per unit of time. For this measurement a Hall Effect wattmeter can be used. The operating principle of such a device, which measures power by use of the Hall Effect, first discovered by E. H. Hall in 1879, is that a conducting material placed in a magnetic field perpendicular to the direction of current flow produces a voltage in a direction perpendicular to both the initial current direction and the magnetic field.

In the early 1950s materials became available to utilize this principle and wattmeters employing this principle provide very reliable power measurements for pilot-scale fermentors.

Accuracy of a Hall Effect wattmeter is ±1% FS. The information obtained, however, represents only the total motor power uptake and does not necessarily reflect the actual amount of energy which is transferred into the liquid via the impellers. The design and scale of the vessel agitator system (i.e., size of shaft, bearings, seals, gears, pulleys, and/or belts) will significantly influence the loss of power between the motor armature and the impeller. A differential measurement is needed in which both the motor power uptake and the definition of power input into the liquid by the impeller can be determined. One source of Hall Effect power measuring devices is Ohio Semitronics Co.

9. Power Input to the Fluid

During the past 10 years there have been reports involving pilot-plant scale fermentors where both dynamometers and strain gauges have been used for the determination of the actual power input to the fluid. Various types of dynamometers and strain gauges were reviewed by Solomons (1969a) and Aiba *et al.* (1973). A detector device utilizing four strain gauges applied directly to the agitator shaft of a 270-liter fermentor was reported by Harmes (1972). The author discusses the degree of care which must be exercised in calibration, assembly, and operation of such a device in order to obtain acceptable process data. Based on such measurements, data were obtained which showed the correlation between the agitation speed and the efficiency of power input to the fluid (ϕ) (ϕ = power measured at the shaft divided by the power measured at the motor armature) in a 270-liter fermentor with 200 liters of working volume. The reported value of ϕ increased with increasing agitation speed. At maximum rpm, almost 30% of the energy supplied to the motor was still lost between the motor and the impeller shaft for the agitator system design used with this unit.

A compromise solution to the problem of obtaining power measurement of the energy imparted to the fermentation fluid is the use of an in-line torque measuring device located external to the vessel. The losses due to bearings and seals will be included in this type of measurement, but by suitable no-load testing it can be effectively eliminated from the reading and a reasonable approximation of power imparted to the fluid can be made. One source of such a torque instrument is Vibrac Corporation.

B. Physicochemical Conditions

The physicochemistry of the culture is related to the culture broth rheology which is a continuously changing condition during the process (Solomons, 1969a; Tuffile and Pinho, 1970). Despite the recognition of the importance of culture broth rheology (Aiba *et al.*, 1973) no systematic work has been performed with respect to development of an on-line measuring device by which the rheology characteristics of the culture broth can be identified. An instrument marketed by Automation Products, Inc. (Dynatrol Viscosity Control Device) might possibly be applied to the fermentation broth rheology if the problem of steam sterilization can be overcome.

Attempts to determine, on-line, the liquid rheology using computed correlations between agitation speed (shear rate) and power uptake (shear stress) in liquids of increasing viscosity (Charles and Toth, 1974) served as an example of utilizing the entire fermentor system as a viscometer. However, this study was confined only to xanthan solutions. Further systematic research is required in this direction to develop these early results into a practical engineering tool.

C. Cellular Physiology

Under the category cellular physiology, the gas-exchange conditions of a culture are of paramount concern. Gas exchange relates to the activity of the aerobically growing cells to consume oxygen and release carbon dioxide. Rates of these activities reflect the cell's overall physiological condition, which is related to the intracellular metabolic activities. Gas exchange or respiratory activity of cells has been investigated in an off-line mode using Warburg respirometers (Umbreit *et al.*, 1957) and washed cells obtained from culture samples. Recently, development of on-line operating gas analyzers has made it possible to analyze respiratory activity in a continuous, real-time mode. In view of the literature which has appeared in this area, one can consider it to be one of the most advanced and practical on-line monitoring facilities for biological process status indication. Instruments which are used to obtain on-line information on gas-exchange conditions are classed in two major categories: (1) liquid phase (dissolved) gas analyzers, and (2) gas phase gas analyzers.

1. Liquid Phase (Dissolved) Gas Analyzer

a. Determination of Dissolved Oxygen. Dissolved oxygen analyzers consist of at least two essential elements: (1) a dissolved oxygen sensor, and (2) a signal amplifier and indicator. Dissolved oxygen sensors of various kinds are described by Solomons (1969b), Aiba *et al.* (1973), and Vincent and Priestley (1975) quoting the relevant original publications. Table II presents several of the currently available commercial dissolved oxygen sensors (probes) usable in aseptic and nonsterile fermentations.

Reviewing the technical background and principles of operation of dissolved oxygen probes, the following factors should be noted:

1. Essentially all of the dissolved oxygen sensors operate on the principle of reduction of oxygen at the surface of a metal electrode as shown by Mackereth (1964):

$$\text{Metal} + \tfrac{1}{2}\,O_2 + H_2O + 2\bar{e} \rightarrow 2\,OH^-$$

At the anode the loss of electrons results in production of ions:

$$\text{Metal} \rightarrow \text{Metal}^2 + 2\,\bar{e}$$

The metal ions combine with the hydroxyl ions to form a metal hydroxide $[ME(OH)_2]$. This interchange occurs between the compounds in the electrolyte surrounding the cathode and anode.

In summary, metal salts will be removed from the anode surface and accumulate in the electrolyte. Consequently, the expendable items in the dissolved oxygen sensor are the metal anode and the electrolyye. Electrons generated by the interaction of oxygen and metals offset the cathode–anode current balance and electric current is generated which is proportional to the amount of O_2 reacted with the metals. Cathodes are generally made of noble metals (most frequently used are platinum or silver). Anodes are generally made of lead.

2. Field experience indicates that oxygen-permeable, membrane-covered sensors have greater stability than a nonmembrane-covered sensor, although the membrane does affect the probe's response time and can be fouled or perforated.

3. Several researchers (Siegel and Gaden, 1962; Taylor *et al.*, 1971) had proved that the current which is produced by the interaction of oxygen and the electrode is a measure of oxygen partial pressure (pO_2) and not the dissolved oxygen concentration. The value of the current (I) from the sensor (at a given temperature and pressure) depends on the electrons transferred (M), the Faraday constant (F), the area of the cathode (A), the permeability coefficient cf membrane (P_m), the thickness of the membrane (b) and the partial pressure of the oxygen (pO_2):

TABLE II. Dissolved Oxygen Sensors

Manufacturer	Type of sensor	DO measurement range	Temperature compensation	Pressure compensation	Response time (sec)[b]
Beckman Instruments	Polarographic	0–9.99 ppm	Yes	No	30
Instrumentation Laboratory, Inc.[a]	Polarographic	0–20 ppm	Yes	Yes	20
New Brunswick Scientific Co., Inc.[a]	Galvanic	0–8 ppm	No	Yes (vented)	45
Transidine General Corp.	Polarographic	0–1999 mm Hg	No	No	—
Yellow Springs Instruments[a]	Galvanic	0–8 ppm	No	Yes (vented)	60–120

[a] Steam sterilizable (min. 121°C).
[b] 90% of reading at 20°C.

$$I = nFA(\mathrm{Pm}/b)\ \mathrm{pO_2} \tag{2}$$

4. The physiological significance of the dissolved oxygen measurement in aerobic cultures can be depicted by the following equation (Aiba *et al.* 1973):

$$d\bar{C}/dt = k_La\,(C^* - \bar{C}) - Q_{O_2}X \tag{3}$$

where C^* = concentration of dissolved oxygen which is in equilibrium with pO_2 in the bulk gas phase ($mM\ O_2$)
$\bar{C}$ = the actual dissolved oxygen concentration in the bulk of the liquid ($mM\ O_2$)
k_La = the volumetric oxygen transfer coefficient ($1/t$)
Q_{O_2} = specific rate of oxygen uptake ($mM\ O_2/t$)
X = cell mass concentration (weight)

On this basis the actual dissolved oxygen level or its concentration [at a given gas (O_2) flow rate, impeller tipspeed, broth temperature, and vessel pressure] reflects conditions which are related both to rheology-dependent O_2 gas transfer rate (k_La), and the oxygen uptake rate (Q_{O_2}) of the culture.

Dissolved oxygen probes have been used to determine k_La and $Q_{O_2}X$ (Lengyel and Nyiri, 1965; Taguchi and Humphrey, 1966; Bandyopadhyay *et al.*, 1967) as well as to control the broth dissolved oxygen concentration (Siegel and Gaden, 1962; Lengyel and Nyiri, 1965; Aiba *et al.*, 1973). Dissolved oxygen control is based on on-line continuous measurement of pO_2. For determination of k_La and/or $Q_{O_2}X$, techniques employing step changes in the air-flow (O_2) rate are used. From the recorded pattern of the changes in dissolved oxygen concentration (as the result of step changes in the air-flow rate) the values of k_La and $Q_{O_2}X$ can be computed (Linek *et al.*, 1973).

b. Determination of Dissolved Carbon Dioxide. One of the end products of complete oxidation of carbon-containing compounds is carbon dioxide (CO_2). The effect of CO_2 on cellular metabolism has been considered as important as that of oxygen (Wimpenny, 1968; Nyiri, 1979). The regulatory effect of CO_2 was successfully applied to the industrial production of α-amylase and protease (Zajic and Liu, 1970; Gandhi and Kjaergaard, 1975), penicillin (Lengyel and Nyiri, 1968), and inosine (Shibai *et al.*, 1973).

Despite its recognized process significance, development of techniques for on-line sensing of dissolved carbon dioxide ($HCO^-{}_3$) are far behind the techniques used for the analysis of dissolved oxygen concentration. On-line dissolved carbon dioxide measurement can be performed using a sensor developed by Severinghaus and Bradley (1958) for human physiology studies.

The operating principle of the instrument is the measurement of the pH change of a $HCO^-_3 + H^+$ solution influenced by HCO^-_3 ions. The sensor itself is a pH probe surrounded by a $HCO^-_3 + H^+$ solution which is separated from the medium with a CO_2 (HCO^-_3)-permeable membrane. Presence of HCO^-_3 in the system makes the sensor sterilizable only by chemical means. In the fermentation industry, this fact has handicapped the widespread application of this otherwise accurately operating sensor. Recently the sensor was adapted by Japanese scientists to fermentors (Ishizaki *et al.*, 1973; Ishizaki and Hirose, 1973) for studies of the effect of CO_2 (HCO^-_3) on various microbial fermentors and the bioengineering aspects of ventilation. Another HCO^-_3 sensor is marketed by Orion Research, Inc. However, its operational principle requires the liberation of HCO^-_3 into CO_2 gas which makes pH adjustment (to pH 5) necessary. This fact makes it difficult to adapt the sensor for on-line analysis of dissolved carbon dioxide.

2. *Gas Phase Gas Analyzers*

a. Determination of Exhaust Gas Oxygen. Instruments which measure O_2 gas concentration in the exhaust gases generally are of the paramagnetic type. Solomons (1969b) discussed in detail their operational principles, which essentially have not changed during the past 10 years. Table III presents a list of several vendors who currently manufacture O_2 gas analyzers which can be utilized for fermentation purposes.

Researchers have also constructed O_2 gas analyzers. Utilizing a nonsterilizable O_2 probe to the exit gas line of a fermentor and connecting the O_2 concentration signal with a gas flowmeter signal, it was possible to electronically calculate the oxygen uptake rate of a yeast culture of a given volume with an accuracy of ± 2 m*M* O_2/liter/hour (Nyiri and Jefferis, 1974). This was among the first "net-effect" sensors which combined more than one on-line operating sensor signal to obtain a third, nondirectly measurable process status indicator, in this particular case the oxygen uptake rate (OXUP or OUR) of the fermentation culture.

b. Determination of Exhaust Gas CO_2. Since CO_2 absorbs in the infrared range, most gaseous CO_2 analyzers commercially developed use this property for analyses. Table IV presents a list of such gas analyzers and their general specifications. Field experience has indicated excellent applicability of infrared devices for CO_2 off-gas analysis. One manufacturer (Infrared) provides an optional "dual channel" design by which another low molecular weight volatile compound which ab-

TABLE III. Oxygen Gas Analyzers

Manufacturer	Model	Ranges (O_2%)	Response time (sec)[a]	Accuracy (% FS)	Sample temperature range (°F)	Measurement Method
Beckman Instruments	755	0–1 to 0–100	7	± 2	− 20–120	Paramagnetic
Mine Safety Appliances	802	0–0.5 to 0–100	80	± 2	32–118	Paramagnetic
Taylor Instrument (Sybron)	OA 184	0–25 to 0–100	7	± 1	14–122	Paramagnetic (both inlet and outlet)

[a] 90% for step change.

TABLE IV. Carbon Dioxide Gas Analyzers

Manufacturer	Model	Range (% CO_2)	Response time (sec)[a]	Accuracy	Drift (zero, span) (per 24 hours)
Anarad, Inc. Cavitron Corp.	AR500	0–30	5	± 1%	± 1%
Beckman Instruments	865	0–5 0–15	0.5	± 1% FS	± 1%
Infrared Industries	IR703 IR702	0–1 0–30	5	± 1% C	1°C
Mine Safety (M.S.A.)	200	0–30	5	± 1%	± 1%

[a] For 90% response time to step change.

sorbs in the infrared region can also be measured (e.g., CH_4, NH_3, acetone, and ethyl alcohol).

Exhaust gas analyzers (either O_2, CO_2, or both) obtain their samples from the exit gas line (after the exit gas filter, foam trap, or dehumidifier). The O_2 and CO_2 concentration of the incoming air is assumed constant at 20.91 and 0.03%, respectively, in order to develop O_2 and CO_2 balances around the fermentor (Aiba *et al.*, 1973). If the incoming air is enriched with O_2 and/or CO_2 the use of an additional gas analyzer in the inlet gas stream is mandatory. The signal output of most commerically available gas analyzers provides data as a percentage of gas concentration. Additional computations are necessary to convert these readings into bioengineering units (e.g., moles) (Fiechter and VonMetenburg, 1963).

Other exhaust gases of potential interest include N_2 (N_2 fixing microbe physiology) and H_2 (*Hydrogenomonas* sp. cultures). N_2 can be determined in the off-gas using gas chromatographs, whereas H_2 in the exit gas can be determined using a thermal conductivity analyzer (Solomons, 1969b).

Though there has been little published to date regarding mass spectroscopic techniques for off-gas analyses, the authors are aware of several laboratories experimenting with this method. As the cost of mass spectrometer instruments has been reduced their competitive position with respect to multiple analyses, such as high-speed gas chromatographs, has improved. Gas chromatographs have been sucessfully applied in several laboratories for the measurement of off-gas O_2, CO_2, H_2O, and other low molecular weight volatile compounds and should be given serious consideration where multiple vessel application for these measurements are needed. Further discussion of the use of both of these methods for off-line analyses will be found in Section III, D, 2, a.

D. Cell Biochemistry

Biochemical measurement of cellular activity is, in fact, measurement of enzymatic activity which, based on our current state of the art, cannot be directly analyzed. On the other hand, the disappearance or appearance of certain compounds (substrates, intermediate products) can indirectly tell the researcher about the biochemical events taking place within the cell. The compounds to be measured must have some physical or chemical characteristics on the basis of which their quantitative determination becomes possible. Typical compound properties used are the ionized forms (H^+, OH^-), molecular weight, volatility (ethanol), variation of functional groups (glucose), or an ability to absorb/scatter light (NAD–$NADH_2$). Based on these physical or chemical characteristics, instruments have been developed for both off-line and on-line determinations of their concentrations. Currently, there is a trend toward development of automatic sampling devices by which the fermentation unit can be connected with an otherwise off-line analytical instrument resulting in "on-line" operation.

1. Direct Reading Sensors

a. pH Measurement. The only steam sterilizable truly direct reading sensor with high selectivity is the pH probe. Of the varieties described by Solomons (1969a), Aiba *et al.* (1973), Vincent and Priestley (1975), we will comment on one sensor which, according to extensive field experience has become a commonly used device in the fermentation industry. This is the combination Ingold pH probe. Advantages of this pH probe can be listed as follows: (1) Combination of the H^+ sensor (half cell) and the reference sensor (half-cell) results in one probe which, therefore, requires only one penetration in the fermentor (bioreactor). (2) Based on widespread field experience, the probe is repeatedly steam sterilizable (more than 100 times at 121°C). (3) Design of the sensor makes it possible to sterilize (one or more models) either in an autoclave or by in-place sterilization techniques. (4) One model (Model 768-35) is designed to be removed under pressure, resterilized, and replaced during process operation. Arrangement of this retractable system is shown in Fig. 5. Connectors and cables make it compatible with most pH analyzers. Recently, another manufacturer, L&N, has come out with a single probe which appears to be comparable to the Ingold unit but there is, as of now, limited market experience with the probe. Table V lists several sources of sterilizable in-place pH electrodes.

b. Specific Ion Measurements. Orion Research and Radiometer have developed on-line sensors for determination of various ions. A list

FIGURE 5. Ingold Model 768-35 steam sterilizable retractable electrode housing for use with Ingold combination pH electrodes.

of specific ion electrodes of interest in biochemical processes is shown in Table VI. It should be noted that none of these probes is steam sterilizable and the two most critical sensors [CO_2 (HCO^-_3) and NH_3 (NH^+_4)] operate in pH ranges (pH 5 for CO_2 and pH 11–13 for NH_3) usually well outside a majority of biological processes. It is felt that the possibilities for measurement of the effects of certain ions on the cell's metabolic activity have not been significantly exploited in either microbial, plant, or animal cell cultures (Kruse and Paterson, 1973; Schroder and Weide, 1974).

c. Redox Potential Measurement. Based on Nernst and other pioneering investigations, several review articles have appeared dealing with the theoretical background and practical application of redox potential in microbiology and biochemical engineering (Hewitt, 1950; Rabotnova, 1963; Kjaergaard, 1977). In this discussion extracellular and intracellular redox potential conditions should be distinguished. Extracellular redox potential (synonyms: rH_2, E_{h_1}, ORP; oxidation–reduction

TABLE V. Steam Sterilizable pH Sensors

Manufacturer	Type	Model	Temperature Range (°C)	Pressure range
Activion Glass, Ltd.	Combination	11-801	10–140	N/A
		11-902	10–140	
Ingold	Combination	761/764 Series	− 10–130	6 bar
Leeds & Northup	Combination	117494	20–130	3.5 bar
		117495		

TABLE VI. Commercially Available Ion Sensors for Biological Processes[a]

Ion species	pH range for direct measurement	Concentration range (moles)
Ammonia	11–13	$1–10^{-6}$
Calcium	6–8	$1–10^{-5}$
Carbon dioxide	5	$10^{-2}–10^{-4}$
Chloride	2–11	$10^{0}–8 \times 10^{-6}$
Magnesium	—	10^{-4} (min)
Nitrate	3–10	$10^{0}–6 \times 10^{-6}$
Nitrate	0–2	$10^{-2}–5 \times 10^{-7}$
Potassium	3–10	$10^{0}–10^{-5}$
Sodium	9–10	Saturation to 10^{-6}
Phosphate	—	5×10^{-3} (min)
Sulfate	—	5×10^{-3} (min)

[a] Based on Orion Research Literature, 8th ed. May, 1977.

potential) is generally measured with two half-cells which compare the standard partial pressure of hydrogen gas (pH_2) or a concentration of two atoms of hydrogen with the amount of hydrogen in the oxidized form $[H^+]^2$:

$$E_h = E_o + \frac{RT}{2F} \ln \frac{[H^+]^2}{[H_2]} \tag{4}$$

where E_h = potential referred to the normal hydrogen electrode
E_0 = standard potential of the system at 25°C when the activities of reactants are unity
F = Faraday constant
R = gas constant
T = absolute temperature

The term $[H^+]^2/[H_2]$ can represent any compound which has the capability to exist in two states, reduced or oxidized respectively, in aqueous solution.

(1) Extracellular Measurement of ORP. The above characteristics represent both advantages and disadvantages for extracellular ORP measurement. According to experimental reports, the hydrogen ion concentration (pH) (Clark and Cohen, 1923), the presence of any inorganic or organic compound with oxidizing and/or reducing capability (including oxygen or metals), as well as temperature, will influence the output potential of a direct sensing ORP probe. This makes the interpretation of extracellular ORP data difficult. Correlations of practical significance were found between ORP and various physiological and biochemical events in microbial cultures (Hewitt, 1950; Rabotnova, 1963). Also, the dependency of mammalian cell growth on the status of redox potential of

the culture media has been demonstrated (Daniels *et al.*, 1970; Toth, 1975). The problems in interpretation of ORP data, however, have handicapped the widespread application of this measurement in fermentation technology. The most probable future use for ORP is in the detection of minute amounts of dissolved oxygen (<1 ppm) in anaerobic processes (between −150 and −450 mV) which are of importance for toxin production and in anaerobic biosynthesis of several organic solvents (acetone, *n*-butanol) where renewed commerical interest has recently occurred due to the energy crisis.

Steam sterilizable ORP sensors are manufactured by various companies, including Ingold, where combined platinum calomel as well as combined pH–redox probes are available. The ORP analyzer is a simple potentiometer with a measuring range of ±500 mV or ±1000 mV and is available from most pH analyzer manufacturers. Redox sensor probes are available from most suppliers of pH electrodes, such as Ingold and Leeds & Northrup. Based on reported field experiences, accuracy of ORP measurement is ±5 mV (FS).

(2) Intracellular Measurement of ORP. An interesting approach was developed by Harrison and Chance (1970) to determine intercellular redox potential by means of measuring NAD–$NADH_2$ conditions. The method involved sending and receiving light through the culture broth in the wavelength range of 360–460 nm. Although the device does have selectivity (Harrison and Harmes, 1972), the scattering of light caused by gas bubbles produces background noise which can only be resolved by the use of laser light. Cost of the equipment, however, has handicapped the commerical application of such a device.

d. Carbohydrate Measurement. Solomons (1969b) discussed a direct reading glucose sensor whose operating principle was based on glucose oxidase enzyme activity (Updike and Hicks, 1967). The reaction results in the formation of hydrogen peroxide which is monitored by electrochemical, amperometric detectors.

Several reactions of this technique have been developed both for off-line and on-line measurement purposes. The significance of these sensors should be emphasized in view of the reportedly developing techniques for biochemical conversion of carbohydrates (e.g., cellulose–glucose, starch–glucose conversions) for a wide range of applications.

Introduction by Leeds & Northrup of an on-line operating glucose analyzer with a sterilizable probe is considered to be a significant development and will be discussed later. Based on current trends (Zabriskie and Humphrey, 1978) it is anticipated that dialysis membrane-based continuous culture broth sampling will be one of the solutions to overcoming the difficulty of sample removal and treatment prior to wet chemical analyses in fermentation technology (cf Section III, D, 2, c).

e. Cell Mass Measurement. Automatic determination of fermentation broth cell mass is still among the most difficult of problems to be resolved. The most common technique is the measurement of light transmittance through the broth. The basis for the method used is the assumption that the intensity of the transmitted light is proportional to the turbidity, which in turn is related to the number of cells present. Devices developed during the past 10 years (e.g., Nephelostat, Peraino and Eisler, 1973; Turbidostat, New Brunswick Scientific Co.) all have similar limitations which can be enumerated as follows: (1) Only discrete "spherical" objects can be measured; i.e., devices are applicable for certain bacteria, yeast, or animal cell cultures, but not for filamentous fungi. (2) The technique is only applicable with clear solutions of soluble culture media, i.e., not applicable for industrial culture media which may contain insoluble and opaque materials, such as starch, corn steep liquor, molasses, or calcium carbonate. (3) The relationship between the cell concentration and light transmittance usually becomes nonlinear beyond 5 gm/liter cell concentration. These problems limit the applicability of such equipment for general scientific investigation. In those cases where application is possible, separate analyses must be made to determine the correlation between the transmitted light intensity and the actual cell population or mass present. The literature generally gives data either in cell numbers (e.g., 10^3 cells/ml) or dry weight (DW) (e.g., 10 gm/liter cells). From an engineering point of view this latter dimension is preferred since on this basis a material balance and conversion yield can be calculated. Automatic and successful direct determination of dry cell mass for general application has not been reported in the literature to date.

2. *Off-Line Automatic Analyzers*

Difficulties in the development of steam sterilizable on-line reading sensors for detection of various metabolites have resulted in the use of off-line analytical devices utilizing repetitive and/or continuous sampling of exhaust gases to monitor specific parameters. Sampling devices usually consist of a pipeline (sterilizable) through which gases released by the culture are connected to the particular analyzer. The following paragraphs discuss several approaches which have passed the rigor of preliminary field experience and will probably be used in the future for quasi on-line chemical analysis of compounds of interest in biochemical processes.

a. Gas Chromatography. Off-line determination of various volatile compounds (e.g., methane, acetaldehyde, ethanol) using gas chromatographs was successfully tested in biochemical processes by several researchers (Hamer, 1967; George and Gaudy, 1973; Abbott *et al.*, 1973).

One of the most elegant applications was reported by O'Brien and Cecchini (1970), who analyzed the metabolic patterns of bacteria by determining their end products (alcohols, di- and tricarboxylic acids) and using this "finger print" for taxonomic identification. It is interesting to note that this approach might also have practical significance for the detection of certain types of contamination in pure culture, though such application has not yet been reported in the literature.

Recent developments in gas chromatograph (GC) technology have made it possible to connect them to the exit gas lines of fermentors. As an example, there is Hewlett-Packard's Series 5830A, Reporting Gas Chromatograph. This instrument can be connected to a maximum of 16 sample lines with an automatic sampler. It can be programmed to control oven temperature, has a dual thermal conductivity detector, and automatically integrates the data and reports them in alphanumeric and/or graphical formats. Signals can also be obtained in binary coded decimal (BCD) form for computer interfacing. An HP5830A GC has been used in the verification of the effects of C/N feed ratios on protein and ethanol biosynthesis for yeast grown on molasses. Figure 6 (G. M. Toth and L. K. Nyiri, unpublished observation), shows the GC profile of a *C. utilis* culture sampled at 15-minute intervals for N_2, CO_2, H_2O, acetaldehyde, and ethanol. Note that the total sample retention time was about 3 minutes for molecular weights up to ethanol.

The advantage of gas chromatography is the ability for automatic determination of a relatively wide range of compounds. Its disadvantages are (1) the discontinuous signal produced and the time length between signals, since the reporting frequency is controlled by the retention time and the number of compounds being analyzed; (2) the relatively high price which, however, can be partially offset if several fermentors are to be monitored; and (3) the practical problem which requires investigation for each scale of use—to what extent do the volatile compound concentrations in the exit gas reflect the actual concentration in the culture broth? For the example shown (acetaldehyde and ethanol) the liquid–gas distribution ratio could be determined for a given gas flow rate.

b. Mass Spectrometer. In recent years reports have appeared about the use of mass spectrometry for monitoring low-molecular-weight compounds in physiological fluids, e.g., ethyl alcohol and dimethylsul-

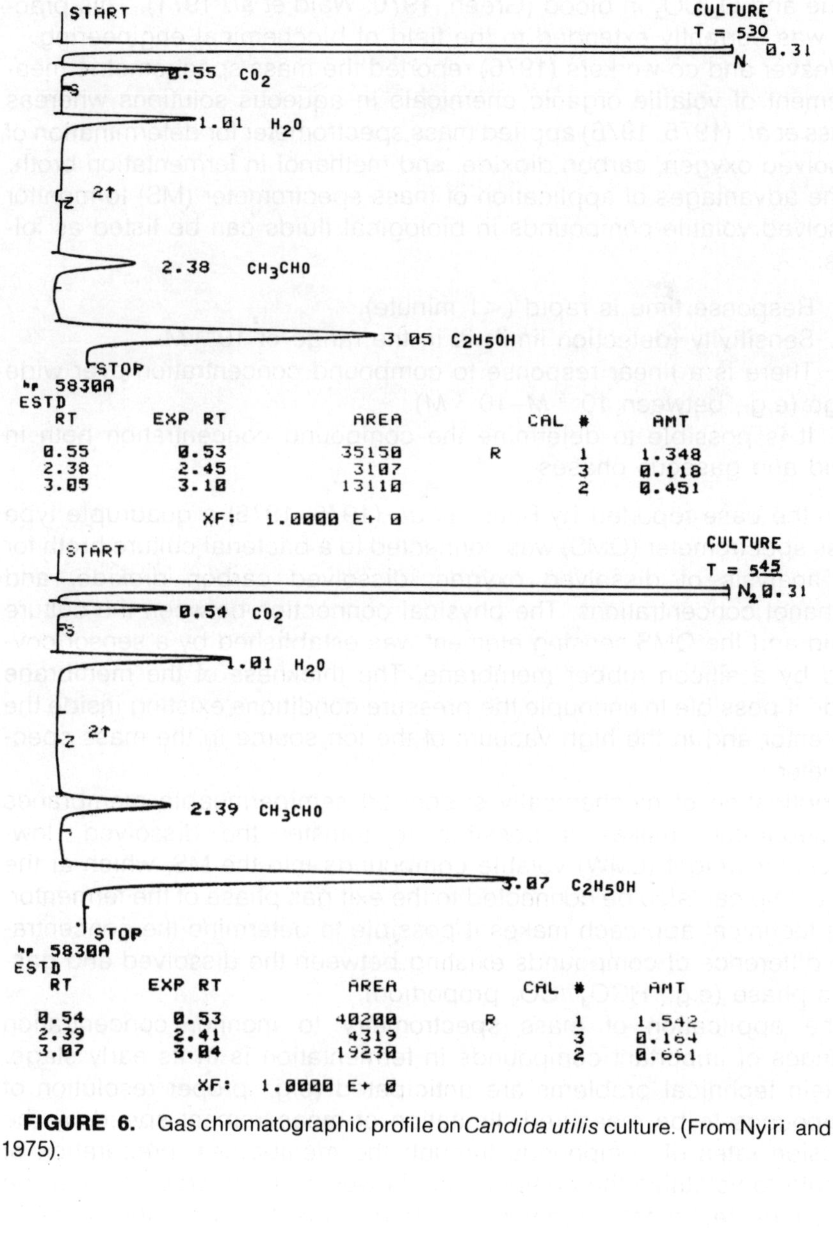

FIGURE 6. Gas chromatographic profile on *Candida utilis* culture. (From Nyiri and Toth 1975).

foxide and O_2/CO_2 in blood (Green, 1970; Wald *et al.* 1971). This practice was currently extended to the field of biochemical engineering.

Weaver and co-workers (1976) reported the mass spectrometric measurement of volatile organic chemicals in aqueous solutions whereas Reuss *et al.* (1975, 1976) applied mass spectrometer for determination of dissolved oxygen, carbon dioxide, and methanol in fermentation broth.

The advantages of application of mass spectrometer (MS) to monitor dissolved volatile compounds in biological fluids can be listed as follows:

1. Response time is rapid (<1 minute).
2. Sensitivity (detection limit) is in the range of 10^{-5} *M*.
3. There is a linear response to compound concentration over wide range (e.g., between 10^{-5} *M*–10^{-1} *M*).
4. It is possible to determine the compound concentration both in liquid and gaseous phases.

In the case reported by Reuss *et al.* (1975, 1976) a quadruple type mass spectrometer (QMS) was connected to a bacterial culture broth for the analysis of dissolved oxygen, dissolved carbon dioxide, and methanol concentrations. The physical connection between the culture liquid and the QMS sensing element was established by a sensor covered by a silicon rubber membrane. The thickness of the membrane made it possible to uncouple the pressure conditions existing inside the fermentor and in the high vacuum of the ion source in the mass spectrometer.

Application of mechanically supported semipermeable membranes as separators makes it possible to transfer the dissolved, low-molecular-weight (LMW) volatile compounds into the MS, which at the same time can also be connected to the exit gas phase of the fermentor. This technical approach makes it possible to determine the concentration difference of compounds existing between the dissolved and gaseous phase (e.g., HCO_3^-/CO_2 proportion).

The application of mass spectrometry to monitor concentration changes of important compounds in fermentation is in its early stage. Certain technical problems are anticipated (e.g., proper resolution of compounds to be measured, limitation of measurement speed by the diffusion rates of compounds through the membranes, preparation of sample to volatilize the compounds); however, these problems can be solved by technical improvements (e.g., selection of suitable membranes, application of selected ion-peak monitoring systems for resolution, and microprocessors for calibration—peak holding as well as tuning).

In view of the fact that more than one fermentor can be connected to one MS, the relatively high unit cost of the mass spectrometer is eliminated. In addition, it is anticipated that with the development of the methodology less expensive and more versatile mass spectrometers will appear on the market, bringing this analytical technique into the realm of widespread applicability.

c. Dialysis Systems. Development of plastic membranes with various molecular weight cutoff values has opened up the possibility for selective separation of molecules from the culture broth (on-line) and analysis off-line in readily available equipment. This same principle has been applied in mass spectrometers. In a recent study Zabriske and Humphrey (1978) described a dialysis system which is attached to a modified baffle in the fermentor. Compounds with low molecular weight are continuously dialyzed from the culture fluid into the water–buffer solution circulating in the dialysis tube. Glucose was measured by a glucose analyzer with a response time of 4–7 minutes to step changes in the culture broth's glucose content. The system is a modified version of a device developed by Glaxo Ltd. (England) for continuous determination of penicillin.

Advantages of the dialysis approach are (1) selectivity for the compound of interest, (2) continuous signal production (analog signal), and (3) relative simplicity of the sampling system. The potential disadvantage of the approach is the relatively long response time which is a function of the distance between the dialysis system and the analyzer and the assay procedure selected.

d. Automatic Wet Chemical Analyzers. A general purpose and sophisticated device which can automatically and semicontinuously determine a variety of organic and inorganic constituents of a culture broth is the Technicon AutoAnalyzer. This instrument consists of a train of modules (Fig. 7) which can pretreat, dilute, sample, react, and analyze the sample broth by a specific quantitative detection device (colorimeter, flame photometer, flowmeter, potentiometer, etc.) for the compound in question. Results of the analysis can be either recorded and/or sent to an interface in the form of a digital signal for further data processing. Typical of the assay procedures for fermentation processes which have been reported for the AutoAnalyzer are mono- and disaccharides, inorganic and organic nitrogens, phosphorus, amino acids, vitamins, and certain antibiotics. Although the AutoAnalyzer is considered an off-line device because of the relatively long elapsed time (typically 10–30 minutes) between the sampling and availability of results, it will find increased "on-line" application because of its accuracy, reliability, and versatility of available measurements. Process information obtained by

FIGURE 7. Technicon dual channel AutoAnalyzer II Continuous-flow modular analytical system for wet-chemical analyses at rates of up to 60 samples per hour.

this method has the potential for process monitoring and possible control response as well as postexperimental comparison of data obtained with on-line operating sensors and other wet-chemical analyzers.

IV. ELECTRONIC ANALYSIS OF PROCESS DATA

The instruments previously described produce information on process status, equipment performance, substrate uptake, and product formation. Sensors, which reflect the process status, generally detect an end result of several metabolic activities and not the metabolic activity itself (e.g., pH is the result of the accumulation of many compounds of acidic or alkaline characteristics as a result of composite metabolic activity).

Furthermore, the specific broth concentration measurements reflect a condition which can be meaningful only if it is placed in the proper time dimension to describe the trend of the process. Consequently, concentration changes in time dimensions (rates) are more useful for experimental and process-control purposes than an absolute concentration value. For instance, oxygen uptake rate gives more information about the biological process conditions than the percentage of O_2 in the exit gas. Finally, as the number of instrumentation signals increases, the sheer management and timely reporting of process data in a usable fashion becomes more difficult.

Based on these considerations, starting in the early 1970s, attempts were made to (1) combine individual process sensor signals to obtain otherwise inaccessible process information, (2) place the direct and combined data outputs in a real-time dimension, and (3) store this data for later retrieval and data reduction (Nyiri, 1971, 1972; Humphrey, 1971; Nyiri and Humphrey, 1972). For these purposes real-time operating digital computers were interfaced with the fermentor's instrumentation. The computer was programmed to acquire the data and combine them to produce new process information. Signals from the sensors (called "gateway sensors," Aiba *et al.*, 1973) are mathematically combined according to the theory of graphs. Nodes of the graph represent a series of computations resulting in a new, previously inaccessible process status indicator. As an example, determination of the respiratory quotient [RQ = CO_2 produced (moles)/O_2 consumed (moles)] is used. O_2 and CO_2 gas concentrations are measured in percentage of the effluent gas. In order to convert these raw signals into mole concentrations, the gas flow rate, temperature, and pressure data of the fermentor as well as the gas constants are required (Fig. 8).

Based on a program containing all these inputs, a computer acquires the sensor data from the instruments and performs the computations resulting in this new information, respiratory quotient. Problems with respect to the use of RQ information and its applicability for identification of the physiological status of a *C. utilis* culture were discussed by Nyiri *et al.* (1975a). Computer determinations of RQ were made by several investigators with promising results using on-line optimization and process control (Nyiri and Krishnaswami, 1974; Aiba *et al.*, 1976). In addition, both material and heat balances have also been successfully computed by this technique (Cooney *et al.*, 1975).

Fermentation data processing hardware requirements are shown in a typical schematic, Fig. 9. The hardware requirements will change in the future as further developments in the electronic state of the art occur. Essentially, however, the following system elements will be required:

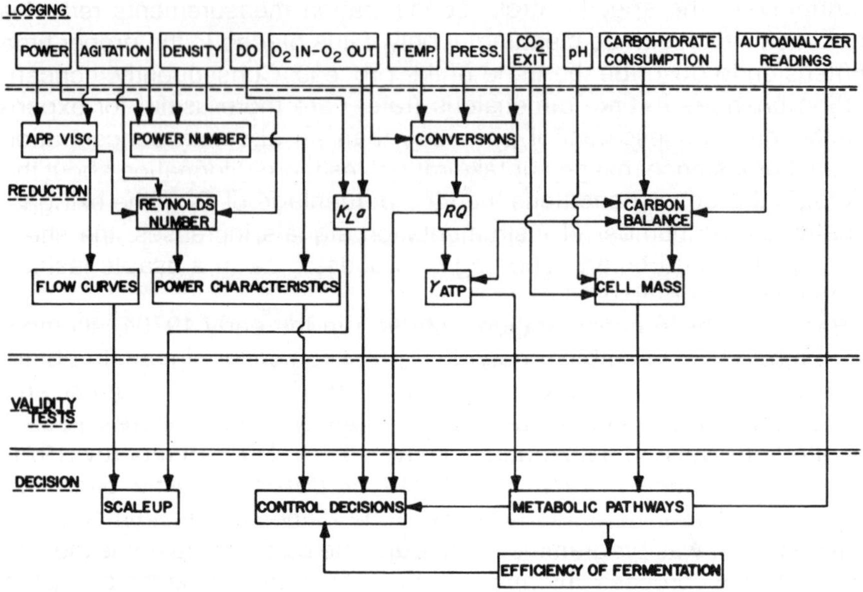

FIGURE 8. Schematic of on-line, real-time data processing in fermentation. Y_{ATP}, organic energy yield factor.

1. Sensor
2. Amplifier
3. Signal conditioning
4. Analog-to-digital converter (for production of BCD signal generally 16-bit word length)
5. Electronic interface
6. Central processing unit (CPU)
7. Real-time clock
8. Data and/or program storage
9. Communication link [teletype (TTY)], video terminal with possible hard-copy capability
10. Data plotter (graphical display of data)

Experience has shown that total execution time for a data acquisition and analysis cycle for a fermentor with 42 separate signal pickups and 61 information outputs required about 1 second, while printing the information on a high-speed printer required about 16 seconds (Nyiri *et al.*, 1975b). These processing times make it possible to acquire data in 1-minute intervals which, for most biological processes, can be considered "continuous" signal pickup and data processing. Metabolic responses to external frequency pulses usually exhibit themselves be-

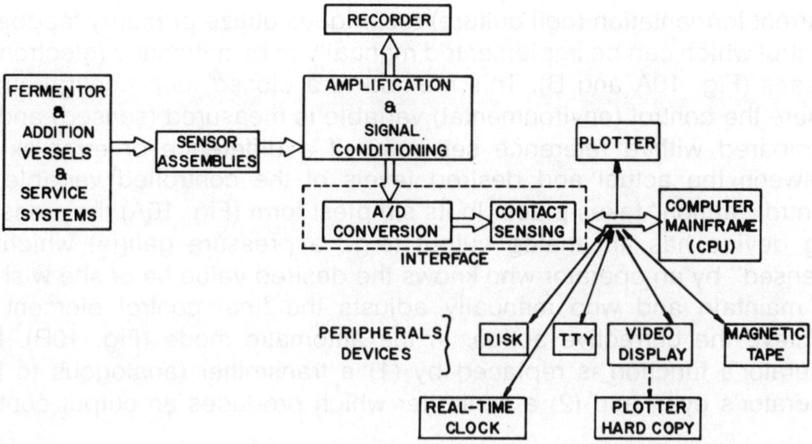

FIGURE 9. Schematics of a computer-aided fermentation system (data acquisition, data processing).

tween 2 and 5 after the step change and (depending on the environmental conditions) can last more than 20 minutes (Nyiri and Toth, 1975). These observations are of importance when process control is to be implemented based on physiological analysis of the culture.

Rapid development in microelectronics has resulted in drastic reduction in the cost per bit of computer memory (Noyce, 1977) and has made it possible to develop microprocessors and microelectronic memories of high density. As a result, low-cost programmable data processors with all of the capabilities needed to perform data acquisition and analysis in real-time and store this information will soon be available for biological processes.

V. EVALUATION OF CONTROL CONCEPTS IN BIOLOGICAL PROCESSES

As was pointed out previously the objective of process control in biological processes is to create a suitable environment for the culture to express the maximum or desired metabolic activity. This primary objective is accomplished by a secondary series of control objectives which maintain a series of set-point conditions in individual control loops. The control problem is complicated further if one considers the fact that the set points may be changed during the process due to changes in physiological conditions of the culture.

Over the past years several control strategies were developed for fermentation processes. These have been discussed by Aiba (1966).

Current fermentation (cell culture) techniques utilize primarily feedback control which can be implemented manually or by automatic (electronic) means (Fig. 10A and B). This method is a closed loop configuration, where the control (environmental) variable is measured (sensed) and is compared with a reference set point. If a difference or error exists between the actual and desired levels of the controlled variable, a control "action" takes place. In its simplest form (Fig. 10A) the measuring device has an analog output (e.g., a pressure gauge) which is "sensed" by an operator who knows the desired value he or she wishes to maintain and who manually adjusts the final control element to achieve the corrective action. In the automatic mode (Fig. 10B), the operator's function is replaced by (1) a transmitter (analogous to the operator's eye), and (2) a controller which produces an output control

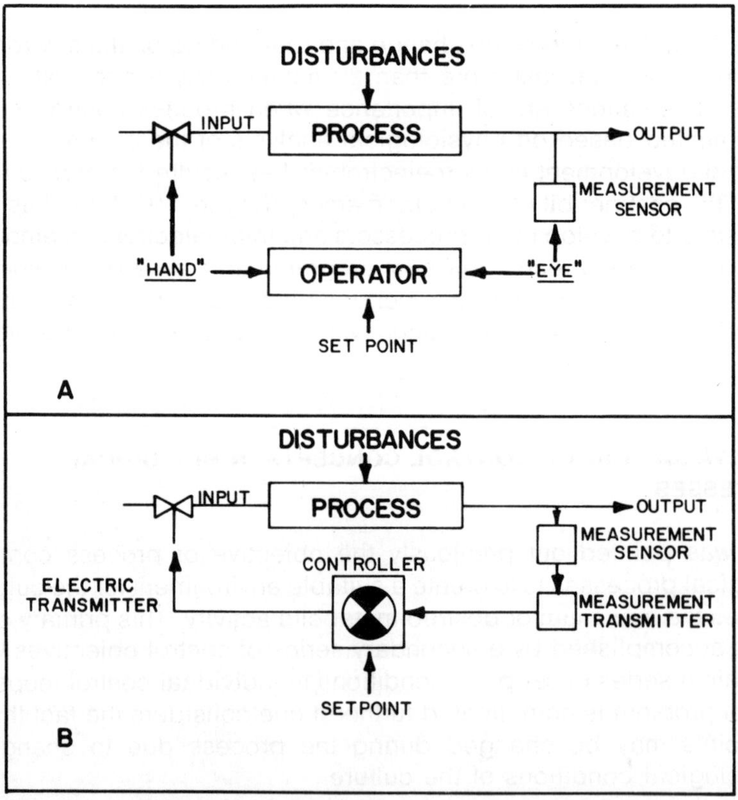

FIGURE 10. Control schematics for biochemical processes. (a) Manual control of a process variable; (b) automatic control of a process variable (feedback control).

signal to the final control element which is a function of the input error signal into the transmitter.

With respect to control variables normally found in most fermentation processes it can usually be stated that (1) the optimum value of the controlled variable (e.g., temperature) was chosen by trial and error over a protracted period of time; (2) the set-point value is generally held constant over the course of the process independent of the physiological stage of the culture (this is mainly due to the lack of information about correlations between the environment and cell physiology); and (3) the sensor accuracy usually exceeds the accuracy of the controller's performance, and to date no evidence is yet available to define how accurate the environmental control must be to maximize cell metabolic activity.

Recent investigations indicate that the optimum environmental conditions change over the course of the process and that certain environmental variable "profiles" need to be developed in an attempt to adjust the environment to cope with the demands of changing cell physiology and enzyme activity (Constantinides *et al.*, 1970 a,b; Rai and Constantinides, 1973; E. B. Lynch and W. F. Ramirez, personal communication, 1975). Though publication of such information has not been made, the authors are aware of several commerical processes which have used this approach to optimize a particular facet of a culture's performance. In addition, changes in one environmental variable may require readjustment of other controlled variables due to the complex interactions between the cells and the environment (Nyiri *et al.*, 1973; Unden and Heden, 1973). Because of the complexity of the biological process, the use of computers to search out optima and apply this knowledge to establish proper environmental conditions for cells of changing physiology is an inevitable step in the future development of both instrumentation and process analysis and control techniques in the fermentation industry (Fuld, 1960; Humphrey, 1971; Nyiri, 1971, 1972).

The first known application of computers to control fermentation process variables was Dista's effort in a penicillin fermentation plant (Grayson, 1969). In this instance, direct digital control was employed where the computer replaced the controller's comparative function in feedback control loops by determining the signal error and computing the necessary final control element action for the manipulated variable. The advantage of this approach was a more accurate set-point control of the individual process variables. However, one potential disadvantage was the loss of process control in the event of computer failure. The overall strategy of control was based on individual set-point control and did not take into account the interactions between the culture and its environment.

Recognition of interactions between the environment and the process culture made it necessary to attempt to develop control modes where more than one process variable was changed simultaneously and where the degree of change was based on (a) the condition of one process status indicator, and (b) the condition of other controlled variables which might interactively influence the process status. The first example (on a bench-scale system) was the interactive control of dissolved oxygen on the basis of computed values of k_La using agitation speed and/or air-flow rate for control purposes (Nyiri *et al.*, 1974) (Fig. 11). Computer linearization of this control system resulted in more stable control performance (Jefferis *et al.*, 1973).

Based on experience and current developments in analog control techniques a modified version of cascade (supervisory) control strategy was recently proposed for biological processes (Nyiri and Jefferis, 1974; Wilson *et al.*, 1975; Nyiri *et al.*, 1975b). Figure 12 illustrates the concept. Environmental variables (e.g., temperature, pressure, and pH) are controlled by individual analog controllers. The control loop consists of conventional elements including sensor, transmitters, controller, and final control element. The set point of every controller, however, is generated by a digital computer. Signals originating from the sensors that indicate the process controller's performance (e.g., temperature reading) as well as signals reflecting biological process conditions, e.g., $O_2\%/CO_2\%$ in the exit gas, are introduced into the computer which (a) reduces the individual signals to process status indicators [e.g., rate of

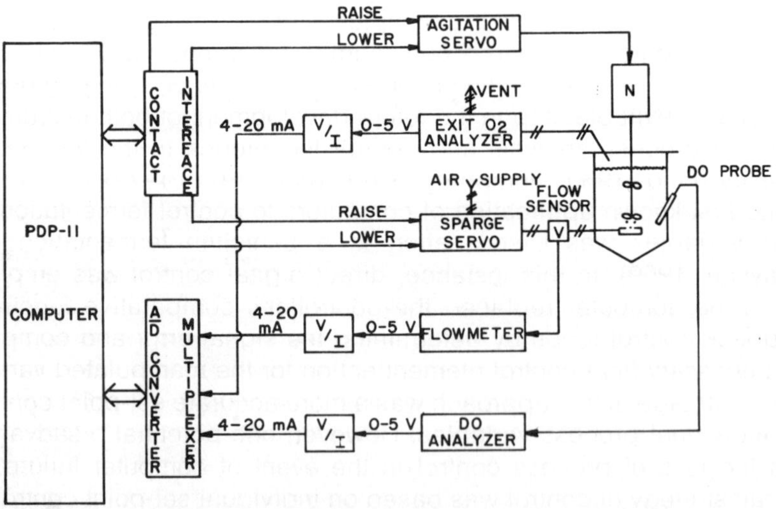

FIGURE 11. Computer control of dissolved oxygen. *N*, agitation; *V/I*, voltage–current signal conditioning; and *V*, volumetric gas flow rate.

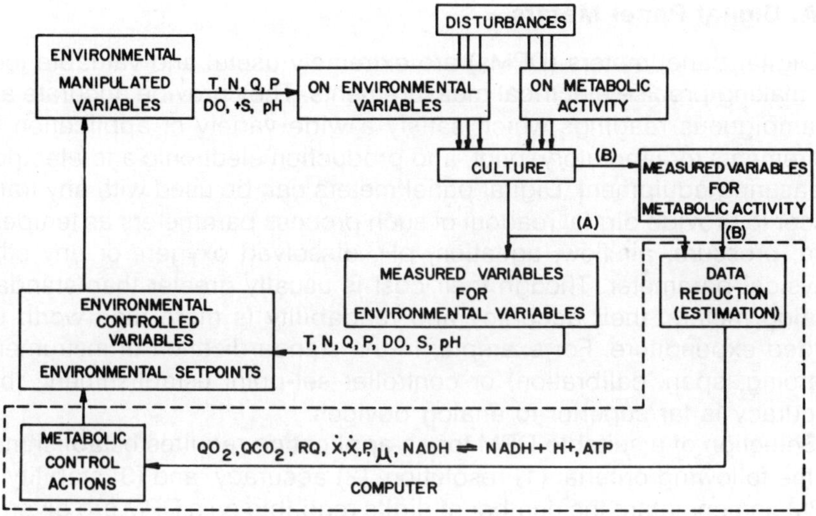

FIGURE 12. Cascade control of fermentation process variables. *DO*, dissolved oxygen concentration; *N*, agitation speed; *P*, pressure, product; *Q*, gas flow rate; *S*, substrate; +*S*, substrate addition rate; *T*, temperature; $\dot{X}$, cell mass formation rate; and μ, specific growth rate.

oxygen uptake (Q_{O_2}) and respiratory quotient (RQ)] and (b), based on experimentally defined interactions between the environment and the cells, alters the set point of one or more individual controllers in order to keep the environmental conditions at the desired optimum.

The individual controllers can be substituted by a digital computer, thereby creating a direct digital control structure (Jefferis, 1973). The decision to rely fully on computers or to use analog controllers interfaced to the computer is at the discretion of the user.

In another reported version, the control of individual environmental variables is performed by microprocessors and the definition of the biological process status (data acquisition, data reduction) is implemented by a real-time digital computer with large memory capacity (Jefferis, 1973). Development of such a structure, however, requires a wealth of information on the culture–environment interactions which, at this moment, is still very much in the exploratory state.

VI. AUXILIARY INSTRUMENTATION

In this category we list those devices which are necessary to complete an instrumentation package, which include digital panel meters and recorders.

A. Digital Panel Meters

Digital panel meters (DPMs) are extremely useful and valuable tools for making precise electrical measurements. They provide accurate and unambiguous readings which satisfy a wide variety of application requirements for laboratory, pilot, and production electronic and electrical measuring equipment. Digital panel meters can be used with any transducer to provide digital readout of such process parameters as temperature, pressure, air flow, agitation, pH, dissolved oxygen, or any other physical parameter. Though their cost is usually greater than standard analog meters, their accuracy and readability is more than worth the added expenditure. For example, in the standardization of instruments (zeroing, span, calibration) or controller set-point establishment, their accuracy is far superior to analog devices.

Selection of a suitable DPM for an application requires establishment of the following criteria: (1) resolution, (2) accuracy, and (3) stability. In DPM nomenclature the number of digits readable on a DPM are given as shown in the tabulation below (excluding decimal point).

DPM digits	Full-scale maximum count
$2\frac{1}{2}$	199
3	999
$3\frac{1}{2}$	1999
$3\frac{3}{4}$	3999
4	9999
$4\frac{1}{2}$	19999
$4\frac{3}{4}$	3999

Sources of digital panel meters are Data Technology, Analog Devices, Canyon Electronics, Datel Systems, Analogic Corp., and Fairchild Instrumentation.

From our own experience, despite the fact that DPMs have output signals directly acceptable by some computer interfaces, the natural oscillation of most DPM input signals will cause "floating" of the last output digit, thereby creating latching problems at the moment of acquisition by the computer. It is therefore preferred to separate the signals (from the DPM) for interfacing. Hence, one of the benefits of DPMs, the fact that they provide their own analog-to-digital (A/D) conversion, is not considered by us to be a general plus if computer coupling is the objective in the electronic system design.

B. Recorders

For permanent record keeping of acquired signals recorders are usually used with most of the various instrumentation systems employed for process monitoring and control. Recorders can be purchased starting with single-point recording units up to recorders which will accept 30 points. Printing can be either by traditional ink–pen systems or newer inkless pressure or thermal recording styli. Essentially all instrument manufacturers are a source of recorders (e.g., Honeywell, L&N, Foxboro, Taylor).

The rationale for recording any variable which is monitored depends to a large extent on the organization doing the work, the need for permanent records (such as in production operations to meet government requirements for product process history), and the nature and type of process work being performed. For several reasons, one of which is the difficulty of simple storage of printed data in standard files, strip-type chart recorders are preferred over circular ones. Also, multipoint recording devices provide a convenient method of recording several relatable variables on the same time axis, making visual data analyses much easier. However, the number of simultaneous recordings on a single chart must be limited to between three and six, depending on the variables involved, if one wishes to utilize full recorder scale width for each variable range. Field experience has shown that beyond six variables on a single recorder chart, regardless of recording chart width, there tends to be added confusion of analyses and not the improvement which is desirable.

Based on the foregoing, it is recommended that multipoint strip chart-type recorders be considered for process data recording. Newer designs in recording equipment have resulted in smaller units which can be easily panel mounted. In addition, several manufacturers are beginning to include microprocessor devices in their recorder which allows the selection and frequency of recording points, as well as variable expansion of recording ranges.

VII. SUMMARY

We have attempted to provide a very broad overview of currently available instrumentation systems based on a generic description of the areas of their application in fermentation. Unfortunately, in a review such as this, it is always possible to overlook both current research work and possibly newer vendor developments. We have also tended to "take a stand" on those classes of instruments which we feel are better suited for use in fermentation work. Some readers will accept this, and others will certainly have their own "best solutions" to champion.

Because of the very rapid developments occurring in the electronics industry, newer techniques and equipment methods are probably already coming to market even during publication of this review. If fermentation processing is to advance, newer instrumentation and, more specifically, newer sensors must be pursued and developed to meet the critical needs of the biological processes with which we must deal. Until the time comes when such equipment is readily available to all, we must rely on the ingenuity and determined efforts of process development pioneers to continue to adapt what is available to the "advancement of the art."

VIII. APPENDIX: LIST OF EQUIPMENT MANUFACTURERS

Activion Glass, Ltd., Halstead, Essex, England
Analog Devices, Norwood, Massachusetts
Analogic Corp., Wakefield, Massachusetts
Anarad, Inc., Santa Barbara, California
Automation Products, Inc., Houston, Texas
Beckman Instruments, Fullerton, California
BLH Electronics, Waltham, Massachusetts
Brooks Instruments Div., Hatfield, Pennsylvania
Canyon Electronics, Phoenix, Arizona
Chemcon, Inc., Medfield, Massachusetts
Data Technology, Santa Ana, California
Datel Systems, Canton, Massachusetts
Electro Corp., Sarasota, Florida
Fairchild Instrumentation, San Jose, California
Fisher and Porter Co., Warminster, Pennsylvania
Foxboro Co., Foxboro, Massachusetts
Gulton Industries, Costa Mesa, California
Hewlett Packard Corp., Santa Clara, California
Honeywell, Process Control Div., Ft. Washington, Pennsylvania
Infrared Industries, Santa Barbara, California
Ingold Electrodes, Inc., Lexington, Massachusetts
Instrumentation Laboratory, Inc., Lexington, Massachusetts
Interpace Corp. (Lapp), Rochester, New York
Leeds and Northrop, North Wales, Pennsylvania
LFE Corp., Waltham, Massachusetts
Meriam Instrument Div. (Scott & Fetzer), Cleveland, Ohio
Micro-strain, Inc., Spring City, Pennsylvania
Milton Roy Co., St. Petersburg, Florida
Mine Safety Appliance Co., Pittsburgh, Pennsylvania
New Brunswick Scientific Co., Edison, New Jersey
Ohio Semitronics Co., Columbus, Ohio
Orion Research, Inc., Cambridge, Massachusetts
Radiometer, Copenhagen, Denmark
Ranco Controls, Columbus, Ohio
Robertshaw Controls Co., Richmond, Virginia
Rosemount, Inc., Minneapolis, Minnesota
Schaevitz Engineering, Pennsauken, New Jersey
Schutte and Koerting, Cornwells Heights, Pennsylvania
A. O. Smith Corp., Erie, Pennsylvania
Taylor Instrument Co., Rochester, New York

Technicon Instrument Corp., Tarrytown, New York
Transducers, Inc., Whittier, California
Vibrac Corporation, Chelmsford, Massachusetts
Yellow Springs Instrument Co., Yellow Springs, Ohio

REFERENCES

Abbot, B. J., Laskin, A. I., and McCoy, C. J. (1973). *Appl. Microbiol.* **25,** 787.
Aiba, S. (1966). *Proc. Int. Cong. Microbiol., 9th, 1966.*
Aiba, S., Humphrey, A. E., and Millis, N. F. (1973). "Biochemical Engineering," 2nd ed. Academic Press, New York.
Aiba, S., Nagai, S., and Nishizawa, Y. (1977). *Biotechnol. Bioeng.* **18,** 1001.
Bandyopadhyay, B., Humphrey, A. E., and Taguchi, H. (1967). *Biotechnol. Bioeng.* **3,** 533.
Charles, M., and Toth, G. M. (1974). *Ann. Meeting Am. Chem. Soc. 168th.*
Clark, W. M., and Cohen, B. (1923). *Public Health Rep.* **38,** 666.
Considine, D. M., ed. (1957). "Process Instruments and Controls Handbook." McGraw-Hill, New York.
Constantinides, A., Spencer, J. L., and Gaden, G. L., Jr. (1970a). *Biotechnol. Bioeng.* **12,** 803.
Constantinides, A., Spencer, J. L., and Gaden, G. L., Jr. (1970b). *Biotechnol. Bioeng.* **12,** 1081.
Cooney, C. L., Wang, H., and Wang, D. I. C. (1975). *U.S./U.S.S.R. Joint Symp. Data Acquisition Process.*
Daniels, W. F., Garcia, L. M., and Rosenstahl, J. F. (1970). *Biotechnol. Bioeng.* **12,** 403.
Fiechter, A., and von Meyenburg, K. (1963). *Biotechnol. Bioeng.* **10,** 535.
Fuld, G. J. (1960). *Adv. Appl. Microbiol.* **2,** 351.
Gandhi, A. P., and Kjaergaard, L. (1975). *Biotechnol. Bioeng.* **17,** 1103.
George, T. K., and Gaudy, A. F. (1973). *Appl. Microbiol.* **26,** 736.
Grayson, P. (1963). *Process Biochem.* **4,** 43.
Green, D. W. (1970). *Inter-Science Chem. Reports* **4,** 211.
Hamer, G. (1967). *Biotechnol. Bioeng.* **8,** 433.
Harmes, C. S., III (1972). *Dev. Ind. Microbiol.* **13,** 146.
Harrison, D. E. F., and Chance, B. (1970). *J. Appl. Microbiol.* **19,** 446.
Harrison, D. E. F., and Harmes, C. S., III (1972). *Process Biochem.* **13,** 1.
Hewitt, L. V. (1950). "Oxidation—Reduction Potentials in Bacteriology and Biochemistry." Livingstone, Edinburgh.
Humphrey, A. E. (1971). *Proc. Labex Symp. Comput. Control Ferment. Process.*
Ishizaki, A., and Hirose, Y. (1973). *Agric. Biol. Chem.* **37,** 1235.
Ishizaki, A., Shibai, H., Hirose, Y., and Shiro, T. (1973). *Agric. Biol. Chem.* **37,** 33.
Jefferis, R. P., III (1973). *Proc. Eur. Conf. Comput. Process Control Ferment., 1st.*
Jefferis, R. P., III, Biela, J. A., and Nyiri, L. K. (1973). *75th Natl. AIChE Meet.*
Kjaergaard, L. (1977). *Adv. in Biochem. Eng.* **7,** 131.
Kruse, P. F., Jr., and Patterson, M. K., eds. (1973). "Tissue Culture: Methods and Applications." Academic Press, New York.
Lengyel, Z. L., and Nyiri, L. K. (1965a). *Biotechnol. Bioeng.* **7,** 31.
Lengyel, Z. L., and Nyiri, L. K. (1965b). *Biotechnol. Bioeng.* **7,** 91.
Lengyel, Z. L., and Nyiri, L. K. (1966). *Biotechnol. Bioeng.* **8,** 337.
Linek, V., Sobotka, M., and Prokop, A. (1973). *Biotechnol. Bioeng.* **4,** 423.
Mackereth, F. J. H. (1964). *J. Sev. Instrum.* **41,** 38.
Metzner, A. B., and Taylor, J. S. (1962). *AIChE J.* **6,** 103.

Noyce, R. N. (1977). *Sci. Am.* **237,** 62.
Nyiri, L. K. (1971). *Proc. Labex. Symp. Comput. Control Ferment. Process.*
Nyiri, L. K. (1972). *Adv. Biochem. Eng.* **2,** 49.
Nyiri, L. K. (1978). *Adv. Biochem. Eng.* **8** (in press)
Nyiri, L. K., and Humphrey, A. E. (1972). *Ferment. Technol. Today, Proc. Ferment. Int. Symp. 4th, 1972.*
Nyiri, L. K., and Jefferis, R. P., III (1974). *In* "Single-Cell Protein II" (S. R. Tannenbaum and D. I. C. Wang, eds.), p. 105, MIT Press, Cambridge, Massachusetts.
Nyiri, L. K., and Krishnaswami, C. S. (1974). *75th Meet. Am. Soc. Microbiol.*
Nyiri, L. K., and Lengyel, Z. L. (1968). *Biotechnol. Bioeng.* **10,** 133.
Nyiri, L. K., and Toth, G. M. (1975). *Annu. SIM Meet. 1975.*
Nyiri, L. K., Jefferis, R. P., III, and Humphrey, A. E. (1973). *Abstr. Pap., 166th Meet., Am. Chem. Soc.*
Nyiri, L. K., Jefferis, R. P., III, and Humphrey, A. E. (1974). *Biotechnol. Bioeng. Symp.* **4,** 613.
Nyiri, L. K., Wilson, J. D., Humphrey, A. E., and Harmes, C. S., III (1975a). U.S. Patent 3,326,738.
Nyiri, L. K., Toth, G. M., and Charles, M. (1975b). *Biotechnol. Bioeng.* **17,** 1663.
O'Brien, R. T., and Cecchini, G. L. (1970). *Dev. Ind. Microbiol.* **11,** 99.
Peraino, C., and Eisler, W. J. Jr. (1973). *In* "Tissue Culture. Methods and Applications" (P. F. Kruse, Jr. and M. K. Patterson, eds.), p. 351. Academic Press, New York.
Rabotnowa, I. L. (1963). "Die Bedeutung physikalisch-chemisher Faktoren (pH und rH_2) für die Lebenstätigkeit der Mikroorganismen." Fischer, Jena.
Rai, V. R., and Constantinides, A. (1973). *AIChE Symp. Ser.* **23,** 114.
Reuss, M., Piehl, H., and Wagner, F. (1975). *Europ. J. Appl. Microbiol.* **1,** 323.
Reuss, M., Piehl, H., and Wagner, F. (1976). *Abstr. Int. Ferment. Symp., 5th, 1976,* p. 24.
Schroder, K. D., and Weide, H. (1974). *Biotechnol. Bioeng. Symp.* **4,** 713.
Severinghaus, J. W., and Bradley, A. F. (1958). *J. Appl. Physiol.* **13,** 515.
Shibai, H., Ishizaki, A., Mizuno, H., and Hirose, Y. (1973). *Agric. Biol. Chem.* **37,** 91.
Siegel, S. O., and Gaden, E. I., Jr. (1962). *Biotechnol. Bioeng.* **4,** 345.
Solomons, G. L. (1969a). "Materials and Methods in Fermentation." Academic Press, New York.
Solomons, G. L. (1969b) *Chem. Eng. (London)* **215,** p. 55.
Taguchi, H., and Humphrey, A. E. (1966). *J. Ferment. Technol.* **44,** 881.
Taylor, G. W., Kondig, J. P., Nagle, S. C., and Higuchi, K. (1971). *Appl. Microbiol.* **21,** 928.
Toth, G. M. (1975). *Proc. Cell Cult. Congr.*
Tuffile, C. M., and Pinho, F. (1970). *Biotechnol. Bioeng.* **12,** 843.
Unden, G. A., and Hedén, C. G. (1973). *Proc. Eur. Conf. Comput. Process Control. Ferment., 1st.*
Umbreit, W. W., Stauffer, R. H., and Stauffer, J. R. (1957). "Manometric Techniques." Burgess, Minneapolis, Minnesota.
Updike, S. J., and Hicks, G. P. (1967). *Nature (London)* **214,** 986.
Vincent, W. A., and Priestley, G. (1975). "Handbook of Enzyme Biotechnology." Academic Press, New York.
Wald, A., Hass, W. K., and Ransohoff, J. (1971). *J. Assoc. Adv. Med. Inst.* **5,** 325.
Weaver, J. C., Mason, M. K., Jarrel, J. A., and Peterson, J. W. (1976). *Biochim. Biophys. Acta.* **438,** 296.
Wilson, D. J., Nyiri, L. K., Humphrey, A. E., and Harmes, C. S., III (1975). U.S. Patent 3,326,737.
Wimpenny, J. W. T. (1968). *Symp. Soc. Gen. Microbiol.* **19.**
Zabriskie, D. W., and Humphrey, A. E. (1978). *Biotechnol. Bioeng.* **20** (in press).
Zajic, J. E., and Liu, F. S. (1970). *Dev. Ind. Microbiol.* **11,** 350.

Chapter 15

Computer Applications in Fermentation Technology

WILLIAM B. ARMIGER
ARTHUR E. HUMPHREY

I. INTRODUCTION

Recent advances in biochemistry and microbiology have resulted in a more complete understanding of fermentations, particularly with respect to substrate metabolism, growth requirements, and enzyme regulation. These developments are enabling the fermentation technologist to describe the dynamic behavior of fermentation processes in terms of kinetic models.

MICROBIAL TECHNOLOGY, 2nd ed., VOL.II
 ISBN 0-12-551502-2

With this understanding, it is now possible to expand the control of fermentation processes beyond independent closed-loop feedback control of culture conditions, such as temperature, pH, and dissolved oxygen, into the more sophisticated strategies of adaptive or interactive control. Eventually, on-line optimization of a fermentation based upon model reference control of a process will be possible. This accomplishment will mean that a feed-forward, interactive control strategy will be utilized. In order to reach this capability, it is necessary to develop a system which encompasses both hardware and software needed to generate the raw data, analyze the resulting information, determine an optimal solution, and implement a control decision. A highly instrumented computer-coupled fermentor is capable of meeting this task.

The accomplishment of on-line optimization of a fermentation is an evolutionary process. The process began with the achievement of fermentation data acquisition by a computer system. Most fermentation companies now routinely utilize computers in data acquisition. This development involved on-line process monitoring and scanning, data logging, and generation of alarm signals. In many cases, this information was collected in a form that was transferred to a computer system for off-line analysis. With the advent of inexpensive and reliable minicomputers, it has become economical to utilize a computer system on-line. With this capability, it has been possible to move rapidly to a computer applications level which has involved partial process control based upon using predetermined set points for controlling primary fermentation parameters, such as temperature, pH, and dissolved oxygen. It has also been possible to utilize the information generated on-line to study the process and to elaborate the process kinetics. Today, on-line computer systems are being utilized for reducing the time required for determining the dynamics of fermentation processes, for testing control strategies, and for modeling the process.

With the development of more powerful minicomputer operating system software, it is possible to operate the computer system using both foreground/background tasks, i.e., simultaneously performing the on-line data acquisition and analysis tasks and simultaneously doing on-line process simulation. This allows the fermentation technologist to quickly develop and refine mathematical models and to identify the model parameters so that methods can be developed for predicting the interactive relationships between primary fermentation parameters. Once significant operating confidence is obtained with process kinetic models, computer systems can advance to the final stage of automatic process control, i.e., on-line optimization based upon model reference control.

II. HISTORY OF COMPUTER APPLICATIONS

Computer applications in the chemical process industry began in the late 1950s. By the late 1960s, publications on computer applications in batch process control were fairly common. However, it was the widespread availability of minicomputers in the 1970s that led to an abundance of process-control applications.

Yamashita and Murao (1967), and Koga *et al.* (1967) published the first papers on the use of digital computers specifically for modeling fermentation processes. Yamashita *et al.* (1969), Grayson (1969), and Harrison *et al.* (1971) published the first real description of a computer-coupled fermentation system. Since that time, numerous articles dealing with system design and software utilization have been published. (These are listed in the Supplementary Reading Section.)

Due to corporate secrecy policies, it is not clear which companies were the first to utilize computers in fermentation processes. However, there are some publications dealing with industrial applications. In 1966, The Ajinomoto Company in Japan utilized a YODIC-500 computer for direct digital control of their glutamic acid fermentation, and Dista Products utilized an ARCH 102 computer for direct digital control of 114 loops in a new fermentation plant. In 1969, Glaxo Laboratories, Ltd., converted an existing fermentation plant from analog to direct digital control. Since the early 1970s, many research-oriented pilot-plant systems have been developed for on-line data acquisition, analysis, and digital set-point control. The University of Pennsylvania (Armiger and Humphrey 1974; Armiger *et al.*, 1975, 1976; Harrison *et al.*, 1971; Harrison and Harmes, 1972), E. R. Squibb (Moes *et al.*, 1971; Young and Koplove, 1972), Lord Rank Research Center (Flynn, 1974), Station de-Gene Microbiologique (Blachere 1973; Corrieu *et al.*, 1974; Lane, 1973), Karolinska Institute (Unden and Heden, 1973), Gesselschaft Fur Biotechnologische Forschung MBH (Jefferis, 1975), and many others (Cooney *et al.*, 1975; Jefferis, 1973; Nyiri *et al.*, 1975; Nagai *et al.*, 1976) have interfaced computers with their fermentation systems.

Computer control and computer applications in fermentation cover a broad spectrum of uses depending upon the needs of a particular process. The requirements for a pilot plant are quite different from those for a production plant. The degree of sophistication of the computer software depends upon the use of the system. Moving from data logging to data analysis, process modeling, process control, and finally to process optimization requires an increasingly more complex system in terms of both the hardware and the software. All of these considerations along with future trends for computer applications will be discussed in the following sections.

III. GENERAL APPLICATIONS

The application of computers to fermentation technology is very dependent upon the end use of the computer system. For example, requirements for a pilot-plant system are quite different from those for a production plant. In a pilot plant, the computer is used primarily as a research tool. It is necessary that the system be designed to be as flexible as possible. It should be particularly useful in obtaining kinetic data necessary for modeling the process and for developing control strategies aimed at process optimization. Pilot-plant computer-coupled systems should be structured to accommodate software development and software testing.

There are two levels in which a computer system can be utilized in a pilot plant. The first level involves operator-oriented programs which can be utilized by a technician. To run these programs, it is necessary for a technician to know only the computer command necessary to start them. The programs are designed to prompt the operator with the responses necessary for implementation. These programs include such functions as data logging, data analysis, and process control. By using operator-oriented routines, the technician can generate the necessary documentation and procedures for system operation and measurements. It is also helpful if the system contains programs for generating standard reports and graphical output.

The second level of computer utilization requires an engineer with programming ability to develop and insert into the operator-oriented programs new routines to be run by the technician. Also, on this level the engineer interacts with the computer to develop control strategies and process optimization routines. The engineer can interact with the system to quickly access large volumes of data. The easy access to this information frees him or her to develop and test control strategies and process optimization routines. Thus, the computer-coupled system is a research tool which allows the engineer to spend most of his or her time interpreting data rather than merely collecting and plotting it.

For a full-scale production plant, the requirements for a computer-coupled system are quite different. The primary objective is to produce a product in an optimal manner. As a result, the computer is used for process control and optimization aimed at maintaining quality control and product uniformity. Here again, it is necessary to interact with the computer system on two levels. The first level involves an operator-oriented command language allowing easy access to the entire process. The command language will contain routines for generating process status information, calculating process variables, and scheduling process operations.

Plant status routines are used to send alarm messages to the operator and to supply him or her with information on the overall operation of the process. Raw material inputs, flow rates, and utility requirements in terms of steam, cooling water, and electric power demands are just a few of the process parameters and variables that are routinely monitored. Scheduling routines provide the operator with information for sequencing operations. This is particularly valuable for a batch process where there are strict protocols for cleaning, media sterilization, vessels charging, and product harvesting. The sequencing routines provide information to step the operator through the process in order to start up or to shut down the plant. With a sequencing routine, it is possible for the computer to verify a command by the operator and to prevent him or her from moving to the next sequence of events until the previous request has been satisfied. In this way, the computer system can be utilized to ensure that the same procedures are followed for each batch run which helps to assure product quality and product uniformity. These routines also monitor raw material inventories in order to maintain proper deliveries, and they assist the operator in planning maintenance schedules for major equipment. Finally, the command language generates specific process information regarding the status of each vessel and automatically produces reports and documentation at the end of each run.

The second level on which the plant-scale computer system can be utilized is similar to that in the pilot plant. At this level, an engineer is required for making changes in the process-control or monitoring routines. These changes are required as process improvements are developed in the pilot plant and transferred to plant-scale operations.

IV. SPECIFIC APPLICATIONS

A. Data Logging

Table I lists variables that are commonly monitored during a fermentation. In data logging applications the raw data are fed into a computer system where they are compared with calibration curves and correction factors in order to obtain values in standard engineering units. The data are usually printed out on a hard-copy device and written onto a mass storage device for a more detailed analysis by a larger computer system at a later time.

TABLE I. Variables Logged and Controlled

Measured parameters
Inlet gas flow rate
Inlet gas pressure
Inlet gas temperature
Inlet gas carbon dioxide
Inlet gas oxygen
Cooling jacket flow rate
Cooling jacket temperature differences
Agitator speed
Power to shaft
Vessel pressure
Glucose concentration
Culture turbidity
Culture fluorescence
Exit gas carbon dioxide
Exit gas oxygen
Controlled variables
Dissolved oxygen
Vessel temperature
pH
Foam
Substrate addition rate
Acid addition rate
Base addition rate

B. Data Analysis

Numerous papers have been published dealing with the analysis of fermentation data (Yamashita and Murao, 1967; Yamashita *et al.*, 1969; Humphrey, 1971; Nyiri, 1971; Jefferis *et al.*, 1972; Wilson, 1975; Wilson *et al.*, 1975; Wang, 1976; Zabriskie, 1976). Many of the analytical techniques depend upon the particular process for which they were developed. However, there are certain calculations that are common for most fermentations, These include computations for the oxygen uptake rate, carbon dioxide evolution rate, respiratory quotient, overall mass transfer coefficient, heat balance, power requirements, cell mass, and cell growth rate. Table II is a summary of some of these calculations.

These calculations require combining inputs from a number of measured variables. As a result, a computer system is a valuable tool for obtaining information on-line concerning the status of the fermentation process. Often, it is not practical to obtain these parameters by any other means. A good example of this is the measurement of biomass concentration and growth rate. Several papers have appeared in the literature concerning the indirect measurement of cell mass concentration, cell growth rate, and substrate concentration (Jefferis *et al.*, 1972; Zabriskie, 1976; Wang, 1976; Armiger, 1977; Harima, 1977). Utilization of com-

TABLE II. Summary of Calculations

Oxygen uptake rate
Carbon dioxide evolution rate
Respiratory quotient
Total oxygen utilized
Total carbon dioxide evolved
Overall oxygen mass transfer rate
Heat balance
Metabolic heat evolution rate
Cell mass concentration
Cell mass growth rate
Specific growth rate
Cell yield
Nitrogen utilization rate
Glucose utilization rate
Carbon balance
Nitrogen balance
Oxygen balance
Apparent viscosity

puters for indirect measurement to give on-line biomass and growth rate data will have a major impact on the fermentation industry.

Two of these techniques have been derived by performing material balances around the fermentor. The first approach selects a specific molecular component and uses a kinetic model relating cell growth with the material balance. Figure 1 shows a balance around a fermentor for component *X*. Zabriskie (1976) estimated biomass concentration by a material balance for oxygen. He corrected the balance for secondary metabolite formation by also monitoring the carbon dioxide evaluation rate. The basic relationship is shown as Eq. (1) (see Fig. 2), where the

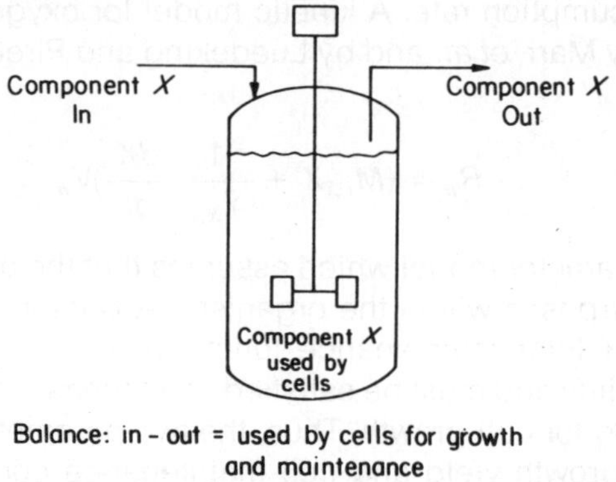

FIGURE 1. Material balance around a fermentor.

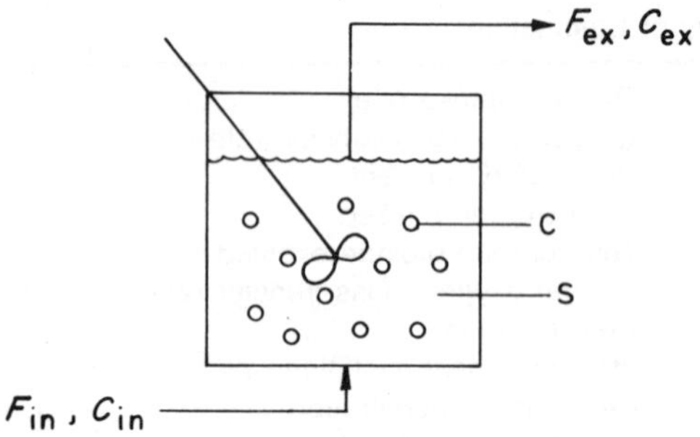

FIGURE 2. Oxygen balance around a fermentor. (From Zabriskie, 1976.)

rate of change of component X with time is equal to the difference in input and output calculations plus an accumulation term.

Gas phase balance: $$\frac{d}{dt}(V_g C)\,dt = F_{in}C_{in} - F_{ex}C_{ex} - J_{g\to a} \quad (1)$$

Utilizing molecular oxygen for the material balance has several advantages over other components. First, there is a close relationship between oxygen and the energetics of growth. Second, oxygen concentration can be measured continuously in both the gaseous and the aqueous phase. Finally the solubility of oxygen in the aqueous phase is so low that a quasi-steady state may be assumed for the concentration of oxygen in the liquid phase. This simplifies the material balance equation. Thus, for oxygen, the uptake rate can be simply expressed as the difference between the inlet and outlet streams [See Eq. (2)].

$$F_{in}C_{in} - F_{ex}C_{ex} - R_a \simeq 0 \quad (2)$$

If one assumes the fermentor is essentially in steady state conditions with respect to oxygen utilization, then the oxygen uptake rate is equal to the oxygen consumption rate. A kinetic model for oxygen consumption, first proposed by Marr, *et al.* and by Luedeking and Piret (1959) is shown in Eq. (3).

Model: $$R_a = (M_{i/X}X + \frac{1}{Y_{X/i}}\frac{dX}{dt})V_a \quad (3)$$

This is a two-parameter model which assumes that the oxygen is utilized for two basic purposes within the organism. A portion of the oxygen is consumed for cellular maintenance functions which are necessary to sustain cell viability and must be satisfied at all times. The balance of the oxygen is utilized for cell growth, Thus, the two parameters in the model are termed the growth yield and cell maintenance constants.

By combining Eqs. (1) and (3), a first-order linear differential equation is obtained [Eqs. (4a) and (4b)].

$$\frac{F_{in}C_{in} - F_{ex}C_{ex}}{V_a} = M_{i/X}X + \frac{1}{Y_{X/i}}\frac{dX}{dt} \quad (4a)$$

In the case of oxygen: $$\text{OUR} = M_{O_2/X}X + \frac{1}{Y_{X/O_2}}\frac{dX}{dt} \quad (4b)$$

Equations (5) and (6) are analytical solutions of this equation for the cell mass concentration and growth rate, respectively:

$$X(t) = \exp(-M_{O_2/X}Y_{X/O_2}t)\left[\int_0^t Y_{X/O_2}\exp(M_{O_2/X}Y_{X/O_2}t)\right.$$
$$\left.\text{OUR}(t)\,dt + X_0\right] \quad (5)$$

$$\frac{dX}{dt} = Y_{X/O_2}(\text{OUR}) - M_{O_2/X}Y_{X/O_2}X \quad (6)$$

To solve this equation only a knowledge of the yield and maintenance parameters plus some way of estimating the initial biomass concentration are needed. Computer results for batch culture of *Thermoactinomyces* sp. and *Saccharomyces cerevisiae* using these correlations based upon real-time oxygen uptake rate data are given in Fig. 3 and 4. For the *Thermoactinomyces* sp. there is good agreement between the on-line estimates of biomass and the laboratory data, especially before the beginning of the stationary phase.

The results are not as good for the experiment involving *S. cerevisiae* due to the diauxic growth of the organism. This observation points to one of the limitations of the first model which assumed that the growth yield and cell maintenance parameters are constant. Another limitation of the basic model is that it does not account for the formation of a secondary product. For example, in order to apply this technique to an antibiotic fermentation, a third term involving a product yield parameter must be included.

In order to deal with the problem of diauxic growth, Zabriskie (1976) introduced a metabolic correction factor (β) into Eq. (5) which is a function of the respiratory quotient. Figure 5 shows the estimates of the biomass concentration for the yeast incorporating the metabolic correction function into the model. It is obvious that a significant improvement in the estimation technique is obtained.

A biomass estimation procedure analogous to the oxygen balancing may be followed using, carbon dioxide. This method was also examined

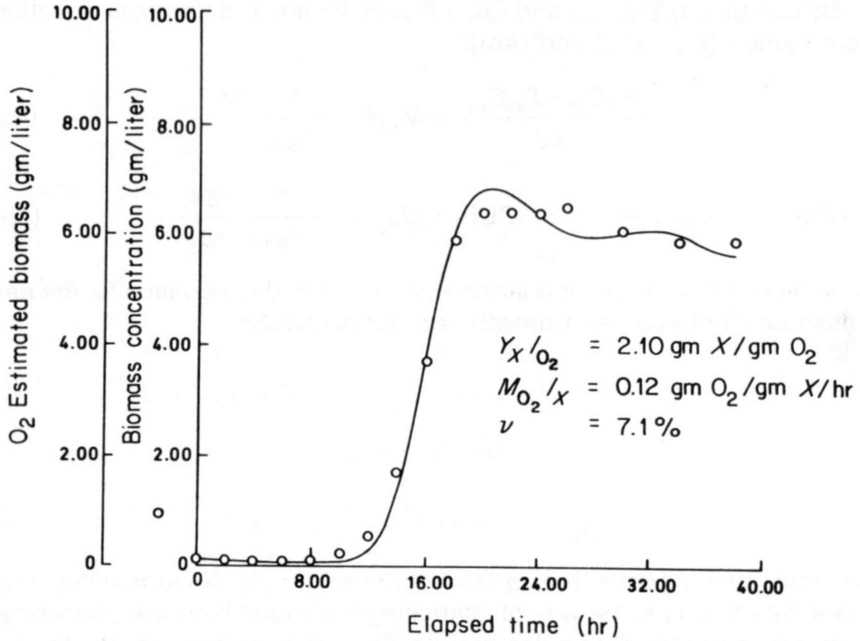

FIGURE 3. *Thermoactinomyces* sp. biomass concentration data and correlation results. (From Zabriskie, 1976.)

by Zabriskie (1976). However, the results show that this technque was much less accurate than oxygen balancing. This is probably due to the fact that carbon dioxide is not directly linked to oxidative phosphorylation so that the relationship between the metabolic correction factor and the respiratory quotient is not as simple.

The second method involving material balancing is based upon the principle of conservation of mass. It does not rely upon determining conversion yields or metabolic rates of activity. However, it does assume that the elemental compositions of the cells and extracellular products are known. The fundamental relationship states that biomass and extracellular products are produced in stoichiometric ratios. The substrate, along with oxygen and ammonia, is converted to biomass, product, carbon dioxide, and water, i.e.,

Substrate + oxygen + ammonia
= biomass + product + carbon dioxide
+ water (carbon source)

Once the molecular composition of the biomass and the product have been determined, it is possible to utilize the stoichiometric relationships along with monitoring the oxygen uptake rate, and the substrate or ammonia consumption rate, to continuously monitor the biomass con-

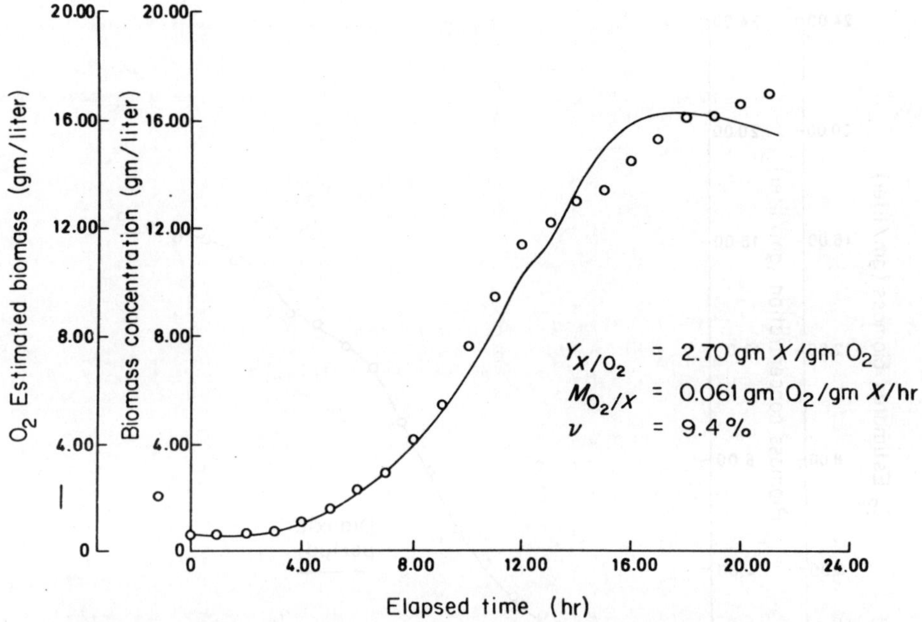

FIGURE 4. *Saccharomyces cerevisiae* biomass concentration data and correlation results. (From Zabriskie, 1976.)

centration and growth rate. The results of Wang (1976) are shown in Fig. 6.

The method of Zabriskie for determining biomass by oxygen uptake has been combined with Wang's technique for material balancing to derive a method of estimating the substrate concentration in a fermentation process that is impossible to measure on-line, i.e., cellulose (Armiger 1977).

This method assumes that there are two stoichiometric equations for the mass balance. One equation is for the growth of cells and production of product:

Ammonia + carbohydrate + oxygen
= cells + product + carbon dioxide + water

The other equation is used for cell maintenance:

Carbohydrate + oxygen = carbon dioxide + water

This technique was demonstrated by growing *Thermoactinomyces* sp. on cellulose where the product formed was extracellular protein. To balance these equations, compositions for the cell mass and protein were determined along with a product yield constant. The stoichiometric coefficients for oxygen and for carbon dioxide were measured experimentally and then the remainder of the coefficients were determined by

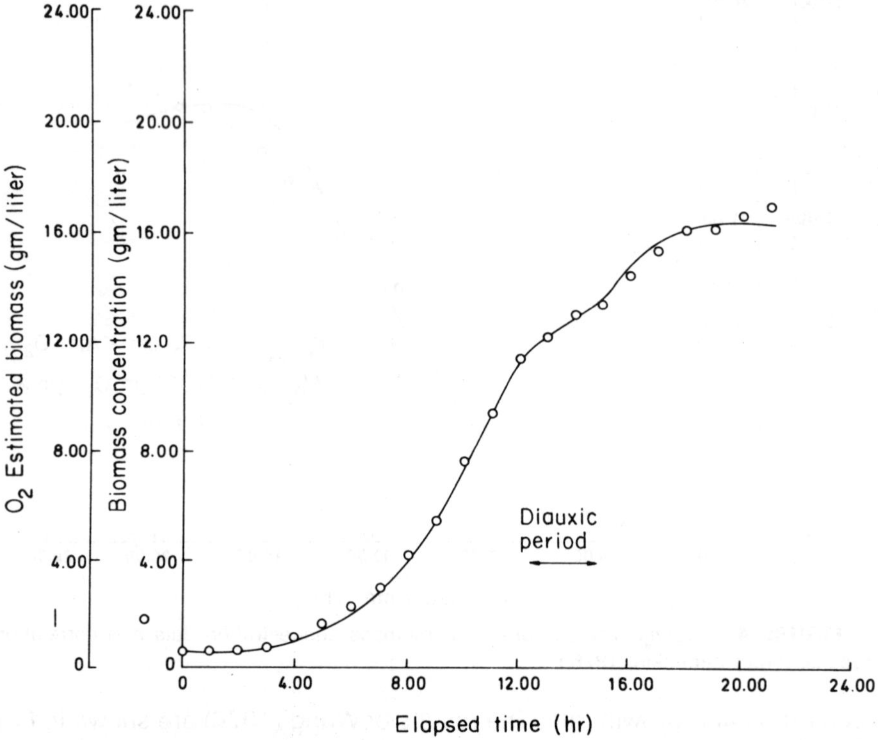

FIGURE 5. *Saccharomyces cerevisiae* biomass concentration data and correlation results with correction function.

balancing the equation. The cellulose concentration was calculated in the following manner: From a given inoculum size and oxygen uptake rate at any particular time, the biomass concentration was estimated. This estimate was then used in the stoichiometric equation to estimate the carbohydrate utilization and hence the concentration of cellulose.

Figure 7 illustrates the success of this technique by comparing laboratory analyses for cellulose and biomass with those values determined on-line. For checking the consistency of the stoichiometry, the on-line analyses for carbon dioxide and nitrogen were compared with the predicted results (Figs. 8 and 9).

Three previous examples all illustrate the value of an on-line computer system for the indirect estimate of important fermentation process parameters. Without the development of these and similar techniques, it would not be possible to evolve more sophisticated control strategies or to realize the on-line optimization of a fermentation.

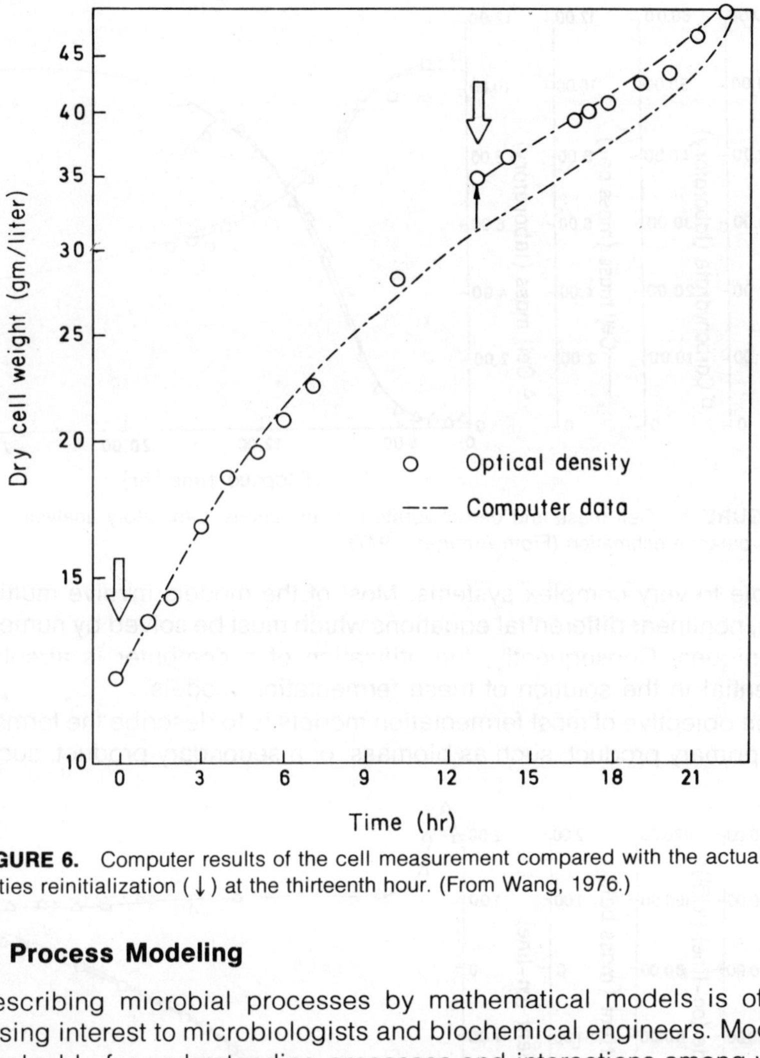

FIGURE 6. Computer results of the cell measurement compared with the actual cell densities reinitialization (↓) at the thirteenth hour. (From Wang, 1976.)

C. Process Modeling

Describing microbial processes by mathematical models is of increasing interest to microbiologists and biochemical engineers. Models are valuable for understanding processes and interactions among variables. In addition, models are necessary to provide the control engineer with insight into possible control strategies. Further, if the model can predict the dynamic behavior of a process, it can be useful for process optimization. Dynamic models are valuable for simulating experiments and testing control strategies. They can be utilized to eliminate many expensive and time-consuming experiments and to design key experiments to test a hypothesis. A large number of models for microbial processes have been proposed in the literature. These range from very

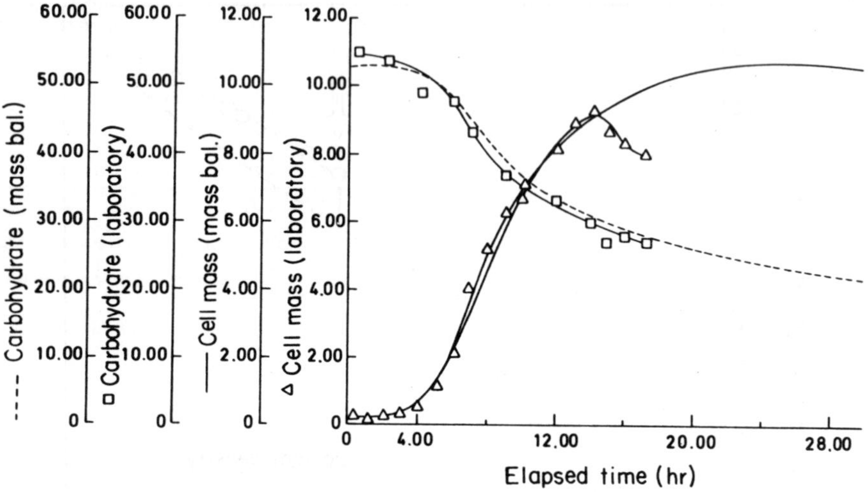

FIGURE 7. Cell mass and carbohydrate concentrations. Laboratory analysis versus mass balance estimation (From Armiger, 1977).

simple to very complex systems. Most of the models involve multivariable, nonlinear differential equations which must be solved by numerical techniques. Consequently, the utilization of a computer is absolutely essential in the solution of these fermentation models.

The objective of most fermentation models is to describe the formation of a primary product, such as biomass, or a secondary product, such as

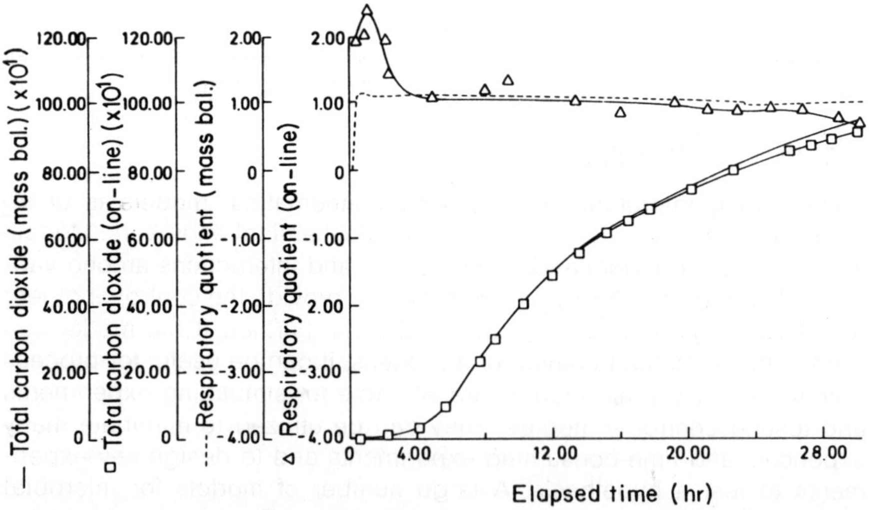

FIGURE 8. Total carbon dioxide (grams) and respiratory quotient on-line analysis versus mass balance estimate. (From Armiger, 1977.)

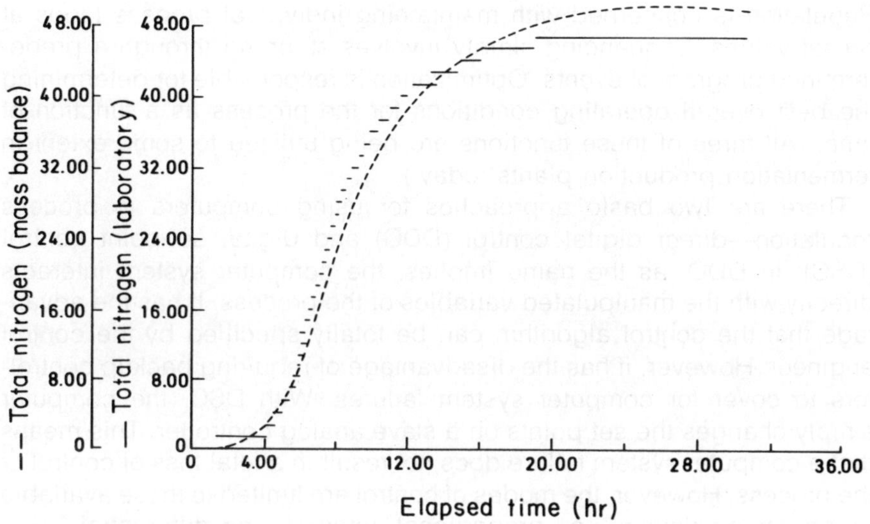

FIGURE 9. Nitrogen utilized (grams). Laboratory analysis versus mass balance estimation. (From Armiger, 1977.)

an antibiotic. These two situations have been modeled for batch, fed-batch, and continuous fermentations. There are two fundamental approaches to modeling systems. These involve the use of structured and unstructured models. Most fermentation models described unstructured systems. They do not account for individual differences between cells but examine variables that are characteristics of the macroscopic environment. Structured models do describe differences between cells and cell components; for example, one might describe the gross cell biomass, an enzyme component of that biomass, and the genetic mechanism activity. Unfortunately, the components of the structured model are difficult to measure; hence, they have not had the utility of the unstructured model. Tsuchiya *et al.* (1966), Fredrickson *et al.* (1970), and Yoshida and Taguchi (1977) all provide reviews of these approaches.

D. Process Control and Optimization

The application of computer control to the fermentation industry has lagged behind similar advances in the chemical process industry by about 10 years. When the complexity of the kinetics of microbial systems is compared to the kinetics of chemical systems, the reasons for this lag become obvious. In spite of these difficulties, process control of fermentations can be implemented. The ways in which control has been implemented include regulation, sequencing, and optimization methods.

Regulation is concerned with maintaining individual process loops at preset values. Sequencing simply involves stepping through a predetermined program of events. Optimization is responsible for determining the best overall operating conditions for the process as a function of time. (All three of these functions are being utilized to some extent in fermentation production plants today.)

There are two basic approaches for using computers in process regulation—direct digital control (DDC) and digital set-point control (DSC). In DDC, as the name implies, the computer system interacts directly with the manipulated variables of the process. It has the advantage that the control algorithm can be totally specified by the control engineer. However, it has the disadvantage of requiring backup controllers to cover for computer system failures. With DSC, the computer simply changes the set points on a slave analog controller. This means that a computer system failure does not result in a total loss of control of the process. However, the modes of control are limited to those available on an analog device, i.e., proportional, integral, and differential.

In selecting a control system for a fermentation plant, the number of control loops required will determine which is the most economical approach. From a functional standpoint, DSC offers some distinct advantages over DDC. For DSC, the computer system only has to calculate set-point changes which are implemented at low frequency so the system is available for more meaningful or complex process calculations. A good example would involve running programs for calculation of operating parameter profiles associated with process optimization.

The regulation of individual process variables and the sequencing of plant operations is not unique to fermentation process control. However, the strategy utilized for manipulating and sequencing variables based upon process models is the area where the biochemical engineer can have the greatest impact. Yamashita *et al.* (1969) discuss these control functions as they relate to the production of glutamic acid. Several papers applying both discrete and continuous maximum principles for optimizing fermentation process have appeared (Constantinides and Vishva, 1974). These studies determined optimum profiles of operating conditions, such as temperature, pH, and dissolved oxygen. However, the optimization of the fermentation step exclusive of the entire process is of limited value. The real objective function of an optimization must involve the entire process from raw materials through product recovery. Blanch and Rogers (1972) approached this problem for the continuous production of gramicidin S. Okabe and Aiba (1973) have approached the problem in a more general manner for antibiotic production. Okabe and Aiba (1974a,b, 1975a,b) have split apart the process into a series of steps that are optimized individually. These parts were recombined to arrive at a global optimization for the process. This more fundamental approach to process optimization is the direction in which the field must move if it is going to have an important impact on fermentation processes. However, these advances will not be realized until more accurate dynamic models of systems are developed.

V. SYSTEM CONFIGURATION

A highly instrumented fermentation system coupled to a digital computer is capable of generating on-line information for conceptualizing kinetic models and implementing strategies for process control and optimization. The process of utilizing the signal produced by an instrument and integrating it into a computer-controlled system involves both hardware and software development. With the advent of inexpensive but very powerful minicomputers, the problem of hardware development is basically one of selecting electronically compatible modules. The development of techniques and software for data analysis and process control is the area in which further research is needed. Configuration of system hardware and software must be structured to operate as an integrated unit. It must also be flexible enough to accommodate growth and expansion of both hardware and software as new products and techniques become available.

A. System Hardware Configuration

The system hardware consists of three main segments: (1) sensors and instrumentation, (2) computer interface, and (3) computer system. In general, the weakest segment in any system involves the sensors. As shown in Table I the number of measurements is very limited. Most of the sensors are measuring variables in the macroscopic environment of the fermentation broth, while the key factors controlling growth and product formation involve the microscopic environment internal to the cell. The development of sensors and techniques for monitoring levels and activities of cellular constituents is only beginning. The physical problem of measuring fermentation parameters is further complicated by the sterilization requirements which severely restrict the choice of materials of construction of the sensors and often limit their sensitivity and accuracy.

The hardware for the computer interface is readily available. With sufficient attention to detail, it will reliably function. However, the architecture of the interface depends upon the application. Analog scanning is the most economical method of interfacing if there is a large number (over 25) of channels involved. However, digital multiplexing, which is cost competitive on a small system (under 25 channels), offers many advantages. First, it allows each channel to function as a separate unit, and it provides the operator with a visual display of all the channels at one time. Second, it allows for all of the data to be logged simultaneously rather than sequentially so that calculations can be made on data from a specific instant in time. For large systems, the advantages of digital multiplexing are difficult to justify because of cost. Therefore, in most plant or pilot-plant applications involving many vessels, an analog scanning interface is selected.

Computer interfacing involves a choice between increased flexibility and increased software requirements. This includes trade-offs such as random access scanning and increased hardware costs with less flexibility but lower software cost. Again, the number of channels involved in the interface will determine if the cost savings are to be gained by using more sophisticated hardware or software.

The selection of a computer system usually starts with the choice between a dedicated computer and a time-sharing system. With the availability of inexpensive but very powerful minicomputers, a dedicated system can usually be justified very easily on the basis of lower operating costs and increased system reliability. The details of selecting the size of computer memory and the peripheral devices are dependent upon the specific application. A disk-based system with sufficient memory to allow for foreground/background operation and some graphics capability is a good starting point. In many cases, communications with a large computer installation can be very beneficial for accessing special software or performing complex calculations.

Figure 10 is a hardware schematic and signal flow diagram for a typical installation. The analog signals from the various sensors are fed to the interface where the analog voltage is converted to a binary coded decimal (BCD) word. The BCD output of the interface is transmitted to the computer under control of the computer software which specifies the scanning rate and sequence. Once the data reach the computer, they are stored on the system disk and analyzed by the system programs. If the results of the analysis are to be used for process control, signals from the computer are then transmitted back to the fermentor to manipulate set points on analog controllers (DSC) or to manipulate variables directly (DDC). The results generated can also be output on a hard-copy device, stored on the disk or tapes, plotted on a graphical device, or transmitted to another computer system.

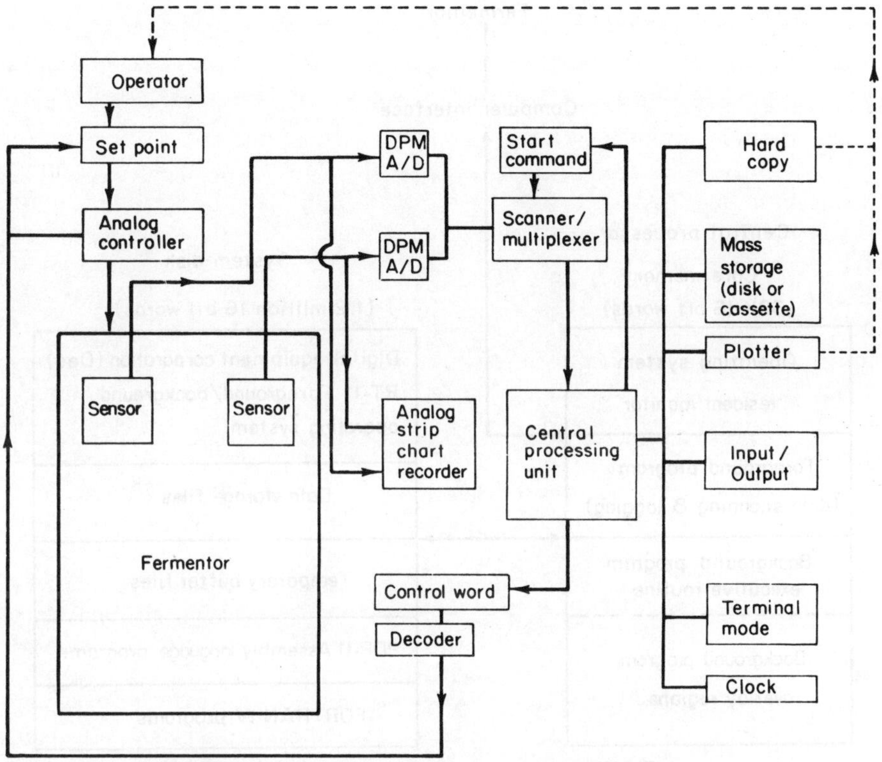

FIGURE 10. Schematic of physical configuration of system.

B. System Software Configuration

The first step in system software configuration is the selection of the proper computer operating system. An operating system with foreground/background, multiprogramming, and multitasking capabilities and a high-level language (i.e., FORTRAN) is preferred. The man–machine interface should be as simple as possible in order to make the computer operator as effective and reliable as possible.

An overview of the system software organization is illustrated in Fig. 11. The computer system can be conceptualized as two parts, the central processor, which contains the core memory; and the system disk. All other computer equipment are really peripheral devices to support these two systems. The central processor contains the computer's instructional logic and a relatively small region of core memory which is allocated into three sections. The first part contains the operating system resident monitor, which is the software enabling the operator to issue commands from the terminal to load programs into core, execute them, and control

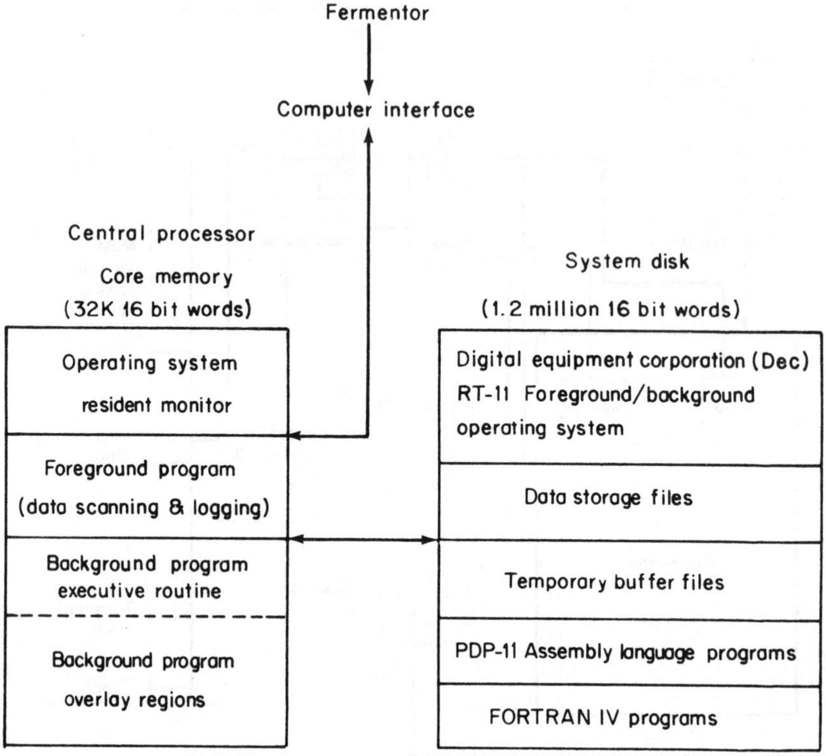

FIGURE 11. Overview of system software.

and monitor information transmitted to or from all system devices. The second portion contains the foreground program, which is responsible for sending the start command at the proper time to the scanner and logging the data on the disk. The third section is for the background program, which is divided into many smaller regions. The first region contains the executive routine for managing the sequence in which different program modules enter, execute their function, and leave the overlay regions.

Since computer core is limited in minicomputers, only data and program modules required at a particular point in time reside in an overlay region. The bulk of the software remains on the disk. As time progresses and certain tasks are completed, data are transferred to a file on the disk and a new program module is read from the disk and overlaid on top of a module whose function is complete. The system disk has no computational power; it is a storage device only. It contains vast amounts of information consisting of the operating system modules required by the resident monitor, data storage files, temporary buffer files, FORTRAN programs, and assembly language routines.

In a foreground/background operating system, tasks in the foreground have system priority. When the foreground has nothing to perform, control is automatically switched to the background to execute its task. Therefore, software must be divided into foreground and background tasks. The foreground functions include starting the scanner, logging data on the disk, and generating alarm messages. The benefits to this arrangement are manifest. The most time-dependent functions are performed first.

Clearly, it would be unacceptable if data acquisition had to be delayed because the computer was occupied by producing a report which could take as long as an hour. Typically, the tasks performed in the foreground can be executed very rapidly (seconds). Thus, without foreground/background capability, all of the time between the end of the foreground function and the generation of a successive data set by the scanner (1–10 minutes) would be wasted. However, with this arrangement, the system can be utilized automatically between scans for many different functions from analyzing laboratory data, simulating the fermentation, producing graphs, or developing new programs.

VI. FUTURE TRENDS

Computer applications in fermentation technology will increase. The availability of microprocessors and software to support these systems will change the hardware configuration of a system and greatly reduce the costs. Soon the computer associated costs will be much less than those related to the sensor systems. However, the real future lies in developing improved mathematical models and strategies for process control and optimization. At this time, the capabilities of computer hardware greatly exceed the capabilities of the software that has been developed. Therefore, the greatest challenges for future application of computers to fermentation are involved with interpreting results rather than physically implementing control decisions. The time has arrived for the biochemical and control engineers to utilize their skills to bring computer applications in fermentation technology up to the level of the available computer technology.

VII. SUMMARY

Computers have a variety of applications in the fermentation industry ranging from data logging and on-line analysis to process modeling, control, and optimization. Only a small fraction of the potential has been realized. This is mainly in the areas of data logging and on-line analysis. Many groups are working on modeling, control, and optimization, but there has not yet been significant developments in terms of interactive control. One of the major problems in developing mathematical models and control strategies lies in the inability to measure on-line many of the important process parameters. Significant improvements in existing sensors and the development of new sensors is needed. Finally, fermentation-related software development needs to be brought to the high level that will fully utilize computer hardware developments, particularly in the microprocessor area.

REFERENCES

Armiger, W. B. (1977). Ph.D. Thesis, University of Pennsylvania, Philadelphia.

Armiger W. B., and Humphrey, A. E. (1974). *Proc. Intersect. Congr. IAMS, 1975 1st,* **5,** 99–119.

Armiger, W. B., Zabriskie, D. W., Humphrey, A. E., Lee S. E., and Jolly, G. (1975). *AIChE Symp. Ser. No. 158* **72,** 77–85.

Armiger, W. B., Zabriskie, D. W., and Humphrey, A. E. (1976). *Proc. Int. Ferment. Symp., 5th, 1976* p. 33.

Blachere, H. T. (1973). *Proc. Eur. Conf. Comput. Process Control Ferment., 1st, 1973,* Session 1, Paper 2.

Blanch, H. W., and Rogers, P. L. (1972). *Biotechnol. Bioeng.* **14,** 151–171.

Constantinides, A., and Vishva, R. R. (1974). *Biotechnol. Bioeng. Symp.* **4,** 663–680.

Cooney, C. L., Wang, H., and Wang, D. I. C. (1975). *Pap., US/U.S.S.R. Semin. Meas. Ferment. Processes.*

Corrieu, G., Blachere, H., and Geranton, A. (1974). *Biotechnol. Bioeng. Symp.* **4,** 607.

Flynn, D. S. (1974). *Biotechnol. Bioeng. Symp.* **4,** 597–605.

Fredrickson, A. G. Megee R.D., and Tsuchiya, H. M. (1970). *Adv. Appl. Microbiol.* **13,** 15–30.

Grayson, P. (1969). *Process Biochem.* **4,** 43–44, 61.

Harima, T. (1977). Ph.D. Thesis, University of Pennsylvania, Philadelphia.

Harrison, D. E. F., and Harmes, C. S. (1972). *Process Biochem.* **7**(4), 13–16.

Harrison, D. E. F. (1971). *Proc. Int. Congr. Microbiol., 10th, 1970.*

Humphrey, A. E. (1971). *Proc. LABEX Symp. Comput. Control Ferment. Processes, 1971* p. 1.

Jefferis, R. (1973). *Proc. Eur. Conf. Comput. Process Control Ferment. 1st, 1973,* Session 1, Paper 2.

Jefferis, R. (1975). *Process Biochem.* **10,** 15.

Jefferis, R., Humphrey, A. E., and Nyiri, L. K. (1972). *164th Annu. Meet., Am. Chem. Soc.*

Koga, S., Berg, C. R., and Humphrey, A. E. *Appl. Microbiol.* **15,** 683.

Lane, A. (1973). *Proc. Eur. Conf. Comput. Process Control Ferment. 1st, 1973,* Session 2, Paper 2.

Luedeking, R., and Pirt, E. L. (1959). *J. Biochem. Microbiol. Technol. Eng.* **1,** 393.

Marr, A. G., Nilsson, E. H., and Clark, D. J. (1962). *Ann. N.Y. Acad. Sci.* **102,** Art. 3, 536.

Moes, H., Ryu, D. Y., Brannick, L., and Cohan, M. (1971). *162nd Natl. Meet. Am. Chem. Soc.*

Nagai, S., Nishizawa, Y., and Yamagata, T. (1976). *Proc. Int. Ferment. Symp., 5th, 1976,* p. 30.

Nyiri, L. K. (1971). *Proc. LABEX Symp. Comput. Control Ferment. Process,* 1972 **13,** 13.

Nyiri, L. K., Toth G. M., and Charles, M. (1975). *Biotechnol. Bioeng.* **17,** 1663
Okabe, M., Aiba, S., Okado, M. (1973). *J. Ferment. Technol.* **51,** 594–605.
Okabe, M., and Aiba, S. (1974a). *J. Ferment. Technol.* **52,** 279–292.
Okabe, M., and Aiba, S. (1974b). *J. Ferment. Technol.* **52,** 759–777.
Okabe, M., and Aiba, S. (1975a). *J. Ferment. Technol.* **53,** 230–240.
Okabe, M., and Aiba, S. (1975b). *J. Ferment. Technol.* **53,** 730–743.
Tsuchiya, H. M., Fredrickson A. G., and Aris, R. (1966). *Adv. Chem. Eng.* **6,** 125–206.
Unden, A. G., and Heden, C. G. (1973). *Proc. Eur. Conf. Comput. Process Control Ferment. 1st, 1973,* Session 2, Paper 1.
Wang, H. Y. (1976). Ph.D. Thesis, MIT, Cambridge. Massachusetts.
Wilson, J. D. (1975). U.S. Patent 3,926,738.
Wilson, J. D., Nyiri, L. K., Humphrey, A. E., and Harmes, C. S. (1975). U. S. Patent 3,926,737.
Yamashita, S. and Murao, C. (1967). *J. Soc. Instrum. Control Eng.* **6,** 735.
Yamashita, S., Hoshi, H. and Inagaki, T. (1969). *Ferment. Adv. Pap. Int. Ferment. Symp., 1968* pp. 441–463.
Yoshida, T., and Taguchi, H. (1977). *In* "Workshop on Computer Applications in Fermentation Technology 1976" (R. P. Jefferson ed.); GBF Monogr. Ser. No. 3, pp. 93–106. Verlag Chemie, Weinheim.
Young, T. B., and Koplove, H. M. (1972). *Ferment. Technol. Today, Proc. Int. Ferment. Symp., 4th, 1972.*
Zabriskie, D. W. (1976). Ph.D. Thesis, University of Pennsylvania, Philadelphia.

SUPPLEMENTARY READING

General Papers and Reviews

Armiger, W. B., ed. (1979). "Computer Applications in Fermentation Technology," *Vol. 9. Biotechnol. Bioeng. Symp. Ser.,* Wiley, New York.
Dolby, D. D. and Jost, J. L. (1977). Computer applications to fermentation operations. *In* "Annual Reports on Fermentation Processes" (D. Perlman, ed.), pp. 114. Academic Press, New York. Vol 1.
Fuld, G. J. (1960). Control applications in fermentation processes. *Adv. Appl. Microbiol.* **2,** 351–360.
Humphrey, A. E. (1971). Present limitations to the control and understanding of a fermentation process. *Proc. LABEX Symp. Comput. Control Ferment. Processes, 1971* p. 1.
Humphrey, A. E. (1973). Rationale and economics of computer process control *Proc. Eur. Conf. Comput. Process Control Ferment. 1st, 1973* Session 1, paper 1.
Humphrey. A. E. (1974). Current developments in fermentation *Chem. Eng.* p. 98.
Humphrey, A. E. (1975). Rationale and problems in the use of computer-coupled fermentation systems. *Pap., US/USSR Semin. Meas. Ferment. Processes.*
Humphrey, A. E., and Jefferis R. P. (1973). Optimization of batch fermentation processes. *Pap. GIAM Meet. 4th 1973.*
Humphrey, A. E., and Jefferis R. P. (1977). Computer-assisted fermentation developments. *Dev. Ind. Microbiol.* **18,** 58–70.
Khosrovi, B., and Topiwala, H. H. (1974). Mathematical modeling and computer control in fermentation processes. *Annu. Rep. Prog. Appl. Chem.* **59,** 357–365.

Mueller, H., and Mueller, F. (1973). Examples and studies of continuous fermentation optimization by computer. *Pap., GIAM Meet., 4th, 1973.*

Nyiri, L. K. (1972). A philosophy of data acquisition, analysis, and computer control of fermentation processes. *Dev. Ind. Microbiol.* **13,** 136–145.

Nyiri. L. K. (1973). Application of computers in biochemical engineering. *Adv. Biochem. Eng.* **2,** 49–93.

Nyiri, L. K. (1974). Strategies and problems related to computer coupled fermentation systems. *Proc. Intersect. Congr. IAMS 1st. 1975* Vol. 5; pp. 120–136.

Nyiri. L. K., and Jefferis, R. P. (1973). Process control aspects of single-cell production in submerged cultures. *Proc. Int. Conf. Single-Cell Protein, 2nd, 1973,* p. 105.

Nyiri, L. K., Jefferis, R. P., and Humphrey, A. E. (1974). Applications of computers to the analysis and control of microbiological processes. *Biotechnol. Bioeng. Symp.* **4,** 613.

Ryu, D. D. V., and Humphrey, A. E. (1973). Examples of computer-aided fermentation systems. *J. Appl. Chem. Biotechnol.* **25,** 283–295. 1973.

Trambouze, P., and Mueller, H. (1973). Computers in fermentation control. *Process Biochem.* **8,** 7–10.

System Design and Application

Geranton, A. (1973). Process control in fermentation—design and construction of software *Proc. Eur. Conf. Comput. Process Control Ferment., 1st, 1973, Session 3, Paper 2.*

Huang, I., and Sonn, M. (1972). Computer control of batch processes. *Br. Chem. Eng. & Process Technol.* **17,** 507–512.

Jefferis, R. P. (1977). Software and file structures for computer-coupled pilot fermentation plants. *In* "Workshop on Computer Applications in Fermentation Technology 1976", (R. P. Jefferis, ed.), GBF Mono. Ser. No. 3, pp. 21–36. Verlag Chemie, Weinheim.

Lundell, R., and Laiho, P. (1976). Engineering of fermentation plants. *Process Biochem.* **11,** 13–17.

Meskanen, A., Lundell R., and Laiho, P. (1976). Engineering of fermentation plants. Part 3. *Process Biochem.* **11,** 31–36.

Shave, J. D. (1976). The computer in the brewhouse. *Process Biochem.* **11,** 10–12.

Spruytenburg, R., Dang, A. D. P., Dunn, I. J., Mart, J. R., Einsele, A., Fiechter, A., and Bourne, J. R. (1976). Experience with a computer-coupled bioreactor. *Chem. Eng. (London)* **310,** 447.

Wilson, J. (1975). Planning a computer-coupled fermentation production plant. Pap., US/USSR Semin. Meas. Ferment. Processes, 1975.

Wilson, J. D., Humphrey, A. E., Harmes, C. S., and Nyiri, L. K. (1974). Procède et appariel pour controler les réactions biochimiques. French Patent 2,184,035,

Data Analysis and Modeling

Armiger, W. B., Moreira, A. R., Phillips, J. A., and Humphery, A. E. (1977). Modeling cellulose digestion for single cell protein *PACHEC, 2nd 1977* pp. 196–205.

Blanch, H. W., and Rogers, P. L. (1977). Production of Gramicidin S. in batch and continuous culture. *Biotechnol. Bioeng.* **13,** 843–864.

Blanch, H. W., and Dunn, I. J. (1974). Modeling and simulation in biochemical engineering. Adv. Biochem. Eng. **3,** 127–165.

Brown, D. E., and Vass, R. C. (1973). Maturity and product formation in cultures of microorganisms. *Biotechnol. Bioeng.* **15,** 321–330.

Cooney, C. L., Wang, H., and Wang, D. I. C. (1975). Computer-aided fermentation monitoring and diagnostics. *Pap., US/USSR Semin. Meas. Ferment. Processes, 1975.*

Cooney, C. L., Wang, H., and Wang, D. I. C. (1977).Computer-aided material balancing for prediction of fermentation processes. *Biotechnol. Bioeng.* **19,** 55–69.

Dang, N. D. P. (1976). The building and use of dynamic models for biological studies. *J. Ferment. Technol.* **54,** 396–405.

Dang, N. D. P., Dunn, I. J., and Mor, J-R. (1975). Modeling of dynamic biological processes with empirical transfer functions. *J. Ferment. Technol.* **53,** 885–894.

Edwards, V. H., and Wilke, C. R., (1968) Mathematical representation of batch culture data. *Biotechnol. Bioeng.* **10,** 205–232.

Erickson, L. E., Nakahara, T., Gutierrez, J. R., MacLean, G. T., and Fan, L. T. (1975). Modeling and characterization of hydrocarbon fermentations. *Pap., US/USSR Joint Semin. Univ. Pennsylvania, 1975,* p. 75.

Fiechter, A., and von Meyenburg, K. (1968). Automatic analysis of gas exchange in microbial systems. *Biotechnol. Bioeng.* **10,** 535–549.

Fishman, V. M., and Biryukov, V. V. (1974). Kinetic model of secondary metabolite production and its use in computation of optimal conditions. *Biotechnol. Bioeng. Symp.* **4,** 647–662.

Imanaka, T., Kaiedo T., Soto K., and Taguchi H. (1972). Optimization of alpha-galactosidase production by mold. *J. Ferment. Technol.* **50,** 633–646.

Imanaka, T., Kaiedo, T., and Taguchi H. (1973). Unsteady-state analysis of a kinetic model for cell growth and alpha-galactosidase production in mold. *J. Ferment. Technol.* **51,** 423–430.

Johnson, D. B., and Berthouex, P. M. (1975). Using multiresponse data to estimate biokinetic parameters. *Biotechnol. Bioeng.* **17,** 571–583.

Khan, A. H., and Ghose T. K. (1973). Kinetics of citric acid fermentation by *Aspergillus Niger. J. Ferment. Technol.* **51,** 734–741.

Kishimoto, M., Yamane, T., and Yoshida, F. (1976). Sensitivity analysis for exponential fed-batch culture. *J. Ferment. Technol.* **54,** 891–901.

Kono, T., and Asai, T. (1969a). Kinetics of continuous cultivation. *Biotechnol. Bioeng.* **11,** 19–36.

Kono, T., and Asai, T. (1969b). Kinetics of fermentation processes. *Biotechnol. Bioeng.* **11,** 293–321.

Kremen, A. (1974). Possible control mechanisms in tetracycline Production: A kinetic model. *Biotechnol. Bioeng. Symp.* **4,** 105–113.

Matsumura, M., Muhotaka, S., Yoshetomo, H., and Kobayashi J. (1973). Analysis of cell behavior in transient state of continuous culture with an apparent age model. *J. Ferment. Technol.* **51,** 904–916.

Minkevich, I. C., and Eroshin, V. K. (1974). Theoretical calculations of mass balance during the cultivation of microorganisms. *Biotechnol. Bioeng. Symp.* **4,** 21.

Moreira, A. R. (1977) Model cellulose degradation by *Thermoactinomyces 168th Annu. Meet. Am. Chem. Soc.*

Nagatani, M., Shoda, M. and Aiba S. (1968). Kinetics of product inhibition in alcohol fermentation. *J. Ferment. Technol.* **46,** 241–248.

Nyiri, L. K., Toth, G. M., Krishnaswami, C. S., and Parmenter, D. V. (1977). On-line analysis and control of fermentation processes. *In* "Workshop on Computer Applications in Fermentation Technology 1976" (R. P. Jefferis, ed.), GBF Monogr. Ser. No. 3, pp. 37–46. Verlag Chemie, Weinheim.

Paynter, M. J. B., and Bungay, H. R. (1970). Responses in continuous culture of lysogenic. *Escherichia coli* following induction. *Biotechnol. Bioeng.* **12,** 347–351. 1970.

Peringer, P. Blachere, H., Borrieu, G., and Lane, A. G. (1973). Mathematical model of the kinetics of growth of *Saccharomyces cerevisiae.* Biotech, Bioeng. Symp. **4,** 27–42. 1973.

Perringer, P., Blachere, H., Borrieu, G., and Lane, A. G. (1974). A generalized mathematical model for the growth kinetics of *Saccharomyces cerevisiae* with experimental determination of parameters. *Biotechnol. Bioeng.* **16,** 431–454. 1974.

Ribot, D. (1977). Polynomial identification methods *In* "Workshop on computer applications in Fermentation Technology 1976" (R. P. Jefferis, ed.), GBF Monogr. Ser. No. 3, pp. 125–140. Verlag Chemie, Weinheim.

Sundstrom, H. E., Klei, H. E., and Brookman, G. T. (1976). Response of biological reactors to sinusoidal variations of substrate concentration. *Biotechnol. Bioeng.* **18,** 1–14.

Svrcek, W. Y., Elliott R. F., and Zajic J. E. (1974). The extended Kalman filter applied to a continuous culture model. *Biotechnol. Bioeng.* **16,** 827–846. 1974.

Swartz, J. R., Wang, H., Cooney, C. L., and Wang, D. I. C. (1976). Computer-aided yeast fermentation. *Proc. Int. Ferment. Symp., 5th, 1976,* pp. 29–63.

Van Dedem, G., and Moo-Young, M. (1975). A model for diauxic growth. *Biotechnol. Bioeng.* **17,** 1301.

Wang, D. I. C. (1975). Modeling and control on the utilization of mixed substrates by mixed cultures. *Pap., US/USSR Joint Semin. Data Acquisition Process. Lab. Ind. Meas. Ferment. Processes, 1975.*

Wang, H., Cooney, C. L., and Wang, D. I. C. (1977). Computer-aided baker's yeast fermentation. *Biotechnol. Bioeng.* **19,** 69–86.

Yamane, T., and Hirano, S. (1977). Semi-batch culture of microorganisms with constant feed of substrate. *J. Ferment. Technol.* **55,** 156–165. 1977.

Yoshida, T., and Taguchi, H. (1977). The use of models in fermentation control. *In* "Workshop on Computer Applications in Fermentation Technology 1976" (R. P. Jefferis, ed.), GBF Monogr. Ser. No. 3, pp. 93–106. Verlag Chemie, Weinheim.

Young, T. B., and Koplove, H. M. (1972). A systems approach to design and control of antibiotic fermentations. *Ferment. Technol. Today, Proc. Int. Ferment. Symp., 4th, 1972.*

Zabriskie, D. W. (1977). Real-time estimation of fermentation biomass concentration using component balancing techniques. *Proc. US/USSR Joint Semin. 1975.*

Zabriskie, D. W., Armiger, W. B., and Humphrey, A. E. (1977). Applications of computers to the indirect measurement of biomass concentration and growth rate by component balancing. *In* "Workshop on Computer Applications in Fermentation Technology 1976." (R. P. Jefferis, ed.) GBF Monogr. Ser. No. 3, pp. 59–72. Verlag Chemie, Weinheim.

Zines, D. O., and Rogers, P. L. 1971). A chemostat study of ethanol inhibition. *Biotechnol. Bioeng.* **13,** 293–308. 1971.

Process Control and Optimization

Aiba, S., Nagai, S., and Nishizawa, Y. (1976). Fed batch culture of *Saccharomyces cerevisiae*—a perspective of computer control to enhance the productivity in baker's yeast cultivation. *Biotechnol. Bioeng.* **18,** 1001–1016.

Blachere, H. T., Peringer P., and Corrieu, G. V. (1977). Optimization of fermentation plants. *In* "Workshop on Computer Applications in Fermentation Technology 1976." (R. P. Jefferis, ed.), GBF Monogr. Ser. No. 3, pp. 1–10. Verlag Chemie, Weinheim.

Bourdaud, D., and Foulard, C. (1973). Identification and optimization of batch culture fermentation processes. *Proc. Eur. Conf. Comput. Process Control 1st, 1973* Session 4, Paper 1.

Chen, H. T., and Fan, L. T. (1975), Principles of on-line dynamic optimization in fermentation systems. *Pap., US/USSR Joint Semin. Data Acquisition Process. Lab. Ind. Meas. Ferment. Processes, 1975.*

Constantinides, A., Spencer, J. L., and Gaden, E. L. (1970a). Optimization of batch fermentation processes. I. Development of mathematical models for batch penicillin fermentations. *Biotechnol. Bioeng.* **12,** 803.

Constantinides, A., Spencer, J. L., and Gaden, E. L. (1970b). Optimization of batch fermentation processes. II. Optimum temperature profiles for batch penicillin fermentations. *Biotechnol. Bioeng.* **12,** 1081.

Gaudy, A. F. (1975). The transient response to pH and temperature shock loading of fermentation systems. *Biotechnol. Bioeng.* **17,** 1051–1064.

Humphrey, A. E., and Jefferis, R. P. (1973). Optimization of batch fermentation processes *Proc. GIAM Meet. 4th, 1973.*

Imanaka, T., Kaieda, Takeji and Taguchi, Hesahaw (1973). Optimization of alpha-galactosidase production in multi-stage continuous culture of mold. *J. Ferment. Technol.* **51,** 431–439.

Lane, A. C., and Blachere, H. (1973). Computer control and optimization in the cultivation of yeast. *Proc. GIAM Meet. 4th, 1973.*

Muzychenko, L. A., Mascheva, L. A., and Yakovleva, G. V. (1974). Algorithm for optimal control of microbiological synthesis. *Biotechnol. Bioeng. Symp.* **4,** 629–645.

Paschold, H., and Verstoep, N. D. L. (1977). Computer based control of fermentation processes. *Soc. Chem. Ind. Meet.*

Perringer, P., Blum, J., and Blachere, H. (1974). Optimal single-cell production from yeast in a continuous fermentation process. *Proc. Intersect. Meet. IAMS,* **5,** pp. 82–98.

Ryu, D. Y., and Humphrey, A. E. (1972). A reassessment of oxygen-transfer rates in antibiotic fermentations. *J. Ferment. Technol.* **50,** 424–431.

Spitzer, D. W. (1976). Maximization of steady-state bacterial production in a chemostat with pH and substrate control. *Biotechnol. Bioeng.* **18,** 167–178.

Wang, D. I. C., Wilcox, R. P., and Evans, L. B. (1975). Modeling and control on the utilization of mixed substrates by mixed cultures *Pap., Joint US/USSR Semin., Univ. Pennsylvania, 1975.*

Chapter 16

General Procedures for Isolation of Fermentation Products

PAUL A. BELTER

I. INTRODUCTION

The subject, recovery and isolation of products produced by fermentations, is a branch of a much larger topic, the separation of natural

MICROBIAL TECHNOLOGY, 2nd ed., VOL.II
 ISBN 0-12-551502-2

products, which in itself is a subdivision of the overall subject of separation science. Books have been written on the subject of separation, and the techniques evolved to obtain "pure" compounds. Scientists and technologists have spent an appreciable part of their technical lives devising methods to separate the complex mixtures that nature produces and, according to the second law of thermodynamics, prefers. This effort requires a great deal of knowledge, ingenuity, and creativity to devise better separation tools and, subsequently, integrated optimal isolation procedures. In the fermentation industry the isolation operations often require more energy, equipment, and labor than the fermentation per se. The recovery plant usually represents a major investment and the isolation costs are a substantial fraction of the cost of the product. It is, therefore, important that the isolation processes be well conceived and the plants well designed.

The isolation of products produced by fermentation is a specialized branch of separation science because of the unique character of the materials being handled rather than any separation science peculiarity. Many of the techniques useful for carbohydrate purification, oil extraction, and protein fractionation are applicable. The wide diversity of product types, and the fact that often living dynamic organisms and/or their enzyme systems are involved, complicate the situation. In spite of this, general principles, theories, and models for separation operations have been developed that are applicable to the isolation of fermentation products (Karger *et al.*, 1973). Many of these fundamentals have been covered in detail in the literature, especially the chemical engineering literature (Holland, 1975), but usually are illustrated by examples that are not related to the fermentation industry. For example, the fermentation of various carbohydrate materials to make ethyl alcohol and the subsequent recovery of "spirits" by distillation was practiced long before fractional distillation theory was developed. In fact, Sorel's (1893) original work was conducted to understand the empirically practiced distillations of the time. Today, after much improvement and refinement, distillation theory forms the backbone of the petroleum industry and is equally applicable in certain areas of the fermentation industry.

King (1971) divided separation processes into three categories depending on their fundamental controlling phenomenon: mechanical, equilibrium and rate-controlled processes. Thirty of the forty-two processes listed have been or are being used to isolate fermentation products. The spectrum of product types discussed in this book covers widely varying categories ranging from relatively simple amino and organic acids to complex mixtures of antibiotics and enzyme preparations. In practice, operational scales range from the laboratory to commercial plants. Batch and continuous operations, both concurrent and countercurrent, are used. In addition, various modes of operation, such as single-stage, multiple discrete-stage, and differential contacting techniques, are employed. Consequently, both steady state and transient state considerations are involved. For the application of these techniques knowledge of the pertinent material and energy balances, equilibrium relationships, and rate phenomena is essential. Unfortunately, both the physical and chemical information for fermentation products are rarely available and must be determined before separation theory and models can be used. This has led in the past to a high degree of empiricism that it is hoped will be minimized in the future. It is obvious from the foregoing that the proposed subject is an extremely diverse one which can only be covered superficially in this chapter. Consequently, the format of this chapter is the discussion of segments of recovery processes, their underlying theory, and general utility for all products rather than to give specific processing details for individual products. In addition, we concentrate on developments that have achieved commercial importance, while laboratory and small-scale techniques are mentioned only briefly.

The basic problems that confront the separation technologists, whether they be chemists or engineers, are the complexity of the mixture and the relatively low concentration of the desired product in the starting material or feed to the isolation process, the harvested fermentation broth or "beer." This heterogeneous mixture contains the intact organisms or mycelium, or fragments thereof, plus other insoluble fractions of medium ingredients, residual substrates, and, in some cases, product. The soluble portions of residual substrates, metabolic pathway intermediates, and usually the desired product are present in the liquid part of this mixture. Although not always, the product usually is present in a relatively low concentration compared to the concentration of other materials present.

Questions that must be answered to devise an isolation process for a given product are listed below.

1. What is the value of this product?
2. What is an acceptable product quality for the proposed end use?
3. Where is the product in the complex mixture?
4. Where are the impurities in the complex mixture?
5. What are the physical and chemical properties of the product and the impurities?
6. What are the economics of various isolation possibilities?

Careful consideration of these questions will provide answers that will enable the isolation technologist to reach the goals of adequate product quality and high recovery with minimum effort. Obviously, the last two items affect the final cost of the product appreciably.

II. GENERAL PROCESSING

Although isolation processes are usually similar for a given class of compound, they vary appreciably between classes. Obviously, one would not use the same isolation techniques for volatile organic solvents that one would use to isolate sensitive and unstable enzymes or antibiotics.

Fortunately, an overall similarity exists in the processing sequence of all of the recovery schemes. In general, conventional isolation processes can be divided into the following four segments.

1. *Removal of insolubles.* Filtration, centrifugation, and decantation are the usual unit operations used in this segment. Relatively little concentration of product occurs during these steps.
2. *Primary isolation of product.* These processes are relatively nonselective in performance and only separate materials of widely divergent polarities. Appreciable concentration and quality increases usually occur during these sequences. Typical operations are sorption (physical and/or ion exchange), solvent extraction, precipitation, and ultrafiltration.
3. *Purification.* Processing techniques which are highly selective for the product and remove impurities of similar chemical functionality and physical properties are used in this segment. Typical approaches are fractional precipitation, chromatography of various types (adsorption, partition, ion exchange, and affinity), chemical derivatization, and decolorization.
4. *Final product isolation.* The desired characteristics of the product for its ultimate use dictate the final sequence utilized at this point. Crystallization followed by centrifugation or filtration and drying are typical steps used for high-quality products. Other products may be drum or spray dried directly from solution if the solvent is water, or desolventized if an organic solvent has been used.

A hypothetical though reasonably typical processing profile for a product is given in Table I.

Several isolation processes have been developed in recent years that simplify the older conventional sequence by combining the first two steps. These developments, whole-broth extraction and fluidized ion-exchange processes, are discussed in Section III, D.

TABLE I. Typical Processing Profile

Step	Product	
	Conc. (gm/liter)	Quality (%)
Harvest broth	0.1–5	0.1–1.0
Filtration	0.1–5	0.1–2.0
Primary isolation	5–10	1–10
Purification	50–200	50–80
Crystallization		90–100

III. SOLIDS REMOVAL

The conventional first step in the isolation of a fermentation product is the separation of the heterogeneous fermentation broth into two fractions, "solubles" and "insolubles." This division usually simplifies or makes possible the application of other techniques to the stream that contains the desired material that would not be possible otherwise; i.e., the use of sorption or ion exchange columns in the fixed-bed mode that would be inoperative unless the feed solution was devoid of particulate matter, the selective precipitation of crude product with salt forms or flocculants, and/or the prevention of excessive emulsification during extraction with an organic solvent. In the cases where the product resides in the insoluble fraction, an appreciable reduction in the volume of material to be processed occurs simultaneously with the removal of most of the soluble impurities.

A. Filtration

In general, fermentation broths are difficult to filter for one or more of a variety of reasons. The nature, sliminess, morphology, and size of the microorganisms; pH and viscosity of the broth; temperature history of the broth; contamination by undesired microorganisms; and the nature of the insoluble portion of residual substrates are a few of the factors that contribute to the filtration difficulties.

Special techniques have been developed to study the filtration characteristics of these broths in the laboratory and to simulate in the laboratory the operation of continuous rotary vacuum filters that are used for filtering these intractable mixtures on a production scale.

The rotary vacuum precoat filter is probably the most widely used filter in the fermentation industry. The drum, which rotates in a trough of fermentation broth, is covered with a layer of precoat, usually diatomaceous earth, prior to filtration. A small quantity of admix is added and thoroughly mixed with the broth before it is pumped into the filter trough. As the drum under vacuum rotates, a thin layer of admix and mycelium adhere to the drum. The layer of solids is dewatered during its passage on the drum to the discharge point, where a knife blade cuts off a thin layer of filter aid precoat and fermentation solids. Usually the cake is washed with water several times during its passage over the drum. These washes serve to displace the broth solution and solubles that are held in the cake voids.

A second type of continuous vacuum filter having a string discharge has been useful for mold fermentations, such as penicillin, where the mold mycelium interlaces upon itself on the drum to form a mat which is subsequently lifted from the drum by the filter strings.

It should be evident that the nature of the insolubles affects the type of filter that can be used for a given application.

Theory

The important variables for a constant pressure filtration can be correlated in a simple fashion by modification of the Poiseuille equation. Integration of that relationship gives Eq. (1).

$$\Theta/V = \mu\alpha CV/2\Delta PA^2 + \mu R_m/\Delta PA \qquad (1)$$

where Θ is time, V is the volume of filtrate, μ is the viscosity of the filtrate, C is the volume of cake deposited per volume of filtrate, A is the filtration area, R_m is the initial resistance of filter medium, ΔP is the pressure drop across the bed, and α is the average specific cake resistance.

A plot of Θ/V versus V results in a straight line having a slope of $\mu\alpha C/2\Delta PA^2$, for a given set of filtration conditions if the particles are noncompressible (α is also constant). Cakes from fermentation broths are seldom noncompressible and experimental measurements of filtration rates are necessary for design. Since filtration theories deal only with attainable rates, no answers are given for important questions, such as filtrate clarity, tenacity of cake to adhere to the drum, ease and efficiency of washing, and dryness of the resultant cake. However, much insight can be gained by the proper interpretation of filter leaf data obtained in the laboratory. Dahlstrom and Pruchas (1957) have described correlating methods for the various rates involved in the application of continuous filters. In addition, the selection of the proper filter aid for a given application requires that a compromise be made between filtration rate and filtrate clarity.

The ability of aminoglycoside antibiotics to bind to diatomaceous earths and cellulosic filter aids has been reported by Wagman *et al.*

(1975). This phenomenon results in an additional source of loss and must also be considered in the selection of a filter aid.

B. Centrifugation

Although numerous types of centrifuges have been developed over the years that can separate liquid phases and minor amounts of suspended solids, the design of relatively high-gravity machines that will handle appreciable volumes of particulate matter is relatively recent. Only in special cases will these units give the clarity of effluent or the "dryness" of cake that can be obtained with a precoat filter. Consequently, these factors plus the existence of plants based on filtration technology have slowed down their acceptance as a tool to remove the insolubles portion of a fermentation broth. As pointed out by Gray *et al.* (1972), the exception appears to be in the continuous processing of enzymes, where centrifugation plays a significant role. Recent operating data for the recovery of microorganisms by centrifugation is given in *Chemical Processing* (Anonymous 1977). The theoretical basis for the scale-up of centrifugation operations have been covered by Aiba *et al.* (1973).

C. Coagulation and Flocculation

Coagulation, defined as the conversion of dispersed colloids into small floc using simple electrolytes, and flocculation, defined as the agglomeration of these small flocs into larger settleable particles using polyelectrolytes, are useful approaches that can improve filtration and centrifugation rates and effluent clarity. Bacterial cells and proteinaceous material in the broth are hydrophilic colloids that usually have negatively charged surfaces. Hydrophobic colloids, such as oil droplets, that lack colloidial stability in the presence of simple electrolytes, are present also.

The simple electrolytes, such as acids and bases, salts, and multivalent ions, are relatively cheap but in general less effective than the polyelectrolytes. Certain finely divided solids, such as clays, activated silica, or carbon, which serve as absorbents and nucleation sites, can also serve as coagulant aids. Polyelectrolytes are high molecular weight, water-soluble organic compounds that are available in forms that give either nonionic, anionic, or cationic functionality.

Each fermentation broth must be studied empirically to establish the usefulness of this approach. Parameters that need to be investigated are the concentrations of colloid, coagulant, and/or flocculant; the effect of severity of agitation; and the kinetics of the binding and settling phenomena. Consequently, subjective responses are flocculation rate, floc size, sedimentation rate, and clarity of the supernatant liquor.

Design approaches based on Stokes' and Newton's laws have been used to predict sedimentation rates for flocced particles. However, complications relating to the presence of nonspherical rather than spherical particles, varying surface charges of polyelectrolytes, and the physical considerations of convection currents, particle–particle interactions (hindered settling), and particle flotation leave much to be desired. These techniques have been developed for waste treatment applications (Shuster and Wang, 1977) but are slowly being introduced into processes for fermentation broths (Treweek and Morgan, 1977).

D. Whole-Broth Processing

In the late 1950s technology was reported (Bartels *et al.*, 1958) for the sorption of a basic antibiotic, streptomycin, on a cationic ion-exchange resin directly from the unfiltered fermentation broth. This approach circumvented the problems and expense of conventional techniques for removing the insolubles. Various investigators recognized the advantages of this type of processing for suspensions of nonfermentation materials as well and have published methods for predicting performance (Prout and Fernandez, 1961; Marchello and Davis, 1963). Belter *et al.* (1973) have described a periodic countercurrent whole-broth process for the recovery of novobiocin. The whole broth, direct from the fermentor, is passed over a high-capacity vibrating screen to remove large particulate matter that would accumulate in the resin bed if not removed. The resulting screened broth is then contacted with an anionic-exchange resin in a series of specially designed, well-mixed columns. Each of the columns is equipped with a special head screen which retains the resin particles but permits passage of the mycelia and other insolubles. After a predetermined time of operation the resin in the lead column is loaded and the column is removed from the train, washed free of insolubles, and eluted with a methanolic ammonium chloride solution. This operation simultaneously removes the novobiocin and restores the resin to the chloride form. The resulting eluates are processed subsequently to crystals by conventional methods. Periodic countercurrent operation is obtained by advancing each of the partially loaded columns one position and adding a freshly eluted and regenerated column in the trail position of the train. At this point the sorption cycle is repeated. It is important to recognize that the operation described is transient in nature during an individual sorption cycle but approaches a pseudo-steady state process when the column positions are changed in the sequence outlined.

A general mathematical model was developed and used to predict the performance of such a system. The model consists of a suitable continuity equation for each stage coupled with the prevailing mass transfer and equilibrium relationships in each stage.

Assuming complete mixing in the column the continuity equation is given for the *n*th column in the train by Eq. (2).

$$V_c \frac{dC_n}{dt} = FC_{n-1} - FC_n - V_R \frac{dq_n}{dt} \tag{2}$$

where V_c is the liquid volume in column *n*, C_n is the antibiotic concentration in the bulk effluent from the column, C_{n-1} is the antibiotic concentration in the bulk feed to column *n*, *F* is the volumetric flow rate, V_R is the volume of resin in the column, q_n is the antibiotic concentration in the resin phase, and *t* is time.

The antibiotic transfer rate is expressed in Eq. (3) as the product of an overall mass transfer coefficient and a concentration difference or driving force.

$$\frac{dq_n}{dt} = K_a(C_n - C_n^*) \tag{3}$$

Equilibrium is expressed by Eq. (4) as a Freundlich-type relationship.

$$C_n^* = b(q_n)^a \tag{4}$$

where C^* is the antibiotic concentration in the bulk liquid that is in equilibrium with the resin phase and *a* and *b* are empirical constants.

The mass transfer coefficient, K_a, can be expressed as a function of resin loading by the empirical correlation given in Eq. (5).

$$K_a = Ae^{-B} \frac{q}{q_\infty} + Ce^{-D} \frac{q}{q_\infty} \tag{5}$$

where q_∞ is the ultimate resin loading and constants *A, B, C,* and *D* determined by laboratory sorption data.

This set of relationships, Eqs. (2)–(5), describes mathematically the time-dependent performance of each column in the sorption train. Modeling of the sorption sequence consists of solving this set of simultaneous equations for each of the columns in the system. Obviously, a computer is required to solve these nonlinear equations for a multiple-stage system. Such calculations permit the determination of effluent concentrations as a function of time (broth loss curves) and concentration on the resin as a function of time (resin loading curves). Subsequent integration of the broth loss profiles can establish total or percentage loss for process evaluation studies.

FIGURE 1. Production-scale columns. (Courtesy of The Upjohn Co.)

It should be realized that these equations apply to all constituents of the broth as well as to the antibiotic; and with additional assay information, resin loading and loss curves could be established for other components and, if present, interactions between competing components evaluated.

Production-scale columns for this type of processing are shown in Fig. 1. This method of primary recovery has general utility for materials that have ionic functionality.

IV. PRIMARY SEPARATIONS

After separation of the fermentation broth into soluble and insoluble fractions, the location of the desired product governs the next step in the sequence. In the case of certain conversions, such as steroid dehydrogenations, intracellular enzymes, and several antibiotics, the desired product remains with the insoluble fraction and the clarified broth can be discarded. In cases where the product is merely insoluble, it can be leached from the mycelia. Standard engineering approaches (Brian, 1972) are available for evaluation and design of this operation. In the cases of intracellular enzymes and bound or associated antibiotics some method of disrupting the cell matrix or membranes and releasing the desired product is needed prior to leaching. Heat, pH adjustment, chemical and/or mechanical disintegration, and osmotic shock are methods that have been used successfully in the laboratory to release the bound product.

However, according to Dunnill (1972) the disruption method that appears to have the greatest commercial potential is the mechanical approach (homogenization). Hetterington *et al.* (1971) and Gray *et al.* (1972) have studied the disintegration of bakers' yeast and *E. coli* cells, respectively. These authors have developed and used Eq. (6), which correlates the release of enzymes with the number of passes through the machine that the cell suspension was given.

$$\log \frac{R_m}{R_m - R} = KNP^{2.9} \tag{6}$$

where R_m is the maximum soluble protein available for release, R is the amount of soluble protein released after N cycles, K is a temperature-dependent rate constant, and P is the homogenizer operating pressure.

After the cell has been lysed and the product released into the liquid, one or more of the approaches discussed in Section III are suitable to remove the cell debris. In most cases the remaining steps in the sequence will be similar to those described for products which are originally extracellular.

However, in the case of enzyme separation, the clear extract contains numerous heat-labile components and the relatively harsh techniques used for extracellular products may not be applicable. Milder fractionating techniques that have been used for these proteinaceous products are ultracentrifugation, ultrafiltration, and various types of chromatography. These techniques are typical ones used for the fractionation of proteins and peptides regardless of their source.

Most materials, antibiotics, extracellular enzymes, organic acids, amino acids, and solvents, are present originally in the liquid portion of the broth. The remainder of this section covers the relatively nonselective separation approaches that are in general use for the primary separation steps.

A. Liquid Extraction

When exposed to a two-phase immiscible liquid system, an organic compound will be distributed in a definite ratio between these two phases. This ratio has been given the name of the distribution coefficient and is defined in Eq. (7).

$$K \equiv C_L/C_H \tag{7}$$

where C_L and C_H are the concentration of the organic compound in the light and heavy phase solvents, respectively. It is a characteristic property of the compound as are melting point, optical rotation, or light absorption properties. It is this property that is exploited in the extraction of clarified fermentation broths with an organic solvent. Choice of the extracting solvent depends on a number of factors, such as distribution properties, volatility, ease of recovery, and cost. The selectivity of a solvent system is characterized by the ratio of distribution coefficients of components under consideration and is designated as the β value, a parameter synonymous to the α value used for distillation calculations.

For the cases where the distribution coefficient for a given compound is linear, independent of concentration, material balance considerations used to predict and evaluate the performance of extraction systems are relatively simple. The definition of an extraction factor, E, is given in Eq. (8).

$$E \equiv LK/H \tag{8}$$

where L is the volumetric flow rate of the phase with the lesser density and H is the volumetric flow rate of the more dense phase. It can be derived that for a single equilibrium contact the fraction unextracted, X/X_0, is equal to $1/1 + E$ where X is the final concentration and X_0 is the initial concentration. Also, this type of reasoning can be extended to predict the performance of multiple-stage contacting. If a heavy phase liquid containing solute is subjected to multiple countercurrent contacts with a light immiscible solvent, each contact attaining equilibrium, it can be derived by a series of material balances that the rejection ratio, R, is given by Eq. (9).

$$R = \frac{E(E^n - 1)}{E - 1} \tag{9}$$

where R is the ratio of the weight of solute leaving in the light phase to the weight of solute leaving in the heavy phase and n is the number of equilibrium stages.

It follows from the definition of R that the fraction of solute extracted into the light phase is $R/(R + 1)$ or $R/(R + 1) \times 100$ to give the yield of desired product. The plot in Fig. 2 shows graphically the relationship between extracted solute, extraction factor, and number of stages used in a countercurrent extraction unit.

When the situation is reversed and the solute enters the system in the light phase, the rejection ratio is that given by Eq. (10).

$$R = \frac{E^m(E - 1)}{E^m - 1} \tag{10}$$

where m equals the number of stages in the unit.

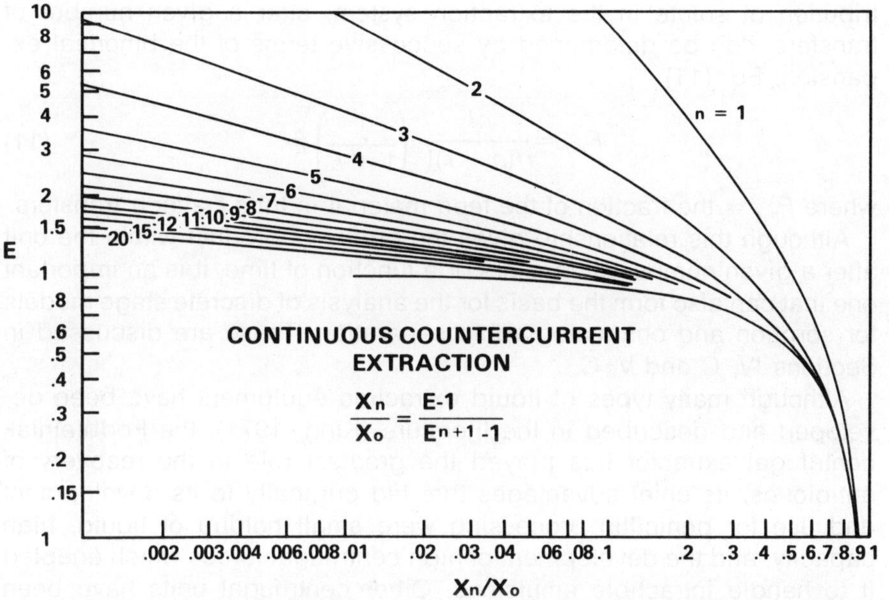

FIGURE 2. Relationship between unextracted solute, extraction factor, and number of stages in continuous countercurrent extraction. (Craig and Craig, 1956; reproduced by permission of Interscience Publishers, Inc.)

It should be remembered that the applicable rejection ratio relationship holds for all components present in the filtered broth and the success of this approach depends on choosing a sufficiently selective solvent to recover the desired product relatively devoid of impurities. Unfortunately, the few solvents that have the proper handling characteristics are not sufficiently selective to permit appreciable purification during the extraction step.

Historically, Craig and Craig (1956) developed this technique and demonstrated its usefulness in the laboratory. The countercurrent distribution system developed by Craig and Post consisted of a series of specially designed tubes, which are equivalent to a series of separatory funnels, connected so that the upper phase solvent can be transferred to successive tubes in the train. The operation of the series of tubes is automated so that a complete cycle consists of agitation to mix the two phases and simultaneous distribution of the solute between the two immiscible phases, a time delay period to allow the equilibrated phases to settle, and subsequent transfer of the upper phase into the next tube leaving the lower phase in the same tube. Simultaneously, a solute-free portion of the upper phase is added to the first tube of the extraction train. If the extraction factor is constant throughout the operation (no change of intrinsic distribution coefficient or solvent volumes), the distribution of solute in the extraction system, after a given number of transfers, can be determined by successive terms of the binomial expansion, Eq. (11).

$$F_{n,r} = \frac{n!}{r!(n-r)!} \left(\frac{1}{1+E}\right)^n E^r \tag{11}$$

where $F_{n,r}$ = the fraction of the feed material in tube r after n transfers.

Although this relationship gives the profile of product within the unit after a given number of transfers or a function of time, it is an important one that can also form the basis for the analysis of discrete stage models for sorption and chromatographic separation. These are discussed in Sections IV, C and V, C.

Although many types of liquid extraction equipment have been developed and described in the literature (King, 1971), the Podbielniak centrifugal extractor has played the greatest role in the recovery of antibiotics. Its chief advantages that led originally to its development and use for penicillin processing were small holdup of liquid, high capacity, and the development of high centrifugal forces which enabled it to handle intractible emulsions. Other centrifugal units have been developed and are used extensively to extract clarified fermentation broths.

B. Dissociation Extraction

A number of products to be recovered from fermentation broths are either weak acids or weak bases. In general, if a compound is to be extracted by an organic solvent, pH conditions are selected so that the material is largely in the un-ionized form. Thereby, weak bases are extracted at relatively high pH values and weak acids under low pH situations. These conditions result in rapid and complete extraction but little separation of impurities of a similar nature. Usually the β values, the ratio of the distribution coefficients, remain relatively constant at extreme pH levels. As discussed in the previous section on liquid extraction the ease of separation depends on this ratio. In some cases the β value will change appreciably with a pH shift and, consequently, the ease of separation will be improved. A typical situation of this nature is shown in Fig. 3, where the distribution properties of penicillins F, G, and K are compared with other major acidic impurities of unknown composition. Unfortunately, the solvent system is not given but it serves to illustrate the phenomenon under discussion.

If the assumption is made that at extreme pH values the ionized species is not soluble in the organic phase and the un-ionized species is soluble in either phase, the mathematical concepts covered in Section

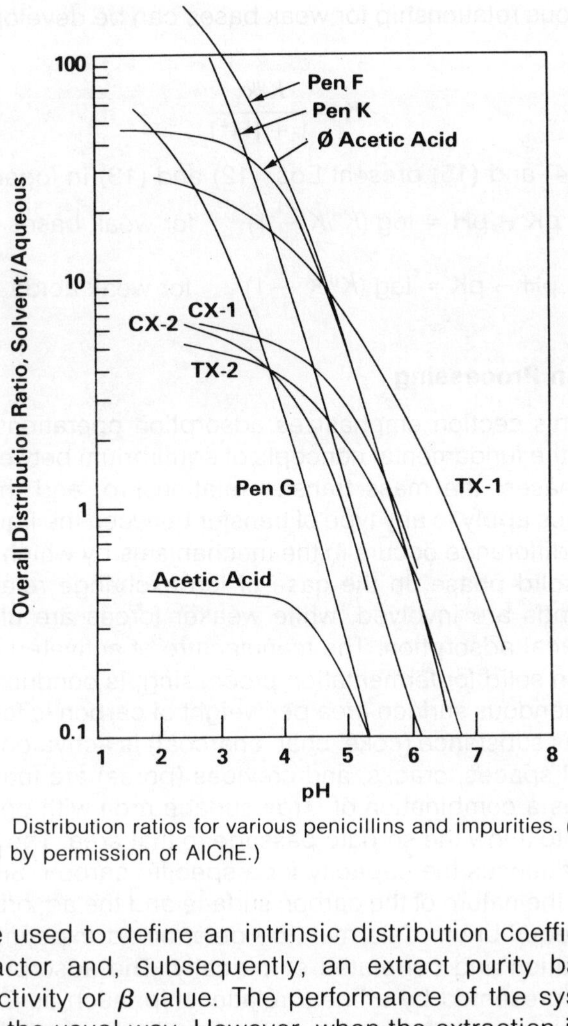

FIGURE 3. Distribution ratios for various penicillins and impurities. (Sounders *et al.*, 1970; reprinted by permission of AIChE.)

IV, A can be used to define an intrinsic distribution coefficient and an extraction factor and, subsequently, an extract purity based on the solvent selectivity or β value. The performance of the system can be evaluated in the usual way. However, when the extraction is conducted at intermediate pH values where the materials are partially ionized, evaluation and design of the extraction process is considerably more complicated. Sounders *et al.* (1970) have given the following relationship shown as Eq. (12) for weak acids.

$$K = \frac{K^0[H^+]}{[H^+] + k_1} \tag{12}$$

where K is the apparent distribution ratio for the weak acid, K^0 is the intrinsic distribution ratio for the un-ionized species, and k_1 is the ionization equilibrium constant for the weak acid.

An analogous relationship for weak bases can be developed as given by Eq. (13).

$$K = \frac{K^0 k_1}{1 + [H^+]} \tag{13}$$

Equations (14) and (15) present Eqs. (12) and (13) in logarithmic form.

$$pK - pH = \log (K^0/K - 1) \quad \text{for weak bases} \tag{14}$$

$$pH - pK = \log (K^0/K - 1) \quad \text{for weak acids} \tag{15}$$

C. Sorption Processing

Although this section emphasizes adsorption operations, it must be realized that the fundamental concepts of equilibrium between the liquid and solid phases, the mass transfer relationship, and the prevailing hydrodynamics apply to any type of transfer between the liquid and solid phases. The difference occurs in the mechanisms by which the solute is held in the solid phase. In the case of ion exchange relatively strong chemical bonds are involved, while weaker forces are utilized in the case of physical adsorption. The manufacture of activated carbons, the usual sorption solid for fermentation processing, is conducted in such a way that tremendous surface area per weight of carbon is formed. As the carbonaceous substance (coke, char, charcoal) is activated a network of large internal spaces, cracks, and crevices (pores) are formed. A good adsorbent has a combination of large surface area with openings sufficiently sized to allow the sorbate passage to that area. The nature of the adsorbate influences the capacity for a specific carbon. Since the dual influences of the nature of the carbon surface and the adsorbate are little understood, equilibrium data must be determined experimentally for a given application and correlated empirically. The adsorption data can usually be represented by a linear relationship, such as $C = mq$, over short ranges of concentrations or by the familiar Freundlich-type equation, $C = mq^n$, over extended concentration ranges. In these relationships, C and q refer to the concentration of the material in the liquid and solid phases, respectively. The m and n symbols are empirically determined constants.

The methods of contacting the liquid and solid phases vary from simple batch contact and multiple-batch contacts followed by filtration, to the usual fixed-bed approach. Prediction of the results of batch operation is relatively simple, involving only material balance and equilibrium considerations. However, prediction of the performance of a fixed bed, which is the most efficient way to use a given amount of carbon, is more difficult because of its inherent transient state nature and the resulting mathematical complexity of two independent variables, time and position.

Fixed-bed contacting has been modeled adequately by establishing the continuity equation across a differential height of solid adsorbant during an increment of time, *dt*. The resulting partial differential equation is given by Eq. (16).

$$\frac{U}{\epsilon}\frac{dC}{dZ} + \frac{dC}{dt} = \frac{-\rho}{\epsilon}\frac{dq}{dt} \tag{16}$$

where C is the concentration of sorbed material in the solution surrounding the sorbent, Z is the height in the column, t is time, ϵ is the porosity of the bed, U is the superficial velocity of the solution, ρ is the bulk density of the sorbent, and q is the concentration of sorbed material in the solid phase.

If it is also assumed that only mass transfer through the liquid film surrounding the particle and the subsequent diffusion of solute into the solid particle form the major resistances to that transfer, it is possible to write a relationship that describes the transfer rate in terms of an overall mass transfer coefficient. Such a relationship is given by Eq. (17).

$$dq/dt = K_a(C - C^*) \tag{17}$$

where C^* is the concentration of solute in equilibrium with q and K_a is the overall mass transfer coefficient.

In general, K_a can be expressed as a function of q for a given particle size and contacting conditions. The actual relationship must be determined experimentally for each application since it is dependent on the composition of the feed solution, the inherent properties of the solute and sorbent, and the liquid and solid phase diffusion mechanisms.

The continuity equation (a partial differential equation), the rate equation for sorption, the mass transfer relationship, and the equilibrium relationship in combination constitute the mathematical model which can be used to describe quantitatively the performance of a fixed-bed sorption process.

Chen *et al.* (1968) have used the approach described above with the help of large-scale computers to study and design fixed-bed sorption processes. This attack had been used by others to predict the performance of fixed-bed columns and processes in other industries.

However, the approach described previously makes the basic assumption that no back-mixing occurs during the contacting operation. In practice, additional zone spreading occurs for a variety of reasons, such as, nonequilibrium contact, axial and radial diffusion which results from molecular and eddy disturbances, irregular flow profiles, thermal effects, and channeling due to nonuniform packing characteristics. Historically, the overall result of these nonidealities has been tackled by introducing a "dispersion" coefficient into the analysis. Tracer techniques and correlations have been established to measure and predict the magnitude of the additional dispersion by a number of authors (Chung and Wen 1968; Hashimoto *et al.* 1977). Since it is difficult to separate and measure the effects of the individual nonidealities, Chen *et al.* (1972) circumvented the problem by simulating an actual sorption train as a series of tanks operating under ideal conditions of back-mixing. Use of the set of equations given in Section III, D for ion-exchange operations and the required number of tanks permit an approximation for quantitative evaluation and/or prediction of fixed-bed sorptions.

A third method of approximation for this problem is that of assuming a number of discrete stages operating with fixed distribution parameters. The eluant and solids profile can be determined by summation of the fundamental distribution for each of the feed increments and its subsequent distributions as given by Eq. (18).

$$\sum_{n=1}^{n=f(\Theta)} \sum_{r=1}^{r=N} \frac{n!}{r!\,(n-r)!} \left(\frac{1}{1+E}\right)^{n} E^{r} \tag{18}$$

where n is the number of feed increments and, consequently, the number of equilibrations; r is the stage number under consideration; E is the sorption parameter for the liquid and solid distribution; and N is the total number of stages.

Use of Eq. (18) permits the calculation of solute profiles throughout the fixed bed in both liquid and solid phases as well as solute concentration in the emerging liquid as a function of feed increments (or time).

Physical sorption is a processing approach that has been used for the recovery of numerous fermentation products, particularly antibiotics.

D. Precipitation

The inherent lack of solubility for salts of various products can be used advantageously to recover selectively and/or purify the desired product. Numerous examples, such as the lactate and oxalate salts of erythromycin, calcium salt of citric acid, and sodium dodecal sulfonate salt of gentamicin, are given in the antibiotic literature. However, its greatest use is in the elimination of proteinaceous impurities and/or the purification of enzymes. Typically, this is accomplished by adding inorganic salts to give relatively high ionic strengths, by reducing the solubility with the addition of organic solvents to change the dielectric strength of the solvent system, or by a combination of both. The last technique is used to precipitate and fractionate dextrans into useful materials of narrow molecular weight ranges.

The underlying principles for this separation approach have been covered by Aiba *et al.* (1973).

V. PURIFICATION OPERATIONS

The unit operations described previously produce mixtures of materials from the fermentation broth that are similar in physicochemical properties to the desired product and therefore require the use of additional separation techniques that are more powerful than those presented, to obtain high-quality products. These techniques gain their separation power because they consist of multiple stages assembled in a cascade so that both product quality and yield can be obtained.

A. Fractional Liquid–Liquid Extraction

The simple concept and basic definitions for this unit operation have been covered in Section IV, A.

A mixture can be separated into only two streams during continuous operation with a single column. A cascade of contacting stages can be assembled as shown in Fig. 4 with feed introduced at or near the center of the column. The section above the feed stage functions as an enriching section, removing impurities from the product solute. The lower section below the feed stage acts to strip the desired product from the impurities. Consequently, the goals of high-quality and improved yields can be met.

If no solvent is added to the column with the feed mixture, the rejection ratio can be shown to be that given in Eq. (19).

$$R = \frac{E^m(E^n - 1)}{E^m - 1} \tag{19}$$

where R is the rejection ratio, E is the extraction factor, m is the number of enriching stages plus 1, and n is the number of stripping stages plus 1. Simplification of this relationship can be made in special cases as shown in Table II. More complex relationships, which deal with the case where the solute mixture is added to the column dissolved in either the upper or the lower phase, may be found in the literature (Brian, 1972). Belter (1967) has published a method by which the separation perfor-

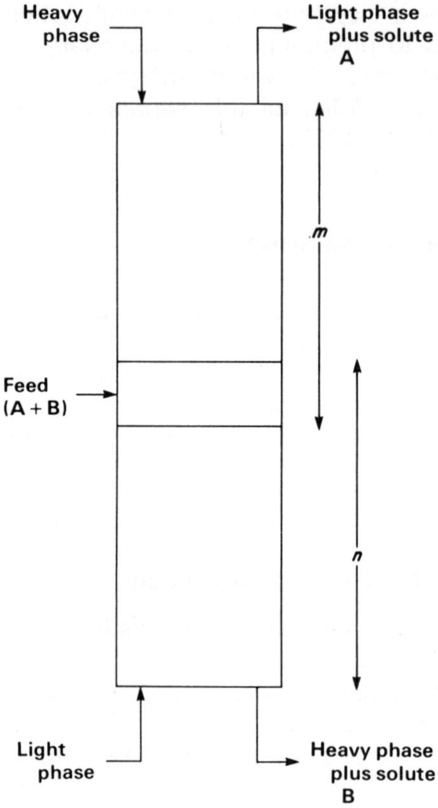

FIGURE 4. Multistage extraction column with center feed.

mance of a continuous column can be simulated in a normal Craig distribution.

Numerous mechanical devices have been conceived for multiple-stage contacting of two liquids. Most of these units have been described and discussed by King (1971). Bartels and Kleiman (1949) described and presented a simplified mathematical prediction approach for the fractionation of streptomycins using a cascade of disk- and bowl-type centrifuges.

B. Chromatography

Chromatography is defined as a method of separation in which the flow of a fluid promotes the fractionation of mixtures by differential migration in a porous sorptive medium (usually a solid) from a narrow initial feed zone. The operation is a time-dependent one but is equilib-

TABLE II. Simplified Forms of Equation (19)

Extraction factor E	Rejection ratio
1	n/m
$>>1$	E^n
$<<1$	E^m
1 (and center feed with $n = m$)	E^n

rium controlled. A schematic illustration of the separation of these components by a differential migration is shown in Fig. 5. Three components are added to the leading edge of the chromatography system prior to elution as shown in Fig. 5a. As the flow of fluid passes through the system, the components separate. Figures 5b–d show the position and separation of the three components at increasing time intervals of fluid flow. Figure 5e shows the concentration of each of the components in the system effluent versus time. A similar plot, which could have the order of components reversed, can be obtained for concentrations of components prior to migration from the system, such as shown in Fig. 5d.

Useful types of chromatography for a separation and purification of fermentation products are adsorption, partition, ion exchange, and gel filtration. Also, dry column chromatography, affinity, and ion exclusion have been advocated. These operations are similar in their mathematical developments but differ considerably in the predominant mechanism by which the differential migration occurs. Separation occurs in adsorption chromatography because of the different strengths of van der Waal's forces binding the solutes to the solid phase. It is accomplished in partition chromatography by the different distribution coefficients of solutes between the mobile and stagnant adsorbed liquid phase. The governing mechanism for ion-exchange chromatography is the varying strengths of the chemical bonds formed with the ionized solutes and the ion-exchange functionality. Gel filtration depends on the ability of molecules of different size and shape to penetrate the matrix of the column material. These mechanisms overlap to some extent and it should not be construed that any chromatographic separation is dependent on only one mechanism.

The basic theory for chromatographic separations parallels those described previously for fixed-bed operations with the mathematical exception that different boundary conditions are employed in the solution of the partial differential equations. Since the concentration of an individual solute is dependent on location within the column and time, it is apparent that the continuity equation will be a partial differential equation (as given in Section IV, C for sorption processes). The transfer rate

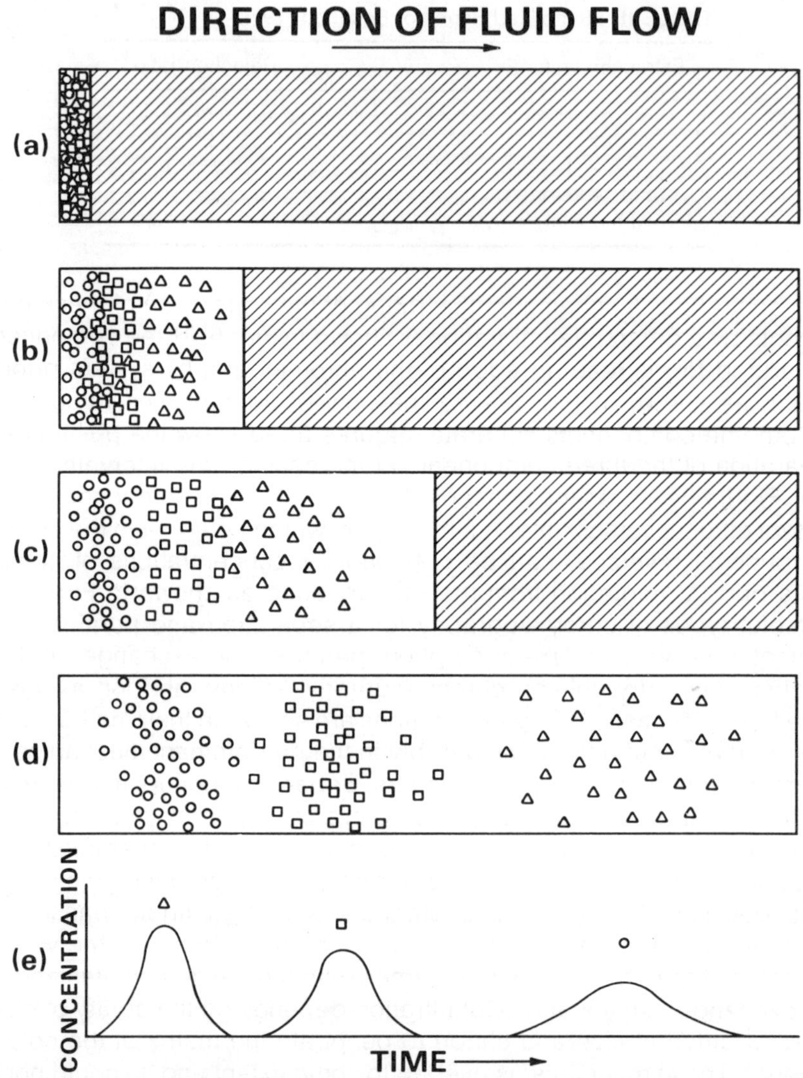

FIGURE 5. Schematic illustration of separation by the differential migration mode. (Karger *et al.*, 1973; reproduced by permission of Wiley, New York.)

and equilibrium relationships can be established experimentally as before.

The fundamental theory for determining concentration profiles during chromatographic separations has been discussed and evaluated by Lightfoot *et al.* (1962). Examples are given for cases of increasing complexity ranging from the simple situation of equilibrium, negligible dispersion and linear sorption isotherms to those cases where none of these ideal conditions exists.

C. Carbon Decolorization

The presence of color bodies and their precursors in the final solutions can seriously affect the quality of the ultimate product and is of concern in the preparation of high-quality products. The usual approach to their removal is by adsorption on activated carbon, either powdered or granular, using techniques and modes of operation previously described.

D. Crystallization

Usually crystallization is the final purification step in the processes that must produce high-quality products. It is an effective and efficient method of purification and possesses some advantages over competitive methods. Relatively low-temperature operations minimize undesirable degradation of heat-sensitive materials. Operations are usually conducted at high concentrations and, consequently, unit costs are low. Separation factors are usually high, which makes this operation an effective fractionation tool.

Various aspects of crystallization phenomena, such as nucleation of different types and growth kinetics, for simple systems have been explored and reported in the literature (Mullins, 1972). Separation or fractionation by crystallization is of special concern but, unfortunately, the prerequisites for prediction—knowledge of phase diagrams for a given mixture, knowledge of true equilibrium states between the crystals and the surrounding solution, and knowledge of the pertinent kinetics—are rarely available. Consequently, determination of optimal crystallization conditions for a given problem becomes a matter of empirical experimentation. Obviously, crystallization is not necessary for all products and in several isolated instances, such as neomycin and certain enzymes, crystallization of the product has never been accomplished.

VI. PRODUCT ISOLATION

Recovery of the product in its final form depends on the prior processing and its ultimate use. Basically, the operations can be divided into two areas; first, processing of crystalline products and, second, drying of products directly from solution.

A. Crystal Processing

High-quality crystals are recovered in conventional batch Nutsche-type filters or centrifugal filters. In the first of these techniques the same theory that was covered in Section III, A for constant pressure filtrations applies to the filtration of crystalline materials, and considerable simplification can be made. If it is assumed that the crystal slurry is uniform and, therefore, the filtrate volume is proportional to the crystal weight, the crystals are noncompressible and that the resistance of filter medium is negligible, Eq. (1) can be simplified to give Eq. (20).

$$dV/d\theta = K(\Delta PA/T) \tag{20}$$

where T is the thickness of crystal layer and the other symbols are as defined for Eq. (1). Upon substitution of $W \cong V$ and integration one obtains Eq. (21).

$$\Theta = K\,(W^2/\Delta PA^2) \tag{21}$$

This relationship can be used to predict filtration and washing rates after the permability factor, K, has been determined from laboratory tests.

A typical centrifugal filter is shown in Fig. 6. The following relationship, Eq. (22), for the centrifugation rate has been developed by Valleroy and Maloney (1960).

$$\frac{dV}{d\theta} = \frac{(2\pi N)^2 \rho L (r_0^2 - r_L^2)}{2\mu(\alpha W/A_m^2 + R_M/A_0)} \tag{22}$$

where N is the speed, μ is the viscosity, α is the specific resistance of the cake, W is the solids concentration in the cake, A_m is the log mean area, and R_M is the medium resistance.

It is readily apparent that this relationship is equivalent to that used previously with the necessary modifications to account for the different geometry and gravitational driving force.

Since the crystals deposited in the basket are subsequently washed to remove adhering mother liquor, the nature and size of the crystals affect the centrifugation and washing rates. Consequently, the crystallization and recovery steps are interrelated and these factors should be considered in the optimization of the crystallization sequence.

After washing the crystals are spun dry and discharged from the basket to trays for subsequent drying or desolventizing, depending on the nature of the solvent employed for crystallization and washing.

B. Drying

The removal of liquid (water or an organic solvent) from the purified product stream, either as liquid-wet crystals or from a solution, usually

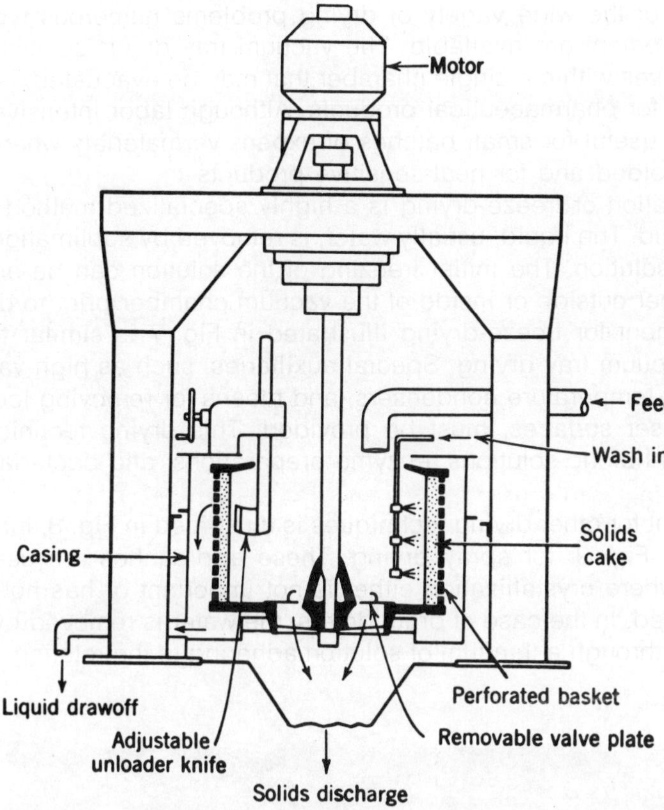

FIGURE 6. Top suspended basket-type centrifugal filter. (Jordan, 1968a; reproduced by permission of Interscience Publishers.)

constitutes the final step in the manufacture of a given product. The underlying principles of drying rates, which are governed by the physicochemical properties of the material to be processed, the psychrometric relationships of the drying medium, and the heat and mass transfer characteristics of both the material and the drying equipment used, have been covered elsewhere (Nonhebel and Moss, 1971). The physical nature of the product, its heat sensitivity, and, again, the physical characteristics of the final product that are acceptable for its ultimate use dictate the drying approach selected. Thirty-four parameters affecting drying have been listed by Nonhebel and Moss (1971). These parameters were divided into four classes according to whether they affect heat transfer, properties of the drying environment, physical properties of the solid–liquid system, or the inherent properties of the solid. Equilibrium relationships between solvent and solid, energy, and mass balances must be considered in the design of optimum drying processes and systems.

Because of the wide variety of drying problems numerous types of drying equipment are available. The vacuum tray dryer, consisting of heated shelves within a single chamber that may be evacuated, is used extensively for pharmaceutical products. Although labor intensive, it is particularly useful for small batches of expensive materials where loss must be avoided and for heat-sensitive products.

Lyophilization or freeze-drying is a highly specialized method of removing liquid. The liquid, usually water, is removed by sublimation from the frozen solution. The initial freezing of the solution can be accomplished either outside or inside of the vacuum chamber prior to drying. The equipment for freeze-drying illustrated in Fig. 7 is similar to that used for vacuum tray drying. Special auxiliaries, such as high-vacuum pumps, low-temperature condensers, and means for removing ice from the condenser surfaces, must be provided. This drying technique is useful for antibiotic solutions, enzyme preparations, and bacterial suspensions.

Equipment for other drying techniques is illustrated in Fig. 8, for drum dryers, and Fig. 9, for spray drying. These approaches are used for situations where crystallization either is not important or has not been accomplished. In the case of drum dryers, the water is removed by heat conduction through a thin film of solution adhering to the rotating steam

FIGURE 7. Freeze–drying system. (Courtesy of The Upjohn Co.)

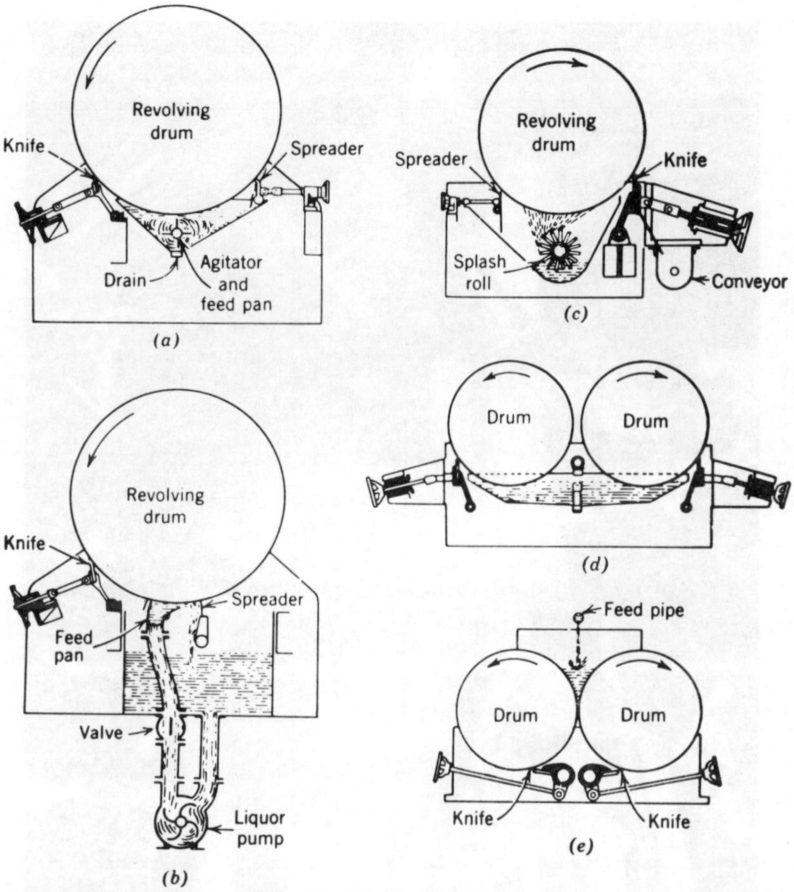

FIGURE 8. Drum dryers. Arrangements for feed input and product take-off. (a) Single drum, dip-feed; (b) single drum, pan-feed; (c) single drum, splash-feed; (d) double drum, dip-feed; (e) double drum, top-feed. (Jordan, 1968b, reproduced by permission of Interscience Publishers.)

heated drum. As the drum rotates, the product is dried and scraped from the drum at the discharge point.

In spray drying the feed solution or slurry is atomized into a chamber through which heated air passes. The sensible heat of the hot gas supplies the latent heat requirements for vaporization of the liquid, leaving small particles of dry powder. Subsequently, the dry product must be separated from the exit air. The solution is dispersed by one of three methods, pressure nozzles, two fluid nozzles, and rotating disks. The product usually dries as small spheres and consequently has a relatively low bulk density. Although relatively expensive because of

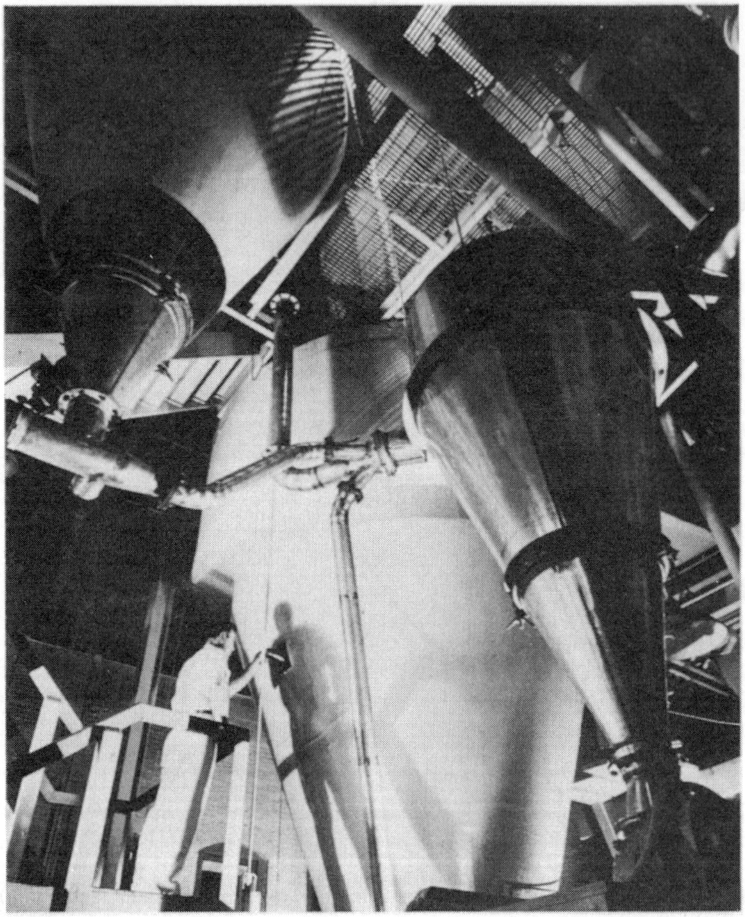

FIGURE 9. Spray-drying system. (Courtesy of The Upjohn Co.)

poor thermal efficiency, the technique is preferred for heat-sensitive materials.

VII. ANCILLARY OPERATIONS

Several operations exist that are essential to an integrated process for the manufacture of a fermentation product but serve only supporting roles. Typical of such operations are the recovery of residual organic solvents from extracted spent broths and the subsequent treatment of the spent broths.

The nature and concentration of materials in the residual fermentation broths depend on their source but, in general, these waste streams contain extremely high levels of biodegradable compounds, resulting in high biological oxygen demand (BOD) requirements and nitrification demands when discharged. The need to maintain and improve the quality of our water resources and our environment has led to extensive studies on ways of handling the solids in these streams.

For economic reasons and since residual organic solvents would contribute appreciably to the BOD load, residual extraction solvents are removed from the liquid streams prior to treatment or disposal. The solvent recovery procedures are reasonably routine stripping operations and can be evaluated and designed by distillation techniques covered in the literature (King, 1971; Brown, 1950).

Numerous methods of waste treatment and disposal for industrial waste treatment have been studied and reported. Approaches that have been studied for spent broths include processes based on activated sludge and trickling filter aerobic digestion, anaerobic filters, spray irrigation, oxidation ponds, deep well injection, and incineration. Comprehensive descriptions of these techniques and examples from the pharmaceutical industry and for other fermentation products are given by Struzeski (1975) and Nemerow (1963).

Use of concentrated spent broths for liquid supplement feeding of ruminants and for the recovery of solids to be used in animal feeds for their nutritive value and the presence of unidentified growth factors constitute another approach to the problem of disposal.

REFERENCES

Aiba, S., Humphrey, A. E., and Millis, N. F. (1973). "Biochemical Engineering," 2nd ed. Academic Press, New York.

Anonymous (1977). *Chem. Process.* **40,** 10, 154.

Bartels, C. R., and Kleiman, G. (1949). *Chem. Eng. Prog.* **45,** 589–594.

Bartels, C. R., Kleiman, G., Korzun, J. N., and Irish, D. B. (1958). *Chem. Eng. Prog.* **54,** 49.

Belter, P. A. (1967). *Ind. Eng. Chem.* **59,** 3.

Belter, P. A., Cunningham, F. L., and Chen, J. W. (1973). *Biotechnol. Bioeng.* **15,** 533–549.

Brian, P. L. T. (1972). "Staged Cascades in Chemical Processing," Prentice-Hall, Englewood Cliffs, New Jersey.

Brown, G. G. (1950). "Unit Operations," Wiley, New York.

Chen, J. W., Buege, J. A., Cunningham, F. L., and Northam, J. I. (1968). *Ind. Eng. Chem., Process Des. Dev.* **7,** 26.

Chen. J. W., Cunningham, F. L., and Buege, J. A. (1972). *Ind. Eng. Chem., Process Des. Dev.* **11,** 430.

Chung, S. F., and Wen, C. Y. (1968). *AIChE J.* **14,** 857.

Craig, L. C., and Craig, D. (1956). *In* "Technique of Organic Chemistry" (A. Weissberger, ed.), 2nd ed., Vol. 3, Part 1. Wiley (Interscience), New York.

Dahlstrom, D. A., and Pruchas, D. B. (1957). "Scaleup Methods for Continuous Filtration Equipment," Inst. Chem. Eng., London.
Dunnill, P. (1972). *Fermen. Technol. Today, Proc. Int. Fermen. Symp. 4th, 1972* pp. 186–194.
Gray, P. P., Dunnill, P., and Lilly, M. D. (1972). *Fermen. Technol. Today, Proc. Int. Ferment. Symp. 4th, 1972* pp. 347–351.
Hashimoto, K., Minura, K., and Tsukano, M. (1977). *J. Chem. Eng. Jpn.* **10,** 1.
Hetherington, P. J., Follows, M., Dunnill, P., and Lilly, M. D. (1971). *Trans. Inst. Chem. Eng.* **49,** 142.
Holland, C. D. (1975). "Fundamentals and Modeling of Separation Processes: Absorption, Distillation, Evaporation and Extraction," Prentice-Hall, Englewood Cliffs, New Jersey.
Jordan, D. G. (1968a). *Chem. Process Dev.* Part 2, p. 846.
Jordan, D. G. (1968b). *Chem. Process Dev.* Part 2, p. 910.
Karger, B. L., Snyder, L. R., and Horvath, C. (1973). "An Introduction to Separation Science," Wiley, New York.
King, C. J. (1971). "Separation Processes," McGraw-Hill, New York.
Lightfoot, E. N., Sanchez-Palma, R. J., and Edwards, D. O. (1962). *In* "Chromatography and Fixed-Bed Separations" (H. M. Schoen, ed.), pp. 99–181. Wiley, New York.
Marchello, J. M., and Davis, M. W., Jr. (1963). *Ind. Eng. Chem.* **2,** 27.
Mullins, J. W. (1972). "Crystallization," CRC Press, Cleveland, Ohio.
Nemerow, N. L. (1963). "Theories and Practices of Industrial Waste Treatment," Addison-Wesley, Reading, Massachusetts.
Nonhebel, G., and Moss, A. A. H. (1971). "Drying of Solids in the Chemical Industry." CRC Press, Cleveland, Ohio.
Prout, W. E., and Fernandez, L. P. (1961). *Ind. Eng. Chem.* **53,** 449.
Shuster, W. W., and Wang, L. K. (1977). *Sep. Purif. Methods* **6,** 153–187.
Sorel, M. (1893). "La rectification de l'alcool." Paris.
Sounders, M., Pierotti, G. J., and Dunn, C. L. (1970). *Chem. Eng. Prog. Symp. Ser.* **66,** 40–42.
Struzeski, E. J., Jr. (1975). "Waste Treatment and Disposal Methods for the Pharmaceutical Industry," EPA 330/1-75-001. EPA Office of Enforcement, Washington, D.C.
Treweek, G. P., and Morgan, J. J. (1977). *J. Colloid Interface Sci.* **60,** 2.
Valleroy, V. V., and Maloney, J. O. (1960). *AIChE J.* **6,** 382.
Wagman, G. H., Bailey, J. V., and Weinstein, M. J. (1975). *Antimicrob. Agents Chemother.* pp. 316–319.

Chapter 17

Use of Immobilized Cell Systems to Prepare Fine Chemicals

ICHIRO CHIBATA
TETSUYA TOSA
TADASHI SATO

MICROBIAL TECHNOLOGY, 2nd ed., VOL.II
 ISBN 0-12-551502-2

I. INTRODUCTION

Since 1960, a number of papers have been published on preparations, properties, and potential utilizations of immobilized enzymes. In 1969, the authors (Tosa *et al.*, 1966, 1967; Chibata *et al.*, 1972) first succeeded in the industrial application of immobilized enzyme, i.e., immobilized aminoacylase, for continuous production of L-amino acids from acyl-DL-amino acids. Production of 6-aminopenicillanic acid and fructose by immobilized penicillin amidase and by immobilized glucose isomerase, respectively, soon followed in the United States, Europe, and Japan.

At present, enzymes derived from microorganisms are used advantageously for industrial purposes. Microbial enzymes are classified into two forms. One is the extracellular enzyme excreted from microbial cells during culture, while the other is the intracellular enzyme which remains in the cells. For the immobilization of the latter, the enzymes must be extracted from cells by some technique, and in some cases further purification is necessary. Extracted intracellular enzymes are, in many cases, unstable. If whole microbial cells can be immobilized directly without extracting enzyme, the immobilized cells can be used as a solid catalyst. In such immobilized microbial cells, the procedure for extraction of enzyme can be omitted, and loss of enzyme activity is expected to be kept to a minimum.

In addition, if the microbial cells having multienzyme systems are immobilized and can be used as a solid catalyst, the possibility exists that fermentative methods involving multienzyme reactions may be replaced by the continuous enzyme reaction using immobilized cells.

More recently, on the basis of these ideas, the studies on immobilization of whole microbial cells have been the subject of increased interest. However, there are fewer reports on immobilized cells than on immobilized enzymes. Chibata *et al.* (1974e) and Tosa *et al.* (1974) studied the immobilization of whole microbial cells, i.e., *Escherichia coli*, having high aspartase activity and in 1973 succeeded in the industrial application of immobilized cells to the continuous production of L-aspartic acid from ammonium fumarate. This was the first international industrial application of immobilized microbial cells.

In 1974 Chibata *et al.* (1975c) and Yamamoto *et al.* (1976, 1977) succeeded in applying industrial production of L-malic acid from fumarate using immobilized *Brevibacterium ammoniagenes* having high fumarase activity.

In this chapter the examples mentioned above and some other applications of immobilized cell systems for preparation of fine chemicals are reviewed.

II. IMMOBILIZATION OF MICROBIAL CELLS

At present, many papers on immobilization of microbial cells have been published (Table I).

Methods for immobilization of microbial cells can be classified into three categories analogous to the immobilization of enzymes, that is, carrier-binding, cross-linking, and entrapping methods. Of these methods, the entrapping method has been most extensively investigated.

A. Carrier-Binding Method

The carrier-binding method is based on direct binding of microbial cells to water-insoluble carriers. Microbial cells are ionically bound to water-insoluble carriers containing ion-exchange residues. Hattori and Furusaka (1960, 1961) bound *E. coli* cells and *Azotobacter agile* cells to Dowex-1 (Cl form). They observed that these immobilized cells showed oxidation activity for glucose and succinic acid.

However, this method is considered not to be advantageous because enzymes and cells may leak out from the carrier due to autolysis during the enzyme reaction.

B. Cross-Linking Method

Microbial cells can be immobilized by cross-linking with bi- or multifunctional reagents such as glutaraldehyde or toluenediisocyanate. Chibata *et al.* (1974e) investigated immobilization of *E. coli* cells having aspartase activity by cross-linking with glutaraldehyde as the bifunctional reagent. The immobilized cells showed aspartase activity corresponding to 34.2% of that of intact cells.

So far very few papers have appeared on this method. However, there is a possibility that suitable cross-linking reagents for immobilization of cells will be found in the future.

C. Entrapping Method

So far the method of directly entrapping microbial cells into polymer matrices has been most extensively investigated for immobilization of cells. For this method, the following matrices are employed: (1) collagen, (2) gelatin, (3) agar, (4) alginate, (5) carrageenan, (6) cellulose triacetate, (7) polyacrylamide, and (8) polystyrene. Among these matrices, polyacrylamide gel has been extensively used for immobilization of many kinds of microbial cells, as shown in Table I. This polyacrylamide

TABLE I. Preparations and Applications of Immobilized Microbial Cells

Enzyme reaction and application[a]	Microorganism	Immobilization method	Reference
[I] Oxidoreductase reaction			
L-Sorbose $\xrightarrow{\text{L-sorbose dehydrogenase}}$ L-sorbosone	*Gluconobacter melanotenus*	Polyacrylamide	Martin and Perlman (1976a)
Cortisone $\xrightarrow{\text{steroid } 11\beta\text{-hydroxylase}}$ cortisol	*Curvularia lunata*	Polyacrylamide	Mosbach and Larsson (1970)
[II] Transferase reaction			
NAD + ATP $\xrightarrow{\text{NAD kinase}}$ NADP + ADP	*Achromobacter aceris*	Polyacrylamide	Chibata *et al.* (1975b) Uchida *et al.* (1977)
2 Glucose + 2 *p*-nitrophenyl phosphate $\xrightarrow{\text{acid phosphatase}}$ glucose 6-phosphate + glucose 1-phosphate + 2 *p*-nitrophenol	*Escherichia freundii*	Polyacrylamide	Saif *et al.* (1975)
[III] Hydrolase reaction			
Triacetin or tributyrin + H_2O $\xrightarrow{\text{lipase}}$ glycerol + acetic acid or butyric acid	*Pseudomonas mephitica* var. *lipolytica*	Heat treatment	Kosugi and Suzuki (1973)
dl-1-Menthol ester + H_2O $\xrightarrow{l\text{-menthol ester hydrolase}}$ *l*-menthol + *d*-menthol ester + organic acid	*Alginomonas nonfermentas*	Polyacrylamide	Nonomura *et al.* (1976)

Lactose + $H_2O \xrightarrow{\beta\text{-galactosidase}}$ D-glucose + D-galactose	*Lactobacillus bulgaricus*	Agar	Miyata and Kikuchi (1976)
		Polyacrylamide	Ohmiya *et al.* (1977)
	Saccharomyces lactis	Cellulose triacetate	Dinelli (1972)
		Polyacrylamide	Ohmiya *et al.* (1977)
	Escherichia coli	Polyacrylamide	Ohmiya *et al.* (1977)
Sucrose + $H_2O \xrightarrow{\text{invertase}}$ D-glucose + D-fructose	Fungal spores	Ion-exchange cellulose	Johnson and Ciegler (1969)
	Saccharomyces pastorianus	Agar	Toda and Shoda (1975)
	Yeast	Freeze-dried	Nanba and Matsuo (1970)
Raffinose + $H_2O \xrightarrow{\text{raffinase}}$ D-glucose + D-. . . galactose + D-fructose	Mold	Glutaraldehyde	Nishimaru *et al.* (1975)
Penicillin G + $H_2O \xrightarrow{\text{penicillin amidase}}$ 6-APA + phenylacetic acid	*Escherichia coli*	Polyacrylamide	Sato *et al.* (1976)
		Cellulose triacetate	Dinelli (1972)
6-APA + D-phenylglycine methyl ester $\xrightarrow{\text{penicillin amidase}}$ ampicillin + methanol	*Bacillus megaterium*	DEAE–cellulose	Fujii *et al.* (1973)
	Achromobacter sp.	DEAE–cellulose	Fujii *et al.* (1973)
	Escherichia coli	Cellulose triacetate	Marconi *et al.* (1975)

(*continued*)

TABLE I (*Continued*)

Enzyme reaction and application[a]	Microorganism	Immobilization method	Reference
7-ADCA + D-phenylglycine methyl ester $\xrightarrow{\text{penicillin amidase}}$ cephalexin + methanol	*Achromobacter* sp.	DEAE–cellulose	Fujii *et al.* (1974)
Acetyl-DL-methionine + H_2O $\xrightarrow{\text{aminoacylase}}$ L-methionine + acetyl-D-methionine + acetic acid	*Aspergillus ochraceus*	Egg albumin and glutaraldehyde	Hirano *et al.* (1977)
ϵ-Aminocaproic acid cyclic dimer + H_2O ϵ-aminocaproic acid $\xrightarrow{\text{cyclic dimer hydrolase}}$ ϵ-aminocaproic acid	*Achromobacter guttatus*	Polyacrylamide	Kinoshita *et al.* (1975)
L-Arginine + H_2O $\xrightarrow{\text{L-Arginine deiminase}}$ L-citrulline + NH_3	*Pseudomonas putida*	Polyacrylamide	Yamamoto *et al.* (1974a)
[IV] Lyase reaction			
Diaminopimelic acid $\xrightarrow{\text{diaminopimelic acid decarboxylase}}$ L-lysine + CO_2	*Microbacterium ammoniaphilum*	Polyacrylamide	Kanemitsu (1975)
L-Aspartic acid $\xrightarrow{\text{aspartate 4-decarboxylase}}$ L-alanine + CO_2	*Pseudomonas dacunhae*	Polyacrylamide	Chibata *et al.* (1975a)
Pyruvate + phenol + NH_3 $\xrightarrow{\beta\text{-tyrosinase}}$ L-tyrosine	*Erwinia herbicola*	Collagen and dialdehyde starch	Yamada *et al.* (1975)
Pyruvate + pyrocatechol + NH_3 $\xrightarrow{\beta\text{-tyrosinase}}$ L-DOPA	*Erwinia herbicola*	Collagen and dialdehyde starch	Kumagai *et al.* (1976)

Fumaric acid + $H_2O \xrightarrow{\text{fumarase}}$ L-malic acid	*Brevibacterium ammoniagenes*	Polyacrylamide	Yamamoto *et al.* (1976)
	Brevibacterium flavum	Carrageenan	Tosa *et al.* (1977)
Fumaric acid + $NH_3 \xrightarrow{\text{aspartase}}$ L-aspartic acid	*Escherichia coli*	Carrageenan	Tosa *et al.* (1977)
		Glutaraldehyde	Chibata *et al.* (1974e)
		Polyacrylamide	Chibata *et al.* (1974e)
Indole + L-serine $\xrightarrow{\text{tryptophan synthetase}}$ L-. . . . tryptophan	*Escherichia coli*	Polyacrylamide	Chibata *et al.* (1974c)
5-Hydroxyindole + L-serine $\xrightarrow{\text{tryptophan synthetase}}$ 5-hydroxy-L-tryptophan	*Escherichia coli*	Polyacrylamide	Chibata *et al.* (1974b)
L-Histidine $\xrightarrow{\text{L-histidine ammonia-lyase}}$ urocanic acid + NH_3	*Achromobacter liquidum*	Polyacrylamide	Yamamoto *et al.* (1974b)
[V] Isomerase reaction			
D-Glucose $\xrightarrow{\text{glucose isomerase}}$ D-fructose	*Actinomyces*	Anion-exchange resin	Ishimatsu *et al.* (1976)
		Drying after dipping in organic acid	Tsumura and Kasumi (1977)
		Gelatin and glutaraldehyde	Yuta *et al.* (1975)
		Heat treatment	Takasaki and Kanbayashi (1969)
		β-Ray irradiation	Tsumura (1969)

(*continued*)

TABLE I *(Continued)*

Enzyme reaction and application[a]	Microorganism	Immobilization method	Reference
	Bacillus coagulans	Glutaraldehyde	Novo Industri (1976a,b)
	Streptomyces griseus	Polyacrylamide	Chibata *et al.* (1974d)
	Streptomyces phaeochromogenes	Carrageenan	Tosa *et al.* (1977)
		Cellulose triacetate	Kolarik *et al.* (1974)
		Chitosan	Tsumura *et al.* (1976)
		Collagen and glutaraldehyde	Vieth *et al.* (1973)
			Wang *et al.* (1973)
	Streptomyces venezuelae		Venkatasubramanian *et al.* (1974)
[VI] Multienzyme reaction			
Oxidation of glucose and succinic acid	*Azotobacter agile*	Dowex-1	Hattori and Furusaka (1961)
	Escherichia coli	Dowex-1	Hattori and Furusaka (1960)
Survival of microorganism in polymer lattice	*Tetrahymena pyriformis*	Polyacrylamide	Updike *et al.* (1969)
Reduction of nitrate and nitrite $NO_3^- \longrightarrow NO_2^- \longrightarrow NH_4^+$	*Micrococcus denitrificans*	Liquid membrane	Mohan and Li (1975)
Production of hydrogen from glucose (hydrogen battery)	*Clostridium butyricum*	Polyacrylamide	Karube *et al.* (1976a)

Photosynthetic regeneration of ATP	*Rhodospirillum rubrum* chromatophores	Polyacrylamide	Yang *et al.* (1976)
Degradation of phenol Phenol $\longrightarrow$ $CO_2 + H_2O$	*Candida tropicalis*	Aluminium alginate	Hackel *et al.* (1975)
		Polystyrene	Hackel *et al.* (1975)
		Polyacrylamide	Hackel *et al.* (1975)
Conversion of L-sorbose to 2-keto-L-gulonic acid L-Sorbose $\xrightarrow{\text{L-sorbose dehydrogenase}}$ L-sorbosone $\xrightarrow{\text{L-sorbosone oxidase}}$ 2-keto-L-gulonic acid	*Gluconobacter melanogenus* and *Pseudomonas syringae*	Polyacrylamide	Martin and Perlman (1976b)
Preparation of orcinol and orcinol monomethyl ester from depside evernic acid by the esterase and decarboxylase activities of the cells	*Umbilicaria pastulate*	Polyacrylamide	Mosbach and Mosbach (1966)
Preparation of coenzyme A from pantothenic acid, L-cysteine, and ATP	*Brevibacterium ammoniagenes*	Polyacrylamide	Shimizu *et al.* (1975)
Catabolism of L-arginine	*Streptococcus faecalis*	Polyacrylamide	Franks (1971)
Production of lactic acid from 1,2-propanediol	*Arthrobacter oxydans*	Polyacrylamide	Yagi *et al.* (1976)
Production of L-glutamic acid from glucose	*Corynebacterium glutamicum*	Collagen	Brownstein *et al.* (1974)
		Polyacrylamide	Slowinski and Charm (1973)
Production of ethanol from glucose	*Saccharomyces cerevisiae*	Calcium alginate	Kierstan and Bucke (1977)
Production of ethanol from inulin	*Kluyveromyces marxianus*	Calcium alginate	Kierstan and Bucke (1977)

(continued)

TABLE I *(Continued)*

Enzyme reaction and application[a]	Microorganism	Immobilization Method	Reference
Production of beer from wort	Yeast	Polyvinyl chloride and porous bricks	Corrieu *et al.* (1976)
Bread making	Bakers' yeast	Alkaline metal salts of alginic acid	Chibata *et al.* (1974a)
BOD sensor	Soil bacteria	Collagen and glutaraldehyde	Karube *et al.* (1976b)
		Polyacrylamide	Karube *et al.* (1977)

[a] Key to abbreviations: 7-ADCA, 7-aminodesacetoxycephalosporanic acid; 6-APA, 6-aminopenicillanic acid; ATP, adenosine triphosphate; BOD, biochemical oxygen demand; L-DOPA, 3,4-dihydroxy-L-phenylalanine; NAD, nicotinamide adenine dinucleotide.

gel method was first used for immobilization of enzymes by Bernfeld and Wan (1963). This technique was later applied to immobilization of lichen cells, *Umbilicaria pustulata,* by Mosbach and Mosbach (1966). Chibata *et al.* (1974e) then investigated the conditions of immobilizing bacterial cells, such as *E. coli,* with aspartase activity, by the same method for industrial use.

In order to most efficiently prepare immobilized *E. coli* cells, the type and the concentration of bifunctional reagents for cross-linking and the concentration of acrylamide monomer were investigated. The following bifunctional reagents were used for lattice formation: *N,N'*-methylene bisacrylamide, *N,N'*-propylene bisacrylamide, diacrylamide dimethyl ether, 1, 2-diacrylamide ethyleneglycol, *N,N'*-diallyl tartardiamide, ethyleneurea bisacrylamide, and 1,3,5-triacryloyl hexahydro-*s*-triazine. The activities of the immobilized *E. coli* obtained were almost the same except for those with ethyleneurea bisacrylamide and 1,3,5-triacryloyl hexahydro-*s*-triazine. *N,N'*-Methylene bisacrylamide was chosen since it was commercially available at low cost.

Further, the concentrations of acrylamide monomer and *N,N'*-methylene bisacrylamide and the amount of cells to be entrapped were investigated. As a result, the optimum conditions were established for immobilization of *E. coli* cells.

Escherichia coli cells (10 gm, wet weight) collected from cultured broth were suspended in 40 ml of physiological saline. To this suspension were added 7.5 gm of acrylamide monomer and 0.4 gm of *N,N'*-methylene bisacrylamide at 8°C, and the total was mixed. To the mixture, 5 ml of 5% β-dimethylaminopropionitrile, an accelerator of polymerization, and 5 ml of 2.5% potassium persulfate, an initiator of polymerization, were added. The reaction mixture was allowed to stand at below 40°C for 10–15 minutes to give a stiff gel. The resultant stiff gel was granulated to a suitable size and was thoroughly washed with physiological saline.

The authors succeeded in immobilizing many kinds of microbial cells using this polyacrylamide gel method. The authors also found (Tosa *et al.*, 1977) that carrageenan, a polysaccharide widely used as a food additive, is a more advantageous matrix for immobilization of microbial cells because the procedure is easy and can be carried out under relatively mild conditions. Consequently, the obtained immobilized cells have higher activity than those prepared by the polyacrylamide gel method and, moreover, the immobilized cells are more stable. Microbial cells are suspended in physiological saline and carrageenan is dissolved in the same saline. After mixing both solutions, gelation occurs easily and rapidly by cooling or contacting with an aqueous solution containing cations, such as K^+, NH_4^+, Ca^{2+}, Cu^{2+}, Mg^{2+}, Mn^{2+}, Fe^{3+}, or amines. For example, 5 gm of whole microbial cells are suspended in 5 ml of physiological saline at 45°–50°C, and 1.7 gm of carrageenan are dissolved in 34 ml of the same saline at 45°–60°C. Both are mixed, and the mixture is cooled at around 10°C. In order to increase the gel strength, the resultant gel is soaked in cold 0.3 *M* potassium chloride solution. After this treatment, the stiff gel which is obtained is granulated to a suitable particle size. If the operational stability of immobilized cells is not satisfied, the immobilized cells are treated with hardening agents such as tannin, glutaraldehyde, or glutaraldehyde and hexamethylenediamine. As a result, more stable immobilized cells can be obtained.

By this method, the authors successfully immobilized whole cells of *E. coli* having aspartase activity, *Brevibacterium flavum* having fumarase activity and *Streptomyces phaeochromogenes* having glucose isomerase activity, respectively. As shown in Table II, the enzyme activities of these immobilized cells were higher than those of the preparations obtained by the polyacrylamide gel method. Moreover, these immobilized cells were stable, and a column packed with them could be used for continuous reaction for long periods. Operational stabilities of immobilized *E. coli* and immobilized *S. phaeochromogenes* were markedly increased by hardening with glutaraldehyde and hexamethylenediamine.

TABLE II. Comparison of Enzyme Activities and Stabilities of Immobilized Microbial Cells Prepared by Carrageenan and Polyacrylamide Gel Methods

		Enzyme activity (μmole/hour/gm cells)		Half-life (days)	
Microorganism	Enzyme	Polyacrylamide	Carrageenan	Polyacrylamide	Carrageenan
Eschericha coli	Aspartase	19,000 (29.2%)[b]	30,400 (46.8%)	120	50
			21,400[a] (32.9%)[b]		686[a]
Streptomyces phaeochromogenes	Glucose isomerase	3,880 (49.1%)[b]	4,280 (54.1%)	53	53
			4,310[a] (54.5%)[b]		289[a]
Brevibacterium flavum	Fumarase	6,680 (34.0%)[b]	9,920 (50.5%)	72	70
			9,120[a] (46.4%)[b]		120[a]

[a] Value after hardening with glutaraldehyde and hexamethylenediamine.
[b] Yield of activity after immobilization or after hardening shown in parentheses.

The carrageenan method is applicable for immobilization of many kinds of microbial cells and is considered to be more advantageous for industrial purposes than the polyacrylamide gel method.

Immobilization of microbial cells by collagen was carried out using the same technique for immobilization of enzymes developed by Vieth *et al.* (1973). This method is suitable for preparation of a membrane type of immobilized cells. After immobilization, various hardening treatments are carried out, since, in this case, the strength of collagen membrane is weak and enzymes easily leak out from the matrix.

Immobilization of microbial cells into cellulose fiber was developed by Dinelli (1972), and the enzyme activity of the immobilized cells prepared by this method is said to be very stable.

Besides entrapping methods described above, agar (Miyata and Kikuchi, 1976), alginate, and polystyrene (Hackel *et al.*, 1975) have been reported as matrices for immobilization of microbial cells. However, the detailed conditions for immobilization of microbial cells and the characteristics of resulting immobilized cells have not been clarified.

More recently, immobilization of cells by liquid membrane has been reported (Mohan and Li, 1975). In this method, a buffer solution containing cells is dispersed and emulsified in an oil phase composed of surfactants, various additives, and hydrocarbon solvent to form encapsulated cells with a liquid membrane.

D. Miscellaneous Immobilization Methods

Fixation of enzyme within microbial cells has been carried out by heat treatment at a temperature which does not inactivate the desired enzyme. This method, however, does not always result in immobilization of microbial cells (Takasaki and Kanbayashi, 1969). When whole cells of *Streptomyces* sp. having glucose isomerase activity were heated at 60°–85°C for 10 minutes, the enzyme was fixed inside the cells and did not leak out from the cells when the cells were incubated under the conditions for enzyme reaction. These immobilized cells were used for industrial production of high-fructose syrup from D-glucose. Glucose isomerase was also fixed inside of *S. phaeochromogenes* cells by β-ray irradiation (Tsumura, 1969).

A method for immobilization of *S. phaeochromogenes* cells by complex formation with chitosan was reported (Tsumura *et al.*, 1976). When the cells were added to chitosan solution dissolved in acetic acid and the mixture adjusted to pH 6.0, the complex of cells and chitosan was formed and precipitated by aggregation.

III. CHARACTERISTICS OF IMMOBILIZED MICROBIAL CELLS

Some enzymatic properties of immobilized microbial cells have been investigated, as in the case of immobilized enzymes, and have been compared with those of intact cells or native enzymes.

A. Optimum pH

A number of papers investigating the effect of pH on enzyme reaction by immobilized microbial cells have been published, and their optimum pHs have been compared with those of intact cells or native enzymes.

Among 15 cases listed in Table III, the shift of optimum pH was observed in 5 cases, and 10 cases did not show the shift of optimum pH.

In *E. coli* and *A. agile,* immobilized using Dowex-1 (C1-form) (Hattori and Furusaka, 1960, 1961) or *B. ammoniagenes* immobilized by entrapping with polyacrylamide gel (Shimizu *et al.*, 1975), the optimum pH for oxidation of glucose or succinate or for synthesis of coenzyme A shifted 1–2 pH units to the alkaline range in comparison with those of intact cells. Further, the optimum pH of *β*-galactosidase activity of *E. coli* immobilized by entrapping with polyacrylamide gel also shifted 0.5 pH units to the alkaline side in comparison with that of native enzyme (Ohmiya *et al.*, 1977). The shift of optimum pH to the acid range was observed in aspartase activity of *E. coli* by the same entrapping method (Chibata *et al.*, 1974e).

B. Optimum Temperature

Effects of temperature on enzyme reaction using immobilized microbial cells have been investigated, and the optimum temperature of immobilized cells has been compared with that of intact cells or native enzymes.

As shown in Table III, the optimum temperature of the immobilized microbial cells was similar to that of intact cells or native enzymes except for L-arginine deiminase activity of *Pseudomonas putida* immobilized by entrapping with polyacrylamide gel (Yamamoto *et al.*, 1974a).

C. Stability

The most important factor in the large-scale use of immobilized microbial cells is their stability. A number of papers on thermal stability of immobilized cells have been published as in the case of immobilized enzymes. In many cases so far reported, thermal stability of microbial

TABLE III. Comparison of Optimum pH and Temperature of Intact and Immobilized Microbial Cells on Enzyme Reaction

Microorganism	Enzyme	Optimum pH		Optimum temp. (°C)		Reference
		Intact	Immobilized	Intact	Immobilized	
Achromobacter aceris	NAD kinase	7.0	7.0	45	45	Uchida *et al.* (1977)
Achromobacter liquidum	L-Histidine ammonia-lyase	9.0[a]	9.0[a]	60[a]	60[a]	Yamamoto *et al.* (1974b)
Aspergillus ochraceus	Aminoacylase	[8.0][b]	8.0	—	—	Hirano *et al.* (1977)
Azotobacter agile	Oxidation of glucose or succinate	6.0	7.0–8.0	—	—	Hattori and Furusaka (1961)
Brevibacterium ammoniagenes	Fumarase	[7.0–7.5][b]	7.0–7.5	[60][b]	60	Yamamoto *et al.* (1976)
	Synthesis of coenzyme A	6.5	7.5	—	—	Shimizu *et al.* (1975)
Escherichia coli	Oxidation of glucose or succinate	6.0–7.0	7.0–8.0	—	—	Hattori and Furusaka (1960)
	Aspartase	10.5	8.5	50	50	Chibata *et al.* (1974e)
	β-Galactosidase	[7.5][b]	8.0	[55][b]	55	Ohmiya *et al.* (1977)
	Penicillin amidase	8.5	8.5	40	40	Sato *et al.* (1976)
Gluconobacter melanogenus	L-Sorbose dehydrogenase	6.3	6.3	50	50	Martin and Perlman (1976a)
Kluyveromyces lactis	β-Galactosidase	[6.3][b]	6.3	[37][b]	37	Ohmiya *et al.* (1977)
Lactobacillus bulgaricus	β-Galactosidase	[5.5][b]	5.5	[55][b]	55	Ohmiya *et al.* (1977)
Pseudomonas putida	L-Arginine deiminase	5.5–6.0[a]	5.5–6.0[a]	37[a]	55[a]	Yamamoto *et al.* (1974a)
Rhodospirillum rubrum chromatophores	Photophosphorylating activity	7.5–8.0	7.5–8.0	43	43	Yang *et al.* (1976)

[a] In the presence of cetylpyridinium chloride.
[b] Data for native enzyme in brackets.

cells was increased by immobilization. A significant reduction in thermal stability was observed only for L-sorbose dehydrogenase of *Gluconobacter melanogenus* immobilized by entrapping with polyacrylamide gel (Martin and Perlman, 1976a). When immobilized *G. melanogenus* cells were incubated in saline for 30 minutes at temperatures between 45° and 60°C, the immobilized cells lost activity more rapidly than the intact cells. However, when L-sorbose was added during this heat treatment, the activity of immobilized cells remained stable. At 50° and 55°C an increase in activity was observed.

In the presence of certain bivalent cations, several immobilized cells showed improved resistance to thermal inactivation. Chibata *et al.* (1974e) revealed that bivalent metal ions, such as Ba^{2+}, Ca^{2+}, Mg^{2+}, Mn^{2+}, and Sr^{2+}, helped to prevent thermal inactivation of aspartase activity of immobilized *E. coli* cells. It was also found that these metal ions enhanced operational stability of aspartase activity of immobilized *E. coli* cells during the continuous enzyme reaction.

Besides metal ions, it has been reported that the addition of antibiotics, such as neomycin, ampicillin, chloramphenicol, and tetracycline, increased the stability of L-sorbose dehydrogenase activity of immobilized *G. melanogenus* cells (Martin and Perlman, 1976a).

D. Activation of Immobilized Microbial Cells

It has been observed that the enzyme activities of immobilized microbial cells increase when autolysis of the cells within matrices occurs or when cells are treated with surfactants.

The authors (Chibata *et al.*, 1974e) at first found that when the immobilized *E. coli* cells prepared by the polyacrylamide gel method were incubated at 37°C for 24–48 hours in a substrate solution, i.e., 1 *M* ammonium fumarate solution (pH 8.5) containing 1m*M* Mg^{2+}, the aspartase activity increased nine to ten times. This phenomenon was considered to be either an adaptive formation of aspartase–protein in the presence of substrate or an increase of membrane permeability for substrate and/or product due to autolysis of *E. coli* cells in the gel lattice.

In order to investigate the adaptive formation of the enzyme, immobilized cells were incubated in 1 *M* ammonium fumarate solution for 48 hours at 37°C in the absence or presence of chloramphenicol, an inhibitor of protein synthesis, at concentrations that completely inhibited protein synthesis. As a result, the enzyme activity increased, even in the presence of chloramphenicol. Therefore, this activation was not considered to be the result of protein synthesis of *E. coli* cells in the gel lattice. This was also confirmed by electron micrographs of immobilized *E. coli* cells. Electron micrographs taken after activation indicated that lysis of cells had occurred. Martin and Perlman (1976a) also reported lysis of immobilized *G. melanogenus* leading to an increase of L-sorbose dehydrogenase activity during early stages of incubation. Whole cells of *Streptococcus faecalis* immobilized by entrapping with polyacrylamide catalyze the degradative conversion of L-arginine to putrescine, and the degradative activity of the immobilized cells was reported to be increased by treating with lysozyme (Franks, 1971).

Enzyme activities of immobilized cells were increased by treatment with surfactants. Yamamoto *et al.* (1976) revealed that fumarase activity of *B. ammoniagenes* immobilized by entrapping with polyacrylamide gel was markedly increased by treating with surfactants. These immobilized cells form L-malic acid as a main product with succinic acid as a by-product. Separation of succinic acid from L-malic acid is very difficult. The formation of succinic acid was found to be suppressed by the treatment with surfactants, such as deoxycholate, bile acid, and bile extract. When immobilized *B. ammoniagenes* cells prepared by polyacrylamide gel method were treated with sodium laurylsulfate, the formation of coenzyme A from pantothenic acid, cysteine, and ATP was also reported to be increased (Shimizu *et al.*, 1975).

Without the treatments described above, Yamamoto *et al.* (1974a) revealed that, in some cases, the permeability of substrate and/or product through the cell membrane is increased by immobilization of microbial cells with polyacrylamide gel. That is, formation of L-citrulline from L-arginine by L-arginine deiminase activity is rarely observed in the case of intact cells of *P. putida* in the absence of surfactant (cetyltrimethylammonium bromide), while the formation occurs without the surfactant by immobilized cells. This phenomenon was also observed in formation of urocanic acid from L-histidine by using intact and immobilized cells of *Achromobacter liquidum* having L-histidine ammonia-lyase activity (Yamamoto *et al.*, 1974b).

IV. PRODUCTION OF CHEMICALS BY IMMOBILIZED MICROBIAL CELLS

Studies so far reported on chemical processes using immobilized microbial cells are listed in Table I. Enzyme reactions are classified according to enzyme reactions recommended in 1972 by the enzyme commission of the International Union of Pure and Applied Chemistry and the International Union of Biochemistry.

In this section we will describe utilization of immobilized microbial cells for the production of useful chemicals.

A. Oxidoreductase Reaction

As shown in Table I, two examples have been reported on immobilization of microbial cells having oxidoreductase activity and its application for the formation of organic compounds.

For example, whole cells of *G. melanogenus* having L-sorbose dehydrogenase were immobilized by the entrapping method using polyacrylamide gel, and the immobilized cells were used for preparation of L-sorbosone from L-sorbose (Martin and Perlman, 1976a). L-Sorbosone is an intermediate for the synthesis of vitamin C. In this study, a very interesting phenomenon was found; namely, addition of electron acceptors, such as phenazine methosulfate or methylene blue, or addition of antibiotics, such as neomycin, ampicillin, chloramphenicol, or tetracycline, increases the stability of the enzyme activity.

Immobilized cells of *Curvularia lunata* entrapped in polyacrylamide gel are used for production of cortisol by 11β-hydroxylation of Reichstein compound S (Mosbach and Larsson, 1970). The immobilized cells lose part of the 11β-hydroxylase activity during storage, but the lost activity is reactivated by incubating the immobilized cells in a nutrient solution containing cortisone for 16 hours. The reactivation of activity was considered to be probably due to growth of the cells and/or enzyme formation by induction with cortisone during the incubation.

B. Transferase Reaction

As shown in Table I, two papers on the use of immobilized microbial cells for transferase reaction have been published. One is on production of NADP from NAD and ATP by the action of NAD kinase (Chibata *et al.*, 1975b; Uchida *et al.*, 1977). That is, *Achromobacter aceris* having higher NAD kinase activity was immobilized into a polyacrylamide gel lattice, and the immobilized cells were used for continuous production of NADP. Although this microorganism had ATP degradation activity, acid treatment of the immobilized cells at pH 4.0 completely removed the undesired activity and at the same time activated the desired NAD kinase. When the acid-treated immobilized cells were packed into a column and the substrate solution (pH 7.0) containing 2.7 m*M* NAD, 2.7 m*M* ATP, and 3.0 m*M* $MnCl_2$ was passed through the column at a flow rate of space velocity (SV, $hour^{-1}$) 0.1 at 37°C, 76% of the NAD in the substrate solution was converted to NADP. The enzyme activity of the immobilized cells was more stable in comparison with that of intact cells. Their half-lives were calculated to be around 20 days for immobilized cells and 6 days for intact cells at 37°C.

Continuous production of glucose 6-phosphate and glucose 1-phosphate from glucose and *p*-nitrophenyl phosphate by the action of acid phosphatase was achieved by Saif *et al.* (1975). The cells of *Escherichia freundii* having acid phosphatase activity were immobilized by the entrapping method using polyacrylamide gel. When the immobilized cells were packed into a column and the mixture of equal volumes of 100 m*M* glucose and 25 m*M* *p*-nitrophenyl phosphate was passed through the column at a flow rate of SV = 0.128 at 37°C, glucose 6-phosphate and glucose 1-phosphate were isolated from the effluent in high yield. The enzyme activity of the immobilized cells was very stable, and the column retained 50% of the initial activity after 120 days of operation.

C. Hydrolase Reaction

There are many examples of the application of immobilized microbial cells for formation of useful organic compounds by hydrolase reaction. Yamamoto *et al.* (1974a) studied continuous production of L-citrulline, which is used medicinally, from L-arginine by using immobilized *P. putida* having high L-arginine deiminase activity. The cells, after entrapment in a polyacrylamide gel lattice, exhibited 56% of the arginine deiminase activity of that of intact cells. In the case of intact cells, the addition of surfactant, such as cetyltrimethylammonium bromide, was essential for a high rate of L-citrulline formation, but in the case of immobilized cells, the surfactant had no effect on L-citrulline formation.

Recently, many studies have been carried out on the continuous production of 6-aminopenicillanic acid (6-APA) from penicillin by using a column packed with immobilized penicillin amidase. 6-APA is used as an intermediate in the production of synthetic penicillin. Some of the immobilized systems are said to be used in industry in the United States and Europe.

Sato *et al.* (1976) studied an efficient continuous method for the production of 6-APA from penicillin G by using immobilized *E. coli* having high penicillin amidase activity. Whole cells of *E. coli* were immobilized by the entrapping method using polyacrylamide gel. These microbial cells have penicillinase activity, decomposing both penicillin and 6-APA. Specific inactivation of penicillinase activity was very difficult. However, penicillinase activity was much lower than penicillin amidase activity. Therefore, optimum conditions for the continuous production of 6-APA without removing the penicillinase activity were investigated. By choosing suitable flow rate, that is, when 50 m*M* penicillin solution (pH 8.5) was passed through a column packed with the immobilized cells at a flow rate of SV = 0.24 at 40°C, 6-APA was found to be efficiently produced in 78% yield.

Immobilized microbial cells having penicillin amidase activity were also used to synthesize ampicillin from 6-APA and D-phenylglycine methyl ester (Fujii *et al.*, 1973). Whole cells of *Achromobacter* sp. or *Bacillus megaterium* were immobilized by ionic binding to DEAE–cellulose. The immobilized cells were used not only to synthesize penicillins but also to acylate the cephalosporin nucleus. However, commercial development of this technique is retarded by low production yield because of product or substrate inhibition and reversibility of the reaction.

A number of papers have been published on the immobilization of microbial cells having invertase activity and their applications for production of invert sugar. This system will compete with immobilized enzyme system.

For the production of L-amino acids, Tosa *et al.* (1966, 1967) and Chibata *et al.* (1972) succeeded in the industrial application of immobilized aminoacylase for continuous production of L-amino acids from acyl-DL-amino acids. Instead of this system, *Aspergillus ochraceus* having aminoacylase activity was immobilized by cross-linking with egg albumin and glutaraldehyde, and the continuous optical resolution of acetyl-DL-methionine was investigated by using these immobilized cells (Hirano *et al.*, 1977).

Further, *Alginomonas nonfermentas* having *l*-menthol ester hydrolase activity was immobilized by the entrapping method using polyacrylamide gel, and preparation of *l*-menthol from synthetic *dl*-menthol ester was investigated by Nonomura *et al.* (1976).

Besides the processes described above, continuous hydrolysis of ε-aminocaproic acid cyclic dimer contained in waste water from nylon factories was investigated by immobilizing *Achromobacter guttatus* by the entrapping method using polyacrylamide gel (Kinoshita *et al.*, 1975).

In the food industry, immobilized microbial cells having β-galactosidase and raffinase activities were used for hydrolysis of lactose in milk and of raffinose interfering with crystallization of sugar from beet, as shown in Table I. These systems will also compete with immobilized enzyme systems in the future.

D. Lyase Reaction

Lyase is an enzyme cleaving C—C, C—O, C—N, or other bonds by means other than hydrolysis or oxidation. A number of studies of the immobilization of lyases and their applications have been published. Many kinds of microbial cells having lyase activity can be immobilized, as shown in Table I, and the immobilized cells have been used for production of useful organic compounds.

1. Carbon–Carbon Lyase Reaction

In the carbon–carbon lyase reaction, carboxyl groups are eliminated from organic compounds by the action of decarboxylase of immobilized microbial cells.

Chibata *et al.* (1975a) immobilized whole cells of *Pseudomonas dacunhae* having high aspartate 4-decarboxylase activity and investigated the conditions for continuous production of L-alanine from L-aspartic acid by using immobilized cells. In this system, additions of pyridoxal phosphate and Co^{2+} to the substrate solution were effective as a cofactor for enzyme reaction and as an enzyme stabilizer, respectively.

Further, *Microbacterium ammoniaphilum* having diaminopimelic acid decarboxylase activity was immobilized into a polyacrylamide gel lattice, and the immobilized cells were used for production of L-lysine from LL-2,6-diaminopimelic acid (Kanemitsu, 1975). When the immobilized cells (amounts corresponding to 200 mg dry cells) were incubated with 100 ml of 0.2 *M* diaminopimelic acid solution (pH 8.0) at 40°C for 8 hours, 280 mg of L-lysine was obtained from the reaction mixture.

In addition to the decarboxylation reactions described above, *Erwinia herbicola* having β-tyrosinase activity was immobilized by cross-linking with collagen and dialdehyde starch, and the immobilized cells were used for production of L-tyrosine from phenol, pyruvate, and ammonia (Yamada *et al.*, 1975), and production of L-DOPA from pyrocatechol, pyruvate, and ammonia (Kumagai *et al.*, 1976), respectively.

2. Carbon–Oxygen Lyase Reaction

The carbon–oxygen lyase reaction has been studied for use in the production of L-malic acid from fumaric acid (Chibata *et al.*, 1975c; Yamamoto *et al.*, 1976), L-tryptophan from indole and DL-serine (Chibata *et al.*, 1974c) and 5-hydroxy-L-tryptophan from 5-hydroxyindole and DL-serine (Chibata *et al.*, 1974b).

L-Malic acid is used in the pharmaceutical field as an antidote for hyperammoniemia and a component of amino acid infusion. This acid is industrially produced from fumaric acid by the action of fumarase in batch process using cultured broth of microbial cells.

For the continuous production of L-malic acid, Chibata *et al.* (1975c) studied the immobilization of whole cells of *B. ammoniagenes* having high fumarase activity by the polyacrylamide gel method. However, such immobilized cells formed succinic acid as a by-product, the separation of which from L-malic acid is very difficult. Therefore, the successful utilization of this system for industrial production of pure L-malic acid is to prevent succinic acid formation during the enzyme reaction. Various treatments of immobilized cells were carried out, as described in previous sections. The treatment with detergent, such as deoxycholic acid, bile acid, and bile extract, was found to be very effective for suppression of succinic acid formation. These detergents also markedly enhanced the formation of L-malic acid by the immobilized cells. For example, when the immobilized cells were allowed to stand in 1 *M* sodium fumarate (pH 7.5) containing 0.3% bile extract at 37°C for 20 hours, succinic acid formation by the immobilized cells was completely suppressed and its fumarase activity was enhanced at about 10 times in comparison with that of untreated cells.

When 1 *M* sodium fumarate solution (pH 7.0) was passed through the column packed with the bile extract-treated immobilized cells at 37°C at an SV = 0.23, the reaction reached equilibrium with about 80% conversion of fumaric acid to L-malic acid. Fumaric acid was removed by acidification from the effluent and pure L-malic acid was separated in about 70% yield from fumaric acid. The fumarase activity of the immobilized cells was very stable and its half-life was calculated to be 53.5 days at 37°C.

Under the conditions described above, 15.4 metric tons of theoretical yield of L-malic acid could be produced per month, using a 1000-liter column. This production system has been utilized industrially since 1974 by Tanabe Seiyaku Co. Ltd., Japan.

For production of L-tryptophan from indole and DL-serine, the authors immobilized whole cells of *E. coli* having tryptophan synthetase activity into a polyacrylamide gel lattice. When the immobilized cells (30 gm) were incubated with 100 ml of substrate solution containing 1 gm of indole and 2 gm of DL-serine at 30°C for 24 hours, 1.5 gm of L-tryptophan was isolated from the reaction mixture. From a similar reaction mixture containing 1 gm of 5-hydroxyindole in place of indole, 1.1 gm of crystalline 5-hydroxytryptophan was obtained.

3. Carbon–Nitrogen Lyase Reaction

Two studies by the authors on production of L-aspartic acid from fumaric acid and ammonia, and production of urocanic acid from L-histidine are examples of the carbon–nitrogen lyase reaction.

L-Aspartic acid is widely used as a medicine and food additive. The acid had been industrially produced by fermentative and enzymatic methods from fumaric acid and ammonia by the action of aspartase.

Since this reaction had been carried out in batch process, there were some disadvantages for industrial purposes. To overcome the disadvantages, Tosa *et al.* (1973) investigated the immobilization of aspartase. The active immobilized aspartase was obtained by entrapment of extracted aspartase in a polyacrylamide gel lattice. However, this immobilized enzyme was relatively unstable; i.e., its half-life was 27 days at 37°C. Therefore, this immobilized aspartase was not considered satisfactory for the industrial production of L-aspartic acid. The authors thus considered that if the microbial cells could be immobilized directly, these disadvantages might be overcome. The immobilization of whole microbial cells was studied from these points of view, and this technique was incorporated into industrial use in 1973 (Chibata *et al.*, 1973, 1974e; Tosa *et al.*, 1974; Sato *et al.*, 1975). This was considered the first industrial application of immobilized microbial cells in the world.

Whole cells of *E. coli* having high aspartase activity were immobilized into a polyacrylamide gel lattice, and the immobilized cells were activated by soaking in 1 *M* ammonium fumarate (pH 8.5) containing 1 m*M* Mg^{2+} at 37°C for 24–48 hours as described in Section III, D.

By using a column packed with the activated immobilized *E. coli* cells, the conditions for the continuous production of L-aspartic acid were investigated. When an aqueous solution of 1 *M* ammonium fumarate (pH 8.5) containing 1 m*M* Mg^{2+} was passed through the column at a flow rate of SV = 0.6 at 37°C, the reaction proceeded to completion. The effluent, 2400 liters, was adjusted to pH 2.8 with 60% H_2SO_4 at 90°C and then cooled at 15°C for 2 hours. The L-aspartic acid that crystallized out was collected by centrifugation and washed with water. This product was pure without the need for recrystallization, and the yield was 304.8 kg (98% of theoretical). The immobilized cell column was very stable and its half-life was estimated to be 120 days at 37°C. The overall production cost of this system was reduced to about 60% of the conventional batch process using intact cells due to marked reduction of the cost of preparing the catalysts and to reduction of labor cost by automation.

Whole cells of *Achromobacter liquidum* having higher L-histidine ammonia-lyase activity were immobilized into a polyacrylamide gel lattice, and the immobilized cells were used for production of urocanic acid from L-histidine (Yamamoto *et al.*, 1974b). Urocanic acid is used as a sun-screening agent in the pharmaceutical and cosmetic fields. Although the microorganism has urocanase activity which decomposes urocanic acid to imidazolone propionic acid, this activity was removed by simple heat treatment (70°C, 30 minutes) of the cells before immobilization.

By using a column packed with the immobilized heat-treated cells, the conditions for the continuous production of urocanic acid were investigated. When an aqueous solution of 0.25 *M* L-histidine (pH 9.0) containing 1 m*M* Mg^{2+} was passed through the column at flow rates of SV below 0.06, L-histidine was completely converted to urocanic acid. Pure urocanic acid was crystallized in a high yield from the effluent by merely adjusting the pH of the effluent to 4.7. The enzyme activity of the column was very stable in the presence of Mg^{2+}, and its half-life was estimated to be about 180 days at 37°C.

E. Isomerase Reaction

As shown in Table I, many papers have been published on the isomerization of D-glucose to D-fructose by immobilized microbial cells having glucose isomerase activity.

Takasaki and Kanbayashi (1969) investigated the use of heat-treated *Streptomyces* sp. having higher glucose isomerase activity for conversion of D-glucose to D-fructose. Into a column packed with the heat-treated cells, 40% glucose solution (pH 8.0) containing 5 m*M* Mg^{2+} and 1 m*M* Co^{2+} was fed by the method of upward flow at 70°C. The continuous reaction could be carried out for 15 days at the average isomerization rate of 40%. However, there are a number of problems yet to be solved for industrial application of this column technique. The industrial production of fructose from glucose is presently accomplished by a batch reaction process, reusing the heat-treated cells two or three times. The preparation of improved stabilized immobilized cells and the development of a continuous reactor system suitable for industrial uses are expected in the future.

Vieth *et al.* (1973) immobilized *S. phaeochromogenes* cells having glucose isomerase activity by the collagen membrane method after heat treatment of the cells. The cell–collagen–membrane was tanned by dipping in alkaline formaldehyde or glutaraldehyde solution, and the tanned cell–collagen–membrane was cut into small chips. When 1 *M* D-glucose solution (pH 7.0) containing 10 m*M* Mg^{2+} was passed through a column packed with the chips at flow rate of 11–12 ml/hour at 70°C for 40 days, the continuous reaction could be carried out at the isomerization rate of 40% for 15 days. After that the activity of the column gradually decreased.

The authors (Chibata *et al.*, 1974d) immobilized whole cells of *Streptomyces griseus* having glucose isomerase activity into a polyacrylamide gel lattice and investigated the use of this preparation for continuous conversion of D-glucose to D-fructose. In this system, Mg^{2+} was necessary for the activation of enzyme reaction and Co^{2+} was useful for the stabilization of enzyme activity.

The industrial isomerization of D-glucose to D-fructose by immobilized glucose isomerase has been carried out in Japan, the United States, and in Europe, and this immobilized microbial cell system is expected to compete with the immobilized enzyme system.

F. Multienzyme Reaction

Applications for enzyme reactions using immobilized microbial cells are carried out mainly by the action of a single enzyme. However, many useful chemicals can be produced by multistep reactions with the action of several kinds of enzymes, especially in fermentative methods. The immobilization of microbial cells for multienzyme reactions has been attempted, as shown in Table I.

Slowinski and Charm (1973) immobilized whole cells of *Corynebacterium glutamicum* (a glutamic acid-producing bacteria) by the entrapping method using polyacrylamide gel and investigated the formation of L-glutamic acid from glucose by using these cells. The immobilized *C. glutamicum* cells corresponding to 2.8 mg (dry weight) of the cells were suspended in 40 ml of fermentation medium containing glucose as the carbon source, inorganic ammonium salts as the nitrogen source, and several kinds of metal salts. The suspension was incubated at 30°C under stirring. After 144 hours of processing, 15 mg/ml of L-glutamic acid were accumulated in the reaction mixture. In this case, productivity of L-glutamic acid by immobilized cells was said to be superior in comparison with that of intact cells. From this work they estimated that the immobilized cells probably can be used in a column with continuous processing.

In addition, continuous production of coenzyme A from pantothenic acid, L-cysteine, and ATP was investigated by immobilized *B. ammoniagenes* cells (Shimizu *et al.*, 1975). The reaction consisted of the following five steps: panthothenic acid → phosphopantothenic acid → phosphopanthothenoylcysteine → phosphopantetheine → dephosphocoenzyme A → coenzyme A. When a substrate solution (pH 6.5) containing sodium pantothenate, cysteine, ATP, $MgSO_4$, and potassium phosphate was passed through a column (0.9 × 30 cm) packed with the immobilized cells at a flow rate of SV = 0.1–0.2 at 37°C, 500 μg/ml of coenzyme A were produced in the effluent. The half-life of the immobilized cell column was 5 days at 37°C.

Recently whole cells of yeast bound to fragments of polyvinyl chloride and porous bricks were used for continuous production of beer from wort (Corrieu *et al.*, 1976). The immobilization of yeast cells was carried out by directly circulating cultured broth into a column packed with polyvinyl chloride and porous bricks. For continuous production of beer, two immobilized cell columns were prepared. One column was used to change wort to beer and another column was used for the maturation of beer. The quality of the beer produced is said to be no different from ordinary beer.

In addition, more recently continuous ethanol production from glucose by *Saccharomyces cerevisiae* cells immobilized in calcium alginate gel was reported (Kierstan and Bucke, 1977). When a solution of 10% (w/v) glucose containing 0.05 *M* $CaCl_2$ was passed through a column (0.9 × 8 cm) packed with the immobilized cells at a flow rate of 50 ml/24 hr, the maximum efficiency of glucose to ethanol conversion was 90% of the theoretical maximum yield. This system has a half-life of about 10 days and has been used for the efficient continuous production of ethanol over a total period of 24 days.

As described above, the examples pose the possibilities that fermentative methods involving multienzyme reactions may be replaced by the continuous enzyme reaction using a column packed with immobilized microbial cells. However, there are a number of difficulties to be overcome, especially the supply of air or oxygen into the column or reactor. If these difficulties can be overcome and energy regeneration can be efficiently carried out, immobilized microbial cell systems are expected to become advantageous bioreactors or catalysts for industrial production of many useful chemicals in the future.

V. CONCLUSION

As described above, many kinds of microbial cells having an enzyme of higher activity can be easily immobilized. Continuous enzyme processing by the immobilized microbial cells will be advantageous in the following cases: (1) when the enzymes are intracellular, (2) when the enzymes extracted from microbial cells are unstable, (3) when the enzymes are unstable during and after immobilization, (4) when the microorganism contains no other enzymes that catalyze interfering side reactions or when those interfering enzymes can be readily inactivated or removed, and (5) when the substrates and products are not high molecular compounds.

Another aspect to be considered is the volume of liquid to be produced. For the unit production of a desired compound, the volume of fermentation broth is much smaller in the case of the continuous method using immobilized cells than in the case of conventional batch fermentative methods. The indication is that the former method is very advantageous from the point of view of water pollution control in plants. Moreover, in the case of the continuous reaction using immobilized cells, it requires less space—mass production becomes possible by using much smaller enzyme reactors than conventional batch fermenters.

In addition, taking into consideration the problems concerning new treatment of fermentative waste for prevention of water pollution, cost of energy, and others, this technique is very promising and is expected to develop along with immobilized enzyme systems in the future.

REFERENCES

Bernfeld, P., and Wan, J. (1963). *Science* **142,** 678–679.

Brownstein, A. M., Vieth, W. R., and Constantinides, A. (1974). *168th Meet., Am. Chem. Soc. Meet.*

Chibata, I., Tosa, T., Sato, T., Mori, T., and Matuo, Y. (1972). *Ferment. Technol. Today, Proc. Int. Ferment. Tec. Symp., 4th, 1972* pp. 383–389.

Chibata, I., Tosa, T., Sato, T., Mori, T., and Yamamoto, K. (1973). *Enzyme Eng.* **2,** 303–313.

Chibata, I., Tosa, T., and Mori, T. (1974a). Japanese Patent Kokai 74/30,582.

Chibata, I., Kakimoto, T., and Nabe, K. (1974b). Japanese Patent Kokai 74/81,590.

Chibata, I., Kakimoto, T., and Nabe, K. (1974c). Japanese Patent Kokai 74/81,591.

Chibata, I., Tosa, T., and Sato, T. (1974d). Japanese Patent Kokai 74/132,290.

Chibata, I., Tosa, T., and Sato, T. (1974e). *Appl. Microbiol.* **27,** 878–885.

Chibata, I., Tosa, T., Sato, T., and Yamamoto, K. (1975a). Japanese Patent Kokai 75/100,289.

Chibata, I., Kato, J., Watanabe, T., and Uchida, T. (1975b). Japanese Patent Kokai 75/135,290.

Chibata, I., Tosa, T., and Yamamoto, K. (1975c). *Enzyme Eng.* **3,** 463–468.

Corrieu, G., Blachere, H., Ramirez, A., Navarro, J. M., Durand, G., Toulouse, I. N. S. A., Duteurtre, B., and Moll, M. (1976). *Proc. Int. Ferment. Symp. 5th, 1976* p. 294.

Dinelli, D. (1972). *Process Biochem.* **7**(8), 9–12.

Franks, N. E. (1971). *Biochim. Biophys. Acta* **252,** 246–254.

Fujii, T., Hanamitsu, K., Izumi, R., Yamada, T., and Watanabe, T. (1973). Japanese Patent Kokai 73/99,393.

Fujii, T., Matsumoto, K., Shibuya, Y., Hanamitsu, K., Yamaguchi, T., Watanabe, T., and Abe, S. (1974). British Patent 1,347,665.

Hackel, U., Klein, J., Megnet, R., and Wagner, F. (1975). *Eur. J. Appl. Microbiol.* **1,** 291–293.

Hattori, T., and Furusaka, C. (1960). *J. Biochem. (Tokyo)* **48,** 831–837.

Hattori, T., and Furusaka, C. (1961). *J. Biochem. (Tokyo)* **50,** 312–315.

Hirano, K., Karube, I., and Suzuki, S. (1977). *Biotechnol. Bioeng.* **19,** 311–321.

Ishimatsu, Y., Shigesada, S., and Kimura, S. (1976). Japanese Patent Kokai 76/44,688.

Johnson, D. E., and Ciegler, A. (1969). *Arch. Biochem. Biophys.* **130,** 384–388.

Kanemitsu, O. (1975). Japanese Patent Kokai 75/132,181.

Karube, I., Matsunaga, T., Tsuru, S., and Suzuki, S. (1976a). *Biochim. Biophys. Acta* **144,** 338–343.

Karube, I., Mitsuda, S., and Suzuki, S. (1976b). *Annu. Meet. Soc. Ferment. Technol. Jpn, 28th, 1976* p. 127.

Karube, I., Mitsuda, S., Matsunaga, T., and Suzuki, S. (1977). *J. Ferment. Technol.* **55,** 243–248.

Kierstan, M., and Bucke, C. (1977). *Biotechnol. Bioeng.* **19,** 387–397.

Kinoshita, S., Muranaka, M., and Okada, H. (1975). *J. Ferment. Technol.* **53,** 223–229.

Kolarik, M. J., Chen, B. J., Emergy, A. H., Jr., and Lim, H. C. (1974). *In* "Immobilized Enzymes in Food and Microbial Processes" (A.C. Olson and C.L. Cooney, eds.), pp. 71–83. Plenum, New York.

Kosugi, Y., and Suzuki, H. (1973). *J. Ferment. Technol.* **51,** 895–903.
Kumagai, H., Sezima, S., Yamada, H., Hino, T., and Okamura, S. (1976). *Annu. Meet. Agric. Chem. Soc. Jpn., 52nd, 1976* p. 233.
Marconi, W., Bartoli, F., Cerere, F., Galli, G., and Morisi, F. (1975). *Agric. Biol. Chem.* **39,** 277–279.
Martin, C. K. A., and Perlman, D. (1976a). *Biotechnol. Bioeng.* **18,** 217–237.
Martin, C. K. A., and Perlman, D. (1976b). *Eur. J. Appl. Microbiol.* **3,** 91–95.
Miyata, N., and Kikuchi, T. (1976). Japanese Patent Kokai 76/133,484.
Mohan, R. R., and Li, N. N. (1975). *Biotechnol. Bioeng.* **17,** 1137–1156.
Mosbach, K., and Larsson, P. O. (1970). *Biotechnol. Bioeng.* **12,** 19–27.
Mosbach, K., and Mosbach, R. (1966). *Acta Chem. Scand.* **20,** 2807–2810.
Nanba, A., and Matuo, Y. (1970). *Annu. Meet. Agric. Chem. Soc. Jpn, 45th, 1970* p. 251.
Nishimaru, H., Izumi, C., Narita, S., and Yamada, K. (1975). Japanese Patent Kokai 75/140,680.
Nonomura, S., Watanabe, M., and Inagaki, T. (1976). Japanese Patent Kokai 76/48,488.
Novo Industri (1976a). Japanese Patent Kokai 76/51,580.
Novo Industri (1976b). U.S. Patent Application No. 501,292.
Ohmiya, K., Ohashi, H., Kobayashi, T., and Shimizu, S. (1977). *Appl. Environ. Microbiol.* **33,** 137–146.
Saif, S. R., Tani, Y., and Ogata, K. (1975). *J. Ferment. Technol.* **53,** 380–385.
Sato, T., Mori, T., Tosa, T., Chibata, I., Furui, M., Yamashita, K., and Sumi, A. (1975). *Biotechnol. Bioeng.* **17,** 1797–1804.
Sato, T., Tosa, T., and Chibata, I. (1976). *Eur. J. Appl. Microbiol.* **2,** 153–160.
Shimizu, S., Morioka, H., Tani, Y., and Ogata, K. (1975). *J. Ferment. Technol.* **53,** 77–83.
Slowinski, W., and Charm, S. E. (1973). *Biotechnol. Bioeng.* **15,** 973–979.
Takasaki, Y., and Kanbayashi, A. (1969). *Kogyo Gijutsuin Biseibutsu Kogyo Gijutsu Kenkyusho Kenkyu Kokoku* **37,** 31–37; *Chem. Abstr.* **74,** 139,538 (1970).
Toda, K., and Shoda, M. (1975). *Biotechnol. Bioeng.* **17,** 481–497.
Tosa, T., Mori, T., Fuse, N., and Chibata, I. (1966). *Enzymologia* **31,** 214–224.
Tosa, T., Mori, T., Fuse, N., and Chibata, I. (1967). *Biotechnol. Bioeng.* **9,** 603–615.
Tosa, T., Sato, T., Mori, T., Matuo, Y., and Chibata, I. (1973). *Biotechnol. Bioeng.* **15,** 69–84.
Tosa, T., Sato, T., Mori, T., and Chibata, I. (1974). *Appl. Microbiol.* **27,** 886–889.
Tosa, T., Sato, T., Yamamoto, K., Takata, I., Nishida, Y., and Chibata, I. (1977). *Int. Cong. Pure Appl. Chem.* **26,** 267.
Tsumura, N. (1969). *Annu. Meet. Soc. Ferment. Technol., Jpn, 21st, 1969* p. 81.
Tsumura, N., and Kasumi, T. (1977). Japanese Patent Kokai 77/44,285.
Tsumura, N., Kasumi, T., and Ishikawa, M. (1976). *Rep. Natl. Food. Res. Inst. (Tokyo)* **31,** 71–75 (in Japanese).
Uchida, T., Watanabe, T., Kato, J., and Chibata, I. (1978). *Biotechnol. Bioeng.* **20,** 255–266.
Updike, S. J., Harris, D. R., and Shrago, E. (1969). *Nature (London)* **224,** 1122–1123.
Venkatasubramanian, K., Saini, R., and Vieth, W. R. (1974). *J. Ferment. Technol.* **52,** 268–278.
Vieth, W. R., Wang, S. S., and Saini, R. (1973). *Biotechnol. Bioeng.* **15,** 565–569.
Wang, S. S., Vieth, W. R., and Constantinides, A. (1973). *Enzyme Eng.* **2,** 123–129.
Yagi, S., Toda, Y., and Minoda, T. (1976). *Annu. Meet. Agric. Chem. Soc. Jpn, 51st, 1976* p. 414.
Yamada, H., Yamada, M., Nakazawa, E., Kumagai, H., Hino, T., and Okamura, S. (1975). *Annu. Meet. Agric. Chem. Soc. Jpn, 50th, 1975* p. 336.
Yamamoto, K., Sato, T., Tosa, T., and Chibata, I. (1974a). *Biotechnol. Bioeng.* **16,** 1589–1599.

Yamamoto, K., Sato, T., Tosa, T., and Chibata, I. (1974b). *Biotechnol. Bioeng.* **16,** 1601–1610.

Yamamoto, K., Tosa, T., Yamashita, K., and Chibata, I. (1976). *Eur. J. Appl. Microbiol.* **3,** 169–183.

Yamamoto, K., Tosa, T., Yamashita, K., and Chibata, I. (1977). *Biotechnol. Bioeng.* **19,** 1101–1114.

Yang, H. S., Leung, K.-H., and Archer, M. C, (1976). *Biotechnol. Bioeng.* **18,** 1425–1432.

Yuta, S., Bhatt, R. R., Yoshida, T., and Taguchi, K. (1975). *Annu. Meet. Soc. Ferment. Technol., Jpn, 27th, 1975* p. 250.

Chapter 18

Economics of Fermentation Processes

WILLIAM H. BARTHOLOMEW
HAROLD B. REISMAN

I. INTRODUCTION

Fermentation is one of the oldest known chemical processes. Fermented foods and alcoholic beverages have been produced for thousands of years, but it is only in the last three to four decades that

MICROBIAL TECHNOLOGY, 2nd ed., VOL.II
 ISBN 0-12-551502-2

biotechnology and chemical engineering have been applied on a large scale to bring to the world fermentation products such as antibiotics, flavoring agents, amino and organic acids, and vitamins.

Fermentation processes, and hence the economics of fermentation, depend upon specificity of enzymatic reactions to produce complex products too costly to produce by direct synthesis. Relatively high efficiency, coupled with availability and cost of agricultural raw materials, allows use of biological systems to produce needed chemical products. In recent years, production of biocatalysts themselves has become an important industry. Large-scale conversion of glucose to high-fructose corn syrup requires equally large-scale production of the needed enzyme—glucose isomerase.

Prior to commercialization, extensive economic evaluation is necessary. Unique characteristics of fermentation processes, including use of agricultural commodities, high-energy use per unit of product, and need for aseptic operation, require involvement of life scientists, chemical engineers, and other engineering disciplines needed for plant construction and startup. If it is assumed that a totally new facility is to be built, site selection is an early requirement. The usual criteria for chemical processes are used to narrow site selection, but there are certain unique requirements which should be considered in fermentation plant design. Process design follows both conventional engineering needs as well as those derived from unique biochemical inputs from laboratory and pilot plant data. Instrumentation and mechanical design follow, again referenced to certain unique control and cleanliness requirements. It is best to consider immediately from the outset that the plant will most likely fall under jurisdiction of one or another governmental regulatory body since a food or pharmaceutical product may be manufactured. Moreover the increasingly complex environmental and regulatory requirements are often recognized late in the design and construction periods, or worse, they sometimes are recognized in the course of operation, particularly those resulting from increasing environmental pressures. Costly correction or modification may then be necessary.

The capital and gross operating cost of a "conventional" fermentation plant is roughly independent of the product produced. At current prices, as a first approximation for a grass-roots fermentation and finishing plant cost, one can use a figure within the range of \$20 to \$50 per liter of installed fermentation capacity. The value within this range depends upon scale and complexity. The economics of scale-up of capital costs will likely have an exponent of 0.75 for the fermentation section of the plant relative to the 0.6 exponent commonly used for a first approximation in processing plants. The fermentation operating costs per unit volume per unit time will vary somewhat but will hold generally within a reasonably narrow range. That is, barring use of an exotic raw material, the cost of labor, utilities, and materials would not vary greatly if put on a unit volume and unit cycle time basis. Also, processing or extraction can vary in complexity but, by and large, they too will hold to a reasonably narrow cost range. The capital and operating costs of all auxiliary operations for fermentation and extraction per se also exert a fly wheel effect relative to total installation cost and cost of operation.

Thus, the unit cost of the bulk product is much a function of fermentation yield and fermentation cycle time. This is illustrated in Table I, presenting certain yield, price, and sales parameters over the 20-year period 1950–1970 for various products. During this period the price of penicillin dropped to one twenty-fifth of the 1950 value, reflecting the substantial increase in yield through strain selection and mutation as well as sales volume increase without substantial new investment. In the overall period from the early 1940s to today this product has enjoyed approximately a 1000-fold improvement in titer. In products like antibiotics and vitamin B_{12}, the organism converts a substantial part of the carbohydrate to cells and CO_2, with the product synthesized only as an incidental by-product of cell growth. In contrast, for production of such commodities as glutamic, gluconic, and citric acids, the organism converts 60–95% of the carbohydrate to the end product on a weight basis. Process improvement by strain selection, engineering, and medium changes, as with antibiotics, has resulted in rate and yield gain. However, the total genetic improvement such as evidenced in antibiotics has necessarily been limited by the nature of these processes which yield such high conversions from the start.

Monosodium glutamate was not produced in any quantity by fermentation until after 1960. Upon changeover from the natural product extraction, the cost of production was reduced substantially, reflected in a selling price reduction to the range $1.75–$2.00 per kilogram in 1961. Over the ensuing years most of the process improvements and increased volume of operation have barely held the line against inflation. Citric acid as a mature product has not been able to hold this line, despite significant process changes, and it has increased in cost.

In the discussion which follows, general principles in both cost estimation and fermentation design will be reviewed and specific details elaborated. Certain tradeoffs will be reviewed and alternatives given to achieve a specific result. Although both cost estimation and biotechnology have advanced to become more or less rigorous sciences, economics has not yet reached that same level of exactitude. In the realm of fermentation economics, therefore, a certain amount of intuition and perhaps even emotion can do a great deal to reach the optimum least cost design.

TABLE I. Cycle Time, Yield, and Sales Volume Parameters

	Bulk price $/kg		U.S. sales kg/year		Approximate cycle time and yield, 1970	
	1950	1970	1950	1970	Time (days)	gm/liter
Vitamin B_{12}	450,000	8,000	20	1,000	5	0.03–0.08
Penicillin G	580	23	160,000	1,250,000	7	10–15
Monosodium glutamate	4[a]	1	4,300,000	22,000,000	1.5	80–100
Citric acid	0.54	0.76	19,000,000	57,000,000	4–5	120–150

[a] Extracted from natural products; dropped to below $2.00 per kg in 1961 with startup of fermentation processes.

II. PLANT DESIGN

A. Fermentors

Fermentors for submerged, aerobic fermentations are generally stirred, gas–liquid contactors; i.e., a closed cylindrical vessel with a diameter/height ratio between 1:1 and 4:1, having concave heads, an agitator, internal baffles, internal heat-exchange coils, and/or a jacket. Steam sparging, automatic temperature, air flow, pressure, pH, and foam controls, internal bearings, and entry methods are almost always included. The main design criteria concern sterility or, perhaps more exactly, asepsis. Design criteria must ensure that foreign microorganisms cannot enter to disrupt normal transformations within the fermentor. Contamination is due to one or more breakdowns in the following steps or precautions: (1) the vessel itself or charged materials are insufficiently sterilized; (2) feed streams (including defoamer) are not properly sterilized; (3) foreign organisms enter through seals or any entry connecting to nonsterile environment; (4) process upsets or mechanical failures.

Fermentor volume may be any convenient value but is normally limited by construction/transportation constraints and potential loss due to contamination of a large broth volume. In recent years, vessel volume has increased for aerobic processes, and there is a report of over 800-m^3 vessel size. When construction information is given, 230- to 380-m^3 vessels are often noted. Internal finish is very important and the relatively small saving achieved by avoiding internal polishing is usually lost many times over by greater effort on cleaning and contamination loss.

The fermentor shaft seal very often creates sterility problems. A stuffing box, while less costly, allows leakage and creates sanitation problems. In recent years, double mechanical seals have become common. Sufficiently high temperatures must be maintained, using pressurized steam, at or near the seal face to prevent entry of foreign organisms.

As to construction specifics, field construction is usually more expensive and has more problems than fabrication in a vendor's shop. Thus, off-site fabrication is a normal choice where possible. A 4- to 4.5-m diameter tank is the usual limiting range for rail or road transport. At the earlier day design of 50- .to 75-m^3 volume fermentation vessels, this resulted in a configuration ranging 1–1.5:1 height/diameter ratio for liquid volume. As the requirement for larger vessels developed, they were increased in height and in some cases shipped in sections for final field connection. This resulted in vessels on the order of 3 or 4:1 height diameter configuration. Such design calls for multiple agitators and usually the air sparging has been limited to the bottom impeller. Thus a constraint on fermentor field erection caused an evolution to tall, multiple-impeller vessels. Such vertical configuration creates a requirement for higher pressure air compressors which are more costly. In certain instances this has resulted in lengthened vessels designed to lie on their side so lower cost air blowers can be used with multiple agitators and air spargers. This configuration can be satisfactory if the system is not extremely aerobic and has no inherent foaming problem. Surface superificial air velocity will increase rapidly with operating volumes above 50% in such configuration. Certain unique designs have been presented (Heden, 1976).

The tall vertical fermentor can be satisfactory in comparatively lengthy fermentations lasting several days. In such cases, where usually less than 5–10% of the oxygen is removed, the air volume can be maintained at a comparatively low practical level to keep foaming under control and still ensure adequate molal supply of oxygen. Generally one is dealing with a viscous non-Newtonian broth, where the controlling factor is getting dissolved oxygen from the bulk liquid to the cell and not from the gas into the liquid. In this system a relatively greater amount of the energy can be put into mixing and pumping to enhance dissolved oxygen distribution within cell clumps and less into the air introduction and transfer of oxygen to the liquid system.

In a highly aerobic, low-viscosity fermentation, lasting on the order of a day or so, as much as 50% of the oxygen can be removed from comparatively large volumes of air passing through the fermentor. Such a system can require as much energy for supplying the air as is used by the agitator for mixing and pumping. Thus a squat 1–1.5:1 height/diameter ratio is more satisfactory for easier foam control and good oxygen transfer from gas to liquid. One can concentrate the energy at a single sparged impeller to obtain the maximum oxygen transfer into the liquid for the energy expended in balance with the requirements for sufficient turnover to ensure uniformity. If one desires a large vessel for the fermentation, consideration must be given to the added problems and cost of field erection versus the long-term advantages of such a squat fermentor over a tall one.

Another consideration in overall plant design must be the number of fermentors. Assuming requirements are for 750 m^3 of installed capacity, one normally would not erect one 750-m^3 fermentor. Either three 250- or four 190-m^3 vessels would be more rational. Each design will have to consider a balance of support facilities, such as utilities, the holding tank size, and size of the initial separation steps to handle the volume from the whole culture storage vessel within a time constraint. Tailored into the consideration is the optimum size of the operating crew needed to run the fermentor system.

Materials of construction are normally rather limited. In some instances a case may possibly be made on paper for utilization of carbon steel fermentors. However, the greater maintenance cost, shorter life, and greater problems in maintaining sterile operation render such construction completely undesirable for pharmaceutical and food operations in particular. In a choice between 304 and 316 stainless steel, consideration for an even longer life favors 316 stainless steel at least for the coils, since 316 does not pit to the extent of 304. Life of coils can be affected particularly in high chloride water when the fermentor is sterilized empty, concentrating chloride ion in residual water on the inside of the coil.

A basic well-designed fermentor system will survive many process and product changes through the years. Even though the plant may be justified for a specific process and product, it is well for the long-term economics of the plant to provide flexible and enduring fermentation equipment.

Engineers in particular like to think in terms of continuous operation to maximize production rates. Such a system can be practical and economic when one is dealing with a product associated with growth of the organism cells themselves or an intracellular product. Also low sensitivity to contamination, such as with a thermophile or production at a lowered pH, is another criterion. It is impractical to consider anything but batch fermentation when the product is an extracellular metabolite that is sensitive to culture and medium balance. Similarly batch fermentation is preferred when the culture can undergo mutation at extended operating time and is sensitive to contamination. Other design criteria are discussed by Lundell and Laiho (1976), Meskanen *et al.* (1976), and Ovaskainen *et al.* (1976).

B. Utilities

Utilities for a fermentation and extraction process represent a major capital cost and operating expense. Steam is required for sterilization of vessels and medium, maintenance of sterility (steam seals), water removal by evaporation, and drying of product. Fermentation processes are most often highly exothermic and considerable amounts of cooling water are needed. Further, medium and feed streams must be cooled after sterilization. Even with recuperation, there is need for large volumes of cooling water either from wells or produced via towers or mechanical means. Process water needs are also high (media preparation, crystallization, cake washing, dilution, equipment washing). Electricity demand is high not only for fermentor drives but for a profusion of rotating machinery in extraction, such as pumps, vacuum systems, agitators, filter, and dryers. With energy costs rising so rapidly it is essential that cost estimates be updated frequently. Various estimates are given below (Table II) for early 1977 but these values must be used with caution even as a base for escalation.

Agitation–aeration requirements are critical for most fermentation processes and both capital and operating costs are appreciable. An article (Ryu and Oldshue, 1977) discusses impact of changes in agitator design and its appreciable effect on overall costs. Capital cost for mixers and blowers, including installation, is given at approximately $900/kW.

Alternatives for consideration in sterile air delivery will affect initial investment and long-term operating costs. An air system can be designed utilizing high-pressure (about 4.5 atm) oil-free compressors (Bartholomew *et al.*, 1974) with exhaust temperature in excess of 220°C. This procedure, combined with a pressure holding tank to maintain adequate retention time, provides sterile air without the need for intermittent inspection and changeout of packed fibrous or membrane filters that must function in the micron interception range. Impingement filters are susceptible to bypass or breakage which would permit entry of nonsterile air. In general, the hot air system requires lower initial investment and reduced maintenance. The choice of systems will vary with the individual plant. One must balance the extra energy cost of somewhat less efficient compression which results in an adequate sterilization temperature, against continued higher labor and maintenace cost for the

TABLE II. Utilities Operating Costs and Range of Energy Prices

Cooling Water		
Well	0.3–1.5 cents/m³	
Cooling tower	0.7–2.2 cents/m³	
Mechanical refrigeration	20–25 cents/m³	
Steam		
At $2.40/therm condensate recovery cost is approximately $6.59–$9.00/1000 kg		
Electricity		
Range is 25–35 mils/kWh		
	Electricity (mils/kWh)	*No. 2 fuel oil ($/therm)*
Eastern United States	30–35	2.40[a]
Europe	25–30	2.60–2.80

[a] Coal cost is estimated at $1.50–$2/per therm with minimal transportation.

more complex filtration system which has the increased probability of failure—resulting in contamination.

C. Pumps and Piping

Standard pumps and piping may be utilized for utilities and services when nonsterile service is concerned. Since construction material is very often stainless steel or clad steel, pipe connections, bolts, and flanges must be correctly chosen to prevent corrosion or galling. Where sterile service is involved, lines must be arranged for (1) complete drainage, (2) sterilization, and (3) ease of cleaning. Pockets and "dead spots" are to be avoided since solids deposition occurs, heat transfer is affected adversely, and an inoculum source for undesirable organisms is created. Valves must be full-flow and closely coupled; need for cleanliness and temperature service (sterile side) means that expensive valves must be used. Materials of construction are usually stainless steel, Teflon, and high-temperature resistant rubber or silicone materials. Normal piping factors must be increased for appropriate costing. All-welded construction is frequently used with further internal polishing of pipe and fittings. Welds should be tested ultrasonically and radiographically.

Sterile transfers are often made by use of sterile air or steam pressure above the material to be transferred. However, special pumps may be used for a low rate feeding of speciality materials (precursors, defoamers, sterility control agents). Diaphragm pumps, which may be steam sterilized, can be used as can various peristaltic pumps. Design criteria are the same as detailed above.

It is vital to decrease turnaround times when dealing with a short-time fermentation. There is a need for rapid transfer of the completed batch to the harvest tank. Sizing of the bottom outlet and the transfer pump combined with a positive pressure in the vessel must be such that the fermentor can be emptied within 15–20 minutes. This may seem axiomatic but it is surprising how such a simple item can be overlooked in the initial design.

D. Heat Exchange

Heat exchange in the normally exothermic biochemical reactions occurring within a fermentor can occur within the vessel or in external exchangers with broth recycle. If heat exchange occurs within a vessel using coils, a balance (or tradeoff) must be maintained between process result versus heat removal. There are two general configurations shown in the tabulation below:

A Best process result	B Best heat transfer
Vessel wall	Vessel wall
Cooling coils	Baffles
(1.0–1.5 coil diam.	Cooling coils
from wall)	Internal volume
Baffles	
Internal volume	

In the authors' experience, a design with configuration A will yield a superior process result. Coils close to the wall leave less annular space where liquor is "hidden" from the impeller. With highly viscous non-Newtonian broths, as much as twice the power input is required by a vessel with configuration B to reach the optimum process plateau compared to configuration A. However, configuration B will provide overall heat transfer coefficients about 40% greater than the former.

Use of broth cycling and external heat exchange offers some advantages but oxygen deprivation may be a problem as well as added contamination potential. Coupling external heat exchange with external aeration (as in a continuous loop design) has been reported for specific applications.

Sterilization is another critical unit operation in fermentation processes affecting economics. Nature of medium, length of cycle, and process conditions will determine whether batch or continuous sterilization is used.

Some processes require a certain minimum cooking time for best results and some may be nearly independent of sterilization up to an hour or more at 121°C. Batch sterilization is perfectly satisfactory in such instances, particularly if the process runs for several days so an 8–10 hour turnaround is not severely limiting.

Short fermentations, 1–1.5 days, almost demand a continuous sterilization system to shorten the turnaround time to the order of 3 hours to maximize fermentor productivity.

Regardless of fermentation time, high concentration of carbohydrate utilized in conjunction with soluble proteins frequently results in a requirement for short interaction. This can be accomplished batchwise by separate sterilization of components. Such a process leads to a cumbersome system of satellite vessels and may not always be economical. Because of the well-known fact that protein denaturation in cellular material occurs at a high activation energy level compared to usual chemical reactions, the high temperature–short time operation in a continuous unit is the more economic and practical solution for sterilization of systems sensitive to chemical changes. If an economizer heat exchanger is used to heat the raw medium or makeup water while cooling the sterilized medium, the steam requirement for sterilization can be reduced at least 50%. Thus, economics of energy utilization favor continuous sterilization.

The choice of sterilization system can affect the design and cost of accessory equipment on the fermentor. In a continuous system, the fermentor is sterilized separately. A single-speed motor and agitator is satisfactory. The design can be made to utilize full power at the air flow selected since the agitator need never be used except in a flowing air system. Batch sterilization calls for a slightly more expensive alternate. Optimum agitator selection is achieved by use of a two-speed motor or a variable-speed drive. Thus one can sterilize the batch at a lower speed in the ungassed condition without overloading the motor, and the higher speed can be used in the gassed condition to draw maximum power. Response times to changes during cooling are vastly different in batch versus continuous sterilization. The vacuum breaker system to avoid tank collapse must be considerably larger for a tank sterilized empty than for one full of liquid. The initial cost and maintenance of this breaker system is thus greater for the continuous sterilization operation.

The cooling system to remove the heat of fermentation and, in the case of batch sterilization, to cool the batch after sterilization can vary considerably in design and economics. Long-duration fermentations present relatively little problem for heat removal. Although more costly to fabricate, jacketed fermentors or utilization of plate coils as baffles can provide adequate surface area. In a low-temperature fermentation, it may be necessary to provide mechanical cooling if adequate chilled well or lake water is not available. This availability of cool water can have considerable influence on overall economics and be a prime consideration in site selection.

When one is not so fortunate as to have access to cold water and must utilize cooling towers, the approach temperature may not allow too much leeway. In such situations, considerable energy can be utilized for pumping large volumes of cooling water. Sufficient volumes often cannot be introduced without excessive pressure drop in the cooling coils. To accommodate the pressure drop and approach temperature problems, it may be necessary to have multiple segments of coils connected in parallel. This all increases the installation cost and influences the economics. It adds another point bearing on overall selection of site as well as the research approach to the process. As noted before, the microbiological approach adds many constraints not required for a strictly chemical process, but there are some advantages. The engineer can only go so far, but the microbial geneticist can extend the degrees of freedom by organism manipulation. Thus when the preliminary design indicates a real problem, such as stated above for the cooling water, a program to develop the fermentation to operate at a somewhat higher temperature allows a desirable alternate not always available in ordinary chemical processing.

Heat exchangers in fermentation applications can be of any type, but preferred configurations are the spiral type and plate type. In these designs, high turbulence is maintained (there may be solids present in suspension), even heating/cooling is achieved, and ease of cleaning is assured. For reasons of cleanliness and corrosion control, stainless steel is a preferred construction material. Direct steam injection, followed by a retention section, is another convenient mode for continuous sterilization. Sterilization, as well as other design criteria, is well covered by Blakebrough (1967, 1969) and Yamada *et al.* (1972).

E. Instrumentation and Control

Fermentation processes are highly instrumented. Instrumentation costs—initial purchase, installation, and maintenance—are therefore relatively high for such processes. Complexity is also high since a centralized control room is desirable (one for all fermentors; one or more for extraction and purification) and local readouts are usually essential. Costs of micro- and minicomputers and even more complex systems have fallen to levels which permit computer control of many variables. Either individual readout or cyclic sensing of control variables is in use. All systems require careful review for degree of interlocking design (example is agitator power interlocked with positive air flow), number and types of alarms, panel arrangement (local and centralized), type of cabling, wiring design, and need for permanent record. Programmable systems must be considered for certain repetitive or sequential functions, such as batching and sterilization.

Table III lists estimates and ranges for instrumentation costs (early 1977). Electronic systems (as pH) include pneumatic converters. A prepiped and prewired panel would cost an additional $600 to $1000 per running foot.

Process computers have been applied to fermentation plants. Requirements and some applications effort are detailed in a number of reports (Humphrey, 1977; Jefferis, 1977; Nyiri, 1972; Nyiri and Charles, 1977). The advantages of computer control are: (1) increased throughput, more consistent quality; (2) increased operating consistency, reduced contamination losses; (3) decreased material and energy re-

TABLE III. Cost of Instrumentation (Pneumatic Systems)[a]

Instrument	Cost ($)
Controllers	
Air-flow control	1,600–1,700 (includes valve)
Pressure control	1,700–1,850 (includes valve)
pH control	4,000–4,500 (includes valve)
Temperature control (dual range)	2,800–3,000 (valves excluded)
Dissolved O_2 control	3,000 (valve excluded)
Defoamer probe	2,600–3,000(includes valve)
Level (diaphragm)	1,000–1,100 (no valve)
Flow (magmeter)	4,500–5,000 (for 2-inch unit including valve)
Analyzers	
Oxygen analyzer	2,000
Carbon dioxide analyzer	3,500
AutoAnalyzer single channel	25,000
pH meter	3,000
Indicators	
Rpm	300
Alarms	50 per point (plus $25/point per pressure switch)
Utility monitor	50 –100 (plus $400 for transmitter)
Integrator	400

[a] Add $300 for recorder function and $200 for second pen. Installation cost factor: 30–40% (exclude laboratory equipment installation for bench equipment as pH meter and Auto Analyzer) Costs as of early 1977.

quirement, greater ease in process development; (4) decreased manpower requirements; and (5) improved data storage.

Interactive control sequences have given higher yields and significantly improved process control. Costs of such systems (including developmental time) are quite high and a clear understanding of potential benefits must be developed. While new computer systems have been developed for pilot-plant use in fermentation research, plant-scale systems have usually been installed in an existing facility rather than in a grass-roots plant. Computer control concepts have appeared in the patent literature (Wilson, 1975; Wilson *et al.*, 1975; and this volume, Chapter 15).

F. Foam Control

Foam generation is a normal accompaniment to highly aerated aqueous suspensions containing proteins and other surface-active materials. Foam formation results in product loss, lowered aeration capacity (reduced broth density), potential for contamination, and need for lowered working volume. Chemical means of foam control are common but problems of toxicity and reduction in oxygen transfer rate must be overcome or at least controlled. Commonly used defoamers include higher carbon alcohols, fatty acids and derivatives, polyglycols, and silicones. Defoamers may be diluted prior to use and are normally added via sterile overpressure (air or steam) or with a controlled flow pump; addition may be on a timed basis, on demand, or by use of a combination of methods. Alarm systems to prevent overaddition are usually included and add to the cost. Defoamer must be sterilized prior to use. Batch methods are common but circulating aseptic systems (common headers) are also utilized.

Mechanical foam breakers are used in conjunction with chemical systems to reduce cost and associated problems. Most such mechanical systems are internal (paddles, ultrasonic devices, high-speed disk or centrifuge bowl) but some recirculating devices are known.

G. Design Checklist

At some point in the design phase, a completely detailed design and construction checklist is needed. The following list is needed not only for project engineering but for investment analysis as well (Weaver and Bauman, 1973). The design and construction detail will bring together process parameters and site construction information which will permit rational meshing of schedules. All details will receive attention commensurate with the degree of sophistication required by design needs.

Design and Construction Checklist

1. *Process*
   Plant and process streams (on process flow diagram)
   Material and energy balance
   Physical/biochemical data
   Equipment performance
   Operating conditions (average, peak)
   Materials of construction
   Relief/blowdown systems
   Raw material availability
   Raw material impurity levels
   Impurity buildup (recycle)
   Risk analysis
   Relate capacity and sales forecast
   Useful life (major elements)
   Temperature, pressure, pH
   Yield, rate, fermentation cycle time
   Batch turnaround time

Data on specific gravity, viscosity, heat of vaporization
Sterilization requirements
Site availability and requirements
Emergency control plan

2. *Design*
Annual operating factor
Expansion capability
Process flexibility (alternate processes)
Instrumentation—need and complexity
In-process storage, spare parts, warehousing

a. *Buildings*
Floors
Utilities distribution
Pressurization
Type of construction
Sanitation
Fire protection and safety
Corrosion resistance
Access (stairways, platforms, elevators)
Cranes or monorails
Foundations
Ground water and soil
Communications
Regulatory and local codes

b. *Equipment*
Design conditions (including access and safety)
Need for stress relieving
Davits or monorails
Inspection/cleaning openings
Pressure/vacuum design
Internal specifications (baffling, demisters, piping, agitators, etc.)
Skirt/support leg designs
Paint and insulation
Code requirements
Process review
Heat exchangers (code and TEMA class, type of bundle, expansion joints)
Need for special gas purges (or blanketing)
Heat recovery

c. *Compressors, blowers*
Type and specification
Driver
Space for auxiliaries and rotor removal
Local/central control panels
Auxilliary lubrication/seal oil pumps
Relief and safety devices
Filters, separators, knockout drum design
Location of air intake
Ambient condition variation
Degree of overdesign
Mode of air sterilization
Heat recovery

d. Pumps, piping

Type and number
Driver and horsepower
Seal types
Seal fluid requirements
Temperature, pressure, material, schedule
Proper arrangement (access and safety)
Underground piping (fire lines, water, sewer, drains)
Special fabrication (rubber lines, plastic, glass)
Ambient temperatures
Tracing requirements
Off-sites (pipe bridges, supports, buried or overhead with clearances)
Shielding, jacketed lines, extreme conditions

e. Fired heaters

Source, availability, price of primary and secondary fuel sources
Standby fuel storage
Design pressure and temperature
Tube and fitting material
Spare burners, tubes, instrumentation
Space for erection and tube pulling
Waste heat generator
Stack height and need for support
Fireproofing of main support
Need for water pretreatment (fresh or recycle)
Emergency alarms and control plan

f. Electricity and instrumentation

Electrical classification by area
Motor list
Voltage variation
Extremes of moisture, temperature, vibration
Grounding requirements
Emergency power requirements
Type of instrumentation
Redundancy and type of alarms

g. Utilities

Need for new facilities
Physical conditions (flow, pressure, temperature, losses)

Steam	Potable water
Condensate	Utility air
Cooling water	Instrument air
Chilled water	Inert gas
Deionized water	Refrigeration

Define all tie-ins and recycle lines

h. Nonprocess requirements

Climatic and geological conditions
Roads, pavings, sprinklers
Sewer systems (storm, sanitary, chemical)
Walks, parking, rail spurs, fencing
Special buildings
  Change houses for both sexes

Maintenance
Guard house
Office
Control room
Lunch room
Laboratories (include inoculum, in-process, product testing, technical support, and quality control)
Garage
Stores
Utilities
Warehouse with associated equipment
Waste treatment and pollution control
Regulatory requirements and plans
Treatment methods
Need for sludge disposal
Need for effluent holding
Plant emissions and compliance
Plant compliance with OSHA
Dust
Fumes
Vapors
Heat
Noise
Failsafe design
Biological hazards
Laboratory (testing) requirements

j. Tank Farm
Location
Dikes
Truck access
Utility requirements
Venting
Lighting
Loading and unloading
Cleaning/sanitation
Inventory monitoring

k. Loading and unloading
Material receipts and shipments
Frequency of tank truck and tank car movement
Spotting
Number of shifts
Weather protection
Office space
Tie-ins (drainage, fire protection, sanitation, communications)
Process and utility requirements
Need for weighing
Turning clearance
Tank car holding area
Railroad standards and responsibilities
Docks and wharves
Pilings
Dredging
Derricks
Offices
Utilities
Telephone
Gangplanks
Fire fighting
Lighting
Drinking water
Car puller
Special warehouse equipment
Warehousing and loading must fulfill regulatory requirements
Need for retention samples and coding

III. PROCESS DESIGN

The assessment of profitability of a process or product is an on-going exercise. It should start in the research stage when technical and commercial data are limited. Further evaluation is made at intervals when justified by new information. Decisions after such evaluations can range from increased effort to abandonment of the project. Fermentation processes often can be improved vastly by microbial manipulation, more so than usually possible in chemical conversions. Therefore, it behooves the microbial scientist or engineer closely involved with a project to make frequent reassessments for evaluation of alternatives during the course of a project. The evaluation will help identify where the data are imprecise, what is important in terms of profitability, and where to concentrate the effort. It will help also to clarify objectives of the necessary program and reveal the supplementary commercial information and market research requirements. Such process evaluation is frequently complex and difficult. A key factor in industry is profitability regardless of technical achievement. Objective procedures to aid such assessment are based on the return on investment (ROI) as the criterion.

This need for objectivity is paramount. A potentially good project or process can be terminated too soon by pessimistic or overly conservative estimates; in contrast, optimistic estimates based on "pie in the sky" thinking can lead to unwarranted project expenditures and, if continued through to a final plant, can lead to economic disaster.

The accuracy of an estimate and the resultant basis for a project decision are no better than the least reliable data. The probability for commercial success lies in achieving projected costs, the target sales and marketing goals, and in overcoming all the other problems, such as FDA and USDA needs and the other regulatory requirements. As the probability for success for any one of these categories approaches zero, the project has little chance for economic success even if the others approach 100%.

The development of a profitability assessment at the product research and early development stages must necessarily be made without a firm cost estimate prepared from detailed design. Nevertheless, an indication of the construction cost is one of the first requirements. This predesign or order of magnitude estimate is usually prepared by a scientist and/or engineer closely concerned with the process. It can vary widely in dependability and in amount of time spent to develop the necessary data.

Preparation of a reasonable estimate may be impeded by lack of typical cost data for the installation of the process and related auxiliary facilities. Chilton (1949, 1960), Aries and Newton (1955), Bauman

(1964), Guthrie (1969), Pikulik and Diaz (1977), and Zimmerman and Lavine (1962) have published cost estimating data for the process industries. Throughout the past quarter century such journals as *Chemical Engineering* and *Cost Engineering* have had a series of articles updating various facets of equipment and process costing. A particularly valuable summation of the entire procedure for cost and profitability estimation is that of Weaver and Bauman (1973). Their chapter in the "Chemical Engineers Handbook" also contains many references.

Chilton (1949) was one of the first to organize data for a systematic estimation of costs for new plants. It is based on developing ratios to installed major process equipment costs as means for arriving at a method for rapid estimation of construction costs. His equipment cost data are keyed to the *Engineering News-Record* (ENR) Construction Cost Index at 400 to facilitate correction for inflation. Since 1949, this ratio has advanced to about 2600 in 1977! Correcting his equipment cost information by such a ratio could be misleading, but his principles and the procedure for selecting ratio factors as percentages of installed major process equipment plus others expressed as percentages of total physical cost can still be applied. Utilizing this procedure provides a means for rapid and comparatively inexpensive preliminary capital cost estimation.

The introductory phases involve preparation of a flow diagram and the selection of processing steps in recovery operations. A generalized flow diagram is given in Fig. 1 and possible unit operations are given for various processes in Table IV. It should be understood that not all steps

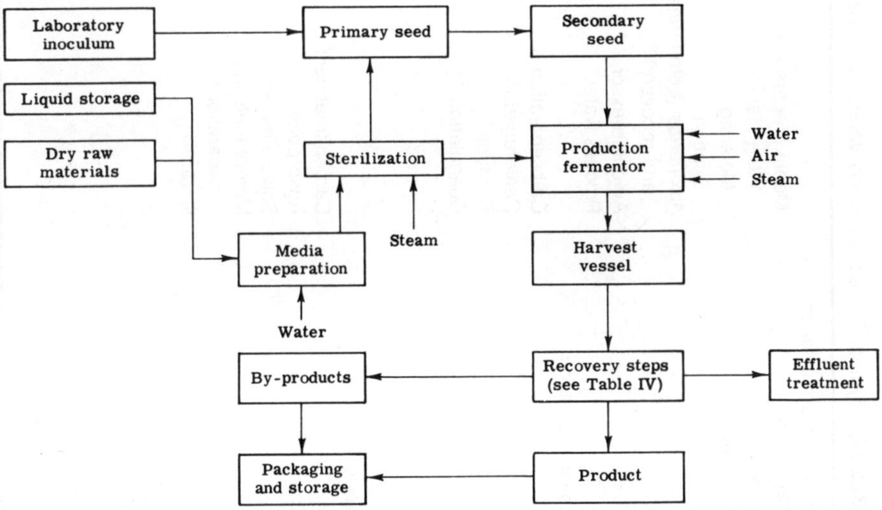

FIGURE 1. Generalized flow sheet.

TABLE IV. Typical Process Showing Various Unit Operations

Submerged fermentation				
Storage tanks	Media preparation	Sterilization	Seed fermentor	Harvest tank
Warehouse	Weigh tanks	Batch	Production fermentor	
(raw materials)	Metering	Continuous	Agitation	
	Dilution	Air sterilization	Aeration	
	Auxiliaries (both fermentation and recovery)			
	Air compressor	Cooling water	Steam	Laboratories
	Process water	Refrigeration	Electricity	
Recovery operation				
	Cell separation	Precipitation	Separation	Reslurry
	Centrifugation	pH/salt adjustment	Ion exchange	Wash
	Filtration	Heat treatment	Rotary filter	Concentration
	Coagulation		Centrifugation	Evaporation
			Solvent extraction	
			Ion exclusion	
			Membrane processes	
Crude separation		*Purification*		
	Carbon treatment	Electrodialysis	Product separation	
	Adsorption	Crystallization	Filter	
	Solvent extraction	Ion exchange	Centrifuge	
	Membrane/osmotic separations			
Drying	*Storage*	*Packaging*	*Warehouse and shipping*	
Freeze				
Fluid bed				
Spray				
Vacuum				
Rotary				
Drum				

need be included nor all possible operations run; selection will be based upon laboratory and pilot-plant data for the individual process.

As an example of considerations in fermentation process design, an outline of information needed is given in Table V. Selection is based upon economics, reaction type, laboratory data, and information developed in the course of scaleup. Once the information is selected and

TABLE V. Process Design Criteria for Fermentation

1.	Substrate (energy, maintenance, conversion)	
	Carbohydrate	Fats and oils
	Molasses	Waste liquors
	Grains	Hydrocarbons
	Spent grains	Starch (and hydrolysates)
2.	Other raw materials	
	Nitrogen sources (ammonia, ammonium salts, urea, proteins)	
	Cations (Fe, Mg, Mn, Zn, Cu, Na, K, Ca, as examples)	
	Anionic constituents (phosphates, for example)	
	Vitamins and cofactors	
	Growth factors (usually from natural products)	
3.	Type of reaction (including rate and yield)	
	Oxidation	Hydrolysis
	Reduction	Hydroxylation
	Decarboxylation	Esterification
	Deamination	Amination
4.	Principal products and by-products	
5.	Organism utilized (including inoculum development)	
	Bacteria	Animal, plant cells
	Mold	Algae
	Actinomycete	Yeast
6.	Agitation–aeration requirements	
	Dissolved oxygen level	Air flow
	Oxygen transfer rate	Power input
	Viscosity effect	Transfer resistances
7.	Special operations	
	Semibatch	Feeding methods
	Continuous	Defoamer type and addition procedure
8.	Special equipment or process specification	
	Membrane	Corrosion (material of construction)
	Recycle	Horizontal fermentor
	Autolysis	Air lift
	Cleaning	Sterilization requirements
9.	Control requirements	
	Temperature	Media concentration
	pH	pCO_2
	Pressure	Redox potential

combined with a flow sheet, plant design and operating cost estimation can begin.

Two detailed and complete reports covering capital and operating cost calculations for numerous fermentation processes have been completed by Fong (1975a,b). Other methods of optimization and cost calculations in fermentation processes can be found in two articles by Richards (1968), in procedure by Whitaker and Walker (1973) for classroom instruction, and in a design report by Schepers (1974).

IV. CASE STUDY—PROJECT EVALUATION

With the preceding information in mind, it is the intention of the authors to develop a project model and to present some data and procedures we have found useful for such preliminary process evaluations. There is no intent to get into the more sophisticated corporate procedures that apply and are necessary as a project develops into the final stages of a capital appropriation request.

A. Selected Equipment Costs

Table VI presents approximate costs for some equipment items frequently used in the fermentation industry. These are derived from recent quotations corrected to current cost indices where necessary and by current contact with some of the manufacturers. It is by no means complete but indicates a sampling of typical equipment. Information of this nature should be kept current by those interested in rapid evaluation techniques. Likely sources are recent quotes, recent company capital appropriation requests, and the corporate engineering department files in addition to the continuing stream of cost update articles in the journals.

B. Production Cost Factors

Table VII contains various approximations for rapid production cost estimation. Percentage ranges of the order indicated have been expressed in many articles, including those referred to above, and in general are confirmed by the authors' experiences.

C. Assumptions and Materials Usage

As an example for our case study, we will assume a scenario that research has discovered, partially developed, and conducted prelimi-

TABLE VI. Approximate Cost of Equipment Frequently Used in the Fermentation Industry

Indices		
Engineering and News-Record (ENR) Construction Cost Index at 2600 (1913 = 100)		
Marshall & Swift Equipment Cost Index at 500 (1926 = 100)		
Agitators and drives		
Standard (304 stainless steel unit with stuffing box)		
Range:	7.5–375 kW	
Scale factor exponent:	0.75	
75 kW unit cost:	$41,000	
Deluxe (316 stainless steel unit with mechanical seal)		
7.5 kW unit cost:	$13,000	
75 kW unit cost:	$58,000	
(Data courtesy of J. A. Mason, Mixing Equipment Company)		
Centrifuges		
Centrifugal desludgers (316 stainless steel)		
100–200 liter/minute unit with accessories:		$125,000
50–100 liter/minute unit with accessories:		$ 75,000
(Rates can vary considerably, depending on feed material properties and clarity required)		
Basket centrifuge (304 stainless steel automated batch)		
76 × 122 cm basket:		$ 75,000 (installed $200,000)

		Per kg of water evaporated per hour
Evaporators (stainless steel)		
Double effect:		$18–$26
Double effect with steam recompression:		$20–$29
Triple effect:		$26–$40
Quadruple effect:		$31–$44
Between 10,000 and 35,000 kg/hour range		
Boiling range:	93°C – 50°C	
Final viscosity:	Up to 200 cP	
(Data courtesy of D. F. Dinnage, APV Company, Inc.)		
Fermentors		
15 m³:	$ 35,000	
113 m³:	$ 95,000	
225 m³:	$165,000	
(304 stainless steel with coils, baffles, 2.75 atm rating, standard openings and connections)		
Holding tank (304 stainless steel)		
190-m³ silo:	$37,000	
Installed with pvf, insts. on pad	$60,000	
Spray dryers (304 stainless steel)		
Nominal water removal kg/hour	*Installed cost*	
1800	$ 600,000	
3200	$1,250,000	
4500	$1,600,000	
(Complete with fans, ducts, air filters, power lines, steam preheater, gas burner, dryer chamber, cyclones, wet collection system, bagging system, buildings, etc.)		

TABLE VII. Approximations for Production Cost Estimation

Production costs	
Depreciation: 6.67–10% of capital (10- to 15-year straight line depreciation)	
Taxes and Insurance: 2–6% of capital (varies with location)	
All salaries except maintenance supervision: 70–100% of direct labor (administration, supervisory, clerical, technical, engineering)	
Indirect Labor: 30–50% of direct labor (includes control and quality control labs, yard, materials handling, cleanup, etc.)	
Associated payroll costs:	20–40% of payroll
Supplies:	20–30% of direct labor
Other plant overhead:	50–70% of direct labor
(telephone, outside services, rentals, tools, travel expenses, demurrage, freight, dues, donations, legal, medical and other professional services, sewer charges, etc.)	
Maintenance: 4–8% of capital	
(salaries, labor, and associated payroll costs are approximately 50% of total)	
General cost factors	
Sales and Administration plus Research and Development expense:	10–15% of net sales
Provision for taxes:	50% of profit from operations
Startup expenses:	5–10% of capital
Working capital:	20–30% of net sales

nary field tests on a new bacterial control agent for a major United States crop of 5,000,000 ha planted annually. Furthermore, there are indications of favorable patent coverage. Field trials with standardized units of activity in 1 kg of product indicate treatment of 3 kg/ha is required. Requirements for control may range from four to eight treatments per crop season with six treatments being the weighted average over the entire country.

Preliminary market research, based on these results, indicates a possible market penetration at a level of 15–25% of the crop and at a \$2.50/kg of product cost to the grower.

This results in a potential market for between 4.5 and 7.5 million ha-treatments which reduces to 13.5–22.5 million kg of product per year. Thus a preliminary evaluation near the midrange level is in order. Assuming a fermentor with an easily manageable capacity of 225 m^3, the bases determined by pilot data and further assumptions are listed in Table VIII. Raw material costs are also calculated for the indicated production level of 17 million kg, a rate fairly near the average of the projected market range.

D. Process Flow Diagram, Material Balance, and Utilities

Simultaneous with development of these numbers is the requirement for a process flow diagram (Fig. 2) and a material balance flow sheet

TABLE VIII. Materials Usage and Cost

Basis
Volume of production fermentors: 225 m^3
Operating volume: 180 m^3
Number of fermentors: 4
Stream factor: 97%
Cycle time: 40 hours (includes 4-hour turnaround)
Production/batch: 20,000 kg (at a standardized activity per kg)
Batches/year: 850
Production/year: 17,000 kg

Raw materials	*kg/batch*	*1000 kg/year*	*Unit cost ($/kg)*	*Cost ($1000)*
Fermentation				
Molasses	11475	9754	0.07	683
Adjuvant solids	3668	3118	0.30	935
Misc. salts, etc.	922	784	0.70 (avg.)	549
Subtotal				2167
Formulation				
Diluent	9100	7735	0.30	2320
Stabilizers	100	85	2.00	170
Sticker	400	340	0.50	170
Anticaking agent	400	340	0.24	82
Subtotal				2742
Packaging	25 kg/bag	680,000 bags	0.40/bag	272
				5181

(Fig. 3). These augment the assumptions and process requirements as well. From these, the nature of the equipment can be resolved and sizing determined. A utilities usage (Table IX) can be calculated based on the equipment selected and the process requirements.

E. Capital Investment, Direct Labor Requirements, and Production Cost

Next a capital investment can be calculated in the manner outlined in Table X, followed by an estimation of the operating labor (Table XI), and finally by the production cost estimate (Table XII). There are many ways to assemble the information and calculations merely by individual preference, but the scheme presented in Figs. 2 and 3 and Tables VIII–XII is fairly straightforward and presents the assumptions conveniently. It outlines these assumptions, the thought process, and calculations in sufficient detail to permit ready cross-checking or easy substitutions of alternatives. The presentation, done in this manner, requires little verbal explanation for easy understanding.

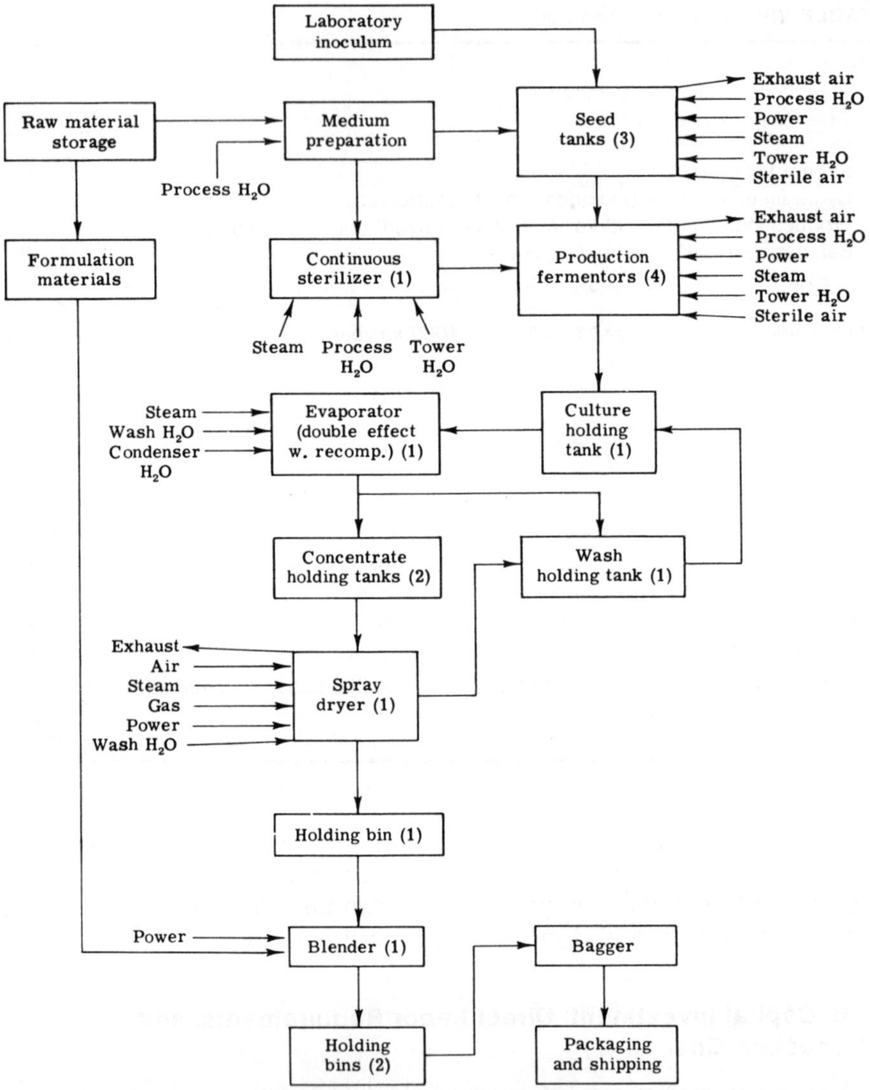

FIGURE 2. Process flow diagram.

This model outlines a simple process of fermentation, evaporation, drying, and blending. If the activity is contained as an intracellular product, then there is an alternative which would substitute centrifugation to separate cells followed by cell drying rather than the concentration step. A calculation exercise for the alternate process could explore the advantages of reduced energy cost versus the inherent problem of waste disposal created by the centrifugation step. If the anticaking agent

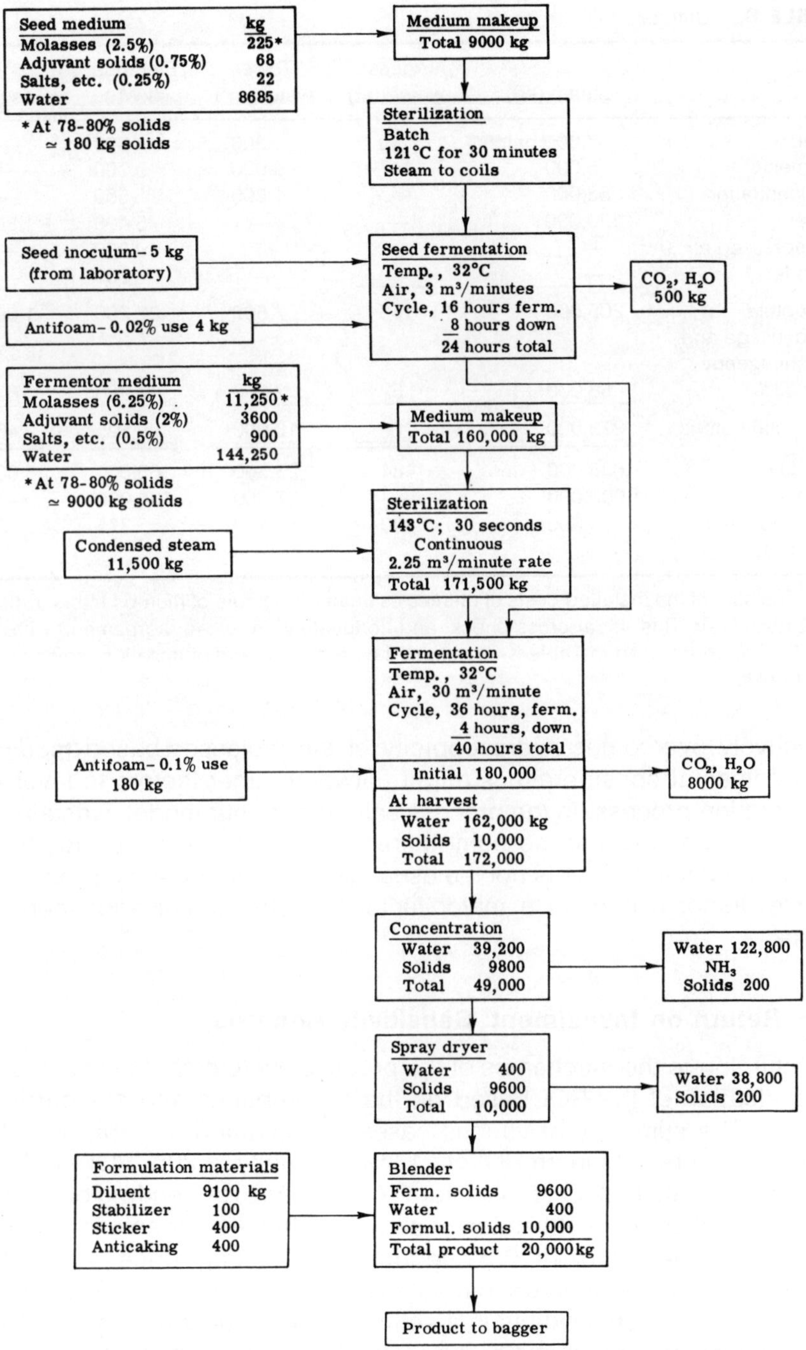

FIGURE 3. Material balance flow sheet (per batch).

TABLE IX. Utilities

	Steam (kg)	Process water (m^3)	Tower water (m^3)	Electrical (kWh)	Gas (m^3)
Seed	7,000/batch	9	300	300	—
Fermentor	25,000	147	6,000	5,700	—
Concentration	53,000	—	1,200	650	—
Dryer	120,000	—	—	5,400	1,800
Compressed air	—	—	—	9,250	—
Blender	—	—	—	900	—
Subtotal	205,000	156	7,500	22,200	1,800
Misc. usage and contingency at 33%	68,000	52	2,500	7,325	600
Total/batch	273,000	208	10,000	29,525	2,400
Per day	636,000	484	23,300	68,800	5,600
Installed	1,000,000	—	27,000	100,000	—
Installed cost[a] ($1000)	600	100	450	375	25

[a] The sum of the installed costs of utilities estimated from rule of thumb factors adds up to $1,550,000. This is a cross-check and fortunately in close agreement with the $1,590,000 figure used in Table X as obtained by arbitrary ratio of installed major equipment costs.

effectively overcomes hygroscopicity of the molasses-based medium, environmental considerations might outweigh other factors to favor the evaporation process. In the process selected for our model, waste problems are minimized to the point there would be no need for additional disposal facilities. This is not the usual case for fermentation processes. Waste disposal can be a major factor in both capital and operating costs.

F. Return on Investment: Sensitivity Analysis

Returning to the mechanics of the process evaluation, the next step is in evaluation of the ROI based on the assumptions and standardized factors. The summary of various cases is contained in Table XIII. The base case A results in an ROI of 21.2%. Normally, a 20% ROI would be considered satisfactory for a venture of this nature, depending on the risk relative to alternative utilization of capital within a company. In cases B, C, and D, the effect of 20% reduction in selling price and sales/production volume or a 20% capital overrun are considered. The indication of greatest sensitivity is in a selling price reduction, although ROIs for all cases are substantially reduced. Thus, the effect of wrong market judgment considerably outweighs the effect of a poor capital projection. If there were concern about selling price holding at the design level but more apt to hold at the lowest volume projected, then case E can be calculated readily to show the effect of the conservative basis of a smaller plant. Quantifiable conceptual information of this nature, developed at reasonably early stages of the project, supplies management with a satisfactory basis of judgment for further action. This is about the simplest form for presenting profitability data. It must be recognized that it is only for a broad look at profitability and a useful tool for consideration of alternatives.

If the project progresses to a capital request stage, by then the process and the plant design should be firm. All major equipment should be specified and quotes obtained with plant capacity firm and process flow sheets, energy and material balances optimized, plus process and instrumentation diagrams nearly complete. Such an estimate for major capital funding can take a 2- to 4-month period to prepare. The financial information is presented in year to year breakdown for the construction year(s) plus 10 years of the project (or whatever is standard for a given company). In such an analysis, the effect of inflation escalators and other capital additions through the years are considered item by item in the cost development. The ROI, cash flows, and payout times are calculated, and usually a present value return analysis is made at the standardized discount rate (or rates) of the company. Some companies will have other financial judgement criteria, but the items mentioned above are basic requirements for a finalized capital request.

In summation, it will be readily apparent that the project model selected is somewhat hypothetical. The authors are not aware of any microbial crop-control agents of such potential in the literature. The only reasonable success story for such a product is that of *Bacillus thuringiensis* which in a 15-year period from 1962 to 1977 has grown from sales in the low hundreds of thousands of dollars to a reported sales volume of approximately $13 million shared by three companies. Satisfactory profitability can be obtained only in multiple-use, shared fermentation facilities at such low volumes. Moreover, this type of bacterial control agent generally will be quite specific relative to chemical control agents. This creates a limited market. Since the cost of research and clearance is nearly as much for a low-dollar volume product as for a large-volume one, it is quite apparent why major fermentation and chemical companies have limited their investments in this direction, although from an ecology viewpoint, it is the desirable approach.

The key point is that such commodity items are sometimes difficult to justify from an ROI basis. For example, SCP processes from hydrocarbons—or for that matter any waste substrate—are often not

TABLE X. Capital Investment Estimate

			Cost (in thousands)		
	Number	Size	Unit cost	Extended cost	Installed cost
Fermentation					
Formulation tanks	2	38 m^3	$ 15	$ 30	
Agitator	2	3.75 kW	4	8	
Continuous sterilizer: 304 Stainless Steel	1	2.25 m^3/min	100	100	
Compressors	2	85 m^3/min	225	450	
Antifoam system			100	100	
Seed tanks: 304 Stainless Steel	3	15 m^3	35	105	
Agitator	3	19 kW	14	42	
Production fermentors: 304 Stainless Steel	4	225 m^3	165	660	
Agitator	4	150 kW	70	280	
Holding tanks: 304 Stainless Steel	2	190 m^3	40	80	
Pumps				150	
Separators and special equipment				100	
Subtotal: Major fermentation equipment				$2,105	
Installed cost: 2.5 × major fermentation equipment[a]					$5,263
Instrumentation: 10% (avg. to complex installation)[b]					526
Process piping: 20% (avg. installation)[b]					1,053
Buildings: Mixed outdoor and indoor const., 40%[b]					2,105
Total: Fermentation capital					$8,947
Processing					
Evaporator: 304 Stainless Steel	1	18,000 kg H_2O/hr	$400	$ 400	$ 600

Holding tanks: 304 Stainless Steel	3	75 m^3	24	72	180
Holding bins: 304 Stainless Steel	3	75 m^3	24	72	180
Blender: 304 Stainless Steel	1	15 m^3			75
Vibrating screen	1	4 m^2-3 deck			50
Subtotal: Process equipment					$1,085
Instrumentation: 10%[b]					109
Process piping: 20%[b]					217
Buildings: Indoor construction, 75%[b]					814
Subtotal					$2,225
Spray dryer: 304 Stainless Steel	1	4,500 kg H_2O/hour	$1070	$1070	1,600
Total: Process capital					$3,825
Utilities and Tankage					
Utilities: (major addition at existing site 20%[b] of installed major equipment)					
0.2 (5263 + 1085 + 1600)	—	—	—	—	$ 1,590
Molasses storage tanks	2	2800 m^3	—	—	270
Subtotal					$ 1,860
Total fixed capital					$14,632
Engineering and construction: 20% straightforward[c]					2,926
Contingency; average, 20%; process subject to change[c]					2,926
Total capital investment (exclusive of land)					$20,484
Use					$20,500

[a] Authors' experience.
[b] Chilton's factors on installed equipment costs (1949).
[c] Chilton's factors on total physical cost (1949).

TABLE XI. Direct Labor Requirements

	Per Shift
Fermentation	3
Evaporator	1
Dryer	1
Blending	1
Packing and shipping	2
Total/shift	8
Total/4 shifts[a]	32
Wages/operator	$16,000 per year[a]
Total direct labor	$512,000

[a] 168 hours/week ÷ 4 = 42 hours/week/operator. Each operator thus has the equivalent of 43 hours straight time pay built into the schedule.

TABLE XII. Production Cost Estimate

	Cost (in thousands)	
Raw materials		
Fermentation	$ 2167	
Formulation	2742	
Packaging	272	
Subtotal		$ 5,181
Payroll Charges		
Direct labor	$ 512	
Indirect labor (40% of direct labor)	205	
Salaried payroll (85% of direct labor)	435	
Associated payroll costs (30% of payroll)	346	
Maintenance salaries, labor and assoc. (3% of capital)	615	
Subtotal		$ 2,113
Supplies and expenses		
Maintenance (3% of capital)	$ 615	
Supplies (25% of direct labor)	128	
Subtotal		$ 743
Utilities		
Steam: $8.50/1000 kg × 273 × 850	$ 1972	
Electricity: $0.035/kWH × 30,000 × 850	893	
Water: $0.008/m³ × 10,000 × 850	68	
Gas: 2400 m³ × $0.09/m³ × 850	184	
Subtotal		$ 3,117
Other Costs		
Depreciation (12 years)	$ 1708	
Taxes and insurance (4% of capital)	820	
Other plant overhead (60% of direct labor)	307	
Subtotal		$ 2,835
Total		$13.989
		$ 0. 82 per kg

TABLE XIII. Return on Investment—10-Year Basis

	Value (dollars in thousands)				
	A Base case 17 million kg/Year at $2.50/kg	B 20% Reduced selling price at $2.00/kg	C 20% Reduced volume at 13.6 kg/Year	D 20% Capital overrun	E 13.6 kg /Year plant design
1. Sales to growers	42,500	34,000	34,000	42,500	34,000
2. Dealer discounts, distribution, freight (30%)	(12,750)	(10,200)	(10,200)	(12,750)	(10,200)
3. Net sales	29,750	23,800	23,800	29,750	23,800
4. Cost of goods sold	(13,989)	(13,989)	(12,210)[b]	(14,743)[b]	(11,818)[b]
5. Gross profit	15,761	9,811	11,590	15,007	11,982
6. Sales and Administration plus Research and Development (12.5% of net sales)	(3,719)	(2,975)	(2,975)	(3,719)	(2,975)
7. Profit from operations (2–10 years)	12,042	6,836	8,615	11,288	9,007
8. Startup costs, first year (10% of capital)	(2,050)	(2,050)	(2,050)	(2,460)	(1,770)
9. Profit from operations (first year)	9,992	4,786	6,565	8,828	7,237
10. Net earnings after taxes (50%)—first year	4,996	2,393	3,283	4,414	3,619
11. Net earnings after taxes (50%)—2–10 years	6,021	3,418	4,308	5,644	4,504
12. Working capital (25% of net sales)	7,438	5,950	5,950	7,438	5,950
13. Original fixed capital investment	20,500	20.500	20,500	24,600	17,700[c]
14. Return on investment (%)[a]	21.2	12.5	15.9	17.2	18.7

[a] $\text{ROI} = \frac{(\text{line } 10 + 9 \times \text{Line } 11)/10}{\text{line } 13 + \text{line } 12}$

[b] Variable costs of materials and utilities × 0.8. Semifixed costs of labor related items × 0.94 (direct labor to 30 men from 32). Capital related costs follow their percentage of capital investment.

[c] Calculated based on $(17/13.6)^{0.67}$ ratio of plant size to obtain new capital cost. The 0.67 factor is a balance between the 0.75 for fermentation section scale factor and the often used 0.6 for process plant scale factor.

viable in the United States in the context of our low-cost competitive protein products from agriculture.

Thus, it is not surprising that about three-quarters of the dollar volume of aerobic microbial products resides in high-markup pharmaceuticals represented by antibiotics and conversion products.

V. REFERENCES

Aries, R., and Newton, R. (1955). "Chemical Engineering Cost Estimation." McGraw-Hill, New York.

Bartholomew, W., Engstrom, D., Goodman, N., O'Toole, A., Shelton, J., and Tannen, L. (1974). *Biotechnol. Bioeng.* **16,** 1005.

Bauman, H. (1964). "Fundamentals of Cost Engineering in the Chemical Industry." Von Nostrand-Reinhold, Princeton, New Jersey.

Blakebrough, N., ed. (1967). "Biochemical and Biological Engineering Science," Vol. I. Academic Press, New York.

Blakebrough, N., ed. (1969). "Biochemical and Biological Engineering Science," Vol. 2. Academic Press, New York.

Chilton, C. (1949). *Chem. Eng. (N.Y.)* **56** (6), 97.

Chilton, C. (1960). "Cost Engineering in the Process Industries." McGraw-Hill, New York.

Fong, W. (1975a). "Proteins from Hydrocarbons," Rep. No. 60A Suppl. Process Econ. Program, Stanford Res. Inst., Menlo Park, California.

Fong, W. (1975b). "Fermentation Processes," Rep. No. 95. Process Econ. Program, Stanford Res. Inst., Menlo Park, California.

Guthrie, K. (1969). *Chem. Eng.* (N.Y.) **46**(6), 114.

Heden, C.-G. (1976). U.S. Patent 3,997,400.

Humphrey, A. (1977). *Process Biochem.* **12,** 19.

Jefferis, R., ed. (1977). "Workshop—Computer Applications in Fermentation Technology, 1976," Ges. Biotechnol. Forsch. mbH. Verlag Chemie, Weinheim.

Lundell, R., and Laiho, P. (1976). *Process Biochem.* **11,** 13.

Meskanen, A., Lundell, R., and Laiho, P. (1976). *Process Biochem.* **11,** 31.

Nyiri, L. (1972). "Advances in Biochemical Engineering", Vol. 2, p. 49. Springer-Verlag, New York.

Nyiri, L., and Charles, M. (1977). "Annual Reports on Fermentation Processes" (D. Perlman, ed.), Vol. 1, p. 365. Academic Press, New York.

Ovaskainen, P., Lundell, R., and Laiho, P. (1976). *Process Biochem.* **11,** 37.

Pikulik, A., and Diaz, H. (1977). *Chem. Eng.* **84** (21), 107.

Richards, J. (1968). *Process Biochem.* **3,** 56.

Ryu, D., and Oldshue, J. (1977). *Biotechnol. Bioeng.* **19,** 621.

Schepers, E. (1974). *Natl. Chem. Eng. Conf. Process Ind. Aust.—Impact Growth, 1974* p. 60.

Weaver, J. and Bauman, H. (1973). *In* "Chemical Engineers Handbook" (R. Perry and C. Chilton, eds.), 5th ed., p. 25–13. McGraw-Hill, New York.

Whitaker, A., and Walker, J. (1973). *J. Biol. Educ.* **7,** 37.

Wilson, J. (1975). U.S. Patent 3,926,738.

Wilson, J., Nyiri, L., Humphrey, A., and Harms, C. (1975). U.S. Patent 3,926,737.

Yamada, K., Kinoshita, S., Tsunoda, T., and Aida, K., eds. (1972). "The Microbial Production of Amino Acids." Kodansha, Ltd., Tokyo.

Zimmerman, O., and Lavine, I. (1962). *Cost Eng.* **4,** 7.

Chapter 19

Fermentation Processes and Products: Problems in Patenting

IRVING MARCUS

I. INTRODUCTION

A patent combines the mix of a scientific document with a legal document. The patent stems from our Constitutional provision:

. . . to promote the progress of Science and Useful Arts, by securing for limited times

MICROBIAL TECHNOLOGY, 2nd ed., VOL.II
 ISBN 0-12-551502-2

to Authors and Inventors the exclusive right to their respective writings and discoveries. (U.S. Constitution, Article I, Section 8)

A. The Patent Statute: Title 35—Patents

Section 101, Inventions patentable, provides

> Whoever invents any new and useful process machine, manufacture or composition of matter on any new or useful improvement thereof may obtain a patent therefor, subject to the conditions and requirements of this title.

The basic conditions for patentability are set forth in the negative. A person is entitled to a patent *unless* the invention was not new (Sec. 102) and if not identically disclosed if it would have been obvious (Sec. 103). Thus, there are, in the positive, three main conditions for obtaining a patent. The invention must be new, must be useful, and must be unobvious.

B. Requirements of a Patent Disclosure

In order, however, to obtain a patent there must be an application for a patent by the inventor with certain prescribed conditions, one of which is set forth in Section 112 entitled "Specification":

> The specification shall contain a written description of the invention, and of the manner and process of making and using it, in such full, clear, concise, and exact terms as to enable any person skilled in the art to which it pertains, or with which it is most nearly connected, to make and use the same, and shall set forth the best mode contemplated by the inventor of carrying out his invention.

These are basics which are required for an understanding of patents that deal with microbiological inventions. The principal microbiological invention is, of course, the fermentation process and processes involving fermentations are in many parts of the patent system.

The microorganism may be referred to as a starting material in process inventions, it may be considered a catalyst, or it may itself be an end product, the result of a man-made variation of an available microorganism.

Thus, there is a mixture of science and the law embodied in a patent. The subject matter is within the realm of science, while the claiming of the subject matter with supportive disclosure is the legal aspect. The legal problems in disclosing and claiming inventions dealing with the action of microorganisms have been in existence for almost 30 years. These problems come to light during the examination of the application and the search of the prior art (patents, domestic and foreign, and the literature).

II. BACKGROUND

A. Applications for Patent Involving the Action Of Microorganisms

For many years inventions dealing with microorganisms received only sporadic attention in the patent area. The cases that came to light involved esoteric legal points that were generally *sui generis*, of first impression. There was little, if any, uniformity either in the patent practice by the examiner in the United States Patent and Trademark Office (PTO) or in the decisions affecting such practice. Case by case, we struggled with the problems that arose. At this time, we are far more knowledgeable in the subject from a legal sense. Meanwhile, the subject began receiving the attention of those working in the patent field and various articles appeared covering the aspects of the application of patent laws to the technology (Marcus, 1975; Levitt, 1976). One aspect of the problem will be covered more fully below with particular attention to the recent diplomatic activity which is now in treaty form. Most recently, the subject attained notoriety with the publicity concerning genetic research and recombinant DNA.

We will concentrate on the practice in patent law dealing with inventions in microbiology from the viewpoint of that law as practiced in the United States. Some mention of foreign practices will, however, be reviewed when the aforementioned treaty is discussed, but basically it is with United States patent laws that we will be dealing. The reasons are that the scope of patent protection permitted by U.S. patent laws is of greater breadth than those of certain of the foreign countries, each of which may be part of a convention under which reciprocal rights are exchanged, all subject, however, to national laws as to the type of patent protection granted.

B. Classification in the United States Patent and Trademark Office (PTO)

In the PTO, there is an entire classification, a single class, that deals with the chemistry of fermentation, Class 195 (Manual of Classification, 1936). The classification is ancient but still applicable, although currently under revision. Most of the inventions are in the fields of fermentation to yield useful products, and the class also provides for enzymes and compositions. However, there are other classifications that deal with microbiological processes, e.g., processes of preparing vaccines and sera using animals as an essential part of the process. Products prepared by microbiological processes are generally classified and examined in the PTO according to their chemical characteristics. Microorganisms can be used in many chemical reactions as the catalytic agent, examples of which include oxidation, reduction, and hydrolysis reactions.

My first contact with the subject in the PTO was as classification examiner charged with the duties of maintaining the classification of chemical patents. With the advent of antibacterials, mostly known as antibiotics, the field of microbiology became very active in patent filings. Each newly discovered organism was carefully cultured by researchers in their search for improvements over penicillin. Soil samples from all over the world were carried to the researchers in the field, and I am certain that many research dollars were spent on the culturing of new organisms that had interesting antibacterial properties but could not be reproduced outside of the laboratory. Although there were many that failed, there were some notorious successes, namely, tetracycline (Conover, 1955), oxytetracycline (Terramycin) (Sobin *et al.*, 1950), chlortetracycline (Aureomycin) (Duggar, 1949), streptomycin (Waksman and Schatz, 1948), erythromycin (Bunch and McGuire, 1953), and fumagillin (Hanson and Eble, 1953).

C. Examination in the United States Patent and Trademark Office

The patent applications filed thus covered many aspects of the inventive concepts. There were process applications for using a newly discovered microorganism to prepare a new product as well as known products. There were applications for improvements in process conditions, in media used, and in the recovery steps. In some cases the recovery was purely chemical and began not with the fermentation itself but with the broth end product, which included the undesirable components. There were also claims for the products themselves. In some cases, the products were defined not by chemical formula but only in terms of characteristics and properties. In some cases, claims were in the form of the product as prepared by the claimed process. There were patents for derivatives of the antibiotics already produced. These are all illustrated by the sample claims provided in Appendixes 1 through 4.

The Classification Division of the PTO made the decision as to which patent examining division each type of claim should be submitted. Inventions in fermentation chemistry, organic chemistry, and medicines were being examined in separate and distinct examining divisions. Where the product was medicinal, and its structure undetermined, the medicine division was charged with the examination. When the structure was elucidated, the burden of examination was shifted to the examiner in organic chemistry. The working arrangement satisfied the parties but had little effect on the patent law at that time. Although this may seem to be an unusual arrangement, it is based on actual experience. No matter how uniform an examining system, how consistent the guidelines, and how highly skilled the examiners, each is a distinct personality. The individual treatment by an examiner in organic chemistry reviewing a product claim might be different from that of an examiner in the medicine arts looking at a product claim that has no known chemical composition but has medicinal properties. Thus, differences in handling applications did occur in spite of PTO efforts to minimize if not prevent all such occurrences. The reason for this background information will become apparent as we discuss how the practice evolved to its current status. Some of this early treatment, however, is discussed by Levy and Wendt (1955). Their review covered deposit of cultures of microorganisms as well as how to claim the products of the fermentation, particularly when the structure is unknown and use is made of infrared absorption spectrum, which is incorporated into the application by a formal drawing in lieu of the structure. These infrared absorption spectra are characterized as "finger prints" for identification purposes insofar as the product is concerned, although they are combined with other known characteristics, including elemental analysis, melting point, molecular weight, X-ray diffraction patterns and optical rotation as well as biological activity against microorganisms. Many other characteristics are used as can be seen by the examples of claims in the appendixes to this chapter.

A real problem occurred in examination when two or more applicants claimed the same product but each used different identifying data and often named the microorganism by their own "catchword" (Reynolds, 1967). The examiners were astute enough to discover this and did not issue patents to two applicants for the same product. In some cases one application defined the product in terms of properties and the other by an elucidated formula. Since the inventions were the same, and only one patent could be granted, it was the examiner's burden to make the discovery of identity and set in motion an interference, which is a contest to determine who was the first inventor. This interference is in the PTO and the final decisions there are appealable.

Thus during the trying years of handling patent applications involving the use of microorganisms which could not be identified, there was no "official" Patent Office policy that instructed the examiners as to the handling of the application and in effect all decisions were those of the individual examiners and their immediate supervisors in their respective "art" areas. So while there was a rule as to which type of application would be examined in the respective examining areas there was no policy to provide uniformity. As case law developed, decisions were made. This meant that examination techniques were based on case law by courts unfamiliar with the technology about microorganisms.

D. Statutory Proposals

It was not until the patent reform legislation was being proposed that statutory provision was suggested. In 1971, S.643 provided:

Section 112—Specification

(d) When the invention relates to a process involving the action of a microorganism not already known and available to the public or to a product of such a process, the written description required by subsection (a) of this section shall be sufficient as to said microorganism, if—

(1) not later than the date that the United States application is filed, an approved deposit of a culture of the microorganism is made by or on behalf of the applicant or his predecessor in title, and

(2) the written description includes the name of the depository and its designation of the approved deposit and, taken as a whole, is in such descriptive terms as to enable any person skilled in the art to which the invention pertains to make and use the same.

An approved deposit shall be a deposit which—

(1) is made in any public depository in the United States which shall be designated for such deposits by the Commissioner of Patents by publication, and

(2) is available, except as otherwise prohibited by law, in accordance with such regulations as may be prescribed,

(a) to the public upon issuance of a United States patent to the applicant or his predecessor or successor in title which refers to such deposit, and

(b) prior to issuance of said patent, under the conditions specified in Section 122.

It is to be noted, however, that S.643 required that the approved deposit be made in "any public depository in the United States." In some cases, this would be impossible for foreign inventors to do since certain pathogenic organisms are subject to import control by the Department of Agriculture as well as other government agencies.

Meanwhile, patent reform legislation was having a stormy passage through Congress and none has yet been approved. However, in the 94th Congress, S.2255 was passed by the Senate on February 26, 1976. Section 112(d) became Section 112(f) and under this was provided:

(g) For the purpose of subsection (f) (1) of this section, an approved deposit shall be a deposit which—

(1) is made in any public depository which shall have been designated for such deposits by the Commissioner by publication, and

(2) is available, except as otherwise prohibited by law, in accordance with such regulations as the Commissioner may prescribe.

Thus the deposit can be made anywhere in the world subject to the guidelines issued by the PTO now in effect. As discussed below the Budapest Treaty provides for the recognition of an International Depository Authority where a deposit of culture of a microorganism may be made in connection with a patent application filed in the patent offices of the respective contracting countries or in designated regional patent offices. The treaty refers to these as Industrial Property Offices. The European Patent Office is a regional patent office.

This discussion is necessitated by the fact that in most of these applications, the subject matter deals with a microorganism which is new, with no known prior description, but which is useful to produce a valuable result, be it in a process or to produce a new chemical product.

How does one meet the statutory provisions of 35 U.S.C 112 by adequately describing such a microorganism? The deposit of a culture in a public depository became the answer. This is a major variation from a conventional patent disclosure and, of course, led to many contentious problems until it was finally resolved in 1971. Since then, the proposed legislation and treaty have been suggested to officially sanction the practice.

It is apparent from this discussion that a microorganism cannot be prepared readily since in most cases it is a naturally occurring living material, and even in those cases where it is prepared or altered by mutation, it is questionable whether it is readily and uniformily reproducible.

E. Patent and Trademark Office Guidelines

The problems in this field leading to the Guidelines for Deposit of Microorganisms (1971) were set forth in an earlier report (Marcus, 1975). Most important was the fact that the idea was conceived by applicants for their own benefit and, for a long time, the contract was under no control by anyone outside of the depositor–depository relationship. As a result, there was confusion about statutory basis and it was not until judicial decisions were made that the PTO took steps to resolve the problems, as is discussed later.

III. PATENT PRACTICE AND PROBLEMS

A. Review of Case Law

To set the stage for a discussion of the patent practice, I will start with a review of some important decisions involving microbiological inventions and then consider current practices based on these decisions keeping in mind the chronology of these decisions. *Guaranty Trust Company of New York et al. v. Union Solvents Corporation* (1931) involved a suit on an invention of Weizmann for "Production of Acetone and Alcohol by Bacteriological Processes" (1919). The 1931 decision of the Delaware District Court found Claims 1 and 3 of the patent valid and infringed. These are reproduced:

> 1. The process of producing acetone and butyl alcohol by the fermentation of liquids containing natural substances rich in starch by means of the herein described bacteria which are capable unaided by converting sterile fermentable grain starch substantially into acetone and butyl alcohol, and also liquefying gelatin.
>
> 3. The process of producing acetone and butyl alcohol by the inoculation of a cereal composition with the herein described bacteria which are capable unaided of converting sterile fermentable cereals substantially into acetone and butyl alcohol.

The question of patent disclosure of the bacteria was considered and effectively handled by the court by acceptance of the plaintiff's Exhibit 12 by plaintiff's expert and by evidence that the selection of the amount of inoculum was left to the judgement and experience of the operator to determine. The specification had referred to the well-defined sources from which Weizmann obtained his bacteria and the precise procedure. The court, therefore, concluded that anyone following this procedure could obtain the bacteria and would have considerable success.

The contention of the defendant that the invention in the Weizmann patent was for the life cycle of a living organism was overcome by pointing out that it is not the bacteria per se that is being claimed but a fermentation process using such bacteria, which the inventor discovered, under the disclosed operating conditions. Thus the inventive concept for Weizmann's discovery was for the use of such bacteria in a process to obtain a very useful known product. The Court of Appeals sustained this decision in 1932.

The Court of Customs and Patent Appeals (CCPA) held the following claim unpatentable in *In re Arzberger* (1940):

> 1. Bacteria herein described and designated as clostridium saccharobutyl-acetonicum-liquefaciens.

The holding was based in the proposition that this bacterium was not statutory subject matter under the plant patent section of the patent laws (Patents for Plants, 1954):

> Sec. 161. Whoever invents or discovers and asexually reproduces any distinct and new variety of plant, including cultivated sports, mutants, hybrids, and newly found seedlings, other than a tuber propagated plant or a plant found in an uncultivated state, may obtain a patent therefor, subject to the conditions and requirements of title. (Amended September 3, 1954, 68 Stat. 1190.)

The provisions of this title relating to patents for inventions shall apply to patents for plants, except as otherwise provided.

In the common language of the people, "plant" as used by Congress did not encompass bacteria even though its reproduction is asexual, by binary fission. The CCPA never decided whether it was encompassed by the statutes as a composition of matter (as will be later discussed).

Next in time was *Ex parte Kropp* (1949), a decision of the PTO Board of Appeals in 1959. The central issue involved the deposit of the culture of the microorganism used in the claimed process. The case held that the hitherto unidentified species of microorganism set forth in the claimed process:

> 4. A process for producing jm-57h which comprises cultivating the organism *Streptomyces* sp. JM-57h in an aqueous nutrient medium containing a source of protein and a source of carbohydrate under aerobic conditions until antibacterial activity is produced in said medium

was insufficiently disclosed if the microorganism had not been deposited in a recognized depository *prior* to filing the application. An offer to deposit was made but the decision held that the deposit, which would have been sufficient to overcome the rejection, had not been made by the time of filing of the application and the subsequent deposit or offer to deposit could not overcome this defect in disclosure.

An interesting variation of this concept was subsequently decided by the CCPA in 1971. The case *In re Argoudelis et al.* (1970) held that the deposit of the microorganism, made no later than the date of filing, could be restricted so that no disclosure of it would be available until the patent would issue. The PTO had held that the Statute (Sec. 112) required a written description, and in the absence of a written description reference to something outside the specification must be a reference to something in the public domain, hence no availability of the "starting material." Since the depositor restricted access to the microorganism only to nominees of the applicant or the PTO, the organism was not adequately described. This PTO position was reversed by the CCPA. Thus a restricted deposit was an enabling disclosure, under 35 U.S.C 112 during the pendency of the application, with the deposit becoming available to the public after grant of the patent as the quid for the patent grant. This decision caused the PTO to publish "Guidelines for Deposit of Microorganisms" (1971), which are still in effect.

Since the publication of these guidelines, problems concerning deposit of microorganisms have been minimized. Without any legislation to the contrary, the PTO will accept a deposit anywhere in the world, if proper access is given to that culture deposit when the United States patent issues.

This is the current practice in the United States and, of course, the practice approved by the Budapest Treaty. The Treaty also provides for access to the organism when there is a publication of the application.

This precise point was decided by the CCPA in *Feldman v. Aunstrup* (1975). The respective applicants were contesting an interference as to a fermentation process. The sole count reads:

> A process for the preparation of a milk-coagulating enzyme which comprises cultivating a milk-coagulating enzyme producing strain of *Mucor miehei* Cooney et Emerson or a natural or artificial variant or mutant thereof in a suitable nutrient medium, and thereafter recovering the milk-coagulating enzyme from the medium.

The microorganism is the essence of the invention. Feldman deposited his culture with the Northern Regional Laboratory of the ARS, and Aunstrup with the Centraalbureau voor Schimmel cultures (CBS) (Central Bureau for Mold Cultures) in Baarn, Netherlands (Table I). The latter deposit was timely, prior to his convention foreign application which was granted priority for his corresponding United States application. The CBS, at the time a private institution, is now part of the Royal Netherlands Academy of Sciences and Arts, a governmental organization. The deposit, at one time restricted, was removed from restriction in 1969. The CBS is a qualified culture collection and distributes samples. The CCPA found the deposit in CBS to satisfy the conditions set forth in *In re Argoudelis et al.* (1970) and in the Patent Office Guidelines. Not only were samples available to the public at the proper time, but the PTO could also obtain access to the CBS sample at any time during pendency of the patent application. The Budapest Treaty carries forward this principle by the contracting states. Thus Aunstrup's application was held to find adequate support in his foreign deposit in accordance with 35 U.S.C. 119 and he won the interference contest. Feldman carried his appeal of this decision to the Supreme Court but certiorari was denied.

This principle was also approved in a decision of the German Supreme Court in 1975 in the so-called bakers' yeast case (Bäckerhefe,

TABLE I. Culture Collection Reviews

A. Culture Collections of Microorganisms (from "Proceedings of the International Conference on Culture Collections." Tokyo, Japan, 1968).
   1. The American Type Culture Collection (W. A. Clark, pp. 3–5).
   2. The Centraalbureau voor Schimmel cultures (J. A. von Ark, pp. 7–9).
   3. History, Policy and Significance of the ARS Culture Collection (C. W. Hesseltine, W. C. Haynes, L. J. Wickersham, and J. J. Ellis).

B. Culture Collections and Patent Depositions [T. G. Pridham and C. W. Hesseltine, *Adv. Appl. Microbiol.* **19,** 1–23 (1975)].

1975). The German Court held that a deposit of a microorganism in a culture collection of an approved depository in a foreign country would satisfy the German Patent Laws if the deposit was made prior to the actual filing date of the application and if based on a convention application, prior to the home country filing. The applicant must also make an irrevocable undertaking to permit access to that deposit by the German Patent Office for examination procedure and to the interested public 18 months after the date of actual filing or the date of the Paris Convention application.

The court noted that the applicant may require third parties obtaining the sample to not only identify themselves but agree not to pass samples to others or to export the samples. This protection is due to the fact that at the time of this publication, enforceable patent rights are not yet in existence. The Budapest Treaty also provides for this notification to the depositor by the depository of those who request samples of microorganisms.

In another important decision in the field, the U.S. Supreme Court (1948) in *Funk Bros. Seed Co. v. Kalo Inoculant Co.* held the following claim invalid:

> An inoculant for leguminous plants comprising a plurality of selected mutually non-inhibitive strains of different species of bacteria of the genus *Rhizobium*, said strains being unaffected by each other in respect to their ability to fix nitrogen in the leguminous plant for which they are specific.

The patent also contains process claims.

The six well-recognized species of bacteria and the corresponding groups (cross-inoculation groups) of leguminous plants are shown in the tabulation below:

Rhizobium trifolii	Red clover, crimson clover, mammoth clover, alsike clover
Rhizobium meliloti	Alfalfa, white or yellow sweet clovers
Rhizobium phaseoli	Garden beans
Rhizobium leguminosarum	Garden peas and vetch
Rhizobium lupini	Lupines
Rhizobium japonicum	Soy beans

In spite of the ease of use provided by the invention, the Supreme Court found a lack of invention in the discovery because no new use of each individual bacteria was discovered:

> Discovery of the fact that certain strains of each species of these bacteria can be mixed without harmful effect to the properties of either is a discovery of their qualities of non-inhibition. It is no more than the discovery of some of the handiwork of nature and hence is not patentable. The aggregation of select strains of the several species

> into one product is an application of that newly-discovered natural principle. But however ingenious the discovery of that natural principle may have been, the application of it is hardly more than an advance in the packaging of the inoculants. Each of the species of root-nodule bacteria contained in the package infects the same group of leguminous plants which it always infected. No species acquires a different use. The combination of species produces no new bacteria, no change in the six species of bacteria, and no enlargement of the range of their utility. Each species has the same effect it always had. The bacteria perform in their natural way. Their use in combination does not improve in any way their natural functioning. They serve the ends nature orginally provided and act quite independently of any effort to the patentee.

This case is of importance today because there was an appeal pending to the CCPA from a Decision of the Board of Appeals of the PTO holding unpatentable as nonstatutory. The claim:

> A biologically pure culture of the microorganism *Streptomyces vellosus* having the identifying characteristics of NRRL 8037, said culture being capable of producing the antibiotic lincomycin in a recoverable quantity upon fermentation in an aqueous nutrient medium containing assimilable sources of carbon, nitrogen and inorganic substances.

This case, known as *In re Bergy et al.* (1977), was argued in March and decided on October 6, 1977; but the question as to whether microorganisms are patentable subject matter may still be in question, in view of a further case (*In re Chakrabarty**) also pending in the same court and, at the same time, a request by the PTO for a rehearing of *In re Bergy et al.**

As noted in *In re Arzberger,* bacteria are not "plants" within the meaning of the plant statute 35 U.S.C. 161: Patents for Plants:

> Whoever invents or discovers and asexually reproduces any distinct and new variety of plant, other than a tuber propagated plant, may obtain a patent therefor, subject to the conditions of this title.

For the most part, microorganisms are products of nature and thus are not novel. The question in the *Bergy et al.* case argued by the applicants was that their culture is modified and not the microorganism as found in nature. This is supported by affidavits of well-known microbiologists including Alma Dietz who deposed that impure cultures of *S. vellosus* gave taxonomic results different from those of the claimed culture. In essence they argue the claimed culture is the product of a microbiologist. The applicants also point to many patents in which a living organism is part of the claimed invention.

On the other hand, the Patent Office Solicitor argued that the present patent laws do not encompass living organisms as statutory subject

* Reargued November 1978 in the CCPA after grant of certiorari by the U.S. Supreme Court.

matter even under the statutory provisions of "manufacture" or "a composition of matter." Several writers disagree and have strongly set forth their views. One, Harold C. Wegner (1974), poses the query "Must something be dead to be patentable?"

Judge Rich in *In re Mancy et al.* (1974), reversing the rejection of

> 1. Process for the production of daunorubicin which comprises aerobically cultivating *Streptomyces bifurcus* strain DS 23,219 (NRRL 3539), of [sic, or] a daunorubicin-producing mutant thereof, using an aqueous nutrient medium containing assimilable sources of carbon, nitrogen and inorganic substances, and separating daunorubicin formed during the culture.
>
> The four dependent claims specify certain further conditions of aerobic cultivation such as pH, temperature, and aeration rate, although apparently the conditions specified are conventional.

included the following statement:

> The Nature of This Invention: A Process of Producing Known Antibiotic Using a New Strain of Microorganism
>
> . . . Here appellants not only have no allowed claim to the novel strain of *Streptomyces* used in their process but would, we presume (without deciding), be unable to obtain such a claim because the strain, while new in the sense that it is not shown by any art of record, is, as we understand it, a 'product of nature.' However, it is not required for unobviousness of the method-of-use claims that the new starting material be patentable, *In re Schneider, supra;* 35 U.S.C. 103.
>
> [4] The process of using the new *Streptomyces* strain which appellants have discovered is clearly within 35 U.S.C. 101, and we note the ready willingness of the board here, as in *Ex parte Arzberger* and *Ex parte Kropp,* to allow appellants to claim it if they show unexpected results from the use of this new strain. Indeed, the public interest appears to be well served by encouraging the patenting of such inventions. While the patent will grant appellants a limited right to exclude others from producing daunorubicin by the use of *Streptomyces bifurcus,* the public receives not only the knowledge of appellants' discovery but also access to *Streptomyces bifurcus* through its deposit with the Department of Agriculture. See *In re Argoudelis, supra.*

The importance of the question of the patentability of microorganisms per se was also set forth in the brief of the Patent Office solicitor in *In re Bergy et al.* (1977):

> The age of "genetic engineering" . . . creation of new living organisms by combining genetic material from different life forms . . . is now upon us.

This concept will be further discussed but is mentioned here in its relation to the patentability of living organisms.

The decision in the *Bergy* case was by a divided court. The majority opinion was supported by a judge from the court of claims sitting by designation who agreed with the majority opinion but indicated he felt it set forth "an extremely limited holding." Only the future can tell whether this is true.

Judge Rich, speaking for the majority, thinks it is in the public interest to include microorganisms within the terms "manufacture" and "composition of matter" as set forth in 35 U.S.C. 101, quoted above. Judge Rich also points out that his views quoted above from *In re Mancy* (1974), even if considered dictum, were ill-considered since it was apparent that no claim for the microorganism could have been obtained in the Mancy case since it lacked novelty. In the *Bergy* case, however, the microorganism is found to be in its purified form and is found to meet the statutory criteria even if it is a living thing.

At this point we cannot speculate how this decision will affect cases arising from genetic engineering discussed below. Judge Miller, however, speaking for the minority, feels that microorganisms are distinct from inanimate chemical compositions and that such subject matter was never intended to be included within the present patent statute. He also refers to the Plant Variety Protection Act (1972) which he thinks further supports this conclusion that Congress did not intend organisms to be included within the scope of "manufacture" or "composition of matter." The fact that the claimed culture of the microorganism is an industrial product ignores such congressional intent. In effect the dissenting judges feel that the CCPA usurped the congressional role in reversing the rejection and that the only clear way that claims such as claim 5 of *Bergy et al.* could be found patentable would be by express statutory provision.*

*On March 2, 1978 in Appeal No. 77–535 the CCPA reversed the decision of the PTO Board of Appeals in *In re Ananda M. Chakrabarty* for an invention of which the following is an illustrative claim:

> 7. A bacterium from the genus *Pseudomonas* containing therein at least two stable energy-generating plasmids, each of said plasmids providing a separate hydrocarbon degradative pathway.

The Court states: "Appellant's invention involves the creation of a new strain of bacteria by the incorporation of a *single* cell, by transmission thereinto of a plurality of compatible plasmids, of a capacity for simultaneously degrading several different components of crude oil with the result that degradation occurs more rapidly. To make this non-technical description somewhat more intelligible we quote from the specification but two of its many definitions:

> *Extrachromosomal element* . . . a hereditary unit that is physically separate from the chromosome of the cell; the terms 'extrachromosomal element' and 'plasmid' are synonymous; when physically separated from the chromosome, some plasmids can be transmitted at high frequency to other cells, the transfer being without associated chromosomal transfer.
>
> *Degradative pathway* . . . a sequence of enzymatic reactions (e.g., 5 to 10 enzymes are produced by the microbe) converting the primary substrate [i.e., oil] to some simple common metabolite, a normal food substance for microorganisms."

and further notes that Appellant's assignee has been granted British Patent 1,436,573 containing the above and other claims for the bacterium.

On June 26, 1978, the U.S. Supreme Court on petition of the Commissioner of Patents granted certiorari (*Parker v. Bergy,* 1978), vacated the decision of the Court of Customs and Patent Appeals, and remanded the case back to the latter court in view of their earlier decision in a case involving computer software (*Parker v. Flook,* 1978) as being nonstatutory subject matter.†

I have briefly touched some of the types of inventions in the field of microbiology. Over the past 30 years, the most significant and commercially important inventions were in the field of antibiotics. The search for antibacterials since the development of penicillin and the "sulfa" drugs became an industry pastime. Everyone brought soil samples back from his vacation trip. Cultures were made and promising microorganisms were more extensively studied. As noted earlier, many never left the laboratory and many microorganisms produced interesting products which could not be made outside of the laboratory in commercial amounts. There were, of course, some highly successful developments, the most significant of which commercially were the tetracyclines. The litigation in which the tetracycline patent has been involved for the past 20 years has cost many millions of dollars to many companies, many states, and the United States government.

The most significant factor in claiming microbiological inventions occurs when the microorganism used to produce the invention, whether it be for a new product or in the process, is new and never before described in the literature. The aforementioned Section 112 of the Patent Statute requires a written description of the invention. The PTO went through some trying years until a definite practice was established in 1971. As explained by Levy and Wendt (1955) there were two serious problems: (1) How to describe unknown microorganisms; and (2) how to describe a chemical product produced in a microbiological process where there is no recognized chemical structure available. The authors relate the PTO solutions. The first was to require a deposit of the microorganism in a depository which is recognized as providing for such deposits and, complementary, the accession number given by such depository with as much of the taxonomic description as the inventors can supply at the time.

The attendant problems were discussed by Marcus (1975). First, there must be a depository that can handle the microorganism and preserve it for at least the life of the patent. Second, the PTO is not privy to the

---

The bacteria in the instant case are modified for the purpose of controlling oil spills by what they refer to as genetic engineering as the term is in the specification.

† On March 29, 1979, the CCPA decided together both the *Bergy* case and the *Chakrabarty* case. In doing so, they reaffirmed their original decision reversing the rejection from the PTO. Once again, it was a divided court. However, Judge Baldwin, who dissented in the original opinion, now joins the majority and writes a concurring opinion.

contract between the depositor and the depository so it has no voice in the deposit, yet its description and availability is a necessary adjunct of the patent application. Finally, can the depositor impose such restriction on availability of the microorganism as to preclude its release until at least the time of patenting unless they give permission for earlier release? Of course, the Commissioner of Patents must always have access to the deposit because it is part of the patent application filed with the PTO.

It was not until the *In re Argoudelis et al.* decision that the practice in the PTO was settled. The PTO issued guidelines in 1971 which are now in full use and have really minimized problems occurring in the United States. However, even with these guidelines, problems arose which, I believe, are now fairly well settled. In drafting Patent Reform Legislation, the drafters provided for the deposit of microorganisms in a public depository as designated by the Commissioner of Patents. The original legislation, as noted *supra,* also required a deposit in the United States. This caused consternation in the profession. Even though other countries at present require such a deposit under their national patent laws in a depository in that country, the prospect of that occurring in the United States appeared to have earthshaking repercussions.

In at least one instance known to me, the microorganism already deposited in a public depository in Europe could not be imported into the United States due to customs and agriculture regulations because of its pathogenicity. The following problem then arose: Can the United States grant a patent for an invention when the public in the United States can never practice the invention?

B. Depositories: Public and International

There was another problem. The deposit of a culture in a public depository, under the guidelines, was required to be open to the public once the United States patent issued. The question, however, was the duration of availability of that culture. The fees the depositor paid theoretically stopped when the patent issued and certainly the deposit could be maintained forever. The "quid" for a patent was the giving of the invention to the public when the patent monopoly expired. But did the inventor or his assigns worry about the status of the deposit when their patent rights expired? These were serious problems since the public would have nothing when the patent expired if the "deposit" was no longer available.

Then, too, what were the controls on depositories? Some were arms of the national government, some were in universities, some were private institutions subsidized by grants, and some were company managed.

These problems became the impetus for the convening of a meeting to consider the legal and scientific aspects of an international system for the deposit of microorganisms for patent purposes. The meeting was under the auspice of World Intellectual Property Organization (WIPO). In the 3 years since the first meeting in 1974, a treaty has been approved and is now being considered for ratification. The Budapest Treaty was approved in June 1977 and is presented in Appendix 5.

The requirements of the national laws of the various countries were considered as well as the various aspects establishing international depositories or, at the least, national depositories recognized by all the participating countries. The establishment of one or more centralized international depositories was considered less advisable by the committee of experts than the recognition of national depositories whose existence would be guaranteed by the signatory nation. These recognized depositories would accept deposits for patent purposes. A deposit in any nationally recognized depository would be recognized for establishing priority rights and the single deposit in such a recognized depository would render the disclosure of a patent application "enabling" for disclosure purposes in the patent system of any of the participating nations where the application is filed.

The treaty made provisions for the status of an International Depository Authority, with assurances by the contracting state guaranteeing its existence and that the depository has the necessary staff and facilities. Of course, there are many other conditions. The requirement of a receipt of the depositor, the maintaining of the deposit, the requirement for secrecy, the ability to supply a sample when properly requested, and the ability to maintain the deposit for a designated period beyond that of the patent grant. This period is at least 30 years after the deposit, or 5 years after the last request.

There are other conditions. The virility of the deposit is checked periodically and if necessary new samples can be requested of the depositor. The depository must supply a sample of any deposited microorganism at the request of the Industrial Property Office (Patent Office) of any contracting state where an application for patent has been filed. In addition the depository will supply samples to those authorized to receive as (1) either designated by the depositor or (2) on evidence that the patent application has been published in accordance with the laws of the country where the application was filed.

Furnishing samples is made interesting. The depositor will be notified by the depository of the furnishing of samples on request, and no charge is made for furnishing a sample to a patent office, although fees may be charged either to the depositor or the requestor as per the conditions set forth in the deposit contract.

A list of recognized depository authorities will be published by the International Bureau under the treaty and from time to time each patent office will furnish the depositories with lists of publications or patents issued in that country which refer to accession numbers in the depositories.

The signatories to the treaty constitute a union (Budapest Union) and members thereof are limited to states that are members of the Paris Union.

Thus a procedure of limiting the number of deposits an applicant for patent need make is being provided by treaty. In other words, a deposit in any signatory country will suffice for a priority claim for an application relying on that deposit for its disclosure filed in that country. And when applications are filed in other countries that deposit will suffice if the proviso that it will be made available after publication of the application is complied with (Behr, 1975). Of course, there might well be national laws that will require a subsequent deposit in their country in connection with a patent application filed under its national laws. However, this later deposit in all probability need not be made before filing of the application but at any time during its pendency so that nationals of that country need not seek to obtain the culture from a depository located abroad. At present time, this is the practice in Japan. An application filed in Japan in connection with an earlier convention application in the United States relies on the United States deposit for its disclosure support. Japan will accept the application for filing and examination but will require a deposit in Japan before the patent is granted there.

Other deposits under national law may occur, but basically the treaty provisions will serve the needs of the international patent community.

We have discussed the practice about depositing and we have shown the international recognition and control of depositories. Now, what are public depositories? Basically they are culture collections with their own curators and are many in number in the United States. Most are private or for company use. As far as I know there are only two public depositories that serve the patent community at present in the United States. There is the ARS Culture Collection of the Agricultural Research Service, U.S. Department of Agriculture, located at the Northern Regional Research Laboratory, Peoria, Illinois. As reported by Hesseltine *et al.* (Table I), the culture collection came into being with that laboratory about 1940 and began with cultures from Dr. Charles Thom's collection. The NRRL services yeasts, molds, bacteria, and actinotimycetes. The designation of deposits in that culture collection follow the letters NRRL. No charge is made for the deposits or their maintenance, or for their distribution "to bona fide scientists in agriculture, in industry, and in education in the

United States and also to those outside the United States as a gesture of international cooperation and good will."

This culture collection does not publish a catalog or list of its cultures. Of course, cultures in connection with patent applications became publicly available when the U.S. patent issues and identifies the microorganism used as being in deposit with the ARS. While the application is pending, the usual secrecy and restriction of accessibility is provided.

The collection at the American Type Culture Collection, located in Rockville, Maryland, was started in 1925 (Table I). At present there is a staff of about 100. The designation of deposits follows the letters ATCC, and the precautions of secrecy and restricted access to deposits in connection with patent applications is maintained. The ATCC publishes a catalog from time to time. Provision is made at ATCC for storage of cultures of bacteria, yeasts, fungi, plant and animal virus, Rickettsiae, protozoa, and animal cell cultures (cell lines). The ATCC charges a fee for deposits and no fee after the patent issues, but requestors of culture samples must pay a service charge.

The ATCC, while funded in part from government grants, is not a governmental arm. In 1976 the ATCC issued a new regulation that in connection with patent applications in the United States and other countries, there is a flat one-time fee of $475 for 25 years of maintenance. At this time we do not know if this will be changed in view of the treaty requirements for at least 30 years of maintenance. The ATCC also has a special fee for notification of the depositors as to who requests samples. Presumably a new fee schedule will be provided when the treaty comes into force.

As far as is now known, the treaty will not convert the "guaranteed" international depositories into governmental agencies where such relationship does not now exist. The independence of the scientific organization serving the scientific as well as the patent community remains. Insofar as that international depository must serve the worldwide patent community, however, a government guarantees their existence. The overall result of nations agreeing to sponsor such institutions not only for international good will but for the protection of industrial property by the patent grant to advance the useful arts brought this treaty to fruitation in a relatively short period of time. The core of the problem—an international depository—that will safely store the culture deposit and make it available for a period of 30 years from the date of deposit or for at least 5 years after the last request is guaranteed by the signatories to the treaty.

In so doing, each state that has an Industrial Property Office—a Patent Office—thus becomes not only a guarantor of the depository's existence but a guarantor of its patent system to see that on patenting or publication the deposit will become available to the public. True, the national laws of the respective nations may vary in the way the deposit is to be made, but overall one deposit is all that is necessary at a great saving of time and money to the users of the patent systems.

While no patent reform legislation is now pending in Congress, the last of the bills, S.2255—passed by the U.S. Senate in 1976—as mentioned, *supra,* had statutory provisions for deposit of microorganisms as providing a description of the invention required by 35 U.S.C. 112, where the microorganism is not available publicly and not otherwise capable of being described in writing [Sections (f) and (g)]. Also proposed is a companion section providing for the grant of priority to an application first applied for in a foreign country which acknowledged therein a deposit in a public depository [Section 119 (d)].

The proposed European Patent Treaty, to be in operation in 1978, by the establishment of a European Patent Office has a Rule 28 (Appendix 6) dealing with the deposit of microorganisms. The conditions are quite similar to those of the Budapest Treaty.

With respect to export of samples of the culture deposits from the United States to Europe, it is my understanding that the ATCC has obtained permission by appropriate export license from the Department of Commerce to transport samples of microorganisms to the West German Patent Office in accordance with the requirements as set forth in the aforementioned bakers' yeast decision.

C. Patentability of Microorganisms

The "burning question" in patent law today is the status of claims for microorganisms per se.

The subject has been introduced by reference to some case law, *supra.* As a product, the microorganism per se is generally found in nature and hence is unpatentable. But what about variations of the microorganism? There may be mutations some of which occur naturally, on storage and some of which are induced. In either case, is the new microorganism a natural product?

In the discussion above, it has been pointed out that most inventions are for the use of bacteria as components or intermediates of microbiological processes and as ingredients in compositions. Many patents have issued on both uses of such microorganisms and in which microorganisms are components of compositions.

In addition, animal and plant cells illustrated by deposits as cell lines in the ATCC have been used as diagnostic tools. Since these cells reproduce themselves, they have been used in mass tissue cultures in the field of testing for diseases such as malignancies, in testing for effectiveness of medicines, and in producing biological substances such as hormones. An example is the testing for the compatibility of a donor by the combining of the white cells of the donor with the white cells of the recipient.

Genetic engineering is the use of DNA to combine with a bacteria host cell and become part of its genetic complement. The ventures in recombinant DNA have been given much notoriety in recent years and the effect on patent practice must also be discussed.

The new breed of writers favors the patentability of microorganisms. Irons and Sears (1975) point out that the "great preponderance of practically useful microorganisms, however, are not natural products." Thus the application of external forces to change their characteristics results in a microorganism different from the product as formed in nature. The writers also suggest that 35 U.S.C. 101 does not limit inventions to "inanimate subject matter." This same thought is expressed by Wegner (1974). Daus *et al.* (1966) are in agreement that as a composition of matter, microorganisms are statutory subject matter and that *In re Arzberger et al.* would probably be reversed today by the CCPA. All these conjectures, however, are no substitute for statutory authority and the thought raised by the "burning question" will be the effect of the decision in the *In re Bergy et al.* case.

It would appear that we have avoided taking sides on the statutory standard for this important subject, while there is statutory basis for plant patents and certificates for plant varieties. It is noted that the statutory basis for deposits of microorganisms is still lacking while we have provided, through case law and PTO Guidelines, the basis for the practice. This is a strange situation.

As I pointed out above, the refusal in the *Arzberger* case to find bacteria within the purview of the plant patent statute as not within the congressional intent as to the word "plant," this being used in the common language of the people and not in any more specific sense.

The refusal to allow claims to microorganisms is further noted as dictum by Judge Rich's language in the *Mancy* case where his position was based on the product by nature theory. But consider his own statement in the *Bergy et al.* case as to his "ill-considered language" in the *Mancy et al.* case. Further, microorganisms are known to be unstable and it raises a question as to the scope of a claim for such a product and how it would be interpreted in the case law if it is subject to test in an infringement action. Would the same mutation or change occur on each application of the process conditions? This is a real problem in an area of such technical complexity, particularly since a living organism is involved.

The most interesting discussion of the problem, however, is by

Hayhurst (1971). The author not only poses the questions as to the patentability of organisms but as to man-made processes for subjecting microorganisms to treatments which result in new mutations. If the new mutant is useful, can the process be protected? which is aside from the question as to whether the product is itself patentable subject matter from a statutory viewpoint. The further question is whether the instructions for the preparation work always, invariably, or occasionally, and if they do work, will a patent issue?

The discussion takes on added importance since many countries did not allow patents for medicines, and most countries did not allow patents for chemical products per se. The Federal Republic of Germany adopted guidelines for a product system (Richtlinien, 1968). Japan followed as of June 1, 1976. In spite of that, however, claims for microorganisms are still in an ethereal state. The importance of product claims is apparent. They dominate all methods for their preparation and all uses later discovered. Processes can be altered to avoid infringement, but if the product used is patented, the patent owner dominates its uses and need not undertake to prove that the process used by others infringes his patented process.

While the CCPA's decision in *In re Argoudelis* prodded the U.S. PTO into the guidelines, and there is statutory provision proposed for microorganisms deposits if the Patent Reform Legislation is enacted, what is the road product claims for microorganisms must take? Would it be better for the PTO to propose that the statute either provide for such subject matter or preclude it and save the unnecessary expense to applicants that such cases engender?

The concern of the PTO is not without merit. If living microorganisms are held to be statutorily acceptable, would the PTO be inundated with applications for patents for newly discovered microorganisms? Of course each would have to be found useful and there would have to be a disclosed process of preparing it—since if it were in naturally occurring form it would not be new under 35 U.S.C. 101.*

*In a concurring opinion in *Chakrabarty*, in a footnote. Chief Judge Markey states:

> As with Fulton's steamboat "folly" and Bell's telephone "toy," new technologies have historically encountered resistance. But if our patent laws are to achieve their objective, extra-legal efforts to restrict wholly new technologies to the technological parameters of the past must be eschewed. Administrative difficulties, in finding and training Patent and Trademark Office examiners in new technologies, should not frustrate the constitutional and statutory intent of encouraging invention disclosures, whether those disclosures be in familiar arts or in areas on the forefront of science and technology.

D. Genetic Engineering

Another concern as to patentability of microorganisms is due to the advent of "genetic engineering"—the rearrangement of the basic genetic material of living things.† Rearrangement of the atoms of DNA, which is known to control the growth and reproduction of living cells, leads to a new form of DNA, known as recombinant DNA. These new molecules can enter a host cell, such as a bacterium, and become a part of its permanent genetic structure. New living microorganisms are formed. However, gene exchange among microorganisms and viruses probably occurs spontaneously in nature. At present, attempts to control recombinant DNA research are being made in Congress. The PTO issued guidelines for accelerated examination of patent applications in this field and then withdrew the practice after a joint meeting between the Secretaries of Commerce and HEW (Appendixes 7 and 8).

The scientific community is alarmed by the potential intervention of government into research activity. Are the dangers and hazards such as to warrant government intervention into the pursuit of scientific knowledge to protect the public welfare? Is there a possibility of antibiotic-resistant genes being produced? Michael Dukakis (1977), governor of Massachusetts, spoke out in *Time* magazine:

> Genetic manipulation to create new forms of life places biologists at a threshold similar to that which physicists reached when they first split the atom. I think it is fair to say that the gene is out of the bottle.

Rifkin (1977) commented that "General Electric is already out in front with the announcement that it has applied for a patent on a tiny microorganism that can eat up oil spills." This article even mentions the companies that are involved in recombinant DNA research almost as if a big

†In one of the Dissents, in *Chakrabarty*, Judge Baldwin notes:

> . . . a modified natural product does not become statutory subject matter until its essential nature has been substantially altered. The issue in the present case becomes whether the modification effected by appellant altered the essential nature of the starting material.
>
> . . . I believe that the essential nature of the unpatentable organism with which applicant started was its animateness or life. Appellant has not changed this essential nature; he has not created a new life. Rather, he has merely genetically grafted an extra plasmid on to the organism and, thereby, made the organism better in cleaning up oil spills. While this improvement in oil digesting ability does exclude the new organism from classification as a mere product of nature, like the borax-impreganted orange which was a better commercial product because it had a longer shelf life, this improvement in the utility for which the unpatentable starting material was already suited does not change the essential nature of the starting material and does not make the modified thing statutory subject matter.

secret is let out of the bag—nobody wanted to tell—nobody wanted the government to know they are in the field.

These are the new problems. At the end, the patent system will be the recipient of the research if (1) patent protection is permitted, and (2) if patent protection is sought. Whether patents will be allowed depends on the first two caveats, since without patent applications, there can be no patents.

Then too, the ultimate decision in the *Bergy et al.* case will decide if microorganisms are patentable subject matter—at least until there is statutory provision for the subject matter. This is the status of the "burning question" stated at the start of this section.

Research in inventions involving the action of microorganisms has had a peripheral problem. Unlike chemical compounds, which are readily reproducible once disclosed, microorganisms which are on deposit in a depository until a patent issues are not readily reproducible. Thus, industrial espionage, the stealing of cultures, puts someone else in possession of the "knowhow" not ordinarily available until the patent issues. The culture can even be taken abroad and used to manufacture valuable pharmaceuticals in countries where patent protection for the original owner was not available. So activity in the field of microorganisms presents an additional factor not present in ordinary research. This problem puts a burden on research organizations, the public depositories, and those who are permitted to obtain samples of microorganism cultures during the pendency of the patent application. Thus, this appears to be a reason why the depositories are asked to notify the depositor who seeks samples of the microorganisms even after the patent is obtained.

The industrial espionage problem came to light by the complaint by *American Cyanamid Co.* (1964) against a former employee who was charged with removing cultures of microorganisms from the research facilities.

The case points up the importance of microorganism deposits and the reason for some sort of controls on their availability. They are not like chemicals, purchasable by the barrel or carload. They could impose danger if improperly handled, but their importance and value to industry and science is tremendous.

IV. SUMMARY

The merging of legal principles and a scientific subject presents many problems. Microbiology presented such problems when it became the subject of valuable patent applications. The discussion above brings together these two disciplines and shows how an attempt was made domestically, and on an international basis, to bring order to the treatment of the subject matter.

APPENDIX 1

2,449,886

STREPTOMYCIN AND PROCESS OF PREPARATION

Selman A. Waksman, New Brunswick, and Albert Schatz, Passaic, N. J., assignors to Rutgers Research and Endowment Foundation, a nonprofit corporation of New Jersey

No Drawing. Application February 9, 1945, Serial No. 577,136

13 Claims. (Cl. 260—236.5)

10. A process for the production of streptomycin that comprises growing a culture of a streptomycin-producing strain of *Actinomyces griseus* in a medium containing corn steep liquor, at a suitable incubation temperature and for a suitable period of cultivation, to form streptomycin in the culture broth, separating the culture broth from the organism growth, and recovering streptomycin from the broth.

11. A process for the production of streptomycin that comprises growing a culture of a streptomycin-producing strain of *Actinomyces griseus* at a suitable incubation temperature and for a suitable period of cultivation, to form streptomycin in the culture broth, separating the culture broth from the organism growth, adsorbing streptomycin from the broth, and recovering the adsorbed streptomycin.

12. A process for the production of streptomycin that comprises growing a culture of a streptomycin-producing strain of *Actinomyces griseus* at an incubation temperature of 22–28° C. for a time of the order of 6–12 days for stationary cultivation and 2–4 days for submerged aerobic cultivation, to form streptomycin in the culture broth, separating the culture broth from the organism growth, and recovering streptomycin from the broth.

13. Streptomycin.

SELMAN A. WAKSMAN

ALBERT SCHATZ

APPENDIX 2

2,482,055

AUREOMYCIN AND PREPARATION OF SAME

B. M. Duggar

Application February 11, 1948, Serial No. 7,592

8 Claims. (167–65)

I claim:

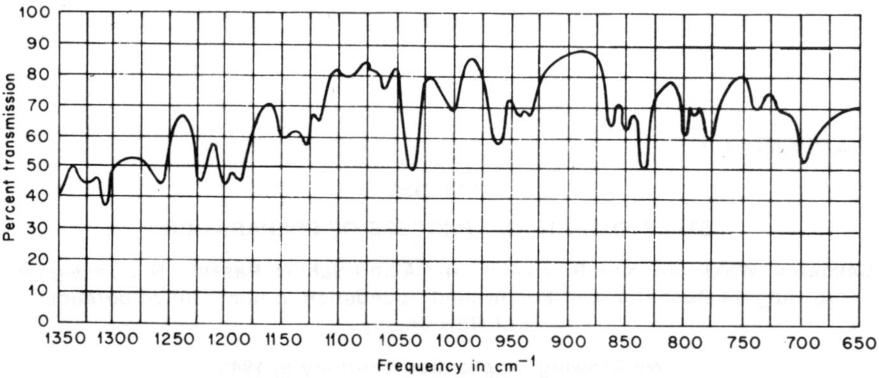

FIGURE 1. Infrared absorption spectrum of HCl, salt of antibiotic.

1. Substances effective in inhibiting the growth of Gram positive and Gram negative bacteria selected from the group consisting of a substance capable of forming salts with acids, containing the elements carbon, hydrogen, nitrogen, chlorine, and oxygen, being very soluble in pyridine, soluble in methanol and in acetone and being slightly soluble in ethanol and in water, its crystals having a refractive index parallel to elongation between about 1.674 and 1.694, and exhibiting characteristic absorption bands in the infra red region of the spectrum when suspended in a hydrocarbon oil in solid form at the following frequencies expressed reciprocal centimeters: 3420, 1643, 1609, 1580, 1523, 1302, 1231, 1209, 1121, 1080, 1050, 969, 943, 867, 844, 825, 805, 794, 788, 733, 713 and the acid salts of said substance.

2. A substance effective in inhibiting the growth of Gram positive and Gram negative bacteria which is capable of forming salts with acids, containing the elements carbon, hydrogen, nitrogen, chlorine, and oxygen, being very soluble in pyridine, soluble in methanol and in acetone and being slightly soluble in ethanol and in water, its crystals having a refractive index parallel to elongation between about 1.674 and 1.694, and exhibiting characteristic absorption bands in the infra red region of the spectrum when suspended in a hydrocarbon oil in solid form at the following frequencies expressed in reciprocal centimeters: 3420, 1643, 1609, 1580, 1523, 1302, 1231, 1209, 1121, 1080, 1050, 969, 943, 867, 844, 825, 805, 794, 788, 733, and 713.

3. The monohydrochloride salt of an organic substance said salt being effective against Gram positive and Gram negative bacteria, being soluble in water and in methanol and very slightly soluble in acetone, containing the elements carbon, hydrogen, nitrogen, chlorine, and oxygen, the crystals of which have refractive indices of

$\alpha = 1.633 \pm 0.005$; $\beta = 1.705 \pm 0.005$; and

$\gamma = 1.730 \pm 0.005$, and exhibiting characteristic absorption bands in the infra red region of the spectrum when measured in the solid state while suspended in hydrocarbon oil at the following frequencies expressed in reciprocal centimeters: 3295, 3050, 1665, 1615, 1575, 1523, 1310, 1263, 1225, 1044, 1009, 969, 863, 851, 840, 800, 770, and 698.

4. A process for the production of aureomycin which comprises growing under aerobic conditions a culture of *Streptomyces aureofaciens* in an aqueous medium having a pH between 4 and 7 and containing a soluble carbohydrate, a source of assimilable nitrogen and essential mineral salts at temperatures within the range 20°C. to 35°C. for a period of time of about twenty-four to forty-eight hours whereby the aqueous medium is fermented and aureomycin is produced.

5. A method of producing aureomycin which comprises the step of introducing a culture of *Streptomyces aureofaciens* into an aqueous nutrient liquor having a pH between about 6 and 7 and containing fermentable carbonaceous and nitrogenous substances and mineral salts and fermenting said liquor aerobically until the pH of the liquor has dropped to below about 5.

6. A process of producing aureomycin, which comprises the steps of introducing spores of the fungus *Streptomyces aureofaciens* in an aqueous liquor containing as nutrient material a usable carbonaceous substance, a source of assimilable nitrogen, and mineral substances essential for the growth of the fungus and after said liquor has been fermented by the said fungus at a temperature within the range 20° to 35°C. at a pH between about 4 and 7 separating the insoluble mycelia from the aqueous solution.

7. A process which comprises the steps of aerobically fermenting an aqueous nutrient liquor at a temperature within the range 20° to 35°C. at a pH between about 4 and 7 with the fungus *Streptomyces aureofaciens* whereby aureomycin is produced.

8. A method which comprises the steps of growing the fungus *Streptomyces aureofaciens* in an aqueous solution containing 0.1% to 5.0% by weight of a nitrogenous substance, 0.5% to 5.0% by weight of a usable carbonaceous material, and having a pH between about 4 and 7 and a temperature of between 20°C. and 35° C., while aerating the liquor, whereby aureomycin is produced.

BENJAMIN M. DUGGAR

APPENDIX 3

2,699,054

TETRACYCLINE

Lloyd H. Conover, Oakdale, Conn.

No drawing. Application October 9, 1953, Serial No. 385,041

18 Claims. (Cl. 260–559)

What is claimed is:

1. A compound chosen from the group consisting of tetracycline, the mineral acid salts of tetracycline, the alkali metal salts of tetracycline and the alkaline earth metal salts of tetracycline.

2. Tetracycline.

3. Mineral acid salts of tetracycline.

4. Alkali metal salts of tetracycline.
5. Alkaline earth metal salts of tetracycline.
6. Tetracycline hydrochloride.
7. A process for the preparation of tetracycline hydrochloride which comprises contacting an organic solvent solution of chlortetracycline with hydrogen in the presence of a catalyst selected from the group consisting of palladium, platinum and Raney nickel until approximately one mole of hydrogen has reacted with each mole of chlortetracycline used.
8. A process for the preparation of tetracycline hydrochloride which comprises contacting a solution of chlortetracycline in an inert polar organic solvent with hydrogen in the presence of a palladium catalyst at about room temperature and under an elevated pressure up to about 200 p. s. i. until approximately one mole of hydrogen has reacted with each mole of chlortetracycline used.
9. A process for the preparation of tetracycline which comprises reacting one mole of chlortetracycline in an inert organic solvent with approximately one mole of hydrogen in the presence of palladium as a catalyst and recovering the tetracycline thus formed.
10. A process as claimed in claim 9, wherein the reaction is performed in the presence of a basic, acid binding agent.
11. A sodium salt of tetracycline.
12. A calcium salt of tetracycline.
13. A solid antibiotic preparation essentially comprising tetracycline hydrochloride.
14. Crystalline tetracycline melting at 170–175° C. with decomposition and having an optical rotation of $[\alpha]_D^{25} = 239°$, when dissolved in methanol at a concentration of 1%.
15. An antibiotic composition useful in the treatment of various infections in animals, whose antibiotic activity is due primarily to the presence of tetracycline.

APPENDIX 4

2,931,798

KANAMYCIN AND THE PROCESSES FOR THE PREPARATION THEREOF

Hamao Umezawa, Kenji Maeda, and Masahiro Ueda, Tokyo, Japan

Application December 16, 1957, Serial No. 703,096

Claims priority, application Japan September 5, 1956

10 Claims. (Cl. 260–210)

We claim:

1. An antibiotic substance effective in inhibiting the growth of Gram-positive bacteria, Gram-negative bacteria and mycobacteria selected from the group consisting of kanamycin A and kanamycin B and acid addition salts thereof, each of said kanamycins being a substance which is soluble in water and substantially insoluble in n-butanol, ethyl acetate, butyl acetate, ether, chloroform and benzene, which forms salts with acids, which exhibits no absorpotion of untraviolet light from 220 mμ to 400 mμ, which gives a positive reaction to ninhydrin reagent dissolved in pyridine and positive Molisch and Elson-Morgan reactions and negative Tollens, Sakaguchi, Fehling, maltol and Seliwanoff reactions, which contains only the elements carbon, hydrogen, oxygen and nitrogen, which contains free primary amino groups, which is dextro-rotatory in water and which exhibits characteristic absorption bands in the infra-red region of the spectrum when pelleted in the form of the free base in potassium bromide at the following wave lengths in microns: 2.96, 6.35, 6.48,

6.85, 7.25, 7.45, 7.86, 8.08, 9.65 and 10.4; further properties of said kanamycins being that kanamycin A base has the empirical formula $C_{18}H_{34-36}N_4O_{11}$, exhibits $[\alpha]_D^{24}$ + 146° (c = 1, 0.1 NH_2SO_4), gives a salicylidene derivative melting at 272–274° C. with decomposition, gives desoxystreptamine on strong acid hydrolysis, gives a product with an ultraviolet absorption spectrum identical to that of furfural on treatment with 40% sulfuric acid for 100 minutes at 100° C. and exhibits additional characteristic absorption bands in the infra-red region of the spectrum when pelleted in the form of the free base in potassium bromide at the following wave lengths in microns: 2.86, 2.93, 3.01, 3.03, 3.15, 3.22, 3.67, 6.23, 6.75, 6.93, 7.10, 7.33, 7.57, 7.63, 7.75, 7.82, 7.90, 7.95, 8.19, 8.43, 8.55, 8.68, 8.79, 8.90, 9.00, 9.20, 9.47, 9.85, 10.08, 10.63, 10.95, 11.25, 11.35, 11.50, 11.73, 11.95, 12.28, 12.48, 12.77 and 13.05; and that kanamycin B base exhibits $[\alpha]_D$ + 135° (c = 0.63 in water), gives a salicylidene derivative which decomposes without melting at 155–165° C., fails to give a product without melting at 255–265°C., fails to give a product with an ultraviolet absorption spectrum identical to that of furfural on treatment with 40% sulfuric acid for 100 minutes at 100° C. and exhibits additional characteristic absorption bands in the infra-red region of the spectrum when pelleted in the form of the free base in potassium bromide at the following wave lengths in microns: 3.44, 6.74, 8.28, 8.76, 9.55 and 11.15.

2. Kanamycin A free base, as defined in claim 1.
3. Kanamycin B free base, as defined in claim 1.
4. An acid addition salt of kanamycin A, as defined in claim 1.
5. An acid addition salt of kanamycin B, as defined in claim 1.
6. Kanamycin A sulfate, as defined in claim 1.
7. Kanamycin B sulfate, as defined in claim 1.
8. A process for the production of a fermentation broth containing the antibiotic described in claim 1 which comprises cultivating a strain of *Streptomyces kanamyceticus* in an aqueous carbohydrate solution containing a nitrogenous nutrient under submerged aerobic conditions until substantial antibacterial activity is imparted to said solution and then recovering said antibiotic from said solution.
9. A process according to claim 8 wherein the antibiotic is separated from the fermentation broth by adsorption on a cation exchange resin and subsequent elution therefrom.
10. A process according to claim 8 wherein the antibiotic is separated from the fermentation broth by adsorption on a cation exchange resin of the carboxylic acid type and subsequent elution therefrom by an acid.

APPENDIX 5

BUDAPEST MICROORGANISM TREATY

The Diplomatic Conference for the Conclusion of a Treaty on the International Recognition of the Deposit of Microorganisms for the Purposes of Patent Procedure was held in Budapest, Hungary from April 14 to 28, 1977. Some 31 States, one intergovernmental organization and 11 international non-governmental organizations were represented in the Diplomatic Conference. Following two weeks of negotiations, the Treaty was adopted by the Diplomatic Conference on April 27, 1976 and signed the following day by 13 States, including the United States. The Treaty will enter into force three months afrer ratification or accession by five States.

The ratification of the Treaty by the United States would give certain advantages to an inventor of a microbiological invention seeking patent protection in a number of countries. An inventor would be able to rely on a single deposit of the microorganism in an approved depositary to satisfy the disclosure requirements of all the member countries of the Union established by the Treaty. Each country will be able to nominate depositaries in its territory for approval, if the depositaries meet the scientific standards required by the Treaty. The public release of a deposit will be governed by the patent laws of the country or countries under which the deposit was made.

We are seeking the views of the public on the advisability of ratification by the United States. Views on the ratification of the Treaty should be submitted to this Office prior to October 14, 1977. If you have any questions or wish further information, please write or call Mr. Michael K. Kirk, Director, Office of Legislation and International Affairs, c/o The Commissioner of Patents and Trademarks, Washington, D.C. 20231 (telephone (703) 557-3065).

C. MARSHALL DANN

August 1, 1977 *Commissioner of Patents and Trademarks*

(The Treaty and Regulations below are reproduced from the May, 1977 issue of *Industrial Property*, published by the World Intellectual Property Organization, Geneva Switzerland.)

Budapest Treaty on the International Recognition of the Deposit of Microorganisms for the Purposes of Patent Procedure*

Done at Budapest on April 28, 1977

TABLE OF CONTENTS**

**Official English title.*

Source: International Bureau of WIPO.

Note: This treaty was signed on April 28, 1977, at Budapest, by the following States: Bulgaria, Denmark, Finland, France, Germany (Federal Republic of), Hungary, Italy, Netherlands, Norway, Spain, Switzerland, United Kingdom, United States of America; it will remain open for signature at Budapest until December 31, 1977.

**This Table of Contents is added for the convenience of the reader (*Editor's Note*).

Article 17: Denunciation of the Treaty
Article 18: Signature and Languages of the Treaty
Article 19: Deposit of the Treaty; Transmittal of Copies; Registration of the Treaty
Article 20: Notifications

APPENDIX 6

Rule 28

Requirements of European patent application relating to micro-organisms

(1) If an invention concerns a microbiological process or the product thereof and involves the use of a micro-organism which is not available to the public, the European patent application and the resulting European patent shall only be regarded as disclosing the invention in a manner sufficiently clear and complete for it to be carried out by a person skilled in the art if:

(a) a culture of the micro-organism has been deposited in a culture collection not later than the date of filing of the application;

(b) the application as filed gives such relevant information as is available to the applicant on the characteristics of the micro-organism;

(c) the culture collection, the date when the culture was deposited and the file number of the deposit are given in the application.

(2) The information referred to in paragraph 1(c) may be submitted within a period of two months after the filing of the application. The communication of this information shall be considered as constituting the culture deposited being made available to the public in accordance with this Rule.

(3) The culture deposited shall be available to any person upon request from the date of publication of the application. The request shall be addressed to the culture collection and shall be deemed to have been made only if it contains:

(a) the name and adddress of the person making the request;

(b) an undertaking vis-à-vis the applicant or proprietor not to make the culture available to any other person;

(c) where the request is made before the date of publication of the mention of the grant of the patent, an undertaking vis-à-vis the applicant to use the culture for experimental purposes only.

(4) A copy of the request shall be communicated to the applicant or proprietor.

(5) The undertaking provided for in paragraph 3(b) shall cease if the application is refused or withdrawn or is demeed to be withdrawn or, if a patent is granted, on the expiry of the patent in the designated State in which it last expires.

(6) The undertaking provided for in paragraph 3(c) shall cease if the application is refused or withdrawn or is deemed to be withdrawn or, if a patent is granted, on the date of publication of the mention of the grant of the patent.

(7) The undertaking under paragraph 3(c) is not applicable insofar as the person making the request is using the culture under a compulsory licence. The term "compulsory licence" shall be construed as including ex officio licences and the right to use patented inventions in the public interest.

(8) The President of the European Patent Office shall publish in the Official Journal of the European Patent Office the culture collections which will be recognised for the purpose of this Rule and shall conclude agreements with them, in particular in respect of the deposit, storage and availability of cultures.

Article 53—Exceptions to Patentability—European patents shall not be granted in respect of: (b) plant or animal varieties or essentially biological processes for the production of plants or animals; this provision does not apply to microbiological processes or the products thereof.

APPENDIX 7

PATENT AND TRADEMARK OFFICE NOTICES

Recombinant DNA

Accelerated Processing of Patent Applications for Inventions

In recent years revolutionary genetic research has been conducted involving recombinant deoxyribonucleic acid ("recombinant DNA"). Recombinant DNA research appears to have extraordinary potential benefit for mankind. It has been suggested, for example, that research in this field might lead to ways of controlling or treating cancer and hereditary defects. The technology also has possible applications in agriculture and industry. It has been likened in importance to the discovery of nuclear fission and fusion. At the same time concern has been expressed over the safety of this type of research. The National Institutes of Health (NIH) has released guidelines for the conduct of research concerning recombinant DNA. "Guidelines for Research Involving Recombinant DNA Molecules," published in the Federal Register of July 7, 1976, 41 F.R. 27902–27943. NIH is sponsoring experimental work to identity possible hazards and safety practices and procedures.

In view of the exceptional importance of recombinant DNA and the desirability of prompt disclosure of developments in the field, the Assistant Secretary of Commerce for Science and Technology has requested that the Patent and Trademark Office accord "special" status to patent applications involving recombinant DNA. Upon appropriate request, the Office will make special patent applications for inventions relating to recombinant DNA, including those that contribute to safety of research in the field. Requests for special status should be written, should identify the application by serial number and filing date, and should be accompanied by affidavits or declarations under 37 CFR 1.102 by the applicant, attorney or agent explaining the relationship of the invention to recombinant DNA research. Requests also must include a statement that the NIH guidelines cited above or as amended in the future are being followed in an experimentation in this field, except that the statement may include an explanation of any deviations considered essential to avoid disclosure of proprietary information or loss of patent rights. The requests will be handled in the same manner as requests to make applications special that relate to energy or environmental quality. See Manual of Patent Examining Procedure 708.02.

Dated: Jan. 7, 1977.

C. MARSHALL DANN
Commissioner of Patents and Trademarks

Approved: January 10, 1977.
BETSY ANCKER-JOHNSON,
Assistant Secretary for Science and Technology.
[FR Doc. 77–1155; Filed 1–12–77; 8:45 am]

APPENDIX 8

PATENT AND TRADEMARK OFFICE NOTICES

Recombinant DNA

Suspension of Accelerated Processing of Patent Applications for Recombinant DNA Research Inventions

On January 10, 1977, the Patent and Trademark Office issued a notice, published in the Federal Register of January 13, 1977, 42 FR 2712–2713, which provided for the accelerated processing of patent applications for inventions relating to Recombinant DNA, including those that contribute to safety of research in the field.

In order that the Federal Interagency Committee for Recombinant DNA Research may consider recommendations concerning research conducted by the private sector in this field, that part of the referenced notice dealing with accelerated processing of patent applications for Recombinant DNA research inventions is suspended until further notice.

That part of the referenced notice dealing with accelerated processing of patent applications relating to safety of research in this field will remain in force.

Dated: March 3, 1977.

RENE D. TEGTMEYER
Acting Commissioner of Patents and Trademarks

Approved: March 3, 1977.
BETSY ANCKER-JOHNSON,
Assistant Secretary for Science and Technology.
[FR Doc. 77–6908; Filed 3–8–77; 8:45 am]

REFERENCES

American Cyanamid Co. v. Fox (1964). 140 U.S.P.Q. 199 New York Supreme Court, New York County.
In re Argoudelis, De Boer, Eble and Herr (1970). 168 U.S.P.Q. 99 (C.C.P.A.–1970).
In re Cornelius F. Arzberger (1940). 46 U.S.P.Q. 32 (C.C.P.A.–1940).
Bäckerhefe (1975). GRUR 1975, p. 430; German Supreme Court.
Behr, O. M. (1975). *J. Pat. Off. Soc.* **57,** 28–45.
In re Bergy, Coats and Malik (1977). 195 U.S.P.Q. 344 (C.C.P.A. 1977).
Bunch, R. L., and McGuire, J. M. (1953). U.S. Patent 2,653,899.
In re Ananda M. Chakrabarty (1978). 197 U.S.P.Q. 72. (CCPA—1978).
Conover, L. H. (1955). U.S. Patent 2,699,054.
Daus, D., Bond, R. T., and Rose, S. K. (1966). *Idea* **10,** 87–100.
Duggar, B. M. (1949). U.S. Patent 2,482,055.
Dukakis, M. (1977). *Time* April 18, p. 32.
Feldman v. Aunstrup (1975). 186 U.S.P.Q. 108 (C.C.P.A.—1975).
Funk Bros. Seed Co. v. Kalo Inoculant Co. (1948). 333 U.S. 127; 76 U.S.P.Q. 280.
Guaranty Trust Company of New York et al. v. Union Solvents Corporation (1931). 12 U.S.P.Q. 47 (Dist. Ct. Delaware, 1931).
Guidelines for Deposit of Microorganisms (1971). April 29; 886 O.G. Pat. Off. 638.
Hanson, F. R., and Eble, T. E. (1953). U.S. Patent 2,652,356.
Hayhurst, W. L. (1971). *Ind. Property* July, pp. 189–198.
Irons, E. S., and Sears, M. H. (1975). *Annu. Rev. Microbiol.* **29,** 319–332.
Ex parte Eleanor J. Kropp (1959). 143 U.S.P.Q. 148 (PTO Board of Appeals 1959).

Levitt, J. (1976). "Current Problems in Patent Law," pp. 49–95. Practising Law Institute.
Levy, D., and Wendt, L. B. (1955). *J. Pat. Off. Soc.* **37,** 855.
In re Mancy, Florent and Preud'Homme. (1974). 182 U.S.P.Q. 303 (C.C.P.A. 1974).
Manual of Classification (1936). U.S. Patent and Trademark Office, Class 195—Chemistry, Fermentation Original Classification.
Marcus, I. (1975). *Adv. Appl. Microbiol.* **19,** 77–83.
Parker v. Bergy (1978). 198 U.S.P.Q. 257 (U.S. Supreme Court—1978).
Parker v. Flook (1978). 198 U.S.P.Q. 193 (U.S. Supreme Court—1978).
Patents for Plants (1954). 35 U.S.C. 161; amended Sept. 3, 1954; 68 Stat. 1190.
Plant Variety Protection Act (1972). 7 U.S.C. 2321.
Reynolds, E. L. (1967). Interference Av. Bv. C; 159 U.S.P.Q. 538.
Richtlinien (1968). Guidelines German Patent Office, Jan. 2 (in Blatt für Patents—, Muster,—und Zeichenwesen).
Rifkin, J. (1977). *Mother Jones Mag.* Feb/March, pp. 23–26 and 39.
Sobin, B. A., Finlay, A. C., and Kane, J. H. (1950). U.S. Patent 2,516,080.
Umezawa, H., Maeda, K., and Ueda, M. (1960). U.S. Patent 2,931,798.
Waksman, S. A., and Schatz, A. (1948). U.S. Patent 2,449,866.
Wegner, H. C. (1974). 5 IIC 285–291.
Weizmann, C. (1919). U.S. Patent 1,315,585.

Subject Index

D

E

Microbial Technology

SECOND EDITION/VOLUME I

Second Edition/Volume I

Microbial Technology

Microbial Processes

Edited by

H. J. PEPPLER
Universal Foods Corporation
Milwaukee, Wisconsin

D. PERLMAN
School of Pharmacy
University of Wisconsin
Madison, Wisconsin

ELSEVIER

AMSTERDAM • BOSTON • HEIDELBERG • LONDON
NEW DELHI • NEW YORK • OXFORD • PARIS • SAN DIEGO
SAN FRANCISCO • SINGAPORE • SYDNEY • TOKYO
Academic Press is an imprint of Elsevier

Microbial Technology: Microbal Processes/2ed Vol. I
Peppler and Perlman

ACADEMIC PRESS
An Imprint of Elsevier
111 Fifth Avenue, New York 10003

Original ISBN: 0-12-551501-4

First Printed in India 2004
Reprinted 2006
Reprinted 2008
Reprinted 2009
Reprinted 2012

Indian Reprint ISBN: 978-81-8147-495-7

Published by Elsevier, a division of Reed Elsevier India Private Limited.

Registered Office: 305, Rohit House, 3 Tolstoy Marg, New Delhi-110 001.
Corporate Office: 14th Floor, Building No. 10B, DLF Cyber City, Phase-II, Gurgaon-122002, Haryana, India.

Printed and bound at Rajkamal Electric Press, Kundli, Haryana.

Contents

List of Contributors

Numbers in parentheses indicate the pages on which the authors' contributions begin.

Ralph F. Anderson* (1), Velsicol Chemical Corporation, Chicago, Illinois 60611

Otto Andresen (281), NOVO Industri A/S, Bagsvaerd, Denmark

Knud Aunstrup (281), NOVO Industri A/S, Bagsvaerd, Denmark

Joe C. Burton (29), Research Department, The Nitragin Company, Milwaukee, Wisconsin 53209

F. Constabel (389), Prairie Regional Laboratory, National Research Council of Canada, Saskatoon, Saskatchewan S7N OW9, Canada

I. W. Cottrell (417), Research and Development Department, Kelco Division, Merck & Company, Inc., San Diego, California 92123

Edvard A. Falch (281), NOVO Industri A/S, Bagsvaerd, Denmark

J. Florent (497), Centre Nicolas Grillet, Rhône-Poulenc Industries, Vitry-sur-Seine, France

Yoshio Hirose (211), Biochemical Department, Central Research Laboratories of Ajinomoto Company, Ltd., Kawasaki-ku, Kawasaki 210, Japan

Carlo M. Ignoffo (1), U.S. Department of Agriculture, Science and Education Administration, Federal Research, Columbia, Missouri 65205

K. S. Kang (417), Biochemical Development, Kelco Division, Merck & Company, Inc., San Diego, California 92123

W. G. W. Kurz (389), Prairie Regional Laboratory, National Research Council of Canada, Saskatoon, Saskatchewan S7N OW9, Canada

John H. Litchfield (93), Battelle, Columbus Laboratories, Columbus, Ohio 43201

Lewis B. Lockwood (355), Biology Department, Western Kentucky University, Bowling Green, Kentucky 42101

* Deceased

J. L. Martin* (187), International Minerals & Chemical Corporation, Terre Haute, Indiana 47808

Yoshio Nakao (311), Microbiological Research Laboratories, Central Research Division, Takeda Chemical Industries, Ltd., Jusohonmachi, Yodogawa-ku, Osaka, Japan

Tage Kjaer Nielsen (281), NOVO Industri A/S, Bagsvaerd, Denmark

L. Ninet (497, 529), Centre Nicolas Grillet, Rhône-Poulenc Industries, Vitry-sur-Seine, France

Hiroshi Okada (211), Biochemical Department, Central Research Laboratories of Ajinomoto Company, Ltd., Kawasaki-ku, Kawasaki 210, Japan

H. J. Peppler† (157), Universal Foods Corporation, Milwaukee, Wisconsin 53201

D. Perlman (241, 483, 521), School of Pharmacy, University of Wisconsin, Madison, Wisconsin 53706

Randolph S. Porubcan (59), Chr. Hansen's Laboratory, Inc., Milwaukee, Wisconsin 53214

J. Renaut (529), Centre Nicolas Grillet, Rhône-Poulenc Industries, Vitry-sur-Seine, France

O. K. Sebek (483), Infectious Diseases Research, The Upjohn Company, Kalamazoo, Michigan 49001

Robert L. Sellars (59), Chr. Hansen's Laboratory, Inc., Milwaukee, Wisconsin 53214

M. T. Walton** (187), International Minerals & Chemical Corporation, Terre Haute, Indiana 47808

* Present address: 1334 S. Center Street, Terre Haute, Indiana 47802.
† Present address: 5157 N. Shoreland Avenue, Whitefish Bay, Wisconsin 53217.
** Present address: R.R. 4, Loogootee, Indiana 47553.

Preface

In the decade since the first edition of "Microbial Technology" appeared, applied microbiology has changed, expanded, and diversified. As new products were introduced in this period, and greater demand for some of the old ones developed, the total fermentation capacity increased at about the same rate noted in the previous 25 years. The number of fermentation products, it is estimated, has quadrupled, while the volume of products manufactured has increased tenfold. This growth has prompted publication of a second edition, completely revised and enlarged.

To accomplish a worldwide survey of industrial microbiology and to describe its contributions to agriculture, industry, medicine, and environmental control, the editors are indebted to 57 willing and expert contributors. Their comprehensive reviews of traditional fermentations and propagations, as well as newly developed microbe-dependent processes and products, are presented in a two-volume set.

Volume I, subtitled "Microbial Processes," describes the production and uses of economic bacteria, yeast, molds, and viruses, and reviews the technologies associated with products of microbial metabolism.

Volume II, subtitled "Fermentation Technology," deals principally with fermentations and modifications of plant and animal products for foods, beverages, and feeds, while reviewing salient aspects of microbial technology: general principles, culture selection, laboratory methods, instrumentation, computer control, product isolation, immobilized cell usage, economics, and microbial patents.

H. J. Peppler
D. Perlman

Contents of Volume II

FERMENTATION TECHNOLOGY

Microbial Technology

SECOND EDITION/VOLUME I

Chapter 1

Bioinsecticides

CARLO M. IGNOFFO
RALPH F. ANDERSON*

I. INTRODUCTION

Control of insect pests costs more than one-half billion dollars annually in the United States. Yet less than 5% of this is spent for microbial insecticides, despite the fact that entomopathogens (pathogens of insects) have been suggested as controlling agents of insect pests for over a century. Synthetic organic chemical insecticides in vogue today were not even available until about 50 years after control of an insect pest was demonstrated by using an entomopathogen. Obviously, the development of microbial insecticides has been pitifully slow. However, within

* In Memorium: Ralph F. Anderson (1924–1978).

MICROBIAL TECHNOLOGY, 2nd ed., Vol. I
 ISBN 0-12-551501-4

the last decade there has been a significant revival. This was largely brought about by recent successes in developing entomopathogenic viruses and bacteria and was further fortified by the public's increased awareness of the impact of toxic, broad-spectrum chemical insecticides.

Of the nearly 1 million species of known insects only about 15,000 species are considered pests and only about 300 are sufficiently destructive to require some control. Fortunately, most insect pests have pathogenic microorganisms associated with them. About 1500 entomopathogens belonging to the fungi, viruses, protozoa, or bacteria are known. Of these, bacteria and viruses, because of their known effectiveness and relative lack of toxicity or pathogenicity to nontarget animals and plants, have been developed into commercial products in the United States.

A candidate entomopathogen, whether it is a bacterium, virus, fungus, or protozoan, must meet certain technical prerequisites before it can be developed into a microbial insecticide. Three of the most important technical prerequisites are: (1) availability or feasibility of a systematic, continuous production technology; (2) minimal or no toxicity or pathogenicity to man, nontarget animals, and plants; and (3) proved effectiveness against the intended target pest. However, from a commercial, nontechnical viewpoint the ultimate and most important question is whether this development will result in a reasonable profit to the industrial developer. Obviously, then, the decision whether or not to develop a candidate entomopathogen into a microbial insecticide is based upon favorable answers to both technical and nontechnical questions.

II. HISTORICAL BACKGROUND

Aristotle's description of diseases of the honey bee in *Historia animalium* was probably the first documentation of an insect disease. Observations of insect maladies, particularly those of bees and silkworms, were reported in early Greek and Roman literature as well as in the works of various poets and naturalists of the sixteenth to nineteenth centuries. For example, Kirby and Spence (1826) included a chapter called "Diseases of Insects" in their classical text "*Introduction to Entomology*."

Insect pathology per se probably had its beginning in the nineteenth century under the stimulus of Bassi (1835) and Pasteur (1870). Bassi was the first to demonstrate that a microorganism, the fungus *Beauveria bassiana,* could cause an infectious disease in an animal (*the silkworm*).

Bassi not only contributed to man's understanding of infectious diseases but also implied that infectious diseases could be used to control insects. Pasteur's investigations were concerned not with insect control but with control of diseases in populations of the beneficial silkworm and honey bee. These studies drew attention to the impact of diseases on insect populations and their feasibility for use as microbial insecticides.

A significant contribution to microbial control of insects was made by Metchnikoff (1879) and Krassilstschik (1888). They were the first to document that an entomopathogen, a muscardine fungus, *Metarrhizium anisopliae,* could be mass produced and applied as a microbial insecticide to control the grain weevil and the sugar beet curculio. These early efforts inspired Giard (1892), Forbes (1898), and Snow (1891) to widely use another muscardine fungus (*B. bassiana*) in an unsuccessful attempt to control chinch bugs.

Control of insect pests with bacteria was probably first attempted by d'Herelle (1914) approximately 35 years after Pasteur's description of silkworm diseases. Apparently, control was not consistent and therefore interest in bacterial pathogens was muted. However, after a lag of nearly 30 years, White and Dutky (1940) succeeded in demonstrating control of the Japanese beetle by distributing spores of the milky disease bacterium, *Bacillus popilliae.* Undoubtedly, this success stimulated other investigators to reinvestigate bacteria, and literature began appearing on the effectiveness of *Bacillus thuringiensis.* Issuance of eight patents between the years 1960 and 1963 for *B. thuringiensis* further attested to the revived interest in bacterial insecticides (Briggs, 1964). A more detailed history of the use of bacteria to control insects is presented by Steinhaus (1964), Heimpel (1965), and DeBach (1964).

Use of viruses to control insect pests was stimulated by the studies of Balch and Bird (1944) and Steinhaus and Thompson (1949) during the years immediately following World War II. This initial interest, which lagged after the first successful demonstration, is presently having a rebirth, as evidenced by the recent registration of the first viral pesticide (Ignoffo, 1973a, b,c) in the United States by the Environmental Protection Agency (EPA).

Documentation of the historical development and use of entomopathogens as microbial insecticides is an empty exercise without noting the contributions of an inspiring, enthusiastic giant, Edward A. Steinhaus, who died in 1969 (Linsley and Smith, 1970). Dr. Steinhaus not only developed the first university curriculum of insect pathology and laboratory of pathology in the United States but also wrote texts, reference sources, and numerous articles on invertebrate pathology. He was responsible for organizing the Society of Invertebrate Pathology and starting the *Journal of Invertebrate Pathology*. His guiding principles and vision still inspire and influence invertebrate pathologists. Through his efforts there are laboratories throughout the world devoted to both basic and applied research in invertebrate pathology and a flourishing society and a journal.

III. CANDIDATE MICROBIAL INSECTICIDES

A. Entomopathogenic Bacteria

From a stockpile of nearly a hundred species associated with insects, only three bacteria (*B. popilliae, B. thuringiensis,* and *B. moritai*) were developed into commercial microbial insecticides (Table I). *In vitro* and *in vivo* production processes both are used (cf. Section V,A). The bacterium *B. popilliae*, produced only in larvae of Japanese beetles, is formulated into dust (Dutky, 1963). In contrast, *B. thuringiensis* and *B. moritai* are produced by conventional fermentation techniques. It is significant that the bacterial species that have been most useful for insect control have been sporeformers, perhaps because they can be

TABLE I. Trade Names of Commercial or Experimental Preparations of Microbial Insecticides Formulated from Bacteria

Bacterial species	Trade name	Producer
Bacillus moritai	Rabirusu	Sumitomo, Japan
Bacillus popilliae	Doom, Japidemic	Fairfax Biological Labs. (U.S.A.)
Bacillus thuringiensis		
β-Exotoxin	Biotoxksybacillin	All Union Inst. Agr. Microbiol. (U.S.S.R.)
	Eksotoksin	Glavmikrobioprom (U.S.S.R.)
	Toxobakterin	Glavmikrobioprom (U.S.S.R.)
δ-Endotoxin	Agritrol	Merck & Co. (U.S.A.)
	Bakthane	Rohm & Haas (U.S.A.)
	Bactospeine	Roger Bellon (France)
	Bathurin	Chemapol-Biokrma (Czechoslovakia)
	Biospor	Farbwerke Hoechst (Germany)
	Biotrol BTB	Nutrilite Prod (U.S.A.)
	Dendrobacillin	Glavmikrobioprom (U.S.S.R.)
	Dipel	Abbott Labs (U.S.A.)
	Entobacterin	Glavmikrobioprom (U.S.S.R.)
	Insektin	Glavmikrobioprom (U.S.S.R.)
	Parasporin	Grain Proc. Lab (U.S.A.)
	Sporeine	LIBEC Laboratoire (France)
	Thuricide	Sandoz-Inc. (U.S.A.)

easily mass produced and are stable enough to be handled in commerce A relatively large number of entomopathogenic spore-forming bacteria were listed by Steinhaus (1959). However, none of the nonsporeformers, (which also can be mass cultured) have been commercialized, probably because they are not easily stabilized for field use. Safety to humans, other vertebrates, and plants was demonstrated for *B. popilliae, B. thuringiensis,* and *B. moritai* but has not been established for other candidate bacterial pathogens of insects (cf. Section V,B). Surprisingly, 19 proprietary names are reported for *B. thuringiensis* preparations (Ignoffo, 1967, 1975). Thirteen formulations are based upon the δ-endotoxin and three on the β-exotoxin. Two trade-name products are formulated from *B. popilliae* and one from *B. moritai*.

B. Entomopathogenic Viruses

Many insect pathologists believe viruses have the greatest potential for development into microbial insecticides. Entomopathogenic viruses are specific and are active against many economically important insects (Ignoffo, 1967, 1968, 1975). About 650 insect viruses are described (Martignoni and Iwai, 1977), and more are being discovered each year. In fact, there has been a phenomenal increase (over 250%) in the number of new viruses reported within the decade covered by this revision (Anderson and Ignoffo, 1967).

Just as the more useful candidate bacteria are protected by spores, the most useful insect viruses are protected by a proteinaceous inclusion body. Those without this protective body are called noninclusion viruses. The inclusion body of insect viruses varies in shape from the irregular or regular polyhedrons (PIB) characteristic of the polyhedrosis and entomopoxviruses (EPV) to ovoid bodies representative of the granulosis viruses (GIV). The site of viral replication also is commonly used to describe insect viruses, e.g., nuclear polyhedrosis (NPV) or cytoplasmic polyhedrosis (CPV). Entomopathogenic viruses are currently placed in seven genera based upon the morphology and physiobiochemical characteristics of the virion (Ignoffo, 1973a; Vago *et al.*, 1974). Latinized binomials are used to name the viruses and the suffix-*virus* is used for all the generic names, e.g., *Baculovirus, Entomopoxvirus, Iridovirus*.

The NPV and GIV, because of their specificity, safety, virulence, and stability, are probably the most promising candidates for consideration as viral insecticides. The CPV and EPV are less virulent, but they have the other characteristics of a good microbial pesticide. Noninclusion viruses, although virulent, are not currently considered likely candidates because less is known about their safety, production feasibility, and efficacy. Considerable information has been obtained on the safety and specificity of entomopathogenic viruses within the last decade (cf. Section V,B,3).

Because of their specificity, insect viruses are more difficult to produce on an industrial scale than bacteria or fungi. They are obligate parasites and consequently must be mass produced *in vivo,* i.e., in living hosts (cf. Section V,A,3). In spite of this inherent difficulty, insect viruses representing every major group have been produced and used in the field as viral insecticides (Ignoffo, 1967, 1975). In addition, more than a dozen commercial or experimental preparations of entomopathogenic viruses have been evaluated sufficiently enough to warrant a trade name designation (Table II). As of 1977, two insect viruses have been commercialized: the *Heliothis* NPV, trade-named Elcar (USA)*; and a CPV of *Dendrolimus* called Matsukemin in Japan. The U.S. Forest Service recently registered an NPV of the tussock moth, TM-BioControl-T (Martignoni and Iwai, 1978), and the gypsy moth NPV, Gypchek (F. B. Lewis, personal communication). Data are also being collated for submission of a registration of another NPV, i.e., *Autographa,* by the Agricultural Research Service. Insect viruses continue to be attractive candidates for microbial insecticides. Within the last decade production feasibility has been established, safety and specificity to nontarget organisms has been confirmed, and effectiveness against target pests has been demonstrated.

C. Entomopathogenic Fungi

No fungi are presently available as commercial microbial insecticides in the United States, but several were produced and used on a large scale in the Soviet Union (Table III). All classes are represented among the more than 500 entomopathogenic fungi (Ignoffo, 1967). Most of the promising fungi are in the class Phycomycetes and Deuteromycetes. Fungi associated with insects are considered effective pathogens if and when field conditions are optimal for spore germination, invasion, and growth. Because these conditions are not predictable, the usefulness of fungi for insect control is often questioned (Yendol and Roberts, 1970). Some mycologists (Bucher, 1964; Latch, 1965) contend that entomogenous fungi are ubiquitous and that field application for insect control is

* This paper reports the results of research only. Mention of a pesticide in this paper does not constitute a recommendation for use by the USDA, nor does it imply registration under the Federal Insecticide, Fungicide and Rodenticide Act (FIFRA) as amended. Also, mention of a commercial product in this paper does not constitute an endorsement of this product by the USDA.

TABLE II. Trade Names of Commercial or Experimental Preparations of Microbial Insecticides Formulated from Viruses

Viral type and host	Trade name	Producer
Cytoplasmic polyhedrosis		
Dendrolimus	Matsukemin	Chugai Pharmaceutical Co., Ltd. (Japan)
		Kumiai Chemical Industry Co., Ltd. (Japan)
Nuclear polyhedrosis		
Heliothis	Biotrol VHZ	Nutrilite Prod. (U.S.A.)
	Elcar	Sandoz, Inc. (U.S.A.)
	Virex	Hays-Sammons (U.S.A.)
	Viron/H	Int. Minerals & Chem. Corp. (U.S.A.)
Lymantria	Virin-ENSH	Glavmikrobioprom (U.S.S.R.)
Mamestra	Virin-EX	Glavmikrobioprom (U.S.S.R.)
Neodiprion	Polyvirocide	Indiana Farm Bureau (U.S.A.)
Orgyia	TM BioControl-1	Forest Service (U.S.A.)
Pieris	Virin GKB	Latvian Agr. Acad. (U.S.S.R.)
Prodenia	Biotrol VPO	Nutrilite Prod. (U.S.A.)
	Viron/P	Int. Minerals & Chem. Corp. (U.S.A.)
Spodoptera	Viron/S	Int. Minerals & Chem. Corp. (U.S.A.)
Trichoplusia	Biotrol VTN	Nutrilite Prod. (U.S.A.)
	Viron/T	Int. Minerals & Chem. Corp. (U.S.A.)

unnecessary. Others (Dunn and Mechalas, 1963; Müller-Kögler, 1965; Roberts and Yendol, 1971; McCoy *et al.*, 1976) are convinced that fungal insecticides are effective if properly used and that their usefulness can be increased when combined with chemical insecticides (Pristavko, 1963).

Fungal mycelium, spores, and toxins can be mass produced, either *in vitro* or *in vivo* (cf. Section V,A). Obviously, the *in vitro* method, either via submerged or surface fermentation, is preferred (Ignoffo, 1967; Dulmage and Rhodes, 1971; Roberts and Yendol, 1971). Some of the fungi that have been successful in controlling insects are *Aschersonia aleyrodis,*

TABLE III. Trade Names of Commercial or Experimental Preparations of Microbial Insecticides Formulated from Fungi

Fungal species	Trade name	Producer
Aschersonia aleyrodis	Aseronija	All Union Inst. Agr. Microbiol. (U.S.S.R.)
Beauveria bassiana	Biotrol FBB	Nutrilite Products (U.S.A.)
	Boverin	Glavmikrobioprom (U.S.S.R.)
Hirsutella thompsonii	ABG-6065	Abbott Labs (U.S.A.)
Metarrhizium anisopliae	Biotrol FMA	Nutrilite Products (U.S.A.)

B. bassiana, M. anisopliae, Entomophthora thaxteriana, and *Nomuraea rileyi* (Ignoffo, 1975).

One advantage of entomopathogenic fungi is that they are generally less specific than other groups of entomopathogens and consequently can be used against different types of insect pests. For example, species of fungi were successful in controlling scale insects, whitefly, plant hoppers, aphids, chinch bugs, phytophagous beetles, flies, caterpillars, and stored-products pests. More extensive development of fungal microbial insecticides, however, must await additional evidence of field efficacy (cf. Section V,C, 2) and safety (cf. Section V,B,3). Fortunately, there has been considerable effort in both these areas within the last decade, which increases the probability that entomopathogenic fungi can be commercially available within the next decade.

D. Entomopathogenic Protozoa

Protozoa of insects never have been industrially produced or sold as microbial insecticides, largely because they are low in virulence, unstable, and difficult to produce. Currently, production can be accomplished only *in vivo* (cf. Section V,A,3). In spite of this difficulty, at least 15 different protozoa were propagated for experimental use against insects (Ignoffo, 1967). Protozoa are promising candidates because many pest insects that are not attacked by other entomopathogens are susceptible to at least one of the 300 described species.

Only five species of protozoa are currently being seriously investigated for development into microbial insecticides. The protozoan *Nosema locustae* has been evaluated for production feasibility and safety and is currently being evaluated for efficacy against rangeland grasshoppers (Henry, 1976); *Mattesia trogodermae* is being evaluated for use against stored-products pests (Schwalbe *et al.*, 1973); *Nosema algerae* is a possible controlling agent of mosquitoes (Undeen and Maddox, 1973); *Vairimorpha necatrix* was suggested for control of caterpillar pests of soybeans (W. M. Brooks, North Carolina State University, personal communication) and *Nosema* (*Perezia*) pyrausta is being considered for control of corn borers (Frye and Olson, 1974; Lewis, 1975). Largely because of problems discussed earlier, none of these have been widely applied (Decker, 1960; Ignoffo, 1967; McLaughlin, 1973); however, successful uses were recorded against caterpillars (Weiser and Veber, 1956; Weiser, 1957; Wilson and Kaupp, 1975), grasshoppers (Taylor and King, 1937; Henry, 1976), and beetles (McLaughlin, 1966; Schwalbe *et al.*, 1973). More recently, favorable information on safety (cf. Section V,B,3), ease of production, environmental stability, and efficacy (cf. Section V,C,2) have encouraged the reconsideration of protozoa as microbial insecticides.

IV. DEVELOPMENTAL PHASES OF A MICROBIAL INSECTICIDE

The initial impetus for development of a candidate microbial insecticide or any other pesticide is the demonstration of its activity against a target pest(s). Fortunately, most of the screening of candidate microbial insecticides has been done by nature, which has relieved developers of the inherent high cost of finding one potential candidate among thousands that would have to be screened. Approximately 10% of the cost of developing a chemical insecticide is that of synthesis and screening (Anonymous, 1969). Actual development toward commercialization of a microbial insecticide usually begins with a concept in the laboratory that progresses through a pilot-plant phase and attains technical realization as a registered product (Ignoffo, 1973a, c). Three technical parameters, i.e., production feasibility, safety (man, vertebrates, beneficial insects, and plants), and effectiveness against important pests (cf. Section I), are evaluated during each phase of development (Fig. 1). These technical parameters, broad generalizations during the laboratory phase, are more closely examined at each successive step as more data are accumulated. The questions asked at each phase are: Can technical success be attained? Can technical success be translated into commercial success? Technical success is attained when an entomopathogen can be continuously produced, is safe, and will effectively control a pest.

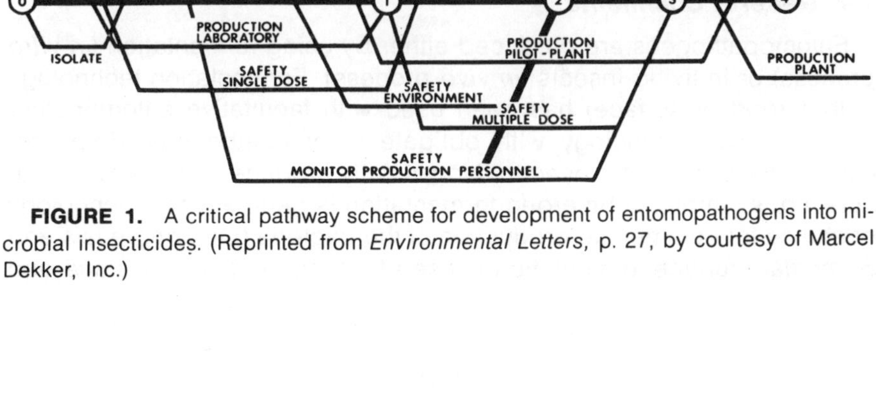

FIGURE 1. A critical pathway scheme for development of entomopathogens into microbial insecticides. (Reprinted from *Environmental Letters,* p. 27, by courtesy of Marcel Dekker, Inc.)

Commercial success is achieved when the product is produced, is effective, and can be used at costs that will return a reasonable profit to the producer. Generally, it takes 5–7 years to develop a known entomopathogen into a microbial insecticide at a cost two to five times lower than for a chemical insecticide. However, if the entomopathogen is similar to those previously investigated (*B. thuringiensis, B. heliothis*) and no problems of safety are encountered, costs may be less, and the entomopathogen may be available within 5 years. A previous success facilitates development of new agents because a framework has been established for constructive dialogue and exchange of information between persons responsible for development and registration.

The time required for development will undoubtedly vary for specific cases and for a variety of reasons, but the sequence (Fig. 1) is applicable to many candidate entomopathogens. Estimated dollar cost and research team years required to develop a candidate microbial insecticide during the first 3 years is about $1–$3 million (Ignoffo, 1975). Most technical questions concerning characteristics of the entomopathogen (biological identity, specificity, mode of action, stability, compatibility, etc.) can be resolved during the first year. Questions of production feasibility, safety, and effectiveness are resolved within the first 2 years. By the end of the third year, all major problems, both technical and nontechnical, should be defined or resolved. Therefore, shortly after the start of the fourth year, sufficient information should be available to decide whether to proceed toward commercial production, marketing, and sales. Optimistically, a product then could be available during the field season of the fourth year of development.

V. TECHNICAL PARAMETERS: MICROBIAL INSECTICIDES

A. Production Feasibility

1. General Comments

Entomopathogens are produced either by using fermentation (*in vitro* process) or in living insects (*in vivo* process). Fermentation technology (submerged or surface) has been used with facultative entomopathogens; *in vivo* technology with obligate entomopathogens. Both processes have been employed to successfully produce commercial products. For example, submerged fermentation is generally the technology of choice for commercial production of the bacteria *B. thuringiensis* and *B. moritai;* surface fermentation is used for fungi, e.g., *B. bassiana* and

Hirsutella thompsonii (McCoy and Kanavel, 1969). The *in vivo* approach is used exclusively to produce insect viruses and protozoa. There are, of course, some notable exceptions; e.g., Doom, a formulation of spores of the bacterium *B. popilliae* can only be produced in living beetle larvae.

Regardless of the process, a large quantity of the desired isolate is first produced and stabilized with a suspending medium. Then, small subsamples of this mix are dispensed into glass vials. The samples are lyophilized, sealed under vacuum, and stored under refrigeration until needed for production. In the production of bacterial entomopathogens test tube slants prepared from these vials are used as the initial inoculum for a stepwise increase in volume of inoculum through a series of two shake flasks and a seed fermentor. This procedure provides a large population of clean cells of the same stage and thereby increases the probability of synchronous growth in the production fermentor. The sequential steps from initial inoculum to seed tank for surface fermentation are similar to those for the submerged fermentation process, but production and recovery are significantly different.

Various sources of carbon (molasses, dextrose, grain mash, grain by-products), nitrogen (cottonseed meal, fish meal, soybean meal, whey, yeast, or casein hydrolysates) are used for fermentation. In submerged fermentation, the total volume of solids is generally limited to about 5%. Essential minerals and growth factors are also included, when necessary.

Production costs of surface fermentation were estimated at $0.50 per lb of technical product and submerged fermentation at $1.69 per lb (Dulmage, 1971). Formulated, wettable powder products of *B. thuringiensis* retail for about $9.00 per lb; use rates, depending upon the pest species, range from ca. ⅛ to 1 lb/acre.

Patents for production of *B. thuringiensis* with submerged fermentation were issued to Bonnefoi (1960), Megna (1963), Drake and Smythe (1963), and Shell International (1969). Production of *B. thuringiensis* using surface fermentation was patented by Mechalas (1963). Production of *B thuringiensis* by submerged or surface fermentation, based upon the patents of Megna (1963) and Mechalas (1963), respectively, are used here as examples of both fermentation processes (Fig. 2). Further details of production are described by Briggs (1963), Fisher (1965) and Dulmage and Rhodes (1971).

2. *In Vitro* Process

a. Submerged Fermentation. In submerged fermentation, the inoculum from slants is incubated in nutrient broth of the first flask for 24 hours.

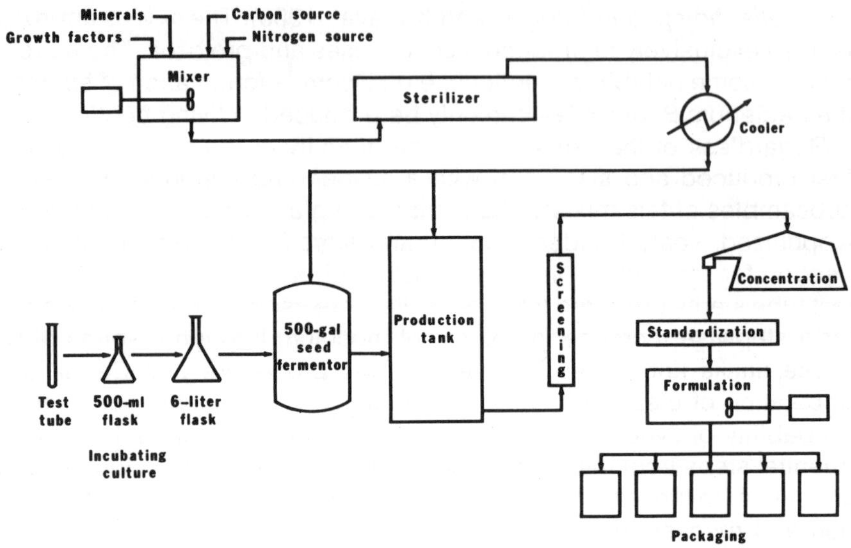

FIGURE 2. Submerged fermentation pathway used to produce the entomopathogenic bacterium *Bacillus thuringiensis*.

Medium for the second flask and the seed fermentor contains beet molasses (1%), corn steep solids (0.85%), and calcium carbonate (0.1%). The production media contain beet molasses (1.86%), corn steep solids (1.7%), cottonseed flour (1.4%), and calcium carbonate (0.1%). Development to sporulation and crystal formation with a viable spore count of $2–5 \times 10^9$ spores/ml is attained within 28–32 hours. Drake and Smythe (1963) obtained spore concentrations five times greater than Megna (1963) by using a production medium containing corn starch (6.8%), corn steep (4.7%), casein (1.94%), sucrose (0.64%), yeast (0.6%), and phosphate buffer (0.6%).

Safety of each fermentation batch is confirmed by intraperitoneal administration to white mice (Harvey, 1960; Anonymous, 1973). The insect-active ingredients of *B. thuringiensis* are recovered by centrifugation, filtration, precipitation, spray-drying, or combinations of these methods and processed into a flowable liquid, wettable powder, dust, or granular products, depending upon the type of fermentation, economics of the process, and the need for a specific formulation. Commercial products of *B. thuringiensis* are available in all four formulations (Fig. 3). Products of fermentation prior to and after blending with adjuvants or concentrated active ingredients are characterized by a bioassay of insecticidal activity and a viable spore count.

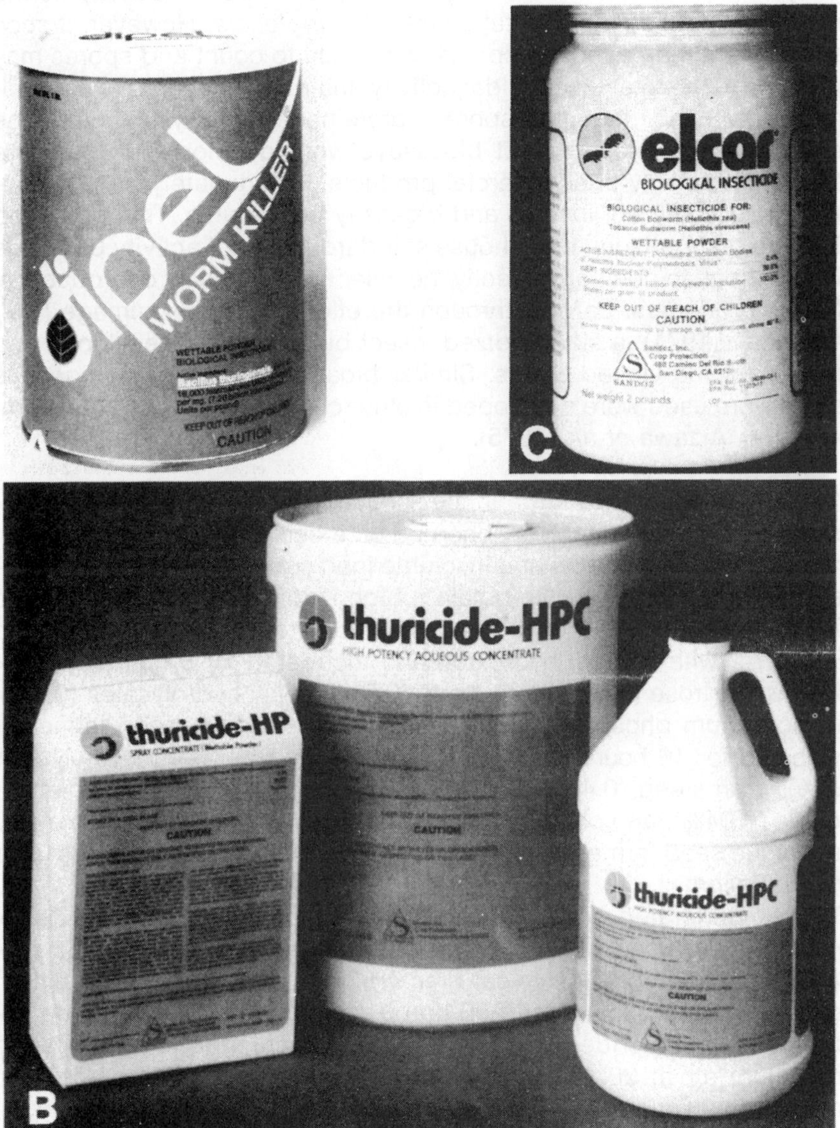

FIGURE 3. (A) Dipel, a wettable powder formulation of *Bacillus thuringiensis* (courtesy of Abbott Laboratories, Inc.). (B) Thuricide, flowable liquid and wettable powder formulations of *Bacillus thuringiensis* (courtesy of Sandoz, Inc.). (C) Elcar, the commercial wettable powder formulation of *Baculovirus heliothis* (courtesy of Sandoz, Inc.).

Only the viable spore count was used as an index of activity during early phases of commercialization of *B. thuringiensis*. However, it soon was apparent that, although spores were easy to count and spores may contribute to the total insecticidal activity, the primary source of caterpillar activity was the parasporal, proteinaceous crystal, i.e., the δ-endotoxin. Therefore, insect bioassays were developed to measure insecticidal activity of commercial products. Unfortunately, at first, producers used different insects and bioassay procedures. This permitted adequate comparisons for in-house standardization of activity but made it very difficult to compare directly the insecticidal activity of products of different producers. Largely through the efforts of H. T. Dulmage (Dulmage *et al.*, 1971), a standardized insect bioassay was developed and adopted for the United States. Similar bioassays in which indigenous insects were used were developed in other countries (Burges and Thomson, 1971; Aizawa *et al.*, 1975).

b. Surface Fermentation. In the surface fermentation process, the inoculum from the seed tank is absorbed on a carrier, usually a mixture of grain or grain by-products and insoluble inert substances (e.g., vermiculite, diatomaceous earth), to obtain a high surface area to volume ratio (Mechalas, 1963). The growth medium in the stepup flasks (adjusted to pH 7.0–7.2 with sodium hydroxide and incubated for 8 hours at 30°C) contains dextrose (1.5%), corn steep (0.5%), yeast hydrolysates (0.5%), and potassium phosphate (0.4%). The contents of the seed fermentor (incubated for 16 hours at 30°C) are dextrose (1.0%), yeast hydrolysates (0.6%), corn steep (0.45%), potassium phosphate (0.35%), sodium hydroxide (0.04%), and calcium chloride (0.01%). An inoculum ratio of two parts of the seed fermentor to five parts of the semisolid medium is used for production.

Large, perforated-bottom metal bins containing about 500 lb of semisolids are used instead of fermentation tanks. The semisolid production medium contains wheat bran (1800 parts), perlite (1200 parts), water (500 parts), soy meal (200 parts), dextrose (120 parts), lime (12 parts), sodium chloride (3 parts), and calcium chloride (1 part). During incubation, air at 30°–34°C and 95–100% relative humidity is forced through the medium via the perforated bottom of each bin for 36 hours. The moisture content during incubation drops from about 60 to 50% while the pH rises from 6.9 to 7.5. Moisture accumulated during incubation is reduced by forcing hot (50°–55°C) dry air through the bins for another 36 hours to achieve a moisture content of about 4%. The final product is then harvested, milled, passed through screens, characterized, blended with various adjuvants to achieve a standard insecticidal activity and spore count, and then packaged as either a dust or a granular formulation.

3. In Vivo Process

Techniques for producing entomopathogens in living hosts, primarily caterpillars, were developed over the last decade and are currently operational (Ignoffo and Hink, 1971). For example, annual production of 1–2 million European pine sawfly larvae, 2–10 million gypsy moth larvae, 15–45 million bollworm larvae, 35–110 million cabbage looper larvae, and 350 million mosquito larvae can be achieved (Ignoffo, 1975). Alternate hosts that are more easily reared and produce more spores than the primary hosts can also be used, e.g., caterpillars for production of a boll weevil protozoan or a mosquito protozoan (Ignoffo and Garcia, 1965; Undeen and Maddox, 1973).

A process to produce insect viruses in yeast and bacteria by using submerged fermentation was patented by Wells (1970) but never commercialized, and laboratory-scale production of insect viruses and protozoa in lines of insect cells both are possible, Virus produced in cell lines was as effective under field conditions as virus obtained from caterpillars (Ignoffo *et al.*, 1974; Jaques, 1977). Although these latter methods have potential as production techniques, the living insect is still the only feasible substrate for commercial production of an obligate entomopathogen.

The production process of the bollworm–budworm NPV, *Baculovirus heliothis* (Ignoffo, 1966a, 1973a) is diagrammed in Fig. 4 as a typical example of an *in vivo* process. As with production of other microorganisms, a large quantity of the selective isolate is first produced. For viruses, the quantity should be sufficient to provide inoculum for several years of production. Subsamples from the first large-scale passage, stabilized, lyophilized, sealed in individual glass vials under vacuum, and stored at −20°C, are maintained as reference standards.

Production of entomopathogenic viruses is often done in several isolated facilities. One facility is devoted to preparation of the diet used to rear the living insect which is used as the substrate for virus production. Another facility, commonly called the insectary, is used to rear the insects, generally caterpillars of 0.1–0.5 ml capacity. The insectary may be further divided so that one section is used solely for maintaining a standard, disease-free, thriving culture of the living insect and the other section is used for mass producing insects to be used in production of virus. A third facility, the actual virus production unit, is used to infect the living insect. Another facility is used for final recovery and formulation of the product. Other functions associated with production, such as quality

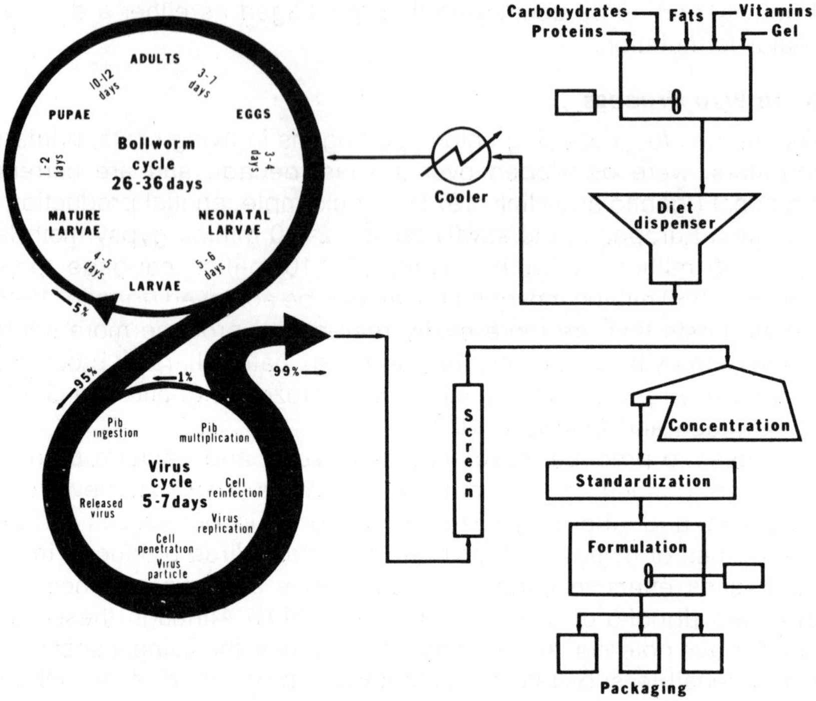

FIGURE 4. The *in vivo* process used to produce the nucleopolyhedrosis virus, *Baculovirus heliothis,* in larvae of species of *Heliothis.*

control bioassay, and standardizations are routinely done in an analytical laboratory.

The semisynthetic diet used to rear bollworm larvae for production of *B. heliothis* is a water-based mixture of casein, sucrose, wheat germ, yeast, Wesson salts, and growth factors (Ignoffo, 1966a). Alphacel is used as a filler, and agar is used to solidify the diet. Antimicrobial substances (e.g., formalin, sorbic acid, methyl parahydroxybenzoate, aureomycin) are often added to inhibit growth of secondary contaminants, such as bacteria, yeast, or fungi. Technology of the food industry was used to automate the virus production process.

Hot (70°–80°C) liquid diet, prepared in a large tank (Fig. 5A), is piped

---

FIGURE 5. (A) Mixing tank for preparing the semisynthetic diet used to rear *Heliothis larvae.* (B) Tray-filling machine that automatically dispenses a uniform volume of semisynthetic diet into individual cubicles. (C) Cages used to obtain eggs from *Heliothis* moths to be reared into larvae for virus production. (D) Mobile racks used to move trays containing diet and larvae. (E) Spray tower used to infect larvae and diet with virus. (F) Recovery of the technical virus product by acetone precipitation and filtration.

A
D
B
E
C
F

to a filling machine that automatically dispenses a uniform volume of diet (Fig. 5B). Plastic trays of individual cells containing the semisynthetic diet are used as the basic larval-rearing and virus-propagation unit (Ignoffo and Boening, 1970). Trays of cells are formed from plastic sheets, and the cells are filled with diet and sealed automatically on a continuous production line. Diet dispensing and cell sealing may be done at the rate of up to 200 trays per hour.

Neonatal caterpillars obtained from eggs laid in the insectary (Fig. 5C) are individually placed in the cell of each tray. Hundreds of these trays are thus semiautomatically infested with larvae and stacked on mobile racks for incubation in controlled environmental rooms (30°C) for 5–7 days (Fig. 5D). Approximately 95% of all larvae are used for virus production, while the other 5% are used to perpetuate the insect culture.

A standard volume of virus is sprayed on the diet of each cell (Fig. 5E). During the next 5–7 days (26°C) (Fig. 4), the virus replicates within the caterpillars and produces 5000–10,000 times more virus than that originally used. Bollworms weigh about 400–600 mg when they die. One caterpillar can produce as much as 36 billion PIB, ca. 30% of its dry weight.

The recovery of virus begins with collection of virus-killed larvae from individual cells by use of a suction tube. The larvae are then slurried with water, and the suspension is filtered through a screen to remove large particles. Further processing to concentrate PIB is done by centrifugation, precipitation, filtration (Fig. 5F), or spray-drying. As with other microbial insecticides, safety and quality control of each batch of virus is confirmed by a mouse injection test, evaluation of the types and levels of secondary microbial contaminants (Duggan, 1970), and bioassays.

A bioassay (mortality of neonatal bollworms) is used to determine the insecticidal activity of each virus batch (Ignoffo, 1966b). Several modes of administration, i.e., surface treatment, diet incorporation, and feeding may be used. The easiest large-scale, analytical method, however, is to incorporate the virus into the diet and feed the diet to neonatal or 1-day-old larvae. All preparations are compared against a reference standard, and their activity is related to this standard. The standard is prepared according to a designated protocol and stored under conditions of optimum stability. Insecticidal activity, measured as an LD_{50} or LC_{50} with 95% fiducial limits, is recorded as the number of insecticidal units per unit weight of virus product. The number of PIB per unit weight of virus product (counted under phase microscopy) is also used to characterize viral preparations.

Significant reductions in the cost of producing viruses and other obligate entomopathogens are anticipated since present techniques are still new. Current cost of producing caterpillars is about $0.1 to $0.4 per insect. The product from 1 to 100 insects is generally used to treat 1 acre. The current commercial formulation of *B. heliothis* (ELCAR) is a wettable powder mixed with several adjuvants to increase stability and efficacy (Fig. 3C). The cost of ELCAR to the grower is about $1.55 per oz. Recommended rates for control of *Heliothis* on cotton are from 2 to 4 oz/acre.

The major problem in virus production has been the time required to infest each cell of a tray with a newly hatched larva. If the process from tray forming to larval infestation were completely automated, production of 10,000 larvae per hour would be possible. Pilot-plant production capacity is about 1 million larvae per month, an increase over laboratory production of approximately 20-fold with no significant increase in production personnel. Maximum capacity for one semiautomatic production line operating continuously is estimated at about 7 million larvae per month.

B. Safety of Entomopathogens

1. General Comments

Because entomopathogens are infectious, replicating, living entities, there has always been a justifiable concern about their specificity, that is, their possible infectivity to nontarget organisms. Most previous evidence of specificity was indirect, such as the lack of reported cases of toxicity or pathogenicity in laboratory and field personnel, the ubiquitous presence of entomopathogens on food products, and the fact that entomopathogens are a natural part of our environment. This indirect evidence that microbial insecticides pose no serious hazard to vertebrates is generally supported by direct evidence (Heimpel, 1971; Ignoffo, 1973b). Nevertheless, safety, as discussed previously (Ignoffo, 1967), cannot be absolutely guaranteed for all entomopathogens in all living systems for all time. Therefore, it is necessary that inferences of lack of safety be corroborated by tests specifically designed to uncover possible risks to man and the environment. Within the last decade, guidelines have been developed to evaluate this impact (Ignoffo, 1973b; Engler, 1974). The establishment of these guidelines is a highly significant achievement not only because they are available but because they result in a regulatory procedure for a systematic review of the safety of many potential microbial insecticides.

2. Types of Tests

Many types of tests and test organisms have been used to evaluate the safety of entomopathogens (Table IV) but not all of those listed are presently required (Engler, 1974). Initially, only acute studies were conducted to evaluate possible toxicity–pathogenicity to nontarget species. More extensive testing began after each entomopathogen was seriously considered a viable candidate. For example, susceptibility was evaluated at all levels, such as in cells, tissues, organs, and organisms. Test organisms also were challenged with both purified entomopathogens (spores, cells, mycelium, crystals, virions, PIB) and with formulations of the entomopathogen. Microbial insecticides known to be relatively nontoxic or nonpathogenic were generally tested in animals at 10–100 times the average rate per acre by using a conversion ratio of weight of test animal to the body weight of man (Ignoffo, 1973b). Animals are generally sacrificed at the end of the test and carefully examined. Criteria used to evaluate the effects of entomopathogens *in vivo* included clinical examinations, histology, serology, and necroscopy. The following paragraphs give a brief summary of the kinds of tests that have been used. Detailed discussion of the protocols and results are presented elsewhere (Ignoffo, 1973b).

a. Acute Toxicity–Pathogenicity. Feeding, inhalation exposure, and dermal toxicity are used to evaluate acute toxicity or pathogenicity of candidate entomopathogens for both invertebrates and vertebrates. In these tests, the entomopathogen is generally administered once and the animal is observed for a short period (generally less than 3 weeks). However, observations may extend for 8 weeks when serological examinations are conducted. Vertebrates, both laboratory-reared and wildlife, are used in these studies.

b. Subacute Toxicity–Pathogenicity. Subacute or chronic studies are initiated after toxicity–pathogenicity has been established. The objective is to detect subtle abnormalities in the test species that are not apparent in acute studies. Since most entomopathogens are nontoxic, large doses are generally administered in the diet or by inhalation, subcutaneous injection, or skin contact throughout the test period, which may extend from 2 to 13 weeks. Two species of animals, one of which is a nonrodent, usually are used. Serological evaluations are done at regular intervals, generally every 4–5 weeks.

c. Primary Irritation. The skin and mucous membrane of the eye frequently are exposed to pesticides; therefore, both systems are used to

TABLE IV. Types of Tests and Organisms Used To Evaluate Safety to Invertebrates, Vertebrates, and Plants[a]

Test type	Organisms used to test:	
	Baculovirus heliothis	*Bacillus thuringiensis*
Acute toxicity–pathogenicity		
Per os, diet	Rat, mouse, birds, fishes, oyster, shrimps, human	Rat, mouse, guinea pig, birds, fishes, swine, human
Inhalation	Rat, guinea pig	Mouse, human
Dermal, skin	Rat, rabbit, guinea pig, human	Mouse, guinea pig, rabbit
Intraperitoneal	Rat, mouse	Chicken, mouse, rat, guinea pig, rabbit
Intravenous	Rat, mouse	
Intracerebral	Mouse	
Eye	Rabbit	Rabbit
Subacute toxicity–pathogenicity		
Diet	Monkey, dog, rat, mouse	Chicken, rat, human
Inhalation	Monkey, dog, rat, mouse	Rat, human
Subcutaneous	Monkey, dog, rat, mouse	Guinea pig, human
Teratogenicity	Rat, mouse	
Carcinogenicity	Rat, mouse	
In vitro replication	Man, primate	
Phytotoxicity	Agricultural crops	Agricultural crops
Invertebrate specificity	Beneficial and other arthropods	Beneficial arthropods

[a] From Ignoffo (1973b).

evaluate the potential of a candidate material to induce irritation. Rabbits are the animals of choice, but rats, mice, guinea pigs, and man have also been used to assess entomopathogens. Tests generally are continued for less than 1 week (eye) or 3 weeks (skin).

d. Teratogenicity. Teratological tests have been used to evaluate several NPV but probably will not be required in the future unless toxicity is observed. The NPV was repeatedly administered (per os) to pregnant dams to ensure maximum effect during organogenesis. Rats and mice are usually the animals of choice, and sufficient levels are used to obtain responses that vary from no effect to a definite effect.

e. Carcinogenicity. Carcinogenic potential has been studied in only a few entomopathogenic viruses. As with the teratological studies, it probably only will be required if toxicity is observed. Strains of rats or mice bred for susceptibility to carcinogens are used as experimental animals. Tests are initiated when the animals are young and are terminated 18 months to 2 years later.

f. *In Vitro* Development. Tests are made to evaluate the potential for entomopathogens, primarily viruses, to replicate in a nonhomologous host (established vertebrate diploid cell lines, including human). Cytopathic effects, inclusion body formation, hemadsorption, transformation, and viral interference are used as criteria to evaluate this potential.

g. Invertebrate Specificity. Acute toxicity studies against beneficial arthropods and invertebrates are also included in an evaluation of the safety and host spectrum of potential microbial insecticides. The scope of the evaluation is based on a knowledge of the organisms that may be present in the ecosystem where the entomopathogen will be used. Representative members of major orders of insects and species of freshwater, estuarine, or marine invertebrates, such as daphnia, dragonfly nymph, oyster, crab, and shrimp, are included. Entomopathogens are commonly tested by topical and per os administration of up to 100 times the anticipated exposure rate.

h. Phytotoxicity–Pathogenicity. Phytotoxicity of formulations of microbial insecticides are included as part of the overall evaluation of safety. Usually, greenhouse-grown plants of species that represent different families of economic plants are selected. Tests are also performed on field-grown species of the crop to be treated as well as on other economic crops.

3. Results of Tests

Extensive tests in the United States have demonstrated that at least four entomopathogenic viruses (*Heliothis, Orgyia, Lymantria, Autographa*); two bacteria (*B. popilliae, B. thuringiensis*); two fungi (*B. bassiana, H. thompsonii*); and three protozoa (*N. locustae, N. algerae, M. trogodermae*) are not toxic or pathogenic to man, other animals, or plants (Heimpel, 1971; Ignoffo, 1973b). Also, sufficient testing has been conducted to demonstrate the safeness of a fungal preparation (Boverin, *B. bassiana*) produced in the Soviet Union and a virus (Matsukemin, *Dendrolimus* CPV) and a bacterium (Rabirusu, *B. moritai*) produced in Japan (Ignoffo, 1973b).

Not one of about 50 entomopathogenic viruses tested *in vivo* was toxic or pathogenic to vertebrates. In fact, all but two of a dozen insect viruses were unable to replicate in vertebrate embryos or cell lines derived from birds, fishes, amphibians, or mammals and the possible replication of virus or development of cytopathic effects reported in the two tests has not been confirmed by other investigators. No deleterious effects at normal rates were reported in tests with *B. thuringiensis* endotoxin, *B. popilliae*, and *B. moritai* (Ignoffo, 1973b).

Presence of allergens may be encountered among fungi (*Beauveria, Entomophthora, Hirsutella, Metarrhizium, Nomuraea*) currently under consideration. Results of recent studies, however, indicate that *Beauveria, Entomophthora, Nomuraea,* and *Hirsutella,* are nontoxic or noninfectious to vertebrates (Ignoffo, 1973b). Only three entomopathogenic protozoa, one from grasshoppers (*N. locustae*), one from mosquitoes (*N. algerae*), and one from beetles (*M. trogodermae*), have been evaluated for safety to nontarget organisms. Initial *in vivo* tests with all three indicated no apparent risk to vertebrates.

Entomopathogens under consideration as microbial insecticides must be evaluated for possible deleterious effects to man, other animals, and plants. Risks cannot be assumed from studies of closely related groups. Undoubtedly, all tests described (Ignoffo, 1973b) will not be required for all entomopathogens, but each candidate should be broadly evaluated prior to large-scale field use. Because infectious microorganisms are being used, studies of risks should be carefully designed to detect presence of the microorganisms in addition to any possible effects induced by the microorganisms. In the final analysis, the risks of field use must be balanced against the benefits obtained. Prudent evaluations against vertebrates and other organisms will, in the long run, hasten development and ensure successful use of safe, effective entomopathogens as microbial insecticides.

C. Effectiveness of Entomopathogens

1. General Comments

Effectiveness of an insecticide is generally interpreted as its ability to kill pests as quickly as possible. However it is better also to consider protection of yield or quality balanced against risks to man and the environment. Initially, a novel microbial insecticide may not provide immediate control equal to that of a highly toxic, broad-spectrum chemical insecticide. Continual use of selective microbial insecticides may, however, preserve beneficial parasites and predators and ultimately result in yields comparable to those obtained with chemical insecticides.

2. Examples of Effectiveness

Entomopathogens belonging to each of the major groups have been successfully used to control at least 100 serious arthropod pests (Ignoffo, 1967; Anderson and Ignoffo, 1967). For example, about 60 insect viruses have been used to control caterpillar pests of agricultural crops, forests, and stored products, phytophagous mites, and beetles. Species of fungi have successfully controlled scale insects, whitefly, plant hoppers, aphids, chinch bugs, phytophagous beetles, flies, caterpillars, and stored-products pests. Bacteria other than the commercially available *Bacillus* preparations have been used to control grasshoppers, termites, caterpillar pests of forests and agricultural crops, and some species of beetles. Applications of entomopathogenic protozoa have suppressed pest populations of boll weevils, mosquitoes, grasshoppers, and about a half-dozen species of caterpillars.

3. Use Patterns and Impact

Microbial insecticides are most often used like chemical insecticides. Need we be so restrictive? There is no question that the use of spray technology is an important approach, but is it the only one available (Ignoffo, 1978)? For example, entomopathogens may be successfully introduced and established to provide long-term seasonal control of pest populations (White and Dutky, 1940; Balch and Bird, 1944; Hall and Dunn, 1958; Henry, 1971). Insects themselves can be used to disseminate entomopathogens (Elmore and Howland, 1964).

An epizootic of virus and fungi might be induced to occur earlier than normal in order to control a target pest (Morris, 1965; Ignoffo *et al.*, 1976). It also may be possible to manipulate the environment to create conditions wherein naturally occurring diseases would exert their greatest effect (Burleigh, 1975). Some of these approaches should provide levels of control equal to or better than that currently obtained with spray technology. However, they should be further explored and additional research should be devoted to evaluating their full potential. At least some of these approaches can be immediately applied, and more may be utilized in the near future.

The potential environmental impact of the substitution of microbial insecticides for some chemical insecticides may be visualized from the following examples. Commercial utilization of the virus, *B. heliothis*, as a control for bollworms and budworms on cotton alone, could decrease the use of at least 17 million lb of chemical insecticides annually. Use of the fungus *H. thompsonii* in Florida against the citrus rust mite and of the bacterium *B. thuringiensis* for cabbage looper control on vegetables could annually eliminate the need for about 4–16 and 1–3 million lb of chemical insecticides, respectively. Development of *N. locustae* for use against western-range grasshoppers and *Entomophthora thaxteriana* against the green peach aphid on potatoes in Maine could annually replace 1–2 or 0.2–0.7 million lb of chemical insecticides, respectively.

VI. CONCLUSIONS

There have been some notable developments in the area of microbial insecticides since publication of our last edition (Anderson and Ignoffo, 1967). Guidelines for the evaluation of the infectivity of entomopathogens to nontarget organisms have been formulated by the Environmental Protection Agency (EPA). The *Heliothis* NPV has been registered in December 1975 by the EPA as the first viral pesticide. Shortly thereafter (September 1976), another insect virus (*Orgyia NPV*) was approved by the EPA after development by the U.S. Forest Service. Two other viruses, the *Lymantria* NPV and the *Autographa* NPV, have been granted a temporary permit for further testing leading to possible registration. A bacterium (*B. moritai*) and a virus (*Dendrolimus CPV*) were registered in Japan for control of flies in feces of livestock pests and a forest caterpillar pest, respectively. A new more active strain of *B. thuringiensis* was produced which has increased the performance and acceptance of commercial products and broadened its use against other insect pests. Data on a fungus (*H. thompsonii*) are being collated by industry in anticipation of an experimental temporary permit for control of mites of citrus, and two other fungi (*E. thaxteriana* and *N. rileyi*) are being intensively investigated for control of aphids and soybean caterpillars, respectively. The protozoan (*N. locustae*) is being extensively field tested against grasshoppers on range crops under a temporary permit issued by the EPA. Currently, about four dozen trade-named products formulated from about 20 entomopathogens have been proposed as either experimental or commercial microbial insecticides (Tables I, II, and III).

Development of entomopathogens or their by-products into microbial insecticides is a reality. Safe, effective entomopathogens formulated as insecticides are being produced by industry and are being effectively used by growers. The technical successes reported herein undoubtedly will continue to be translated into commercial successes and will in turn stimulate additional research on use of entomopathogens as safe, selective insecticides.

REFERENCES

Aizawa, K., Fujiyoshi, N., Ohba, M., and Yoshikawa, N. (1975). *Proc. Int. Congr. International Association of Microbiological Societies* 1st, 1972 Vol. 2, pp. 577–606.

Anderson, R. F., and Ignoffo, C. M. (1967). *In* "Microbial Technology" (H. J. Peppler, ed.), pp. 172–181. Van Nostrand-Reinhold, Princeton, New Jersey.

Anonymous (1969). *Chem. Week* **104,** 36–38.

Anonymous (1973). *Code Fed. Regul.* **38,** 10643.

Balch, R. E., and Bird, F. T. (1944). *Sci. Agric.* **25,** 65–80.

Bassi, A. (1835). *Orcesi Lodi.*

Bonnefoi, A. (1960). French Patent 1,225,179.

Briggs, J. D. (1963). *In* "Insect Pathology" (E. A. Steinhaus, ed.), Vol. 2, pp. 519–548. Academic Press, New York.

Briggs, J. D. (1964). *Bull. W. H. O.* **31,** 495.

Bucher, G. E. (1964). *Ann. Soc. Entomol. Que.* **9,** 30.

Burges, H. D., and Thomson, E. M. (1971). *In* "Microbial Control of Insects and Mites" (H. D. Burges and N. W. Hussey, eds.), pp. 591–622. Academic Press, New York.

Burleigh, T. G. (1975). *Environ. Entomol.* **4,** 574–576.

DeBach, P. (1964). "Biological Control of Insect Pests and Weeds." Van Nostrand-Reinhold, Princeton, New Jersey.

Decker, G. C. (1960). *Agric. Chem.* **93,** 30–33.

d'Herelle, F. (1914). *Ann. Inst. Pasteur,* Paris **28,** 280.

Drake, B. B., and Smythe, C. V. (1963). U.S. Patent 3,087,865.

Duggan, R. E. (1970). *Fed. Regist.* **35,** 18690.

Dulmage, H. T. (1971). *In* "Microbial Control of Insects and Mites" (H. D. Burges and N. W. Hussey, eds.), pp. 581–590. Academic Press, New York.

Dulmage, H. T., and Rhodes, R. A. (1971). *In* "Microbial Control of Insects and Mites" (H. D. Burges and N. W. Hussey, eds.), pp. 507–540. Academic Press, New York.

Dulmage, H. T., Boening, O. P., Rehnborg, C. S., and Hansen, G. D. (1971). *J. Invertebr. Pathol.* **18,** 240–245.

Dunn, P. H., and Mechalas, B. J. (1963). *J. Insect Pathol.* **5,** 451.
Dutky, S. R. (1963). *In* "Insect Pathology" (E. A. Steinhaus, ed.), Vol. 2, pp. 75–115. Academic Press, New York.
Elmore, J. C., and Howland, A. F. (1964). *J. Insect Pathol.* **6,** 430–438.
Engler, R. (1974). *Dev. Ind. Microbiol.* **15,** 200–207.
Fisher, R. A. (1965). *Int. Pest Control* **7,** 4.
Forbes, S. A., II (1898). *16th Rep. State Entomol. Noxious Beneficial Insects State Ill.*
Frye, R. D., and Olson, L. C. (1974). *Farm Res.* September.
Giard, A. (1892). *Bull. Sci. Fr. Belg.* **24,** 1.
Hall, I. M., and Dunn, P. H. (1958). *J. Econ. Entomol.* **51,** 341–344.
Harvey, A. M. (1960). *Fed. Regist.* **25,** 3207–08.
Heimpel, A. M. (1965). *World Rev. Pest Control* **4,** 150.
Heimpel, A. M. (1971). *In* "Microbial Control of Insects and Mites" (H. D. Burges and N. W. Hussey, eds.), pp. 469–489. Academic Press, New York.
Henry, J. E. (1971). *J. Invertebr. Pathol.* **18,** 389–394.
Henry, J. E. (1976). *Agric. Res.* **24,** 5.
Ignoffo, C. M. (1966a). *In* "Insect Colonization and Mass Production" (C. N. Smith, ed.), pp. 501–530, Academic Press, New York.
Ignoffo, C. M. (1966b). *J. Invertebr. Pathol.* **8,** 547–548.
Ignoffo, C. M. (1967). *In* "Insect Pathology and Microbial Control" (P. A. van der Laan, ed.), pp. 91–117. North-Holland Publ., Amsterdam.
Ignoffo, C. M. (1968). *Curr. Top. Microbiol. Immunol.* **42,** 129–167.
Ignoffo, C. M. (1973a). *Exp. Parasitol.* **33,** 380–406.
Ignoffo, C. M. (1973b). *Ann. N. Y. Acad. Sci.* **217,** 141–172.
Ignoffo, C. M. (1973c). *Misc. Publ. Entomol. Soc. Am.* **9,** 57–61.
Ignoffo, C. M. (1975). *Environ. Lett.* **8,** 23–40.
Ignoffo, C. M. (1978). *J. Invertebr. Pathol.* **31,** 1–3.
Ignoffo, C. M., and Boening, O. P. (1970). *J. Econ. Entomol.* **63,** 1696–1697.
Ignoffo, C. M., and Garcia, C. (1965). *J. Invertebr. Pathol.* **7,** 260–262.
Ignoffo, C. M., and Hink, W. F. (1971). *In* "Microbial Control of Insects and Mites" (H. D. Burges and N. W. Hussey, eds.), pp. 541–580. Academic Press, New York.
Ignoffo, C. M., Hostetter, D. L., and Shapiro, M. (1974). *J. Invertebr. Pathol.* **24,** 184–187.
Ignoffo, C. M., Marston, N. L., Hostetter, D. L., Puttler, B., and Bell, J. V. (1976). *J. Invertebr. Pathol.* **27,** 191–198.
Jaques, R. P. (1977). *J. Econ. Entomol.* **70,** 111–118.
Kirby, W., and Spence, W. (1826). "An Introduction to Entomology: or Elements of the Natural History of Insects," p. 197, Longman, Rees, Orme, Brown, & Green, London.
Krassilstschik, I. M. (1888). *Bull. Sci. Fr. Belg.* **19,** 461.
Latch, G. C. M. (1965). *N. Z. J. Agric. Res.* **8,** 384.
Lewis, L. C. (1975). *Iowa State J. Res.* **49,** 435–445.
Linsley, E. G., and Smith, R. F. (1970). *J. Econ. Entomol.* **63,** 689–691.
McCoy, C. W., and Kanavel, R. F. (1969). *J. Invertebr. Pathol.* **14,** 386–390.
McCoy, C. W., Brooks, R. F., Allen, J. C., and Selhime, A. G. (1976). *Proc. Tall Timbers Conf.* **6,** 1–17.
McLaughlin, R. E. (1966). *J. Econ. Entomol.* **59,** 909–910.
McLaughlin, R. E. (1973). *Misc. Publ. Entomol. Soc. Am.* **9,** 95–98.
Martignoni, M. E., and Iwai, P. J. (1977). *U.S.D.A., For. Serv. Gen. Tech. Rep.* PNW 40-28.
Martignoni, M. E., and Iwai, P. J. (1978). *U.S.D.A. For. Serv., Tech. Rep.* (in press).
Mechalas, B. J. (1963). *U.S. Patent* 3,076,992.
Megna, J. C. (1963). *U.S. Patent* 3,073,749.

Metchnikoff, E. (1879). "Issue III. The Grain Weevil" *Commission attached to the Odessa Zemstvo Office*, Odessa.
Morris, O. N. (1965). *J. Insect Pathol.* **5,** 401–414.
Müller-Kögler, E. (1965). "Pilzkrankheiten bei insekten." Parey, Berlin.
Pasteur, L. (1870). "Etudes sur la maladie des vers à soie," Parts I and II. Gauthier-Villars, Paris.
Pristavko, V. P. (1963). *Sb.: 5, Soveshck. Vses. Entomol. Obshck.* p. 110.
Roberts, D. W., and Yendol, W. G. (1971). *In* "Microbial Control of Insects and Mites" (H. D. Burges and N. W. Hussey, eds.), pp. 125–149. Academic Press, New York.
Schwalbe, C. P., Boush, G. M., and Burkholder, W. E. (1973). *J. Invertebr. Pathol.* **21,** 176–182.
Shell International Research Maatschappij, N. V. (1969). Netherlands Patent 68,163,097.
Snow, F. H. (1891). *Annu. Rep. Entomol. Soc. Ont.* **21,** 93.
Steinhaus, E. A. (1959). *Trans. Int. Conf. Insect Pathol. Biol. Control, 1st, 1958* p. 37.
Steinhaus, E. A. (1964). *Proc. Int. Conf. Insect Pathol. Biol. Control, 2nd, 1962* p. 7.
Steinhaus, E. A., and Thompson, C. G. (1949). *J. Econ. Entomol.* **42,** 301.
Taylor, A. B., and King, R. L. (1937). *Trans. Am. Microsc. Soc.* **56,** 172.
Undeen, A. H., and Maddox, J. V. (1973). *J. Invertebr. Pathol.* **22,** 258.
Vago, C., Aizawa, K., Ignoffo, C., Martignoni, M. E., Tarasevitch, L., and Tinsley, T. W. (1974). *J. Invertebr. Pathol.* **23,** 133–134.
Weiser, J., III (1957). *Cesk. Parazitol.* **4,** 359.
Weiser, J., and Veber, J. (1956). *Z. Angew. Entomol.* **40,** 55.
Wells, F. E. (1970). Republic of South Africa Patent 700,831.
White, R. T., and Dutky, S. R. (1940). *J. Econ. Entomol.* **33,** 306–309.
Wilson, G. G., and Kaupp, W. J. (1975). *Can. For. Serv. Inf. Rep.* **IP-X-11,** 28.
Yendol, W. G., and Roberts, D. W. (1970). *Proc. Int. Colloq. Insect Pathol., 4th, 1970* pp. 28–42.

Chapter 2

Rhizobium Species

JOE C. BURTON

INTRODUCTION

Man lives in an atmosphere dominated by nitrogen gas, but he is unable to use this element in its molecular form. Air nitrogen (N_2) must first be "fixed," that is, incorporated into compounds which animals and plants can utilize as food.

N_2 can be fixed chemically or biologically. However, a large amount of fossil fuel energy is required to manufacture nitrogen fertilizers chemically. The recent shortage of fossil fuel and resulting high cost of N fertilizers has focused greater attention to biological processes.

Certain bacteria, *Rhizobium* spp., have the ability to infect roots of leguminous plants, form nodules, and work symbiotically with their host in fixing molecular N. It has been estimated (Hardy and Holsten, 1972)

MICROBIAL TECHNOLOGY, 2nd ed., Vol I
 ISBN 0-12-551501-4

that nodule bacteria (rhizobia), in association with leguminous hosts, fix at least 90 × 10^6 metric tons of N annually. This is more than twice the amount of N in chemical fertilizers and more than one-half the total amount of this element fixed biologically each year Nonetheless, with the rapidly increasing world population and the acute shortage of fossil fuel to manufacture nitrogen fertilizer, large increases in biological N_2 fixation are mandatory. The *Rhizobium*–leguminous plant association offers the greatest promise of all systems for providing the nutritious protein food which will be needed in the years ahead.

Culture of *Rhizobium* species (rhizobia) in the laboratory for use in agriculture began shortly after Hellriegel (1886) revealed his discovery of the magic of nodulated leguminous plants in 1886. Beijernick, a Dutch microbiologist isolated the causal organisms, rhizobia, from the leguminous nodules in 1888. Culture of *Rhizobium* species in the laboratory to increase legume crop production began soon afterward, but early attempts to use laboratory-produced cultures of the organism for improvement of leguminous crops met with little success. This should not be too surprising considering the many factors involved. The perplexing problems encountered during this early period are described in detail by Fred *et al.* (1932).

It was recognized early that leguminous plants have preferences for rhizobia and vice versa, but the full implication of this was not understood. The selection of effective strains of rhizobia for specific leguminous species was one of the key developments and continues to be one of the major responsibilities of the inoculant manufacturer today. Other key developments in understanding the *Rhizobium*–legume association and in development of the inoculant industry are described by Burton (1967).

II. *Rhizobium*: LEGUMINOUS PLANT ASSOCIATIONS

A. Plant Groups and *Rhizobium* Species

The family, Leguminosae, comprises more than 12,000 species of leguminous plants although fewer than 100 of these are currently used for food production. These are chiefly in the subfamily, Papilionatae, which consists predominantly of nodulating species with the ability to work symbiotically with rhizobia to utilize atmospheric nitrogen.

The nodule bacteria (*Rhizobium* spp.) are one of two genera in the family Rhizobiaceae and are characterized by their ability to infect and induce nodule formation on roots of leguminous plants (Buchanan and Gibbons, 1974). *Rhizobium* species are differentiated by the kinds of plants they nodulate. Plants mutually susceptible to nodulation by a particular kind of rhizobia constitute a cross-inoculation group. The bacteria capable of nodulating the plants in one of these groups are considered a *Rhizobium* species.

The six recognized *Rhizobium* species and their respective leguminous hosts are shown in the tabulation below.

Rhizobium meliloti	*Medicago, Melilotus, and Trigonella* spp.
Rhizobium trifolii	*Trifolium* spp.
Rhizobium leguminosarum	*Pisum, Vicia, Lathyrus*, and *Lens* spp.
Rhizobium phaseoli	*Phaseolus vulgaris* and *P. coccineus*
Rhizobium japonicum	*Glycine max*
Rhizobium lupini	*Lupinus* and *Ornithopus* spp.

The designation of new species of *Rhizobium* on the basis of cross-inoculation groups was discontinued about 1939 because the many unexplainable, incongruous plant and bacteria reactions cast doubt on the validity of the concept. Rhizobial taxonomists have now focused major attention on speed of growth, acid production, serology, DNA base ratios, numerical taxonomy, DNA hybridization, and phage susceptibilities. Graham (1976) and Vincent (1974) discuss these systems in detail and point out the inconsistencies. The difficulty is evidenced by the fact that no new *Rhizobium* species have been added for almost four decades. The practice today is to identify the nodule bacteria strain with its parent host plant, e.g., lotus rhizobia from *Lotus* species or baptisia rhizobia from *Baptisia* species, when there is no *Rhizobium* species name.

The grouping of leguminous plants on the basis of susceptibility to nodulation by a particular species or kind of rhizobia has been useful to inoculant manufacturers. With leguminous species, it is generally accepted that nodulation is essential for symbiotic N_2 fixation. The tests can be limited to strains capable of nodulating the particular plant and the best N_2-fixing bacteria can be selected from these.

B. *Rhizobium* Strain Selection

Rhizobium strain selection is one of the most important responsibilities of the inoculant manufacturer. *Rhizobium* strains are selected for the following qualities: (a) They must form N_2-fixing nodules on the particular plant and provide an adequate supply of nitrogen; (b) they must be competitive as evidenced by colonization of the root hairs and nodulation of the host in the presence of other highly infective rhizobia;

(c) nodulation must occur over a range of temperatures under which the particular plant will grow; (d) the *Rhizobium* strain must be able to grow well in culture medium, in the carrier medium, and in the soil after the seed is planted; and (e) the organism should be able to survive in the soil from one season to another.

1. Nitrogen-Fixing Ability

A prime consideration in selecting *Rhizobium* strains is nitrogen-fixing ability. The initial tests are generally made in a greenhouse or growth chamber (Vincent, 1970; Wacek and Brill, 1976; Date, 1976). Surface-sterilized seeds are planted and inoculated with the various strains of rhizobia. *Rhizobium* strains vary widely in ability to fix N_2 and stimulate growth (Fig. 1).

Through the years, investigators have sought simple, quick tests for measuring N_2-fixing efficiency, but the only way of knowing with certainty that a strain of rhizobia will fix N_2 with a particular host is to put them together under a favorable environment for growth and measure the N_2 fixed. More recently the discovery that nitrogenase, the enzyme in the legume nodule responsible for reducing N_2 to NH_3, also reduces

FIGURE 1. Growth of white clover, *Trifolium repens,* inoculated with various strains of *Rhizobium trifolii.* (A) Noninoculated; (B) and (D) inoculated with effective strains; (C) and (E) inoculated with ineffective strains.

acetylene to ethylene has provided a quicker method of evaluating *Rhizobium*–plant associations and weeding out inferior strains. Hardy and Holsten (1977) discuss the merits of acetylene reduction and other measurements of dinitrogen fixation in legumes.

Rhizobium strains which are effective on a wide range of host plants are preferred over those which are effective on only a few. The responses of three *Vicia* spp. to seven strains of rhizobia are shown in Fig. 2. Strains 175G9, 175G10, and 128C2 were highly effective on *V. villosa* but very ineffective on *V. grandiflora*. In contrast, strains 128C53 and 128C56 were very effective on all three host plants and would thus be more desirable in an inoculum for *Vicia* species.

Certain leguminous hosts are very promiscuous and have the ability to work effectively with a wide range of rhizobia. Siratro (*Macroptilium purpureum*) and the cowpea, *Vigna unguiculata*, are examples. Other plants are highly specific and work only with a special *Rhizobium* strain. The Texas bluebonnet, *Lupinus subcarnosus*, and Kura clover, *Trifolium ambiguum*, exemplify this group.

In studying N_2 fixation by numerous species of legumes in association with many strains of rhizobia, it has been noted that within cross-inoculation groups, certain plant species tend to respond similarly to inoculation with strains of rhizobia. This information has led to a subdivision of plants on the basis of N_2 fixation predictability. These subdivisions are referred to as "effectiveness groupings" and constitute the basis for multiple host inocula (Table I).

It should be remembered that all plants within an effectiveness group may not respond similarly to all strains of rhizobia. However, there is

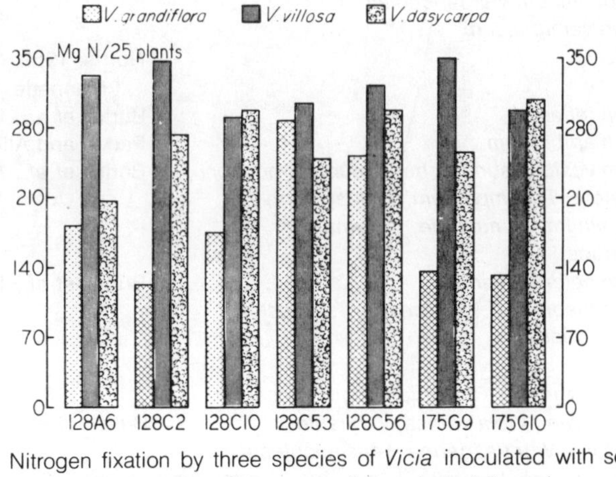

FIGURE 2. Nitrogen fixation by three species of *Vicia* inoculated with seven strains of rhizobia in growth chamber studies; plants were harvested at 6 weeks.

TABLE I. Effectiveness Groupings of Leguminous Species within Cross-Inoculation Groups

Cross-inoculation group	Reference
Alfalfa subgroups/*Rhizobium meliloti*	
A. *Medicago sativa, M. falcata, M. minima,*	Burton and Erdman, 1940
M. tribuloides, Melilotus denticulata	Burton and Wilson, 1939
M. alba, M. officinalis, M. indica	Brockwell and Hely, 1961
B. *Medicago arabica, M. hispida, M. lupulina,*	Brockwell and Hely, 1961
M. orbicularis, M. praecox, M. truncatula	
M. scutellata, M. polymorpha, M. rotata,	Kassim, 1976
M. rigidula, Trigonella foenum-graecum	
C. *Medicago laciniata*	Brockwell and Hely, 1961
D. *Medicago rugosa*	Brockwell, 1971
Clover subgroups/*Rhizobium trifolii*	
A. *Trifolium incarnatum, T. subterraneum,*	Burton and Briggeman, 1948
T. alexandrinum, T. hirtum, T. glomeratum,	Vincent, 1945;
T. arvense, T. angustifolium	Purchase and Vincent, 1949
	Strong, 1937
B. *Trifolium pratense, T. repens, T. hybridum*	Erdman and Means, 1946
T. fragiferum, T. procumbens, T. nigrescens	
T. glomeratum	Burton *et al.*, 1977
C. *Trifolium rueppellianum, T. tembense,*	Norris, 1959, 1965; Norris and 't Mannetje, 1964
T. usambarense, T. baccarinii, T. steudneri,	
T. burchellianum var. *burchellianum*	
T. burchellianum var. *johnstonii*	
T. africanum, T. pseudostriatum	
D. *Trifolium semipilosum* var. *kilimanjaricum*	Norris, 1959; Norris and 't Mannetje, 1964
T. masaiense, T. cheranganiense,	
T. rueppellianum var. *lanceolatum*	
E. *Trifolium semipilosum*	
F. *T. masiense*	Norris, 1959; Norris and 't Mannetje, 1964
G. *Trifolium reflexum*	Burton *et al.*, 1977
H. *Trifolium ambiguum*	Parker and Allen, 1949
I. *Trifolium vesiculosum, T. berytheum, T. bocconi*	Burton *et al.*, 1977
T. boissiere, T. compactum, T. dasyurum,	
T. leucanthum, T. mutabile, T. vernum,	
T. physodes	
J. *Trifolium heldreichianum*	Burton *et al.*, 1977
K. *Trifolium medium, T. sarosience, T. alpestre*	
L. *Trifolium rubens*	
Pea and vetch subgroups/Rhizobium leguminosarum	
A. *Pisum sativum, Vicia villosa, V. hirsuta,*	Burton *et al.*, 1977
V. tenuifolia, V. tetrasperma, Lens esculenta,	
Lathyrus aphaca, L. cicera, L. hirsutus, L. odoratus	
B. *Vicia faba, V. narbonensis*	Burton *et al.*, 1977
C. *Vicia sativa, V. amphicarpa*	Burton *et al.*, 1977

(*continued*)

TABLE I (*Continued*)

Cross-inoculation group	Reference
D. *Lathyrus sativus, L. clymenum, L. tingitanus*	Burton *et al.*, 1977
E. *Lathyrus sylvestris*	Burton *et al.*, 1977
F. *Lathyrus ochrus, L. tuberosus, L. szenitzii*	Burton *et al.*, 1977
Cowpea subgroups/*Rhizobium* sp.	
A. *Vigna unguiculata, V. sesquipedalis, V. luteola, V. cylindrica, V. angularis, V. radiata, V. mungo, Desmodium* sp., *Alysicarpus vaginalis, Crotalaria* sp., *Macroptilium lathyroides, M. atropurpureus, Psophocarpus* sp., *Lespedeza striata, L. stipulaceae*	Burton, 1952 Burton *et al.*, 1977
B. *Phaseolus limensis, P. lunatus, P. aconitifolius Canavalia ensiformis, Canavalia lineata*	Burton *et al.*, 1977
C. *Arachis hypogaea, A. glabrata, Cyamopsis tetragonoloba, Lespedeza sericea, L. japonica*	Burton *et al.*, 1977
D. *Centrosema pubescens*	Bowen, 1959
E. *Lotononis bainesii*	Norris, 1958
Lotus subgroups/*Rhizobium* sp.	
A. *Lotus corniculatus, L. tenuis, L. angustissimus*	Erdman and Means, 1949
L. tetragonolobus, Dorycnium hirsutum	Burton, 1964
L. caucasieus, L. crassifolius, L. creticus, L. edulus, L. frondosus, L. subpinnatus, L. weilleri, Anthyllis vulneria, A. latoides	Burton *et al.*, 1977
B. *Lotus uliginosus, L. americanus, L. scoparius, L. angustissimus, L. pedunculatus, L. strictus, L. strigosus*	Burton *et al.*, 1977
Ornithopus sativum	Jensen, 1967
Lupine subgroups/*Rhizobium lupini*	
A. *Lupinus albicaulis, L. albifrons, L. albus, L. angustifolius, L. arboreus, L. argenteus, L. benthami, L. formosus, L. luteus, L. micranthus, L. perrenis, L. sericeus*	Burton *et al.*, 1977
B. *Lupinus densiflorus, L. vallicola*	Burton *et al.*, 1977
C. *Lupinus*	Burton *et al.*, 1977
D. *Lupinus polyphyllus*	Burton *et al.*, 1977
E. *Lupinus subcarnosus*	Burton *et al.*, 1977
F. *Lupinus succulentus*	Burton *et al.*, 1977
Coronilla subgroups/*Rhizobium* sp.	
A. *Coronilla varia, Onobrychis viciaefolia, Petalostemum purpureum, P. candidum, P. microphyllum, P. multiflorum, P. villosum Leucaena leucocephala, Albizia julibrissin*	Burton *et al.*, 1977

sufficient congruity within a group to justify this categorization until the precise *Rhizobium* requirements for each host genotype can be determined.

2. Competitiveness

Competitiveness implies aggressiveness and the ability to produce nodules on the roots of a particular leguminous host growing in a soil which contains other highly infective rhizobia. A *Rhizobium* strain is considered competitive when it produces a large proportion of the nodules on a particular leguminous plant growing in a soil or substrate heavily infested with other infective strains. The competitive nature of a strain of *Rhizobium* is difficult to assess but its importance is not questioned. Date (1976) asserts that controlled-environment facilities are not usually considered suitable for measuring competitiveness or other characteristics, such as pH and pesticide tolerance or longevity in culture or on seed. He considers field tests imperative to measure these qualities.

Johnson *et al.* (1965) have used serological techniques very successfully in studying competitiveness of strains of *R. japonicum* under greenhouse conditions and in field soils free of *R. japonicum.* However, this technique is too laborious and time-consuming to permit screening of large numbers of *R. japonicum* strains for competitiveness in the field, particularly where soybeans have been grown for several years. Only a few strains could possibly be studied.

The writer (unpublished results) has used a growth chamber technique in screening strains of *R. trifolii* for use on soils infested with numerous highly infective non-N_2-fixing rhizobia. A peat-base inoculum containing the native strains is added to the sterile sand in Leonard jars. Each strain of *R. trifolii* being tested is prepared as a peat-base inoculum which is applied to the seeds. The inoculated seeds are planted in Leonard jars containing sterile sand as a control. In a parallel series, the inoculated seeds are planted in the Leonard jars containing sand inoculated with the native ineffective rhizobia. Growth and nitrogen fixation obtained in the latter series is a good measure of strain competitiveness (Fig. 3).

Often effective strains of rhizobia, as judged by growth in the sterile sand, are ineffective in the presence of native rhizobia. These are considered noncompetitive strains. Results from growth chamber tests have been tested and verified under field conditions.

3. Other Qualities of Rhizobia

Rhizobium strains equally effective in N_2 fixation in the greenhouse often show marked differences when tested in the field. The reasons for

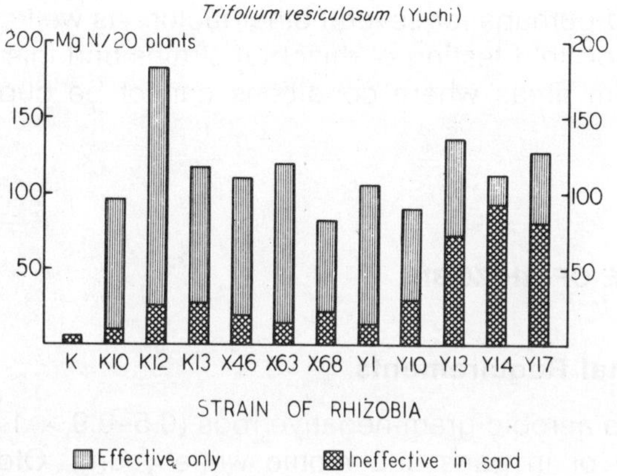

FIGURE 3. Competitiveness of effective rhizobia on arrowleaf clover (*Trifolium vesiculosum* Yuchi) as influenced by the presence of highly infective, non-nitrogen-fixing rhizobia in the substrate. Cross-hatched portion of bars show competitiveness of *Rhizobium* strains.

this are not completely understood, but certain factors have been suggested (a) adaptability of the strain of rhizobia to the fertility level or the pH of the soil, (b) inherent variation in N_2-fixing potential of the *Rhizobium* strain which is manifested only under the nitrogen stress of a mature fruiting plant, or (c) variation in ability to tolerate an antagonistic soil microflora

The question of whether rhizobial strains respond differently to levels of mineral nutrients and soil pH has not been answered adequately. One reason for this is that it has not been possible to separate effects on the bacteria from those on the plant. Also, information on acid tolerances of many of our important legumes is lacking.

Soil acidity influences nutrient solubility and growth of the host plant as well as growth and effectiveness of the rhizobia. With a low pH, nodulation failure may result from poor survival or inability of the rhizobia to colonize the root. Even if nodulation occurs at a low pH, this does not assure N_2 fixation. Optimal nodule function may require higher calcium and pH than is needed for nodulation (Munns, 1976, 1977).

Under acid conditions, aluminum and manganese are more soluble and may inhibit N_2 fixation simply by reducing growth of the host plant. *Rhizobium* strains tolerant to high concentrations of aluminum and manganese would be of little benefit unless the host plant were also tolerant to aluminum and manganese. Acid-tolerant strains of *R. meliloti* have been reported by Rice (1976) and by L. M. Bordeleau (personal communication), but these have been selected by screening under field conditions and perhaps for several other factors as well as pH. There is no substitute for field testing of rhizobial strains and this is particularly true in problem areas where conditions cannot be duplicated in the glasshouse.

III. CULTURE OF RHIZOBIA

A. Nutritional Requirements

Rhizobia are aerobic gram-negative rods (0.5–0.9 × 1.2–3.0μm), occurring singly or in pairs and motile while young. Older cells stain unevenly in bands because of granules or polymeric β-hydroxybutyrate (PHB) which does not stain.

Rhizobia are generally divided into what is called "fast growers" and "slow growers." The term, "fast grower" is assigned to cultures of *R. meliloti, R. trifolii, R. leguminosarum, R. phaseoli, Lotus* (birdsfoot type), *Astragalus, Sesbania, Caragana,* and various other varieties which produce turbidity in broth or visible colonies on solid media in 3–5 days at 28°C. The slow growers, *R. japonicum, R. lupini,* cowpea, *Lotus* (big trefoil type), and other varieties, require 8–10 days to attain an equal amount of growth. This division is not absolute. Differences in rates or amounts of growth among strains may be reduced, but not eliminated by supplementing the media with different forms of carbohydrate, nitrogenous compounds, and organic extracts.

The extensive literature on nutrition of rhizobia is reviewed by Wilson (1940), Allen and Allen (1950), and Vincent (1974, 1977). Rhizobia utilize monosaccharides and disaccharides readily, and to a lesser extent trisaccharides, alcohols, and acids. Pentoses, such as arabinose and xylose, are the preferred source of carbon for *R. japonicum* and other slow growers (Wilson and Umbreit, 1940; Graham and Parker, 1964). Graham and Parker reported that strains of rhizobia differ in their abilities to utilize carbohydrates. None of 16 different carbohydrates proved completely satisfactory for all strains of particular species of rhizobia. This should not be surprising, however, when consideration is given to the host preferences of the rhizobia. Also, the ability of rhizobia to utilize carbohydrates depends to some extent upon the basal medium, oxidation–reduction potential, size of inoculum, method of culture, and possibly other factors.

The composition of culture media used in industrial propagation of rhizobia is given in Table II. Sucrose is the most commonly used carbon

TABLE II. Media Used in Culturing Rhizobia

Ingredient (gm/liter)	Wright[a] (1925)	Bond (1940)	Van Schreven (1953)	Burton (1967)	Date (1976)
Sucrose	—	10.00	15.00	10.00	—
Mannitol	10.00	—	—	2.00	10.00
K_3PO_4	—	—	—	0.10	—
K_2HPO_4	0.50	9.50	0.50	—	0.50
KH_2PO_4	—	—	—	0.37	—
$MgSO_4 \cdot 7\ H_2O$	0.20	0.20	0.20	0.18	0.80
NaCl	0.10	0.10	—	0.06	0.20
$(NH_4)_2HPO_4$	—	—	—	0.10	—
$CaSO_4 \cdot 2\ H_2O$	—	—	—	0.04	—
$Ca(NO_3)_2 \cdot 4\ H_2O$	—	—	—	—	—
$CaCO_3$	3.00	1.00	2.00	0.25	—
Ca gluconate	—	1.50	—	—	—
$FeCl_3 \cdot 6\ H_2O$	—	—	—	—	0.10
Yeast water (ml)	100	50	100.0	—	100.0
Autolysed yeast	—	—	—	1.00	—
Water (ml)	900	950	900	1000	900

[a] Commonly referred to as Fred and Waksman Medium No. 79.

source, but mannitol and glycerol are sometimes used for the slow growers.

Yeast extract or autolyzate is used almost universally as a source of growth factors and inorganic nitrogen. While some strains of rhizobia can use nitrate or ammonium as a nitrogen source, better growth is usually obtained when low molecular weight amino acids are added. Yeast extract, hydrolyzed casein, and corn steep solids serve as sources of growth factors and amino acids.

The balance of amino acids appears to be important also (Holding *et al.*, 1960; Hamdi, 1965, 1968a, b, 1969; Badawy, 1965; Strijdöm, 1963; Strijdöm and Allen, 1965, 1967, 1969). Rhizobia cultured through several transfers on media enriched with glycine, alanine, and certain D-forms of amino acids lost their ability to fix nitrogen. Rhizobial strain efficiency was maintained when the organisms were cultured on media containing only the regular complement of yeast extract consisting predominantly of the naturally occurring L-isomers of amino acids. The composition of various organic autolyzates used in culturing rhizobia is given in Table III.

The growth factor requirements of *Rhizobium* spp. are simple and vary with species and strain. Strains of *R. leguminosarum*, *R. trifolii*, and *R.*

TABLE III. Amino Acid and Vitamin Content of Various Medium Supplements

Component	Difco yeast extract[a]	Hydrolyzed protein				Corn steep solids[c]
		Yeast autolysate Amberex 1003[b]	Casein[b] Amber EHC	Soy-Peptone Sheffield	Cottonseed[b] Pharmamedia	
Nitrogen (%)						
Total N	9.5	8.0	13.5	9.2	9.0	7.3
Amino N	73.0	50.0	30.0	20.0	50.0	37.0
Amino acid (%)						
Alanine	—	—	2.6	2.3	—	5.6
Arginine	1.0	3.7	3.7	4.6	7.1	5.6
Aspartic acid	5.0	—	5.7	5.3	7.5	0.2
Cystine	—	0.8	0.3	0.6	1.4	—
Glutamic acid	6.5	8.5	20.1	12.1	17.4	0.7
Glycine	2.5	5.0	1.0	2.8	3.7	1.6
Histidine	1.0	2.3	2.0	1.8	2.2	1.2
Isoleucine	3.0	4.2	4.8	0.7	2.9	1.3
Leucine	3.5	6.0	9.4	3.8	5.5	4.7
Lysine	4.0	6.0	6.8	3.7	3.0	3.3
Methionine	1.0	2.0	2.8	0.8	1.7	1.0
Phenylalanine	2.0	4.0	5.5	3.3	4.9	1.5
Threonine	3.5	3.2	4.3	1.9	2.9	1.2
Tyrosine	0.5	3.0	4.4	1.9	3.1	—
Valine	3.5	4.0	6.2	2.0	4.0	2.4
Vitamins (μg/gm)						
Biotin	1.0	2.0	—	—	0.6	—
Niacin	280.0	550.0	—	—	74.8	—
Pyridoxine	20.0	25.0	—	—	15.0	—
Riboflavin	20.0	35.0	—	—	13.6	—
Thiamin	3.0	50.0	—	—	20.0	—
Calcium pantothenate	—	100.0	—	—	58.0	—

[a] Analysis by Difco.
[b] Analysis by Amber Laboratories, Milwaukee, Wisconsin.
[c] D. D. Christianson *et al.* (1965).

phaseoli may require one, all, or any combination of biotin, thiamin, and calcium pantothenate (Wilson, 1940; Allen and Allen, 1950; Graham, 1963). Only certain strains of *R. meliloti, R. lupini, R. japonicum,* and the cowpea rhizobia require any vitamins and biotin is the only one needed. The vitamin needs of rhizobia from the many legumes have not been determined, but the majority grow well in yeast extract–mannitol—mineral salts medium.

Our knowledge of the role of inorganic salts in nutrition of rhizobia is scanty. Plant extracts have generally been used in culturing rhizobia. The mineral content of these have probably been adequate for most rhizobia. A synthetic medium is generally necessary to demonstrate mineral deficiencies. Yeast extract contains nutritionally significant concentrations of iron, calcium, magnesium, strontium, sodium, potassium, barium, manganese, copper, lead, aluminum, and vanadium (Steinberg, 1938). It is generally conceded that rhizobia need small amounts of iron, magnesium, calcium, potassium, manganese, zinc, and cobalt (Vincent, 1974, 1977). It appears that fast growers need more calcium than slow growers (Badawy and Allen, 1963). Since rhizobia in the nodules are nourished by the leguminous host, it is reasonable to assume that their mineral nutrient requirements are somewhat similar.

B. Growing Rhizobia

Unlike most microorganisms, *Rhizobium* spp. are not grown for their by-products or protein. The product envisaged is an inoculum, a live culture of rhizobia for application to leguminous seeds or to the soil where seeds are to be planted—a culture teeming with billions of effective rhizobia capable of producing nodules and fixing N_2 on the plant or plants for which it is prepared. Cell yield or biomass is not important; the main concern in formulating a medium for growing rhizobia is retention of the nitrogen-fixing capabilities of the organisms. It is desirable, of course, to have high viable counts in the broth culture and in the inoculant to prepare a strong inoculum and to assure early nodulation. The steps in large-scale growth of *Rhizobium* species are given in Fig. 4.

The fermentors used in growing rhizobia vary from small bottles which can be sterilized in an autoclave to large, jacketed stainless steel vessels built for high-pressure steam and equipped with mechanical agitators and various control devices. Regardless of the size of the operation, small and intermediate size fermentors which can be used either for production or starter cultures are desirable.

Rhizobium species are aerobic. The fermentors must be equipped to supply sterile air through spargers. Large vessels must have mechanical

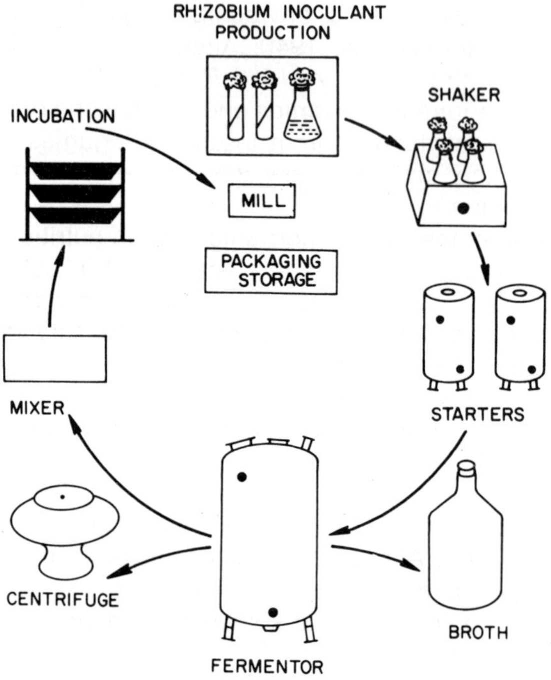

FIGURE 4. Steps in production of *Rhizobium* inoculants.

agitation to facilitate heat exchange and avoid overcooking of the medium and to increase aeration efficiency. The essentials are described by Date and Roughley (1977).

While rhizobia are considered obligate aerobes, excellent growth is obtained at oxygen partial pressures as low as 0.01 atm (Wilson, 1940). An oxygen partial pressure of 0.15 atm is optimum for respiration. As a rule of thumb, 1 liter of air per minute should be supplied for every 20 liters of medium.

Rhizobia grow best in the range of 28°–30°C, with the exception of *R. meliloti* which prefers a temperature of 35°C. With a good medium and adequate aeration, a cell population of 4 or 5 $\times$ 10^9 can be attained in 96 hours with a 1.0% inoculum. The incubation time can be reduced substantially by increasing the size of the inoculum.

Norris (1965) claims that the fast growing rhizobia produce acids when mannitol is broken down. However, Laird and West (1938), working with strains selected especially for their ability to produce acid, obtained a maximum of 2–3 ml of 0.1*N* acid per 100 ml of medium in 48 hours.

Even with a weakly buffered medium, *R. meliloti, R. trifolii, R. phaseoli,* and *R. leguminosarum* cultured in mass in fermentors produce no appreciable change in pH in 96 hours. When incubated for 144 hours, *R. meliloti* may lower the pH from 6.8 to 6.4; other fast growers produce no change in pH. An acid reaction is usually indicative of contamination usually by an aerobic thermophilic spore former which has survived the sterilization. Broth culture should contain a minimum of 1×10^9 viable rhizobia per milliliter and should be completely free of contamination to be satisfactory for use in preparing inoculants.

C. Carrier Media

The qualities of a good carrier medium for rhizobia are: (1) nontoxic to *Rhizobium* species, (2) good absorption qualities, (3) easily pulverized and sterilized, (4) good adhesion to seeds, and (5) readily available at moderate cost. Peats from different areas of the same bog as well as peats from diverse areas differ greatly in their suitability as carriers for nodule bacteria. Chemical analysis is helpful, but other hidden qualities can be more important. In reality, the only way the suitability of a particular peat can be determined is by putting rhizobia into the carrier and monitoring their growth. Rhizobia should be able to increase in numbers by tenfold or more in 3–5 weeks.

Good-quality peat is not available in many countries. Roughley (1968, 1970, 1975) cites problems with variation in quality of peat from the same bog, high salt concentration, and in using excessive temperatures for drying. Strijdöm and Deschodt (1975) had good success with a mixture of coal dust, bentonite, and lucerne powder as a carrier for rhizobia. Bagasse was satisfactory when the sugar was removed before grinding. Corby (1975) had good success with a compost of corn cobs, 1000 kg; limestone, 27 kg; single superphosphate, 9 kg; and ammonium nitrate, 12 kg; but 6 months were needed for drying and grinding. Dube and Mandeo (1973) used finely pulverized lignite.

The composition of Demilco peat, the one used most commonly in the United States, is given in Table IV. After the wet peat is mined and screened to remove stones and roots, it is shredded and fed into a revolving drum through which very hot air is continuously passing. The peat is flash-dried. Moisture content drops from 60% to 8 or 9% in a matter of seconds. While still hot, the peat is pulverized in a hammer mill to the desired fineness and packaged in multiwalled paper bags until used.

TABLE IV. Analysis of Sedge Peat, a Carrier for *Rhizobium* spp.[a]

Sedge peat contents	*Amounts*
Total nitrogen	1.62 (%)
Organic matter	86.80 (%)
Ash (500°)	13.20 (%)
Exchangeable potassium	62 ppm
Nitrogen as NH_4 and NO_3	94 ppm
Available phosphorus	12 ppm
pH	4.5–5%
Moisture	7–8%
Analysis of ash	*Amount (%)*
Potassium	1.12
Phosphorus	0.33
Calcium	5.21
Magnesium	1.14
Iron	2.10
Silicon	28.00
Aluminum	6.32
Sodium	0.52
Sieve analysis	*Amount (%)*
Powder inoculant	
Through 200 mesh (ASTM)	80–90
On 200 mesh	18–8
On 50 mesh	Less than 1
Granular inoculant	
On 16 mesh (ASTM)	0
On 40 mesh	70–85
On 50 mesh	15–30
Through 50 mesh	Less than 4

[a] Peat mined and processed by Demilco, Waukesha, Wisconsin.

IV. INOCULANT MANUFACTURE

A. Types of Products

Based on use, there are two kinds of rhizobial inoculants: (1) those designed for application to leguminous seeds, and (2) those designed for direct application to the soil. Seed inoculants are more common.

1. Seed Inoculants

The practice of applying rhizobia to leguminous seeds, inoculation, was started soon after these bacteria were first cultured in the laboratory. Rhizobia were applied to seeds because this was an easy convenient way of implanting the bacteria in the root zone of the developing seedling. The laborious task of spreading tons of soil to provide a few rhizobia could thus be bypassed.

Currently, six forms of culture or inoculum are marketed: (1) bottle or agar slant, (2) peat-base, (3) liquid or broth, (4) lyophilized or freeze-dried, (5) frozen concentrates, and (6) oil-dried culture. Generally, the peat-base inocula are considered more dependable because of the protection the peat affords the bacteria in the package and on the seed (Burton, 1964, 1967; Vincent, 1974). The liquid culture from the fermentor is used in preparation of all forms except the agar slant. Concentration of bacterial cells by centrifugation is a second step in preparation of the frozen concentrates, the lyophilized, and the oil-dried forms.

Two different approaches are used in the large-scale manufacture of peat-base inoculants. The method commonly used in Australia and in Europe is to grow the rhizobia in broth and then inject this broth culture either into a pliable film package containing sterilized peat or into small bottles containing sterile peat enriched with alfalfa meal. The proper amount of broth culture to add is predetermined to provide optimum moisture in the package to ensure growth. The rhizobia proliferate on the substrate in the individual packages. This method has the advantage of eliminating or greatly reducing contaminants, but it is not readily adaptable to large-scale operations because of its excessive demands on labor and time. Also, it imposes difficulties in carrying out an effective quality-control program.

The method favored in the United States is to increase the rhizobia in broth to a minimum of 1×10^9 ml. A predetermined amount of extremely fine $CaCO_3$ is then added to the broth culture to neutralize the acid and assure attaining a pH of 6.8 in the peat-base inoculum. The broth–$CaCO_3$ suspension is then sprayed onto the pulverized heat-treated sedge peat while it is being agitated in a horizontal paddle-or ribbon-type mixer. The finished inoculant should have a moisture content of about 38% (wet basis).

After mixing, the inoculant is placed in this layers in roller tubs and allowed to incubate for a period of 48–72 hours at 26°–28°C. During this time, the heat of wetting is dissipated and the moisture is absorbed uniformly. The inoculant is then milled to break up lumps and is packaged. Viable rhizobia reach their peak in numbers 2 or 3 weeks with the fast-growing rhizobia and in 4–5 weeks for the slow-growing rhizobia (Fig. 5). Viable rhizobia decline in numbers after 5 or 6 weeks unless the inoculant is refrigerated. The decline is directly related to storage temperature. Inoculants kept at 4°C or lower will retain a high count of viable rhizobia for 12 months or longer.

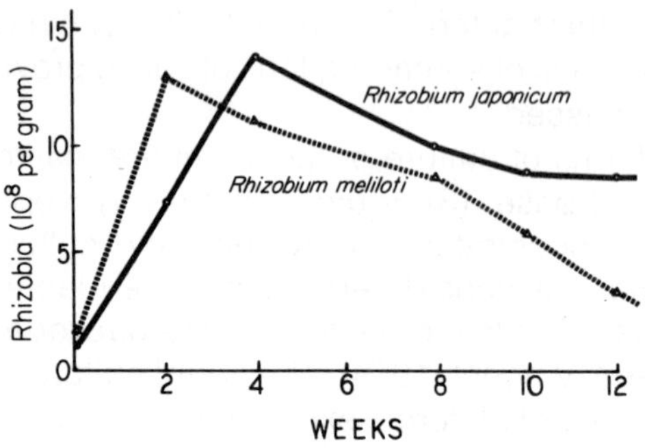

FIGURE 5. Growth of *Rhizobium meliloti* and *Rhizobium japonicum* in a peat-base carrier at 25°C for 12 weeks following preparation.

2. *Soil Inoculants*

Under some conditions inoculation of seeds is not practical. Seeds coated with toxic fungicides or insecticides cannot be successfully inoculated with rhizobia. Wetting the coats of fragile seeds, such as peanuts, with an inoculant slurry may reduce germination. Also, seeds with any particulate matter, such as peat-base inoculant or powdery chemicals, cannot be sown with the new type of air planters which depend upon a smooth seed coat for proper operation. Under these circumstances, the inoculant must be distributed separately from the seed.

Soil inoculants are of three types. In the granular peat-base type called "Soil Implant," the rhizobia actually grow in the peat particles. It is designed for application through an attachment on the drill which will meter out as little as 4 or 5 kg/ha of a granular material and place it in the furrow with the seed at planting time. The inoculant must consist of fine granules and must flow freely (Fig. 6). A second type, comprised of inert cores coated with peat-base inoculants and absorbant clays, was introduced by Taylor (1972) for mixing and sowing directly with seeds.

A third type of soil inoculant is prepared by centrifuging the broth culture to produce a creamy liquid of *Rhizobium* cells which is kept frozen until used by the farmer. At planting time the frozen concentrate is thawed and diluted with sufficient water to permit distribution in the seed furrow using the liquid fertilizer attachment on the seed drill (Scudder, 1975; Leffler, 1976). From 15 to 30 liters of bacterial suspension is usually applied per hectare.

Soil inoculants, both the granular and liquid types, have the following advantages: (1) They enable application of greater numbers of rhizobia than can be applied to most seed. This can help to overcome a flora of

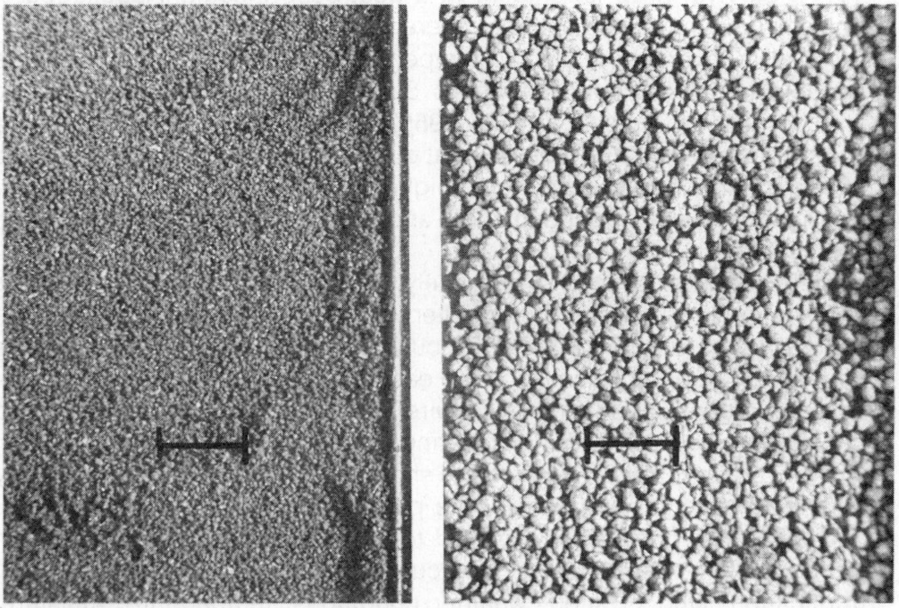

FIGURE 6. *Rhizobium* inoculants, Left: conventional seed type; right: granular type applied directly to soil by drill attachment. Space = 1 cm.

dominantly ineffective rhizobia in the soil. (2) Soil inoculants do not interfere with use of chemical protectants on the seed. Inoculants applied separately from the seed are not adversely affected by toxic chemicals on the seed coat. (3) Soil inoculants are easy to apply. (4) Soil inoculants can be used to effect nodulation on growing plants which for some reason are not effectively nodulated.

B. Longevity

The prime objective in inoculating leguminous seeds is to provide a sufficient number of highly effective rhizobia to nodulate the host and keep it well supplied with nitrogen. Longevity of *Rhizobium* species is essential for success. The effective life of an inoculant is therefore a primary concern of the inoculant manufacturer, the seedsman, and the farmer.

The peat-base inoculant is the predominant form because it provides better protection for the rhizobia, both in the package and on the seed (Burton, 1964, 1967; Vincent, 1974; Roughley, 1975; Date and Roughley,

1977). Survival of *R. japonicum* in a peat-base inoculant packaged in polyethylene incubated at three different temperatures is shown in Fig. 7. Survival is inversely related to temperature. At 43°C only 0.01% of the rhizobia lived for as long as 2 weeks. Survival was good at 21 and 32°C.

According to Vincent (1958, 1965) lyophilized inoculants have a slightly longer shelf-life than do peat cultures at 28°–30°C, but death rate after application to seed is much higher. In the United States, commercial mixtures of lyophilized rhizobia and talc have not proved acceptable (Schall *et al.*, 1975).

Broth cultures are used successfully in Europe but the seed inoculated with the culture is planted soon afterward (Van Schreven, 1958). In the United States, liquid or broth inoculants for soybeans have become moderately popular because of the ease of application. However, under field conditions, good results require very large inocula and the seed must be planted soon after inoculating (Burton and Curley, 1965). Broth, lyophilized, and agar slant cultures share a common disadvantage; they lack the protection afforded by the peat on the seed and in the soil.

The number of rhizobia required to bring about effective nodulation depends upon age and form of inoculant, strain of organism, method of application, kind of seed, weather conditions, temperature, moisture, type of soil, seed bed preparation, and possibly other factors. However, as field data accumulate, the evidence is growing that far greater numbers of rhizobia are required for successful inoculation than was formerly suspected. Australian and Canadian workers now consider 1000 rhizobia per seed as a minimum (Date, 1976; Stevenson, 1976). Soybeans planted under field conditions require a minimum of 10^5 rhizobia per seed for effective nodulation (Burton and Curley, 1965).

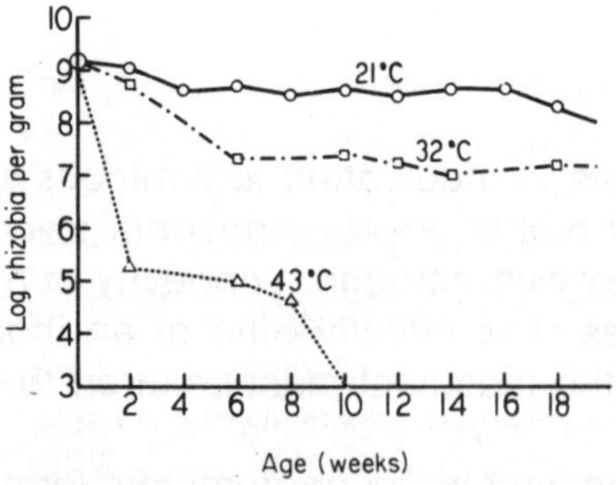

FIGURE 7. Effect of temperature on survival of *Rhizobium japonicum* in a peat-base inoculant packaged in polyethylene to prevent loss of water.

C. Quality Control

The quality of *Rhizobium* products (inoculants and preinoculated seeds) is monitored in only a few countries. The standards and methods of testing vary.

In Australia, a central organization (AIRCS, Australian Inoculant Research and Control Service) assures the quality of Australian inoculants. University or Department of Agriculture personnel (a) test and select rhizobial strains, (b) maintain and issue stock cultures of the chosen strains, (c) check quality of broth culture before it is mixed with peat, and (d) collect samples of the finished product from various distribution points and test them for rhizobia content. A minimum viable rhizobia count of 10^8 per gram up to the expiration date is considered satisfactory. The previous standard was 10^6 per gram. Samples of preinoculated seeds are also collected. A minimum of 10^3 viable rhizobia per seed is considered satisfactory, but few samples meet this standard (Brockwell *et al.*, 1975).

Quality control in the United States is in sharp contrast to the Australian system. Manufacturers are responsible for selecting their own strains of rhizobia and for setting their own standards in manufacture. The generally poor quality of inoculants during the 1920s and early 1930s led the United States Department of Agriculture and several of the states to set up testing departments for legume inoculants, but the tests were not difficult. The minimal greenhouse test used served only to caution the farmer regarding use of questionable brands the following year. It did not function as a real measure of inoculant quality.

The introduction and widespread promotion of preinoculated seed in the United States has created a new interest in regulatory testing. In general, seeds inoculated longer than 2 or 3 weeks in advance of planting have not proved to be adequately inoculated when planted in soils devoid of rhizobia.

Without refrigeration, neither legume inoculants nor preinoculated seed will maintain high levels of viable rhizobia for long periods of time. It is necessary to specify the effective shelf-life. In Australia, inoculants are dated to expire 6 months after preparation as compared to 1 year in the United States, but the effective use period does not generally overlap two planting seasons. The net effect is almost the same, but the Australian system is more realistic.

The rule regarding preinoculated seed has not yet been specified. Certain states require that the date of inoculation be stamped on the seed tag and also the date beyond which the manufacturer does not consider the seed effectively inoculated.

V. USE OF *Rhizobium* INOCULANTS

Seed Application

1. Conventional Methods

Inoculants are generally designed for application to seeds, because this is a convenient practical way of placing the bacteria in the soil zone where the young seedling will develop. When the inoculant is mixed with water to form a slurry and then applied, the seed are more uniformly coated and the inoculant adheres to the seed firmly (Fig. 8). Seeds should be inoculated just before planting.

Inoculants have been introduced for dry application but the results have not been good. This can best be observed in a field devoid of effective rhizobia rather than in a field which has previously grown the particular legume.

Leguminous seeds differ greatly in size, shape, color, nature of seed coat, and many other characteristics. The amounts of inoculant and particularly water which can be applied safely depend upon the kind of seed. Caution must be exercised in inoculating not to get seeds so moist

FIGURE 8. Chickpea (*Cicer arietinum*) seeds. Left: noninoculated; right: inoculated with peat-base inoculant.

or sticky that they will not flow freely through a mechanical seeder. Small seeds can absorb more water than large seeds (Table V).

In the United States the normal rate of application is 4.4 gm of inoculant per kilogram of seed regardless of seed size, but the amount of water is greatly reduced for the larger seeds. The big seeds naturally receive a larger number of rhizobia per seed, but according to Burton and Curley (1965) they need more rhizobia for effective nodulation. Regulatory standards often specify a minimum of 10^3 viable rhizobia per seed regardless of size. An inoculant with 10^6 rhizobia per gram would meet this standard with peas, but with white clover, one would have to apply 400 times as much inoculant or an inoculant with 400 times as many rhizobia per gram. The minimum standard is much more difficult to attain with the small seeds.

2. *"Custom Inoculation"*

The modern farmer is extremely busy at planting time and prefers to have his seed inoculated in advance. Seedsmen who are equipped with treaters can apply the inoculant mechanically when the farmer is ready to plant. This practice is called "custom inoculation" (Carter, 1963). This

TABLE V. Amounts of Inoculant and Water Commonly Used in Inoculating Leguminous Seeds[a]

Seed type	Inoculant (gm/kg seed)	Water (ml/kg seed)	Rhizobia (per seed)[b]	Seeding rate (kg/ha)	Inoculant[c] (gm/ha)
White clover (*T. repens*)	4.4	22	500	2.2	10
Birdsfoot trefoil (L. corniculatus)	7.4	25	1,800	3.4	25
Red clover (*T. pratense*)	4.4	22	1,500	10.0	44
Alfalfa (*M. sativa*)	4.4	22	2,000	17.0	75
Vetch (*V. atropurpureal*)	4.4	13	40,000	43.0	190
Soybean (*Glycine max*)	4.4	13	150,000	65.0	285
Lupine blue (*L. angustifolius*)	4.4	13	160,000	82.0	360
Peas (edible) (*P. sativum*)	4.4	6	200,000	260.0	1,150

[a] Data from The Nitragin Co. The amounts of inoculant per gram of seed vary slightly with manufacturers.

[b] Based on rhizobia content of 200×10^6/gm inoculant.

[c] At specified seeding rate.

is in contrast to "preinoculation," the term used to describe inoculation of leguminous seeds months in advance of planting and before the seed are offered for sale.

Both peat-base and liquid cultures are used in "custom inoculation." Additives consisting of gums and sugars are often included in the inoculant slurries to improve adhesion to the seed and improve rhizobia survival on the seed.

3. *Survival on Seeds*

As pointed out earlier, inoculants are applied to seed because of convenience. Seeds generally provide a poor substrate for sustaining rhizobia (Burton, 1952, 1964, 1967, 1975; Vincent, 1958, 1965; Thompson *et al.*, 1975). Like other nonsporulating gram-negative rods, rhizobia die rapidly under drying conditions, whether on the seed or in the soil. The death rate will depend upon form, age, size of inoculum, temperature, relative humidity, and to some extent *Rhizobium* species. Subterranean clover, white clover, and some alfalfa seeds have water-soluble antibiotics in their seed coats which adversely affect rhizobia (Bowen, 1961; Thompson, 1960, 1961).

Rhizobia in a peat carrier are able to survive in greater numbers on seed than the rhizobia in liquid or lyophilized inoculants. Strains of *R. japonicum, R. lupini,* and the cowpea rhizobia appear to die more rapidly than *R. meliloti, R. trifolii, R. leguminosarum,* or *R. phaseoli* (Burton and Curley, 1965; Carter, 1963). Scanty gum production has been offered as an explanation, but the soybean and cowpea rhizobia survive in soil for long periods (Reid *et al.*, 1935).

4. *Special Inoculation Techniques*

Ordinary methods of inoculating seeds are adequate under many conditions, particularly when a good, quality peat inoculum slurry is applied just before planting. However, there are many other circumstances where special techniques are mandatory.

a. Lime Pellet. When leguminous seeds are to be mixed with acid fertilizers or planted in highly acidic soils, a lime coating is essential to protect the rhizobia (Roughley *et al.*, 1966). An adhesive mixture containing peat inoculant is first applied to the seed which are then rolled in a finely pulverized limestone until all the seed are thoroughly coated. The large amount of gum slurry required to wet the seed, 5 liters/50 kg seed, permits application of a very large inoculum which gives added insurance of success. Two types of stickers, the synthetic substituted cellulose gums and plant resins, acacia or gum arabic, are used. The coating is usually done in a rotating drum equipped with paddles (Brockwell, 1972; Date, 1970).

Lime pelleting is not recommended for lupins, trefoils, and many of the tropical legumes, because the *Rhizobium* spp. for these plants will not tolerate the high pH of lime. Powdered rock phosphate or some other neutral powder is more suitable.

b. Inoculant Pellet. When leguminous seeds are to be sown on neutral or slightly acid soils and these soils are heavily infested with native ineffective rhizobia, it is very desirable to add as many of the effective inoculant rhizobia as possible to the seed. The seeds are first mixed with a gum arabic inoculant slurry (1 liter to 15 kg seed). The seed are then dried by quickly mixing with another 800 gm of the powdered inoculant (Burton, 1975). Plants from seeds inoculated this way are effectively inoculated even with a highly competitive native flora of rhizobia in the soil. Also seed inoculated by this method can better withstand the hazards of dry soils and high temperature for as long as 2 or 3 weeks following sowing.

c. Seed Coatings. Seed coatings under proprietary brand labels such as Prillcote, CelPril, Rhizo-Kote, Nu-Kote, and Gold-Coat, have recently been introduced. Special adhesives are used to stick various combinations of particulate substances such as clay, charcoal, gypsum, $CaCO_3$ micronutrients, and pesticides to the seed coat and give the seed a uniform smooth covering. Peat-base inoculants are often included. Claims are made for better stands with fewer seed, more available nutrients, beneficial growth factors, healthier seedlings, more effective nodulation, and other benefits, but whether these treatments are really superior to the conventional lime and phosphate coatings remains to be proved. The limited evidence is that rhizobia do not survive any longer than they do on regular lime-pelleted seed (Brockwell *et al.*, 1975). Coated seed is considerably more expensive.

VI. OUTLOOK

A world population of 7 billion people is predicted by the year 2000. With the diminishing supply of fossil fuels and the resulting increase in cost of nitrogen fertilizers, biological N_2 fixation is destined to supply a larger share of food protein. Of all the biological systems, the *Rhizobium*–leguminous plant association appears the most promising. Nonetheless, formidable barriers cloud the picture.

Grain legumes are our main source of plant protein for human consumption, but world production of grain legumes is only about 10% that of cereal grains, although they are planted on 15% of the total area (Hardy, 1976). Grain legumes, even with their nitrogen-fixing nodules, are not as productive as the cereal grains.

According to Hardy, about one-third of the total grain legume production comes from the soybean. Despite the fact that soybeans work symbiotically with rhizobia and fix almost one-half of the nitrogen they need, yields have increased only moderately during the past 25 years. Soybean production has increased fivefold in the past 25 years, but this has resulted largely from a fourfold increase in area used to grow soybeans.

Agronomists are puzzled and rhizobiologists are embarrassed that strains of *R. japonicum* cannot supply 75% or more of the soybean plants' need for nitrogen. Hardy and Havelka (1973, 1975) suggest that the plant is the limiting symbiont and that plants with greater photosynthetic capability would result in more nitrogen fixation and higher yields. Soybean physiologists are working on this problem.

Apparently the other grain legumes are not as productive as the soybean. Although the soybean occupies only about 30% of the total area planted to grain legumes (Table VI), it produces about one-half the food protein in addition to substantial yield of oils.

The greatest challenge to the rhizobiologist is to develop more effective *Rhizobium* strains for dry beans, chickpeas, peanuts, and dry peas. The limited data available indicate that only a small percentage of the

TABLE VI. World Areas Planted to Grain Legumes[a]

Leguminous crop	Hectares planted × 10^3
Soybeans (*Glycine max*)	37,500
Peanuts (*Arachis hypogaea*)	18,000
Dry beans (*Phaseolus vulgaris*)	22,279
Chickpeas (*Cicer arietinum*)	10,543
Dry peas (*Pisum sativum*)	9,264
Cowpeas (*Vigna unguiculata*)	4,953
Broadbeans (*Vicia faba*)	4,683
Pigeon peas (*Cajanus cajan*)	2,587
Vetches (*Vicia* sp.)	1,879
Lentils (*Lens esculenta*)	1,717
Lupins (*Lupinus* sp.)	1,042
Other	6,324
Total	120,771

[a] From Food and Agricultural Organization (FAO), 1966.

seed is inoculated and that the benefits when obtained are of very low magnitude. The benefits or lack of them which have been noted could be attributed to poor nitrogen-fixing strains or a method of inoculating which could not possibly allow colonization of the roots and nodulation by the applied rhizobia. Working with soybeans, Johnson *et al.* (1965) found that an inoculum 800 to 1000 times the usual one had to be applied to attain even a moderate percentage of nodules from the applied rhizobia. Ecological problems in inoculating legumes have not received adequate attention. Certainly this is a fruitful area for research.

Data on world usage of *Rhizobium* inoculants are scanty. In the United States, *R. japonicum* inoculants constitute about 80% of the total. It is estimated that about 40% of the legume seeds planted are inoculated. Brazil, Australia, India, and most of the European nations manufacture *Rhizobium* inoculants.

There is good reason to suspect that *Rhizobium* inoculants will continue to be used and to increase in importance as the world population grows. Plant and *Rhizobium* geneticists will have to unite in purpose and work symbiotically to develop more efficient host plants and more efficient strains of rhizobia. Together they can accomplish what neither can achieve alone. Also, it will still be necessary to develop methods of inoculation which can assure nodulation by the improved strains.

REFERENCES

Allen, E. K., and Allen, O. N. (1950). *Bacteriol. Rev.* **14,** 273–330.
Badawy, F. H. (1965). Ph. D. Thesis, University of Wisconsin, Madison.
Badawy, F. H., and Allen, O. N. (1963). *Agron. Abst.* p. 31.
Beijerinck, M. W. (1888). *Bot. Ztg.* **46,** 726–804.
Bond, V. S. (1940). U.S. Patent No. 2,200,532.
Bowen, G. D. (1959). *Queens. J. Agric. Sci.* **16,** 267–282.
Bowen, G. D. (1961). *Plant. Soil*, **15n,** 155–165.
Brockwell, J. (1971). *CSIRO Div. Field St. Rec.* No. 10, pp. 51–58.
Brockwell, J. (1972). *Aust. Seed Rev.* **2,** 10–13.
Brockwell, J., and Hely, F. W. (1961). *Aust. J. Agric. Res.* **12,** 630–643.
Brockwell, J., Herridge, D. F., Roughley, R. J., Thompson, J. A., and Gault, R. R. (1975). *Aust. J. Exp. Agric. Anim. Husb.* **15,** 780–787.
Buchanan, R. E., and Gibbons, N. E. (eds.) (1974). "Bergey's Manual of Determinative Bacteriology," 8th ed. Williams & Wilkins, Baltimore, Maryland.
Burton, J. C. (1952). *Soil Sci. Soc. Am. Proc.* **16,** 356–358.
Burton, J. C. (1964). *In* "Microbiology and Soil Fertility" (C. M. Gilmour and O. N. Allen, eds.), pp. 107–134. Oregon State Univ. Press, Corrallis.
Burton, J. C. (1967). *In* "Microbial Technology" (H. J. Peppler, ed), pp. 1–33. Van Nostrand-Reinhold, Princeton, New Jersey

Burton, J. C. (1975). *In* "Symbiotic Nitrogen Fixation in Plants" (P. S. Nutman, ed.), IBP Synth. Vol. pp. 175–189. Cambridge Univ. Press, London and New York.
Burton, J. C., and Briggeman, D. S. (1948). *Soil Sci. Soc. Am. Proc.*, **13,** 275–278.
Burton, J. C., and Curley, R. L. (1965). *Agron. J.* **57,** 379–381.
Burton, J. C., and Erdman, L. W. (1940). *J. Am. Soc. Agron.* **32,** 439–450.
Burton, J. C., and Wilson, P. W. (1939). *Soil Sci.* **47,** 293–303.
Burton, J. C., Martinex, C. J., and Curley, R. L. (1977). "Rhizobia Inoculants for Various Leguminous Species," Inf. Bull. 101. Nitragin Co., Milwaukee, Wisconsin.
Carter, A. S. (1963). *Soybean News* **14,** 4.
Christianson, D. D., Cavins, J. F. C., and Wall, J. S. (1965). *J. Agric. Food Chem.* **13,** 277–280.
Corby, H. D. L. (1975). *In* "Symbiotic Nitrogen Fixation in Plants" (P. S. Nutman, ed.), pp. 169–173. Cambridge Univ. Press, London and New York.
Date, R. A. (1970). *Plant Soil* **32,** 703–725.
Date, R. A. (1976). *In* "Symbiotic Nitrogen Fixation in Plants" (P. S. Nutman, ed.), Cambridge Univ. Press, London and New York.
Date, R. A., and Roughley, R. J. (1977). *In* "A Treatise on Dinitrogen Fixation" (R. W. F. Hardy and A. H. Gibson, eds.), Sect. IV. pp. 243–276. Wiley, New York.
Dube, J. N., and Mandeo, S. L. (1973). *Res. Industry,* **18,** 94.
Erdman, L. W., and Means, U. M. (1946). *Soil Sci. Soc. Am. Proc.* **11,** 255–259.
Erdman, L. W., and Means, U. M. (1949). *Soil Sci. Soc. Am. Proc.* **14,** 170–175.
Fred, E. B., Baldwin, I. L., and McCoy, E. (1932). "Root Nodule Bacteria and Leguminous Plants." Univ. of Wisconsin Press, Madison.
Graham, P. H. (1963). *J. Gen. Microbiol.* **30,** 245–248.
Graham, P. H. (1976). *In* "Symbiotic Nitrogen Fixation in Plants" (P. S. Nutman, ed.), pp. 99–112. Cambridge Univ. Press, London and New York.
Graham, P. H., and Parker, C. A. (1964). *Plant Soil* **20,** 383–395.
Hamdi, Y. A. (1965). Ph.D. Thesis, University of Wisconsin, Madison.
Hamdi, Y. A. (1968a). *Acta Microbiol. Pol.* **17,** 277–278.
Hamdi, Y. A. (1968b). *Arch. Microbiol.* **63,** 227–231.
Hamdi, Y. A. (1969). *Plant Soil* **31,** 111–121.
Hardy, R. W. F. (1976). *Proc. Int. Symp. Nitrogen Fixation*, Vol. 2, pp. 693–717.
Hardy, R. W. F., and Havelka, U. D. (1973). *Plant Physiol.* Suppl. **48,** 35.
Hardy, R. W. F., and Havelka, U. D. (1975). *Science* **188,** 633–643.
Hardy, R. W. F., and Holsten, R. D. (1972). "The Aquatic Environment: Microbial Transformations and Water Quality Management Implications." Environmental Protection Agency, Washington D.C.
Hardy, R. W. F., and Holsten, R. D. (1977). *In* "A Treatise on Dinitrogen Fixation" (R. W. F. Hardy and A. H. Gibson, eds.), Sect. IV, pp. 451–486. Wiley, New York.
Hellriegel, H., (1886). *Ztschr. Ver.* Rübenzucker Industrie Deutschen Reichs, **36,** 863–877.
Holding, A. J., Tilo, S. N., and Allen, O. N. (1961). *Trans. Int. Congr. Soil Sci., 7th, 1960* Vol. II, pp. 608–616.
Jensen, H. L. (1967). *Arch. Microbiol.* **59,** 174–179.
Johnson, H. W., Means, U. M. and Weber, C. R. (1965). *Agron. J.* **57,** 179–185.
Kassim, M. (1976). M.Sc. Thesis, University of Missouri, Columbia.
Laird, D. G., and West, P. M. (1938). *Can. J. Res. Sect. C* **16,** 347–353.
Leffler, W. A. (1976) U.S. Patent 3,976,017.
Munns, D. N. (1976). "Proceeding of the NifTAL, Workshop on Nitrogen Fixation by Tropical Legumes," pp. 211–236. Kahului, Maui, Hawaii.

Munns, D. N. (1977). *In* "A Treatise on Dinitrogen Fixation" (R. W. F. Hardy and A. H. Gibson, eds.), Sect. IV, pp. 353–392, Wiley, New York.
Norris, D. O. (1958). *Aust. J. Agric. Res.* **9,** 629–632.
Norris, D. O. (1959). *Emp. J. Exp. Agric.* **27,** 87–97.
Norris, D. O. (1965). *Plant Soil* **22,** (2), 143–166.
Norris, D. O. (1965). *Proc. Int. Grassl. Congr., 9th, 1965* pp. 1087–1092.
Norris, D. O., and Date, R. A. (1976). *In* "Tropical Pasture Research, Principles and Methods" (N. H. Shaw and W. W. Bryan, eds.) pp. 134–174. Commonw. Agric. Bur., Harley Berkshire, England.
Norris, D. O., and 't Mannetje, L. (1964). *East Afr. Agric. For. J.* **29,** No. 3.
Parker, D. T., Allen, O. N., and Algren, H. L. (1949). *Crops Soils.* **1,** 10–11.
Purchase, H. F., and Vincent, J. M. (1949). *Proc. Linn. Soc. N. S. W.* **74,** 227-236.
Reid, J. J., Fred, E. B., and Baldwin, I. L. (1935). *J. Bacteriol.* **29,** 75–76.
Rice, W. A. (1976). "Alfalfa Production in the Peace River Region," pp. D1-D12. Can. Res. Stn. Beaverlodge, Alberta.
Roughley, R. J. (1968). *J. Appl. Bacteriol.* **31,** 259–265.
Roughley, R. J. (1970). *Plant Soil* **32,** 675–701.
Roughley, R. J. (1975). *In* "Symbiotic Nitrogen Fixation in Plants" (P. S. Nutman, ed.), IBP Vol. 7, pp. 125–136. Cambridge Univ. Press, London and New York.
Roughley, R. J., and Vincent, J. M. (1967). *J. Appl. Bacteriol* **30,** 362–376.
Roughley, R. J., Date, R. A., and Walker, M. H. (1966). *N.S.W. Dep. Agric., Div. Sci. Serv. Plant Ind. Bull.* **S-67,** 5.
Schall, E. D., Schenberger, L. C., and Swope, A. (1975). *Indiana* Agric. Exp. St., **106.**
Scudder, W. T. (1975). *Soybean Dig.* **35,** 16.
Steinberg, R. A. (1938). *J. Agric. Res.* **57,** 461–476.
Stevenson, C. L. (1976). *Legume Inoculant Insp. Program, Agric. Can.* **T-4-1.**
Strijdöm, B. W. (1963). Ph.D. Thesis, University of Wisconsin, Madison.
Strijdöm, B. W., and Allen, O. N. (1965). *Can. J. Microbiol.* **12,** 275–283.
Strijdöm, B. W., and Allen, O. N. (1967). *S. Afr. J. Agric. Sci.* **10,** 623–630.
Strijdöm, B. W., and Allen, O. N. (1969). *Phytophylatica* **1,** 147–152.
Strijdöm, B. W., and Deschodt, C. C. (1975). *In* "Symbiotic Nitrogen Fixation in Plant" (P. S. Nutman, ed.), pp. 151–168. Cambridge Univ. Press, London and New York.
Strong, T. H. (1937). *J. Counc. Sci. Ind. Res. Aust.* **10,** 12–16.
Taylor, G. G. (1972). U.S. Patent 3,672945.
Thompson, J. A. (1960). *Nature (London)* **187,** 619–620.
Thompson, J. A. (1961). *Aust. J. Agric. Res.* **12,** 578–592.
Thompson, J. A., Brockwell, J., and Roughley, R. J. (1975). *J. Aust. Inst. Agric. Sci.* **41,** 253–254.
Van Schreven, D. A. (1958). *In* "Nutrition of the Legumes" (E. G. Hallsworth, ed.), pp. 328–333. Butterworth, London.
Van Schreven, D. A., Otzen, D., and Lindenbergh, D. J. (1954). *Antonie van Leeuwenhoek* **20,** 33–57.
Vincent, J. M. (1945). *J. Aust. Inst. Agric. Sci.* **11,** 121–127.
Vincent, J. M. (1958). *In* "Nutrition of the Legumes" (E. G. Hallsworth, ed.), pp. 328–333, Butterworth, London.
Vincent, J. M. (1965). *Amr. Soc. Agron. Monogr.,* **10,** 384–435.
Vincent, J. M. (1970). *IBP Handb.* 15.
Vincent, J. M. (1974). *In* "Biology of Nitrogen Fixation" (A. Quispel, ed.), pp. 265–341. Am. Elsevier, New York.

Vincent, J. M. (1977). *In* "A Treatise on Dinitrogen Fixation" (R. W. F. Hardy and W. S. Silver, eds.), Sect. III, pp. 277–366. Wiley, New York.
Wacek, T. J., and Brill, W. J. (1976). *Crop Sci.* **16,** 519–523.
Wilson, J. B., and Umbreit, W. W. (1940). *Soil Sci. Soc. Am. Proc.* **5,** 262–263.
Wilson, P. W. (1940) "Biochemistry of Symbiotic Nitrogen Fixation," Univ. of Wisconsin Press, Madison.
Wright, W. H. (1925). *Soil Sci.* **20,** 95–129.

Chapter 3

Lactic Starter Culture Concentrates

RANDOLPH S. PORUBCAN
ROBERT L. SELLARS

I. INTRODUCTION

Commercial lactic starter culture concentrates have gained immense popularity in the United States food industry in the last 10 years. In particular, liquid nitrogen frozen concentrates have become the dominant form used by the dairy and cheese industry. Sold as either "bulk sets" or "direct sets" in approximately 70–500 ml quantities, such cultures are capable of inoculating from 300 to 1500 gal (1136–5678 liters) of milk directly. Subsequent incubation thereby produces either intermediate bulk starter when "bulk sets" are used, or the final product when "direct sets" are used. Cheddar cheese, cottage cheese, sour cream, buttermilk, and yogurt are some examples of these final products.

Other food industries also employ liquid nitrogen frozen lactic starter

MICROBIAL TECHNOLOGY, 2nd ed., Vol. I
 ISBN 0-12-551501-4

culture concentrates; namely, the meat, bakery, and fermented vegetable industries. Respective examples of cultured products produced in these industries are semidry sausage, sourdough French bread, and pickled cucumbers.

Freeze-dried and spray-dried concentrates of lactic starter cultures are also commercially available in the food industry. These have the advantageous feature of not requiring cryogenic* shipment and storage. However, performance limitations are usually experienced when such dry cultures are used directly without first making one or two intermediate propagations.

To clarify the terminology, lactic starter cultures or lactic acid starter cultures encompass several important genera of bacteria: Streptococci of group N, such as *Streptococcus lactis, Streptococcus cremoris, Streptococcus lactis* subsp. *diacetylactis,* and *Streptococcus thermophilus*†; Lactobacilli, such as *Lactobacillus bulgaricus, Lactobacillus acidophilus, Lactobacillus plantarum,* and *Lactobacillus helveticus;* and Pediococci, such as *Pediococcus cerevisiae.*

This chapter describes specifically the large-scale bioprocessing of concentrated lactic acid bacteria with associated emphasis on the commercial use of the resulting cultures. In the first edition of "Microbial Technology" Sellars (1967) discussed much of the general background information complementary to this chapter. Information on propagating lactic seed cultures, classification of cultures used in cheese manufacture, cultural and physiological characteristics of typical lactic cultures, starter metabolism, and growth curve data are given in this earlier chapter, all of which is still valid for most intents and purposes and, therefore, will not be reiterated here.

II. INDUSTRIAL BIOMASS PRODUCTION

The successful large-scale bioprocessing of lactic acid starter cultures depends on three interrelated and equally important elements: (1) processing equipment; (2) support facilities; and (3) operating procedures. We will limit our discussion in this section to processing where the desired end result is a highly concentrated starter culture destined to be used in preparing fermented food products.

Conventional fermentation equipment, such as that used by the phar-

* This term is used loosely here to indicate temperatures at or colder than −40°C.

† This organism may be more closely related to the group D streptococci (Buchanan and Gibbons, 1974).

maceutical industry and standard dairy processing equipment, do not satisfy the unique requirements for starter culture production. A compromise that hybridizes the pharmaceutical industry's ability to steam sterilize and maintain aseptic vessels with modern dairy ultrahigh-temperature (UHT) and clean-in-place (CIP) technology is required. Such a balance can be achieved by making appropriate modifications on conventional but modern dairy processing equipment. These modifications vary in their degree of sophistication and will be discussed in more detail later.

Figure 1 presents a general flow scheme in semiblock diagram form for the major components involved in the large-scale production of concentrated lactic acid starter cultures. The scheme begins at points A and B with media reconstitution followed by ultrahigh-temperature pasteurization (step C) and subsequent aseptic transfer to the fermentor (H). Here the sterile medium is inoculated and incubated; concomitantly, various growth parameters are controlled if required. The goal is to reach maximum biological concentration and activity in the fermentor prior to any further processing.

Once fermentation is complete the culture or fermentate is cooled and pumped into a "cold wall" storage tank (Fig. 1, item J). From here a variety of physical methods of concentration may be selected varying from lyophilization to centrifugal separation. Ultimately, the concen-

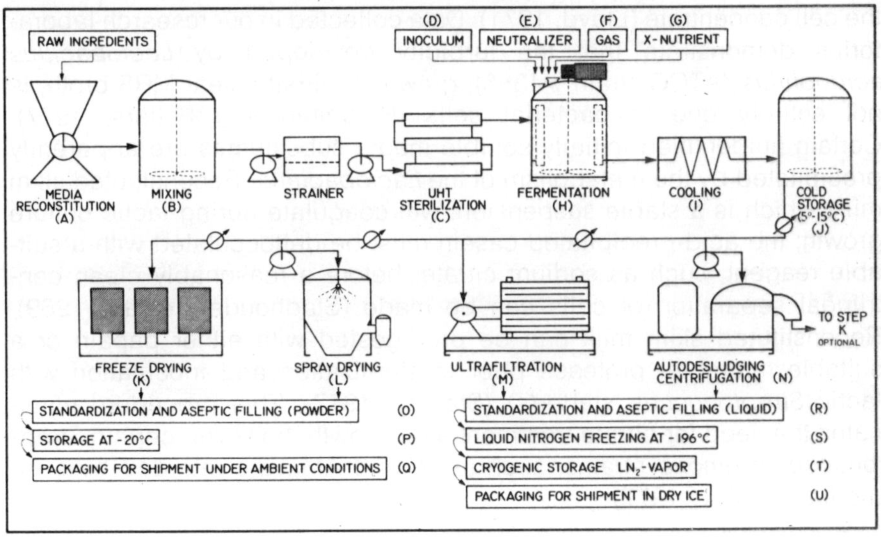

FIGURE 1. Flow scheme for the large-scale production of lactic starter culture concentrates.

trated starter culture* is marketed as either a powder or a frozen liquid concentrate. With this brief overview of the bioprocess we will proceed to a more detailed evaluation.

A. Growth Media

1. Formulation

Basal media ingredients for starter culture biomass production should be either soluble or able to remain in homogeneous suspension. With only a few exceptions this is generally the rule, particularly when centrifugal or membrane procedures are used to physically concentrate the starter bacteria after fermentation. Insoluble, nondispersible ingredients will be recovered with the cell concentrate in these procedures, consequently limiting the maximum level of cells attainable per unit volume (Stadhouders *et al.*, 1969). This problem may be circumvented in certain cases by employing enzymatic predigestion of the otherwise insoluble ingredient(s). More specifically, cheese whey upon proteolytic digestion no longer precipitates during heat sterilization and is thus rendered more useful in preparing basal media for lactic *Streptococcus* starters (Hansen, 1969; Jost and Monti, 1977).

The use of totally soluble and/or dispersible ingredients does not in itself eliminate the possibility of recovering residual media solids with the cell concentrate (Lloyd, 1971). Data collected in our research laboratories demonstrate that the turbidity developed by *Lactobacillus acidophilus* (ATCC strain #4356), grown in initially clear MRS broth, is not entirely due to bacterial cells (Porubcan and Sellars, 1977). Certain unidentified initially soluble media substituents are apparently precipitated by the metabolism of the *Lactobacillus*. Reconstituted skim milk, which is a stable suspension, will coagulate during lactic culture growth; the acid-precipitated casein must be deflocculated with a suitable reagent, such as sodium citrate, before a reasonably clean centrifugal separation of cells can be made (Stadhouders *et al.*, 1969). Reconstituted skim milk can be predigested with either papain or a suitable microbial protease prior to sterilization and inoculation with lactic *Streptococcus* starter(s) (Stanley, 1977). This procedure eliminates the need for citrate treatment after growth; however, certain *Streptococcus cremoris* strains, such as P_2 and AM_2, are retarded by such media (Stanley, 1977).

* The terms "starter cultures" and "cultures" are used as synonyms to "lactic acid starter cultures" in this chapter.

Insoluble, nondispersible ingredients, although often desirable from a price standpoint, have the additional drawback of being less available to the cell. Often the degree of "insolubility" is related to the heat treatment parameters employed. For example, when reconstituted sweet whey is ultrapasteurized or UHT heated at 290°F (143.3°C) for 4 seconds, it remains in suspension; however, when it is batch heat treated at 90°C (194°F) for 40 minutes, it precipitates dramatically and subsequently fails to support the prolific growth of lactic streptococci.

A variety of ingredients are in general use for the bulk cultivation of lactic acid bacteria. They can be classified into four broad groups, as listed and characterized in Table I.

To achieve maximum growth in batch fermentations with any particular strain or mixture of strains of lactic acid bacteria, it is necessary to formulate an advantageous basal medium. It may be as "simple" as reconstituted skim milk or a very complex mixture of several ingredients from each group in Table I. The importance of well-controlled fermentation research in establishing a correct formulation cannot be overemphasized. Most of the useful strains of lactobacilli, group N streptococci, and pediococci have rather fastidious growth requirements (Lawrence *et al.*, 1976; Etchells and Costilow, 1968; Thompson, 1976). Indeed, our technology has not advanced to the point where we can chemically define the growth media for these lactic bacteria and at the same time achieve maximum cell density and activity. Thus, we must continue to use such complex nutrients as skim milk, whey, yeast extract, peptones, and the like.

Unfortunately, complex nutrients such as those listed in group I of Table I are often subject to a great deal of variance from one manufacturer to another and even between different production lots from the same manufacturer. It is sensible to pretest samples of such ingredients prior to purchasing large quantities. In some cases where lot-to-lot uniformity is irregular it may be helpful to use two or more commercial brands simultaneously. For example, if a given basal medium formulation calls for 0.5% yeast extract, it may prove beneficial to use 0.25% from each of two different commercial brands, say brands A and B. Thus, the formulation would become more "poised" in that it would be less sensitive to a loss in quality in brand A than if brand A were used alone. The "poising" of a medium formulation so that it becomes less sensitive to lot-to-lot variations in the major ingredients is vital to obtaining uniform fermentations. Sometimes such "poising" can be effected by using an adjuvant ingredient(s) not directly related to the ingredient in question. For example, it has been shown that the unsatisfactory clotting of springtime milk by various strains of lactic streptococci can be over-

TABLE I. Ingredients Used in the Large-Scale Fermentation of Lactic Starter Cultures

Group I. Complex nutrients

A. Soluble (forming clear 1% solutions)[a]
   1. Yeast extracts or autolysates
   2. Casein hydrolysates
   3. Soy protein hydrolysates
   4. Peptones of animal origin
   5. Corn steep liquor
   6. Liver extracts
   7. Whey protein hydrolysates

B. Insoluble (forming stable aqueous suspensions)[b]
   1. Skim milk powder
   2. Sweet whey powder (from Swiss or mozzarella cheese)
   3. Acid whey powder (from Cheddar or cottage cheese)
   4. Whey protein concentrates
   5. Hydrolyzed cereal solids, dextrins, malto-dextrins
   6. Demineralized whey powder

Group II. Simple nutrients

A. Organic
   1. Glucose, fructose, and other monosaccharides
   2. Lactose, maltose, sucrose, and other disaccharides
   3. Tweens and related compounds
   4. Vitamins

B. Inorganic
   1. Mineral salts of Ca, Mn, Mg, Mo, Fe, Co, etc.

Group III. Buffers

A. Organic
   1. K, Na, or ammonium salts of citrate,[c] glycerol phosphate, and acetate

B. Inorganic
   1. Mainly the orthophosphates of K, Na, or ammonium; examples, monoammonium phosphate, disodium phosphate

Group IV. Defoamers

A. Silicone types
   1. Dow Corning FG-10
   2. Hansen's defoamer

B. Organic types
   1. Tweens (i.e., Tween 80)
   2. Acetylated monoglycerides

[a] Crude or unrefined varieties of the examples listed may be insoluble or may contain insoluble components.

[b] Stable suspensions may be dependent upon the method of preparation—see text.

[c] Citrate can also be a fermentable substrate and/or a deflocculating agent.

come by the addition of seven different amino acids at 50 ppm each (Erzinkyan *et al.*, 1966).

2. Reconstitution

Thorough and complete reconstitution of the various media substituents prior to sterilization or heat treatment is essential for two reasons: (1) incompletely hydrated proteins and peptones may heat denature differently compared to their hydrated counterparts (Porubcan, 1975; Babel 1976), and (2) nondispersed semidry aggregates of protein and/or carbohydrate may not sterilize during heat treatment due to inadequate heat penetration. This is particularly true when UHT processing is used where the thermal exposure time is so brief (Carlsen, 1977).

Proper reconstitution can be achieved by using modern dairy-type powder-to-liquid blenders, such as the Ladish Triblender or the Lanco Liquifier. Generally, the blender is connected to an agitated holding tank so that thorough recirculation back to the blender can be achieved during powder addition. (Refer to Fig. 1, parts A and B, for a diagrammatic representation of this process.)

3. Sterilization

By definition, sterilization is the complete destruction of all microorganisms, including mycotic and bacterial spores, by either wet steam at 120°C and 15 psig for 15 minutes or by dry heat at 160°–180°C for 3 hrs (Lawrence and Block, 1968). Sterilization per se cannot be achieved in complex media, such as milk or whey, by these conventional techniques without the associative destruction of certain growth-promoting properties, (Thomas and Turner, 1977; Webb, 1965). It is common knowledge, for example, that autoclaved (120°C, 15 minutes) skim milk is degraded to where it will not support active lactic culture growth. In the past, batch heat treatment at 85°–95°C for 40–120 minutes was used for such heat-sensitive media (Christensen, 1967). It was recognized, however, that although vegetative cells were destroyed under these conditions, it was not possible to guarantee the destruction of bacterial and mycotic spores. Thus, a condition less than sterile had to be accepted. The process of tyndallization (or sporulation), which employs the alternating application of batch heat treatment with appropriate incubation periods, can be effective but is not practical in large-scale bioprocessing.

The best way to effect the sterilization of complex media without inducing a concomitant loss in growth-supporting properties is through UHT* pasteurization. In our laboratory we have confirmed that UHT

* Also referred to as higher heat, shorter time (HHST) processing.

processing at 140°–145°C for 4–8 seconds will render 10% reconstituted skim milk (or sweet whey) sterile without any accompanying deleterious effects, such as caramelization. Such media support the prolific growth of lactic streptococci and popular lactobacilli such as *L. bulgaricus* and *L. acidophilus*. Understandably, chemical changes take place in milk and whey when they are UHT pastuerized (Hansen and Melo, 1977). Specifically, casein appears to be more resistant to structural alteration than β-lactoglobulin when raw skim milk is heated at 143° ± 1°C for 8–10 seconds (Hansen and Melo, 1977). These changes in the whey protein are potentially stimulatory to lactic acid bacteria (Webb, 1965; Woods, 1975), and current data from our laboratory support this contention.

Various types of industrial UHT pasteurizers are available. They can be classified as either indirect or direct referring to the method by which high-pressure steam heats the product. Of the indirect UHT pasteurizers, the tubular models are better engineered to withstand the high pressures and they do not encounter the gasketing problems that always plague the plate-type pasteurizers. The direct steam-injected UHT pasteurizers are not particularly suited for processing culture media. Direct injection of steam creates turbulence and physical stress that may induce media instability (Woods, 1975), not to mention the requirement for absolutely pure steam. In addition, moderate whey protein denaturation is enhanced by indirect UHT pasteurization, a potentially beneficial condition as previously discussed.

When skim milk is used as a substrate for lactic acid bacteria, the type of heat treatment to which it is subjected determines whether growth stimulation or inhibition will occur. For example, heating milk at 143°–160°F (62°–71°C) for 30–40 minutes stimulates the growth of *Streptococcus* starter cultures, whereas heating at 160°F (71°C) for 45 minutes or at 178°F (81°C) for 10–45 minutes inhibits growth (Webb, 1965; Woods, 1975). This "temperature conditioning" of milk and whey and other complex media should be treated as a separate entity apart from the time/temperature profile required to effect sterilization. A heat treatment that can optimize the growth response and also effect sterilization is the goal to achieve with any new medium formulation. Indirect tubular UHT processing systems, such as the Cherry-Burrell Spiratherm, can accomplish this difficult task. Large-scale Spiratherm systems with capacities over 2000 gal/hour regenerate input calories to between 55 and 65% depending on the unit.

As shown in Fig. 1 (C) reconstituted medium is pumped through a strainer to a ballast tank (not shown) associated with the UHT unit. This is usually done with automatic level control. The medium in the ballast tank is then fed to the UHT via a high-pressure pump, such as a homogenizer. After media sterilization, careful attention must be focused on the method of transfer to the presterilized fermentor(s) (Fig. 1, H). Here it is necessary that steam-locked sanitary pumps and valves be used and that the transfer piping be steam sterilized after conventional CIP.

B. Fermentation

1. Fermentor Design

The large-scale fermentation of lactic acid starter cultures requires a round dairy processor with some rather sophisticated modifications if both security against bacteriophage infection and maximum biomass production are to be achieved. An ideal processor/fermentor is diagrammatically indicated as H in Fig. 1; special design features include those listed below:

1. An ability to hermetically seal and pressurize the vessel at 15 psig (working) to allow in-place sterilization at 121°C when required.

2. A full pressure-rated jacket with zone controls for in-place fermentor sterilization control and temperature control during incubation. It should be possible to hold the fermentor contents at ± 0.5°C during incubation cycles.

3. The "round processor" design where the vessel's functional depth is about twice the inside diameter.

4. A variable-speed agitator with a full bottom sweep impeller. Mechanical speed reducing transmissions, such as a Reeves drive, are effective up to about 10 HP. Above 10 HP variable frequency AC control is recommended. Teflon agitator shaft seals are desirable.

5. Type 316 stainless steel on all product contact surfaces; type 304 stainless steel is acceptable for exterior surfaces.

6. Two, three, or four baffles installed at angles of 180°, 120°, or 90°, respectively, and extending above the liquid level perpendicular to the side wall with a separate CIP spray ball assembly for each baffle. Proper baffling ensures top-to-bottom turnover during agitation.

7. A high-temperature CIP system able to deploy a chlorinated alkaline detergent at 185°F (85°C) with subsequent water rinse, acid, and chlorination cycles. Single-use CIP systems, where fresh cleaning solutions are used for each independent operation, are preferred as cross-contamination possibilities are virtually eliminated by such systems. Economics Laboratory at Beloit, Wisconsin manufactures and installs an excellent system of this type.

8. An absolute 0.2–0.35 μm filter(s) on the headplate to assure that all air entering the headspace is bacteria free.* Filter units are attached to large sanitary butterfly valves that can be closed during hermetic (and CIP) operations. The filter(s) should be sized such that it will provide sufficient air displacement capacity to keep up with the unloading pump.

9. A vacuum/pressure safety relief valve on the headplate along with a separate rupture disk if double protection against vacuum collapse is desired.

10. Standard sanitary inlet ports for all transfer operations; one is usually a "no-foam" inlet for media fill (on headplate) and another can be steam jacketed for aseptic inoculation. Additional ports may be required on the headplate for various additives, such as neutralizer, gas, extra nutrient, and defoamer (see Fig. 1, E, F, and G). Ports are also required for electrodes such as pH electrodes. Thermistor or thermocouple probes for temperature sensing and control are generally installed in sidewall-mounted hermetically sealed wells that extend into the product by 6–10 inches.

11. A manhole with sight glass and a light fixture at 180°; both should be mounted on the headplate. Visual assessment of agitator rpm for a given starter culture is necessary to ensure optimization of this parameter. In addition, the manhole should be "safety interlocked" with the CIP system so that it must remain open with a louvered door in place during CIP.

12. Remote manometer readout of vessel volume.

The foregoing special design features truly convert an otherwise standard round processor into a functional fermentor for lactic acid bacteria. In contrast to typical pharmaceutical-type fermentors the parts and fittings here are of sanitary design; the agitator is low speed, usually 5–60 rpm on 1000–8000-gal vessels; electrodes are mounted through the headplate, not through the sidewall, and all connection fittings are sanitary welded and are, therefore, without threads.

2. Fermentation Process.

Illustration of the general fermentation process for lactic acid bacteria is best developed by referring once again to the example shown in Fig. 1. Ultrahigh-temperature sterilized growth media are aseptically transferred into a presterilized fermentor (Fig. 1, H) through a no-foam inlet in the headplate. Alternatively, the growth media can be batch heat treated

* Generally, this will be a safe filter for viruses such as bacteriophage since most airborne phage is associated with macroscopic dust (Jespersen, 1977). If phage contamination develops, incinerated air must be used instead of the filter.

at 85°–95°C for 40–120 minutes in the fermentor; however, UHT processing is usually the desired course. Subsequently, gentle agitation is started and the temperature is adjusted via the jacket controls until the exact incubation temperature is reached. The fermentor is then inoculated (usually at 1–2%) with fresh mother culture (Fig. 1, D) via an aseptic centrifugal pump* or by pressurized air unloading using sterile air. Fermentation is then commenced and various parameters are automatically controlled as required to achieve maximum biomass populations (Blaine, 1972; Gilliland, 1977; Efstathiou *et al.*, 1975). These controls include agitation, temperature, pH, and in some instances redox potential (ORP) and nutrient concentration. After the fermentation is complete, usually by late log or early stationary phase, the culture or fermentate is cooled rapidly to 5°–15°C by pumping it through a tubular heat exchanger where 1°–2°C water is the cooling agent (Fig. 1, I). Slow cooling can be effected in the fermentor by circulating 1°–2°C water through the jacket. This may be a more advantageous approach with some cultures. The cooled culture can then be held in a cold wall dairy storage tank (Fig. 1, J) or in the actual fermentor if desired. From here it is further processed by one or more of four physical methods of concentration (Fig. 1, K, L, M, and N). These will be detailed in a following section.

The pH control function is of particular importance. Generally, "tight control" to within ± 0.1 pH unit of the set-point pH is most advantageous (Porubcan and Sellars, 1977; Richardson *et al.*, 1977). This can be effected by using an automatic pH controller (or autotitrator) coupled with precise agitation control. Neutralizer, such as concentrated ammonium hydroxide, should be admitted into the fermentor headspace at 180° to the pH electrode(s) (Porubcan and Sellars, 1977). Providing that fermentor agitation is adequate in terms of top-to-bottom turnover, the pH electrode(s) can be submerged in the top 10–15% of the fermentor's working volume. Combination electrodes are preferred over separate glass and calomel electrodes, and pH signal preamplification up to approximately 1 V is highly desirable (Richardson *et al.*, 1977). Preamplified pH signals are more stable and can be conveyed over longer distances. "Saw-toothed type" recordings of the pH control function (pH versus time) are important and should be obtained routinely.

The foregoing hypothetical descriptions depict a purely batch-type process. Continuous and/or dialysis culture fermentations with lactic streptococci have been studied (Osborne, 1977; Keen, 1972). From an industrial-scale standpoint, however, these are not practical. For example, Osborne's work with dialysis of lactic *Streptococcus* cultures

* A sanitary centrifugal pump with steam-jacketed shaft seals should be used.

against sterile media requires a medium-to-culture ratio of 30:1 to be effective. In our laboratory we have tried ultrafiltration as a means of removing metabolic waste products during the active log phase pH-controlled growth of lactic streptococci. When sterile medium was added to the fermentor at fixed intervals during ultrafiltration to compensate for the discarded permeate, the end result, with the cultures studied, was not an increase in biomass but an increase in the rate of lactic acid production in the fermentor (Porubcan and Sellars, 1977). This apparent increased rate of acid production during fermentation was not carried over by subsequent subcultures. No practical utility was demonstrated by these experiments.

An equation for batch bacterial growth discussed by Konak (1975) defines eight distinct growth phases: (1) lag phase, (2) accelerated growth, (3) logarithmic growth, (4) decelerated growth, (5) stationary phase, (6) accelerated death, (7) logarithmic death, and (8) survival phase. The overall result of the first five phases is a sigmoidal growth curve. The general equation for logarithmic growth is:

$$dN/dt = kN \tag{1}$$

where k is the specific growth rate, N is cell numbers per unit volume, and t is unit time. A straight line relationship is maintained during true log phase growth, but at the onset of decelerated growth dN/dt starts decreasing due to three growth-limiting factors: (1) limitations in nutrient supply, (2) accumulation of toxic metabolic waste products; and, (3) overcrowding—the "bump coefficient."

In phases 2 and 3 the rate of growth is directly proportional to N or the concentration of bacteria. In phases 4 and 5 collective retardation forces set in due to all or some of the above factors, and the growth rate becomes directly proportional to $(N_s - N)$ at any given time t (Konak, 1975). N_s is the concentration of bacteria per unit volume in the stationary phase. Assuming that the relationships are governed by exponential law we can write:

$$\frac{1}{N_s^{a+b}} \frac{dN}{dt} = k \left(\frac{N}{N_s}\right)^a \left(1 - \frac{N}{N_s}\right)^b \tag{2}$$

where a, b, and k are constants and the sum $a + b$ may be considered the "overall order" of the bacterial growth process. In the special case where $a = 1$ and $b = 0$, Eq. (2) indicates logarithmic growth.

If we now define $F_P = N/N_s$ as the fractional increase in population density, we can write:

$$\frac{dF_P}{dt} = k\, N_s^{a+b-1}\, F_P^a\, (1 - F_P)^b \tag{3}$$

Maximum growth rate occurs when $F_P = a/(a + b)$; therefore, from the inflection point on the curve of F_P versus t (or N versus t) we can obtain one simple relationship between the constants a and b. This can be determined along with N_s and k in laboratory experiments where initial nutrient concentration, initial bacterial concentration, bacterial species, temperature, and hydrodynamics are all controlled. Equation (3), which in general form has proven validity in crystallization and precipitation applications, can be used here to guide scale-up operations for batch-type fermentations (Konak, 1974; Nielsen, 1964).

With lactic acid bacteria, particularly with group N streptococci and many lactobacilli, the most significant of the three growth-limiting factors seems to be the accumulation of toxic metabolic waste products. dN/dt for most *Streptococcus cremoris* strains grown at 22°C in 12% reconstituted skim milk under pH control at pH 5.40 (NH_4OH neutralization) will start to decrease once the ammonium lactate concentration reaches about 2–3% (w/w) (Porubcan and Sellars, 1977). There is still a surplus of nutrient (carbohydrate and protein) at this time and overcrowding is not a particularly significant issue under these specific growth conditions (viable plate counts run 5×10^9/ml or less after 16 hours of growth).

C. Physical Concentration

1. Centrifugal Separation

The concentration of bacteria by centrifugal separation depends on two general criteria: (1) the properties of the bacterial culture to be separated, and (2) the efficiency of the centrifuge. There are two bacterial properties that are directly proportional to separation efficiency: the cell diameter squared and the density difference between the cell and the liquid medium. The viscosity of the supernatant is inversely proportional to separation efficiency (Elsworth, 1962; Webb, 1964). Obviously, then, large, dense cells that are suspended in a low-viscosity medium are most easily separated; the composite effect of these properties determines the overall "centrifugability" of the particular strain(s). It is the cellular size and shape that most significantly influence separation because of the square function. Spherical cells separate more readily than rod-shaped cells (Elsworth, 1962). Thus, by this theory, a yeast cell having a diameter of 5 μm will separate 25 times more readily than a bacterial coccus cell having a diameter of 1 μm (everything else equal). Another way of considering this difference is that the feed rate to an industrial centrifuge can be 25 times faster for the yeast culture.

Temperature also affects the efficiency of separation; higher temperatures benefit separation but exert a negative effect on cell viability and activity. Thus, a compromise is required; usually most lactic cultures are held somewhere between 5° and 15°C during separation. The highest temperature that can be tolerated by a given strain should be determined; this temperature will depend on cell sensitivity and centrifuge design.

The efficiency of a centrifuge is often mistakenly confused with the relative centrifugal force (RCF)* or gravitational effect that can be generated. The sigma number for a particular machine is the correct function that defines efficiency. Sigma is proportional to w^2r^2/S, where w is the bowl rpm, r is the bowl radius, and S is the maximum distance that a particle must travel to reach the bowl wall or a collecting surface. In effect, sigma is the surface area of the theoretical gravity tank that is required to effect the same degree of separation as the machine (Elsworth, 1962).

For a number of years the only industrial centrifuges used for separating lactic acid bacteria were the Sharples and Carl-Padberg (CEPA) type tubular rotor machines. These machines spin at high rpm, exerting correspondingly high RCFs. The largest CEPA machine, the model Z-101, spins at 14,000 rpm, exerts 17,760 RCFs, and has a rotor sludge capacity of 10 liters. The major drawback of these machines is their manual nature. The rotor (or bowl) must be stopped, dismantled from the machine, and opened before the bacterial concentrate can be scraped out. Even when waxed parchment or polypropylene inserts are used as removable rotor liners, the process is not practical and is subject to contamination.

The development of modern automatic desludging CIP separators made the large-scale bioprocessing of concentrated lactic acid bacteria realistic. These disk–bowl type separators are very efficient machines because of their large effective surface areas (small S values in the sigma expression) and large bowl diameters. They spin at relatively low rpm (4500–6000) but nevertheless have sigma values comparable to the tubular rotor machines. There are two major types of automatic desludging CIP separators: first, there are those where a closed bowl is intermittently opened during operation to eject the concentrate; second, there are the nozzle machines that have continuously open ports in the bowl periphery for continuous concentrate ejection. Examples of the first type are De Laval's MRPX-213 and MRPX-217 and Westfalia's SA-60, SA-80, and SB-80. Of the second type, or the nozzle machines, the best example is Alfa-Laval's Hermetic Bactofuge. According to the manufacturer's

* RCF = $g \cdot w^2 \cdot r$ where g is the gravitational constant, w is the bowl rpm, and r is the bowl radius in centimeters.

specifications the Bactofuge Type D-3187M will remove 90% of the bacteria in raw milk when operating at 65°C.

Insofar as the concentration of lactic acid bacteria is concerned, the intermittent desludgers afford better control over concentration. With these machines the bowl can be precisely opened and closed whenever desired to effect a particular degree of concentration. With the nozzle-type machines the feed rate to the machine can be varied within certain limits (the nozzles must be kept satisfied) but beyond that it is a "take what you get" situation. In both cases, the machines require certain modifications for lactic acid bacteria processing. The conical disk plates should be highly polished and somewhat closer together than standard and they should have grooved channeling. The centripetal pump should be constructed with a smooth bore for low-capacity feed rates, and provisions for aseptic sludge removal should be considered. It is common to run these separators at 1/10 to 1/20 of their specified feed capacity when processing lactic acid bacteria. Variable-speed sanitary positive pumps are preferred over any other product-to-separator delivery system. With the intermittent automatic desludgers it is possible to concentrate certain lactic streptococci to between 20 and 100 times their initial concentration (Porubcan and Sellars, 1977).

A recent innovation pioneered by Chr. Hansen's Laboratory, Inc. in Milwaukee, Wisconsin, involves cooling the spinning separator bowl (Westfalia SA-80-06 autodesludger) with liquid nitrogen. Actually, liquid nitrogen per se does not contact the bowl, but the resulting supercold vapor makes the contact. The liquid nitrogen is sprayed onto the top of the spinning bowl; it is delivered through a cryogenic solenoid valve that is automatically opened and closed as a function of the supernatant discharge temperature. By cooling with liquid nitrogen in this manner it is possible to exactly counterbalance the frictional heat of the machine such that the supernatant and concentrate discharge at a temperature exactly equal to the fermentate input temperature. This degree of cooling efficiency cannot be achieved by using a water-cooled jacket on the centrifuge cover bonnet.

2. *Ultrafiltration*

The industrial-scale concentration of lactic *Streptococcus* starters by ultrafiltration was reported by Porubcan and Sellars (1976). In this work a skim milk culture of *S. lactis* mixed with *S. cremoris* (Hansen's #70) was controlled between pH 5.40 and 6.00 with concentrated ammonium hydroxide during a 16-hour incubation period at 24°C. Following this the culture was cooled to 12°C and concentrated on an Aqua Chem Ultra-Sep ultrafilter (150 ft^2 of membrane) for 2 hours with inlet and outlet pressures of 60 psig* and 30 psig, respectively. The result was a threefold increase in bacterial concentration with a proportional increase in lactic acid-producing activity.

Ultrafiltration (UF) is different from reverse osmosis in that the membrane is permeable to both water and soluble low molecular weight compounds rather than to water alone. Consequently, lower operating pressures of almost an order of magnitude are characteristic of UF systems (Payne *et al.*, 1973). In the concentration of bacteria it is advantageous to operate under conditions of lower pressure and lower osmolarity; thus, UF is the membrane process of choice. Depending on the nature of the bacterial culture and on the type of UF machine, the extent to which a culture can be concentrated varies from two to about twelvefold. Beyond this level of concentration the viscosity of the concentrate limits the attainable velocity across the membrane and thus the permeate flux through the membrane. A point of diminishing returns is ultimately reached. In addition, frictional heat buildup damages the cells at these higher concentrations. In any event, an efficient method of cooling the bacterial concentrate during ultrafiltration is a vital requirement.

To be of value in the concentration of starter bacteria an ultrafiltration system must be able to tolerate pH 13 caustic cleaning (CIP) and subsequent chemical sterilization. Aqua Chem's Ultra-Sep machines can be supplied with either polyvinyl or Teflon membranes. Both types of membranes tolerate pH 13 caustic; the former can be sterilized with either H_2O_2 or formaldehyde and the latter with hypochlorite as well. These Ultra-Sep machines are of a rectangular plate "sandwich" design. The degree of compression on this sandwich of membranes and plastic spacers is of critical importance. Hydraulic compression to between 5000 and 6000 psig is required for optimum performance on units having 150–600 ft^2 of membrane surface (Schantz, 1976). Another competitive UF machine of a plate-and-frame design is manufactured by DDS in Denmark; Chr. Hansen's Laboratory in Copenhagen has concentrated lactic streptococci with a pilot-scale (Als de Danske Sukkerfabrikker DDS) unit having 4 m^2 of membrane surface (Jespersen, 1974a).

3. *Spray-Drying*

Spray-drying is the transformation of a solution or suspension into a dry powder in a single operation. The feed liquid is atomized into a fine

* psig = pounds per square inch gauge—not absolute pressure, but gauge pressure.

spray which immediately contacts a flowing stream of hot air. The spray-drying of yogurt and related cultures was reported by Porubcan and Sellars (1975b). A mixture of *L. bulgaricus* and *S. thermophilus* (Hansen's CH-3) was grown in 12% reconstituted skim milk for 16 hours at 37°C. A Niro Mobile Minor laboratory-scale spray-drier was then used to dry the culture at an outlet temperature of 145°F (63°C), a culture feed rate of 12–15 ml/minute, and a nozzle velocity of 3.5 kg/cm^2. The resulting powder had a plate count on Hansen's yogurt agar (Porubcan and Sellars, 1973) of 12 × 10^8 viable colony-forming units (CFU) per gram and a moisture content of 4.5%.

The successful spray-drying of yogurt cultures is dependent upon a chemical stabilization process developed by Porubcan and Sellars (1975a). The cooled fermented culture is mixed with ascorbate and glutamate followed by pH adjustment to 6.10–6.60. This combination induces a chemical synergism that imparts partial protection to the cells during the drying process and significant stabilization to subsequent storage at ambient temperatures. The chemical mechanism involved is linked to the antioxidant properties of the ascorbate with glutamate (although aspartate works equally well), moderating the reaction by controlling the rate of ascorbate oxidation. After the ascorbate expires, the stabilization effect is no longer realized and the dry culture deteriorates. Air, moisture, and light accelerate the oxidation of ascorbate and, thus, should be minimized whenever possible.

Spray-drying per se is a concentration process by virtue of water removal; however, significant cellular destruction occurs even when the ascorbate/glutamate treatment is used. In fact, losses in viable plate count (in CFU) of about 70% are normal. The major advantage of the ascorbate/glutamate treatment process is the enhanced stability of the spray-dried powder to storage at 21°–23°C. Specifically, untreated spray-dried *L. acidophilus* will decay by more than 1000-fold (from 1.6 × 10^9 to 1.0 × 10^6 per gram) after 1 month at 21°C, whereas treated culture decays by only twofold (10.7 × 10^9 to 5.7 × 10^9 per gram) under the same conditions (Porubcan and Sellars, 1975a).

Two parameters are critical to the success of large-scale spray-drying of yogurt and related cultures: (1) The outlet temperature of the main chamber should be controlled between 63° and 68°C (145°–155°F) and (2) the relative humidity of the intake air should not exceed 60%. Different strains of *L. bulgaricus, L. acidophilus,* and *L. helveticus* exhibit different degrees of sensitivity to the spray-drying process. Acceptable results are achieved with most of these thermoduric lactics; however, mesophilic lactics are very sensitive to spray-drying.

4. Lyophilization

Batch freeze-drying of lactic starter cultures has been practiced commercially for over two decades; both tray- and cylinder shell-type driers have been used. Generally, the freeze-drying process is less destructive to starter culture cells than the spray-drying process. Mesophilic lactic streptococci can be freeze-dried with rather satisfactory results; losses in viable cell populations can run as high as 80%, but 50% is about normal. The ascorbate/glutamate treatment process (Porubcan and Sellars, 1975a) is also of significant value when used in conjunction with freeze-drying. As with spray-drying the stability of the resulting dry culture to storage at ambient temperatures is the most pronounced attribute gained by using this chemical treatment process.

Procedurally, the stabilized fermented culture is frozen in a thin film of 1-cm thickness or less, typically by slow rotation in a horizontally positioned stainless steel (10-liter) cylinder that is submersed in either liquid nitrogen (−196°C) or dry ice/alcohol (−65°C). Two liters of liquid culture containing 12–18% solids by weight require about 5–12 minutes for complete freezing in the dry ice/alcohol mixture. Subsequently, each cylinder is clamped to a manifold assembly where it is subjected to a strong vacuum that should reach 25 μm of mercury in less than 3 hours. The cylinders are maintained at 20°–24°C during vacuum drying. The vacuum is held for 12–16 hours and the condensate is collected at −65°C. The resulting crusted material is gently ground to a uniform powder while appropriate aseptic procedures are maintained. The moisture content in this culture powder will be less than 1% if proper procedures have been followed; this level may be difficult to achieve with thermoduric yogurt-type cultures.

For liquid capacities of 50 liters or more the tray-type driers, such as those manufactured by Stokes or Hull, are more appropriate than the cylinder shell driers. Here liquid nitrogen and/or its vapor is the preferred freezing agent. Compared to spray-drying, freeze-drying is considerably more expensive due to greater energy requirements.

D. Packaging

1. Containers

Seamless extruded aluminum cans in 70 to 360-ml capacities are ideal containers for liquid starter culture concentrates, particularly when liquid nitrogen freezing is employed (Rinfret and LaSalle, 1975). Extruded plastic containers, such as those used for frozen sherbet, and cylindrical cardboard containers are also used by some commercial starter culture suppliers. These containers cannot be subjected to submersion in liquid nitrogen, however.

Spray-dried or freeze-dried cultures are preferably packaged in moisture-proof pouches of a laminated foil–plastic construction. Larger quantities of 1–50 kg are commonly packaged in heavy guage plastic bags contained in fiber drums. Since oxygen, moisture, and light are deleterious to dry cultures, it is desirable to pack them under dry inert gas or under vacuum in opaque enclosures.

2. *Liquid Nitrogen Freezing*

The optimum method of freezing liquid starter culture concentrates is by direct immersion in liquid nitrogen or its vapor. A 360-ml aluminum can full of liquid culture concentrate at 12°C will freeze solid after about 3–5 minutes in liquid nitrogen at −196°C; complete freezing to −196°C will take about 12 minutes and will require from 1 to 1.3 liters of liquid nitrogen (Porubcan and Sellars, 1977; Reddy *et al.*, 1974).

The initial freezing rate for 350 ml of skim milk contained in a 360-ml aluminum can is 0.4°F/second in liquid nitrogen (complete immersion) and 0.04°F/second in static liquid nitrogen vapor (Shuster, 1976). Because of viscosity effects and altered colligative properties these rates are lowered anywhere from three- to fivefold when concentrated starter cultures containing over 15% solids (dry wt. basis) are frozen.

A newly developed process for preparing frozen pellets of concentrated starter culture is currently under investigation at Chr. Hansen's Laboratory in Milwaukee. By a proprietary process, droplets of liquid starter concentrate are "rained" into an agitated bath of LN_2. The resulting pellets range from 1–2 mm in diameter and, as such, permit a free-flowing bulk-packaged product to be prepared.

Generally, most lactic acid bacteria freeze well in liquid nitrogen (LN_2) and do not require special cryoprotective compounds such as glycerol. A satisfactory diluent for starter culture concentrates destined for LN_2 freezing is reconstituted skim milk (Porubcan and Sellars, 1977; Peebles *et al.*, 1969). Most lactic acid bacteria have an indefinite shelf-life when stored in liquid nitrogen (−196°C) or its vapor (−150°C). We have had Redi-Set brand bulk set cultures (*S. cremoris* and *S. lactis*) in test storage at −150° to −196°C for over 8 years without any measurable loss in viable plate count or lactic acid-producing activity. Storage at less optimum temperatures leads to completely different results. For example, these same bulk set cultures that last 8 years at −150° to −196°C will only keep 4–6 weeks at −40°C before showing a measurable loss in viability. In addition, Leach and Sandine (1976) have shown that strain dominance in mixed strain lactic streptococcus starters can change in prolonged storage (8 months) at −20°C. The data of Reddy *et al.* (1974) clearly indicate that LN_2 freezing and storage is the best procedure to maintain component balance in mixed strain lactic cultures.

Frozen starter cultures should be thawed just prior to use by direct submersion in water at 24°C for 3–20 minutes depending on the amount of agitation employed. Once thawed, the general rule is to use the culture immediately whenever possible; however, most thawed cultures will keep for 1–2 hours at 5°–10°C without a measurable loss in viability and/or activity. The particular manufacturer's instructions should be followed in all cases as these recommendations may vary from one supplier to another.

Liquid nitrogen frozen cultures can be shipped from manufacturer to user in special portable liquid nitrogen refrigerators, such as the Linde Model 30-10. It is possible to transport 88 × 70 ml size culture units in such a container which, in addition, contains enough liquid nitrogen to hold the cultures for about 28–30 days. More commonly, however, cultures are shipped in styrofoam boxes packed with dry ice (about −60°C effective temperature). Such containers retain the frozen cultures for 72–96 hours, depending on ambient conditions. Upon arrival the user immediately transfers the dry ice refrigerated cultures to a mechanical deep freeze at −40°C or below or to LN_2 vapor storage.

E. Building Facility

1. Design Concepts

A detailed discussion on the construction and design of the basic building facility for starter culture production is beyond the scope of this chapter. There are some important design concepts, however, that should be discussed here. Figure 2 illustrates a circular flow pattern for the production of frozen concentrated starter cultures. Raw materials enter the sequence at pie section 1 (media reconstitution) and process flow proceeds counterclockwise ending up at pie section 8 (cryogenic storage of frozen cultures). Warehouse facilities for the storage of raw materials are not shown. At the central core is the CIP system and the pipe chase for all major utilities. Although this circular design has some very desirable attributes, in actuality modifications are required to facilitate construction and enhance floor space economy.

Sanitary construction typical of modern dairy plants is an essential element here. In addition, various design sophistications are required, about ten of which encompass the most critical design aspects. For purposes of brevity these ten are listed below without attempts to cite and contrast all of the various alternatives.

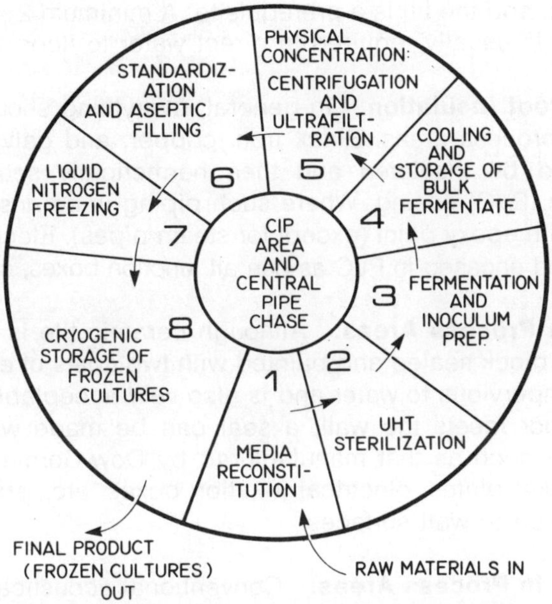

FIGURE 2. Circular design concept for the layout of processing equipment for the production of frozen lactic starter culture concentrates.

a. Air Handling System. The air supply to process rooms (Fig. 2, pie sections 2 through 6) should be HEPA* filtered at 0.35 μm. Both Flanders and Cambridge manufacture such filters. Laminar flow HEPA filtered air is desirable in the aseptic filling areas and over fermentors (Austin, 1970). In these specific areas a Class 100 environment should be achieved; by definition this means that there are to be fewer than 100 particles of less than 0.5 μm in length per cubic foot of air (Anonymous, 1973). Filtered air is not required in the media reconstitution room since raw ingredients routinely contaminate this area; personnel working in this room should be prohibited from routinely entering process areas 2 through 6. Air lock vestibules are recommended where aseptic sections (Fig. 2, pie sections 2 through 6) access into warehouses and other nonfiltered areas.

b. Floors in Process Areas. The floors in the process areas (all pie sections, Fig. 2) should be constructed of quarry tile of at least 0.75-inch thickness. An antiskid grit finish is desirable and flush epoxy grout and epoxy bedding is recommended. A waterproof membrane between the

* HEPA = high efficiency particulate air filters.

structural slab and the fill is a prerequisite. A minimum 2% pitch on the finished floor is usually required to direct water to floor drains.

c. Waterproof Insulation. In general, everything should be waterproof in the process areas. Black iron, copper, and galvanized utility piping should be insulated and then hermetically sealed in polyvinyl chloride (PVC) casing. Where such piping is not insulated it can be painted with epoxy paint (except for steam pipes). Electrical conduit can be ordered encased in PVC as can all junction boxes, couplings, etc.

d. Walls in Process Areas. Although ceramic tile is very acceptable, cement block sealed and painted with two coats of epoxy paint is completely impervious to water and is also very acceptable. Where the quarry tile floor meets the wall, a seal can be made with "10-year" silicone caulk such as that manufactured by Dow-Corning or General Electric. Bracket plates, electrical junction boxes, etc., should also be silicone caulked to wall surfaces.

e. Ceilings in Process Areas. Conventional acoustical ceiling tile, if used, should be treated with a nonleaching bacteriostatic/fungistatic compound, such as Steridex manufactured by Pentagon Plastics of Springfield, Virginia.

f. Process Water. In localities where it is not desirable to use the municipal water supply directly in processing, the best compromise treatment procedure is reverse osmosis (RO). Deionized water is expensive and very difficult to maintain free of bacterial contamination. Reverse osmosis water should be collected in a large reservoir (generally a reinforced fiberglass tank is used). In our operation we continuously circulate the collected RO water through ultraviolet sterilizing equipment. Milwaukee water has a hardness of 8 grains and commercial RO systems, such as those from Continental Water, Inc., are capable of reducing this to less than 1 grain in a single pass. Such water is very acceptable for culture media preparation (Porubcan and Sellars, 1977).

g. Process Cooling System. All process cooling should be performed with refrigerated (1°C) water wherever possible. A central ice bank reservoir maintained by tandem screw compressors is generally used. Since the jackets of the various process vessels have multipurpose functions where steam, hot water, municipal cold water, and refrigerated water all share common channels at times, it is clear that glycol-type coolants would not be compatible.

h. Process Steam. Pure culinary steam generated by a "drinking water boiler" is most desirable, particularly if steam is to be directly injected into product, sanitary piping, or even cleaning solutions. If this type of boiler is not practical, then shell-and-tube-type heat exchangers and other indirect methods of heating can be used. Steam containing various boiler treatment compounds should not be allowed to chemically contaminate product contact surfaces.

i. Compressed Air. Instrument-grade compressed air should be used throughout the plant. At Chr. Hansen's Laboratories we define such air as having a dew point maximum of 35°F (1°–2°C) with absolute filtration of all particles larger than 0.5 μm.

j. Liquid Nitrogen (LN_2). A central liquid nitrogen dispensing system should be laid out such that distances from the source are absolutely minimized. Vacuum-insulated stainless steel transfer lines are best. If gaseous nitrogen is required for process applications, it can be obtained from the central LN_2 system. Both Minnesota Valley Engineering (MVE) Inc., Minneapolis, Minnesota and Linde Division of Union Carbide, Inc. can provide total cryogenic system engineering.

III. SPECIFIC STARTER CULTURES

A. Lactic Streptococci

1. Commercial Applications

Buttermilk, sour cream, hard and soft cheeses, cottage cheese, and cultured butter are now routinely produced using commercially available frozen concentrates of lactic streptococci. Available as either "bulk sets" or "direct sets," these concentrates are used to prepare intermediate bulk starter or the final product, respectively. In the former case the culture units range in size from 70 to 130 ml and contain from 5×10^9 to 50×10^9 CFU/ml. In the latter case the cultures are packaged in 130 to 500-ml units and contain from 50×10^9 to 400×10^9 CFU/ml. More specifically, a typical DVS (direct-vat-set; such as, Hansen's DVS #950 or DVS #960) culture for Cheddar cheese will contain 360 ml of concentrate containing 200×10^9 CFU/ml of *S. cremoris* (two to eight different strains) and will inoculate up to 5000 lb (2273 kg) of cheese milk. This represents an addition of 14.4×10^{13} CFU to a 10,000-lb vat which is approximately equivalent to using 1% of bulk starter that has a viable plate count of 4.0×10^9 CFU/ml.

Preparation of bulk starter from "bulk-set" frozen concentrates is rather straightforward. Reconstituted skim milk or phosphate-buffered starter medium is batch heat treated at 85°–95°C (alternatively, 230°F for 12–16 seconds can be used) for 40–60 minutes and subsequently cooled to 21°–28°C (for mesophilic starters) and inoculated with a defrosted culture. Typically, 70 ml of "bulk set" culture (such as Hansen's Redi-Set #253 or #70) will inoculate up to 300 gal. Incubation then proceeds at 21°–28°C for 12–18 hours or until the starter has developed sufficiently, i.e., when the pH reaches 4.90–5.00 for maximum activity. At this time the starter is cooled to 5°–10°C and held cold until used; it is not a good idea to hold "ripe" starter for more than 24 hours.

Richardson *et al.* (1977), Chen and Richardson (1977), and Ausavanodon *et al.* (1977) have proposed an improved method for preparing bulk starter. The procedure calls upon the culture user to prepare bulk starter utilizing a whey-based bacteriophage inhibitory medium and pH control. Fresh Cheddar cheese whey is first fortified with 1.0% monoammonium phosphate, 1.9% disodium phosphate, and 0.32% yeast extract and is then heated at 89°–95°C for 45 minutes. Inoculation follows with a suitable commercial lactic *Streptococcus* culture and incubation is carried out at 21°–27°C with pH control via anhydrous ammonia injection at pH 6.0 ± 0.1 pH unit. After about 16 hours of pH-controlled growth, a significant decrease in acid production is observed (by a decrease in neutralizer consumption) and the starter is essentially ready to use. Advantages claimed for this type of starter preparation are increased activity at reduced costs.

There are many disadvantages inherent to bulk starter systems in general, including the improved version just described, that are not encountered when liquid nitrogen frozen direct-vat-set (DVS) concentrates are used. First, frozen direct-vat-set cultures eliminate any need to cultivate starters in the cheese factory. By doing so there is a real savings in energy, labor, and raw material costs to the culture user. Second, frozen direct-vat-set cultures allow the cheese maker to rotate or use different starters, for purposes of bacteriophage control, on a vat-to-vat basis; or different cheese varieties can be made on a vat-to-vat basis. Third, DVS concentrates can be prepared with specific control over strain balance. For example, the culture manufacturer can grow select single strains of lactic streptococci in separate fermentors and combine them after subsequent concentration (Farr, 1969). Alternatively, the culture manufacturer can inoculate his large-scale fermentors with discreet proportions of compatible strains which would "meet each other" for only

10–12 generations before they must meet again in the cheese vat. Such procedures require the technological coordination and control of a sophisticated bacteriological laboratory. Fourth, DVS concentrates can be pretested for activity (ability to produce lactic acid), strain balance, bacteriophage, purity, and starter bacteria concentration. This is not at all possible with bulk starter systems. Pretesting helps to ensure positive daily performance and uniformity.

In addition to these four inherent advantages of DVS over bulk starter systems there are other potential advantages presently under investigation. For example, improved cheese yields in Cheddar cheese manufacturing and improved cheese quality have been reported (Rasmussen, 1977). More data is required in these areas, however, in order to ascertain statistical significance.

2. Strain Selection

One of the key criteria for selecting single strains of either *S. cremoris* or *S. lactis* destined for direct-vat-set use is "industrial centrifugability." It is important that whey-based cultures of lactic streptococci separate efficiently on industrial auto-desludging separators. Data from our laboratory indicate that "centrifugability" is largely an intrinsic characteristic of the particular strain, although growth media composition and various chemical treatments can influence this parameter.

A second important selection factor is activity or the ability to produce lactic acid. A potentially fast strain is one that will reduce litmus milk with acid coagulation within 16 hours at 21°C at a 1% inoculation or less (Babel, 1976).

Finally, a third selection factor is resistance to bacteriophage. Here, it now seems important to find strains that are not "readily lysogenic" or that do not spontaneously produce temperate phage and are induced to do so only with difficulty, if at all (Lawrence, 1977). Such a strain is not necessarily phage resistant but stands a better chance of being compatible with other lactic streptococci. Potentially good sources for lactic streptococci include commercial starter cultures and natural Cheddar cheese (Babel, 1976).

The above selection criteria do not necessarily hold true for "bulk set" culture strains. Some suppliers have allowed "natural selection" to govern the absolute composition of these starters. Here pseudolysogenic* and/or lysogenic strains perpetuate in such a way that variants to parent strains develop spontaneously. Although it is true that such strain com-

* Defined as a phage-carrying strain where the bacteria and phage can be separated.

binations are not "scientifically defined," it remains a fact of many years experience in the United States that such cultures have a definite resistance to bacteriophage infection. Since "bulk set" concentration levels can often be achieved by fermentation alone, such as fermentation in skim milk with pH control, "centrifugability" is not necessarily a concern (Jespersen, 1974). In addition, since "bulk set" starters are further propagated (often under less than ideal conditions) in the cheese factory for another 10–14 generations prior to use in the cheese vat, inherent stability to bacteriophage is of primary concern and may supersede the importance of scientifically defining the starter's composition in terms of individual strain percentage.

3. Concentrate Preparation

The preparation of concentrated *S. cremoris* and *S. lactis* cultures has been studied by Stadhouders *et al.* (1969), Peebles *et al.* (1969), Blaine (1972), Stanley (1977), Cogan *et al.* (1971), and Efstathiou *et al.* (1975). A recent general review on cheese starters has been prepared by Lawrence *et al.* (1976).

There is some disagreement regarding the optimum growth parameters for the lactic streptococci. In part, it stems from whether one is discussing the simultaneous propagation (in one fermentor) of different species and/or different genera (e.g., leuconostocs combined with streptococci) or whether the discussion relates to the propagation of single strains or simple "defined" mixtures of compatible strains (same species). Basically, the following information is valid, in a general sense, for most strains of lactic streptococci when they are grown separately or in simple compatible mixtures: An orthophosphate-buffered, yeast extract-fortified, sweet whey (preferably from Swiss cheese) based medium containing from 6 to 12% total solids by weight is a good general medium. A small amount of added skim milk powder (1–2%) would be of benefit if not for the problems encountered during separation of the cells as discussed previously. Ultrahigh-temperature and batch heat treatment options have already been discussed. Optimum incubation parameters are 25°–30°C for 8–18 hours depending on the strain(s). Generally, it is a good idea to automatically pH control the fermentation; optimum control is achieved between pH 5.40 and pH 6.30, with pH 6.00 most often preferred. Control should be maintained at ±0.1 pH unit. Neutralization is best effected with ammonium hydroxide or anhydrous ammonia, which are preferred over sodium, potassium, and calcium hydroxides (Blaine, 1972). Slow, constant agitation is generally the rule; nitrogen or CO_2, when sparged into the fermentor headspace, do not benefit the fermentation (Gilliland, 1977). At a temperature of 25°C with pH control at pH 6.00 (with ammonium hydroxide) the balance of component strains in a mixed strain culture is not altered appreciably (Efstathiou *et al.*, 1975; Lamprech and Foster, 1963).

Keen (1972) has reported that inhibitory hydrogen peroxide develops during fermentation with some lactic streptococci and that catalase treatment can reverse the inhibition. Yeast extract has a reasonable amount of pseudocatalase-type activity and its presence obviates the need to add catalase per se (Porubcan and Sellars, 1977). Gilliland and Speck (1968) have demonstrated that D-leucine accumulates during fermentation and acts as an autoinhibitor. Data from our laboratory unequivocally point to ammonium lactate accumulation (due to the pH-control process) as the number one factor responsible for the slowing down of the growth rate. Again, the virtuous yeast extract seems to exert a therapeutic effect on the cells in that it somewhat increases resistance to ammonium lactate. This effect often varies considerably from one commercial yeast extract brand to another.

Subsequent to fermentation the cell crop is cooled to between 5° and 15°C and is treated with a deflocculant, such as sodium citrate; from 2 to 4% by weight is usually required. The pH is then increased to about 6.6 and the cells are harvested by centrifugation or ultrafiltration as previously discussed. The cell concentrate can be suspended in a skim milk diluent or in a 7.5% lactose solution prior to freezing. The lactose is of particular benefit if the freezing is carried out at −37°C versus −196°C (Gilliland, 1977).

B. Lactobacilli

1. Commercial Applications

The genus *Lactobacillus* encompasses many species, only a few of which have significant commercial value at this time. In focusing our attention on these few it is appropriate to begin with *Lactobacillus bulgaricus*. In most commercial applications in the United States this organism is grown in symbiotic association with *Streptococcus thermophilus*. Concentrated liquid nitrogen frozen cultures containing these two organisms are used in the direct preparation of yogurt and in the preparation of bulk starter for Italian cheese varieties such as mozzarella, parmesan, provolone and romano. Direct-vat-set concentrates for these Italian cheeses are presently under development at Chr. Hansen's Laboratories.

Bulk starter for Italian cheese is prepared directly from frozen concentrates. Typically, a 70-ml unit of starter concentrate (such as Hansen's

Redi-Set R1 or CH-3) will inoculate up to 300 gal of either reconstituted skim milk (10–12% solids by weight) or thermophilic-type bulk starter media. Both types of growth media are batch heat-treated at 85°–95°C for 40–60 minutes and are subsequently cooled to between 37° and 45°C before inoculation and incubation. Incubation time is regulated accordingly; for example, at 37°C usually 12–14 hours are required, whereas at 45°C, 6 hours may be sufficient time to effect proper acid development (pH 4.90 or less). In general, the higher incubation temperatures somewhat favor *L. bulgaricus* over *S. thermophilus*. Since bacteriophages exist for both of these thermoduric lactics (Sozzi *et al.*, 1976) proper precautions with regard to sanitation and culture rotation should be followed just as for the mesophilic lactics.

Yogurt can be prepared directly from the same "bulk set" cultures that are used for Italian cheese bulk starter. A 70-ml unit, once again, is sufficient inoculum for up to 300 gal of milk. Milk for the manufacture of yogurt contains from 1.0 to 3.5% fat. The serum solids content of the milk is increased to 10.5–11.5% by addition of nonfat dry milk or condensed skim milk. The milk is then homogenized and heated to 185°–195°F (85°–90.5°C); holding time is 30–60 minutes. A high degree of heat treatment is essential because it improves the body of the product by limiting whey expulsion (Sellars and Babel, 1970). Following heat treatment, the milk is cooled to 104°–113°F (40°–45°C) and inoculated as stated above. Usually 8 hours are required at 104°F to effect proper acid development and coagulation. Yogurt which has a pH of 4.0 is very acceptable to the consumer. To attain this pH, cooling should be started when the pH is 4.3–4.4 (Sellars and Babel, 1970).

Lactobacillus acidophilus, because of its ability to implant the gastrointestinal tract, has recently become very popular in the form of Sweet Acidophilus. Here, cold, pasteurized low-fat milk is inoculated with a cell concentrate of *L. acidophilus* such that a final viable population of about 3×10^6 CFU/ml results in the milk. No incubation cycle is permitted and the inoculated milk is held refrigerated at 45°F (7°C) or below. The product tastes exactly like low-fat milk. Since incubated milk cultures of *L. acidophilus* have an "off flavor" due to the metabolism of the organism, this Sweet Acidophilus product has met with more general acceptance. Liquid nitrogen frozen concentrates of *L. acidophilus* in 70 to 130-ml units containing from 40×10^9 to 50×10^9 CFU/ml are now commercially available. One 130-ml unit, when added to 500 gal of cold milk, will effect a viable population of about 3×10^6 CFU/ml in the milk.

Various thermoduric lactobacilli, such as *L. acidophilus, L. bulgaricus,* and *L. lactis,* have been fed to livestock for the purpose of controlling disease, such as scours (Metchnikoff, 1907; Sandine *et al.*, 1972; Mitchell and Kenworthy, 1976; Muralidhara *et al.*, 1977). Renewed interest in this old concept has developed due to the general concern over the widespread prophylactic use of broad-spectrum antibiotics in animal feed. In addition, the present availability of commercial frozen concentrates of these lactobacilli has improved the functionality of the entire concept. Spray-dried thermoduric lactobacilli are also being evaluated commercially in this regard (Porubcan *et al.*, 1977). In terms of human health and nutrition there is a growing general consensus that these lactobacilli may perform a beneficial function (Metchnikoff, 1908; Meyers, 1931; Mitsuoka, 1971; Sandine *et al.*, 1972; Speck, 1976).

The use of frozen concentrates of *Lactobacillus plantarum* (such as Hansen's LP-1) in the fermented vegetable industry is gaining in popularity. A process for the pure culture fermentation of cucumbers was patented by Etchells and Bell (1968). This process produces a firm textured, fermented cucumber, whereas previous state-of-the art processing often resulted in bloated and/or softened cucumbers. Extraneous fermentations involving nonlactic acid-producing bacteria and/or yeasts and molds would often cause these aforementioned defects. The Etchells and Bell process involves several steps; (1) heat blanching of the cucumbers to destroy surface asporogenous microorganisms and inactivate any deleterious heat-labile enzyme systems that may be present; (2) packing of the cucumbers aseptically into sanitized containers and covering with a preacidified (pH 4.6–4.2) brine that gives an equilibrated brine strength (% NaCl by wt.) of about 1.6–10%; (3) incubation of the containers with a suitable pure culture, such as *L. plantarum* or *Pediococcus cerevisiae,* at a favorable temperature until the fermentation has progressed to completion. If 1 billion viable cells are added per quart, about 72 hours at 90°F will be required.

For practical reasons this process is more or less limited to small containers such as No. 10 tin cans and quart or gallon glass jars. Etchells (1973) reported an adapted process suitable for the controlled fermentation of cucumbers and other vegetables brined in bulk. Here chemical (chlorine) sanitation of the washed green cucumbers replaced the heat shocking step; suitable buffering (to ensure complete fermentation of all sugar) and nitrogen purging (to reduce the CO_2 content in the brine) steps were added. When the 25° salometer (6.6% NaCl) brining procedure is followed, the fermentation reaches completion in 7–12 days when brine temperatures are held between 78° and 85°F (26–29°C). A single 130-ml liquid nitrogen frozen concentrated culture unit (Hansen's LP-1) of *L. plantarum* contains over 6.5×10^{12} viable cells. When added to 2000 gal of brined cucumbers it will provide 3–5 billion viable organisms per gallon, a very adequate inoculum level.

2. Concentrate Preparation

Published articles on the preparation of concentrates of lactobacilli suitable for commercial use are not as numerous as for the lactic streptococci. It has been our experience that pH-controlled milk cultures of thermoduric lactobacilli such as *L. bulgaricus, L. lactis, L. acidophilus,* and *L. helveticus* are difficult to concentrate effectively by industrial centrifugation even when copious amounts (2–6% by wt.) of sodium citrate and/or other casein deflocculating reagents are added after growth. Clarified sweet whey fortified with yeast extract is a satisfactory medium for some of these lactobacilli but not for others. Centrifugal separation of the denatured whey proteins along with the cells is usually a severe problem if unclarified whey is used, and, once again, treatment with sodium citrate does not usually remedy the problem (Porubcan and Sellars, 1977).

Various broth-type growth formulas are commercially available that can, with some modification, be rather effective. BBL's (Baltimore Biological Laboratory Division, Bioquest, Inc., Baltimore, Maryland) LBS broth and MRS broth are good examples of these. Etchells and Costilow (1968) have patented an improved version of the LBS broth formula originally developed by Rogosa (1951). Their improvement involves an upward adjustment of the final pH of the medium. *Lactobacillus plantarum* grows very well in LBS broth modified as such. In spite of the strong acetate buffer capacity of LBS it may be desirable to control the pH during growth for some strains of *L. plantarum. Lactobacillus plantarum* strain 442 (Fleming, N.C. State), for example, does very well at 30°C in LBS broth pH controlled at 5.60 with 50% NaOH. A total incubation time of 18 hours produces a viable plate count of between 50 and 100×10^8 cells per ml in the fermentor. Subsequent concentration by either centrifugation or ultrafiltration produces a very acceptable concentrate (Porubcan and Sellars, 1977).

Lactobacillus acidophilus (ATCC 4356) grows well in commercial MRS broth at 37°C. The acetate level in LBS broth is too high for prolific *L. acidophilus* growth. Control at pH 6.0 (±0.1 pH unit) with either ammonium hydroxide or sodium hydroxide improves the cell yield. On an industrial scale it is necessary to adapt the MRS formulation by substituting industrial peptones and yeast extracts for the laboratory types given. For example, Humko Sheffield's Pepticase (tryptic digest of casein) at approximately one-fourth the cost per pound is a very acceptable substitute for BBL's Trypticase.

Some thermoduric lactobacilli are particularly sensitive to changes in the oxidation–reduction potential (ORP) of the media. This effect may or may not be related to dissolved oxygen or fermentor headspace gas composition. An anaerobic headspace, such as a N_2 purged headspace, will stimulate certain strains of *L. bulgaricus* (Porubcan and Sellars, 1977).

C. Pediococci

1. Commercial Applications

The use of *Pediococcus cerevisiae* as a starter culture for fermented sausage was reviewed by Niven *et al.* (1955). Their process was patented (Niven *et al.*, 1959).

Previous state-of-the-art processing relied on certain microorganisms, usually indigenous to the sausage ingredients, to produce satisfactory levels of lactic acid during fermentation. Because of the unpredictability of this "natural process," spoilage often resulted.

Frozen concentrates of *P. cerevisiae,* (such as Hansen's PC-1 and PC-2) in quantities of 70–130ml provide sufficient inoculum for up to 300 lb of the meat–sodium chloride–nitrite–seasonings–dextrose mixture, also known as the emulsion. Viable plate counts on either BBL's APT or BBL's LBS agar run as high as 100×10^9 CFU/ml in these commercial concentrates.

A general processing schedule for all sausage products cannot be given; however, the starter culture technique reduces traditional production times by about 80% for thuringer, cervelat, or summer sausage (Everson *et al.*, 1970). At controlled smokehouse temperatures of 100°–110°F, freshly stuffed casings of inoculated sausage emulsion will usually reach pH 4.80 in 24 hours or less depending upon the kind of sausage and the activity of the particular starter strain. Nitrite levels should not exceed ⅛ oz/cwt of emulsion; and dextrose additions should be at least 0.75 lb/cwt (Niven *et al.*, 1958).

2. Concentrate Preparation

Pediococcus cerevisiae is a particularly desirable microorganism for sausage fermentation because of its salt tolerance and ability to autoinactivate around pH 4.80, a very desirable end-point pH for many types of fermented sausage. The commercial production of this organism can be effected in fermentors charged with industrially modified versions of APT, MRS, or LBS broth. pH control with either concentrated ammonium hydroxide or 50% sodium hydroxide solution (between pH 5.40 and 6.00) significantly benefits biomass yield in the fermentor. Incremental additions of fermentable carbohydrate(s) may also encourage higher biomass yields. Optimum incubation conditions are 16–18 hours at 25°–30°C, depending on the particular strain of *P. cerevisiae* being processed. Bacterial concentrations in the fermentor should average between 5×10^9 and 15×10^9 CFU/ml after about 16–18 hours of incubation (Porubcan and Sellars, 1977).

The appropriate method of physical concentration subsequent to fermentation is ultrafiltration. With a fermentate concentration of 10×10^9 CFU/ml it is only necessary to reduce the fermentate volume by four-fifths to achieve the desired level of 50×10^9 CFU/ml in the commercial concentrate. The "broth-based" fermentation substrate facilitates the ultrafiltration process. Culture temperatures should be held between 10° and 15°C during concentration. Once concentrated, the culture can be dispensed into 70- or 130-ml aluminum cans and subsequently flash frozen in liquid nitrogen at −196°C. Essentially 100% of the cells survive this method of preservation and, when cryogenic freezing as such is employed, no glycerol or other cryoprotectants need be added (Porubcan and Sellars, 1977).

REFERENCES

Anonymous (1973). *Dairy & Ice Cream Field,* July.

Ausavanodom, N., White, R. S., Young, G., and Richardson, G. H. (1977). *J. Dairy Sci.* **60,** 1245–1251.

Austin, P. R. (1970). "Design and Operation of Clean Rooms." Business News Publ. Co., Detroit, Michigan.

Babel, F. J. (1976). Purdue University (personal communication).

Blaine, J. W. (1972). Ph.D. Thesis, Oregon State University, Corvallis.

Buchanan, R. E., and Gibbons, N. W., eds. (1974). "Bergey's Manual of Determinative Bacteriology," 8th ed. Williams & Wilkins, Baltimore, Maryland.

Carlsen, V. R. (1977). Cherry-Burrell, Inc. (personal communication).

Chen, Y. L., and Richardson, G. H. (1977). *J. Dairy Sci.* **60,** 1252–1255.

Christensen, V. W. (1967). U.S. Patent 3,354,049.

Cogan, T. M., Buckley, D. J., and Condon, S. (1971). *J. Appl. Bacteriol.* **34,** 403.

Efstathiou, J. D., McKay, L. L., Morris, H. A., and Zottola, E. A. (1975). *J. Milk Food Technol.* **38**(8), 444–448.

Elsworth, R. (1962). "Phase Separation of Bacterial Suspensions by Centrifuge." New Brunswick Sci. Co., New Jersey.

Erzinkyan, L. A., Pakhlevanyan, M. Sh., and Akopyan, L. G. (1966). *Dairy Sci. Abstr.* **21,** 386.

Etchells, J. L. (1973). *Pickle Pak. Sci.* **3,** 4–14.

Etchells, J. L., and Bell, T. A. (1968). U.S. Patent 3,403,032.

Etchells, J. L., and Costilow, R. N. (1968). U.S. Patent 3,410,755.

Everson, C. W., Danner, W. E., and Hammes, P. A. (1970). *J. Agric. Food. Chem.* **18,** (4), 570.
Farr, S. M. (1969). U.S. Patent 3,420,742.
Gilliland, S. E. (1977). *J. Dairy Sci.* **60** (5), 805–809.
Gilliland, S. E., and Speck, M. L. (1968). *J. Dairy Sci.* **51,** 1573.
Hansen, A. P., and Melo, T. S. (1977). *J. Dairy Sci.* **60,** 1368–1373.
Hansen, R. (1969). *Nord. Mejeri-tidsskr.* **9,** 192–194.
Jespersen, N. J. T. (1974). Chr. Hansen's Laboratory, Inc. (personal communication).
Jespersen, N. J. T. (1974). *S. Afr. J. Dairy Technol.* **6,** 63.
Jespersen, N. J. T. (1977). Chr. Hansen's Laboratory, Inc. (personal communication).
Jost, R., and Monti, J. C. (1977). *J. Dairy Sci.* **60,** 1387–1393.
Keen, A. R. (1972). *J. Dairy Res.* **39,** 151–159.
Konak, A. R. (1974). *Krist. Tech.* **3,** 243.
Konak, A. R. (1975). *Biotechnol. Bioeng.* **17,** 271–272.
Lamprech, E. D., and Foster, E. M. (1963). *J. Appl. Bacteriol.* **26,** 359.
Lawrence, C. A., and Block, S. S. (1968). "Disinfection, Sterilization, and Preservation," Lea & Febiger, Philadelphia, Pennsylvania.
Lawrence, R. C. (1977). "Cheddar Cheese Starters." A paper presented at the annual meeting of the American Dairy Science Association.
Lawrence, R. C., Thomas, T. D., and Terraghi, B. E. (1976). *J. Dairy Res.* **43,** 141–193.
Leach, R. D., and Sandine, W. E. (1976). *J. Dairy Sci.* **59** (8), 1293–1397.
Lloyd, G. T. (1971). *Dairy Sci. Abstr.* **33,** 411.
Lyster, R. L. (1971). *J. Dairy Res.* **38,** 403.
Metchnikoff, E. (1907). "Quelques remarques sur le lait aigni." Putnam, New York.
Metchnikoff, E. (1908). "Prolongation of Life." Putnam, New York.
Meyers, R. P. (1931). *Am. J. Public Health,* **21,** 867–872.
Mitchell, I., and Kenworthy, R. (1976). *J. Appl. Bacteriol.* **41,** 163–174.
Mitsuoka, T. (1971). *Proc. Int. Symp. Conversion & Manuf. Food Stuffs Microorg., Kyoto, Japan.*
Muralidhara, K. S., Sheggeby, G. G., Elliker, P. R., England, D. C., and Sandine, W. E. (1977). *J. Food Prot.* **40** (5), 288–295.
Nielsen, A. E. (1964). "Kinetics of Precipitation." Pergamon, Oxford.
Niven, C. F., Jr., Deibel, R. H., and Wilson, G. D. (1955). "The Use of Pure Culture Starters in the Manufacture of Summer Sausage," Annu. Meet. Am. Meat Inst., Chicago, Illinois.
Niven, C. F., Jr., Deibel, R. H., and Wilson, G. D. (1958). *Am. Meat Inst. Found. Cir.* **41.**
Niven, C. F., Jr., Deibel, R. H., and Wilson, G. D. (1959). U.S. Patent 2,907,661.
Osborne, R. J. W. (1977). *J. Soc. Dairy Technol.* **30** (1), 40–44.
Payne, R. E., Hill, C. G., Jr., and Amundson, C. H. (1973). *J. Milk Food Technol.* **36** (7), 359–363.
Peebles, M. M., Gilliland, S. E., and Speck, M. L. (1969). *Appl. Microbiol.* **17,** 805.
Porubcan, R. S. (1975). M.S. Thesis, Dept. of Chemistry, University of Wisconsin, Milwaukee.
Porubcan, R. S., and Sellars, R. L. (1973). *J. Dairy Sci.* **56,** 634.
Porubcan, R. S., and Sellars, R. L. (1975a). U.S. Patent 3,897,307.
Porubcan, R. S., and Sellars, R. L. (1975b). *J. Dairy Sci.* **58,** 787.
Porubcan, R. S., and Sellars, R. L. (1976). *Pap. ADSA Meet., 1976.*
Porubcan, R. S., and Sellars, R. L. (1977). Chr. Hansen's Laboratory, Inc. (unpublished research).
Porubcan, R. S., Sellars, R. L., Daines, T., and Rovics, J. (1977). Unpublished studies on the feeding of *L. acidophilus* to calves. (Data collected at FAR, Inc., Juneau, Wisconsin, supported by Chr. Hansen's Laboratory, Inc.)

Rasmussen, H. (1977). *Dairy Field* Sept., p. 70H.
Reddy, M. S., Vedamuthu, E. R., Washam, C. J., and Reinbold, G. W. (1974). *J. Dairy Sci.* **57** (1), 124–127.
Richardson, G. H., Cheng, C. T., and Young, R. (1977). *J. Dairy Sci.* **60** (3), 378–386.
Rinfret, A. P., and LaSalle, B. (1975). "Round Table Conference on the Cryogenic Preservation of Cell Cultures." Natl. Acad. Sci., Washington, D.C.
Rogosa (1951). *J. Bacteriol.* **62,** 132–133.
Sandine, W. E., Muralidhara, K. S., Elliker, P. R., and England, D. C. (1972). *J. Milk Food Technol.* **35,** 691–702.
Schantz, R. N. (1976). Aqua Chem, Milwaukee, Wisconsin (personal communication).
Sellars, R. L. (1967). *In* "Microbial Technology" (H. J. Peppler, ed.), pp. 34–75. Van Nostrand-Reinhold, Princeton, New Jersey.
Sellars, R. L., and Babel, F. J. (1970). "Cultures for the Manufacture of Dairy Products." Chr. Hansen's Laboratory, Inc., Milwaukee, Wisconsin.
Shuster, J. M. (1976). Minnesota Valley Engineering, Inc., Minneapolis (personal communication).
Sozzi, T., Maret, R., and Poulin, J. M. (1976). *Appl. Environ. Microbiol.* **32,** (1), 131–137.
Speck, M. L. (1976). *J. Dairy Sci.* **59,** 338–343.
Stadhouders, J., Jansen, L. A., and Hup, G. (1969). *Neth. Milk Dairy J.* **23,** 182–199.
Stanley, G. (1977). *J. Soc. Dairy Technol.* **30** (1), 36–39.
Thomas, T. D., and Turner, K. W. (1977). *N. Z. J. Dairy Sci. Technol.* **12,** 15–21.
Thompson, J. (1976). *J. Bacteriol.* **127** (2), 719–730.
Webb, B. H. (1965). "Fundamentals of Dairy Chemistry," Avi Publ. Co., Westport, Connecticut.
Webb, F. C. (1964). "Biochemical Engineering," Van Nostrand-Reinhold, Princeton, New Jersey.
Woods, W. C. (1975). "Trends in High Temperature Sterilization of Dairy Products—A Symposium," Annu. Conf. Am. Cultured Dairy Products Inst.

Chapter 4

Production of Single-Cell Protein for Use in Food or Feed

JOHN H. LITCHFIELD

I. INTRODUCTION

What are single-cell proteins and what are their present and potential uses for human food or animal feed? In this chapter, we discuss the various groups of microorganisms that have been considered for food or feed use, including algae, bacteria, yeasts, molds, and higher fungi. The dried cells of these organisms are collectively referred to as "single-cell proteins" (SCP). This term was coined by C. L. Wilson at the Massachu-

MICROBIAL TECHNOLOGY, 2nd ed., Vol. I
 ISBN 0-12-551501-4

setts Institute of Technology in 1966 and has been used widely since that time in referring to microbial cells grown for food or feed applications (Scrimshaw, 1968).

People have eaten certain microorganisms as a portion of their diet since ancient times. Top-fermenting yeast (*Saccharomyces* species) was recovered as a leavening agent for bread as early as 2500 B.C. (Frey, 1930). Bread and yeast-leavened baked products contain 1–4% of bakers' yeast based on the weight of flour used (Reed and Peppler, 1973).

Fermented milks and cheeses produced by lactic acid bacteria of the genera *Streptococcus* and *Lactobacillus* were consumed by the early Egyptians and Greeks and reached a higher state of development during the Roman era (100–50 B.C.) (Kosikowski, 1977). These cultured dairy products contain from 10^7 to 10^{10} lactic acid bacteria per gram (Lampert, 1965; Pederson, 1971).

Wild mushrooms were prized as a delicacy by the pharaohs of Egypt. During the First Century B.C., the epicures of the Roman Empire held wild mushrooms in high esteem, as indicated in the verse of the poet Horace. The Romans also regulated the grading and selling of mushrooms. In addition, they recognized the danger resulting from consumption of poisonous wild mushrooms (Seelig, 1959). Mushrooms were also eaten in ancient times in the Orient. The current practices utilized in cultivating the various species used in China, Japan, Malaya, and the Philippines undoubtedly originated from ancient lore.

Bluegreen algae of the genus *Spirulina* grow in alkaline lakes in certain regions of the world. People in the Lake Chad region of Africa have eaten *Spirulina* species for many generations. Also the Aztecs in Mexico were consuming *Spirulina* species as a major source of protein at the time of arrival of the Spanish explorers in the sixteenth century (Clement, 1968; Ad Hoc Panel, 1975).

The purposeful cultivation of microorganisms for direct use in human food or animal feed is a fairly recent development. Bakers' yeast, (*Saccharomyces cerevisiae*) was grown in an aerated molasses–ammonium salts medium for food use in Germany during World War I (1914–1918). Aerobic yeasts, particularly, *Candida utilis* (torula yeast) were produced in Germany for food and feed use during World War II (1939–1945) (Prescott and Dunn, 1959; White, 1954).

Considerable effort has been expended since World War II to develop processes for mass cultivation of microbial cells (SCP). Numerous reviews and symposia describe the development of processes for utilizing raw materials, including simple sugars, starches, cellulose, agricultural and food-processing wastes, and hydrocarbons by bacteria, yeasts, molds, and higher fungi, and photosynthetic conversions of wastes by algae (Davis, 1974; Gutcho, 1973; Humphrey, 1974; Kihlberg, 1972; Lipinsky and Litchfield, 1970, 1974; Litchfield, 1974, 1977a,b,c, 1978; Mateles and Tannenbaum, 1968; Rockwell, 1976; Tannenbaum and Wang, 1975).

II. PRODUCTION

A. Algae

Algae can be grown either photosynthetically and autotrophically using either artificial lighting or sunlight, or heterotrophically in the dark with organic carbon and energy sources. Most of the research on mass cultivation of algae has been conducted under photosynthetic conditions particularly for waste treatment in sewage oxidation ponds using sunlight (Oswald, 1969; Oswald and Golueke, 1968) or using artificial illumination in conjunction with the development of bioregenerative systems for use in life-support systems for extended space exploration (Casey *et al.*, 1963; Lachance, 1968; Litchfield, 1967d).

On the other hand, Japanese investigators have conducted research on growth of algae under heterotrophic conditions in the dark using organic carbon and energy sources. The goal of this research has been to identify algal strains that have higher growth rates under heterotrophic conditions than under photosynthetic–autotrophic conditions (Endo and Shirota, 1972).

1. Growth Conditions

Table I summarizes the results of studies on the growth of selected species of algae of interest in SCP production. More important strains include *Chlorella sorokiniana, Scenedesmus acutus, Scenedesmus quadricauda, Scenedesmus obliquus, Spirulina maxima,* and *Spirulina platensis.* Limited studies have been conducted with *Oocystis polymorpha. Chlorella regularis* has been grown under both autotrophic and heterotrophic conditions.

In open-circulation systems, particularly sewage oxidation ponds, mixed cultures of algae tend to predominate rather than a single strain. For example, a mixed culture of *Chlorella ellipsoidea* and *S. obliquus* developed in open pond systems in Japan. Mixed cultures of algae dominated by *S. quadricauda* was used in sewage oxidation pond studies at the University of California experiment station at Richmond (Oswald, 1969).

TABLE I. Growth of Selected Algae

Organism	Carbon and energy source	Scale	temperature (°C)	pH	Specific growth (hr^{-1})	Culture density	Yield (dry wt. basis)	References
Chlorella pyrenoidosa 71105	CO_2(0.7% in air), urea (0.4 gm/liter) fluorescent lamp, 15,000 lm	600-gal tank	39	4.5–5.8	0.0625	0.28 v/v%	1 g/day	Casey *et al.* (1963)
C. pyrenoidosa	CO_2(10%), fluorescent lamp, 52,000 lm	2.7 liters	38	6.4–6.5	—	8.98 gm/liter	36.5 gm/day	Matthern and Koch (1964)
Chlorella regularis	(1) Autotrophic: CO_2 (5% in air), fluorescent lamp, 40 klux	1 liter	36	6.5	0.26	6 ml/liter	—	Endo and Shirota (1972)
	(2) Heterotrophic (dark): acetic acid (100 ml/liter)	20-liter jar fermentor (13 liters medium)	36	6.8	0.28	15 ml/liter	0.48 gm/gm acetate utilized	
Chlorella sorokiniana	CO_2 (5% in air), 70 ml/min, 20mmol M KNO_3/liter, fluorescent lamps	405-ml annular chemostat	39	—	—	—	1.59 gm/day	Richardson, *et al.* (1969)
Oocystis polymorpha	CO_2 (5% in air), 70 ml/min, fluorescent lamps	405-ml annular chemostat	39	—	—	—	1.25 gm/day	Richardson, *et al.* (1969)
Scenedesmus acutus 276-3A	CO_2, sunlight,	shallow tanks 55m², 30cm deep	Ambient (30)	7–8	— —	—	20gm/m²/day	Anonymous (1970); Becker and Venkataraman (1978)
Scenedesmus quadricauda	SO_2, sunlight	10^6 liter pond	25–35	8.5–9.5	—	1 gm/liter	20 tons/acre/year (12.6 gm/m²/day)	Oswald and Golueke (1968)
S. quadricauda and Scenedesmus obliquus	CO_2 (0.5%), sunlight	54,000 liters (900 m²)	34	6.5–7.0	—	1.5–2.0 gm/liter	10 gm/m²/day	Enebo (1969)
Spirulina maxima	CO_2 (0.5%) combustion gases, bicarbonate, sunlight	700-m² pond	Ambient	—	—	—	15 gm/m²/day	Durand-Chastel and Clement (1972); Clement (1975)

S. maxima	CO_2 (combustion gas), sunlight	20 × 5 × 0.1 m basin	Ambient	—	—	—	12 gm/m²/day	Clement (1968); Clement *et al.* (1967)
S. maxima	CO_2-enriched air, 250-ml Erlenmeyer flasks 100 ml medium) 4000 lux, 2 v/v/min;							
	(a) synthetic medium,		30	9.5		2 gm/liter (12 days)	—	Kosaric *et al.* (1974)
	(b) sewage effluent		30	9.5		0.77 gm/liter (9 days)		
Spirulina platensis	CO_2, 10-liter jar fermentor (20 cm diameter), 10 klux		35–37	8–10.5	0.0189	4.2 gm/liter		Ogawa *et al.* (1972)

The limiting factor in the growth of algae on a large scale is illumination. In outdoor pond cultivation length of day and night, cloud cover, and the fixed nature of the spectrum and intensity of sunlight limit algal cultivation to latitudes between 35° north and south.

Estimated energy requirements for algal growth at a 20% efficiency indicate a requirement of 35kWh for producing 1 kg of algae (Enebo, 1969). In studies with *C. sorokiniana* (*pyrenoidosa* 71105), Shuler and Affens (1970) obtained maximum efficiency of conversion of light energy to chemical energy of 12% in an annular cylindrical culture unit illuminated with artificial light at 15,700 lm (36,000 fc). In other studies with the same organism, Matthern *et al.* (1969) obtained cultured densities in the range of 25.5 gm/liter (dry weight basis) at an illumination of 300,000 lm which is 30 times the intensity of sunlight as measured at the Earth's surface. Taking into account projected future costs of electricity, it is clear that artificial illumination for the production of algae as a source of SCP is not feasible and the outdoor culture in open ponds is the only possible alternative.

Carbon dioxide is the source of carbon for photosynthetic–autotrophic growth of algae. Air contains only 0.03% CO_2. Accordingly, additional CO_2 must be supplied to the culture. Combustion gas was used as a source of CO_2 for growing *Spirulina maxima* by the Institute Français du Petrole. In regions where naturally alkaline lakes are present, such as Lake Texcoco in Mexico, *S. maxima* can be grown without enrichment with CO_2 because of the naturally high concentrations of sodium carbonate present in these waters (Clement, 1975).

In sewage ponds, algal growth is limited by the extent of liberation of CO_2 and ammonia by bacterial action. Slow and uniform liberation of CO_2 is necessary to provide a uniform supply of the carbon source to the growing culture (Oswald and Golueke, 1968).

A heterotrophic culture of algae has been investigated in Japan to determine means of producing algae as a source of SCP from organic carbon sources without light (Endo and Shirota, 1972; Nakayama *et al.*, 1974). Specific growth rates in the range of 0.3 per hr are obtained with *Chlorella regularis* strain S-50 using glucose as a substrate in the dark (Endo and Shirota, 1972).

Nitrogen sources suitable for algal growth include ammonium salts or nitrates. The nitrogen sources together with phosphorus and mineral nutrients may be readily available in sewage ponds. However, in other water supplies where synthetic media are used in culturing algae these nutrients may have to be supplied.

A number of pond and tank systems for large-scale algal cultivation have been developed in various countries including Japan, Czechoslovakia, West Germany, India, Algeria, and Mexico (Anonymous, 1970; Becker and Venkataraman, 1978; Clement, 1975; Enebo, 1969). In the United States extensive studies have been conducted on a pilot scale by Oswald and his co-workers at the University of California, Richmond, for a number of years.

Aseptic conditions are not maintained during the large-scale growth of algae in ponds. However, the potential contamination of algal cultures grown in sewage oxidation ponds by enteric pathogenic bacteria and viruses must be given serious consideration (Cooper, 1962). *Chlorella* cultures also excrete water-soluble substances into the growth medium that provide nutrients for bacterial growth (Ward *et al.*, 1965; Vela and Guerra, 1966).

The key factors in large-scale algal cultivation are agitation to maintain cells in suspension and prevent sedimentation and adjusting flow rates so that the detention period in the pond will exceed the generation time of the particular algal culture to allow maximum population development.

2. Yields

Yields of algal cultures given in Table I that were obtained under artificial light conditions are of only academic interest since sunlight is the only economically feasible energy source for algal cultivation. The yields in the range 12–15 gm/m^2/day (dry weight basis) obtained with *Spirulina maxima* and *Scenedesmus quadricauda* are typical of those that can be obtained in outdoor pond conditions (Clement, 1968, 1975; Durand-Chastel and Clement, 1972; Oswald and Golueke, 1968).

3. Cell Recovery

A major problem in algal culture is harvesting the culture. Experience with *Chlorella*–Scenesdesmus cultures indicates that large volumes of water must be handled because of the low cell density in the range of 1 to 2 gm (dry weight) per liter obtained with these organisms. Also these cultures settle out slowly. Cells must be recovered by concentration, dewatering, and drying. Inorganic compounds, such as aluminum sulfate and calcium hydroxide, and cationic polymers have been investigated as flocculants for algae (Oswald and Golueke, 1968). However, these flocculants cannot be separated from the harvested cells, which makes this method unsatisfactory for recovering algae for food or feed applications. Algae may autoflocculate in shallow ponds at pH 9.5 or above without flocculants (Oswald and Golueke, 1968).

Ion-exchange resins effect a good separation of algae in the pH range 2.8–3.5 (Golueke and Oswald, 1970). However, the cost of the hydrochloric or sulfuric acid regenerants are sufficiently great that this process would not be economically feasible. The cost of separating and concentrating algae by centrifugation, flocculation, or a combination of these steps appears to be economically prohibitive at the present time.

Spirulina maxima floats on the surface in clumps when maximum growth is attained. It can be harvested by skimming at much lower cost than would be incurred by centrifugation (Durand-Chastel and Clement, 1972).

B. Bacteria and Actinomycetes

Bacteria are of interest for use in SCP production because of their high growth rate (20–30 minutes generation time) as compared to 2–3 hours for yeast and 16 hours or more for algae, molds, and higher fungi. Various species of bacteria utilize a wide range of carbon and energy sources including those from renewable resources, such as carbohydrates (sugars, starch, cellulose) in either pure form or in agricultural and forest product wastes, or those from nonrenewable resources, such as hydrocarbons and petrochemicals. Certain species of actinomycetes have growth rates and substrate utilization patterns similar to those of bacteria and consequently are of interest for use in SCP production.

1. Growth Conditions

A wide range of bacterial species has been considered for SCP production. Important considerations in selecting bacterial cultures suitable for use in SCP production include specific growth rate, yield on a given substrate, pH and temperature tolerance, aeration requirement, genetic stability, and freedom from associated bacteriophages. The strain also should not be pathogenic to plants, animals, or humans either by direct infection or by excretion of toxins. Also certain bacterial members of the Enterobacteriaceae may contain endotoxins which may effect man or animals adversely even after the cells are killed by heat. Another important consideration is that the strain must not have the potential for mating with known pathogenic bacteria to yield potentially pathogenic hybrids. This type of mating can occur quite readily within the Enterobacteriaceae.

Substrate and nitrogen concentrations are important factors affecting growth of bacteria. Substrate concentrations for both cultures are in the 1–5% range but may be considerably lower for continuous cultures. Carbon-to-nitrogen ratios in the growth medium should be maintained in the range of 10 : 1 or less to favor high protein contents in the cells and prevent accumulation of lipids or storage substances, such as poly-β-

hydroxybutyrate. Anhydrous ammonia or ammonium salts are suitable nitrogen sources for SCP production by bacteria and actinomycetes. Phosphorus is required by bacteria and should be supplied as a feed-grade phosphoric acid to avoid contamination of the product with arsenic or fluoride that is found in crude industrial phosphoric acid. Generally water supplies provide adequate quantities of mineral salts for bacterial growth; however, some natural waters may require supplemental minerals, such as iron, magnesium, and manganese, to make up deficiencies. In such cases minerals should be added as sulfates or hydroxides rather than as chlorides to avoid corrosion problems in stainless steel fermentation equipment.

During growth, pH may be controlled in the 5–7 range preferred by most bacteria by the rate of ammonia and phosphoric acid addition to the medium. In continuous processes, instrumentation can be designed to effect feedback control of pH and nitrogen content of the medium.

Temperature tolerance is an important characteristic of bacterial strains for SCP production from hydrocarbons or alcohols. For example, in the production of bacterial SCP from methane, heat production is in the order of 10 kcal/gm of cells at a yield coefficient of 1.0 gm of cells per gram of methane (Hammer *et al.*, 1975). Therefore, cooling cost for bacterial SCP production can be significant unless high-temperature-tolerant strains in the range of 35°–45°C can be used. Holve (1976) and Wang *et al.* (1976) discuss methods for calculating the heat production associated with bacterial growth on hydrocarbon and carbohydrate substrates based on the elemental composition of the cells.

Oxygen transfer to growing cells is also an important factor in bacterial SCP production under aerobic conditions, especially from hydrocarbons, methanol, or ethanol. Mateles (1971) presented an empirical equation for estimating oxygen requirements for cell production based on the elemental compositions of the carbon source and cells, the molecular weight of the carbon source, and the yield of cells based on the carbon source consumed. Also, cooling requirements can be gauged from the estimated oxygen requirement by use of the factor 0.11 kcal/mmole of oxygen consumed, developed by Cooney *et al.* (1969).

Maintaining sterility is important in the mass cultivation of bacteria. Undesirable contaminants such as enteric pathogens may grow equally as well as the desired strain since most bacterial SCP processes operate in a pH range of 5–7. Metabolic excretion products, such as amino acids, may provide sufficient nutrients for the growth of contaminants. Both growth media and equipment should be sterilized in bacterial SCP production processes and aseptic conditions must be maintained throughout the growth.

Table II presents growth characteristics of selected bacteria and actinomycetes having potential utility for SCP production. Some of the most important processes will be discussed here.

During the 1960s there was considerable interest in the potential for producing SCP from bacteria using gaseous and liquid hydrocarbons and chemicals derived from them, such as methanol and ethanol. In particular, there was interest in utilizing methane in regions remote from markets where this gas would otherwise be flared at the wells. The option also existed for manufacturing methanol from the methane and shipping the methanol to the site of SCP production (Rosenzweig and Ushio, 1974).

Research on a pilot plant-scale process for producing bacterial SCP from methane using *Methylococcus capsulatus* (Harrison *et al.*, 1972; Hamer *et al.*, 1975), or a mixed culture of *Pseudomonas* sp., *Hyphomicrobium* sp., *Acinetobacter* sp., and *Flavobacterium* sp. (Wilkinson *et al.*, 1974; Harrison, 1976) has been conducted at Shell Research Limited in the United Kingdom. High productivities and yield coefficients can be obtained in these processes. However, the productivity of the processes for continuous production of bacterial SCP for methane is limited by transfer of oxygen and methane from the gas phase to the bacterial cells, occurrence of single or double gaseous substrate limitation, high heat production necessitating cooling, and possible formation of inhibitory products (Hamer *et al.*, 1975; Harwood and Pirt, 1972).

In addition, potential explosive hazards require operation below 12.1% by volume of oxygen (Hamer *et al.*, 1967). Therefore capital costs for production of bacterial SCP from methane may be significantly higher than from other substrates.

Kyowa Hakko Kogyo Company, Limited in Japan has developed several processes for the production of bacterial SCP from gaseous hydrocarbons. Methane, ethane, propane, *n*-butane, isobutane, propylene, butylene, or mixtures of these hydrocarbons were utilized by *Brevibacterium ketoglutamicum* ATCC No. 15, 587 (Tanaka *et al.*, 1972).

Other bacteria that have been investigated for growth on gaseous hydrocarbons substrates include various mixed cultures on methane (Wolnak *et al.*, 1972; Sheehan and Johnson, 1971), *Arthrobacter simplex* on propane and butane (Orgel *et al.*, 1971); and *Corynebacterium hydrocarbonoclastus* on propane (Akiba *et al.*, 1973). The actinomycete *Nocardia paraffinica* utilizes *n*-butane readily (Sugimoto *et al.*, 1972).

Liquid *n*-alkanes are suitable carbon and energy sources for a wide range of bacteria and actinomycetes. Only a few examples are cited here.

Acinetobacter cerificans has been investigated on a small pilot plant-scale for the production of SCP from purified normal alkanes by Exxon

Corporation in a joint venture with Nestle Alimentana, S.A. (Perkins and Furlong, 1967; Guenther and Perkins, 1968). The Chinese Petroleum Corporation, Taiwan has developed a pilot-scale process for producing *Achromobacter delvacvate* from diesel oil (Ko *et al.*, 1964) and *Pseudomonas* No. 5401 from fuel oil (Ko and Yu, 1968). However, none of these processes has been practiced on a full commercial scale.

Actinomycetes, including *Mycobacterium phlei* and *Nocardia* sp., have been grown on liquid C_{10}–C_{20} hydrocarbons on a laboratory scale (Wagner *et al.*, 1969). However, no further development of these processes has occurred.

Bacterial SCP production from methanol is an interesting recent development. Methanol has the advantages of high solubility in water, lack of explosive hazards, freedom from traces of hydrocarbons, and ease of removal from the cell product. Pathways of C_1 compound metabolism, microorganisms, and the utilization of methanol by bacteria are discussed in recent reviews (Cooney and Levine, 1972; Kosaric and Zajic, 1974) and in a symposium (Organizing Committee, 1975). Bacteria of interest in producing SCP from methanol include *Methylomonas methanolica* (Dostalek *et al.*, 1972; Dostalek and Molin, 1975), *Methylophilus (Pseudomonas) methylotrophus* (MacLennan *et al.*, 1973; Young, 1973; Gow *et al.*, 1975; Anonymous, 1976b, 1977a), *Pseudomonas* sp. (Goldberg *et al.*, 1976; Mateles *et al.*, 1976), *Pseudomonas utilus* and *P. inaudita* (Yoshikawa *et al.*, 1975), and a mixed culture consisting of *Methylomonas methylovora* and species of *Flavobacterium, Pseudomonas,* and *Xanthomonas* (Cremieux *et al.*, 1977; Ballerini *et al.*, 1977).

The most advanced process for the production of SCP from methanol has been developed by Imperial Chemical Industries Limited. The organism used is *Methylophilus (Pseudomonas) methylotrophus* (MacLennan *et al.*, 1976; Anonymous, 1976b, 1977a). Gow *et al.* (1975) have described the novel "pressure cycle" airlift fermentor used in this process. The advantages claimed for this system include high oxygen transfer rates without oxygen limitation, ready removal of heat liberated during growth at high productivity, a homogeneous liquid phase, and lack of contamination through drive shafts and mechanical seals encountered with conventional fermentors.

Initial studies were accomplished on a 1000 metric tons/year pilot plant over a 3-year period. Construction has started on a 50,000–75,000 metric ton annual capacity plant, which is scheduled for startup in late 1979 (Anonymous, 1976b, 1977a).

In Japan, Mitsubishi Gas Chemical Company has investigated the growth of bacteria of the genus *Pseudomonas* that will grow on methanol at 38°C (Nagai, 1973). Also, Mitsubishi Petrochemical Company, Ltd.,

TABLE II. Growth of Selected Bacteria and Actinomycetes on Various Substrates

Organism	Carbon and energy source	Scale aeration, agitation	Temperature (°C)	pH	Specific growth rate or dilution rate (D) (hr^{-1})	Culture density (gm/liter) (dry wt. basis)	Yield (dry wt. basis) (gm/gm substrate used)	References
Achromobacter delvacvate	Diesel oil	6000-liter fermentor, 3000 liters of medium, 1 vol/vol/min	35–36	7.0–7.2	—	10–15 (48 hours)	—	Ko *et al.* (1964)
Acinetobacter (Micrococcus) cerificans	(a) Gas oil	7.5-liter fermentor,	30	7.0	(a) 0.4–1.0	8–10	0.10–0.12	Ertola *et al.* (1969)
	(b) *n*-Hexadecane	4.5 liters of medium, 1 vol/vol/min 500-rpm,			(b) 1.1–2.0	8–10	0.80–0.90	
	(c) *n*-Hexadecane	7.5-liter fermentor 3.5–7.0 mMO_2/liter/min	30	6.8	(c) 1.33	—	1.20	Guenther and Perkins (1968)
Bacillus megaterium	Collagen meat packing waste	2.3 liter working volume fermentor, 3 v/v/min (continuous), 400 rpm	34	7.0	0.25 (*D*)	2.3	—	Bough *et al.* (1972)
Brevibacterium sp.	Mesquite wood	14-liter fermentor, 1.0–1.5 v/v/min, 1350–1500 rpm	30–37	6.45–7.2	—	—	0.444	Fu and Thayer (1975); Thayer *et al.* (1975)
Cellulomonas sp. Alcaligenes	Bagasse,	7-,14-,530-liter fermentors, batch and continuous	34	6.6–6.8	0.2–0.29 0.08–0.1 (*D*)	16 (batch) 10 (continuous)	0.44–0.50	Han *et al.* (1971)
Methylococcus capsulatus	Methane	2.8 liters of medium, 6.7% CH_4, 17.1–19.4% O_2 v/v, 38.6–49.5 ml/min, 1450-rpm (continuous)	37	6.9	0.14	0.4	1.00–1.03	Harwood and Pirt (1972)

Methylomonas clara	methanol	1000-m^3 reactor volume (continuous)	39	6.8	0.5 0.3–0.5(*D*)	—	0.50	Faust *et al.* (1977)
Methylomonas methanolica	Methanol	4-liter working volume fermentor, 300–400 mM O_2/liter/min 1.5 v/v/min, 1200–1500 rpm, continuous	30	6.0	0.24 (*D*)	9.6	0.48	Dostalek *et al.* (1972); Dostalek and Molin (1975)
Methylophilus (Pseudomonas) methylotrophus	Methanol	Pressure cycle air-lift fermentor, continuous	35–40	—	0.38–0.5	30	0.5	MacLennan *et al.* (1973); Young (1973); Gow *et al.* (1975); Anonymous (1976, 1977)
Nocardia sp. NBZ-23	*n*-Alkanes	7.5-liter, fermentor, 6–8 liters of medium, 1500 rpm	30	6.8	1.25	14.7	0.98	Wagner *et al.* (1969)
Pseudomonas sp. No. 5401	Fuel oil	6000-liter fermentor: (a) batch (b) continuous	36–38	7.0	0.16 0.12 (*D*) 0.25 (*D*)	16 (24–26 hr) 10 8	1.00 — —	Ko and Yu (1968)
Pseudomonas sp. *Hyphomicrobium* sp. *Acinetobacter* sp. *Flavobacterium* sp. *(mixed culture)*	Methane	1.0-liter fermentor, 0.9 liters of medium 1 atm; 470 rpm	32	5.7	0.06 (*D*)	0.8	0.99	Wilkinson *et al.* (1974)
Rhodopseudomonas gelatinosa	Bicarbonate, wheat bran	4-liter working volume fermentor, 75-W incandescent lamp, 1100 rpm	40	0.2	(a) batch, 0.31 (b) continuous, 0.028 (*D*)	4.33 3.15	— —	Shipman *et al.* (1975)
Thermomonospora fusca	Cellulose pulping fines	10-liter fermentor, 3 liter/min, 60 rpm	55	7.4	—	—	0.35–0.40	Crawford *et al.* (1973)

has developed a bacterial SCP process based on the continuous culture system utilizing methanol in a 500-metric tons/year pilot plant (Masuda and Yoshikawa, 1973; Tung, 1975).

Hoechst/Uhde has developed a pilot-plant-scale process for growing *Methylomonas clara* on methanol. A 1000 metric tons/year unit has been constructed near Frankfurt, West Germany (Faust *et al.*, 1977).

Actinomycetes, particularly *Streptomyces* sp., will also grow on methanol as a substrate (Kato *et al.*, 1974). However, only laboratory-scale studies have been reported.

Ethanol has characteristics similar to methanol insofar as SCP production is concerned. Exxon Corporation has conducted small-scale pilot-plant studies on the growth of *Acinetobacter calcoaceticus* on ethanol (Laskin, 1975).

Recent price increases of petroleum and refined petroleum products have made hydrocarbons and chemicals derived from them, such as methanol and ethanol, less attractive as raw materials for SCP production than renewable resources, such as agricultural waste or by-products. In particular, sugars and starches are readily available from tropical crops, such as sugar cane or cassava, respectively. Also, cellulose is present in agricultural or forest products residues and food-processing waste. Certain chemical industry wastes, such as those from maleic anhydride and phthalic anhydride production, may provide utilizable substrates for bacterial growth (Edwards *et al.*, 1972). Some of the bacteria and corresponding substrates that have been investigated for utilizing waste material for SCP production include *Bacillus megaterium* on collagen meat packing waste (Bough *et al.*, 1972), *Brevibacterium* sp. and *Cytophaga* sp. on mesquite wood (Chang and Thayer, 1974; Thayer *et al.*, 1975; Fu and Thayer, 1975), *Cellulomonas* sp. on bagasse (Han *et al.*, 1971; Dunlap, 1975a, b; Srinivasen *et al.*, 1977), *Pseudomonas dentrificans* on sulfite waste liquor (Camhi and Rogers, 1976), sequential cultures of *Bacillus* sp. and *Lactobacillus* sp. on collagen- and starch-containing wastes (Busta *et al.*, 1977), and mixed cultures of lactic acid bacteria and yeasts on swine waste–corn mixtures (Weiner, 1977a, b).

Shipman *et al.* (1975) have grown the photosynthetic bacterium *Rhodospeudomonas gelatinosa* in continuous culture using wheat bran infusion as the substrate. Although the cells contained significant quantities of protein and vitamins, productivities did not appear sufficiently attractive to justify further development of this process.

Celluloytic, thermophilic actinomycetes offer an opportunity to produce SCP from cellulosic wastes without requiring extensive cooling systems for fermentors. Laboratory studies on *Thermomonospora fusca* indicated that 60–65% degradation of paper mill fines could be achieved in 96 hours, yielding a 30% protein product (Crawford *et al.*, 1973). General Electric Company constructed a small pilot plant at Casa Grande, Arizona for producing SCP from feed lot wastes using thermophilic actinomycetes (Bellamy, 1974, 1975). However, this pilot plant is not now in operation because of adverse economic conditions.

2. *Yield*

A number of factors affect growth rate, yield, and productivity (dry weight of cells per unit volume-unit time) of bacteria used for SCP production. Oxygen transfer, mass transfer of substrate to the cell, and heat production are important limiting factors for bacterial growth on either carbohydrates or hydrocarbons as carbon and energy sources. Also, inhibitory substances may be produced during growth particularly in continuous cultures.

Table II shows that typical yields of bacteria and actinomycetes grown on hydrocarbons range from 0.80 to 1.20 gm dry matter/gm substrate utilized. The maximum theoretical yields of bacteria from methane assuming the involvement of a mixed function oxidase is 0.91 gm dry cells/gm substrate according to Van Dijken and Harder (1975).

Harwood and Pirt (1972) attained experimental yields for *Methylococcus capsulatus* of 1.00–1.03 gm (dry wt.)/gm methane. These differences could be accounted for by the difficulties in determining carbon balances with gaseous substrates in continuous culture systems.

Experimental values of yields of bacteria from methanol given in Table II range from 0.48 to 0.50 gm. In studies with various *Pseudomonas* sp. Goldberg *et al.* (1976) obtained yields from methanol of 0.38–0.54 gm dry cells/gm substrate. The theoretical values calculated by Van Dijken and Harder (1975) for the yield of bacterial cells from methanol was 0.73 gm (dry wt.)/gm substrate. They assumed a yield of adenosine triphosphate (Y_{ATP}) from 3-phosphoglycerate as a key intermediate of 10.5 and a cell composition of C_4H_8ON for their calculations.

Table II shows yields of bacteria and actinomycetes from complex carbohydrates of 0.35–0.50 gm(dry wt.)/gm substrate. A typical yield value for *Pseudomonas fluorescens* grown aerobically on glucose is 0.51 gm (dry wt.)/gm substrate (Mennett and Nakayama, 1971).

During the 1960s there was interest in bioregenerative systems based on hydrogen-fixing bacteria for recycling expired carbon dioxide and urinary excretion products during extended space missions. Breathable oxygen would be produced by the electrolysis of water. The hydrogen also produced by electrolysis, along with a portion of the oxygen, the expired carbon dioxide, and urea, could be used by *Alcaligenes (Hy-*

drogenomonas) species to produce a microbial cell product to be utilized as food. Foster and Litchfield (1964, 1967, 1968a,b, 1969) designed a continuous culture system for *Alcaligenes eutrophus* which utilizes hydrogen, oxygen, and carbon dioxide in a 6 : 2 : 1 ratio. Productivities were in the range of 2.0 gm (dry wt.)/liter/hour (Litchfield, 1972). Yields of 5.2 gm (dry wt.) of *A. eutrophus* cells per gram-mole of hydrogen at a specific growth rate of 0.42 have been obtained (Bongers, 1969, 1970).

3. Cell Recovery

A number of problems arise in recovering bacterial cells from the growth medium. In most SCP processes, cell densities are in the order of 10–20 gm (dry wt.)/liter. Consequently, large volumes of water must be handled. Also, bacterial cells have a smaller size (1–2 μm) than other microorganisms and bacterial cell densities are very close to that of water (1.003 gm/cm^3). Taking into account these factors, the cost of separating bacterial cells by centrifugation would be prohibitively high.

Wang (1968a,b, 1969) estimated that total costs for centrifugal separation of bacteria from the growth medium would be approximately four times as great as for yeasts.

Labuza (1975) pointed out that plate-and-frame filter presses are not amenable to continuous processing as used in SCP production. Also, in vacuum filters, the cell size and packing density of bacteria results in compression on screens and filtration ceases because the void volume decreases to nearly zero. The use of filter aids and flocculants would contaminate the cell product and make it undesirable for food or feed applications.

As a result of the high cost of centrifugation consideration has been given to alternative methods of bacterial cell recovery in SCP production.

Acinetobacter cerificans can be concentrated by a two-zone froth-flotation process according to Perkins and Furlong (1967). Concentration ratios of bacterial cells in the froth were significantly greater than those in the slurry.

Imperial Chemical Industries, Ltd. have developed a proprietary process for separating *Methylophilus methylotrophus* grown on methanol from the growth medium without using flocculating agents. A dewatered cell product 25 gm (dry wt.)/liter or greater is then concentrated by conventional decanter-type centrifuges (Gow *et al.*, 1975; MacLennan *et al.*, 1976). The resulting concentrated product is then dried.

Hoechst/Uhde have operated a pilot-plant-scale process for separating *Methylomonas clara* grown on methanol from the culture medium based on electrochemical coagulation and centrifugation. The cell product is then spray-dried (Faust *et al.*, 1977).

There is a lack of information in the literature on initial concentration of bacterial cells by evaporation prior to spray-drying or drum-drying. Also, information is not readily available on flow characteristics of bacterial cell suspensions or of drying characteristics of initially dewatered bacterial cells of interest in SCP production. A further discussion of dewatering and drying of cells will be given subsequently in the section on yeasts.

Cell recovery processes from SCP production result in wastes from the spent growth medium and cell wash water. To the extent possible, spent media should be recycled since their biochemical oxygen demands (BOD) are substantial. In the case of methane, the excess substrate can be recycled readily and the spent medium and cell washings contain primarily inorganic salts and small amounts of cells. These residues must be treated by conventional waste treatment processes.

C. Yeasts

The technology of yeast production has developed since the nineteenth century. The reader should consult Reed and Peppler (1973) and Chapter 5, on yeast production, in this volume for information on the historical background and the current status of bakers' yeast production processes. This section covers the current status of food and fodder yeast production.

1. Growth Conditions

Table III lists some of the yeasts and substrates that are of potential interest for use in SCP production other than those in the genus *Saccharomyces,* such as *S. cerevisiae* and *S. carlsbergensis (S. uvarum),* which can be grown as primary yeast from cane or beet molasses as substrates.

Bakers' yeast, *S. cerevisiae,* was used for food and fodder yeast in Germany during World War I (Prescott and Dunn, 1959). However, this yeast does not utilize pentoses and requires supplementation with amino acids and B vitamins, mainly thiamin, niacin, pyridoxine, pantothenic acid, and inositol, which are usually furnished by blends of cane and beet molasses.

During World War II, *Candida utilis* was produced as a source of feed or food yeast in Germany from wood hydrolyzates which contain pentoses. This yeast does not require amino acids or B vitamins for growth and utilizes ammonium salts as a source of nitrogen. This development

TABLE III. Selected Yeasts and Substrates of Interest for SCP Production

Yeast	Substrate	References
Candida spp.	*n*-Alkanes	Kanazawa (1975); Sonoda *et al.* (1973)
Candida boidinii	Methanol	Reuss *et al.* (1975)
Candida ethanothermophilum	Ethanol	Masuda *et al.* (1975)
Candida guilliermondii	Diesel oil	
Candida kofuensis	*n*-Alkanes	Ueno *et al.* (1974, a, b, c)
Candida lipolytica	*n*-Alkanes	Evans and Shennan, (1974)
	Gas oil	Champagnat and Filosa (1971); Laine and du Chaffaut (1975)
	Oxanone wastewater	Wiken (1972)
Candida methanolica	Methanol	Goto *et al.* (1976)
Candida pseudotropicalis	Waste polyethylene	Brown *et al.* (1975)
Candida rigida	Liquefied petroleum gas	Imada *et al.* (1972)
Candida steatolytica	Brewery wastes	Shannon and Stevenson (1975 a, b)
Candida tropicalis	*n-Alkanes*	*Cooper et al.* (1975)
	Ethanol	Rychtera *et al.* (1977)
	Sulfite waste liquor	Rychtera *et al.* (1977)
	Starch	Spencer-Martins and Van Uden (1977)
Candida utilis	Ethanol	Ridgeway *et al.* (1975)
	Sulfite waste liquor	Holderby and Moggio (1960); Anderson *et al.* (1974); Camhi and Rogers (1976)
	Anaerobic digester supernatant	Irgens and Clarke 1976)
	Molasses	Mian *et al.* (1976)
	Tapioca starch	Thanh and Wu (1975)
Candida utilis and *Endomycopsis fibuligera*	Potato starch waste	Wiken (1972)
Debaryomyces kloeckeri	Soybean spent solubles	Sugimoto (1974)
Hansenula polymorpha	Methanol	Cooney and Levine (1975); Cooney *et al.* (1975)
Kluyveromyces fragilis	Cheese whey	Wasserman *et al.* (1961); Bernstein and Everson (1973); Moulin *et al.* (1976)
	Crude lactose	Vananuvat and Kinsella (1975 a, b)
	Lactose permate, ultrafiltrate	Delaney *et al.* (1975); Lane (1977)

(*continued*)

TABLE III (*Continued*)

Yeast	Substrate	References
Rhodotorula glutinis	Domestic sewage	Fustier and Simard (1976)
Rhodotorula rubra	Potato hydrolyzate	Tong *et al.* (1973a, b)
Torulopsis methanosorba	Methanol	Yokote *et al.* (1974)
Torulopsis methanolovescens	Methanol	Goto *et al.* (1976)
Trichosporon cutaneum	Cheese whey	Akin *et al.*(1967)
	Oxanone wastewater	Wiken (1972)

led to the postwar advances and present-day practices in SCP production processes utilizing a variety of yeasts and substrates.

Many of the same characteristics cited for bacteria are important criteria in selecting yeasts for use in SCP production, particularly specific growth rate, yield on a given substrate, pH and temperature tolerance, aeration requirements, and genetic stability. Also, the yeast strain must not be pathogenic to plants, animals, or humans.

Substrate and nitrogen concentrations for yeast growth should be adjusted to provide a C : N ratio in the range of 7 : 1 to 10 : 1 to favor high protein contents. Concentrations of carbohydrates in batch cultures range from 1 to 5%. In continuous cultures, with hydrocarbons or alcohols, lower concentrations are used. At higher C : N ratios many yeasts, particularly those of the genus *Rhodotorula*, accumulate a substantial portion of the cell weight in the form of lipids. Again, anhydrous ammonia or ammonium salts are suitable nitrogen sources for yeasts of interest in SCP production.

Anhydrous ammonia can be used in combination with phosphoric acid to effect pH control. It is desirable to select a yeast having a high specific growth rate and yield in the pH range of 3.5–4.5 to minimize the possibility of growth of any bacterial contaminants.

Carbohydrate substrates are soluble in aqueous media. Mass transfer is not limited by solubility. The same situation applies to methanol which is completely miscible with water. However, C_{10}–C_{18} hydrocarbons are soluble in water only to the extent of 6–16 mg/ml. The uptake of hydrocarbons by yeasts such as *C. lipolytica* may take place from either the liquid or the gas phase. These substrates are believed to penetrate into the cells by passive diffusion (Prokop and Sobotka, 1975).

As with bacteria, heat is liberated by yeasts growing on carbohydrates, hydrocarbons, or alcohols. For example, *C. utilis,* growing on sulfite waste liquor, which is representative of carbohydrate substrates, generates approximately 3740–8000 kcal/kg dry weight (Holderby and

Moggio, 1960). On the other hand, for *Candida* species growing on hydrocarbons, calculated values of heat liberated range from 4400–8000 kcal/kg for yields of 1.2–0.9 gm cells (dry wt.)/gm substrate utilized, respectively (Kanazawa, 1975; Laine and du Chaffaut, 1975). Actual observed values of heat liberated during growth are in this range. A value of 7600 kcal/kg dry weight of yeast is cited by Bennett *et al.* (1969) for *Candida* sp. growing on *n*-alkanes. Cooney *et al.* (1975) estimated a heat load of 37 kcal/liter/hour for production of yeast on methanol assuming a productivity of 5 gm/liter/hour, a yield of 0.5 gm cells per gram oxygen and using a factor of 0.12 kcal/mmole of oxygen.

It is clear that heat tolerance is a desirable attribute for a yeast to be used in SCP production. Most yeasts have the highest specific growth rates in the range 30°–34°C. A yeast capable of growing well in the range 40°–45°C would be of considerable interest. *Hansenula polymorpha* ATCC 26012 has an optimum growth temperature range of 37°–42°C (Cooney and Levine, 1975). This organism gives good cell yields from methanol at 42°C.

In most regions where SCP production facilities are located, cooling water is rarely available at a temperature below 20°C. Mechanical refrigeration would be required for cooling fermentors to maximum desired temperatures for growth.

The growth rate of yeasts under aerobic conditions depends upon the rate of mass transfer of oxygen and substrates to and across the cell surface. For growth on carbohydrates, the oxygen requirement is 1 gm or less per gram (dry weight) of cells (Reed and Peppler, 1973). On *n*-alkanes the oxygen requirement is approximately 2 gm of oxygen per gram (dry weight) of cells (Johnson, 1967).

According to Laine and du Chaffaut (1975) yeast grown on hydrocarbons requires an oxygen transfer coefficient of 10–15 kg/m^3/hour. A Waldhof-type agitated and aerated fermentor has been used extensively for achieving desired levels of aeration in food and fodder yeast production from both carbohydrate and hydrocarbon substrates (Prokop and Sobotka, 1975). However, in recent years, air-lift-type fermentors of the type developed by LeFrançois (1955) have been used to an increasing extent to achieve increased oxygen transfer rates and/or decreased power requirements for aeration and agitation, particularly in the production of yeast SCP from hydrocarbons (Laine, 1972; Hatch, 1975; Kanazawa, 1975).

Yeast SCP production may or may not take place under sterile conditions. In either batch or long-term continuous yeast production, one must balance the need for contamination control, by maintaining sterile conditions, with the capital and operating costs of the equipment required. In producing *C. utilis* (torula yeast) from sulfite waste liquor, the growth medium is passed through heat exchangers to effect sterilization and then charged into clean Waldhof-type fermentors. However, sterile conditions are not maintained during growth and contamination control depends upon maintaining a pH below 6.0, aerating with a sterile air supply, and maintaining a large population of yeast in the fermentor (Holderby and Moggio, 1960; Anderson *et al.*, 1974).

In the case of yeasts grown on hydrocarbons, clean but nonsterile conditions were used by British Petroleum Company, Limited in an air-lift fermentor pilot plant for producing *Candida* sp. from gas oil at Lavera, France. However, in producing *Candida* sp. from purified *n*-alkanes in a conventional agitated–aerated fermentor at Grangemouth, Scotland, British Petroleum used a completely sterile system and maintained sterile conditions throughout operations (Evans, 1968; Bennett *et al.*, 1969).

Table IV summarizes the growth characteristics of selected yeasts of importance in SCP production. Only the more important current processes will be discussed here.

Production of *C. utilis* dates from the 1940s. As mentioned previously, sulfite waste liquor from the paper industry was used as the substrate to convert to a salable product what would otherwise be a difficult to treat waste (Harris *et al.*, 1951; Holderby and Moggio, 1960).

A 10 million lb/year capacity torula yeast plant has been constructed and operated by Boise-Cascade Corporation at Salem, Oregon, to treat 85,000 tons of sulfite liquor from an ammonia-base pulp process that would otherwise be burned as fuel. Both food-grade and feed-grade yeasts are produced. Two continuous agitated and aerated 60,000-gal (227,124-liter) Waldhof fermentors are used to process 48,000–50,000 gal (181,700–189,270 liters) of medium (Anderson *et al.*, 1974).

The temperature is maintained at 36°–37°C by passing the medium through a shell and tube heat exchanger. The yeast is harvested by centrifugation, washed three times, pasteurized in a plate heat exchanger at 95°C, and spray-dried.

Candida utilis has been produced from a wide range of other substrates, as indicated in Table III. Ethanol, produced from ethylene, is a suitable substrate for this yeast. Amoco Foods Company has constructed a 10,000,000-lb (454,000-kg) capacity plant for continuous production of *C. utilis* from ethanol at Hutchinson, Minnesota. All liquid streams are sterilized at 149°C. Ammonia is sterilized by filtration and is used for pH control. The cells are harvested continuously, washed, and spray-dried to give a food-grade product (Anonymous, 1977b; Ridgeway *et al.*, 1975).

TABLE IV. Growth of Selected Yeasts on Various Substrates

Organism	Carbon and energy source	Scale, aeration, agitation	Temperature (°C)	pH	Specific growth rate or dilution rate (*D*) (hr^{-1})	Culture density (gm/liter) (dry wt. basis)	Yield (dry wt. basis) (gm/gm substrate used)	References
Candida ethanothermophilum ATCC 20380	Ethanol	30-liter fermentor, 17 liters of medium, 17 liters/min, 500 rpm (continuous)	40	3.5	0.20 (*D*)	8.0	0.95	Masuda *et al.* (1975)
Candida lipolytica	(a) *n*-Alkanes	1800-liter working volume fermentor, 1.5 v/v/min, continuous	32	5.5	0.16 (*D*)	23.6	0.88	Evans and Shennan (1974)
	(b) Gas oil	12-m^3 fermentor, 2400 liters of medium, 1 v/v/min, continuous	30	4.0	—	25	0.18	Champagnat and Filosa (1971) Laine and du Chaffauf (1975)
Candida tropicalis	*n*-Alkanes	Air-lift 50,000-liter fermentor	30	3.0	0.15–0.24	10–30	1.0–1.1	Silver and Cooper (1972a, b), Cooper *et al.* (1975)
Candida utilis	(a) Sulfite waste liquor	Walhof fermentor 0.50 cfm	32	5.0	0.5	—	0.39	Harris *et al.* (1951)
	(b) Ethanol	100–140 m*M* O_2/liter/hour (continuous)	30	4.6	0.3	6–7	—	Ridgeway *et al.* (1975)
Hansenula polymorpha ATCC 26012	Methanol	1-liter fermentor, 375 ml of medium	37–42	4.5–5.5	0.22 (0.13,*D*)	1.2	0.36	Cooney and Levine (1975); Cooney *et al.* (1975)
Kluyveromyces (Saccharomyces) fragilis	Cheese whey	1600-gal (6057-liter) fermentor, 1.5 m*M* O_2/gal/min	32	5.5	0.66	14.6	0.55	Wasserman *et al.* (1961)

Rhodotorula glutinis	Domestic sewage	14-liter fermentor, 0.7 v/v/min, 300 rpm, continuous	28	7.4	0.158–0.211	0.08	—	Fustier and Simard (1976)
Saccharomyces cerevisiae	Cane molasses	(a) 75–225 m^3 fermentors, 2 m*M* O_2/liter/min (l v/v/min)	30	4.5–5.0	0.20–0.25	40–45	0.50	Reed and Peppler (1973)
		(b) 5,900-gal (34,068-liter fermentors, continuous	30	4.5–5.0	0.14	70	—	Olson (1961)

Candida utilis has also been grown on ethanol in a 1000-metric ton/year pilot plant at Kojetin, Czechoslovakia. Scaleup is planned to a 60,000 ton/year scale (Wells, 1975).

In Japan, Mitsubishi Petrochemical Company, Limited has conducted research on a pilot-plant-scale process for growing *Candida acidothermophilum* and *Candida ethanothermophilum* on ethanol. These organisms are claimed to tolerate temperatures up to 40°C (Masuda *et al.*, 1975).

Over 25 × 10^9 lb (11.4 × 10^9 kg) of sweet and acid cheese whey are produced in the United States each year. Early research on the utilization of this whey, which contains 69–77% lactose (dry basis), by *Kluyveromyces (Saccharomyces) fragilis* was conducted at the Eastern Regional Research Center of the U.S. Department of Agriculture (Wasserman *et al.*, 1961).

Dry food-grade *Kluyveromyces fragilis* yeast has been produced in the United States for more than a decade. A 5000 short ton/year plant for producing either yeast aerobically or ethanol anaerobically from cheese whey is operated by Amber Laboratories at Juneau, Wisconsin (Bernstein and Everson, 1973; Bernstein *et al.*, 1975). Also, research has been conducted on the growth of *K. fragilis* on lactose permeates and ultrafiltrates from whey protein recovery (Delaney *et al.*, 1975; Lane, 1977).

The Symba process developed by the Swedish Sugar Corporation utilizes two yeasts symbiotically for the treatment of wastes containing starch. *Endomycopsis fibuligera* produces α- and β-amylases which hydrolyze starch to lower saccharides, predominantly glucose and maltose, and *C. utilis* utilizes these sugars for growth. The process was originally designed to treat potato-processing wastes but can be applied to other wastes that contain starch, such as rice-processing waste.

Data obtained on a pilot plant providing for the production of 40–100 kg dry yeast per day were used to design a plant to handle 126 m^3 of waste per hour having a biochemical oxygen demand (BOD) of 15,000 mg/liter. BOD reductions are estimated to be 85% or greater (Wiken, 1972).

Research at the Dutch State Mines demonstrated that *C. lipolytica* or *Trichosporon cutaneum* can be used to treat oxanone water, a waste mixture of organic acids resulting from the manufacture of caprolactam used as the raw material for producing Nylon 6. Special strains of these yeasts have high resistance to the toxic effects of organic acids at pH 6.5. A yeast production of 4–4.5 kg/m^3/hour and 80% removal of organic acids are obtained (Wiken, 1972).

During the early 1960s, British Petroleum Company, Limited introduced the concept of using *C. lipolytica* and related yeasts for the production of SCP from hydrocarbon fractions including gas oil and purified *n*-alkanes prepared by molecular sieve purification (Champagnat *et al.*, 1963; Champagnat and Filosa, 1965, 1971; Evans, 1968; Bennett *et al.*, 1969; Evans and Shennan, 1974; Laine and du Chaffaut, 1975). In the case of gas oil, the objective was to utilize the *n*-alkane fraction, effect dewaxing, and concurrently produce SCP. About 10% of the gas oil is utilized by the yeast and the remaining 90% is returned to an adjacent refinery.

British Petroleum constructed two pilot plants for producing *Candida* sp. from hydrocarbons. A 16,000 metric ton/year plant at Lavera, France was designed to utilize gas oil containing 25% C_{15}–C_{30} *n*-alkanes having a boiling range of 300°–380°C. A continuous air-lift-type fermentor was used under clean but nonsterile operating conditions. This plant ceased operation in 1976. A 4000 metric ton/year pilot plant has been constructed and operated at Grangemouth, Scotland. It utilized purified *n*-alkanes prepared by a molecular sieve process (97.5–99% C_{10}–C_{23}, boiling range 175°–300°C). A continuous aerated and agitated fermentor was operated under aseptic conditions. The mineral nutrients were sterilized by heating, while air and ammonia are sterilized by filtration (Evans, 1968; BP Proteins Ltd., 1974).

The British Petroleum processes led to the development of similar yeast-based SCP processes utilizing hydrocarbon substrates by petroleum and chemical industries throughout the world. These processes use the following yeasts: *Candida* sp. (Kanegafuchi Chemical Industry Company, Limited; Takata, 1969; Kanazawa, 1975; Kanazawa *et al.*, 1975), *Candida kofuensis* (Mitsui-Toatsu Chemicals, Inc.; Ueno *et al.*, 1974a,b,c), *Candida novellus* (Liquichimica Biosintesi, S.p.a.; Giacobbe *et al.*, 1975), and *Candida tropicalis* (Gulf Research and Development Company; Cooper *et al.*, 1975).

A 100,000 metric ton/year yeast SCP plant constructed in Sardinia by British Petroleum and Ente Nazionale Idrocarburi has not yet been placed into operation because of a dispute with Italian governmental authorities over hydrocarbon residues and product quality (Anonymous, 1976a, 1978a,b). Another 100,000 metric ton yeast SCP plant constructed by Liquichimica Biosintesi S.p.a. at Saline, di Montebello, Italy using the Kanegafuchi Chemical Industry process has not been operated for the same reason (Anonymous, 1978a,b).

Other pilot-plant yeast SCP processes have been operated by Institute Français du Petrole, Soleige, France (Decerle *et al.*, 1969), Sovnaft, Kojotin, Czechoslovakia (Anonymous, 1973b), and All-Union Research Institute of Biosynthesis of Protein Substances, Bashkira and Gorki, USSR (Anonymous, 1974b). Moo-Young (1976) surveyed existing or planned SCP production facilities throughout the world. His compilation should be consulted for further information.

2. Yield

Factors affecting the growth rate, yield, and productivity of yeasts used for SCP production from various substrates are similar to those for bacteria.

Table IV presents yields of selected yeasts grown on various substrates. Experimental yields on carbohydrates range from 0.39 to 0.55 gm cells (dry wt.)/gm substrate utilized. These values are similar to the 0.51 gm (dry wt.)/gm substrate reported by Hernandez and Johnson (1967) for *C. utilis* grown on glucose. Data on yields reported in the literature should be evaluated carefully to determine the extent to which nonsugar carbon sources in complex natural products and wastes contribute to yield.

In the case of *C. utilis* grown on ethanol, Rychtera *et al.* (1977) obtained a yield of 0.75 gm (dry wt.)/gm substrate. They cited a previous published yield value in the range of 0.68–0.72 gm (dry wt.)/gm ethanol. The 0.95 gm (dry wt.)/gm ethanol yield reported for *Candida ethanothermophilum* by Masuda *et al.* (1975) appears to be unusually high and the reasons for these differences are not readily discernible from the information given in their patent.

The yield of *Hansenula polymorpha* grown on methanol of 0.36 gm (dry wt.)/gm substrate obtained by Cooney and Levine (1975) is lower than a calculated theoretical yield of 0.6 gm (dry wt.)/gm substrate based on the number of electrons available for transfer to oxygen. They point out that this discrepancy could probably be explained when the pathway of methanol oxidation and incorporation into cell mass is better understood. They cite other published yield data for yeasts grown on methanol ranging from 0.29 to 0.45 gm (dry wt.)/gm substrate.

The yields of various *Candida* species grown on purified *n*-alkanes range from 0.88 to 1.1gm (dry wt.)/gm substrate. These yields were obtained in pilot-plant-scale operations and are representative of the values attainable in large-scale production facilities. These values agree with the calculations of Guenther (1965), which indicate that yields greater than 1.0 gm (dry wt.)/gm substrate are theoretically possible from hydrocarbons on an energy basis.

Various equations have been developed for showing the stoichiometry of yeast growth on carbohydrates or hydrocarbons (Darlington, 1964; Guenther, 1965; Bennett *et al.*, 1969; Kanazawa, 1975). Typical equations are given by Bennett *et al.* (1969) for carbohydrates [Eq. (1)] and hydrocarbons [Eq. (2)]:

$$8n\ CH_2O + 0.8n\ O_2 + 0.19n\ NH_4 + \text{essential elements} \rightarrow$$
$$n(CH_{1.7}O_{0.5}N_{0.19}Ash) + 0.8n\ CO_2 + 1.3n\ H_2O + 80{,}000n\ \text{kcal} \qquad (1)$$

$$2n\ CH_2 + 2n\ O_2 + 0.19n\ NH_4 + \text{essential elements} \rightarrow$$
$$n(CH_{1.7}O_{0.5}N_{0.19}Ash) + n\ CO_2 + 1.5\ H_2O + 200{,}000\ n\ \text{kcal} \qquad (2)$$

These equations will vary with the composition of the yeast cell product and the same considerations apply to other microorganisms used for SCP production, including algae, bacteria, molds, or higher fungi.

Holve (1976) presented a series of equations for describing the stoichiometry of conversion of carbohydrates, hydrocarbons, alcohols, or organic acids to SCP by microorganisms. His model depends upon assumed chemical compositions of microorganisms based on literature values. In practice, actual analyses of the composition of the cells being grown should be used in such calculations.

3. Cell Recovery

Yeast cells range in size from 5 to 8 μm and have a density of 1.04–1.09 gm/cm^3. They can be recovered readily from the growth medium by continuous centrifugation. Usually, the cells are centrifuged in a first stage to initially dewater, yielding a yeast cream. This is followed by two subsequent washing centrifugations. The final washed yeast cream usually contains 15–20% solids. This procedure is typical of that used in the recovery of *C. utilis* after growth on sulfite waste liquor (Holderby and Moggio, 1960; Anderson *et al.*, 1974).

In the case of *Candida* sp. grown on *n*-alkanes, a wash with a surfactant may be required to remove traces of hydrocarbon following initial centrifugation (Wang, 1968; Labuza, 1975). With gas oil, a much more complex separation and cleanup procedure must be used to remove residual hydrocarbons and lipids. Decantation, phase separation with solvents, washing with sufactants, and solvent extraction have been used alone or in combination. Also hydrocarbons and lipids can be depleted by allowing the yeast to remain in contact with the growth medium in absence of added substrate for varying periods. Solvents can be removed by steam (Litchfield, 1977b).

For recovering *Kluyveromyces fragilis* after growth on cheese whey, the entire growth medium can be passed through a three-stage evaporation to concentrate the solids from 8 to 27% to give a feed-grade product. If a high-protein–low-ash product is desired, multistage centrifugation is used (Anonymous, 1975; Bernstein and Everson, 1973).

Wang (1968b, 1969) has analyzed methods and costs for recovering microbial cells including yeasts. He points out that the cost of cell separation could be reduced if the size of the cells could be increased or if flocculants compatible with a food- or feed-grade product could be developed.

The separated cells can either be drum- or spray-dried. According to Labuza (1975) drum-drying brings about two to five logarithmic cycles of kill of yeast cells. Associated bacteria having similar thermal resistances would be killed to the same extent. However, drum-drying, if not carefully controlled, may result in excessive darkening of the product.

Under the temperature and time conditions used in spray-drying, 5–8 logarithmic cycles of kill of yeast are obtained and cells of pathogenic bacteria, such as *Salmonella* and *Staphylococcus,* should be killed to the same extent. The color and finished characteristics of the final yeast product prepared by spray-drying are superior to those obtained after drum-drying (Labuza, 1975).

D. Molds and Higher Fungi

As mentioned earlier in this chapter, certain higher fungi, the mushrooms, have been used as human food since ancient times. Also, many Asian fermented food products are produced by mold fermentations.

Molds and higher fungi grow at much slower rates than bacteria and yeasts. Doubling times may range from 6 to 16 hours, or longer.

Mushrooms are usually grown as the fruiting body in manure or synthetic compost beds (see Lambert, 1938; Stoller, 1954; Litchfield, 1967a; Volume II, Chapter 8, for reviews of mushroom cultivation practices).

In the United States, the basidomycete *Agaricus campestris* (*A. bisporus*) is the sole species produced. In Japan, the Shiitake mushroom, *Cortinellus berkelyanus* Ito and Inci, is produced by a 2-year incubation of logs of certain trees of the Fagaceae and Betulaceae inoculated with spore suspensions. The Chinese padistraw mushroom, *Volvaria volvacea* (Bull.) Fr., is cultured by inoculating moistened rice straw with spawn (Lambert, 1938).

The Perigord truffle, *Tuber melanosporum* Vitt, or black truffle, which belongs to the family Ascomycetes, is produced in southern France by taking soil from the vicinity of truffle-bearing oaks and raking this soil beneath newly planted oak trees (truffieres or truffle orchards). About 5–10 years after the oaks have been planted, truffles appear and yields range up to 100 lb/acre (Seelig, 1959).

This section covers production of mold or fungal mycelium for food or feed applications, since the mycelium of these organisms can be grown in submerged culture in a manner similar to growing yeasts.

1. Growth Conditions

Fungal mycelium was grown in submerged culture in Germany during World War II for use as food. Numerous species have been investigated for their potential utility as food or feed since that time and the reader is referred to previous reviews for background information (Litchfield, 1967a,b, 1968; Gray, 1970).

The same general considerations for selecting yeast cultures for SCP production apply to molds and higher fungi, including specific growth rate, yield, pH and temperature tolerance, aeration requirements, and genetic stability. Lack of pathogenicity, or toxigenicity, is very important. Strains of some species of molds such as *Aspergillus fumigatus* and *Fusarium graminearum* are very hazardous to humans (Wilson, 1971; Sherwood and Peberdy, 1974). Yet these organisms have been proposed for SCP production (Anderson *et al.*, 1975; Gregory *et al.*, 1976; Khor *et al.*, 1976). The use of organisms of this type should be avoided even though the particular strain selected may be judged "safe" after preliminary toxicological evaluation.

Numerous species of molds and higher fungi have been evaluated for production as SCP or as food flavor additives on carbohydrates or on agricultural, food-processing, or forest products wastes that contain carbohydrates. Table V presents some typical examples of both organisms and substrates.

The concentrations of carbohydrate substrates used are in the 1–10% range. Carbon-to-nitrogen (C : N) ratio in some species, when as high as 20 : 1, have given cells with satisfactory protein contents, but in most cases C : N values of 5 : 1 to 15 : 1 are preferred.

Anhydrous ammonia or ammonium salts can be used as nitrogen sources in continuous cultures and at the same time effect pH control. In many waste treatment applications, fungal cultures are grown in either a batch or semicontinuous mode and ammonium phosphate, ammonium sulfate or ammonium nitrate are added to the waste as supplemental nitrogen sources.

Phosphoric acid is suitable as a source of phosphorus. Mineral nutrients, such as potassium, sulfur, magnesium, calcium, iron, manganese, zinc, copper, and cobalt, are required for growth of most fungi. However, the concentrations of specific inorganic constituents in the growth medium may vary considerably when complex organic materials of variable composition are used as substrates. In the case of mushroom mycelia produced for food flavoring purposes, the concentrations of inorganic salts in the medium must be limited to levels actually required for growth to prevent the development of a bitter flavor in the mycelium (Litchfield, 1967c).

The mycelium of certain mushrooms will not grow in synthetic media unless supplemented with vitamins. Some strains of *Coprinus comatus* have an absolute requirement for thiamin, and the addition of thiamin also enhances growth of other mushrooms. *Agaricus* spp. and *Morchella* spp. will grow in a simple synthetic medium. However, the rate of growth of the latter group of organisms is enhanced by supplements, such as

TABLE V. Growth of Selected Molds and Higher Fungi on Various Carbon and Energy Sources

Organism	Carbon and energy source	Scale, aeration, agitation	Temperature (°C)	pH	Specfic growth rate or dilution rate (*D*)(hr^{-1})	Culture density (gm/liter)	Yield of Mycelium gm/100 gm substrate (dry wt. basis) Supplied	Used	References
Agaricus blazei	Glucose, citrus press water, orange juice	250-ml Erlenmeyer shake flask, 2-inch stroke, 80 cpm, 1-liter, 40-liter bottle	Ambient	3.5–7.5	—	15.2 18.8 20.6	30 31 34	42 46 41	Block *et al.* (1953)
Agaricus campestris	Glucose	20-liter fermentor, air flow, 2–3 v/v/min, 400 rpm	25	4.5		20	—	—	Humfeld and Sugihara (1949)
	Malt syrup–cane molasses	2/liter fermentor, 13 m*M* O_2/liter/hour, 400 rpm	27	5.0–5.5	—	7.2	15.8	44.6	Moustafa (1960)
Aspergillus niger	Carob bean extract	3000-liter fermentor (total capacity)	30–36	3.4	0.25	31.5	45	—	Imrie and Vlitos (1975)
Aspergillus oryzae	Coffee waste water	5000-gal (18,927-liter tank, air flow, 1 v/v/min	28	4.0–4.5	0.037 (*D*)	—	—	45	deCabrera *et al.* (1974)
Boletus edulis	Glucose	300-ml Erlenmeyer flask, 100-ml medium, 150 rpm	25	4.5–5.5	0.00166	2.5	—	25	Van Eybergen and Scheffers (1972)
Calvatia gigantea	Brewery waste: (a) Grain press (b) Trub press liquor	250-ml Erlenmeyer flasks, 100-ml medium, 200 rpm	25	6.0	— —	(a) 6.25 (b)27.72	— —	24.8 74.9	Shannon and Stevenson (1975b)
Fusarium moniliforme	Carob bean extract	14-liter fermentor (8.5 liter working volume), (semicontinuous), 400–700 rpm, 0.25 v/v/min	30	5.5–6.5	0.18	8.8	—	0.384	Drouliscos *et al.* (1976)
Geotrichium sp.	Corn and pea waste	Aeration pool 10,000 gal (37,854 liter), continuous, 10,000-gal ditch	Ambient	3.7	—	0.75–1.0	—	—	Church *et al.* (1973)

Gliocladium deliquescens	Corn net milling waste	50,000 gal (189,270 liter continuous)	Ambient	4.6	—	1.2	—	—	Church *et al.* (1973)
	Soy whey	18-liter fermentor	Ambient	4.6	—	3.2–3.5	—	—	Church *et al.* (1972)
Lentinus edodes	Glucose	30-liter fermentor, 2 v/v/min, 400–600 rpm	25	5.5	0.12	7.5	—	26	Teramoto *et al.* (1966)
Morchella crassipes	Glucose,	10-liter carboy, 7 liters of medium, 0.08 m*M* O_2/liter/min	25	6.5	—	8.02	33.6	48.6	Litchfield *et al.* (1963a)
	matose,					3.55	31.4	47.8	
	lactose,					3.30	29.2	46.3	
	cheese whey,					2.38	5.96	32.7	Litchfield and Overbeck (1966)
	corn canning waste,					0.75	13.9	27.8	
	pumpkin canning waste,					5.51	20.0	42.6	
	sulfite liquor (NH_3)	5-gal carboy, 16 liters of medium, 0.25 v/v/min	Ambient	5.0–7.0	—	2.0	—	78	Kosaric *et al.* (1973)
Morchella deliciosa	Sulfite waste liquor	5-gal carboy, 16 liters of medium, 0.25 v/v/min	Ambient	6.0	—	1.0	—	32	Kosaric *et al.* (1973)
Morchella esculenta	Glucose,	10-liter carboy, 7 liters of medium, 0.08 m*M* O_2/liter/min	25	6.5	—	7.85	32.8	48.1	Litchfield *et al.* (1963a);
	maltose,					3.40	30.1	47.5	
	lactose,					1.65	14.6	43.4	
	cheese whey,					1.28	3.21	32.1	Litchfield and Overbeck (1966)
	corn canning waste,					0.85	15.3	33.3	
	pumpkin canning waste,					7.92	28.8	50.4	
	sulfite waste liquor (NH_3)	5-gal carboy, 16 liters of medium, 0.25 v/v/min	Ambient	5.0–6.0	—	1.9	—	65	Kosaric *et al.* (1973)
Morchella hortensis	Glucose,	10-liter carboy, 7 liters of medium, 0.08 m*M* O_2/liter/min	25	6.5	—	8.20	34.3	48.8	Litchfield *et al.* (1963a)
	maltose,					3.75	33.2	49.0	
	lactose,					3.60	31.8	47.9	
	cheese whey,					8.65	23.5	43.6	
	corn canning waste,					1.23	22.3	33.5	Litchfield and Overbeck (1966)
	pumpkin canning waste					8.25	27.5	48.1	

(continued)

TABLE V *(Continued)*

Organism	Carbon and energy source	Scale, aeration, agitation	Temperature (°C)	pH	Specific growth rate of dilution rate(*D*)(hr^{-1})	Culture density (gm/liter)	Yield of Mycelium gm/100 gm substrate (dry wt. basis) Supplied	Used	References
Paecilomyces varioti (Pekilo)	Spent sullite liquor	360-m^3 fermentor	—	—	0.2(*D*)	17	—	55	Romantschuk (1975); Romantschuk and Lehtomaki (1978)
Trichoderma harzianum	Coffee waste	5000 (18,927-liter) 7500-gal (28,391-liters) tanks (continuous), 2 v/v/min	Ambient	3.5	0.19–0.10	—	—	31–47	Aquirre *et al.* (1976); Espinosa *et al.* (1975)
Trichoderma viride	Corn and pea waste	Aeration pool 10,000 gal (37,854 liters), continuous 11,000-gal ditch (41,640 liter)	Ambient	4.6	—	1.2	—	—	Church *et al* (1973)
T. viride	Ball-milled cellulose	14-liter fermentor, 6.1 liters of medium, 0.3 v/v/min, 350 rpm (continuous)	30	5.0	0.033–0.08(*D*)	4.0	—	70	Peiterson (1977)
Tricholoma nudum	(a) Glucose	3-liter fermentor, 1 liter/min, 400 rpm	25	5.0	—	22.6	—	37.6	Reusser *et al.* (1958b)
	(b) Sulfite waste liquor	30-liter fermentor, 8.2 liter/min, 300 rpm	25	5.5	—	10.6	—	64.0	Reusser *et al.* (1958b)

corn steep liquor, which are rich in trace nutrients including vitamins (Litchfield *et al.*, 1963a).

A pH range of 3.0–7.0 is suitable for most fungi. It is generally desirable to operate at a pH of 5 or below to minimize bacterial contamination problems, particularly in growing fungi on wastes under nonaseptic conditions. In growing mycelia of morel mushrooms (*Morchella* sp.) a pH of 6.0–6.5 is used to maximize growth rate, yield, and flavor. This requires that the culture be grown under aseptic conditions (Litchfield, 1964).

Most fungi grow best in the range 25–36°C. Growth rates decrease markedly above this range. Mushroom mycelia appear to be less tolerant of temperatures above 30°C than other fungi. Other fungi grow well at higher temperatures. For example, *Aspergillus niger* grown on carob bean extract grows well up to 36°C (Imrie and Vlitos, 1975).

Cooling must be provided to control the heat generated by aerobic growth of fungal cultures, and the same considerations noted previously for bacteria and yeasts apply here.

The oxygen requirements for growth of fungal cultures are quite complex. Fungal mycelium generally grows either in a filamentous or a pellet form in submerged culture. A recent review discussed the factors affecting the growth of molds in pellet form (Metz and Kossen, 1977).

The formation of pellets is largely determined by the extent of agitation. In general, pellet formation is favored by low agitation and aeration rates. For example, with *Morchella* spp. sparging to give aeration rates of 0.08–0.15 mmole oxygen/liter/minute gave highest yields in the pellet form. Above 0.15 m*M* oxygen/liter/minute growth was filamentous and yields decreased (Litchfield *et al.*, 1963a). This decrease in yield with increased aeration is not always observed with other fungi and in many cases higher yields are obtained when growth is in the filamentous form.

In many cases, the mycelium will develop around the impeller in agitated vessels. In such instances, it is preferable to use fermentors in which the aeration system also provides agitation. In most reports of mass cultivation of fungal mycelium, aeration rates are given as air flow per volume of medium per minute, rather than as oxygen absorption rates, which makes it difficult to compare the conditions used in different studies.

Only a few reports of growth of fungi on noncarbohydrate substrates have appeared in the literature. A strain of *Graphium* grows on natural gas (methane and ethane), with ethane being the preferred substrate (Volesky and Zajic, 1971). *Lentinus edodes, Pleurotus ostreatus,* and *Schizophyllum* sp. grow on a variety of carbon sources, particularly C_{10}–C_{16} hydrocarbons, methanol, and ethanol (Sugimori *et al.*, 1971).

Only selected fungal SCP processes listed in Table V will be discussed here. Of particular interest are processes for utilizing wastes, particularly those containing cellulose and those for producing edible fungi.

Edible mushroom mycelium is primarily of interest as a food flavoring rather than a primary source of protein. Laboratory and small pilot plant-scale studies have been conducted on species of *Agaricus, Boletus, Calvatia, Cantharellus, Collybia, Lentinus, Morchella, Tricholoma,* and *Xylaria* (Block, 1960; Litchfield, 1967a, 1968; Reusser *et al.*, 1958a,b; Sugihara and Humfeld, 1954).

Morchella mycelium has been produced on a commercial scale for sale as a food flavoring (Heinemann, 1963; Klis, 1963; Litchfield, 1964, 1967a,b,c). Typical growth conditions are shown in Table V for *Morchella crassipes, Morchella esculenta* and *Morchella hortensis.*

In a large pilot-plant run, *Morchella* sp. were grown in a 2000-gal (7570-liter) tank containing 1500 gal (5678 liters) of medium. Approximately 1.5–2 short tons of fresh mycelium were obtained containing 90% moisture or 24–30 gm (dry wt.)/liter (Robinson and Davidson, 1959).

Carob beans are widely grown in the Mediterranean region for edible gum production. The waste pods contain sugars which can be extracted and used as a substrate for growth of fungi. A process for utilizing the carbohydrates in carob bean extract using *Aspergillus niger* has been developed by Tate and Lyle, Limited in the United Kingdom. Cell densities of 31.5 gm/liter were obtained in 3000-liter capacity fermentors (Imrie and Vlitos, 1975). In other work in Greece, *Fusarium moniliforme* has been grown on carob bean extract in a semicontinuous laboratory fermentor system. A harvest of 8.8 gm/liter/day was obtained (Droulisos *et al.*, 1976).

Corn and pea canning wastes have been treated in aeration pools of 10,000 gal (37,854 liters), or in a 11,000-gal (41,639-liter) oxidation ditch using *Trichoderma viride* as the inoculum. However, a *Geotrichium* sp. that was naturally present tended to predominate over the season in corn wastes apparently because of an oversupply of nitrogen and phosphorus, while the pea waste was dominated by a *Fusarium* sp. Biochemical oxygen demand reductions of 96% and 95% were obtained for corn and pea wastes, respectively (Church *et al.*, 1973).

In other studies, *Gliocladium deliquescens* was used to treat corn wet milling wastes in a 50,000-gal aerated pilot unit and soy-processing waste in a laboratory fermentor scale study (Church *et al.*, 1972, 1973). A 87% reduction in chemical oxygen demand (COD) of the corn wet milling waste and greater than a 97% reduction in the BOD of soy whey waste were obtained.

Ranks, Hovis, McDougall, Limited in the United Kingdom have performed extensive studies on a wide range of fungi as a potential source of edible protein (Anderson *et al.*, 1975). *Fusarium graminearum* was selected as the organism for process development (Solomons and Scammell, 1976). The strain used is claimed to be nontoxic (Towersey *et al.*, 1976, 1977). Pilot plant-scale studies are in progress but detailed operating data have not been released.

ICAITI in Guatemala has investigated growth of various fungi on coffee-processing wastes (deCabrera *et al.*, 1974; Rolz, 1975; Aguirre *et al.*, 1976). *Trichoderma harzianum* was used to treat wastes from coffee processing in El Salvador in a system consisting of a 2000-gal (7571-liter) equalization tank, two 5000-gal (18,927-liter) fermentors, and two 450-gal (1734-liter) seed fermentors. Aeration was supplied by sparging and the fermentation was run under nonaseptic conditions at pH 3.5. A 73% reduction of COD was observed in 24 hours with net gain of 3.2 gm/liter of solids. Approximately 56% of the dry solids was protein.

Large quantities of cellulose are present in agricultural wastes and in pulp and paper industry wastes. The conversion of cellulose in bagasse from sugar cane operations by bacteria and in animal feed lot wastes by thermoactinomycetes to SCP was covered previously.

Research at the U.S. Army Natick Laboratories has shown that *Trichoderma viride* has a complete cellulase system that catalyzes the conversion of cellulose to cellobiose, and cellobiose in turn to glucose (Mandels and Sternberg, 1976). Cellulose must be pretreated to achieve satisfactory yields of glucose by the *T. viride* cellulase system because its insolubility, crystallinity, and intimate association with lignin limit access of this enzyme to glycosidic bonds. Alkali treatment and/or ball milling to a fine particle size (70–100 μm) enhances enzyme action.

Alkali or photochemical treatment, electron irradiation, and hydrolysis were investigated for increasing the biodegradability of cellulose by various fungi, including *A. fumigatus, Penicillium* sp., *T. viride, Chaetomium* sp., and *Geotrichium candidum* (Rogers *et al.*, 1972). Best results were obtained by preheating the cellulose with dilute solutions of sodium nitrite followed by irradiation of the cellulose nitrite slurries with ultraviolet light at 3650 Å for 1–5 days.

Either *T. viride* alone or a mixed culture of *T. viride* and *C. utilis* converted barley straw to a protein product containing 21–26% protein (Peitersen, 1975a,b). The time required for maximum cellulase and cell production was less when the combined cultures were used. In subsequent studies *T. viride* was grown on ball-milled cellulose in a 14-liter continuous culture system; cellulose was consumed to the extent of 50–75% (Peitersen, 1977).

According to Eriksson and Larsson (1975) the rot fungus *Sporotrichum pulverulentum* utilized waste mechanical fibers from a newsprint mill only to a limited extent. The protein content of these fibers after treatment was only 14%.

Moo-Young *et al.* (1977) found that *Chaetomium cellulyticum* gave 50–100% greater growth rates and 80% more cellular protein than *T. viride* when it was grown on a medium containing 1% of a purified form of amorphous cellulose (Solka-floc) or partially delignified sawdust. However, *T. viride* produced greater quantities of free cellulases at higher rates and degraded greater amounts of both substrates.

Fungal degradation of cellulose and concomitant SCP production have not yet been practiced successfully on a large pilot-plant or commercial scale. Pretreatment of cellulose by ball milling or other physical or chemical means to make it amenable to utilization by fungi appears to be excessively costly at the present time.

The Pekilo process developed by the Finnish Pulp and Paper Research Institute utilizes *Paecilomyces varioti* for converting sulfite waste liquor to SCP. A 10,000-metric ton capacity plant has been constructed (Romantschuk, 1975; Romantschuk and Lehtomäki, 1978). A 360-m^3 continuous fermentor system is used, having a retention time of 4.5–5 hours. The mycelial concentration in fermentors reaches 17 gm (dry wt.)/liter. Productivity is in the range 2.7–2.8 kg/m^3/hr, and daily production ranges from 15 to 16.5 metric tons/24hr. The process is designed to be installed between the cooking operation and the evaporation plant of the pulp mill to reduce pollutional loads.

2. Yield

Table V presents yield data on selected molds and higher fungi grown on various substrates. Yields of mycelium vary widely, depending upon organism and substrate. As would be expected, the highest yields were obtained in laboratory-scale studies. Lower conversions of substrate to cell mass are often obtained with many of the fungi listed than with bacteria or yeasts.

A yield of 0.45 gm (dry wt.)/gm substrate for *Aspergillus niger* grown on carob bean extract is typical for molds and yields of 0.32–0.50 gm (dry wt.)/gm substrate is typical for mushroom mycelium such as the *Morchella* sp.

Factors affecting growth rate and yield productivity of molds and higher fungi are similar to those of bacteria and yeasts. The problem of oxygen transfer to filamentous or pellet mycelium, discussed previously, is a major limiting factor in improving yields of fungi in commercial-scale production facilities.

3. Cell Recovery

Molds and higher fungi are more easily recovered from the growth medium because of the filamentous or pellet form of the mycelium which permits the use of screens or filter presses. Basket centrifuges, rotary vacuum filters, and inclined screens have been evaluated for their utility in recovery of mycelium of *Morchella* sp.

In one study a Tolhurst basket centrifuge operated at 1179 rpm for 10 minutes yielded a 2-inch-thick *Morchella* mycelial cake containing 74% moisture. The pellets are ruptured, but this is not a problem when the mycelium is to be dried for food use into powdered form. Dewatering with a Byrd-Young vacuum filter was unsuccessful because the pellets did not cling to the filter and the spaces between the pellets were large enough to make vacuum operation ineffective (Robinson and Davidson, 1959). Inclined screens effectively separate mycelial pellets from the spent medium and aid in washing the mycelium.

In studies with corn and pea canning wastes, the mycelium was collected on vibratory 120-mesh screen. A rotary vacuum filter was used successfully in collecting *G. deliquescens* mycelium after growth on corn wet milling wastes with solid contents ranging up to 22% (Church *et al.*, 1973).

In the Pekilo process for producing *Paecilomyces* sp. on sulfite waste liquor, the mycelium is collected and washed on a drum filter and the spent liquor passes to the evaporator at the pulp mill. The mycelium is mechanically dewatered to 35–45% dry weight content (Romantschuk, 1975).

Aspergillus niger grown on carob bean extract is in a filamentous rather than a pellet form. Rotary vacuum filters can be used to separate the mycelium from the medium without a precoat. In contrast to *Morchella* mycelium, the porosity of the mycelium permits washing in the filter without resuspending it, and the cake can be scraped from the rotary drum and disrupted mechanically (Imrie and Vlitos, 1975).

Fungal mycelium can be dried in steam-tube rotary, tray, or belt driers. Mushroom mycelium for food use must be washed thoroughly to remove excess growth medium which may contribute undesirable flavor to the dried product. Drying temperatures must be controlled carefully to avoid damaging heat-sensitive flavor components and darkening the product excessively.

III. PRODUCT QUALITY AND SAFETY

Three aspects of SCP product quality will be considered here: (1) nutritional value of SCP products for use in human food or animal feeds,

(2) safety of SCP products for human or animal feeding, and (3) production of functional protein concentrates and isolates free from nucleic acids and toxic factors. Although the primary use of SCP products is for their protein content, cells of microorganisms also contain carbohydrates, lipids, vitamins, and minerals that contribute to nutritional value. On the other hand, microbial cells also contain nonprotein nitrogen materials, such as nucleic acids, that have harmful effects when consumed by humans or nonruminant animals in quantities above certain limits. This problem is discussed subsequently.

A. Nutritional Value

There is considerable information available on the composition of microbial cells, including protein, amino acid, vitamin, and mineral contents (Waslien, 1975). Table VI summarizes proximate analysis of selected microorganisms of interest in SCP production. Crude protein values are based on nitrogen values multiplied by the factor 6.25. Protein content computed in this manner is used for formulating protein products into feeds. However, it should be emphasized that these values do not reflect the true protein content of microorganisms since nitrogen or nonprotein nitrogenous substances, including nucleic acids, is erroneously included as protein.

Protein and lipid contents of microorganisms reflect the composition of the medium and growth conditions. Yeasts, molds, and higher fungi have higher cellular lipid contents and lower nitrogen (and protein contents) when grown in media having high amounts of available carbon as energy source and a deficiency of nitrogen. The importance of C : N ratio in the medium for producing various microorganisms has been discussed previously.

Ash contents of SCP products also vary depending upon the mineral content of the growth medium. Also, procedures used in product separation, including washing, can result in large variations in ash contents of a given organism. Table VII presents amino acid contents of selected microorganisms of interest in SCP production as compared with the Food and Agricultural Organization (FAO) reference values. In general, only a few microorganisms, particularly the bacteria, have amino acid profiles that compare favorably with the FAO values as regards methionine content. With yeasts, in addition to methionine deficiencies, lysine/arginine ratios are wider than desirable and must be adjusted in poultry rations (Shacklady and Gatumel, 1973). Also, in algae, glycine and methionine deficiencies have been observed in chick feeding studies (Leveille *et al.*, 1962).

However, actual performance of SCP products as determined in feeding studies is the most important measure of nutritional value. For human food applications, protein digestibility and protein efficiency ratio (PER), biological value (BV), or net protein utilization (NPU) determined in rats are used as measures of the utility of an SCP product. For animal feed applications metabolizable energy, protein digestibility, and feed conversion ratio (weight of ration consumer/weight gain) are usual measures of performance in broiler chickens, swine, and calves (Shacklady and Gatumel, 1973; Smith and Palmer, 1976), while egg production is used for laying hens (Shannon *et al.*, 1976; Waldroup and Hazen, 1975).

Table VIII shows typical values obtained from animal feeding studies of SCP products. Supplementation with methionine is necessary to obtain a PER or BV equivalent to animal proteins, such as casein, for most microorganisms. (Shacklady and Gatumel, 1973; Waldroup and Flynn, 1975). This amino acid, or its hydroxy analog, is available commercially at low enough cost to permit supplementation of feedstuffs.

The results of feeding studies with broiler chickens, using *Methylophilus methylotrophus* (Gow *et al.*, 1975), *Pseudomonas* sp. grown on methanol (Waldroup and Payne, 1974), or *Aspergillus niger* grown on carob bean extract (Imrie and Vlitos, 1975), do not show significant differences between experimental and control diets at the 10% level of SCP in the diet. Similar results were also observed with broiler chickens in feeding studies with the yeasts *C. lipolytica* (Shacklady and Gatumel, 1973) and *C. tropicalis* (Waldroup *et al.*, 1971) up to the 15% level in the diet. In contrast, an SCP product grown on methanol gave greater and more efficient gains in broiler chickens at the 15% level in pelleted diets as compared with soybean meal (White and Balloun, 1977). In swine feeding studies with yeast grown on *n*-alkanes feed conversion ratios tend to decrease as percentages of SCP in the diet increase above about 10% (Litchfield, 1977b). For bacteria grown on methanol, efficiencies of feed conversion were equivalent to those of whitefish meal at levels up to 20% of the diet of pigs (Whittemore *et al.*, 1976). When yeast SCP grown on hydrocarbons is used as a milk replacer for calves and steers, the limit appears to be approximately 7.5% in the ration (Shacklady and Gatumel, 1973).

The older literature on animal feeding studies with molds and higher fungi has been summarized previously (Litchfield, 1968). More recent studies, summarized in Table VIII, confirm earlier observations that supplementation with DL-methionine is necessary to obtain satisfactory performance with molds and higher fungi. In the United States, the Food and Drug Administration regulations allow the use of the dried yeasts *S. cerevisiae, K. (Saccharomyces) fragilis,* and *C. utilis* in foods provided

TABLE VI. Proximate Composition of Selected Microorganisms of Interest in SCP Production

Organism	Substrate	Composition (gm/100 gm dry wt)							References
		Nitrogen	Protein	Fat	Total CHO	Crude fiber	Ash	Energy (kcal/gm)	
Algae									
Chlorella sorokiniana	CO_2	9.6	60	8	22	3	9	5.2	Lubitz (1963)
Chorella regularis S-50	CO_2	9.3	58	16	—	4.4	6.7	—	Endo and Shirota (1972)
Spirulina maxima, synthetic medium	CO_2	10	62	3	—	—	—	—	Clement *et al.* (1967)
Spirulina maxima, sewage	CO_2	8.5	53	4.8	28	—	15	—	Kosaric *et al.* (1974); Nguyen *et al.* (1974)
Bacteria and actinomycetes									
Acinetobacter (Micrococcus) cerificans	Hexadecane	11	72	—	—	—	—	—	Guenther and Perkins (1968)
Cellulomonas sp.	Bagasse	14	87	8	—	—	7	—	Han *et al.* (1971)
Methalomonas clara	Methanol	12–13	80–85	8–10	—	—	8–12	—	Faust *et al.* (1977)
Methylophilus (Pseudomonas) methylotrophus	Methanol	13	83	7	—	<0.05	8.6	3.0	Gow *et al.* (1975)
Thermomonospora fusca	Pulping fines	4.8–5.6	30–35	—	—	—	—	—	Crawford *et al.* (1973)
Yeasts									
Candida lipolytica (Toprina)	*n*-Alkanes	10	65	8.1	—	—	6	—	Shacklady (1972)
Candida lipolytica (Toprina)	Gas oil	11	69	1.5	—	—	8	—	Shacklady (1972)
Candida utilis	Ethanol	8.3	52	7	—	5	8	—	Amoco Foods Co. (1974)
Candida utilis	Sulfite	9	55	5	—	—	8	—	Peppler (1968)
Hansenula polymorpha	Methanol	—	50	—	—	—	—	—	Cooney and Levine (1975)
Kluyveromyces (Saccharomyces) fragilis	Cheese whey	9	54	1	—	—	9	—	Peppler (1968)
Saccharomyces cerevisiae	Molasses	8.4	53	6.3	—	—	7.3	—	Peppler (1968)
Trichosporon cutaneum	Oxanone wastes	8.6	54	8	31	—	7	11	Wiken (1972)

Molds and higher fungi									
Agaricus campestris white variety	Glucose	—	36	3	49	6.9	4.5	—	Humfeld and Sugihara (1949)
Agaricus campestris, brown variety	Glucose	—	45	—	—	—	5.2	—	Humfeld and Sugihara (1949)
Aspergillus niger	Molasses	7.7	50	—	—	—	—	—	Anderson *et al.* (1975)
Fusarium graminearum	Starch	8.7	54	—	—	—	—	—	Anderson *et al.* (1975)
Morchella crassipes	Glucose	5.0	31	3.1	—	—	—	—	Litchfield *et al.* (1963b)
Morchella crassipes	Sulfite waste liquor	4.1	26	4.4	39	—	5.9	—	Kosaric *et al.* (1973)
Morchella esculenta	Glucose	5.0	31	1.9	—	—	—	—	Litchfield *et al.* (1963b)
Morchella hortensis	Glucose	5.4	34	1.4	—	—	—	—	Litchfield *et al.* (1963b)
Paecilomyces varioti (Pekilo)	Sulfite waste liquor	8.8	55	1.3	25	7	6	—	Romantschuk and Lehtomaki (1978)
Trichoderma viride	Starch	10.2	64	—	—	—	—	—	Anderson *et al.* (1975)

TABLE VII. Amino Acid Content of Selected Organisms of Interest for SCP Production[a, b]

Organism	Substrate	Ala	Arg	Asp	Cys	Glu	Gly	His	Ile	Leu	Lys	Met	Phe	Pro	Ser	Thr	Try	Tyr	Val	References
Algae																				
Chlorella sorokiniana	CO_2	5.9	5.6	5.9	—	9.3	4.8	1.4	3.4	4.0	7.8	1.8	2.7	4.0	2.2	3.2	1.4	2.7	5.1	Lubitz (1963)
Chlorella regularis S-50 Autotrophic	CO_2	7.3	5.8	8.8	0.7	11.8	5.4	1.8	4.2	8.1	7.7	1.3	5.1	4.3	3.0	3.6	1.5	2.6	5.9	Endo and Shirota (1972)
S-50 heterotrophic	Glucose	7.4	10.2	7.7	0.9	9.9	4.9	3.0	3.4	7.0	9.4	1.8	3.2	2.2	3.4	4.0	1.4	3.0	5.1	Endo and Shirota (1972)
Spirulina maxima	CO_2	6.5	6.6	8.6	0.4	13.6	4.6	1.7	5.8	7.8	4.8	1.5	4.6	3.8	4.3	4.6	1.3	3.9	6.3	Clement *et al.* (1967)
Bacteria and Actinomycetes																				
Acinetobacter (Micrococcus) cerificans	*n*-Alkanes	6.7	4.7	8.0	—	10.7	5.0	1.7	4.3	6.5	5.2	1.8	3.6	3.1	2.8	4.1	2.8	1.3	5.4	Ertola *et al.* (1971)
Cellulomonas–Alcaligenes	Bagasse	—	6.5	—	—	—	—	7.8	5.4	7.4	7.6	2.0	4.7	—	—	5.5	—	—	7.1	Han *et al.* (1971)
Methylococcus capsulatus	Methane	—	6.2	—	0.6	—	5.1	2.2	4.3	8.1	5.7	2.7	4.6	—	—	4.6	—	3.8	6.5	Hamer *et al.* (1975)
Methylophilus (Pseudomonas) methylotrophus	Methanol	6.8	4.5	8.5	0.6	9.6	4.9	1.8	4.3	6.8	5.9	2.4	3.4	3.0	3.4	4.6	0.9	3.1	5.2	Gow *et al.* (1975)
Thermomonospora fusca	Cellulose pulping fines	13.9	5.6	6.7	0.4	18.0	4.4	2.0	3.2	6.1	3.6	2.0	2.6	6.1	2.6	4.0	—	1.9	13.0	Crawford *et al.* (1973)
Yeasts																				
Candida lipolytica (Toprina)	*n*-Alkanes	7.4	4.8	10.2	1.1	11.3	4.8	2.0	4.5	7.0	7.0	1.8	4.4	4.4	4.8	4.9	1.4	3.5	5.4	Evans (1968)
Candida lipolytica	Gas oil	5.8	5.0	10.0	0.9	12.1	4.5	2.1	5.3	7.8	7.8	1.6	4.8	3.7	5.1	5.4	1.3	4.0	5.8	Evans (1968)
Candida utilis	Sulfite waste	5.8	5.4	9.2	—	15.6	3.6	1.2	3.8	7.6	4.8	1.1	8.6	6.0	5.0	5.4	2.4	6.2	3.8	Peppler (1965)
Candida utilis	Ethanol	5.5	5.4	8.8	0.4	14.6	4.5	2.1	4.5	7.1	6.6	1.4	4.1	3.4	4.7	5.5	1.2	3.3	5.7	Amoco Foods Co. (1974)
Hansenula polymorpha ATCC 26012	Methanol	6.1	5.6	10.5	0.7	14.4	5.2	2.4	5.1	8.3	8.1	1.5	5.0	—	—	5.2	—	4.8	6.2	Levin and Cooney (1973)
Kluyveromyces (Saccharomyces) fragilis	Cheese whey	—	—	—	—	—	—	2.1	4.0	6.1	6.9	1.9	2.8	—	—	5.8	1.4	2.4	5.4	Bernstein and Everson (1973)

Rhodotorula glutinis	Domestic sewage	5.1	4.5	6.0	—	9.3	3.5	2.2	3.3	5.4	7.1	—	3.2	3.0	3.3	3.1	—	3.1	3.6	Fustier and Simard (1976)
Saccharomyces cerevisiae	Molasses	—	5.0	—	1.6	—	—	4.0	5.5	7.9	8.2	2.5	4.5	—	—	4.8	1.2	5.0	5.5	Reed and Peppler (1973)
Fungi																				
Aspergillus niger	Carob bean extract	—	—	—	1.1	—	—	—	4.2	5.7	5.9	2.6	3.8	—	—	5.0	2.1	3.2	5.2	Imrie and Vlitos (1975)
Fusarium graminearum	Starch	6.6	6.8	10.4	1.2	11.0	6.5	3.3	4.0	6.0	6.9	1.8	4.0	4.8	5.3	4.6	—	3.8	6.6	Anderson *et al.* (1975)
Gliocladium deliquescens	Waste starch	6.1	4.8	7.9	1.0	8.6	4.7	2.3	3.8	6.2	5.3	1.3	3.0	4.2	3.9	4.3	1.4	3.6	4.7	Smith *et al.* (1975)
Morchella crassipes	Glucose	6.2	3.0	4.8	0.4	14.1	3.1	2.0	2.9	5.6	3.5	1.0	1.9	5.1	3.1	3.0	1.5	1.7	3.0	Litchfield *et al.* (1963b)
	Sulfite waste liquor (NH_3)	5.3	3.9	5.9	0.6	8.3	3.8	2.0	2.9	6.1	5.6	1.2	5.8	4.0	4.0	4.1	—	—	4.1	LeDuy *et al.* (1974)
Morchella esculenta	Glucose	4.8	8.0	5.0	0.3	14.8	2.9	2.1	2.7	5.1	3.8	0.9	2.5	4.2	3.1	3.0	0.9	1.7	3.4	Litchfield *et al.* (1963b)
Morchella hortensis	Glucose	4.5	4.0	4.6	0.4	15.4	3.0	1.9	2.4	5.0	3.0	0.7	2.3	4.5	2.8	2.7	1.0	1.9	2.9	Litchfield *et al.* (1963b)
Paecilomyces varioti	Spent sulfite liquor	—	—	—	1.1	—	—	—	4.3	6.9	6.4	1.5	3.7	—	—	4.6	1.2	3.4	5.1	Romantschuk (1975)
Trichoderma harzianum	Coffee-processing wastes	—	—	—	—	—	—	—	4.0	5.8	4.7	1.1	3.5	—	—	3.8	—	—	4.9	Aguirre *et al.* (1976)
Trichoderma viride	Barley straw	4.8	4.0	7.8	1.5	8.0	4.5	1.9	3.5	5.8	4.4	1.4	8.7	3.8	4.2	4.9	—	3.3	4.4	Peitersen (1975a)
Trichosporon cutaneum	Oxanone wastewater	—	—	—	1.0	—	—	—	5.2	7.1	7.3	1.6	4.5	—	—	5.5	1.4	4.2	6.1	Wiken (1972)
FAO reference		—	—	—	—	—	—	—	4.2	4.8	4.2	2.2	2.8	—	—	—	1.4	—	4.2	Food and Agricultural Organization (1957)

[a] Amino acid content, gm/16 gm N.

[b] Key to amino acid abbreviations:

Ala, alanine; Arg, arginine; Asp, aspartic acid; Cys, cystine; Glu, glutamic acid; Gly, glycine; His, histidine; Ile, isoleucine; Leu, leucine; Lys, lysine; Met, methionine; Phe, phenylanine; Pro, proline; Ser, serine; Thr, threonine; Try, tryptophan; Tyr, tyrosine; Val, valine.

TABLE VIII. Performance of Selected Single-Cell Protein Products in Animal Feeding Studies

Single-cell protein and substrate	Treatment and percent in diet	Animal	Performance: Protein digestibility (%)	Performance: Protein efficiency ratio (PER)	Performance: Biological value (BV)	Feed conversion ratio (kg/kg weight gain)	References
Algae							
Chlorella sorokiniana	(1) Dried, 10	Rat	86	2.19	—	—	Lubitz (1963)
	(2) Dried + 0.2 L-methionine; 10	Rat	86	2.90	—	—	—
Spirulina maxima	Dried, 10	Rat	84	2.3–2.6	72	—	Bourges *et al.* (1972); Durand-Chastel and Clement (1972)
Bacteria							
Acinetobacter (Micrococcus) cerificans (*n*-hexadecane)	Dried	Rat	83.4	—	67	—	Ertola *et al.* (1969)
Bacillus megaterium (collagen waste)	Dried	Rat	—	1.88[a]	—	—	Bough *et al.* (1972)
Methylophilus methylotrophus (methanol)	(1) Dried; 9.8	Broiler chicken	— / —	— / —	— / —	2.30 (2.33)[b]	Gow *et al.* (1975)
	(2) Dried; 6.7	Pig	—	—	—	3.13 (3.34)[b]	
Pseudomonas sp. (methanol)	Dried; 10	Broiler chicken	— / —	— / —	— / —	1.74 (1.64)[b]	Waldroup and Payne (1974)

Yeasts							
Candida lipolytica (*n*-alkanes)	(1) Dried	Rat	96	—	61	—	Shacklady and Gatumel (1973)
	(2) Dried + 0.3 DL-methionine	Rat	96	—	91	—	
	(3) Dried; 10	Broiler chicken	88	—	—	2.58	
			—	—	—	(2.68)[a]	
	(4) Dried; 7.5	Pig	92	—	—	3.04 (3.11)	
C. lipolytica (gas and oil)	(1) Dried	Rat	94	—	54	—	Shacklady and Gatumel (1973)
	(2) Dried + 0.3 DL-methionine	Rat	95	—	96	—	
		—	—	—	—	—	
Candida tropicalis (*n*-alkanes)	Dried; 15	Broiler chicken	—	—	—	2.03	Waldroup *et al.* (1971)
			—	—	—	(2.06)[b]	
Candida utilis (sulfite waste liquor)	(1) Dried	Rat	85–88	0.9–1.4	32–48	—	Bressani (1968)
	Dried + 0.5	—	—	—	—	—	
	(2) DL-methionine	Rat	90	2.0–2.3	88	—	
Kluyveromyers fragilis (cheese whey)	Dried	Rat	—	—	—	1.5	Booth *et al.* (1962)
Saccharomyces cerevisiae (molasses)	Dried	Rat	80–90	1.7	58–69	—	Bressani (1968)
	—	—	—	—	—	—	

(*continued*)

TABLE VII *(Continued)*

Single-cell protein and substrate	Treatment and percent in diet	Animal	Performance: Protein digestibility (%)	Performance: Protein efficiency ratio (PER)	Performance: Biological value (BV)	Feed conversion ratio (kg/kg weight gain)	References
Fungi							
Aspergillus niger (carob bean extract)	Dried; 20	Broiler chicken	—	2.50	—	2.02	Imrie and Vlitos (1975)
			—	—	—	(2.12)[a]	
Fusarium graminearum	(1) Dried; 10.5	Rat	—	1.89	—	—	Duthie (1975)
	(2) Dried; 40	Pig	—	—	—	1.76 (2.04)[a]	
Fusarium moniliforme (carob bean extract)	(1) Dried	Rat	—	1.15	—	—	Drouliscos *et al.* (1976)
	(2) Dried + 6.3 DL-methionine	Rat	—	2.38	—	—	
			—	—	—	—	
Gliocladium deliquescens (waste starch)	Dried	Rat	—	—	49	—	Smith *et al.* (1975)
	—	—	—	—	—	—	
Trichoderma viride (waste starch)	Dried	Rat	—	—	48	—	Smith *et al.* (1975)
	—	—	—	—	—	—	

[a] Adjusted to casein PER of 2.5.
[b] Control ration value.

that the folic acid content of the yeast does not exceed 0.04 mg/gm (0.008 mg pteroyl glutamic acid/gm) (Code of Federal Regulations, 1978). Use of dried yeast in foods has been limited to flavor enhancement and B vitamin supplementation. High levels of *C. utilis* in foods have led to poor acceptance (Klapka *et al.*, 1958).

Bressani (1968) discussed earlier studies on supplementation of foods with yeasts and on their value in human nutrition. More recently Young and Scrimshaw (1975) point out that yeast and algae have low protein digestibility values that can be improved considerably if these SCP products are processed suitably. However, nucleic contents are a limiting factor in the extent to which SCP products can be used in human foods. The Protein Advisory Group (PAG) of the United Nations (1970a,b,c, 1971) has developed guidelines for preclinical testing of novel protein sources and supplementary foods which should be followed in any development program on SCP products for human food use.

B. Safety

There are several key problem areas in using SCP products for human foods. These include the following: (1) high nucleic acid contents of many microorganisms that would result in kidney stone formation or gout if consumed as a significant portion of the protein intake; (2) poor digestibility and possible adverse gastrointestinal and skin reactions; and (3) possible presence of toxic or carcinogenic compounds from residues of substrates, from biosynthesis by the organism, or from chemical reactions during processing and drying.

Calloway (1974) pointed out that adverse responses may be observed in human subjects fed algae, bacteria, yeast, or fungal SCP products even though favorable results were obtained previously in animal studies. Published values of nucleic acid contents on a dry weight basis of algae are in the range of 4–6% (Waslien *et al.*, 1970; Durand-Chastel and Clement, 1972), bacteria, up to 16% (Senez, 1973; Gow *et al.*, 1975); yeasts, 6–11% (Trevelyan, 1975; Waslien *et al.*, 1970); and molds, 2.5–6% (Smith *et al.*, 1975; Trevelyan, 1975). Scrimshaw (1975) states that a human intake of 2 gm of yeast nucleic acid per day would be within safe limits. However, there would be a significant risk of kidney stone formation or gout at intakes greater than 3 gm/day.

Gastrointestinal disturbances, including nausea and vomiting, have been encountered in human feeding studies with algae (Powell *et al.*, 1961; Dam *et al.*, 1965; Lee *et al.*, 1967); with bacteria, including *A. eutrophus (H. eutropha)* and *Klebsiella (Aerobacter aerogenes)* (Waslien *et al.*, 1970); and with yeast grown on ethanol (Scrimshaw, 1973). Also, skin lesions were observed in human volunteers who consumed *C. utilis* grown on sulfite waste but not with glucose-grown cells (Scrimshaw, 1975). In bacteria the toxic materials were either intracellular or bound to the cell, since washed cells were used in feeding studies (Calloway, 1974).

It has been mentioned previously that strains of some molds, particularly *S. fumigatus* and *F. graminearum,* may produce substances that are toxic to humans and animals.

From the standpoint of animal feeding, high nucleic acid contents in SCP are not a problem with ruminants and are better tolerated by nonruminant livestock than by humans. De Groot (1974) discusses minimal evaluations for establishing the safety of SCP. Taylor *et al.* (1974) present criteria for the evaluation of SCP for feeding nonruminants.

Questions have arisen on possible hazards from the presence of traces of residual *n*-alkanes, odd-carbon fatty acids, or polycyclic aromatic hydrocarbons in yeast SCP grown on hydrocarbon substrates. The results of extensive studies have shown that the polycyclic hydrocarbon contents of *Candida* sp. grown on hydrocarbons were no greater than those in other commercial yeast products or other representative foods. Also no carcinogenic, mutagenic, or teratogenic effects were noted in appropriately designed studies (Engel, 1973).

The PAG Ad Hoc Working Group on Single-Cell Proteins concluded that low levels of residual *n*-alkanes and presence of odd-carbon fatty acids do not constitute a hazard (Protein Advisory Group, 1976a,b). The PAG has developed guidelines for production of SCP for human consumption and on the nutritional and safety aspects of novel proteins for animal feeding (Protein Advisory Group, 1972, 1974).

C. Functional SCP Products

As a result of high nucleic acid contents and possible presence of toxic factors in microbial cells, there is an alternative of processing the cells to remove cell walls, toxic factors, and nucleic acids. To be useful as functional food additives, the resulting SCP concentrates or isolates should have desirable physical–chemical functional properties, including water and fat binding, emulsion stability, dispersibility, gel formation, whippability, and thickening.

Both chemical and enzymatic processes can be used to reduce nucleic acid contents of SCP to acceptable levels of 2–3% or less. Chemical processes include acid precipitation at pH 5 (Hedenskog and Mogren, 1973) and acid or alkaline hydrolysis (Daly and Ruiz, 1975; Newell *et al.*, 1975). Enzyme processes are based on heat shock and incubation or on incubation alone to degrade nucleic acids by endonucleases within the cells (Sinskey and Tannenbaum, 1975; Robbins, 1976) or by autolysis (Trevelyan, 1976).

Methods for disrupting microbial cells include mechanical means, such as ball mills or homogenization at elevated pressures, autolysis, or acid, alkaline, or enzyme hydrolysis (Litchfield, 1977a). Concentrates and isolates can be prepared by precipitation from acid or alkaline solutions (Kalina and Nicholas, 1974; Mogren *et al.*, 1974; Newell *et al.*, 1975; Robbins, 1976). A preliminary attempt to improve the functional characteristics of *C. utilis* protein concentrates by succinylation was not successful (McElwain *et al.*, 1975).

Both bacterial protein and *C. utilis* protein can be spun into fibers (Heden *et al.*, 1971; Huang and Rha, 1971). Fibers can be formed by extrusion of the disrupted cell extract into an acid coagulating medium such as acetate buffer, pH 4.5 (Akin, 1973, 1974; Ridgeway, 1974; Litchfield, 1977a). These fibers would then be used for forming textured protein-type products.

Bakers' yeast protein concentrate is now approved by the Food and Drug Administration for use as a functional food additive (Code of Federal Regulations, 1978). In addition, it is reasonable to expect that other yeast concentrates or isolates will be approved in the future when their safety and utility are demonstrated to the satisfaction of regulatory authorities. Extensive evaluations will be required to establish the safety of concentrates or isolates from algal, bacterial, or fungal cells where there is no past history of safe use for human consumption.

IV. ECONOMIC AND ENERGY CONSIDERATIONS

Economic and energy considerations in SCP production are intimately linked together. Single-cell protein production processes are capital and energy intensive. Marketing considerations are also important aspects of the overall economics of SCP production.

A. Manufacturing Costs

Manufacturing costs include costs of the raw materials (carbon and energy source, nitrogen source, mineral nutrients, and other chemicals) materials and supplies, utilities, labor including fringe benefits, maintenance, taxes and insurance, working capital and interest, depreciation, and general administration. Several cost comparisons of various SCP processes have been published (Litchfield, 1977b,c; Moo-Young, 1977).

Manufacturing costs of SCP are very dependent upon the cost of the carbon and energy source to the producer of SCP products. Various estimates for the costs of hydrocarbon substrates, as a percentage of total manufacturing cost, range from 13 to 57.5% depending upon the process and the year of the estimate (Litchfield, 1977b). SCP production costs from methanol show a much greater dependency upon increases in substrate costs than in the case of *n*-paraffins (Dimmling and Seipenbusch, 1978).

Rising costs of petroleum and products derived from this raw material have made the economics of SCP processes based on hydrocarbons tenuous. This has already been reflected in a lack of announcements of new production facilities, suspension of operations on existing facilities, and abandonment of announced plans for new facilities (Anonymous, 1978a,b).

For waste-type materials, raw materials costs are a lower percentage of total costs than for hydrocarbons, methanol, or ethanol. Moo-Young (1977) cites a range of 17–26% of total costs for sulfite waste liquor and bagasse.

Waste materials must be generated in large quantities in close proximity to the SCP production plant over most of the year to be economically attractive. Otherwise, the costs associated with collecting and transporting wastes may be excessive (Burch *et al.*, 1963).

Next to materials costs, utilities are the second most important element of production costs for most SCP processes. Utilities required include water, electricity, and steam. Ratledge (1975) estimated the following water requirements in millions of liters for 100,000 metric tons/year for plants of different processes: yeasts, *n*-alkanes, 18.2; bacteria, methanol, 45.5; and bacteria, methane, 18.2. Energy in the form of electricity and steam is used for sterilization, aeration, cooling, product separation, and drying. For example, electric power requirements for producing yeasts from *n*-alkanes on a 100,000-metric ton/year scale range from 500 to 3620 kWh/metric ton of SCP; and steam, from 4600 to 10,600 kg/metric ton of SCP (Litchfield, 1977b). Energy requirements for cooling are considerable for hydrocarbon- and methanol-based processes. British Petroleum estimates a cooling requirement of 100×10^6 kcal/hour for producing yeast from purified *n*-alkanes (BP Proteins, Ltd., 1974). The requirements for carbohydrate-based processes would be significantly lower as discussed previously in the sections on various microorganisms used in SCP production.

It is difficult to compare manufacturing cost estimates given in the literature for various processes because differing cost bases are used for raw materials, utilities, labor, maintenance, and depreciation. In some instances, cost estimates do not include working capital requirements or realistic interest charges.

Location is an all-important factor in determining manufacturing costs for SCP products. Raw materials availability and costs, labor, and utilities costs vary widely in different geographical regions. Also proximity to markets is an important factor in plant location.

Lewis (1976) considered both indirect and direct energy inputs for SCP processes, including energy used for carbohydrate substrate cultivation and transportation, for production and construction of capital equipment, and for production and processing of SCP and treatment of wastes. Typical gross energy requirements in megajoules (MJ) per kilogram of protein for selected organisms and substrates were: (1) *Candida* sp.; methanol, ethanol, or gas oil, 190; *n*-paraffins, 185; distilling industry wastes, 155; molasses, 75; (2) Pseudomonads; methane, 165; methanol or ethanol, 145; and (3) *Aspergillus niger*: solid agricultural waste, 80; agricultural process effluent, 30.

He concluded that processes based on hydrocarbons compared very unfavorably to those based on carbohydrates. The energy requirements for SCP production compared with those for equivalent protein production by conventional agriculture indicate limited applicability of SCP processes to less developed countries.

B. Capital Costs

Costs for constructing SCP production facilities that require aseptic operation, complex product separation, and cleaning, as is the case of hydrocarbons and methanol, have increased markedly. Even SCP plants based on carbohydrates now require clean operations to meet the requirements of regulatory agencies for food- or feed-grade materials; and costs for stainless steel equipment, suitable centrifuges, etc., to meet these requirements are also increasing.

Estimates of fixed capital costs in 1975 for 100,000-metric tons/year SCP plants (U.S. Gulf Coast) were $90.5 million and $66.0 million for processes based on *n*-alkanes and methanol, respectively (Brownstein and Constantinides, 1975). For a 50,000–75,000 metric tons/year SCP plant based on methanol, ICI estimated a 1976 cost of $70 million (Anonymous, 1976b). Future plants in advanced countries will be even more costly because of rapid increases in construction costs.

C. Market Considerations

At the present time, food-grade yeasts *(S. cerevisiae, C. utilis,* and *K. fragilis)* have a position in food ingredient markets for their flavor enhancement characteristics in foods. Soybean concentrates (70–72% protein) and isolates (90–95% protein) are the major competitive products to food-grade SCP products for either functional or nutritional uses at the present. Most SCP products and soybean products have lower methionine contents than are desirable in human nutrition and would require supplementation to make a balanced source of protein for human nutrition.

The SCP isolates and concentrates developed to date do not have the wide range of functional properties exhibited by soy protein concentrates and isolates. However, some yeast SCP products, particularly *C. utilis,* do have interesting functional effects in baked and pasta products. Any new SCP products will have to compete on cost and functional equivalent bases with soy protein products and perhaps additional protein concentrates or isolates from plant protein sources, such as cottonseed and peanuts, or animal proteins such as fish, milk, or meat.

Food industry and consumer acceptance of SCP products in foods will be an important factor in their future success. Stone and Sidel (1974) describe a consumer survey on SCP acceptance.

Algae have a bitter lingering flavor and impart a dark undesirable color to foods (Cook *et al.*, 1963). Further processing, such as solvent extraction, or formulating into other foods may minimize flavor problems, but the color is difficult to mask. Also, further processing is necessary to improve digestibility of algae.

There is a lack of published information on human sensory studies with bacterial SCP, although bland colorless powders can be made. The primary focus of bacterial SCP has been toward feedstuff applications.

The reports of adverse gastrointestinal reactions in human feeding studies with bacteria cited previously will limit further development of food-grade bacterial SCP products. A process for removing endotoxins from bacterial SCP has been described (Dasinger and Nasland, 1972) but its effectiveness has not been verified in published toxicological studies.

The flavor characteristics of food-grade yeasts are well known. Dried food yeasts and yeast autolysates have been used for many years (Reed and Peppler, 1973).

Dried mycelia of certain higher fungi, the mushrooms, have pleasant flavors. These products can be used in soups, sauces, or gravy formulations as replacements for dried cultured mushrooms. The high costs of producing dried morel mushroom mycelium made this product uncompetitive with imported dried mushrooms in the United States.

Similar considerations apply to SCP products in feedstuff markets. Soybean meal and fish meal are the major competitive feedstuffs. It has been pointed out previously that wide fluctuations in the prices of these feedstuffs may make feed-grade SCP products noncompetitive for substantial periods in Western Europe and Japan because of the high fixed capital and operating costs of SCP plants (Lipinsky and Litchfield, 1974; Litchfield, 1977c).

The producer of formulated feedstuffs uses price, acceptability, and nutritional performance as factors in deciding which protein ingredients to use. Least cost linear programming is now used in the feedstuff industry to determine the lowest cost protein that will give the desired performance. Feed-grade SCP products will have to compete with other feed proteins on this basis.

V. CONCLUSIONS

Processes are available for producing a wide range of microorganisms as protein sources for food or feed applications. In general, most algae have serious shortcomings for either food or feed use, including poor digestibility and sensory characteristics, high costs of harvesting, and ease of contamination with pathogenic organisms. *Spirulina* species are claimed to have more desirable characteristics than many algae species, but their requirements for alkaline conditions for growth limit production to a few geographical regions, such as Mexico and Central Africa.

Bacteria and actinomycetes have many desirable characteristics for SCP production, including high growth rates and protein contents and favorable amino acid profiles, as compared with other microorganisms. However, the possible presence of toxic substances in bacterial cells limits human food applications. Capital and energy costs for bacterial SCP production are also high. Although substantial experimental animal feeding studies have been conducted on bacterial SCP produced from methanol, the potential place of this product in feedstuff markets will not be established until after the one plant now under construction comes into production.

Certain species of yeasts have had established positions in food and feed markets. Increasingly higher costs of hydrocarbons make processes based on these raw materials appear unattractive in the future. Carbohydrates and wastes that contain carbohydrates are the lowest cost substrates for yeast SCP production. Future growth of markets for food and feed yeast and for yeast protein concentrates and isolates will depend upon their economic competitiveness with alternative proteins from plant or animal sources.

Molds and higher fungi have relatively slow growth rates and are easily contaminated by other microorganisms unless aseptic conditions are maintained during growth. However, they are easily harvested by filtration or screening, which is an advantage over other microorganisms. Use of molds or higher fungi as food or feed will be governed by the same economic, market, nutritional, and safety factors that effect the acceptability of other SCP products.

REFERENCES

Ad Hoc Panel (1975). "Underexploited Tropical Plants with Promising Economic Value." Nat. Acad. Sci., Washington, D.C.

Aguirre, F., Maldonado, O., Rolz, P., Menchu, J. F., Espinosa, R., and deCabrera, S. (1976). *ChemTech* **6**(10) 636–642.

Akiba, T., Kajiyama, S., and Fukimbara, T. (1973). *J. Ferment. Technol.* **51,** 343–347.

Akin, C. (1973). U.S. Patent 3,781,264.

Akin, C. (1974). U.S. Patent 3,833,552.

Akin, C., Witter, L. D., and Ordal, Z. J. (1967). *Appl. Microbiol.* **15,** 1339–1344.

Amoco Foods Co. (1974). "Torutein Product Bulletin." Chicago, Illinois.

Anderson, C., Longton, J., Maddix, C., and Scammel, G. W. (1975). *In* "Single Cell Protein II" (S. R. Tannenbaum and D. I. C. Wang, eds.), pp. 314–329. MIT Press, Cambridge, Massachusetts.

Anderson, R., Weisbaum, R. B., and Robe, K. (1974). *Food Process.* **35**(7), 58–59.

Anonymous (1970). *Chem. Eng. (N.Y.)* **77**(14), 30.

Anonymous (1973a). *Eur. Chem. News* **23** (573), 14.

Anonymous (1973b). *Eur. Chem. News* **24** (591), 12.

Anonymous (1974a). *Eur. Chem. News* **25** (622), 18.

Anonymous (1974b). *Chem. & Eng. News* **52** (33), 30.

Anonymous (1975). *Chem. Eng. (N.Y.)* **82**(6), 36–37.

Anonymous (1976a). *Eur. Chem. News* **28** (720), 6.

Anonymous (1976b). *Chem. Ind. (London)* **20,** 859.

Anonymous (1977a). *Process Biochem.* **12**(1), 30.

Anonymous (1977b). *Food Eng.* **49**(6), 95–97.

Anonymous (1978a). *Chem. Eng. News* **56**(12), 10.

Anonymous (1978b). *Chem. Eng. News* **56**(38), 12–13.

Ballerini, D., Parlouar, D., Lapeyronnie, M., and Sri, K. (1977). *Eur. J. Appl. Microbiol.* **4,** 11–19.

Becker, E. W., and Venkataraman, L. V. (1978). "A Manual on the Cultivation and Processing of Algae as a Source of Single Cell Protein." Central Food Technological Research Institute, Mysure, India.

Bellamy, W. D. (1974). *Biotechnol. Bioeng.* **16,** 869–880.

Bellamy, W. D. (1975). *In* "Single Cell Protein II" (S. K. Tannenbaum and D. I. C. Wang, eds.), pp. 263–372. MIT Press, Cambridge, Massachusetts.

Bennett, I. C., Hondermarck, J. D., and Todd, J. R. (1969). *Hydrocarbon Process.* **48**(3), 104–108.

Bernstein, S., and Everson, T. C. (1973). *Proc. Natl. Symp. Food Process. Wastes, 4th, 1973* EPA 660/2–73–031.

Bernstein, S., Tzeng, C. H., and Sisson, D. (1975). Abstract BMPC 68, *Chem. Congr. North Am. Continent, 1st, 1975,* pp. 151–166.

Block, S. S. (1960), *J. Biochem. Microbiol. Technol. Eng.*, **2**, 243–252.
Block, S. S., Stearns, T. W., Stephens, R. L., and McCandless, R. F. J. (1953). *J. Agric. Food Chem.* **1,** 890–893.
Bongers, L. (1969). *Dev. Ind. Microbiol.* **11,** 241–255.
Bongers, L. (1970). *J. Bacteriol.* **104,** 145–151.
Booth, A. N., Robinson, D. J., and Wasserman, A. E. (1962). *J. Dairy Sci.* **45,** 1106–1107.
Bough, W. A., Brown, W. L., Porsche, J. D., and Doty, D. M. (1972). *Appl. Microbiol.* **24,** 226–235.
Bourges, H., Sotomayor, A., and Mendoza, E. (1972). *Abstr. Int. Congr. Nutr., 9th, 1972* p. 85.
BP Proteins, Ltd. (1974). "The Toprina Cycle." London, England.
Bressani, R. (1968). *In* "Single Cell Protein" (R. I. Mateles and S. R. Tannenbaum, eds.), pp. 90–121. MIT Press, Cambridge, Massachusetts.
Brown, B. S., Jones, J. C., and Hulse, J. M. (1975). *Process Biochem.* **10,** 3–7.
Brownstein, A. M., and Constantinides, A. (1975). *169th Natl. Meet., Am. Chem. Soc.*, Paper No. 86.
Burch, J. E., Lipinsky, E. S., and Litchfield, J. H. (1963). *Food Technol.* **17**(10), 54–60.
Busta, F. F., Schmidt, B. E., and McKay, L. L. (1977). U.S. Patent 4,018,650.
Calloway, D. H. (1974). *In* "Single Cell Protein" (P. Davis, ed.), pp. 129–146. Academic Press, New York.
Camhi, J. D., and Rogers, P. L. (1976). *J. Ferment. Technol.* **54,** 437–449 and 450–458.
Casey, R. P., Lubitz, J. A., Benoit, R. J., Weissman, B. J., and Chau, H. (1963). *Food Technol.* **17,** 1039–1043.
Champagnat, A., and Filosa, J. A. (1965). U.S. Patent 3,193,390.
Champagnat, A., and Filosa, J. A. (1971). U.S. Patent 3,560,341.
Champagnat, A., Vernet C., Laine, B., and Filosa, J. (1963). *Nature (London)* **197,** 13–14.
Chang, W. T. M., and Thayer, D. W. (1974). *Dev. Ind. Microbiol.* **16,** 456–464.
Church, B. D., Nash, H., and Brosz, W. (1972). *Dev. Ind. Microbiol.* **13,** 30–46.
Church, B. D., Erickson, E. E., and Widmer, C. M. (1973). *Food Technol.* **27**(2), 36–42.
Clement, G. (1968). *In* "Single-Cell Protein" (R. I. Mateles and S. R. Tannenbaum, eds.), pp. 306–308. MIT Press, Cambridge, Massachusetts.
Clement, G. (1975). *In* "Single-Cell Protein II" (S. R. Tannenbaum and D. I. C. Wang, eds.), pp. 467–474. MIT Press, Cambridge, Massachusetts.
Clement, G., Giddey, C., and Menzi, R. (1967). *J. Sci. Food Agric.* **18,** 497–501.
Code of Federal Regulations (1978). "Title 21, Food and Drugs; 172.325, 172.896, Bakers Yeast Protein; Dried Yeasts." U.S. Govt. Printing Office, Washington, D.C.
Cook, B. B., Lau, E. W., and Bailey, B. M. (1963). *J. Nutr.* **81,** 23–29.
Cooney, C. L., and Levine, D. W. (1972). *Adv. Appl. Microbiol.* **15,** 337.
Cooney, C. L., and Levine, D. W. (1975). *In* "Single-Cell Protein II" (S. R. Tannenbaum and D. I. C. Wang, eds.), pp. 402–423. MIT Press, Cambridge, Massachusetts.
Cooney, C. L., Wang, D. I. C., and Mateles, R. I. (1969). *Biotechnol. Bioeng.* **11,** 269–281.
Cooney, C. L., Levine, D. W., and Snedecor, B. (1975). *Food Technol.* **29**(2), 33–42.
Cooper, P. G., Silver, R. S., and Boyle, J. P. (1975). *In* "Single-Cell Protein II" (S. R. Tannenbaum and D. I. C. Wang, eds.), pp. 454–466. MIT Press, Cambridge, Massachusetts.
Cooper, R. C. (1962). *Am. J. Public Health* **52,** 252–257.
Crawford, D. L., McCoy, E., Harkin, J. M., and Jones, P. (1973). *Biotechnol. Bioeng.* **13,** 833–843.
Cremieux, A., Chevalier, J., Combet, M., Dumenil, G., Parlouar, D., and Ballerini, D. (1977). *Eur. J. Appl. Microbiol.* **4,** 1–9.

Daly, W. H., and Ruiz, L. P., Jr. (1975). "Fabrication of Single Cell Protein from Cellulosic Wastes, EPA" Rep. No. EPA-6702–75–032.
Dam, R., Lee, S., Fry, P. C., and Fox, H. (1965). *J. Nutr.* **86,** 376–382.
Darlington, W. A. (1964). *Biotechnol. Bioeng.* **6,** 241–242.
Dasinger, B. L., and Nasland, L. A. (1972). U.S. Patent 3,644,175.
Davis, P., ed. (1974). "Single-Cell Protein." Academic Press, New York.
deCabrera, S., Mayorga, H., Espinosa, R., and Rolz, C. (1976). *Proc. Int. Congr. Food Sci. Technol., 4th, 1974* Vol. 4, pp. 296–301.
Decerle, C., Franckowiak, S., and Gatellier, C. (1969). *Hydrocarbon Process.* **48**(3), 109–112.
DeGroot, A. P. (1974). *In* "Single-Cell Protein" (P. Davis, ed.), pp. 75–92. Academic Press, New York.
Delaney, R. A. M., Kennedy, R., and Walley, B. D. (1975). *J. Sci. Food Agric.* **26,** 1177–1186.
Dimmling, W., and Seipenbusch, R., *Process Biochem.* **13**(3), 9–12, 14, 15, 34.
Dostalek, M., and Molin, N. (1975). *In* "Single-Cell Protein II" (S. R. Tannenbaum and D. I. C. Wang, eds.), pp. 385–401. MIT Press, Cambridge, Massachusetts.
Dostalek, M., Haggström, L., and Molin, N. (1972). *Ferment. Technol. Today, Proc. Int. Ferment. Symp., 4th, 1972,* pp. 497–501.
Drouliscos, N. J., Macris, B. J., and Kokke, R. (1976). *Appl. Environ. Microbiol.* **31,** 691–694.
Dunlap, C. E. (1975a). *In* "Single-Cell Protein II" (S. R. Tannenbaum and D. I. C. Wang, eds.), pp. 244–262. MIT Press, Cambridge, Massachusetts.
Dunlap, C. E. (1975b). *Food Technol.* **29**(12), 62–67.
Durand-Chastel, H., and Clement, G. (1972). *Int. Congr. Nutr., 9th, 1972* Paper C-3-6.
Duthie, I. F. (1975). *In* "Single-Cell Protein II" (S. R. Tannenbaum and D. I. C. Wang, eds.), pp. 505–544. MIT Press, Cambridge, Massachusetts.
Edwards, V. H., Kinsella, J. E., and Sholiton, D. B. (1972). *Biotechnol. Bioeng.* **14,** 123–147.
Endo, H., and Shirota, M. (1972). *Ferment. Technol. Today, Proc. Int. Ferment. Symp., 4th, 1972* pp. 533–541.
Enebo, L. (1969). *Chem. Eng. Prog., Symp. Ser.* **65**(93), 60–65.
Engel, C. (1973). *In* "Proteins from Hydrocarbons" (H. Gounelle de Pontanel, ed.), pp. 53–81. Academic Press, New York.
Eriksson, K. E., and Larsson, K. (1975). *Biotechnol. Bioeng.* **17,** 327–348.
Ertola, R. J., Mazza, L. A., Balatti, A. P., and Sanahuja, J. (1969), *Biotechnol. Bioeng.* **11,** 409–416.
Ertola, R. J., Segovia, R. F., Gambaruto, M., Monosiglio, J. C., and Artuso, C. (1971). *Dev. Ind. Microbiol.* **12,** 72–76.
Espinosa, R., Maldonado, O., Menchu, J. F., and Rolz, C. (1975). *Chem. Congr. North Am. Continent, 1st, 1975* Abstract BMPC 65.
Evans, G. H. (1968). *In* "Single-Cell Protein" (R. I. Mateles and S. R. Tannenbaum, eds.), pp. 243–254. MIT Press, Cambridge, Massachusetts.
Evans, G. H., and Shennan, J. L. (1974). U.S. Patent 3,846,238.
Faust, U., Präve, P., and Sukatsch, D. A. (1977). *J. Ferment. Technol* **55,** 609–614.
Food and Agricultural Organization (1957). *FAO Nutr. Stud.* No. 16.
Foster, J. F., and Litchfield, J. H. (1964). *Biotechnol. Bioeng.* **6,** 441–456.
Foster, J. F., and Litchfield, J. H. (1967). *NASA Spec. Publ. SP*–134, 201–212.
Foster, J. F., and Litchfield, J. H. (1968a). *NASA Spec. Publ.* **165,** 67–74.
Foster, J. F., and Litchfield, J. H. (1968b). *SAE Trans.* 2639–2646.
Foster, J. F., and Litchfield, J. H. (1969). *NASA Contract. Rep.* **NASA CR-1296.**

Frey, C. N. (1930). *Ind. Eng. Chem.* **22,** 1154–1162.
Fu, T. T., and Thayer, D. W. (1975). *Biotechnol. Bioeng.* **17,** 1749–1960.
Fustier, P., and Simard, R. E. (1976). *Can. Inst. Food Sci. Technol.* **9,** 182–185.
Giacobbe, F., Puglisi, P., and Longbardi, G. (1975). *169th Natl. Meet., Am. Chem. Soc.* Paper No. 74.
Goldberg, I., Rock, J. S., Ben-Bassat, A., and Mateles, R. I. (1976). *Biotechnol. Bioeng.* **18,** 1657–1668.
Golueke, C. G., Jr., and Oswald, W. J. (1970). *J. Water Pollut. Control Fed.* **42,** R304.
Goto, S., Okamoto, R., Kuwajima, J., and Takemetsu, A. (1976). *J. Ferment. Technol.* **54,** 213–224.
Gow, J. S., Littlehailes, J. D., Smith, S. R. L., and Walter, R. B. (1975). *In* "Single Cell Protein II" (S. R. Tannenbaum and D. I. C. Wang, eds.), pp. 370–384. MIT Press, Cambridge, Massachusetts.
Gray, W. D. (1970). *Crit. Rev. Food Technol.* **1,** 225–229.
Gregory, K. F., Reade, A. E., Khor, G. L., Alexander, J. C., Lumsden, F. H., and Losos, G. (1976). *Food Technol.* **30**(3), 30–35.
Guenther, K. R. (1965). *Biotechnol. Bioeng.* **7,** 445–446.
Guenther, K. R., and Perkins, M. B. (1968). U.S. Patent 3,384,491.
Gutcho, S. (1973). "Proteins from Hydrocarbons." Noyes Data Corp., Park Ridge, New Jersey.
Hamer, G., Heden, C. G., and Carenberg, C. D. (1967). *Biotechnol. Bioeng.* **9,** 499–514.
Hamer, G., Harrison, D. E. F., Harwood, J. H., and Topiwala, H. H. (1975). *In* "Single-Cell Protein II" (S. R. Tannenbaum and D. I. C. Wang, eds.), pp. 357–369. MIT Press, Cambridge, Massachusetts.
Han, Y. W., Dunlap, C. E., and Callihan, C. D. (1971). *Food Technol.* **25,** 130–133 and 154.
Harris, E. E., Hajny, G. J., and Johnson, M. C. (1951). *Ind. Eng. Chem.* **43,** 1593–1596.
Harrison, D. E. F. (1976). *ChemTech,* **6,** 570–574.
Harrison, D. E. F., Topiwala, H. H., and Hamer, G. (1972). *Ferment. Technol. Today, Proc. Int. Ferment. Symp., 4th, 1972* pp. 491–495.
Harwood, J. H., and Pirt, S. J. (1972). *J. Appl. Bacteriol.* **35,** 597–607.
Hatch, R. T. (1975). *In* "Single-Cell Protein II" (S. R. Tannenbaum and D. I. C. Wang, eds.), pp. 46–68. MIT Press, Cambridge, Massachusetts.
Heden, C. G., Molin, N., Olsson, U., and Rupprecht, A. (1971). *Biotechnol. Bioeng.* **13,** 147–150.
Hedenskog, G., and Mogren, H. (1973). *Biotechnol. Bioeng.* **15,** 129–142.
Heinemann, B. (1963). U.S. Patent 3,086,320.
Hernandez, E., and Johnson, M. J. (1967). *J. Bacteriol.* **94,** 991–1001.
Holderby, J. M., and Moggio, W. A. (1960). *J. Water Pollut. Control Fed.* **32,** 171–181.
Holve, W. A. (1976). *Process Biochem.* **11**(10), 2–7.
Huang, F., and Rha, C. K. (1971). *J. Food Sci.* **36,** 1131–1134.
Humfeld, H., and Sugihara, T. F. (1949). *Food Technol.* **3,** 355–356.
Humphrey, A. E. (1974). *Chem. Eng. (N.Y.)* **81**(26), 98–112.
Imada, O., Hoshiai, K., and Tanaka, M. (1972). U.S. Patent 3,635,796.
Imrie, F. K. E., and Vlitos, A. J. (1975). *In* "Single-Cell Protein II" (S. R. Tannenbaum and D. I. C. Wang, eds.), pp. 223–243. MIT Press, Cambridge, Massachusetts.
Irgens, R. L., and Clarke, J. D. (1976). *Eur. J. Appl. Microbiol.* **2,** 231–241.
Johnson, M. J. (1967). *Science* **155,** 1515–1519.
Kalina, V., and Nicholas, P. (1974). U.S. Patent 3,821,080.
Kanazawa, M. (1975). *In* "Single-Cell Protein II" (S. R. Tannenbaum and D. I. C. Wang, eds.), p. 438–453. MIT Press, Cambridge, Massachusetts.

Kanazawa, M., Maruyama, T., and Katoh, S. (1975). *169th Natl. Meet., Am. Chem. Soc.* Paper No. 85.
Kato, N., Tsuji, K., Tani, Y., and Ogata, K. (1974). *J. Ferment. Technol.* **52,** 917–920.
Khor, G. L., Alexander, J. C., Santos-Nuñez, J., Reade, A. E., and Gregory, K. F. (1976). *Can. Inst. Food Sci. Technol. J.* **9**(3), 139–143.
Kihlberg, R. (1972). *Annu. Rev. Microbiol.* **26,** 427–466.
Klapka, M. R., Duby, G. A., and Pavcek, P. L. (1958). *J. Am. Diet. Assoc.* **34,** 1317–1320.
Klis, J. (1963). *Food Process.* **24**(9), 99.
Knecht, R., Prave, P., Seipenbusch, R., and Sukatsch, D. A. (1977). *Process Biochem.* **12**(4), 11–14.
Ko, P. C., and Yu, Y. (1968). *In* "Single-Cell Protein" (R. I. Mateles and S. R. Tannenbaum, eds.), pp. 255–262. MIT Press, Cambridge, Massachusetts.
Ko, P. C., Yu, Y., and Li, C. S. (1964). "Protein from Petroleum by Fermentation Process." Chinese Solvent Works, Chinese Petroleum Corp., Taiwan.
Kosaric, N., and Zajic, J. E. (1974). *Adv. Biochem. Eng.* **3,** 89–125.
Kosaric, N., LeDu, A., and Zajic, J. E. (1973). *Can. J. Chem. Eng.* **51,** 186–190.
Kosaric, N., Nguyen, H. T., and Bergougnou, M. A. (1974). *Biotechnol. Bioeng.* **16,** 881–896.
Kosikowski, F. (1977). "Cheese and Fermented Milk Foods," 2nd ed. Edwards, Ann Arbor, Michigan.
Labuza, T. P. (1975). *In* "Single-Cell Protein II" (S. R. Tannenbaum and D. I. C. Wang, eds.), pp. 69–104. MIT Press, Cambridge, Massachusetts.
Lachance, P. A. (1968). *In* "Single-Cell Protein" (R. I. Mateles and S. R. Tannenbaum, eds.), pp. 122–152. MIT Press, Cambridge, Massachusetts.
Laine, B. M. (1972). *Can. J. Chem. Eng.* **50,** 154–156.
Laine, B. M., and du Chaffaut, J. (1975). *In* "Single-Cell Protein II" (S. R. Tannenbaum and D. I. C. Wang, eds.), pp. 424–437. MIT Press, Cambridge, Massachusetts.
Lambert, E. B. (1938). *Bot. Rev.* **4,** 397–426.
Lampert, L. M. (1965). "Modern Dairy Products," Chapter 15, p. 20. Chem. Publ. Co., New York.
Lane, A. G. (1977). *J. Appl. Chem. Biotechnol.* **27,** 165–169.
Laskin, A. I. (1975). Abstract BMPC39. *Chem. Congr. North Am. Continent, 1st, 1975*
LeDuy, A., Kosaric, N., and Zajic, J. E. (1974). *Can. Inst. Food Sci. Technol. J.* **7,** 44–49.
Lee, S. K., Fox, H. M., Kies, C., and Dam, R. (1967). *J. Nutr.* **92,** 281–285.
LeFrançois, L., Marillen, C. G., and Mejane, T. V. (1955). French Patent 1,102,200.
Leveille, G. A., Sauberlich, H. E., and Shockley, J. W. (1962). *J. Nutr.* **76,** 423–428.
Levine, D. W., and Cooney, C. L. (1973), *Appl. Microbiol.* **26,** 982–990.
Lewis, C. W. (1976). *J. Appl. Chem. Biotechnol.* **26,** 568–575.
Lipinsky, E. S., and Litchfield, J. H. (1970). *Crit. Rev. Food Technol.* **1,** 580–618.
Lipinsky, E. S., and Litchfield, J. H. (1974). *Food Technol.* **28**(5), 16–22 and 40.
Litchfield, J. H. (1964). *In* "Global Impacts of Applied Microbiology" (M. P. Starr, ed.), pp. 327–337. Wiley, New York.
Litchfield, J. H. (1967a). *In* "Microbial Technology" (H. J. Peppler, ed.), pp. 107–144. Van Nostrand-Reinhold, New Jersey.
Litchfield, J. H. (1967b). *Food Technol.* **21**(2), 55–57.
Litchfield, J. H. (1967c). *Biotechnol. Bioeng.* **9,** 289–304.
Litchfield, J. H. (1967d). *Proc. Int. Congr. Nutr. 7th, 1966* Vol. 4, pp. 1068–1074.
Litchfield, J. H. (1968). *In* "Single-Cell Protein" (R. I. Mateles and S. R. Tannenbaum, eds.), pp. 309–329. MIT Press, Cambridge, Massachusetts.
Litchfield, J. H. (1972). *Dev. Ind. Microbiol.* **13,** 317–331.

Litchfield, J. H. (1974). *Chem. Process. (London)* **20**(9), 11–18.
Litchfield, J. H. (1977a). *Food Technol.* **31**(5), 175–179.
Litchfield, J. H. (1977b). *Biotechnol. Bioeng. Symp.* **7,** 77–90.
Litchfield, J. H. (1977c). *Adv. Appl. Microbiol.* **22,** 267–305.
Litchfield, J. H. (1978). *Chem Tech* **8,** 218–223.
Litchfield, J. H., and Overbeck, R. C. (1966). *Food Sci. Technol. Proc. Int. Congr. 1st, 1962,* Vol. 2, pp. 511–520.
Litchfield, J. H., Overbeck, R. C., and Davidson, R. S. (1963a). *J. Agric. Food Chem.* **11,** 158–162.
Litchfield, J. H., Vely, V. G., and Overbeck, R. C. (1963b). *J. Food Sci.* **28,** 741–743.
Lubitz, J. A. (1963). *J. Food Sci.* **28,** 229–232.
McElwain, M. D., Richardson, T., and Amundson, C. H. (1975), *J. Milk Food Technol.* **28,** 521–526.
MacLennan, D. G., Gow, J. S., and Stringer, D. A. (1973). *Process Biochem.* **8**(6), 22–24.
MacLennan, D. G., Ousby, J. C., Owen, T. R., and Steer, D. C. (1976). U.S. Patent 3,989,594.
Mandels, M., and Sternberg, D. (1976). *J. Ferment. Technol.* **54,** 267–286.
Masuda, Y., and Yoshikawa, K. (1973). *Expert Work. Group Meet. Manuf. Proteins Hydrocarbons, U. N. Ind. Dev. Organ.* Paper ID/WG 164/20.
Masuda, Y., Kato, K., Takayama, Y., Kida, K., and Nakanishi, M. (1975). U.S. Patent 3,868,305.
Mateles, R. I. (1971). *Biotechnol. Bioeng.* **13,** 581–582.
Mateles, R. I., and Tannenbaum, S. R., eds. (1968). "Single-Cell Protein." MIT Press, Cambridge, Massachusetts.
Mateles, R. I., Goldberg, I., and Battat, E. (1976). U.S. Patent 3,989,595.
Matthern, R. O., and Koch, R. B. (1964). *Food Technol.* **18**(5), 58–62 and 64–65.
Matthern, R. O., Kostick, J. A., and Okada, I. (1969). *Biotechnol. Bioeng.* **11,** 863–874.
Mennett, R. H., and Nakayama, T. O. M. (1971). *Appl. Microbiol.* **22,** 772–776.
Metz, B., and Kossen, N. W. F. (1977). *Biotechnol. Bioeng.* **19,** 781–799.
Mian, F. A., Ajdery, A., and Fazeli, A. (1976). *J. Ferment. Technol.* **54,** 76–81.
Mogren, H. L., Hedenskog, G. O., and Enebo, L. E. (1974). U.S. Patent 3,848,812.
Moo-Young, M. (1976). *Process Biochem.* **11**(10), 32–34.
Moo-Young, M. (1977). *Process Biochem.* **12**(4), 6–10.
Moo-Young, M., Chahal, D. S., Swan, J. E., and Robinson, C. W. (1977). *Biotechnol. Bioeng.* **19,** 527–538.
Moulin, G., Galzy, P. and Joux, J. L. (1976). *Food Sci. Technol., Proc. Int. Congr., 4th, 1974* Vol. 3, pp. 47–53.
Moustafa, A. M. (1960). *Appl. Microbiol.* **8,** 667.
Nagai, I. (1973). *Expert Work. Group Meet. Manuf. Proteins Hydrocarbons, U. N. Ind. Dev. Organ* Paper ID/WG 164/28.
Nakayama, O., Ueno, T., and Tsuchiya, F. (1974). *J. Ferment. Technol.* **52,** 225–232.
Newell, J. A., Robbins, E. A., and Seeley, R. D. (1975). U.S. Patent 3,867,555.
Nguyen, H. T., Kosaric, N., and Bergougnou, M. A. (1974). *Can. Inst. Food Sci. Technol. J.* **7,** 114–116.
Ogawa, T., Kozasa, H., and Terui, G. (1972). *J. Ferment. Technol.* **50,** 143–149.
Olson, A. J. C. (1961). *SCI Monogr.* **12,** 81–93.
Organizing Committee (1975). "Microbial Growth on C-1 Compounds." Microbiol., Washington, D.C.
Orgel, G., Pietrusza, E. W., and Joris, G. G. (1971). U.S. Patent 3,622,465.
Oswald, W. J. (1969). *Chem. Eng. Prog., Symp. Ser.* **65**(93), 87–92.

Oswald, W. J., and Golueke, C. G. (1968). *In* "Single-Cell Proteins" (R. I. Mateles and S. R. Tannenbaum, eds.), pp. 271–305. MIT Press, Cambridge, Massachusetts.
Pederson, C. S. (1971). Microbiology of Food Fermentations." Avi Publ. Co., Westport, Connecticut.
Peitersen, N. (1975a). *Biotechnol. Bioeng.* **17,** 361–374.
Peitersen, N. (1975b). *Biotechnol. Bioeng.* **17,** 1291–1299.
Peitersen, N. (1977). *Biotechnol. Bioeng.* **19,** 337–348.
Peppler, H. J. (1965). *J. Agric. Food Chem.* **13,** 34–36.
Peppler, H. J. (1968). *In* "Single-Cell Protein" (R. I. Mateles and S. R. Tannenbaum, eds.), pp. 229–242. MIT Press, Cambridge, Massachusetts.
Perkins, M. B., and Furlong, L. E. (1967). U.S. Patent 3,355,296.
Powell, R. C., Nevels, E. M., and McDowell, M. E. (1961). *J. Nutr.* **75,** 7–12.
Prescott, S. C., and Dunn, C. G. (1959). "Industrial Microbiology," 3rd ed., Chapter 3. McGraw-Hill, New York.
Prokop, A., and Sobotka, M. (1975). *In* "Single-Cell Protein II" (S. R. Tannenbaum and D. I. C. Wang, eds.), pp. 127–157. MIT Press, Cambridge, Massachusetts.
Protein Advisory Group (1970a). "PAG Statement No. 4 on Single Cell Protein." FAO/WHO/UNICEF, United Nations, New York.
Protein Advisory Group (1970b). "PAG Guideline No. 6 for Pre-Clinical Testing of Novel Sources of Protein." FAO/WHO/UNICEF, United Nations, New York.
Protein Advisory Group (1970c). "PAG Guideline No. 7 for Human Testing of Supplementary Food Mixtures." FAO/WHO/UNICEF, United Nations, New York.
Protein Advisory Group (1971). "PAG Guideline No. 8 on Protein-Rich Mixtures for Use as Weaning Foods." FAO/WHO/UNICEF, United Nations, New York.
Protein Advisory Group (1972). "PAG Guideline No. 12 on the Production of Single Cell Protein for Human Consumption." FAO/WHO/UNICEF, United Nations, New York.
Protein Advisory Group (1974). "PAG Guideline No. 15 on Nutritional and Safety Aspects of Novel Protein Sources for Animal Feeding." FAO/WHO/UNICEF, United Nations, New York.
Protein Advisory Group (1976a). *PAG Bull.* **6**(2), 1–6.
Protein Advisory Group (1976b). *PAG Bull.* **6**(2), 6–9.
Ratledge, C. (1975). *Chem. Ind. (London)* **21,** 918–920.
Reed, G., and Peppler, H. J. (1973). "Yeast Technology." Avi Publ. Co., Westport, Connecticut.
Reuss, M., Gniesser, J., Reng, H. G., and Wagner, F. (1975). *Eur. J. Appl. Microbiol.* **1,** 295–305.
Reusser, F., Spencer, J. F. T., and Sallans, H. R. (1958a). *Appl. Microbiol.* **6,** 1–4.
Reusser, F., Spencer, J. F. T., and Sallans, H. R. (1958b). *Appl. Microbiol.* **6,** 5–8.
Richardson, B., Orcutt, D. M., Schwertner, H. A., Martinez, C. L., and Wickline, H. E. (1969). *Appl. Microbiol.* **18,** 245–250.
Ridgeway, J. A., Jr. (1974). U.S. Patent 3,843,807.
Ridgeway, J. A., Jr., Lappin, T. A., Benjamin, B. M., Corns, J. B., and Akin, C. (1975). U.S. Patent 3,865,691.
Robbins, E. A. (1976). U.S. Patent 3,991,215.
Robinson, R. F., and Davidson, R. S. (1959). *Adv. Appl. Microbiol.* **1,** 261–278.
Rockwell, P. J. (1976). "Single Cell Proteins from Cellulose and Hydrocarbons, 1976." Noyes Data Corp., Park Ridge, New Jersey.
Rogers, C. J., Coleman, E., Spino, D. F., and Purcell, T. C. (1972). *Environ. Sci. Tecnol.* **6,** 715–718.
Rolz, C. (1975). *In* "Single-Cell Protein II" (S. R. Tannenbaum and D. I. C. Wang, eds.), pp. 273–313. MIT Press, Cambridge, Massachusetts.

Romantschuk, H. (1975). *In* "Single Cell Protein II' (S. R. Tannenbaum and D. I. C. Wang, eds.), pp. 344–355. MIT Press, Cambridge, Massachusetts.
Romantschuk, H., and Lehtomäki, M. (1978). *Process Biochem.* **13**(3), 16, 17, 23.
Rosenzweig, M., and Ushio, S. (1974). *Chem. Eng. (N.Y.)* **81**(1), 62–63.
Rychtera, M., Barta, J., Feichter, A., and Einselle, A. A. (1977). *Process Biochem.* **12**(2), 26–30.
Scrimshaw, N. S. (1968). *In* "Single-Cell Protein" (R. I. Mateles and S. R. Tannenbaum, eds.), pp. 3–7. MIT Press, Cambridge, Massachusetts.
Scrimshaw, N. S. (1973). *In* "Proteins from Hydrocarbons" (H. Gounelle de Pontanel, ed.), pp. 189–213. Academic Press, New York.
Scrimshaw, N. S. (1975). *In* "Single-Cell Protein II" (S. R. Tannenbaum and D. I. C. Wang, eds.), pp. 24–45. MIT Press, Cambridge, Massachusetts.
Seelig, R. A. (1959). "Cultivated Mushrooms," 3rd ed. United Fresh Fruit and Vegetable Association, Washington, D. C.
Senez, J. (1973). *In* "Proteins from Hydrocarbons" (H. Gounelle de Pontanel, ed.), pp. 3–26. Academic Press, New York.
Shacklady, C. A. (1972). *World Rev. Nutr. Diet.* **14,** 154–179.
Shacklady, C. A., and Gatumel, E. (1973). *In* "Proteins from Hydrocarbons." (H. Gounelle de Pontanel, ed.), pp. 27–52. Academic Press, New York.
Shannon, D. W. F., McNab, J. M., and Anderson, G. B. (1976). *J. Sci. Food Agric.* **27,** 471–476.
Shannon, L. J., and Stevenson, K. E. (1975a). *J. Food Sci.* **40,** 826–826.
Shannon, L. J., and Stevenson, K. E. (1975b). *J. Food Sci.* **40,** 830–832.
Sheehan, B. T., and Johnson, M. J. (1971). *Appl. Microbiol.* **21,** 511–515.
Sherwood, R. F., and Peberdy, J. F. (1974). *J. Sci. Food Agric.* **25,** 1081–1087.
Shipman, R. H., Kao, I. C., and Fan, L. T. (1975). *Biotechnol. Bioeng.* **17,** 1561–1570.
Shuler, R. L., and Affens, W. A. (1970). *Appl. Microbiol.* **19,** 76–86.
Silver, R. S., and Cooper, P. G. (1972a). *72nd Natl. Meet., AIChE, 1972* Paper 16d.
Silver, R. S., and Cooper, P. G. (1972b). *Pap., Div. Pet. Chem. 164th Natl. Meet., Am. Chem. Soc.* Paper No. 31.
Sinskey, A. J., and Tannenbaum, S. R. (1975). *In* "Single-Cell Protein II" (S. R. Tannenbaum and D. I. C. Wang, eds.), pp. 158–178. MIT Press, Cambridge, Massachusetts.
Smith, R. H., and Palmer, R. (1976). *J. Sci. Food Agric.* **27,** 763–770.
Smith, R. H., Palmer, R., and Reade, A. E. (1975). *J. Sci. Food Agric.* **26,** 785–795.
Solomons, G. L., and Scammell, G. W. (1976). U.S. Patent 3,937,654.
Sonoda, Y., Someya, J., Futai, N., Tagaya, N., and Murakami, T. (1973). *J. Ferment. Technol.* **51,** 479–483.
Spencer-Martins, I., and Van Uden, N. (1977). *Eur. J. Appl. Microbiol.* **4,** 29–35.
Srinivasan, V. R., Fleenor, M. B., and Summers, R. J. (1977). *Biotechnol. Bioeng.* **19,** 153–154.
Stoller, B. B. (1954). *Econ. Bot.* **8,** 48–95.
Stone, H., and Sidel, J. (1974). *In* "Single-Cell Protein" (P. Davis, ed.), pp. 147–160. Academic Press, New York.
Sugihara, T. F., and Humfeld, H. (1954). *Appl. Microbiol.* **2,** 170–172.
Sugimori, T., Oyama, Y., and Omichi, T. (1971). *J. Ferment. Technol.* **49,** 435–446.
Sugimoto, H. (1974). *J. Food Sci.* **39,** 934–938.
Sugimoto, M., Yokoo, S., and Imada, O. (1972). *Ferment. Technol. Today, Proc. Int. Ferment. Symp., 4th, 1972* pp. 503–507.
Takata, T. (1969). *Hydrocarbon Process.* **48**(3), 99–103.
Tanaka, K., Ohshima, K., and Yamamoto, M. (1972). U.S. Patent 3,639,210.

Tannenbaum, S. R., and Wang, D. I. C., eds. (1975). "Single-Cell Protein II." MIT Press, Cambridge, Massachusetts.
Taylor, J. C., Lucas, E. W., Gable, D. A., and Graber, G. (1974). *In* "Single Cell Protein" (P. Davis, ed.), pp. 179–186. Academic Press, New York.
Teramoto, S., Taguchi, H., Veda, R., and Yoshida, T. V. (1966). *Proc. Int. Congr. Food Sci. Technol. 2nd, 1966* Paper A 3.9.
Thanh, N. C., and Wu, J. S. (1975). *Can. Inst. Food Sci. Technol. J.* **8,** 202–205.
Thayer, D. W., Yang, S. P., Key, A. B., Hang, H. H., and Barker, J. W. (1975). *Dev. Ind. Microbiol.* **16,** 465–474.
Tong, P. Q., Riel, R. R., and Simard, R. E. (1973a). *Can. Inst. Food Sci. Technol. J.* **6,** 212–215.
Tong, P. Q., Simard, R. E., and Riel, R. R. (1973b). *Can. Inst. Food Sci. Technol. J.* **6,** 239–243.
Towersey, P. J., Longton, J., and Cockram, G. N. (1976). U.S. Patent 3,937,693.
Towersey, P. J., Longton, J., and Cockram, G. N. (1977). U.S. Patent 4,041,089.
Trevelyan, W. E. (1975). *J. Sci. Food Agric.* **26,** 1673–1680.
Trevelyan, W. E. (1976). *J. Sci. Food Agric.* **27,** 753–762.
Tung, T. (1975). *Food Eng.* **47**(4), 24.
Ueno, K., Asai, Y., Shimada, M., and Goto, S. (1974a). *J. Ferment. Technol.* **52,** 861–866.
Ueno, K., Asai, Y., Shimada, M., and Kametani, T. (1974b). *J. Ferment. Technol.* **52,** 867–872.
Ueno, K., Asai, Y., Yonemura, H., and Kametani, T. (1974c). *J. Ferment. Technol.* **52,** 873–877.
Vananuvat, P., and Kinsella, J. E. (1975a). *J. Food Sci.* **40,** 336–342.
Vananuvat, P., and Kinsella, J. E. (1975b). *J. Food Sci.* **40,** 823–825.
Van Dijken, J. P., and Harder, W. (1975). *Biotechnol. Bioeng.* **17,** 15–30.
Van Eybergen, G. C., and Scheffers, W. A. (1972). *Antonie van Leeuwenhoek* **38,** 448–450.
Vela, G. R., and Guerra, C. N. (1966). *J. Gen. Microbiol.* **42,** 123.
Volesky, B., and Zajic, J. E. (1971). *Appl. Microbiol.* **21,** 614–622.
Wagner, F., Kleeman, T., and Zahn, W. (1969). *Biotechnol. Bioeng.* **11,** 393–408.
Waldroup, P. W., and Flynn, N. W. (1975). *Poult. Sci.* **54,** 1129–1133.
Waldroup, P. W., and Hazen, K. K. (1975). *Poult. Sci.* **54,** 635, 637.
Waldroup, P. W., and Payne, J. R. (1974). *Poult. Sci.* **53,** 1039–1042.
Waldroup, P. W., Hillard, C. M., and Mitchell, R. J. (1971). *Poult. Sci.* **50,** 1022–1029.
Wang, D. I. C. (1968a). *Chem. Eng. (N.Y.)* **75**(18), 99.
Wang, D. I. C. (1968b). *In* "Single-Cell Protein" (R. I. Mateles and S. R. Tannenbaum, eds.), pp. 217–228. MIT Press, Cambridge, Massachusetts.
Wang, D. I. C. (1969). *Chem. Eng. Prog., Symp.* Ser. **65**(93), 66.
Wang, H. Y., Mou, D.-G., and Swartz, J. R. (1976). *Biotechnol. Bioeng.* **18,** 1811–1814.
Ward, C. H., Moyer, J. E., and Vela, G. R. (1965). *Dev. Ind. Microbiol.* **6,** 213–222.
Waslien, C. I. (1975). *Crit. Rev. Food Sci. Nutr.* **6,** 77–151.
Waslien, C. I., Calloway, D. H., Margen, S., and Costa, F. (1970). *J. Food Sci.* **35,** 294–298.
Wasserman, A. E., Hampson, J. W., and Alvare, N. F. (1961). *J. Water Pollut. Control Fed.* **33,** 1090–1094.
Weiner, B. A. (1977a). *Eur. J. Appl. Microbiol.* **4,** 51–57.
Weiner, B. A. (1977b). *Eur. J. Appl. Microbiol.* **4,** 59–65.
Wells, J. (1975). *169th Natl. Meet., Am. Chem. Soc.* Paper No. 72.
White, J. (1954). "Yeast Technology," Chapter 2, p. 23. Wiley, New York.
White, W. B., and Balloun, S. L. (1977). *Poultry Sci.* **56,** 266–273.
Whittemore, D. J., Moffat, I. W., and Taylor, A. B. (1976). *J. Sci. Food Agric.* **27,** 1163–1170.

Wiken, T. O. (1972). *Ferment. Technol. Today, Proc. Int. Ferment. Symp., 4th, 1972* pp. 569–596.

Wilkinson, T. G., Topiwala, H. H., and Hamer, G. (1974). *Biotechnol. Bioeng.* **16,** 41–59.

Wilson, B. J. (1971). *Microb. Toxins* **6,** 207–295.

Wolnak, B., Andreen, B., and Wen, C. V. (1972). U.S. Patent 3,649,459.

Yokote, Y., Suginoto, M., and Abes, S. (1974). *J. Ferment. Technol.* **52,** 201–209.

Yoshikawa, J. Katsur, T., Fukita, Y., Wada, H., and Tanigawa, Y. (1975). U.S. Patent 3,901,762.

Young, R. J. (1973). *Expert Work. Group Meet. Manuf. Proteins Hydrocarbons, U. N. Ind. Dev. Organ.* Paper ID/WG 164/9.

Young, V. R., and Scrimshaw, N. S. (1975). *In* "Single-Cell Protein II" (S. R. Tannenbaum and D. I. C. Wang, eds.), pp. 564–586. MIT Press, Cambridge, Massachusetts.

Chapter 5

Production of Yeasts and Yeast Products

H. J. PEPPLER

I. INTRODUCTION

Yeast technology comprises those propagations involving several types of yeast used in the alcoholic fermentation of foods, beverages, and feeds as well as those processes which enrich, alter, or extract concentrates of living yeast cells. Among the derived products are nutritional supplements, flavor carriers, enzymes, nucleosides, nucleotides, proteins, and polysaccharides. Yeasts and their derivatives are manufactured on every continent. The most common product, bakers' yeast, is grown by at least 73 companies in 32 countries, excluding (for lack of data) the eastern European nations (Perlman, 1977). The market for yeast products sold to the baking industry in the United States is close to $120 million annually; in 1976 about 15.7 billion lb of yeast-

MICROBIAL TECHNOLOGY, 2nd ed., VOL.I

ISBN 0-12-551501-4

leavened bread-type products were shipped (*Bakery Trends*, 1977). By 1980 shipments are expected to increase 4.7% in this category of products which includes white pan bread, hearth bread and rolls, variety breads, rolls (hamburger, wiener, brown and serve, Kaiser, Parkerhouse, and other types), English muffins, bread stuffing, and crumbs.

Of the 349 species of yeast described by taxonomists only a few are of economic importance (Lodder, 1970). The more commonly used species and the processes and products usually associated with them are listed in Table I. Outstanding in this group of yeasts are the versatile strains of *Saccharomyces cerevisiae* used universally in food-related fermentations and beverage production. For protein biosynthesis, however, *Candida utilis* production is growing relatively rapidly, mainly to provide dried biomass (nonfermentative) as grain-sparing supplements for formulated livestock feeds. How these yeasts are grown and how the cells are utilized are reviewed in this chapter.

II. YEAST PRODUCTION

Yeast biomass is grown on a variety of energy-yielding carbon compounds and phosphate and nitrogen sources in aerobic batch and continuous culture systems. During yeast growth in the fed-batch process, the addition of sugar-containing media, usually diluted molasses, is

TABLE I. Principal Industrial Yeasts and Their Usage

Yeast species	Products/applications
Saccharomyces cerevisiae	Bread-type products, beer brewing, wine making, distilled beverages, ethanol, cider, food yeast, feed yeast, yeast-derived products, (autolysates, hydrolysates, protein, biochemicals), invertase
Saccharomyces uvarum (syn. *S. carlsbergensis*)	Beer brewing
Saccharomyces saké	Sake brewing
Saccharomyces bayanus	Sparkling wines
Saccharomyces lactis	Lactase
Kluyveromyces fragilis (syn. *S. fragilis*)	Food yeast, feed yeast, ethanol
Candida utilis	Food yeast, nucleic acids, feed yeast
Candida tropicalis	Food yeast, feed yeast
Candida pseudotropicalis	Food yeast, feed yeast
Candida lipolytica	Feed yeast

metered incrementally to minimize ethanol accumulation. It is a flexible system, suitable for small and large fermentors [up to 60,000 gal (225 m^3)], and readily permits switching to alternate substrates and yeast cultures. The fed-batch process is the principal method of producing bakers' yeast, some food yeasts, and biomass for the extraction of enzymes and the preparation of autolysates.

In continuous and semicontinuous yeast propagating systems at steady state, the input of nutrients and sugar source (usually molasses, wood sugars, or spent sulfite liquor) and the withdrawal of yeast emulsion are in balance. Modified Waldhof fermentors with capacities of around 100,000 gal (375 m^3) provide the aeration needed by *Candida* species for efficient assimilation of hexoses, pentoses and other carbonyls. While lacking some of the flexibility of the fed-batch process, the continuous culture systems developed for *Candida utilis, C. tropicalis,* and other species have made possible the marked increases in annual tonnage of fodder yeast and food yeast. Estimates of yeast production throughout the world in 1977 are shown in Table II.

Changes in yeasting practices for leavening and beverage fermentations have influenced advances in industrial microbiology since the 1850s. Among the many contributions are (1) Pasteur's discovery in 1874 that yeast utilizes ammonia, and yeast growth is stimulated by aeration which represses ethanol formation; (2) the pure-culture concept and methods introduced by Hansen in 1896; (3) the fed-batch (Zulauf) process invented by Hayduck in 1919; and (4) the replacement of grain mashes with molasses media in the 1920s. The historical and modern eras of yeast technology have been reviewed often: Kiby (1912) published a production manual for bakers' compressed yeast; Walter (1953)

TABLE II. Estimated Annual Yeast Production, 1977 (Dry Tons)[a]

Location	Bakers' yeast	Dried yeast[b]
Europe	74,000	160,000
North America	73,000	53,000
The Orient	15,000	25,000
United Kingdom	15,500	[d]
South America	7,500	2,000
Africa	2,700	2,500

[a] Based in part on surveys by Moo-Young (1976); World Wide Survey of Fermentation Industries, 1967 (1971); and private sources.

[b] Dried yeast includes food and fodder yeasts; data for petroleum-grown yeasts are not available.

[c] Production figures for USSR not reported.

[d] None reported.

and Irvin (1954) outlined the traditional Vienna process and the productive bakers' yeast modifications developed from 1870 to 1915; and advances in modern yeast making were published by Frey (1930, 1957), White (1954), Pyke (1958), Butschek and Kautzmann (1962), Peppler (1967, 1970, 1978), Burrows (1970), Harrison (1971), Reed and Peppler (1973), Rosen (1977), and Hoogerheide (1977). Descriptive information covering over 160 yeast-related patents granted in the United States since 1970 was compiled by Johnson (1977).

A. Active Yeasts

Viable biomass, the primary product of all yeast propagations, is obtained for use as (1) active yeast for a variety of food fermentations, (2) active yeast for autolysis or enzyme extraction, or (3) inactive (nonfermentative) dried yeast (95–96% solids). Furthermore, the yeast crop may be used directly, without separation of cells and medium, as starter yeast in breweries, distilleries, and wineries. The cells can also be removed from the growing medium, washed, and concentrated. Such concentrates, known generally as yeast cream, may be (1) dewatered to form press or filter cake (as bakers' compressed yeast), (2) dehydrated further to produce active dry yeast (ADY) for bread making and other fermentations, (3) extracted to recover invertase and other cell constituents, (4) autolysed to prepare flavor and nutritional concentrates, or (5) pasteurized and desiccated by drum- or spray-drying processes to obtain inactive yeast for use as food and feed ingredients, and as raw material for extraction of ribonucleic acid and other cellular components.

The major process streams for separated yeast, shown in Fig. 1, outline the pathway to three major end products: compressed yeast and active dry yeast, forms of active yeast; and primary-grown food yeast, a common type of inactive yeast.

1. Bakers' Yeast

a. Cultures. The production of bakers' yeast is a multistage, aerobic, fed-batch process. It begins in the laboratory with pure cultures of *Saccharomyces cerevisiae* carried on agar media. The naturally occurring strains have been selected for their vigorous growth, high yields on molasses, storage stability, and functional properties in doughs and the breads made from them. Stock cultures of the proved strains are maintained in the diploid state. Yeast hybrids for leavening have been patented, but none has come into general use (Burrows and Fowell, 1961; Langejan and Khoudokormoff, 1976).

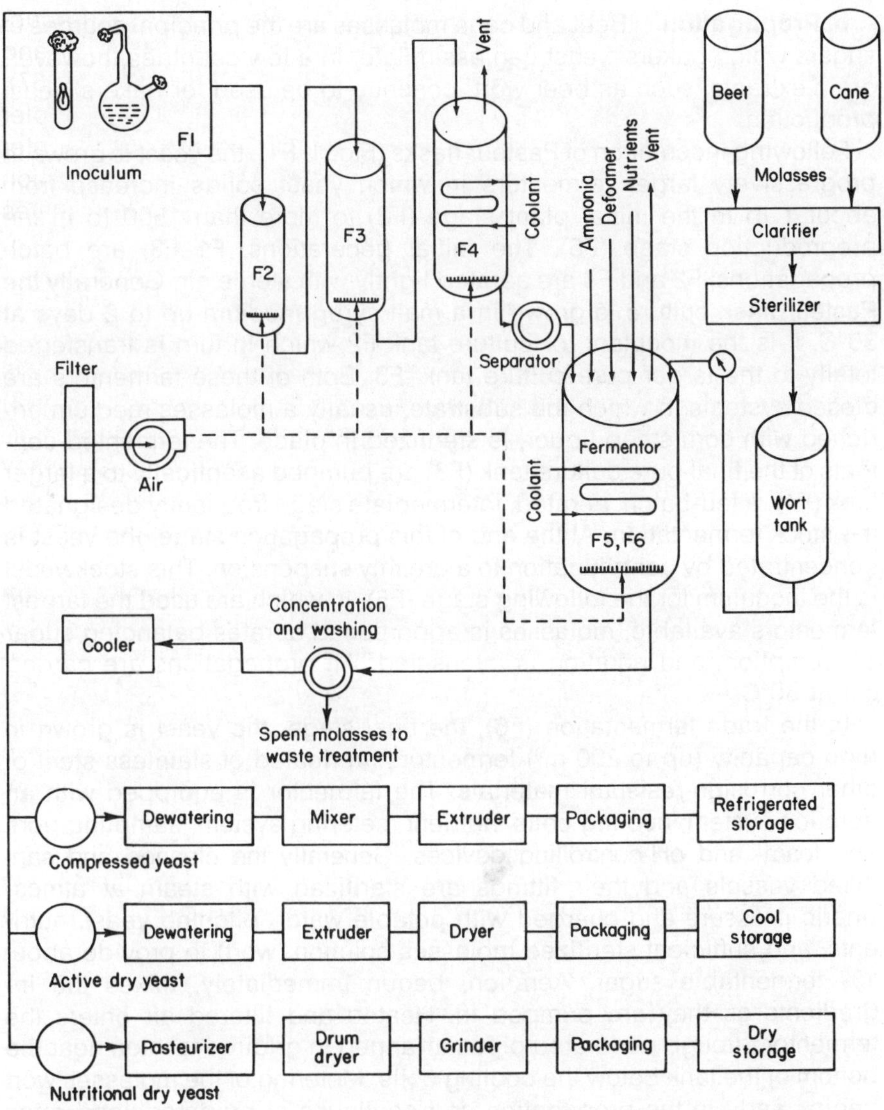

FIGURE 1. Process steps in the production of bakers' yeast and food yeast. (After a sketch by W. H. Chan, Universal Foods Corp., Milwaukee, Wisconsin.)

b. Propagation. Beet and cane molasses are the principal sources of sugars which bakers' yeast can assimilate. In a few countries, however, grain extracts, such as beer wort, continue to be used for bakers' yeast production.

Following inoculation of Pasteur flasks (Fig. 1, F1), the yeast is grown in progressively larger fermentors in which yeast solids increase from about 1 lb in the initial plant stage (F2) to more than 1500 lb in the preproduction stage (F5). The initial generations, F1–F3, are batch propagations; F2 and F3 are aerated lightly with sterile air. Generally the Pasteur flask culture is grown in a malt syrup medium up to 2 days at 30°C. It is the inoculant for culture tank F2 which in turn is transferred totally to the larger pure-culture tank, F3. Both of these fermentors are closed systems in which the substrate, usually a molasses medium enriched with corn steep liquor, is sterilized in place. The fermented contents of the final pure culture tank (F3) are pumped aseptically to a larger tank (F4), a fed-batch, aerated, intermediate stage frequently designated the stock fermentation. At the end of this propagation stage, the yeast is concentrated by centrifugation to a creamy suspension. This stock yeast is the inoculum for the following stage (F5), in which are used the largest fermentors available, molasses is apportioned at rates balancing sugar consumption, and aeration is intensified. All propagations are carried out at 30°C.

In the trade fermentation (F6), the final stage, the yeast is grown in high-capacity (up to 200 m^3) fermentors fabricated of stainless steel or other corrosion-resistant materials. The fermentor is equipped with an aeration system, cooling coils, nutrient metering system, sampling port, and foam- and pH-controlling devices. Generally the cleaned and sanitized vessels and their fittings are sterilized with steam at atmospheric pressure and charged with potable water, pitching yeast, nutrients, and sufficient sterilized molasses solution (wort) to provide about 1% fermentable sugar. Aeration, begun immediately, mixes the ingredients as they are pumped in. Heated and filtered air enters the fermentor through perforated pipes arranged in gridiron fashion near the bottom of the tank below the cooling coils. Metering of the molasses wort begins early in the propagation and continues at predetermined rates throughout most of the schedule. Efficient biosynthesis demands careful management of sugar and air inputs to avoid excess carbon losses. Overfeeding is avoided by controlling the available sugar concentration in the fermentor well below 0.1%, and by adjusting aeration to provide a minimum oxygen absorption rate of 0.2 m moles O_2/liter/min. Maximum yeast yield is obtained when the increase in yeast solids in the fermentor is limited to 20–25% per hour. Phosphate and ammonia are added in the early hours of the schedule in amounts according to the yeast crop expected, and ammonia or other alkali subsequently to hold the pH between 4 and 6.

(1) Nutrients. Beet molasses and cane molasses (blackstrap) supply fermentable sugars, essential minerals (potassium, phosphorus, magnesium, zinc, iron, and copper), amino nitrogen (asparagine, aspartic acid, alanine, glutamic acid, and glycine), and vitamins (biotin, thiamin, pyridoxine, pantothenic acid, and inositol). These nutrients vary according to the type of molasses, its geographical origin, management of the sugar crop, and efficiency of sugar mill operations. In general, beet molasses is fivefold higher in organic nitrogen than cane, but half of it is betaine which is not assimilated by *Saccharomyces.* Cane molasses is substantially richer in biotin, pantothenic acid, thiamin, magnesium, and calcium. Both types of molasses are nearly equal in their content of fermentable sugar, potassium, trace minerals, niacin, pyridoxine, and inositol. For bakers' yeast production, beet molasses is mixed with at least 20% cane molasses to assure an adequate supply of biotin.

To prepare the mixed molasses solutions, usually called wort, beet and cane are processed separately; i.e., diluted, pH adjusted, heated, clarified, sterilized, and blended.

Molasses contains only a small portion of the mineral nutrients required to grow bakers' yeast of desired composition. All propagations must be supplemented heavily with phosphorus and nitrogen to sustain optimum growth, maximum yield, and quality. Phosphoric acid, its ammonium salts, anhydrous ammonia, and ammonium sulfate are the usual additives. Other supplements may be added to assure highest yields, performance, and stability of bakers' yeasts. These additions include magnesium sulfate, potassium chloride, zinc sulfate, and thiamin. Where urea is used, biotin supplements may be required to obtain maximum yields of yeast.

(2) Aeration. Of the numerous aeration systems that have been developed for bakers' yeast propagations, the sparger type is most practical and in general use. It is a simple, horizontal network (grid) of parallel pipes, with numerous drilled openings (0.8–1.5 mm diameter), attached to a large, central, vertical air conduit. The assembly is secured near the bottom of the fermentor. With proper design, this aeration system easily supplies the high air inputs required for yeast biosynthesis (1 v air/1 v broth/min). In a new Danish yeast plant, which began operations in 1973, the sparger system in the trade fermentors (150 m^3) has 24 side pipes containing about 30,000 holes of 1.5-mm diameter (Rosen, 1977).

Alternate systems with increased air utilization efficiencies have come into commercial use. They combine mechanical agitation and novel air distributors and are described as agitator–aerator system, deep jet aerator (Vogelbusch), self-priming aerator (Frings), and turbine aerator system (Chemapec). The operating characteristics of such systems were discussed by Reed and Peppler (1973).

(3) Process Control. Most parameters of yeast propagation can be regulated automatically, generally by sensors which signal the activation of pumps and pneumatic devices. Temperature, pH, defoamer demand, and nutrient rate may be regulated automatically. Molasses feed rate may be controlled by monitoring the ethanol concentration in the effluent air (Rosenquist and Egnell, 1966; Rungaldier and Braun, 1961; Bach *et al.*, 1978). Yeast growth may be regulated by combining an on-line computer with process sensors which determine indirectly cell concentration, yield, and growth rate (Wang *et al.*, 1977). An alternate system for optimizing yeast production, described by Nyiri *et al.* (1977), regulates the ratio of carbon : nitrogen in the nutrient feed on the basis of the respiratory quotient (*RQ*) which is calculated from determinations of rates of CO_2 evolution and O_2 consumption of the culture. For maximal yeast productivity the RQ ($= dCO_2/dO_2$) should be around 1.0.

In the modern bakers' yeast plant in Denmark, an electronic control system automatically monitors startup, propagation, and harvesting (Rosen, 1977). The problems of process measurements in computer-coupled fermentation systems, based on experience with a yeast-producing process, have been discussed by Swartz and Cooney (1978); also see Armiger and Humphrey, Volume II, Chapter 15.

c. Harvesting Yeast. Depending upon the amount of pitching yeast used, the length of trade propagations vary from 10 to 20 hours, and the cell solids from 3.5 to 5% in the fermentor broth. In some European plants, bakers' yeast production generally passes through more stages than described above and requires more time to reach the trade stage (Suomalainen, 1963; Butschek and Kautzmann, 1962). In the new Danish bakers' yeast plant described by Rosen (1977), a trade propagation of about 10 hours is used. At the end of the trade propagation, the broth is cooled as yeast cells are recovered by concentration in centrifugal separators, diluted with water, and reseparated. The washed suspension of cells, known in the trade as yeast cream, usually contains 17–19% solids. Some nozzle-type centrifuges with special bowl design can concentrate yeast to 23% solids. Yeast cream is chilled and stored at 0°C until processed further, either as bakers' compressed yeast, active dry yeast, or dried inactive yeast. For bakers' compressed yeast and active dry yeasts, the cream is dewatered in custom-modified plate-and-frame filter presses. A friable cake of 27–30% solids, known as press cake, is generally obtained. Some plants use rotary vacuum filters to dewater yeast cream intended for the compressed yeast market.

d. Compressed Yeast. This perishable product may be packaged in a variety of market forms: foil-wrapped blocks (3/5 to 5/8 oz) and waxed paper wrapped and sealed blocks (1–2 oz) for household use; larger blocks (0.5, 1, and 5 lb) wrapped in waxed paper and 25- to 50-lb bulk yeast (crumbled) in sealed, polyethylene-lined, multiwall bags (Schuldt and Seeley, 1966).

Except for the bulk (crumbled) product, compressed yeast is prepared for packaging by mixing press cake with a small amount of water containing emulsifiers to facilitate smooth extrusion, even cutting, a standard consistency, and uniform plasticity in storage. The preferred emulsifiers bring about a whiter yeast cake and reduce surface water spotting due to water migration. Certain processed vegetable oils, glyceryl monostearates, and lactylated monoglycerides are suitable for this purpose. Further shelf-life enhancement of the household yeast products is provided by the incorporation of cereal starch and mold inhibitors; both ingredients are declared on the label.

Crumbled yeast is prepared from press cake which is broken into small, irregular pieces by forcing it through a suitable screen or perforated plate.

Domestic brands of compressed yeast contain from 8 to 9% nitrogen, 1–1.4% phosphorus, and 29–31% solids. In other countries, compressed yeast of 27% solids are common but may reach 30% in some trade areas. All compressed yeast products are generally stored at 4°C or lower.

e. Active Dry Yeast. Dehydrated or active dry yeast is a viable product produced from yeast press cake extruded directly through a perforated plate, forming spaghetti-like strands which are broken into short pieces and broadcast to uniform depth on a continuous, perforated, stainless steel belt. Dehydration is by stages over a 6-hour period (Thorn and Reed, 1959). Warm air (25°–45°C) is forced at moderate velocities through the bed of cylindrical fragments as they move from chamber to chamber in a tunnel dryer. The air humidity is so regulated that yeast moisture can equilibrate at about 8%. The light-tan, porous granules obtained are friable, have a mild odor, and may vary in moisture from 7.5 to 8.3%, and contain about 6.5–7.2% nitrogen. The performance characteristics of active dry yeast and its uses in baking are discussed in reviews by Thorn and Reed (1959) and Reed and Peppler (1973).

Domestic active dry yeast (ADY) is marketed in a variety of forms and packages. As unground, irregular cylinders, it is air-packed in fiber drums, cartons, and multiwall bags containing 10–300 lb product. Coarsely ground, free-flowing ADY is vacuum-packed in glass jars (4 oz), metal cans (8 oz, 2 lb), 25-lb nitrogen-pack cans, and also in pouches of paper–foil–pliofilm construction (5–42 gm ADY) packed in nitrogen or in air-pack protected with butylated hydroxyanisole. The smallest units in each category are generally for household use and for insertion in packages of prepared mixes (bread, hot rolls, coffee cake, pizza). In Europe active dry yeast is also vacuum-packed with flexible film in 500-gm, 1-kg, and 10-kg blocks.

Active dry yeast produced as described above may be packaged in nitrogen or other oxygen-free atmosphere to prolong shelf-life. When antioxidants and nonionic surfactants are emulsified with the yeast cream before dehydration, an active dry yeast of enhanced storage stability is obtained (Chen and Cooper, 1962). This means of protecting active dry yeast permits its use in bakery product prepared mixes of yeast-leavened cereal flours in which the yeast is an integral part of the ingredients (Chen *et al.*, 1966).

Numerous methods have been developed to dehydrate yeast press cake with high viability (95–99%). One alternate batch process, the rotolouver dryer, yields semihard, smooth-surfaced pellets. This old process, still in limited use, consists of tumbling the fragments of extruded press cake in a double-walled, rotating drum while warm air streams through the fluted inner shell. The light-tan, firm ovals formed vary in size, usually 0.5–3 mm in diameter and 1–4 mm in length.

Fluidized bed drying methods for producing active dry yeast have been described by Burrows (1976), Taylor (1975), and Trevelyan (1973). A spray-drying process for viable yeast was developed by Fantozzi and Trevelyan (1971).

2. Brewing Yeasts

For the production of lager beer and ale, selected and adapted strains of *S. cerevisiae* and *S. carlsbergensis (S. uvarum)*, respectively, are used (Thorne, 1975). Pure cultures are commonly employed. After brewing, most of the yeast is recovered and prepared to pitch another fermentor or is processed and dried for use as food or feed supplements.

Yeast handling in a modern brewery involves the preparation and keeping in readiness of pure yeast cultures for plant use at all times, as well as the recovery of yeast from finished fermentations, storing it, and determining its suitability for pitching new fermentations. Following a weekly schedule, according to Strandskov (1965), pooled colonies of the selected yeast strain are grown in tubes of hopped wort and transferred through successive stages until a mass culture is obtained in 10 liters of wort. The yeast crop from this fermentation is transferred to a final pure-culture vessel of 6-barrel (715-liter) capacity and diluted to a count of 6 million cells per ml with sterile plant wort. This process is repeated until the working capacity of the fermentor is reached. Then the buildup begins, also stagewise, to supply pitching yeast to the plant. Fermentors are usually pitched at the rate of 1–2 lb yeast slurry (15–17% yeast solids) per barrel of wort, or 10–14 million yeast/ml wort (see Volume II, Chapter 1).

Thorne (1970) described a yeasting system for brewing yeast which begins with a single cell in a drop of wort and ends, after numerous cultivations, in progressively increased volumes of wort, with enough pure yeast to pitch a 100-hl fermentor twice every week.

Yeast recovered from beer may be recycled after washing, cooled (0°–2°C), and held overnight. After about 20 recyclings, replacement with pure culture, as described above, is recommended (Strandskov, 1965). Before recycling plant yeast, its bacterial contaminants may be inactivated by washing the yeast with phosphoric acid solution (pH 2.8), by suspension in 0.75% ammonium persulfate solution, or by a combination of both treatments (Bruch *et al.*, 1964).

Methods of determining brewing yeast vigor and other aspects of its behavior during fermentation have been discussed by Gilliland (1971), Reed and Peppler (1973), Thorne (1975), and MacLeod (1977).

3. Distillers' Yeasts

The variety of yeasts selected for brewery, distillery, and winery operations function primarily as catalysts to effect a maximum ethanol yield efficiently and to impart distinctive and uniform characteristics to the final product. A wide range of practices are found. For example, distillers' yeast may be produced in modern closed systems with meticulous control to deliver the inoculum free of interfering microorganisms (Coulter, 1964; also see Volume II, Chapter 3; Lyons and Rose, 1977). A new mash may be inoculated with a previously finished yeast mash, thereby reducing the preparation of a laboratory culture to weekly intervals. For spirits production, the mash may be seeded directly with bakers' yeast (Stark, 1954). For the latter practice, special strains in the form of active dry yeast are produced by bakers' yeast manufacturers.

In the modern yeasting system described by Coulter (1964), plant cultures are started daily from purity-checked wort agar slants of proved strains of *S. cerevisiae.* After building up the yeast concentration in three laboratory stages of sterilized malt syrup (25-ml, 500-ml, and 2-gal volume), the yeasting continues in sterilized mash through three successive plant stages of 45-, 450-, and 9000-gal batches. A rye–barley malt medium is soured with *Lactobacillus delbrückii* to facilitate sterilization. When a yeast population of approximately 100 million viable cells per ml is reached, the culture is used to inoculate plant fermentors at the rate of 3% by volume. Following fermentation and distillation, the yeast is recovered in combination with the grain residues (see Volume II, Chapter 3; Harrison and Graham, 1970).

When barley malt and rye were replaced with less expensive corn or milo and the mash was treated with glucamylase (produced with *Aspergillus awamori*), Van Lanen *et al.* (1975) obtained higher than usual cell counts and longer retention of viability of the stored seed yeast.

4. Wine Yeasts

The wine maker selects the strain of yeast which will exhibit the alcoholic fermentation capacity and flavor properties judged to be proper for the kind of grapes used. Although many cultures with characteristic names (Burgundy, Tokay, Montrachet, Epernay, etc.) are traditional in wine making, they are predominantly strains of *S. cerevisiae* (Kunkee and Amerine, 1970; also see Volume II, Chapter 5). Suitable pure cultures may be obtained from commercial laboratories serving the wine industry and from culture collections maintained by government agencies in wine-producing areas. The production of yeast starter has been described by Thoukis *et al.* (1963) and DeSoto (1955). When the crushing season begins, a sample of pasteurized or sterilized must (about 1–5 liters grape juice) is inoculated from a laboratory pure culture slant and built up to bulk fermentation size by serial transfers to increased volumes of fresh must treated with bisulfite or pasteurized to destroy unwanted yeast and bacteria. During the cultivations, the yeast grows to a peak population near 25 million cells per ml, and adapts itself to sulfur dioxide and to ethanol. The vigorous master culture, or starter, is charged to fresh batches of must at the rate of 2–5% by volume. At the end of fermentation, the yeast settles with the lees and cannot be recovered economically for reuse.

Three species of wine yeasts are produced in compressed and active dry form and marketed in North America: *Saccharomyces cerevisiae* (Montrachet strain) for table wines, *S. bayanus* for secondary fermentations of sparkling wines, and *S. beticus* for sherry fermentations. These products, available mainly as bulk, active dry wine yeast, are produced by propagating the selected yeast strain aerobically in molasses wort and employing equipment and procedures used in the manufacture of bakers' yeast (Thoukis *et al.*, 1963; Reed and Peppler, 1973).

5. *Sake Yeast*

Yeast starters, called *moto*, are prepared for the sake brewing process from a mash of steamed rice, mold stock (*koji*) and water. Two procedures are in general use: a natural sequential bacteria and yeast fermentation at 10°C, which requires about 30 days to complete (Ohwaki and Lewis, 1970), and a pure-culture modification, which is ready for pitching the main fermentation (*moromi*) in about 9 days. In the shorter yeasting process, the mash is acidified with lactic acid and inoculated with a pure culture of sake yeast, a selected strain of *S. cerevisiae* (Kodama, 1970).

6. *Yeast Culture for Livestock Feeds*

Yeast culture is one of eight feed yeast products defined by the Association of American Feed Control Officials (AAFCO, 1977):

> Section 96.8 Yeast Culture is the dried product composed of yeast and the media in which it is grown, dried in such a manner as to preserve the fermenting activity of the yeast. The media must be stated on the label.

In general, the production of yeast culture begins with a culture medium, such as diluted molasses, which is inoculated with a suitable yeast, usually a selected bakers' yeast. Following several hours of fermentation, part of the brew is mixed with a mixture of moistened cereal grains, incubated for varying periods, and dried. Mild drying conditions are used to avoid damage to the yeast, enzymes, and other heat-sensitive products formed during the fermentation stages. Yeast culture is fed to livestock and poultry primarily as a digestive aid (Peppler and Stone, 1976). Rations for ruminants are generally fortified with 1–1.5% yeast culture, 1.5–2% in hog feeds, and 2–3% in poultry rations. The addition of 0.5% yeast culture to corn silage at ensiling markedly stimulated the natural lactic acid fermentation (Lane *et al.*, 1974). Increased lactic acid contents of silage media were also noted by Woolford (1976) when unidentified yeasts isolated from various silages were added with the inoculum used.

B. Inactive Yeasts

Inactive (heat-killed, nonfermentative) yeasts, generally designated "dried yeast," are obtained either by recovery of residual brewers' yeast or by harvesting yeast grown primarily for the purpose of supplementing human diets and animal rations as sources of protein, flavor, and vitamins of the B complex. Such food and feed (fodder) yeasts are grown aerobically to high nitrogen contents (above 8%) under conditions comparable to those used in bakers' yeast propagations. However, a greater variety of carbon substrates may be used, notably molasses, sulfite mill waste liquor, wood hydrolysate, whey, hydrocarbons, and ethanol (Moo-Young, 1976; Litchfield, 1977a,b).

Food and feed yeasts are known by many synonyms: dried yeast, inactive dried yeast, dry yeast, dry inactive yeast, levure-aliment sèche, dried torula yeast, sulfite yeast, wood sugar yeast, xylose yeast, levadura alimenticia, la levure alimentieri, Nahrungshefe, and Saccharomyces Siccum, single-cell protein (SCP), and microbial protein. Since 1968 the term single-cell protein (SCP), introduced by M.I.T. scientists (Scrimshaw, 1968), has been identified worldwide with the efforts of microbial technology to provide an abundance of low-cost protein. Such programs initially concentrated on growing yeast on hydrocarbons, but they soon expanded rapidly to include the propagation of many species of bacteria, fungi, algae, and yeasts on a wide variety of agricultural and industrial by-products and waste materials (Laskin, 1977). The developments in this field are discussed by Litchfield in this volume, Chapter 4, and in numerous reviews: GIAM V (1978, also abstracted by Behrman, 1978), O'Sullivan (1978), Litchfield (1977a,b), Moo-Young (1976, 1977), Rockwell (1976), and Lipinsky and Litchfield (1974).

1. Primary-Grown Yeasts

The principal dried yeast products in the United States are comprised of two species of *Saccharomyces*, *Candida utilis*, and *Kluyveromyces (Sacch.) fragilis* (Table I): *S. cerevisiae*, molasses-grown strains of bakers' yeast, and recovered brewing yeast strains of this species; *K. fragilis* cultured on cheese whey; *C. utilis* propagated on sulfite mill waste liquor and ethanol; and *S. uvarum (S. carlsbergensis)* recovered from beer and ale. In some countries *Candida tropicalis* is grown on wood sugars and sulfite waste liquor (Butschek, 1962), *Candida pseudotropicalis* is cultured on whey, and *C. utilis* is propagated on molasses (Chien, 1960).

Quality guidelines for dried yeast grown for human use have been established by governmental agencies, professional associations, and food processors. Food yeast must conform to specifications of yeast type, color, flavor, microbial content, chemical composition, and vitamin potency as published by the International Union of Pure and Applied Chemistry (IUPAC), The British Pharmaceutical Codex, the National Formulary (N.F. XIII) of the American Pharmaceutical Association, and the food additive regulations of the Food and Drug Administration (FDA), U.S. Department of Health, Education and Welfare. Some food manufacturers impose additional chemical and bacteriological specifications for dried yeast added to products for thermal processing and special formulations.

As defined by N.F. XIII, dried yeast is "the dried cells of any suitable strain of *Saccharomyces cerevisiae* . . . or *Candida utilis* . . . grown in media other than those required for beer production. . . . Such yeasts, properly designated as to species, are commonly known as Primary Dried Yeast . . . and Torula Dried Yeast." "Brewer's dried yeast" is defined as "a by-product from the brewing of beer . . . washed free of beer prior to drying." When the washing step includes one or more alkaline treatments to remove hop resins adsorbed on the yeast cells, the product is designated "debittered brewer's dried yeast." Each type of dried yeast, N.F. XIII, must be free of fillers (starch, corn meal, etc.) and contain a minimum of 45% protein (N × 6.25); not more than 7% moisture and 8% ash; and, in each gram, the equivalent of not less than 120 μg thiamin hydrochloride, 40 μg riboflavin, and 300 μg niacin. Live bacterial and mold count maxima are 7500 and 50/gm respectively. In addition, the U.S. Food and Drug Administration has established a zero tolerance for *Salmonella*.

Dietary usage of food yeast is further restricted by the FDA on the basis of its folic acid content. When added to foods as a flavor ingredient, dried yeast may be used to the extent that the folic acid content it contributes does not exceed 40 μg/gm yeast [about 8 μg of free folic acid (pteroylglutamic acid)/gm]. Additions to foods in special dietary usage are limited to 400 μg total folic acid per day (100 μg free folic acid/adult). Analyses of primary dried yeast on the market show both total and free folic acid levels to be well within the limits prescribed by the FDA. For each type of food yeast, the average total folic acid content was found to be 12 μg/gm. The free folic acid content of both dried yeasts was consistently less than 1 μg/gm yeast (Peppler, 1965).

Standards and methods specified by IUPAC in 1966, describe dried yeast as "the whole organism of one individual yeast or a mixture of several yeasts belonging to the family Saccharomycetaceae . . . and to the family Cryptococcaceae . . . obtained either as a by-product of fermentation processes or by special culture and conforming to such standards as may be laid down." Specifically excluded are yeasts which have been extracted, those containing more than 20% fat, and yeasts which carry inert fillers or substances that are not incorporated components of normal yeasts cells. Nine standards are set: Upper limits are specified for moisture (10%), ash (10%), lead (5 μg/gm), arsenic (5 μg/gm), live bacteria (7500/gm), and molds (50/gm). Minimum levels are set for nitrogen (7.2%, equivalent to 45% protein; i.e., N × 6.25), thiamin (10 μg/gm), riboflavin (30 μg/gm), and niacin (300 μg/gm). Dried yeast must also be free of starch and bacteria of the genus *Salmonella*.

The British Pharmaceutical Codex in 1949 stated that food yeast be dried by "a process which avoids decomposition of the vitamins present" and in 1954 required, in each gram, not less than 100 μg thiamin, 40 μg riboflavin, and 300 μg niacin (Pyke, 1958). The Swedish Pharmacopoeia in 1948, according to Pyke (1958), specified only the minimum content for thiamin (150 μg/gm).

Color, flavor, granulation, and packaging of food yeast are not specified in official standards. In the trade, however, these properties assume economic and esthetic importance according to the requirements of dietetic preparations, pharmaceutical products, and food-processing applications. The National Formulary (N.F. XIII) merely states, "Dried Yeast occurs as yellowish white to weak yellowish orange flakes, granules or powder, with an odor and taste characteristic of the type." Dried torula yeast is described in identical terms but "with a characteristic odor and bland taste."

Food yeast marketed in the United States generally conforms to the specifications prescribed in N.F. XIII, the food additive orders of the FDA, and the requirements of food processors.

Dried yeasts are readily compounded with a variety of spices, extracts, and imitation food flavors. One such flavor carrier, hickory-smoked food yeast, is used to impart a distinctive baconlike flavor and aroma to foods, especially snack-type products. Special processing and roasting of brewers' yeast produces a novel cocoa powder replacement (Andres, 1978).

Official specifications for dried yeasts used to supplement animal feeds are published by the Association of American Feed Control Officials (AAFCO, 1977). Crude protein (N × 6.25) values of 40% (minimum) are specified for primary dried yeast, grain distillers' dried yeast, molasses distillers' dried yeast, and torula dried yeast. The minimum for brewers' dried yeast is 35%. Most of the feed yeast available comes from two sources: breweries and fermentation of pulp mill sulfite liquor (torula dried yeast). Average values, in percentage, for major components are as follows (torula dried values in parentheses): crude protein 44 (47), crude fat 0.9 (1.6), crude fiber 2.7 (2.5), ash 6.6 (8.1), calcium 0.1 (0.5), phosphorus 1.4 (1.6). The vitamin values (in μg/gm) are: thiamin 90 (6), riboflavin 30 (40), niacin 440 (550), pantothenic acid 110 (65), choline 3700 (2800), folic acid 9 (20), and biotin 0.9 (1.0) (Peppler and Stone, 1976).

a. Molasses-Grown Dried Yeast. Dried yeast may be produced on molasses in bakers' yeast plants using procedures described above (Fig. 1), or in factories designed expressly for food and feed yeast production, as in Taiwan's facility for continuous propagation of *C. utilis* in a battery of six Waldhof fermentors (Chien, 1960). In both aerobic systems, the yeasts are grown under sugar-limiting conditions to a nitrogen content above 8% (dry basis), separated, chilled and stored, or processed directly. After enrichment with vitamins (if needed to meet customer specifications), the yeast cream is pasteurized, usually drum-dried to about 95% dry matter, pulverized, and packed in plastic-lined fiber drums. In addition to the chemical and bacteriological standards discussed above, food processors may specify color and particle size. A detailed review of dried yeast composition was published by Peppler (1970).

Propagations of food and feed yeasts may also be fed selected minerals to enrich dried yeast with organically bound essential trace elements, notably zinc, selenium, and chromium. Such custom-grown yeast products are being developed in the wake of recent dietary trace element research (Mertz, 1969, 1975, 1977; Toepfer *et al.*, 1973; Schwarz, 1974, 1976; Levander, 1975; Frost and Lish, 1975; Ammerman and Miller, 1975; Sakurai and Tsuchiya, 1975; Tuman *et al.*, 1978).

A fluidized bed system for drying inactive yeast has been developed by Bardot (1972). Press or filter cake is mixed with dried yeast to a solids content which can be pelleted and is dried by fluidization with air or nitrogen at 150°C. Large-scale production at costs below those for spray-drying and improved digestibility of the product are claimed. The costs of removing water from yeast cells were compared by Labuza (1975).

b. Wood Sugar-Grown Dried Yeast. *Candida utilis* and *C. tropicalis* assimilate a variety of organic acids, aldehydes, and sugars (2–3% pentoses and hexoses) occurring in sulfite liquor. In the production of *C. utilis* in the United States, as described by Kaiser and Jacobs (1957), a blend of sulfite liquor from several pulp digesters is stripped of sulfur dioxide, adjusted to pH 4.5 with ammonia, sampled for sugar and phosphorus analysis, and supplemented with the required phosphorus and potassium. No biotin is added since both species, unlike most yeasts, synthesize this vitamin. Continuous, automatic, on-stream monitoring of sugar and phosphorus contents assures accurate and economical proportioning of added phosphate based on the sugar input. Ammonia additions and pH adjustments are also made and recorded automatically.

At the rate of 100 gal/min, the fortified liquor is introduced at the top of the draft tube of a 60,000-gal (225-m^3) modified Waldhof fermentor. Rapid mixing of liquor and yeasty emulsion occurs at the bottom of the draft tube as air (about 2300 ft^3/minute) is distributed by the rotor (about 390 rpm). Fermentor temperature is maintained near 32°C by continuously circulating the emulsion through an external cooler coupled with a refrigeration system. Yeast is harvested by withdrawing emulsion at a rate to balance the substrate input, defoamed mechanically, separated, washed, concentrated to about 15% solids, pasteurized, and dried to less than 6% moisture on atmospheric, steam-heated, double-drum dryers.

For production of food yeast, the yeast cream is washed several times to remove most of the residual lignosulfonates. This results in a high-nitrogen (above 8%) dried yeast unusually low in sodium content (less than 0.002%).

c. Whey-Grown Dried Yeast. As whey is drawn from the cheese vat it contains about 4% lactose, 1% protein, 1% ash, and a small amount of lactic acid. It will support the growth of the lactose-fermenting yeasts *Saccharomyces lactis, K. fragilis* and *C. pseudotropicalis*. Only *K. fragilis* is used in the United States for the production of dried yeast.

In the process reported by Bernstein and Plantz (1977), diluted whey is supplemented with ammonia, phosphoric acid, minerals, and yeast extract. After the medium is adjusted to pH 4.5, it is pasteurized (80°C for 45 minutes), cooled to 30°C, and transferred to a 15,000-gal (55-m^3) stainless steel fermentor equipped with an aeration system and jacketed for cooling. Following inoculation with *K. fragilis*, the cell count is increased to 1×10^9/ml in a period of 8–12 hours before the continuous yeast propagating phase is started. The addition of diluted whey and nutrients (1250 gal/hour) is monitored automatically over the 5- to 7-day growing period. Rapid analysis of broth samples for lactose and glucose concentration aids in maintaining the desired fermentation efficiency and in balancing whey input with yeast withdrawal. The yeast is recovered either as feed yeast or food yeast. For feed yeast the broth and cells are concentrated in a three-stage evaporator and spray-dried; the product contains about 45% crude protein. For the production of food yeast (50% protein), the yeast is harvested by centrifugation and the cream is spray-dried. Ethanol is recovered from the evaporator condensate and the supernatant streams from the centrifuges. Annual production of dried yeast is estimated at 5000 tons.

d. Ethanol-Grown Dried Yeast. Since 1975 *C. utilis* has been produced commercially in a solution of inorganic nutrients containing 100–500 ppm ethanol (Ridgway *et al.*, 1975). With two fermentors in parallel, operation of the closed propagating system is continuous, aseptic, fully automated, and computer-monitored. Fermentor broth is withdrawn continuously and separated by centrifugation. After pasteurization, the yeast cream is spray-dried. Modifications in substrate composition and cream processing yield four yeast products which differ in emulsion capacity, water binding, nutritional value, and flavor (Andres, 1977). The entire annual production, about 7000 tons, is processed for use in the food industry.

Weatherholtz and Holsing (1976) evaluated the acceptance of test meals prepared with these materials (20 gm/day) and found no major differences, as compared with soybean and potato supplements in the control meals consumed by the student test group.

e. Hydrocarbon-Grown Dried Yeast. Since the early 1960s, the demand and subsequent research for new sources of protein led to the development of numerous processes for producing high-protein (above 50%) microbial biomass on a variety of carbonaceous raw materials. Bioconversion of petroleum fractions for use in feeds and foods was most attractive while supplies of crude oil were cheap and plentiful and the cost of protein from agricultural sources was climbing. Since yeasts and bacteria proved to be superior protein synthesizers, processes for both were the first to reach pilot stage and advance to commercial-scale development.

Petroprotein from current production, principally in the form of dried yeast (*Candida* species), is used directly in animal feeds. It partially replaces protein which is in short supply and is needed for human consumption. Direct human use of hydrocarbon-grown yeasts and bacteria awaits the completion of clinical studies aimed at reconfirming the safety of microbial protein products. Until the public health issue of residual hydrocarbons is fully resolved, two new yeast factories in Italy remain on standby until authorities grant operating permits (O'Sullivan, 1978; Behrman, 1978). Each plant expects to produce 100,000 metric tons of yeast annually in aseptic systems. The facility at Sarroch (Sardinia) is designed to grow *C. lipolytica* on *n*-paraffins with a protein content of more than 60% in the dried yeast. The other production complex, located at Saline di Montebello (Calabria), will grow *Candida maltosa*.

Although the petroprotein industry may be set back by the present high cost of hydrocarbon feedstocks and the low price of soybeans, the evaluation of SCP processes on a variety of substrates continues. These studies include, for example, biomass production on cellulosic materials, starch, molasses, paper mill wastes, and the by-product and waste streams of food plants, livestock feeder lots, municipal garbage, and industrial plants (Gillies, 1978; Han, 1978). This worldwide interest in the SCP field has been the subject of many excellent recent reviews: Chapter 4 in this volume and Litchfield (1977a,b); global activity (GIAM V, 1978); United States patents issued since 1970 (Johnson, 1977); update of developments and problems (Laskin, 1977); survey of manufacturers and economics (Moo-Young, 1976, 1977); utilization of cellulose and hydrocarbons (Rockwell, 1976); symposium (Tannenbaum and Wang, 1975); and developments in France (Senez, 1975).

f. Dried Yeast Produced on Methanol. In addition to the raw materials discussed above, microbial protein production on methanol and methane has received the most attention. Processes with yeasts and bacteria have reached the pilot stage (Laskin, 1977). In the most advanced yeast-based process, *Candida boidinii* is grown on methanol in continuous culture (Reuss *et al.*, 1975; Laskin, 1977).

2. *By-Product Yeast*

Yeast grown during beer brewing exceeds the amounts required for pitching fresh wort. This surplus yeast may be separated from beer by centrifugation. The concentrated yeast cream may be pasteurized and drum-dried as feed yeast, mixed with spent brewers' grains for fodder, debittered and dried as food yeast, or converted to products used for nutritional and flavoring purposes (Andres, 1978).

The bitter components (hop resins and tannin materials) adsorbed on brewers' yeast are removed by making the slurry slightly alkaline and recovering the yeast by centrifuging. After one or two more alkaline washes, if needed, the final suspension of debittered yeast is acidified (about pH 5.5) and dried (East *et al.*, 1966).

III. YEAST-DERIVED PRODUCTS

Debittered brewers' yeast and primary-grown species of *Saccharomyces, Candida,* and *Kluyveromyces* are the raw materials for the manufacture of yeast autolysates, hydrolysates, invertase, lactase, ribonucleic acid, isolated protein, and yeast glycan. These products meet specific applications in the food, beverage, pharmaceutical, and fermentation industries.

A. Autolysates and Hydrolysates

Active and inactive dried yeasts can be partially solubilized and converted to nutritional products and condiments in the form of syrups (pastes) and powders. These products are generally known as yeast extract paste, dried yeast extract, dried yeast autolysate, yeast hydrolysate paste, and yeast hydrolysate powder. The principal raw materials are debittered brewers' yeast and *S. cerevisiae* grown on molasses and whey.

1. Autolysed Yeast

Autolysis is a process of self-digestion in which the cellular enzymes of viable yeast cells are induced to ferment some of the glycogen and solubilize a portion of the yeast protein. Fermentation and hydrolysis begin rather quickly in an agitated slurry of viable yeast (15–18% yeast solids) as it is slowly warmed to 50°C. Initiation of autolysis is stimulated and accelerated by small amounts (3–5%) of sodium chloride, ethyl acetate, or chloroform (not permitted in the United States). The reaction continues at about 45°C, with moderate agitation, for 12–24 hours. When the desired concentration of soluble nitrogen is reached, the autolysate may be pasteurized and drum-dried without removing the yeast cell walls, glycogen, and other undigested constituents of the cell. In the usual practice, however, a clear extract is obtained by centrifugation and then concentrated by evaporation to a thick syrup or paste (80% solids). It is used in this form, or spray-dried to form hygroscopic, water-soluble powders (East *et al.*, 1966).

Autolysed dried yeast extracts differ in composition. When the autolysis of primary-grown yeast is induced with NaCl, the solids (96%) contain 6.8% nitrogen, 5% monosodium glutamate, and 37% NaCl. Autolysates prepared with organic solvents as modifiers of cell wall permeability generally contain 97% solids, 11% ash, 5% monosodium glutamate, 2.5% NaCl, and 8% nitrogen (about 30% as α-amino nitrogen). Some low-salt autolysates can be drum-dried without separation of undigested yeast cell matter. Such nonhygroscopic powders are only 50% soluble and contain 92% solids, 7–9% nitrogen, 6–8% ash, and 2–3% free amino acids.

The water-soluble, spray-dried, autolysed yeast extracts are important food ingredients, contributing a meaty flavor and aroma to soups, sausages, meats, gravies, sauces, seafoods, and vegetables.

2. Hydrolysed Yeast

Hydrolysis by controlled cooking of yeast in a strong acid solution is the most efficient method of manufacturing yeast extracts. Dried yeast is reslurried with water to a solids content of 65–80% and concentrated hydrochloric acid is added. The mixture is heated to 100°C in a wiped-film evaporator with reflux condenser. Hydrolysis continues until the desired level of amino nitrogen is reached, generally 6–12 hours for converting 50–60% of the total nitrogen to amino nitrogen, depending upon the initial acidity (Ziemba, 1967). Neutralization of the hydrolysate to pH 5–6, usually with sodium hydroxide, is followed by filtration, decolorization, and concentration to a syrup (45% solids) or paste (85% solids), or spray-dried (95% solids). Yeast hydrolysate syrups contain about 42% solids, 18% NaCl, 2.5–3% nitrogen, and 3.5% free glutamic acid.

Liquid hydrolysates diluted with two parts of yeast cream can be dried successfully on drum-dryers. While hydrolysis results in the highest yields of yeast extract, there is some loss of vitamins, protein, and flavor. The high final salt content of hydrolysates limits their use in foods.

To meet the special needs of food processors and health food formulators, yeast extracts and yeast autolysates may be fortified with monosodium glutamate and 5′-nucleotides, combined with extracts of vegetable or animal proteins, or blended with dried yeast. One such dietary product combines, prior to drying, equal weights of yeast autolysate and enzyme-hydrolysed lactalbumin. This high-nitrogen (9.5–10.5%) powder is also high in α-amino nitrogen (35–40% of the total nitrogen).

B. Enzymes

Numerous enzymes and coenzymes can be extracted from yeasts commercially. Those obtained in substantial quantities include invertase, lactase, coenzyme A, and nicotinamide adenine dinucleotide (NAD). Two of these, invertase and lactase, are produced in the largest volume and primarily for food use.

1. Invertase

Invertase (sucrase, saccharase, E.C. 3.2.1.26) is a β-fructofuranosidase whose action converts sucrose to fructose and glucose, and raffinose is hydrolysed to fructose and melibiose. Although the enzyme is present in many yeasts, *S. cerevisiae,* as bakers' yeast, is the principal commerical source (Meister, 1965).

Two commercial forms are produced: an invertase-rich dry yeast used to prepare high-test molasses from sugar cane juice, and a clear liquid concentrate extracted from autolysed yeast for the candy industry. The liquid product is the concentrate obtained by evaporation, under vacuum, of filtrates or extracts obtained from bakers' yeast autolysates. An invertase-enriched dry product is obtained by separating the yeast cells, after a short period of autolysis, dewatering the cream, extruding the press cake, and drying the pellets as is done in dehydrating active dry yeast (Peppler and Thorn, 1960).

Suomalainen *et al.* (1967) found invertase activity of bakers' yeast was nearly tripled when grown with mannose as carbon source instead of glucose. Highly purified invertase for biochemical research use is obtained by processing *C. utilis* cells. This yeast is also the preferred source for uricase and 6-phosphogluconate dehydrogenase.

2. Lactase

Lactase (β-galactosidase, β-D-galactopyranosidase, β-D-galactosidegalactohydrolase E.C. 3.2.1.23) catalyzes the hydrolysis of lactose to glucose and galactose. The hydrolytic reaction is used to alter the chemical and physical properties of milk, milk products, and whey (Bouvy, 1975; Holsinger, 1978; Guy and Bingham, 1978). Lactase is present in a number of common yeasts, including *S. lactis, K. fragilis* and *C. pseudotropicalis.* The commercial product, an odorless, tasteless, high-activity [40,000 oNPG (ortho-nitrophenyl-β-D-galactoside) units/gm] white powder, is extracted from *S. lactis* (Bouvy, 1975).

The process, described by Mahoney *et al.* (1975), compared four strains of *K. fragilis* grown aerobically in a lactose–salts medium. The harvested cells were autolysed at 37°C in 1 *M* potassium phosphate buffer (pH 7.0) containing manganese chloride (0.1 m*M*) and magnesium sulfate (0.5 m*M*). Lactase was isolated from the extract, after concentration, by precipitation with acetone. Stable preparations were obtained, but the lactase activity was low (860–1220 oNPG units/gm, dry wt.). Commercial, whey-grown, spray-dried yeast (*K. fragilis*) could also be used for lactase extraction. The highest enzyme yield (about 500 oNPG units/gm dry yeast) was also obtained with the phosphate buffer extraction procedure used for wet cells.

C. Ribonucleic Acid

Ribonucleic acid (RNA) is present in all living organisms, and microorganisms are exceptionally rich in nucleic acids. Yeasts, mainly species of *Candida* grown aerobically on molasses or sulfite waste liquor, are the preferred industrial raw material for the extraction of RNA. Potassium-sensitive strains of *Candida,* especially *C. utilis,* contain 10–15% RNA and are low in deoxyribonucleic acid (DNA) (Miyata, 1976; Akiyama *et al.*, 1975). Modifications in culture media composition (additions of zinc, antibiotics, etc.) and growing conditions influence the RNA biosynthesis (consult this volume, Chapter 10).

RNA is recovered from dried yeast (*C. utilis*) by extracting slurries (10% solids) with hot sodium hydroxide solution containing potassium chloride (about pH 11). After centrifuging, the supernate is treated with acid or ethanol and the crude precipitate is dried. Bakers' yeast may be extracted by a similar procedure using hot KCl–NaOH (Kuninaka *et al.*, 1976).

The crude RNA powder is converted enzymatically to the flavor-potentiating nucleotides 5′-guanosine monophosphate (5′-GMP) and 5′-inosine monophosphate (5′-IMP). For use in food applications the salts of these flavor accentuators are generally mixed, e.g., a 1 : 1 blend of disodium inosinate and disodium guanylate, with 5% of this mixture diluted in monosodium glutamate.

Japan is the principal producer of yeast nucleic acids, RNA, and its derivatives (Yamada, 1977).

D. Other Products from Yeasts

Candida utilis, and bakers' and brewers' yeasts, are the principal sources of biocatalysts, biochemicals, and metabolic intermediates used primarily for analytical and biochemical research, and in medical applications. Some of these are alcohol dehydrogenase, hexokinase, L-lactate dehydrogenase, glucose-6-phosphate dehydrogenase, glyceraldehyde-3-phosphate dehydrogenase, inorganic pyrophosphatase, the phosphoric acid esters of adenosine (AMP, ADP, ATP), the cytochromes, adenine, cytidine, guanosine, uridine, amino acids, glutathione (Horie *et al.,* 1976), vitamins, and enzymes of the citric acid cycle. Most of these products are isolated and purified in small quantities; their preparation has been reviewed by Harrison (1968, 1970).

As the predominant source of microbial protein (SCP), the isolation and characterization of protein from yeasts has attracted universal attention (Chen and Peppler, 1978). However, the commercial production of yeast protein concentrates is uneconomical at present (see Litchfield in this volume, Chapter 4). A recently developed process, described below, indicates the technology and feasibility of protein isolation from bakers' yeast.

1. Isolated Yeast Protein

A bakers' yeast cream containing 7–10% solids is passed through an industrial homogenizer (8000 psig) three times (Newell *et al.*, 1975; Robbins *et al.*, 1975). The homogenization ruptures more than 90% of the yeast cells and releases about 90% of the protein. After adjustment to pH 9.5, the homogenate is centrifuged. Two streams are collected: One is an extract of soluble protein and the other is a slurry of cell wall fragments. The soluble extract, following acidification, is incubated to initiate yeast nuclease activity and to precipitate the protein (Robbins, 1976). The protein is concentrated by centrifugation and spray-dried (Seeley, 1977). The final product, a free-flowing white powder, contains 75% protein and less than 1% nucleic acid. Its functional and nutritional properties have been described by Seeley (1977) and Robbins and Seeley (1977).

The proteins of *C. utilis* grown on ethanol have been isolated and modified for extrusion by Akin (1973, 1974). Vananuvat and Kinsella (1975) extracted protein from *K. fragilis* and studied its functional properties.

TABLE III. Principal United States Manufacturers of Yeasts and Yeast Products

	Bakers' yeast	Dried yeast	Wine Distillers, etc.	Extracts, lysates, etc.
Fleischmann Division Standard Brands, Inc. New York, New York	1	1	1	1
Red Star Yeast Division Universal Foods Corp. Milwaukee, Wisconsin	1	1	1	1
Anheuser-Busch, Inc. St. Louis, Missouri	1	2		2
Federal Yeast Division Diamond-Shamrock Corp. Baltimore, Maryland	1			
Dixie Yeast Corp. Gastonia, North Carolina	1			
Yeast Products, Inc. Clifton, New Jersey		1, 2		2
Amber Laboratories Juneau, Wisconsin		2, 4		2
Lake States Yeast Division St. Regis Paper Company Rhinelander, Wisconsin		3		
Boise-Cascade, Inc. Salem, Oregon		3		
Pure Culture Products, Inc. Hutchinson, Minnesota		5		
Stauffer Chemical Co. Westport, Connecticut		4		4
Kraft Inc. Chicago, Illinois		4		

[a] Numbers identify carbon source on which yeast was grown: (1), molasses; (2), beer; (3), sulfite liquor; (4), whey; (5), ethanol.

2. *Isolated Yeast Glycan*

Yeast glycan is the concentrated, insoluble material separated from slurries of mechanically ruptured yeast cells (Newell *et al.*, 1975). After it is spray dried, the free-flowing white powder (6% moisture) contains about 87% carbohydrate, 11% protein, less than 0.5% nucleic acid, and 1% ash. In aqueous food systems yeast glycan functions as a thickener (Seeley, 1977).

IV. OUTLOOK

The market for yeast and yeast products sold to the food industry is relatively mature. It is expected to maintain a unit growth of 2–4% annually in the next 3–5 years. Production of yeast as a protein source for animal feeds will increase as plant proteins become more expensive and the safety of yeasts grown on uncommon substrates is established. Table III lists the manufacturers of yeast and yeast products operating in the United States.

REFERENCES

AAFCO (1977). Official Publication. Association of American Feed Control Officials, Inc., Baton Rouge, Louisiana.

Akin, C. (1973). U.S. Patent 3,781,264.

Akin, C. (1974). U.S. Patent 3,833,552.

Akiyama, S., Doï, M., Arai, Y., Nakao, Y., and Fukuda, H. (1975). U.S. Patent 3,909,352.

Ammerman, C. B., and Miller, S. M. (1975). *J. Dairy Sci.* **58,** 1561–1577.

Andres, C. (1977). *Food Process.* **38**(11), 87–88.

Andres, C. (1978). *Food Process.* **39**(11), 22–23.

Bach, H. P., Woehrer, W., and Roehr, M. (1978). *Biotechnol. Biochem.* **20,** 799–807.

Bakery Trends (1977). *Bakery Production and Marketing,* June 1977, pp. 79–100. Gorman Publ. Co., Chicago, Illinois.

Bardot, P. (1972). U.S. Patent 3,660,908.

Behrman, D. (1978). *ASM News* **44,** 102–106.

Bernstein, S., and Plantz, P. E. (1977). *Food Eng.* **49**(11), 74–75.

Bouvy, F. A. M. (1975). *Food Prod. Dev.* **9**(2), 10–14.

Bruch, C. W., Hoffman, A., Gosine, R. M., and Brenner, M. W. (1964). *J. Inst. Brew.* **70,** 242.

Burrows, S. (1970). *In* "The Yeasts" (A. H. Rose and J. S. Harrison, eds.), Vol. 3, pp. 349–420. Academic Press, New York.

Burrows, S. (1976). U.S. Patent 3,962,467.

Burrows, S., and Fowell, R. R. (1961). British Patents 868,133 and 868,621.

Butschek, G. (1962). *In* "Die Hefen" (F. Reiff *et al.*, eds.), Vol. II, pp. 611–664. Verlag Hans Carl, Nürnburg.

Butschek, G., and Kautzmann, R. (1962). *In* "Die Hefen" (F. Reiff *et al.,* eds.), Vol. II, pp. 501–610. Verlag Hans Carl, Nürnburg.

Chen, S. L., and Cooper, E. J. (1962). U.S. Patent 3,041,249.
Chen, S. L., and Peppler, H. J. (1978). *Dev. Ind. Microbiol.* **19,** 79–94.
Chen, S. L., Cooper, E. J., and Gutmanis, F. (1966). *Food Technol.* **20,** 1585.
Chien, H. C. (1960). *Food Eng., Feb.,* pp. 92–95.
Coulter, C. J. (1964). *Proc. Am. Soc. Brew. Chem.* pp. 64–69.
DeSoto, R. T. (1955). *Am. J. Enol. Vitic.* **6,** 26–30.
East, E., Smith, B. J., and Borsden, D. G. (1966). *Food Manuf.* Sept.
Fantozzi, E., and Trevelyan, W. E. (1971). U.S. Patent 3,615,685.
Frey, C. N. (1930). *Ind. Eng. Chem.* **22,** 1154–1162.
Frey, C. N. (1957). *In* "Yeast—Its Characteristics, Growth and Function in Baked Products—a Symposium" (C. S. McWilliams and M. S. Peterson, eds.), pp. 7–33. Quartermaster Food and Container Institute for the Armed Forces, Chicago, Illinois.
Frost, D. V., and Lish, P. M. (1975). *Annu. Rev. Pharmacol.* **15,** 259–284.
GIAM V (1978). "Global Impacts of Applied Microbiology" (P. Matangkasombut, ed.), *Proc. Int. Conf., 5th, 1977,* Mahidol University, Bangkok, Thailand. John Wiley, New York.
Gillies, M. T. (1978). "Animal Feeds from Waste Materials." Noyes Data Corp., Park Ridge, New Jersey.
Gilliland, R. B. (1971). *In* "Modern Brewing Technology" (W. P. K. Findlay, ed.), pp. 108–128. Macmillan, New York.
Guy, E. J., and Bingham, E. W. (1978). *J. Dairy Sci.* **61,** 147–151.
Han, Y. W. (1978). *Adv. Appl. Microbiol.* **23,** 119–153.
Hansen, E. C. (1896). "Practical Studies in Fermentation." C. Griffin & Co., London.
Harrison, J. S. (1968). *Process Biochem.* **3**(8), 59–62.
Harrison, J. S. (1970). *In* "The Yeasts" (A. H. Rose and J. S. Harrison, eds.), Vol. 3, pp. 529–545. Academic Press, New York.
Harrison, J. S. (1971). *Prog. Ind. Microbiol.* **10,** 129–177.
Harrison, J. S., and Graham, J. C. J. (1970). *In* "The Yeasts" (A. H. Rose and J. S. Harrison, eds.), Vol. 3, pp. 283–348. Academic Press, New York.
Hayduck, F. (1919). German Patents 300,662 and 303,221.
Holsinger, V. H. (1978). *Food Technol.* **32**(3), 35–40.
Hoogerheide, J. C. (1977). *CHEMTECH* **7**(2), 94–97.
Horie, Y., Katayama, H., Uchida, T., and Kuroiwa, Y. (1976). Japan Kokai 76/76,483.
Irvin, R. (1954). *In* "Industrial Fermentations" (L. A. Underkofler and R. J. Hickey, eds.), pp. 273–306. Chem. Publ. Co., New York.
I.U.P.A.C. World Wide Survey of Fermentation Industries 1963 (1966). *Pure Appl. Chem.* **13,** 405–417.
Johnson, J. C. (1977). "Yeasts for Food and Other Purposes." Noyes Data Corp., Park Ridge, New Jersey.
Kaiser, W., and Jacobs, W. (1957). *Food Process.* **1,** 22 and 58.
Kiby, W. (1912). "Handbuch für Presshefenfabrikation." Friedrich, Viehweg & Son, Braunschweig, West Germany.
Kodama, K. (1970). *In* "The Yeasts" (A. H. Rose and J. S. Harrison, eds.), Vol. 3, pp. 225–282. Academic Press, New York.
Kuninaka, A., Fujimoto, M., and Yoshino, H. (1976). Japan Kokai 76/86,188.
Kunkee, R. E., and Amerine, M. A. (1970). *In* "The Yeasts" (A. H. Rose and J. S. Harrison, eds.), Vol. 3, pp. 5–71. Academic Press, New York.
Labuza, T. P. (1975). *In* "Single Cell Protein II" (S. R. Tannenbaum and D. I. C. Wang, eds.), pp. 69–104. MIT Press, Cambridge, Massachusetts.
Lane, G. T., Spiegel, R. E., and Botts, R. L. (1974). *J. Dairy Sci.* **57,** 135.
Langejan, A., and Khoudokormoff, B. (1976). U.S. Patent 3,993,783.
Laskin, A. I. (1977). *Annu. Rep. Ferment. Processes* **1,** 151–175.

Levander, O. A. (1975). *J. Am. Diet. Assoc.* **66,** 338–344.
Lipinsky, E. S., and Litchfield, J. H. (1974). *Food Technol.* **28**(5), 16.
Litchfield, J. H. (1977a). *Food Technol.* **31**(5), 175–179.
Litchfield, J. H. (1977b). *Adv. Appl. Microbiol.* **22,** 267–305.
Lodder, J. (1970). "The Yeasts, a Taxonomic Study," 2nd ed. North-Holland Publ., Amsterdam.
Lyons, T. P., and Rose, A. H. (1977). *In* "Alcoholic Beverages" (A. H. Rose, ed.), Vol. 1, p. 635–692. Academic Press, New York.
MacLeod, A. M. (1977). *In* "Alcoholic Beverages" (A. H. Rose, ed.), Vol. 1, pp. 43–137. Academic Press, New York.
Mahoney, R. R., Nickerson, T. A., and Whitaker, J. R. (1975). *J. Dairy Sci.* **58,** 1620–1629.
Meister, H. (1965). *Wallerstein Lab. Commun.* **28,** 7–15.
Mertz, W. (1969). *Physiol. Rev.* **49,** 163–239.
Mertz, W. (1975). *Nutr. Rev.* **33,** 129–135.
Mertz, W. (1977). *Food Prod. Dev.* **11**(6), 62 and 67–69.
Miyata, S. (1976). Japan Kokai 76/54,975.
Moo-Young, M. (1976). *Process Biochem.* **11**(10), 32–34.
Moo-Young, M. (1977). *Process Biochem.* **12**(4), 6–10.
Newell, J. A., Robbins, E. A., and Seeley, R. D. (1975). U.S. Patent 3,867,555.
Nyiri, L. K., Toth, G. M., Parmenter, D. V., and Krishnaswami, C. S. (1977). U.S. Patent 4,064,015.
Ohwaki, K., and Lewis, M. J. (1970). *Wallerstein Lab. Commun.* **33,** 105–116.
O'Sullivan, D. A. (1978). *Chem. & Eng. News* **56**(12), 12; 56(38), 12–13.
Pasteur, L. (1874). *Compt. Rend.* **78,** 213.
Peppler, H. J. (1965). *J. Agr. Food Chem.* **13,** 34–36.
Peppler, H. J. (1967). *In* "Microbial Technology" (H. J. Peppler, ed.), pp. 145–171. VanNostrand-Reinhold, Princeton, New Jersey.
Peppler, H. J. (1970). *In* "The Yeasts" (A. H. Rose and J. S. Harrison, eds.), Vol. 3, pp. 421–462. Academic Press, New York.
Peppler, H. J. (1978). *Annu. Rep. Ferment. Processes* **2,** 191–202.
Peppler, H. J., and Stone, C. W. (1976) *Feed Manag.* **27**(8), 17–18.
Peppler, H. J., and Thorn, J. A. (1960). U.S. Patent 2,992,748.
Perlman, D. (1977). *CHEMTECH* **7,** 434–443.
Pomper, S., and Akerman, E. (1971). U.S. Patent 3,617,306.
Pyke, M. (1958). *In* "The Chemistry and Biology of Yeasts" (A. H. Cook, ed.), pp. 535–586. Academic Press, New York.
Reed, G., and Peppler, H. J. (1973). "Yeast Technology," Avi Publ. Co., Westport, Connecticut.
Reuss, M., Gneiser, J., Reng, H. G., and Wagner, F. (1975). *Eur. J. Appl. Microbiol.* **1,** 295–305.
Ridgway, J. A., Jr., Lappin, T. A., Benjamin, B. M., Corns, J. B., and Akin, C. (1975). U.S. Patent 3,865,691.
Robbins, E. A. (1976). U.S. Patent 3,991,215.
Robbins, E. A., Sucher, R. W., Schuldt, E. H., Sidoti, D. R., Seeley, R. D., and Newell, J. A. (1975). U.S. Patent 3,887,431.
Rockwell, P. J. (1976). "Single Cell Proteins from Cellulose and Hydrocarbons." Noyes Data Corp., Park Ridge, New Jersey.
Rosen, K. (1977). *Process Biochem.* **12**(3), 10–12.
Rosenquist, S. O., and Egnell, E. R. (1966). Swedish Patent 209,600.
Rungaldier, K., and Braun, E. (1961). U.S. Patent 3,002,894.

Sakurai, H., and Tsuchiya, K. (1975). *Environ. Physiol. & Biochem.* **5,** 107–118.
Schuldt, E. H., and Seeley, R. D. (1966). *Baker's Dig.* **40**(2), 42.
Schwarz, K. (1974). *Fed. Proc., Fed. Am. Soc. Exp. Biol.* **33,** 1748–1757.
Schwarz, K. (1976). *Med. Clin. North Am.* **60**(4), 745–758.
Scrimshaw, N. S. (1968). *In* "Single-Cell Protein" (R. I. Mateles and S. R. Tannenbaum, eds.), pp. 3–7. MIT Press, Cambridge, Massachusetts.
Seeley, R. D. (1975). *Food Prod. Dev.* **9**(7), 45.
Seeley, R. D. (1977). *MBAA Tech. Q.* **14**(1), 35–39.
Senez, J. C. (1975). "Producing Food Protein from Hydrocarbons," pp. 29–33. CNRS, Paris.
Stark, W. H. (1954). *In* "Industrial Fermentations" (L. A. Underkofler and R. J. Hickey, eds.), Vol. I, pp. 17–72. Chem. Publ. Co., New York.
Strandskov, F. B. (1965). *Wallerstein Lab. Commun.* **28,** 29–33.
Suomalainen, H. (1963). *Pure Appl. Chem.* **7,** 639.
Suomalainen, H., Christiansen, V., and Oura, E. (1967). *Suom. Kemistil. B* **40,** 286–287.
Swartz, J. R., and Cooney, C. L. (1978). *Process Biochem.* **13**(2), 3–7 and 24.
Tannenbaum, S. R., and Wang, D. I. C., eds. (1975). "Single-Cell Protein II." MIT Press, Cambridge, Massachusetts.
Taylor, R. (1975). U.S. Patent 3,885,049.
Thorn, J. A., and Reed, G. (1959). *Cereal Sci. Today* **4,** 198.
Thorne, R. S. W. (1970). *Process Biochem.* **5**(4), 15–16 and 22.
Thorne, R. S. W. (1975). *Process Biochem.* **10**(7), 17–20 and 25–28.
Thoukis, G., Reed, G., and Bouthilet, J. R. (1963). *Am. J. Enol. Vitic.* **14,** 148–154.
Toepfer, E. W., Mertz, W., Roginski, E. E., and Polanski, M. M. (1973). *J. Agric. Food Chem.* **21,** 69–73.
Trevelyan, W. E. (1973). U.S. Patent 3,780,181.
Tuman, R. W., Bilbo, J. T., and Doisy, R. J. (1978). *Diabetes* **27,** 49–56.
Vananuvat, P., and Kinsella, J. E. (1975). *J. Agric. Food Chem.* **23,** 216–221 and 613–616.
Van Lanen, J. M., Smith, M. B., and Maisch, W. F. (1975). U.S. Patent 3,868,307.
Walter, F. G. (1953). "The Manufacture of Compressed Yeast," 2nd ed. Chapman & Hall, London.
Wang, H. Y., Cooney, C. L., and Wang, D. I. C. (1977). *Biotechnol. Bioeng.* **19,** 69–86.
Weatherholtz, W. M., and Holsing, G. C. (1976). *Ecol. Food Nutr.* **5**(3), 153–159.
White, J. (1954). "Yeast Technology." Chapman & Hall, London.
Woolford, M. K. (1976). *J. Appl. Bacteriol.* **41,** 29–36.
World Wide Survey of Fermentation Industries, 1967 (1971). *I.U.P.A.C. Inf. Bull.* No. 3.
Yamada, K. (1977). "Industrial Fermentation." Int. Tech. Inf. Inst., Tokyo.
Ziemba, J. V. (1967). *Food Eng.* **19**(1), 82–85.

Chapter 6

Production of Butanol–Acetone by Fermentation

M. T. WALTON
J. L. MARTIN

MICROBIAL TECHNOLOGY, 2nd ed., VOL. I
 ISBN 0-12-551501-4

I. INTRODUCTION

From its start as a wartime venture in England in 1916 until the establishment of the synthetic butanol process in the late 1930s, the fermentation process for the manufacture of butanol was the major method for its production. The plants were large operations in comparison to other fermentations then or now. A plant at Peoria, Illinois, contained ninety-six 50,000-gal fermentors and was capable of grinding 25,000 bu of corn per day for use as raw material.

Because of the unfavorable economics in comparison to the synthetic process using petroleum-based feed stocks, the fermentation process ceased to operate in the early 1960s in the United States and in most other countries. However, as price and availability of ethylene and propylene as feed stocks for the synthetic processes have become subjects of concern, there is renewed interest in examining the fermentation processes as a means of producing all or a portion of our future needs of butanol and acetone.

This chapter outlines the procedures used in the fermentation processes and presents data and suggestions which should be considered in evaluating the economics of these processes, based on replenishable carbohydrate raw materials, and of the synthetic processes which use diminishing hydrocarbon feed stocks.

II. HISTORY

The interesting history of the butanol–acetone process has been described by several authors and will be only briefly summarized here (Gabriel, 1928; Gabriel and Crawford, 1930; McCutchan and Hickey, 1954).

The original observation that bacteria produce butanol was made by Pasteur in 1861. The interest in commercializing the process was accelerated in 1909 in England primarily as a means to obtain butadiene as raw material for synthetic rubber. In 1912 Dr. Chaim Weizmann broke from the group that was developing the process but continued to study the fermentation. After about 2 years he found an organism, which he called *Clostridium acetobutylicum,* which successfully fermented starchy grains to produce acetone, butanol, and ethanol. He applied for a patent on the process and British Patent 4845 was issued to him (Weizmann, 1915).

With the outbreak of World War I in 1914, the production of acetone was of great interest to England for the manufacture of cordite, the explosive used in naval warfare. A plant was built at Kings Lynn, England, in 1914 for the production of acetone by fermentation, but the operation was a failure until Dr. Weizmann was placed in charge of the plant and installed his process. Because of the shortage of corn in England, the process was transferred in August 1916 to Canada where it was operated until November 1918. To supplement the output of acetone from Canada, a butanol–acetone plant was built in Terre Haute, Indiana, which operated from May to November of 1918.

While the Terre Haute plant was operating during World War I, there was no use for butanol and it was stored. Shortly after the end of the war, E. I. DuPont de Nemours & Company developed nitrocellulose lacquers for the automobile industry, and it was found that butyl acetate was the solvent of choice for this coating system. In order to supply butanol for conversion to butyl acetate, the butanol–acetone (ABE) fermentation unit at Terre Haute was started in 1920 as a private venture by Commercial Solvents Corporation, which had an exclusive license under the United States Weizmann patent (Weizmann, 1919).

The validity of the patents covering the Weizmann fermentation process was challenged in England in 1926 and in the United States in 1931. In landmark decisions, British Patent 4845 and U.S. Patent 1,315,585 were held to be valid and infringed. One of the key issues in these suits was whether the patent specifications gave proper directions for the isolation of *Clostridium acetobutylicum* (BY). In his decision on the English suit Lord Romer wrote,

> . . . the description in the Specification is intended for the guidance of those who are desirous of obtaining BY and not for people who wish to avoid getting it. (Reports of Patent Design and Trademark Cases, 1926; *U.S. Patent Quarterly,* 1932.)

After the expiration of the Weizmann patent in 1936, new ABE fermentation plants were built in Philadelphia, Pennsylvania; Baltimore, Maryland; Puerto Rico; and Japan.

Many early attempts were made to produce acetone and butanol by the fermentation of molasses since it furnished carbohydrates at a cost lower than did corn. However, it was not until 1938, when new microorganisms were discovered, that molasses was used successfully as a substrate for this fermentation (Arzberger, 1938; Carnarius and McCutchan, 1938; Legg and Tarvin, 1938; Woodruff *et al.*, 1937; Müller, 1938).

III. VERSATILITY OF THE PROCESSES

A. Solvent Ratios

Acetone and butanol are produced from starchy materials such as corn by use of the species *C. acetobutylicum* with a solvents yield of about 26.5% based on dry corn. The solvent ratio usually runs 30% acetone, 60% butanol, and 10% ethanol (Gabriel, 1928; Gabriel and Crawford, 1930; McCutchan and Hickey, 1954). An unusual variant of *Clostridium saccharo-butyl-acetonicum-liquefaciens* (Code C-12) ferments molasses rather poorly but gives excellent yields on corn with a high butanol ratio. The solvent yield is 25% based on dry corn, with a ratio of 19% acetone, 78% butanol, and 3% ethanol.

The species of microorganism most widely used to ferment molasses is named *C. saccharo-butyl-acetonicum-liquefaciens* (Arzberger, 1938; Carnarius and McCutchan, 1938). With the exception of the variant (Code C-12), discussed in the preceding paragraph, microorganisms of this species give poor results on starchy grains. Selected variants of the foregoing species give solvent yields of 30–33% based on sucrose, with solvent ratios which vary from 20 to 35% acetone, 76–61% butanol, and 4% ethanol. (Carnarius and McCutchan, 1938; Legg and Tarvin, 1928; Woodruff *et al.*, 1937). Baba (1954a,b) reported data obtained by fermenting with *Clostridium toanum* a medium containing 0.3% defatted rice bran, 0.3% $(NH_4)_2SO_4$, and 7.0–7.8% sugar derived from cane juice, or blackstrap molasses, or a mixture of the two. The yield of solvents was about 30% based on glucose with an average solvents ratio of 52.92% butanol, 42.47% 2-propanol, 3.20% ethanol, and 1.41% acetone. Baba (1953) stated that the volume of gas evolved from the fermentation of a 6% sucrose solution with *C. toanum* was about 26 times the volume of mash and that the composition of the gas was about 50% CO_2–50% H_2.

Variations in solvent ratios have been accomplished largely by isolation of different cultures. Little success has been achieved by genetic manipulation of the cultures, but no work with butanol cultures has been reported in the last few years using newer techniques that have been successfully employed with other species of bacteria.

The ratio of solvents can be modified to some extent by manipulating the temperature of the mash and the composition of the nutrients in it. Tarvin (1941) reported that the addition of nitrates instead of ammonia to a molasses fermentation increased the amount of acetone at the expense of the butanol, e.g., in a typical experiment the addition of $NaNO_3$ at a concentration of 4.5 gm/liter increased the acetone percentage from 30.7 to 40.9. Hongo (1958) stated that the addition of neutral red in a concentration of 6 m*M* to a *C. acetobutylicum* fermentation of either xylose or glucose increased the butanol ratio at the expense of acetone, giving about 10–15% more butanol. Carnarius (1940) found that the butanol ratio could be increased 3–5% at the expense of acetone by cooling a molasses mash from 30°C to 24.5°C 16 hours after inoculation.

B. Raw Materials

Most of the acetone and butanol made by fermentation has used one or more of three sources of carbohydrates, i.e., corn, blackstrap molasses,* or high-test molasses.* Other common sources of carbohydrates such as wheat, rice, horse chestnuts, Jerusalem artichokes, potatoes, and beet molasses can be readily used. However, when we consider these fermentation processes as possible replacements for those based on hydrocarbon feed stocks, we must look carefully to carbohydrates derived from sources now considered to be waste products. These include hydrolyzed wood, hydrolyzed corncobs, hydrolyzed garbage, whey, sulfite liquor, and hydrol, a by-product from the manufacture of glucose from corn.

Leonard and Peterson (1947) found that sugar solutions prepared by the hydrolysis of maple, oak, and fir could be fermented with *C. acetobutylicum* to give a solvents yield of 24.5–38.5% based on the sugar utilized if the sugar solutions were first steam stripped of furfural and other impurities and then adjusted with lime to pH 6.5. The sugar utilization was about 80–85% with a solvent ratio of approximately 72% butanol, 25% acetone, and 3% ethanol. Further work needs to be done on this raw material since yields fell off when the initial sugar concentration of the mash increased much beyond 3% (Tsuchiya *et al.*, 1949).

The production of butanol and acetone from the sugars present in waste sulfite liquor was studied by Wiley *et al.* (1941), who treated the liquor first with lime to remove sulfites, then with additional lime to precipitate lignin, followed by removal of excess calcium salts as the sulfate. After the pH was adjusted to 5.8, the following nutrients were added: 0.05% $(NH_4)_2HPO_4$, 0.1% molasses, and 0.1% $CaCO_3$. In the fermentation about 80% of the sugars were utilized with a yield of 25–30% based on the sugars used. The ratio of solvents was 75% butanol, 20% acetone, and 5% ethanol. This process suffers from the same deficiency as that described above in that the sugar concentration of the sulfite liquor runs from 1 to 3%.

Langlykke *et al.* (1948) studied the production of butanol and acetone by fermenting xylose and glucose derived from the acid hydrolysis of corncobs. Their sugar solution, which was made in a bronze hydrolysis unit, contained about 18 ppm of copper, and this toxic element was removed by treating the liquor with powdered iron. The purified hy-

* Blackstrap molasses is the concentrated mother liquor from the crystallization of sucrose from sugar cane juice and contains 50–55% sucrose. High-test molasses is the concentrate of sugar cane juice in which 67% of the sucrose has been inverted to glucose and fructose to prevent crystallization and to avoid sugar import duties.

drolysate with an initial sugar concentration of 5.03% gave a sugar utilization of 90% and a 30.6% yield of solvents based on the sugar used. The ratio of solvents was butanol 61.7%, acetone 31.8%, and ethanol 6.5%. The hydrolysis work showed that from 2000 lb of corncobs one could obtain 606 lb of glucose and 484 lb of xylose (Dunning and Lathrop, 1945). It may be noted that the toxic effect of copper could be eliminated by a proper choice of materials of construction for the hydrolyzer. The problem of furfural toxicity would yield to steam stripping of the hydrolysate since furfural forms an azeotrope with water (Horsley, 1973).

Some recent results on the butanol fermentation of pentoses are exhibited in Table I.

IV. FERMENTATION PROCEDURES AND EQUIPMENT

This section discusses procedures geared to a plant equipped with fermentors of 50,000-gal capacity.

A. Overview of Procedures and Results

1. Corn and Other Grains

The fermentation of corn with *C. acetobutylicum* does not require any additives. Since corn oil is a valuable product and is not beneficial to the fermentation, it is recovered by steeping the corn followed by grinding with roller mills which flatten the germ so that it can be removed by screens. The oil is extracted from the germ and the degerminated meal used for the fermentation. Depending upon the value of the oil-cake meal as livestock feed, it is either sold for this purpose or is added with the degerminated meal to the fermentation mash. The starch portion of the oil-cake meal is used by the bacteria and the protein portion enhances the feed value of the recovered solids.

The corn meal is added to 60% water and 40% stillage (fermented liquor from which solvent has been distilled) to a calculated concentration of 8.5% of original dried corn. The concentration of solvents in the finished fermentation liquor (beer), after 50–56 hours, is about 22.5 gm/liter, with a solvent ratio of 30% acetone, 60% butanol, and 10% ethanol. About 0.25–0.26 lb of solvents are produced per pound of dry corn equivalent.

In order to obtain complete utilization of the starch, its starting concentration in the mash must be chosen so the butanol concentration in the final beer is no higher than 13 gm/liter. Depending on the butanol ratio

TABLE I. Recent Reports on the A.B.E. Fermentation of Hydrolysates of Corncobs, Cornstalks, and Wood

Culture	Substrate	Time (hours)	Yield (%)	Solvent Composition (%) B	A	E	References[a]
C. acetobutylicum S_{25}	Hydrolysate of corncobs, 7%		26.2				(1)
	Hydrolysate of sawdust, 7%		22.2				(1)
Butyl culture	One part hydrolysate of cornstalks Three parts molasses	50–55	31–37				(2)
Butyl culture	Wood and plant hydrolysates, 8%		35.5	62	32	6	(3)
C. acetobutylicum 314	Corncob hydrolysate, 40–60%, (3.7% total) Molasses, 60–40% (sugar)	48–72	40	67.5			(4)
Butyl culture	Pentoses, 13.5%		25.4	67	33		(5)

[a] Key to references: (1) Taha *et al.* (1974); (2) Nakhamonivich *et al.* (1965); (3) Tornescher Hefe, G.m.b.H. (1954); (4) Nakhmanovich *et al.* (1960); (5) Moldenhauer and Lechner (1951).

characteristic of the culture being used, the required initial concentration of starch will be in the range of 6–8% on dry corn equivalent.

Differing greatly from the Weizmann organism, the Code C-12 culture has poor proteolytic powers and the medium must be fortified during the fermentation with ammonia, adding about 1.0% on dry corn equivalent to maintain a pH of 5.5–5.8. The fermentation is slower, requiring 72–80 hours, giving yields of about 25–25.5% on dry corn equivalent with a ratio of 19% acetone, 78% butanol, and 3% ethanol.

2. Molasses

As in the corn fermentation, a butanol concentration of 13 gm/liter in the final fermentation will limit the fermentation. Thus, the sugar concentration used in the starting mash varies from 5.5 to 7.5% depending upon the solvent ratio produced by the culture selected. The mash is augmented by the addition of superphosphate and ammonia. The amount of phosphate varies from 0.05 to 0.2% of P_2O_5 based on sugar. Blackstrap molasses requires the lesser amount and high-test molasses requires amounts in the higher range. The exact amount is determined by laboratory fermentation tests. The ammonia is added to high-test molasses mashes in hourly increments as needed to maintain the pH at 5.6–6.0, about 1.2–1.3% NH_3 based on sugar concentration being required. Because of the higher buffering properties of blackstrap molasses, all or nearly all of the ammonia can be added to mashes of this molasses before cooking.

3. Procedures Common to Fermentation of Both Types of Raw Material

a. Slopback. *Slopback* is the term used for the procedure of adding hot still residue from the solvent stripping column to the mash makeup. This procedure is used in all of the butyl fermentation processes. The amount of slopback varies from 25 to 50% of the total mash, the percentage used being the highest tolerated by the culture without giving reduced yields. The advantages are an increase in yield of solvents, a decrease in the amount of nutrients required, a saving in heat required in cooking, a saving in cooling water, and a substantial saving in steam required to evaporate the residue.

b. Sterilization. The fermentation equipment (tanks and lines) is sterilized 4–6 hours at 110°–115°C and is then cooled before use while maintaining it under a pressure of 15 psig (pounds per square inch gauge) with sterile fermentor gas.

Sterile fermentor gas is preferred to air to replace steam during the cooling of the fermentors to ensure an anaerobic atmosphere during the start of the fermentation in the plant. The absence of air in the fermentor gas reduces the cleanup problem when the gas is to be used for manufacture of methanol, ammonia, or Dry Ice. If the gas is not to be used for further processes, sterile air may be used for pressure maintenance.

The gas is sterilized by sending it through a tower of raschig rings in which a 5% solution of cresol in water is circulating. Other satisfactory methods, such as the use of glass wool filters, can be employed.

B. Culture Maintenance

Since the butanol–acetone organisms produce true spores, the cultures are maintained as spores, preferably with soil as a carrier. Each fermentor is started from the spore stock. The method of preparing the spore stock is as follows.

Sand, soil, volcanic ash, or a mixture of these, is screened and about 250 gm is placed in a 500-ml Erlenmeyer flask. Sufficient water is added to moisten the material. The flasks are plugged with cotton and sterilized four times at 24-hour intervals for 3 hours at 15 psig steam pressure. Excess moisture is removed in an oven.

Sufficient second- or third-generation culture (a generation for this discussion is a 24-hour transfer) is allowed to incubate until the fermentation is complete (about 72 hours) and then is added aseptically to the dried sterile soil until it is moderately wet. The flasks are then placed in a desiccator containing $CaCl_2$, the pressure is reduced, and they are allowed to remain until dry (about 3–4 days).

To check for contamination, a liberal amount of the soil–spore mixture is added to a molasses or corn medium, incubated, and plated daily for 3 days. The incubating flasks are plated aerobically and if no contaminating organisms appear, the cultures are then tested for solvent yield. If yields and solvent ratios are satisfactory, plant tests of several fermentors are made before the culture is adopted for general use.

C. Laboratory Development of Cultures

The first, second, and third generations are carried in the laboratory. For molasses fermentations and aided corn fermentations using *C. saccharo-butyl-acetonicum-liquefaciens*, the first generation is carried in 15 ml of peeled potato medium. This medium is prepared by blending in a Waring Blendor 250 gm wet weight of potatoes, 5 gm glucose, and 2 gm $CaCO_3$ with water to make 1000 ml of medium. This is distributed 15 ml to a tube and the tubes are plugged with cotton and autoclaved 1 hour at 15 psig steam pressure. The tubes are inoculated with about 1 gm of stock culture. The inoculated tubes are immediately heated for 1 minute in boiling water or in a steamer to inactivate vegetative cells and to "shock" the spores and are immediately cooled to inoculating temperature. The tubes are then incubated at 31°C for 24 hours.

For unaided corn fermentations (*C. acetobutylicum*), the first-generation medium is made with 5% dry corn equivalent in water. No additives are necessary. The medium is sterilized for 2 hours at 20 psig steam pressure. The inoculation procedure is the same as above, but the incubation is carried out at 37°C.

One first-generation tube is used to inoculate 300 ml of medium in a 500-ml flask. The inoculum medium for aided molasses contains 5.6% sucrose, 0.3% each of $(NH_4)_2SO_4$ and $CaCO_3$, and 0.01% P_2O_5 as superphosphate. The medium for aided corn fermentations is similar except that corn is substituted for molasses. The inoculated medium is incubated at 31°C. Medium for the unaided corn process is made with 5.5% dry corn equivalent and is incubated at 37°C.

After 24 hours, the one second-generation flask is used to inoculate two 4-liter flasks containing 2900 ml of the same medium as that for the second generation. One of these flasks is used to inoculate the first vessel in the plant. Since the fermentors in the plant are set in groups of four or more, at least one more flask is developed than needed to allow a choice of the flasks to be used. The flasks used should be gassing actively, have a pH of about 5.4, and pass a microscopic examination showing actively growing bacteria. The carbohydrate should be about one-half fermented when the flask is used.

D. Seed Stage

The first plant stage is usually a 1000-gal tank, although with certain cultures an intermediate 80-gal tank is used before the 1000-gal tank. The tanks are filled to 75% capacity with the desired medium. For the molasses fermentation cultures, the medium is 5.5% sucrose as molasses, 0.3% each of $(NH_4)_2SO_4$ and $CaCO_3$, and 0.02% P_2O_5 as superphosphate. The aided corn fermentation uses the same medium, except that corn on a dried equivalent basis is substituted for the molasses. Unaided corn seeds are made to 6.5% dry corn equivalent concentration and require no nutrients.

The medium is generally cooked in a batch cooker. The molasses medium is cooked at 118°C for 60 minutes and the corn mash is cooked at 133°C for 90 minutes. Either one is pumped hot through steam-sterilized lines to previously sterilized tanks still under steam and held a few minutes before being cooled to fermenting temperature by water in the jackets of the tanks.

After the tanks are cooled to the desired temperature, each is inoculated with the contents of one 4-liter flask. In the molasses fermentation, pH and Brix measurements are made on seeds immediately after inoculation and again after 24 hours to determine activity. The pH should be about 6.0 at the start and fall to about 5.4 at 24 hours. The Brix will drop about 1.5 points in 24 hours. In seeds with unaided corn, the activity is measured by titratable acidity. Gas rates measured on samples taken at 24 hours of age, using a Smith gas-rate tube, should show a displacement of at least 3 inches in 60 minutes in all three types of media (Salle, 1939). Aerobic contamination tests are started at 12 hours of age and read before the seeds are used at 26–28 hours. If these tests are negative, the seeds are used. An extra seed or two is usually carried with the group and if all are good, any extras are added to the fermentors.

The seeds are fermented under 15 psig gas pressure. Before the seeds generate sufficient gas, the gas pressure is maintained with fermentor gas from the sterile system, or with sterile air if gas is not available.

E. Fermentors

The final liquid volume in a 50,000-gal fermentor is in the range 45,000–47,500 gal, this volume being largely dependent on the extent of the foaming encountered.

After the fermentor has been washed free of residue from the previous fermentation, it is sterilized 4–5 hours with steam at 15 psig. It is then cooled, using sterile fermentor gas to hold a pressure of 10–15 psig. The molasses medium ranges from 5.5 to 7.5% sugar as molasses, 0.0025–0.015% P_2O_5 as superphosphate, and 0.07–0.1% NH_3 depending upon the culture used and on whether the medium contains blackstrap or high-test molasses. Aided corn mash consists of 6.8% dry corn equivalent, 0.014% P_2O_5 as superphosphate, and about 0.07% NH_3. The ammonia is added during the fermentation in hourly increments to maintain the pH in the range 5.5–6.0.

The ingredients, including 25–50% "slopback," are added to horizontal batch cookers with horizontal shafts to which are attached vertical rakes. Steam is added directly to the mash through several check valves in the bottom of the cooker. The molasses medium is cooked 60 minutes at 107°C; the corn medium is cooked 90 minutes at 133°C. After cooking, the mash is pumped through sterilized lines and cooled with jacketed water coolers to the fermenting temperature. Four cooks are required to fill one fermentor. Two seeds are added to the fermentor early in the filling period. A slightly positive pressure of sterile fermentor gas is maintained on the fermentors during the filling period (2–3 psig) and this is increased to 10–15 psig after filling and maintained at this value during fermentation by the gas evolved by the fermentation.

Samples are withdrawn from the fermentor at frequent intervals to check the pH of the aided medium and the titratable acidity of unaided corn medium. About every 8 hours samples are streaked on agar plates and incubated aerobically to check for contaminants. The molasses fermentation is completed in 40–45 hours. The unaided corn fermentation is completed in 50–60 hours, and the aided corn requires 70–80 hours.

F. Recovery of Solvents

After fermentation, the fermented liquor (beer) is sent to a beer well about twice as large as a fermentor at intervals sufficient to allow for a constant flow to the beer still. This is a continuous still, containing about 30 perforated plates, which concentrate the solvents to a 50% solution in water. The mixed solvents can then be separated by batch fractionation in the following steps: (1) Charge a kettle equipped with steam scrolls and a 40-plate column; (2) first take off the acetone fraction and collect it into the acetone running tank; (3) after a 5°C rise in outflow temperature, turn the flow into the ethyl fraction tank; (4) when the distillate starts to separate into two layers, turn the top layer to the crude butyl tank and return the bottom layer to the kettle. When there is no longer a top layer in the distillate, the steam is shut off and the liquid in the kettle is discarded.

The various crude fractions are further fractionated to remove traces of the contaminating solvents. The butanol fraction is dried by taking off the distillate through a decanter, returning the top layer to the kettle and the lower layer to the crude solvent fraction tank. When there is no longer separation and the butanol tests water-free, it is collected in the finished butanol tank. The kettle residue containing higher alcohols and esters is generally esterified and sold separately.

Of course, suitable reflux ratios are used throughout the separation process. A continuous fractionation system can be used, but its lack of flexibility might limit the choice of fermentation raw materials and the choice of cultures producing different solvent ratios.

G. Production Controls

Sugar in molasses is determined by means of Fehling's solution by any one of several common procedures. The sugar analysis of the cargo is used to calculate the amount of molasses needed. A hydrometer calibrated in degrees Brix is used to check the sugar content of the mash at the start of the fermentation and at the end to determine completeness of fermentation.

The starch content of the meal is determined by converting to glucose and analyzing by Fehling's solution. The moisture of the meal is obtained by determining loss in weight of a sample under desiccation in a drying oven. The presence of starch in the fermentation residue is determined by the iodine reaction.

Since the fermentations produce gas, the rate of gas evolution is a good measure of the fermentation rate. This rate is checked at intervals of about 8 hours using a Smith gas-rate tube (Salle, 1939). Since the harmful bacterial contaminants are facultative and since the butanol–acetone producers are strict anaerobes, the presence of growth on a nutritive agar plate streaked with a fermenting sample indicates the presence of contaminants. This tool is used in all fermentation stages.

A pH determination is used to indicate the need for ammonia addition during fermentation, and a large drop in pH indicates the presence of contaminants. The titratable acidity in the unaided corn fermentation is used to follow the course of that fermentation.

The concentration of solvents in the beer is estimated by comparing the specific gravity of the distillate from a measured sample of beer with a known amount of water solution containing solvents of about the same ratio. The solvent ratio can be determined by either mass spectrometry or gas chromatography. If this type of equipment is not available, an older method can be used in which the first step is the determination of acetone in the distillate by changing it to iodoform with hypoiodite. A fresh sample of distillate is salted out to obtain an anhydrous solution of solvents. This anhydrous solution is adjusted to a 25% acetone content by the addition of either acetone or butanol. By titrating the adjusted anhydrous solution to a turbidity point with water and comparing the result with prepared tables, a value for the ethanol content is obtained. The concentration of butanol is obtained by difference (McCutchan and Hickey, 1954).

V. STERILITY AND CONTAMINATION

A. Sterilization and Cleanliness

The butanol–acetone fermentation is less subject to contamination problems than are many others because it operates in the absence of air, thus discouraging aerobic organisms. Steam is the general means of maintaining sterile conditions. To make the steam heat effective, it is necessary to operate without building up organic products in the system.

The primary purpose of cooking the medium is to sterilize it. However, in the case of corn it is also necessary to gelatinize the starch, which makes it necessary to cook under more extreme conditions.

The mash can be cooked in either a batch cooker or a continuous cooker with high temperatures and short residence times. The latter is more adaptable to molasses mashes, since the medium is essentially soluble.

B. Bacterial Contamination

Although the butanol–acetone fermentation is less subject to contamination than are most aerobic processes, it must be operated essentially aseptically. The lactobacilli are particularly troublesome in the corn fermentation and to a lesser extent in the molasses fermenting processes. Samples of the fermenting medium from the laboratory through the final fermentors must be checked for bacterial contamination on aerobically incubated agar plates.

C. Bacteriophage

Problems with bacteriophage have been reported wherever the butanol–acetone process has been operated. It is detected when the fermentation slows as evidenced by low gas rate, nonacceptance of usual ammonia additions in the case of molasses fermentations, and nondisappearance of carbohydrates. A fermentation infected with a phage may stop without apparent cause. Although the fermentation may start again after 24–48 hours, it will probably be slow and give low yields. In order to recover values from a fermentor which has been infected with a phage, some workers have added an inoculum of *Candida utilis* to complete the utilization of carbohydrates (Kinoshita *et al.*, 1954).

The bacteriophage can be separated from its bacterial host by passing through a filter such as Berkefeld or a sintered glass, which retains the bacteria but allows the bacteriophage to pass through. The presence of phage is confirmed when the filtrate interferes with fermentations by its host in laboratory tests.

The culture can be made phage resistant by repeated transfers in the presence of the phagic filtrate. A small amount of the filtrate is added after each transfer. Legg (1928), when using the *C. acetobutylicum* culture, allowed each transfer to incubate to completion and "heat shocked" the culture after each transfer for 3 minutes at 100°C just prior to addition of the phagic filtrate. Legg and Walton (1938), in working with molasses fermenting cultures, found it preferable to make each transfer as soon as the culture recovered, adding fresh phagic filtrate at the same time. McCoy (1946) employed a special medium consisting of liver infusion, glucose, and Bactotryptone for carrying the resistant culture.

The development of a phage-resistant culture is not a complete solution of the problem. The resistant cultures contain particles of phage added to develop the property of resistance, and unless these particles are removed, they will accompany the culture in a new fermentation. The phage will then increase in virulence and in many instances the fermentation will again be affected within a short time (Hongo and Murata, 1966). Physical or chemical methods of inactivation are seldom successful because the tolerance of the phage and bacteria are quite similar.

In 1937 M. T. Walton in hitherto unpublished work discovered a method, based on antiphagic serum, by which he could inactivate the phage particles associated with the resistant bacteria. This method is carried out by injecting 1.0–5.0 ml of the phagic filtrate intravenously in rabbits every other day for 20–30 days. The development of the antiphagic property is confirmed when a sample of phagic filtrate to which the serum is added fails to inhibit the susceptible culture. The serum is judged acceptable when 1 ml inactivates all the phagic particles in 10 ml of filtrate. Two to five milliliters of the serum is then added to the phage-resistant culture immediately after the heat shocking step. The second generation is used to make soil cultures.

The butyl cultures may be attacked by more than one bacteriophage and the culture made resistant to one phage may be susceptible to another. It would, in that event, be necessary to repeat the foregoing process using filtrates of the new phage. Some phages are specific to a given strain of butanol–acetone-producing cultures; others may attack more than one strain, usually to a varying degree.

VI. BY-PRODUCTS

A. Gases

The gas produced from a molasses culture giving solvents with a butanol content of around 70% consists of 67% CO_2 and 33% H_2. Cultures giving a lower butanol percentage produce gases with a higher content of hydrogen. About 4.7 ft^3 of gas is produced per pound of sucrose. About 5.9 ft^3 of gas is produced per pound of dry corn from the

C. acetobutylicum and the gas composition is about 40% hydrogen and 60% CO_2. The gases can be used to produce methanol or ammonia. When the two gases are separated, the hydrogen can be used in chemical synthesis or burned for energy, and the CO_2 solidified into Dry Ice.

The gases leaving the fermentors carry appreciable amounts of solvents with them. The solvents may be recovered by passing the gases through beds of activated carbon or by a water scrubbing system.

B. Solid Residues

The liquid effluent from the bottom of the beer still contains solids which must be processed because of environmental considerations. The amount of solids and the biological oxygen demand (BOD) will vary depending on the raw material fermented and the amount of slopback, but the solids are normally in the range of 2.0–4.5 gm/100 ml and the BOD runs from 20,000 to 40,000 ppm.

In the late 1930s, Miner (1940) found that the bottoms from the beer still contained riboflavin and other vitamins. The bottoms from the beer still are concentrated in multiple-effect evaporators and the concentrate is dried with either drum or spray driers to give solids which can be incorporated in poultry and swine feeds. The dried residue from the fermentation of high-test molasses contains from 60 to 100 μg riboflavin/gm and from 28 to 30% protein.

Because of its high solids content, blackstrap molasses produces more dried residue than do other commercial raw materials. However, the disposal of the residue from a fermentation of blackstrap molasses is a problem because the high salt content leads to a hygroscopic product and it produces diarrhea in poultry.

The foregoing discovery by Miner became the foundation of riboflavin production by fermentation. Arzberger (1943) showed that the fermentation by *C. acetobutylicum* of corn mashes in vessels made of glass or aluminum gave a solid residue which assayed about 4000 μg of riboflavin per gram of solids. Walton (1945) discovered that the fermentation of a mash containing ground whole rice produced a solid which contained up to 5500 μg of riboflavin per gram of solids.

Meade *et al.* (1945, 1947) reported the fermentation by *C. acetobutylicum* of whey or skim milk to produce a beer with a riboflavin concentration of 25–50 μg/ml, while with a substrate of whey plus xylose he obtained riboflavin concentrations in the range 70–97 μg/ml.

Imai (1955) tested a large number of adjuvants in a study of riboflavin formation by *C. acetobutylicum* but the highest concentration of the vitamin obtained was 15 μg/ml.

Oguni (1955) assayed the stillage from a butanol–acetone plant and found B_{12} in a concentration of 11–27 μg/liter.

VII. PRODUCTION OF BUTANOL–ACETONE IN COUNTRIES OTHER THAN ENGLAND AND THE UNITED STATES

Most of the work on butanol–acetone published in the last 20 years has originated in countries other than England and the United States. A considerable portion of this work is summarized in Table II.

Several reports from Russian microbiologists deal with the production of butanol–acetone by continuous fermentation methods, and it is stated the use of this procedure allows the rate of solvents production from a given fermentor volume to be increased from 20 to 60% over the rate obtainable by the batch method (Yarovenko, 1964a,b). Furthermore, it is claimed the continuous method results in a 5–17% increase in the percentage of butanol in the solvent mixture (Yarovenko *et al.*, 1966).

Yarovenko (1975) has recently published a book on the continuous fermentation procedure applied to butanol–acetone production.

VIII. FUTURE PROSPECTS FOR THE BUTANOL–ACETONE FERMENTATION

As the prices of raw materials based on petroleum increase and as the supplies diminish, the cost of producing butanol–acetone by the synthetic route will increase substantially. At some time in the next several years, this price increase may make the production of butanol–acetone by fermentation the process of choice. Whether cereal grains and molasses will then be available is not certain since there is mounting pressure to use these products for human food. However, there is an abundance of waste carbohydrates which could be transformed into a substrate for the butanol–acetone fermentation. In the United States alone we produce nearly 1 billion tons per year of solid waste, about one-half of which is cellulosic material (Humphrey, 1974). Cellulose (hexosans) and pentosans can be hydrolyzed to their constituent sugars, glucose and xylose, each of which is a suitable raw material for butanol and acetone.

In the next few paragraphs we list a number of problems associated with the use of waste carbohydrates as raw materials for fermentations. If one or more of these problems can be solved, it would greatly enhance the competitive position of the fermentation process for butanol–acetone.

TABLE II. Production of Butanol–Acetone in Countries Other Than England and the United States

Culture	Substrate	Time (hours)	Yield (%)	Solvent composition (%)				References[c]
				B	A	E	I[a]	
SES-4	Molasses sugar conc. 5–6%		33	56.8	33.5	9.6		(1)
SES-5	Temperature, 33°C		33	56.2	29.9	13.8		(1)
Culture I (100,000-liter tank)	Molasses [6.37% sugar in mash plus rice bran 0.3%, $(NH_4)_2SO_4$ 0.3%], and $CaCO_3$ 0.4%; temperature 37.5°–27°C	104	33.6[b]					(2)
SES-5 (100,000-liter tank)	Molasses [6.28% sugar in mash plus rice bran 0.3%, $(NH_4)_2SO_4$ 0.3%], and $CaCO_3$ 0.4%; temperature 37°–35°C	90	32.0[b]					(2)
Butyl culture	By addition of calcium acetate and acetic acid to fermentation yield of 40% is claimed		40					(3)
C. acetobutylicum	Fermentation of 12% sugar solution in presence of 6% active carbon; carbon could be reused		36	66	33			(4)
Butyl culture	Fermentation of starch and starch plus beet molasses							(5)
	Starch, 100%		37.1					
	Starch, 90%; beet molasses, 10%		36.7					
	Starch, 50%; beet molasses, 50%		38.6					
	Starch, 37%; beet molasses, 63%		37.9					
C. acetobutylicum isolated from sugar cane roots	Mash ingredients in 1 liter: 250 ml sugar cane juice 20°Brix. 1.0 gm ground *Vicia sativa*, 2.5 gm KH_2PO_4, 4 gm $CaCO_3$		32	73	19–23	3–4		(6)
C. acetobutylicum	Molasses plus rye or wheat meal		26.5	66	31	3		(7)
C. acetobutylicum	Addition of 1.0% animal charcoal increases yield of solvents							(8)

C. saccharoperbutylacetonicum ATCC 13564	4% sugar solution with $(NH_4)_2SO_4$ 0.2%, Ca superphosphate 0.1%, and $CaCO_3$ 0.3%	60	28.5					(9)
Butyl culture	In a medium of rye flour, sterile air blown in for 2–3 min every hour; butanol output increased 3.4–9.1%							(10)
C. aurianticum	Organism used glucose, fructose, sucrose and starch; nitrogen sources used: $(NH_4)_2SO_4$, $NaNO_3$, peptone, urea		30–40	60.5	0.5	7.4	25.6	(11)
C. saccharoperbutylacetonicum N-I-41	Molasses plus $(NH_4)_2SO_4$, $CaCO_3$, and Ca superphosphate	60	None given	82.1	15	2.9		(12)
C. saccharoperbutylacetonicum	Plant-scale test using blackstrap molasses (5.14 gm sugar/100 ml mash); great deal of trouble with phage	55	34.2	75.4	20.2	4.4		(13)
C. saccharo-butylacetonicum-liquefaciens	Molasses (5% sugar in mash), protein nutrients, and an alkaline buffer added to mash	36–48	30	60	30	10		(14)
C. acetobutylicum S_{25}	5% Millet		34.4					(1)
	5% Corn		33.9					
	5% Sweet potatoes		33.7					
	7% Corncobs		26.2					
	12% Molasses		24.9					
	6% Corn, 4% molasses		24.3					
	7% Sawdust		22.2					

[a] B, butanol; A, acetone; E, ethanol; I, isopropyl alcohol.

[b] The yields in this reference have been recalculated to conform to the conventional formula:

$$\text{Yield} = \frac{\text{weight of solvents}}{\text{weight of sugar added}} \times 100\%.$$

[c] Key to references: (1) Bah *et al.* (1950); (2) Lee and Leu (1951); (3) Karsch and Schoedler (1956); (4) Hongo and Nagata (1958); (5) Zalesskaya *et al.* (1958); (6) Perdomo (1958); (7) Logotkin and Zaritskii (1959); (8) Bahadûr and Saroj (1960); (9) Hongo (1960); (10) Nakhmanovich and Kochkina (1960); (11) Suto *et al.* (1960); (12) Motoe (1961); (13) Hongo *et al.* (1965); (14) Srivastava (1968).

If the rate of hydrolysis of hexosans and pentosans by enzymes could be substantially increased and if excessive pretreatment costs were not encountered, this method could be cheaper than acid hydrolysis. The enzyme system would be expected to form fewer products toxic to the butanol–acetone organisms.

At present, acid hydrolysis is the best method of hydrolyzing hexosans and pentosans, but better data are needed on the hydrolysis conditions for each type of waste product so as to get the highest conversion to sugar consistent with a product which does not contain reversion compounds harmful to the bacterium being used for the fermentation.

Some of the waste products, e.g., corncobs and cornstalks, suffer from seasonal output and from the difficulty of collecting and delivering them cheaply to a central location. If these products are to be used, the formidable logistics problems must be solved. There are, however, numerous products such as bagasse, sawdust, other wood by-products, manure, and garbage which are already collected and which could be delivered to a plant.

Fermentation tests are needed for each candidate substrate in order to find the optimum conditions for the highest yield and the best ratio of butanol to acetone.

With the microorganisms now available, fermentation activity is severely depressed when the concentration of butanol in the mash goes above 13 gm/liter. If an organism could be found, either by isolation from natural sources or by genetic manipulation, which would function in mashes of higher butanol concentration, the operating costs of the fermentation process would be reduced considerably.

Hongo and Nagata (1958) report a method for doubling the amount of butanol which can be tolerated by the *C. acetobutylicum* organism. Their method involved the addition of 6% activated carbon to a 12% sugar solution, which was then fermented in good yields to give a calculated final butanol concentration of 28.5 gm/liter. This method should be tested on the substrates from waste carbohydrates to see whether it gives an economic advantage over the conventional fermentation process.

The continuous fermentation procedure should be tested with the substrates derived from waste carbohydrates to determine whether this principle is applicable to these products. If these tests prove that the increased productivity from a given fermentor volume and the increased ratio of butanol to acetone claimed in the literature can be verified, they would show a significant improvement in the economics of the fermentation processes.

By short hourly periods of aeration in a fermenting mash of rye flour,

Nakhmanovich and Kochkina (1960) were able to increase the amount of butanol in the range of 3.4–9.1%, with a concomitant sharp reduction in the oxidation–reduction potential. This experiment should be repeated with substrates from waste carbohydrates to determine whether aeration increases the ratio of butanol to acetone in the fermentation of mashes containing them.

As indicated earlier, present technology does not permit butanol made by fermentation to compete with butanol made by the oxo process using propylene as a feed stock. If the price of propylene rises to the point where present fermentation technology becomes competitive, the price of butanol may be so high that it will become a specialty product with a limited market. We believe that the future of producing butanol–acetone on a large scale by the fermentation process depends on the successful solution of the foregoing problems relating to the use of waste carbohydrates.

REFERENCES

Arzberger, C. F. (1938). U.S. Patent 2,139,108.

Arzberger, C. F. (1943). U.S. Patent 2,326,425.

Baba, T. (1953). *Bull. Fac. Eng., Hiroshima Univ.* **2,** 19–32. *Chem. Abstr.* **48,** 1624 (1954).

Baba, T. (1954a) a. *Bull. Fac. Eng., Hiroshima Univ.* **3,** 147–152; *Chem. Abstr.* **48,** 10289 (1954).

Baba, T. (1954b) b. *Bull. Fac. Eng., Hiroshima Univ.* **3,** 229–235; *Chem. Abstr.* **49,** 4227 (1955).

Bah, H., Lee, H., and Leu, C. (1950). *Rep. Taiwan Sugar Exp. Stn.* **6,** 193–206; *Chem. Abstr.* **49,** 14263 (1955).

Bahadur, K., and Saroj, K. K. (1960). *Jpn. J. Microbiol.* **4,** 43–51; *Chem. Abstr.* **55,** 8753 (1961).

Carnarius, E. H. (1940). U.S. Patent 2,198,104.

Carnarius, E. H., and McCutchan, W. N. (1938). U.S. Patent 2,139,111.

Dunning, J. W., and Lathrop, E. C. (1945). *Ind. Eng. Chem.* **37,** 24–29.

Gabriel, C. L. (1928). *Ind. Eng. Chem.* **20,** 1063–1067.

Gabriel, C. L., and Crawford, F. M. (1930). *Ind. Eng. Chem.* **22,** 1163–1165.

Hongo, M. (1958). *Nippon Nogei Kagaku Kaishi* **32,** 219–223; *Chem. Abstr.* **57,** 5123 (1962).

Hongo, M. (1960). U.S. Patent 2,945,786; *Chem. Abstr.* **54,** 21632 (1960).

Hongo, M., and Murata, A. (1966). *Hakko Kyokaishi* **24,** 354–358; *Chem. Abstr.* **66,** 1565 (1967).

Hongo, M., and Nagata, K. (1958). *Nippon Nogei Kagaku Kaishi* **32,** 585–594, 679–689, 758–770, and 855–858; *Chem. Abstr.* **55,** 19126 (1961).

Hongo, M., Harada, K., Akahoshi, K., Nagata, K., and Takahashi, S. (1965). *Nippon Nogei Kagaku Kaishi* **39,** 247–256; *Chem. Abstr.* **63,** 15505 (1965).

Horsley, L. H. (1973). *In* "Azeotropic Data-III," p. 24. Am. Chem. Soc., Washington, D.C.

Humphrey, A. E. (1974). *Chem. Eng. (N.Y.)* **81,** 98–112.

Imai, K. (1955). *Mem. Coll. Agric., Kyoto Univ., Chem. Ser.* **29,** 1–31; *Chem. Abstr.* **50,** 13363 (1956).
Karsch, W., and Schoedler, K. (1956). German Patents 941,184 and 941,185; *Chem. Abstr.* **52,** 19014 (1958).
Kinoshita, S., Shiga, A., and Okumura, T. (1954). *Nippon Nogei Kagaku Kaishi* **28,** 83–87; *Chem. Abstr.* **51,** 6937 (1957).
Langlykke, A. F., Van Lanen, J. M., and Fraser, D. R. (1948). *Ind. Eng. Chem.* **40,** 1716–1719.
Lee, H., and Leu, C. (1951). *Rep. Taiwan Sugar Exp. Stn.* **7,** 149–156; *Chem. Abstr.* **50,** 2115 (1956).
Legg, D. A. (1928). U.S. Patent 1,668,814.
Legg, D. A., and Tarvin, N. R. (1938). U.S. Patent 2,107,262.
Legg, D. A., and Walton, M. T. (1938). U.S. Patent 2,132,358.
Leonard, R. H., and Peterson, W. H. (1947). *Ind. Eng. Chem.* **39,** 1443–1445.
Logotkin, I. S., and Zaritskii, I. M. (1959). *Spirt. Prom'st.* **25,** 14–20; *Chem. Abstr.* **53,** 9562 (1959).
McCoy, E. F. (1946). U.S. Patent 2,398,837.
McCutchan, W. N., and Hickey, R. J. (1954). "Industrial Fermentations" (L. A. Underkofler and R. J. Hickey, eds.), Vol. I, pp. 347–352 and 364. Chem. Publ. Co., New York.
Meade, R. E., Pollard, H. L., and Rodgers, N. E. (1945). U.S. Patent 2,369,680.
Meade, R. E., Rodgers, N. E., and Pollard, H. L. (1947). U.S. Patent 2,433,232.
Miner, C. S. (1940). U.S. Patent 2,202,161.
Moldenhauer, O., and Lechner, R. (1951). German Patent 824,932; *Chem. Abstr.* **49,** 1354 (1955).
Motoe, M. (1961). Japanese Patent 61/5797; *Chem. Abstr.* **56,** 7806 (1962).
Müller, J. (1938). U.S. Patent 2,123,078.
Nakhmanovich, B. M., and Kochkina, L. V. (1960). *Tr. Tsentr. Nauchno-Issled. Inst. Spirt. Likero-Vodochn. Prom'sti* **9,** 147–153; *Chem. Abstr.* **58,** 9594 (1963).
Nakhmanovich, B. M., Senkevich, V. V., Shcheblykina, N. A., and Lipshits, V. V. (1960). *Tr. Inst. Lesokhoz. Probl. Khim. Drev., Akad. Nauk Latv. SSR* **21,** 47–53; *Chem. Abstr.* **56,** 15960 (1962).
Nakhmanovich, B. M., Lipshits, V. V., and Pavlovich, L. A. (1965). *Prikl. Biokhim. Mikrobiol.* **1,** 635–639; *Chem. Abstr.* **64,** 13351 (1966).
Oguni, Y. (1955). *Vitamins* **9,** 227–231; *Chem. Abstr.* **50,** 15023 (1956).
Perdomo, E. V. (1958). *Inst. Cubano Invest. Tecnol. Ser. Estud. Trab. Invest.* **3,** 1; *Chem. Abstr.* **52,** 20871 (1958).
Reports of Patent Design and Trademark Cases (1926). Vol. XLIII, No. 7, pp. 185–238. (Great Britain.)
Salle, A. J. (1939). *In* "Fundamental Principles of Bacteriology," pp. 247–251. McGraw-Hill, New York.
Srivastava, S. B. (1968). *Chem. Age India* **19,** 295–296; *Chem. Abstr.* **69,** 42777 (1968).
Suto, T., Namba, Y., Kurashima, K., Matsuki, M., and Furusaka, C. (1960). *Sci. Rep. Res. Inst., Tohoku Univ. Ser. D* **11,** 115–127 and 129–137; *Chem. Abstr.* **55,** 14582 and 14583 (1961).
Taha, S. M., Mahmoud, S. A. Z., Ishac, Y. Z., El-Sawy, M., and El-Demerdash, M. E. (1974). *Egypt. J. Microbiol.* **8,** 15–27; *Chem. Abstr.* **83,** 145, 747 (1975).
Tarvin, N. B. (1941). U.S. Patent 2,260,126.
Tornescher Hefe, G.m.b.H. (1954). German Patent 920,724; *Chem. Abstr.* **53,** 4649 (1959).
Tsuchiya, H. M., Van Lanen, J. M., and Langlykke, A. F. (1949). U.S. Patent 2,481,263.
U.S. Patent Quarterly (1932a). **12,** 47–57.
U.S. Patent Quarterly (1932b). **15,** 237.

Walton, M. T. (1945). U.S. Patent 2,368,074.
Weizmann, C. (1915). British Patent 4845.
Weizmann, C. (1919). U.S. Patent 1,315,585.
Wiley, A. J., Johnson, M. J., McCoy, E., and Peterson, W. H. (1941). *Ind. Eng. Chem.* **33,** 606–610.
Woodruff, J. C., Stiles, H. R., and Legg, D. A. (1937). U.S. Patent 2,089,522.
Yarovenko, V. L. (1964a). *Continuous Cultiv. Microorganisms, Proc. Symp., 2nd, 1962* Vol. 2, pp. 205–219; *Chem. Abstr.* **63,** 1193 (1965).
Yarovenko, V. L. (1964b). *Tr. Vses. Nauchno-Issled. Inst. Fermentn. Spirt. Prom'sti.* **15,** 105–124; *Chem. Abstr.* **63,** 15,504 (1965).
Yarovenko, V. L. (1975). *Pishch. Prom'st. SSSR* pp. 1–102; *Chem. Abstr.* **84,** 134121 (1976).
Yarovenko, V. L., Nakhmanovich, B. M., Shcheblykin, N. F., and Shcheblykina, N. A. (1966). *Prikl. Biokhim. Mikrobiol.* **2,** 344–347; *Chem. Abstr.* **65,** 6257 (1966).
Zalesskaya, M. I., Logotkin, I. S., Marima, A. M., Guskova, N. P., and Chekasina, E. V. (1958). *Tr. Tsentr. Nauchno-Issled. Inst. Spirt. Likero-Vodochn. Prom'sti* **8,** 52–60; *Chem. Abstr.* **55,** 3918 (1961).

Chapter 7

Microbial Production of Amino Acids

YOSHIO HIROSE
HIROSHI OKADA

I. INTRODUCTION

Konbu (kelplike seaweed) has been widely used as an important traditional seasoning source in Japan. In 1908, the taste of konbu was identified as being due to L-glutamic acid (Ikeda, 1908). Based on this discovery, the industrial production of monosodium L-glutamate was initiated by Ajinomoto Co. in 1909. At that time, L-glutamic acid was produced by the acid hydrolysis of wheat gluten or soybean protein. A half-century after this discovery, it was reported that considerable quantities of L-glutamic acid accumulated in bacterial cultures (Kinoshita *et al.*, 1957a). The research and development, carried out mainly in Japan, resulted in the successful and economical production of L-glutamic acid

MICROBIAL TECHNOLOGY, 2nd ed., VOL. I
 ISBN 0-12-551501-4

by the fermentative process. The significance of the establishment of microbial production of L-glutamic acid cannot be overestimated. Such essential metabolites as amino acids or nucleotides were considered not to be accumulated in microbial cultures due to regulatory mechanisms in the cell. The discovery of L-glutamic acid fermentation stimulated a wide variety of research aimed at the isolation of wild strains and the genetic derivation of mutants which could accumulate large amounts of primary metabolites. Fortunately, fundamental knowledge about biosynthetic pathways of amino acids and their regulatory mechanisms had already been elucidated to some extent.

Amino acids are constituents of proteins and play various important roles in living organisms. There have been increased demands for amino acids for use in the areas of food and feed additives and drug manufacturing.

Amino acids are used in foods as flavor enhancers, seasonings, and nutritional additives. They are also used to improve bread quality. Monosodium L-glutamate, the most important commercial amino acid, is widely used as a flavor enhancer. L-Alanine and glycine are used as flavoring agents. The essential amino acids which are often present only in insufficient amount in cereals are L-lysine, L-threonine, and L-methionine. L-Lysine and DL-methionine are now used practically to increase the efficiency of protein utilization of animal feeds. The reductive power of L-cysteine is utilized to improve bread quality in the baking process.

In the field of medicine, amino acids are used for infusions and as therapeutic agents. In the past, based on their nutritional values, essential amino acids were the principal components of the infusion mixtures, only glycine among nonessential amino acids being used to increase nitrogen content. However, it has recently been disclosed that nonessential amino acids also have nutritional significance (Breuer *et al.*, 1964) and the development of amino acid infusions has been stimulated. Infusion formulas now contain various dispensable amino acids as well as essential ones. When the condition of the patients does not allow them to take required amounts of protein in a normal manner, for example, in a postoperative situation, an infusion formula is used.

Based on the specific physiological properties, some amino acids are also being used therapeutically. For example, L-arginine is effective in the alleviation of hyperammonemia and liver disorders by increasing arginase activity in the liver. L-Cysteine is used for the treatment of bronchitis or nasal catarrh, and L-glutamine and L-histidine for gastric ulcers.

Amino acid derivatives are also used in the chemical industry. They are applied in cosmetics, synthetic leathers, surface-active agents, fungicides, and pesticides. One typical example is use of poly-γ-methyl glutamate in the manufacture of synthetic leather.

To meet the diversified demands mentioned above, almost all of the amino acids are now commercially produced. The production methods developed so far may be summarized as follows: (1) protein hydrolysis methods; (2) chemical synthesis methods; (3) microbiological methods, including (a) direct production of amino acids from carbon source such as glucose; (b) precursor addition methods; and (c) enzymatic methods.

Table I presents production methods, uses, annual outputs, and representative manufactures of each amino acid now produced industrially. In the following, the microbial methods of amino acid production are reviewed with special emphasis on L-glutamic acid and L-lysine, which are the ones most abundantly produced. Production of other amino acids is referred to only very briefly.

Amino acid production methods include (1) biosynthesis and excretion of amino acids from carbohydrate, as in the case of L-glutamic acid and L-lysine accumulation from glucose; (2) salvage synthesis of amino acids from intermediate metabolites, for example, L-tryptophan accumulation from anthranilic acid; and (3) enzymatic production of amino acids from their substrates, for example, L-aspartic acid from fumaric acid and ammonia by aspartase of *Escherichia coli*. With selection of a suitable strain, careful determination of culture medium, and optimization of cultural condition, amino acid production can be precisely controlled and enhanced.

Since there have been many excellent reviews on amino acid fermentations exemplified by those of Kinoshita(1963), Yamada *et al.*(1972), Tsunoda and Okumura(1972), and Nakayama(1976), the authors' efforts are focused here on two essential aspects of the microbial production of amino acids: the strains and the process control.

II. MICROBIAL STRAINS EMPLOYED IN AMINO ACID PRODUCTION

A great variety of microorganisms has been isolated or induced for amino acid production. They are classified into wild strains, auxotrophic mutants, and regulatory mutants.

A. Direct Production of Amino Acids from Carbon Sources

The ultimate goal of microbiological amino acid production may be the direct accumulation of amino acid from cheap carbon sources.

TABLE I. Production Methods, Annual Outputs, and Representative Manufacturers of Amino Acids

Amino acids	Industrial production methods[a]	Annual output (tons/year)	Main uses	Representative manufacturers
L-Ala	C-3	10–50	Flavoring	Ajinomoto Kyowa Tanabe
DL-Ala	B C-1	150–200	Flavoring	Ajinomoto Tanabe
L-Arg	C-1 A	200–300	Infusion, therapeutic	Ajinomoto Kyowa
L-Asp	C-3	500–1000	Therapeutic flavoring	Ajinomoto Kyowa Tanabe
L-Asn	B	10–50	Therapeutic	Ajinomoto Tanabe
L-CysH	A	100–200	Improver of bread quality, antioxidant	Ajinomoto Kyowa Kyowa
L-Glu	C-1	200,000	Seasoning	Ajinomoto Kyowa Stauffer, Orsan
L-Gln	C-1	300	Therapeutic	Ajinomoto Kyowa
Gly	B	5000–6000	Sweetner	Ajinomoto Kyowa Tanabe
L-His	C-1	100–200	Therapeutic	Ajinomoto Tanabe
L-Ile	C-1	10–50	Infusion	Ajinomoto Kyowa Tanabe
L-Leu	A	50–100	Infusion	Ajinomoto Kyowa Tanabe
L-Lys	C-1	15,000–20,000	Feed additive	Ajinomoto Kyowa Tanabe
DL-Met	B	70,000	Feed additive	Ajinomoto Sumitomo Chemical Degussa
L-Phe	B	50–100	Infusion	Ajinomoto Tanaba
L-Pro	C-1	10–50	Infusion	Ajinomoto Tanabe
L-Ser	C-2	10–50	Cosmetic	Ajinomoto Kyowa Tanabe
L-Thr	C-1, B	50–100	Food additive	Ajinomoto Kyowa Tanabe

(*continued*)

TABLE I (*Continued*)

Amino acids	Industrial production methods[a]	Annual output (tons/year)	Main uses	Representative manufacturers
L-Trp	B	50–100	Infusion	Ajinomoto Tanabe
L-Tyr	A	50–100	Infusion	Ajinomoto Kyowa Tanaba
L-Val	C-1	50–100	Infusion	Ajinomoto Tanabe

[a] Key to methods: A, protein hydrolysis method; B, chemical synthesis method; C, methods employing microbial activities: (C-1) direct production from glucose or other carbon sources; (C-2) precursor addition methods; (C-3) enzymatic methods.

Therefore, much effort has been made to isolate wild strains and to derive genetic mutants, auxotrophs and regulatory mutants, which can accumulate amino acids directly from glucose. Wild strains are used in L-glutamic acid and L-glutamine fermentations. Amino acids which are excreted by auxotrophic mutants include L-citrulline, L-leucine, L-lysine, L-ornithine, L-proline, L-threonine, and L-tyrosine. Regulatory mutants are more widely used for the accumulation of L-arginine, L-histidine, L-isoleucine, L-leucine, L-lysine, L-methionine, L-phenylalanine, L-threonine, L-tryptophan, L-tyrosine, and L-valine.

1. Wild Strains

A large number of L-glutamic acid-producing bacteria have been isolated. Representative strains are shown in the following tabulation:

Genus *Corynebacterium:*	*C. glutamicum* (Kinoshita *et al.*, 1957b); *C. lilium* (Lee and Good, 1963); *C. callunae* (Lee and Good, 1963); and *C. herculis* (Distillers Co., Ltd., 1965)
Genus *Brevibacterium:*	*B. divaricatum* (Su and Yamada, 1960); *B. aminogenes* (Ota and Tanaka, 1959); *B. flavum* (Okumura *et al.*, 1962a); *B. lactofermentum* (Okumura *et al.*, 1962a); *B. saccharolyticum* (Okumura *et al.*, 1962a); *B. roseum* (Okumura *et al.*, 1962a); *B. immariophilum* (Okumura *et al.*, 1962a); *B. ammoniagenes* (Oishi and Aida, 1965); *B. alanicum* (Ogawa *et al.*, 1959); and *B. thiogenitalis* (Kanzaki *et al.*, 1967)
Genus *Microbacterium:*	*M. salicinovolum* (Doi and Kaneko, 1960); *M. flavum* var. *glutamicum* (Masuo *et al.*, 1962); and *M. ammoniaphilum* (Miyai *et al.*, 1964)
Genus *Arthrobacter:*	*A. globiformis* (Veldkamp *et al.*, 1963); and *A. aminofaciens* (Mogi *et al.*, 1967)

Most of these L-glutamic acid-producing bacteria are gram-positive, nonspore-forming, and nonmotile and require biotin for growth (Oishi, 1967). Among various components of the culture medium, the concentration of biotin is the most decisive factor in controlling the fermentation.

A great variety of investigations have clarified the role of biotin in the L-glutamic acid production. These works may be summarized as follows.

1. Product formation is greatest when the biotin concentration is suboptimal for the maximum growth. Excess biotin in the medium supports abundant growth but seriously inhibits product formation (Tanaka *et al.*, 1960).
2. Even in the presence of excess biotin, addition of penicillin at the growth phase permits the cells to accumulate L-glutamic acid. Other antibiotics which inhibit cell wall synthesis, such as cephalosporin C, could replace penicillin (Sommerson and Phillips, 1962).
3. The addition of C_{16}–C_{18} saturated fatty acids or their esters with hydrophilic polyalcohols also permits the cells to accumulate L-glutamic acid even in the medium of excess biotin (Takinami *et al.*, 1963, 1964, 1965).
4. Oleic acid could replace biotin (Okumura *et al.*, 1962b). An oleic acid-requiring mutant of *B. thiogenitalis* could accumulate L-glutamic acid even in the presence of excess biotin. The amount of L-glutamic acid was maximal when the oleic acid supply was limited (Kanzaki *et al.*, 1967).

Discovery of penicillin and saturated fatty acids as antibiotin agents allowed industrial utilization of such biotin-rich raw materials as cane and beet molasses for L-glutamic acid production. It is physiologically and industrially important that these chemical agents be added to the medium during a specific phase of the cell growth.

The excretion of L-glutamic acid is closely related to the cell permeability, which is associated with both chemical and physical constituents of the cell membrane. It is due not to the lack of biosynthesizing activity but to the lack of L-glutamic acid permeability that the biotin-rich cells could not accumulate L-glutamic acid, as demonstrated by Shiio *et al.* (1963). The glutamic acid-producing cells, e.g., biotin-limited cells, penicillin- or Tween 60-treated cells, and oleic acid-limiting cells, excreted intracellular L-glutamic acid when they were washed with phosphate buffer. On the other hand, biotin-rich cells did not excrete endogeneous L-glutamic acid when washed with the buffer. It was clarified that the cells, whose biotin content was greater than 0.5 ng/mg of cell, biosynthesized sufficient amounts of oleic acid, which resulted in a high content of phospholipid in the cell membrane. These cells excreted endogenous L-glutamic acid poorly, as demonstrated by Takinami *et al.* (1966, 1967). Oleic acid-dependent mutants and C_{16}–C_{18} saturated fatty acid-treated cells, as well as the cells with biotin content less than 0.5 ng/mg of cell, synthesized insufficient amounts of phospholipid. Phospholipid-poor cells excreted endogenous L-glutamic acid. These results are in good agreement with the reports of Izumi *et al.* (1973) and Kamiryo *et al.* (1976) that acetyl-CoA carboxylase, which is a biotin enzyme, is involved in the biosynthesis of oleic acid and other fatty acids and that C_{16}–C_{18} saturated fatty acids inhibited oleic acid synthesis by repressing acetyl-CoA carboxylase (Fig. 1).

The effect of penicillin on L-glutamic acid permeability was of particular interest. The cells grown in the presence of penicillin or C_{16}–C_{18} saturated fatty acids showed enhanced permeability toward L-glutamic acid. Penicillin inhibited cell wall synthesis, resulting in elongated or swelled cells. Saturated C_{16}–C_{18} fatty acids did not have as drastic an effect on the cell structure. The effects of penicillin and C_{16}–C_{18} saturated fatty acids were compared, in connection with osmotic pressures of the

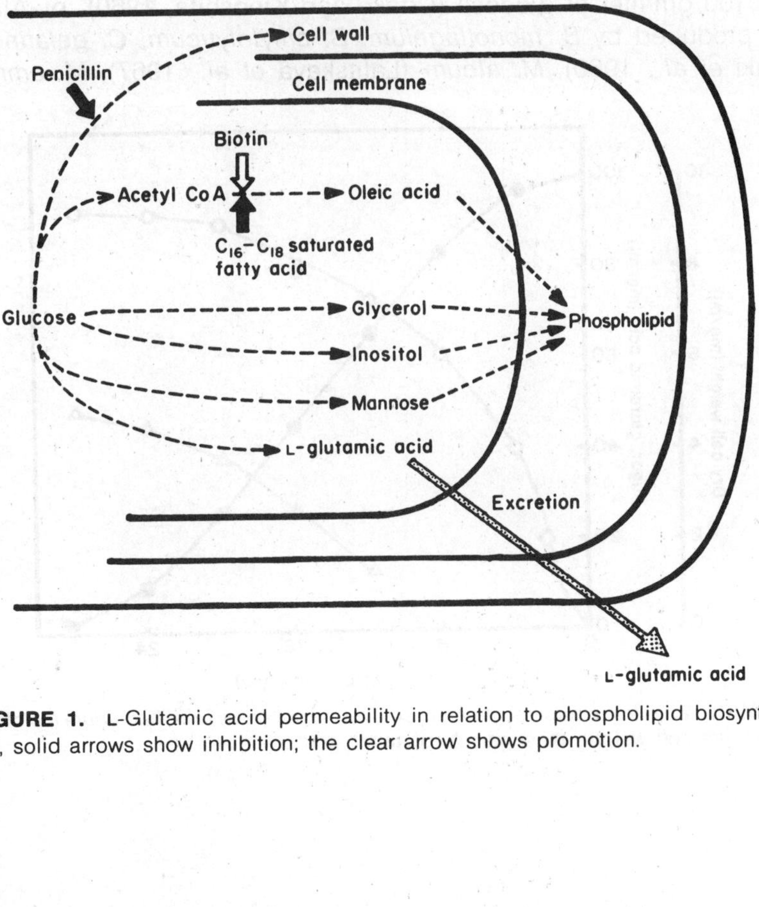

FIGURE 1. L-Glutamic acid permeability in relation to phospholipid biosynthesis. Filled, solid arrows show inhibition; the clear arrow shows promotion.

medium, by Shibukawa *et al.* (1968). The addition of polyoxyethylene monopalmitate enhanced L-glutamic acid permeability in a biotin-rich medium independently of the osmotic pressures. On the other hand, penicillin promoted the excretion of L-glutamic acid only in a medium of low osmotic pressure and was ineffective in a medium of high osmotic pressure. The cell structure was changed by penicillin only at low osmotic pressure. These results led to the conclusion that penicillin had a secondary effect on the membrane function. It primarily inhibited cell wall synthesis leaving the cell membrane unprotected, thus breaking the permeability barrier causing physical damage to the cell membrane.

An example of time course curves of L-glutamic acid fermentation is shown in Fig. 2. L-Glutamic acid-producing bacteria assimilated 100–200 gm/liter of sugar, giving about 50% of the production yield. Usable raw materials included carbon sources, such as a hydrolysate of starch, cane, or beet molasses, and nitrogen sources, such as urea or gaseous ammonia.

Some other amino acids are also extracellularly accumulated by wild strains. *Paracolobacterium coliform* accumulated 15 gm/liter of L-valine from 100 gm/liter of glucose (Udaka and Kinoshita, 1960). DL-Alanine was produced by *B. monoflagelum, B. amylolyticum, C. gelatinosum* (Ozaki *et al.*, 1960), *M. album* (Lafitskaya *et al.*, 1967), *M. ammoni-*

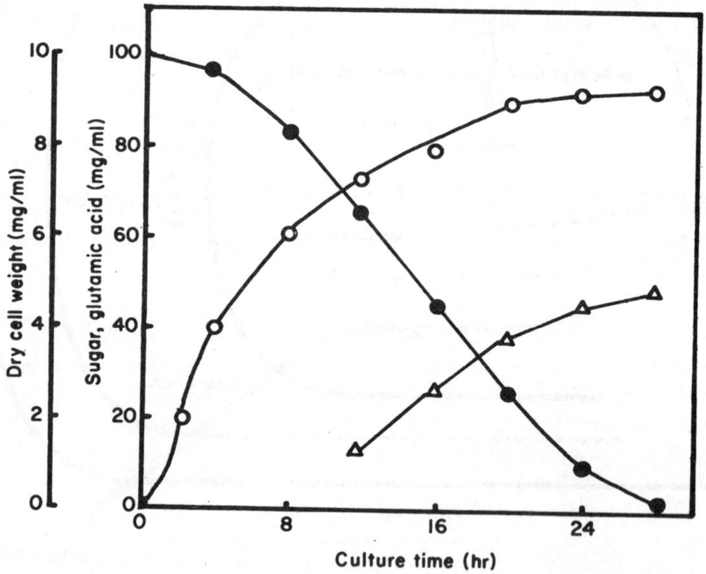

FIGURE 2. Time course of glutamic acid fermentation by *Brevibacterium lactofermentum.* ○, dry cell weight; ●, sugar; △, glutamic acid.

aphilum (Yoshida and Nakayama, 1975), and *Bacillus subtilis* (Matsumoto *et al.*, 1967). L-Glutamine fermentation by *B. flavum* or *C. glutamicum* was of industrial significance. L-Glutamic acid fermentation was converted to L-glutamine fermentation by altering the composition of the culture medium and the pH. In this conversion, *N*-acetyl-L-glutamine accumulated as a by-product. A high concentration of ammonium chloride, weakly acidic pH, and the presence of zinc ions favored the L-glutamine production at the expense of L-glutamic acid or *N*-acetyl-L-glutamine accumulation, as was demonstrated by Nakanishi *et al.* (1975). The maximum accumulation of L-glutamine by this wild strain reached 40 gm/liter at the production yield of 30%. However, regulatory mutants were found more potent. A sulfaguanidine-resistant mutant of *B. flavum* accumulated 39 gm/liter of L-glutamine at the production yield of 39% (Yoshinaga *et al.*, 1976).

2. *Auxotrophic Mutants*

Among auxotrophic mutants derived from *B. flavum* and *C. glutamicum,* several amino acids are produced, including L-citrulline, L-leucine, L-lysine, L-ornithine, L-proline, L-threonine, and L-tyrosine. The application of a homoserine (or L-threonine plus L-methionine)- requiring auxotrophic mutant of *C. glutamicum to* L-lysine production is one example (Kinoshita *et al.*, 1958; Nakayama *et al.*, 1961, 1966). For the effective production of L-lysine, sufficient amounts of biotin must be added to prevent the excretion of L-glutamic acid. Furthermore, the concentration of homoserine must be suboptimal for the growth, because the limitation of the feedback inhibition of aspartate kinase is necessary. Figure 3 shows the biosynthetic pathway of L-lysine and its metabolic controls in *C. glutamicum.* L-Lysine is synthesized through the diaminopimelic acid pathway. The key enzyme, aspartate kinase, is under the influence of the concerted feedback inhibition by L-threonine and L-lysine; excess homoserine increases the L-threonine pool within the cells, and aspartate kinase is inhibited.

Table II shows amino acids which are accumulated by auxotrophic mutants.

3. *Regulatory Mutants*

Knowledge of regulatory mechanisms of amino acid biosynthesis and the role of amino acid analogs has been applied to the breeding of regulatory mutants for amino acid production. Regulatory mutants possess a feedback-insensitive key enzyme and can be expected to accumulate the related amino acid in large amounts. A typical example of the application of regulatory mutants to amino acid fermentation was the

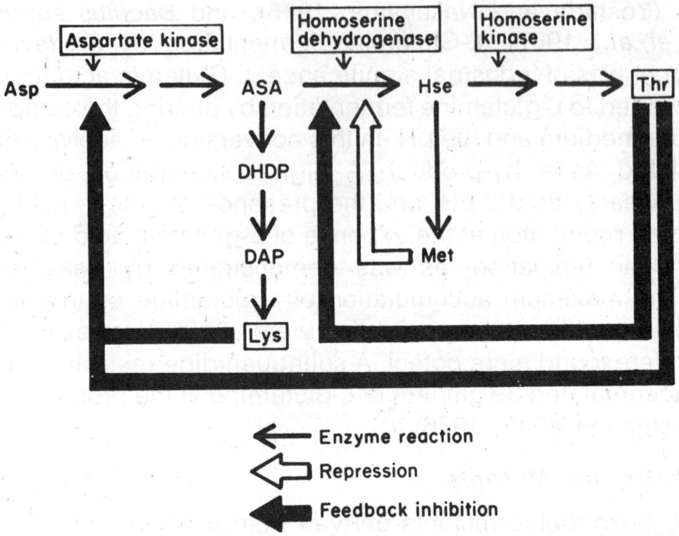

FIGURE 3. Regulation in lysine biosynthesis in *Brevibacterium flavum* and *Corynebacterium glutamicum*. ASA, aspartate semialdehyde; DADP, dihydrodipicolinate; Hse, hemoserine; DAP, diaminopimelate.

derivation by Sano and Shiio (1970, 1971) of a potent L-lysine producer. Unlike the regulatory mechanism in *E. coli* (Fig. 4), the regulation of L-lysine synthesis by *B. flavum* was rather simple: Only aspartate kinase was sensitive to feedback inhibition. This implied that *B. flavum* was a suitable parent strain from which to obtain regulatory mutants for L-lysine production. Regulatory mutants were obtained by the isolation of L-lysine analog-resistant mutants whose growth was not inhibited by an L-lysine analog, which behaved as a feedback inhibitor on aspartate kinase and which did not allow growth of the parent strain. Some of the mutants resistant to the analog were expected to be regulatory mutants with aspartate kinase desensitized to concerted feedback inhibition. Regulatory mutants might also be obtained by the induction of revertants from aspartate kinase-deficient mutants. Various changes in the enzyme structure might take place and some of them were expected to desensitize the regulatory enzyme.

S-(2-Aminoethyl)-L-cysteine (AEC) was employed for L-lysine production.

$$CH_2(NH_2)-CH_2-CH_2-CH_2-CH(NH_2)COOH$$
L-Lysine
$$(NH_2)-CH_2-CH_2-S-CH_2-CH(NH_2)COOH$$
S-(2-Aminoethyl)-L-cysteine (AEC)

TABLE II. Amino Acids Accumulated by Auxotrophic Mutants

Amino acids	Microorganism	Genetic markers	Accumulation and product yield (gm/liter) (yield, %)	References
L-Cit	*Bacillus subtilis*	*Arg*$^-$	16.5 (12.7)	Okumura *et al.* (1964)
	Corynebacterium glutamicum	*Arg*$^-$	10.7 (10.0)	Nakayama and Hagino (1966)
L-Lys	*Corynebacterium glutamicum*	*Homoserine*$^-$	44 (22)	Kinoshita *et al.* (1958)
L-Orn	*Corynebacterium glutamicum*	*Arg*$^-$	26 (26)	Kinoshita *et al.* (1957b)
	Arthrobacter citreus	*Thiamine*$^-$, *Arg*$^-$	27 (27)	Shibuya *et al.* (1967)
L-Pro	*Brevibacterium flavum*	*Ile*$^-$, *Sulfa-guanidine*^{r}	35 (35)	Yoshinaga *et al.* (1966, 1976)
L-Tyr	*Corynebacterium glutamicum*	*Phe*$^-$, *Purine*$^-$	15.1 (7.5)	Hagino *et al.* (1973)
L-Leu	*Corynebacterium glutamicum*	*Phe*$^-$, *His*$^-$	16 (13.3)	Araki *et al.* (1974a)
L-Thr	*Escherichia coli*	*Diaminopimelate*$^-$ *Met*$^-$, *Ile revertant*	13.8 (18.4)	Kase *et al.* (1971)

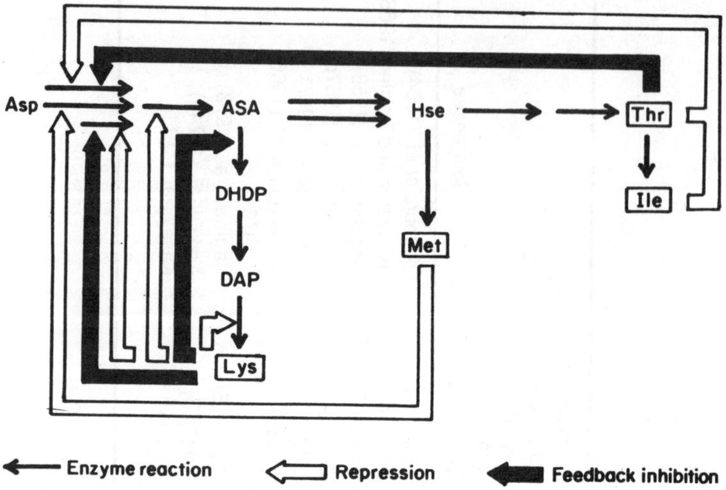

FIGURE 4. Regulation in lysine biosynthesis in *Escherichia coli.*

Figure 5 shows that the combined dose of AEC and L-threonine showed strong inhibitory effect on the cell growth of *B. flavum.* This inhibition was recovered by the addition of L-lysine. Among AEC-resistant mutants which could grow in the presence of AEC and L-threonine, there appeared some potent L-lysine producers. The best producer, FA 1-30, produced 57 gm/liter of L-lysine from 130 gm/liter of glucose. The regulatory mutant exhibited the following characteristics:

1. Its growth was no longer influenced by AEC, which inhibited growth of the parent strain with false feedback inhibition.
2. It produced large amounts of L-lysine.
3. L-Lysine production by the mutant was not restricted by exogenous L-threonine and L-lysine.
4. Activity of its aspartate kinase was not inhibited by L-threonine and L-lysine.

Similar techniques of breeding regulatory mutants were successfully applied to obtain other amino acid producers, examples of which are given in Table III.

B. Precursor Addition Methods

For some amino acids, e.g., L-tryptophan and L-serine, direct fermentation was very difficult to establish, perhaps due to the difficulty in avoiding metabolic control. Instead, the precursor addition method was

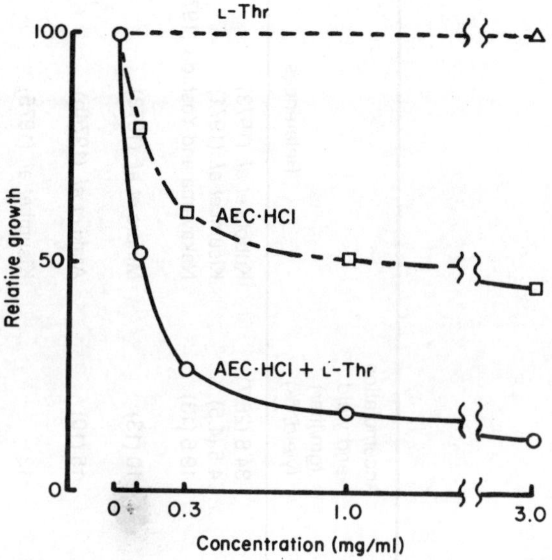

FIGURE 5. Effect of L-threonine and AEC [*S*-(2-aminoethyl)-L-cysteine] on cell growth of *Brevibacterium flavum*.

developed for these amino acids. Some adequate intermediate metabolites in the biosynthetic pathway of these amino acids were converted to the corresponding amino acids by microorganisms. Amino acid biosynthesis was accompanied by some other side reactions and an ATP supply was required. Therefore, glucose or some other carbon source was also supplied in a medium.

Terui and Niitsu (1969) isolated *Hansenula anomala,* which accumulated only traces of L-tryptophan from glucose; however, significant amounts of the amino acid where accumulated in the medium when anthranilic acid was added as a precursor. Furthermore, indole-resistant mutants of *H. anomala* were derived. They had the following favorable characteristics for L-tryptophan production: (1) Anthranilic acid and L-tryptophan decomposing activities were very weak; and (2) the enzymes concerned with L-tryptophan synthesis were not subject to repression and inhibition by the end product. Glucose, ethanol, or glycerol was used as a carbon source, and anthranilic acid or indole was used as a precursor. With stepwise feeding of the mixed solution of glycerol and anthranilic acid, 3 gm/liter of L-tryptophan was accumulated at the production yield of about 90%.

L-Serine was also produced by *Corynebacterium glycinophilum* (Kubota *et al.*, 1971, 1972). About 20 gm/liter of glycine were converted into 10 gm/liter of L-serine per 100 gm/liter of glucose.

TABLE III. Amino Acids Accumulated by Regulatory Mutants

Amino acids	Microorganisms	Genetic markers[a]	Accumulation and yield (gm/liter) (yield, %)	References
L-Arg	*Brevibacterium flavum*	TA^r	34.8 (26.7)	Kubota *et al.* (1973)
	Bacillus subtilis	$ArgHX^r$	4.5 (4.5)	Kisumi *et al.* (1971)
	Corynebacterium glutamicum	$D\text{-}Ser^r$, $D\text{-}Arg^r$, $ArgHX^r$, *Ile* revertant	19.6 (13)	Nakayama and Yoshida (1972)
L-His	*Brevibacterium flavum*	TA^r, SG^r, AHV^r, ETH^r, ABT^r	10 (13)	Mihara *et al.* (1973)
	Corynebacterium glutamicum	TRA^r, MG^r, AG^r, TU^r, MP^r, $5MT^r$	15 (10)	Araki *et al.* (1974b)
	Serratia marcescens	Histidase$^-$, TRA^r, MH^r, TA^r	13	Kisumi *et al.* (1975)
L-Ile	*Brevibacterium flavum*	AHV^r, OMT^r	14.5 (14.5)	Shiio *et al.* (1973)
	Serratia marcescens	$IleHX^r$, ABA^r	12 (12)	Komatsubara *et al.* (1973)
	Corynebacterium glutamicum	Met^-, AHV^r, AEC^r, THI^r, ETH^r, AL^r, ABA^r	10.6 (10.6)	Kato and Nakayama (1976)
L-Leu	*Brevibacterium lactofermentum*	TA^r, MET^-, Ile^-	28 (22)	Tsuchida *et al.* (1975a)
	Serratia marcescens	ABA^r, *Ile* revertant	13.5 (11)	Kisumi *et al.* (1973)

L-Lys	*Brevibacterium flavum*	*AEC^r*	57 (44)	Sano and Shiio (1970)
L-Met	*Corynebacterium glutamicum*	*Thr$^-$, ETH^r, SM^r, MetHX^r*	2 (2)	Kase and Nakayama (1975)
L-Phe	*Brevibacterium flavum*	*MFP^r*	2.2 (2.2)	Sugimoto *et al.* (1973)
	Corynebacterium glutamicum	*Tyr$^-$, PEP^r, PAP^r*	9.5 (9.5)	Hagine and Nakayama (1974)
L-Trp	*Brevibacterium flavum*	*Phe$^-$, 5FT^r, 3FP^r*	6.2 (6.2)	Shiio *et al.* (1975)
	Corynebacterium glutamicum	*5MT^r, TrpHX^r, 6FT^r, 4MT^r, PEP^r, PAP^r, TyrHX^r, PheHX^r*	12 (12)	Hagino and Nakayama (1975)
L-Tyr	*Brevibacterium flavum*	*MFP^r*	1.9 (1.9)	Sugimoto *et al.* (1973)
	Corynebacterium glutamicum	*AT^r, PAP^r, PEP^r, TyrHX^r, Phe$^-$*	17.6 (17.6)	Hagino and Nakayama (1973)
L-Thr	*Brevibacterium flavum*	*AHV^r, Met$^-$*	18 (18)	Nakamora and Shiio (1972)
	Brevibacterium flavum	*AHV^r*	13 (13)	Shiio and Nakamori (1970)
	Corynebacterium glutamicum	*Met$^-$, AHV^r, AEC^r*	14 (14)	Kase and Nakayama (1972)
L-Val	*Brevibacterium lactofermentum*	*TA^r*	31 (31)	Tsuchida *et al.* (1975b)

[a] TA, 2-thiazole alanine; ArgHX, arginine hydroxamate; SG, sulfaguanidine; AHV, α-amino-β-hydroxyvaleric acid; ETH, ethionine: ABT, 2-aminobenzothiazole: TRA, 1,2,4-triazole alanine: MG, 6-mercaptoguanine; AG, 8-azaguanine; TU, 4-thiuracil; MP, 6-mercaptopurine; 5MT, 5-methyltryptophan; OMT, *o*-methylthreonine; ABA, α-aminobutylic acid; AEC, *S*-(2-aminoethyl)-L-cysteine; SM, selenomethionine; MetHX, methionine hydroxamate; MH, 2-methylhistidine; PEP, *p*-fluorophenylalanine; PAP, *p*-aminophenylalanine; MFP, *m*-fluorophenylalanine; 5FT, 5-fluorotryptophan; 3FP, 3-fluoro-phenylalanine; TrpHX, tryptophan hydroxamate; PheHX, phenylalanine hydroxamate; PEP, *p*-ethylphenylalanine; TyrHX, tyrosine hydroxamate; AT, 3-aminotyrosine; IleHX, isoleucine hydroxamate; AL, 4-azaleucine; THI, thiaisoleucine.

C. Enzymatic Methods

Another microbial method of amino acid production is that of enzymatic conversion of a substrate to the corresponding amino acid. Amino acids which are effectively produced by this method include L-aspartic acid, L-alanine, L-tyrosine, L-cysteine, L-tryptophan, L-phenylalanine, and L-lysine. Table IV lists the amino acids produced by this method and shows substrate, enzymatic reaction, enzyme source, and productivity of each method. Some amino acids are produced with a single enzyme, for example, L-aspartic acid from fumaric acid and ammonia by aspartase of *E. coli.* On the other hand, the enzymatic conversion of chemically synthesized compounds into amino acids often requires several types of enzymes. For example, the conversion of DL-2-amino-thiazoline-4-carboxylic acid (DL-ATC) into L-cysteine requires three enzymes of *Pseudomonas thiazolinophilum:* ATC racemase, L-ATC hydrolase, and *S*-carbamoyl-L-cysteine (SCC) hydrolase. The reactions are as follows:

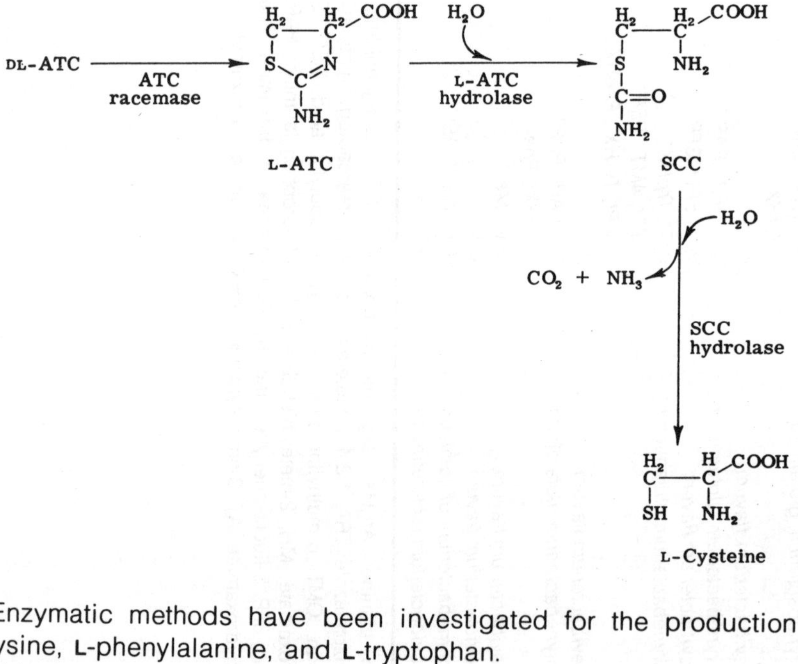

Enzymatic methods have been investigated for the production of L-lysine, L-phenylalanine, and L-tryptophan.

Fukumura (1977) established an enzymatic process for producing L-lysine using DL-α-amino-ϵ-caprolactam (DL-aminolactam) as the starting material. This process consisted of a racemization of D-aminolactam and the hydrolyzation of L-aminolactam to form L-lysine, as shown in the reactions below. *Achromobacter obae* nov. sp. and *Cryptococcus laurentti* were selected as the enzyme sources for D-aminolactam racemase and L-aminolactam hydrolase. DL-Aminolactam at a concentration of 100 gm/liter was converted into L-lysine in 25 hours. The production yield was 99.8% (moles product per mole of substrate).

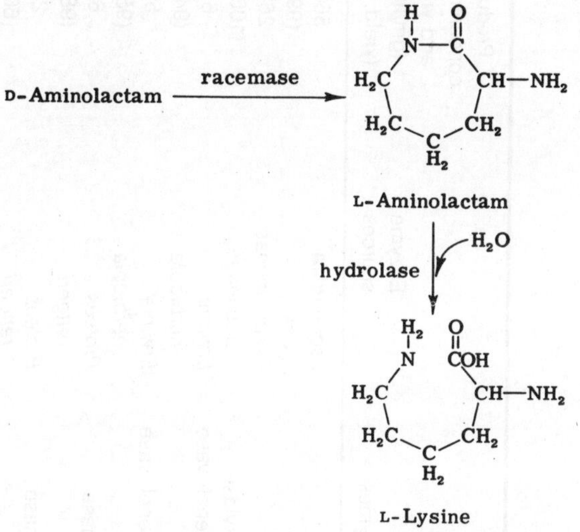

III. PROCESS CONTROL IN AMINO ACID FERMENTATION

During the course of industrialization of penicillin fermentation, pure-culture technology was developed in large-scale submerged fermentors. It included (1) maintenance of the pure-culture condition (sterilization of equipment, media, and air); (2) control of pH, temperature, and foaming; and (3) aeration and agitation. All these advances in biochemical engineering were successfully applied to industrialization of amino acid fermentation. It would be out of place, however, to attempt to cover all these diverse advances in this chapter. Instead, a few aspects which have particular relation to the large-scale production of amino acids are described.

A. Maintenance of Pure-Culture Conditions

In amino acid fermentation, contamination is controlled according to the procedures which have been established in penicillin fermentation.

TABLE IV. Amino Acids Produced by Enzymatic Methods

Amino acids	Reaction[a]	Enzymes	Enzyme sources	Product concentration and yield (gm/liter (yield, %)	References
L-Asp	Fumaric acid + NH_4^+ → L-Asp	Aspartase	*Escherichia coli*	560 (99)	Kitahara *et al.* (1959)
L-Ala	L-Asp → L-Ala + CO_2	Aspartate decarboxylase	*Pseudomonas dacunhae*	268 (100)	Chibata *et al.* (1965)
L-Tyr	Phenol + pyruvate + NH_4^+ → L-Tyr + H_2O	Tyrosinephenol-lyase	*Erwinia herbicola*	61 (94)	Enei *et al.* (1973)
L-DOPA	Pyrocatechol + pyruvate + NH_4^+ → L-DOPA + H_2O	Tyrosinephenol-lyase	*Erwinia herbicola*	59 (95)	Enei *et al.* (1973)
L-Trp	Indole + pyruvate + NH_4^+ → L-Trp + H_2O	Tryptophanase	*Proteus rettgeri*	63 (98)	Nakazawa *et al.* (1972)
5-(OH)-L-Trp	5-(OH)-Indole + pyruvate + NH_4^+ → 5-(OH)-L-Trp + H_2O	Tryptophanase	*Proteus rettgeri*	28 (60)	Nakazawa *et al.* (1972)
L-Trp	DL-Trp hydantoin + 2 H_2O → L-Trp + CO_2 + NH_3	D-Trp hydantoin racemase, L-Trp hydantoin hydrolase, *N*-carbamoyl-Trp hydrolase	*Flavobacterium aminogenes*	50 (100)	Yokozeki *et al.* (1976)

L-CysH	DL-ATC + 2 H_2O → L-CysH + CO_2 + NH_3	D-ATC racemase, L-ATC hydrolase, *S*-carbamoyl-L-CysH hydrolase	*Pseudomonas thiazolinophilum*	30 (95)	Sano *et al.* (1976)
L-CysH	β-Chloro-L-Ala + Na_2S → L-CysH + NaCl + NaOH	Cysteine desulfhydrase	*Aerobacter aerogenes*	49 (89)	Ogishi *et al.* (1977)
L-Phe	DL-Phe hydantoin + 2 H_2O → L-Phe + CO_2 + NH_3	D-Phe hydantoin racemase, L-Phe hydantoin hydrolase, *N*-carbamoyl-L-Phe hydrolase	*Flavobacterium aminogenes*	50 (100)	Yokozeki *et al.* (1976)
L-Lys	DL-Aminocaprolactam + H_2O → L-Lys	D-Caprolactam racemase	*Achromobacter obae*	100	Fukumura (1974)
		L-Caprolactam hydrolase	*Cryptococcus laurentii*	(100)	

[a] L-DOPA, 3-(3,4-dihydroxphenyl)-L-alanine; DL-Trp hydantoin; DL-5-indolylmethyl-hydantoin; DL-ATC; DL-2-aminothiazoline-4-carboxylic acid; DL-Phe hydantoin; DL-5-benzyl-hydantoin; DL-aminocaprolactam; DL-α-amino-ε-caprolactam.

Unlike penicillin fermentation, in which penicillinase-producing microorganisms often inactivate penicillin, no specific contaminants have been observed in amino acid fermentations. However, amino acids have a nutritional value to many microorganisms, and the culture conditions of amino acid fermentations are very favorable for many undesirable microbes. To ensure the maximum productivity, therefore, the maintenance of pure-culture conditions is essential.

In order to eliminate foreign microbes in the media, heat sterilization is most commonly used. For large volumes of the media in the commercial operation, continuous rather than batchwise sterilization has proved to be more successful. The following advantages, pointed out by Humphrey and Deindoerfer (1959), have been substantiated in amino acid fermentations: (1) energy saving, (2) better quality control, and (3) improved productivity. Figure 6 shows an example of a continuous sterilization system with a plate exchanger employed for L-glutamic acid fermentation.

The problem of producing a large quantity of sterile air is not peculiar (but is no less important) to amino acid fermentation. In laboratories, cotton plugs are satisfactory, and the oxygen transfer characteristics of various types of cotton plugs have been quantitatively elucidated in relation to the productivity of L-glutamic acid (Hirose *et al.*, 1966a). In pilot-plant jar fermentors and large-scale commercial fermentors, air sterilization is performed with glass-wool filters. Developments in the theoretical basis of air filtration were very effective in the design and construction of air filters (Aiba *et al.*, 1973).

B. Automatic Control of Amino Acid Fermentation

Figure 7 shows a typical structure of a commercial fermentor for amino acid fermentation. Although most of the microorganisms for amino acid production are bacteria and less power input may be needed for agitation than is used in antibiotic fermentations, it should be noted that the oxygen requirement and heat evolution per unit volume of fluids and time are large because of the high rate of sugar consumption and respiration.

Utilization of gaseous ammonia is a great advantage in amino acid fermentation. Since the molecules of amino acids contain much nitrogen, the selection of an adequate nitrogen source and its sufficient supply are essential for the performance of this type of fermentation. The use of gaseous ammonia as a nitrogen source and as a pH-controlling agent solves various technological problems. It avoids undesired dilution of the fermentation broths, controls the pH more precisely, and makes the equipment and operation more simple.

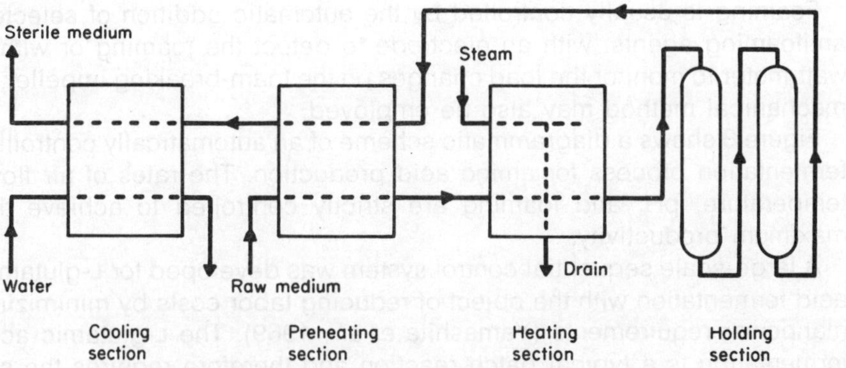

FIGURE 6. Plate–exchanger continuous sterilizer.

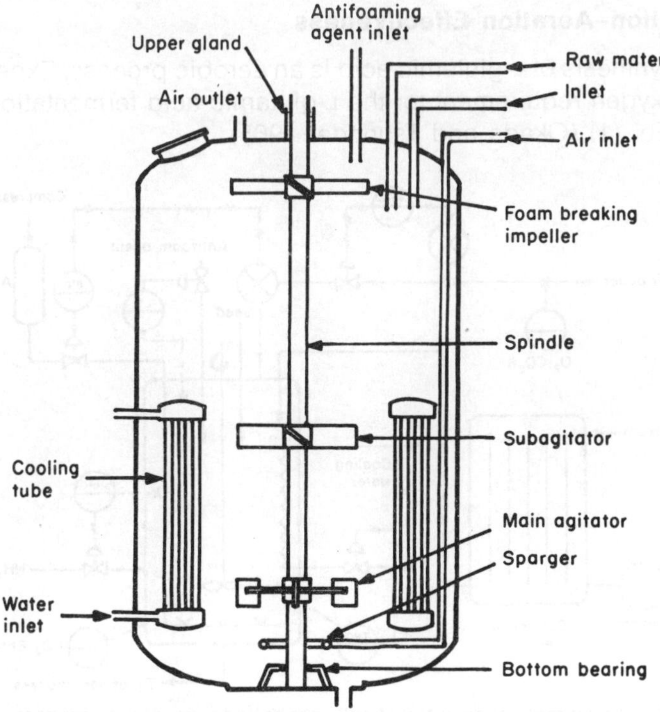

FIGURE 7. Structure of fermentor for an aerobic fermentation process.

Foaming is usually controlled by the automatic addition of selected antifoaming agents, with an electrode to detect the foaming or with a watt-meter to monitor the load changes on the foam-breaking impeller. A mechanical method may also be employed.

Figure 8 shows a diagrammatic scheme of an automatically controlled fermentation process for amino acid production. The rates of air flow, temperature, pH, and foaming are strictly controlled to achieve the maximum productivity.

A large-scale sequential control system was developed for L-glutamic acid fermentation with the object of reducing labor costs by minimizing manpower requirements (Yamashita *et al.*, 1969). The L-glutamic acid fermentation is a typical batch reaction and therefore requires the sequential operation of many valves to sterilize the system, and then to introduce the medium to the fermentor. During the fermentation, conventional analog control instruments regulate the sterilized air flow, temperature, pressure, and pH. Some of these analog controllers are programmed to execute a desired profile in time.

C. Agitation–Aeration Effectiveness

The biosynthesis of L-glutamic acid is an aerobic process. Experimentally, the oxygen requirement for the L-glutamic acid fermentation is as shown in Eq. (1) (Okada and Tsunoda, 1965).

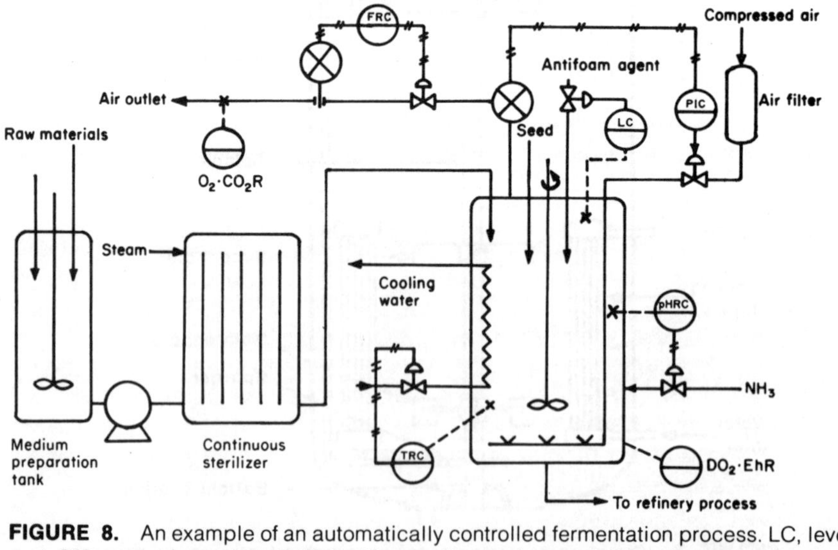

FIGURE 8. An example of an automatically controlled fermentation process. LC, level control; PIC, pressure control; PHRC, pH recording and control; FRC, flow rate recording and control; TRC, temperature recording and control; EhR, redox potentials recording.

$$C_6H_{12}O_6 + 2.33\ O_2 \longrightarrow 0.82\ C_5H_9O_4N + 1.94\ CO_2 \qquad (1)$$

Oxygen is consumed in large amounts. Therefore, quantitative discussion on oxygen transfer during the fermentation is indispensable for both the optimization of operation and the scaleup.

The following equations are the simplest expression of the basis for discussing oxygen transfer during the fermentation [Eq. (2)]:

$$\begin{aligned} r_{ab} = K_d(P_B - P_L) &= K_rM \quad (P_L \geqslant P_{Lcrit}) \\ &< KM \quad (P_L < P_{Lcrit}) \end{aligned} \qquad (2)$$

where r_{ab} is the rate of cell respiration (moles of O_2/ml/min); K_d is the oxygen absorption coefficient (mole/ml/min/atm); P_B is the gas phase oxygen tension (atm); P_L is the liquid phase oxygen tension (atm); K_r is the cell's oxygen demand (mole/min/gm of cell); M is the cell density (gm/ml); and P_{Lcrit} is the critical level of liquid phase oxygen for cell respiration (atm). Rate of cell respiration (r_{ab}) equals K_rM *at* P_L levels above P_{Lcrit}; in other words, cell respiration is satisfied when oxygen supply is sufficient to maintain P_L above P_{Lcrit}.

Values of P_{Lcrit} are usually less than 0.01 atm, too low to be determined with conventional membrane-coated oxygen electrodes. The determination of extremely low P_{Lcrit} values was indispensable; therefore, a new method was developed in which both P_L and the redox potentials of culture medium were measured simultaneously (Shibai *et al.*, 1974). This value was 0.0002 atm for *B. flavum* (Akashi *et al.*, 1977a).

Lack of quantitative analysis on oxygen transfer sometimes leads to inaccurate conclusions. For example, L-glutamic acid-producing bacteria are known to excrete large amounts of lactic acid in a biotin-rich medium. However, the analysis of oxygen transfer clarified the true factor: It was due not to excess biotin but to oxygen deficiency that lactic acid became the dominant product in a biotin-rich medium (Hirose *et al.*, 1968). In biotin-rich media cell density became high, which resulted in a high oxygen demand of culture broth. Consequently, the amount of oxygen was insufficient. In fact, the cell did not accumulate lactic acid even in biotin-rich media when sufficient oxygen was supplied (Fig. 9).

In L-glutamic acid fermentation, the P_L level is closely related to the product formation, as shown in Fig. 10 (Hirose *et al.*, 1966b). The production yield markedly decreased when the P_L level was zero as measured by conventional membrane-coated oxygen electrodes. This implies that aeration and agitation must be controlled so as to keep P_L slightly above zero for the effective production of L-glutamic acid. In these cultures of high productivity, cell respiration was satisfied. Limited oxygen supply resulted in lactic acid or succinic acid accumulation at the expense of L-glutamic acid formation (Okada *et al.*, 1961). Oxygen tension in the

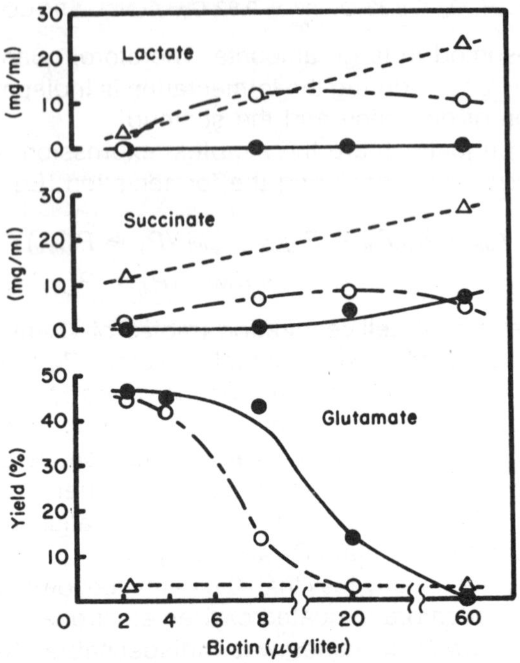

FIGURE 9. Accumulation of glutamate, succinate, and lactate vs. biotin concentration in L-glutamic acid fermentation. ●, P_L controlled at 0.15 atm; ○, sulfite r_{ab} = 10.5 × 10^{-7} mole/ml/minute; △, sulfite r_{ab} = 1.2 × 10^{-7} mole/ml/minute.

growth phase also seriously influenced the product formation (Hirose *et al.*, 1967). Oxygen deficiency in the growth phase did not bring about any marked change in L-glutamic acid formation, whereas a high aerating condition seriously reduced the L-glutamic acid-producing capability of the cell.

Although the productivity of other amino acids was also influenced by oxygen supply, the extent or the mode of the influence was diverse according to the characteristics of biosynthetic pathways of each amino acid. Figure 11 shows the relationship between the degree of oxygen satisfaction (r_{ab}/K_rM) and the relative productivity of each amino acid. For example, oxygen shortage caused inhibition of both L-glutamic acid and L-lysine production, but the extent of the inhibition was more serious in L-glutamic acid fermentation than in L-lysine fermentation (Akashi *et al.*, 1977b). Moreover, the maximum production of L-leucine was observed when the cell respiration was slightly repressed. Aerobic microorganisms require oxygen to reoxidize $NADH_2$ to form NAD and to generate ATP for growth and metabolism. L-Glutamic acid is biosynthesized

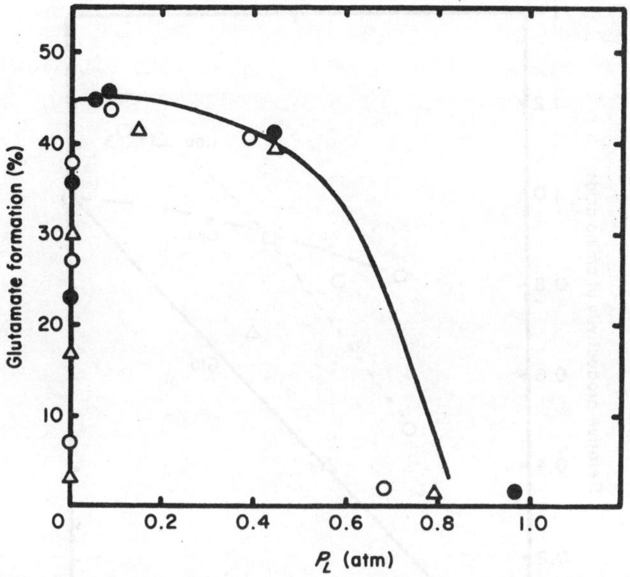

FIGURE 10. Effect of dissolved oxygen level, P_L, on the production of L-glutamic acid. P_L was controlled by changing gas-phase oxygen tension. ●, 20 ml working volume in 500-ml flask; ○, 40 ml working volume in 500-ml flask; △, 80-ml working volume in 500-ml flask.

by way of pyruvic acid, citric acid, and α-ketoglutaric acid. Formation of 5¹/₃ moles of $NADH_2$ and generation of 2 moles of ATP are accompanied with assimilation of 1 mole of glucose. On the other hand, L-lysine is biosynthesized by way of pyruvic acid, citric acid, fumaric acid, and L-aspartic acid, forming 2⅔ moles of $NADH_2$ and ⅔ moles of ATP from 1 mole of glucose. In case of L-leucine biosynthesis, assimilation of 1 mole of glucose accompanies generation of 2 moles of ATP and $NADH_2$. Table V summarizes ATP consumption and $NADH_2$ formation in relation to the biosynthesis of these amino acids. The amount of $NADH_2$ formed in L-glutamic acid fermentation is more than that in L-lysine fermentation by 2⅔ moles. This suggests that the inhibition of product formation due to oxygen shortage was more serious in L-glutamic acid fermentation. L-Leucine is biosynthesized not through the TCA cycle but by way of pyruvic acid. Therefore, the maximum production might occur when aerobic degradation of glucose was inhibited due to oxygen shortage. However, extremely deficient condition of oxygen might inhibit the cell to reoxidize 2 moles of $NADH_2$ and lead to the accumulation of lactic acid in place of L-leucine biosynthesis. Therefore, it is reasonable that the

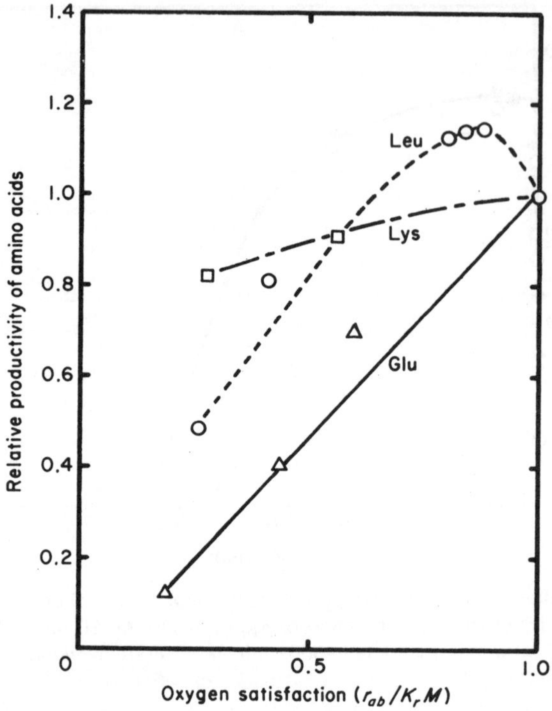

FIGURE 11. Relationship between relative productivity of amino acids and degree of oxygen satisfaction.

maximum production of L-leucine was observed when the cell respiration was slightly inhibited.

IV. CONCLUSION

The demands for amino acids are still increasing in the fields of foods, animal feeds, pharmaceuticals, chemicals, etc. To meet these increas-

TABLE V. ATP Consumption and $NADH_2$ Formation in the Biosynthesis of L-Glutamic acid, L-Lysine, and L-Leucine

Amino acids	Formation of $NADH_2$ (moles)	Consumption of ATP (moles)
L-Glu	$5\frac{1}{3}$	−2
L-Lys	$2\frac{2}{3}$	$-\frac{2}{3}$
L-Leu	2	−2

ing and diversified demands, almost all amino acids are now being produced commercially. Most of them are fermentation products. Large amounts of L-glutamic acid and L-lysine, the most important commercial amino acids, are available at a very low price, but the productivity of some other amino acids, such as L-tryptophan and L-histidine, is still insufficient and, consequently, their selling prices are still high. Every effort to supply each amino acid in sufficient quantities at moderate prices will surely continue. Promising approaches for improving the productivity of amino acid fermentations in the future include: (1) selection and improvement of cultures, (2) selection of raw or starting materials, and (3) physical and metabolic controls of fermentation. Remarkable progress in microbial genetics should accelerate the effective breeding of potent producers of amino acids. In addition to auxotrophs and regulatory mutants, which are now widely employed for industrial production of amino acids, other types of mutants induced by gene engineering technique are expected to be developed. Pursuits of more prospective raw materials are endless. Progress in organic chemistry may supply the fermentation industry with new starting materials. A combination of the synthetic techniques of organic chemistry with enzymatic techniques of microbiology will surely bring about new industrial methods for amino acid production. The productivity of amino acid fermentations will be increased also by better physical and metabolic controls in growing cultures or reaction mixtures. Advances in biochemical engineering will contribute most to successful improvement. An understanding of the microbial physiology and biosynthetic mechanisms involved in the fermentation process, as well as quantitative approaches to the activity of microorganisms, is usually fundamental to the successful control of amino acid fermentations.

REFERENCES

Aiba, S., Humphrey, A. E., and Mills, N. F. (1973). "Biochemical Engineering," 2nd ed., p. 270. Academic Press, New York.

Akashi, K., Shibai, H., and Hirose, Y. (1977a). *Biotechnol. Bioeng.* **20,** 27 (1978).

Akashi, K., Shibai, H., and Hirose, Y. (1977b). *Proc. Annu. Meet. Agric. Chem. Soc. Jpn.* p. 211.

Araki, K., Ueda, H., and Saigusa, S. (1974a). *Agric. Biol. Chem.* **38,** 565.

Araki, K., Shimijo, S., and Nakayama, K. (1974b). *Agric. Biol. Chem.* **38,** 837.

Breuer, L. H., Pond, W. G., Warner, R. G., and Loosli, J. K. (1964). *J. Nutr.* **82,** 499.

Chibata, I., Kamimoto, T., and Kato, J. (1965). *Appl. Microbiol.* **13,** 638.

Distillers Co., Ltd. (1965). British Patent 996,906.

Doi, S., and Kaneko, Y. (1960). *Hakko To Taisha* **2,** 26.

Enei, H., Nakazawa, H., Okumura, S., and Yamada, H. (1973). *Agric. Biol. Chem.* **37,** 725.

Fukumura, T. (1974). *Proc. Int. Congr. IAMS, 1st, 1974, IX, Tokyo.* Abstract Vol. V, p. 495.
Hagino, H., and Nakayama, K. (1973). *Agric. Biol. Chem.* **37,** 2013.
Hagino, H., and Nakayama, K. (1974). *Agric. Biol. Chem.* **38,** 157.
Hagino, H., and Nakayama, K. (1975). *Agric. Biol. Chem.* **39,** 343.
Hagino, H., Yoshida, H., Kato, F., Arai, T., Katsumata, R., and Nakayama, K. (1973). *Agric. Biol. Chem.* **37,** 2001.
Hirose, Y., Sonoda, H., Kinoshita, K., and Okada, H. (1966a). *Agric. Biol. Chem.* **30,** 49.
Hirose, Y., Sonoda, H., Kinoshita, K., and Okada, H. (1966b). *Agric. Biol. Chem.* **30,** 585.
Hirose, Y., Sonoda, H., Kinoshita, K., and Okada, H. (1967). *Agric. Biol. Chem.* **31,** 1210.
Hirose, Y., Sonoda, H., Kinoshita, K., and Okada, H. (1968). *Agric. Biol. Chem.* **32,** 855.
Humphrey, A. E., and Deindoerfer, F. H. (1959). *Appl. Microbiol.* **7,** 256.
Ikeda, K. (1908). *J. Tokyo Chem. Soc.* **30,** 820.
Izumi, Y., Tani, Y., and Ogata, K. (1973). *Biochim. Biophys. Acta* **326,** 485.
Kamiryo, T., Parthasarathy, S., and Numa, S. (1976). *Proc. Natl. Acad. Sci. U.S.A.* **73,** 386.
Kanzaki, T., Isobe, K., Okazaki, H., Mochizuki, K., and Fukuda, H. (1967). *Agric. Biol. Chem.* **31,** 1416.
Kase, H., and Nakayama, K. (1972). *Agric. Biol. Chem.* **36,** 1611.
Kase, H., and Nakayama, K. (1975). *Agric. Biol. Chem.* **39,** 153.
Kase, H., Tanaka, H., and Nakayama, K. (1971). *Agric. Biol. Chem.* **35,** 2089.
Kato, H., and Nakayama, K. (1976). *Proc. Annu. Meet. Agric. Chem. Soc. Jpn.* p. 127.
Kinoshita, S. (1963). *Recent Prog. Microbiol., Symp. Int. Congr. Microbiol., 8th, 1962* p. 317.
Kinoshita, S., Udaka, S., and Shimono, M. (1957a). *J. Gen. Appl. Microbiol.* **3,** 193.
Kinoshita, S., Nakayama, K., and Udaka, S. (1957b). *J. Gen. Appl. Microbiol.* **3,** 276.
Kinoshita, S., Nakayama, K., and Kitada, S. (1958). *J. Gen. Appl. Microbiol.* **4,** 128.
Kisumi, M., Kato, J., Sugiyama, M., and Chibata, I. (1971). *Appl. Microbiol.* **22,** 987.
Kisumi, M., Komatsubara, S., and Chibata, I. (1973). *J. Biochem. (Tokyo)* **73,** 107.
Kisumi, M., Nakazawa, N., Takagi, T., and Chibata, I. (1975). *Proc. Annu. Meet. Agric. Chem. Soc. Jpn.* p. 30.
Kitahara, K., Fukui, S., and Misawa, M. (1959). *J. Gen. Appl. Microbiol.* **5,** 74.
Komatsubara, S., Kisumi, M., and Chibata, I. (1973). *Proc. Annu. Meet. Soc. Ferment. Technol., Jpn* p. 98.
Kubota, K., Kageyama, K., Shiro, T., and Okumura, S. (1971). *J. Gen. Appl. Microbiol.* **17,** 167.
Kubota, K., Kageyama, K., Maeyashiki, I., Yamada, K., and Okumura, S. (1972). *J. Gen. Appl. Microbiol.* **18,** 365.
Kubota, K., Onoda, T., Kamijo, H., Yoshinaga, F., and Okumura, S. (1973). *J. Gen. Appl. Microbiol.* **19,** 339.
Lafitskaya, T. N., and Aseeva, I. V. (1967). *Mikrobiologia* **36,** 568.
Lee, W. H., and Good, R. C. (1963). U. S. Patent 3,087,863.
Masuo, E., Wakisaka, Y., and Hattori, M. (1962). Japanese Patent 37-6342.
Matsumoto, T., Yano, T., and Yamada, K. (1967). *Agric. Biol. Chem.* **31,** 1381.
Mihara, O., Kamijo, H., and Kubota, K. (1973). *Proc. Annu. Meet. Agric. Chem. Soc. Jpn.* p. 112.
Miyai, K., Osawa, T., and Tsuruo, I. (1964). Japanese Patent 39-2988.
Mogi, M., Sugizaki, Z., and Mizusawa, K. (1967). Japanese Patent 42-502.
Nakamori, S., and Shiio, I. (1972). *Agric. Biol. Chem.* **36,** 1675.
Nakanishi, T., Nakajima, J., and Kanda, K. (1975). *J. Ferment. Technol.* **53,** 543.
Nakayama, K. (1976). *Process Biochem.* **12,** 4.
Nakayama, K., and Hagino, H. (1966). *J. Agric. Chem. Soc. Jpn.* **40,** 377.

Nakayama, K., and Yoshida, H. (1972). *Agric. Biol. Chem.* **36,** 1675.
Nakayama, K., Kitada, S., and Kinoshita, S. (1961). *J. Gen. Appl. Microbiol.* **7,** 145.
Nakayama, K., Tanaka, H., Hagino, H., and Kinoshita, S. (1966). *Agric. Biol. Chem.* **30,** 611.
Nakazawa, H., Enei, H., Okumura, S., Yoshida, H., and Yamada, Y. (1972). *FEBS Lett.* **25,** 43.
Ogawa, C., Ode, M., and Midorikawa, Y. (1959). *Hakko To Taisha* **1,** 45.
Ogishi, N., Nishikawa, D., Kumagai, H., and Yamada, H. (1977). *Proc. Annu. Meet. Agric. Chem. Soc. Jpn.* p. 90.
Oishi, K. (1967). *J. Agric. Chem. Soc. Jpn.* **41,** R35.
Oishi, K., and Aida, K. (1965). *Agric. Biol. Chem.* **29,** 83.
Okada, H., and Tsunoda, T. (1965). *Agric. Biol. Chem.* **29,** 923.
Okada, H., Kameyama, I., Okumura, S., and Tsunoda, T. (1961). *J. Gen. Appl. Microbiol.* **7,** 177.
Okumura, S., Tsugawa, R., Tsunoda, T., Kono, K., Matsui, T., and Miyachi, N. (1962a). *J. Agric. Chem. Soc. Jpn* **36,** 141.
Okumura, S., Tsugawa, R., Tsunoda, T., and Motozaki, S. (1962b). *J. Agric. Chem. Soc. Jpn.* **36,** 204.
Okumura, S., Shibuya, M., Konishi, S., Ishida, M., and Shiro, T. (1964). *Agric. Biol. Chem.* **28,** 742.
Ota, S., and Tanaka, M. (1959). *J. Ferment. Technol.* **37,** 261.
Ozaki, S., Kono, K., Okumura, S., Okada, H., and Sakaguchi, K. (1960). *Proc. Symp. Amino Acids Ferment., 2nd, 1960,* p. 4.
Sano, K. (1976). *J. Agric. Chem. Soc. Jpn.* **50,** R201.
Sano, K., and Shiio, I. (1970). *J. Gen. Appl. Microbiol.* **16,** 373.
Sano, K., and Shiio, I. (1971). *J. Gen. Appl. Microbiol.* **17,** 97.
Sano, K., Matsuda, K., Yokazeki, K., Tamura, F., Yasuda, N., Noda, I., and Mitsugi, K. (1976). *Proc. Annu. Meet. Agric. Chem. Soc. Jpn.* p. 238.
Shibai, H., Ishizaki, A., and Hirose, Y. (1974). *Proc. Int. Congr. IAMS, 1st, 1974, IX, Tokyo.* Abstract Vol. V, p. 154.
Shibukawa, M., Kurima, M., Okabe, S., and Osawa, T. (1968). *Hakko To Taisha* **17,** 61.
Shibuya, M., Yoshinaga, F., and Okumura, S. (1967). *Proc. Annu. Meet. Agric. Chem. Soc. Jpn.* p. 178.
Shiio, I., and Nakamori, S. (1970). *Agric. Biol. Chem.* **34,** 448.
Shiio, I., Otsuka, S., and Katsuya, N. (1963). *J. Biochem. (Tokyo)* **53,** 333.
Shiio, I., Nakamori, S., and Sasaki, A. (1973). *Proc. Annu. Meet. Agric. Chem. Soc. Jpn.* p. 113.
Shiio, I., Sugimoto, S., and Nakagawa, M. (1975). *Agric. Biol. Chem.* **39,** 627.
Sommerson, N. L., and Phillips, T. (1962). U. S. Patent 3,080,297.
Su, Y. C., and Yamada, K. (1960). *Bull. Agric. Chem. Soc. Jpn.* **24,** 69.
Sugimoto, S., Nakagawa, M., Tsuchida, T., and Shiio, I. (1973). *Agric. Biol. Chem.* **37,** 2327.
Takinami, K., Okada, H., and Tsunoda, T. (1963). *Agric. Biol. Chem.* **27,** 858.
Takinami, K., Okada, H., and Tsunoda, T. (1964). *Agric. Biol. Chem.* **28,** 114.
Takinami, K., Yoshii, H., Tsuri, H., and Okada, H. (1965). *Agric. Biol. Chem.* **29,** 351.
Takinami, K., Yamada, K., and Okada, H. (1966). *Agric. Biol. Chem.* **30,** 674.
Takinami, K., Yamada, K., and Okada, H. (1967). *Agric. Biol. Chem.* **31,** 223.
Tanaka, K., Iwasaki, T., and Kinoshita, S. (1960). *J. Agric. Chem. Soc. Jpn.* **34,** 593.
Terui, G., and Niitsu, H. (1969). *Biotechnol. Bioeng. Symp.* **1,** 33.
Tsuchida, T., Yoshinaga, F., Kubota, K., Momose, H., and Okumura, S. (1975a). *Agric. Biol. Chem.* **39,** 1149.

Tsuchida, T., Yoshinaga, F., Kubota, K., and Momose, H. (1975b). *Agric. Biol. Chem.* **39,** 1319.

Tsunoda, T., and Okumura, S. (1972). *Proc. Int. Symp. Convers. Manuf. Foodst. Microorg.*, p. 229.

Udaka, S., and Kinoshita, S. (1960). *J. Gen. Appl. Microbiol.* **5,** 159.

Veldkamp, H., Berg, G. V. D., and Zevenhuizen, L. P. T. M. (1963). *Antonie van Leeuwenhoek* **29,** 35.

Yamada, K., Kinoshita, S., Tsunoda, T., and Aida, K. (1972). "Microbial Production of Amino Acids." Kodansha, Tokyo.

Yamashita, S., Hoshi, H., and Inagaki, T. (1969). *In* "Fermentation Advances" (D. Perlman, ed.), p. 444. Academic Press, New York.

Yokozeki, K., Sano, K., Eguchi, T., Yasuda, N., Noda, I., and Mitsugi, K. (1976). *Proc. Annu. Meet. Agric. Chem. Soc. Jpn.* p. 238.

Yoshida, H., and Nakayama, K. (1975). *J. Agric. Chem. Soc. Jpn.* **49,** 527.

Yoshinaga, F., Konishi, S., Okumura, S., and Katsuya, N. (1966). *Proc. Symp. Amino Acids Nucleic Acids Ferment.* p. 4.

Yoshinaga, F., Kikuchi, K., Tsuchida, T., and Okumura, S. (1976). Japanese Patent 51-44196.

Chapter 8

Microbial Production of Antibiotics

D. PERLMAN

I. INTRODUCTION

The production of antibiotics by microorganisms is similar in many respects to the synthesis of other compounds of extreme molecular complexity. In general, the special reactions required for synthesis are superimposed upon the metabolic systems responsible for growth and reproduction. These special steps may be often regarded as distortions or elaborations of normal cellular mechanisms in that they give such structural peculiarities as fused rings, D-amino acid residues, and rare saccharides.

MICROBIAL TECHNOLOGY, 2nd ed., VOL. I
 ISBN 0-12-551501-4

In modern industrial microbial processes, the products may account for as much as 30% by weight of the nutrients supplied, so that what was previously a minor metabolic route represents a substantial portion of the metabolic activity of the organism. The biosynthesis of antibiotics may be regarded as the result of a series of "inborn errors of metabolism." These errors may be exaggerated by subjecting the original microorganism to mutagenic influences. The high yields of commercially important antibiotics owe much to the selection of such mutant strains. For example, the yield of benzylpenicillin was improved to well over 20 mg/ml (equivalent to 33,000 Oxford units/ml) (Chain, 1964) by the use of strains selected from *Penicillium chrysogenum* NRRL 1951, a culture producing only 120 μg/ml (Backus and Stauffer, 1955; Coghill and Koch, 1945).

In a period of a little more than 35 years, over 4000 antibiotics have been discovered, and the industry producing them has become highly competitive. World production of antibiotics exceeds 100,000 tons annually, and in the United States alone, sales of antibiotics for all purposes are over $1,000,000,000 per year. Of the thousands of antibiotics known, however, only a mere handful—the 90 listed in Table I—are produced on a commercial scale. In addition, 46, including 23 penicillins, 13 cephalosporins, dihydrostreptomycin, clindamycin, 6 tetracyclines, and 2 rifamycins, are prepared by a combination of microbial synthesis and chemical modification. Only chloramphenicol and pyrrolnitrin are manufactured on a commercial scale by chemical synthesis. The total chemical syntheses for penicillins, cephalosporins, polymyxins, gramicidin, tetracyclines, kanamycins, griseofulvin, and cycloheximide have been completed on a laboratory scale.

The cephalosporins group accounts for the largest sales volume (greater than $800,000,000 annually) with the Tetracycline group next in importance and leading in tonnage. Some clinical uses of antibiotics are listed in Table II. A number of antibiotics which were produced on a commercial scale have been withdrawn from large-scale distribution. These include stendomycin, ristocetin, and fumagillin. Those which are currently used for agricultural purposes include the antifungals blasticidin S and cycloheximide; the antibacterial kasugamycin; the coccidiostats monensin, lasalocid, and salinomycin; and the peptides nosiheptin, siomycin, and thiopeptin (all useful as growth promotants in poultry and animals). Antibiotics used only in veterinary medicine include hygromycin B, thiostrepton, and tylosin. Nisin, a polypeptide antibiotic from *Streptococcus cremoris,* is used as a food preservative in England.

II. GENERAL PRODUCTION METHODS

Since all of the antibiotics listed in Table I are made by aerobic fermentations, there are several similarities in the processes used for their production. The industry has been rather reluctant to publish details of the production methodology, but the general outline is fairly well known. Useful general accounts were described by Hastings (1955) and Jackson (1958), and particulars can be pieced together from the patent literature.

The culture vessels may have a capacity of as much as 100,000 gal. They usually measure two or three times as deep as they are wide. Stainless steel or nickel–chrome alloys are normally used in constructing these vessels, though in the past years less expensive grades of steel have been widely used. The vessels are fitted with aerating and agitating devices and with cooling coils and/or jackets for accurate temperature control.

The medium may be sterilized in a separate cooker and then pumped into the sterile culture vessels, or it may be sterilized *in situ*. Strict attention is paid to sterility in the antibiotic industry, and it is claimed that fewer than 2% of the batches are lost as a result of infections with antibiotic-resistant microorganisms or with phages. Since fermentation operations are costly, efforts are usually directed toward running as short a fermentation operation as economically possible. Large inoculum levels, e.g., as much as 10% of the volume of the fermentation, have been reported; careful choice and control of incubation temperatures, pH control, and continuous addition of energy sources and precursors, and other techniques are usually helpful in keeping the fermentation process operating at maximum economic efficiency.

In addition to concern with an efficient economic process from a fermentation viewpoint, there is concern with the problems of extraction and sometimes recovery of the antibiotics. An antibiotic is often present in the culture medium at concentrations of less than 1 gm/liter, and recovery of a pure antibiotic from the fermented medium (containing as much as 40 gm of dissolved solids per liter) is a very difficult assignment. At times a change of fermentation medium composition, e.g., use of invert cane molasses as a replacement for glucose solutions, which is economically attractive, introduces such complications into the recovery process for purifying the antibiotic that the economic advantage is lost.

Each of the 50 commercially operated antibiotic fermentations has its own peculiarities, and space limitations prevent a detailed analysis of

TABLE I. Some Antibiotics Produced on a Commercial Scale

		Antibiotic spectrum[b]								
Antibiotic	Microbial source	G+	G−	My	AF	AT	Other	Chemical type	Therapeutic or other use	Some commercial source[c]
Amphomycin	*Streptomyces canus*	+						Peptide	Topical	42
Amphotericin B	*Streptomyces nodosus*				+			Polyene	Oral or parenteral	58
Avoparcin[a]	*Streptomyces candidus*	+	+					Glycopeptide	Animal growth promotant	4
Azalomycin F[a]	*Streptomyces hygroscopicus*	+	+		+				Topical (AF)	54
Bacitracin	*Bacillus subtilis*	+						Peptide	Topical; also animal growth promotant	7, 33, 42, 45, 47, 48
Bambermycins	*Streptomyces bambergenesis*	+						Phosphoglycolipid	Animal growth promotant	31
Bicyclomycin[a]	*Streptomyces sapporonensis*	+	+						Topical	27
Blasticidin S[a]	*Streptomyces griseochromogenes*				+			Nucleoside	Agricultural (AF)	35
Bleomycins	*Streptomyces verticillus*	+	+			+		Peptide	Parenteral (AT)	44a
Cactinomycin[a]	*Streptomyces chrysomallus*	+				+		Peptide	Parenteral (AT)	22
Candicidin B	*Streptomyces griseus*				+			Polyene	Topical	42, 47
Candidin[a]	*Streptomyces viridoflavus*				+			Polyene	Topical	47
Capreomycin	*Streptomyces capreolus*			+				Peptide	Parenteral	21

Cephalosporins	Cephalosporin C is produced by *Cephalosporium acremonium* and converted to 7-ACA, which is used for prep. of semisynthetic cephalosporins	+	+					Peptide	Oral and parenteral	5, 12, 14, 18, 23, 26, 27, 29, 41, 45, 49, 52, 58, 59, 60
Chloramphenicol	*Streptomyces venezuelae;* commercial manufacture is by chemical synthesis	+	+				Rickettsia		Oral or parenteral	46
Chromomycin A_3[a]	*Streptomyces griseus*	+				+			Parenteral (AT)	59
Colistin	*Bacillus colistinus*		+					Peptide	Parenteral	8, 37, 39, 52
Cycloheximide	*Streptomyces griseus*				+				Agricultural (AF)	35, 61
Cycloserine	*Streptomyces orchidaceus*			+				Amino acid	Parenteral (TB)	5, 33
Dactinomycin	*Streptomyces antibioticus*					+		Peptide	Parenteral (AT)	44
Daunorubicin	*Streptomyces peucetius*					+			Parenteral (AT)	23, 52
Doxorubicin	*Streptomyces peucetius*					+			Parenteral (AT)	23
Enduracidin[a]	*Streptomyces fungicidus*	+		+				Peptide	Animal growth promotant	59
Erythromycin	*Streptomyces erythreus*	+						Macrolide	Oral and parenteral; animal growth promotant	1, 5, 13, 14, 18, 19, 21, 41, 48, 49, 52, 61
Fortimicins[a]	*Micromonospora olivoasterospora*	+	+					Aminoglycoside	Parenteral	39

(*continued*)

TABLE I (*Continued*)

Antibiotic	Microbial source	Antibiotic spectrum[b] G+	G−	My	AF	AT	Other	Chemical type	Therapeutic or other use	Some commercial source[c]
Fungimycin[a]	*Streptomyces coelicolor* var. *aminophilus*				+			Polyene	Topical	42
Fusidic Acid[a]	*Fusidium coelcineum*	+						Steroid	Parenteral	40
Gentamicins	*Micromonospora purpurea*	+	+					Aminoglycoside	Parenteral	14, 15, 56
Gramicidin A	*Bacillus brevis*	+						Peptide	Topical	42, 47, 62
Gramicidin J[a](S)	*Bacillus brevis*	+						Peptide	Topical	43
Griseofulvin	*Penicillium griseofulvum*				+			Spirolactone	Oral	14, 29, 34, 40, 44_n, 59
Hygromycin B	*Streptomyces hygroscopicus*	+	+				Helminths	Aminoglycoside	Animal feed suppl.	11, 41, 59
Josamycin[a]	*Streptomyces narbonesis*	+						Macrolide	Oral and parenteral	55
Kanamycins	*Streptomyces kanamyceticus*	+	+	+				Aminoglycosides	Parenteral	12, 14, 43, 44a, 52
Kasugamycin[a]	*Streptomyces kasugaensis*	+	+					Aminoglycoside	Agricultural antibacterial	8, 43, 55
Kitasatamycin[a]	*Streptomyces kitasatoensis*	+						Macrolide	Oral and parenteral	60
Lasalocid	*Streptomyces hazelensis*	+					Coccidia	Polyether	Agricultural use as coccidiostat and growth promotant	22
Lincomycin	*Streptomyces lincolnensis*	+							Oral and parenteral	61
Lividomycin[a]		+	+							36, 52

Macarbomycins[a]	*Streptomyces phaechromogenes*	–						Phosphoglycolipid	Animal growth promotant	43
Mepartricin[a]					+			Polyene	Topical	57
Midecamycin[a]		–						Macrolide	Oral and topical	43
Mikamycins[a]	*Streptomyces mitakaensis*	+						Peptide	Animal growth promotant	8
Mithramycin	*Streptomyces* species	+		+		+			Parenteral (AT)	48
Mitomycin C	*Streptomyces caespitosus*	+	+		+				Parenteral	12, 39
Mocimycin[a]		+							Animal growth promotant	28
Monensin	*Streptomyces cinnamonensis*	+					Coccidia	Polyether	Animal growth promotant	41
Myxin	*Chromobacterium iodinum* plus chemical modification				+			Phenazine	Topical in veterinary use	32
Neocarzinostatin	*Streptomyces carzinostaticus*	+				+		Peptide	Parenteral (AT)	
Neomycins	*Streptomyces fradiae*	+	+					Aminoglycoside	Oral and topical	6, 7, 11, 14, 23, 42, 44a, 47, 48, 52, 53, 58, 59, 61
Nosiheptide[a]	*Streptomyces actinosus*	+						Peptide	Animal growth promotant	52
Nisin	*Streptococcus cremoris*	+						Peptide	Food preservative	
Novobiocin	*Streptomyces niveus*	+							Oral and topical	43, 52, 61
Nystatin	*Streptomyces noursei*				+			Polyene	Oral and topical	4, 14, 15, 52, 58
Oleandomycin	*Streptomyces antibioticus*	+						Macrolide	Oral and parenteral	48
Paromomycin	*Streptomyces rimosus*	+	+				Protozoa	Aminoglycoside	Oral	46

(*continued*)

TABLE I *(Continued)*

Antibiotic	Microbial source	Antibiotic spectrum[b] G+	G−	My	AF	AT	Other	Chemical type	Therapeutic or other use	Some commercial source[c]
Penicillin G[d]	*Penicillium chrysogenum*	+						Peptide	Oral and parenteral; also as animal growth promotant	2, 3, 6, 8, 9, 10, 12, 14, 22, 25, 28, 29, 31, 41, 43, 44, 45, 48, 52, 58, 60, 63
Penicillin V[d]	*Penicillium chrysogenum*	+						Peptide	Oral	1, 2, 3, 8, 10, 11, 12, 21, 25, 28, 29, 31, 41, 43, 48, 52, 58, 60, 63
Pimaricin	*Streptomyces natalensis*						+	Polyene	Topical; also used for food preservation	28
Polymyxin B	*Bacillus polymyxa*				+			Peptide	Parenteral	45, 48
Polyoxins[a]	*Streptomyces cacaoi* var. *asoensis*								Agriculture (AF)	35, 55
Pristinamycins[a]	*Streptomyces pristinaspiralis*	+						Peptide	Parenteral	52
Quebemycin[a]	*Streptomyces viridans*	+						Phosphoglycolipid	Animal growth promotant	39, 52
Ribostamycin[a]	*Streptomyces ribosidificus*	+	+					Aminoglycoside	Parenteral	43
Rifamycin SV[a]	*Nocardia mediterranei*			+				Anasamycin	Parenteral	30, 26
Ristocetin	*Nocardia lurida*	+						Glycopeptide	Parenteral	42

Sagamycin[a]	*Micromonospora sagamiensis*	+	+				Aminoglycoside	Parenteral	39
Salinomycin[a]	*Streptomyces albus*	+				Coccidia	Polyether	Veterinary use	35
Siccanin[a]	*Helminthosporium siccans*				+			Veterinary use	54
Siomycin	*Streptomyces sioyaensis*	+					Peptide	Animal growth promotant	56a
Sisomicin	*Micromonospora inyoensis*	+	+				Aminoglycoside	Parenteral	22, 56
Spectinomycin	*Streptomyces spectabilis*	+					Aminocyclitol	Parenteral	1, 35, 61
Spiramycin[a]	*Streptomyces ambofaciens*	+					Macrolide	Parenteral and oral	39, 52
Streptomycin	*Streptomyces griseus*	+	+	+			Aminoglycoside	Parenteral; use in agriculture to control bacteria	6, 14, 28, 29, 43, 44, 48, 52, 58, 63
Dihydrostreptomycin	*Streptomyces humidus* (Also chemical reduction of streptomycin)	+	+	+			Aminoglycoside	Parenteral	6, 14, 28, 43, 45, 48, 52, 63
Tetracyclines									
Chlortetracycline	*Streptomyces auerofaciens*	+	+			Rickettsia	Tetracycline	Parenteral and oral; animal growth promotant	4, 5, 14, 20, 23, 49, 52, 50, 57,
6-Demethyl-7-chlortetra-cycline	*Streptomyces aureofaciens*	+	+			Rickettsia	Tetracycline	Parenteral and oral	4, 18, 49, 52,
5-Hydroxy-tetracycline	*Streptomyces rimosus*	+	+			Rickettsia	Tetracycline	Parenteral and oral animal growth promotant	5, 11, 14, 15, 18, 28, 34, 38, 48, 49, 50
Tetranactin[a]	*Streptomyces flaveolus*	+				Insects	Macrotetralide	Insecticide	17

(*continued*)

TABLE I (*Continued*)

Antibiotic	Microbial source	Antibiotic spectrum[b] G+	G−	My	AF	AT	Other	Chemical type	Therapeutic or other use	Some commercial source[c]
Thiopeptin[a]	*Streptomyces tateyamensis*	+						Peptide	Animal growth promotant	27, 35
Thiostrepton	*Streptomyces azureus*	+						Peptide	Animal growth promotant	58
Tobramycin[a]	*Streptomyces tenebrarius*	+	+					Aminoglycoside	Parenteral	11, 41
Trichomycin[a]	*Streptomyces hachijoensis*				+		Trichomonas	Polyene	Topical	27
Tylosin	*Streptomyces fradiae*	+					PPLO	Macrolide	Veterinary; animal growth promotant	21, 41
Tyrothricin	*Bacillus brevis*	+	+					Peptide	Topical	10, 42, 47, 62
Tyrocidine	*Bacillus brevis*	+	+					Peptide	Topical	42
Validamycin[a]	*Streptomyces hygroscopicus* var. *limonesis*	+	+					Aminoglycoside	Parenteral	59
Vancomycin	*Streptomyces orientalis*	+						Glycopeptide	Parenteral	41
Variotin[a]	*Paecilomyces varioti*				+				Topical	40, 44a
Viomycin	*Streptomyces floridae*	+	+	+				Peptide	Parenteral	48, 60
Virginiamycin	*Streptomyces virginiae*	+						Peptide	Animal growth promotant	51

[a] Not distributed in the United States (1978).

[b] G+, gram-positive bacteria; G−, gram-negative bacteria; My, mycobacteria; AF, antifungal; AT, antitumor.

[c] Key to sources (from Perlman, 1977):

1. Abbott Laboratories (U.S.A.)
2. Aktiebolaget Astra (Sweden)
3. Aktiebolaget Fermenta (Sweden)
4. American Cyanamid (U.S.A.)
5. Ankerfarm S.p.A. (Italy)
6. Antibioticos S.A. (Spain)
7. Apothekernes Laboratorium fur Specialpraeparater A/S (Norway)
8. Banyu Pharmaceutical Company (Japan)
9. Beecham Pharmaceutical Company (England)
10. Biochemie GmbH (Austria)
11. Biogal (Hungary)
12. Bristol-Myers Company (U.S.A.)
13. Chemibiotic Ltd. (Ireland)
14. China National Chemicals Import and Export Corporation (China)
15. Chinoin (Hungary)
16. Chong-Kun-Dong Corporation (S. Korea)
17. Chugai Pharmaceutical Company (Japan)
18. Companhia Industrial Produtora de Antibioticos S.A.R.L. (CIPAN) (Portugal)
19. Compania Espanola de la Penicillina y Antibioticos S.A. (Spain)
20. Diaspa S.p.A. (Italy)
21. Dista Products Ltd. (England)
22. Farbenfabriken Bayer AG (W. Germany)
23. Farmitalia S.p.A. (Italy)
24. Fermentfarma S.p.A. (Italy)
25. Fermion Oy (Finland)
26. Fervet S.p.A. (Italy)
27. Fujisawa Pharmaceutical Company (Japan)
28. Gist-Brocades n.v. (Holland)
29. Glaxo Laboratories Ltd. (England)
30. Gruppo Lepetit S.p.A. (Italy)
31. Hoechst AG (W. Germany)
32. Hoffmann-LaRoche Inc. (U.S.A.)
33. IMC Chemicals Group, Inc. (U.S.A.)
34. Imperial Chemical Industries Ltd. (England)
35. Kaken Chemical Company (Japan)
36. Kowa Company (Japan)
37. Kayaku Antibiotics Research Company Ltd. (Japan)
38. Krakow Pharmaceutical Works ("Polfa") (Poland)
39. Kyowa Hakko Kogyo Company (Japan)
40. Leo Pharmaceutical Products (Denmark)
41. Eli Lilly Company (U.S.A.)
42. H. Lundbeck and Company (Denmark)
43. Meiji Seika Kaisha Ltd. (Japan)
44. Merck and Company, Inc. (U.S.A.)

44a. Nihon Kayaku Company (Japan)

45. Novo Industri A/S (Denmark)
46. Parke, Davis and Company (U.S.A.)
47. S. B. Penick and Company (U.S.A.)
48. Pfizer, Inc. (U.S.A.)
49. Proter S.p.A. (Italy)
50. Rachelle Laboratories Inc. (U.S.A.)
51. Recherche et Industrie Therapeutique (Belgium)
52. Rhone-Poulenc S.A. (France)
53. Roussel-UCLAF (France)
54. Sankyo Company Ltd. (Japan)
55. Sanraku Ocean Company, Ltd. (Japan)
56. Schering Corporation (U.S.A.)

56a. Shionogi & Co., Ltd. (Japan)

57. Societa Prodotti Antibiotici S.p.A. (Italy)
58. E. R. Squibb and Sons (U.S.A.)
59. Takeda Chemical Industries Ltd. (Japan)
60. Toyo Jozo Company Ltd. (Japan)
61. The Upjohn Company (U.S.A.)
62. Wallerstein Laboratories, Inc. (a division of G. B. Fermentation Industries, Inc., USA)
63. Wyeth Laboratories (U.S.A.)

[d] Both penicillins G and V are used as sources of 6-aminopenicillanic acid, an intermediate in the synthesis of semisynthetic penicillins. Also, penicillin V is used as a starting material for the chemical synthesis of cephalexin.

TABLE II. Some Antibiotics Useful in Treatment of Infections

Infective microorganism	Disease	Suggested Antibiotics
I. Gram-positive cocci		
Staphylococcus aureus	Abscess *Penicillin G-sensitive:* Bacteremia Endocarditis *Penicillin G-resistant:* Pneumonia Meningitis *Methicillin-resistant:* Osteomyelitis	Penicillin G; gentamicins; erythromycin; lincomycin; cephalosporins; clindamycin Penicillinase resistant penicillin Vancomycin; cephalosporin
Streptococcus pyogenes (groups (A, B, C)	Pharyngitis Scarlet fever Otitis media Pneumonia Cellulitis	Penicillin G and V; cephalosporin; clindamycin; erythromycin; tetracycline
Streptococcus viridans	Dental infections Subacute bacterial endocarditis Urinary tract infections	Penicillin G ± streptomycin; cephalosporin; erythromycin; vancomycin
Streptococcus faecalis (enterococcus)	Endocarditis Urinary tract infections Bacteremia Meningitis Brain abscess	Penicillin + aminoglycoside Ampicillin; vancomycin; penicillin G + aminoglycoside; erthromycin ± aminoglycoside
Streptococcus (anaerobic)	Bacteremia Endocarditis Brain abscess and other abscesses	Penicillin G; chloramphenicol; clindamycin; erythromycin; tetracycline
Streptococcus pneumoniae (diplococcus)	Pneumonia Meningitis Endocarditis Arthritis	Penicillin G; cephalosporin; chloramphenicol; lincomycin; clindamycin; erythromycin
II. Gram-negative cocci		
Neisseria gonorrhoeae (gonococcus)	Genital infections Arthritis Meningitis Endocarditis	Penicillin G; ampicillin; spectinomycin; erythromycin; tetracycline
Neisseria meningitidis	Meningitis Bacteremia Carrier state	Penicillin G; chloramphenicol; erythromycin Rifampin
III. Gram-positive bacilli		
Clostridium perfringens	Gas gangrene	Penicillin G; erythromycin; tetracycline; cephalosporin
Clostridium tetani	Tetanus	Penicillin G; tetracycline; erythromycin

(*continued*)

TABLE II (*Continued*)

IV. Gram-negative bacilli		
Escherichia coli	Urinary tract infection	Ampicillin; cephalosporin; carbenicillin; tetracycline; chloramphenicol
	Other infections	Ampicillin; gentamicin; cephalosporin; chloramphenicol; tetracycline; colistin; kanamycin; polymyxin B
Enterobacter aerogenes	Urinary tract infection	Gentamicin; carbenicillin; chloramphenicol; kanamycin; tetracycline; colistin
Alcaligenes faecalis	Urinary tract infection	Chloramphenicol; tetracycline; colistin; polymyxin B; kanamycin; gentamicin
Proteus mirabilis	Urinary tract infection	Ampicillin; cephalosporin; gentamicin; kanamycin; chloramphenicol; tetracycline
Pseudomonas aeruginosa	Urinary tract infection	Carbenicillin; gentamicin; colistin; polymyxin B
Salmonella sp.	Typhoid	Ampicillin; chloramphenicol
Shigella sp.	Gastroenteritis	Ampicillin; chloramphenicol; polymyxin B; kanamycin; tetracycline
Vibrio cholerae	Cholera	Tetracycline; chloramphenicol; erthromycin
V. Acid-fast Bacilli		
Mycobacterium tuberculosis	Tuberculosis	Streptomycin + synthetic drugs; rifampin; cycloserine
Atypical *Mycobacterium*	Pulmonary infection	Streptomycin + synthetic drugs; erythromycin; rifampin; cycloserine
VI. Miscellaneous Agents		
Mycoplasma pneumonia	Atypical viral pneumonia	Erythromycin; tetracycline
Rickettsia	Typhus fever Rocky Mountain spotted fever Q fever	Tetracycline; chloramphenicol
VII. Fungi		
Candida albicans	Skin lesions	Amphotericin B; nystatin
Candida albicans	Pneumonia	Amphotericin B
Cryptococcus	Meningitis	
Aspergillus	Bone lesions Disseminated disease	
Microsporon	Skin lesions	Griseofulvin
Trichophyton	Hair lesions	
Epidermophyton	Nail lesions	

the literature on each. However, the techniques used in production of penicillins, tetracyclines, and bacitracin are fairly typical of those used for all antibiotics (the reader is directed to the patent literature for details of processes for producing other antibiotics).

III. PROCESSES USED FOR PENICILLIN PRODUCTION

A. Historical Background

Penicillin is the name applied by Fleming (1929) to the bacteriostatic principle produced by a mold later identified as *Penicillium notatum.* Since Fleming's discovery in 1929, it has been found that penicillin is produced by a variety of molds belonging to other species and genera, and also that there is a series of closely related penicillins, all of which show approximately the same antibiotic characteristics. The naturally occurring penicillins differ from each other in the R group as follows:

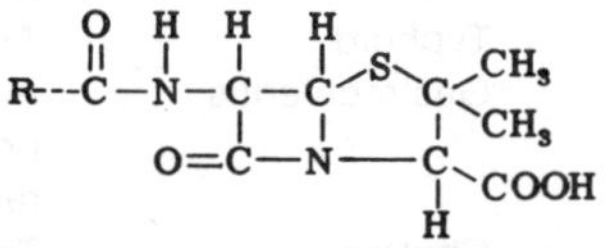

R group	Chemical name	Trivial name
$CH_3CH_2CH{=}CHCH_2—$	6-(2-Hexenamido)penicillin	Penicillin F
$CH_3(CH_2)_3CH_2—$	6-(Hexanamido)penicillin	Dihydropenicillin F
$CH_3(CH_2)_5CH_2—$	6-(Heptamido)penicillin	Penicillin K
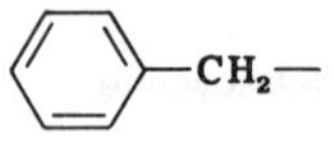	6-(Phenylacetamido)penicillin	Penicillin G
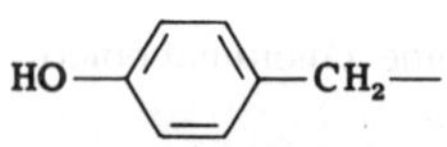	6-(*p*-Hydroxyphenylacetamido) penicillin	Penicillin X

Following Fleming's report, Clutterbuck *et al.* (1932) undertook to study the chemistry of penicillin. Their work showed that penicillin was an organic acid which was extractable into organic solvents from aqueous solutions of low pH, and that it was extremely labile to hydrogen ion and heat. In nearly all of their experiments the bioactivity disappeared upon evaporation of a solution to dryness. In view of this great lability and the very low fermentation yields (less than 0.1% of current practice), work on penicillin was abandoned.

In 1937, stimulated by an interest in antiinfective agents, Chain *et al.*

(1940) undertook a reinvestigation of penicillin. They cultured Fleming's organism in surface culture in what amounted to a small pilot-plant scale. By keeping a low temperature during their extraction procedure, they were able to concentrate the penicillin 1000-fold and produce a dry powder in the form of a salt of penicillin, which had a reasonable stability on storage. This represented the first great advance in penicillin production and furnished material used to demonstrate the phenomenal curative properties of this antibiotic (Abraham *et al.*, 1941).

Because of the virtual impossibility of quickly producing a significant amount of penicillin under wartime conditions in England, Drs. Florey and Heatley came to the United States in July 1941 in order to enlist the aid of the American government and pharmaceutical industry. The story of their success in this direction has been told before and there is no need to repeat it here (Chain *et al.*, 1940; Abraham *et al.*, 1949; Coghill, 1944; Stewart, 1961). Suffice it to say that by the time of the invasion of France in 1944, adequate amounts of penicillin were available on all fronts to effect a tremendous saving of life among the wounded. Penicillin was most certainly one of the few benefits accruing from World War II. Had it not been for the urgent need for such a curative agent, the wartime ease of securing government money for such research projects, and the excess profits tax which made it possible for industry to take a terrific gamble with fifteen-cent dollars, we might not have penicillin even today.

B. Selection of Culture

The mold described by Fleming (1929) and subsequently identified as a strain of *Penicillium notatum* was used in the early developmental stages of the production of penicillin. The yields with this culture were poor, and considerable work was carried out in several laboratories in an attempt to obtain a higher yielding strain. Several strains, including a natural variant of Fleming's culture (given the registry number NRRL 1249B21), were found to produce higher yields of penicillin (Moyer and Coghill, 1946). These new strains, together with other improvements in the process, combined to give yields of penicillin in the range of 200 Oxford units/ml (1 Oxford unit = 0.6 μg) by the surface culture method; none of these cultures performed well in submerged culture.

Many strains of *P. notatum* and the related *P. chrysogenum* were tested in an effort to find one which would produce good yields in submerged culture processes. One of these, *P. chrysogenum* NRRL 1951, was found to be a "superior producer." By plating NRRL 1951, followed by selective isolation, strain NRRL 1951B25 was obtained. This strain was significantly better than the parent. Demerec (at the Carnegie Institution) subjected NRRL 1951B25 to X-ray irradiation, and one of the survivors, X-1612, was found to produce twice as much penicillin (it gave yields of 300 units/ml) (Demerec, 1948). This culture was .subjected to ultraviolet irradiation by Backus and Stauffer (Coghill and Koch, 1945) at the University of Wisconsin, and a strain designated as Wisconsin (Wis.) Q176 (producing 1200 units/ml) (Foster *et al.,* 1946) was selected. This strain was adopted by most of the penicillin manufacturers, and monospore isolates from it are the parent cultures of those now in use. The genealogy of this family of strains is given in Fig. 16 (Volume II, Chapter 12). Descendents of these cultures have many properties not found in the parents. They produce 10 times as much penicillin [titers of 30,000 units are reported (Chain, 1964)] and they do not produce the yellow pigment which hampered early efforts at isolating pure, colorless benzylpenicillin. The practice of improvement of antibiotic production by selection of "better" cultures first tested in the penicillin manufacturing program has been widely applied to other antibiotic processes, and the role of the microbial geneticist in the fermentation industry has increased in importance.

The penicillin-producing molds are characterized by unusual variability; in general, the greater the productivity of the strain, the less stable it is. Stock cultures can be maintained on agar slants, in dry soil, in lyophilized form, or as spore or cell suspensions stored in liquid nitrogen. Stocks carried on agar slants are perhaps most liable to variation since successive transfer of the organisms provides greater opportunity for variation to occur. Frequent transfers tend to propagate selectively those portions of the culture population that sporulate most readily. If agar slants are used, transfers should be relatively infrequent. The wide variation in penicillin-producing ability of single-spore isolates from *P. chrysogenum* Wis. Q176 was described by Reese *et al.* (1949) and others (Coghill, 1944; Haas *et al.,* 1956). Similar variation has been noted in the cultures producing streptomycin (Perlman *et al.*, 1954).

C. Fermentation Medium and Conditions

Surface culture of *P. notatum* for penicillin production was first carried out in simple chemically defined media, such as that of Czapek and Dox. Later it was found that the culture grown in this medium supplemented with organic materials, e.g., casein digest, yeast extract, or meat products, resulted in higher antibiotic titers. Lactose was found to be a suitable replacement for the sucrose or glucose of the chemically defined media (probably due to its slow metabolism by the fungus and resultant steady carbohydrate supply over a protracted incubation period). Increased yields were also obtained with more complex nitrogen sources, such as extracted oilseeds, e.g., cottonseed, peanut meal, and soybean meal (Chain, 1966; Moyer and Coghill, 1946; Perlman, 1949).

A great step forward was brought about by the use of corn steep liquor which, together with a change to *P. chrysogenum* as the source organism, resulted in increasing yields from about 20 units/ml to about 100 units/ml (Perlman, 1950). Apart from the supply of utilizable nitrogenous materials in the corn steep liquor, the corn steep contained phenylalanine and phenethylamine which we now know act as precursors for the formation of benzylpenicillin.

At the same time, efforts were made to devise chemically defined media with a similar potential. The first reasonably successful one was that of Jarvis and Johnson (1947) which contained lactose, mineral salts, ammonium lactate and acetate, and phenylacetic acid as a precursor for benzylpenicillin. Another useful medium was that of Calam and Hockenhull (1949) which supported the production of antibiotic by a large number of strains of *P. chrysogenum*. A comparison of this medium with corn steep liquor (Table III) shows a close relationship, chemically speaking. Inclusion of vegetable oils, first used as antifoams, had a beneficial effect on yield (Davey and Johnson, 1953; Perlman, 1950; and E. R. Squibb and Sons, Inc., 1952). The effect of these lipids was to increase both the amount of mycelium formed and the antibiotic yield.

TABLE III. Comparison of Calam and Hockenhull's Medium with Corn Steep Liquor Medium

	Corn steep liquor medium (%)	Calam and Hockenhull's medium (%)
Main carbohydrate	Lactose, 3.0–4.0	Lactose, 3.0
Other carbohydrates	Glucose, 0–0.5 nonreducing, mainly polysaccharides[a]	Glucose, 1.0 Starch, 1.5
Organic acids	Acetic, ca. 0.05[a] Lactic, ca. 0.5[a]	Acetic, 0.25 Citric, 1.0
Special precursors	Phenethylamine and other precursors[a]	Phenylacetic acid, 0.05
Main nitrogen source	Amino acids, peptides, amines[a]	Ammonium sulfate, 0.5
Other nitrogen sources	Ammonia[a]	Ethylamine, 0.3
Total solids	8.0–9.0	8.5
Total N	0.15–0.2	0.2

[a] Concentration varies from one batch of corn steep liquor to next. Phenethylamine, found in many batches of corn steep liquor, was often increased in these media by addition of pure phenethylamine or phenylacetic acid supplements.

Most of the recent developmental work on the penicillin fermentation process has been concerned with increasing production and decreasing manufacturing costs. Replacement of the expensive lactose component of the medium with continuous addition of glucose or sucrose did not change antibiotic production (Davey and Johnson, 1953; Freaney, 1958; Hosler and Johnson, 1953; and Soltero and Johnson, 1953) and resulted in some financial advantages. Increased antibiotic production was also obtained by controlling the pH of the fermentation between pH 6.8 and 7.4 (Brown and Peterson, 1950; Singh and Johnson, 1948). This was accomplished by adding buffering agents to the medium (both $CaCO_3$ and phosphates have been used) and by additions of sterile H_2SO_4 and NaOH as needed. The availability of glass electrodes which withstand repeated autoclavings has made the pH-controlled fermentation possible on a routine basis (Deindoerfer and Wilker, 1957; Gualandi *et al.*, 1960; West *et al.*, 1961).

Aeration and agitation are fundamental to the penicillin process, and the theory of oxygen transfer in large-scale fermentations has been studied in many laboratories. Applications to the penicillin process were studied by Bartholomew *et al.* (1950a), Brown and Peterson, (1950), Calam *et al.* (1951), and Chain and Gulandi (1954). Many parameters have been measured, and in general it is agreed that high levels of aeration and agitation are required to maintain the desired dissolved oxygen levels. Temperature of incubation is also very important in the penicillin production process, and most operations are conducted at 25° ± 0.5°C. Owen and Johnson (1955) showed that slightly higher incubation temperatures resulted in faster mycelium formation, but best antibiotic production took place at 25°C (with their strain of *P. chrysogenum*).

The first moiety of the penicillin molecule to be chemically identified was the acyl portion, which differs among the naturally occurring penicillins. The most important of the penicillins in current production is penicillin G, where the acyl moiety is phenylacetyl. This was originally derived from the phenethylamine in the corn steep liquor (Mead and Stack, 1948; Moyer and Coghill, 1947; Smith and Bide, 1948), and inclusion of this liquor in the fermentation medium not only increased antibiotic production but also shifted the major penicillin produced from penicillin F to penicillin G. Addition of phenylacetic acid and phenylacetamide to penicillin-producing fermentations was investigated as a means of increasing production of penicillin G (Behrens *et al.*, 1948; Moyer and Coghill, 1947). When large amounts of phenylacetic acid, e.g., 1 gm/liter, were included in the fermentation medium at time of the inoculation, some inhibition of mycelium growth was observed as well as inefficient conversion to penicillin G (Moyer and Coghill, 1946). These difficulties were overcome by supplying the fermentation with compounds which were slowly metabolized to phenylacetic acid by the particular strain of *P. chrysogenum* used. Among the compounds found useful for this purpose were *N*-phenacetylethanolamine (Behrens *et al.*, 1948), *N*-methylphenylacetamide (Behrens *et al.*, 1948), esters of phenylacetic acid (Tabenkin *et al.*, 1952), and esters of phenethanol (Perlman *et al.*, 1954). As the bioengineering technology became more sophisticated, continuous addition of phenylacetic acid to the growing cultures during the incubation period was evaluated and found to be economically desirable (Pan and Perlman, 1954; Singh and Johnson, 1948).

A flow sheet of a typical fermentation process for production of procaine penicillin G is shown in Fig. 1. Spores of *P. chrysogenum* strain are used to inoculate 100 ml of medium in a 500-ml Erlenmeyer flask, and the inoculated flasks are placed on a rotary shaker (250 rpm, 2-inch displacement) located in a 25°C incubator. After 4 days' incubation, the contents of the flask are transferred to 2 liters of medium (in a 4-liter flask). This second flask fermentation is incubated on the shaker for 2 days and the contents are transferred to 500 liters of medium in an 800-liter stainless steel tank. This tank is equipped with air spargers, agitators, cooling coils for temperature control, and antifoam addition devices. After 3 days' incubation, the contents of the tank are used to inoculate 180,000 liters of fermentation medium in a 250,000-liter fermentor. This larger fermentor is equipped with devices for continuous addition of sterile glucose syrups, pH control (automatic addition of NaOH and H_2SO_4 to keep the pH at a preset value), foam sensing devices to activate automatic addition of antifoams (e.g., animal or vegetable oils), and metering pumps for continuous addition of sterile

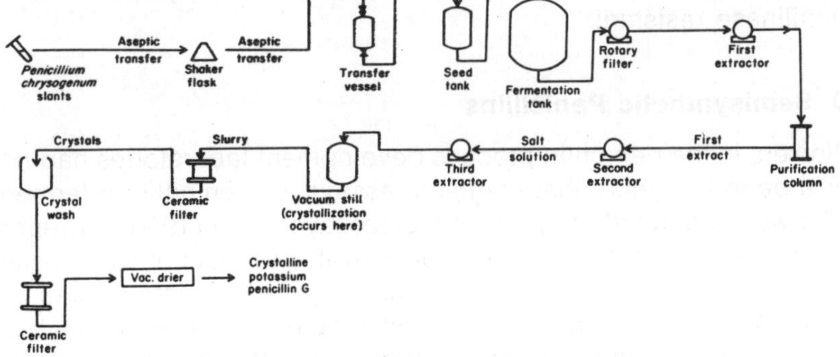

FIGURE 1. Penicillin fermentation process flow chart.

phenylacetic acid. After 5–6 days' incubation, the contents of the fermentor are filtered (all the penicillin is in the cell-free liquid), and the filtrate is passed through a series of Podbelniak extractors where the penicillin is extracted into amyl or butyl acetate. The penicillin in the amyl acetate extract is transferred back into an aqueous solvent by extraction with phosphate buffer, and the potassium penicillin G is crystallized from a butanol–water mixture. The potassium penicillin G may be further purified and used as such, or it can be converted to procaine penicillin G for clinical use.

Most penicillin-producing fungi are capable of incorporating a variety of compounds into the acyl portion of the penicillin molecule when the appropriate compounds are added to the growing culture. More than 1000 compounds have been tested to determine whether they would be incorporated and nearly 100 penicillins have been prepared in relatively pure form (Behrens *et al.*, 1948; Soper *et al.*, 1948). Among the objectives of this research program have been the preparation of penicillins with all of the "good" attributes of penicillin G and additional advantages. These latter include: (1) acid stability so that the penicillin can be administered orally, (2) lowered allergenicity since nearly 8% of the population is allergic to penicillin G, and (3) resistance to the penicillin-inactivating enzyme, penicillinase (Abraham and Chain, 1940; Pollock, 1965; Pollock *et al.*, 1956) found in many bacteria. Phenoxymethylpenicillin (also known as penicillin V), prepared by addition of phenoxyacetic acid to fermentations (Behrens *et al.*, 1948), has the desired acid stability characteristics and has found a place in clinical practice as an orally absorbed penicillin. Patients allergic to penicillin G were found to be less sensitive to allylmercaptomethylpenicillin (also known as penicillin O), a penicillin produced biosynthetically by addition of allylmercaptoacetic acid to *P. chrysogenum* cultures (Ford *et al.*, 1953). None of the penicillins prepared biosynthetically appeared to be penicillinase resistant.

D. Semisynthetic Penicillins

Workers in the penicillin process development laboratories had noted over a period of years that chemical assays for penicillin in fermented media were frequently higher than bioassays. Kato (1953a,b) reported finding in his fermented media a considerable amount of the substance responsible for the discrepancy and that more of this material was found in fermented media which had not been supplemented with phenylacetic acid. Kato postulated that this material was the penicillin molecule "without" the R group. This hypothesis was confirmed in part by Demain (1953), who found that the substance did not form penicillin in the presence of washed, resting *P. chrysogenum* cells and phenylacetic acid precursor. The identification of the substance as 6-aminopenicillanic acid was reported by Batchelor *et al.* (1959), who showed that it was sensitive to the enzyme pencillinase (which opened the β-lactam ring of penicillin) and also that it could be converted to penicillin G by treatment with phenacetyl chloride.

Since the isolation of the acid from fermented media was different than for penicillin G (because of its hydrophilic nature), other methods were sought for the production. Rolinson *et al.* (1960) noted that certain strains of gram-negative bacteria and streptomycetes converted penicillin G to 6-aminopenicillanic acid (a reaction first proposed by Sakaguchi and Murao (1950; Murao, 1955). Similar results were obtained by Claridge *et al.* (1960) and Huang *et al.* (1960, 1963). The distribution of this hydrolytic enzyme (called penicillin amidase or benzylpenicillin acylase) was studied in many laboratories (Batchelor *et al.*, 1961; Chain, 1964; Huang *et al.*, 1963; Kaufmann and Bauer, 1960) with the results shown in Table IV.

Since 6-aminopenicillanic acid can be readily acylated by chemical means, it has been possible to prepare a new series of semisynthetic penicillins. Tens of thousands of compounds have been prepared (Price, 1969, 1977), and a number have been found to be markedly superior to penicillin G in oral absorption, acid stability, and resistance to penicillin-inactivating enzymes. The structures of some of these which have been of clinical importance are shown in Fig. 2 and their special properties summarized in Table V.

IV. CEPHALOSPORINS

The production of antibiotic activity by a particular strain of *Cephalosporium* was first noted by Brotzu (1948). Several antibiotics with differing

TABLE IV. Distribution of Penicillin Amidase

Group of microorganisms	Genera in which penicillin amidase was reported
Bacteria	*Pseudomonas, Xanthomonas, Alcaligenes, Flavobacterium, Escherichia, Aerobacter, Erwinia, Serratia, Proteus, Bordetella, Micrococcus, Sarcina, Corynebacterium, Cellomonas, Nocardia, Bacillus*
Molds	*Alternaria, Aspergillus, Epicoccum, Fusarium, Mucor, Penicillium, Phoma, Trichoderma*
Yeasts	*Cryptococcus, Saccharomyces, Trichosporon*
Actinomycetes	*Streptomyces*

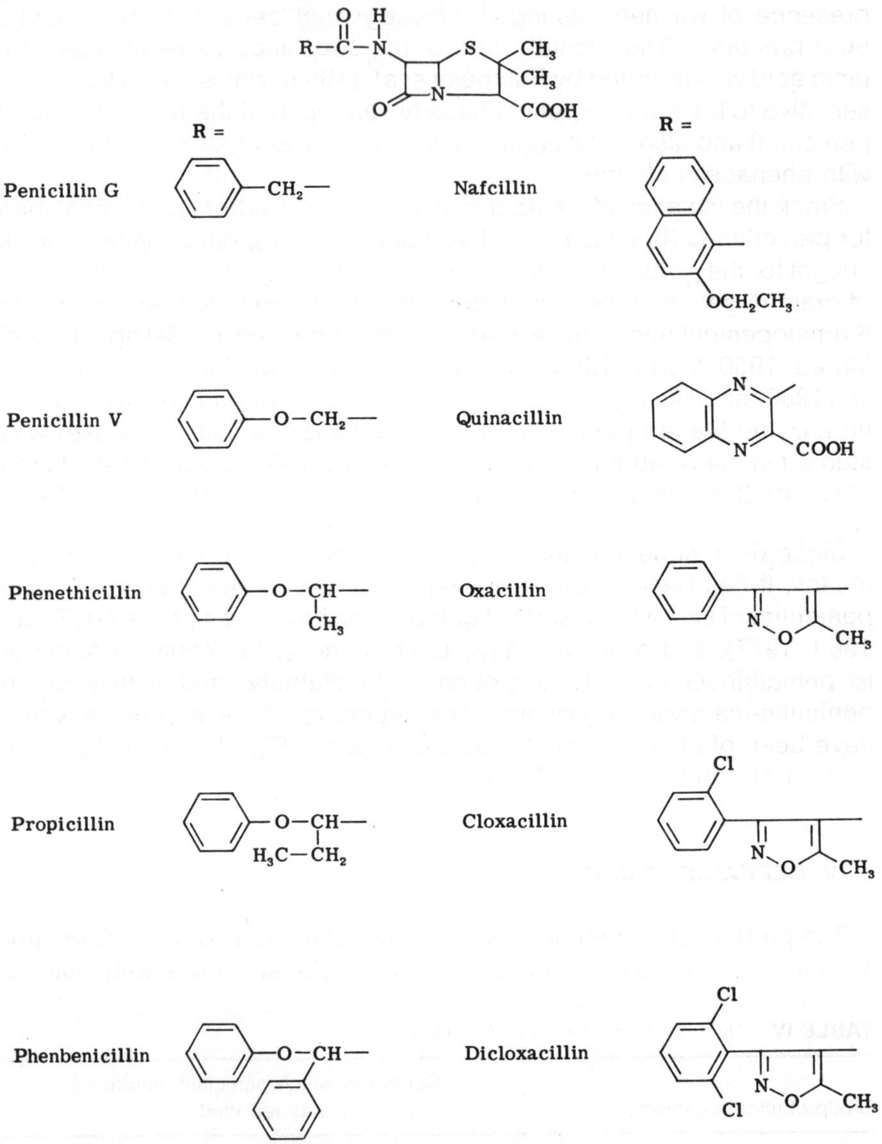

FIGURE 2. Structures of some clinically useful penicillins.

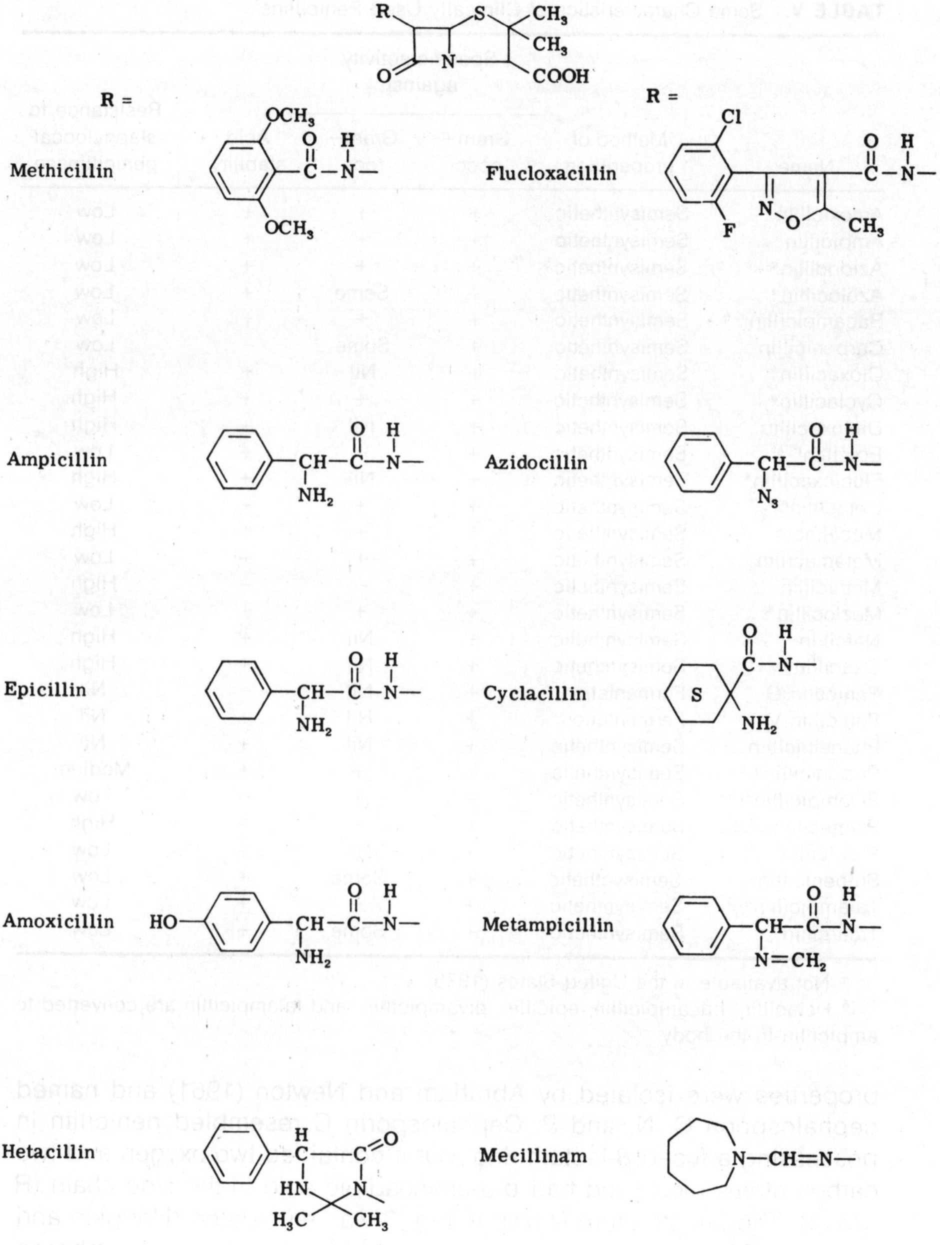

FIGURE 2. (*Continued.*)

TABLE V. Some Characteristics of Clinically Used Penicillins

Name	Method of preparation	Specific activity against: Gram+ cocci	Gram− rods	Acid stability	Resistance to staphyloccal penicillinase
Amoxicillin	Semisynthetic	+	+	+	Low
Ampicillin	Semisynthetic	+	+	+	Low
Azidocillin[a]	Semisynthetic	+	+	+	Low
Azolocillin[a]	Semisynthetic	+	Some	+	Low
Bacampicillin[a,b]	Semisynthetic	+	+	+	Low
Carbenicillin	Semisynthetic	+	Some	−	Low
Cloxacillin	Semisynthetic	+	Nil	+	High
Cyclacillin[a]	Semisynthetic	+	+	+	High
Dicloxacillin	Semisynthetic	+	Nil	+	High
Epicillin[a,b]	Semisynthetic	+	+	+	Low
Flucloxacillin[a]	Semisynthetic	+	Nil	+	High
Hetacillin[b]	Semisynthetic	+	+	+	Low
Mecillinam	Semisynthetic	+	+	+	High
Metamacillin	Semisynthetic	+	+	+	Low
Methicillin	Semisynthetic	+	−	−	High
Mezlocillin[a]	Semisynthetic	+	+	+	Low
Nafcillin	Semisynthetic	+	Nil	+	High
Oxacillin	Semisynthetic	+	Nil	+	High
Penicillin G	Fermentation	+	Nil	−	Nil
Penicillin V	Fermentation	+	Nil	+	Nil
Phenethicillin	Semisynthetic	+	Nil	+	Nil
Pirbenicillin[a]	Semisynthetic	+	+	+	Medium
Pivampicillin[a,b]	Semisynthetic	+	+	+	Low
Pivmecillinam[a]	Semisynthetic	+	+	+	High
Propicillin[a]	Semisynthetic	+	Nil	+	Low
Sulbenicillin[a]	Semisynthetic	+	Some	+	Low
Talampicillin[a,b]	Semisynthetic	+	+	+	Low
Ticarcillin	Semisynthetic	+	Some	+	Low

[a] Not available in the United States (1978).

[b] Hetacillin, bacampicillin, epicillin, pivampicillin, and talampicillin are converted to ampicillin in the body.

properties were isolated by Abraham and Newton (1961) and named cephalosporin C, N, and P. Cephalosporin C resembled penicillin in possessing a fused β-lactam ring, but it contained two oxygen and two carbon atoms more, and had D-α-aminoadipic acid in the side chain (R group). The full structure shown in Fig. 3 was elucidated (Hodgkin and Maslen, 1961) and the nucleus shown to consist of a six-membered dihydrothiazine ring (in place of the five-membered thiazolidine ring of the penicillins) fused to a β-lactam ring. This nucleus (7-amino-

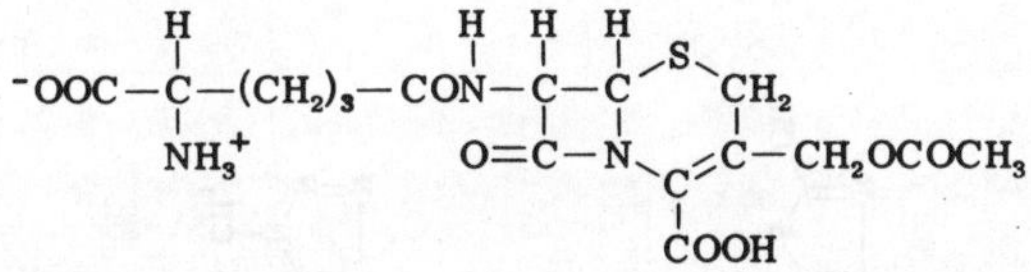

FIGURE 3. Structure of cephalosporin C, an antibacterial agent produced by *Cephalosporium acremonium.*

cephalosporanic acid) contains structures which are present in 6-aminopenicillanic acid, including a peptide linkage (CO : NH) vulnerable to hydrolysis by amidase (Holt and Stewart, 1964; Walton, 1964a,b) and the β-lactam grouping susceptible to enzymatic hydrolysis by lactamase (Fleming *et al.*, 1963; Sabath and Abraham, 1966). Cephalosporins are in general less susceptible than penicillin G or ampicillin to hydrolysis by β-lactamase from staphylococci or *Bacillus cereus,* but they are readily hydrolyzed by enzymes from gram-negative bacteria. The aminoadipic acid side chain (Fig. 3) lessens susceptibility to β-lactamase from certain bacteria.

The antistaphylococcus activity of cephalosporin C is relatively low, but it has some action against streptococci, and also against certain gram-negative bacilli, and is remarkably nontoxic in experimental animals. In view of similarity in structure to penicillin, it is not surprising to find sensitivity in subjects who are hypersensitive to penicillins.

Microbiological processes for production of cephalosporin C resemble in many respects those used for penicillin production. Special strains of *Cephalosporium* have been selected which produce more cephalosporin C and less cephalosporin N than the parent culture, and the growth of these in certain special fermentation media has resulted in higher antibiotic titers. Even with these modifications in processing, the antibiotic yields are much lower than those reported for the penicillins. Manufacturing operations on a commercial scale are carried out by many companies (see Table I).

The possibilities of preparing derivatives of cephalosporin C with new attributes were investigated by Abraham and Newton, who found that the pyridine-treated material had higher antibacterial activity especially against staphylococci (Abraham, 1962; Loder *et al.*, 1961). Profiting from the experience with semisynthetic penicillins, scientists at the Eli Lilly Company laboratories, and at other companies, used the 7-aminocephalosporanic acid prepared by chemical degradation of cephalosporin C (Morin *et al.*, 1962) to prepare thousands of semisynthetic cephalosporins (Webber and Ott, 1977). The structures of some of the clinically useful group are shown in Fig. 4 and some of their therapeutic uses are mentioned in Table VI.

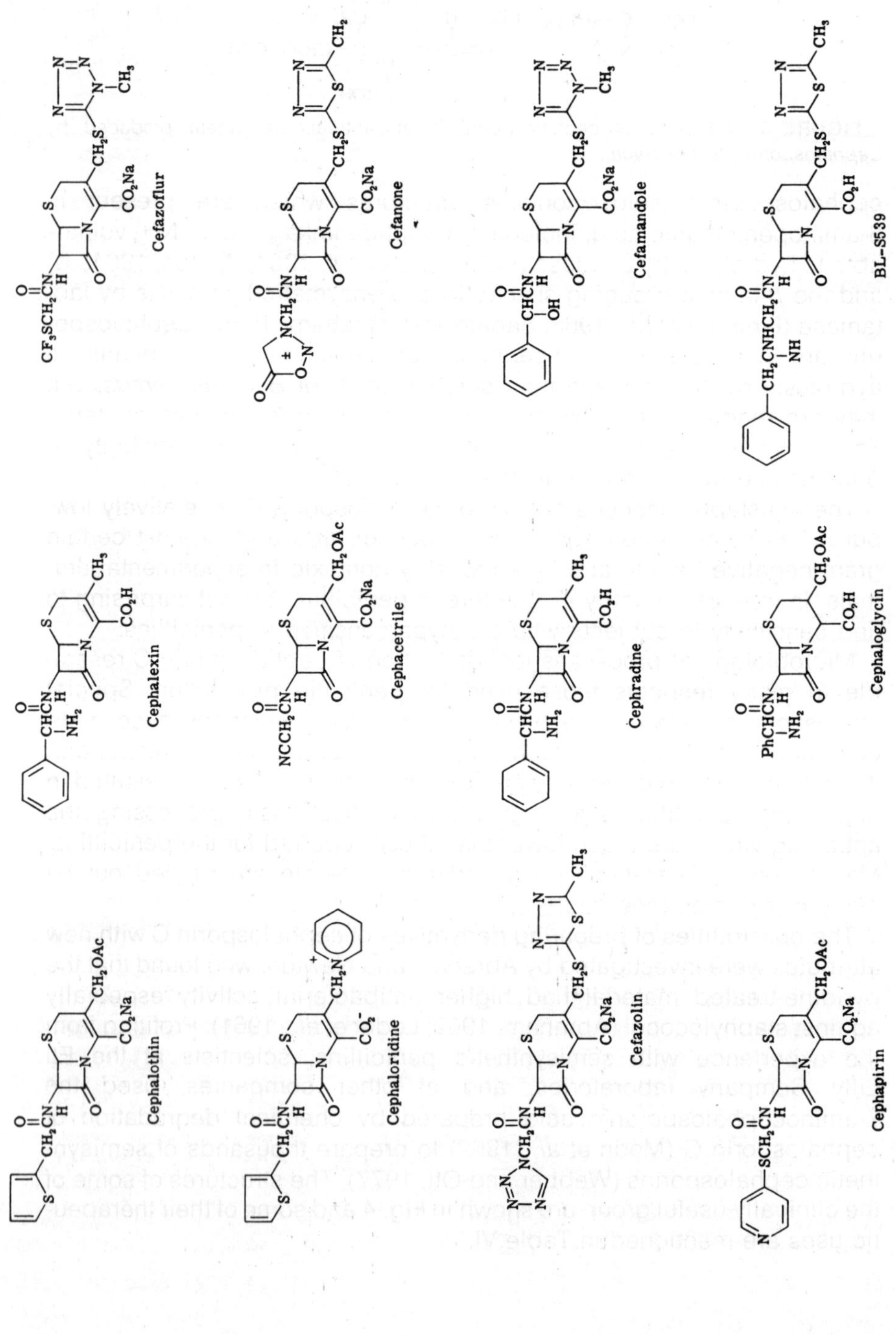

FIGURE 4. Structures of some clinically useful cephalosporins.

TABLE VI. Some Characteristics of Cephalosporins of Clinical Interest

Name	Method of Administration		Specific activity (minimum inhibitory concentration, ug/ml) against:		
	Injection	Oral	*S. aureus*	*Pseudomonas*	*E. coli*
Cefanone[a]	+	−	0.5	0.7	>200
Cefamandole	+	−	0.9	0.8	>200
Cefazolin	+	−	0.8	2	>200
Cefoxitin	+	−	2	6	>200
Cefuroxime[a]	+	−	0.7	8	>200
Cefazaflur	+	−	0.6	0.7	>200
Cefaparole[a]	+	−	4	4	>200
Cefatrizine[a]	+	−	3	3	>200
Ceftezol[a]	+	−	0.6	3	>200
Cephalexin	−	+	5	9	>200
Cephaloglycin	−	+	2	2	>200
Cephaloridine	+	−	4	5	>200
Cephalothin	+	−	0.4	17	>200
Cephapirin	+	−	0.6	13	>200
Cephacetrile	+	−	2	8	>200
Cephradine	−	+	5	20	>200
BL-S339[a]	+	−	0.6	3	>200
BL-S217[a]	+	−	0.7	6	>200
SCE-129[a]	+	−	6	58	0.9

[a] Not approved for use in the United States (1978).

V. PRODUCTION OF TETRACYCLINES

The tetracyclines are a group of crystalline antibiotics possessing a common hydronaphthacene skeleton. Tetracycline can be considered the parent compound of the group and is 4-dimethylamin-1, 4, 4α, 5, 5α, 6, 11, 12α-octahydro-3, 6, 10, 12, 12α-pentahydroxy-6-methyl-1, 11-dioxo-2-naphthacenecarboxamide (Stephens *et al.*, 1954; Waller *et al.*, 1952) (see Fig. 5 for a diagrammatic presentation). Chlortetracycline (Aureomycin) is 7-chloro, and oxytetracycline (Terramycin) the 5-hydroxy derivative. 7-Bromotetracycline (Lepetit S.p.A. 1957a,b) and 6-demethyl-7-chlortetracycline (Declomycin) have also been of clinical interest. All of these antibiotics are manufactured by fermentation processes using species of *Streptomyces* and, in addition, tetracycline is also manufactured by hydrogenolysis of chlortetracycline (Boothe *et al.*, 1953; Forbath, 1957).

The tetracyclines are active *in vivo* against numerous gram-positive and gram-negative organisms, and some of the pathogenic rickettsiae

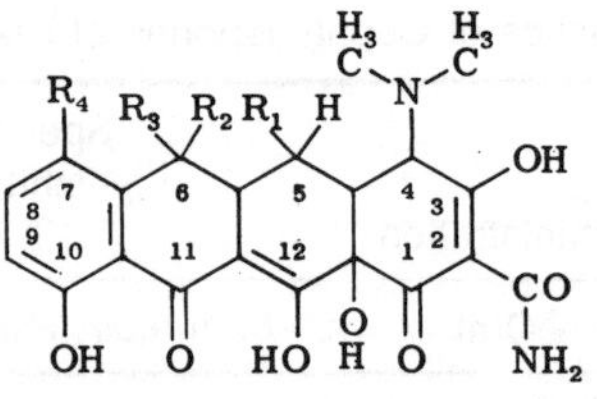

FIGURE 5. Structures of clinically important tetracyclines.

	R_1	R_2	R_3	R_4	Manufactured by
Tetracycline	H	OH	CH_3	H	Fermentation
7-Chlortetracycline	H	OH	CH_3	Cl	Fermentation
5-Oxytetracycline	OH	OH	CH_3	H	Fermentation
6-Demethyl-7-chlortetracycline	H	OH	H	Cl	Fermentation
6-Deoxy-5-oxytetracycline	OH	H	CH_3	H	Semisynthetic
7-Dimethylamino-6-deoxy-6-demethyl-tetracycline	H	H	H	$N(CH_3)_2$	Semisynthetic

and large viruses. Some infections where tetracyclines are therapeutic agents of choice are listed in Table II. The antibiotics are primarily bacteriostatic in action and in higher concentrations do have a bactericidal effect.

The systemic administration of tetracyclines may be carried out utilizing either oral or intravenous dosage forms; the intramuscular route is not feasible due to considerable pain and irritation at the site of injection. For veterinary use, tetracyclines are given by intravenous injection in bovines and equines, except that oral routes are used for young nonruminating calves.

In connection with work on the "animal protein factor," it was found that crude chlortetracycline fermentation mash (containing some vitamin B_{12}) gave growth responses in animals well above those obtained with supraoptimal levels of vitamin B_{12}. At the present time, both chlortetracycline and oxytetracycline are being used extensively for growth stimulation and improvement of feed efficiency in poultry and hogs, and for the reduction of losses from certain disease conditions. Purified antibiotics as well as dried fermentation residues and the mycelium of *Streptomyces aureofaciens* grown on shredded barley (Belik *et al.*, 1957; Herold *et al.*, 1958; Herold and Necasek, 1959) are used for these purposes. The action of antibiotics in increasing poultry growth is apparently confined to their effect on the bacteria within the intestinal tract. Chlortetracycline and oxytetracycline have also been used in the preservation of fish and poultry.

A. Selection of Culture

Chlortetracycline, the first of the group to be investigated (Broschard *et al.*, 1949), is prepared from *S. aureofaciens* fermentations. The original strain, isolated by Duggar from a soil sample from Missouri (Duggar, 1948, 1949), shows extreme variability in microscopic, macroscopic, and physiological characteristics, and a variety of natural and artificially induced mutants have been examined (Backus *et al.*, 1954; Duggar *et al.*, 1954). Van Dyck and deSomer (1952), in attempts to correlate mycelium pigmentation and chlortetracycline production of their mutant strains, concluded that the best producers were the highly yellow pigmented strains.

Many investigators have observed that in the absence of special precautions, degeneration of antibiotic-producing ability took place in selected strains (Haas *et al.*, 1956). Storage of spores in lyophilized form has been used in some laboratories, while in others the spores have

TABLE VII. Streptomycetes Species Producing Tetracyclines

Name	Products[a]	References
S. alboflavus	CTC, TC, OTC, actinomycin	Villax (1963)
S. antibioticus	OTC, TC	Villax (1963)
S. aureofaciens	CTC, TC	Duggar (1948, 1949) Minieri *et al.* (1954, 1956)
S. aureus	TC	Villax 1963)
S. californicus	TC, actinomycin	Villax (1963)
S. cellulosae	OTC, actinomycin	Villax (1963)
S. feofaciens	TC	Lepetit S.p.A. (1957a)
S. flaveolus	OTC, TC, actinomycin	Villax (1963)
S. flavus	CTC, TC, OTC, actinomycin	Villax (1963)
S. fuscofaciens	OTC, quinocycline	Chas. Pfizer and Co. (1958)
S. lusitanus	CTC, TC	Villax (1963)
S. parvus	OTC, TC, actinomycin	Villas (1963)
S. platensis	OTC	McGuire (1954)
S. rimosus	OTC, TC, rimocidin	Finlay *et al.* (1950); Perlman *et al.* (1960); Sobin *et al.* (1950)
S. sayamaensis	TC, CTC	Arishima and Sekizawa, (1959)
S. vendargensis	OTC, vengacide	N.V. Koniklijke Nederlandsche Gist-en Spiritus-Fabriek (1956)
S. viridofaciens	CTC, TC	Lepetit S.p.A. (1957a)

[a] CTC, 7-chlortetracycline; TC, tetracycline; OTC, oxytetracycline.

been stored in sterile, desiccated soil in a refrigerator. Among the complicating factors in the preservation and maintenance of high antibiotic-producing strains is the possible contamination with actinophages. Although phage-resistant strains can be obtained without too much difficulty, some of these may be lysogenic and potentially dangerous.

In the 20 years since Duggar's isolation of *S. aureofaciens,* a number of other streptomycetes producing various tetracyclines, together with other antibiotics, have been described. The names given to some of these and the fermentation products are presented in Table VII. In a number of instances, the nature of the antibiotics produced by these cultures was determined by paper chromatographic methods (Martin *et al.*, 1955), while in others (e.g., *S. aureofaciens, S. feofaciens, S. fuscofaciens, S. lusitanus, S. platensis, S. rimosus, S. sayamaensis, S. vendargensis,* and *S. viridofaciens*) the products were isolated in pure form.

As with penicillin production, certain cultures of the tetracycline-producing streptomycetes produce much more antibiotic activity than others. Intensive screening of *S. aureofaciens* mutants (Backus *et al.*, 1954; Duggar *et al.,* 1954) showed wide variation in antibiotic production, and some were obtained which produced as much as 50 times the amount of chlortetracycline as the original culture. The original *S. aureofaciens* produced both chlortetracycline and tetracycline when grown in media where chloride ion was limiting (Demain, 1953; Federal Trade Commission, 1958; Minieri *et al.*, 1954, 1956).

Mutants with impaired halogen metabolism derived from this culture are now available and produce only tetracycline when grown in media containing large amounts of chloride (Doerschuk *et al.*, 1956, 1959). Certain other cultures are nonselective in choice of halogen, accepting bromide in place of chloride, and thus producing 7-bromotetracycline (Lepetit S.p.A., 1957a.b). *Streptomyces rimosus,* the oxytetracycline-producing culture, synthesizes traces of tetracycline (Perlman *et al.*, 1960) but apparently will not synthesize the chloro analog of 5-oxytetracycline. Mutants from *S. aureofaciens* with impaired methylation systems synthesize 6-demethyltetracyclines (McCormick *et al.*, 1957). These antibiotics, which are chemically more stable than the tetracycline analogs, are widely used clinically.

B. Fermentation Medium and Conditions

Most of the available information on processes for the fermentation production of chlortetracycline (Duggar, 1949; Niedercorn, 1952), oxytetracycline (Sobin *et al.*, 1950), and tetracycline (Minieri *et al.*, 1956; Lepetit S.p.A., 1957a,b; Gourevitch and Lein, 1955; Hatch *et al.*, 1956; McGhee and Megna, 1957) is in the patent literature. Although different media are used for each of the fermentation processes, the general operations appear to be the same for all: Spores of the streptomycete taken from stock slants or soil stocks are used to inoculate agar plates. The plates are incubated for several days, and the second crop of spores is used to inoculate flasks of liquid media. These flasks are placed on shakers, and when vegetative growth is at a maximum, they are used to inoculate aerated bottles or small fermentors. Several small fermentors are sometimes used to inoculate larger tanks, and eventually the final fermentation stage (20,000–100,000 liters) is seeded with about 5% of inoculum. Among the media used for the agar slants is Duggar's agar (Duggar, 1948; Duggar *et al.*, 1954) which contains 2% beef extract, 0.05% asparagine, 1% glucose, 0.5% K_2HPO_4, and 1.3% agar. The shaken flasks and aerated bottles may contain a variety of media. A typical one contains 2% corn steep liquor (50% solids), 3% sucrose, and 0.5% $CaCO_3$.

The medium described by Duggar (1949) for the fermentation phase contains 1% sucrose, 1% corn steep liquor, 0.2% $(NH_4)_2HPO_4$, 0.2% KH_2PO_4, 0.1% $CaCO_3$, 0.025% $MgSO_4 \cdot 7\ H_2O$, 0.005% $ZnSO_4 \cdot 7\ H_2O$, 0.00033% $CuSO_4 \cdot 5\ H_2O$, and 0.00033% $MnCl_2 \cdot 4\ H_2O$. Chlortetracycline yields of 170 μg/ml were obtained by Duggar (1949) with this medium. Van Dyck and deSomer (1952) used a medium containing sucrose, peanut oil meal, corn steep liquor, molasses, sucrose, $(NH_4)_2SO_4$, $CaCO_3$, and NaCl and, with selected cultures of *S. aureofaciens,* obtained yields of 1250 μg/ml of chlortetracycline. A yield of 2000 μg/ml was reported by McCormick *et al.* (1959) when they grew selected cultures in chemically defined media, showing that no special "stimulating materials" found in seed meals and other nitrogenous substances are needed for chlortetracycline production. However, they did find (McCormick *et al.*, 1959, 1960, 1961) that a cosynthetic factor produced by certain streptomycetes stimulated production of the tetracyclines by mutants of *S. aureofaciens.* Yields obtained in the other tetracycline-producing fermentations range from 1000 μg/ml for oxytetracycline (Regna *et al.*, 1951; Sobin *et al.*, 1950) to 2000–9500 μg/ml for tetracycline (Laboratories Pro-Ter, 1964; Rolland and Sensi, 1955; Sensi *et al.*, 1955; Villax, 1965).

The direct microbial production of tetracycline is the practical alternative to the chemical dechlorination of chlortetracycline for the manufacture of tetracycline (Forbath, 1957). When *S. aureofaciens* (strain NRRL 2209, the Duggar culture) is grown on Duggar's medium (Duggar, 1949) a mixture of chlortetracycline and tetracycline is produced (Federal Trade Commission, 1958). Growth of this and other chlortetracycline-producing organisms in media containing biologically insignificant amounts of chloride results in biosynthesis of tetracycline instead of chlortetracycline (Federal Trade Commission, 1958; Gourevitch and Lein, 1955; McGhee and Megna, 1957; Minieri *et al.*, 1954, 1956; Petty and Matrishin, 1950; Rolland and Sensi, 1955; Sensi *et al.*, 1955). Since corn steep liquor and other nitrogenous materials of natural origin contain significant amounts of inorganic chloride, this must be removed before these substances can be used in media for a tetracycline-producing fermentation. Among the methods used for removal of the chloride is absorption by ion-exchange resins (applied to corn steep liquor) (Hatch *et al.*, 1956) and precipitation of the chloride as the insoluble AgCl (Gourevitch and Lein, 1955). Chemically defined media, where only chloride-free reagents are used, have also been successfully prepared (Darken *et al.*, 1960).

A biochemical alternative to the use of chloride-free media for the production of tetracycline by *S. aureofaciens* is the inclusion in the media of compounds which inhibit the incorporation of chloride into the antibiotic. Among the more useful are the halides, bromide (Gourevitch *et al.*, 1956; Lepetit S.p.A., 1957a,b) and fluoride (Szumski, 1959), and a variety of organic compounds, including 2-mercapto-2-thiazoline, 2-mercapto-4,5-dimethylthiazole, 2-benzoxazolethiol, 2-mercaptobenzimidazole (Lein *et al.*, 1959), thiadiazoles, oxadiazoles, triazoles, and benzthiazoles (Goodman *et al.*, 1959). The use of bromide, which blocks chloride utilization in certain cultures (Gourevitch *et al.*, 1956), results in biosynthesis of bromotetracycline with others (Doerschuk *et al.*, 1956, 1959; Lepetit S.p.A., 1957a,b).

The addition of antimetabolites to tetracycline-producing fermentations has resulted in modifying the biosynthetic pathways, with "new" antibiotic entities being produced. For example, addition of sulfonamides (Goodman and Miller, 1963; Perlman *et al.*, 1961; Perlman and Heuser, 1962) to chlortetracycline-producing fermentations interfered with some of the methylation steps in the biosynthesis, resulting in biosynthesis of 7-chlor-6-demethyltetracycline. A similar result was obtained when aminopterin and ethionine were added (Neidleman, 1962; Neidleman *et al.*, 1963), while addition of ethionine to an oxytetracycline fermentation resulted in biosynthesis of 4-*N*-ethylmethyl-5-oxytetracycline (Goodman and Miller, 1963). There is some variation from strain to strain in response to these antimetabolites, and what is one strain's "poison" is the next strain's nutrient.

VI. PRODUCTION OF BACITRACIN

Bacitracin is a polypeptide antibiotic active against many gram-positive and a few gram-negative species of bacteria. It is not assimilated significantly into circulating body fluids following oral ingestion, but it exerts its activity locally in the gastrointestinal tract. Although it is highly active parenterally, its systemic use has so far been considerably restricted because of the appearance of kidney damage when it is administered under some conditions. In addition to its considerable success in topical applications, bacitracin is used effectively in the treatment of surgical wounds. Among the important infectious problems of animals in which bacitracin has been used with reasonable success are bovine mastitis, infected wounds, hemorrhagic septicemia, otitis, enteritis (necrotic), infectious keratitis, dermatitis, and some types of dysentery. Feed supplements containing bacitracin in various forms have proved to be quite effective and economical in animal and poultry nutrition, increasing feed efficiency and reducing infectious diseases.

The original bacitracin-producing organism was obtained from debrided tissue taken from a compound fracture of the tibia of 7-year-old Margaret Tracy (Johnson *et al.*, 1945; Meleney *et al.*, 1949). The culture, first identified as *Bacillus subtilis,* was later classified as *Bacillus licheniformis*. Several cultures of *B. licheniformis* were later found to produce antibiotics, and one of these produced the antibiotic ayfivin, a peptide antibiotic shown to be chemically in the bacitracin group of antibiotics (Arrigada *et al.*, 1949; Sharp *et al.*, 1949). Bacitracin has also been identified in *B. subtilis* cultures (Aida, 1962; Freaney and Allen, 1958; Keko *et al.*, 1953; Snoke, 1960).

The various products generally referred to as "bacitracin" are mixtures of related polypeptides. They are generally known as bacitracins A, A′, B, C, D, E, F, F_1, F_2, F_3, and G. Bacitracin A is of primary importance and is the most potent, antibiotically speaking. Bacitracins B and C are less potent antibiotics than A, and the rest have very little antibacterial activity. Commercial products contain mostly bacitracin A, but some preparations have had as much as 20% bacitracin F (Craig and Konigsberg, 1957).

A unit of bacitracin is the amount which, in a dilution of 1 : 1024 based on serial twofold dilutions in beef infusion broth, inhibits the growth of a particular group A hemolytic streptococcus under defined conditions (Jackson, 1958). An international standard has been set up and has an activity of 55 units/mg (Craig and Konigsberg, 1957). Products with potencies of the order of 80 units/mg have been reported in the literature (Hickey, 1964).

The amino acid sequence of bacitracin A is:

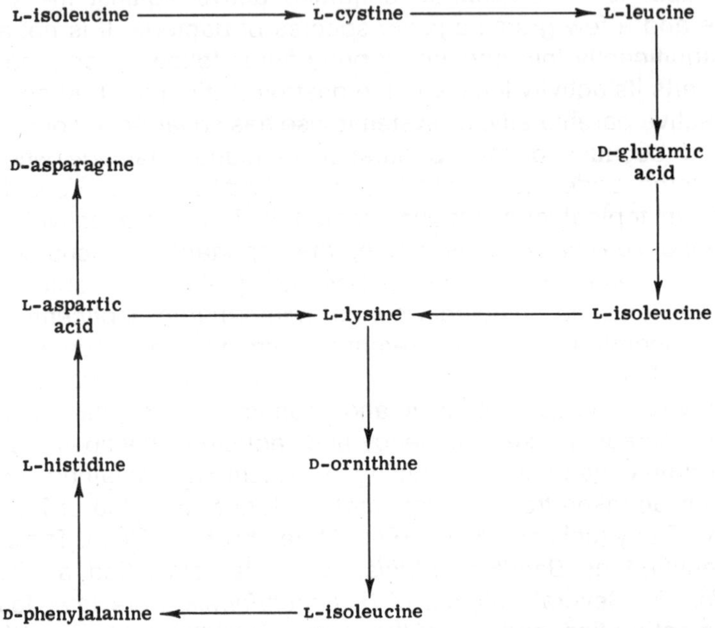

Methods for biosynthesis of bacitracin evolved from static procedures employing flasks and bottles and commercially impractical laboratory media to deep tank aerated operations employing low-cost media for industrial production. Flask and bottle surface cultures were used on an industrial scale in 1946 at the Ben Venue Laboratories to produce enough bacitracin for clinical and pharmacological trials. As the medical potential developed and more material was required to meet the demand, the submerged culture process was developed. Some of the characteristics of processes mentioned in the patent literature (and probably used on an industrial scale) are given in Table VIII.

A flow sheet describing the large-scale manufacture of bacitracin at Terre Haute, Indiana, plant of IMC Chemicals Group, Inc. (formerly Commercial Solvents Corporation) is shown in Fig. 6 (Inskeep *et al.*, 1953). The bacitracin-producing strains of *B. licheniformis* were carried on agar slants or in the spore form dispersed in dry, sterile soil. For culture development, a suitable small amount of soil culture was used to inoculate a 4-liter Erlenmeyer flask of tryptone or peptone broth. This was shaken on a reciprocating shaker located in a 37°C incubator. After

TABLE VIII. Characteristics of Processes Used in the Production of Bacitracin

Medium composition	Aeration method	Incubation period	Bacitracin yield (units/ml)	References
Tryptone, meat infusion, other broths	Surface culture	3–5 days	10–26	Jackson (1958)
Soybean meal, starch, calcium lactate	Surface culture	7 days	88	Darker (1951)
Soybean meal, calcium lactate, $CaCO_3$, starch	Submerged culture	26 hours	80–90	Keko *et al.* (1953)
Cottonseed and soybean meals, $CaCO_3$, dextrin	Submerged culture	24 hours	125	Cohen (1957)
Soybean meal, starch, $CaCO_3$	Submerged culture	24 hours	325	Freaney and Allen (1958)
Soybean grits, sucrose	Submerged culture	30 hours	400	Flickinger and Perlman (1979)

18–24 hours' incubation, the contents of the flask were used to inoculate 150 gal of medium contained in a 200-gal stainless steel tank. The culture was aerated vigorously for 6 hours (temperature maintained at 37°C), and this cell suspension was used to inoculate 800 gal of medium in a seed tank. The seed tank is made of stainless steel and is operated in a manner similar to the 150-gal stage, except that 800 gal of medium are prepared in a separate 1200-gal carbon steel cooker. This medium is blown hot to the seed tank, where it is cooled. A typical seed medium contained 4% soybean meal, 0.5% $CaCO_3$, and 0.5% starch. The seed culture is aerated and agitated after inoculation, and when the bacterial culture is in the log phase of growth, this inoculum is blown with sterile air via sterile pipe lines to a fermentor of cooled, sterile medium.

The fermentors are vertical, cylindrical, stainless steel vessels of 24,000-gal total capacity, having somewhat dished bottoms. The vessels are equipped with sterile air spargers, stainless steel tubing acting as cooling coils, and antifoam addition devices.

At the end of the incubation period, the bacitracin is recovered from the fermented medium by extraction with *n*-butanol. The bacitracin in the rich butanol is extracted with buffers, and the aqueous phase is concentrated (see Fig. 6). Ion-exchange resins have been used to effect further purifications (Gollaher and Honohan, 1956). Zinc bacitracin is used as a high-potency product for certain pharmaceutical preparations since its

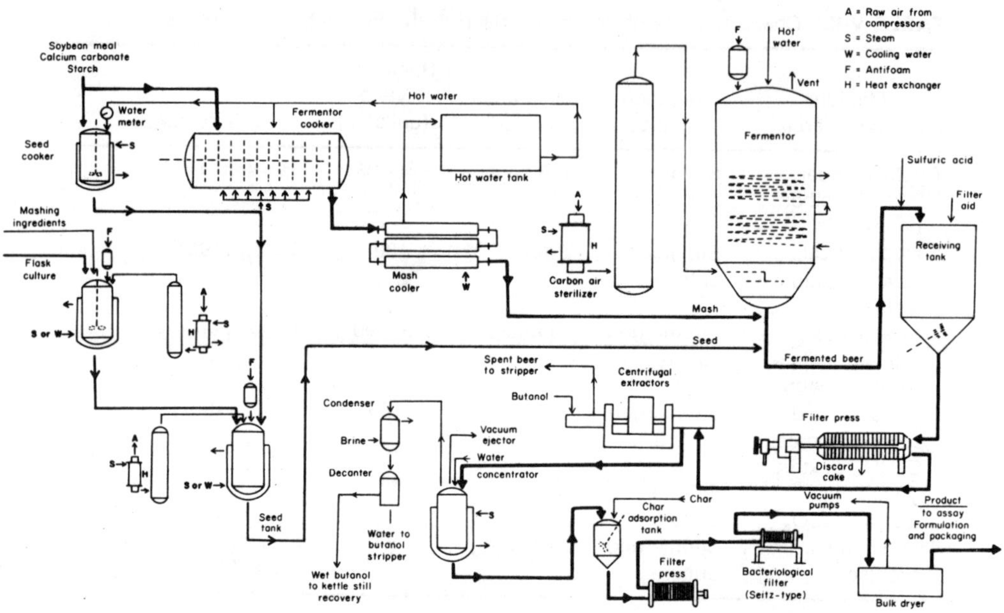

FIGURE 6. Bacitracin fermentation process flow chart.

storage stability is better than that of bacitracin itself. For animal feed purposes, the processing is quite simple, involving evaporation of the whole fermentation broth, drying, and blending with other animal feed supplements. Zinc bacitracin and bacitracin–methylene–disalicylate are two forms used for some feed formulations.

REFERENCES

Abraham, E. P. (1962). *Pharmacol. Rev.* **14,** 473.

Abraham, E. P., and Chain, E. (1940). *Nature (London)* **146,** 837.

Abraham, E. P., and Newton, G. G. F. (1961). *Biochem. J.* **79,** 393.

Abraham, E. P., Chain, E., Fletcher, C. M., Florey, H. W., Gardner, A. D., Heatley, N. G., and Jennings, M. A. (1941). *Lancet* **2,** 177.

Abraham, E. P., Chain, E., Florey, H. W., Florey, M. E., Heatley, N. G., Jennings, M. H., and Sanders, A. G. (1949). *In* "The Antibiotics" (H. W. Florey, ed.), p. 631. Oxford Univ. Press, London and New York.

Aida, T. (1962). *J. Agric. Chem. Soc. Jpn.* **36,** 724.

Arishima, M., and Sekizawa, Y. (1959). Japanese Patent 9,497.

Arrigada, A., Florey, H. W., Florey, M. E., Jennings, M. A., and Wallmark, I. G. (1949). *Br. J. Exp. Pathol.* **30,** 458.

Backus, E. J., Duggar, B. M., and Campbell, T. H. (1954). *Ann. N.Y. Acad. Sci.* **60,** 86.

Backus, M. P., and Stauffer, J. F. (1955). *Mycologia* **47,** 429.

Bartholomew, W. H., Karow, E. O., Stat, M. R., and Wilhelm, R. H. (1950a) *Ind. Eng. Chem.* **42,** 1801.
Bartholomew, W. H., Karow, E. O., Sfat, M. R., and Wilhelm, R. H. (1950b). *Ind. Eng.Chem.* **42,** 1810.
Batchelor, F. R., Doyle, F. P., Naylor, J. H. C., and Rolinson, G. N. (1959). *Nature (London)* **183,** 257.
Batchelor, F. R., Chain, E. B., Richards, M., and Rolinson, G. N. (1961). *Proc. R. Soc. London, Ser. B* **154,** 522.
Behrens, O. K., Corse, J., Edwards, J. P., Garrison, L., Jones, R. G., Soper, Q. F., van Abeele, F. R., and Whitehead, C. W. (1948). *J. Biol. Chem.* **175,** 793.
Belik, E., Herold, M., and Doskocil, J. (1957). *Folia Biol. (Prague)* **3,** 229.
Boothe, J. H., Morton, J., Petisi, J. P., Wilkinson, R. G., and Williams, J. H. (1953). *J. Am. Chem. Soc.* **75,** 4621.
Broschard, R. W., Dornbush, A. C., Gordon, S., Hutchings, B. L., Kohler, A. R., Krupa, G., Kushner, S., Lefemine, D. W., and Pidacks, C. (1949). *Science* **109,** 199.
Brotzu, G. (1948). *Lav. Ist. Ig. Cagliari*
Brown, W. E., and Peterson, W. H. (1950). *Ind. Eng. Chem.* **42,** 1769.
Calam, C. T., and Hockenhull, D. J. D. (1949). *J. Gen. Microbiol.* **3,** 19.
Calam, C. T., Driver, N., and Bowers, R. H. (1951). *J. Appl. Chem.* **1,** 209.
Chain, E. B. (1964). *In* "New Perspectives in Biology" (M. Sela, ed.), p. 205. Elsevier, Amsterdam.
Chain, E. B. (1966). *Antimicrob. Agents Chemother.* p. 1.
Chain, E., and Gulandi, G. (1954). *Rend. Ist. Super. Sanita* **17,** 5.
Chain, E., Florey, M. W., Gardner, A. D., Heatley, N. G., Jennings, M. A., Orr-Ewing, J., and Sanders, A. G. (1940). *Lancet* **2,** 226.
Chas. Pfizer and Co. (1958). German Patent 1,028,744.
Claridge, C. A., Gourvitch, A., and Lein, J. (1960). *Nature (London)* **187,** 237.
Clutterbuck, P. W., Lovell, R., and Raistrick, H. (1932). *Biochem. J.* **26,** 1907.
Coghill, R. D. (1944). *Chem. & Eng. News* **22,** 588.
Coghill, R. D., and Koch, R. D. (1945). *Chem. & Eng. News* **23,** 2310.
Cohen, I. R. (1957). U.S. Patent 2,789,941.
Craig, L. C., and Konigsberg, W. (1957). *J. Org. Chem.* **22,** 1345.
Darken, M. A., Berenson, H., Shirk, R. J., and Sjölander, N. O. (1960). *Appl. Microbiol.* **8,** 46.
Darker, G. D. (1951). U.S. Patent 2,567,698.
Davey, V. F., and Johnson, M. J. (1953). *Appl. Microbiol* **1,** 208.
Deindoerfer, F. H., and Wilker, B. L. (1957). *Ind. Eng. Chem.* **49,** 1223.
Demain, A. L. (1953). *Adv. Appl. Microbiol.* **1,** 23.
Demerec, M. (1948) U.S. Patent 2,445,748.
Doerschuk, A. P., McCormick, J. R. D., Goodman, J. J., Szumski, S. A., Growich, J. A., Miller, P. A., Bitler, B. A., Jensen, E. R., Petty, M. A., and Phelps, A. S. (1956). *J. Am. Chem Soc.* **78,** 1058.
Doerschuk, A. P., McCormick, J. R. S., Goodman, J. J., Szumski, S. A., Growich, J. A., Miller, P. A., Bitler, B. A., Jensen, E. R., Matrishin, M., Petty, M. A., and Phelps, A. S. (1959). *J. Am. Chem. Soc.* **81,** 3069.
Duggar, B. M. (1948). *Ann. N.Y. Acad. Sci.* **51,** 177.
Duggar, B. M. (1949). U.S. Patent 2,482,055.
Duggar, B. M., Backus, E. J., and Campbell, T. H. (1954). *Ann. N.Y. Acad. Sci.* **60,** 71.
E. R. Squibb and Sons, Inc. (1952). British Patent 679,087.
Federal Trade Commission (1958). "Economic Report on Antibiotic Manufacture," US Gov. Printing Office, Washington, D.C.

Finlay, A. C., Hobby, G. L., P'an, S. Y., Regna, P. P., Routien, J. B., Seeley, D. B., Shull, G. M., Sobin, B. A., Solomons, I. A., Vinson, J. W., and Kane, J. H. (1950). *Science* **111,** 85.

Fleming, A. (1929). *Br. J. Exp. Pathol.* **10,** 226.

Fleming, P. C., Goldner, H., and Glass, D. G. (1963). *Lancet* **1,** 1399.

Flickinger, M. C., and Perlman, D. (1979). *Appl. Environ. Microbiol.* **15,** 222.

Forbath, T. P. (1957). *Chem. Eng. (N.Y.)* **64,** 228.

Ford, J. H., Churchill, B. W., and Colingsworth, D. R. (1953). *Antibiot. Chemother. (Washington, D.C.)* **3,** 1149.

Foster, J. W., Woodruff, M. B., Perlman, D., McDaniel, L. E., Wilker, B. L., and Hendlin, D. (1946). *J. Bacteriol.* **51,** 695.

Freaney, T. E. (1958). U.S. Patent 2,830,934.

Freaney, T. E., and Allen, L. P. (1958). U.S. Patent 2,828,246.

Gollaher, M. G., and Honohan, E. J. (1956). U.S. Patent 2,763,590.

Goodman, J. J., and Miller, P. A. (1963). *Biotechnol. Bioeng.* **4,** 391.

Goodman, J. J., Matrishin, M. M., Young, R. W., and McCormick, J. R. D. (1959). *J. Bacteriol.* **78,** 492.

Gourevitch, A., and Lein, J. (1955). U.S. Patent 2,712,517.

Gourevitch, A., Misiek, M., and Lein, J. (1956). *Antibiot. Chemother. (Washington, D.C.)* **5,** 448.

Gualandi, G., Caldarola, E., and Chain, E. B. (1960). *Rend. Ist. Super. Sanita* **23,** 869.

Haas, F. L., Puglisi, T. A., Moses, A. J., and Lein, J. (1956). *Appl. Microbiol.* **4,** 187.

Hastings, J. J. H. (1955). *Ind. Chem. Manuf.* **31,** 494.

Hatch, A. B., Hunt, G. A., and Lein, J. (1956). U.S. Patent 2,763,591.

Herold, M., and Necasek, J. (1959). *Adv. Appl. Microbiol* **1,** 1.

Herold, M., Matelova, V., Necasek, J., and Hostalek, Z. (1958). *Cesk. Mikrobiol.* **3,** 313.

Hickey, R. J., (1964). *Prog. Ind. Microbiol.* **5,** 93.

Hodgkin, D., and Maslen, E. N. (1961). *Biochem. J.* **79,** 393.

Holt, R. J., and Stewart, G. T. (1964). *J. Gen. Microbiol.* **36,** 203.

Hosler, P., and Johnson, M. J. (1953). *Ind. Eng. Chem.* **45,** 871.

Huang, H. T., English, A. R., Seto, T. A., Shull, G. M., and Sobin, B. A. (1960). *J. Am. Chem. Soc.* **82,** 3790.

Huang, H. T., Seto, T. A., and Shull, G. M. (1963). *Appl. Microbiol.* **11,** 1.

Inskeep, G. C., Bennett, R. E., Dudley, J. F., and Shepard, M. W. (1953). *Ind. Eng. Chem.* **43,** 1488.

Jackson, T. (1958). *In* "Biochemical Engineering" (R. Steel, ed.), p. 185. Macmillan, New York.

Jarvis, F. G., and Johnson, M. J. (1947). *J. Am. Chem. Soc.* **69,** 3010.

Johnson, B. A., Anker, H., and Meleney, F. (1945). *Science* **102,** 376.

Kato, K. (1953a). *J. Antibiot. Ser. A* **6,** 130.

Kato, K. (1953b). *J. Antibiot. Ser. A* **6,** 184.

Kaufmann, W., and Bauer, K. (1960). *Naturwissenschaften* **47,** 474.

Keko, W. L., Bennett, R. E., and Arzberger, C. F. (1953). U.S. Patent 2,627,494.

Laboratories Pro-Ter (1964). Belgian Patent 649,537.

Lein, J., Sawmiller, L. F., and Cheney, L. E. (1959). *Appl. Microbiol.* **7,** 149.

Lepetit S.p.A. (1957a). British Patent 772,149.

Lepetit S.p.A. (1957b). British Patent 775,139.

Loder, B., Newton, G. G. F., and Abraham, E. P. (1961). *Biochem. J.* **79,** 408.

McCormick, J. R. D., Sjölander, N. O., Hirsch, U., Jensen, E. R., and Doerschuk, A. P. (1957). *J. Am. Chem. Soc.* **79,** 4561.

McCormick, J. R. D., Sjölander, N. O., Johnson, S., and Doerschuk, A. P. (1959). *J. Bacteriol.* **77,** 475.
McCormick, J. R. D., Hirsch, U., Sjölander, N. O., and Doerschuk, A. P. (1960). *J. Am. Chem. Soc.* **82,** 5006.
McCormick, J. R., Sjölander, N. O., and Hirsch, U. (1961). U.S. Patent 2,998,352.
McGhee, W. J., and Megna, J. C. (1957). U.S. Patent 2,776,243.
McGuire, J. M. (1954). British Patent 713,795.
Martin, J. H., Shay, A. J., Pruess, L. M., Porter, J. N., Mowat, J. H., and Bohonos, N. (1955). *Antibiot. Annu.* p. 1020.
Mead, T. M., and Stack, M. V. (1948). *Biochem. J.* **42,** XVIII.
Meleney, F., and Johnson, B. A. (1949). *Am. J. Med.* **7,** 794.
Minieri, P. P., Firman, M. C., Mistretta, A. G., Abbey, A., Bricker, C. E., Rigler, N. E., and Sokol, H. (1954). *Antibiot. Annu.* p. 81.
Minieri, P. P., Sokol, H., and Firman, M. C. (1956). U.S. Patent 2,734,018.
Morin, R. B., Jackson, B. G., Flynn, E. H., and Roeske, R. W. (1962). *J. Am. Chem. Soc.* **84,** 3400.
Moyer, A. J., and Coghill, R. D. (1946). *J. Bacteriol.* **51,** 57.
Moyer, A. J., and Coghill, R. D. (1947). *J. Bacteriol* **53,** 329.
Murao, S. (1955). *J. Agric. Chem. Soc. Jpn.* **29,** 400.
Neidleman, S. L. (1962). U.S. Patent 3,061,522.
Neidleman, S. L., Bienstock, E., and Bennett, R. E. (1963). *Biochim. Biophys. Acta* **71,** 199.
Niedercorn, J. G. (1952). U.S. Patent 2,609,329.
N. V. Koninklijke Nederlandsche Gist-en Spiritus-Fabriek (1956). British Patent 764,193.
Owen, S. P., and Johnson, M. J. (1955). *Appl. Microbiol.* **3,** 375.
Pan, S. C., and Perlman, D. (1954). *Anal. Chem.* **26,** 1432.
Perlman, D. (1949). *Bull. Torrey Bot. Club.* **76,** 79.
Perlman, D. (1950). British Patent 700,316.
Perlman, D. (1977). *CHEM TECH* **7,** 434.
Perlman, D., and Heuser, L. J. (1962). U.S. Patent 3,028,311.
Perlman, D., and O'Brien, E. (1954). *Arch. Biochem. Biophys.* **51,** 266.
Perlman, D., Greenfield, R. B., and O'Brien, E. (1954). *Appl. Microbiol.* **2,** 199.
Perlman, D., Heuser, L. J., Dutcher, J. D., Barrett, J. M., and Boska, J. A. (1960). *J. Bacteriol.* **80,** 419.
Perlman, D., Heuser, L. J., Semar, J. B., Frazier, W. R., and Boska, J. A. (1961). *J. Am. Chem. Soc.* **83,** 4481.
Petty, M. A., and Matrishin, M. (1950). *Abstr. 118th Meet., Am. Chem. Soc.* p. 18A.
Pollock, M. R. (1965). *Biochem. J.* **94,** 666.
Pollock, M. R., Torriani, A. M., and Tridgell, E. J. (1956). *Biochem. J.* **62,** 387.
Price, K. E. (1969). *Adv. Appl. Microbiol.* **11,** 17.
Price, K. E. (1977). *In* "Structure-Activity Relationships among the Semisynthetic Antibiotics" (D. Perlman, ed.), p. 61. Academic Press, New York.
Reese, E. T., Sanderson, K., Woodward, R., and Eisenberg, G. M. (1949). *J. Bacteriol.* **57,** 15.
Regna, P. P., Solomons, I. A., Murai, K., Timreck, A. E., Brunnings, K. J., and Lazier, W. A. (1951). *J. Am. Chem. Soc.* **73,** 4211.
Rolinson, G. R., Batchelor, F. R., Butterworth, D., Cameron-Wood, J., Cole, M., Eustale, G. C., Hart, M. W., Richards, M., and Chain, E. B. (1960). *Nature (London)* **187,** 236.
Rolland, G., and Sensi, P. (1955). *Farmaco, Ed. Sci.* **10,** 37.
Sabath, L. D., and Abraham, E. P. (1966). *Antimicrob. Agents Chemother.* p. 392.
Sakaguchi, K., and Murao, S. (1950). *J. Agric. Chem. Soc. Jpn.* **23,** 411.

Sensi, P., de Ferrari, G. A., Gallo, G. G., and Rolland, G. (1955). *Il Farmaco, Ed. Sci.* **10,** 337.
Sharp. V. E., Arrigada, A., Newton, G. G. F., and Abraham, E. P. (1949). *Br. J. Exp. Pathol.* **30,** 444.
Singh, K., and Johnson, M. J. (1948). *J. Bacteriol.* **50,** 339.
Smith, E. L., and Bide, A. E. (1948). *Biochem. J.* **42,** XVII.
Snoke, J. E. (1960). *J. Bacteriol.* **80,** 552.
Sobin, B. A., Finlay, A. C., and Kane, J. H. (1950). U.S. Patent 2,516,080.
Soltero, F. V., and Johnson, M. J. (1953). *Appl. Microbiol.* **1,** 52.
Soper, Q. F., Whitehead, C. W., Behrens, O. K., Corse, J. J., and Jones, R. G. (1948). *J. Am. Chem. Soc.* **70,** 2849.
Stephens, C. R., Conover, L. H., Hochstein, F. A., Moreland, W. T., Regna, P. P., Pilgrim, F. J., Brunings, K. J., and Woodward, R. B. (1954). *J. Am. Chem. Soc.* **76,** 3568.
Stewart, G. T. (1961). *Lancet* **1,** 509.
Szumski, S. A. (1959). U.S. Patent 2,871,167.
Tabenkin, B., Lehr, H., Wayman, A. C., and Goldberg, M. W. (1952). *Arch. Biochem. Biophys.* **38,** 43.
Van Dyck, P., and deSomer, P. (1952). *Antibiot. Chemother. (Washington, D.C.)* **2,** 184.
Villax, I. (1963). *Antimicrob. Agents Chemother.* p. 661.
Villax, I. (1965). Eire Patent 49565.
Waller, C. W., Hutchings, B. L., Wolf, C. F., Goldman, A. A., Broschard, R. W., and Williams, J. H. (1952). *J. Am. Chem. Soc.* **74,** 4981.
Walton, R. B. (1964a). *Dev. Ind. Microbiol.* **5,** 349.
Walton, R. B. (1964b) French Patent 1,357,977.
Webber, J. A., and Ott, J. L. (1977). *In* "Structure-Activity Relationships among the Semisynthetic Antibiotics" (D. Perlman, ed.), p. 161. Academic Press, New York.
West, J. M., Stickle, G. P., Walter, K. D., and Brown, W. E. (1961). *J. Biochem. Microbiol. Technol. Eng.* **3,** 125.

Chapter 9

Production of Microbial Enzymes

KNUD AUNSTRUP
OTTO ANDRESEN
EDVARD A. FALCH
TAGE KJAER NIELSEN

MICROBIAL TECHNOLOGY, 2nd ed., VOL. I
 ISBN 0-12-551501-4

I. INTRODUCTION

Production and use of microbial enzymes is an ancient art. By experience and empirical methods it has developed to a highly sophisticated state; the preparation of fermented oriental food is an example of this.

The combination of microbiological and biochemical science with modern technology has made possible the development of a large enzyme industry. This industry was founded well over 50 years ago, but the economically important developments have taken place only within the last decade.

A. Recent Developments

The regrowth of the enzyme industry began in the early 1960s with the introduction of glucoamylase in starch hydrolysis. The enzyme method was substituted for the traditional acid hydrolysis and resulted in an increased dextrose yield.

Enzyme-containing detergents have been known since 1913, but their use was limited because the enzymes available were unstable in detergent formulations. Around 1965 a new, stable protease produced by *Bacillus licheniformis* was introduced, which was very successful and found wide application for use in detergents.

In 1968–1969, it was discovered that some workers handling enzyme concentrates were experiencing allergic reactions. This caused a strong public reaction against the use of enzymes. Consequently, enzymes were removed from most detergents used in the United States. An official investigation showed that there was no risk involved for the detergent user. After the introduction of dust-free, encapsulated enzyme products, the risk involved in handling enzyme concentrates was also eliminated. Since that time the use of detergent enzymes has grown steadily.

The substitution of microbial enzymes for calf rennet is an old dream, and numerous microbial proteases have been tested for this purpose. In the period 1965–1970 three microbial rennets were successfully introduced, and they are now being widely used in the rennet market.

Due to the high cost of sucrose and the availability of large amounts of inexpensive dextrose, the development of an enzymatic process which would isomerize glucose to fructose has been very attractive. The first commercial process was in use around 1970, and since then it has grown to become one of the most important enzyme applications. The use of immobilized enzymes for this purpose has made it possible to keep the conversion costs to a low level.

The brewing industry has traditionally used malt as an enzyme source.

The partial substitution of barley for malt and the inclusion of microbial amylase, protease, and β-glucanase has avoided some of the difficulties caused by variation of the malt quality. Furthermore, this change has been economically advantageous. The distilling industry, another traditional malt consumer, has in many countries substituted the microbial enzymes, α-amylase and glucoamylase, for malt.

In the pharmaceutical industry, the use of enzymes has been limited. The most important application is for digestive aids, in which a mixture of pancreatic and microbial enzymes is used. Another important application is the use of penicillin acylase in the manufacture of semisynthetic penicillins.

Apart from these large-scale industrial applications of enzymes, there has been a substantial growth in the preparation of microbial enzymes for diagnostic, scientific, and analytical purposes. These enzymes are usually prepared under laboratory-type conditions and are not dealt with here.

B. Present Position

A rough estimate is that the present world market for industrial microbial enzymes represents a sales value of $150–175 million. The economically most important enzymes are listed in Table I, together with an estimate of the relative value of the enzymes and the amount of enzyme produced per year, calculated as pure enzyme protein.

The major producers of industrial enzymes are listed in Table II. NOVO and Gist-Brocades have by far the largest share of the market.

TABLE I. Production of Industrial Enzymes

Enzyme	Amounts produced per year (tons of pure enzyme protein)	Relative sales value (%)
Bacillus protease	500	40
Glucoamylase	300	14
Bacillus amylase	300	12
Glucose isomerase	50	12
Microbial rennet	10	7
Fungal amylase	10	3
Pectinases	10	10
Fungal protease	10	1
Other	—	1

TABLE II. Companies Producing Microbial Enzymes

Denmark	Grindstedvaerket
	NOVO Industri
France	Sté. Rapidase
Germany	Boehringer Ingelheim
	Miles-Kalichemie
	Röhm
Holland	Gist Brocades (KNGS)
Japan	Amano
	Daiwa Kasei
	Godo Shusei
	Kyowa Hakko Kogyo
	Meito Sangyo
	Nagase
	Sankyo
	Shin Nihon
Switzerland	Swiss Ferment
United Kingdom	ABM
	Glaxo
United States	CPC
	G. B. Fermentation Industries, Inc.
	Miles
	Novo Biochemical Industries
	Pfizer
	Rohm & Haas
	Standard Brands

II. DEVELOPMENT OF NEW ENZYMES

Fewer than 50 microbial enzymes are of industrial importance today, but patents have been applied for on more than a thousand different enzymes. This reflects the increasing interest in developing new enzyme products and shows that it is easier to find a new enzyme than to find a profitable application.

A new enzyme product becomes a commercial success only if a demand exists and if the product possesses properties which satisfy the technical and economic requirements of the process. Most new enzymes fail in at least one of these respects. Demands for new enzymes arise from the development of new processes or from the unsatisfactory performance of known enzymes in established processes.

The properties of enzymes can to some extent be changed by chemical modification of the molecule. The scientific literature on the subject is extensive, but the methods have so far not been used on technical enzymes, presumably because the results have not been economically or technically attractive. The search for new enzymes has, therefore, been a search for new microorganisms. The basic principles which are guidelines in this search are described below.

A. Strain Selection

1. Enzyme Properties and Taxonomy

It is generally held that macromolecules, such as nucleic acids and proteins, are specific for the species and that the phylogenetic development which has given rise to the microbiological variation basically has been caused by variation in these molecules. As a consequence, it is to be expected that enzyme types, such as protease, α-amylase, and lactase, which are found in several species, will have properties which vary as much as the other properties of the organism. One usually finds that closely related organisms have enzymes with closely related properties, while unrelated organisms have enzyme systems which differ widely.

The protease subtilisin Carlsberg from *Bacillus licheniformis* is closely related to the protease subtilisin NOVO from *Bacillus amyloliquefaciens,* just as the species are closely related. Yet, the differences are distinct, and it appears that these enzymes are specific for the species. Unfortunately, many incorrect classifications of microorganisms in the biochemical literature make the establishment of such rules as this difficult.

2. Enzyme Properties and the Environment

Extracellular enzymes must work in the environment of the microorganism. One may therefore rightly assume that they have their optimum activity and stability close to the optimum conditions for growth of the microorganism. Numerous examples of this may be cited.

Proteases from alkalophilic *Bacillus* species have pH optima several units higher than that of the protease from the "neutralophilic" *B. licheniformis.* The α-amylase from the thermophile *Bacillus coagulans* can be used at 10°C higher temperatures than that of the mesophile *B. amyloliquefaciens.* There are known exceptions. The α-amylase of the mesophile *B. licheniformis* is more thermostable than the amylase of *B.*

coagulans. Several *Aspergillus* species, which have growth optima at about pH 4, produce an alkaline protease, with pH optimum about 10, in addition to an acid protease.

The properties of intracellular enzymes are usually not too dependent on environmental conditions. The internal pH of the cells is close to neutrality and appears to be little influenced by external conditions. Thermostable enzymes, however, are usually found in thermophilic organisms. The thermostable lactase of *Bacillus stearothermophilus* is such an example. Thermostability of an isolated enzyme may be lower than the growth optimum of the strain would indicate, because the enzyme is especially stabilized in the intact cell.

3. Enzyme Properties and the Presence of Substrate

Microorganisms are responsible for the majority of the mineralizations of organic material in nature. It is claimed that all organic material can be microbiologically decomposed. Microorganisms which will produce enzymes for the degradation of certain materials are, therefore, usually found where these substances are abundant. For example, producers of lignolytic enzymes, pentosanases, and cellulases are found in forest soil, and uric acid decomposers in chicken pens. Microorganisms capable of metabolizing a new compound are often quickly established in sewage plants and effluent streams.

4. Methods for the Isolation of Enzyme-Producing Organisms

Traditional microbiological methods are most often used for isolation. Especially important is the use of enrichment cultures and selective media. In the ideal system, the substrate for the enzyme serves as the sole source of one or more vital elements. For example, a uric acid–salts medium is used in the isolation of urate oxidase-producing microorganisms.

Agar diffusion tests are used whenever possible for the detection of enzyme activity. The principle is valuable as a qualitative or very rough quantitative test but the results are often misleading, particularly if the enzyme substrate is incorporated in the growth medium for the microorganism.

In the design of isolation methods, the imagination of the researcher is essential for success. A combination of a good source of microorganisms and a specific isolation method, often nontraditional, will lead to a useful result. An example is the search for alkali-stable proteases, where enrichment and selection on media with 0.1 *M* sodium sesquicarbonate and with casein as a protease indicator led to the discovery of a large group of *Bacillus* species that produce alkaline proteases.

5. Safety Measures

Microorganisms may be pathogenic or producers of toxic materials, and this must constantly be borne in mind when working with unknown organisms. Until they are identified, therefore, cultures and products should be handled with the proper precautions. This is especially important for fungal isolates since fungal spores spread easily.

B. Strain Development

1. Desired Properties of Production Strains

The organism should grow on an inexpensive medium and give a constant, high yield of enzyme in a short time. Secondary enzyme activities and the content of metabolites in the fermented broth should be minimal. Furthermore, recovery of the enzyme should be simple and inexpensive and lead to a stable product which can be handled safely and which has an acceptable appearance. Last but not least, the process must be safe to the personnel in the production plant, and the effluents from the plant should not disturb the environment.

The fulfillment of these objectives requires a combined optimization of the strain properties and process parameters. Optimization of strain properties is very attractive because this, as a rule, offers an inexpensive and permanent solution to the problem. An example is the development of a constitutive mutant which eliminates the need for an expensive inducer, a problem which has been relevant in the production of the xylose-induced glucose isomerase.

Health and safety problems can often be alleviated by strain development. The viable counts of the products, for example, can be reduced by using asporogenic mutants of *Bacillus* strains. Objectionable by-products, such as antibiotics, can be eliminated by using mutants.

Wild strains will often have undesirable enzymatic side activities which have to be removed in the purification process. This is the case with the transglucosidase in glucoamylase. A better method is to develop a mutant which does not produce the side activity.

2. Methods for Strain Development

The exciting progress in microbial genetics offers many attractive methods to the industrial researcher. The use of most of these methods, however, is hampered by the facts that the enzyme-producing organism usually is genetically unknown and that the efforts and especially the time which can be spent on a single organism is limited. The method of choice, therefore, will, as a rule, be straightforward mutagenization and selection. The mutagens used are those commonly applied in microbiology. In the development of selection methods, experience and personal skill are important.

The success of the mutation work is of great economic value to the enzyme industry. The methods used are therefore carefully guarded and very little information about them is published.

3. Strain Maintenance

The highly developed production strains must be protected against degeneration, contamination, and loss of viability. The most convenient way to secure this is to store the strains lyophilized or at the temperature of liquid nitrogen. It is possible to store the strains almost indefinitely without any changes using these methods.

III. THE FERMENTATION PROCESS

Traditionally, microbial enzymes were produced by surface cultures, i.e., cultures of microorganisms in thin layers of liquid or moist, solid media. This technique is still used for a few products, primarily of fungal origin, such as *Aspergillus* amylase, proteases from *Aspergillus* and *Mucor* species, pectinases from *Penicillium* and *Aspergillus* species, and also cellulases. Originally, the cultures grew in trays that were handled manually, but today mechanical systems for cleaning, filling, and emptying the trays are used. Systems where the semisolid culture is tumbled in a rotating drum also have been developed. Control of infections and also uniform control of temperature, humidity, and aeration present difficulties. Few developments of this technique have been published in the last decade.

Submerged culture methods today dominate in the production of enzymes because handling costs and the risk of infection are reduced, and because modern methods of control are more easily adapted to these processes. Yields are also generally higher. The production methods discussed below refer to submerged fermentation.

A. Inoculum Preparation

Highly mutated strains are increasingly used for enzyme production. An important requirement for a propagation technique, therefore, is that the production capability of the strain be preserved. The technique should also minimize the risk of contamination. Numerous recultivations of laboratory cultures and multiple propagation steps in the factory should be avoided. It is generally possible to develop a method where the seed flask is inoculated directly from a lyophilized culture and where only one seed tank is used.

In a few cases, the enzyme yield or the mycelial form of the culture depends on a specific propagation method (Meyrath *et al.*, 1973); but this is the exception. In general, the process pattern is unaffected by rather wide variations in the propagation technique. Agar media may often be used in the seed flask. These cultures are easier to handle and have a longer period of applicability than shake-flask cultures. Cells or spores from the agar culture are suspended in a sterile solution before being transferred to the seed tank.

The medium in the seed tank often resembles the production medium. Excessive heat sterilization of the medium retards the growth of the inoculum. The volume of the seed tank usually constitutes 3–10% of the volume of the production fermentor.

The propagation time in the seed fermentor varies from 10 to 80 hours, depending on the process. Examples of parameters used as applicability criteria of the seed culture are change of pH, relative mycelial volume, development of carbon dioxide, or an easily detectable enzyme activity.

B. Medium Formulation and Preparation

The medium should provide the energy source for the process. Furthermore, it should include nutrients providing carbon and nitrogen sources and also special growth requirements, such as essential amino acids. Stimulating factors may be added to reduce the lag time or to increase the growth rate.

However, enzyme production may be negligible in a medium designed for good growth. The enzyme production often depends on the presence of an inducer in the medium, or it may be repressed by a component of the medium. Catabolite repression often controls enzyme production (Paigen and Williams, 1970). The strongest repression is seen in media containing glucose. Repression may be avoided by replacing glucose by slowly fermentable carbohydrates or by partly hydrolyzed starch. In a process for pullulanase production, the use of liquefied starch with a dextrose equivalent less than 42 is recommended (Heady, 1971).

In order to increase the productivity, highly concentrated media are used. In a patented process for the production of glucoamylase, concentrations of liquefied starch as high as 25% on a dry weight basis are used (Smith and Frankiewicz, 1978). Often a concentrated medium will result in catabolite repression, in retardation caused by the high osmotic pressure, or in a culture which is difficult to aerate. These difficulties may be overcome by incremental or continuous feeding of a concentrated medium to the culture (Kalaboklas, 1971; Hulme, 1973). Typical protein and carbohydrate components of industrial media for enzyme production are listed in Table III. Salts are often added as supplementary sources of nitrogen, phosphorus, sulfur, or calcium. Trace metals are normally present in sufficient amounts in the tap water or in the main raw materials. Strong pH buffers, mostly phosphate buffers, are still widely used to reduce pH fluctuations during the fermentation. Proper balance of carbon and nitrogen sources is also important to the pH pattern of the process if pH control is not applied.

Economy is very important in medium formulation. Typically, raw materials account for 60–80% of the variable costs of an enzyme fermentation process. Much development work is directed toward the replacement of costly ingredients with components available in large quantities at low cost. An enzyme fermentation medium normally has a high content of solids, such as ground whole grains or flakes.

Certain raw materials must be avoided in enzyme fermentations because the quality is too variable. Either a food quality is not available or they have an adverse effect on enzyme recovery, product quality, or waste disposal.

Most media are still sterilized batchwise in the fermentor. Continuous sterilization methods, however, are gaining wider application, particularly for the feeding medium. In some cases the feeding medium may be sterilized by filtration. Continuous heat sterilization is performed as a high-temperature/short-time process which results in improved preservation of growth factors and less development of color. Recovery of heat may be an additional advantage. Fouling of heat-exchange surfaces and clogging of pipes by grainy material in the medium should be considered when the sterilization system is designed. Direct steam injection may, in some cases, solve this problem.

TABLE III. Typical Medium Constituents for Enzyme Fermentation

Carbohydrates	Starch hydrolysate, molasses, saccharose, corn, barley, wheat
Proteins	Soybean meal, cotton seed meal, peanut meal, corn steep liquor, yeast hydrolysate, whey, gluten

C. Process Conditions and Equipment

Since microbial enzymes are relatively low-volume products, it has been difficult to justify the development and construction of specialized equipment for submerged fermentation. Equipment and techniques are most often adapted from antibiotics fermentations. Tall cylindrical fermentors of stainless steel with capacities of 10–100 tons and furnished with strong mechanical agitators and air spargers are typical. The advantage of this traditional setup is flexibility. It is easy to switch between products.

A schematic diagram of a typical enzyme fermentation process is shown in Fig. 1. The diagram indicates the extent of auxiliary equipment and control systems required for the process. As in many other fermentation processes, control of an enzyme fermentation is hampered by the fact that neither growth nor product formation may be determined rapidly enough to be of value for control. Primary physical variables, such as temperature, air flow, and pressure, are controlled within narrow limits. Other measurable variables, such as pH, oxygen tension, or oxygen consumption, are often also applied in process control. Foaming is normally controlled by automatic oil addition.

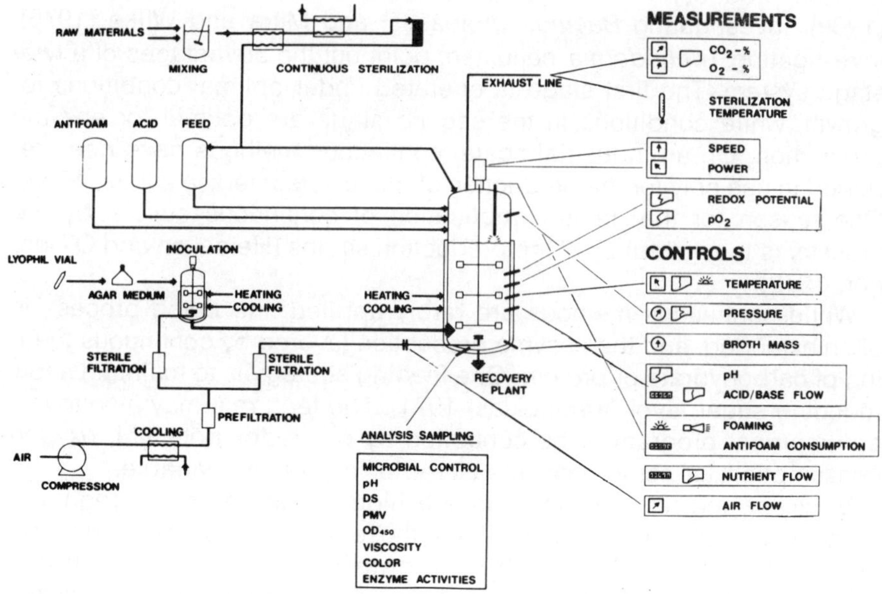

FIGURE 1. Schematic diagram of a typical enzyme fermentation process. Examples of useful measurements and controls are indicated.

Enzyme fermentations are especially vulnerable to microbial contamination. Rich media with a neutral pH value are typical, and the protection afforded by antibiotic activity is normally lacking. Infection generally means loss of a batch. Strict attention must be given to the contamination risk in the design and construction of fermentor, pipes, and auxiliary equipment. A fully welded system is recommended. Ports and valves should be steam-sealed, and all transfers of cultures and media should be done by compressed sterile air. A positive pressure must be maintained in the aseptic system. Mechanical agitator seals are normally used.

As a rule, extracellular enzymes are produced by batch processes which last from 30 to 150 hours. The optimal harvesting time falls between the point of maximum productivity and the point of maximum enzyme activity. Relative costs of raw materials, utilities, and recovery, as well as utilization of the plant capacity, determine the optimum. Often the process is terminated before or carried on beyond the optimum in order to obtain a broth with properties which facilitate product recovery. Finally, the scheduling in the plant restricts the choice of fermentation time. The optimal harvesting time is often signaled by a characteristic change in pH, oxygen tension, or some other parameter which is easy to determine.

Continuous culture techniques have been applied in several studies of enzyme production on a laboratory scale. Jensen (1972) and Fabian (1969), investigating *Bacillus* proteases, and Mitra and Wilke (1975), investigating *Trichoderma* cellulase, point out the advantages of a two-stage system. The first stage is operated under optimal conditions for growth, while conditions in the second stage are optimal for enzyme production. On an industrial scale, continuous methods have been reported in use only for the production of glucose isomerase (Diers, 1976). One reason for the limited application of continuous culture by the industry is the instability of the production strains (Heineken and O'Connor, 1972).

While continuous methods are rarely applied, the batch process is often extended, and the enzyme production favored by continuous feeding of carbohydrate or protein. One feeding strategy is to maintain a low reducing sugar level (Kalaboklas, 1971). The feed rate may also follow an empirical program or be controlled by pH, redox potential, oxygen consumption, or some other measured or calculated variable.

Most enzyme fermentations have a high oxygen demand, requiring aeration and agitation rates similar to those used in antibiotic fermentations. Very viscous media with non-Newtonian behavior are often employed; in other cases heavy mycelial growth complicating the oxygen transfer develops. In the glucoamylase process the productivity is limited by the oxygen transfer rate (Aunstrup, 1978). However, high oxygen tension may, in certain processes, inhibit enzyme formation. Diers (1976) reports that glucose isomerase production by an atypical variety of *Bacillus coagulans* is optimal during oxygen limitation and simultaneous addition of glucose at such a rate that the concentration in the broth is infinitesimal.

During process development much attention must be given to the properties of the broth in product recovery and to the quality of the final product. Since enzymes used for technical purposes are marketed as rather crude protein solutions or precipitates, the quality is strongly influenced by the fermentation method. Choice of raw materials, sterilization method, foam-control method, aeration and agitation rates, and fermentation time affect not only the product recovery yield and cost but also color, smell, stability, powder properties, and similar quality parameters.

A general kinetic model of enzyme synthesis has not been developed, but the regulation of the formation of many specific enzymes has been studied. A few enzymes used commercially are formed during exponential growth, but most are formed in the postexponential growth phase. As mentioned in the preceding sections, many factors other than growth control enzyme formation. A stoichiometric relationship for the process cannot be expressed. The yield of useful enzyme protein may reach 1–5% of the initial medium dry substance. The cell yield in a typical enzyme fermentation may be 2–10% on a similar basis. Residual nutrients and metabolites usually constitute 5–10% of the broth at the end of a fermentation.

IV. RECOVERY AND FINISHING

Methods for the recovery of enzymes for use commercially are simple unit operations such as centrifugation, filtration, vacuum evaporation, and precipitation of proteins. The complications in the processes arise from the character of the fermented broth. The broth has a variable, unspecified composition, a high content of colloidal material, and often a high viscosity.

The most significant development in recent years has taken place in the finishing processes for solid enzymes. Ten years ago, commercial products were dusty powders, but today the majority of solid enzymes are supplied as dustless granulates or as immobilized enzymes.

A. Cooling and Pretreatment of the Broth

After termination of the fermentation, the broth is rapidly cooled from the fermentation temperature of 30°–50°C to about 5°C. The culture broth is an excellent medium for a wide variety of microorganisms. Thorough cooling of the broth and strict hygienic measures in the plant are the most important methods to control infections during the recovery process. The addition of preservatives is normally unacceptable because many enzyme products are used in food processing. Also, many preservatives may effect the biological treatment of the wastewater from the recovery plant.

The refrigerated broth is normally pretreated before the separation processes—filtration or centrifugation. The character of pretreatment depends on the type of broth and on the equipment used for recovery.

Broth fermented by filamentous organisms can often be filtered or centrifuged directly after adjustment of pH to the stability optimum of the enzyme.

Broth from bacterial fermentations is more difficult to process. Fermentations on rich media and fermentations that leave large amounts of colloidal particles in the broth present the largest problem to the recovery plant. In these cases pretreatment with a coagulating or flocculating agent is needed. Inorganic salts, e.g., calcium phosphate precipitated in the broth from soluble phosphate and calcium salts, may be used to enclose cells and colloids in a precipitate. More efficient is a flocculation using synthetic polyelectrolytes which are available in several types, ranging from strongly cationic over nonionic to anionic. Each type can be used alone or in combination with another type or with inorganic chemicals. The synthetic polyelectrolytes are large molecules often based on polymerized ethyleneimine, acrylates, or acrylamide.

A special pretreatment step prior to a filtration may, for instance, be the addition of 2–4% diatomaceous earth to the broth as a body feed.

B. Separation of Solids

Most enzymes of industrial interest are extracellular, i.e., excreted from the cells into the liquid. The first task in recovery of an extracellular enzyme is to remove cells and other suspended material from the broth.

During the past decade there have been few developments in equipment used for this purpose. The main alternatives are still vacuum drum filters and centrifuges. The disk-type centrifuge with a self-cleaning bowl is the one mainly used. The use of vacuum drum filters precoated with diatomaceous earth or expanded perlite are particularly widespread in the industry.

C. Purification

The purification of the enzyme after the initial separation step may be more or less extensive, depending on the intended use of the product.

Figure 2 shows examples of purification steps leading to commercial-grade enzyme products.

After a polishing filtration, the solution is concentrated by vacuum evaporation at low temperature or by ultrafiltration. Ultrafiltration, although a relatively new technique, has been developed to a degree which makes it suitable for large-scale application. Compared with vacuum evaporation, ultrafiltration has the advantage that it removes smaller molecules, e.g., molecules of a MW below 10,000, from the concentrate. Consequently it is often possible to make a more concentrated enzyme solution by ultrafiltration than by vacuum evaporation. One drawback, however, is that the membranes are easily clogged by precipitates which may form during the concentration of the solution.

The concentrated enzyme solution can be clarified by a polishing filtration, and the remaining germs can be removed by a sterile filtration on cellulose-containing filter pads.

Stabilizers or preservatives may be added before or after the sterile filtration. Typical stabilizers are calcium salts, proteins, starch hydrolyzates, and sugar alcohols. Microbial stability of a liquid product may be secured by addition of 18–20% NaCl or by the use of a food-grade preservative, such as benzoate, parabens, or sorbate.

There is a preference of liquid rather than solid enzyme products, because the liquid products are cheaper to produce and are also safer and more convenient to apply. In some cases, however, only a solid preparation can be used. Solid enzyme concentrates can be obtained by precipitation or by direct spray-drying of the enzyme solution. It is often impossible to spray-dry a solution concentrated by evaporation, but it can usually be done on an ultrafiltered solution.

A higher degree of purity is obtained when the enzyme is precipitated with acetone, alcohols, or inorganic salts, such as ammonium sulfate or sodium sulfate. Fractional precipitation yields a higher purity than a one-step precipitation. In large-scale operations salts are preferable to organic solvents to eliminate the possibility of explosion. The precipitate may be dried by freeze-drying, vacuum-drying, or spray-drying depending on the heat stability of the enzyme.

D. Intracellular Enzymes

Recovery of intracellular enzymes is more difficult. The yield of enzyme on a total broth basis is often less than for extracellular enzymes because the cell mass is limiting.

To harvest the cells, centrifugation is usually preferred to filtration to avoid mixing the cells with diatomaceous earth. In industry the cells are often used without extraction of the enzyme, thereby saving activity and

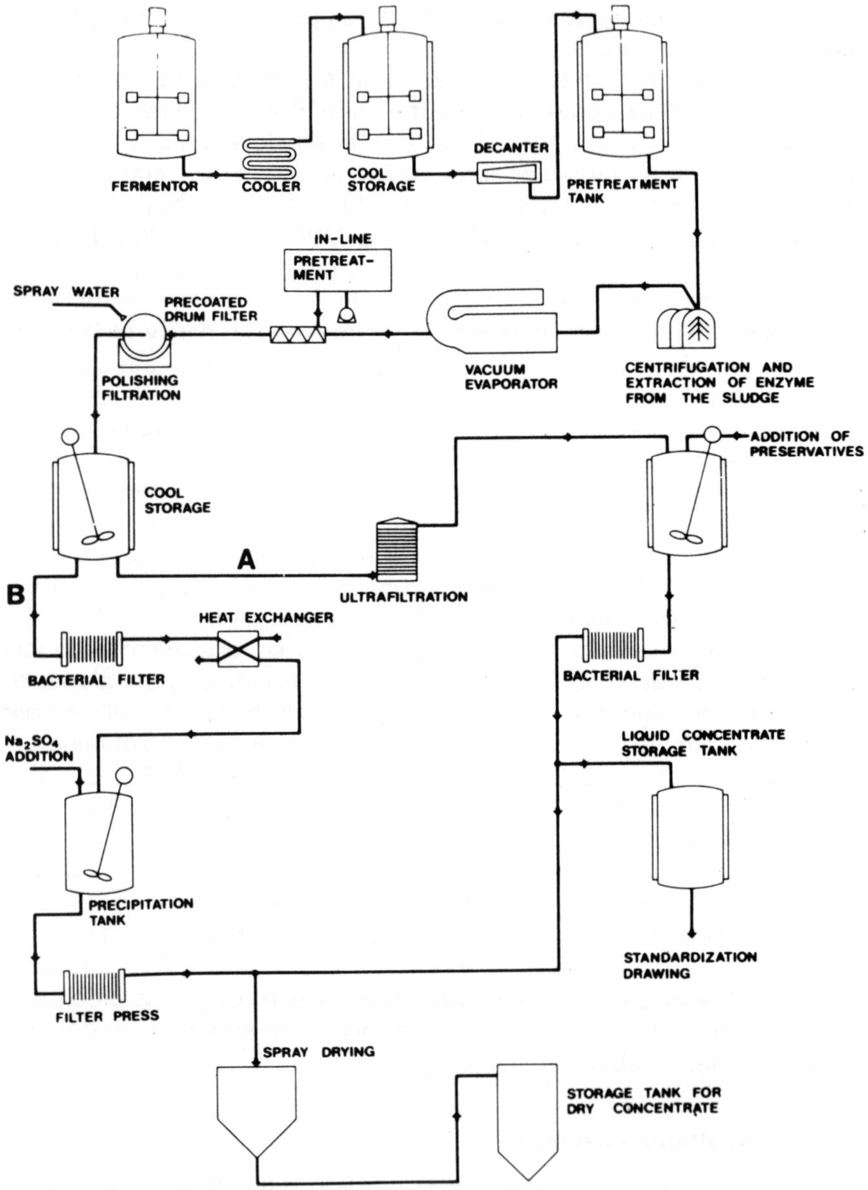

FIGURE 2. Examples of enzyme recovery processes.

expensive purification steps. If the enzymes must be extracted, autolysis may be used, but on an industrial scale physical methods are preferred, e.g., disruption of the cells by means of a homogenizer or a bead mill (Melling and Phillips, 1975). When the cells have been broken, the enzyme can be purified as are extracellular enzymes, but the process is usually more difficult due to the content of cell debris and nucleic acids from the broken cells.

E. Finishing of Solid Enzymes

A few years ago, all solid enzymes were sold as dusty powders with a small particle size. A significant proportion of the particles were smaller than 10 μm. This type of product requires very careful handling in order to avoid exposure to enzyme dust. Problems in handling and admixture of the powders, especially in the detergent industry, led to the development of methods to convert the enzyme concentrate to dustless granulates. An additional advantage of the granulates is often improved storage stability.

In the field of detergent enzyme granulation several methods have been developed, but on an industrial scale only a few are applied. A common method is to embed the enzyme into spheres of a waxy material consisting of a nonionic surfactant by means of a spray-cooling or prilling process. The "marumerizer" process is another important method by which the enzyme is mixed with a filler, a binder, and water; then extruded; and subsequently formed into spheres in a so-called marumerizer. After drying the spheres are coated with a layer of waxy material (see Fig. 3). Enzyme granules produced by these methods are rigid and have a very low dust level. The particles have an average diameter of about 500 μm, and practically no particles are smaller than 200 μm. Furthermore, the granules may be coated with an inert film. Today most enzymes used in the detergent industry are granulated.

In other applications, where the enzymes normally are used in industrial processes, the handling properties are less critical, and simple powders are still used. One reason is that the additives used in the granulates for the detergent industry are unacceptable in many other applications. A significant improvement in handling properties may be obtained, however, by a relatively simple granulation step, such as a fluid bed agglomeration.

F. Immobilization of Enzymes

Methods for the immobilization of enzymes have been known for several years, but the technique has gained industrial importance only

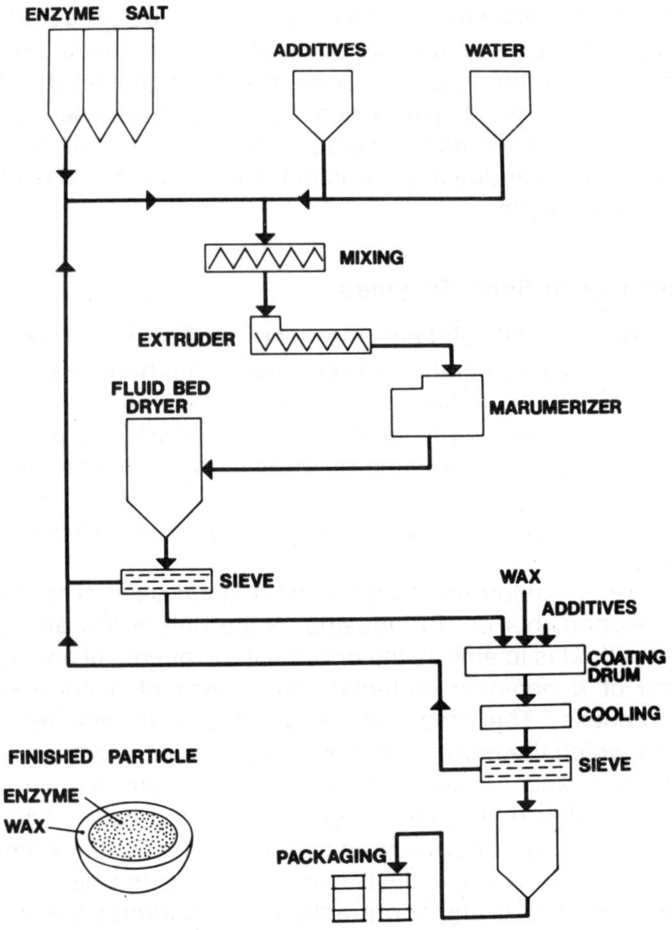

FIGURE 3. Typical flow sheet for the preparation of a dust-free enzyme product.

within the last few years. By far the most significant immobilized enzyme product at present is glucose isomerase, an intracellular enzyme produced by various microorganisms. Several manufacturers use the enzyme-containing cells for immobilization without further purification.

Numerous immobilization methods have been published (Barker and Kay, 1975). In some methods of industrial importance, glutaraldehyde is used to cross-link the enzyme or the supporting protein in which the cells are suspended. According to a method described by Van Velsen (1971), the enzymes or the cells are suspended in an aqueous solution of a gelling protein. The protein is suspended in an organic water-immiscible liquid to produce enzyme-containing droplets. The gelling protein is forced to gel. Subsequently, the particles are cross-linked with glutaraldehyde. Amotz *et al.* (1976) describe a method in which the cells are harvested, concentrated, and homogenized. Glutaraldehyde is added, and the resulting gel is shaped into granules which are dried.

Another attractive immobilization method is the use of polyelectrolytes to flocculate the enzyme-containing cells. After filtration the filter cake is shaped to particles of the desired size and dried (Long, 1974). Other methods are based on the fixation of the extracted enzyme through chemical reaction or adsorption to solid supports, such as ion exchangers, glass beads, or cellulose fibers.

The particles of an immobilized enzyme must be porous, uniform in size, and physically strong in order to perform satisfactorily in a column over a long period of time. Glucose isomerase, for example, is usually used more than 1000 hours at temperatures between 60° and 65°C.

V. REGULATIONS AND SPECIFICATIONS

Because microbial enzymes are natural products, and in view of their traditional use for food purposes in the orient, they were regarded as inherently harmless substances until about 15 years ago. The increasing knowledge of mycotoxins and the allergenic properties of enzymes has made everybody concerned more cautious. Microbial enzyme products must now meet strict specifications with regard to toxicity and other safety aspects.

A. Toxicity

As a rule, enzymes are completely nontoxic. A well-known exception is phospholipase C, which is formed by several pathogenic bacteria. Toxic materials may, however, be present in the raw materials, and theoretically they may be formed during the manufacturing process by the metabolism of the enzyme-generating organism or contaminants. Since most enzyme products undergo a relatively crude purification process, they contain all components of the fermentation broth in small or large quantities. It is necessary, therefore, to ensure that no toxic products are formed at any stage of the process. This is done by prescribing food- or feed-grade raw materials throughout the process, by regularly testing for microbial contamination, and by a thorough investigation of the enzyme-generating organism and of the enzyme product.

This investigation also includes a complete survey of the scientific literature on the organism and closely related organisms, followed by extensive feeding studies on several species of animals. The feeding studies are usually made with a mixture of several batches, e.g., ten batches of enzyme concentrate prepared according to the actual production process. The amounts used are chosen to ensure the highest possible safety factor. Usually the limit is set by the amount of solids that the animals are able to ingest. In some cases a high content of salts, such as NaCl, Na_2SO_4, and phosphates, will set an unacceptably low limit, necessitating desalting of the enzyme preparation before it is fed to the animals.

The preparations will always be tested for aflatoxin. They will be tested also for other mycotoxins if their presence is theoretically possible.

B. Allergenicity

Like all proteins, enzymes are antigenic and may thus cause allergy. The antigenicity varies from one enzyme to another, but no general rules can be given. The antigenicity of the enzyme as such cannot be changed, but the formulation can be made so that exposure of the user of the enzyme is minimized. The most important factor is the level of dust formed by handling the enzyme. To keep it low, the enzyme may be encapsulated in inert material. An encapsulated product may contain less than 0.5 μg of pure enzyme dust per kilogram as determined according to Harris and Rose (1972). Whenever possible, liquid products are preferred because they can easily be handled without exposing personnel to enzyme dust.

C. Microbial Safety

The enzyme-generating strain must be a harmless saprophyte. This is ensured by a literature survey and by animal pathogenicity tests, in which the organism must be administered both orally and by injection. In addition, the organism, as a rule, will be completely removed from the preparations, a precaution that serves the additional purpose of preventing competitors from obtaining the strain.

During the fermentation process contamination tests are regularly performed, and contaminated batches discarded. During recovery and handling of the finished product the hygienic standard must be as high as in the food industry. Nevertheless a certain degree of contamination is unavoidable. Regular controls of the microbial standard of the finished product are therefore necessary.

D. Regulations

Responsibility for the safety of an enzyme product remains with the manufacturer. In most countries, though, a new enzyme product for use in food requires approval by the appropriate authorities. In the United States, approval of the Federal Drug Administration is required. National and international bodies have worked out recommendations and specifications for a number of known enzyme products (Food Chemicals Codex, 1975; World Health Organization, 1972, 1975).

E. Specifications

All technical enzyme preparations are sold on an activity basis. This means that an analytical method must be specified by the manufacturer. Although recommendations for enzyme analysis have been given by the Commission on Enzyme Nomenclature, the different enzyme manufacturers generally use their own method of analysis. Even when activities are given in so-called international units, they should be interpreted with caution, because small differences in reaction conditions may have an important effect on the result. The manufacturer will do his best to reduce the deviation of the analytical method. For most preparations it is possible to give the activity with a standard deviation better than 5%.

Usually the manufacturer will guarantee a certain storage stability under specified conditions. For example, less than 10% loss per 6 months at 25°C for liquid products or less than 10% loss per year for solid preparations. In addition, the gross chemical composition may be given as well as the kind and amount of preservatives added. For special enzymes, such as detergent proteases, the specifications usually include limits for enzyme dust, particle size, bulk density, etc.

Immobilized enzymes present special problems, since performance cannot be judged on activity alone. The activity test must be performed in a column operating over a considerable period of time, and other parameters, such as particle size, particle strength, and color formation, become important.

The customer expects uniform, reliable, and safe enzyme products, and the manufacturer must prepare specifications which ensure that these expectations are met.

VI. SURVEY OF ENZYMES AND APPLICATIONS

A list of the enzymes used commercially is given in Table IV All the enzymes with the exception of two are hydrolases. It can thus be said

TABLE IV. Microbial Enzymes for Industrial Use

Enzyme	Microbial source
α-Amylase	*Aspergillus oryzae*
	Bacillus amyloliquefaciens
	Bacillus licheniformis
β-Amylase	*Bacillus cereus*
	Bacillus megaterium
	Bacillus polymyxa
β-Glucanase	*Aspergillus niger*
	Bacillus amyloliquefaciens
Cellulase	*Aspergillus niger*
	Trichoderma reesei
Dextranase	*Penicillium* sp.
	Trichoderma sp.
Glucoamylase	*Aspergillus awamori*
	Aspergillus niger
	Rhizopus sp.
Glucose isomerase	*Actinoplanes* sp.
	Arthrobacter sp.
	Bacillus sp. (thermophilic)
	Streptomyces sp.
Glucose oxidase	*Aspergillus niger*
Invertase	*Saccharomyces* sp.
Lactase	*Aspergillus niger*
	Kluyveromyces fragilis
	Kluyveromyces lactis
Lipase	*Aspergillus* sp.
	Candida lipolytica
	Geotrichum candidum
	Rhizopus sp.
Pectinase	*Aspergillus niger*
	Aspergillus sp.
Penicillin acylase	*Bacillus megaterium*
	Erwinia carotovorum
	Escherichia coli
Pentosanase	*Aspergillus* sp.
	Bacillus amyloliquefaciens
Protease, acid	*Aspergillus* sp.
Protease, neutral	*Aspergillus oryzae*
	Bacillus amyloliquefaciens
	Bacillus thermoproteolyticus
Protease, alkaline	*Aspergillus oryzae*
	Bacillus amyloliquefaciens
	Bacillus licheniformis
	Bacillus sp. (alkalophilic)
	Streptomyces griseus
Pullulanase	*Klebsiella aerogenes*
Rennet	*Endothia parasitica*
	Mucor miehei
	Mucor pusillus

that enzyme technology so far has been concentrated on the relatively simple hydrolytic processes, whereas the more complicated enzymatic reactions have not been possible on a technical scale. Most of the more complicated systems require co-enzymes, and this has deterred their use since an economical way of applying them has not been found. The explanation may be that there is no demand for a large-scale application of such complicated systems.

A. Microbial Sources of Technical Enzymes

The majority of industrially produced enzymes are derived from microorganisms of two genera, *Aspergillus* and *Bacillus*. It is characteristic of these microorganisms that they are common, harmless saprophytes; they grow rapidly, are metabolically very active, and they are highly variable. They secrete a number of extracellular enzymes in relatively large quantities. Usually several enzymes are produced at the same time.

Bacillus amyloliquefaciens produces α-amylase, β-glucanase, and neutral and alkaline protease. *Bacillus licheniformis* normally produces an almost pure serine protease with minimal amounts of α-amylase. However, special strains that produce high amounts of α-amylase together with the protease have been developed. Although the taxonomic distinction between the two species seems pretty clear, many manufacturers and scientists unfortunately still prefer to name both strains *B. subtilis*.

Aspergillus oryzae, when grown in semisolid culture, will produce a large number of enzymes, primarily α-amylase, lactase, glucoamylase, and protease. In submerged culture the α-amylase formation is increased, whereas the formation of other enzymes becomes minimal. *Aspergillus niger* strains are used in the production of glucoamylase, pectinase and protease. Special strains are developed for each of these purposes, but it is characteristic that all strains produce most of the enzymes in varying amounts. The nomenclature of the *Aspergilli* is complicated and unclear. *Aspergillus niger* and *A. oryzae* should here be understood as representing the groups of black and green *Aspergillus* species, respectively.

Members of the family *Mucoraceae* have frequently been used for enzyme production, e.g., the use of *Rhizopus* sp. for the manufacture of α-amylase, glucoamylase, protease, and lipase. These organisms are still important enzyme sources in the manufacture of oriental foods but their use in the production of enzymes for technical use is limited. The two thermophilic *Mucor* species, *M. miehei* and *M. pusillus*, are important in the manufacture of microbial rennet.

The *Actinomycetales* have become important sources of glucose isomerase which is produced commercially using several *Streptomyces* species and one *Actinoplanes* species. Few other microbial species are used for commercial enzyme production (Table IV).

B. Applications

Enzymes offer an attractive solution to many catalytic problems, especially within the food industry, because of their specificity, nontoxicity, and mild reaction conditions. Only the most important processes are described here. A list of selected publications on the application of enzymes is given at the end of this chapter.

1. Enzymes for the Detergent Industry

Proteinaceous dirt will often precipitate on soiled clothes, or it may coagulate during the normal washing process. Furthermore, proteins make dirt adhere to the textile fibers. Such stains, which are otherwise difficult to remove, can be dissolved easily by addition of proteolytic enzymes to the detergent.

The alkaline serine protease of *B. licheniformis*, otherwise known as subtilisin Carlsberg, is the preferred protease for this purpose. The enzyme is well suited for this application. It will attack many peptide bonds and therefore easily dissolve proteins. It may be used in the presence of most nonionic and anionic detergents at temperatures up to 65°C, and its pH optimum is close to 9, the pH normally used in washing fluids. Furthermore, extensive tests have proved the enzyme harmless to the user.

A few other proteases are used in detergents. The serine protease of *B. amyloliquefaciens* has found some application, presumably because the preparations usually have a substantial content of α-amylase. For some applications this may be an advantage. Proteases with improved stability and protein solubilizing properties under washing conditions have been found in alkalophilic bacilli (Aunstrup *et al.*, 1972). These enzymes are particularly advantageous in detergents with low phosphate content.

The success of the detergent proteases has inspired numerous attempts to use enzymes for removal of other stubborn stains, such as fruit colors and lipids, especially in low-temperature washing. So far no satisfactory solution to these problems has been found.

2. Enzymes for the Starch Industry

Until around 1960, dextrose and glucose syrups were prepared from starch exclusively by acid hydrolysis. The process was corrosive, and the dextrose yield low. It was, therefore, a great step forward when pure glucoamylase in combination with bacterial α-amylase made possible a complete enzymatic hydrolysis of starch to dextrose. Today several enzymatic processes are used in the industry. A schematic summary is given in Table V. Most of the processes are applied in combination, starting with a liquefaction followed by a more or less extensive saccharification depending on the desired end product.

Prior to liquefaction the starch should be gelatinized by heat treatment. For the widely used maize starch, temperatures above 100°C are normally applied in order to secure complete gelatinization. It is consequently an advantage to use the extremely heat-stable amylase of *B. licheniformis* which can function at temperatures up to 110°C. Normally high concentrations (30–50% dry substance) are used in the saccharification process. The equilibrium concentrations of the saccharides formed by resynthesis limit the maximum degree of hydrolysis obtainable, e.g., 97% dextrose at 30% dry matter. Since the activity of glucoamylase toward branching points (α-1,6 bonds) is low, it may be an advantage to use a debranching enzyme, such as pullulanase, early in the hydrolysis process. In the preparation of fructose syrup, a fructose concentration of minimum 42% (on solids) is normally sought. Since this is close to the equilibrium concentration (around 50%), it is important that the initial glucose syrup be as pure as possible, i.e., that the hydrolysis be complete. Glucose isomerase is used in immobilized form, and a major share of the fructose syrup production is made by continuous column processes (Hilmer Nielsen *et al.*, 1976).

Since the processes described here are directly dependent on the enzyme activities, it is important that the enzymes be pure, i.e., free from undesired enzyme activities, such as transglucosidase or protease, and that the activity be accurately standardized. Liquid preparations of amylase and glucoamylase are used, whereas the immobilized glucose isomerase comes in various forms, from frozen cells to dried granules. For column operation, mechanical strength, flow properties, and active lifetime are important parameters.

3. Enzymes for the Dairy Industry

In the dairy industry microbial rennet has to a large extent replaced the expensive rennet from calves. The microbial rennets are acid aspartate

TABLE V. Examples of Enzymatic Processes in the Starch Industry

Process	Enzyme	Source	Reaction conditions pH	Reaction conditions Temperature °C	Product
Liquefaction	α-Amylase	*B. amyloliquefaciens*	5.5–7	90	Maltodextrins
		B. licheniformis	5.5–9	110	DE 10–20
Debranching	Pullulanase	*K. aerogenes*	6–7	50–60	Intermediate process in the manufacture of dextrose
Saccharification	α-Amylase	*A. oryzae*	5–7	50–55	High-maltose syrup, high-DE syrup,
Saccharification	Glucoamylase	*A. niger* (*Rhizopus* sp.)	4–5	55–60	High-DE syrup, cryst. dextrose
Isomerization	Glucose isomerase	(*Streptomyces* sp.), *B. coagulans*, *Actinoplanes* sp.	6.5–8.5	60–65	Fructose syrup

proteases. The reaction mechanism with casein is closely related to that of calf rennet. However, minor differences in the enzyme properties, such as dependency on Ca^{2+}, temperature, and pH, necessitate small adjustments of the cheese-making procedure according to the enzyme used. When the proper procedures are applied, the quality of cheese prepared with microbial rennet equals that of cheese prepared with traditional rennet.

Lactase for the hydrolysis of lactose in whey or milk is gaining increasing importance. Lactase from *Kluyveromyces fragilis* or *Kluyveromyces lactis* is usually used, although the properties of these enzymes are not ideal because of a narrow pH optimum and a low maximum temperature.

Lipase for flavor development in special cheeses (romano, provolone) is an interesting application, and microbial esterases for this purpose are presently being developed.

4. Other Applications

In the textile industry, α-amylase has been used for many years to remove starch sizes. This application is still an important outlet for bacterial α-amylase. An improved, rapid, high-temperature process is possible with the thermostable amylase from *B. licheniformis*.

Bacterial α-amylase is also used in the brewing industry for the liquefaction of brewing adjuncts. The substitution of barley for malt requires three enzymes, α-amylase, β-glucanase, and protease. All are produced by *B. amyloliquefaciens*. It is noteworthy that only the neutral protease is of importance because the alkaline protease is inhibited by an inhibitor in barley. β-Glucanase is often used alone to alleviate filtration problems due to poor malt quality or precipitation of glucans during the fermentation.

Substitution of amylase and glucoamylase for malt in the distilling industry makes high-temperature liquefaction possible and renders the process more reproducible.

The use of pectic enzymes in the wine and juice industry improves the yield and the quality of the products. Empirical methods and experience are especially important in the production and use of pectinase products. They contain the enzymes pectinase, pectin esterase, transeliminase, and polygalacturonidase. In addition, several nonpectolytic enzymes, such as protease, pentosanase, and glucoamylase, are present in small quantities. The combined action of several of these enzymes often decides its application.

Alkaline protease from alkalophilic bacteria may be used for dehairing of hides. The process functions well and is considerably more pleasant, both to workers and to the environment, than the usual sulfide process.

The method is little used, partly because it requires strict control of the process conditions, and partly for economic reasons.

VII. CONCLUSION

The industrial use of enzymes is insignificant and the processes are unimportant when compared with the overwhelming magnitude and importance of the enzymatic processes performed in nature. Some of these processes, such as lignocellulose hydrolysis, are relatively simple and seem to be technologically possible within a few years. Others, such as nitrogen fixation, are extremely complex, and their industrial realization is unthinkable with present technology.

Solution to such important problems as these will require systems of a higher degree of complexity than are so far known in the enzyme technology. The development will include multienzyme systems, free or immobilized, and in most cases coupled to cofactors. The possibilities are legion but the research effort necessary is some orders of magnitude larger than that for the simple problems so far attempted by the industry.

There is no doubt that industrial evolution will favor the use of enzymes. Their specificity, mild reaction conditions, and nontoxicity make them ideal catalysts in a world which becomes more and more conscious of pollution and energy waste.

REFERENCES

Amotz, S., Nielsen, T. K., and Thiesen, N. O. (1976). U.S. Patent 3,980,521.

Aunstrup, K. (1978). *In* "Biotechnology and Fungal Differentiation" (J. Meyrath and J. D. Bu'lock, eds), p. 157. Academic Press, New York.

Aunstrup, K., Outtrup, H., Andresen, O., and Dambmann, C. (1972). *Ferment. Technol. Today, Proc. Int. Ferment. Symp., 4th, 1972* pp. 299–305.

Barker, S. A., and Kay, I. (1975). *In* "Handbook of Biotechnology" (A. Wiseman, ed.), pp. 89–95. Halsted Press, New York.

Diers, I. (1976). *In* "Continuous Culture 6: Applications and New Fields" (A. C. R. Dean *et al.*, eds.), pp. 208–225. Halsted Press, New York.

Fabian, J. (1969). *In* "Continuous Cultivation of Microorganisms" (I. Málek, ed), pp. 489–495. Academia, Prague.

Food Chemicals Codex (1975). 2nd ed. Natl. Acad. Sci., Washington, D.C.

Harris, R. G., and Rose, T. J. (1972). British Patent 1,343,963.

Heady, R. E. (1971). British Patient 1,232,130.

Heineken, F. G., and O'Connor, R. J. (1972). *J. Gen. Microbiol.* **73,** 35–44.

Hilmer Nielsen, M., Zittan, L., and Hemmingsen, S. H. (1976). *In* "Chemical Engineering in a Changing World" (W. T. Koetsier, ed.), pp. 183–198. Elsevier, Amsterdam.

Hulme, M. A. (1973). U.S. Patent 3,734,831.

Jensen, D. E. (1972). *Biotechnol. Bioeng.* **14,** 647–662.
Kalaboklas, G. (1971). U.S. Patent 3,623,956.
Long, M. E. (1974). U.S. Patent 3,935,069.
Melling, J., and Phillips, B. W. (1975). *In* "Handbook of Enzyme Biotechnology" (A. Wiseman, ed.), pp. 63–64. Halsted Press, New York.
Meyrath, J., Volavsek, G., and Stahl, U. (1973). *Biotechnol. Bioeng. Symp.* **4,** 257.
Mitra, G., and Wilke, C. R. (1975). *Biotechnol. Bioeng.* **17,** 1–13.
Paigen, K., and Williams, B. (1970). *Adv. Microb. Physiol.* **4,** 251–324.
Smith, J. A., and Frankiewicz, J. R. (1978). British Patent 1,526,237.
Van Velsen, A. G. (1971). U.S. Patent 3,838,007.
World Health Organization (1972). "Food Additives Series," No. 2. WHO, Geneva.
World Health Organization (1975). "Food Additives Series," No. 6, WHO, Geneva.

GENERAL REFERENCES

de Becze, G. I. (1970). *Crit. Rev. Food Technol.* **1,** 479–518.
Meltzer, Y. L. (1973). "Encyclopedia of Enzyme Technology." Future Stochastic Dynamics Inc., New York.
Reed, G. (1975). "Enzymes in Food Processing," 2nd ed. Academic Press, New York.
Spencer, B., ed. (1974). "Industrial Aspects of Biochemistry." Vol. 30, Part I. Am. Elsevier, New York.
Whitaker, J. R. (1974). "Food related Enzymes," Adv. Chem. Ser. No. 136. Am. Chem. Soc., Washington, D.C.
Wiseman, A., ed. (1975). "Handbook of Enzyme Biotechnology." Halsted Press, New York.

Chapter 10

Microbial Production of Nucleosides and Nucleotides

YOSHIO NAKAO

MICROBIAL TECHNOLOGY, 2nd ed., VOL. I

ISBN 0-12-551501-4

I. INTRODUCTION

In Japan, *konbu* (kelplike seaweed) and *katsuobushi* (dried bonito) have been widely used as food seasoning substances from the early seventeenth century. The flavor-enhancing material in konbu was identified as sodium glutamate by Ikeda in 1908, and that of katsuobushi, as histidine salt of inosinic acid (IMP) by Kodama in 1913. The industrial manufacturing method for sodium glutamate was established immediately following the discovery by Ikeda. First, sodium glutamate was produced by the acid hydrolysis of plant proteins. In 1957, a process for the fermentative production of L-glutamic acid was developed in Japan by Kinoshita *et al.* L-Glutamic acid is now industrially produced, mainly by the fermentation process, worldwide.

In the 1950s, a wide range of research was initiated for the development of methods for industrial production of 5′-IMP when knowledge of the biosynthesis of nucleic acids, its regulation, and enzymatic degradation of nucleic acids was accumulating and analytical methods for nucleosides, nucleotides, and nucleic acids were becoming established. At that time applied microbiologists were stimulated by the success in glutamic acid fermentation and were aware that they could contribute to the development of processes for the fermentative production of other useful substances.

Techniques for the derivation of various mutants were being developed. Mutants useful for the industrial production of primary metabolites had to possess the following properties: high productivity of metabolites, lack of specific enzyme in the biosynthetic pathway, release from metabolic regulation, and altered membrane permeability.

Attempts to produce 5′-IMP and 5′-guanylic acid (GMP) were borne on the background of growing concepts and techniques in applied microbiology and increasing knowledge and methods in nucleic acid biochemistry and biochemical genetics.

In 1960, the manufacturing method of 5′-IMP and 5′-GMP by the enzymatic hydrolysis of ribonucleic acid (RNA) was established in Japan. Several years later, a process for fermentative production of 5′-IMP and 5′-GMP was developed. The industrial production of 5′-IMP and 5′-GMP is at present carried out by both hydrolysis of yeast RNA and direct fermentation.

Observation of the synergy of flavor between sodium glutamate and the sodium salt of 5′-IMP or 5′-GMP has resulted in a rapid enlargement of the market size of complex food seasonings (a mixture of sodium glutamate, 5′-IMP and 5′-GMP) in which 5′-IMP and 5′-GMP are usually added at levels of 8–12% or 1.5–2.0%.

Microbial technology established for the production of 5′-IMP and 5′-GMP has been applied to develop processes for the production of nucleic acid-related substances used for medicinal purposes. Amounts and uses of nucleic acid-related substances produced in Japan are shown in Table I (Yamada, 1976).

In this chapter, production methods of nucleosides and nucleotides are described with particular emphasis on the industrial manufacturing of 5′-IMP and 5′-GMP. Preparation of adenosine triphosphate (ATP), nicotinamide adenine dinucleotide (NAD), flavin adenine dinucleotide (FAD), coenzyme A (CoA), adenosine 3′,5′-cyclic phosphate (cAMP), and guanosine 5′-triphosphate 3′-diphosphate (pppGpp) is also described.

II. CLASSIFICATION OF METHODS FOR PRODUCTION OF 5′-IMP AND 5′-GMP

Nucleotides are the building units of polymerized nucleic acids which are present in all living organisms: viruses, microorganisms, and plant and animal cells. They also occur in the nucleotide pool as free or coenzyme forms. Therefore, nucleotides can be produced by the methods listed below:

1. Enzymatic or chemical degradation of nucleic acids contained in organisms
2. Direct fermentation using mutants which are genetically blocked in

TABLE I. Outputs and Utilization of Nucleic Acid-Related Compounds[a]

Compound	Amount (tons/year)	Use
5′-IMP	2000	Seasonings
5′-GMP	1000	Seasonings
Inosine	25	Drug (coronary insufficiency, myocardial infarction)
ATP	6	Drug (metabolic disorders, muscular dystrophy)
CDP-choline	2	Drug (clouding of consciousness, accompanying head trauma or cerebral surgery)
FAD	Trace	Drug (hepatic or renal diseases)
NAD	Trace	Drug (hepatic or renal diseases)
5′-AMP	?	Drug (circulatory disorders, rheumatism)
Orotic acid	?	Drug (hepatic diseases)

[a] From Yamada (1976)

the biosynthetic pathway of nucleic acids and which are released from feedback regulation by end products

3. Extraction of free nucleotides existing in the pool

Method (1) is broken down into three procedures: (a) Hydrolysis of yeast RNA by microbial enzymes; (b) degradation of microbial endogenous RNA by endogenous enzymes; and (c) chemical hydrolysis of yeast RNA into nucleosides and its chemical phosphorylation.

Method (2) is broken down into four procedures: (a) Fermentative production of nucleosides and their phosphorylation; (b) fermentative production of 5′-nucleotides; (c) salvage synthesis of nucleosides from bases and phosphorylation of the nucleosides; and (d) salvage synthesis of 5′-nucleotides from bases.

Methods (1a), (1c), (2a), and (2b) are now industrially used for the production of 5′-IMP and 5′-GMP and are described in detail in the sections below.

III. PRODUCTION OF 5′-IMP AND 5′-GMP BY HYDROLYSIS OF RNA

A. Production of 5′-IMP and 5′-GMP by Hydrolysis of RNA with Microbial Enzymes

Industrial production of 5′-IMP and 5′-GMP started in Japan in 1961 using the method of hydrolysis of RNA by microbial enzymes. The manufacturing process includes (1) production of yeast cells with high RNA content, (2) extraction of RNA from yeast cells, (3) production of RNA-hydrolyzing enzymes, (4) enzymatic hydrolysis of RNA, and (5) recovery and purification of 5′-IMP and 5′-GMP from the enzymatic digest. Discovery of microbial nucleolytic enzymes which yield 5′-mononucleotides as a degradation product of RNA was an important factor which stimulated the establishment of this manufacturing method. The properties of nucleolytic enzymes and their producing organisms are described in the next section.

1. Nucleolytic Enzymes and Their Producing Microorganisms

As shown in Table II nucleolytic enzymes are classified on the basis of the substrates and degradation products of the enzymes. Figure 1 shows the scheme of production of 5′-IMP and 5′-GMP from RNA. Enzymatic formation of 5′-IMP and 5′-GMP requires two reactions: Hydrolysis of RNA into 5′-mononucleotides and deamination of 5′-adenylic acid

TABLE II. Classification of Nucleolytic Enzymes

Enzyme				Substrate	Product
Nucleolytic enzyme	(A) Nuclease	Endonuclease	RNase	RNA	Oligonucleotide with 3′-terminal phosphate; Oligonucleotide with 5′-terminal phosphate
			Nuclease	RNA, DNA	Oligonucleotide with 3′-terminal phosphate; Oligonucleotide with 5′-terminal phosphate
			DNase	DNA	Oligonucleotide with 3′-terminal phosphate; Oligonucleotide with 5′-terminal phosphate
		Exonuclease		RNA, DNA	3′-Mononucleotide; 5′-Mononucleotide
				DNA	5′-Mononucleotide
	(B) Polynucleotide phosphorylase			RNA	Nucleoside diphosphate

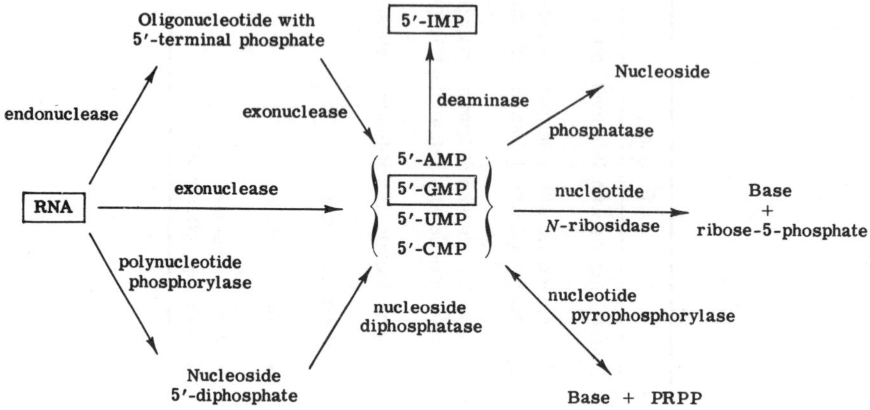

FIGURE 1. Enzymes participating in the production of IMP and GMP by RNA hydrolysis.

(AMP) to 5′-IMP. Therefore, microorganisms capable of producing these two activities should be screened first. In the 1950s, phosphodiesterases of snake venom (Cohn and Volkin, 1953) and calf intestinal mucosa (Cohn and Volkin, 1951) were known to form 5′-mononucleotides from RNA. These enzyme sources, of course, were not applicable for industrial purposes. Then, Kuninaka and his co-workers (1959, 1961) and Ogata and his co-workers (1963) searched for microbial enzymes which hydrolyze RNA to 5′-mononucleotides. Fungi, such as *Penicillium citrinum*; actinomycetes, such as *Streptomyces aureus;* bacteria; and yeasts were found to produce enzymes fitting their purposes. Table III lists those microorganisms capable of hydrolyzing RNA to 5′-nucleotides. The enzymological properties of these enzymes have been extensively studied and the enzymes have been classified according to the type of hydrolysis (Table IV).

2. Production of RNA

Economical production of RNA has been pursued since 1960, when the demand for RNA as raw material for 5′-mononucleotide production increased. RNA is contained in all living organisms and microorganisms would be considered to be the most suitable source of RNA. The RNA content of microorganisms is 5–25% in bacteria, 2.5–15% in yeasts, and 0.7–28% in fungi (Wade and Morgan, 1957; Ogata and Imada, 1962; Belozersky and Spirin, 1960; Belozersky, 1947; Di Carlo and Schultz, 1948; Vendrely, 1955; Kuroiwa and Horie, 1955; Miwa, 1959). Among microorganisms, yeasts have been considered to be the best RNA

TABLE III. Microorganisms Capable of Hydrolyzing RNA to 5′-Nucleotides

Microorganism	References
Mold	
Ascomycetes	
Aspergillus	Ogata *et al.* (1963); Nakao and Ogata (1963)
Chaetomidium	Ogata *et al.* (1963)
Glomerella	Ogata *et al.* (1963)
Monascus	Saruno *et al.* (1964); Soeda *et al.* (1968)
Neurospora	Ogata *et al.* (1963)
Penicillium	Kuninaka *et al.* (1959, 1961)
Sordaria	Ogata *et al.* (1963)
Fungi imperfecti	
Acrocylindrium	Ogata *et al.* (1963)
Fusarium	Ogata *et al.* (1963)
Gliomastix	Ogata *et al.* (1963)
Helminthosporium	Ogata *et al.* (1963)
Phoma	Tone *et al.* (1964); Tone and Ozaki (1968)
Verticillium	Ogata *et al.* (1963)
Basidiomycetes	
Pellicularia	Hasegawa *et al.* (1964); Fujimura *et al.* (1964, 1967)
Actinomycetaceae	
Streptomyces	Ogata *et al.* (1963); Sugimoto *et al.* (1964a,b)
Bacteria	
Bacillus	Ogata *et al.* (1963)
Micrococcus	Berry and Campbell (1967a,b)
Yeast	
Rhodotorula	Nakao *et al.* (1963)

source because their nucleic acids consist mainly of RNA, with very limited deoxyribonucleic acid (DNA) content.

Moreover, yeast cells are easily collected by yeast separators and many strains of the yeasts can grow on cheap carbon sources, such as molasses or sulfite waste liquor. *Candida utilis,* among the yeasts, is most widely used as an RNA source on the industrial scale. The RNA content of yeast cells can be increased by optimization of culture condition. Mutants with high RNA content were also isolated (Akiyama *et al.*, 1975). As in bacteria, the RNA content of yeasts is high at their logarithmic growth phase with active protein biosynthesis (Katchman and Fetty, 1955; Chayen *et al.*, 1955). RNA content in bakers' yeasts increased in media containing higher ammonium and phosphate con-

TABLE IV. Microorganisms Producing 5′-P-Forming Nucleolytic Enzyme[a]

Nucleolytic enzyme				
Enzyme	Substrate	Product	Microorganism	References
RNase	RNA	Oligonucleotides with 5′-terminal phosphate	*Bacillus* AU2	Jacobsen and Rodwell (1972)
			Escherichia coli	Spahr and Schlessinger (1963); Spahr (1964)
			Lactobacillus plantarum	Logan and Singer (1968)
			Lactobacillus casei	Keir *et al.* (1964)
			Physarum polycephalum	Hiramaru *et al.* (1969a,b)
Endonuclease	RNA DNA	Oligonucleotides with 5′-terminal phosphate	*Azotobacter agilis*	Stevens and Hilmoe 1960a,b)
			Serratia marcescens	Nestles and Roberts (1969a,b)
			Streptomyces aureus	Yoneda (1964a,b)
			Saccharomyces fragilis	Nakao *et al.* (1968): Lee *et al.* (1968)
			Acrocylindrium sp.	Suhara and Yoneda (1973); Suhara (1973)
			Aspergillus oryzae	Uozumi *et al.* (1968a,b,c, 1969a,b,c, 1972); Hino *et al.* (1971); Uozumi and Arima (1974)
			Neurospora crassa	Linn and Lehman (1965a,b, 1966)
Exonuclease	RNA Oligonucleotide	5′-Mononucleotide	*Micrococcus sodonensis*	Berry and Campbell (1967a,b)
			Streptomyces aureus	Nakao (1976)
			Streptomyces sp. No. 41	Sugimoto *et al.* (1964a,b)
			Aspergillus oryzae	Ando (1966) Shishido and Ikeda (1970, 1971a,b) Shishido and Ando (1972)
			Aspergillus quercinus	Ohta and Ueda (1968)
			Acrocylindrium sp.	Suhara (1974)
			Monascus purpureus	Saruno *et al.* (1964); Soeda *et al.* (1968, 1969)
			Penicillium citrinum	Fujimoto *et al.* (1969, 1974a—d, 1975a,b); Kuninaka *et al.* (1975a,b)
			Phoma cucurbitacearum	Tone *et al.* (1964); Tone and Ozaki (1968)
			Physarum polycephalum	Hiramaru *et al.* (1969a,b)

[a] Enzymes hydrolyzing nucleic acid to mono- and/or oligonucleotides terminated by 5′-phosphate.

centrations (Di Carlo and Schultz, 1948). Addition of copper, zinc, and ferrous ions to the culture medium greatly stimulated RNA formation in *C. utilis*; the RNA content reached about 11% in media containing more than 0.25 ppm zinc ion (Yamashita *et al.*, 1974). Several antibiotics, such as blasticidin S (Hashimoto and Okauchi, 1965) and anisomycin (Watanabe, 1976), were observed to increase RNA content in *C. utilis* and *Saccharomyces cerevisiae.* The Waldhof-type fermentor was devised in Germany during World War II in order to economically produce fodder yeast on a large scale. The Waldhof-type fermentor has been used for cultivation of yeasts as an RNA source. Recently, the air-lift-type fermentor (Kanazawa, 1974) was found to be effective for cultivation of yeasts. In Japan, yeasts are now industrially cultivated on sulfite waste liquor or molasses by continuous culture systems. RNA content is about 10 and 15% on the respective carbon sources. Yeasts are separated by yeast separators and dried, and then the RNA is extracted with hot dilute alkaline saline. After yeast cells are removed, RNA in the extract is precipitated by the addition of acid or ethanol. Crude RNA powder is obtained by dehydration of the precipitate. For the production of 5′-IMP and 5′-GMP, such RNA powder is subjected to enzymatic hydrolysis. Fukuda *et al.* (1963) devised the method of direct enzymatic hydrolysis of dried yeast without RNA extraction.

3. Production of Nucleolytic Enzymes

Penicillium citrinum and *S. aureus* are industrially used as the enzyme producers. Each organism has been improved by repeated mutation and selection. The improved mutant of *P. citrinum,* which has lost the ability to form pigment and has acquired the ability to form abundant nuclease, is cultivated on wet wheat bran at 30°C for 5 days. Wheat bran with grown mycelia is extracted with water. The extract contains at least two enzymes: exonuclease, which hydrolyzes RNA into 5′-mononucleotides (Fujimoto *et al.*, 1974a–d), and phosphatase, which dephosphorylates 5′-mononucleotides to nucleosides. Phosphatase should be selectively inactivated by heat treatment before the extract is subjected to hydrolysis of RNA.

In the case of *S. aureus,* improved mutants which contain higher endonuclease and exonuclease and lower phosphatase activities have been isolated. The selected mutant is cultivated at 28°C for 30 hours in a liquid medium with agitation and aeration. The liquid medium contains (%): enzymatic hydrolyzate of starch, 3.0; soybean meal, 2.0; corn steep liquor, 1.0; $(NH_4)_2SO_4$, 0.1; $MgSO_4 \cdot 7\ H_2O$, 0.05; $CaCO_3$, 0.5; soybean oil, 0.005. The culture filtrate contains endonuclease (Yoneda, 1964a,b), exonuclease, adenylic deaminase, and phosphatase (Nakao, 1976).

4. Enzymatic Hydrolysis of RNA and Recovery of 5′-IMP and 5′-GMP

Phosphatase contained in enzyme solutions dephosphorylates 5′-mononucleotides into nucleosides. Therefore, it is highly important to destroy or suppress the activity of phosphatase. Heat treatment of enzyme solutions greatly reduces the phosphatase activity, but complete selective inactivation of phosphatase is difficult. Enzyme solution obtained from wheat bran culture of *P. citrinum* contains exonuclease, which acts optimally at 65°C, and phosphatase, which acts optimally at 45°C. Digestion of RNA with *P. citrinum* enzyme solution at 65°C for about 4 hours at pH 5.0 yields 5′-mononucleotides with little concomitant formation of nucleosides. Since this enzyme solution does not contain adenylic deaminase, the degradation products of RNA are 5′-AMP, 5′-GMP, 5′-cytidylic acid (CMP), and 5′-uridylic acid (UMP). The 5′-AMP in the digest is deaminated into 5′-IMP with adenylic deaminase of *Aspergillus oryzae.*

In the case of hydrolysis with the enzyme solution of *S. aureus,* RNA is hydrolyzed into 5′-mononucleotides by the cooperative action of endo- and exonucleases. The 5′-AMP, once formed, is further deaminated to 5′-IMP by deaminase present in the same enzyme solution. Therefore, as the hydrolysis reaction with a multienzyme system proceeds, suitable conditions for enzyme reaction should be set to produce 5′-IMP and 5′-GMP in maximum yield. Industrial production of 5′-IMP and 5′-GMP by hydrolysis of RNA with the enzyme solution of *S. aureus* is carried out at 0.5–1.0% of RNA and at 42°–65°C for about 10 hours in a pH region of 7.0–8.0.

The 5′-mononucleotides in the digest are recovered by adsorption on activated charcoal at acidic pH, followed by elution with dilute methanol–ammonia solution. 5′-IMP and 5′-GMP can be separated from 5′-CMP and 5′-UMP by activated charcoal column (Tanaka *et al.*, 1964a,b). Each nucleotide can be fractionated with anion- or cation-exchange resins (Tanaka *et al.*, 1963; Ueno *et al.*, 1965). Sanno *et al.* (1964) and Ishibashi *et al.* (1965) proposed a fractional precipitation method for the separation of 5′-purine nucleotides from 5′-pyrimidine nucleotides. Industrially, 5′-IMP and 5′-GMP are recovered by the treatment of the enzymatic hydrolysate with ion-exchange resin and charcoal, followed by the fractional precipitation with a lower alcohol.

B. Formation of 5′-Nucleotides by Degradation of Cellular RNA by Endogenous Enzymes

In the processes described in the previous section, yeast RNA was hydrolyzed with enzymes obtained from other microorganisms. An idea to simplify the process is to digest microbial RNA with endogenous nucleolytic enzymes. Several bacteria and yeasts were found to digest their cellular RNA by their endogenous enzymes and excrete 5′-mononucleotides (Imada, 1976). Such a process has not been adopted for industrial production of 5′-IMP and 5′-GMP. If economical mass production of single-cell protein (SCP) is to be practiced to cover the worldwide shortage of protein sources, it would be necessary to reduce nucleic acid content in single-cell protein. In such a case, the autodigestive process would be useful to achieve both the reduction of nucleic acid content and the production of 5′-nucleotides.

C. Production of Nucleosides by Chemical Hydrolysis of RNA and Their Chemical Phosphorylation

It is well known that 3′(2′)-nucleotides are formed by alkaline hydrolysis of RNA, but 5′-mononucleotides have never been directly formed from RNA by chemical methods. As a result of extensive investigation, production of nucleosides from RNA by chemical processes followed by their phosphorylation have been applied for the industrial manufacturing of 5′-IMP and 5′-GMP.

Production of nucleosides from RNA is composed of two reactions: Formation of 3′(2′)-nucleotides by alkaline hydrolysis of RNA and hydrolysis of the phosphomonoester linkage. These two reactions, however, can be accomplished by a single process. RNA is quantitatively converted into nucleosides by heating RNA at 130°C for 3–4 hours in calcium hydroxide solution (Koshiro *et al.*, 1972). Guanosine in the hydrolysate is easily crystallized due to its low solubility in water. Adenosine, cytidine, and uridine are separated with anion-exchange resin (Numata *et al.*, 1969), and each nucleoside is crystallized by concentration of each fractionated eluate. Adenosine is converted into inosine by deamination with nitrous acid. Inosine and guanosine are phosphorylated to 5′-IMP and 5′-GMP by the chemical method described in later sections.

IV. PRODUCTION OF 5′-IMP AND 5′-GMP BY FERMENTATION

After establishment of the industrial production of 5′-IMP and 5′-GMP by enzymatic hydrolysis of RNA, applied microbiologists in Japan actively explored the fermentative production of these 5′-nucleotides. In order to produce purine nucleotides by direct fermentation, the biosynthetic pathway of purine nucleotides and the regulation involved in it should be taken into account.

A. Biosynthesis of Purine Nucleotides and Its Regulation

Many reviews have appeared on the biosynthesis of purine nucleotides (Buchanan and Hartman, 1959; Moat and Friedman, 1960; Magasanik, 1962; Hartman, 1970; Blakley and Vitols, 1968; Momose, 1968; Gots, 1971). Therefore only a summary is presented here.

1. De novo Biosynthesis of Purine Nucleotides and Its Regulation

a. Precursors of Purine Nucleus. Isotope incorporation studies (Buchanan and Hartman, 1959) proved that nine atoms of purine nucleus come from five precursor molecules, as shown in Fig. 2. For the synthesis of 1 mole of 5′-IMP, 1 mole each of phosphoribosyl pyrophosphate (PRPP), aspartic acid, glycine, and CO_2 and 2 moles each of formic acid and glutamine are required as substrate. Eight moles of ATP are necessary as energy donor.

b. Biosynthetic Pathway. The biosynthetic pathway of purine nucleotides was first elucidated in pigeon liver homogenate by Buchanan and Hartman (1959). The pathway in microorganisms was later found to be identical with that in pigeon liver. As shown in Fig. 3, 5′-IMP is synthesized through 11 reactions starting from ribose 5-phosphate. 5′-AMP and 5′-GMP are synthesized through branch pathways from 5′-IMP. The 5′-AMP and 5′-GMP are interconvertible through reactions of the so-called purine nucleotide cycle. The 5′-AMP and 5′-GMP thus formed are used as building units of RNA, DNA, and various coenzymes.

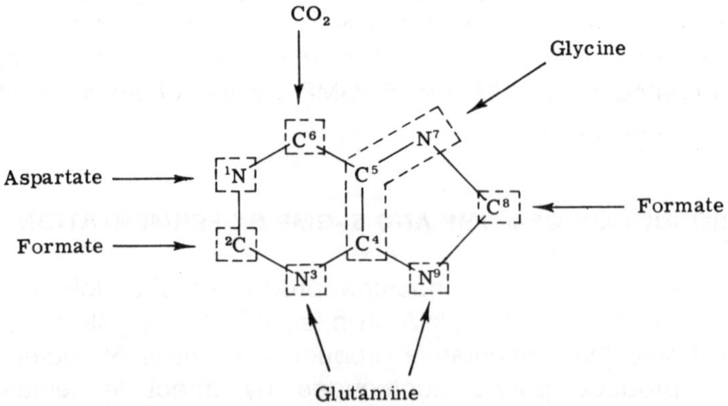

FIGURE 2. Precursors of the purine ring system.

c. Mutants of Purine Nucleotide Biosynthesis. Magasanik (1957) classified purine-requiring mutants into several classes according to their requirements for purine bases and intermediates accumulated by them (Fig. 4). Class 1 mutants required for their growth either adenine, hypoxanthine, xanthine, or guanine. Class 1 mutants, therefore, had an intact purine nucleotide cycle but lacked some enzymes for the synthesis of 5′-IMP. These mutants were classified into Classes 1a, 1b, and 1c: Class 1b mutants accumulated aminoimidazole riboside (AIR) and Class 1c mutants 5-amino-4-imidazole-carboxamide riboside (AICAR).

Class 2 mutants lacked 5′-IMP dehydrogenase and required xanthine or guanine for growth. Class 3 mutants lacked 5′-XMP aminase, required guanine for growth, and accumulated xanthosine in the culture medium. Class 4 mutants showed specific requirement for adenine and were subdivided into two groups: mutants lacking adenylosuccinate (SAMP) lyase and accumulating 5-amino-4-imidazole-*N*-succinocarboxamide riboside (SAICAR), and mutants lacking SAMP synthetase and accumulating inosine. Mutants being industrially used for fermentative production of nucleic acid-related substances have been isolated and improved from various bacterial species, based on the findings described above.

d. Regulation of Biosynthetic Pathway. Complex and fine control of purine biosynthetic pathway has been extensively studied. In this section, information on the control mechanisms which are indispensable for the development of the fermentation process for production of 5′-IMP and 5′-GMP will be summarized. Details have been reviewed by Blakley and Vitols (1968), Momose (1968), and Gots (1971).

(1) Metabolic Regulation of 5′-IMP Biosynthesis. PRPP amidotransferase catalyzes the first reaction of purine nucleotide biosynthesis: formation of PRA from PRPP and glutamine. Wyngaarden and Ashton (1959) and Caskey *et al.* (1964) demonstrated that this enzyme purified from pigeon liver was inhibited by AMP, ADP, ATP, GMP, GDP, or GTP. The inhibition was due to competition of these nucleotides for PRPP. Combination of 6-aminopurine and 6-hydroxypurine nucleotides, such as GMP plus AMP or IMP plus AMP, showed stronger inhibition to the reaction than the combination of two 6-aminopurine nucleotides, such as AMP plus ADP, or two 6-hydroxypurine nucleotides, such as GMP plus IMP. PRPP amidotransferase from *Aerobacter aerogenes* showed properties similar to that from pigeon liver (Nierlich and Magasanik, 1965).

Repression and derepression of PRPP amidotransferase in *A. aerogenes* was also studied by Nierlich and Magasanik (1963). The enzyme synthesis was completely repressed by the simultaneous addition of

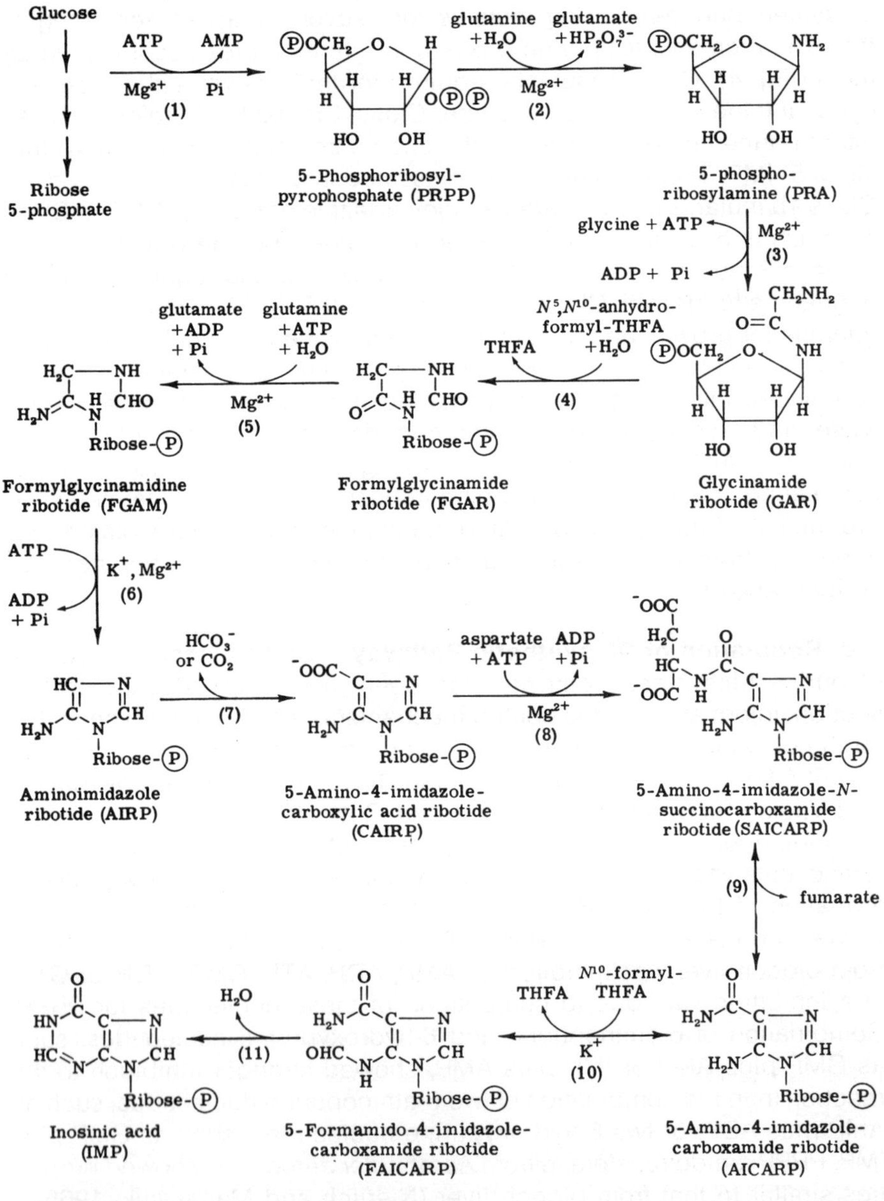

FIGURE 3. Synthesis of IMP. (1) PRPP synthetase, (2) PRPP amidotransferase, (3) phosphoribosylglycinamide synthetase, (4) phosphoribosylglycinamide formyltransferase, (5) phosphoribosylformylglycinamidine synthetase, (6) phosphoribosylaminoimidazole synthetase, (7) phosphoribosylaminoimidazole carboxylase, (8) phosphoribosylamino-imidazole succinocarboxamide synthetase, (9) adenylosuccinate lyase, (10) phosphoribo-

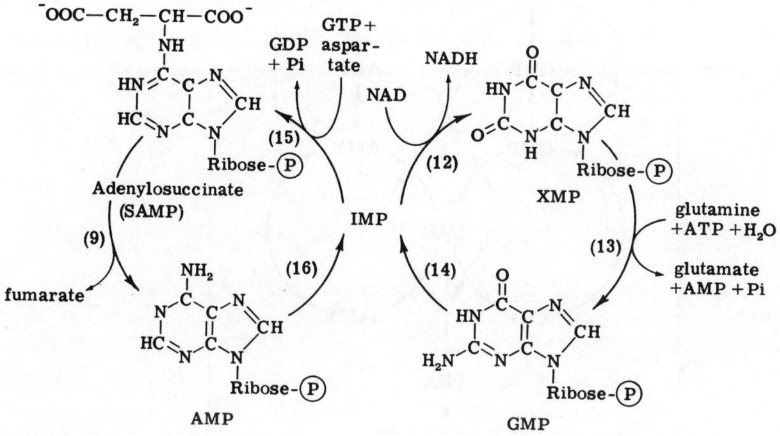

FIGURE 3 (*continued*)
sylaminoimidazole carboxamide formyltransferase, (11) IMP cyclohydrolase, (12) IMP dehydrogenase, (13) XMP aminase, (14) GMP reductase, (15) Adenylosuccinate synthetase, (16) AMP deaminase. (Microbial production of nucleic acid-related substances (1976), p. 26, Fig. 2.2, Kodansha Ltd. Halsted Press.)

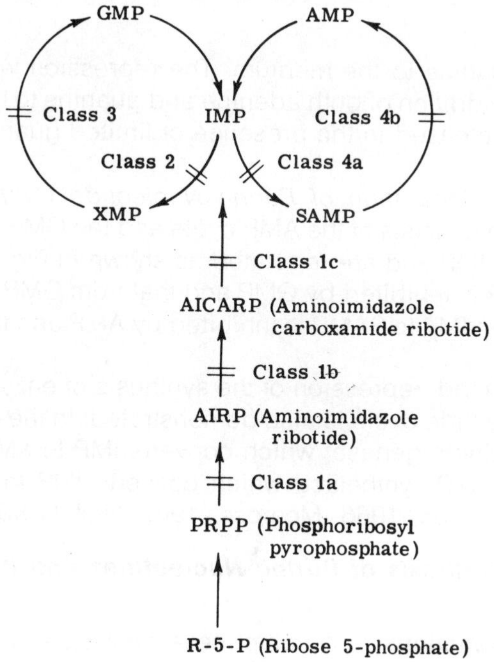

FIGURE 4. Classification of mutants blocked in the purine nucleotide biosynthetic pathway.

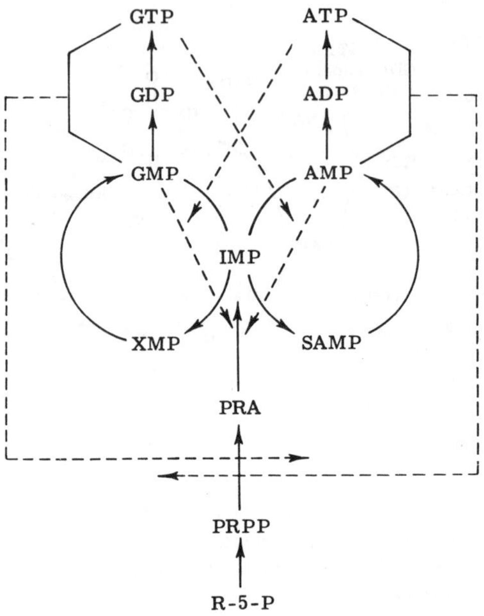

FIGURE 5. Inhibition of enzymes participating in the biosynthesis of purine nucleotides.

adenine and guanine to the medium. The repression was released by limitation of the addition of both adenine and guanine to the medium, but it was partially released in the presence of limited guanine and excess adenine.

(2) Metabolic Regulation of Purine Nucleotide Cycle. The purine nucleotide cycle consists of the AMP cycle and the GMP cycle which are connected at 5′-IMP and are regulated as shown in Fig. 5. The reaction from IMP to XMP is inhibited by GMP and that from GMP to IMP, by ATP. The reaction from IMP to SAMP is inhibited by AMP and that from AMP to IMP, by GTP.

On the other hand, repression of the synthesis of enzymes involved in the purine nucleotide cycle is also demonstrated. In the case of *Bacillus subtilis,* IMP dehydrogenase, which converts IMP to XMP, is repressed by GMP and SAMP synthetase which converts IMP to SAMP by AMP (Momose *et al.*, 1965, 1966; Momose, 1967; Nishikawa *et al.*, 1967).

2. *Salvage Synthesis of Purine Nucleotides and Its Regulation*

a. Salvage Reactions. Microbial cells synthesize purine nucleotides from purine bases present in a medium. Such a reaction is called a salvage synthesis, in contrast to a *de novo* synthesis. Salvage synthesis takes place by either of the two reactions shown in Fig. 6. One is direct conversion of bases to nucleotides catalyzed by nucleotide pyrophosphorylase. The other is a two-step reaction in which bases are first converted to nucleoside by nucleoside phosphorylase, followed by phosphorylation of nucleosides by nucleoside kinase or nucleoside phosphotransferase.

b. Regulation of Salvage Synthesis. Nucleotide pyrophosphorylase of *B. subtilis* is inhibited by purine nucleotides (Berlin and Stadtman, 1966). Nucleotide formation from adenine and guanine was specifically inhibited by the respective 5′-nucleotides formed as the products. On the other hand, conversion of hypoxanthine and xanthine to IMP was inhibited more strongly by GMP and GTP than by IMP and XMP.

B. Production of 5′-IMP

The following methods have been reported for the production of 5′-IMP.

1. Fermentative production and chemical phosphorylation of inosine
2. Fermentative production of 5′-IMP
3. Fermentative production of adenosine or 5′-AMP followed by chemical or enzymatic conversion to 5′-IMP
4. Salvage synthesis by microorganisms of 5′-IMP from chemically synthesized hypoxanthine

Methods (1) and (2) have proved to be useful for economical production of 5′-IMP and are described in detail in this section.

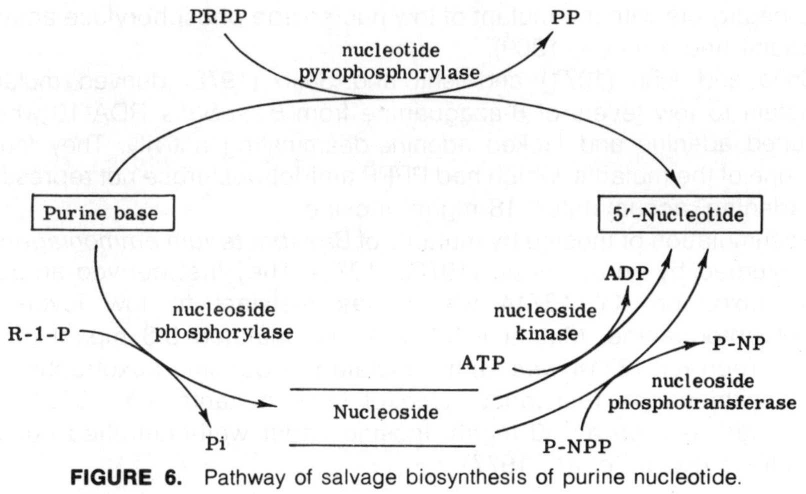

FIGURE 6. Pathway of salvage biosynthesis of purine nucleotide.

1. Production of Inosine by Fermentation and Its Chemical Phosphorylation

a. Fermentative Production of Inosine. *(1) Producing Microorganisms.* Inosine producers should be endowed with the following characteristics: (i) lack of SAMP synthetase, (ii) release from regulation operating in 5′-IMP biosynthesis, and (iii) lack of inosine-degrading enzymes, such as nucleoside phosphorylase and hydrolase.

In 1957, Gots and Gollub and Partridge and Giles (1957) first reported that hypoxanthine was accumulated in the culture broth of an adenine auxotroph of *Escherichia coli* and *Neurospora crassa.* Since the establishment of the manufacturing process for 5′-IMP by enzymatic hydrolysis of RNA, applied microbiologists actively have explored fermentative methods for production of inosine using adenine auxotrophs of species of the genera *Bacillus, Brevibacterium, Corynebacterium, Streptomyces,* and *Saccharomyces.* Mutants of the genera *Bacillus* and *Brevibacterium* have been found to accumulate large amounts of inosine.

In 1961, Uchida *et al.* first reported accumulation of inosine by mutants obtained from species of *Bacillus.* Aoki and co-workers (Aoki, 1963; Aoki *et al.*, 1963a,b, 1968; Yamanoi *et al.*, 1967; Yamanoi and Shiro, 1968) mutated *B. subtilis* and isolated mutant No. 2 which accumulated hypoxanthine with a small amount of inosine. Then mutants requiring various amino acids were isolated from mutant No. 2. One of the isolates, mutant C-30 (*ad⁻his⁻try⁻*), was found to have low inosine-degrading activity and to accumulate 6.3 mg/ml inosine.

Nogami *et al.* (1968) derived a mutant with low nucleoside phosphorylase activity from an inosine-guanosine-producing mutant of *Bacillus* sp. No. 102. They found that concomitant formation of purine bases was negligible with the mutant of low nucleoside phosphorylase activity (Nogami and Yoneda, 1969).

Shiio and Ishii (1971) and Ishii and Shiio (1972) derived mutants resistant to low levels of 8-azaguanine from *B. subtilis* RDA-16 which required adenine and lacked adenine-deaminating activity. They found that one of the mutants, which had PRPP amidotransferase not repressed by adenine, accumulated 18 mg/ml inosine.

Accumulation of inosine by mutants of *Brevibacterium ammoniagenes* was studied by Furuya *et al.* (1970a, 1975). They first derived an adenine auxotroph KY 13714 which was resistant to low levels of 6-mercaptoguanine. Mutant KY 13714 accumulated 9.3 mg/ml of inosine. Then KY 13714 was further mutated to guanine auxotrophy and 6-methylthiopurine resistance. One of such mutants, KY 13761, accumulated as high as 30 mg/ml inosine under well-controlled culture conditions (Kotani *et al.*, 1977).

Industrial production of inosine is mainly carried out by using the *ad⁻his⁻try⁻* series of mutants derived from *B. subtilis* by Aoki *et al.* described above.

(2) Culture Conditions. Representative fermentation media suitable for inosine production are shown in Table V.

Fermentative production of inosine by mutants of *B. subtilis* is industrially carried out under the following conditions (Aoki *et al.*, 1968; Shibai *et al.*, 1973): carbon source, starch hydrolyzate; adenine source, dried yeast or crude RNA; pH, 6.0–6.2; temperature, 30°–34°C; aeration, more than 0.01 atm dissolved oxygen. The aeration rate for maximum inosine production was determined from two parameters: (i) fulfillment of oxygen absorption by the producing microorganism which requires more than 5×10^{-7} mole/ml/min and (ii) avoidance of inhibition of inosine formation by generated carbon dioxide.

Figure 7 shows an example of inosine fermentation with a *B. subtilis* mutant under such manipulated conditions. Figure 8 shows an example of inosine fermentation by *B. ammoniagenes* KY 13761 derived by Kotani *et al.* (1977).

(3) Isolation of Inosine. Culture filtrate is adjusted to pH 11 with sodium hydroxide, and then concentrated *in vacuo* until the concentration of inosine becomes greater than 30 mg/ml. The concentrate is again adjusted to pH 11 and is further concentrated. Crystals of inosine sodium salt are collected and then suspended in water. The suspension is adjusted to pH 5.2 with hydrochloric acid and concentrated *in vacuo.* Crystalline inosine, separated out by concentration, is collected by centrifugation.

b. Chemical Phosphorylation of Inosine. In order to phosphorylate inosine, enzymatic methods (Brawerman and Chargaff, 1955; Mitsugi, 1964; Mitsugi *et al.*, 1964, 1965) as well as chemical methods (Muramatsu and Takenishi, 1965; Yoshikawa and Kato, 1967; Yoshikawa *et al.*, 1967, 1969, 1970; Honjo *et al.*, 1966) were widely tested. Industrially, chemical phosphorylation methods are employed. Yoshikawa *et al.* (1967, 1969) could chemically convert nucleosides to 5′-nucleotides in 68–90% yield by suspending unprotected nucleosides in trimethyl phosphate or triethyl phosphate followed by a reaction with phosphoryl chloride (Table VI). In the reaction system, 2′(3′),5′-diphosphoryl ester was concomitantly formed, but this disadvantageous by-production of diphosphoryl ester could be suppressed by addition of about equimolar water to nucleoside in the reaction mixture. Conversion of inosine or guanosine to 5′-IMP or 5′-GMP reached more than 90% in the improved

TABLE V. Composition of Media Employed for Inosine Accumulation

Compounds	Microorganisms					
	B. subtilis[a,g] ad^-, his^-, tyr^-	*B. subtilis*[b,g] ad^-, try^-, red^-, dea^-, $8AG^r$	*Bacillus* sp.[c,g] ad^-, his^-, red^-, $8AG^r$	*Bacillus* sp.[d,g] ad^-, his^-, thr^-, dea^-	*Brev. ammoniagenes*[e,g] ad^L, $6MG^r$, gua^-	*Brev. ammoniagenes*[f,g] ad^L, $6MG^r$, $6MTP^r$, gua^-
Glucose	60-70 gm/liter	70 gm/liter		100 gm/liter	130 gm/liter	
Maltose			100 gm/liter			
Inverted molasses						75 (0 hours), 100 (22 hours) gm/liter
NH_4Cl	20 gm/liter	15 gm/liter		20 gm/liter		
$(NH_4)_2SO_4$			20 gm/liter			
$(NH_2)_2CO$					4 gm/liter	2 gm/liter
KH_2PO_4	1 gm/liter	1 gm/liter		2 gm/liter	10 gm/liter	1 gm/liter
K_2HPO_4					10 gm/liter	1 gm/liter
$CaHPO_4$			5 gm/liter			
$Ca_3(PO_4)_2$			5 gm/liter			
$MgSO_4 \cdot 7\,H_2O$	0.4 gm/liter	0.4 gm/liter	2 gm/liter	0.5 gm/liter	10 gm/liter	1 gm/liter
$FeSO_4 \cdot 7\,H_2O$	2 ppm (Fe^{2+})	2 ppm (Fe^{2+})			10 mg/liter	
$ZnSO_4 \cdot 7H_2O$					10 mg/liter	
$MnSO_4 \cdot xH_2O$	2 ppm (Mn^{2+})	2 ppm (Mn^{2+})			10 mg/liter	
$MnCl_2 \cdot 4h_2O$				0.01 gm/liter		
Cysteine					20 mg/liter	

Tryptophan		300 mg/liter				
Thiamin					5 mg/liter	
Pantothenate					10 mg/liter	
Biotin			0.2 mg/liter		30 μg/liter	
Nicotinic acid					5 mg/liter	
Mieki	4 gm/liter	2 gm/liter				
Casamino acid				1 gm/liter		
Yeast extract				10 gm/liter		
Meat extract					10 mg/liter	
Dry yeast	14 gm/liters		10 gm/liters			
Adenine		100 mg/liter			100 mg/liter	
Guanine					100 mg/liter	
$CaCO_3$	2%	2.5%	2%	5%		
pH	6.0	7.0	7.6		8.3	
Max. inosine accumulated	10.5 gm/liter	18 gm/liter	14.1 gm/liter	10.8 gm/liter	13.6 gm/liter	30 gm/liter

[a] From Aoki *et al.* (1968).
[b] From Ishii and Shiio (1972).
[c] From Nogami *et al.* (1968).
[d] From Komatsu *et al.* (1972).
[e] From Furuya *et al.* (1975).
[f] From Kotani *et al.* (1977).
[g] Key to Abbreviations; ad^-, adenine-requiring; gua^-, guanine-requiring; his^-, histidine-requiring; tyr^-, tyrosine-requiring; thr^-, threonine-requiring; try^-, tryptophan-requiring; ad^L, leaky adenine mutant; red^-, GMP reductase-deficient; dea^-, AMP deaminase-deficient; $8AG^r$, 8-azaguanine-resistant; $6MG^r$, 6-mercaptoguanine-resistant; $6MTP^r$, 6-methylthiopurine-resistant.

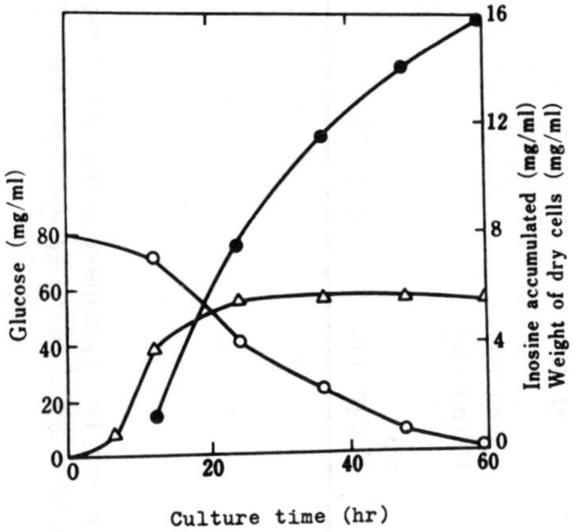

FIGURE 7. Time course of inosine fermentation by a mutant of *B. subtilus* K. (From Shibai *et al.*, 1973.) ●—● Inosine, ○—○ glucose, △—△ weight of dry cells.

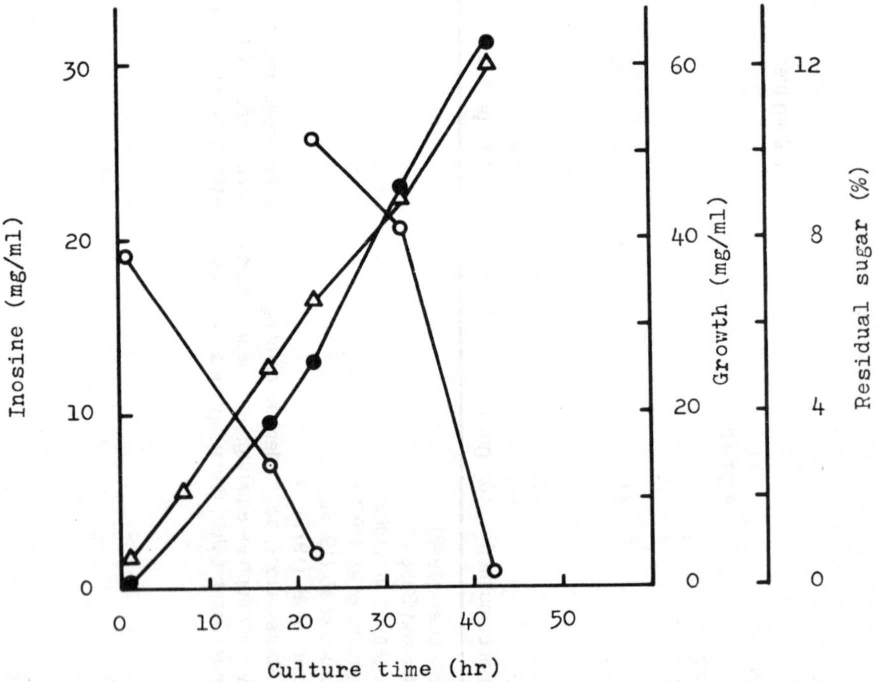

FIGURE 8. Time course of inosine accumulation by *B. ammoniagenes* KY 13761. (From Kotani *et al.*, 1977.) ●—●, Inosine, ○—○, residual sugar, △—△, growth.

TABLE VI. Direct Phosphorylation of Unprotected Nucleosides in Trialkyl Phosphate[a,b]

Nucleoside	R group in $(RO)_3PO$	$POCl_3$ (mmole)	H_2O (mmole)	Time (hours)	Yield of nucleoside-phosphate (mole %) 5′-mono	2′(3′) 5′-di
Inosine	CH_3CH_2	6	0	2	68	18
Inosine	CH_3CH_2	6	2	2	91	8
Guanosine	CH_3	6	0	6	85	9
Guanosine	CH_3	6	2	6	90	5
Adenosine	CH_3CH_2	4	0	6	84	11
Xanthosine	CH_3	4	0	9	80	5
Uridine	CH_3	4	0	12	89	8
Cytidine	CH_3CH_2	4	0	1	88	8
Deoxyinosine	CH_3	4	0	5	73	6
Deoxycytidine	CH_3	4	0.5	6	64	19
AICAR	CH_3	4	0	4	91	5

[a] From Yoshikawa *et al.* (1967, 1969).
[b] Conditions: nucleoside, 2 mmole; trialkyl phosphate, 5 ml; at 0°C.

system. The 5′-IMP was neutralized with sodium hydroxide to 5′-IMP $\cdot Na_2 \cdot 7.5\ H_2O$ and supplied commercially.

2. *Production of 5′-IMP by Fermentation*

a. Producing Microorganisms. Phosphate esters, including 5′-nucleotides, are generally impermeable through cytoplasmic membranes. The following four characteristics are considered to be necessary for a 5′-IMP-producing strain: (1) lack of SAMP synthetase, (2) release from regulation operating in the biosynthesis of 5′-IMP, (3) lack of 5′-IMP-degrading enzymes such as 5′-nucleotidase, and (4) cytoplasmic membrane permeable to 5′-IMP.

In order to obtain mutants lacking nucleotide dephosphorylating enzymes, Momose *et al.* (1964), Fujimoto and Uchida (1965), and Fujiwara *et al.* (1967) isolated mutants which can grow on adenine but not on 5′-AMP from inosine-producing *B. subtilis*. The mutants obtained showed reduced nucleotide-degrading activity but, unexpectedly, did not accumulate considerable amounts of 5′-IMP.

On the assumption that trace levels of nucleotide-degrading activity remained, interfering with the accumulation of 5′-IMP, Fujimoto *et al.* (1965) tried to decrease the nucleotide-degrading activity further and obtained mutant A-1-25. They found that about 0.6 mg/ml of 5′-IMP was accumulated by mutant A-1-25. Akiya *et al.* (1972) examined culture conditions for 5′-IMP formation with mutant A-1-25Z, which is a derivative of the above A-1-25, and succeeded in accumulating about 3.4 mg/ml 5′-IMP.

Nakayama *et al.* (1964, 1965) and Satō *et al.* (1965) mutated a glutamic acid-producing *Corynebacterium glutamicus* to adenine auxotrophy and 6-mercaptopurine resistance. The mutant accumulated 2 mg/ml 5′-IMP. Demain and co-workers (1965, 1966; Demain, 1968) also derived an adenine-requiring and IMP dehydrogenase-lacking mutant of *C. glutamicus* and found that the mutant accumulated 0.8–0.9 mg/ml 5′-IMP. Nara *et al.* (1967) derived an adenine auxotroph, KY 7208, from *B. ammoniagenes* ATCC 6872. They succeeded in accumulating 5 mg/ml 5′-IMP under elaborated culture conditions. Furuya *et al.* (1968) derived a leaky mutant, KY 13102, the growth of which was stimulated by adenine, from *B. ammoniagenes* ATCC 6872. The mutant KY 13102 accumulated 12.8 mg/ml 5′-IMP.

Accumulation of 5′-IMP by *B. ammoniagenes* mutants obtained by Nara *et al.* (1968b), Furuya *et al.* (1968, 1970b), and Komuro *et al.* (1969) was greatly influenced by Mn^{2+} concentration as well as by adenine concentration in the medium. From the above-described KY 13102, Furuya *et al.* (1969) isolated mutant KY 13105 and KY 13171 (Kato *et al.*, 1971) which could accumulate 5′-IMP in the presence of excess Mn^{2+}.

b. Culture Conditions. Nara *et al.* (1967, 1968b, 1969a,b), Furuya *et al.* (1968, 1970b) , and Komuro *et al.* (1969) examined culture conditions for 5′-IMP production by *B. ammoniagenes* mutants. Their results are summarized as follows: (1) optimum concentration of phosphoric acid was as high as 2%; (2) that of $MgSO_4$, about 1%; (3) that of adenine, about 25 mg/liter; (4) that of biotin, 20–30 μg/liter; and (5) that of Mn^{2+}, 10–20 μg/liter. In addition, (6) thiamin and calcium pantothenate were necessary for growth of producer microorganisms and accumulation of IMP and (7) a mixture of amino acids, e.g., histidine plus lysine or casamino acids, stimulated IMP accumulation.

Figure 9 shows the time course of 5′-IMP production by *B. ammoniagenes* KY 13102 under optimal culture conditions (Furuya *et al.*, 1968). A characteristic feature of the fermentation process is that hypoxanthine is first accumulated at an early stage and 5′-IMP is formed in later stages on consumption of hypoxanthine. Nara *et al.* (1969a) proposed a mechanism of IMP accumulation by *B. ammoniagenes* mutants as diagramatically shown in Fig. 10. Later, Furuya *et al.* (1970b) and Kato *et al.* (1971) demonstrated that *de novo* formed 5′-IMP leaked out directly from producer cells in later fermentation stages.

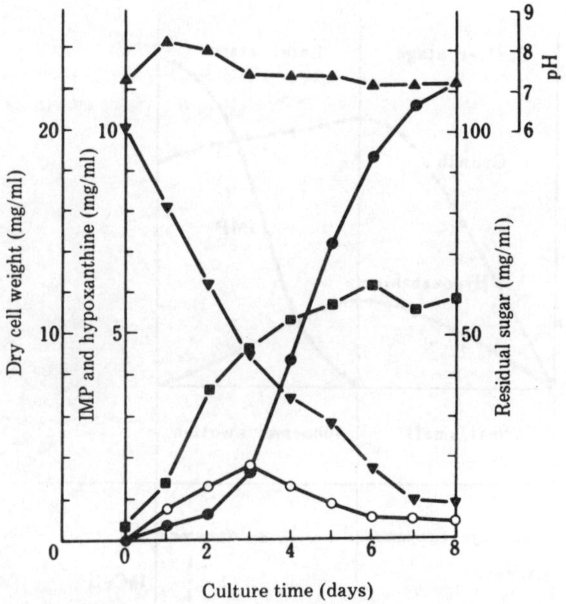

FIGURE 9. Time course of IMP accumulation by *B. ammoniagenes* KY 13102. (From Furuya *et al.*, 1968.) ●—● IMP, ○—○ hypoxanthine, ▲—▲ pH, ▼—▼ residual sugar, ■—■ dry cell weight.

3. Production of Adenosine by Fermentation

Production of 5′-IMP from fermentatively produced adenosine does not seem very practical for economical manufacturing. Adenosine fermentation was first studied for the purpose of producing 5′-IMP, but later studies were made not for 5′-IMP production but for the supply of adenosine or 5′-AMP, which are important as raw materials for synthesis of several valuable medicinary compounds.

a. Producing Microorganisms. Konishi *et al.* (1968) derived an isoleucine auxotroph, which accumulated 2.5 mg/ml adenosine, from *B. subtilis* K. The correlation between isoleucine auxotrophy and adenosine productivity is not clear. Haneda *et al.* (1971) serially mutated *Bacillus* sp. No. 1043 and obtained mutant P53-18, capable of producing 16 mg/ml adenosine. Genetical characteristics of the mutant P53-18 were *his*$^-$*thr*$^-$*xan*$^-$*8AX*^{r} and lack of adenosine deaminase and GMP reductase.

The following characteristics of the producing strain favor abundant accumulation of adenosine: (1) lack of IMP dehydrogenase, (2) release

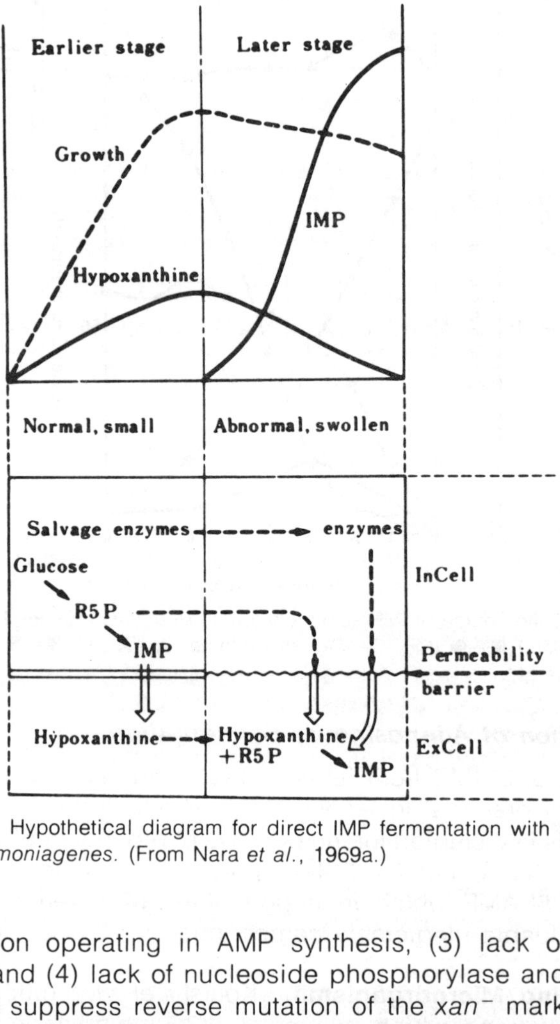

FIGURE 10. Hypothetical diagram for direct IMP fermentation with an adenineless mutant of *B. ammoniagenes*. (From Nara *et al.*, 1969a.)

from regulation operating in AMP synthesis, (3) lack of adenosine deaminase, and (4) lack of nucleoside phosphorylase and hydrolase.

In order to suppress reverse mutation of the *xan*$^-$ marker, a further genetic defect in XMP aminase was effective (Haneda *et al.*, 1972).

b. Culture Conditions. Haneda *et al.* (1971) evaluated culture conditions for adenosine production by the mutant P53-18 and found that concentration of yeast extract, calcium carbonate, and aeration rate were important factors for adenosine production. A process of adenosine fermentation by mutant P53-18 under optimum conditions is shown in Fig. 11.

C. Production of 5′-GMP

Accumulation of large amounts of 5′-GMP in the microbial culture fluid appeared to be very difficult since there exist regulation mechanisms to control the overproduction of 5′-GMP.

It was also presumed that 5′-GMP, once formed, is converted to other nucleotides by the purine nucleotide cycle and that 5′-GMP is degraded into guanosine or guanine by phosphatases or nucleoside phosphorylases, which are active in most microbial cells. Based on these presumptions, various indirect methods have been attempted in order to produce 5′-GMP beside trials to accumulate 5′-GMP directly in the fermentation broth. These methods include: (1) fermentative production of AICAR followed by chemical conversion of it to 5′-GMP; (2) fermentative production of guanosine followed by chemical or biochemical phosphorylation of it; (3) mixed culture of two microorganisms, one accumulating 5′-XMP (or xanthosine) and the other converting 5′-XMP (or xanthosine) to 5′-GMP; and (4) direct fermentation of 5′-GMP. Among these methods, (1) and (2) are now industrially employed.

1. Production of AICAR by Fermentation and Its Chemical Conversion to 5′-GMP

a. Fermentative Production of AICAR. *(1) Producing Microorganisms.* Accumulation of AICA in culture broth of sulfonamide-inhibited *E. coli* was first observed in 1942 (Fox, 1942; Stetten and Fox, 1945; Shive *et al.*, 1947). Since then, the accumulation of AICAR was studied using mainly *E. coli* (i) wildtype strains, (ii) mutants genetically blocked in purine nucleotide biosynthesis pathway, and (iii) wildtype strains metabolically blocked by drugs which inhibits purine biosynthesis. However, the amount of AICAR accumulated by *E. coli* was low. Shiro *et al.* (1962) isolated a purine auxotroph, D-421, from X-ray treated *B. subtilis* No. 2093. The mutant accumulated about 2.6 mg/ml of AICAR in its culture broth. This finding led to the establishment of industrial production of AICAR by fermentation. Later, Kinoshita *et al.* (1967) isolated many auxotrophic mutants of *B. subtilis* IAM 1523, *Bacillus megaterium* IAM 1245, and *Brevibacterium flavum* No. 2247 and examined their productivity of AICAR. AICAR was produced only by purine auxotrophs. They found that a purine auxotroph No. 336 of *B. megaterium* accumulated 15 mg/ml AICAR. Shirafuji *et al.* (1968) derived a mutant requiring adenine plus hypoxanthine or guanine from nonexacting purine auxotroph of *Bacillus pumilus* which formed detectable AICAR. The derived mutant accumulated as high as 15 mg/ml of AICAR. For industrial pro-

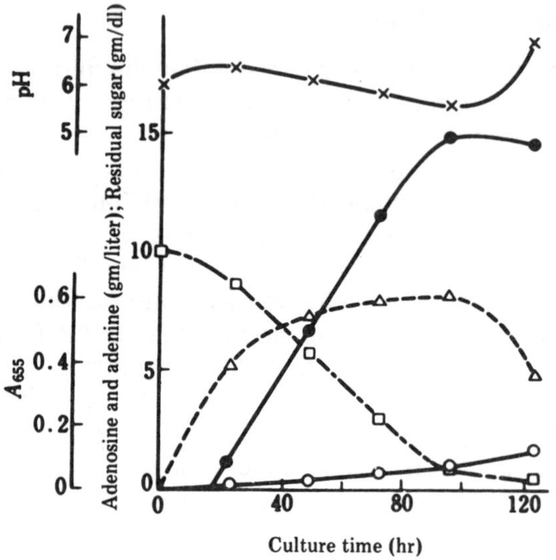

FIGURE 11. Time course of adenosine fermentation by *Bacillus* sp. P53-18. (From Haneda *et al.*, 1971.) ●—● Adenosine, ○—○ adenine, □-—-□ residual sugar, △- - -△ A_{655}, ×—× pH.

duction of AICAR, a purine auxotroph of *B. megaterium* is now being used.

(2) Culture Conditions. Kinoshita *et al.* (1967, 1968, 1969a,b,c, 1973) examined production conditions for AICAR by *B. megaterium* No. 336 and found that suppression of spore formation and back-mutation during fermentation were important factors for the production of AICAR in high yield. The sporulation could be suppressed by the regulation of oxygen supply (Kinoshita *et al.*, 1973) or by the addition of sporulation inhibitors, such as butyric acid (Kinoshita *et al.*, 1969b). Reversion of prototrophy could be controlled by selection of mutants with low reversion rates (Kinoshita *et al.*, 1967) or by the addition of erythromycin, which preferentially inhibits the growth of revertants (Kinoshita *et al.*, 1968). The fermentation medium should be supplemented with purine sources, such as RNA, adenine, or inosine, at a suboptimal level for growth and with suitable amounts of inorganic ions, such as phosphate, potassium, and sulfate, in order to accumulate large amounts of AICAR. Figure 12 shows a typical time course of AICAR fermentation (Kinoshita *et al.*, 1969a).

b. Chemical Synthesis of 5′-GMP from AICAR. AICAR in the fermented broth is recovered in more than 90% yield through several

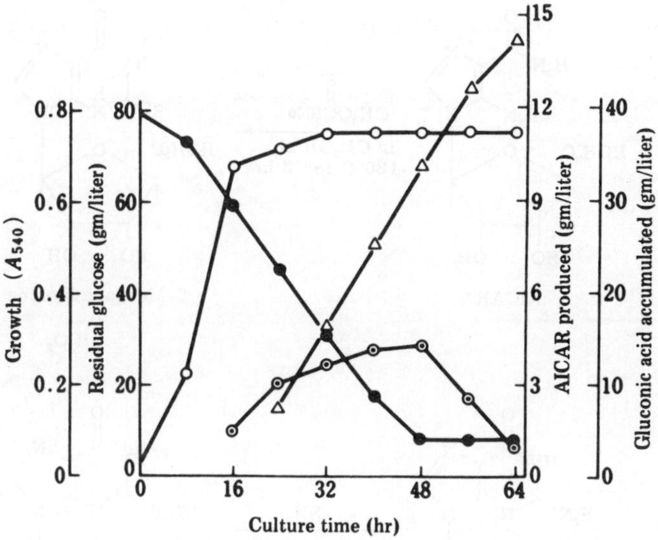

FIGURE 12. Time course of AICAR fermentation by *B. megaterium* MA-336. (From Kinoshita *et al.*, 1969a.) △—△ AICAR, ⊙—⊙ gluconic acid, ○—○ growth, ●—● glucose.

purification steps: adsorption on and elution from cation-exchange resin, concentration, crystallization, and drying. AICAR is chemically converted in 70% yield to guanosine as shown in Fig. 13 and then chemically phosphorylated to 5′-GMP in about 90% yield (Kinoshita *et al.*, 1967; Yamazaki *et al.*, 1967; Kumashiro *et al.*, 1968). The 5′-GMP formed is neutralized with sodium hydroxide and crystalline 5′-GMP · Na_2 · 7 H_2O is prepared as the final product.

2. Production of Guanosine by Fermentation and Its Phosphorylation

a. Fermentative Production of Guanosine. Investigations to develop a process for the direct fermentative production of guanosine have been attempted since the process appeared to be more useful than the above two-step method of guanosine production from AICAR. Accumulation of guanosine in the culture fluid of *B. subtilis, B. pumilus, Bacillus licheniformis, Corynebacterium petrophilum, Corynebacterium guanofaciens, and Streptomyces griseus* have been reported. The yield of guanosine production is high with mutants of *Bacillus* sp. and these mutants are now being used for industrial guanosine production.

Producing Organisms. For the accumulation of guanosine, the producing microorganisms should have the following properties: (i) lack of

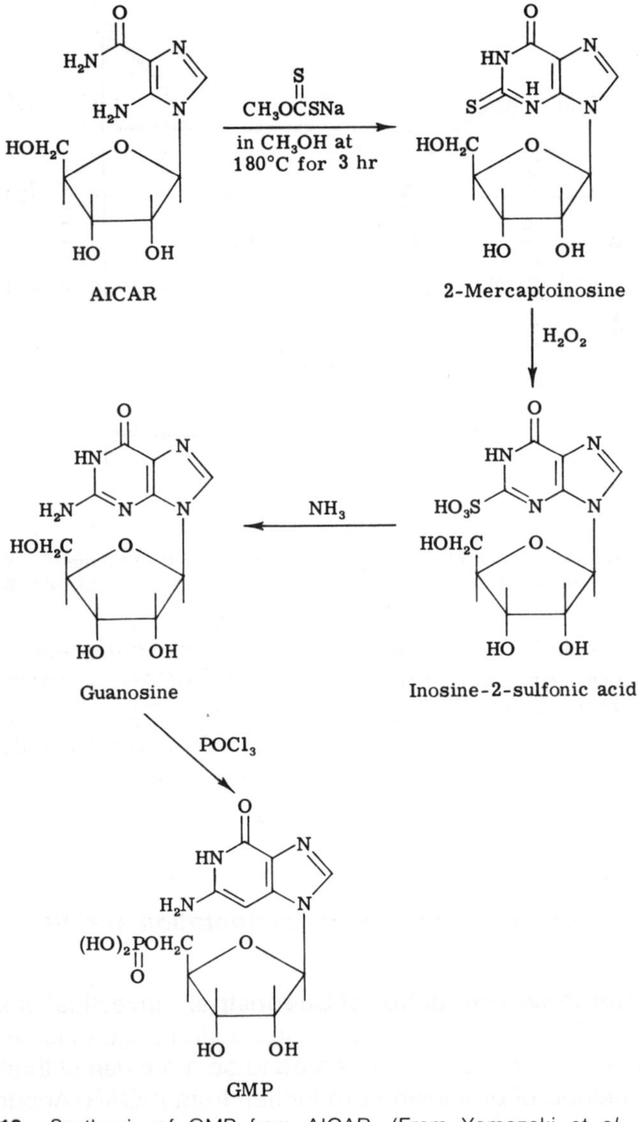

FIGURE 13. Synthesis of GMP from AICAR. (From Yamazaki *et al.*, 1967; and Kumashiro *et al.*, 1968.)

SAMP synthetase and GMP reductase; (ii) lack of guanosine-degrading enzymes, such as purine nucleoside phosphorylase or nucleoside hydrolase; (iii) release of regulation by GMP of GMP-synthesizing enzymes; and (iv) faster formation of 5′-GMP from 5′-IMP than dephosphorylation of 5′-IMP.

Mutants endowed with such favorable characteristics for guanosine formation have been derived from *B. subtilis*. The results are summarized in Table VII.

Konishi and Shiro (1968) obtained mutants resistant to 8-azaguanine from an inosine-producing mutant, AJ-1987, of *B. subtilis*. One of such mutants, AJ-1993, accumulated 4.3 mg/ml guanosine. Strain AJ-1993 possessed a higher level of 5′-IMP dehydrogenase than the original AJ-1987 and showed weak GMP-reductase activity.

Momose and Shiio (1969) isolated many 8-azaxanthine-resistant mutants from two kinds of mutants of *B. subtilis:* an inosine-producing adenine auxotroph K38-3, and a GMP-reductase-lacking adenine and tryptophan double auxotroph No. 30-12, which was derived from mutant K38-3. On examining nucleoside-forming capacities, 57% of 8-azaxanthine-resistant mutants derived from K38-3, which possessed GMP reductase, were found to accumulate inosine and hypoxanthine and the remainder (47%) only accumulated inosine. On the other hand, 70% of 8-azaxanthine resistant mutants derived from GMP reductase-lacking mutant No. 30-12 accumulated guanosine together with inosine, while the remainder (30%) only accumulated inosine. The yield of guanosine by one of the latter mutants, No. 30-12-3, was 5 mg/ml.

Nogami *et al.* (1968) derived a guanosine-producing mutant, designated T-780, from a *Bacillus* sp. No. 102 after multistep mutation (Nogami and Yoneda, 1969). They showed that adenine auxotrophy, lack of GMP reductase, and resistance to adenine or adenosine are essential characteristics for guanosine production. Other genetic characteristics, such as high resistance to streptomycin and lack of nucleoside phosphorylase, were effective for guanosine production.

Komatsu *et al.* (1972) and Komatsu and Kodaira (1973), based on a similar concept, derived a guanosine-producing mutant No. 20, from a *Bacillus* sp. No. 1043. It produced guanine, hypoxanthine, and xanthine together with guanosine. They further derived a mutant GnR-176, which had weak nucleoside-degrading enzyme activity. The yield of guanosine by GnR-176 increased to 10.6 mg/ml.

Matsui *et al.* (1974) isolated a guanosine-producing mutant, No. 14119, which became resistant to methionine sulfoxide, from *B. subtilis* No. 1411, capable of producing inosine. Slight increase in IMP dehydrogenase activity and marked decrease in nucleotidase activity were ob-

TABLE VII. Production of Guanosine by Microorganisms[a]

Microorganism	Genetic character	Guanosine accumulated (gm/liter)	References
Bacillus subtilis	Ad^-, red^-, $8AG^r$	4.3	Konishi and Shiro (1968)
Bacillus subtilis	Ad^-, red^-, $8AX^r$, try^-	5.0	Momose and Shiio (1969)
Bacillus sp.	Ad^-, red^-, $8AG^r$, AAR^r SM^r, NP^-	5.8	Nogami *et al.* (1968); Nogami and Yoneda (1969)
Bacillus sp.	Ad^-, red^-, $8AX^r$, his^- thr^-, $adenase^-$, NS^{pd}	10.6	Komatsu *et al.* (1972); Komatsu and Kodaira (1973)
Bacillus subtilis	Ad^-, red^-, MSO^r, his^- $Psicofuranine^r$, $Decoyinine^r$	16	Enei *et al.* (1976)

[a] Key to abbreviations; ad^-, adenine-requiring; his^-, histidine-requiring; thr^-, threonine-requiring; try^-, tryptophan-requiring; $8AG^r$, 8-azaguanine-resistant; $8AX^r$, 8-azaxanthine-resistant; AAR^r, adenine or adenosine resistant; SM^r, streptomycin resistant; MSO^r, methionine sulfoxide resistant; $psicofuranine^r$, psicofuranine resistant; $decoyinine^r$, decoyinine-resistant; red^-, GMP reductase deficient; NP^-, nucleoside phosphorylase deficient; $adenase^-$, adenase deficient; NS^{pd}, partially deficient purine nucleoside-hydrolyzing activity.

served with the mutant No. 14119. Enei *et al.* (1976) further derived a psicofuranine-resistant mutant Gp-1 from strain 14119. Since Gp-1 formed xanthosine besides guanosine, they mutated it to be decoyinine resistant and obtained mutant MG-1. GMP synthetase in MG-1 was not inhibited by 5′-GMP. The yield of guanosine by MG-1 was as high as 16 mg/ml. Figure 14 shows an example of guanosine fermentation by mutant MG-1.

b. Isolation and Phosphorylation of Guanosine. To recover guanosine, the fermented broth is first adjusted to pH 12 with sodium hydroxide. After cells are removed by centrifugation, the supernatant fluid is concentrated under reduced pressure. Guanosine, precipitated by neutralizing the concentrate, is phosphorylated and crystalline 5′-GMP · Na_2 · 7 H_2O is prepared as described previously.

3. Production of Xanthosine and 5′-XMP by Fermentation and Their Conversion to 5′-GMP

The accumulation of xanthosine or 5′-XMP, an intermediate of 5′-GMP biosynthesis, in microbial culture fluid appears easier than accumulation of guanosine or 5′-GMP as an end product. Methods for production of xanthine derivatives have been investigated together with the conversion of those to 5′-GMP.

Accumulation of xanthosine or 5′-XMP would take place with microorganisms fulfilling the following two characters: (1) lack of XMP aminase and (2) lack of SAMP synthetase. High producers of xanthosine have been found in mutants of *B. subtilis,* and those of 5′-XMP, in mutants of *B. ammoniagenes*.

Ishii and Shiio (1973) mutated an inosine-producing strain of *B. subtilis,* which requires adenine, lacks GMP reductase and AMP deaminase, and is resistant to 8-azaguanine, to guanine auxotrophy. The obtained guanine auxotroph accumulated 17–18 mg/ml xanthosine under elaborated culture conditions. Xanthosine is phosphorylated to 5′-XMP by the chemical method described previously. One-step conversion of xanthosine to 5′-GMP, however, has not yet been reported.

Misawa *et al.* (1969) found that 6.5 mg/ml 5′-XMP was accumulated in the culture fluid of a guanine-requiring mutant of *B. ammoniagenes*. They concluded that the accumulated 5′-XMP was synthesized *de novo* and excreted through the cytoplasmic membrane of the producing microorganism.

In order to convert the accumulated 5′-XMP to 5′-GMP, Furuya *et al.* (1971) used a mutant of *B. ammoniagenes,* which is resistant to psicofranine or decoyinine and has increased XMP aminase activity. By cultivation of the mutant in a medium containing 13 mg/ml 5′-XMP at 30°C for

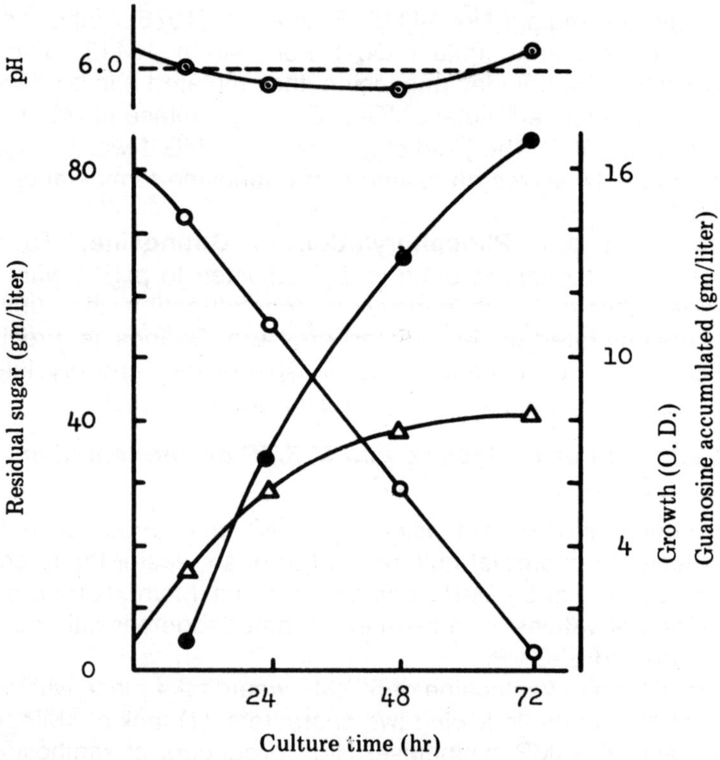

FIGURE 14. Time course of guanosine fermentation by *B. subtilus* MG-1. (From Enei *et al.*, 1976). ●—● Guanosine, ○—○ residual sugar, △—△ growth, ⊙—⊙ pH.

48 hours, about 10 mg/ml of 5′-GMP was produced in the culture fluid. Recently, they succeeded in the development of a single-step fermentative process for 5′-GMP formation by mixed culture of the above two mutants: One accumulates 5′-XMP and the other converts 5′-XMP to 5′-GMP (Furuya *et al.*, 1973). The yield of 5′-GMP was also about 10 mg/ml. Although they demonstrated a possibility that 5′-GMP is produced by fermentation, the processes have not yet been employed in industrial scale due to insufficient yield of 5′-GMP and formation of GDP and GTP as by-products.

4. Production of 5′-GMP by Fermentation

It seems necessary that 5′-GMP producers have the following characteristics: (1) lack of SAMP synthetase, (2) lack of 5′-GMP-degrading activity, (3) insensitivity of GMP-synthesizing enzymes to the product,

5′-GMP, and (4) membrane permeability from the interior to the exterior for 5′-GMP.

Demain *et al.* (1966) mutated a 5′-IMP-forming *Micrococcus glutamicus* ATCC 13761 and derived a mutant MB-1806 which is released from feedback inhibition by GMP with respect to biosynthesis of 5′-GMP. It accumulated 0.5 mg/ml GMP in its culture broth. Many trials followed in order to derive 5′-GMP producers from microorganisms such as *Pseudomonas aeruginosa*, *B. subtilis*, *Brevibacterium helvolum*, *E. coli*, *B. ammoniagenes*, and *Candida tropicalis*. Among these mutants, an adenine auxotroph derived from *B. ammoniagenes* ATCC 6871 showed the highest yield: 5.1 mg/ml 5′-GMP in broth (Abe and Udagawa, 1966a,b,c). This yield, however, is not sufficient for economical production of 5′-GMP on an industrial scale.

V. PRODUCTION OF NUCLEIC ACID-RELATED SUBSTANCES BY FERMENTATION

Investigation to develop a process for fermentative production of nucleic acid-related substances other than those related to 5′-IMP or 5′-GMP production has also been carried out. In this section, production by microorganisms of nucleic acid-related substances which are or will be useful for medicinal purposes is described.

A. ATP

Based on an old finding by Lutwal-Mann and Mann (1935), Tochikura *et al.* (1967) developed an improved process to produce ATP. ATP was obtained in a molar yield of 72% by phosphorylation of adenosine or 5′-AMP with bakers' yeast dried with acetone. Such a method was improved for industrial ATP production with several modifications: treatment of yeast with anionic detergents (Kawai *et al.*, 1974) and mild drying of yeasts at 60°C with a liquid-phase dryer to give 8.5% water content, followed by mild rehydration (30°C, more than 10 minutes) (Saeki *et al.*, 1976). In the latter case, the reaction mixture containing 15 mg/ml yeast cells, 36 mg/ml glucose, 2.03 mg/ml $MgCl_2$, and 3.83 mg/ml AMP in phosphate buffer of pH 7.5 was incubated at 30°C with shaking. The molar yield of ATP from AMP was about 70%.

Direct fermentative production of ATP was also attempted by Tanaka *et al.* (1968) and Nara *et al.* (1968a). Addition of 2 mg/ml adenine to the culture of *B. ammoniagenes* ATCC 6872 followed by an additional 1 day incubation yielded 1.57 mg/ml ATP, 1.59 mg/ml ADP, and 2.16 mg/ml AMP. Addition of guanine instead of adenine yielded GTP, GDP, and GMP.

B. NAD

Takebe and Kitahara (1963) demonstrated that the NAD content is highest in lactic acid bacteria among various microorganisms. Yeast cells grown on a medium supplemented with adenine and nicotinamide, however, accumulated NAD in the cells as high as 40 mg/gm (Sakai *et al.*, 1973b). NAD can be economically extracted from such yeast cells.

Nakayama *et al.* (1968) cultivated *B. ammoniagenes* for 2 days on a usual medium and for an additional 2–3 days on a medium containing adenine and nicotinamide. NAD in the culture fluid reached 2–3 mg/ml.

NAD is generally phosphorylated to NADP by NAD kinase. Tochikura *et al.* (1969) found a unique phosphate-transferring activity in *Proteus mirabilis.* In that transferring reaction, *p*-nitrophenylphosphate or nucleoside monophosphates are used as phosphate donor. The enzyme catalyzing the transferring reaction was also found in species of the genera *Aerobacter, Bacterium, Staphylococcus,* and *Streptomyces* (Kuwahara *et al.*, 1972). Another unique reaction, in which only metaphosphate served as a phosphate donor, was found in *B. ammoniagenes* and several species of the genus *Micrococcus* (Murata *et al.*, 1977). Application of such microorganisms for NADP production has not yet been reported.

C. FAD

Eremothecium ashbyii accumulates much FAD in its mycelia (Kuwata *et al.*, 1954; Kuwata, 1955). Based on that finding, a process to produce FAD was established by extracting it from *E. ashbyii* mycelia. A process to accumulate FAD extracellularly, however, seems to be more favorable than that described above.

Nakamura *et al.* (1968) found that the addition of adenine to the culture of *B. ammoniagenes* resulted in the accumulation of FAD in its culture filtrate. There have been several attempts to accumulate FAD by fermentation (Sakai *et al.*, 1973a). Among them, *Sarcina lutea* showed the highest yield and conditions for FAD production were elaborated. Alteration of cell wall structure by the addition of cycloserine to a culture medium was effective for FAD production. Cultivation of *S. lutea* in a medium containing 100 mg/liter cycloserine for 24 hours, followed by cultivation for 3–4 days more in the presence of adenine and riboflavin resulted in the accumulation of 780 mg/liter FAD (Watanabe *et al.*, 1974). This method may be the most economical way to manufacture FAD.

D. Coenzyme A

Coenzyme A has been industrially produced by extraction from yeast cells (Beinert *et al.*, 1953; Reece *et al.*, 1959). The yield is 8–9 mg/kg of bakers' or brewery yeast. Ogata *et al.* (1972a,b) and Shimizu *et al.* (1972) succeeded in accumulating 2–3 mg/ml CoA in the culture broth by incubating *B. ammoniagenes* IFO 12071 in a medium containing pantothenic acid, cysteine, and ATP as precursors in the presence of anionic detergents. They also improved the method by immobilizing bacterial cells in polyacrylamide gel (Shimizu *et al.*, 1975).

Accumulation of CoA in the culture filtrate is also possible. *B. ammoniagenes* was first cultivated in a medium containing AMP; pantothenic acid, cysteine, and a cationic detergent were then added to the medium. Upon further cultivation, 3–5 mg/ml CoA (as disulfide) were accumulated (Shimizu *et al.*, 1973a). Nishimura *et al.* (1974) demonstrated that *S. lutea* IAM 1099 showed strong activity to synthesize CoA. Shimizu *et al.* (1973b, 1974) have reviewed CoA biosynthesis and its regulation.

E. cAMP

Okabayashi *et al.* (1964) demonstrated the direct fermentative production of cAMP without addition of precursors: *Brevibacterium liquefaciens* ATCC 14929 accumulated 1.2 mg/ml cAMP in an alanine-containing medium. A mutant of *Microbacterium* sp. No. 205, which required biotin and was resistant to 6-mercaptopurine, 8-azaguanine, and methylsulfoxide, accumulated 2 mg/ml cAMP (Ishiyama *et al.*, 1973). It is also known that *Corynebacterium murisepticum* No. 7 and *Microbacterium* sp. No. 205 accumulated 10–100 μg/ml cGMP (Ishiyama, 1975).

However, cAMP can be most effectively produced by salvage synthesis. Ishiyama *et al.* (1974) cultivated mutant ATCC 21376 which was derived from *Microbacterium* sp. No. 205 in nutrient broth and transferred the culture to a medium containing 0.3% hypoxanthine (or 0.5% inosine), glucose, phosphate, peptone, and yeast extract. After cultivation for 40 hours at 28°C, 3–6 mg/ml cAMP was accumulated.

The 3′,5′-cyclic nucleotides can be also prepared by hydrolyzing RNA with a specific RNase followed by incubation in ⅓ *M* phosphate buffer (pH 7.0–8.7) (Yamasaki and Arima, 1967; Mantani *et al.*, 1972).

F. Guanosine 5′-Triphosphate 3′-Diphosphate

Murao and Nishino (1973, 1974) first demonstrated ATP: nucleotide pyrophosphotransferase in microbial culture broth. This enzyme formed magic spot I (ppGpp) or II (pppGpp) (Cashel and Gallant, 1969; Lazzarini *et al.*, 1972) from ATP and GDP or GTP in the presence of Mg^{2+}. Purine nucleoside 5′-mono-, di-, and triphosphate accept pyrophosphate from ATP and the respective 3′-pyrophosphates are formed (Hamagishi *et al.*, 1975). Distribution of this enzyme is widespread: *Streptomyces adephosphoryticus* (Murao and Nishino, 1973, 1974), *Streptomyces hydrogenans* (Betz *et al.*, 1977), *Streptomyces morookaensis, Streptomyces aspergilloides, Streptomyces hachijoensis, Actinomyces violasceus, Streptoverticillium septatum* (Oki *et al.*, 1976), and mutant KY 13510 (decoyinine resistant) derived from *B. ammoniagenes* ATCC 6872 (Sato and Furuya, 1976, 1977). Incubation in a reaction mixture containing 50–150 mg/ml of freeze-dried cells of mutant KY 13510, 30 mg/ml 5′-GMP·Na_2·7 H_2O, glucose, potassium phosphate, $MgSO_4$, and polyoxyethylene stearylamine (pH 7.4) gave 12 mg/ml ppGpp and 5 mg/ml pppGpp.

REFERENCES

Abe, S., and Udagawa, K. (1966a). Japanese Patent 41-12,673.

Abe, S., and Udagawa, K. (1966b). Japanese Patent 41-13,796.

Abe, S., and Udagawa, K. (1966c). Japanese Patent 41-13,797.

Akiya, T., Midorikawa, Y., Kuninaka, A., Yoshino, H., and Ikeda, Y. (1972). *Agric. Biol. Chem.* **36,** 227–233.

Akiyama, S., Doi, M., Arai, Y., Nakao, Y., and Fukuda, H. (1975). U.S. Patent 3,909,352.

Ando, T. (1966). *Biochim. Biophys. Acta* **114,** 158–168.

Aoki, R. (1963). *J. Gen. Appl. Microbiol.* **9,** 397–402.

Aoki, R., Momose, H., Kondō, Y., Muramatsu, N., and Tsuchiya, Y. (1963a). *J. Gen. Appl. Microbiol.* **9,** 387–396.

Aoki, R., Kondō, Y., and Momose, H. (1963b). *J. Gen. Appl. Microbiol.* **9,** 403–408.

Aoki, R., Kondō, Y., Hirose, Y., and Okada, H. (1968). *J. Gen. Appl. Microbiol.* **14,** 411–416.

Beinert, H., Korff, R. W., Green, D. E., Buyske, D. A., Handschumacher, R. E., Higgins, H., and Strong, F. M. (1953). *J. Biol. Chem.* **200,** 385–400.

Belozersky, A. N. (1947). Cold Spring Harbor Symp. Quant. Biol. **12,** 1–6.

Belozersky, A. N., and Spirin, A. S. (1960). *In* "The Nucleic Acids" (E. Chargaff and J. N. Davidson, eds.), Vol. 3, pp. 147–185. Academic Press, New York.

Berlin, R. D., and Stadtman, E. R. (1966). *J. Biol. Chem.* **241,** 2679–2686.

Berry, S. A., and Campbell, J. N. (1967a). *Biochim. Biophys. Acta* **132,** 78–83.

Berry, S. A., and Campbell, J. N. (1967b). *Biochim. Biophys. Acta* **132,** 84–93.

Betz, J. W., Schneider, B. H., and Träger, L. (1977). *Hoppe-Seyler's Z. Physiol. Chem.* **358,** 353–359.

Blakley, R. L., and Vitols, E. (1968). *Annu. Rev. Biochem.* **37,** 201–224.

Brawerman, G., and Chargaff, E. (1955). *Biochim. Biophys. Acta* **16,** 524–532.

Buchanan, J. M., and Hartman, S. C. (1959). *Adv. Enzymol.* **21,** 199–261.
Cashel, M., and Gallant, J. (1969). *Nature (London)* **221,** 838–841.
Caskey, C. T., Ashton, D. M., and Wyngaarden, J. B. (1964). *J. Biol. Chem.* **239,** 2570–2579.
Chayen, R., Chayen, S., and Roberts, E. R. (1955). *Biochim. Biophys. Acta* **16,** 117–126.
Cohn, W. E., and Volkin, E. (1951). *Nature (London)* **167,** 483–484.
Cohn, W. E., and Volkin, E. (1953). *J. Biol. Chem.* **203,** 319–332.
Demain, A. L. (1968). *Biotechnol. Bioeng.* **10,** 291–302.
Demain, A. L., Jackson, M., Vitali, R. A., Hendlin, D., and Jacob, T. A. (1965). *Appl. Microbiol.* **13,** 757–761.
Demain, A. L., Jackson, M., Vitali, R. A., Hendlin, D., and Jacob, T. A. (1966). *Appl. Microbiol.* **14,** 821–825.
Di Carlo, F. J., and Schultz, A. S. (1948). *Arch. Biochem.* **17,** 293–300.
Enei, H., Sato, K., Matsui, H., and Hirose, Y. (1976). *Abstr. Annu. Meet. Agric. Chem. Soc. Jpn.* p. 119.
Fox, C. L., Jr. (1942). *Proc. Soc. Exp. Biol. Med.* **51,** 102–104.
Fujimoto, M., and Uchida, K. (1965). *Agric. Biol. Chem.* **29,** 249–259.
Fujimoto, M., Morozumi, M., Midorikawa, Y., Miyakawa, S., and Uchida, K. (1965). *Agric. Biol. Chem.* **29,** 918–922.
Fujimoto, M., Kuninaka, A., and Yoshino, H. (1969). *Agric. Biol. Chem.* **33,** 1517–1518.
Fujimoto, M., Kuninaka, A., and Yoshino, H. (1974a). *Agric. Biol. Chem.* **38,** 777–783.
Fujimoto, M., Kuninaka, A., and Yoshino, H. (1974b). *Agric. Biol. Chem.* **38,** 785–790.
Fujimoto, M., Kuninaka, A., and Yoshino, H. (1974c). *Agric. Biol. Chem.* **38,** 1555–1561.
Fujimoto, M., Fujiyama, K., Kuninaka, A., and Yoshino, H. (1974d). *Agric. Biol. Chem.* **38,** 2141–2147.
Fujimoto, M. Kuninaka, A., and Yoshino, H. (1975a). *Agric. Biol. Chem.* **39,** 1991–1997.
Fujimoto, M., Kuninaka, A., and Yoshino, H. (1975b). *Agric. Biol. Chem.* **39,** 2145–2148.
Fujimura, Y., Hasegawa, Y., Kaneko, Y., and Doi, S. (1964). *Nippon Nogei Kagaku Kaishi* **38,** 467–471.
Fujimura, Y., Hasegawa, Y., Kaneko, Y., and Doi, S. (1967). *Agric. Biol. Chem.* **31,** 92–100.
Fujiwara, M., Byhovsky, V., Nakamura, H., Tamura, G., and Arima, K. (1967). *J. Gen. Appl. Microbiol.* **13,** 1–14.
Fukuda, H., Yashima, S., Akiyama, S., Fugono, T., Nakazawa, B., and Tada, T. (1963). British Patent 933,829.
Furuya, A., Abe, S., and Kinoshita, S. (1968). *Appl. Microbiol.* **16,** 981–987.
Furuya, A., Abe, S., and Kinoshita, S. (1969). *Appl. Microbiol.* **18,** 977–984.
Furuya, A., Abe, S., and Kinoshita, S. (1970a). *Appl. Microbiol.* **20,** 263–270.
Furuya, A., Abe, S., and Kinoshita, S. (1970b). *Agric. Biol. Chem.* **34,** 210–221.
Furuya, A., Abe, S., and Kinoshita, S. (1971). *Biotechnol. Bioeng.* **13,** 229–240.
Furuya, A., Okachi, R., Takayama, K., and Abe, S. (1973). *Biotechnol. Bioeng.* **15,** 795–803.
Furuya, A., Kato, F., and Nakayama, K. (1975). *Agric. Biol. Chem.* **39,** 767–771.
Gots, J. S. (1971). *Metab. Pathways, 3rd Ed.* **5,** 225–255.
Gots, J. S., and Gollub, E. G. (1957). *Proc. Natl. Acad. Sci. U.S.A.* **43,** 826–834.
Hamagishi, Y., Nishino, T., and Murao, S. (1975). *Agric. Biol. Chem.* **39,** 1015–1023.
Haneda, K., Hirano, A., Kodaira, R., and Ohuchi, S. (1971). *Agric. Biol. Chem.* **35,** 1906–1912.
Haneda, K., Komatsu, K., Kodaira, R., and Ohsawa, H. (1972). *Agric. Biol. Chem.* **36,** 1453–1460.
Hartman, S. C. (1970). *Metab. Pathways, 3rd Ed.* **4,** 1–68.
Hasegawa, Y., Nakai, T., Fujimura, Y., Kaneko, Y., and Doi, S. (1964). *Nippon Nogei Kagaku Kaishi* **38,** 461–466.
Hashimoto, K., and Okauchi, M. (1965). Japanese Patent 40-15,946.

Hino, T., Uozumi, T., Tamura, G., and Arima, K. (1971). *Agric. Biol. Chem.* **35,** 1109–1115.
Hiramaru, M., Uchida, T., and Egami, F. (1969a). *J. Biochem. (Tokyo)* **65,** 693–700.
Hiramaru, M., Uchida, T., and Egami, F. (1969b). *J. Biochem. (Tokyo)* **65,** 701–708.
Honjo, M., Marumoto, R., Kobayashi, K., and Yoshioka, Y. (1966). *Tetrahedron Lett.* **32,** 3851–3856.
Imada, A. (1976). *In* "Microbial Production of Nucleic Acid-related Substances" (K. Ogata *et al.*, eds.), pp. 101–111. Halsted Press, New York.
Ishibashi, J., Kamio, H., and Yoneda, M. (1965). Japanese Patent 40-12,914.
Ishii, K., and Shiio, I. (1972). *Agric. Biol. Chem.* **36,** 1511–1522.
Ishii, K., and Shiio, I. (1973). *Agric. Biol. Chem.* **37,** 287–300.
Ishiyama, J. (1975). *Agric. Biol. Chem.* **39,** 1331–1332.
Ishiyama, J., Yokotsuka, T., Saito, N., Kato, M., and Yoshida, F. (1973). *Abstr. Annu. Meet. Agric. Chem. Soc. Jpn.* p. 306.
Ishiyama, J., Yokotsuka, T., and Saito, N. (1974). *Agric. Biol. Chem.* **38,** 507–514.
Jacobsen, G. B., and Rodwell, V. W. (1972). *J. Biol. Chem.* **247,** 5811–5817.
Kanazawa, M. (1974). *Sekiyu To Biseibutsu (Petroleum and Microorganisms)* **12,** 33–36.
Katchman, B. J., and Fetty, W. O. (1955). *J. Bacteriol.* **69,** 607–615.
Kato, F., Furuya, A., and Abe, S. (1971). *Agric. Biol. Chem.* **35,** 1061–1067.
Kawai, H., Sako, F., and Endo, K. (1974). *Amino Acid Nucleic Acid (Tokyo)* **30,** 62–73.
Keir, H. M., Mathog, R. H., and Carter, C. E. (1964). *Biochemistry* **3,** 1188–1193.
Kinoshita, K., Shiro, T., Yamazaki, A., Kumashiro, I., Takenishi, T., and Tsunoda, T. (1967). *Biotechnol. Bioeng.* **9,** 329–342.
Kinoshita, K., Aoki, M., Yamanoi, A., and Shiro, T. (1968). *Nippon Nogei Kagaku Kaishi* **42,** 529–535.
Kinoshita, K., Tsuri, H., Sakai, S., Yasunaga, M., Okada, H., and Shiro, T. (1969a). *Nippon Nogei Kagaku Kaishi* **43,** 400–403.
Kinoshita, K., Sakai, S., Yasunaga, M., Sasaki, H., and Shiro, T. (1969b). *Nippon Nogei Kagaku Kaishi* **43,** 404–409.
Kinoshita, K., Yasunaga, M., Sakai, S., and Shiro, T. (1969c). *Nippon Nogei Kagaku Kaishi* **43,** 473–477.
Kinoshita, K., Niwa, K., Sasaki, H., Sakai, S., and Hirose, Y. (1973). *Nippon Nogei Kagaku Kaishi* **47,** 793–798.
Komatsu, K., and Kodaira, R. (1973). *J. Gen. Appl. Microbiol.* **19,** 263–271.
Komatsu, K., Haneda, K., Hirano, A., Kodaira, R., and Ohsawa, H. (1972). *J. Gen. Appl. Microbiol.* **18,** 19–27.
Komuro, T., Nara, T., Misawa, M., and Kinoshita, S. (1969). *Agric. Biol. Chem.* **33,** 1018–1029.
Konishi, S., and Shiro, T. (1968). *Agric. Biol. Chem.* **32,** 396–398.
Konishi, S., Kubota, K., Aoki, R., and Shiro, T. (1968). *Amino Acid Nucleic Acid (Tokyo)* **18,** 15–20.
Koshiro, M., Fukuhara, S., Sowa, T., Ohuchi, S. (1972). Japanese Patent 47-23,296.
Kotani, Y., Yamaguchi, T., Kato, T., and Furuya, A. (1977). *Abstr. Annu. Meet. Agric. Chem. Soc. Jpn.* p. 397.
Kumashiro, I., Yamazaki, A., Meguro, T., Takenishi, T., and Tsunoda, T. (1968). *Biotechnol. Bioeng.* **10,** 303–320.
Kuninaka, A., Otsuka, S., Kobayashi, Y., and Sakaguchi, K. (1959). *Bull. Agric. Chem. Soc. Jpn.* **23,** 239–243.
Kuninaka, A., Kibi, M., Yoshino, H., and Sakaguchi, K. (1961). *Agric. Biol. Chem.* **25,** 693–701.
Kuninaka, A., Fujimoto, M., and Yoshino, H. (1975a). *Agric. Biol. Chem.* **39,** 597–601.

Kuninaka, A., Fujimoto, M., and Yoshino, H. (1975b). *Agric. Biol. Chem.* **39,** 603–610.
Kuroiwa, Y., and Horie, Y. (1955). *Bull Agric. Chem. Soc. Jpn.* **19,** 35–42.
Kuwahara, H., Tachiki, T., Tochikura, T., and Ogata, K. (1972). *Agric. Biol. Chem.* **36,** 745–754.
Kuwata, S. (1955). *Vitamins* **9,** 453–462.
Kuwata, S., Masuda, T., Sawa, Y., and Asai, M. (1954). *Vitamins* **7,** 990.
Lazzarini, R. A., Cashel, M., and Gallant, J. (1972). *J. Biol. Chem.* **246,** 4381–4385.
Lee, S. Y., Nakao, Y., and Bock, R. M. (1968). *Biochim. Biophys. Acta* **151,** 126–136.
Linn, S., and Lehman, I. R. (1965a). *J. Biol. Chem.* **240,** 1287–1293.
Linn, S., and Lehman, I. R. (1965b). *J. Biol. Chem.* **240,** 1294–1304.
Linn, S., and Lehman, I. R. (1966). *J. Biol. Chem.* **241,** 2694–2699.
Logan, D. M., and Singer, M. F. (1968). *J. Biol. Chem.* **243,** 6161–6166.
Lutwak-Mann, C., and Mann, T. (1935). *Biochem. Z.* **281,** 140–156.
Magasanik, B. (1957). *Annu. Rev. Microbiol.* **11,** 221–252.
Magasanik, B. (1962). *In* "The Bacteria" (I. C. Gunsalus and R. Y. Stanier, eds.), Vol. 3, pp. 295–334. Academic Press, New York.
Mantani, S., Fukumoto, J., and Yamamoto, T. (1972). *Agric. Biol. Chem.* **36,** 242–248.
Matsui, H., Satou, K., Anzai, Y., Enei, H., and Hirose, Y. (1974). *Proc. Symp. Amino Acid Nucleic Acid 1974,* p. 20.
Misawa, M., Nara, T., Udagawa, K., Abe, S., and Kinoshita, S. (1969). *Agric. Biol. Chem.* **33,** 370–376.
Mitsugi, K. (1964). *Agric. Biol. Chem.* **28,** 669–677.
Mitsugi, K., Komagata, K., Takahashi, M., Iizuka, H., and Katagiri, H. (1964). *Agric. Biol. Chem.* **28,** 586–600.
Mitsugi, K., Nakazawa, E., Okumura, S., Takahashi, M., and Yamada, H. (1965). *Agric. Biol. Chem.* **29,** 1051–1058.
Miwa, M. (1959). *Hakko Kyokaishi* **17,** 535–540.
Moat, A. G., and Friedman, H. (1960). *Bacteriol. Rev.* **24,** 309–339.
Momose, H. (1967). *J. Gen. Appl. Microbiol.* **13,** 39–51.
Momose, H. (1968). *Tanpakushitsu Kakusan Koso* **13,** 781–797.
Momose, H., and Shiio, I. (1969). *J. Gen. Appl. Microbiol.* **15,** 399–411.
Momose, H., Nishikawa, H., and Katsuya, N. (1964). *J. Gen. Appl. Microbiol.* **10,** 343–358.
Momose, H., Nishikawa, H., and Katsuya, N. (1965). *J. Gen. Appl. Microbiol.* **11,** 211–220.
Momose, H., Nishikawa, H., and Shiio, I. (1966). *J. Biochem. (Tokyo)* **59,** 325–331.
Muramatsu, N., and Takenishi, T. (1965). *J. Org. Chem.* **30,** 3211–3212.
Murao, S., and Nishino, T. (1973). *Agric. Biol. Chem.* **37,** 2929–2930.
Murao, S., and Nishino, T. (1974). *Agric. Biol. Chem.* **38,** 2483–2489.
Murata, K., Uchida, T., Kato, J., and Chibata, I. (1977). *Abstr. Annu. Meet. Agric. Chem. Soc. Jpn.* p. 287.
Nakamura, N., Takasawa, S., and Tanaka, M. (1968). *Vitamins* **37,** 622.
Nakao, Y. (1976). *In* "Microbial Production of Nucleic Acid-related Substances" (K. Ogata *et al.*, eds.), pp. 87–100. Halsted Press, New York.
Nakao, Y., and Ogata, K. (1963). *Agric. Biol. Chem.* **27,** 291–301.
Nakao, Y., Nogami, I., and Ogata, K. (1963). *Agric. Biol. Chem.* **27,** 507–517.
Nakao, Y., Lee, S. Y., Halvorson, H. O., and Bock, R. M. (1968). *Biochim. Biophys. Acta* **151,** 114–125.
Nakayama, K., Suzuki, T., Satō, Z., and Kinoshita, S. (1964). *J. Gen. Appl. Microbiol.* **10,** 133–142.
Nakayama, K., Nara, T., Tanaka, H., Satō, Z., Misawa, M., and Kinoshita, S. (1965). *Agric. Biol. Chem.* **29,** 234–238.

Nakayama, K., Satō, Z., Tanaka, H., and Kinoshita, S. (1968). *Agric. Biol. Chem.* **32,** 1331–1336.
Nara, T., Misawa, M., and Kinoshita, S. (1967). *Agric. Biol. Chem.* **31,** 1351–1356.
Nara, T., Misawa, M., and Kinoshita, S. (1968a). *Agric. Biol. Chem.* **32,** 561–567.
Nara, T., Misawa, M., and Kinoshita, S. (1968b). *Agric. Biol. Chem.* **32,** 1153–1161.
Nara, T., Misawa, M., Komuro, T., and Kinoshita, S. (1969a). *Agric. Biol. Chem.* **33,** 358–369.
Nara, T., Komuro, T., Misawa, M., and Kinoshita, S. (1969b). *Agric. Biol. Chem.* **33,** 1030–1036.
Nestle, M., and Roberts, W. K. (1969a). *J. Biol. Chem.* **244,** 5213–5218.
Nestle, M., and Roberts, W. K. (1969b). *J. Biol. Chem.* **244,** 5219–5225.
Nierlich, D. P., and Magasanik, B. (1963). *Fed. Proc., Fed. Am. Soc. Exp. Biol.* **22,** 476.
Nierlich, D. P., and Magasanik, B. (1965). *J. Biol. Chem.* **240,** 358–365.
Nishikawa, H., Momose, H., and Shiio, I. (1967). *J. Biochem. (Tokyo)* **62,** 92–98.
Nishimura, N., Shibatani, T., Kakimoto, T., and Chibata, I. (1974). *Appl. Microbiol.* **28,** 117–123.
Nogami, I., and Yoneda, M. (1969). *Kagaku To Seibutsu* **7,** 371–377.
Nogami, I., Kida, M., Iijima, T., and Yoneda, M. (1968). *Agric. Biol. Chem.* **32,** 144–152.
Numata, T., Kishi, M., Sato, T., Yano, N., Senoo, S., and Hiro, H. (1969). Japanese Patent 44-25,587.
Ogata, K., and Imada, A. (1962). *Annu. Rep. Takeda Res. Lab.* **21,** 31–53.
Ogata, K., Nakao, Y., Igarashi, S., Ohmura, E., Sugino, Y., Yoneda, M., and Suhara, I. (1963). *Agric. Biol. Chem.* **27,** 110–115.
Ogata, K., Shimizu, S., and Tani, Y. (1972a). *Agric. Biol. Chem.* **36,** 84–92.
Ogata, K., Tani, Y., Shimizu, S., and Uno, K. (1972b). *Agric. Biol. Chem.* **36,** 93–100.
Ohta, Y., and Ueda, S. (1968). *Appl. Microbiol.* **16,** 1293–1299.
Okabayashi, T., Ide, M., and Yoshimoto, A. (1964). *Amino Acid Nucleic Acid (Tokyo)* **10,** 117–123.
Oki, T., Yoshimoto, A., Ogasawara, T., Sato, S., and Takamatsu, A. (1976). *Amino Acid Nucleic Acid (Tokyo)* **34,** 52–62.
Partridge, C. W. H., and Giles, N. H. (1957). *Arch. Biochem. Biophys.* **67,** 237–238.
Reece, M. C., Donald, M. B., and Crook, E. M. (1959). *J. Biochem. Microbiol. Technol. Eng.* **1,** 217–228.
Saeki, N., Watanabe, T., Hoda, Y., and Oka, O. (1976). *Abstr. Annu. Meet. Agric. Chem. Soc. Jpn.* p. 117.
Sakai, T., Watanabe, T., and Chibata, I. (1973a). *Agric. Biol. Chem.* **37,** 849–856.
Sakai, T., Uchida, T., and Chibata, I. (1973b). *Agric. Biol. Chem.* **37,** 1041–1048.
Sanno, Y., Nara, K., Minato, S., and Hirose, U. (1964). Japanese Patent 39-15,846.
Saruno, R., Takahira, H., and Fujimoto, M. (1964). *J. Ferment. Technol.* **42,** 475–480.
Sato, A., and Furuya, A. (1976). *Agric. Biol. Chem.* **40,** 465–474.
Sato, A., and Furuya, A. (1977). *Agric. Biol. Chem.* **41,** 641–646.
Satō, Z., Nakayama, K., Tanaka, H., and Kinoshita, S. (1965). *Agric. Biol. Chem.* **29,** 412–418.
Shibai, H., Ishizaki, A., Mizuno, H., and Hirose, Y. (1973). *Agric. Biol. Chem.* **37,** 91–97.
Shiio, I., and Ishii, K. (1971). *J. Biochem. (Tokyo)* **69,** 339–347.
Shimizu, S., Tani, Y., and Ogata, K. (1972). *Agric. Biol. Chem.* **36,** 370–377.
Shimizu, S., Miyata, K., Tani, Y., and Ogata, K. (1973a). *Agric. Biol. Chem.* **37,** 607–613.
Shimizu, S., Kubo, K., Tani, Y., and Ogata, K. (1973b). *Agric. Biol. Chem.* **37,** 2863–2870.
Shimizu, S., Kubo, K., Morioka, H., Tani, Y., and Ogata, K. (1974). *Agric. Biol. Chem.* **38,** 1015–1021.

Shimizu, S., Morioka, H., Tani, Y., and Ogata, K. (1975). *J. Ferment. Technol.* **53,** 77–83.
Shirafuji, H., Imada, A., Yashima, S., and Yoneda, M. (1968). *Agric. Biol. Chem.* **32,** 69–75.
Shiro, T., Yamanoi, A., Konishi, S., Okumura, S., and Takahashi, M. (1962). *Agric. Biol. Chem.* **26,** 785–786.
Shishido, K., and Ando, T. (1972). *Biochim. Biophys. Acta* **287,** 477–484.
Shishido, K., and Ikeda, Y. (1970). *J. Biochem. (Tokyo)* **67,** 759–765.
Shishido, K., and Ikeda, Y. (1971a). *J. Mol. Biol.* **55,** 287–291.
Shishido, K., and Ikeda, Y. (1971b). *Biochem. Biophys. Res. Commun.* **42,** 482–489.
Shive, W., Ackermann, W. W., Gordon, M., Getzendaner, M. E., and Eakin, R. E. (1947). *J. Am. Chem. Soc.* **69,** 725–726.
Soeda, E., Murata, A., Utsu, M., and Saruno, R. (1968). *Seikagaku* **40,** 688.
Soeda, E., Abe, S., Murata, A., and Saruno, R. (1969). *Nippon Nogei Kagaku Kaishi* **43,** N86.
Spahr, P. F. (1964). *J. Biol. Chem.* **239,** 3716–3726.
Spahr, P. F., and Schlessinger, D. (1963). *J. Biol. Chem.* **238,** PC 2251–2253.
Stetten, M. R., and Fox, C. L., Jr. (1945). *J. Biol. Chem.* **161,** 333–349.
Stevens, A., and Hilmoe, R. J. (1960a). *J. Biol. Chem.* **235,** 3016–3022.
Stevens, A., and Hilmoe, R. J. (1960b). *J. Biol. Chem.* **235,** 3023–3027.
Sugimoto, H., Iwasa, T., and Yokotsuka, T. (1964a). *Nippon Nogei Kagaku Kaishi* **38,** 135–143.
Sugimoto, H., Iwasa, T., and Yokotsuka, T. (1964b). *Nippon Nogei Kagaku Kaishi* **38,** 567–575.
Suhara, I. (1973). *J. Biochem. (Tokyo)* **73,** 1023–1032.
Suhara, I. (1974). *J. Biochem. (Tokyo)* **75,** 1135–1141.
Suhara, I., and Yoneda, M. (1973). *J. Biochem. (Tokyo)* **73,** 647–654.
Takebe, I., and Kitahara, K. (1963). *J. Gen. Appl. Microbiol.* **9,** 31–40.
Tanaka, H., Satō, Z., Nakayama, K., and Kinoshita, S. (1968). *Agric. Biol. Chem.* **32,** 721–726.
Tanaka, K., Mizuno, K., Sanno, Y., and Hamuro, Y. (1963). Japanese Patent 38-16,892.
Tanaka, K., Ohmura, E., Honjo, M., Sanno, Y., and Sugino, Y. (1964a). Japanese Patent 39-7779.
Tanaka, K., Ohmura, E., Ogata, K., Sanno, Y., Yoneda, M., and Suhara, I. (1964b). Japanese Patent 39-12,342.
Tochikura, T., Kuwahara, M., Yagi, S., Okamoto, H., Tominaga, Y., Kano, T., and Ogata, K. (1967). *J. Ferment. Technol.* **45,** 511–529.
Tochikura, T., Kuwahara, M., Komatsubara, S., Fujisaki, M., Suga, A., and Ogata, K. (1969). *Agric. Biol. Chem.* **33,** 840–847.
Tone, H., and Ozaki, A. (1968). *Enzymologia* **34,** 101–128.
Tone, H., Umeda, M., Ishikura, T., and Miyachi, N. (1964). *Amino Acid Nucleic Acid (Tokyo)* **10,** 142–149.
Uchida, K., Kuninaka, A., Yoshino, H., and Kibi, M. (1961). *Agric. Biol. Chem.* **25,** 804–805.
Ueno, Y., Wada, S., Imada, I., and Tomoda, K. (1965). Japanese Patent 40-12,515.
Uozumi, T., and Arima, K. (1974). *Agric. Biol. Chem.* **38,** 1739–1740.
Uozumi, T., Tamura, G., and Arima, K. (1968a). *Agric. Biol. Chem.* **32,** 963–968.
Uozumi, T., Tamura, G., and Arima, K. (1968b). *Agric. Biol. Chem.* **32,** 969–974.
Uozumi, T., Tamura, G., and Arima, K. (1968c). *Agric. Biol. Chem.* **32,** 1409–1413.
Uozumi, T., Tamura, G., and Arima, K. (1969a). *Agric. Biol. Chem.* **33,** 25–30.
Uozumi, T., Tamura, G., and Arima, K. (1969b). *Agric. Biol. Chem.* **33,** 636–644.
Uozumi, T., Tamura, G., and Arima, K. (1969c). *Agric. Biol. Chem.* **33,** 645–652.
Uozumi, T., Hino, T., Tamura, G., and Arima, K. (1972). *Agric. Biol. Chem.* **36,** 434–441.

Vendrely, R. (1955). *Int. Rev. Cytol.* **4,** 115–142.
Wade, H. E., and Morgan, D. M. (1957). *Biochem. J.* **65,** 321–331.
Watanabe, K. (1976). *In* "Microbial Production of Nucleic Acid-related Substances" (K. Ogata *et al.*, eds.), pp. 55–65. Halsted Press, New York.
Watanabe, T., Uchida, T., Kato, J., and Chibata, I. (1974). *Appl. Microbiol.* **27,** 531–536.
Wyngaarden, J. B., and Ashton, D. M. (1959). *J. Biol. Chem.* **234,** 1492–1496.
Yamada, K. (1976). *Hakko To Kogyo* **34,** 658–694.
Yamanoi, A., and Shiro, T. (1968). *J. Gen. Appl. Microbiol.* **14,** 1–17.
Yamanoi, A., Konishi, S., and Shiro, T. (1967). *J. Gen. Appl. Microbiol.* **13,** 365–380.
Yamasaki, M., and Arima, K. (1967). *Biochim. Biophys. Acta* **139,** 202–204.
Yamashita, T., Hidaka, T., and Watanabe, K. (1974). *Agric. Biol. Chem.* **38,** 727–734.
Yamazaki, A., Kumashiro, I., and Takenishi, T. (1967). *J. Org. Chem.* **32,** 1825–1828.
Yoneda, M. (1964a). *J. Biochem. (Tokyo)* **55,** 475–480.
Yoneda, M. (1964b). *J. Biochem. (Tokyo)* **55,** 481–485.
Yoshikawa, M., and Kato, T. (1967). *Bull. Chem. Soc. Jpn.* **40,** 2849–2853.
Yoshikawa, M., Kato, T., and Takenishi, T. (1967). *Tetrahedron Lett.* **50,** 5065–5068.
Yoshikawa, M., Kato, T., and Takenishi, T. (1969). *Bull. Chem. Soc. Jpn.* **42,** 3505–3508.
Yoshikawa, M., Sakuraba, M., and Kusashio, K. (1970). *Bull. Chem. Soc. Jpn.* **43,** 456–461.

Chapter 11

Production of Organic Acids by Fermentation

LEWIS B. LOCKWOOD

MICROBIAL TECHNOLOGY, 2nd ed., VOL. I
 ISBN 0-12-551501-4

I. INTRODUCTION

Organic acids discussed here are those which are manufactured in large volume or which offer potential for future development. They are marketed as relatively pure chemicals or their salts. These processes have been developed within the last hundred years, in contrast to some of the major food and beverage fermentations, the origin of which is found in antiquity. Operation of the first plant for the bulk manufacture of a fermentation chemical was started in 1881 in Massachusetts for the production of calcium lactate by bacterial fermentation. Wehmer in 1893 reported that certain fungi of the genus *Citromyces* could accumulate citric acid in liquid cultures. The genus *Citromyces* is now included in the monoverticillate group of the genus *Penicillium*. A plant to produce citric acid using Wehmer's culture was established in Alsace, France and operated for about 10 years (Mial, 1975). These plants provided the starting point from which much modern fermentation technology was developed.

II. CITRIC ACID

Citric acid was recovered in 1869 in England from calcium citrate imported from Italy. Calcium citrate was marketed through an Italian government-controlled cartel, which held the price very high and thus retarded the development of new markets for citric acid. Introduction of citric acid made by a new fermentation process in 1923 broke the power of the cartel, and today's prices are considerably lower than those prior to 1923. The price of citric acid dihydrate in the United States in 1976 was $0.495 per pound in 250-lb drums, and the price of anhydrous citric acid was $0.555–0.605 per pound. The price of calcium citrate was $0.69 per pound. The four citric acid plants of two manufacturers in the United States are in the eastern part of the country at Southport, North Carolina and Groton, Connecticut (Charles Pfizer, Inc.), Dayton, Ohio and Elkhart, Indiana (Miles Laboratories). Prices are about $0.045 per lb higher west of the Rocky Mountains.

The fermentation processes now account for almost all the citric acid marketed in the world today, although a small amount is recovered from citrus fruit in Mexico and in South America. This is estimated to amount to less than 1% of total world production. Citric acid production in the United States and Europe probably amounts to 210,000,000 lb annually. In addition, Japan produces about 10,000,000 lb, and about 20,000,000 lb are produced in Brazil, Israel, Mexico, Colombia, and Argentina.

Three processes are used in the manufacture of citric acid. These are the Japanese koji process, in which special strains of *Aspergillus niger* are used; the liquid surface culture fermentation, in which *A. niger* floats on the surface of a solution; and the submerged fermentation process, in which the fungal mycelium grows throughout a solution in a deep tank.

Scheele (1789) first reported the isolation and crystallization of the sour constituent of lemon juice in 1784. Grimaux and Adams (1880) synthesized citric acid from glycerol.

Wehmer (1893) established the occurrence of citric acid as a microbial metabolite. The next noteworthy step in the development of the citric acid production process was the recognition by Molliard (1922) that citric acid accumulates in cultures of *A. niger* under conditions of nutrient deficiency. He attributed his results to phosphate deficiency in his culture media. Yields were too small to permit commercial exploitation, but his publication lead to further development of the concept of nutrient deficiency especially in the development of the submerged fermentation process. In a culture solution adequate to permit the maximum growth of mycelium or to permit sporulation, little or no citric acid accumulates.

A. Koji Fermentation Process

The Japanese wheat bran process accounts for about one-fifth of Japanese citric acid production annually. Originally the solid residues left after sweet potato starch recovery were the fermentation medium. Later, wheat bran was substituted for the sweet potato material, but data are not available on whether sweet potato wastes are still used to any extent. In terms of process operations, the conditions of fermentation are almost identical with those when fungal amylase and protease of *Aspergillus oryzae* are produced. Prior to sterilization, the pH of the bran is adjusted to between 4 and 5, and additional moisture is picked up during steaming, so that the final water concentration of the mash is 70–80%. When the bran has cooled to 30°–36°C the mass is inoculated with a koji which was made by a special strain of *A. niger* which is probably not as responsive to the presence of ions of iron as the culture strains used in other processes. Hisanaga and Nakamura (1966) and Yamada (1965) reported that the temperature of the mash during fermentation should not exceed 28°C. Addition of 3–7% of filter cake from a glutamic acid fermentation resulted in improvement in yield. Starch originally present in the bran is saccharified by the amylase of *A. niger*. Addition of α-amylase to the bran after cooling was also beneficial. The inoculated bran is spread in trays to a depth of 3–5 cm or spread in windrows on a floor. After 5–8 days, the koji is harvested and placed in percolators, and the citric acid is extracted with water. Purification and recovery follow essentially the same process steps used in the recovery of the acid when either the liquid surface culture process or the submerged fermentation processes is used.

B. Liquid Surface Culture Process

The liquid surface culture process is the one which first supplied citric acid at a low price. It is still used by many manufacturers, but operations are highly secret, and only one detailed description of the process has been published (Mallea, 1950). The inoculated fermentations are dispersed in shallow pans made of high-purity aluminum or of stainless steel. Humidified air is blown over the surface of the solution for 5 or 6 days, after which dry air is used. If the CO_2 content of the air rises as high as 10%, a marked reduction of yield occurs. Spores germinate within 24 hours, and a white, wrinkled mycelium covers the surface of the solution. Eight or 10 days after inoculation, the original sugar concentration of 20–25% has been reduced to the range of 1–3%. At the end of the fermentation, the liquid can be drained off and replaced by fresh culture solution if care is taken to introduce the replacement solution under the mycelium to ensure that the mycelium still floats on the surface. Any portion of the mycelial mat which becomes submerged is inactivated. The average length of the fermentation cycle is reduced since the initial growth period of about 3 days is bypassed. Little citric acid is produced during the growth phase. Such reuse of mycelium has been done in Argentina and in Russia, but it is doubtful if it is done in any large plant today.

The preferred sugar substrate for the fermentation is sugar beet molasses which has not undergone the Steffens process for sucrose recovery. This molasses contains sucrose as the principal carbohydrate, and some glucose as well as proteins, peptides, amino acids, and inorganic ions. It has been subjected to heat during the concentration of the beet extract and sucrose crystallization, so doubtless it contains saccharinic acids and related compounds at least in traces. These may serve to counteract in part the effect of traces of undesirable metallic ions.

Mallea (1950) appreciated the importance of metallic ions in the control of the accumulation of citric acid in solutions for the surface culture of *A. niger*. He developed on a commercial scale a system for removal of metallic ions, or reduction in quantity of undesirable ions in sucrose syrup by adsorption with a combination of $CaCO_3$, colloidal silica, tricalcium phosphate, and starch. Steinberg (1945) had previously removed trace metallic ions from sucrose solutions by heating with $Ca(OH)_2$. It appears probable that some activated carbons might also be effective in trace element removal from liquid surface culture media. Some manufacturers precipitate iron by addition of calcium ferrocyanide.

The pH of the culture solution for citric acid accumulation is usually initially in the range of 5–6 but, on spore germination, rapidly approaches the range of pH 1.5–2.0 as ammonium ions are removed from the solution. If it rises to about 3.5 after the initial change or is adjusted to 3.5 or above, some oxalic acid is formed along with citric acid. The presence of iron or sodium salts also favors the accumulation of oxalic acid, and of yellow or yellow-green pigments in the mycelium. These pigments are sometimes excreted into the culture solution and are difficult to remove during product recovery and purification.

The final yield of citric acid in the liquid surface culture process is in the range of 80–85% of the weight of the carbohydrate initially supplied.

C. Submerged Culture Processes

The advantages of the submerged fermentation process over the liquid surface fermentation processes in terms of investment and operating costs induced investigators to attempt to develop a submerged fermentation process for citric acid manufacture. Amelung (1930), using a black *Aspergillus* which he called *A. niger japonicus* (probably *Aspergillus japonicus* Saito) slowly bubbled a stream of air through a culture solution of 15-cm depth. He obtained subsurface growth and found citric acid in the culture solution. Yields were inferior to those commonly found in liquid surface culture fermentation.

Kluyver and Perquin (1933) described the use of cultures on shakers, a technique for the study of submerged fermentation in small scale. Up to this time, the study of submerged fermentation required large equipment and an air compressor. Perquin (1938) used shaker cultures for the study of the production of organic acids. He published in his doctoral dissertation a laboratory-scale method for the production of citric acid by *A. niger* in which he extended Molliard's phosphate deficiency concept to submerged fermentations. Neither Molliard nor Perquin appears to have recognized the role of trace metallic ions which are commonly occurring impurities in phosphate salts. Following the Molliard–Perquin concepts, Szücs (1944) obtained a patent for the submerged culture production of citric acid by the method of growing *A. niger* on a culture solution which contained KH_2PO_4 and transferring the mycelium to a new culture solution to which he had not added phosphate, or by the method of growing mycelium in a culture solution initially inadequately supplied with phosphate. He assumed that the growing mycelium would remove all the phosphate from the solution, leaving it free of phosphate.

Moyer (1953) found that addition of methanol to his shaker cultures of *A. niger* resulted in improvement in yield of citric acid. Neither this nor the application of the Molliard–Perquin concepts has led to a commercially successful process. Studies conducted by Miles Laboratories have failed to demonstrate any relationship between assimilable phosphorus in the culture solution and citric acid accumulation after the initial mycelial growth period (Lockwood and Snell, 1976).

Shu and Johnson (1947, 1948a,b) demonstrated that when ions of manganese or iron were present in sufficient concentration little or no citric acid accumulates in the culture solutions, but *A. niger* grew much better than when these ions were inadequately supplied. Removal from the culture solution of undesirable metallic ions by ion exchange was demonstrated by Woodward *et al.* (1949) and the use of copper ion as an antagonist to iron, discovered by Schweiger (1957), made possible the present commercial process. Tables I and II show the response of the citric fermentation to ferric ions, and the counteraction of this response by copper ions. Bruchmann (1961) found that in the presence of copper ions, the enzyme aconitase is inhibited. He believed that the action of the zinc ion was at some later point in the destruction of citric acid.

Culture solutions are flash sterilized by passage through the pipes of a

TABLE I. Effect of Iron on the Production of Citric Acid from Decationized Sucrose by *A. niger* in Submerged Culture[a]

Iron[b] (mg/liter)	Weight yield citric acid[c] (%)
0.0	67.0
0.05	71.0
0.5	88.0
0.75	79.0
1.00	76.0
2.00	71.0
5.00	57.0
10.0	39.0

[a] From Schweiger (1957). Medium composition: sucrose purified by mixed resin bed ion exchange from 3,600,000 ohms resistance at 40% concentration, diluted with deionized water at 18,000,000 ohms resistance to 14.2% sugar content, and recrystallized nutrient salts added: KH_2PO_4, 0.014%; $MgSO^4 \cdot 7\,H_2O$, 0.1%; $(NH_4)_2CO_3$, 0.2%; HCl to pH 2.6.

[b] Supplied as $FeCl_3$.

[c] (Grams citric acid produced/ Equivalent grams hexose moiety supplied) × 100.

TABLE II. Effect of Ions of Copper and Iron on the Production of Citric Acid from Glucose by *A. niger* in Submerged Culture[a]

Fe^{3+} (mg/liter)	Cu^{2+} (mg/liter)	Citric acid yield[b] (%)
10	50	77.8
50	50	69.1
100	50	50.7
150	50	14.2
10	100	77.2
50	100	65.4
100	100	53.9
150	100	29.8
10	500	74.0
50	500	65.4
100	500	60.6
150	500	27.6

[a] From Schweiger (1957). Medium composition: glucose (commercial grade), approx. 14%, $(NH_4)_2CO_3$, 0.2%; KH_2PO_4, 0.014%; $MgSO_4 \cdot 7\ H_2O$, 0.1% Fe and Cu supplied as sulfates.

[b] (Grams citric acid produced/grams glucose supplied) × 100.

steam-jacketed heat exchanger and immediately cooled to about 30°C in another heat exchanger. To operate efficiently, it may be necessary to induce turbulent flow through the first heat exchanger. For decationized solutions No. 316 stainless steel is necessary in the construction of the heat exchanger and lines to the fermentor in order to avoid corrosion by the acidic solutions.

Inocula are usually spores of a suitable strain of *A. niger* grown on solid nutrient media. Stock cultures are kept refrigerated in tubes of a good loam soil, or as lyophilized spores. *Aspergillus niger* is an unstable organism, and it is desirable to minimize the number of culture transfer generations from a culture which has proved satisfactory.

The range of copper ion required to be added for process control varies from 0.1 to 50.0 mg/liter, depending on the amount of iron in the culture solution (Table II). Other additives to the culture solution are KH_2PO_4, 0.01–0.3% (but may be as great as 2%), and $MgSO_4 \cdot 7\ H_2O$ to give up to 0.25%. If the content of iron is low enough, iron must be added to give 0.1–0.2 mg/liter. Copper ion may be required to counteract any excess of iron. The pH of the decationized solutions is between 1.5 and 2.0. Before inoculation of decationized solutions, the pH value is adjusted with ammonia to about pH 4.0. In both fermentations, the pH rapidly changes to the range of 1.5–2.0 by rapid uptake of ammonium ion, leaving inorganic acid residues. Little citric acid is formed in young cultures before the pH of 2.0 is reached.

Aeration of the submerged fermentation must be continuous. Air is bubbled through the solution at a rate of 0.5–1.5 v/v solution per minute. Mechanical stirring is not necessary. Batti (1966) has shown that even a brief failure of air supply stops the fermentation and that it will not be reactivated for several days unless the pH is adjusted to permit new growth. At the same time, additional copper ion must be added to ensure that the new growth is of the right biochemical type. The combination of aeration, low pH, and the extreme sensitivity to iron and copper make the use of ordinary iron fermentors impossible unless some impervious coating is applied to them. Glass and plastic coatings have been used, but the common blue glass linings are unsatisfactory because cobalt leaches out of the glass. Antifoam agents are necessary to prevent loss of material due to foamover at any time during the fermentation. Such agents must be free of iron, cobalt, or nickel.

Process controls in the submerged fermentation depend primarily on microscopic examination of mycelia. Schweiger (1957) has described the short, thick, stubby branches and abundant chlamydospores of active mycelia. An experienced technician can estimate the amount of iron or copper to be added to control the rate and efficiency of the fermentation. Figure 1 is a flow sheet of the citric fermentation process.

A series of organic compounds which counteract the effect of iron on

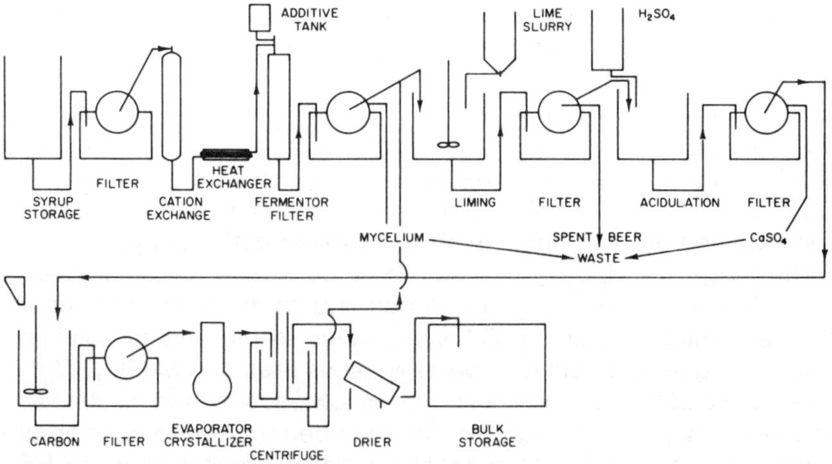

FIGURE 1. Flow sheet of the manufacture of citric acid. (From Lockwood and Schweiger, 1967; reprinted by permission of the Van Nostrand-Reinhold Co., New York.)

citric acid accumulation has been patented (Lockwood and Batti, 1965). Their action appears to be similar to the action of copper ion in that growth and sporulation are inhibited and mycelium produced has the proper enzymatic makeup for citric acid accumulation. The organic compound antagonists are probably not used in the commercial manufacture of citric acid at this time since the copper ion is effective at lower cost.

Clark (1962) described a process in which ferrocyanide was used to remove or complex iron in beet or cane molasses. The initial pH before dilution of the molasses was pH 6 before sterilization but was adjusted to pH 6.5 after sterilization. Ferrocyanide concentrations were in the range of 20–150 mg/liter. Optimal yields were obtained when the initial concentrations of ferrocyanide were 20 mg/liter and were increased to 80 mg/liter after the initial growth period. The removal of trace metallic ions by ion exchange or precipitation by ferrocyanides are probably the only practical methods of purifying culture solutions for use in submerged culture production of citric acid. It is not known how many plants use each of these methods.

Solutions of high-test cane syrup, glucose, or sucrose are suitable carbohydrate substrates for the production of citric acid by either process. High-test syrup is concentrated cane juice in which about 30–35% of the sucrose has been inverted by yeast invertase. When the ferrocyanide treatment is used the substrate can be a solution of any of the above, or beet molasses which has not been treated by the Steffens process. Ion exchange may be too costly for use with the cruder molasses solutions. Yields obtained from decationized refined syrups are generally superior to those obtained from the less pure sugar sources. The initial sugar concentration is about 20–25% and yields of citric acid approximate those obtained in surface culture.

Meyrath (1967) has discussed the energy requirements for growth and citric acid excretion in relation to the acid production by nonproliferating cells. A significant amount of energy is required since the acid is excreted into a solution of much greater concentration. The culture solution concentration of citric acid may be as great as 1 M, while that within the cell probably never exceeds 0.01 M.

Continuous culture techniques are not considered suitable for use in citric acid production. A multitank system would be required for continuous fermentation in any process in which cell growth and metabolic product formation occur primarily at different times. This would require a significantly larger capital expenditure and would probably not be economically competitive.

D. Product Recovery

The filtered culture solution is subjected to a polishing filtration if the solution is hazy due to the presence of residual antifoam agents, mycelium, or oxalate. Calcium citrate is precipitated from the clear solution by addition of a slurry of $Ca(OH)_2$. This must be very low in magnesium content if losses due to the relatively soluble magnesium citrate are to be avoided. Calcium citrate is filtered off on any conventional type of industrial filter, and the filter cake is transferred to a tank where it is treated with sulfuric acid to precipitate calcium sulfate. The dilute filtrate containing the citric acid is purified by treatment with carbon which has been activated by treatment with heat or an hydrochloric acid wash. It is demineralized by successive passages through ion-exchange beds, and the purified solution is evaporated in a circulating granulator or in a circulating evaporator–crystallizer. Crystals are removed by centrifugation, and a predetermined fraction of the mother liquor is returned to the evaporator. The remaining mother liquor is returned to the recovery stream at a point prior to liming and decolorization. Citric acid so prepared may require recrystallization from water in order to meet the standards of the United States Pharmacopoeia. Mother liquors from recrystallization are recycled to a point prior to liming in primary product recovery. Direct crystallization of citric acid from fermented beers has been studied, but quality problems have always precluded the use of this method of recovery.

In the Usines de Melle patented recovery process (Anonymous, 1963), the filtered beer is extracted countercurrently at 10°–30°C with a mixture of 100 parts tri-*n*-butyl phosphate and 5–30 parts *n*-butyl acetate or methyl isobutylketone. The solvent is then extracted with water at 70°–95°C. Citric acid is further concentrated in the hot-water extract, decolorized, and crystallized in the usual manner.

E. Chemistry of the Citric Acid Fermentation

Citric acid is excreted into and accumulates in culture solutions of pH about 1.8–2.0, an acidity too great to be tolerated by many organisms. Data are not available as to the pH within the mitochondrion or even within the cell. Data are available which clarify some aspects of the problem. Blumenthal (1965) has reported on the relative importance of different metabolic pathways in fungi. In *A. niger* 78% of the sugar consumed passes through the Embden–Meyerhof–Parnas (EMP) pathway. Under the special conditions of the citric acid fermentation, this amount may be either reduced or increased, but such data are not available for cultures under the conditions required for citric acid accumulation. Both EMP and hexose monophosphate (HMP) pathways of glycolysis were reported to be active at all times, the greatest activity of the EMP pathway occurred during the vegetative stage, and that of the HMP pathway occurring during conidiation. Relatively little citric acid is found in cultures during the growth period, and most of the citric acid is produced by nonproliferating cells. Since cultures which sporulate accumulate almost no citric acid, it appears that the EMP pathway may play the major role in glycolysis in the citric acid fermentation.

Acetyl-CoA derived by the EMP pathway condenses with oxalacetate to form citrate, the point of entry into the oxidative sequence of an abortive citric acid cycle (Fig. 2). The cycle is aborted by inhibition of the enzymes aconitase and isocitric dehydrogenase either by copper ions, by H_2O_2 (Bruchmann, 1961), or possibly by the action of some constituent of beet molasses. Iron is an essential cofactor for these enzymes in the conversion of either citrate to aconitase or isocitrate to aconitate by isocitrate dehydrogenase. In noninhibited cultures supplied with iron, citric acid is an excellent substrate for growth. La Nauze (1966) has compared activity of aconitase, and NAD- and NADH-linked isocitrate dehydrogenases of culture strains which gave widely different citric acid yields. All three enzymes were present in both culture strains when grown on media in which citric acid did or did not accumulate. Copper ion was supplied to the culture solutions at 0.3 mg/liter and ferrous ion at 2 mg/liter. The ratio of copper ion to iron ion would not permit activity of the three enzymes at pH 2.0, an acidity approximating that of the commercial fermentations. She also found some citric acid inside the mycelium, an observation which I have confirmed. This would indicate that the intracellular pH was in the neighborhood of 2, a value much too

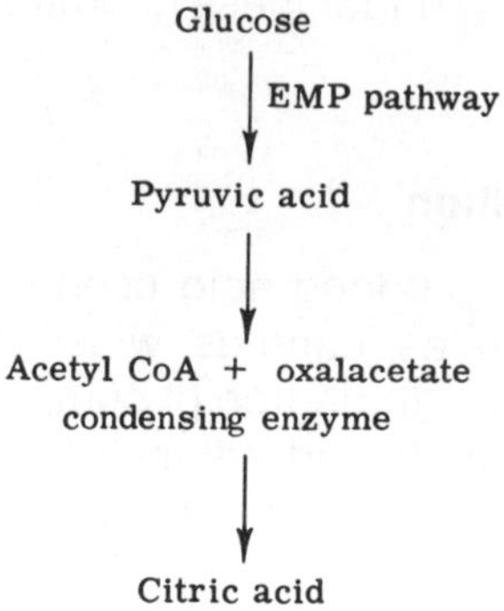

FIGURE 2. Biochemical sequence in citric acid fermentation.

acidic for these enzymes to be active. Thus, citric acid accumulation would result not from failure to produce the citric acid cycle enzymes, but from the inability of the enzymes to function under the physiologically adverse conditions of the fermentation. *Aspergillus niger* and *Penicillium javanicum* convert several intermediates of the citric acid cycle to citric and oxalic acids. However, Cleland and Johnson (1954), in radioactive carbon studies, found no evidence of recycling of carbon in the citric fermentation. Commercial yields of citric acid from D-glucose are great enough to preclude the possibility of significant recycling of carbon.

F. Marketing: Market Forms and Control Procedures

Citric acid is marketed as an anhydrous crystalline chemical, as the crystalline monohydrate, or as the crystalline sodium salt. Most of the free acid meets the specifications of the United States Pharmacopoeia, which sets maxima and minima for ash, heavy metals, sugar, other acids, and materials which cause a dark color when a sample of citric acid is heated in concentrated sulfuric acid (90°C for 1 hour). Anhydrous and monohydrate crystals are available in granular (15–50 mesh), fine, and powder form (35–100 mesh for anhydrous and 60–200 mesh for the monohydrate).

G. Packaging and Stability

Both anhydrous and monohydrate forms of citric acid are supplied in 100-lb multiwall moisture-barrier paper bags and in 100- and 250-lb drums. Caking may occur if citric acid is stored at high temperature and humidity. Storage at 21°C and 50–70% relative humidity is satisfactory. Citric acid does not oxidize under these conditions. Its solutions are only mildly corrosive.

H. Uses and Competition

Citric acid is the principal food acid used in the preparation of soft drinks, desserts, jams, jellies, candies, wines, and frozen fruits. In this area, malic acid offers a small degree of competition. Citric acid is also used in gelatin food products and artificial flavors of dry compounded materials, such as soft drink tablets and powders. Here, fumaric acid offers competition.

In pharmacy, citrates are used in blood transfusion, and the free acid is used in effervescent products. It is rapidly metabolized almost completely in the human body and can serve as a source of energy. Cosmetic uses of citric acid include those in astringent lotions, where it is used to adjust pH and act as a sequestrant and in hair rinses and hair setting fluids.

Citric acid is a superior sequestering agent in the neutral and low pH range, exhibiting some activity at pH values as low as 3.0. This complexing activity permits its wide industrial use in electroplating, in leather tanning, and in the reactivation of old oil wells, where the pores of the sand face have become clogged with iron.

III. ITACONIC ACID

Itaconic acid (methylene succinic acid) accumulates in cultures of *Aspergillus terreus* and to a lesser extent in those of *Aspergillus itaconicus*. The compound is a structurally substituted metharyllic acid.

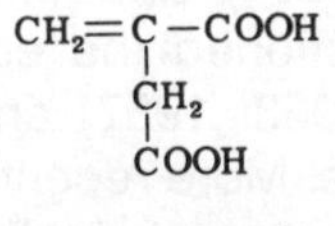

Itaconic acid

Consequently its principal use would be expected to be as a copolymer in acryllic of methacryllic resins. In these resins, inclusion of about 5% itaconate in the polymer imparts the ability to take and hold printing inks. Resins derived by polymerization of methyl itaconate tend to be rather brittle. Copolymerization with other esters makes a less brittle resin. Itaconic acid may also find some use in detergents.

Itaconic acid is marketed in two grades: refined, which is a pale tan to white crystalline solid and sells for \$0.83 per pound in car lots, and the industrial grade, which is darker in color and sells for \$0.22 per pound (U.S. Tariff Commission, 1974). The Charles Pfizer Co. is the sole manufacturer of this substance in the United States at the present time.

Itaconic acid was formerly obtained by pyrolysis of citric acid. Water is lost to produce aconitic acid. On loss of CO_2 from aconitic acid, two isomeric compounds are produced, the anhydrides of itaconic and citraconic acids. The process was not commercially successful. Aconitic acid is present in sugar cane juice and was believed to interfere in sucrose crystallization. Removal of calcium aconitate by sugar refiners, and its conversion to itaconic acid by heating of its solutions, was the principal source of itaconic acid until the fermentation process was developed.

In fermentation solutions, itaconic acid is accompanied by succinic acid and itatartaric acid. These interfere in analysis of the solutions when itaconic acid is measured by titration with alkali but do not interfere in the bromination method of Friedkin (1945).

A. Technological Development

Itaconic acid was first reported as a fungal metabolite by Kinoshita (1931, 1937) who described a new species, *Aspergillus itaconicus,* which produced as its principal water-soluble metabolites mannitol and itaconic acid. Calam *et al.* (1939) found that the compound accumulated in cultures of *A. terreus.* Moyer made a survey of isolates of *A. terreus* and found one strain, NRRL 1960, which was superior to his other strains in either surface or submerged culture. His strain or a mutant derived from it is believed to be the one now used industrially.

In early studies it was observed that sugar concentrations greater than 7% were not fermented efficiently. This value was confirmed by Tsao *et al.* (1962; Tsao and Su, 1964) and by Pfeifer *et al.* (1953, 1957). Kobayashi (1967) has questioned the economics of the process developed by Pfeifer *et al.* (1953, 1957) and considered his continuous fermentation to be preferable. More recently, it was observed that if beet molasses was used in the spore germination medium, it was possible to ferment cane molasses at 15% sugar concentration, with an itaconic acid concentration in the beer of 8.5%, a yield of 85% of the theoretical maximum.

B. Current Manufacturing Process

The current manufacturing process is the one based on the use of molasses. Detailed information on this process has not been published; however, the manufacturer owns the patent of Neubel and Ratajak (1962) which probably provides the best index of current usage.

Stock cultures are maintained on Czapek agar slants or as spores scraped off these slants and suspended in garden soil. Spores for inoculum are grown on either a liquid or solid vegetable medium. Spores are germinated in a solution of the composition shown in the tabulation below.

Component	Amount
	Sugar-beet molasses to give 15% sugar
$ZnSO_4$	1.5 gm
$MgSO_4 \cdot 7\ H_2O$	5.0 gm
$CuSO_4 \cdot 5\ H_2O$	0.02 gm
Soybean oil	0.25 ml
Water to	1 liter

Aeration at one-fourth v/v solution and high-speed agitation are maintained for about 18 hours; the temperature is held at 33°–37°C. Under these conditions spore germination results in the formation of mycelia of a suitable biochemical type. The pH changes from about 7.5 to 4.0.

The fermentation solution is inoculated with one-fifth volume of the suspension of germinated spores. This is an excessively large volume of inoculum but is probably desirable on account of the high temperature of the fermentation (39°–42°C), and to obtain enough beet molasses into the solution to give the benefits of beet molasses. The composition of the fermentation solution is shown below:

Component	Amount
	Cane molasses to give 150 gm sugar
$ZnSO_4$	1 gm
$MgSO_4 \cdot 7\,H_2O$	3.0 gm
$CuSO_4 \cdot 5\,H_2O$	0.01 gm
Water to	1 liter

Air is bubbled through the culture solution, and vigorous agitation is applied. During the first 24 hours, the pH changes from about 5.1 to 3.1. Lime or ammonia is added to readjust the pH to about 3.8, and the fermentation is continued for 2 days more. The final itaconate concentration is about 85 gm/liter. Use of ammonia is preferable to use of lime because calcium itaconate has low solubility in the solution, and much of it adheres to the mycelium during filtration. On the other hand, if ammonia is the neutralizing agent, ammonium itaconate crystallizes out with the itaconic acid on evaporation of the solution. This is not too serious because the major market is for the esters which are distilled off on heating with sulfuric acid.

In another process (Batti and Schweiger, 1964) spore germination and growth of *A. terreus* is restricted by adding $CuSO_4 \cdot 5\,H_2O$ to the decationized syrups. Sugar concentration as great as 30% can be used with final beer concentrations of itaconic acid reaching 180 gm/liter. Ten to fifteen percent of the acidity may be succinic and itatartaric acids. Alkali earth metal salts added to the initial culture solution restrict the accumulation of these acids to trace amounts and result in improvement in itaconic acid yield (Table III).

In several characteristics both processes are similar and resemble the citric fermentation. Copper ion restricts growth and product destruction in both processes. The concentrations of copper ions in the two processes are similar to those used in the citric fermentation. In both processes excessive concentration of iron in the culture solution results in reduction of the accumulation of itaconic acid (Table IV); a similar

TABLE III. Effect of Addition of Alkaline Earth Metals on Itaconic Acid Accumulation by *Aspergillus terreus*[a]

Metallic ion added		
(%) Type	Amount (mg/liter)	Weight yield[b]
None		9
Calcium	337	43
Calcium	2700	59
Magnesium	337	20
Magnesium	2700	48

[a] From Batti and Schweiger (1964).

[b] (Grams itaconic acid produced/grams sugar supplied) × 100. Itaconic acid determined by bromination.

effect is observed in the manufacture of citric acid. Both itaconic fermentations and the citric fermentation require continuous aeration, and stoppages of air of very brief duration may be enough to stop the fermentation. Fermentation will not be rapidly resumed until new growth of mycelium in the presence of copper ion in inhibitory concentration has occurred.

Saccharinic acid-type compounds present in molasses may in part substitute for the copper ion requirement in the present manufacturing process. A similar affect has been observed in the citric fermentation.

In either process whenever the concentration of itaconic acid exceeds 7% the fermentation rate is much reduced. This is not an effect of pH as no significant change occurs at this time. If the excess titratable acidity is neutralized with ammonia or NaOH, the fermentation rate remains constant, and total itaconate and itaconic acid concentrations may reach 15–18%.

TABLE IV. Effect of Iron on the Production of Itaconic Acid by *Aspergillus terreus*[a]

Iron (mg/liter)	Weight yield[b] (%)
0	57
1	25
2	17
4	17

[a] From Batti and Schweiger (1964).

[b] (Grams itaconic acid produced/grams sugar supplied) × 100

C. Product Recovery

If part of the acid has been neutralized with NH_4OH, NaOH, or $Ca(OH)_2$ during the fermentation, the solution must be acidified with an inorganic acid prior to evaporation. Mycelium and any suspended solids are removed by filtration on any conventional industrial precoat filter with a filter aid precoat. If a refined-grade free acid product is made, the hot solution is treated with activated carbon and filtered. This filtrate is then evaporated and cooled to crystallize. A second crop of crystals may be obtained on further evaporation of the mother liquor, but these are industrial grade. When only the industrial grade is made, the carbon treatment may be omitted since this product will be esterified and the ester distilled. The mother liquor obtained from the crystallization steps may be extracted with a solvent, such as *n*-amyl alcohol, which has a favorable partition coefficient. The solvent extracts too much color, however, and further carbon treatment may be required. However, carbon treatment of the solvent may be much less effective than carbon treatment of the aqueous solution. Alternatively the mother liquor may pass through an anion-exchange resin bed, and the itaconic acid eluted with mineral acid. Crude itaconate is suitable for esterification without further purification or may be precipitated from fermentation filtrates by liming and the calcium precipitated with sulfuric acid. The itaconic acid crystallizes on concentration. Crude itaconic acid obtained from mother liquor of recrystallization may be esterified directly or, if a higher purity acid is desired, recrystallized from hot water after decolorization with carbon and filtration. This requires considerable care in temperature control to avoid large losses on the filter. A flow sheet is given in Fig. 3.

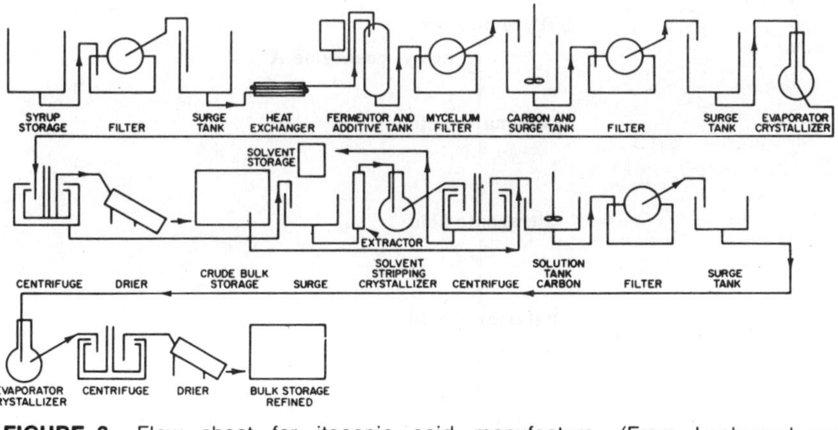

FIGURE 3. Flow sheet for itaconic acid manufacture. (From Lockwood and Schweiger, 1967; reprinted by permission of the Van Nostrand-Reinhold Co., New York.)

D. Chemistry of Itaconic Acid Fermentation

As in the citric acid fermentation, it is very probable that the EMP glycolytic pathway is followed as far as pyruvic acid. Here a variation occurs. Instead of acetyl-CoA combining with oxalacetate to form citric acid, acetyl-CoA condenses with pyruvic acid to form citramalic acid (Jakubowska and Metodiewa, 1974) (Fig. 4). This is then converted to itatartaric acid which is reduced to itaconic acid. It was formerly believed that itaconic acid was formed by loss of CO_2 from *cis*-aconitic acid of the citric acid cycle (Bentley and Theisen, 1957; Casida, 1968; Berry, 1975). *Aspergillus terreus* has the enzyme system for this sequence of reactions when grown in the absence of copper ion, and at a higher pH than that of the itaconic fermentation. If the conversion of citric acid to *cis*-aconitic acid is inhibited by copper ion which inhibits the action of aconitase, it should also result in reduction in itaconic acid yield. Actually addition of copper ion results in an increase in yield. Although either mechanism can result in the accumulation of itaconic acid, it is doubtful if the decarboxylation of *cis*-aconitic acid contributes significantly to itaconic acid accumulation. The role of copper ion must be the inhibition of some enzyme involved in the catabolism of itaconic acid.

As in the citric acid fermentation, aeration requirements are small, but aeration must be continuous. Cessation of aeration for as little as 15

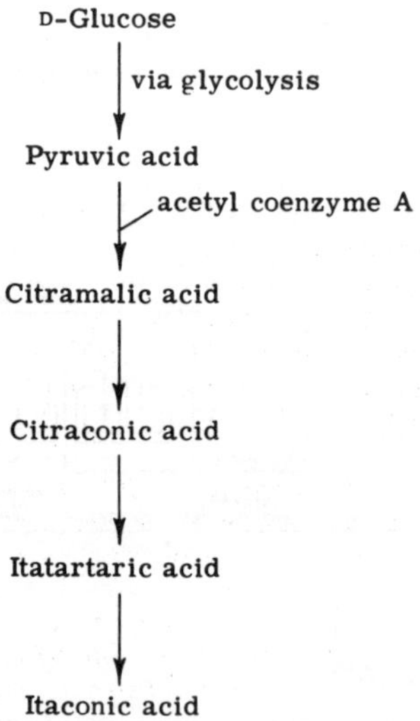

FIGURE 4. Proposed metabolic sequence in the production of itaconic acid.

seconds stops the fermentation and it will not be resumed until new mycelial growth of suitable biochemical type has occurred.

IV. LACTIC ACID

Lactic acid was first isolated by Scheele (1789) from sour milk. Many studies on the physical and chemical properties of the compound established its structure and applications and found that the compound occurred in two isomeric forms and as a racemic mixture of these. It was also found that the acid is produced in muscle when work is done and that the acid is produced by many microorganisms by the fermentation of sugars.

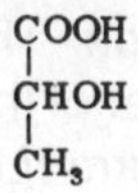

Lactic acid

A. Technological Development

Lactic acid was the first organic acid to be manufactured industrially by fermentation. In 1880 a plant was built in Littleton, Massachusetts, to manufacture calcium lactate by fermentation. Little is known about the process beyond the fact that calcium lactate was manufactured and marketed. The Clinton Processing Co., Clinton, Iowa is the only manufacturer using the fermentation process in the United States.

Numerous types of microorganisms have been isolated which accumulate lactic acid or lactates in their culture solutions. Homofermentative lactobacilli, the ones which produce only cells and lactic acid in significant amounts, are used commercially, although algae, yeasts, and phycomycetous fungi are also known to produce the acid. Yields comparable to those obtained by homofermentative lactic acid bacteria have been obtained only in glucose cultures of the fungus *Rhizopus oryzae*. The homofermentative lactic acid fermentation produces almost 2 moles lactic acid per mole of glucose supplied. The common heterofermentative types have not been used on account of their much lower yields although they occur not infrequently as contaminants.

B. Isomers

Lactic acid of commerce is the L(+)-isomer, D(−)-isomer, or any possible mixture of the two. The entire range of isomers has been found.

The capital letters prefixed to the names of the isomers indicate configuration in relation to the isomers of glyceraldehyde, and the symbols (+) and (−) indicate the direction of rotation of a plane of polarized light. The mixture of isomers is now called DL-lactic acid. Most lactic acid-producing organisms produce only one isomer but a few bacteria, such as *Lactobacillus plantarum,* which are frequent contaminants produce enzymes which convert one isomer to a mixture of the two. These enzymes, called racemases, are now known to be lactic dehydrogenases which maintain an equilibrium between the lactic acid isomers and pyruvic acid (Dennis and Kaplan, 1960). When both dehydrogenase enzymes are present, racemization occurs.

Identification of the isomers by polarimetric methods is made difficult by the low optical activity of the pure isomers and by their spontaneous polymerization in aqueous solution. The dimers, polymers, salts, and esters have rotation reversed to that of the free acid solutions. This is due to the occurrence of an epoxide form (Lockwood *et al.*, 1965). Observed rotations of the free acids are of direction opposite to that of the monomers on account of the presence of the epoxide.

C. Current Manufacturing Process

Thirty years ago five lactic acid fermentation plants operated in the United States. Today only one remains. Much lactic acid is made by hydrolysis of lactonitrile, a by-product of another process.

Peckham (1944) and Schopmeyer (1954) have described the lactic acid fermentation process. Little change has occurred since these publications appeared. Lactic acid is very corrosive to metals. Consequently, wooden fermentors have been used in most plants. These were uncovered or were covered with loosely fitting wooden lids. They were steamed empty before charging, and fermentation solutions were pasteurized or sterilized by passage through steam-jacketed heat exchangers. Contamination of culture solutions by thermophilic clostridia resulted in the production of some butanol and butyric acid contaminating the lactic acid. The product could be sold only to leather tanners for deliming of hides. Such contamination actually forced one manufacturer out of the lactic acid business.

Inoculum cultures of *Lactobacillus delbrueckii* are transferred from test tubes through successively larger culture vessels held at 45°–55°C. Each stage of culture buildup requires 16–18 hours. A slight excess of $CaCO_3$ is present in each stage. Inoculum volume is usually about 5% of fermentation solution volume.

The fermentation solution is usually hydrolyzed starch or dextrose syrups, although D-glucose, maltose, lactose, or sucrose can be fermented. The sugar concentration does not exceed 12%. At a greater sugar concentration, calcium lactate may crystallize in the fermentor as a solid mass which is difficult to handle.

Lactic acid bacteria have complex requirements for B vitamins. This ordinarily is met by enrichment of the culture solution with crude vegetable materials. Malt sprouts, the rootlets produced by germinating barley grains in the malting process, are the most commonly used vegetable material. If they have been overheated in drying, they lose some of their value as a nutrient for lactobacilli. The culture solution contains $CaCO_3$ to neutralize 2 moles lactic acid per mole hexose supplied, or a slight excess. After inoculation the solution is slowly stirred to keep the $CaCO_3$ suspended, but aeration of the solution is not necessary.

Plant fermentation time is usually 5–10 days, varying from batch to batch with uncontrolled conditions. An excess of $CaCO_3$ in the culture solution keeps the pH in the range of 5.5–6.5. The pH necessary varies somewhat with the composition of the culture solution. Control of pH between 6.3 and 6.5 by continuous neutralization with a slurry of $Ca(OH)_2$ will permit the complete fermentation of 12–13% glucose in 72 hours. It is very important that residual sugar be reduced to 0.1% or less during the fermentation because residual sugar makes recovery of the better quality lactic acid more difficult. Commercial fermentation yields are 93–95% of the weight of the glucose supplied. Recovery yields vary with the various recovery processes and product grades.

D. Product Recovery and Grades

The suspended solids and most of the bacteria are removed from the solution on any conventional industrial precoat filter. A polishing filtration may also be required. For technical grade product, calcium is precipitated as $CaSO_4 \cdot 2\ H_2O$ which is filtered off. On evaporation of the filtrate to 35–40% lactic acid content, more $CaSO_4 \cdot 2\ H_2O$ precipitates and is removed by filtration. Further evaporation to 44–45% total acidity gives the technical-grade product.

Food-grade lactic acid is a pale yellow, straw-colored solution of about 50% total acidity. The acid is made by fermenting a higher grade of sugar, with minimal protein content of the culture solution. Calcium is precipitated as $CaSO_4 \cdot 2\ H_2O$ and washed, and the filtrate and wash water are combined. The solution is bleached with activated carbon, evaporated to about 25% solids, then bleached again, further evaporated to about 50% total acidity, and further bleached. If the final product is off-color, it may be treated with ferrocyanide to precipitate iron or copper ions and filtered. An alternative procedure is the evaporation of the fermented beer to recover calcium lactate which is washed with cold water in a centrifuge before decomposing with H_2SO_4. Further processing is as above.

Plastics-grade lactic acid is a colorless product which can be obtained by esterification with methanol after concentration. Another procedure is solvent extraction with isopropyl ether followed by reextraction of the isopropyl ether with water (Anonymous, 1959).

Lactic acid, USP, is a colorless solution of 85% total acidity, 76–78% lactic acid concentration. It contains 2–3.5% volatile acids and 0.5–1.0% ash. Approximately 30% of the acid occurs as polymers, either lactide, linear polymers, or both.

Entry into the market of DL-lactic acid obtained as a by-product altered the economic value of the fermentation process very unfavorably, especially in the paler grades. In the event that a shortage of hydrocarbons occurs, lactonitrile will be in short supply or available only at higher cost. In this instance, fermentation lactic acid will again be economically competitive. However, it appears probable that the production of L(+)-lactic acid by *Rhizopus* will be preferred on account of the much simpler recovery process and shorter fermentation time. Yields of lactic acid obtained by use of homofermentative lactobacilli and *Rhizopus* are identical but the 48-hour fermentation by *Rhizopus* gives an almost colorless fermented solution. The lactic acid on acidulation of the solution with H_2SO_4 and filtration to remove $CaSO_2 \cdot 2\,H_2O$ will crystallize on evaporation. Crystals so obtained are colorless and are not hygroscopic but tend to liquefy by loss of water due to polymerization.

V. GLUCONIC ACID AND GLUCONO-δ-LACTONE

D-Glucono-δ-lactone, the simplest of the direct dehydrogenation products of D-glucose, and its free acid form, gluconic acid, are produced by a large variety of bacteria and fungi. Both types of microorganisms are used commercially. Bacterial accumulation of gluconic acid was observed by Alsburg (1911) in cultures of a bacterial parasite of olive trees now known as *Pseudomonas savastanoi*. Molliard (1922) described its occurrence in cultures of *A. niger*. Fungi which produce D-glucono-δ-lactone and D-gluconic acid are found in the genera *Aspergillus, Penicillium, Gliocladium, Scopulariopsis, Gonatobotrys*, and *Endomycopsis*. Doubtless many other fungi will be found to produce these compounds. Bacteria which produce D-glycono-δ-lactone and D-gluconic acid are found among the genera *Pseudomonas, Vibrio*, and

Gluconobacter. The organisms commonly used in this fermentation are *A. niger* or *Gluconobacter suboxydans*. The larger volume production uses the fungal process. Most of the *Gluconobacter* production is marketed as glucono-δ-lactone for use in baking powders and effervescent products.

The equilibrium of the lactone and the free acid in solution is readily controlled by pH and temperature. The relative amounts of lactone and free acid can be determined approximately by titration as rapidly as possible up to the first end point with phenolphthalein as indicator with standard alkali. An excess of standard alkali is added to the solution heated to at least 53°C and then back-titrated with standard acid. The difference between initial and total alkali titer represents the lactone, while the initial titer represents the free acid. The lactone may be determined more precisely and independently by the hydroxamate method of Hestrin (1949). Some *Pseudomonas* strains produce a lactonase which hydrates the lactone and the lactone does not accumulate in their culture solutions. The lactone may also arise spontaneously in solutions of the free acid, but the fermentation product is the δ-lactone.

A. Fermentation Processes

Only submerged process technology has been used for many years, although at one time at least one manufacturer in the United States made gluconates by a surface liquid culture process. The manufacturers of gluconic acid and gluconates in America use the fungal process. Only one chemical process may be competitive. Although numerous chemically mild oxidizing agents convert D-glucose to D-gluconic acid, they are usually too costly, even though a lesser initial capital investment may be required. Manufacturers of gluconic acid or its salts in the United States are Bristol-Meyers Co., Syracuse, New York; Pfizer Inc., New York, New York; and Premier Malt Products Inc., Milwaukee, Wisconsin.

B. Calcium Gluconate Fermentation

Calcium gluconate is manufactured primarily to meet the pharmaceutical market. The solubility of calcium gluconate in water is about 4% at 30°C (Ward, 1967). Supersaturation of this salt commonly occurs in fermentation solutions up to about 15% or more glucose concentration. Consequently this imposes a limitation on initial glucose concentration to about 13–15% (Moyer *et al.*, 1940). The calcium source must also be sterilized separately from the glucose solution to avoid the Lobry–du Bryn–Van Ekenstein reaction, which alters the conformation of about

30% of the glucose, thus reducing the yield. This reaction occurs slowly at pH 5.5 and at values above pH 7 is almost instantaneous. The neutralizing agent must be added aseptically to the sterile glucose solution at intervals after inoculation. If added before spore germination, spores will not germinate; if added to mycelium in large enough quantity to make the solution alkaline the mycelium will be killed. Addition of $CaCO_3$ slurry must be delayed until enough glucono-δ-lactone has been formed to buffer the solution against the high alkalinity of a heat sterilized slurry of $CaCO_3$. On sterilization of an aqueous suspension of $CaCO_3$ with steam under pressure, considerable $Ca(OH)_2$ is formed. Dry sterilization of $CaCO_3$ in large scale has not proved practical.

Moyer *et al.* (1940) tried to increase the solubility of gluconate by adding borates or boric acid to the fermentation solutions. Calcium borogluconate is formed. Although one company used this procedure for a short time, the borogluconate was found to be deleterious to blood vessels of animals, and the product was withdrawn from the market.

Culture solutions used for this fermentation are D-glucose solutions to which several inorganic salts and corn steep liquor have been added. The solution has not been deionized. Much variation in mineral salt concentration and corn steep liquor concentration can be tolerated with little effect on yield. Most if not all the mineral salt nutrient requirements are supplied by the corn steep liquor.

The beneficial effects of growing the mold under increased air pressure were demonstrated by May *et al.* (1934), who studied the fermentation in submerged aerated cultures at various pressures above atmospheric. Pressures up to 3 atm above sea level pressure gave more rapid fermentations than cultures at 1 atm pressure. Most of the advantage obtained was found at 2 atm gauge pressure, and no further improvement was found between 45 and 105 lb gauge pressure.

C. Sodium Gluconate Fermentation

The major differences between the production of calcium and sodium gluconate are in the concentrations of initial glucose and final concentration of gluconate (Fig. 5) and in pH control. In the calcium gluconate fermentation pH control results from the $CaCO_3$ slurry added to the solution during the fermentation, and in the sodium gluconate fermentation pH control is accomplished by automatic addition of NaOH solution. Sodium gluconate is much more soluble than is calcium gluconate (Table V). Consequently, the controlled automatic addition of NaOH provides an easy and precise means of neutralizing the acid as it is formed. Again, it is important that the pH of the fermentation solution

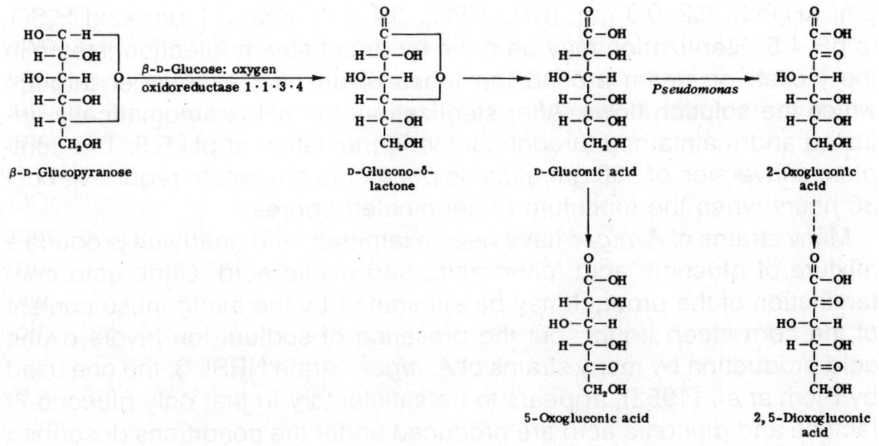

FIGURE 5. Gluconate and oxogluconate production.

should never be less acid than pH 6.5 after inoculation. The method described by Blom *et al.* (1952), developed at the Northern Regional Research Laboratory of the United States Department of Agriculture, appears to be typical of the industrial production of sodium gluconate. Efficient oxygenation of the fermentation solution is ensured by the use of 1.0–1.5 volumes air per volume solution in baffled fermentors equipped with turbomixers. The air is introduced directly beneath the turbomixer blade. A pressure of 30 psig is maintained. The fermentation solution must be cooled by circulation of water through coils inside the fermentor or in the jacket of the tank.

A typical fermentation solution contains, per liter, glucose (anhydrous basis), 250–350 gm; corn steep liquor, 3.7 gm; $MgSO_4 \cdot 7\,H_2O$, 3.5–4.0

TABLE V. Solubility of Sodium Gluconate in Water at Various Temperatures[a]

Temperature (°C)	Sodium gluconate in saturated solution (wt. %)
15	34.6
20	36.7
30	39.6
50	45.9
60	49.9
70	53.0
80	56.9

[a] From Ward (1967).

gm; KH_2PO_4, 0.2–0.3 gm; $(NH_4)_2HPO_4$, 0.4 gm; urea, 0.1 gm; and H_2SO_4 to pH 4.5. Sterilization may be done by direct steam injection, steam in the jacket, or steam around the tubes of an heat exchanger through which the solution flows. After sterilization, the pH is automatically adjusted and maintained throughout the fermentation at pH 6.5. The complete conversion of 300 gm glucose per liter to gluconate requires about 36 hours when the inoculum is germinated spores.

Many strains of *A. niger* have been examined, and nearly all produce a mixture of gluconic acid, citric acid, and oxalic acid. Citric acid contamination of the product may be eliminated by the manganese content of the corn steep liquor, but the presence of sodium ion favors oxalic acid production by many strains of *A. niger*. Strain NRRL 3, the one used by Blom *et al.* (1952), appears to be satisfactory in that only glucono-δ-lactone and gluconic acid are produced under the conditions described above.

Glucono-δ-Lactone Fermentation

In the *Gluconobacter suboxydans* fermentation, a 10% concentration of glucose is converted to glucono-δ-lactone and free gluconic acid in about 3 days in the absence of pH control. Little is known about this fermentation, but it appears that either corn steep liquor or cooked yeast cells provide all the essential nutrients. Approximately 40% of the gluconic acid is in the δ-lactone form. Most of the gluconic acid is converted to the δ-lactone during recovery operations.

It has been informally reported to the writer that one major manufacturer in the United States now recovers the lactone directly and converts it to the desired salt by heating with the hydroxide of the desired cation. No details of the fermentation were reported but it has been known for more than 50 years that such fermentation is possible with *A. niger* or with one of several penicillia as well as by *Gluconobacter*.

D. Product Recovery

1. Calcium Gluconate

Calcium gluconate may be recovered by removal of the mycelium and any other suspended solids on any conventional industrial filter, carbon treatment of the filtrate to decolorize it, filtration to remove carbon, and evaporation to a 15–20% calcium gluconate concentration. On cooling to a temperature just above 0°C, most of the calcium gluconate in the highly supersaturated solution crystallizes. Seeding with calcium gluconate crystals may be necessary in order to obtain rapid crystallization.

The needle crystals tend to form a spongy mass so pressure filtration or centrifugation in a perforate basket centrifuge is necessary to remove most of the mother liquor. Washing should be done with cold water.

2. Sodium Gluconate

Sodium gluconate in the commercial grades may be prepared from filtrates of the sodium gluconate fermentation by concentrating to 42–45% solids, adjusting to pH 7.5 with NaOH, and drum-drying (Ward, 1967). A more refined product may be made by carbon treatment of the hot solution before drying, or by recrystallization from the crude drum-dried product after carbon treatment.

3. Crystalline Lactones of Gluconic Acid and the Crystalline Acid

Aqueous solutions of gluconic acid are an equilibrium mixture of free gluconic acid, glucono-δ-lactone, and glucono-γ-lactone. Crystals separating out of a supersaturated solution at a temperature below 30°C (preferably near 0°C) will be predominantly free gluconic acid; from 30° to 70°C the crystals will be principally glucono-δ-lactone, and at temperatures greater than 70°C they will be principally the γ-lactone. Seeding with the desired compound and close control of conditions are required to obtain crystals of high purity (Ward, 1967).

4. Free Gluconic Acid Solution

The standard 50% solution of gluconic acid may be prepared by treatment of the calcium gluconate solutions with H_2SO_4 and heating to precipitate calcium sulfate, filtration, and concentration of the filtrate to about 50% solids. If any residual sugar or fungal polysaccharide is present, carbon treatment may be necessary.

E. Market Data

Gluconic acid is usually marketed as a 50% technical-grade aqueous solution (w/w). It finds use as a mild acidulant in metal pickling and in foods. Calcium gluconate is a readily available form of soluble calcium widely used for oral and intravenous therapy. Sodium gluconate is an excellent sequestering agent in neutral or alkaline solutions, where it finds use in the cleansing of glassware, such as returnable bottles and dishes, and as a constituent of compounds for washing walls.

Prices of gluconic acid, calcium gluconate, and sodium gluconate in March 1976 varied considerably with market form (U.S. Tariff Commission, 1976). Technical-grade sodium gluconate sold for $0.80 per pound in 100-lb bags. It was sometimes blended with NaOH to make the alkaline product used in bottle washing solutions and was sold at a slightly lower price. Calcium gluconate USP grade in ton lots packed in 100-lb drums brought $0.75 per pound. The price of the free acid sold as a 50% solution and delivered in the United States was $0.33–0.355 per pound. The sodium salt is the principal market form, amounting in 1974 in the United States to 8,169,000 lb (U.S. Tariff Commission, 1974).

1. Specifications

The specifications shown in Table VI are typical of those now prevailing. Calcium gluconate must meet USP standards.

TABLE VI. Specifications[a]

Substance	USP Standard
Sodium gluconate	
Appearance	White to tan powder
Purity	98.8–99.8%
Moisture	0.1–0.2%
Reducing substances	0.1–0.5%
Solubility, 20°C	
In water	Very soluble
In ethanol	Slightly soluble
In ether	Insoluble
Aqueous solution	pH 5.3–6.3
Stability, 0°–100°C	Stable
Glucono-δ-lactone	
Specific rotation	
(freshly prepared) [α]20	+64°
Decomposition point	158°–163°C
Solubility, 20°C	
In water	Very soluble
In ethanol	Slightly soluble
In ether	Insoluble
Appearance	White crystalline powder
Taste	Sweet
Gluconic acid (technical 50%)	
Color	Light yellow
Gluconic acid content	49–51%
Ash	Not over 0.4%
Specific gravity	1.23–1.24
Sulfate	Trace
Chloride	Trace
Reducing substances	Not more than 3% on 100% acid basis
	Nontoxic

[a] From Ward (1967).

2. Toxicity

Storage above 18°C is recommended to prevent crystallization of the acid or the lactone.

3. Uses

The principal market for gluconic acid and its derivatives is for the sodium salt. Hydroxyl groups of the gluconate molecule bind di- and trivalent metallic ions in soluble form, thus preventing precipitation of calcium, magnesium, or iron compounds from solutions used in bottle and dish washing. Alkaline bottle washing compounds which contain 2.5–10% sodium gluconate, depending on the hardness of the water, are used extensively in automatic cleansing equipment. Metal carbonate precipitates are removed by sodium gluconate containing products for the washing of painted walls or metals without inducing appreciable corrosion. When added to cement, gluconates control setting time and increase the strength and water resistance of the cement.

In pharmaceutical use, calcium gluconate has long been used as a preferred source of calcium in cases of calcium deficiency due to diet or pregnancy, and for rapid relief of symptoms of allergies. Many drugs can be administered more effectively when combined with gluconates.

Glucono-δ-lactone finds use as a constituent of baking powders, bread mixes, and other foods where its latent acidogenic properties may be useful.

VI. OXOGLUCONATES

Three oxogluconates are accumulated in glucose culture by bacteria. *Gluconobacter suboxydans* converts D-glucose into a mixture of 2-oxogluconate and 5-oxogluconate. 5-Oxogluconate predominates, and 2-oxogluconate may be found in only traces. Many species of *Pseudomonas* accumulate 2-oxogluconate in D-glucose cultures. Weight yields as great as 90% are not uncommon. Katznelson *et al.* (1953) produced 2,5-dioxogluconate from glucose using *Gluconobacter melanogenum*.

The metabolic scheme (Fig. 5) showing the biochemical mechanism is probably correct. D-Glucono-δ-lactone is an intermediate in the production of D-gluconates by these organisms. D-Gluconate salts are a good substrate for *Pseudomonas* growth and conversion to 2-oxogluconate.

2-Oxogluconate is the only one of the oxogluconates that has been manufactured commercially. Operating conditions are the same as those used in the fungal production of calcium gluconate except that 5 ml of corn steep liquor or 0.3 gm of yeast extract may be added per liter of culture solution. 2-Oxogluconate formation follows quickly on the initial formation of gluconate. Figure 6 illustrates the simultaneous kinetic pattern of the fermentation of 10% glucose medium under 30 lb gauge pressure of air in the presence of an excess of $CaCO_3$ (Stubbs *et al.*, 1940). Yields of calcium 2-oxogluconate equivalent to 90% of theory have been obtained but industrial yields are believed to be slightly less than this. Commercial usage calls for the fermentation of 20% glucose solutions by *Pseudomonas fluorescens* strains acclimatized to high osmotic pressure. In order to maintain acclimatization, stock cultures are maintained on 20% sodium gluconate agar slants, and seed buildup plant stages all require 20% glucose solutions.

In a few individual culture strains, the fermentation does not proceed beyond the gluconate stage. In others, although 2-oxogluconate accumulates as the major metabolite, if the fermentations are prolonged δ-oxoglutaric acid, succinic acid, and pyruvic acid accumulate in considerable quantity (Lockwood and Stodola, 1946).

2-Oxogluconic acid is usually recovered and shipped in the free acid state after centrifugation or filtration to remove cells. Calcium is removed

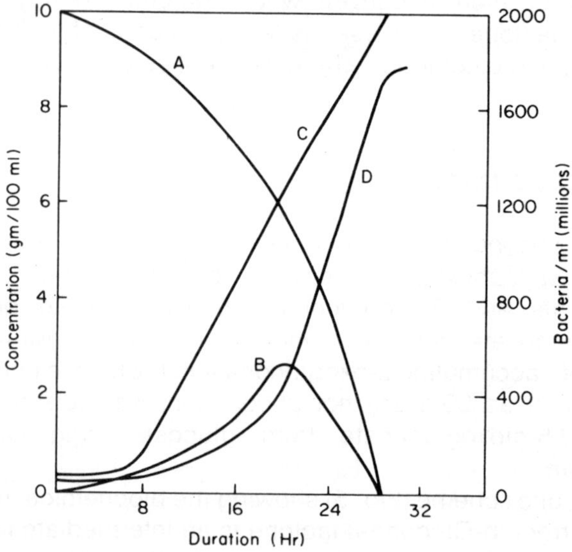

FIGURE 6. Composition of fermenting culture solution at various ages for 2-oxogluconate production based on data of Stubbs *et al.* (1940). Curve A: glucose in solution (gm/100 ml); curve B: glucono-δ-lactone and gluconate in solution (gm/100 ml): curve C: 2-oxogluconate in solution (gm/100 ml); curve D: *Pseudomonas* cells per milliliter (millions).

by precipitation with H_2SO_4. The filtered acid may be shipped as a syrup or as a crystalline material after evaporation under reduced pressure at a temperature which must never exceed 50°C. The material may be shipped as esters since esterification is necessary to conversion to erythorbic acid.

Productions of erythorbic acid, its esters, and its salts are the major markets for 2-oxogluconic acid. They are used as water- or fat-soluble antioxidants to retain color, flavor, and nutritive values in canned fruits and vegetables and in meats and meat products.

Simultaneous analysis of fermenting solutions which contain calcium salts of two acids, gluconic and 2-oxogluconic, and two compounds which reduce alkaline copper sulfate is done by a method developed by G. E. Ward (Stubbs *et al.*, 1940) which distributes the copper-reducing activity on the basis of the optical rotation of the plane of polarized light on passing through a standard layer of the solution.

5-Oxogluconate has not been manufactured on a commercial scale but is of interest as a potential source of L(+)-tartaric acid.

VII. MISCELLANEOUS ACIDS FORMERLY MADE BY FERMENTATION

Examination of the history of lactic acid production in the United States suggests the possibility that lactic acid may someday be transferred to this category. However, only two organic acids of any volume now belong to this group.

The manufacture of fumaric acid, formerly done by fermentation, is now entirely done by oxidation of aromatic hydrocarbons. When the time comes that hydrocarbons are in very short supply, or at too high a price, further consideration of fermentation methods for the production of fumaric and lactic acids is to be expected. Fumaric acid was made by the aerobic fermentation of glucose by selected strains of *Rhizopus* in the presence of $CaCO_3$. It is also found in cultures of *Mucor* and *Cunninghamella*. Most strains of *Rhizopus* produce L(+)-lactic acid under the ordinary fermentation conditions, or a mixture of lactic and fumaric acids. Strain selection offers the promise of improvement in yield. Only inorganic nutrients other than the carbon source are required. Calcium fumarate has low solubility in cold water, so it precipitates among the filaments of mycelium. It can be extracted with hot water by heating the culture solution before filtration.

The other organic acid made by fermentation in the United States about 1916 was gallic acid. Tannase of *Aspergillus wentii* was produced in deep tanks with agitators by growing the fungus on an extract of gallotannins. Aeration was discontinued after the third day, more tannins were added, and agitation was stopped. Excreted tannase hydrolyzed the gallotannins. The gallic acid so obtained was used in blue pigments.

REFERENCES

Alsburg, C. L. (1911). *J. Biol. Chem.* **9,** 1–7.

Amelung, H. (1930). *Chem.-Ztg.* **44,** 118.

Anonymous (1959). *Chem. & Eng. News,* June 15.

Anonymous (1963). *Chem. Week,* Sept. 21.

Batti, M. A. (1966). U.S. Patent 3,290,227.

Batti, M. A., and Schweiger, L. B. (1964). U.S. Patent 3,078,217.

Bentley, R. C., and Theisen, C. (1957). *J. Biol. Chem.* **226,** 703–720.

Berry, D. R. (1975). *In* "The Filamentous Fungi, Industrial Mycology" (J. E. Smith and D. R. Berry, eds.), Vol. I. Arnold, London.

Blom, R. H., Pfeifer, V. F., Moyer, A. J., Traufler, D. H., Conway, D. H., Crocker, H. F., Farison, R., and Hannibal, D. B. (1952). *Ind. Eng. Chem.* **44,** 435.

Blumenthal, H. (1965). *In* "The Fungi: An Advanced Treatise" (G. C. Ainsworth and A. S. Sussman, eds.), Vol. I. Academic Press, New York.

Bruchmann, E. E. (1961). *Biochem. Z.* **335,** 199–211.

Calam, C. J., Offord, A. C., and Raistrick, H. (1939). *Biochem. J.* **33,** 1488–1495.

Casida, L. E. (1968). "Industrial Microbiology." Wiley, New York.

Clark, D. S. (1962). *Biotechnol. Bioeng.* **4,** 17–21.

Cleland, W. W., and Johnson, M. J. (1954). *J. Biol. Chem.* **208,** 679–689.

Dennis, D., and Kaplan, N. (1960). *J. Biol. Chem.* **235,** 810–818.

Friedkin, M. (1945). *Ind. Eng. Chem., Anal. Ed.* **17,** 637–638.

Grimaux, P., and Adams, P. (1880). *C. R. Hebd. Seances Acad. Sci.* **90,** 1252.

Hestrin, S. (1949). *J. Biol. Chem.* **180,** 249–261.

Hisanga, W., and Nakamura, S. (1966). Japanese Patent 16555/66.

Jakubowska, J., and Metodiewa, S. (1974). Final Report, Grant No. FG-P1-244. U.S. Dept. of Agriculture, Lodz, Poland.

Katznelson, H., Tannenbaum, S. W., and Tatum, E. L. (1953). *J. Biol. Chem.* **204,** 43–59.

Kinoshita, K. (1931). *Bot. Mag.* **45,** 45–61.

Kinoshita, K. (1937). *Acta Phytochim.* **9,** 129.

Kluyver, A. J., and Perquin, L. H. C. (1933). *Biochem. Z.* **266,** 668–681.

Kobayashi, T. (1967). *Process Biochem,* Sept., pp. 61–65.

La Nauze, J. M. (1966). *J. Gen. Microbiol.* **43,** 73–81.

Lockwood, L. B., and Batti, M. A. (1965). U.S. Patent 3,189,527.

Lockwood, L. B., and Schweiger, L. B. (1967). *In* "Microbial Technology" (H. J. Peppler, ed.), pp. 183–199. Van Nostrand-Reinhold, Princeton, New Jersey.

Lockwood, L. B., and Snell, R. L. (1976). "Development of Miles Laboratories Process for the Manufacture of Citric Acid," Miles Analecta, Vol. 3, No. 2, pp. 23–26.

Lockwood, L. B., and Stodola, F. H. (1946). *J. Biol. Chem.* **164,** 81–83.

Lockwood, L. B., Yoder, D. E., and Zienty, M. (1965). *Ann. N.Y. Acad. Sci.* **119** 844–867.

Mallea, O. (1950). *Ind. Quim.* **12,** 264–280.

May, O. E., Herrick, H. T., Moyer, A. J., and Wells, P. A. (1934). *Ind. Eng. Chem.* **26,** 575–578.

Meyrath, J. (1967). *Process Biochem.* Oct., pp. 25–28.
Mial, L. M. (1975). *In* "The Filamentous Fungi, Industrial Mycology" (J. E. Smith and D. R. Berry, eds.), Vol. I. Arnold, London.
Molliard, M. (1922). *C. R. Hebd. Seances Acad. Sci.* **174,** 881–883.
Moyer, A. J. (1953). *Appl. Microbiol.* **1,** 1–13.
Moyer, A. J., Umberger, E. J., and Stubbs, J. J. (1940). *Ind. Eng. Chem.* **32,** 1347.
Neubel, R. D., and Ratajak, E. J. (1962). U.S. Patent 3,044,941.
Peckham, G. T. (1944). *Chem. & Eng. News* **22,** 440–443.
Peckham, G. T., and Filachione, E. M. (1962). *Kirk-Othmer Encycl. Chem. Technol., 2nd Ed.,* Vol. 12, pp. 170–188.
Perquin, L. H. C. (1938). Dissertation Technische Hoogeschool, W. D. Mainema, Delft.
Pfeifer, V. F., Nelson, G. E. N., Vojnovich, C., and Lockwood, L. B. (1953). U.S. Patent 2,657,173.
Pfeifer, V. J., Vojnovich, C., and Heger, E. N. (1957). *Ind. Eng. Chem.* **44,** 2975–2980.
Scheele, W. (1789). *Crells Ann.* **2,** 1.
Schopmeyer, H. H. (1954). *In* "Industrial Fermentations" (L. A. Underkofler and R. J. Hickey, eds.), Vol. I, pp. 291–419. Chem. Publ. Co., New York.
Schweiger, L. B. (1957). U.S. Patent 2,970,084.
Scüzs, J. (1944). U.S. Patent 2,353,771.
Shu, P., and Johnson, M. J. (1947). *J. Bacteriol.* **54,** 161.
Shu, P., and Johnson, M. J. (1948a). *J. Bacteriol.* **56,** 577.
Shu, P., and Johnson, M. J. (1948b). *Ind. Eng. Chem.* **40,** 1202.
Steinberg, R. A. (1945). *Soil Sci.* **60,** 185–189.
Stubbs, J. J., Lockwood, L. B., Roe, E. T., Tabenkin, B., and Ward, G. E. (1940). *Ind. Eng. Chem.* **32,** 1626–1630.
Tsao, C. Y., Jr., and Su, S. C. (1964). *Sugar y Azucar* **59,** 36.
Tsao, C. Y., Jr., Huang, J. C., and Su, S. C. (1962). *Bull. Acad. Sci. Sinica* [N. S.] **3,** 26–38 and 48.
United States Tariff Commission (1974). *Chem. Mark. Rep.*
United States Tariff Commission (1976). *Chem. Mark. Rep.*
Ward, G. E. (1967). *In* "Microbial Technology" (H. J. Peppler, ed.), pp. 200–211. Van Nostrand-Reinhold, Princeton, New Jersey.
Wehmer, C. (1893). *Bull. Soc. Chim. Fr.* **9,** 728.
Woodward, J. C., Snell, R. L., and Nichols, R. S. (1949). U.S. Patent 2,492,673.
Yamada, K. (1965). "Science in Japan," p. 401. Am. Assoc. Adv. Sci., Washington, D.C.

Chapter 12

Plant Cell Suspension Cultures and Their Biosynthetic Potential*

W. G. W. KURZ
F. CONSTABEL

I. INTRODUCTION

The development of a plant from germination to flowering and the formation of a fruit is governed by an integrated system of internal and external factors, i.e., genotype, interrelationship of cells, tissues and organs, and habitat. An analysis of developmental processes, the ger-

*NRCC No. 15779.

MICROBIAL TECHNOLOGY, 2nd ed., VOL. I
 ISBN 0-12-551501-4

mination of seeds, the formation of pigments, or the initiation of buds, for instance, confronts botanists with a complex set of factors which generally resists exact identification and computation. The very gradual change of plant morphology from a mere descriptive to an experimental discipline (Goebel, 1908) and the late emergence of morphogenesis as a unit of research focusing on the biochemistry of plant ontogeny (Wardlaw, 1968) reflect the difficulties of an analytical approach to plant development. Progress had to await the introduction of model systems which would display developmental steps in plant ontogeny in a simplified manner, not necessarily identical to *in situ* situations. The search for a model system resulted in organs, tissues, and cells excised from intact plants and grown under controlled environmental conditions *in vitro.* This cell culture system has successfully been employed in numerous morphogenetic analyses, mainly by varying (1) the origin of the cells and (2) the environment, i.e., physical and chemical external factors (Gautheret, 1959; Reinert and Bajaj, 1977; Steward, 1969; Street, 1973, 1974; White, 1963).

Results of morphogenetic investigations coupled with refinement of methods revealed a potential of cell cultures for application in various fields of plant sciences. As soon as cultured cells were found to be amenable to regeneration of entire plants, large-scale cloning was introduced to horticulture (Murashige, 1974). Cell culture, furthermore, offered a unique way to extend studies in plant pathology to causal relationships in tumor formation (Braun, 1968; Nester, 1977; Sacristan and Melchers, 1970), in host–parasite interaction (Hollings, 1965; Takebe, 1975), and in sanitation of pathogen-infected stock of crop plants (Kartha and Gamborg, 1975). Crop improvement through modification of somatic cells, cell hybridization, is entirely based on cell culture methods (Constabel, 1976; Gamborg *et al.*, 1974; Smith, 1974).

Plant cells when cultured as suspensions have a number of properties in common with cultures of microorganisms: (1) they grow in a sterile environment; (2) they are fairly homogeneous in size, rate of metabolism, and stage of development; (3) they have a doubling time which is longer than that of microorganisms but considerably shorter than of cells *in situ;* and (4) they can be grown on a large scale. In contrast to microorganisms, plant cells feature an immensely greater morphogenetic potential. This potential includes the synthesis and accumulation of plant products, terpenoids, glycosides (steroids, phenolics), and alkaloids. A number of these products are indispensable in the pharmaceutical industry. The idea of employing cell suspension cultures for industrial exploitation, therefore, has been widely considered (Becker, 1969; Furuya *et al.*, 1972; Jones, 1974; Puhan and Martin, 1970; Reinhard, 1967; Staba, 1969; Steck and Constabel, 1974; Zenk, 1976). The problem encountered would be to control environmental conditions of cell suspension cultures in such a manner that their morphogenetic potential is expressed and channeled into the production of desirable substances.

This chapter is intended to describe principles, experiments, and results of plant cell culture as they relate to product formation. For an introduction to detailed techniques which would guide students in establishing and maintaining plant cell cultures, excellent publications for laboratory use are available (Brown and Sommer, 1975; Gamborg and Wetter, 1975; Street, 1973; White, 1963).

II. PLANT CELL CULTURE TECHNIQUES

A. Plant Tissues

Seed plants, which occur in an infinite variety of forms as trees, herbs, or grasses, exhibit three basic morphological units or organs: root, stem, and leaf. The anatomy of these organs reveals coherent masses of cells, associated in tissues. Variation in form and function, differences in cell combination, and topographic distribution in the plant body determine the type of tissue. Parenchyma or ground tissue is the most versatile of all types of tissues.

Parenchyma cells are living nucleated cells, capable of growth and division. They vary in shape and are often isodiametric, but some maybe elongate. Their function is concerned with photosynthesis, storage, wound healing, origin of adventitious structures, and excretion. Parenchyma cells may occur in extensive continuous masses in the cortex and pith of stems and roots, in the mesophyll of leaves and fruits, or in the endosperm of seeds. They are also associated with other types of cells in morphologically heterogeneous tissues, such as phloem and xylem. Perpetually young parenchyma, primarily responsible for the formation of new cells, are the meristems. They are found in shoot and root tips as well as in a conical to cylindrical layer, cambium, in perennial stems and roots. In active meristems, part of the daughter cells remains meristematic, while others develop into the various tissue elements. These cells gradually change physiologically, chemically, and morphologically and assume more or less specialized functions through differentiation. The development of plant tissues is characterized by three types of growth: cell division, cell elongation, and cell differentiation.

B. Explants

Any tissue excised from a plant and transferred to nutrient medium *in vitro* is referred to as an explant.

While cell cultures may be obtained from any seed plant, some species can be manipulated more successfully than others. Variation in response to explantation is caused by the culture conditions employed but may also depend on the nature of the explants. Predominantly woody tissues will generally fail to grow; parenchyma rich in tannins may intoxicate the entire explant because of leakage from dead and wounded cells. In general, the selection of a species for establishing cell cultures will be dictated by the project under study and consequently will be directed to those species which feature useful characteristics.

Given a particular species, availability of parenchyma for explantation is a first consideration. Shoot tips appear most promising because during the growing season they harbor actively dividing cells, the shoot meristem. Shoot tip excision, however, requires microsurgery and isolation generally is limited to a low number of explants. Transferred to *in vitro* conditions shoot tips tend to retain their structure and regenerate a shoot and entire plant rather than yield a cell culture. Parenchyma from stems, rhizomes, tubers, and roots, by contrast, is easily accessible and will generally respond quickly to culture conditions *in vitro.* Preference for this material results also from the uniformity of the tissue. Explants from trees and shrubs can be obtained by excising phloem pieces. A method of explantation which applies to the widest range of species requires the germination of seeds and excision of segments from hypocotyls and young roots. It is estimated that the majority of cell strains maintained today originated from segments of seedlings. Recent advances in methodology permit the growth of haploid tissue and in a few cases the regeneration of haploid plants from cultured anthers.

Given the fact that all parenchyma cells of a plant have an identical morphogenetic potential, the origin of an explant, stem, root, or tuber, would be irrelevant. All tissue and cell cultures resulting from an explantation would have an equal potential for organogenesis and for synthesis and accumulation of natural products specific for the species in question. The selection of parenchyma for explantation, therefore, can largely be determined by practical reasons, i.e., accessibility, ease of excision and isolation, rapidity of response to *in vitro* conditions.

C. Culture Media

Successful plant cell culture depends on the choice of nutrient medium. In general, the medium is composed of five groups of ingredients: inorganic nutrients, carbon source, vitamins, growth regulators, and organic supplements.

Inorganic nutrients (mineral salts) are the same as required by intact plants, such as nitrate, sulfate, phosphate, potassium, calcium, magnesium, iron, and trace nutrients, and provide building stones for amino acids, chlorophyll, and vitamins, participate in oxidoreduction systems, and affect changes in plasma colloids. The concentration of mineral salts in media has to be well adjusted in order to prevent intoxication of cultured cells through ion antagonism. A concentration of 1–3 m*M* calcium, magnesium, and sulfate generally is adequate. For most purposes the nutrient should contain at least 25 m*M* each of potassium and nitrate. Replacement of nitrate by ammonium may result in growth inhibition (Gamborg, 1970). With respect to nitrate and ammonium it may be noted that depletion through uptake by the cells or tissues may change the pH of a medium. The required trace nutrients include I, B, Mn, Zn, Cu, Mo, and Co (Gamborg *et al.*, 1976).

Plant cells cultured *in vitro* are heterotrophic with respect to carbohydrates, even when maintained in light. Growth therefore depends on an adequate supply of organic carbon. Sucrose at 2–4% is an efficient carbon source for most cell strains. The capacity of plant tissues to use a variety of carbon sources depends on the species in question. In cell cultures derived from carrots the following series of carbohydrates with decreasing efficiency in growth promotion was established; sucrose, glucose, maltose, raffinose, fructose, galactose, mannose, and starch (Koblitz, 1972).

Although most cell cultures are capable of synthesizing vitamins, the rate of synthesis may be too slow to support rapid growth. *Myo*-inositol may not be essential but is routinely added, since it has been shown to enhance callus growth. The supply of vitamins may become critical for the growth of cells cultured at low densities (Kao and Michayluk, 1975).

As a rule cell cultures are autotrophic with respect to amino acids. They are perfectly able to metabolize nitrate and ammonium; therefore, amino acids generally are not an essential ingredient of culture media. In case inorganic nutrients do not seem to be adequate, the best approach is to add 0.05–0.1 gm/liter casein hydrolysate to initiate and establish a culture. The amino acid mixture may later be replaced by L-glutamine (2–10 m*M*) or the organic nitrogen can be omitted from the medium completely (Gamborg *et al.*, 1976).

Plant cells cultured *in vitro* commonly require media supplemented with growth hormones, auxins, and cytokinins, and sometimes gibberellins, alone or in combination (Street, 1966). Hormones play a dominant role in intercellular regulation of morphogenetic processes. Like all hormones, plant hormones are usually effective in small amounts and sites of synthesis and action may be different. One hormone may have a broad action spectrum and be capable of influencing a number of different processes. The mechanism of action of hormones is not fully understood. In particular the sites of action are still being debated. Assumptions are that they (a) activate all potentially active genes; (b) activate selected genes which in turn will activate other genes; or (c) do not affect genes but regulate one particular metabolic reaction, thus modifying a set of metabolic reactions which may result in an activation of one or several genes. The action spectrum of auxins is extremely wide. They induce cell division in explants, promote cell elongation, stimulate root formation, interact with the development of shoots, and affect the accumulation of secondary metabolites. A natural auxin is indole-3-acetic acid (IAA). Naphthalene-1-acetic acid (NAA) and the very active 2,4-dichlorophenoxyacetic acid (2,4-D) are synthetic products. Cytokinins are known to stimulate cell division and promote shoot formation. The most commonly used cytokinins are kinetin (6-furfurylaminopurine = 6-furfuryladenine) and 6-benzyladenine (6-benzylaminopurine). Cell cultures from monocotyledons may require an exogenous supply of zeatin (γ-methyl-γ-oxymethylallylaminopurine). Among many naturally occurring gibberellins only gibberellic acid (GA_3) has been employed in tissue culture to some extent. GA_3 appears to stimulate differentiation including the accumulation of secondary metabolites (Constabel and Nassif-Makki, 1971).

At times a number of undefined natural products have assisted in establishing plant tissue cultures, the liquid endosperm of coconuts (coconut water), of corn, as well as yeast and malt extract. Even today, a supplement of 10–15% coconut water may be recommended in cases where explants resist to grow despite the presence of auxins and cytokinins (Shantz and Steward, 1952; van Staden and Drewes, 1974; Steward, 1958). The advantages of avoiding the addition of any undefined supplements to nutrient media are evident.

Most water sources for the preparation of media may require demineralization, glass distillation, or both. The chemicals should be of the highest grade available. Growth regulators may require recrystallization before use. The pH is adjusted to a specified value between 5.5 and 6.8 and the medium autoclaved at 120°C for 15–20 minutes (up to 1 liter) or 30–40 minutes (volumes to 10 liters). Most components in the media tolerate autoclaving, although some caution should be exercised (Bragt *et al.*, 1971). Exceptions may be IAA and glutamine. The media can be stored at room temperature, but +10°C is preferable (Gamborg *et al.*, 1976).

Media after Murashige and Skoog (1962), Eriksson (1965), and Gamborg *et al.* (1968) have found widest acceptance in cell culture laboratories (Table I). Murashige and Skoog's medium appears to satisfy the nutritional as well as physiological needs of many plant cell strains.

A distinguishing feature of this medium is its high content of nitrate, potassium, and ammonium relative to other nutrient media. Eriksson's medium is similar to that of Murashige and Skoog but contains twice the amount of phosphate and much lower concentrations of micronutrients. Gamborg's medium has been tested with a wide range of cell cultures.

TABLE I. Culture Media

Nutrients	MS[a] mg/liter		ER[a] mg/liter		B5[a] mg/liter	
		m*M*		m*M*		m*M*
NH_4NO_3	1650	20.6	1200	15.0		
KNO_3	1900	18.8	1900	18.8	2500	25
$CaCl_2 \cdot 2\,H_2O$	440	3.0	440	3.0	150	1.0
$MgSO_4 \cdot 7\,H_2O$	370	1.5	370	1.5	250	1.0
KH_2PO_4	170	1.25	340	2.5		
$(NH_4)_2SO_4$					134	1.0
$NaH_2PO_4 \cdot H_2O$					150	1.1
		μ*M*		μ*M*		μ*M*
KI	0.83	5.0			0.75	4.5
H_3BO_3	6.2	100	0.63	10	3.0	50
$MnSO_4 \cdot 4\,H_2O$	22.3	100	2.23	10		
$MnSO_4 \cdot H_2O$					10	60
$ZnSO_4 \cdot 7I\,H_2O$	8.6	30			2.0	7.0
Zn versenate			15	37		
$Na_2MoO_4 \cdot 2\,H_2O$	0.25	1.0	0.025	0.1	0.25	1.0
$CuSO_4 \cdot 5\,H_2O$	0.025	0.1	0.0025	0.01	0.025	0.1
$CoCl_2 \cdot 6\,H_2O$	0.025	0.1	0.0025	0.01	0.025	0.1
$Na_2 \cdot EDTA$	37.3	100	37.3	100	37.3	100
$FeSO_4 \cdot 7\,H_2O$	27.8	100	27.8	100	27.8	100
Sucrose (gm)	30		40		20	
Inositol	100				100	
Nicotinic Acid	0.5		0.5		1.0	
Pyridoxine·HCl	0.5		0.5		1.0	
Thiamine·HCl	0.1		0.5		10.0	
Glycine	2.0		2.0			
IAA[b]	1–30					
NAA[c]			1.0			
Kinetin	0.04–10		0.02		0.1	
2,4-D[d]					0.1–1.0	
pH	5.7		5.8		5.5	

[a] Media of: MS, Murashige and Skoog (1962); ER, Eriksson (1965); B5, Gamborg *et al.* (1968).

[b] Indoleacetic acid.

[c] Naphthaleneacetic acid.

[d] 2,4-Dichlorophenoxyacetic acid.

This medium contains relatively low amounts of ammonium, a nutrient that may repress growth in batch cultures (Gamborg *et al.*, 1976).

The selection of media presented here should not be interpreted as implying that other media are not suitable. The media are suggested as a starting point. For a more complete discussion of nutrition, several reviews are available (Dougall, 1972; Murashige and Skoog, 1962; Street, 1966; White, 1963).

D. Establishment of Callus and Cell Cultures

1. Explant Culture

A typical explant culture is a 2–5 mm^3 sterile cube or segment excised from a stem, tuber, or root transferred onto 30 ml of nutrient agar medium inside a petri dish or culture tube. The culture is maintained in the dark, in light, or in a regime of light and dark cycles at 20°–30°C, preferably 25°–28°C. Histological analysis reveals that within 2–5 days the explanted tissue responds to the excision and implantation onto a nutrient medium with a wound reaction. Noninjured living nucleated cells adjacent to the wounded cells redifferentiate and reenter mitotic cycles. Given a uniform explant and suitable culture conditions the first division can occur in an almost perfectly synchronous manner (Mitotic Index = 70–80%) (Yoeman, 1974). The presence of hormones, in particular auxins, in the medium will induce further divisions of the daughter cells. Soon the surface of the explant and primarily the part exposed to the air will be covered with a thin layer of newly formed callus. Resumption of cell division by parenchyma cells in an explant may not be restricted to cells near the surface but may also progress toward the interior. As a result the explant will grow and, depending on the intensity of cell proliferation, will burst and become obliterated. Explants which consist of several types of cells show the formation of young tissue only around parenchyma cells and yield irregularly shaped callus. Tanniferous parenchyma cells were observed to undergo divisions and give rise to a sequence of nontanniferous derivatives (Constabel, 1968). Cells which store anthocyanins, chloroplasts, or Ca oxalate crystals were found to lose their pigments or inclusions, when reentering mitotic cycles, a process referred to as dedifferentiation (Buvat, 1944, 1945).

2. Callus Culture

A callus culture begins with the isolation of the newly formed tissue from the explant and its transfer to fresh medium and maintenance under conditions as mentioned before. The transferred callus continues to grow but, in time, may lose its uniformity. Actively dividing cells form a layer of meristem and build a globular mass of nondividing parenchyma or form small meristem islands producing a nodulated irregular callus. Differentiation in the callus may proceed further and result in the formation of tracheids. The homology between a callus meristem and a cambium is evident. As the callus grows the synthesis and accumulation of natural products may be substantial. Tannin, anthocyanin, betalain, alkaloid, and carotenoid production in callus has been reported (Krikorian and Steward, 1969).

After several weeks the size of the callus may permit cutting it into smaller pieces for inoculation of fresh media and establishment of subcultures. Selection of those callus pieces which resulted from optimum growth will be the material of choice for subculturing. This selection may mean the inception of a cell strain characterized by relatively rapid growth. In time this strain may alter its properties further and lose its capacity to produce roots, shoots, or natural products, in other words, lose some of its morphogenetic potential.

Heterogeneity of callus cannot only be found in the appearance of various kinds of tissues and cells but is also reflected in the level of ploidy of the cells. Poly- and aneuploid cells frequently occur in older callus cultures (Kao *et al.*, 1970).

3. Cell Culture

Rapidly growing and soft callus tissue is suitable for transfer to liquid medium and may easily render a cell suspension culture. Typically, an inoculum of about 500 mg fresh weight is suspended in 50 ml nutrient contained in 250-ml Delong flasks. The cultures are maintained on gyratory shakers (120–150 rpm) under light and temperature conditions as before. While growth of a callus culture is more or less linear, cells suspended in liquid medium pass through growth phases similar to those of microorganisms and documented as a sigmoid curve. Cell suspension cultures have several advantages over callus cultures besides growing faster: (1) The suspensions can be pipetted; (2) they are less heterogeneous and differentiation of cells is less pronounced; (3) they can be cultured in volumes up to 1500 liters; (4) they can be subjected to more stringent environmental controls; and (5) they can be manipulated in their production of natural products by feeding precursors.

4. Anther Culture

Anther culture is a method which permits the callusing of pollen and growth of haploid tissue. This material may then be (1) exposed to mutagens to yield mutant cell lines, (2) treated with colchicine to result in homozygous cells, or (3) transferred to conditions conducive to the regeneration of plants, haploids. Anther culture could become a major tool in improving product formation in cell cultures (Gresshoff and Doy, 1972; Kasha, 1974; Zenk, 1976). Success in growing haploid cells depends on recognizing the stage of mononucleated pollen in the development of stamen and on the choice of a nutrient medium (Nitsch, 1972).

Anthers are excised from superficially sterilized flower buds (the exact developmental stage has to be determined by means of microscopic analysis of spores) and transferred to agar medium. Anther culture media appear to require a supplement of amino acids (Nitsch, 1974).

5. Protoplast Culture

The observations that protoplasts, i.e., cells minus cell wall, are able to take up inert as well as biologically active particles and that they can be induced to fuse have led to project somatic cell genetics, including the regeneration of parasexually hybrid plants. The experimental basis for this approach is the isolation and culture of protoplasts (Cocking, 1972; Gamborg *et al.*, 1974).

Excised pieces of parenchyma or cultured cells are incubated in a solution containing cell wall digesting enzymes and osmotic stabilizers. Isolated protoplasts appear as perfectly spherical, wall-less cells. After isolation they have to be washed free of the enzyme solution and can now be subjected to experimental conditions or incubated with nutrient medium. Formation of a new cell wall may begin 20 minutes after removal of enzymes or may take several days. Cell division in protoplasts has been observed as early as 24 hours after isolation. Wall formation and resumption of cell divisions depend as much on the physiological conditions of the protoplast source as on the choice of nutrient medium. Rapidly growing cell suspension cultures have been a particularly suitable protoplast source. In attempts to achieve greatest variation in protoplast species, the mesophyll of young, expanded leaves has been used most extensively as starting material (Gamborg and Wetter, 1975). Prolonged culture and growth of protoplasts proper, i.e., wall-less cells, has not yet been achieved.

6. Plating Cells

Cloning of cell lines requires plating techniques. A simple way would be the dispersal of a mass of cells on the surface of nutrient agar medium by means of a spatula. A more efficient technique calls for the mixture of a cell suspension with agar medium autoclaved and cooled to 45°–40°C, and dispensing the mixture as a thin layer in petri dishes (Bergmann, 1960). Growth of plated cells may depend on the density of the cell population. Low densities require the conditioning of media by either supplementing media of growing cultures or applying feeder layers of inactivated cells (Raveh *et al.*, 1973). The chemical nature of a conditioned medium is still poorly understood. Substances leaked or excreted from cells are a factor in conditioning.

7. Deep-Freeze Preservation

A cell suspension culture derived from seed plants is similar to cultures of yeast and other microorganisms. Storage and maintenance of such cultures is a pressing problem, the resolution of which would mean freedom from the problem of genotypic changes of clones. In a few cases, deep-freeze preservation of cultured cells (carrot and sycamore) has been achieved. After thawing, the cells demonstrated their full morphogenetic potential—carrot cells regenerated entire plants (Nag and Street, 1973; Street, 1975).

E. Regeneration of Organs and Plants

The cell theory (Schleiden, 1838; Schwann, 1839) postulates that cells are the unit of function of any organism. This may be interpreted by saying that living, nucleated cells carry the entire genetic information of the organism from which they are derived. As a conclusion cells cultured *in vitro* would have the morphogenetic potential of the original plant. Expression of this potential would be seen as shoot, root, and embryo formation, as regeneration of a complete plant.

The cell theory was proved to be correct when single carrot cells suspended in liquid medium were found to react like zygotes, forming embryos. These could subsequently be grown to mature, seed-bearing plants (Reinert, 1959; Steward, 1958). Since then embryogenesis has been observed in cell and anther cultures of a number of species (Murashige, 1974). Predictions are that this expression of totipotency of cultured cells will become a general phenomenon, once conducive environmental factors have been clearly defined. As a rule, embryogenesis is triggered by transfer of cells to media without hormones. However, only a fraction of the cells in question may respond to this treatment, and it is still impossible to identify with certainty those cells which are about to undergo the process of embryogenesis (Street and Withers, 1974).

An alternate and much more common route to plant regeneration from cultured cells is via shoot and root formation. The primordia of these organs originate from single cells or small groups of cells. Shoots tend to arise from the periphery, while roots emerge from the interior of a callus. A classical example for the regulation of organogenesis is found in callus derived from tobacco stem pith (Skoog and Miller, 1957). When this tissue is isolated and grown *in vitro* cell division occurs only on a medium containing both an auxin and a cytokinin. A combination of 2 mg/liter IAA and 0.1 mg/liter kinetin results in mere callus growth. If, however, the kinetin concentration is lowered to 0.02 mg/liter roots will develop; 0.5 mg/liter kinetin, conversely, will result in shoot formation. Shoots may root and thus regenerate an entire plant.

The inception of organs and embryos occurs most frequently in recently isolated tissues and this ability decreases with increasing duration of a culture. It is suggested that the expiration of the capacity to express totipotency may be related to the instability of genomes in cultured cells, for instance, to changes in the level of ploidy (Torrey, 1959).

III. CELL SUSPENSION CULTURES

One of the prerequisites for the use of plant cells in the commercial production of natural substances is their adaptability to submerged culture conditions. These conditions not only facilitate mass propagation but allow also stringent control of the culture environment as well as direction toward a high-yield output of the desired metabolite. In the past 25 years the development of various culture techniques has increased our knowledge of the conditions of culture to the extent that it would be now possible to cultivate most plant cells in suspension on a large scale. A cell derived from seed plants differs in its requirements considerably from a microbe and it is therefore not surprising that culture techniques developed for the latter cannot be adapted as such. Cell suspension cultures are generally heterogeneous, with single cells and cell aggregates growing at different rates. The proportion between free cells and aggregates, as well as their size, depends on the species and age of the culture, the composition of the medium, and environmental conditions. It is therefore not surprising that attempts to achieve ideal conditions for cell division and separation to yield a uniform culture of single cells have been numerous but only to a certain degree sucessful (Ganapathy *et al.*, 1964; Kurz, 1973; Street and Henshaw, 1963; Street *et al.*, 1965). The fundamental aspects of plant suspension cultures on a much broader basis have recently been reviewed (Street, 1973).

A. Physical Parameters of Culture

1. Inoculum

The size of the inoculum is important for the initiation of growth in suspension culture. Below a critical ratio of inoculum to volume of medium, cell division ceases. This observation led to the assumption that the medium has to be "conditioned" by factors released from the cells of the inoculum. Growth only resumes after a critical concentration of these factors is attained in the medium (Street and Henshaw, 1963; Street, 1966, pp. 665–680). Ten to fifteen percent v/v of an actively growing suspension culture has been found a satisfactory inoculum for most species (Veliky and Genest, 1972; Verma and van Huystee, 1971; Wang and Staba, 1963). The ratio of free cells to cell aggregates influences the growth rate (Torrey *et al.*, 1962) and in some investigations the inoculum was prefiltered to obtain a more homogeneous cell suspension (Staba and Lamba, 1963; Torrey *et al.*, 1961). Larger inocula may reduce the "conditioning time" of the medium but may lower the relative growth rate of the cells (Caplin, 1963).

2. Temperature

The temperature range within which plant cells are able to grow is slightly higher than for the plant proper and lies for most cultures between 20° and 28°C. Tulecke and Nickell (1960) studied the temperature effect on the growth of five suspension cultures and found optimal ranges between 20° and 21°C for *Lolium perenne* and an extreme temperature of 31°–32°C for *Rosa* sp.

3. Aeration and Agitation

In most suspension cultures effective aeration is essential for optimal growth and a minimum ratio of 1 : 4 (air : medium) was recommended for cultures grown under forced aeration (Routien and Nickell, 1956). However, this should only be taken as a rule of thumb. Aeration should be viewed in the whole context of a particular process as it has been found that forced aeration in some instances (Lamport, 1964; Muir *et al.*, 1958; Verma and van Huystee, 1971) reduces the growth rate. Agitation has been achieved by placing culture vessels on rotary or reciprocating shakers; by passing air through filtered tubes (Wang and Staba, 1963), open-ended tubing on pipes (Kato *et al.*, 1976; Lamport, 1964; Tulecke and Nickell, 1960; Tulecke, 1966; Wilson, 1976), or fritted disks (Graebe and Novelli, 1966; Reinhard, 1967); by compressed air (Kurz, 1973), magnetic stirrers (Bragt *et al.*, 1971; Constabel *et al.*, 1971, 1974), in conventional stirred-jar fermentors (Kato *et al.*, 1972; Veliky and Martin, 1970; Veliky and Genest, 1972; Wang and Staba, 1963); or by using roller devices (Lamport, 1964; Muir *et al.*, 1958; Short *et al.*, 1969).

4. Light

Plant suspension cultures are usually grown in diffuse fluorescent light with light intensities ranging from 15 to 315 μ Einstein m^{-2} sec^{-1} ($\mu\epsilon/m^2$/second) (Hood, 1964; Mitsuoka and Nishi, 1974; Vasil *et al.*, 1964; Venketeswaran, 1964). However, light seems only to be necessary for cultures containing chlorophyll and has no effect on the growth rate (Torrey and Reinert, 1961). A difference in the consistency of *Haplopappus gracilis* suspension culture was reported, showing that cultures grown in light were firm, while those grown in the dark were friable (Blakely and Steward, 1961).

5. pH

The pH range for plant suspension cultures is rather wide, being between 5.2 and 6.8, and the optimum depends on the tissue cultivated (Veliky and Martin, 1970; Verma and van Huystee, 1971). As autoclaving of the medium can change the pH, which might effect the growth of cells in suspension, special consideration should be given to the buffering of media (Holsten *et al.*, 1965).

B. Culture Systems

The culture systems mentioned here were developed solely for the cultivation of plant suspensions and were designed to avoid shearing effects and other mechanical features which would rupture the thin-walled plant cell. Nevertheless microbial fermentor systems may be successfully adapted for cultivation of plant suspension cultures. The purpose of all systems is an even distribution of cells within the culture, thorough mixing of medium and cells, and adequate gas exchange between the liquid and gas phases of the culture.

Early systems were slowly rotating roller-tube devices holding different types of culture vessels (Caplin and Steward, 1949; de Ropp, 1946; White, 1953). One of them, the Auxophyton (Steward *et al.*, 1958), was designed for rapid proliferation of explants in nippled Florence flasks. Through the slow movement of the Auxophyton, part of the medium was distributed as a thin film within the nipples of the culture vessel, which facilitated a high rate of gas exchange. A somewhat similar apparatus was designed by Muir and co-workers (1958). Good growth has also been obtained with rolling and spinning cultures, where glass bottles carrying the culture were rotated at different speeds at 45° to horizontal position (Lamport, 1964; Short *et al.*, 1969). Reciprocal or rotary shakers are often used to grow plant cell suspensions in various types and sizes of flasks and bottles. The approximate speed for rotary shakers is up to 150 rpm and for reciprocal shakers up to 100 oscillations per minute. Large batch-type suspension culture devices have been described by several authors, where agitation and aeration is achieved either solely by forced aeration (Graebe and Novelli, 1966; Kurz, 1973; Tulecke and Nickell, 1960; Tulecke, 1966; Wilson, 1976) or by aeration and a mechanical stirrer (Melchers and Engelmann, 1955; Miller *et al.*, 1968; Veliky and Martin, 1970; Wilson *et al.*, 1971). These devices are more versatile compared with the one described earlier because of the ease with which growth conditions can be controlled through regulation of such parameters as temperature, agitation, aeration, and pH.

Because of the need of more uniform cell material for metabolic and physiological studies, grown under defined, nonlimiting culture conditions in a state of balanced growth, several continuous culture systems for cell suspension cultures have been developed in the past few years.

The system by Kurz (1973) (Fig. 1) employs compressed air for both agitation and aeration by feeding air pulses at regular intervals into the culture vessel, a flat-bottomed glass cylinder. As the compressed air (5–10 psi) enters the cylinder via a central pipe at the base, it expands into a large bubble having the same diameter as the culture vessel. As the bubble slowly moves upward through the vessel the entire culture passes as a thin layer between the fermentor wall and the surface of the air bubble, thus being aerated and stirred effectively. The reduction of air pressure from 5–10 psi to atmospheric pressure and expansion of the bubble causes a mild vibration in the culture, which probably is the main factor in producing a culture consisting of predominantly single cells, in continuous cultivation for long periods of time. This fermentor system was also successfully employed in the induction of partial synchrony of cell division in continuously grown plant cell suspension cultures (Constabel *et al.*, 1974, 1977).

A conical glass V-fermentor was developed by Veliky and Martin (1970) (Fig. 2). In this system the culture is agitated by a Teflon-coated double bar magnet stirrer supported on a short glass rod at the bottom of the fermentor. The air is supplied through a 17-gauge hypodermic needle. This reduces the number of devices inserted into the culture to a minimum and avoids the buildup of cell clusters on them. The V-fermentor is mainly used for batch operations and in semicontinuous culture through intermittent renewal of medium and harvesting of culture.

The Phytostat described by Miller and co-workers (1968) (Fig. 3), as

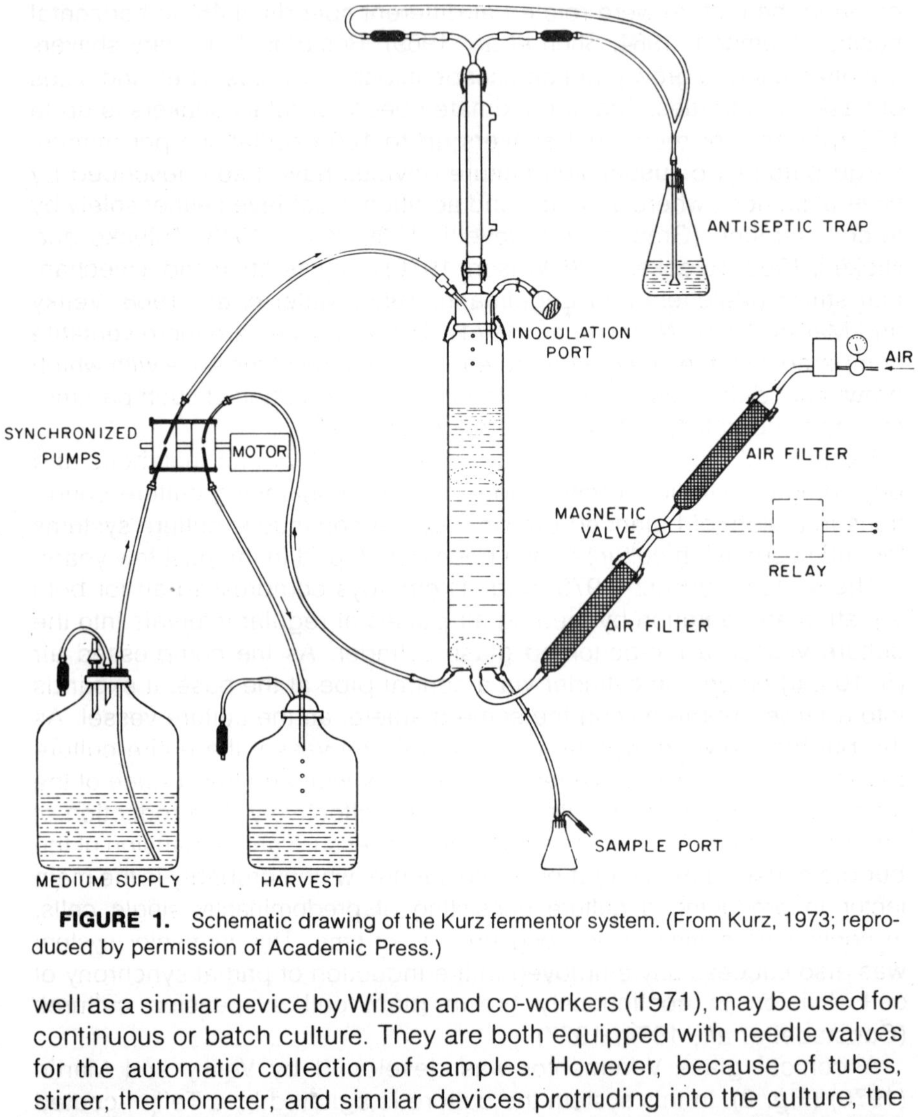

FIGURE 1. Schematic drawing of the Kurz fermentor system. (From Kurz, 1973; reproduced by permission of Academic Press.)

well as a similar device by Wilson and co-workers (1971), may be used for continuous or batch culture. They are both equipped with needle valves for the automatic collection of samples. However, because of tubes, stirrer, thermometer, and similar devices protruding into the culture, the buildup of cell masses on those at the interphase between culture and gas phase poses a problem.

C. Methods of Cultivation

The ideal state for a cell suspension culture is that of morphological, biochemical, and genetic homogeneity grown in a fully controllable

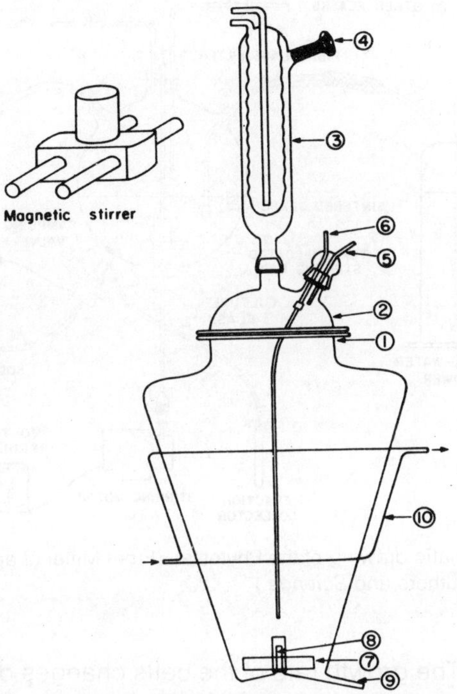

FIGURE 2. Schematic drawing of the V-fermentor: (1) Large-diameter flat-flange joint; (2) multisocket lid; (3) water-cooled condenser; (4) air exhaust with a sterilizing filter; (5) inlet for medium; (6) air inlet; (7) Teflon-coated double-bar magnetic stirrer; (8) short glass rod; (9) sampling outlet; (10) water jacket. (From Veliky and Martin, 1970; reproduced by permission of the National Research Council of Canada.)

environment. Within plant cell research, however, this state is far from being attained. It is not within the scope of this chapter to discuss different growth patterns in cell cultures in detail. This has been done in an excellent review (Street, 1973). Instead we shall try to evaluate different methods of cultivation with respect to production of natural substances.

1. Batch Culture

The majority of investigations involving plant cell suspension cultures have been performed under batch culture conditions. In batch culture, growth takes place in a "closed system"; the cells grow in a limited amount of medium and multiply to form a population in which succeeding generations progressively modify their environment. This produces a sequence of changes in the culture that is often mistakenly referred to as

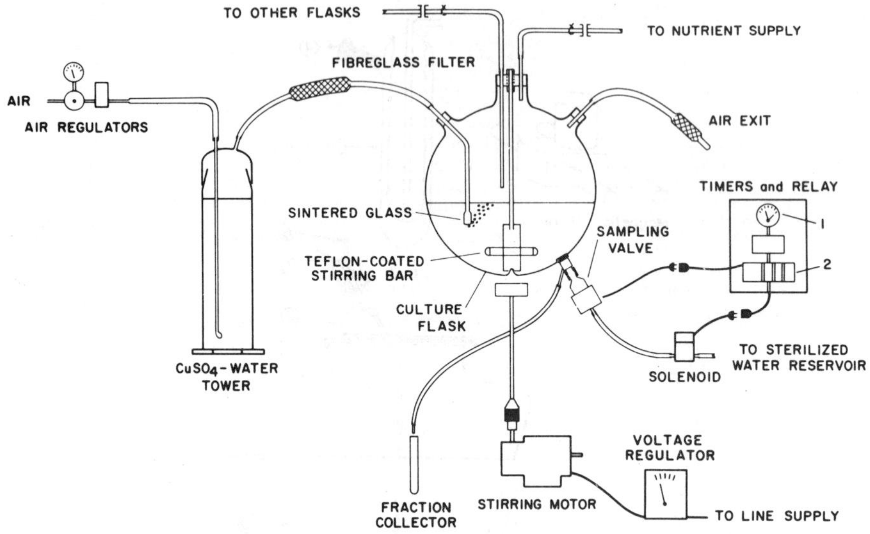

FIGURE 3. Schematic drawing of the Phytostat. (From Miller *et al.*, 1968; reproduced by permission of the authors and *Science*.)

the growth cycle. The growth rate of the cells changes during the period of cultivation so that the time required for the population to double usually increases as the culture progresses. Under these unbalanced conditions the metabolism and composition of the cells are also altered with successive generations and this makes it extremely difficult, if not impossible, to determine any factors responsible for these fluctuations. This notwithstanding, the batch culture technique in plant cell research has its place in the production of biomass and for exploratory research into synthesis of metabolites.

2. Continuous Culture

In continuous culture (chemostat), growth takes place in an "open system" of culture. In this method a continuous feed of medium flowing into a constant volume of growing culture produces a steady-state condition for growth. The environment imposes a constant growth rate on the cells so that the doubling time and overall metabolism of the cells remain constant and thus characteristic of the steady state. It is possible to change the growth rate by altering the flow through the system; in this way the growth of cells may be examined at different growth rates by analysis of the appropriate steady states and metabolic regulation in the production of secondary metabolites is possible. This significant advantage over the batch culture technique in connection with specifically designed cultural environments, such as the controlled feeding of precursors, will lead in the not too distant future to the commercial use of plant cell cultures in the production of secondary metabolites.

3. Continuous Synchronous Culture

In batch and continuous culture the cells divide randomly. In batch culture the doubling time of the successive generations is likely to be changing, but in the chemostat this remains constant. In continuous synchronous growth, an "open system" ideally, all the cells are at the same stage of development in the cell cycle. Thus, instead of an average condition as is obtained in the steady state of a chemostat from the randomly dividing population of cells, there is in the continuous synchronous culture a pattern of change which coincides with the cell cycle and repeats itself with each successive doubling of the cell population. The cell, thereby amplified by the size of the synchronous population, may be examined at any stage of the cell cycle and at any desired growth rate. This has the advantage that enzymes or metabolites occurring only at certain stages of the cell cycle can be obtained at maximal yields, a most useful criterion for consideration in a possible industrial production of such compounds.

The method of continuous synchronous cultivation of plant cells in suspension culture has been developed recently by imposing flushes of nitrogen or ethylene gas at regular intervals upon chemostat cultures at steady state (Constabel *et al.*, 1974, 1977). On plant cells tested so far, this method gave up to 25% (mitotic index) of synchronization of the population over a period of several weeks. Earlier attempts of synchronization in plant suspension cultures (Eriksson, 1966, 1967; King *et al.*, 1973) were based on batch culture conditions and synchrony only persisted for up to six generations.

IV. BIOSYNTHESIS AND ACCUMULATION OF NATURAL PRODUCTS IN PLANT CELL CULTURES

The metabolism of inorganic and organic compounds in plants results in a great number of products which are catabolized as soon as they are formed; these are referred to as intermediary or primary metabolites. At particular stages in the development of plants primary metabolites may be transformed into storage products, fats, starch, and proteins. Secondary metabolites are products which result from biosynthetic transformations of primary metabolites. They occur as terpenoids, glycosides (steroids, phenolics), and alkaloids. Although detected in various groups of animals, secondary metabolites typically are plant products. The diversity and specificity of secondary metabolites has been used as an argument in the classification of plants.

Secondary metabolites appear to originate in cell organelles and may be sequestered into the cytoplasm. As they seem not to be essential for the life of the cell (plant), they are either deposited in vacuoles or excreted from gland cells. Genotype, physiological condition, and state of development as much as location within a given plant determine the formation of metabolites. A meristematic cell, which divides frequently, will rarely exhibit terpenoids, phenolics, and alkaloids; a mature parenchyma cell, however, may be loaded with a whole spectrum of products. This observation can be explained reasonably well on the basis of differential gene activity.

The occurrence of secondary plant products in cell cultures has been well documented (Becker, 1969; Jones, 1974; Krikorian and Steward, 1969; Petiard *et al.*, 1972; Reinhard, 1967; Zenk, 1976). In general the yield is low. It appears that commonly used media favor growth by cell division and not synthesis and accumulation of secondary metabolites. Relatively few cell suspension cultures have shown appreciable amounts of products (Table II). Two kinds of behavior may account for this exceptional result: Cell division and accumulation of products in individual cells do not exclude each other but occur simultaneously [anthocyanin production in *Haplopappus gracilis* cells (Constabel *et al.*, 1971), anthraquinone production in *Morinda citrifolia* cells (Zenk *et al.*, 1975)], or rapidly growing suspension cultures continuously sequester a substantial fraction of the cell population which matures and accumulates metabolites.

Temperature and photoperiod as well as nutrition and changes in the level of hormones may modify, induce, or repress the biosynthetic activity of cultured cells. As these factors can be strictly controlled they offer a chance to manipulate metabolite production.

The effect of various components of nutrient media on the synthesis and accumulation of secondary metabolites has been assessed in only a few cases (Matsumoto *et al.*, 1971, 1973). In cell suspension cultures of soybean, exhaustion of nitrate was thought to cause a switch in cellular metabolism to reactions forming nitrogen-free products, phenylpropanoids for example (Hahlbrock *et al.*, 1974). In *Morinda* cell cultures, nitrate did not affect the biosynthesis of anthraquinone. However, anthraquinone accumulation showed a drastic response to changes in the level of phosphate, 5 gm/liter KH_2PO_4 being the optimum concentration. Iron and calcium ions were necessary for growth and metabolite produc-

TABLE II. Natural Substances Produced by Plant Cell Suspension Cultures.

Products	Species	References
Tobacco alkaloids	*Nicotiana tabacum*	Solt (1957)
		Solt, *et al.* (1960);
	Nicotiana glauca	Speake, *et al.* (1964)
Tropane alkaloids	*Datura stramonium*	Chan and Staba (1965);
		Stohs (1969)
Indole alkaloids	*Vinca minor*	Petiard and Guinebault (1976)
		Boder *et al.* (1964);
	Catharanthus roseus	Patterson and Carew (1969)
		Stöckigt *et al.* (1976);
	Phaseolus vulgaris	Velicky (1972)
Purine alkaloids	*Coffea arabica*	Keller *et al.* (1972)
Quinoline alkaloids	*Ruta graveolens*	Boulanger *et al.* (1973)
Betalains	*Phytolacca americana*	Misawa *et al.* (1973)
Betaines	*Medicago sativa*	Sehti and Carew (1974)
Saponines	*Panax ginseng*	Furuya and Ishii (1973)
	Glycyrrhiza glabra	Tamaki *et al.* (1975)
	Stevia rebaudiana	Komatsu *et al.* (1976)
Cardiac glycosides	*Digitalis purpurea*	Staba and Lamba (1963)
	Apocynum cannabinum	Lee *et al.* (1972)
Glucosinolates	*Reseda lutea*	
	Trapaeolum majus	Kirkland *et al.* (1971)
Furanocoumarins	*Ruta graveolens*	Steck *et al.* (1971)
Anthocyanins	*Populus nigra*	Matsumoto *et al.* (1970)
		Fritsch *et al.* (1971);
	Haplopappus gracilis	Constabel *et al.* (1971)
Quinones	*Nicotiana tabacum*	Ikeda *et al.* (1974)
Anthraquinones	*Morinda citrifolia*	Zenk *et al.* (1975)
Carotenoids	*Daucus carota*	Sugano *et al.* (1971)
Peptides	*Scopolia japonica*	Misawa *et al.* (1973); (1975c)
Antibiotics	*Phytolacca americana*	Misawa *et al.* (1974b)
Plant virus inhibitors	*Phytolacca americana*	Misawa *et al.* (1974c); (1976)
	Agrostemma githago	Misawa and Sakato (1976)
		Misawa *et al.* (1975a)
Coccidiostatic substances	*Catharanthus roseus*	Misawa *et al.* (1975c)
Phosphodiesterase	*Catharanthus roseus*	Furuya *et al.* (1973)
	Phytolacca americana	Ukita *et al.* (1973)
	Nicotiana tabacum	Kubo (1974a)
	Daucus carota	Kubo (1974b)
	Scopolia japonica	
	Zea mays	
Amylases	*Saccharum officinarum*	Maretzki *et al.* (1971)
Glucanases	*Triticum aestivum*	
	Hordeum vulgare	Gamborg and Eveleigh (1968)

tion, while KI appeared to be dispensable. Glucose at 7.5% was superior to all other carbohydrates for anthraquinone production. *Morinda* cells had a definite requirement for organically-bound reduced nitrogen. This requirement was satisfied by supplementing the medium with casein hydrolysate. A dose–response curve showed that 0.2% casein hydrolysate was optimal for growth but suboptimal for anthraquinone production, which in turn was optimal at 0.3% (Zenk *et al.*, 1975).

Synthesis of secondary metabolites in cell cultures can be regulated by the kind and concentration of hormones in the medium in just the same manner as has been demonstrated for the differentiation of roots and shoots. The ability of 2,4-D to suppress and of NAA to enhance anthraquinone accumulation in *Morinda* cell suspension cultures (Zenk *et al.*, 1975), or of increased NAA concentrations to inhibit anthocyanin production in *Haplopappus* cells (Constabel *et al.*, 1971), or of increased 2,4-D concentrations to inhibit the accumulation of polyphenols in rose (Davies, 1972) and *Cassia* cultures (Shah *et al.*, 1976), and of nicotine in tobacco (Tabata *et al.*, 1971) as well as indole alkaloids in *Vinca minor* cultures (Petiard and Guinebault, 1976) are striking examples. Since kinetin is known to stimulate cell division it is reasonable to believe that it might be influencing metabolite production by enhancing cell division in a particular hormonal environment which stimulates the daughter cells to increased synthetic activity.

Morinda cells cultured in media supplemented with NAA not only proved that the biosynthetic potential of this material can exceed that of the original plant, but its biosynthetic capacity may also be taken as an excellent example for the expression of totipotency. Explants isolated from plant parts which normally do not synthesize and accumulate anthraquinones (leaves, stems, fruits) formed callus, and this material gave rise to suspension cultures which showed anthraquinone production (Zenk *et al.*, 1975).

The synthesis of flavonoids in parsley cultures as influenced by light can be traced back to the molecular level. As long as these cultures are kept in darkness, they multiply but do not form flavonoids. Once they are exposed to light the glucoside apiin can be detected in appreciable amounts. This reaction is due to the photocontrol of the synthesis of a key enzyme in flavonoid synthesis, phenylalanine-ammonia lyase (PAL), and of UDP-apiose synthetase (Hahlbrock and Wellmann, 1971). The photocontrol of PAL activity has been subjected to further analysis. In *Haplopappus* cell cultures blue light always leads to an increase of PAL synthesis and subsequent anthocyanin production. Irradiation with red and far red light prior to or after incubation with blue light had no effect on the enzyme synthesis (Gregor and Reinert, 1972). These results illustrate how cells in a culture undergo biochemical modifications in a controllable manner.

Metabolite accumulation cannot succeed when precursors are absent. An example is demonstrated by *Ruta graveolens* cultures producing the alkaloid dictamnine. Supplementing the medium with 4-hydroxy-2-quinolone stimulated the synthesis and accumulation of dictamnine from less than 10^{-4}% to 0.1–0.2% of dry weight (Steck *et al.*, 1973). *o*-succinylbenzoic acid (OSB) is an intermediary product in anthraquinone biosynthesis. Media supplemented with OSB stimulated the accumulation of anthraquinones in *Morinda* cell suspension cultures significantly. After 20 days of culture metabolite levels had increased by 100% (Zenk *et al.*, 1975). The production of alkaloids in cell cultures of *Datura stramonium* was considerably enhanced by feeding DL-tropic acid or tropine (Konoshima *et al.*, 1970); alkaloid production in tobacco cells was affected by methylputrescine (Neumann and Müller, 1971). The supplementation of nutrients with precursors, artificial or natural, therefore, can be a limiting factor in exploiting cultured cells for the production of secondary metabolites.

Aromatic secondary metabolites may not be as stable as their configuration, stage of oxidation and polymerization, and position in biosynthetic pathways would suggest. For many years diurnal or seasonal fluctuations in phenolics and terpenoids have been observed. It is only recently that through sterile systems as cell cultures the degradation of secondary metabolites has been proved (Berlin *et al.*, 1971; Ellis and Towers, 1970). A high rate of O_2 uptake accompanying the more active degradation indicates that it is the rapidly growing cultures which are most effective in degrading secondary metabolites (Constabel and Nassif-Makki, 1971).

Transformations of organic compounds by microorganisms have long been known. The most important are transformations of steroids. Advances in culture methods have permitted the studies of biotransformations of steroids in plant cells. Suspension cultures of *Digitalis lanata* performed a variety of transformations with digitalis glycosides, i.e., glycosylation, acetylation, hydroxylation, and reverse reactions (Reinhard, 1974). Biotransformations of progesterone and pregnenolone were obtained with callus cultures of *Nicotiana tabacum* and *Sophora angustifolia* (Furuya *et al.*, 1972). Tobacco cells which do not contain any compounds of morphinan structure were shown to perform the conversion (demethylation) of thebaine to codeine and morphine (Mothes, 1966).

Apart from secondary metabolites, plant cell cultures have proved to be a source of a variety of proteins, which in the future may become of economic importance, as well as enzymes, peptides, plant virus inhibitors, and antibiotics (Mitsuoka and Nishi, 1974) (Table II). It thus appears that plant cell cultures have infinite possibilities for producing useful natural substances in industry. Again, the remaining difficulty is to control the induction of product synthesis.

At present, attention is focused on employing mutants for high-yield product formation. Mutant cell lines would be characterized by increased enzymatic activities for the biosynthesis of a particular product, or by a defect in enzymes which degrade an accumulated product, or by a defect in enzymes located at a branching point of a metabolic pathway resulting in increased product formation. The culture of haploid cells derived from pollen (anther culture) affords the starting material. Exposure to X-rays or chemicals (ethyl-methanesulfonate, 8-azaguanidine) has proved to result in a variety of mutant cell lines. While the stability of the mutants may be a problem, progress is still hampered by a lack of adequate selection systems for cloning cells with higher than wild-type production of specific metabolites (Zenk, 1974).

Table II lists a variety of products which have been detected in cell suspension cultures. Reports on the occurrence of metabolites in callus cultures have not been included, because this kind of material appears less amenable to projected large-scale industrial operations. It becomes quite obvious from studying the table that substances which are known to exert physiological activities have attracted particular interest. *Vinca* and *Catharanthus* cultures which may produce cancerostatic alkaloids have received most attention. The range of products of cell culture origin would substantiate the feasibility of exploitation of plant cell cultures.

REFERENCES

Becker, H. (1969). *Mitt. Dtsch. Pharm. Ges.* **39,** 273–279.

Bergmann, L. (1960). *J. Gen. Physiol.* **43,** 841–851.

Berlin, J., Barz, W., Harms, H., and Haider, H. (1971). *FEBS Lett.* **16,** 141–146.

Blakley, L. M., and Steward, F. C. (1961). *Am. J. Bot.* **48,** 351–358.

Boder, G. B., Gorman, M., Johnson, I. S., and Simpson, P. J. (1964). *Lloydia* **27,** 328–333.

Boulanger, D., Bailey, B. K., and Steck, W. (1973). *Phytochemistry* **1[illegible],** 2399–2405.

Bragt, J. V., Mossel, D. A. A., Pierik, R. L. M., and Veldstra, H. (1971). Effects of sterilization on components in nutrient media. Misc. Pap. No. 9. Wageningen, Netherlands.

Braun, A. C. (1968). *Results Prob. Cell Differ.* **1,** 128–135.

Brown, C. L., and Sommer, H. S. (1975). "An Atlas of Gymnosperms Cultured in Vitro." Georgia Forest Research Council, Macon.

Buvat, R. (1944). *Ann. Sci. Nat., Bot. Biol. Veg.* **5,** 1–130.

Buvat, R. (1945). *Ann. Sci. Nat., Bot. Biol. Veg.* **6,** 1–119.
Caplin, S. M. (1963). *Am. J. Bot.* **50,** 91–94.
Caplin, S. M., and Steward, F. C. (1949). *Nature (London)* **163,** 920–921.
Chan, W. N., and Staba, E. J. (1965). *Lloydia* **28,** 1–26.
Cocking, E. C. (1972). *Annu. Rev. Plant Physiol.* **23,** 29–50.
Constabel, F. (1968). *Planta Med.* **16,** 241–247.
Constabel, F. (1976). *In Vitro* **12,** 743–747.
Constabel, F., and Nassif-Makki, H. (1971). *Ber. Dtsch. Bot. Ges.* **84,** 629–636.
Constabel, F., Shyluk, J. P., and Gamborg, O. L. (1971). *Planta* **96,** 306–316.
Constabel, F., Kurz, W. G. W., Chatson, B., and Gamborg, O. L. (1974). *Exp. Cell Res.* **85,** 105–110.
Constabel, F., Kurz, W. G. W., Chatson, B., and Kirkpatrick, J. W. (1977). *Exp. Cell Res.* **105,** 263–268.
Davies, M. E. (1972). *Planta* **104,** 50–65.
de Ropp, R. S. (1946). *Science* **104,** 371–373.
Dougall, D. K. (1972). *In* "Growth, Nutrition, and Metabolism of Cells in Culture" (G. H. Rothblat and V. J. Cristofalo, eds.), Vol. 2, pp. 372–406. Academic Press, New York.
Ellis, B. E., and Towers, G. H. N. (1970). *Phytochemistry* **9,** 1457–1461.
Engvild, K. C. (1974). *Physiol. Plant.* **32,** 390–393.
Eriksson, T. (1965). *Physiol. Plant.* **18,** 976–993.
Eriksson, T. (1966). *Physiol. Plant.* **19,** 900–910.
Eriksson, T. (1967). *Physiol. Plant.* **20,** 348–354.
Fritsch, H., Hahlbrock, K., and Grisebach, H. (1971). *Z. Naturforsch. Teil B* **26,** 581–585.
Furuya, T., and Ishii, T. (1973). Japanese Patent (Kokai) 73/31917.
Furuya, T., Syona, U., Kojima, H., Hirotani, M., Ikuta, A., Hikichi, M., Kawaguchi, K., and Matsumoto, K. (1972). *Ferment. Technol. Today, Proc. Int. Ferment. Symp. 4th, 1972,* pp. 705–709.
Furuya, A., Ukita, M., Kotani, Y., Misawa, M., and Tanaka, H. (1973). Japanese Patent (Kokai) 73/33093.
Gamborg, O. L. (1970). *Plant Physiol.* **45,** 372–375.
Gamborg, O. L., and Eveleigh, D. E. (1968). *Can. J. Biochem.* **46,** 417–421.
Gamborg, O. L., and Wetter, L. R. (1975). "Plant Tissue and Culture Methods." National Research Council of Canada, Ottawa.
Gamborg, O. L., Miller, R. A., and Ojima, K. (1968). *Exp. Cell Res.* **50,** 151–158.
Gamborg, O. L., Constabel, F., Fowke, L. C., Kao, K. N., Ohyama, K., Kartha, K. K., and Pelcher, L. E. (1974). *Can. J. Genet. Cytol.* **16,** 737–750.
Gamborg, O. L., Murashige, T., Thorpe, T. A., and Vasil, I. K. (1976). *In Vitro* **12,** 473–478.
Ganapathy, P. S., Hildebrandt, A. C., and Riker, A. J. (1964). *Am. J. Bot.* **51,** 669.
Gautheret, R. J. (1959). "La culture des tissus végétaux." Masson, Paris.
Goebel, K. (1908). "Einleitung in die experimentelle Morphologie der Pflanzen." Teubner, Leipzig.
Graebe, J. E., and Novelli, G. D. (1966). *Exp. Cell Res.* **41,** 509–520.
Gregor, H. D., and Reinert, J. (1972). *Protoplasma* **74,** 307–319.
Gresshoff, P. M., and Doy, C. H. (1972). *Aust. J. Biol. Sci.* **25,** 259–264.
Hahlbrock, K., and Wellmann, E. (1971). *Planta* **94,** 236–239.
Hahlbrock, K., Ebel, J., Oaks, A., Auden, J., and Liersch, M. (1974). *Planta* **118,** 75–84.
Hollings, M. (1965). *Annu. Rev. Phytopathol.* **3,** 367–396.
Holsten, R. D., Sugii, M., and Steward, F. C. (1965). *Nature (London)* **208,** 850.
Hood, K. J. (1964). *Plant Physiol.* **39,** Suppl. 11.

Ikeda, T., Matsumoto, T., Kato, K., and Noguchi, M. (1974). *Agric. Biol. Chem.* **38,** 2297–2298.
Jones, L. H. (1974). *In* "Industrial Aspects of Biochemistry" (B. Spencer, ed.), FEBS, Vol. 30, Part 2, pp. 813–833. Elsevier, Amsterdam.
Kao, K. N., and Michayluk, M. R. (1975). *Planta* **126,** 105–110.
Kao, K. N., Miller, R. A., Gamborg, O. L., and Harvey, B. L. (1970). *Can. J. Genet. Cytol.* **12,** 297–301.
Kartha, K. K., and Gamborg, O. L. (1975). *Phytopathology* **65,** 826–828.
Kasha, K. T., ed. (1974). "Haploids in Higher Plants." University of Guelph, Ontario.
Kato, K., Shiozawa, Y., Yamada, A., Nishida, K., and Noguchi, M. (1972). *Agric. Biol. Chem.* **36,** 889–902.
Kato, A., Kawazoe, S., Ijima, M., and Shinizu, Y. (1976). *J. Ferment. Technol.* **54,** 82–87.
Keller, H., Wanner, H., and Baumann, T. W. (1972). *Planta* **108,** 339–350.
King, P. J., Mansfield, K. J., and Street, H. E. (1973). *Can. J. Bot.* **51,** 1807–1823.
Kirkland, D. F., Matsuo, M., and Underhill, E. W. (1971). *Lloydia* **34,** 195–198.
Koblitz, H. (1972). "Gewebekulturen." Fischer, Jena.
Komatsu, K., Nozaki, W., Takemura, M., and Nakaminami, M. (1976). Japanese Patent (Kokai) 76/19169.
Konoshima, M., Tabata, M., Yamamoto, H., and Hiraoka, N. (1970). *Yakugaku Zasshi* **90,** 370–377.
Krikorian, A. D., and Steward, F. C. (1969). *In* "Plant Physiology" (F. C. Steward, ed.), Vol. 5B pp. 227–326. Academic Press, New York.
Kubo, Y. (1974a). Japanese Patent (Kokai) 74/85286.
Kubo, Y. (1974b). Japanese Patent (Kokai) 74/86589.
Kurz, W. G. W. (1973). *In* "Tissue Culture: Methods and Applications" (P. F. Kruse, Jr. and M. K. Patterson, eds.), p. 359. Academic Press, New York.
Lamport, D. T. A. (1964). *Exp. Cell Res.* **33,** 195–206.
Lee, P. K., Carew, D. P., and Rosazza, J. (1972). *Lloydia* **35,** 150–155.
Maretzki, A., Delacruz, A., and Nickell, L. G. (1971). *Plant Physiol.* **49,** 521–525.
Matsumoto, T., Nishida, K., Noguchi, M., and Tamaki, E. (1970). *Agric. Biol. Chem.* **34,** 1110–1114.
Matsumoto, T., Okunishi, K., Nishida, K., Noguchi, M., and Tamaki, E. (1971). *Agric. Biol. Chem.* **35,** 543–551.
Matsumoto, T., Nishida, K., Noguchi, M., and Tamaki, E. (1973). *Agric. Biol. Chem.* **37,** 561–567.
Melchers, G., and Engelmann, U. (1955). *Naturwissenschaften* **42,** 564–565.
Miller, R. A., Shyluk, J. P., Gamborg, O. L., and Kirkpatrick, J. W. (1968). *Science* **159,** 540–542.
Misawa, M. (1977). *In* "Plant Tissue Culture and Its Bio-Technological Application" (W. Barz, E. Reinhard, M. H. Zenk, eds.), pp. 17–26. Springer, Berlin, Heidelberg, New York
Misawa, M., and Sakato, K. (1976). Japanese Patent Appl. 76/113203.
Misawa, M., Hayashi, M., Nagano, Y., and Kawamoto, T. (1973). Japanese Patent (Kokai) 73/6153.
Misawa, M., Tanaka, H., Chiyo, O., and Mukai, N. (1974a). Japanese Patent (Kokai) 74/7491.
Misawa, M., Sakato, K., and Hayashi, M. (1974b). Japanese Patent (Kokai) 74/126894.
Misawa, M., Hayashi, M., Tanaka, H., Ko, K., and Misato, T. (1974c). *Biotechnol. Bioeng.* **17,** 1335–1347.

Misawa, M., Sakato, K., Hayashi, M., Takayama, S., Tanaka, H., Misato, T., and Ko, K. (1975a). Japanese Patent Appl. 75/9828.
Misawa, M., Sakato, K., and Tanaka, H. (1975b) Japanese Patent (Kokai) 75/5891.
Misawa, M., Hayashi, M., Shimada, K., and Omotani, Y. (1975c). Japanese Patent (Kokai) 75/101510.
Misawa, M., Sakato, K., Tanaka, H., Misato, T., and Ko, K. (1976). Japanese Patent (Kokai) 76/70889.
Mitsuoka, S., and Nishi, T. (1974). Japanese Patent (Kokai) 74/94897.
Mothes, K. (1966). *Naturwissenschaften* **53,** 317–323.
Muir, W. H., Hildebrandt, A. C., and Riker, A. J. (1958). *Am. J. Bot.* **45,** 589–597.
Murashige, T. (1974). *Annu. Rev. Plant Physiol.* **25,** 135–166.
Murashige, T., and Skoog, F. (1962). *Physiol. Plant.* **15,** 473–497.
Nag, K. K., and Street, H. E. (1973). *Nature (London)* **245,** 270–272.
Nester, E. (1977). "Crown Gall". Raven, New York.
Neumann, D., and Müller, E. (1971). *Biochem. Physiol. Pflanz.* **162,** 503–513.
Nitsch, C. (1974). *C. R. Hebd. Seances Acad. Sci.* **278,** 1031–1034.
Nitsch, J. P. (1972). *Z. Pflanzenzuecht.* **67,** 3–18.
Patterson, B. D., and Carew, D. P. (1969). *Lloydia* **32,** 131–140.
Petiard, V., and Guinebault, P. R. (1976). German Federal Republic Patentagency Offenlegungschrift 2,603,588.
Petiard, V., Demarly, Y., and Paris, R. R. (1972). *Plant. Med. Phytother.* **6,** 41–49.
Puhan, Z., and Martin, S. M. (1970). *Prog. Ind. Microbiol.* **9,** 13–39.
Raveh, D., Hubermann, E., and Galun, E. (1973). *In Vitro* **9,** 216–222.
Reinert, J. (1959). *Planta* **53,** 318–333.
Reinert, J., and Bajaj, Y. S. P., eds. (1977). "Applied and Fundamental Aspects of Plant Tissue Culture." Springer, Berlin and New York.
Reinhard, E. (1967). *Dtsch. Apoth.-Ztg.* **107,** 1201–1207.
Reinhard, E. (1974). *In* "Tissue Culture and Plant Science" (H. E. Street, ed.), pp. 433–459. Academic Press, New York.
Routien, J. B., and Nickell, L. G. (1956). U.S. Patent 2,747,334.
Sacristan, M. D., and Melchers, G. (1970). "Die Kulturpflanze," Suppl. 6. Akademie-Verlag, Berlin.
Schleiden, M. J. (1838). *Arch. Anat. Physiol. Physiol. Abt.* **137.**
Schwann, T. (1839). "Mikroskopische Untersuchungen über die Übereinstimmung in der Struktur und dem Wachstum der Tiere und der Pflanzen." Berlin.
Sehti, J. K., and Carew, D. P. (1974). *Phytochemistry* **13,** 321–324.
Shah, R. R., Subbaiah, K. V., and Mehta, A. R. (1976). *Can. J. Bot.* **54,** 1240–1245.
Shantz, E. M., and Steward, F. C. (1952). *J. Am. Chem. Soc.* **74,** 6133–6135.
Short, K. C., Brown, E. G., and Street, H. E. (1969). *J. Exp. Bot.* **20,** 572–578.
Skoog, F., and Miller, C. O. (1957). *Symp. Soc. Exp. Biol.* **11,** 118–130.
Smith, H. H. (1974). *BioScience* **24,** 269–276.
Solt, M. L. (1957). *Plant Physiol.* **32,** 480–484.
Solt, M. L., Dawson, R. F., and Christman, D. R. (1960). *Plant Physiol.* **35,** 887–894.
Speake, T., McCloskey, P., Smith, W. K., Scott, T. A., and Hussey, H. (1964). *Nature (London)* **201,** 614–615.
Staba, E. J. (1969). *Recent Adv. Phytochem.* **2,** 75–106.
Staba, E. J., and Lamba, S. S. (1963). *Lloydia* **26,** 29–35.
Steck, W., and Constabel, F. (1974). *Lloydia* **37,** 185–191.
Steck, W., Bailey, B. K., Shyluk, J. P., and Gamborg, O. L. (1971). *Phytochemistry* **10,** 191–195.

Steck, W., Gamborg, O. L., and Bailey, B. K. (1973). *Lloydia,* **36,** 93–95.
Steward, F. C. (1958). *Am. J. Bot.* **45,** 709–713.
Steward, F. C., ed. (1969). "Plant Physiology," Vol. 5B. Academic Press, New York.
Steward, F. C., Caplin, S. M., and Millar, F. K. (1952). *Am. J. Bot.* **16,** 58–77.
Stöckigt, J., Treimer, J., and Zenk, M. H. (1976). *FEBS Lett.* **70;** 267–270.
Stohs, S. J. (1969). *J. Pharm. Sci.* **58,** 703–705.
Street, H. E. (1966). *In* "Cells and Tissues in Cultures" (E. N. Wilmer, ed.), Vol. 3, pp. 534–629. Academic Press, New York.
Street, H. E. (1973). "Plant Tissue and Cell Culture," Bot. Monogr. Vol. 11. Univ. of California Press, Berkeley and Los Angeles.
Street, H. E., ed. (1974). "Tissue Culture and Plant Science." Academic Press, New York.
Street, H. E. (1975). *IAPTC Newsl.* **15,** 2–4.
Street, H. E., and Henshaw, G. G. (1963). *Symp. Soc. Exp. Biol.* **17,** 234–256.
Street, H. E., and Withers, L. A. (1974). *In* "Tissue Culture and Plant Science" (H. E. Street, ed.), pp. 71–100. Academic Press, New York.
Street, H. E., Henshaw, G. G., and Buiatti, M. C. (1965). *Chem. Ind. (London)* **1,** 27–33.
Sugano, N., Miya, S., and Nishi, A. (1971). *Plant Cell Physiol.* **12,** 525–531.
Tabata, M., Yamamoto, H., Hiraoka, N., Marumoto, Y., and Konoshima, K. (1971). *Phytochemistry* **10,** 723–729.
Takebe, I. (1975). *Annu. Rev. Plant Pathol.* **13,** 105–125.
Tamaki, E., Morishita, K., Nishida, K., Kato, K., and Matsumoto, T. (1975). Japanese Patent 75/16440.
Torrey, J. G. (1959). *In* "Cell, Organism and Milieu" (D. Rudnick, ed.), pp. 189–222. Ronald, New York.
Torrey, J. G., and Reinert, J. (1961). *Plant Physiol.* **36,** 483–491.
Torrey, J. G., Reinert, J., and Merkel, N. (1962). *Am. J. Bot.* **49,** 420–425.
Tulecke, W. (1966). *Ann. N.Y. Acad. Sci.* **139,** 162–175.
Tulecke, W., and Nickell, L. G. (1960). *Trans. N.Y. Acad. Sci.* [2] **22,** 196–206.
Ukita, M., Furuya, A., Tanaka, H., and Misawa, M. (1973). *Agric. Biol. Chem.* **37,** 2849–2854.
van Staden, I., and Drewes, S. E. (1974). *Physiol. Plant.* **32,** 347–352.
Vasil, I. K., Hildebrandt, A. C., and Riker, A. J. (1964). *Am. J. Bot.* **51,** 677.
Veliky, I. A. (1972). *Phytochemistry* **11,** 1405–1406.
Veliky, I. A., and Genest, K. (1972). *Lloydia* **35,** 450–456.
Veliky, I. A., and Martin, S. M. (1970). *Can. J. Microbiol.* **16,** 223–226.
Venketeswaran, S. (1964). *Am. J. Bot.* **51,** 669.
Verma, D. P. S., and van Huystee, R. B. (1971). *Exp. Cell Res.* **69,** 402–408.
Wang, C. J., and Staba, E. J. (1963). *J. Pharm. Sci.* **52,** 1058–1062.
Wardlaw, C. W. (1968). "Morphogenesis in Plants" Methuen, London.
White, P. R. (1953). *Am. J. Bot.* **40,** 517–524.
White, P. R. (1963). "The Cultivation of Animal and Plant Cells," 2nd Ed. Ronald, New York.
Wilson, G. (1976). *Ann. Bot. (London)* [N.S.] **40,** 919–932.
Wilson, S. B., King, P. J., and Street, H. E. (1971). *J. Exp. Bot.* **21,** 177–207.
Yoeman, M. M. (1974). *In* "Tissue Culture and Plant Science" (H. E. Street, ed.), pp. 1–18. Academic Press, New York.
Zenk, M. H. (1974). *In* "Haploids in Higher Plants" (K. J. Kasha, ed.), pp. 339–353. University of Guelph, Ontario.
Zenk, M. H. (1976). "Das physiologische Potential pflanzlicher Zelkulturen." Westdeutscher-Verlag, Düsseldorf.
Zenk, M. H., El-Shagi, H., and Schulte, U. (1975). *Planta Med. Suppl.* pp. 79–101.

Chapter 13

Polysaccharides

K. S. KANG
I. W. COTTRELL

MICROBIAL TECHNOLOGY, 2nd ed., VOL. I
 ISBN 0-12-551501-4

I. INTRODUCTION

The polysaccharides derived from plants and seaweeds have been in use for thousands of years. Gum arabic was a component of embalming fluids used by the ancient Egyptians, and red seaweeds were cooked by the Chinese during the time of Confucius to produce a gel which was then flavored and sweetened.

Large-scale utilization of plant gums and seaweed gums, such as algin, and development of economical extraction processes for the seaweed gums began in the 1920s. Since that time other natural sources have been developed to yield gums such as guar and locust bean. Also, chemical modifications of starch and cellulose have been successful in producing water-soluble gums.

In the continuing search for novel, natural, water-soluble polysaccharides, particular attention has been directed in recent years to the production of extracellular polysaccharides of microorganisms. The first microbial polysaccharide to be commercialized was dextran; it was used as a blood plasma extender and in other applications. This polysaccharide has had only limited success as a commercial water-soluble gum, although it has been the subject of a large amount of research and many attempts at commercialization.

Some other bacteria that have been shown to produce polysaccharides with novel properties are *Alcaligenes faecalis* var. *myxogenes*, *Azotobacter vinelandii,* and *Beijerinckia indica.* These organisms produce curdlan, bioalgin, and PS-7, respectively. The yeast *Aureobasidium pullulans* produces pullulan, a polysaccharide with unique properties; scleroglucan, a fungal polysaccharide, is produced by a species of *Sclerotium*. However, none of these microbial polysaccharides has been produced on a large scale commercially.

At this time the only microbial polysaccharide that has reached and maintained large-scale commercial production is xanthan gum, the polysaccharide produced by the bacterium *Xanthomonas campestris.*

II. NATURE OF MICROBIAL POLYSACCHARIDES

A. Relation to Cell Structure

Microorganisms produce polysaccharides of three distinct types: extracellular, structural, and intracellular storage forms. The extracellular polysaccharides, the main subject of this chapter, can be further classified into two forms: (a) capsules that are integral with the cell wall, as well as structurally demonstrable microcapsules, and (b) loose slime components that accumulate in large quantities outside of the cell wall then diffuse into the culture medium. During growth these extracellular polysaccharides contribute a gummy texture to bacterial colonies on a solid medium or cause increased viscosity in a liquid medium.

The capsular components can be separated readily from the amorphous loose slime by centrifugation. The slime formers may produce slime in large quantities and, in some cases, the viscosity becomes so great that a liquid culture remains in place when the culture flask is inverted. The polysaccharides produced by slime formers have the greatest industrial potential, since their polysaccharides can be recovered in large quantities from the culture fluids.

B. Natural Functions of Polysaccharides

Extracellular polysaccharides probably protect microorganisms against various adverse environmental conditions (Wilkinson, 1958). For example, the remarkably high moisture-holding capacity of the polysaccharide enables bacteria to maintain at least a minimum of moisture in their immediate environment, even after prolonged exposure to low humidity. Also, the thick capsule associated with extracellular polysaccharide production by bacteria provides a definite protection against bacteriophages, simply by serving as a physical barrier against the agent. This fact was substantiated by Maxted (1952) for group A streptococci. The capsules also provide a partial protection against phagocytosis and amoebic attack (Wilkinson, 1958).

A strain of *B. indica* (Lopez and Becking, 1968) was reported to metabolize its own polysaccharide. However, slime-producing bacteria normally are not capable of catabolizing their own extracellular polysaccharides (Wilkinson, 1958). There are indications, however, that fungi degrade and utilize their own capsular polysaccharides (Szaniszlo *et al.*, 1968), and the relatively large amount of starch stored by *Polytomella coeca* during growth serves as an energy reserve (Bourne *et al.*, 1958).

The development of microbial populations in natural environments, such as soil, seawater, lakes, and ponds, occurs most abundantly on the surface of solid particles or films. Capsular polysaccharides and slime may serve to cement the cells and their microcolonies to solid substrata (Corpe, 1970).

C. Environmental Influences

Many polysaccharide producers respond to environmental factors directly. For some microorganisms the carbon source determines both the quantity of polysaccharide formation and the quality of the product synthesized. For example, several strains of *Bacillus polymyxa,* including USDA strain 354, produce a heteropolysaccharide consisting of glucose, mannose, and fructose when grown on a sucrose agar medium; whereas on a medium containing a monosaccharide, such as fructose, glucose, galactose, or arabinose, the polysaccharide formed consists of glucose, mannose, and uronic acid (Forsyth and Webley, 1949). In contrast, another strain of *B. polymyxa* produces a heteropolysaccharide when the organism is cultured in liquid medium with glucose as a carbon source: however, a homopolysaccharide that contains only fructose accumulates when sucrose is used as the carbon source (Ninomiya, 1967). Conversely, the extracellular heteropolysaccharides produced by many microorganisms, such as *X. campestris,* do not vary significantly in composition or molecular weight as a result of changes in the substrate.

Ordinarily the carbon source concentration affects the efficiency of carbon source conversion into polysaccharide. For example, it has been reported that the conversion efficiency of glucose to polymer by *X. campestris* decreased markedly with an increase in glucose concentration (Rogovin *et al.*, 1961).

Although a nitrogen source is necessary for both cell growth and enzyme synthesis for polysaccharide formation, an excess of nitrogen, in general, reduces conversion of carbohydrate substrate to extracellular polysaccharide. Usually, the effects of aeration can be related to the relative efficiencies of energy production during aerobic and anaerobic culturing. Under anaerobic conditions, the "Pasteur effect" takes place due to a low efficiency of energy production. All polysaccharide production of commercial significance requires an aerobic process. Also, in all polysaccharide fermentation, the product yield parallels an increase in the viscosity of the fermentation liquor. Thus, as the liquor thickens, a marked decrease in aeration efficiency occurs, which can be attributed to a much lower mass transfer rate between the air bubbles and the liquor.

Temperature often is critical in polysaccharide synthesis. In general, the optimal temperature for cell growth also is the optimum for product formation. All polysaccharide producers of any commercial significance, such as *X. campestris,* are mesophilic organisms. In order to combat the inefficient mass transfer caused by the high viscosity, large amounts of air that usually are hot and high degrees of agitation are applied during fermentation.

Another important factor affecting polysaccharide production is pH. During the latter stages of acidic polysaccharide fermentations, such as those involving *Xanthomonas* (Moraine and Rogovin, 1966) or *Azotobacter* gums (Cohen and Johnstone, 1963), the pH decreases because polymers and certain organic and inorganic acids form. Nevertheless, the bacterial polysaccharides of possible commercial significance appear to have an optimal pH for synthesis between 6.0 and 7.5. For fungal gum production, the optimal pH lies between 4.0 and 5.5.

Many microorganisms have a strict requirement for certain mineral elements. Among these elements are K, P, Mg, Mn, and Ca. Other elements, such as Mo, Fe, Cu, and Zn, may also be required. However, certain of the minerals that might be added can inhibit product formation. As a result, the mineral requirements for polysaccharide synthesis vary from species to species.

III. MECHANISM OF SYNTHESIS

A. Homopolysaccharides

The mechanism for microbial synthesis of polysaccharides, from monosaccharides as well as from disaccharide donors, has been reviewed by a number of authors (Wilkinson, 1958; Hestrin, 1962; Anderson, 1963; Sutherland, 1972). In general, the synthesis of most polysaccharides involves the addition of a single monomeric unit to the nonreducing end of an existing primer polysaccharide molecule. The general reaction for homopolysaccharide synthesis may be represented by Eq. (1):

$$\text{G-O-X} + [(\text{G-O})_n - \text{G}] \xrightarrow{\text{E}} [(\text{G-O})_{n+1} - \text{G}] + \text{X} \qquad (1)$$

In this equation G-O-X designates a donor system in which G is the glycoside and X is a nonpolymeric product, such as pyrophosphate, uracil diphosphate (UDP), or sugar. The symbol E denotes the transglycosidation enzyme, and *n* is the molar quantity. In mammalian glycogen synthesis the symbol X varies in different systems. X in the equation is UDP. The X also represents UDP in bacterial cellulose synthesis by *Acetobacter xylinum* (Glaser, 1958) and glucan synthesis by *Rhizobium japonicum* (Dedonder and Hassid, 1964).

In the case of starchlike polysaccharides synthesized by a strain of *Neisseria perflava* (Hehre *et al.*, 1947; Hehre and Hamilton, 1948), by *P. coeca* (Lwoff *et al.*, 1950), and by certain species of *Corynebacterium* and streptococci (Carlson and Hehre, 1949), the X in the above equation is phosphate. However, most *Neisseria* strains, except *N. perflava,* have an amylosucrase that catalyzes the polymerization of the glucose moiety of sucrose to yield a starchlike polysaccharide. In this reaction, the intermediary formation of glucose 1-phosphate (G-1-P) is not involved

(Hehre *et al.*, 1947). The production of a starchlike polysaccharide by an amylomaltase obtained from mutants of *Escherichia coli* (Doudoroff *et al.*, 1949; Monod and Torriani, 1950) is another example of transglycosylation. In these cases, maltose serves as the donor substrate.

For dextran and levan synthesized from disaccharide donors, the X in the equation is a single monomeric sugar unit—fructose for the former and glucose for the latter. Ebert and Patat (1962) propose a two-step mechanism for dextran synthesis as shown in Eq. (2): (2a) is the propagation reaction and (2b) is the acceptor reaction.

$$(EP)_n + S \rightarrow ESP_n \rightarrow \left[(EP)_{n+1}\right] + F \tag{2a}$$

$$+$$

$$A \longrightarrow AESP_n \rightarrow ES + APN \tag{2b}$$

Where E is the enzyme, S is sucrose, P_n is growing dextran chain, F is fructose, and ESP_n is a complex. The dissociation of ESP_n controls the overall reaction that results in growth of the polymer chain by one glucose unit with the liberation of one molecule of fructose. It was found that dextran sucrase has no detectable catalytic action on glucose 1-phosphate (G-1-P) (Hehre, 1943), or glucose 6-phosphate (G-6-P), fructose 6-phosphate (F-6-P), fructose 1,6-diphosphate (F-1,6-diP), and adenosine triphosphate (ATP) (Carlson *et al.*, 1953). Likewise, the levansucrases produced by *Aerobacter levanicum* (Hestrin *et al.*, 1955), *B. polymyxa*, and *Bacillus subtilis* (Hestrin, 1944; Hestrin and Shapiro, 1944) are not affected by inorganic phosphate, F-6-P, or F-1,6-diP.

Thus, the evidence indicates that dextransucrase, levansucrase, and amylosucrase directly transfer a single monomeric sugar unit, which originated from the disaccharide donors, Eq. (3):

$$\begin{aligned} &n\ \text{Sucrose} \xrightarrow{\text{dextransucrase and amylosucrase}} (\text{glucose})_n + n\ \text{fructose} \\ &n\ \text{Sucrose} \xrightarrow{\text{levansucrase}} (\text{fructose})_n + n\ \text{glucose} \\ &n\ \text{Maltose} \xrightarrow{\text{amylomaltase}} (\text{glucose})_n + n\ \text{glucose} \end{aligned} \tag{3}$$

The branching of glycogen- and starchlike substances, as well as levan and dextran, results from the action of additional enzymes in the reaction mixture. The Q enzyme, which is capable of converting linear amylose to branched amylopectin, has been found in *N. perflava* (Hehre *et al.*, 1947) and *P. coeca* (Bebbington *et al.*, 1952). A similar branching enzyme operates in dextran synthesis (Bovey, 1959). Ebert and Brosche (1967) showed that branches in dextran molecules are produced by already formed dextran acting as acceptors. When tritium-labeled dextran is added to the reaction mixture for dextran synthesis, the branched dextran formed has unlabeled side chains and a labeled main chain.

B. Heteropolysaccharides

Exocellular heteropolysaccharide synthesis is catalyzed by multienzyme systems, and each enzyme assumes a high degree of specificity. Therefore, heteropolysaccharide synthesis is much more complex than homopolysaccharide synthesis. Information on the mechanism of synthesis of exocellular heteropolysaccharides of commercial significance is still very scant. Most of the information has been obtained from polysaccharide syntheses of immunologically and clinically important organisms.

Sutherland (1977) divided the process of exocellular polysaccharide synthesis into four major sequences: (1) substrate uptake, (2) intermediary metabolism, (3) formation of exopolysaccharide, and (4) modification and extrusion. These sequences are logical and useful in explaining the polysaccharide biosynthesis. Briefly, the substrate enters the cell, usually by active transport and group translocation involving substrate phosphorylation. After entry, the substrate is subjected to either catabolic pathways or pathways leading to the polysaccharide synthesis. The synthetic pathways involve substrate activation by formation of various sugar nucleotides, sugar phosphates, and the interconversion of sugar.

Monosaccharides from sugar nucleotides are added sequentially to growing polysaccharide chains in correct linkages. Isoprenoid alcohol phosphate serves as a carrier.

The role of the isoprenoid lipid, as monosaccharide carrier in polysaccharide synthesis, is well documented in *Streptococcus faecalis* (Umbreit *et al.*, 1972), *Aerobacter aerogenes* (Troy *et al.*, 1971), *Micrococcus lysodeikticus* (Scher *et al.*, 1968), and various other bacteria (Sutherland, 1972). Finally, intermediate or incomplete polymer molecules are modified and extruded from the cell surface to form exocellular polysaccharides.

Bacterial algin exemplifies Sutherland's scheme of exocellular polysaccharide synthesis. This algin (which is a copolymer of D-mannuronic acid and L-guluronic acid) is produced by a strain of *A. vinelandii.* It is partially acetylated (Gorin and Spencer, 1966). Pindar and Bucke (1975) describe an anabolic pathway leading to the formation of alginic acid. More recently, Deavin *et al.* (1977) summarized algin synthesis by *A. vinelandii* as shown in Fig. 1.

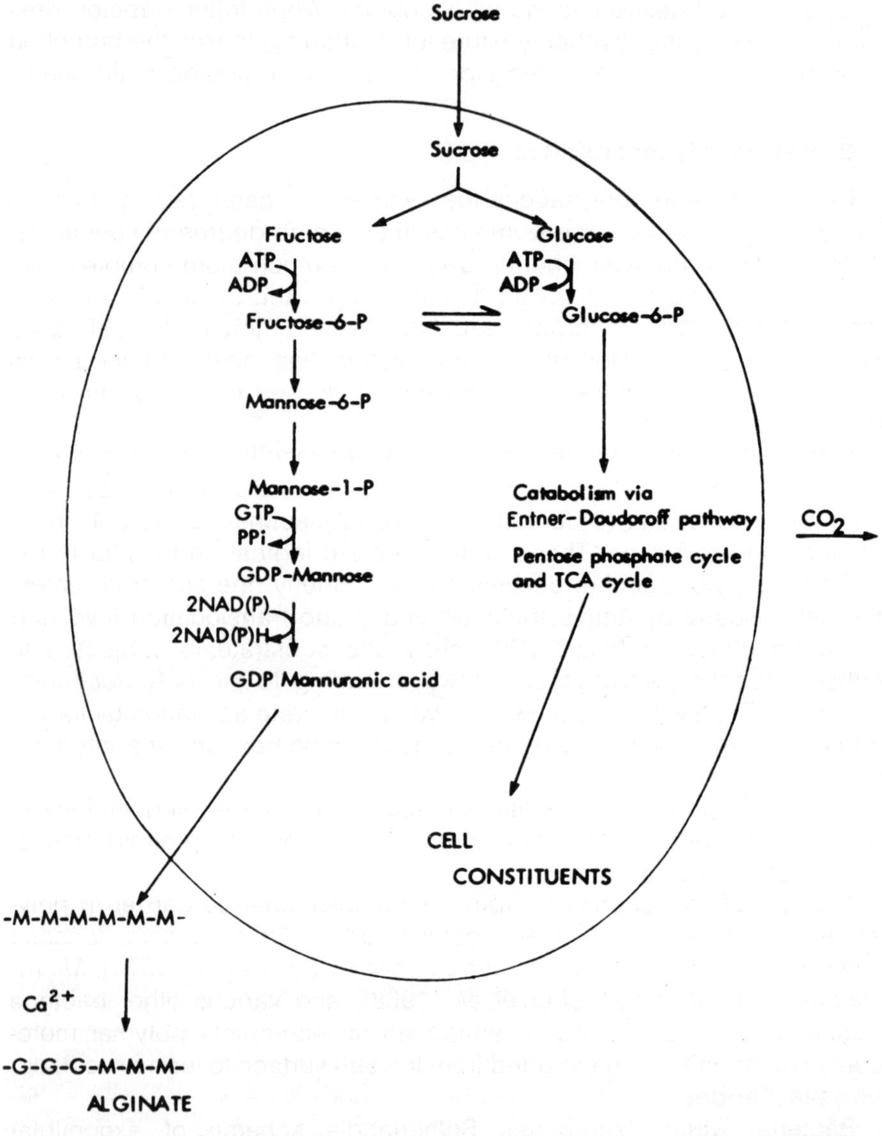

FIGURE 1. Metabolism of *Azotobacter vinelandii* in relation to alginate synthesis.

Sucrose, the carbon source, is transported into the cell and acted on by invertase, and G-6-P and F-6-P are formed by their respective kinases. Fructose 6-phosphate enters the alginate biosynthetic pathway in the sequence shown in Eq (4):

$$\begin{array}{ccc} \text{Mannose 6-phosphate (M-6-P)} \longrightarrow & \text{mannose 6-phosphate (M-1-P)} \\ & \downarrow \\ \text{GDP–mannuronic acid} \longleftarrow & \text{glucose diphosphate (GDP)–mannose} \end{array} \qquad (4)$$

As mannuronic acid residues accumulate, they then polymerize to form polymannuronate and are subsequently extruded from the cell. As Haug and Larsen (1971) demonstrated earlier, a polymannuronic acid C-5-epimerase partially converts D-mannuronic acid residues to L-guluronic acid residues extracellularly to form alginate. A part of the carbon source is converted via the Entner–Duodoroff pathway, pentose phosphate cycle, and tricarboxylic acid cycle for energy and biomass production.

Recently, Sutherland (1977) constructed a biosynthetic pathway for xanthan gum produced by *X. campestris* (Fig. 2) based on information obtained from other bacteria. Although only limited information is avail-

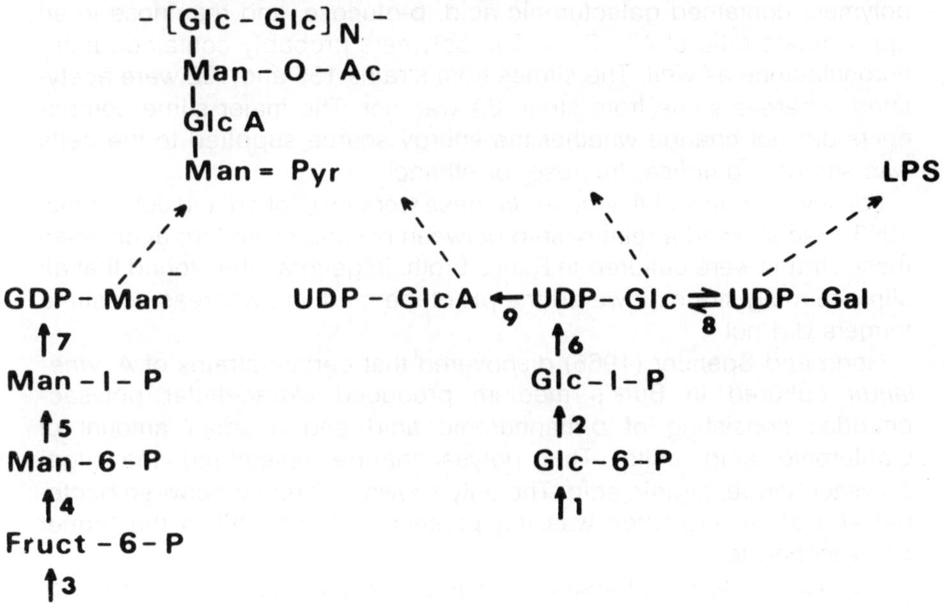

FIGURE 2. Biosynthesis of *Xanthomonas* polysaccharides. (1) Hexokinase, (2) phosphoglucomutase, (3) phosphoglucose isomerase, (4) phosphomannose isomerase, (5) phosphomannomutase, (6) UDP-Glc pyrophosphorylase, (7) GDP-Man pyrophosphorylase, (8) UDP-Gal epimerase, (9) UDP-Glc dehydrogenase.

able, enzyme specificity and the cytoplasmic membrane structure appear to be the factors that determine and control the polysaccharide synthesis.

IV. BACTERIAL POLYSACCHARIDES

A. Azotobacteraceae

1. *Azotobacter Polysaccharides*

All four genera listed in the family Azotobacteraceae (Breed *et al.*, 1957) contain exocellular gum formers. The better known slime formers are *A. vinelandii*, *B. indica*, *Azotomonas agilis*, and *Derxia gummosa*. The nomenclature of microorganisms for this section follows that described in the 8th edition of Bergey's Manual (Buchanan and Gibbons, 1974) in which the species previously classified as *Azotobacter indicus* is described as *B. indica*.

Cohen and Johnstone (1964) isolated extracellular polysaccharides from three strains of *A. vinelandii*. Their analysis revealed that all three polymers contained galacturonic acid, D-glucose, and rhamnose in an approximate ratio of 43 : 2 : 1. The polymers probably contained mannuronolactone as well. The slimes from strains 155 and 102 were acetylated, whereas slime from strain 3A was not. The major slime components did not change whether the energy source supplied to the cells was sucrose, glucose, fructose, or ethanol.

For seven strains of *A. vinelandii* these workers (Cohen and Johnstone, 1963) also showed a relationship between pH and slime formation when these strains were cultured in Burk's broth. In general, they found that all slime-forming strains lowered the pH of the medium, whereas nonslime formers did not.

Gorin and Spencer (1966) discovered that certain strains of *A. vinelandii* cultured in Burk's medium produced extracellular polysaccharides consisting of D-mannuronic acid and a small amount of L-guluronic acid units. This polysaccharide resembled the algal polysaccharide, alginic acid. The only known difference between bacterial and algal aliginates was the presence of *O*-acetyl in the former polysaccharide.

A structural study by Larsen and Haug (1971) that used three strains of *A. vinelandii* revealed that the sequences of uronic acid monomers were distributed in a blockwise manner as previously demonstrated in algal alginate. It was also shown that the D-mannuronic acid-L-guluronic acid (M/G) ratio could be controlled by adjusting the calcium concentration in the fermentation medium; the M/G ratio of the polymer was increased by decreasing calcium ion concentration. A subsequent study (Haug and Larsen, 1971) demonstrated that the polymannuronic acid C-5-epimerase, which catalyzes the transformation of the mannuronic acid residue to the G residue, was operating and the enzyme activity was influenced by Ca^{2+} (Deavin *et al.*, 1977). Couperwhite and McCallum (1974) observed that EDTA in Burk's medium affected the M/G ratio. In the presence of EDTA the M/G ratio of the polymer increased markedly. This increase resulted from alteration of the epimerase, probably occurring when EDTA chelated Ca^{2+}.

Davidson and co-workers (1977) showed by using polyguluronate lyase and polymannuronate lyase that *O*-acetyl groups in the algin elaborated by *A. vinelandii* do not occur regularly in the polymer. They were associated with only one portion of the polymer. From this the authors theorized that the acyl groups protected certain portions of the mannuronic acid residues from epimerization during the formation of bacterial alginate.

Deavin and co-workers (1977) reported a continuous fermentation under phosphate-limited and oxygen-controlled conditions. Under these conditions the maximum yield of alginate (~45% of the sucrose utilized) occurred at the low respiration rates. At higher respiration rates the yield decreased dramatically, because a greater proportion of the substrate was being oxidized to carbon dioxide. The alginate yield in a typical batch culture was only 25% of the sucrose utilized.

2. *Beijerinckia* Polysaccharides

Quinnell and co-workers (1957) reported that the polysaccharide produced by one strain of *B. indica* consisted of glucose, glucuronic acid, and aldoheptose in the ratio of 3 : 2 : 1. On the basis of extensive structural work on the extracellular polysaccharide of another strain of *B. indica*, Parikh and Jones (1963) proposed a linear molecular structure composed of repeating units of D-glucuronic acid, D-glucose, and D-glycero-D-mannoheptose. Lopez and Becking (1968) reported that the polymer of *B. indica* strain Hawai-2 consisted of glucose, galactose, mannose, glucuronic acid, and galacturonic acid, but no heptose.

3. Polysaccharide PS-7

Kang and McNeely (1976a) developed a heteropolysaccharide, PS-7, by using a strain of *B. indica* that they isolated from soil. PS-7 is composed of glucose, rhamnose, and glucuronic acid in an approximate weight ratio of 6.6:1.5:1. PS-7 has an acetyl content of about 9%. Solutions of PS-7 gum are characterized by a high viscosity, a high degree of pseudoplasticity, and an excellent solubility in seawater and even in brine containing 25% salt. PS-7 also exhibits good pH and temperature stability. Figure 3 indicates the high viscosity of this polysaccharide.

Probably the most outstanding property of PS-7 is its exceptionally good suspending ability. Carico (1976a) compared the suspending ability of PS-7 to that of various other gums by means of a standard American Petroleum Institute sand-content tube. Figure 4 is a plot of the settling rates of sand in different polymer fluids. The dotted portion of the lines represents extrapolation to 100%, the time required for complete settling of sand to occur. PS-7 was the most efficient polymer for suspending the sand particles.

The high pseudoplasticity is depicted in Fig. 5 from a viscogram obtained using a Fann viscometer, Model 35. Note that the concentration of PS-7 is only one-half the concentration of xanthan gum or one-quarter of other polymer concentrations. These properties indicate that PS-7

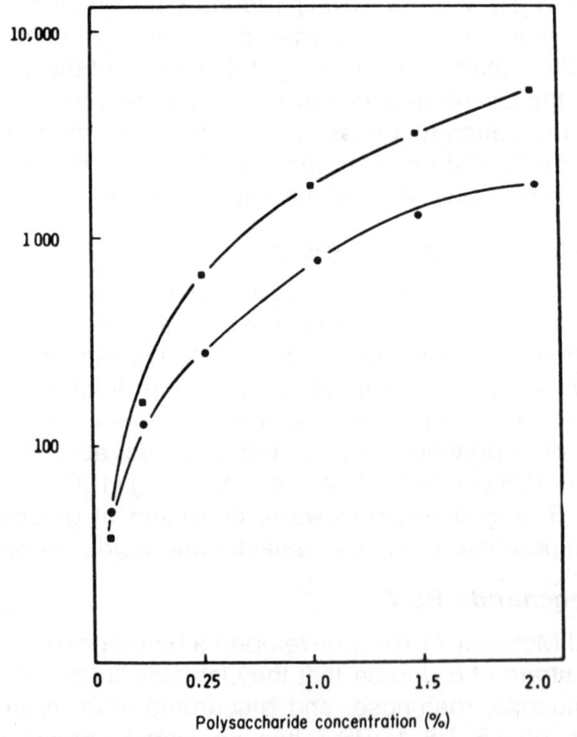

FIGURE 3. The viscosity of PS-7 (■) and xanthan gum (●).

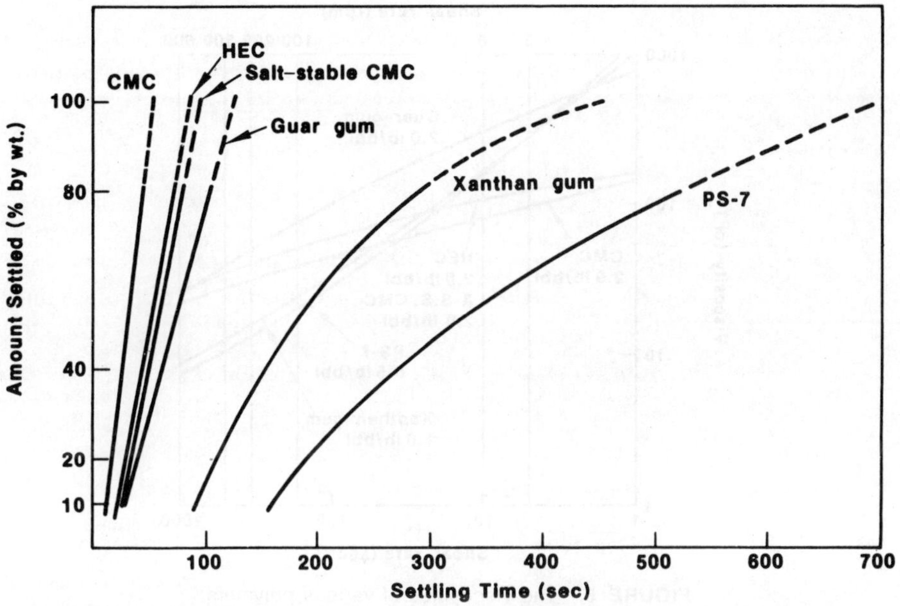

FIGURE 4. The settling rates of sand in different polymer fluids. CMC, carboxymethylcellulose; HEC, hydroxyethylcellulose.

gum will find a variety of applications particularly in oil well drilling (Kang and McNeely, 1976b) and pseudoplastic water-based paint (Kang and McNeely, 1975).

Possible food applications have been also suggested (Kang and Kovacs, 1975). Recently, Kang and McNeely (1977) described production, properties, and possible applications of PS-7.

B. *Alcaligenes* Polysaccharides

Harada (1965) isolated *A. faecalis* var. *myxogenes* 10C3 from soil and found that it produced a unique extracellular polysaccharide when D-glucose or ethyleneglycol was used as a carbon source. The polymer was named succinoglucan 10C3, as it comprised approximately 10% succinic acid, 70–80% D-glucose, and small amounts of galactose and mannose. Methylation and periodate oxidation showed that this gum consisted of β-D-glucose linked through the (1→3), (1→4), and (1→6) positions and a small proportion of β-D-galactose linked through the (1→3) positions (Harada *et al.*, 1969). A high-yielding mutant (strain 22) produced 1.2 gm of the polysaccharide from 4 gm of glucose in 3 days

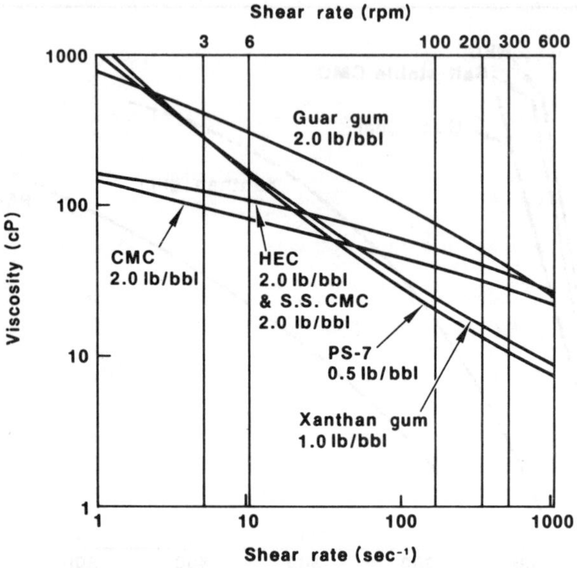

FIGURE 5. The viscosities of various polymers.

(Amemura and Harada, 1971). In 1% aqueous dispersions the free acid and the calcium salt exhibited high viscosities of 1560 and 2500 cP, respectively, as measured by a Cannon-Fenke viscometer at 30°C. It is interesting to note that the viscosity of the sodium salt was extremely low and was increased greatly by addition of many inorganic salts, especially ferric and aluminum ions (Harada and Yoshimura, 1965).

Another mutant derived from *A. faecalis* var. *myxogenes* 10C3 (mutant K) produced a homopolysaccharide known as curdlan (Harada *et al.*, 1968). This polymer consisted of a nonbranched linear chain of D-glucopyranosyl units joined entirely by β-D-(1→3) glycosidic linkages. In a defined or in a semisynthetic medium, the strain converted 50% of the added D-glucose into curdlan after 5 days' incubation at 32°C (Harada *et al.*, 1967). Thus, this polysaccharide, known to be neutral and soluble neither in water nor in acidic solution, dissolves in alkaline solutions, such as 0.5 *N* potassium hydroxide. Curdlan forms a resilient gel when a water suspension is heated above 54°C. The gel strength increases with the increase in temperature. The gel strength is almost constant between pH 2.5 and 10; strength decreases rapidly at pH slightly above 10 and it does not gel at about pH 12. The curdlan gel has properties intermediate between the brittleness of agar gel and the elasticity of gelatin, as shown in Fig. 6 (Harada, 1977).

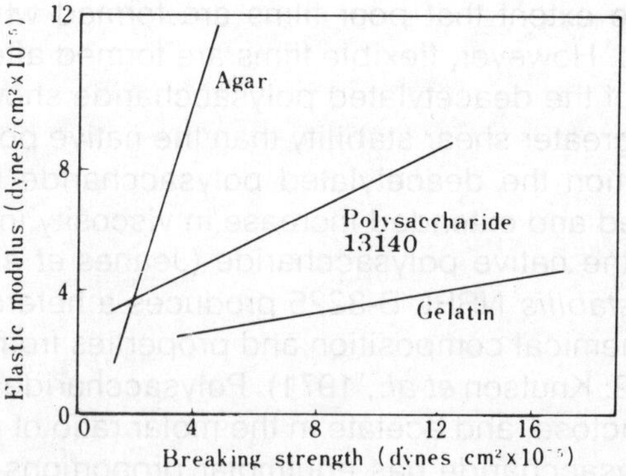

FIGURE 6. Relationship between the breaking strength and elastic modulus of various gels. Samples of gels were sliced into cylindrically shaped pieces 23-mm thick and then measured with Autograph model IM-100.

The curdlan-producing organism was further mutated to obtain strain NTK-u (Kimura *et al.*, 1963). This strain produces a thermogelable polysaccharide 1314) similar to curdlan. Various food applications of this polysaccharide, such as in processing boiled noodles, hamburger, and sausage, have been suggested by Harada (1977).

C. *Arthrobacter* Polysaccharides

Two strains of *Arthrobacter viscosus,* NRRL B-1797 and NRRL B-1973, are known to be extracellular polysaccharide producers (Gasdorf *et al.*, 1965; Knutson *et al.*, 1971). Polysaccharide B-1973 contains D-glucose, D-galactose, and D-mannuronate in the molar ratio of 1 : 1 : 1. The sugar composition of B-1797 polysaccharide is similar to that of B-1973, except that the former gum contains glucuronic acid. In addition, B-1973 gum is highly acetylated (25%), whereas B-1797 polysaccharide is only slightly acetylated. The polysaccharide B-1973 has been subjected to more extensive investigation (Jeanes *et al.*, 1965; Anonymous, 1962; Cadmus *et al.*, 1963; Sloneker *et al.*, 1968). The fermentation medium used for these investigations contained 2–4% glucose, 0.3% hydrolyzed casein, dipotassium acid phosphate adjusted to a pH of 7.0, and traces of magnesium and manganese salts. After clarification the native polysaccharide B-1973 was recovered by precipitation with methanol. Yields of 40–45%, based on glucose, have been obtained in a 4-day aerobic fermentation at 25°C.

These studies indicate that a high proportion of acetylated groups in the native polysaccharide interferes with hydration and molecular alignment to the extent that poor films are formed when aqueous solutions are dried. However, flexible films are formed after deacetylation. Also, solutions of the deacetylated polysaccharide show pseudoplastic properties and greater shear stability than the native polysaccharide. At 0.1% concentration the deacetylated polysaccharide B-1973 shows a more pronounced and extended increase in viscosity in the presence of KCl than does the native polysaccharide (Jeanes *et al.*, 1965).

Arthrobacter stabilis NRRL B-3225 produces a heteropolysaccharide that differs in chemical composition and properties from the *A. viscosus* gums (Gill, 1972; Knutson *et al.*, 1971). Polysaccharide B-3225 consists of glucose, galactose, and acetate in the molar ratio of 4 : 2 : 1 (Jeanes, 1974). This polysaccharide has equimolar proportions of pyruvate and succinate linked as the half ester. Treatment with dilute alkali removes acetyl and succinyl ester substituents, leaving the pyruvic acid ketal as the only acidic group. This polysaccharide is characterized by a good heat stability of its solution viscosity.

Arthrobacter polysaccharides have not been produced commercially. The potential market appears to be small, probably due to the greater salt sensitivity of the polysaccharide, its cross-linking tendencies on heating, as well as a lack of toxicity studies.

D. *Methylomonas* Polysaccharides

Methylomonas mucosa produces a heteropolysaccharide in liquid medium containing methanol as a carbon source. Acid hydrolysis followed by gas chromatographic analysis indicates that this polysaccharide is composed of glucose, galactose, mannose, and pyruvic acid in a molar ratio of (1.5–2.0) : 1 : 1 : (1.5–2.0). The solution viscosity is quite stable to a wide range of NaCl concentrations. However, the gum is sensitive to pH in that the viscosity is at a maximum at pH 4.0 and is slightly lower at pH 7.0. Above pH 7 or below pH 4 the viscosity decreases.

The *Methylomonas* polysaccharide is known to have drag-reducing properties and to be pseudoplastic. On the basis of these properties various industrial applications for this product, such as thickening and suspending agents in foods, cosmetics, paints, oil well drilling, and oil well flooding (Tannahill and Finn, 1975), have been suggested. The crude polymer is negatively charged and has a 1% viscosity of approximately 350 cP as measured by a Brookfield viscometer at 30 rpm and 25°C. However, it has been claimed that the product viscosity can be drastically increased and the fermentation time shortened by the controlled addition of methanol, the carbon source, to the fermentation process near the end of the logarithmic phase of the bacterial growth. Recent studies by Tam and Finn (1977) on the growth kinetics for polymer production indicate that a semicontinuous fermentation is feasible for better reproducibility and a more economic fermentation than a batch process.

E. Miscellaneous Bacterial Polysaccharides

1. Zoogloea

It is well known that species of *Zoogloea* are one of the predominant bacterial groups in activated sludge (Dias and Bhat, 1964; Friedman *et al.*, 1968) and that polysaccharides are a major factor contributing to the physical characteristics of the sludge (Tenney and Stumm, 1965; Unz and Farrah, 1974). Friedman and his co-workers (1968) isolated a bacterium from sewage and identified it as *Zoogloea ramigera.*

Wallen and Davis (1972) reported that a water-soluble heteropolysaccharide was formed by *Z. ramigera* NRRL B-3669M when glucose was used as a carbon source. The polymer consisted of glucose, mannose, and galactose units that were linked through the (1→3) and (1→6) positions. This gum had a viscosity of about 2000 cP at 0.75% polymer concentration in water at 25°C, and the gum solution was quite pseudoplastic. Studies in our laboratory (unpublished results) indicate that this gum has fairly good stability against heat, pH, and mechanical shear.

2. Corynebacterium

Recently, Nagahama *et al.* (1977) isolated *Corynebacterium* species from effluents of the sweet potato starch-manufacturing plant. These isolates produce viscous anionic polysaccharides that contain 44–57% glucuronic acid and 9–12% pyruvic acid, the remaining constituents being mannose, galactose, and glucose in an approximately equimolar basis. These strains convert up to 66% of 3% "mashed potato" into polysaccharide in 4 days at 34°C. Although no detailed physical property study has been made, these polysaccharides appear to have a relatively high viscosity.

Corynebacterium equi. var. *mucilaginosus* ATCC 21521 produced a polysaccharide with *n*-paraffin mixture of C_{12}–C_{17} (Yamatodani and Kanamaru, 1972). This polymer consisted of glucose and mannose, succinic acid, lactic acid, and acetic acid. The polymer yield in a typical fermentation with 7% *n*-paraffin mixture (v/v) was 1.7 gm/100 ml fermentation liquor. This gum was claimed to have a strong adhesive property at low concentrations. *Corynebacterium viscosus* also produced an acidic polysaccharide with an *n*-paraffin mixture of C_{13}–C_{16} (Yamada and Furukawa, 1973).

3. *Rhizobium*

The extrapolysaccharides that have been elaborated by *Rhizobium leguminosarum, Rhizobium phaseoli,* and *Rhizobium trifdii* contained D-glucose, D-glucuronic acid, D-galactose, pyruvic acid, and *O*-acetyl groups in the approximate proportions 5 : 2 : 1 : 2 : 3 (Zenvenhuizen, 1973). These polymers contained residues of (1→3)-linked D-glucose substituted at C-4 and C-6 by pyruvate (13%), (1→4)-linked D-glucose (32%), and (1→4)-linked D-glucuronic acid (20%); (1→4, 1→6)-linked branching residues of D-galactose and/or D-glucose (13%), and terminal D-glucose and/or D-galactose residues substituted at C-4 and C-6 by pyruvate (13%). *Rhizobium meliloti* gum did not contain glucuronate.

4. *Bacillus*

Cadmus *et al.* (1967) reported that *B. polymyxa* could utilize 3% glucose within 72–96 hours with 30–35% conversion. A different strain of *B. polymyxa* was isolated by Ninomiya and Kizaki (1969), which converted 5% glucose into polysaccharide with an efficiency of 36% after a 62-hour fermentation. In our laboratory, fermentation conditions were devised for a strain of *B. polymyxa* in which 3% glucose could be completely utilized within 20–24 hours with approximately 45% conversion efficiency (unpublished data). Therefore, under these rapid fermentation conditions, microbial contamination is not a major problem, since the native bacteria outgrow most contaminants.

Bacillus polymyxa slime is a heteropolysaccharide composed of D-glucose, D-mannose, D-galactose, and D-glucuronic acid in a ratio of 3 : 3 : 1 : 2. The gum exhibits relatively high viscosity and high pseudoplasticity. The viscosity increases sharply in aqueous solution when various salts are added, but it decreases drastically with a small increase in temperature.

Morita and Murao (1974) reported that *B. subtilis* FT-3 produced an acidic polysaccharide consisting of glucose, galactose, fucose, glucuronic acid, and *O*-acetyl in the molar ratio of 2 : 2 : 1 : 1 : (1.4). The mean molecular weight of this polymer was estimated as 10^4–10^6. Since these organisms are spore formers, the polysaccharide should be further processed in most cases either to remove or to kill the spores. This fact has discouraged commercial development of the *Bacillus* polysaccharides.

5. Levan

Levans are polymers of anhydro-β-D-fructose that contain mainly β-2,6-glycosidic linkages. Levan is produced by a number of microorganisms, including species of *Bacillus*, *Pseudomonas*, *Acetobacter*, and *Corynebacterium*. Biosynthesis of levan also can be readily accomplished when cell-free extracts of *B. subtilis* are used with the addition of sucrose.

Like dextran, levan is a branched-chain homopolysaccharide having molecular weights in the range of $1-100 \times 10^6$ (Hestrin, 1962). Despite the high molecular weight, the viscosities of aqueous solutions of levan are considerably lower than those of linear polysaccharides such as algin. Thus, levans appear to be unlikely prospects for development as commercial water-soluble gums.

6. Chromobacterium

Corpe (1960) described the formation of an extracellular polysaccharide by *Chromobacterium violaceum*. In peptone broth this organism converted glucose into a polysaccharide comprised of glucose, uronic acid, and amino sugar in a ratio of 5 : 1 : 1. Later, it was found (Corpe, 1964) that growth and polysaccharide synthesis in a defined medium were markedly enhanced when amino acids were substituted for ammonia as a nitrogen source, and the best polysaccharide yield occurred when the ratio of carbohydrate to nitrogen was 10 : 1. Studies conducted by Corpe (1960) indicate that the polysaccharide exhibits an unusual resistance to the microbial degradation by soil microorganisms and has some interesting rheological properties.

7. Starch- and Celluloselike Polymers

Many strains of *Neisseria* produce starchlike extracellular polysaccharides from sucrose (Carlson and Hehre, 1949). The product is a mixture of a linear α-1,4-linked amylose fraction and a branched-chain portion resembling amylopectin or glycogen. In these biosyntheses the enzyme amylosucrase transfers the D-glucopyranosyl units from the sucrose molecule to the end of the growing polymer chain.

As early as 1886 Brown reported that *Acetobacter xylinium* elaborated tough membranes of cellulose when grown in a suitable nutrient solution. Much later Tarr and Hibbert (1931a) reported cellulose formation by *Acetobacter pasteurianus*, *Acetobacter rancens*, *Sarcina ventriculi*, and *Bacterium xyloides*. The authors established the preferred substrate for *A. xylinium* as 5–10% of a hexose sugar, 0.1% asparagine, 0.5% potassium dihydrogen phosphate, 0.1% sodium chloride, and 0.5% ethanol. Aerobic fermentation of *A. xylinium* for 10 days at 30°C resulted in the formation of extracellular cellulose as a meshwork of intertwined microfibrils (Muhlethaler, 1949), a product identical to plant cellulose (Hestrin and Schramm, 1954).

Recently, Kjosbakken and Colvin (1975) demonstrated the existence of a soluble polymer of glucose in a cell-free particulate enzyme system from *A. xylinium*. This polymer was subsequently shown by King and Colvin (1976) to be an intermediate in bacterial cellulose synthesis. More recently, Colvin and his co-workers (1977) characterized this polymer as being a linear chain of β-(1→4) glucose with single glucose residues branched at position 2 of every third glucose, on the average.

A highly specialized use for bacterial cellulose in membranes for osmometry was reported (Masson *et al.*, 1946). Since starch and cellulose are available from plant sources at low prices and in very large amounts, it is unlikely that the microbial products could develop any significant commercial production.

V. FUNGAL POLYSACCHARIDES

A. Pullulans

Pullulan is the generic name applied to a type of α-D-glucan produced by *Aureobasidium (Pullularia) pullulans*. The polysaccharide was isolated and partially characterized by Bernier (1958) and further characterized by Bender *et al.* (1959).

Typical pullulans are linear polymers consisting of predominant repeating units of maltotriose attached via α-1,6-linkages. Small numbers of α-maltotetrose units are frequently present within the polymer chain (Catley and Whelan, 1971; Taguchi *et al.*, 1973). In addition, the presence of varying amounts of 1,3-linked glucosyl residues in pullulans elaborated by different strains is also reported (Sowa *et al.*, 1963). Pullulanase depolymerizes pullulan by hydrolyzing α-1,6-linkages (Drummond *et al.*, 1969; Taguchi *et al.*, 1973). The enzyme also attacks α-1,6 bonds in starch (Hathaway, 1971).

The molecular weight of a pullulan varies considerably depending on the fermentation conditions and strains used (Zajic and LeDuy, 1973). To meet specific applications Hayashibara Laboratories in Japan is said to be producing pullulan with molecular weights ranging between 1×10^4 and 4×10^5 (Yuen, 1974).

Catley and Whelan (1971) investigated the efficiency of several mono-

and disaccharides, glycerol, and acetate as carbon sources for pullulan elaboration and found that sucrose resulted in the best conversion to pullulan. However, it was reported that product yield over 70% could be obtained by replacing the conventional carbon sources with less expensive starch syrups (Yuen, 1974). The optimal dextrose equivalent of starch hydrolysate for pullulan production was 50–60%.

Yuen (1974) described pullulan as having characteristics similar to those of polyvinyl alcohol, and pullulan-made articles closely resemble those of styrene in transparency, gloss, and hardness, but are much more elastic. Pullulan forms very low oxygen-permeable films (Table I) and high-tensile fibers, the strengths of which are described as comparable to those of nylon.

A number of applications of pullulan and its derivatives have been proposed and patented primarily by Japanese workers. The applications include use as a flocculator of clay slimes in hydrometallurgical processes (Zajic and LeDuy, 1973), use in coating and packing material for foodstuffs and pharmaceuticals, in noncaloric food, in adhesives, and in the manufacture of special fibers and fabrics (Yuen, 1974).

B. *Rhinocladiella* Polysaccharides

The black yeastlike fungus *Rhinocladiella mansonii* NRRL Y-6272 elaborates an extracellular polysaccharide composed of residues of *N*-acetyl-D-glucosamine and *N*-acetyl-D-glucosaminuronic acid in a molar ratio of approximately 2 : 1 (Jeanes *et al.*, 1971). The organism converted 5% glucose to polysaccharide with a conversion efficiency of 13% in 4 days at 25°C. The medium used contained L-asparagine or urea

TABLE I. Oxygen Permeability Comparison of Various Films

Sample[a]	Oxygen permeability (ml/m^2, 24 hours, atm, 25°C)
PS	2.50
PM	2.01
PO	2.15
P_1	1.30
P_2	0.60
Cellophane	4.70
Moisture-proof cellophane	8.58
Polypropylene	1100.00

[a] PS, PM, PO, P_1 and P_2 represent pullulan films of different origins.

and several metal salts. The viscosity of the resulting cell-free polysaccharide in 1% aqueous solution was 9400 cP as measured by a Brookfield viscometer, Model LVF at 30 rpm (Burton *et al.*, 1976).

Another black yeastlike fungus, *Rhinocladiella elatior* Mangenot NRRL YB-4163, isolated from paper mill effluent, also produces a polysaccharide (Watson *et al.*, 1976). This polysaccharide is composed of 2-acetamido-2-deoxy-D-glucuronic acid residues. It is superior to YB-6272 in stabilization of oil–water emulsions and shows a high elasticity; when solutions (1%) of YP-4163 are poured or stirred, the entire solution tends to cling together.

C. Miscellaneous Fungal Polysaccharides

Bouveng *et al.* (1967) reported a water-soluble polysaccharide production when *Armillaria mellae* was grown in a broth containing D-glucitol as carbon source. They were able to fractionate this gum into four fractions by DEAE-cellulose chromatography. The two fractions consisted of a glycogen-type glucan and a β-D-(1→3) glucan, respectively. The third fraction was found to consist of a glucan and a xylomannan, and the last was comprised of D-galactose, D-mannose, fucose, and 3-*O*-methyl-D-galactose in a weight percentage ratio of 42 : 30 : 13 : 15.

Szaniszlo *et al.* (1968) described a capsular polysaccharide derived from *Leptosphaeria albopunctata*, a marine fungus. The gum is composed of "large amounts of glucose and minute amounts of mannose." Polysaccharide synthesis by this organism requires concentrations of salts similar to those in seawater. Furthermore, the synthesis is enhanced by NaCl concentrations above those of artificial seawater. Organisms of this type that produce sufficient gum to have industrial significance have not been reported.

Archer and his co-workers (1977) described an extracellular branched D-glucan produced by *Monilinia fructigena* in glucose–malt medium. More than 90% of this gum is glucose which has 1→4 main linkages, with a lower proportion of 1→2 linkages. The degree of branching varied between every fifth and every tenth residue. The gum contained small amounts of mannose and galactose.

A large number of fungal species are known, and new polysaccharides are frequently discovered. The authors recommend an excellent review article by Gorin and Spencer (1968) for the structural chemistry of yeast and fungal polysaccharides.

Antitumor Activities

Many fungal polysaccharides of the glucan type have been reported to have antitumor activities. A large amount of literature has been published in Japan on this subject. Whistler and his co-workers (1976) reviewed these and other microbial polysaccharides with bacterial, yeast, and plant origins. Several examples of fungal glucans having antitumor activities are given in Table II.

From their studies the authors concluded that the polysaccharides examined so far do not seem eligible for clinical trial in human cancer therapy, although they are quite active in suppressing transplanted tumors. Thus, they suggest further search for new polysaccharides having stronger antitumor activities than those so far discovered.

VI. YEAST POLYSACCHARIDES

A. Phosphomannan

Yeasts of the genus *Hansenula* and related genera produce extracellular phosphomannans from glucose (Slodki *et al.*, 1961). These polysaccharides yield on hydrolysis only D-mannose and D-mannose 6-phosphate. Developmental work was carried out on the phosphomannan produced by *Hansenula holstii* NRRL Y-2448 (Jeanes *et al.*, 1961a) at the Northern Utilization Research and Development Division of the USDA. This phosphomannan was produced in an aerobic 4-day fermentation at 28°C in a medium containing 6% glucose, organic nitrogen, potassium dihydrogen phosphate, and trace elements, with adjustment of pH to 5.0 (Anderson *et al.*, 1960). The product was isolated by alcohol precipitation from the fermentation beer. Yields were 40–55% based on utilization of the glucose substrate. *Hansenula capsulata* Y-1842 also produces a phosphomannan (Slodki, 1963).

Phosphomannan from *H. holstii* Y-2448 contains mannose and phosphorus in a molar ratio of 5 : 1 (Jeanes *et al.*, 1961a). The molecular weight by light scattering is $\sim 16 \times 10^6$. Jeanes and Watson (1962) proposed that the structure of this polymer consists of a repeating unit of 10 mannose residues which are distributed on an average of five between a phosphodiester group that is linked between C-6 of one mannose unit and C-1 of another. The proposed structure of phosphomannan is illustrated in Fig. 7.

A later study by Bretthauer *et al.* (1973) involving a mild acid hydrolysis of the polymer showed that at least 65% of the mannose and phosphate in the native polymer occur as homogeneous monophosphomannopentose units.

Soft, gellike solutions are formed by phosphomannan at concentrations as low as 0.25% in water. The viscosity versus concentration curve is unusual for a water-soluble gum in that the viscosity reaches a

TABLE II. Antitumor Activity of Some Fungal Glucans against Subcutaneously Implanted Sarcoma 180 in Mice.[e]

Glucan	Source	Linkages	Dose (mg/kg × number)	Route of injection	Complete regression	Inhibition ratio (%)
Lentinan[a]	*Lentinus edodes*	(1→3)-β-D	25 × 10	i.p.	2/9	73.0
			5 × 10	i.p.	7/10	97.5
			1 × 10	i.p.	6/10	95.1
			0.2 × 10	i.p.	6/10	78.1
Schizophyllan[b]	*Schizophyllum commune*	(1→3)-β-D, (1→6)-β-D	5 × 10	i.p.	4/10	89
			1 × 10	i.p.	7/10	81
			0.5 × 10	i.p.	7/10	82
			5 × 4	i.v.	5/10	100
			1 × 4	i.v.	4/10	96
			10 × 10	s.c.	4/10	82
			1 × 10	s.c.	0/10	11
Pachymaran[c]	Pachyman from *Poria cocos*	(1→3)-β-D	5 × 10	i.p.	4/9	96
Scleroglucan[d]	*Sclerotium glucanicum*	(1→3)-β-D, (1→6)-β-D	50 × 10	i.p.	2/10	41.2
			5 × 10	i.p.	5/10	88.2
			0.5 × 10	i.p.	7/10	91.3
Fraction LC-1[a]	*Lentinus edodes*	(1→4)-β-D, (1→6)-β-D	30 × 10	i.p.	8/9	96.5
			15 × 10	i.p.	10/10	100
			5 × 10	i.p.	8/10	99.0

[a] From Chihara, G. *et al.* (1969).
[b] From Komatsu, N. *et al.* (1969).
[c] From Chihara, G. *et al.* (1970).
[d] From Singh, P. P. (1974).
[e] Treatment with glucans was started 24 hr after tumor implantation, and the results were recorded after 5 weeks.

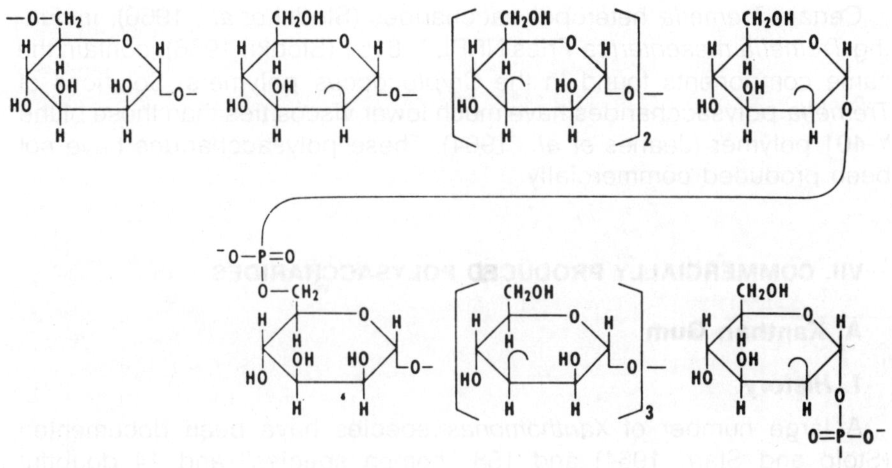

FIGURE 7. Structural types and relative proportions of constituent mannose units in phosphomannan Y-2448. Absence of branching and some details of sequence have not been established.

maximum of about 2500 cP at a 1.5% polysaccharide concentration and a minimum of 1700 cP at 3%, as measured by a Brookfield viscometer. Solutions of approximately 0.65% concentration have the same viscosity as those of 3% concentration. Solutions are thixotropic and are highly sensitive to salts. At 0.5% KCl concentration, solutions having 1.5, 2, and 3% polymer concentrations lose 94, 87, and 63%, respectively, of their viscosity is distilled water. The polymer is also sensitive to vigorous mechanical shearing and heat (Jeanes *et al.*, 1961a). Weak acids split the diester cross-link, causing a loss of viscosity (Anonymous, 1958).

Development of phosphomannan as a commercial water-soluble gum appears to be difficult. The applications are limited primarily because of the sensitivity to salts, shear, and heat, as well as its instability in acid solutions.

B. Polysaccharide Y-1401

This polysaccharide was produced by *Cryptococcus laurentii* var. *flavescens* NRRL Y-1401 in a 6-day aerobic fermentation at 25°C with 5% glucose substrate in the presence of autolyzed brewers' yeast and inorganic salts (Cadmus *et al.*, 1962). The conversion efficiency was 30–35% based on the glucose consumed. Polysaccharide Y-1401 contains D-mannose, D-xylose, D-glucuronic acid, and *O*-acetyl in the molar ratio of 4 : 1 : 1 : (1.5).

Certain *Tremella* heteropolysaccharides (Slodki *et al.*, 1966), including *Tremella mesenterica* Fries NRRL Y-6151 (Slodki, 1966), contain the same components found in the *Cryptococcus* polymers. Solutions of *Tremella* polysaccharides have much lower viscosities than those of the Y-401 polymer (Jeanes *et al.*, 1964). These polysaccharides have not been produced commercially.

VII. COMMERCIALLY PRODUCED POLYSACCHARIDES

A. Xanthan Gum

1. History

A large number of *Xanthomonas* species have been documented (Stolp and Starr, 1964) and 158 "nomen species" and 44 doubtful species are listed in the 8th Edition of Bergey's Manual (Buchanan and Gibbons, 1974). However, only a few species, such as *Xanthomonas campestris, X. phaseoli, X. malvacearum,* and *X. carotae,* were reported as efficient gum formers (Lilly *et al.*, 1958). Recently, a strain of *X. manihotis* was reported to be an efficient polysaccharide producer (Chen and Tsuo, 1976).

The Northern Utilization Research and Development Division of the USDA conducted an extensive screening program for gum producers stored in their large microbial culture collection. Of the various biosynthetic polysaccharides produced in their laboratory polysaccharide B-1459 (xanthan gum), produced by *X. campestris* NRRL B-1459, was found to have characteristics that rendered it very promising as a commercial product (Jeanes *et al.*, 1961b). The Kelco Division of Merck & Co., Inc., San Diego, California, carried out the pilot plant feasibility studies for xanthan gum in 1960, semicommercial production of Kelzan xanthan gum in 1961, and substantial commercial production in early 1964.

2. The Safety and Regulatory Status of Xanthan Gum

With respect to safety properties xanthan gum is one of the most extensively investigated polysaccharides (McNeely and Kovacs, 1975). Short-term acute feeding tests on rats and dogs at the Pharmacology Laboratory of the Western Regional Research Laboratory of the USDA have indicated that xanthan gum causes no acute toxicity or growth-inhibiting activity (Booth *et al.*, 1963). Xanthan gum was also found to be a nonsensitizer (Robbins *et al.,* 1964; Woodard Research Corp., unpublished report, 1973), and it caused no eye or skin irritation (Woodard Research, Corp. unpublished report, 1971).

In short-term subacute toxicological studies of xanthan gum it was not possible to determine acute oral toxicity (LD_{-50}) to rats and dogs. The test animals fed over a 24-hour period—with doses of as much as 45 gm/kg to rats (Woodard Research Corp., unpublished report, 1968a) and 20 gm/kg to dogs (Woodard Research Corp., unpublished report, 1968b)—had no fatalities and showed no signs of toxicity or changes in their internal organs.

The long-term feeding studies on xanthan gum consisted of 2-year studies of rats and dogs and a three-generation reproduction study of the rat (Woodard *et al.*, 1973). No significant effects on growth rate, survival, hematological values, or organ weights were found, and no tumors were detected.

On the basis of the above feeding studies the Food and Drug Administration (FDA) issued a food additive order on March 19, 1969 permitting the use of xanthan gum in food products without any specific quantity limitations (Anonymous, 1969a). In the United States the FDA regulations permit addition of xanthàn gum to many standardized foods, such as cheeses, cheese products, milk and cream products, mellorine (imitation ice cream), food dressings, table syrups, and vegetables in butter sauce (Anonymous, 1977a). In the standards for these products xanthan gum is either mentioned by name or with the phrase "safe and suitable stabilizers" in the listing of approved ingredients. The FDA has also approved xanthan gum as a component of paper and paperboard that is intended for food contact.

USDA regulations also permit the use of xanthan gum in sauces, gravies, and breadings used with meat and poultry products (Anonymous, 1977b). The United States Environmental Protection Agency (EPA) has exempted xanthan gum from tolerance requirements when it is used as an inert ingredient in pesticide formulations (Anonymous, 1976).

The Canadian Governor-in-Council has also formally approved the general use of xanthan gum in foods (Anonymous, 1971). Xanthan gum is on Annex II of the European Economic Community emulsifier/stabilizer list (Anonymous, 1974). The Joint Expert Committee of the Food and Agriculture Organization/World Health Organization of the United Nations (FAO/WHO) has issued an acceptable daily intake (ADI) for xanthan gum (Anonymous, 1975). In addition, many other countries have approved xanthan gum for various food uses (Kovacs and Kang, 1977).

3. Production

Xanthomonas campestris is a bacterium originally isolated from the rutabaga plant; it produces viscid or gummy colonies on agar media (Breed *et al.,* 1957). Lilly and co-workers (1958) reported that *X. campestris* produced substantial quantities of extracellular polysaccharides during the fermentation of a carbohydrate substrate. This fermentation had a close similarity to that of *X. campestris* growth with cabbage extracts as the sole substrate.

Sutton and Williams (1970) determined that the chemical components of xanthan gum are the same whether produced by an industrial-type fermentation or on living cabbage tissues under natural conditions. Furthermore, xanthan gum from both sources produced the same precipitin bands when reacted with an antiserum specific for the polysaccharide. Additional tests demonstrated that the metal ion analysis profile and the rheological properties of both the commercially available food-grade xanthan gum and the xanthan gum produced on a pure cabbage extract are similar (Kang, 1972). These data indicate that the commercially produced xanthan gum is identical to the naturally occurring xanthan gum.

The nutritional requirements for minimum growth of 30 different species of the genus *Xanthomonas* were reported by Starr (1946). A mixture of glucose, ammonium chloride, a phosphate buffer, magnesium sulfate, and certain trace minerals was found to be sufficient for minimum growth. During the fermentation the pH of the medium decreases due to the formation of metabolic acids and xanthan gum, which also contains acidic functions. If the pH reaches a critical point, such as 5.0, the gum production drastically decreases. A nearly neutral pH allows the gum synthesis to continue until all the carbohydrate substrate is exhausted, at which point fermentation is complete. In the commercial process, with maintenance of optimal conditions for gum production, including a high degree of aeration and a constant temperature, the fermentation proceeds at a much faster rate than on the cabbage plant.

In aerobic fermentation at around 28°C, 1–5% glucose concentrations have been found to provide the best xanthan gum yields (Lilly *et al.*, 1958). At higher glucose concentrations an actual decrease in conversion efficiency is observed. Glucose, sucrose, and starch have been found to be efficient in polysaccharide production. Recently, Charles and Radjai (1977) reported the use of acid whey as a carbon source. Cottage cheese whey was hydrolyzed by immobilized lactase. The authors reported that glucose in the whey hydrolysate was used more rapidly than galactose, but both sugars were almost completely consumed without diauxy.

After the completion of fermentation the liquor is pasteurized, and the polysaccharide is recovered by precipitation with isopropyl alcohol. This is followed by drying, milling, testing, and packaging. Meticulous precautions must be taken to meet all microbiological, chemical, and physical specifications. Detailed procedures for culture maintenance, polysaccharide production, purification, and analysis of xanthan gum were recently described by Jeanes *et al.* (1976). A flow sheet diagram for a typical xanthan gum manufacturing process is shown in Fig. 8.

A number of alternate methods have been suggested for the recovery of xanthan gum from the fermentation liquor. Each process starts with the pasteurization of the liquor to kill the bacterial cells. Since xanthomonads are nonspore formers and the vegetative cells are sensitive to elevated temperatures, complete pasteurization can be readily accomplished.

Drum- or spray-drying of the fermentation beer yields a crude polysaccharide product (Rogovin *et al.*, 1965). Long-chain quaternary ammonium salts have been proposed as precipitants (Rogovin and Albrecht, 1964). However, it is doubtful that these methods would be acceptable for the manufacture of food-grade products. Xanthan gum can be recovered also by precipitation with calcium ions followed by washing of the insoluble calcium complex with acid or salt (McNeely and O'Connell, 1966).

A number of continuous fermentation methods have been reported (Silman and Rogovin, 1972; Patton and Lindblom, 1962). More recently, Ellwood and Evans (1977) described a single-stage continuous fermentation of *Xanthomonas juglandis* and *X. campestris*. However, the problem of maintaining sterility even in the batch fermentors indicates that the installation of a continuous fermentation system on a commercial scale would significantly increase the contamination risk.

A clear product can be produced by diluting the fermentation liquor and clarifying by filtration. A process whereby xanthan gum is treated

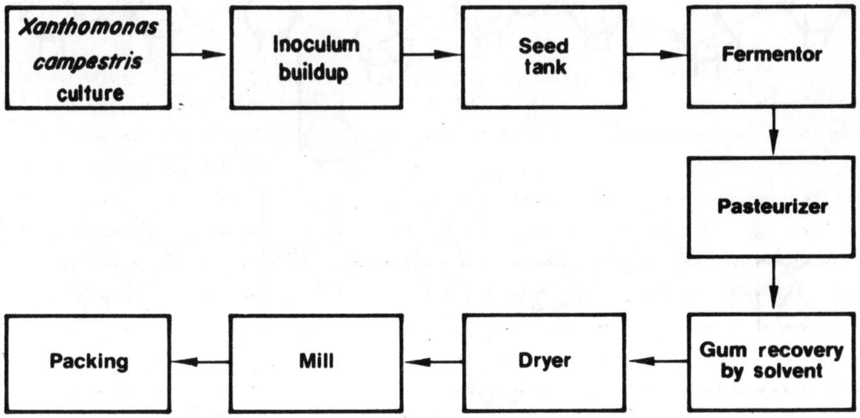

FIGURE 8. Flow sheet diagram for xanthan gum process.

with hypochlorite to produce a clear solution was patented by Colegrove (1970). Another method to clarify xanthan gum utilized high pH at high temperatures (Patton, 1973).

4. Composition, Structure, and Conformation

Xanthan gum is a high molecular weight, natural polysaccharide containing D-glucose (2.8 moles), D-mannose (3.0 moles), D-glucuronic acid (2.0 moles), acetic acid (~4.7%), and pyruvic acid (~3%) (Sloneker and Jeanes, 1962). A primary structure for this polysaccharide was proposed by Sloneker *et al.* (1964) and is shown in Fig. 9. This structure has a 15-sugar repeat that implies a biosynthetic complexity hitherto unknown in the production of extracellular bacterial polysaccharides. The primary structure of xanthan gum was reinvestigated by Jansson *et al.* (1975), using an improved degradative technique (Lindberg *et al.*, 1973) and improved methylation analysis (Bjorndal *et al.*, 1970), and was shown to be composed of pentasaccharide repeating units containing two glucose units, two mannose units, and one glucuronic acid unit (Fig. 10). The main chain of xanthan gum contains β-D-glucose units linked through the 1- and 4-positions; i.e., the chemical structure of the xanthan gum backbone is identical to that of cellulose. The side chain contains the two mannose units and the glucuronic acid unit. The terminal β-D-mannose unit is linked glycosidically to the 4-position of β-D-glucuronic acid, which in turn is glycosidically linked to the 2-position of α-D-mannose. This three-sugar side chain is linked to the 3-position of

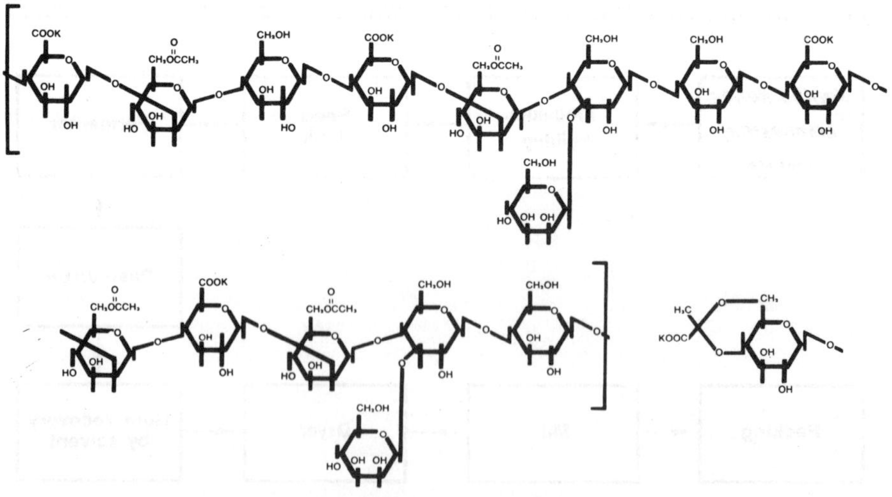

FIGURE 9. Previously accepted structure of xanthan gum.

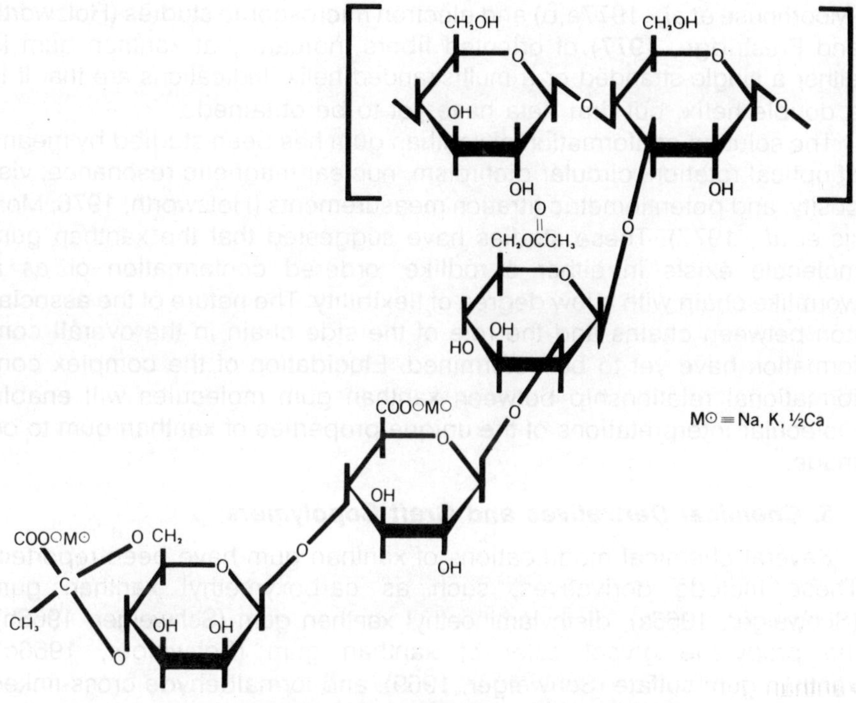

FIGURE 10. Structure of xanthan gum.

every other glucose unit, on an average, in the main chain. The distribution of the side chains has not been determined. The D-mannose unit adjacent to the main chain contains an acetyl group at position 6, and about half of the terminal D-mannose units are ketalically linked through the 4- and 6- positions to pyruvic acid. The distribution of the pyruvate groups has not been elucidated. The molecular weight has been reported to range between 2 and 50 $\times$ 10^6 (or 5 $\times$ 10^7) daltons (Dintzis *et al.*, 1970). This wide range is probably due to an association phenomenon between polymer chains that results in aggregates. The covalent or chemical structure of xanthan gum detailed here only partially describes the xanthan gum molecule. A complete description of the xanthan gum molecule is necessary to determine the shape or conformation in solution. Although most polysaccharides occur as random coils in solution, certain polysaccharides, such as carrageenan, have a special shape in solution (Rees, 1970). The unusual properties of xanthan gum in solution indicate that xanthan gum has a specific conformation in solution. Considerable research has been carried out recently to elucidate the secondary and tertiary structure of xanthan gum. X-Ray diffraction studies

(Moorhouse *et al.*, 1977a,b) and electron microscopic studies (Holzworth and Prestridge, 1977) of oriented fibers indicate that xanthan gum is either a single-stranded or a multistranded helix. Indications are that it is a double helix, but firm data have yet to be obtained.

The solution conformation of xanthan gum has been studied by means of optical rotation, circular dichroism, nuclear magnetic resonance, viscosity, and potentiometric titration measurements (Holzworth, 1976; Morris *et al.*, 1977). These studies have suggested that the xanthan gum molecule exists in either a rodlike, ordered conformation or as a wormlike chain with a low degree of flexibility. The nature of the association between chains and the role of the side chain in the overall conformation have yet to be determined. Elucidation of the complex conformational relationship between xanthan gum molecules will enable molecular interpretations of the unique properties of xanthan gum to be made.

5. Chemical Derivatives and Graft Copolymers

Several chemical modifications of xanthan gum have been reported. These include derivatives, such as carboxymethyl xanthan gum (Schweiger, 1966a), diethylaminoethyl xanthan gum (Schweiger, 1966b), the propylene glycol ester of xanthan gum (Schweiger, 1966c), xanthan gum sulfate (Schweiger, 1969), and formaldehyde cross-linked xanthan gum (Patton, 1962). Also, deacetylated xanthan gum has been prepared (Jeanes and Sloneker, 1961).

Graft copolymers of xanthan gum have been prepared by grafting acrylic monomers onto xanthan gum in the fermentation broth using ceric ion initiation (Pettitt, 1973). Specifically, acrylamide and/or acrylic acid have been grafted onto xanthan gum to produce xanthan gum-g-poly(acrylamide), xanthan gum-g-poly(acrylic acid), or xanthan gum-g-poly(acrylamide-coacrylic acid), respectively. Also, acrylamide and 2-acrylamido-2-methylpropane sulfonic acid have been grafted onto xanthan gum to produce xanthan gum-g-poly(acrylamide-co-2-acrylamido-2-methylpropane sulfonic acid) (Cottrell, 1977; Cottrell *et al.*, 1978). A sulfonic acid-containing graft copolymer has also been prepared by derivatization of xanthan gum-g-poly(acrylamide) using formaldehyde and sodium metabisulfite (Cottrell *et al.*, 1978). Although many derivatives of xanthan gum have been prepared, none has been put into commercial production at this time.

6. Properties

a. Rheological Properties. One of the most important properties of xanthan gum is its ability to control the rheological properties of fluids.

Xanthan gum will dissolve with stirring in either hot or cold water to produce a high-viscosity solution even at low gum concentrations. The relationship between solution viscosity and the concentration of xanthan gum is shown in Fig. 11, as measured using a Brookfield Model LVF viscometer at 60 rpm and 25°C.

Figure 12 shows the effect of salts, such as sodium chloride, on the viscosity of xanthan gum solutions. At low concentrations of xanthan gum, 0.1% or lower, addition of sodium chloride causes a slight reduc-

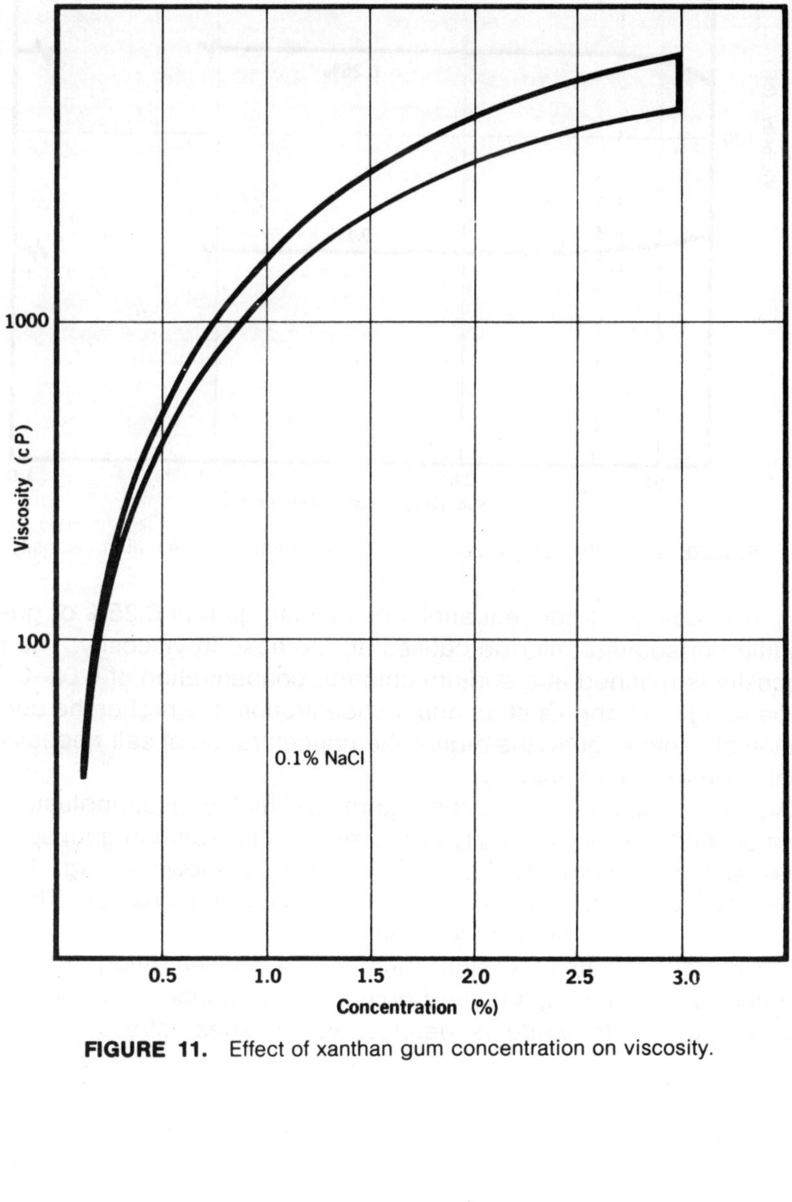

FIGURE 11. Effect of xanthan gum concentration on viscosity.

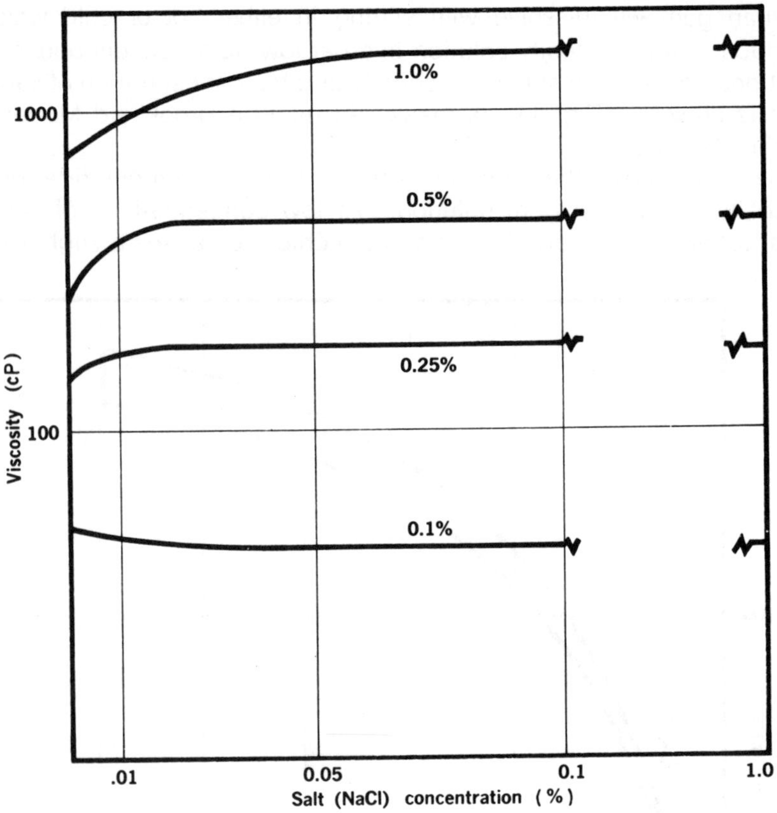

FIGURE 12. Effect of salt concentration on xanthan gum solution viscosity.

tion in viscosity. At concentrations of xanthan gum of 0.25% or greater, addition of sodium chloride causes an increase in viscosity. The peak viscosity is reached at a sodium chloride concentration of 0.02–0.07%, depending upon the xanthan gum concentration; the higher the concentration of xanthan gum, the higher the concentration of salt necessary to achieve maximum viscosity.

Aqueous solutions of xanthan gum are highly pseudoplastic. The relationship between viscosity and shear rate for xanthan gum solutions ranging in concentration from 0.5 to 2.5% is shown in Fig. 13. The viscosity decreases rapidly as the rate of shear is increased. This relationship is instantaneous and reversible.

Aqueous solutions of xanthan gum with a concentration of 0.75% or greater have a rheological yield point. This is demonstrated in Fig. 14. The working yield value is defined as the shear stress required to

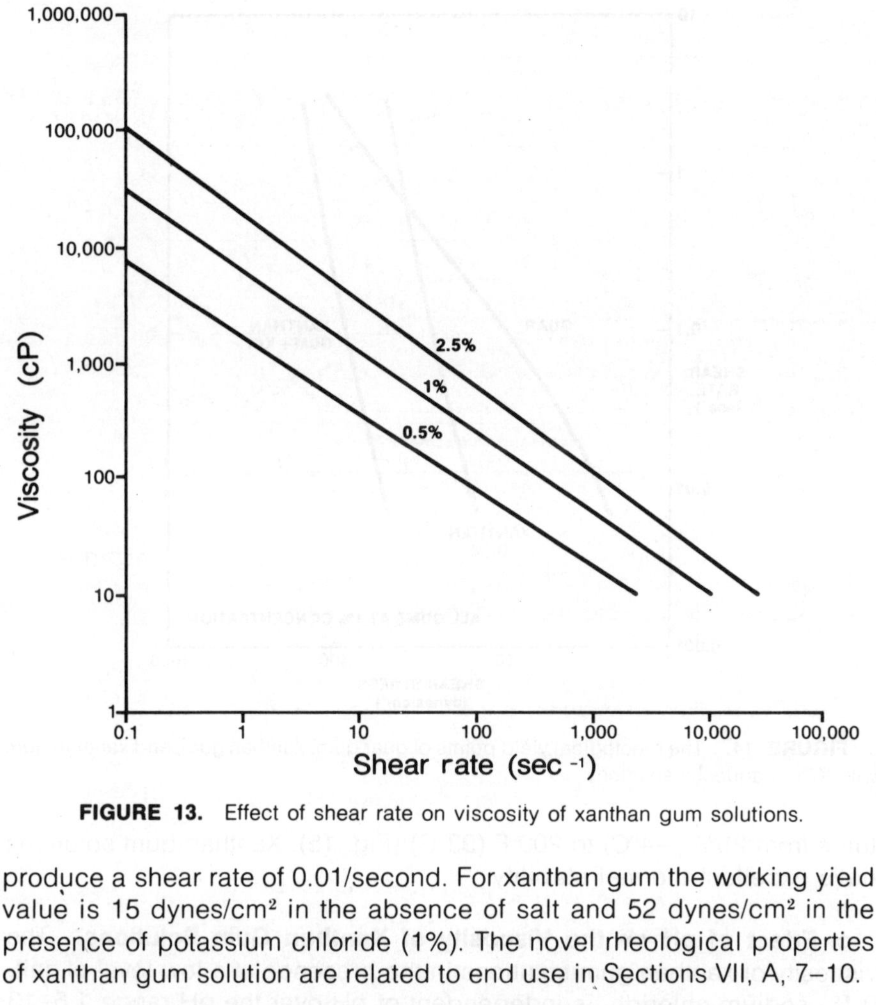

FIGURE 13. Effect of shear rate on viscosity of xanthan gum solutions.

produce a shear rate of 0.01/second. For xanthan gum the working yield value is 15 dynes/cm² in the absence of salt and 52 dynes/cm² in the presence of potassium chloride (1%). The novel rheological properties of xanthan gum solution are related to end use in Sections VII, A, 7–10.

b. Effect of Heat on the Properties of Xanthan Gum Solutions. The viscosity of solutions of most polysaccharides generally decreases when they are heated. Solutions of xanthan gum in a salt-free system, however, increase in viscosity after the initial decrease (Jeanes *et al.*, 1961b). This behavior is consistent with the unwinding of an ordered conformation, such as a double helix, changing into a random coil conformation with a resulting increase in effective hydrodynamic volume of the molecule and, therefore, a resulting increase in viscosity.

In the presence of a small amount of salt, 0.1% sodium chloride, the viscosity of a xanthan gum solution is virtually unaffected by tempera-

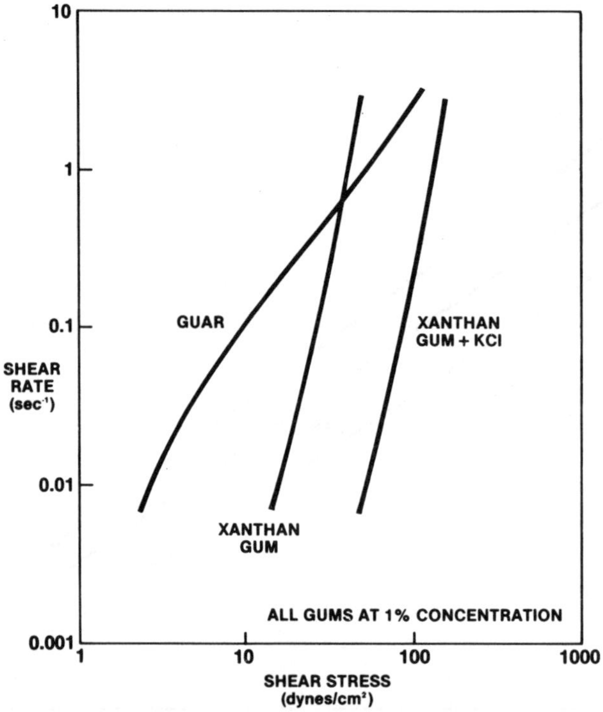

FIGURE 14. The rheological yield points of guar gum, xanthan gum, and xanthan gum with KCl in aqueous solutions.

tures from 25°F (−4°C) to 200°F (93°C) (Fig. 15). Xanthan gum solutions have excellent thermal stability.

c. Effect of pH on the Viscosity of Xanthan Gum Solutions. The viscosity of xanthan gum solutions in the presence of a low level of salt, 0.1% sodium chloride, is independent of pH over the pH range 1.5–13 (Fig. 16).

d. Compatibility of Xanthan Gum Solutions. Xanthan gum will dissolve directly in many acid solutions, such as 5% sulfuric acid, 5% nitric acid, 5% acetic acid, 10% hydrochloric acid, and 25% phosphoric acid. These solutions will be reasonably stable at ambient temperature (~25°C) for several months.

Xanthan gum will also dissolve directly in 5% sodium hydroxide solutions to produce thickened solutions. However, thickened 10–15% sodium hydroxide solutions must be prepared by adding 50% sodium

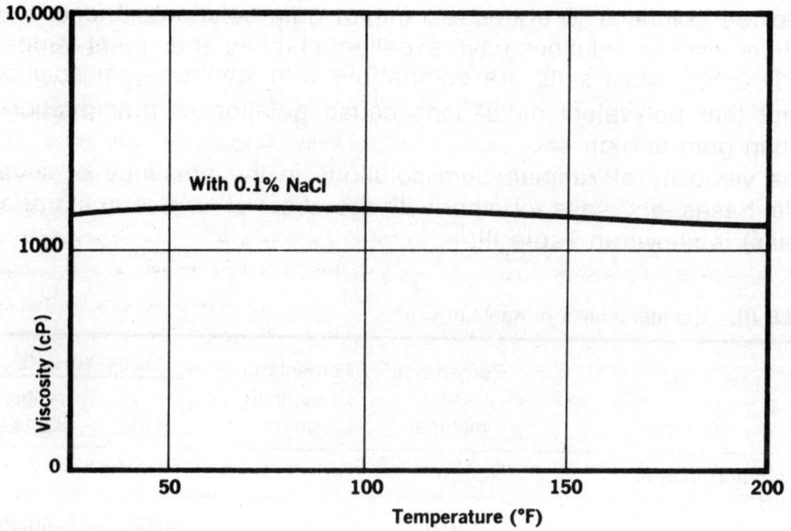

FIGURE 15. Effect of temperature on 1% xanthan gum solution viscosity.

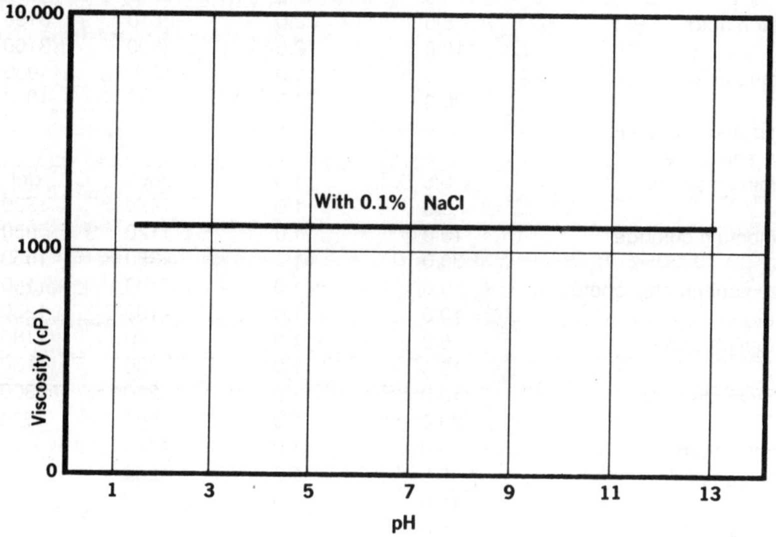

FIGURE 16. Effect of pH on viscosity of 1% xanthan gum solution.

hydroxide solutions to aqueous xanthan gum solutions. These highly alkaline, viscous solutions have excellent stability at ambient temperature (~25°C). Most salts are compatible with xanthan gum solutions, except that polyvalent metal ions cause gelation or precipitation of xanthan gum at high pH.

The viscosity of xanthan gum solutions in the presence of several acids, bases, and salts together with the effect of long-term storage (3 months) is shown in Table III.

TABLE III. Compatibilities of Xanthan Gum

Material	Percentage of material	Percentage of xanthan gum[a]	Viscosity (cP)[b]	
			Initial	After 90 days
Organic and inorganic acids				
Acetic acid	10.0	2.0	4200	4050
	20.0	2.0	4300	4400
Citric acid	10.0	1.0	1200	900
	20.0	1.0	1400	1030
Hydrochloric acid	5.0	2.0	3400	2750
	10.0	2.0	3500	1250
Phosphoric acid	30.0	2.0	4850	5250
	40.0	2.0	5600	7000
Sulfuric acid	5.0	2.0	3610	3150
	10.0	2.0	3800	3100
Tartaric acid	10.0	1.0	1210	900
	20.0	1.0	1430	1010
Organic and inorganic salts and bases				
Aluminum sulfate	0.5	1.0	2860	gel
	5.0	1.0	1460	1330
Ammonium chloride	10.0	1.0	1120	980
	30.0	1.0	1380	1050
Diammonium phosphate	5.0	1.0	1100	1100
	10.0	1.0	1190	1230
Barium chloride	5.0	1.0	1240	1180
	15.0	1.0	1330	1130
Calcium chloride	10.0	1.0	1260	1300
	20.0	1.0	1580	1610
Chrome alum	5.0	1.0	1430	gel
Cobalt chloride	5.0	1.0	1280	1110
	15.0	1.0	1600	1160
Cobalt sulfate	5.0	1.0	1300	1120
	15.0	1.0	1170	1170

(continued)

TABLE III (*Continued*)

Material	Percentage of material	Percentage of xanthan gum[a]	Viscosity (cP)[b] Initial	Viscosity (cP)[b] After 90 days
Cupric chloride	5.0	1.0	1180	970
	15.0	1.0	1500	2000
Ferrous sulfate	5.0	1.0	1410	300
	15.0	1.0	1600	gel
Magnesium chloride	5.0	1.0	1130	1070
	15.0	1.0	1210	1180
Potassium chloride	5.0	1.0	1240	1200
	15.0	1.0	1200	1115
Sodium bisulfite	5.0	1.0	1210	1160
	15.0	1.0	1540	1485
Sodium tetraborate	5.0	1.0	1140	1140
	15.0	1.0	1630	1520
Sodium carbonate	5.0	1.0	1130	1050
Sodium chloride	5.0	1.0	1110	1090
	15.0	1.0	1330	1200
Sodium chromate	5.0	1.0	1620	gel
Sodium citrate	5.0	1.0	1240	1170
	15.0	1.0	1390	1240
Sodium hydroxide	5.0	1.0	1360	810
	10.0	1.0	1390	115
Disodium hydrogen phosphate	5.0	1.0	1310	1100
	10.0	1.0	1190	1230
Sodium sulfate	5.0	1.0	1070	1100
	15.0	1.0	1460	1550
Zinc chloride	5.0	1.0	1200	1090
	15.0	1.0	1460	1320

[a] 0.5% xanthan gum (preserved with 0.1% formaldehyde): initial viscosity, 290 cP; after 90 days, 285 cP; 1.0% xanthan gum (preserved with 0.1% formaldehyde): initial viscosity, 860 cP; after 90 days, 990 cP; 2.0% xanthan gum (preserved with 0.1% formaldehyde): initial viscosity, 2350 cP; after 90 days, 2500 cP.

[b] Brookfield Model LVF Viscometer at 60 rpm, appropriate spindle.

Solutions of xanthan gum are compatible with methanol, ethanol, isopropanol, and acetone up to a final concentration of 50–60%. Concentrations greater than this range cause gelation or precipitation of the xanthan gum. Xanthan gum will not dissolve in most organic solvents, but it is soluble in formamide at 25°C and in glycerol and ethylene glycol at 65°C.

Enzymes such as protease, cellulase, hemicellulase, pectinase, and amylase will not degrade xanthan gum in solution. As with all polysaccharides or gums, xanthan gum is degraded by strong, oxidizing agents, such as peroxides, persulfates, and hypochlorites. This degradation is accelerated at elevated temperatures.

Most of the commercially available thickeners, such as sodium alginate and starch, have excellent compatibility with xanthan gum. Galactomannans, such as locust bean gum and guar, exhibit synergistic viscosity increases with xanthan gum. This "useful incompatibility" (Kovacs, 1973a) is described in the following section.

e. Interaction with Galactomannans. One of the more novel properties of xanthan gum is its ability to react with galactomannans, such as guar (Rocks, 1971) and locust bean gum (Schuppner, 1971). In the case of guar the interaction is made manifest by a synergistic increase in viscosity; i.e., the viscosity of the combination is greater than would be expected from the viscosities of solutions of xanthan gum and guar gum alone. The viscosity increase resulting from the synergistic interaction is shown in Fig. 17. In the case of locust bean gum the interaction is made manifest by a large synergistic viscosity increase at low gum concentrations (Fig. 18), but as the concentration of the gum is increased, a heat-reversible gel is formed (Fig. 19). The utility of the xanthan gum–galactomannan combinations is discussed in Sections VII, A, 7 and 8.

Considerable research has been carried out in recent years to determine the specific interactions between polysaccharides and galactomannans that result in synergistic viscosity increases or gelation (Dea *et al.*, 1972, 1977). The studies have suggested that the interaction of agarose and carrageenan with galactomannans results from the association between the carrageenan and agarose double helices, as well as sequences of unsubstituted mannose residues in the galactomannan. A schematic representation of the galactomannan conformation is shown in Fig. 20, and the double-helical conformation of polysaccharides was discussed in section VII, A, 4. A possible model for the interaction between xanthan gum and locust bean gum is shown in Fig. 21. This model is similar to that suggested by Morris *et al.*. (1977) to explain the interaction between xanthan gum and galactomannans which results in a three-dimensional network of polysaccharide molecules that contain water molecules in the "open" spaces within the network. Hence, the state of the water in the system is modified with either a resultant increase in viscosity or gelation (Rees, 1969).

7. Industrial Applications

Because of its unique properties xanthan gum has utility in a wide range of industrial applications.

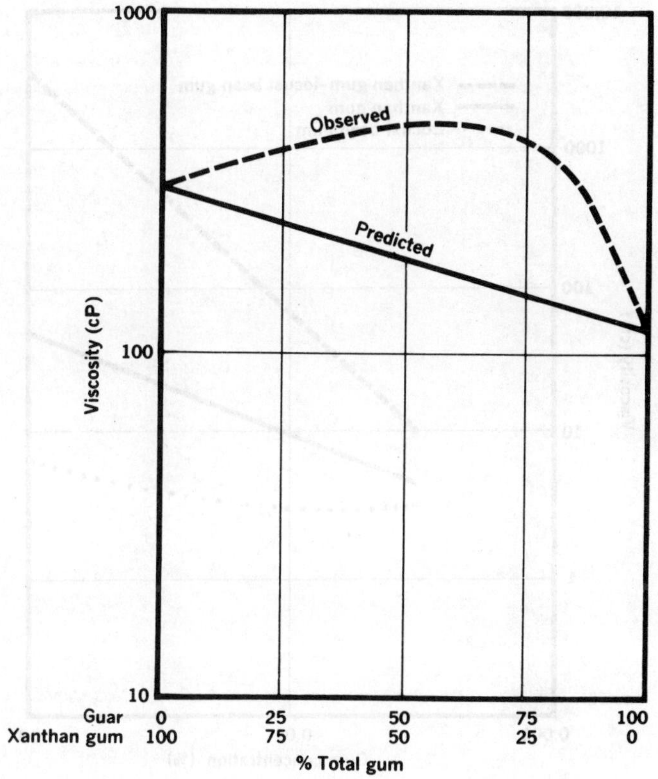

FIGURE 17. Effect of colloid ratio on xanthan gum/guar gum solutions (0.5% total gum).

a. Petroleum. *(1) Oil Well Drilling Fluids.* Xanthan gum is used at low levels, 0.5–1.5 lb/bbl, to viscosify and control the rheological properties of water-based drilling fluids. Specifically, the pseudoplasticity or shear-thinning property provides low viscosity at the drill bit, where the amount of shear is very high, and high viscosity in the annulus, where the amount of shear is low. Therefore, xanthan gum serves a double purpose in allowing faster penetration at the bit because of low viscosity; at the same time its high viscosity suspends the cuttings in the annulus (Carico, 1976b). The excellent compatibility of xanthan gum with salts, such as sodium chloride and calcium chloride, and the resistance to thermal degradation are additional properties that make xanthan gum the ideal drilling fluid additive.

(2) Completion and Workovers. The shear thinning and suspending properties of xanthan gum are utilized to prepare completion and work-

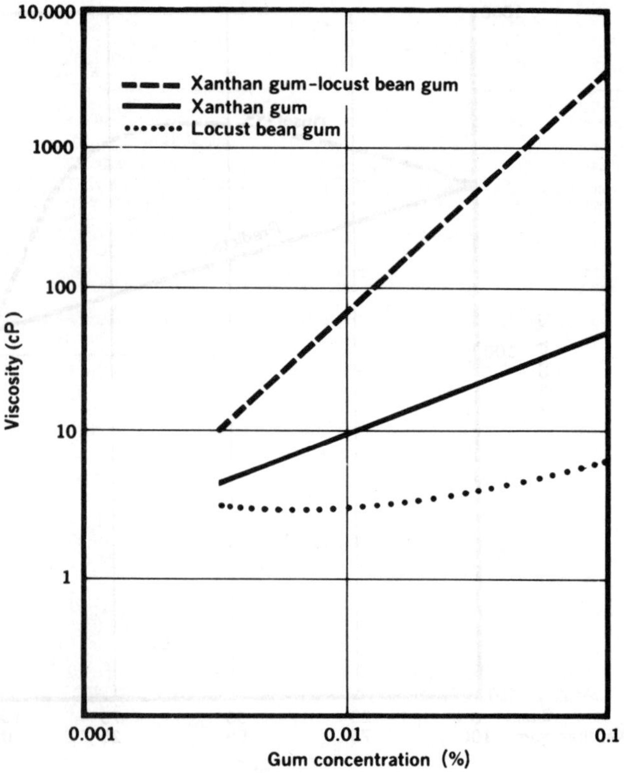

FIGURE 18. Viscosity as a function of colloid concentration; xanthan gum–locust bean gum combinations exhibit a synergistic increase in viscosity. Solutions were prepared by diluting at 140°F from 1% concentration.

over fluids with the desired rheological properties (Lipton and Burnett, 1976). After completion xanthan gum can be completely degraded with "breakers" such as hypochlorite or hypochloric acid. Because of the chemical nature of xanthan gum these fluids cause minimum formation damage. Weighted completion fluids containing high levels of calcium chloride or calcium bromide can also be formulated with xanthan gum because of its excellent salt compatibility.

(3) Fracturing Fluids. The combination of unique properties allows xanthan gum to be used to formulate water-based hydraulic fracturing fluids. Xanthan gum imparts high viscosity at low concentrations, and the shear-thinning property provides ease of pumping under the high-shear conditions, yet suspension of proppants (such as sand) under low-shear

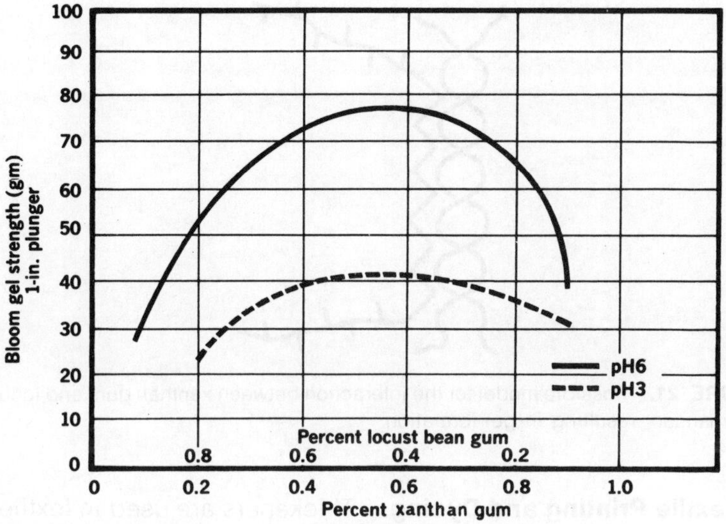

FIGURE 19. Gel strength of 1% xanthan gum–locust bean gum gel as a function of colloid ratio.

conditions. Since xanthan gum contains a very small percentage of insoluble material, formation damage (blockage of the rock material by insoluble material) is kept to a minimum. Xanthan gum can be removed from the formation by adding a controlled amount of hypochlorite or hydrochloric acid to the fluid before injection into the well. The excellent thermal stability, shear stability, and salt compatibility are additional properties that enable xanthan gum to find utility in fracturing fluids.

(4) Enhanced Oil Recovery. Xanthan gum produces viscous pseudoplastic aqueous solutions that provide excellent mobility control for increased oil displacement efficiency in secondary or tertiary flooding processes. As with the other petroleum applications described the thermal stability, shear stability, and salt compatibility are important properties of xanthan gum, because they provide additional functionality to viscous water-flooding fluids. The development of xanthan gum-based water-flooding fluids, which can be injected into formations having low permeability, has been the subject of much research (Lipton, 1974; Burnett, 1975; Colegrove, 1976).

FIGURE 20. Schematic representation of galactomannan conformation.

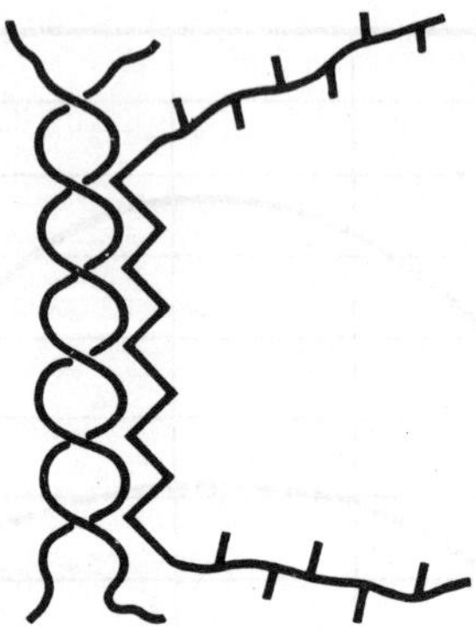

FIGURE 21. Possible model for the interaction between xanthan gum and locust bean galactomannan, resulting in gel formation.

b. Textile Printing and Dyeing. Thickeners are used in textile printing pastes to control the rheological properties of the paste during application and, also, to prevent migration of the dye and thereby produce sharp, clean patterns. Xanthan gum is an excellent thickener for this application because of its high viscosity at low concentrations and its unique rheological properties.

Xanthan gum in combination with guar gum imparts excellent thermal stability and desirable rheological properties to carpet printing pastes. Xanthan gum is compatible with most components of the printing pastes, and the wash-out properties of xanthan gum are excellent (Racciato, 1976).

Xanthan gum is used to thicken dye solutions used in space dyeing. This technique is used to print yarns that are primarily used in multicolor carpet manufacture. The xanthan gum prevents flushing and bleeding of the dyes on the yarn prior to drying.

c. Ceramic Glazes. The unique rheological properties of xanthan gum at low concentration enable the insoluble components of the glaze to be suspended for extended periods of time. Xanthan gum is compatible with the components of the glaze and it also prevents agglomeration of the glaze components during grinding, thereby allowing a reduction in grinding time. Also, the drying time is controlled and the level of imperfections is reduced (McNeely *et al.*, 1966). Ceramic processes based on extrusion are improved by the addition of xanthan gum since its pseudoplastic nature imparts plasticity to the mix.

d. Cleaners. Acid and alkaline cleaners with excellent functionality can be formulated with xanthan gum because it has good viscous stability in strong acids and alkalies. The pseudoplasticity of xanthan gum produces cleaners that have very high viscosity at a very low shear rate; this allows the cleaner to cling to inclined surfaces for extended periods of time and maximizes contact time. Abrasive cleaners can be formulated with xanthan gum so that the abrasive component is uniformly suspended and the flow properties of the cleaner are not impaired.

e. Slurry Explosives. Xanthan gum is used to thicken slurry explosive formulations containing 60–80% sodium, ammonium, and calcium nitrate because of the excellent compatibility of xanthan gum with high levels of salts. The thickened system is reacted with a cross-linking agent to form a gel that is either pumped into fissures in the rock for immediate use or is packed into small-diameter casings. The cross-linked xanthan gum prevents the highly soluble nitrate salts from being leached out of the gel when it is immersed in water (Colegrove, 1967). Also, xanthan gum is used in combination with guar to provide better heat stability.

f. Other Applications. Xanthan gum has also found utility in ink, paint, paper, and wallpaper adhesives and suspension polymerization (Hartnek *et al.*, 1974). A process to prepare a concentrated pumpable suspension of xanthan gum has been developed for use in industrial applications that require metered or automated addition of xanthan gum (Colegrove, 1975).

8. Food Applications

Xanthan gum has found widespread utility in the food industry as a result of its unique properties. These unique properties are summarized as follows: (a) solubility in hot or cold water; (b) high viscosity at low concentrations; (c) little variation in viscosity with changing temperatures; (d) excellent solubility and stability in acid systems; (e) unique rheological properties that provide high viscosity under low shear and low viscosity under high shear; (f) excellent suspending properties due to high yield value; (g) excellent compatibility with a wide range of salts; (h) excellent thermal stability; (i) good emulsifying properties; and (j) freeze-thaw stability.

Some of the application areas where xanthan gum has utility have been reviewed (Rocks, 1971; McNeely and Kang, 1973; Kovacs, 1973b; Kovacs and Kang, 1977· Andrew, 1977; Cottrell and Kang, 1978). Several application areas that illustrate the utility of xanthan gum in foods follow.

a. Pastry Fillings. A typical bakery jelly used to fill pastries or donuts contains a high concentration of sugar, a low level of starch, and small amounts of colors and flavors. Addition of 0.2% xanthan gum to the formulation provides a filling that has a low level of syneresis and, as a result, the filling is not absorbed by the pastry. The pseudoplasticity of xanthan gum, i.e., very high viscosity at very low shear, is responsible for this performance. The pseudoplasticity imparted by the xanthan gum also allows the fillings to be easily pumped during the filling operation because the viscosity is low under high shear.

b. Sauces and Gravies. Spaghetti sauce contains high levels of tomato solids and cheese, together with small amounts of spices and flavors. Addition of xanthan gum to the formulation provides a sauce that has excellent heat stability for extended periods of time and that maintains constant viscosity over wide ranges of temperature. The high viscosity at very low shear rate allows the sauce to cling to the spaghetti during consumption. The sauce also has excellent flavor release and mouthfeel because of the low viscosity under the high-shear conditions in the mouth. These important properties of the sauce result from the pseudoplasticity imparted by the xanthan gum. Cheese sauces and barbecue sauces can also be formulated using xanthan gum. The same properties that are important for sauces are also important for gravies. Addition of a small amount of xanthan gum to the gravy mix provides these properties.

c. Pourable Salad Dressings. Salad dressings are vegetable oil–water emulsions containing vinegar, sugar, egg yolk, and flavors. Addition of 0.25% xanthan gum provides excellent stability to the emulsion for periods up to 1 year. The dressings can be easily pumped during the filling operation and they also flow easily from the bottle during use (O'Connell, 1962). These unique properties are a result of the rheological properties of xanthan gum, specifically the high working yield value and the high degree of pseudoplasticity.

d. Dairy Products. Xanthan gum in combination with locust bean gum and guar gum stabilizes cottage cheese creaming emulsions. In addition to providing stability the xanthan gum blend also emulsifies the fat, prevents whey-off, and improves the cling of the dressing to the curd (Kovacs and Titlow, 1976). A blend of xanthan gum, locust bean gum, and guar gum improves the physical and organoleptic properties of pasteurized processed cheese spread (Kovacs and Igoe, 1976). Blends of xanthan gum, locust bean gum, and guar gum are effective stabilizers in ice cream, ice milk, and milk shakes. At relatively low concentrations in ice cream and ice milk, the water-absorbing characteristics provide excellent mix viscosity control. The frozen products exhibit smooth texture, creamy body, and exceptional tolerance to heat shock. A low-pH dessert gel can be prepared by combining a highly acid fruit juice with milk using xanthan gum in combination with locust bean gum and carboxy methyl cellulose (CMC). The resulting dessert gel has a smooth, creamy consistency and relatively long refrigerated shelf-life. Certain milk-based beverages can be stabilized with xanthan gum and guar gum (Igoe, 1977).

e. Other Applications. Xanthan gum has found utility in canned foods, dry mixes, frozen foods, juice drinks, relishes, spoonable dressings, and syrups. These are only a few examples of the foods that benefit from xanthan gum, and it is to be expected that more applications that use the unique properties of xanthan gum will emerge in the future.

9. Agricultural Applications

a. Animal Feed. Liquid feed supplements often contain ingredients which tend to stratify, settle out, or alter their physical stability with time, such as ground limestone, magnesium oxide, trace minerals, salts, or fat-soluble vitamins. Addition of 1–2 lb xanthan gum per ton of supplement minimizes any stratification or separation of these ingredients during shipping or extended storage (Andrew, 1977; Petrowski, 1978).

Calf milk replacers have traditionally used skimmed milk powder as the primary protein source. It is advantageous, at times, to use vegetable or fermentation proteins, but these proteins are partially or totally insoluble in water and settle out of the milk replacer in a very short time. Addition of 0.03–0.04% xanthan gum suspends the insoluble proteins to give a product having uniform quality (Andrew, 1977).

b. Agricultural Chemicals. Xanthan gum alone or in combination with locust bean gum improves the efficiency of flowable fungicides, herbicides, and insecticides by uniformly suspending the solid components of the formulation in an aqueous system or by stabilizing emulsions and multiphase liquid systems. The unique rheological properties of xanthan gum also improve sprayability, reduce drift, and increase pesticide cling and permanence (Schuppner, 1972; Gibsen and Saddington, 1973).

10. Pharmaceutical and Cosmetic Applications

a. Pharmaceutical Applications. Xanthan gum can be used to stabilize emulsion cream formulations containing pharmaceuticals. The rheological properties of xanthan gum account for the stabilization and ease of application (Felty, 1975). The aluminum salt of xanthan gum may be used to suspend barium sulfate in a radiographic opacifier. Xanthan gum will also suspend pharmaceuticals, such as antibiotics, to provide formulations of uniform dosage.

b. Cosmetic Applications. Denture cleaning compositions containing dilute solutions of phosphoric acid can be formulated with xanthan gum. The gum suspends the diatomaceous earth in the formulation in the presence of the acid (Regan and Regan, 1975). Toothpaste can be formulated with xanthan gum. The unique rheological properties of xanthan gum enable the paste to cling to the brush yet spread over the teeth during brushing. Xanthan gum provides body to creams and lotions but allows ease of application.

B. Dextrans

1. History

Structurally, dextrans are defined today as a large class of α-D-glucans in which α-1,6 linkages are predominant (Jeanes, 1977). Dextrans are the oldest microbial water-soluble gums known to man. In 1822 a report was made that cane sugar juice, after storage, had changed to a thick mucilage and was unfit for use (Brown, 1906). Pasteur, in 1861, noted that the viscous fermentations of sucrose were caused by a microbial action. Scheibler (1869) named the gum "dextran" because of its similarity to dextrin. He found that it yielded only D-glucose on acid hydrolysis. In 1930 Hucker and Pederson classified three distinct species of *Leuconostoc—L. mesenteroïdes, L. dextranicum,* and *L. citrovorum*—as dextran producers. Tarr and Hibbert (1913b) reported optimal conditions for dextran production from sucrose.

Jeanes (1952) of the Northern Utilization Research and Development Division of the USDA published a bibliography of 410 references related to the clinical use of dextran. The same author updated the bibliography, which includes dextran information from 1861 through mid-1976 (Jeanes, 1978). Comprehensive reviews were made by Murphy and Whistler (1973) and by Jeanes (1977).

2. Production

a. Whole-Culture Fermentation. A large number of bacterial strains is capable of producing dextran (Jeanes, 1952; Jeanes *et al.*, 1954). Commercial production of dextran has been carried out with only two of these strains, *L. mesenteroides* and *L. dextranicum.* Jeanes (1965) described methods for the preparation of three types of dextrans from strains NRRL B-512F, B-1146, and B-523. For B-512F dextran, Jeanes recommended a medium for both inoculum buildup and production that contained 10% sucrose, 0.5% dipotassium acid phosphate, 0.25% Difco yeast extract, and 0.02% magnesium sulfate heptahydrate. Dextran production by *Streptococcus bovis* (Bailey and Oxford, 1958) and *Streptococcus* NRRL B-1351 (Rogovin *et al.*, 1960) were also reported.

Many publications and patents disclose media composition and the use of adjuncts for the whole-culture fermentation of dextran and the process used for the commercial production of dextran (Murphy and Whistler, 1973; Jeanes, 1968). In the commercial whole-culture process the inoculum is increased in several stages in a medium similar to that of the final fermentation at 25°C in an agitated fermentor. After the fermentation, dextran is recovered by precipitation with alcohol and purified to make it suitable for clinical use. The use of molasses as an inexpensive source of substrate to produce technical grade dextrans was also reported (Behrens and Wuensche, 1969).

b. Enzymatic Synthesis. This method offers tremendous advantages over the whole-culture method for simpler product purification and greater product uniformity. Hehre (1941) accomplished the enzymatic synthesis of dextran from sucrose in cell-free extracts of *L. mesenteroides.* Hehre (1941) and later Tsuchiya *et al.* (1955) investigated the properties of the transglycosidase enzyme, dextransucrase. Optimal pH values for enzyme production and for dextran synthesis were 6.7 and 5.1, respectively.

Jeanes (1965) described a method of enzymatic dextran synthesis with *L. mesenteroides* NRRL B-512F. The medium for the initial stage of inoculum buildup contained 2% sucrose, 0.5% dipotassium acid phosphate, 0.5% Difco yeast extract, and 0.25% Difco Tryptone. The media for successive stages include small amounts of metal salts and exclude Tryptone. Bacterial cells are removed by centrifugation. Dextran is produced enzymatically by adding sucrose to the cell-free culture liquid at pH 5.0–5.2 and holding the temperature at 25°–30°C. Since dextransucrase incorporates only the glucose portion of sucrose into dextran, approximately 50% of the original weight of sucrose remains in solution as D-fructose after removal of the dextran. Methods for the recovery of this by-product (Koepsel *et al.*, 1956) have been used commercially to produce fructose with dextran as a by-product. A semicontinuous process (Behrens *et al.*, 1965) and fully continuous fermentation were described by Ogino (1973).

3. Derivatives

Like cellulose and starch, dextran contains no charged groups and is readily substituted. A large number of chemically modified dextrans have been reported, including acetates, nitrates, sulfates, phosphates, and various ethers (Murphy and Whistler, 1973). The cross-linked insoluble dextrans and the iron–dextran complex are of major commercial interest.

Products for laboratory use in gel filtrations and as molecular sieves are produced commercially from dextran cross-linked and insolubilized with epichlorohydrin (Anonymous, 1960a,b, 1964; Smiley, 1966). These materials have proved useful for rapid molecular weight determinations on small quantities of polymers. Epichlorohydrin-insolubilized dextran also serves as a base for ion-exchange resins. The diethylaminoethyl, carboxymethyl, and sulfoethyl derivatives of cross-linked dextran are available (Anonymous, 1965). They are effectively used in the purification, separation, and isolation of enzymes and hormones (Kagadel and Akerstroem, 1971).

4. Properties

The molecular weight of dextran ranges from approximately 1.5×10^4 to 2×10^7 or higher, depending on the method of preparation. Solutions of dextran in water are quite low in viscosity for such a high molecular weight polymer. A 2% solution of native dextran has a viscosity of approximately 150 cP. Dextran has relatively good compatibilities with salts, acids, and bases. It is also resistant to degradation at high temperatures.

5. Applications

a. Blood Plasma Extender. A considerable research effort was expended in the period from 1948 to 1955 to develop dextran as a blood plasma extender (Grönwall, 1957). Sterile, pyrogen-free dextran solutions (~6%) having a molecular weight between 50,000 and 100,000 were stockpiled in pint bottles. In emergency situations these solutions could be used to restore blood volume in patients suffering from shock due to blood loss. The United States' military stockpiling program for dextran as a blood plasma extender was discontinued in 1955, and the producing plants were shut down. Since 1955 production of dextran for use as a blood plasma extender has been limited.

b. Oil Well Drilling. Dextran was used as an oil well drilling fluid additive in the 1950s (Owen, 1950). Several patents (Owen, 1959; Sparks, 1962; Cypert and Patton, 1963) have been issued. Because of the low viscosity of dextran in aqueous solutions it was used in relatively high concentrations in drilling fluids to control fluid loss. As a result, dextran was uneconomical, and its use in drilling fluids was discontinued. Recently, a limited number of further studies are being performed (Jeanes, 1977). However, the authors are not aware of a major drilling operation where dextrans or modified dextrans are being used.

c. Iron–Dextran Complex. The complex of iron with dextran has attained limited use as a source of nutritional iron. The iron in this complex was reported to be completely available to form the hemoglobin complex. This source of iron was used in combating anemia (Martin *et al.*, 1955; Cox *et al.*, 1965). The solution contains 5% iron and 20% dextran of molecular weight of 5000.

d. Miscellaneous Uses. Dextran was tested extensively in thickened water flooding for the secondary recovery of oil. It proved to be uneconomical because of its low viscosity in solution. Various applications in foods, metallurgy, photographic emulsions, drug encapsulations, and other applications have been reviewed (Murphy and Whistler, 1973; Jeanes, 1977). However, most of these applications have not reached the practical stage primarily because other, better performing, competitive products are available.

C. *Sclerotium* Polysaccharides

Many species of *Sclerotium* produce glucans (Johnson *et al.*, 1963; Halleck, 1967) in a medium containing suitable carbon sources, such as sucrose and glucose, in the presence of autolyzed yeast and certain salts. Similar polysaccharides are produced by other fungi, such as *Claviceps purpurea* (Perlin and Taber, 1963; Buck *et al.*, 1968), *Stereum sanguinolentum* (Axelsson *et al.*, 1968), *Plectania occidentalis* NRRL 3137 (Wallen *et al.*, 1965; Davis *et al.*, 1966), and *Helotium* sp. NRRL 3129 (Davis *et al.*, 1966). The polysaccharide scleroglucan has been developed by the Pillsbury Company and commercialized under the trade name of Polytran. This polysaccharide was described by Rodgers (1973).

1. Structure

The typical scleroglucan produced by *Sclerotium glucanicum* has a structure consisting of a main chain of β-D-(1→3)-glucopyranose units to which glucopyranose units are joined by β-D-(1→6) linkages to every third unit of the main chain (Johnson *et al.*, 1963). The number of single-unit side chains varies, depending upon the organisms producing glucans: 25% in a gum produced by *C. purpurea* (Perlin and Taber, 1963) and 50% in a glucan elaborated by *P. occidentalis* NRRL 3137 (Axelsson *et al.*, 1968). The degree of polymerization (DP) of the scleroglucan from *S. glucanicum* varies widely between ~110 and 1600. However, the commercial scleroglucan has an average DP of ~800 (Rodgers, 1973).

2. Production

Scleroglucan is produced commercially by aerobic submerged fermentation in a medium containing D-glucose, corn steep liquor, nitrate as the major source of nitrogen, and mineral salts (Rodgers, 1973). Synthesis of scleroglucan proceeds simultaneously with mycelial growth and, as a result of the polysaccharide production, the medium develops a gellike consistency. The pH of the fermentation, initially 4.5, drops to about 2 as a result of the accumulation of oxalic acid.

After fermentation the fermentation liquor is heated to inactivate glucanase enzymes. This is followed by homogenization to detach the glucan from the mycelial tissue. The homogenized liquor is diluted to decrease viscosity and filtered to eliminate mycelia and other particulate matter. The gum in the clear filtrate is alcohol precipitated, dried, and milled. For some industrial applications the homogenized fermentation liquor is spray-dried without purification. Scleroglucan content in this grade is about 30%.

3. Properties

Scleroglucan disperses readily in water upon mechanical stirring to give high-viscosity solutions (Fig. 22). This polysaccharide produces a self-supporting sliceable gel at 25°C when dispersed in water at a concentration of 1.2–1.5%. However, at temperatures below 10°C even very dilute solutions form diffusely structured gels. Such gels tend to shrink and undergo syneresis when left undisturbed for long periods of time. However, these gels disperse quickly when a mild agitation is applied. It forms a stable pH-irreversible gel by slowly complexing with borate in alkaline solutions.

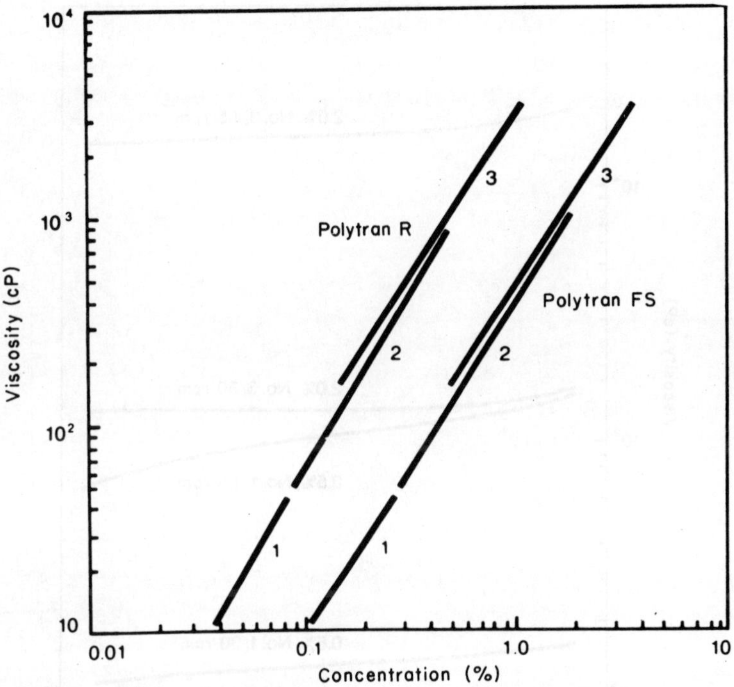

FIGURE 22. Relation of viscosity of solutions of refined (Polytran R) and commercial (Polytran FS) scleroglucans to concentration. Brookfield LVT viscometer; spindles 1, 2, and 3; 30 rpm.

The viscosity of scleroglucan solutions is relatively stable at temperatures between 10° and 90°C, as shown in Fig. 23. The effect of pH on stability of 0.5% solutions of commercial scleroglucan heated at 121°C for 0.25 and 20 hours is given in Fig. 24. The viscosity is stable at the autoclaving temperature (121°C) from initial pH 3 to 11 for 20 hours. The zero-hour curve indicates the viscosity of a solution aged 24 hours, heated to 90°C, cooled, and adjusted to varied pH values. This solution exhibits good stability between pH 1 and 10.

Scleroglucan has good compatibility with various salts. However, aluminum sulfate and magnesium sulfate precipitate the gum at 10% salt concentration. Probably the most outstanding property of scleroglucan is its synergistic viscosity increase when combined with bentonite. The apparent viscosities of 0.15% purified scleroglucan and 5.0% bentonite are near 200 and 300 cP, respectively. A combined viscosity is ~4000 cP as measured by the Brookfield viscometer LVT at 30 rpm with No. 3 spindle. However, the stability of this synergistic viscosity in the application situation, such as in drilling muds, is questionable.

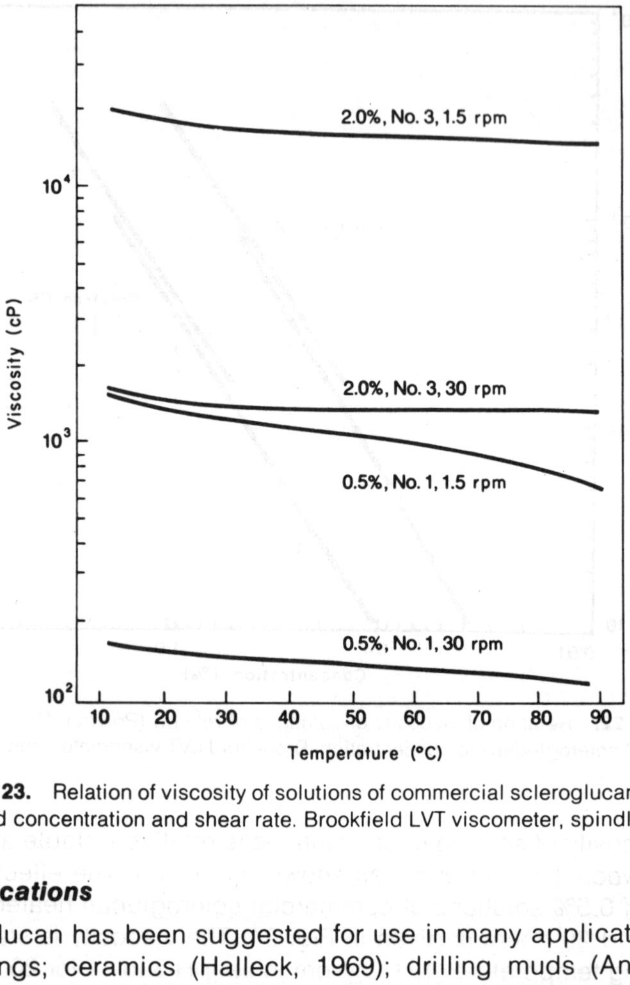

FIGURE 23. Relation of viscosity of solutions of commercial scleroglucan to temperature at varied concentration and shear rate. Brookfield LVT viscometer, spindles 1 and 3.

4. Applications

Scleroglucan has been suggested for use in many applications: inks and coatings; ceramics (Halleck, 1969); drilling muds (Anonymous, 1969a); pharmaceuticals (Sheth and Lachman, 1969); and cosmetics (Rodgers, 1973). Despite these suggested applications scleroglucan has not captured any significant portion of industrial applications with the possible exception of ceramics.

D. Zanflo Polysaccharides

Zanflo is the trademark of a heteropolysaccharide produced by an extensively mutated strain of a bacterium that was isolated from soil. This mutant grows at 30°C only in the presence of added iron. The organism will not grow at 37°C whether or not iron is present in the medium.

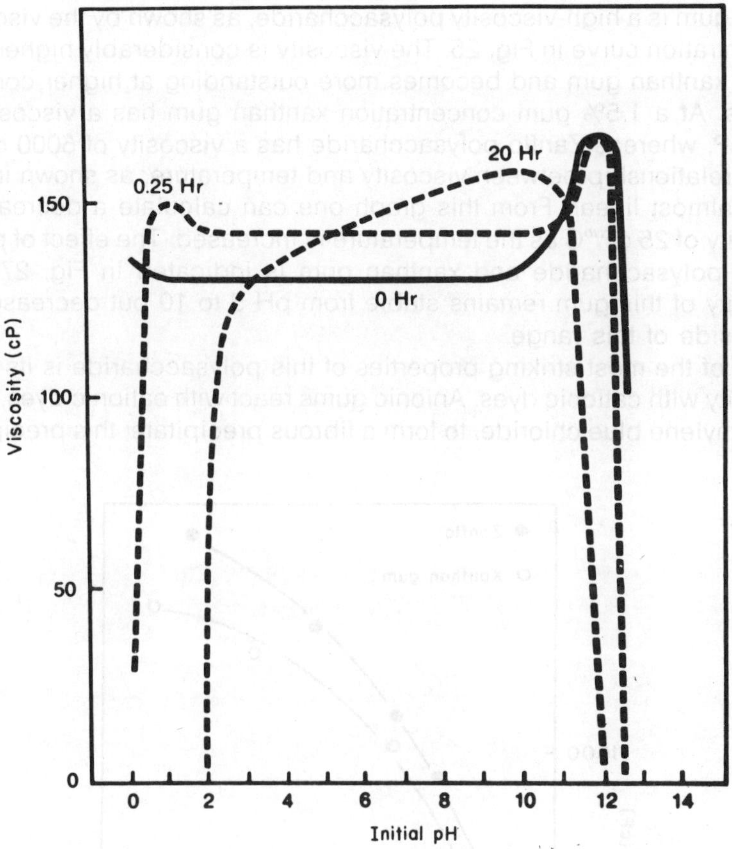

FIGURE 24. Effect of pH on stability of 0.5% solutions of commercial scleroglucan heated at 121°C for 0.25 and 20 hours. Brookfield LVT viscometer, spindle 1, 30 rpm.

Zanflo gum is produced by aerobic submerged fermentation in a medium containing phosphate as a buffering agent, ammonium nitrate and a soy protein product as nitrogen sources, magnesium sulfate, trace minerals, and a carbon source. This organism requires specific carbon sources for optimal polysaccharide production. Lactose and hydrolyzed starch are preferred over glucose, sucrose, or maltose. This polysaccharide contains 97% carbohydrate and 3% protein. The carbohydrate portion was found to contain glucose, galactose, glucuronic acid, and fucose in the molar ratio of 3 : 2 : (1.5) : 1. Uronic acid accounts tor approximately 20% of the polysaccharide on a weight basis. It is noteworthy that fucose is not commonly found as a structural constituent of exocellular bacterial heteropolysaccharides.

This gum is a high-viscosity polysaccharide, as shown by the viscosity concentration curve in Fig. 25. The viscosity is considerably higher than that of xanthan gum and becomes more outstanding at higher concentrations. At a 1.5% gum concentration xanthan gum has a viscosity of 2500 cP, whereas Zanflo polysaccharide has a viscosity of 5000 cP.

The relationship between viscosity and temperature, as shown in Fig. 26, is almost linear. From this graph one can calculate a decrease in viscosity of 25 cP/°C as the temperature is increased. The effect of pH on Zanflo polysaccharide and xanthan gum is indicated in Fig. 27. The viscosity of this gum remains stable from pH 5 to 10 but decreases on either side of this range.

One of the most striking properties of this polysaccharide is its compatibility with cationic dyes. Anionic gums react with cationic dyes, such as methylene blue chloride, to form a fibrous precipitate; this precipitate

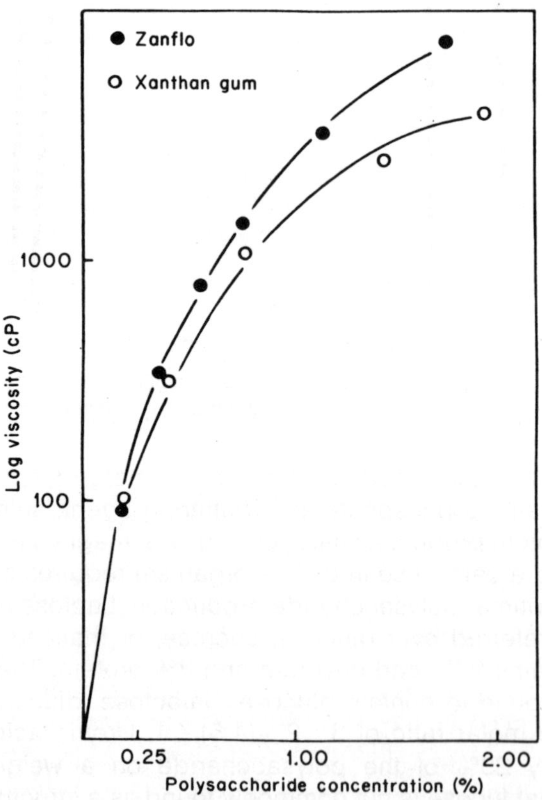

FIGURE 25. Viscosity versus concentration for Zanflo.

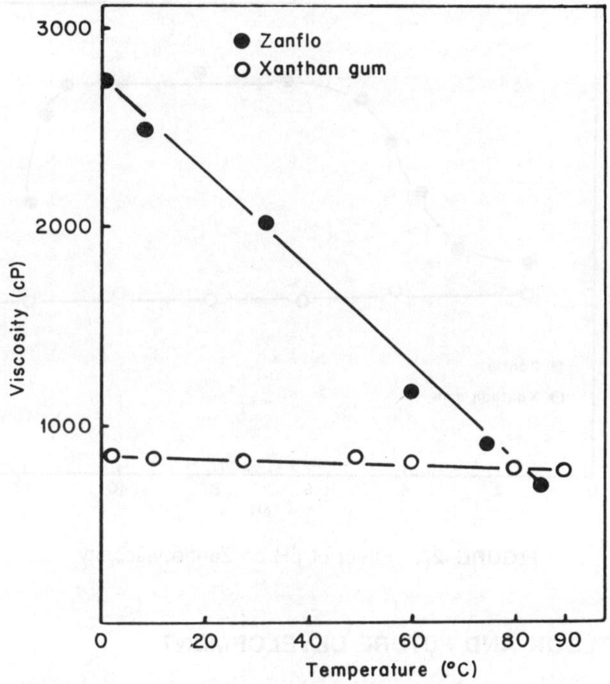

FIGURE 26. Effect of temperature on Zanflo viscosity (1%).

limits their industrial applications. However, Zanflo polysaccharide, even though it contains a substantial amount of uronic acid, does not precipitate with methylene blue chloride at any pH. Zanflo gum has a high resistance to enzymatic attack and has excellent flow and leveling properties. Because of these properties, as well as those described earlier, Zanflo polysaccharide has already found applications in the paint industry.

In our laboratories we developed another extensively mutated strain (tTR-45) of a bacterium that was isolated from a rhizosphere soil sample. This mutant requires thymine or thymidine for growth at 30°C but will not grow at 37°C. This strain produces a heteropolysaccharide consisting of 33% mannose, 29% glucose, 21% galactose, and 17% glucuronic acid. The polysaccharide also contain ~5.7% acetyl and ~4.9% pyruvate. The viscosity of this polymer is even higher than that of Zanflo gum. For example, a 1% solution of this polymer has a viscosity of ~3000 cP as compared to ~2000 for Zanflo gum. The flow and leveling properties of this polymer are as good as those of Zanflo gum.

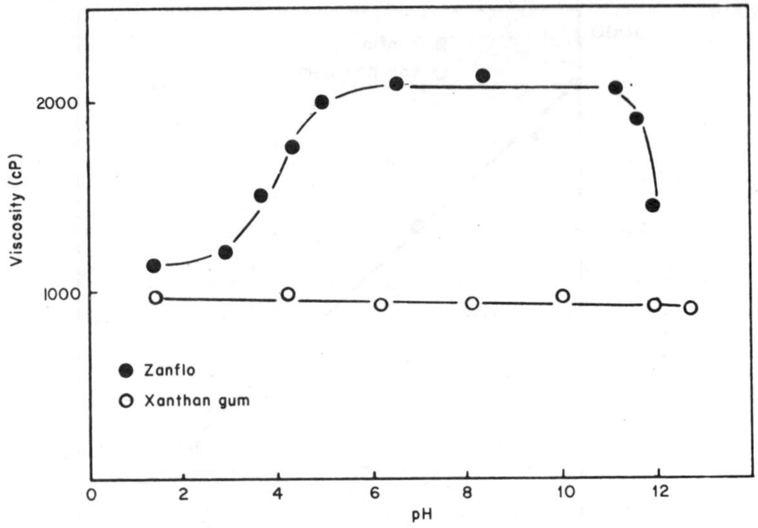

FIGURE 27. Effect of pH on Zanflo viscosity.

VIII. OUTLOOK AND FUTURE DEVELOPMENT

Because world labor rates will continue to climb, it can be expected that the price of the traditional plant gums, such as gum arabic, will increase. Also, because of the lack of uniformity and problems with purity it can be expected that these gums will be replaced in industrial usage.

The number of microbial species that could be polysaccharide producers is large. In view of the encouraging results obtained during the last two decades, it is to be expected that several of the microbial polysaccharides described in this chapter will achieve commercial production on a substantial scale in the future. It is also expected that additional new microbial polysaccharides having novel properties will be discovered. However, the new microbial polysaccharides must have a superior cost/performance compared with existing gums, and they must also comply with all safety and regulatory requirements. Therefore, it must be emphasized that commercialization of new microbial polysaccharides will be extremely arduous and costly.

Further elucidation of metabolic pathways leading to the polysaccharide synthesis should be made to establish control mechanisms of microbial polysaccharides. This sort of information is still lacking, even though it is fundamental for a thorough understanding of polysaccharide production and quality differentiation of microbial polysaccharides through metabolic manipulation and mutation.

REFERENCES

Amemura, A., and Harada, T. (1971). *Hakko Kogaku Zasshi* **49,** pp. 559–564.
Anderson, R. F. (1963). *In* "Biochemistry of Industrial Micro-organisms" (C. Rainbow and A. H. Rose, eds.), pp. 300–316. Academic Press, New York.
Anderson, R. F., Benedict, R. U., Cadmus M. C., and Slodki, M. E. (1960). *Arch. Biochem. Biophys.* **89,** 289.
Andrew, T. R. (1977). *In* "Extracellular Microbial Polysaccharides" (P. A. Sandford and A. Laskin, eds.), ACS Symp. Ser. No. 45, pp. 231–241. Am. Chem. Soc., Washington, D.C.
Anonymous (1958). "Information on Phosphomannan Y-2448," CA-N-7. USDA North. Util. Res. Dev. Div., Peoria, Illinois.
Anonymous (1960a). "Sephadex in Gel Filtration." Pharmacia, Uppsala.
Anonymous (1960b). British Patent 854,715 (Aktiebolaget Pharmacia). (Nov. 23) ibid., 974,054 (Nov. 4, 1964).
Anonymous (1962). "Microbial Polysaccharide B-1973," CA-N-21. USDA Agric. Res. Serv., Peoria, Illinois.
Anonymous (1964). British Patent 974,054 (Aktiebolaget Pharmacia).
Anonymous (1965). "Sephadex Ion Exchangers." Pharmacia, Uppsala.
Anonymous (1969a). 21 CFR 172.695.
Anonymous (1969b). Federal Food and Drug Administration (FDA).
Anonymous (1971). Canadian Governor-in-Council. *Can. Gazette Part II* **105.**
Anonymous (1974). European Economic Community Emulsifier/Stabilizer List, Annex II.
Anonymous (1975). Food and Agriculture Organization/World Health Organization (FAO/WHO) Joint Expert Committee.
Anonymous (1976). U.S. Environmental Protection Agency.
Anonymous (1977a). 21 CFR, Parts 131, 133, 155 and 168.180.
Anonymous (1977b). 9 CFR, 318.7 (c) (4) and 381.147 (f) (3).
Archer, S. A., Clamp, J. R., and Migliore, D. (1977). *J. Gen. Microbiol.* **102,** 157–167.
Axelsson, K., Bjorndal, H., and Ericksson, K. E. (1968). *Acta Chem. Scand.* **22,** 1363–1364.
Bailey, R. W., and Oxford, A. E. (1958). *Nature (London)* **182,** 185.
Bebbington, A., Bourne, E. J., and Wilkinson, J. A. (1952). *J. Chem. Soc.* pp. 246–253.
Behrens, U., and Wuensche, L. (1969). *Cana Azucar* **3,** 39; *Chem. Abstr.* **73,** 5743Y (1970).
Behrens, U., Ringpfeil, M., Bagert, A., and Krueger, K. (1965). East German Patent 35,367; *Chem. Abstr.* **63,** 7624 (1965).
Bender, H., Lehmann, J., and Wallenfels, K. (1959). *Biochim. Biophys. Acta* **36,** 309–316.
Bernier, B. (1958). *Can. J. Microbiol.* **4,** 195–204.
Bjorndal, H., Hellerquist, C. G., Lindberg, B., and Svensson, S. (1970). *Angew. Chem. Int. Ed. Engl.* **9,** 610–619.
Booth, A. N., Deeds, F., and Hendrickson, A. P. (1963). *Toxicol. Appl. Pharmacol.* **5,** 478–484.
Bourne, E. J., Stacey, M., and Wilkinson, I. A., (1950). *J. Chem. Soc.* 2694–2698.
Bouveng, H. O., Fraser, R. L., and Lindberg, B. (1967). *Carbohydr. Res.* **4,** 20–31.
Bovey, F. A. (1959). *J. Polym. Sci.* **35,** 183–190.
Breed, R. S., Murray, E. G. D., and Smith, N. R., Eds. (1957). "Bergey's Manual of Determinative Bacteriology," 7th ed. Williams & Wilkins, Baltimore, Maryland.

Bretthauer, R. K., Kaczorowski, G. J., and Weise, M. J. (1973). *Biochemistry* **12,** 1251–1256.
Brown, A. J. (1886). *J. Chem. Soc.* **49,** 172 and 432.
Brown, C. A., Jr. (1906). *J. Am. Chem. Soc.* **28,** 453.
Buchanan, R. E., and Gibbons, N. E., eds. (1974). "Bergey's Manual of Determinative Bacteriology," 8th Ed. Williams & Wilkins, Baltimore, Maryland.
Buck, K. W., Chain, E. B., Chen, A. W., and Dickerson, A. G. (1968). *J. Gen. Microbiol.* **51,** 377–385.
Burnett, D. B. (1975). *Soc. Pet. Eng. AIME* **SPE 5372.**
Burton, K. A., Cadmus, M. C., Lagoda, A. A., Sandford, P. A., and Watson, P. R. (1976). *Biotechnol. Bioeng.* **18,** 1669–1677.
Cadmus, M. C., Anderson, R. F., and Lagoda, A. A. (1962). *Appl. Microbiol.* **10,** 153.
Cadmus, M. C., Gasdorf, H., Lagoda, A. A., Anderson, R. F., and Jackson, R. W. (1963). *Appl. Microbiol.* **11,** 488–492.
Cadmus, M. C., Burton, K. A., Lagoda, A. A., and Smiley, K. L. (1967). *Bacteriol. Proc.* A102.
Carico, R. D. (1976a). *Oil and Gas J.* **74**(27), 81–84.
Carico, R. D. (1976b). *Soc. Pet. Eng. AIME* **SPE 5870.**
Carlson, A. S., and Hehre, E. J. (1949). *J. Biol. Chem.* **177,** 281.
Carlson, W. W., Rosano, C. L., and Whiteside-Carlson, V. (1953). *J. Bacteriol.* **54,** 136.
Catley, B. J., and Whelan, W. J. (1971). *Arch. Biochem. Biophys.* **143,** 138–142.
Charles, M., and Radjai, M. K. (1977). *In* "Extracellular Microbial Polysaccharides" (P. A. Sandford and A. Laskin, eds.), ACS Symp. Ser. No. 45, pp. 27–39. Am. Chem. Soc., Washington, D.C.
Chen, W. P., and Tsuo, C. H. (1976). *Taiwan Sugar* **23,** 14–16.
Chihara, G. *et al.* (1976). *Nature (London)* **222,** 687–688.
Chihara, G. *et al.* (1970). *Nature (London)* **225,** 943–944.
Cohen, G. H., and Johnstone, D. B. (1963). *Nature (London)* **198,** 211.
Cohen, G. H., and Johnstone, D. B. (1964). *J. Bacteriol.* **88,** 329–338.
Colegrove, G. T. (1967). U.S. Patent 3,326,733.
Colegrove, G. T. (1970). U.S. Patent 3,516,983.
Colegrove, G. T. (1975). U.S. Patent 3,894,879.
Colegrove, G. T. (1976). U.S. Patent 3,966,618.
Colvin, J. R., Chéné, L., Sowden, L. C., and Takai, M. (1977). *Can. J. Biochem.* **55,** 1057–1063.
Corpe, W. A. (1960). *Can. J. Microbiol.* **6,** 153–163.
Corpe, W. A. (1964). *J. Bacteriol.* **88,** 1433–1441.
Corpe, W. A. (1970). *In* "Adhesion in Biological System" (R. S. Manly, ed.), pp. 73–87. Academic Press, New York.
Cottrell, I. W. (1977). German Offen. 2,643,140.
Cottrell, I. W., and Kang, K. S. (1978). *Dev. Ind. Microbiol.* **19,** pp. 117–131.
Cottrell, I. W., Shim, J. L., Best, G. H., and Empey, R. H. (1978). *In* "Carbohydrate Sulfates" (R. G. Schweiger, J. L. Shin, and G. H. Best, eds.), ACS Symp. Ser. No. 77, pp. 193–202. Am. Chem. Soc., Washington, D.C.
Couperwhite, I., and McCallum, M. F. (1974). *Arch. Microbiol.* **97,** 73–80.
Cox, J. S., King, R. E., and Reynolds, G. F. (1965). *Nature (London)* **207,** 1202–1203.
Cypert, J. D., and Patton, J. T. (1963). U.S. Patent 3,084,122; *Chem. Abstr.* **59,** 1422 (1963).
Davidson, I. W., Sutherland, I. W., and Lawson, C. J. (1977). *J. Gen. Microbiol.* **98,** 603–606.
Davis, E. N., Rhodes, R. A., and Schulk, H. R. (1966). *Appl. Microbiol.* **13,** 267–271.
Dea, I. C. M., McKinnon, A. A., and Rees, D. A. (1972). *J. Mol. Biol.* **68,** 153–172.
Dea, I. C. M., Morris, E. R., Rees, D. A., Welsh, E. J., Barnes, H. A., and Price, J. (1977). *Carbohydr. Res.* **57,** 249–272.

Deavin, L., Jarman, T. R., Lawson, C. J., Righaleto, R. C., and Slocombe, S. (1977). *In* "Extracellular Microbial Polysaccharides" (P. A. Sandford and A. Laskin, eds.), ACS Symp. Ser. No. 45, pp. 14–26. Am. Chem. Soc., Washington, D.C.
Dedoner, R. A., and Hassid, W. Z. (1964). *Biochim. Biophys. Acta* **90,** 239.
Dias, F., and Bhat, J. (1964). *Appl. Microbiol.* **12,** 412–417.
Dintzis, F. R., Babcock, G. E., and Tobin, R. (1970). *Carbohydr. Res.* **13,** 257–267.
Doudoroff, M., Hassid, W. A., Putman, E. W., Potter, A. L., and Lederberg, J. (1949). *J. Biol. Chem.* **179,** 921.
Drummond, G. S., Smith, E. E., Whelan, W. G., and Jai, H. (1969). *FEBS Lett.* **5,** 85.
Ebert, K. H., and Brosche, M. (1967). *Biopolymers* **5,** 423–430.
Ebert, K. H., and Patat, F. (1962). *Z. Naturforsch. Teil B* **17,** 738.
Ellwood, D. C., and Evans, G. T. C. (1977). Netherlands Patent 7612-448.
Felty, L. G. (1975). U.S. Patent 3,906,108.
Forsyth, W. G. C., and Webley, D. M. (1949). *Biochemistry* **44,** 455.
Friedman, B., Dugan, P., Pfister, R., and Ramsen, C. (1968). *J. Bacteriol.* **96,** 2144–2153.
Gadsdorf, H. J., Benedict, R. G., Cadmus, M. C., Anderson, R. F., and Jackson, R. W. (1965). *J. Bacteriol.* **90,** 147–150.
Gibsen, K. F., and Saddington, A. W. (1973). U.S. Patent 3,717,452.
Gill, J. W. (1972). U.S. Patent 3,632,570.
Glaser, L. (1958). *J. Biol. Chem.* **232,** 627.
Gorin, P. A. G., and Spencer, J. F. T. (1968). *Adv. Carbohydr. Chem.* **23,** 367.
Gorin, P. A. T., and Spencer, J. F. T. (1966). *Can. J. Chem.* **44,** 993–998.
Grönwall, A. (1957). *In* "Dextran and Its Use in Colloidal Infusion Solutions." Academic Press, New York.
Halleck, F. E. (1967). U.S. Patent 3,301,848.
Halleck, F. E. (1969). U.S. Patent 3,447,940.
Harada, T. (1977). *In* "Extracellular Microbial Polysaccharides" (P. A. Sandford and A. Laskin, eds.), ACS Symp. Ser. No. 45, pp. 265–283. Am. Chem. Soc., Washington, D.C.
Harada, T. (1965). *Arch. Biochem. Biophys.* **112,** 65–69.
Harada, T., and Yoshimura, T. (1965). *Agric. Biol. Chem.* **29,** 1027.
Harada, T., Maeda, I., Hiroshi, S., Masada, M., and Misaki, A. (1967). *Agric. Biol. Chem.* **31,** 1184–1188.
Harada, T., Misaki, A., and Saito, H. (1968). *Arch. Biochem. Biophys.* **124,** 292–298.
Harada, T., Misaki, A., Hiroshi, S., and Toshiko, I. (1969). *Biochemistry* **8,** 4645–4650.
Hartnek, H. G., Empey, R. A., Pettitt, D. J., and Van Winkle, T. L. (1974). U.S. Patent 3,852,257.
Hathaway, R. J. (1971). U.S. Patent 3,556,942.
Haug, A., and Larsen, B. (1971). *Carbohydr. Res.* **17,** 297–308.
Hehre, E. J. (1941). *Science* **93,** 237.
Hehre, E. J. (1943). *Proc. Soc. Exp. Biol. Med.* **54,** 240.
Hehre, E. J., and Hamilton, D. M. (1948). *J. Bacteriol.* **55,** 197.
Hehre, E. J., Carlson, A. E., and Neill, V. M. (1947). *Science* **106,** 523.
Hestrin, S. (1944). *Nature (London)* **154,** 581.
Hestrin, S. (1962). *In* "The Bacteria" (I. C. Gunsalus and R. Y. Stanier, eds.), Vol. 3, pp. 373–388. Academic Press, New York.
Hestrin, S., and Shapiro, S. (1944). *Biochem. J.* **38,** 2.
Hestrin, S., and Shramm, N. (1954). *Biochem. J.* **58,** 345.
Hestrin, S., Feingold, D. S., and Avigad, G. (1955). (*J. Am. Chem. Soc.* **77,** 6710.
Holzworth, G. (1976). *Biochemistry* **15,** 4333–4339.

Holzworth, G., and Prestridge, E. B. (1977). *Science* **197,** 757–759.
Hucker, G. J., and Pederson, C. S. (1930). *N.Y., Agric. Exp. Stn., Geneva, Tech. Bull.* **167,** 3.
Igoe, R. S. (1977). U.S. Patent 4,046,925.
Jansson, P. E., Kenne, L., and Lindberg, B. (1975). *Carbohydr. Res.* **45,** 275–282.
Jeanes, A. (1952). *In* "Dextran, A Selected Bibliography," Bur. Agric. Ind. Chem. AIC-288. USDA, Peoria, Illinois.
Jeanes, A. (1965). *Methods Carbohydr. Chem.* **5,** 118–127.
Jeanes, A. (1968). *Encyl. Poly. Sci. Techno.* (N. M. Bikales, ed.) **4,** 693–711.
Jeanes, A. (1974). *Food Technol. (Chicago)* **28**(5), 34–40.
Jeanes, A. (1978). Dextran Bibliography, *US, Dep. Agric., Agric, Res. Serv.,* Misc. Publ. #1355.
Jeanes, A. (1977). *In* "Extracellular Microbial Polysaccharides" (P. A. Sandford and A. Laskin, eds.), ACS Symp. Ser. No. 45, pp. 284–298. Am. Chem. Soc., Washington, D.C.
Jeanes, A., and Sloneker, J. H. (1961). U.S. Patent 3,000,790.
Jeanes A., and Watson, P. R. (1962). *Can. J. Chem.* **40,** 1318–1325.
Jeanes, A., Haynes, W. C., Wilham, C. A., Rankin, J. C., Melvin, E. H., Austin, M., Cluskey, J. E., Tsuchiya, H. M., and Rist, C. E. (1954). *J. Am. Chem. Soc.* **76,** 5041–5052.
Jeanes, A., Dimler, R. J., Pittsley, J. E., and Watson, P. R. (1961a). *Arch. Biochem. Biophys.* **92,** 343.
Jeanes, A., Pittsley, J. E., and Santi, F. R. (1961b). *J. Appl. Polym. Sci.* **5,** 519–526.
Jeanes, A., Pittsley, J. E., and Watson, P. R. (1964). *J. Appl. Polym. Sci.* **8,** 2775–2787.
Jeanes, A., Knutson, C. A., Pittsley, J. E., and Watson, P. R. (1965). *J. Appl. Polym. Sci.* **9,** 627–638.
Jeanes, A., Burton, K. A., Cadmus, M. C., Knutson, C. A., Rowin, G. L., and Sandford, P. A. (1971). *Nature (London), New Biol.* **233,** 259–260.
Jeanes, A., Cadmus, M. C., Knutson, C. A., Rogovin, P., and Silman, R. W. (1976). *U.S., Agric. Res. Serv., North Cent. Reg.* [*Rep.*] **ARS-NC-51.**
Johnson, J. J., Jr., Kirkwood, S., Misaki, A., Nelson, T. E., Scaletti, J. V., and Smith, F. (1963). *Chem. Ind. (London),* p. 820.
Kagadel, L., and Akerstroem, S. (1971). *Acta Chem. Scand.* **25,** 1855–1859.
Kang, K. S. (1972). Kelco Div. of Merck & Co., Inc., San Diego, California (unpublished data).
Kang, K. S., and Kovacs, P. (1975). *Int. Food Sci. Technol., Proc. Int. Congr., 4th, 1974,* pp. 33–34.
Kang, K. S., and McNeely, W. H. (1975). U.S. Patent 3,894,976.
Kang, K. S., and McNeely, W. H. (1976a). U.S. Patent 3,960,832.
Kang, K. S., and McNeely, W. H. (1976b). U.S. Patent 3,979,303.
Kang, K. S., and McNeely, W. H. (1977). *In* "Extracellular Microbial Polysaccharides" (P. A. Sandford and A. Laskin, eds.), ACS Symp. Ser. No. 45, pp. 220–230. Am. Chem. Soc., Washington, D.C.
Kimura, H., Moritaka, S., and Misaki, M. (1973). *J. Food Sci.* **38,** 668–670.
King, G. G. S., and Colvin, J. R. (1976). *Appl. Polym. Symp.* **28,** 623–636.
Kjosbakken, J., and Colvin, J. R. (1975). *Can. J. Microbiol.* **21,** 111-120.
Knutson, C. A., Pittsley, J. E., and Jeanes, A. (1971). ACS *Abstr. Pap., 161st Meet., Am. Chem. Soc.* CARB. 28.
Koepsell, H. J., Jackson, R. W., and Hoffman, C. A. (1956). U.S. Patent 2,729,587.
Komatsu, N., Okubo, S., Kikumoto, S., Kimura, K., Saito, G., and Sakai, S. (1969). *Gann* **60,** 137–144.
Kovacs, P. (1973a). *Food Technol.* **27,** 26–30.
Kovacs, P. (1973b). *Food Trade Rev.* **11,** 17–22.

Kovacs, P., and Igoe, R. S. (1976). *Food Prod. Dev.* **10,** 32, 34, 36, and 38.
Kovacs, P., and Kang, K. S. (1977). *In* "Food Colloids" (H. D. Graham, ed.), pp. 500–522. Avi Publ. Co., Westport, Connecticut.
Kovacs, P., and Titlow, B. D. (1976). *Am. Dairy Rev.* **4,** 34J–34N.
Larsen, B., and Haug, A. (1971). *Carbohydr. Res.* **17,** 287–296.
Lilly, V. G., Leach, J. G. and Wilson, H. A. (1958). *Appl. Microbiol.* **6,** 105–108.
Lindberg, B. Lörngren, J., and Thompson, J. L. (1973). *Carbohydr. Res.* **28,** 351–357.
Lipton, D. (1974). *Soc. Pet. Eng. AIME* **SPE 5099.**
Lipton, D., and Burnett, D. B. (1976). *Soc. Pet. Eng. AIME* **SPE 5872.**
Lopez, R., and Becking, J. H. (1968). *Microbiol. Expan.* **21,** 1–23.
Lwoff, A., Ionesco, H., and Gutman, A. (1950). *Biochim. Biophys. Acta* **4,** 270.
McNeely, W. H., and Kang, K. S. (1973). *In* "Industrial Gums" (R. L. Whistler and J. N. Be Miller, eds.), 2nd ed., pp. 473–498. Academic Press, New York.
McNeely, W. H., and Kovacs, P. (1975). *In* "Physiological Effects of Food Carbohydrates" (A. Jeanes and J. Hodge, eds.), ACS Symp. Ser. No. 15, pp. 269–281. Am. Chem. Soc., Washington, D.C.
NcNeely, W. H., and O'Connell, J. J. (1966). U.S. Patent 3,232,929.
McNeely, W. H., Fairchild, W. P., and Hunter, A. R. (1966). Canadian Patent 727,071.
Martin, L. E., Bates, C. M., Beresford, C. R., Donaldson, J. D., McDonald, F. F., Dunlop, D., Sheard, P., London, E., and Twigg, G. D. (1955). *Br. J. Pharmacol. Chemother.* **10,** 375.
Masson, C. R., Menzies, R. F., and Cruikshank, L. (1946). *Nature (London)* **157,** 74.
Maxted, W. R. (1952). *Nature (London)* **170,** 1020–1021.
Monod, J., and Torriani, A. M. (1950). *Ann. Inst. Pasteur, Paris* **78,** 65.
Moorhouse, R., Walkinshaw, M. D., and Arnott, S. (1977a). *In* "Extracellular Microbial Polysaccharides" (P. A. Sandford and A. Laskin, eds.), ACS Symp. Ser. No. 45, pp. 90–102. Am. Chem. Soc., Washington, D.C.
Moorhouse, R., Walkinshaw, M. D., Winter, W. T., and Arnott, S. (1977b). *In* "Cellulose Chemistry and Technology" (J. C. Arthur, ed.), ACS Symp. Ser. No. 48, pp. 133–152. Am. Chem. Soc., Washington, D.C.
Moraine, R. A., and Rogovin, S. P. (1966). *Biotechnol. Bioeng.* **8,** 511–524.
Morita, N., and Murao, S. (1974). *J. Ferment. Technol.* **52,** 438–444.
Morris, E. R., Rees, D. A., Young, G., Walkinshaw, M. D., and Darke, A. (1977). *J. Microbiol.* **110,** 1–16.
Muhlethaler, K. (1949). *Biochim. Biophys. Acta* **3,** 527.
Murphy, P. T., and Whistler, R. L. (1973). *In* "Industrial Gums" (R. L. Whistler and J. N. Be Miller, Eds.), 2nd ed., pp. 513–542. Academic Press, New York.
Nagahama, T., Fujimoto, S., and Kanie, M. (1977). *Agric. Biol. Chem.* **41,** 9–16.
Ninomiya, E. (1967). Japanese Patent 42-7600.
Ninomiya, E., and Kizaki, T. (1969). *Angew. Makromol. Chem.* **6,** 179.
O'Connell, J. J. (1961). U.S. Patent 3,067,038.
Ogino, S. (1973). Japanese Patent 48-2800.
Owen, W. E. (1950). *Sugar* **45,** 42.
Owen, W. I. (1959). U.S. Patent 2,868,725.
Parikh, V. M., and Jones, J. K. N. (1963). *Can. J. Chem.* **41,** 2826–2835.
Pasteur, L. (1861). *Bull. Soc. Chim. Fr.,* pp. 30–31.
Patton, J. T. (1962). U.S. Patent 3,020,207.
Patton, J. T. (1973). U.S. Patent 3,729,460.
Patton, J. T., and Lindblom, G. P. (1962). U.S. Patent 3,020,206.
Perlin, A. S., and Taber, W. A. (1963) *Can. J. Chem.* **41,** 2278–2282.
Petrowski, G. E. (1977). *Proc., AFMA Symp.* pp. 57–59.

Pettitt, D. J. (1973). U.S. Patent 3,708,446.
Pindar, D. F., and Bucke, C. (1975). *Biochem. J.* **152,** 617–622.
Quinnell, C. M., Knight, S. G., and Wilson, P. W. (1957). *Can. J. Microbiol.* **3,** 277–288.
Racciato, J. S. (1976). *Am. Dyest. Rep.* **65**(11), 51 and 69.
Rees, D. A. (1969). *Adv. Carbohydr. Chem. Biochem.* **24,** 209.
Rees, D. A. (1970). *Sci. J.* **6**(12), 47–51.
Regan, B. F., and Regan, G. B. (1975). U.S. Patent 3,899,437.
Robbins, D. J., Booth, A. N., and Moulton, J. E. (1964). *Food Cosmet. Toxicol.* **2,** 545–550.
Rocks, J. K. (1971). *Food Technol. (Chicago)* **25,** 22–31.
Rodgers, N. E. (1973). *In* "Industrial Gums" (R. L. Whistler and J. N. Be Miller, eds.), 2nd ed., pp. 499–511. Academic Press, New York.
Rogovin, S. P., and Albrecht, W. J. (1964). U.S. Patent 3,119,812.
Rogovin, S. P., Senti, F. R., Benedict, R. G., Tsuchiya, H. M., Watson, P. R., Tobin, R., Sohns, V. E., and Slodki, M. E. (1960). *J. Biochem. Microbiol. Technol. Eng.* **2,** 381.
Rogovin, S. P., Anderson, R. F., and Cadmus, M. C. (1961). *J. Biochem. Microbiol. Technol. Eng.* **3,** 51.
Rogovin, S. P., Albrecht, W. J., and Sohns, V. (1965). *Biotechnol. Bioeng.* **7,** 161–169.
Scheibler, Z. (1869). *Verh. Dtsch. Zucker-Ind.* **19,** 472.
Scher, M., Lennarz, W. J., and Sweeley, C. C. (1968). *Proc. Natl. Acad. Sci. U.S.A.* **59,** 1313–1320.
Schuppner, H. R. (1971). U.S. Patent 3,557,016.
Schuppner, H. R. (1972). U.S. Patent 3,659,026.
Schweiger, R. G. (1966a). U.S. Patent 3,236,831.
Schweiger, R. G. (1966b). U.S. Patent 3,244,695.
Schweiger, R. G. (1966c). U.S. Patent 3,256,271.
Schweiger, R. G. (1969). U.S. Patent 3,446,796.
Sheth, P., and Lachman, L. (1969). U.S. Patent 3,421,920.
Silman, R. W., and Rogovin, S. P. (1972). *Biotechnol. Bioeng.* **14,** 23–31.
Singh, P. P. (1974). *Carbohydr. Res.* **37,** 245–247.
Slodki, M. E. (1963). *Biochim. Biophys. Acta* **69,** 96.
Slodki, M. E. (1966). *Can. J. Microbiol.* **12,** 495–499.
Slodki, M. E., Cadmus, M. C., and Wickerham, L. J. (1961). *J. Bacteriol.* **82,** 269.
Slodki, M. E., Bandoni, R. J., and Wickerham, L. J. (1966). *Can. J. Microbiol.* **12,** 489–494.
Sloneker, J. H., and Jeanes, A. (1962). *Can. J. Chem.* **40,** 2066–2071.
Sloneker, J. H., Orentas, D. G., and Jeanes, A. (1964). *Can. J. Chem.* **42,** 1261–1269.
Sloneker, J. H., Orentas, D. G., Knutson, C. A., Watson, P. R., and Jeanes, A. (1968). *Can. J. Chem.* **46,** 3353–3361.
Smiley, K. L. (1966). *Food Technol.* **20,** 1206–1210.
Sowa, W., Blackwood, A. C., and Adams, G. A. (1963). *Can. J. Chem.* **41,** 2314–2319.
Sparks, W. J. (1962). U.S. Patent 3,053,765; *Chem. Abstr.* **58,** 8838 (1963).
Starr, M. P. (1946). *J. Bacteriol.* **51,** 131–143.
Stolp, H., and Starr, M. P. (1964). *Phytopathol. Z.* **51,** 442–478.
Sutherland, I. W. (1972). *Adv. Microb. Physiol.* **8,** 143–208.
Sutherland, I. W. (1977). *In* "Extracellular Microbial Polysaccharides" (P. A. Sandford and A. Laskin, eds.), ACS Symp. Ser. No. 45, pp. 40–57. Am. Chem. Soc., Washington, D.C.
Sutton, J. C., and Williams, P. H. (1970). *Can. J. Bot.* **48,** 645–651.
Szaniszlo, P. J., Wirsen, C., Jr., and Mitchell, R. (1968). *J. Bacteriol.* **96,** 1474–1483.
Taguchi, R., Kikuchi, Y., Sakuno, Y., and Kobashi, T. (1973). *Agric. Biol. Chem.* **37,** 1583–1588.

Tam, K. T., and Finn, R. K. (1977). *In* "Extracellular Microbial Polysaccharides" (P. A. Sandford and A. Laskin, eds.), ACS Symp. Ser. No. 45, pp. 58–80. Am. Chem. Soc., Washington, D.C.
Tannahill, A. T., and Finn, R. K. (1975). U.S. Patent 3,878,045.
Tarr, H. L. A., and Hibbert, H. (1931a). *Can J. Res.* **4,** 372.
Tarr, H. L. A., and Hibbert, H. (1931b). *Can. J. Res.* **5,** 414.
Tenney, M., and Stumm, W. (1965). *J. Water Pollut. Control Fed.* **37,** 1370–1388.
Troy, F. A., Frerman, F. E., and Heath, E. C. (1971). *J. Biol. Chem.* **246,** 118–133.
Tsuchiya, H. M., Hellman, N. N., Koepsell, H. J., Corman, J., Stringer, C. S., Rogovin, S. P., Bogard, M. O., Bryant, G., Feger, V. H., Hoffman, C. A., Senti, F. R., and Jackson, R. W. (1955). *J. Am. Chem. Soc.* **77,** 2412–2419.
Umbreit, J. N., Stone, K. J., and Strominger, J. L. (1972). *J. Bacteriol.* **112,** 1302–1305.
Unz, R. F., and Farrah, S. R. (1974). *Envion. Prot. Technol. Ser.* **EPA-670/2-74-018.**
Wallen, L. L., and Davis, E. N. (1972). *Environ. Sci. Technol.* **6,** 161–164.
Wallen, L. L., Rhodes, R. A., and Shulke, H. R. (1965). *Appl. Microbiol.* **13,** 272–278.
Watson, P. R., Sandford, P. A., Burton, K. A., Cadmus, M. C. and Jeanes, A. (1976). *Carbohydr. Res.* **46,** 259–265.
Whistler, R. L., Bushway, A. A., Nakamura, W., Singh, P. P., and Tokuzen, R. (1976). *Adv. Carbohydr. Chem. Biochem.* **32,** 235–275.
Wilkinson, J. F. (1958). *Bacteriol. Rev.* **22,** 46.
Woodard, G., Woodard, M. W., McNeeley, W. H., Kovacs, P., and Cronin, M. T. I. (1973). *Toxicol. Appl. Pharmacol.* **24,** 30–36.
Woodard Research Corp. (1973). Unpublished report. Herndon, VA.
Woodard Research Corp. (1971). Unpublished report. Herndon, VA.
Woodard Research Corp. (1968a). Unpublished report. Herndon, VA.
Woodard Research Corp. (1968b). Unpublished report. Herndon, VA.
Yamada, K., and Furukawa, T. (1973). Japanese Patent 48-33396.
Yamatodani, T., and Kanamaru, T. (1972). U.S. Patent 3,674,642.
Yuen, S. (1974). *Process Biochem.* **9,** 7–9.
Zajic, J. E., and LeDuy, A. (1973). *Appl. Microbiol.* **25,** 628–635.
Zevenhuizen, L. P. T. M. (1973). *Carbohydr. Res.* **26,** 409–419.

Chapter 14

Microbial Transformation of Steroids and Sterols

O. K. SEBEK
D. PERLMAN

I. INTRODUCTION

Naturally occurring steroids possess remarkable hormonal properties which are of critical importance to human well-being. They include hormones of the adrenal cortex (cortisone, cortisol, corticosterone), the progestational hormone (progesterone), the androgens or male hormones (testosterone, dihydrotestosterone), and the female sex hormones (estradiol, estrone) (see Scheme 1).

They are all derived from a tetracyclic hydrocarbon (perhydrocyclopentanophenanthrene) and differ primarily in the number, type, and location of the substituent functional groups and in the number and position of the double bonds (Scheme 2).

In the last 30 years, thousands of derivatives of these compounds have been synthesized and tested for their endocrine and other activities. Many of them were found to be active as antiinflammatory and progestational agents, sedatives, as well as anabolic and antitumor substances. Some are effective in allergic, dermatologic, and ocular diseases, in cardiovascular therapy, and in veterinary products, and some have been

MICROBIAL TECHNOLOGY, 2nd ed., VOL. I
 ISBN 0-12-551501-4

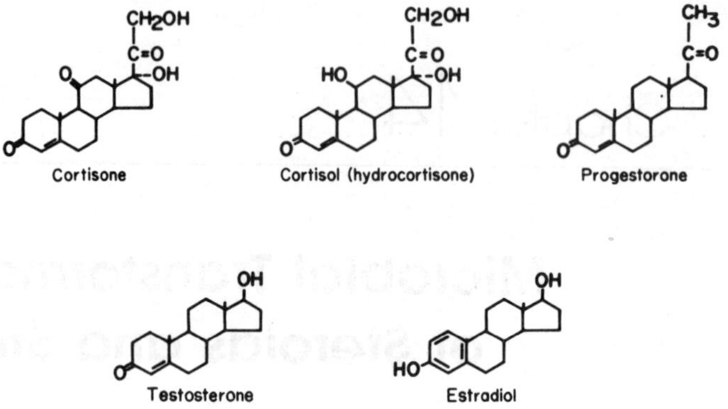

SCHEME 1

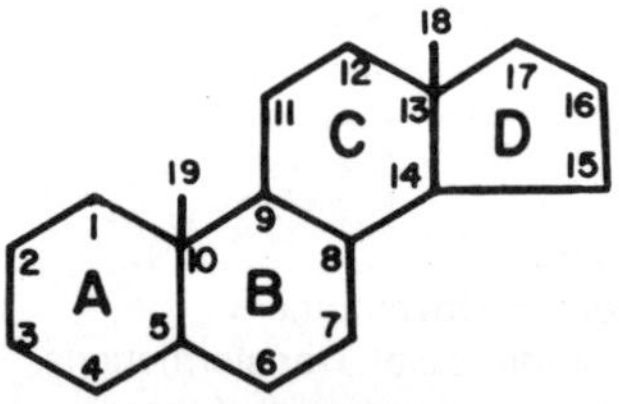

SCHEME 2

outstandingly successful as oral contraceptives (Applezweig, 1962, 1978).

The demonstration of such extraordinary activities had its origin in the discovery by Hench and his associates in 1949 of the remarkable antiinflammatory property of cortisone in rheumatoid arthritis (Hench *et al.*, 1949, 1950).

In the wake of this discovery, the dramatic surge in the demand for cortisone to treat millions of arthritic sufferers went unmet since no adequate method for its manufacture existed. The chemical synthesis from deoxycholic acid (Sarett, 1946) developed at Merck and Company, was workable but complicated and uneconomical: 615 kg of deoxycholic acid had to be processed through 31 chemical steps in order to obtain 1 kg of cortisone acetate. The oxygen shift from C-12 to C-11 alone, which is essential for the activity of cortisone (and of all adrenal corticosteroid hormones), required nine steps* and is shown in Scheme 3.

Economic synthesis of cortisone, cortisol, and other corticosteroids eventually developed from three lines of information: (1) diosgenin from

*The efficiency of this shift was subsequently improved and deoxycholic acid, obtained from beef bile, is at present a substrate in some manufacturing processes.

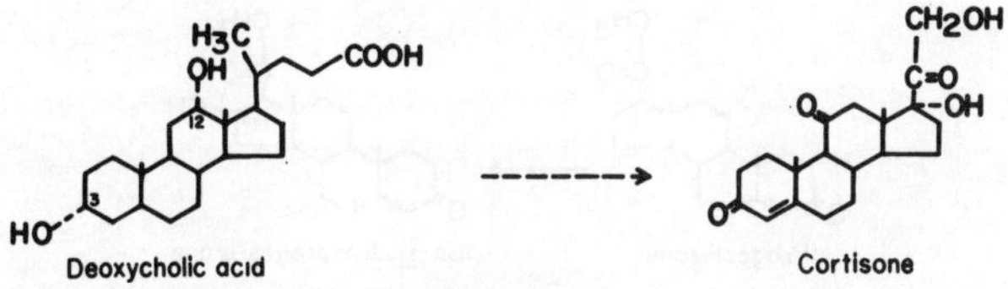

SCHEME 3

the Mexican barbasco plant (*Dioscorea composita*) and stigmasterol from soybeans (*Glycine max*) were found abundant and inexpensive raw materials for chemical conversion to 16-dehydropregnenolone and pregnenolone, respectively; (2) deoxycorticosterone was hydroxylated to corticosterone by perfusion with isolated beef adrenal glands* (Hechter *et al.*, 1949) (see Scheme 4); (3) yeasts and bacteria were reported to

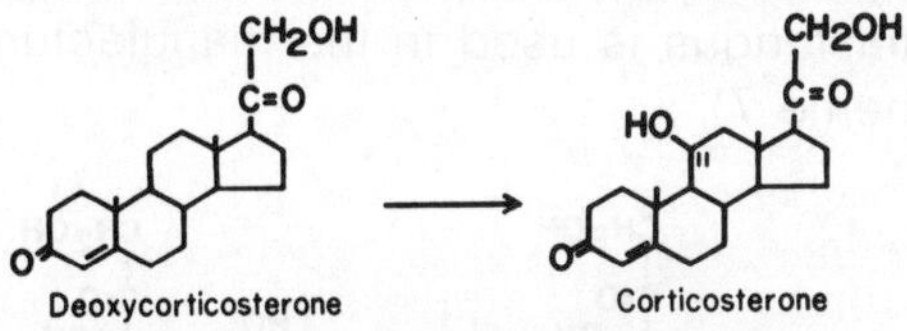

SCHEME 4

carry out nuclear reduction, oxidation, isomerization, and hydrolysis of steroids (Mamoli and Vercellone, 1937a,b).

By applying this information, Peterson and Murray succeeded in 1952 in introducing the essential oxygen at C-11 by oxidizing progesterone to 11α-hydroxyprogesterone by means of first *Rhizopus arrhizus* and then by *Rhizopus nigricans* in practically quantitative yields (>85%) (Peterson and Murray, 1952; see also Murray and Peterson, 1953) (Scheme 5).

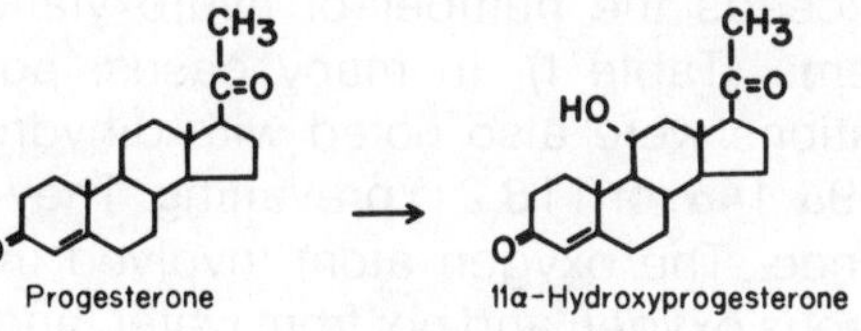

SCHEME 5

Almost simultaneously, *Streptomyces argenteolus* was reported to oxidize progesterone to 16α-hydroxyprogesterone (Perlman *et al.*, 1952) (Scheme 6).

*This process was employed at one time by G. D. Searle and Company; it produced the first reasonably large quantities of cortisol for clinical testing.

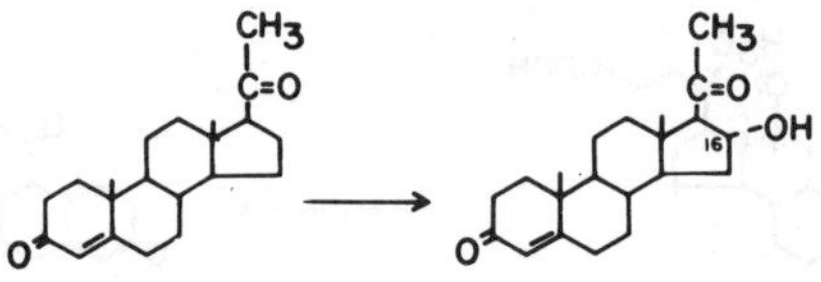

Progesterone 16α-Hydroxprogesterone

SCHEME 6

This reaction became of considerable importance 5 years later in the synthesis of 16α-hydroxy-9α-fluoroprednisolone (triamcinolone) which has greatly improved antiinflammatory activity.

Following these reports, the disclosures of other hydroxylations were made in a rapid sequence: *Cunninghamella blakesleeana* (Hanson *et al.*, 1953) and *Curvularia lunata* (Shull *et al.*, 1953; Shull and Kita, 1955) were found to carry out 11β-hydroxylation at C-11 in good yields (about 60%) and the latter fungus is used in the manufacture of cortisol from compound S (Scheme 7).

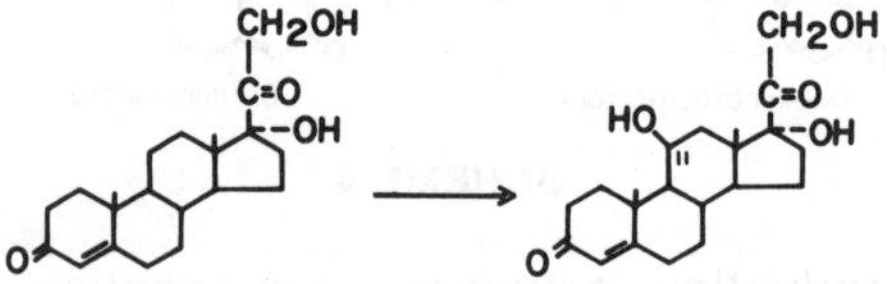

Compound S Cortisol

SCHEME 7

These observations with their highly practical applications stimulated an intensive search for additional microbial modifications of these compounds. In time, all available carbon atoms were found to be hydroxylated by different organisms and by now the number of microbial hydroxylations exceeds the number of hydroxylations carried out by mammalian systems (Table I). In many cases, polyhydroxylations in different combinations were also noted with dihydroxylations (such as 6β,11α; 7α,14α; 9α,14α; or 11β,21) prevailing. They are not, however, of practical importance. The oxygen atom involved in these reactions is derived from gaseous oxygen and not from water, and the hydroxyl group thus formed retains the same configuration as the hydrogen that has been replaced (Hayano *et al.*, 1958a,b).

The therapeutic properties of cortisone and cortisol were further improved by microbial introduction of a 1,2-double bond whereby prednisone (1-dehydrocortisone) and prednisolone (1-dehydrocortisol) were formed, respectively (for example, see Scheme 8).

This 1-dehydrogenation was first described to be carried out by *Cylin-*

TABLE I. Comparison of Hydroxylation of Steroids by Microbial and Mammalian Systems

Microbial systems	Mammalian systems
1α	1α
1β	—
2α	2α
2β	2β
3β	—
5α[a]	—
5β[b]	
	6α
6β	6β
7α	7α
7β	—
9α	—
10β[c]	—
11α	—
11β	11β
12α	12α
12β	—
14α	—
15α	15α
15β	—
16α	16α
16β	—
17α	17α
18	18
19	19
21	21

[a] In A-norprogesterone series.
[b] In cardiac glycosides.
[c] In 19-nortestosterone.

drocarpon radicicola and *Streptomyces lavendulae* (Fried *et al*., 1953), by *Fusarium solani* and *Fusarium caucasicum* (Vischer and Wettstein, 1953), and by *Septomyxa affinis*, which is also used in the preparation of 6α-methylprednisolone (Medrol), a valuable synthetic corticoid which does not cause significant sodium retention (Sebek and Spero, 1959; Spero *et al*., 1956). These organisms also degrade the side chain,

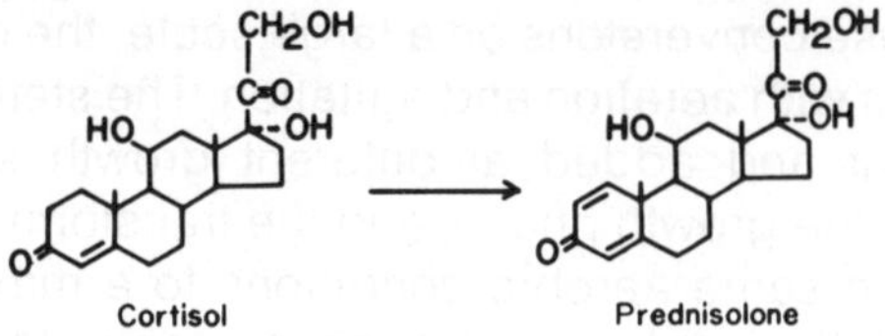

SCHEME 8

oxidize the hydroxyl group at C-17, and bring about expansion of ring D of some steroids (such as progesterone and deoxycorticosterone) (see Scheme 9).

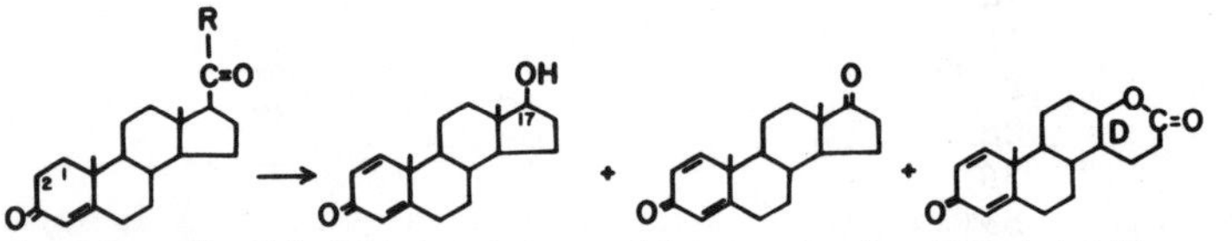

Progesterone ($R = CH_3$) Deoxycorticosterone ($R = CH_2OH$) 1-Dehydrotestosterone 1,4-Androstadienedione 1-Dehydrotestololactone

SCHEME 9

In contrast, *Arthrobacter (Corynebacterium) simplex* which was the first organism used commercially to 1-dehydrogenate cortisol to prednisolone (Nobile *et al.*, 1955) is devoid of these attendant reactions. The same is true of two other 1-dehydrogenating bacteria (*Bacillus sphaericus* and *Bacterium cyclooxydans*) which have also been used in industrial practice.

A process called "pseudo-crystallofermentation" was also described and reported to give high yields of the 1-dehydrogenated product. According to this procedure, *A. simplex* transforms finely powdered solid cortisol in 1–50% concentrations of crystalline prednisolone in high (>93%) yields directly in the fermentation beers within 5 days (Kondō and Masuo, 1961).

II. COMMERCIAL DEVELOPMENT

As these considerations indicate, microbial modifications were directly responsible for the generation of new analogs, which in turn served in many cases as substrates for a still larger number of steroid products. They had also a significant effect on the cost of the desired hormones. Thus the 11α-hydroxylation of progesterone brought the 1949 price of cortisone in a very short time from \$200 to \$6 per gram and, through further improvements, to less than \$1 per gram at present.

To carry out these conversions on a large scale, the culture is grown in fermentation tanks with aeration and agitation. The steroid is dissolved in a suitable solvent and added at different growth stages, preferably toward the end of the growth phase, and the transformation is allowed to proceed under the same aerobic conditions to a maximum. Since steroids are essentially insoluble or sparingly soluble in water, it is of considerable importance that they react in aqueous suspensions and that the reactions are completed in most cases within a reasonable time.

Thus 11α-hydroxylation of 2 gm of progesterone per liter by *R. nigricans* proceeds satisfactorily in 48 hours. The product is released into the medium from which it is recovered by extraction with methylene chloride. With further development (use of surfactants, improved methods of steroid addition) the substrate levels have been increased with the concomitant yield increases of the product. Solid substrate addition also improved the efficiency of 11α-hydroxylation. Thus 2 g of finely ground aqueous suspension of progesterone with 0.01% wetting agent (Tween 80) per 100 ml was converted by *Aspergillus ochraceus* to 11α-hydroxyprogesterone in 90% yields in 3 days with a minimum of a side-product (6β, 11α-dihydroxyprogesterone, Weaver *et al.*, 1960). A study of this hydroxylation indicated that 14% of the substrate progesterone remained unconverted regardless of the concentrations at which it had been added. This was due to the formation of a crystalline structure consisting of 6 parts of the product and 1 part of progesterone, which thus became unavailable for the transformation. Continuous feeding of the substrate overcame this drawback. A similar substrate–product interaction was observed during 1-dehydrogenation by *S. affinis* of 11β,21-dihydroxy-4,17(20)-pregnadien-3-one (dienediol) to 11β,21-dihydroxy-1,4,17(20)-pregnatrien-3-one (trienediol), an intermediary step in the synthesis of 6α-methylprednisolone (see above). This difficulty was overcome by multistage feeding of the substrate (Maxon *et al.*, 1966).

For illustration, some of the steroid transformations of commercial importance are listed in Table II. The fermentation conditions used in laboratory studies of these and related transformations are summarized in Table III (also see Perlman, 1976).

The complexity of industrial processes may be illustrated by the information in Fig. 1, which outlines the conversion of stigmasterol to six steroids of clinical importance.

Twenty-one chemical and microbial steps are involved in the conversion of stigmasterol to fluorometholone, and 10 steps to prednisolone and to cortisone. By continuous refinement of the process operations as high as 10 mole-% conversions of the starting material to final product have been realized.

III. TYPES OF MICROBIAL TRANSFORMATIONS

In addition to the six reactions of commercial importance (see Table II), many other microbial transformations have been discovered. They are listed in Table IV and include introduction of nuclear and side-chain hydroxy groups; cleavage of carbon-to-carbon linkages in the side chain as well as in ring D; introduction of a 1,2-double bond in ring A; oxidation

TABLE II. Some Steroid Transformations of Commercial Importance

Reaction	Substrate → product	Microorganism	Some industrial producers
11α-Hydroxylation	Progesterone → 11α-hydroxyprogesterone	*Rhizopus nigricans*	The Upjohn Company
11β-Hydroxylation	Compound S → cortisol	*Curvularia lunata*	Pfizer, Inc.; Gist-Brocades
16α-Hydroxylation	9α-Fluorocortisol → 9α-fluoro-16α-hydroxycortisol	*Streptomyces roseochromogenus*, *S. argenteolus*	E.R. Squibb and Sons; Lederle Laboratories
1-Dehydrogenation	Cortisol → prednisolone	*Arthrobacter simplex*	Schering Corporation
	Dienediol[a] → trienediol[a]	*Septomyxa affinis*	The Upjohn Company
1-Dehydrogenation, side-chain cleavage, and ring D expansion	Progesterone → 1-dehydrotestololactone	*Cylindrocarpon radicicola*	E.R. Squibb and Sons
Side-chain cleavage	β-Sitosterol → androstadienedione, 9α-hydroxyandrostenedione and/or androstenedione	*Mycobacterium* spp.	G.D. Searle and Company The Upjohn Company

[a] See text for chemical names.

TABLE III. Media and Fermentation Conditions Used in Microbial Transformations of Steroids

Microorganism	Steroid substrate	Steroid product (approx. yields wt. %)	Composition of medium[a]	Length of incubation, temperature, aeration	Reference
Alcaligenes faecalis	Cholic acid	Ketocholic acids (90–100%)	A	2 days (monoketo acid); 4 days (diketo acid); 6 days (triketo acid); 37-39°; surface culture	Schmidt and Hughes (1944)
Corynebacterium mediolanum	21-Acetoxy-3β-hydroxy-5-pregnen-20-one	21-Hydroxy-4-pregnene-3,20-dione (30%)	B	6 days, 36-37°C, pure oxygen with agitation	Mamoli, (1944)
Cunninghamella blakesleeana (H334)	Compound S	Cortisone (19%), cortisol (65%)	C	3 days, 28°C, rotary shaker (250 rpm)	O'Connell, *et al.* (1955)
Cylindrocarpon radicicola (ATCC 11011)	Progesterone	1-Dehydrotestololactone (50%)	D	3 days, 25°C, reciprocating shaker (120 spm)	Fried and Thoma, (1956)
Fusarium solani	Progesterone	1,4-Androstadiene-3,17-dione (85%)	E	4 days 25°C, rotary shaker (100 rpm)	Vischer and Wettstein (1953)
Rhizopus arrhizus ATCC 11145	4-Androstene-3,17-dione	11α-Hydroxy-4-androstene-3,17-dione (25%)	F	4 days, 28°C small aerated tank (6-7 mM O_2/liter/min)	Murray and Peterson (1953)
Streptomyces albus	Estradiol	Estrone (90-95%)	G	6 hours of substrate oxidation with resting cells, 30°C	Heusghem and Welch (1948)
Streptomyces aureus	Progesterone	15α-Hydroxy-4-pregnene-3,20-dione (11%)	H	3 days, 25°C, rotary shaker (280 rpm)	Fried *et al.* (1956)

[a] Composition of the media: (A) 4.7 gm $(NH_4)_2SO_4$, 0.5 gm asparagine, 5 gm NaCl, 0.65 gm NaOH, 2 gm glycerol, inorganic salts, 5 gm cholic acid, distilled water to 1 liter; (B) 60 ml yeast water, 10 ml of 0.2 *M* Na_2HPO_4, 1 gm KH_2PO, 0.2 gm 21-acetoxy-3-hydroxypregnan-20-one; (C) 3 gm $NaNO_3$, 0.5 gm KCl, 30 gm dextrin, 0.01 gm $FeSO_4 \cdot 7H_2O$, 10 mg Tween 80, 0.5 gm $MgSO_4 \cdot 7H_2O$, 1,3 gm $K_2HPO_4 \cdot 3H_2O$, distilled water to 1 liter, pH 7.2; (D) 3 g corn steep solids, 3 gm $NH_4H_2PO_4$, 2.5 gm $CaCO_3$, 2.2 gm soybean oil, 0.5 gm progesterone, distilled water to 1 liter, pH 7.0; (E) 15 gm peptone, 6 ml corn steep liquor, 50 gm glucose, distilled water to 1 liter, pH 6.0, progesterone (0.25 gm) added after 2 days of incubation; (F) 20 gm lactalbumin digest, 5 ml corn steep liquor, 50 gm glucose, tap water to 1 liter, pH 5.5–5.9; androstenedione (0.25 gm) added after 27 hours of incubation; (G) nutrient broth (to grow cells), phosphate buffer (pH 7.0) (to oxidize estradiol); and, (H) 2.2 gm soybean oil, 15 gm soybean meal, 10 gm glucose, 2.5 gm $CaCO_3$, 0.25 gm progesterone, water to 1 liter

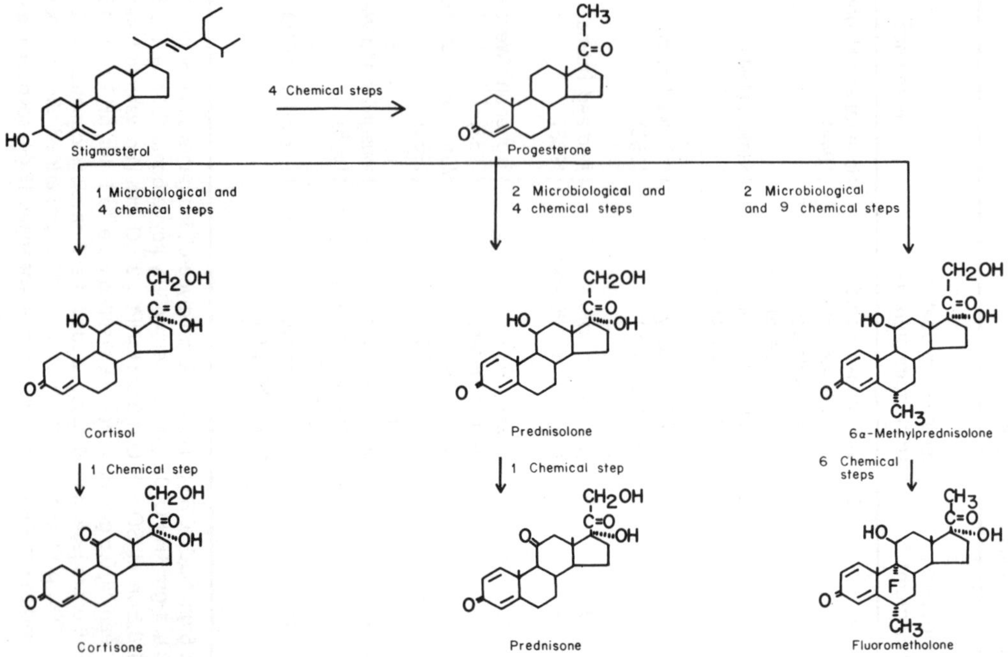

FIGURE 1. Conversion of stigmasterol to six clinically important steroids.

of hydroxyl groups to ketones; reduction of carbonyl groups; and hydrolysis and formation of esters.

The methods generally used in screening microorganisms for the ability to transform steroids are: (1) the microorganism is first grown in suitable media for 1 or 2 days; (2) the steroid, dissolved in a water-soluble solvent (e.g., dimethylformamide, acetone, propylene glycol), is then added; (3) incubation is continued for 1 or 5 days; (4) the transformation products are obtained by extraction of the beer with methylene chloride or nonpolar solvents (chloroform, ethyl acetate) and by purification by column chromatography and other methods; (5) the structure of the transformation products is determined by classical methods of organic chemistry. Most of the fungi and streptomycetes examined have been found to transform some steroids, and sufficient information has accumulated to predict the type of transformation that might be expected if the genus and species of the culture is known or, conversely, to predict the identity of the microorganism if the transformation is known (Čapek *et al.*, 1966, 1976; Charney and Herzog, 1968, 1980; Vézina and Rakhit, 1974).

The transformations can be performed by cells at different stages of growth. Thus the steroid may be added to: (1) the growing culture (either at the time of inoculation or toward the end of the growth phase); (2) the

TABLE IV. Types of Microbial Transformation of Steroids

A. Oxidation
1. Conversion of secondary alcohol to ketone
2. Introduction of primary hydroxyl on steroid side chain
3. Introduction of secondary hydroxyl on steroid nucleus
4. Introduction of tertiary hydroxyl on steroid nucleus
5. Dehydrogenations of ring A of steroid nucleus in positions 1(2) and 4(5)
6. Aromatization of ring A of the steroid nucleus
7. Oxidation of the methylene group to ketone group
8. Cleavage of side chain of pregnane at C-17 to form ketone
9. Cleavage of side chain of pregnane at C-17 and opening of D ring to form testololactone
10. Cleavage of side chain of steroids to form carboxyl group
11. Cleavage of side chain of pregnane steroids at C-17 to form secondary alcohol
12. Formation of epoxides
13. Decarboxylation of acids

B. Reduction
1. Reduction of ketone to secondary alcohol
2. Reduction of aldehyde to primary alcohol
3. Hydrogenation of double bond at position 1(2) of ring A
4. Hydrogenation of double bond at positions 4(5) of ring A and at 5(6) of ring B
5. Elimination of secondary alcohol
6. Formation of homosteroids of the androstane series from pregnane derivatives

C. Hydrolysis
1. Saponification of steroid esters
1. Acetylation

D. Esterification

nonproliferating ("resting") cell suspensions (or cell-free extracts); (3) microbial systems which carry out simultaneously other reactions as the major fermentation operations; (4) the spore suspensions; or (5) the immobilized whole cells and immobilized cell-free enzymes. The last two methods are receiving considerable attention (Abbott, 1976; Vézina *et al.*, 1968) but have not been used on a commercial scale.

IV. SIDE-CHAIN DEGRADATION OF STEROLS

In the past 18 years, the degradation of the side chain of cholesterol and of related plant sterols by microorganisms has been investigated extensively, and the results of these studies have become of considerable economic importance (Martin, 1977). The first data of this kind by Turfitt in 1948 reported on the conversion of cholestenone and bile acids to small amounts of 3-keto-4-androstene-17-carboxylic acid (Turfitt, 1948) and by Whitmarsh in 1964 on the degradation of cholesterol into a mixture of 3-ketobisnor-4-cholenic acid, 3-ketobisnor-1,4-choladienic acid, 4-androstene-3,17-dione (androstenedione), and 1,4-androstadiene-3,17-dione (androstadienedione) in the presence of small amounts of 8-hydroxyquinoline (Whitmarsh, 1964). By this approach, other compounds were found which prevent the degradation of the steroid nucleus while allowing the degradation of the cholesterol side chain (Nagasawa *et al.*, 1969, 1970): iron- or copper-chelating agents (such as quinoline-2-carboxylic acid, 2,2′-bipyridine, 1,10-phenanthroline); divalent ions that replace iron or block -SH functions (Ni^{2+}, Co^{2+}, Pb^{2+}, SeO_3^{2-}, AsO_2^-); or redox dyes. Sih and his collaborators showed that the degradation of the sterol nucleus can be blocked also by the chemical modification of the substrate cholesterol at C-19 (19-hydroxy-, 19-nor-, 6β,19-oxidosterols). The same workers also established that the cholesterol side chain is degraded to the C-17 ketosteroid by a stepwise β-oxidation via the intermediary C-24 and C-22 carboxylic acids (Sih *et al.*, 1968) as shown in Scheme 10.

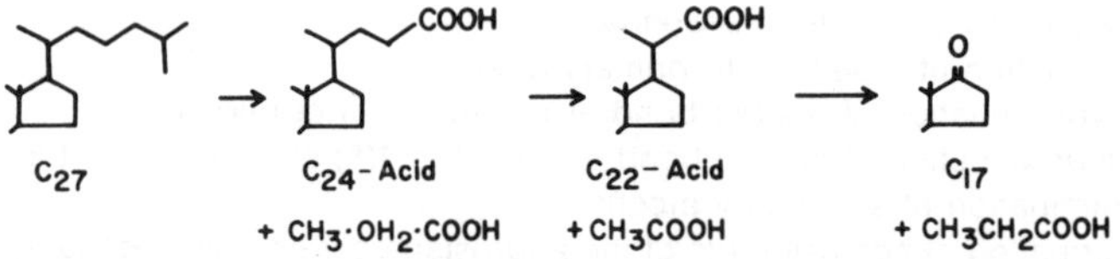

SCHEME 10

When these inhibitors are not present or when the substrate cholesterol is not chemically modified, most of the cholesterol-utilizing microorganisms degrade the nucleus as well. There are two reactions involved in the initial stages of the degradation of the steroid nucleus and the sequence in which they proceed depends on the microorganism used. As seen in Scheme 11, *Pseudomonas* sp. and *Mycobacterium phlei*

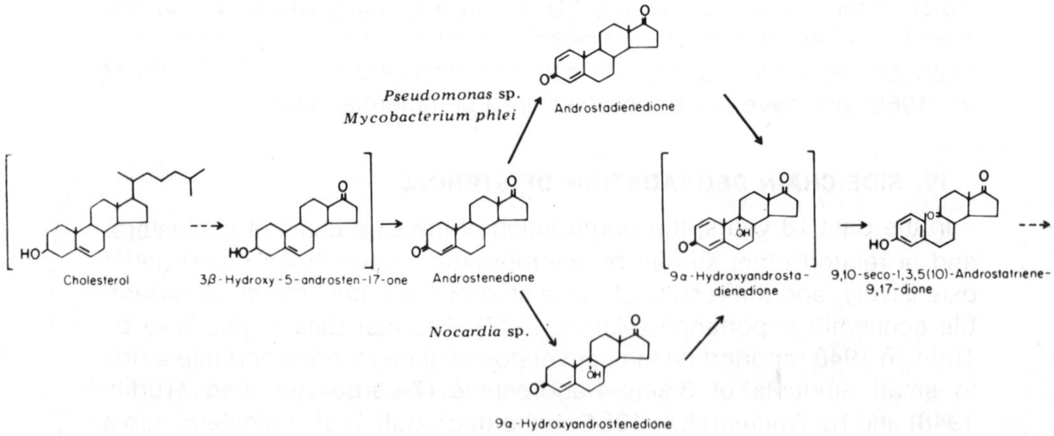

SCHEME 11

carry out first the 1-dehydrogenation which is followed by 9α-hydroxylation, while *Nocardia* sp. performs these reactions in the reverse sequence.

With this information, the problem of accumulating the desired products androstenedione and androstadienedione from cholesterol and related plant sterols in the absence of inhibitors or chemically modified substrates has been solved. Marsheck and his collaborators (1972) prepared a mutant of a soil mycobacterium which converted cholesterol, stigmasterol, and sitosterols to androstadienedione. By additional uv irradiation, a new mutant was selected which accumulated androstenedione as the main product. Both of these compounds (androstadienedione and androstenedione) are convenient substrates for the synthesis of other steroids. This discovery is of signal practical importance. It has made soy sterols (especially β-sitosterol and campesterol) commercially attractive new steroid substrates, since they are readily available in large quantities from soybeans and tall oil (waste product of the pulpwood kraft process; Conner *et al.*, 1976).

REFERENCES

Abbott, B. J. (1976). *Adv. Appl. Microbiol.* **20,** 203.

Applezweig, N. (1962). "Steroid Drugs." McGraw-Hill, New York.

Applezweig, N. (1978). *In* "Crop Resources" (D. S. Seigler, ed.), p. 149. Academic Press, New York.

Čapek, A., Hanč, O., and Tadra, A. (1966). "Microbial Transformations of Steroids." Academia, Prague.

Čapek, A., Fassatiová, O., and Hanč, O. (1976). *Folia Microbiol.* (*Prague*) **21,** 70.

Charney, W., and Herzog, H. (1968). "Microbiological Transformations of Steroids," 1st ed. Academic Press, New York.

Charney, W., and Herzog, H. (1980). "Microbiological Transformations of Steroids," 2nd ed. Academic Press, New York.

Conner, A. H., Nagaoka, M., Rowe, J. W., and Perlman, D. (1976). *Appl. Microbiol.* **32,** 301.

Fried, J., Thoma, R. W., and Klingsberg, A. (1953). *J. Am. Chem. Soc.* **75,** 5764.

Fried, J., and Thoma, R. W.. (1956). U.S. Patent 2,744,120.

Hanson, F. R., Mann, K. M., Nielson, E. D., Anderson, H. V., Brunner, M. P., Karnemaat, J. N., Colingsworth, D. R., and Haines, W. J. (1953). *J. Am. Chem. Soc.* **75,** 5369.

Hayano, M., Gut, M., Dorfman, R. I., Sebek, O. K., and Peterson, D. H. (1958a). *J. Am. Chem. Soc.* **89,** 2336.

Hayano, M., Saito, A., Stone, D., and Dorfman, R. I. (1958b). *Biochim. Biophys. Acta* **21,** 380.

Hechter, O., Jacobsen, R. P., Jeanloz, R., Levy, H., Marshall, C. W., Pincus, G., and Schenker, V. (1949). *J. Am. Chem. Soc.* **71,** 3261.

Hench, P. S., Kendall, E. C., Slocumb, C. H., and Polley, H. F. (1949). *Mayo Clin. Proc.* **24,** 181.

Hench, P. S., Kendall, E. C., Slocumb, C. H., and Polley, H. F. (1950). *Arch. Intern. Med.* **85,** 545.

Heusghem, C., and Welsch, M. (1948). *Bull. Soc. Chim. Biol.* **31,** 282.

Kondō, E., and Masuo, E. (1961). *J. Gen. Appl. Microbiol.* **7,** 113.
Mamoli, L. (1944). U.S. Patent 2,341,110.
Mamoli, L., and Vercellone, A. (1937a). *Ber. Dtsch. Chem. Ges.* **70,** 470 and 2079.
Mamoli, L., and Vercellone, A. (1937b). *Hoppe-Seyler's Z. Physiol. Chem.* **245,** 93.
Marsheck, W. J., Kraychy, S., and Muir, R. D. (1972). *Appl. Microbiol.* **23,** 72.
Martin, C. K. A. (1977). *Adv. Appl. Microbiol.* **22,** 29.
Maxon, W. D., Chen, J. W., and Hanson, F. R. (1966). *Ind. Eng. Chem., Process Des. Dev.* **4,** 421.
Murray, H. C., and Peterson, D. H. (1953). U.S. Patent 2,646,370.
Nagasawa, M., Bae, M., Tamura, G., and Arima, K. (1969). *Agric. Biol. Chem.* **33,** 1644.
Nagasawa, M., Watanabe, N., Hashiba, H., Murakami, M., Bae, M., Tamura, G., and Arima K. (1970). *Agric. Biol. Chem.* **34,** 838.
Nobile, A., Charney, W., Perlman, P. L., Herzog, H. L., Payne, C. C., Tully, M. E., Jevnik, M. A., and Herschberg, E. B. (1955). *J. Am. Chem. Soc.* **77,** 4184.
O'Connell, P. W., Mann, K. M., Nielson, E. D., and Hanson, F. R. (1955). *Appl. Microbiol.* **3,** 16.
Perlman, D. (1976). *In* "Applications of Biochemical Systems in Organic Chemistry" (J. B. Jones, C. J. Sih, and D. Perlman, eds.), Wiley, New York.
Perlman, D., Titus, E., and Fried, J. (1952). *J. Am. Chem. Soc.* **74,** 2126.
Peterson, D. H., and Murray, H. C. (1952). *J. Am. Chem. Soc.* **74,** 1871.
Sarett, L. (1946). *J. Biol. Chem.* **162,** 591.
Schmidt, L. H., and Hughes, H. B. (1944). U.S. Patent 2,360,447.
Sebek, O. K., and Spero, G. B. (1959). U.S. Patent 2,897,218.
Shull, G. M., and Kita, D. A. (1955). *J. Am. Chem. Soc.* **77,** 763.
Shull, G. M., Kita, D. A., and Davisson, J. W. (1953). U.S. Patent 2,658,023.
Sih, C. J., Tai, H. H., Tsong, Y. Y., Lee, S. S., and Coombe, R. G. (1968). *Biochemistry* **7,** 808.
Spero, G. B., Thompson, J. L., Magerlein, B. J., Hanze, A. R., Murray, H. C., Sebek, O. K., and Hogg, J. A. (1956). *J. Am. Chem. Soc.* **78,** 6213.
Turfitt, G. E. (1948). *Biochem. J.* **42,** 376.
Vézina, C., and Rakhit, S. (1974). *In* "CRC Handbook of Microbiology" (A. I. Laskin and H. A. Lechevalier, eds.), Vol. IV, p. 117. CRC Press, Cleveland, Ohio.
Vézina, C., Sehgal, S. N., and Singh, K. (1968). *Adv. Appl. Microbiol.* **10,** 221.
Vischer, E., and Wettstein, A. (1953). *Experientia* **9,** 271.
Weaver, E. A., Kenney, H. E., and Wall, M. E. (1960). *Appl. Microbiol.* **8,** 345.
Whitmarsh, J. M. (1964). *Biochem. J.* **90,** 23P.

Chapter 15

Vitamin B_{12}

J. FLORENT
L. NINET

I. INTRODUCTION

A. Historical Background

Vitamin B_{12}, or cyanocobalamin, is an important biological compound active as an hematopoietic factor in mammals and as a growth factor for many microbial and animal species. Although its existence was suspected in 1925 by Wipple and Robscheit-Robbins when studying the therapeutic effect of beef liver in pernicious anemia, vitamin B_{12} was isolated in 1948 as pure crystals independently by Rickes *et al.* and by Smith. In 1955 Hodgkin *et al.* unraveled its structure by means of X-ray crystallography. Ten years were spent from the first efforts of Woodward and Eschenmoser until a full chemical synthesis was achieved (Krieger,

MICROBIAL TECHNOLOGY, 2nd ed., VOL. I

ISBN 0-12-551501-4

1973; Maugh, 1973). With 70 steps required, this difficult synthesis is, for industrial purposes, of little value and today the vitamin B_{12} group of compounds is obtained by fermentation processes.

B. Structure of Vitamin B_{12}

Vitamin B_{12} belongs to the large family of cobaltocorrinoids exhibiting the general formula presented in Fig. 1.

The X radical for the cobalamins, which are the most interesting cobaltocorrinoids from the medical point of view, is 5,6-dimethylbenzimidazole. The group includes cyanocobalamin, or true vitamin B_{12}; hydroxocobalamin, or vitamin B_{12a}; methylcobalamin, or mecobalamin; and 5′-deoxyadenosylcobalamin, or cobamamide (or coenzyme B_{12} of Barker), where Y is a cyano, hydroxyl, methyl, or 5′-deoxyadenosyl radical, respectively.

Other cobaltocorrinoids, with different heterocyclic bases (substituted benzimidazoles, purines) as the X radical, are also produced by various microorganisms either spontaneously or after the introduction of the corresponding base in the fermentation medium. Pseudovitamin B_{12} and factor III, in which X is adenine and 5-hydroxybenzimidazole, respectively, are the most frequently encountered. Natural analogs designated as "incomplete" are also known, for instance, etiocobalamin, or factor B, which is devoid of the X–ribose–phosphate moiety.

These "complete" or "incomplete" analogs have been reviewed by Mervyn and Smith (1964). Generally, they have low activity as growth factors for vertebrates but are very potent for the microorganisms. The cobalamins, in biological tests, are always the most active.

C. Sources of Vitamin B_{12}

Although vitamin B_{12} is present in small amounts in almost every animal tissue, e.g., 1 mg/kg in beef liver, it originates from microorganisms. Depending on the nature of their nutritional habits and digestive physiology, animals obtain the vitamin from their own intestinal flora or from other animals through their meat diet. An exogenous supply is mandatory for man.

Vitamin B_{12} derived from cultures of microorganisms soon supplanted beef liver as a practical source of the vitamin for therapeutic purposes. Around 1950, materials rich in biomasses, such as activated sludges or broths of antibiotic-producing streptomyces, were used for isolating vitamin B_{12} either in a crude form for animal feeding or in a pure state for a medical use (Noyes, 1969). Later, high-producing bacterial strains were specially selected for commercial production.

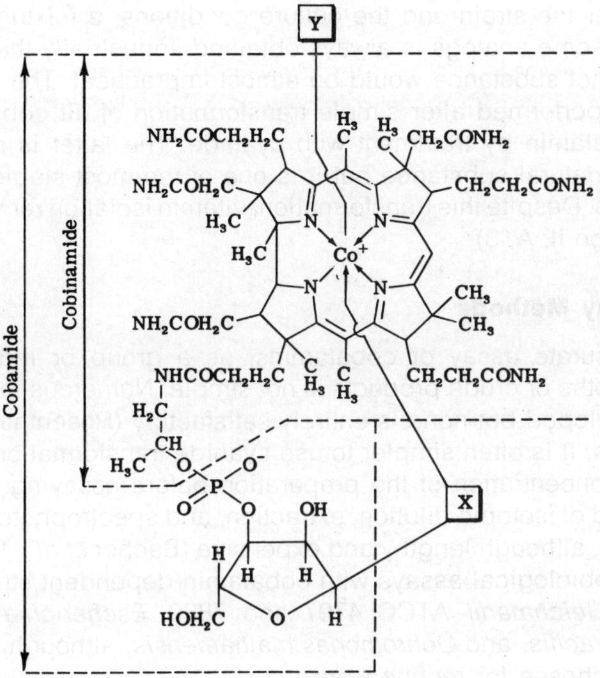

Cobalamins		
X	Y	Trivial name
	—CN	Vitamin B_{12}
	—OH	Vitamin B_{12a}
	—CH_3	Mecobalamin
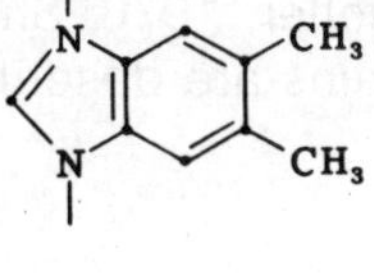	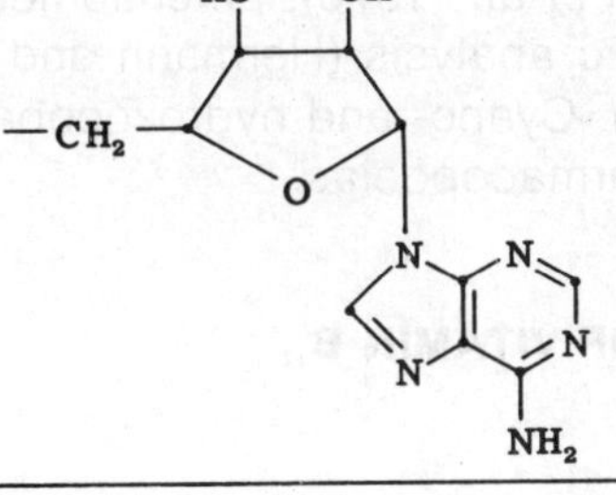	Cobamamide or coenzyme B_{12}

FIGURE 1. General structure of cobaltocorrinoids with references to cobalamins.

Whatever the strain and the culture conditions, a mixture of cobalamins and some analogs is always obtained. Industrially the isolation of each distinct substance would be almost impractical. The extraction is generally performed after simple transformation of all cobalamins into cyanocobalamin by treatment with cyanide. The latter is not the most abundant natural substance but it is one of the most stable and active derivatives. Despite this transformation, vitamin isolation remains tedious (see Section II, A, 3).

D. Assay Methods

The accurate assay of cobalamins, as a group or individually, in culture broths or crude products is not simple. Numerous methods have been developed but none is entirely satisfactory (Rosenthal, 1968). For this reason, it is often simpler to use cyanide transformation with simultaneous concentration of the preparation before assaying. In practice, the method of isotopic dilution, extraction, and spectrophotometry is one of the best, although lengthy and expensive (Bacher *et al.*, 1954). Therefore, microbiological assays with cobalamin-dependent strains as *Lactobacillus leichmanii* ATCC 4797 and 7830, *Escherichia coli* 113-3, *Euglena gracilis,* and *Ochromonas malhamensis,* although nonspecific, are often chosen for routine work.

In practice the purer the preparation, the easier the analysis. Hence it is advisable to carry out preliminary separations by thin-layer or column chromatography (Bilkus and Mervyn, 1970; Tortolani *et al.*, 1970a,b; Vogelmann and Wagner, 1973; Tortolani and Mantovani, 1974; Matsumoto *et al.*, 1974; Lotti, 1975) or by electrophoresis (Kato and Shimizu, 1962; Tortolani and Ferri, 1974). Afterward, several classical methods are available for assaying the isolated fractions or pure products: spectrophotometry in the visible spectrum (Farmaceutici Italia, 1971b; Celletti *et al.*, 1976) or in the IR spectrum (Goldstein *et al.*, 1974), atomic absorption (Whitlock *et al.*, 1976), potentiometry (Goldstein and Duca, 1976), and enzymatic analysis (Hermann and Mueller, 1976; Sheehan and Hercules, 1977). Cyano- and hydroxocobalamins are described in several national pharmacopoeias.

II. PRODUCTION OF VITAMIN B_{12}

A. General

During the last 20 years several accounts on the vitamin B_{12} production processes have appeared (Perlman, 1959, 1967; Prescott and Dunn,

1959; Mervyn and Smith, 1964; Friedmann and Cagen, 1970). The main lines of this research have been: (1) to obtain a better knowledge of the biosynthetic pathway in order to sustain biochemical and genetic studies; (2) to improve the better known strains (propionibacteria and *Pseudomonas*) by mutation; (3) to replace traditional sugars by cheaper nutrients in the media; and (4) to improve extraction and purification techniques.

1. Biogenesis of Vitamin B_{12}

Since the reviews by Wagner (1966) and Friedmann and Cagen (1970), a great deal of attention has been devoted to biosynthesis with *Streptomyces olivaceus* (Sato *et al.*, 1968, 1971) and *Propionibacterium shermanii* (Lowe and Turner, 1970; Mueller *et al.*, 1970, 1971; Renz and Bauer-David, 1972; Russel, 1974; Scott, 1975, 1976; Scott *et al.*, 1976; Battersby and McDonald, 1976; Battersby *et al.*, 1976, 1977a,b; Bykhovsky and Zaitseva, 1976; Ford and Friedmann, 1976; Imfeld *et al.*, 1976).

With few exceptions all the authors agree with the general pathway shown in Fig. 2. According to this pathway, several heme derivatives (cytochromes and catalase) which are vital elements for microbial life can also repress the first steps of vitamin B_{12} biosynthesis (Vorobeva *et al.*, 1967; Simon, 1968; Bykhovsky *et al.*, 1969, 1976; Brazenas and Kanopkaite, 1970a; Miyazaki *et al.*, 1972). Likewise, oxidation–reduction potentials and metal concentrations can play a leading part in the course of vitamin B_{12} fermentations.

2. Producing Strains, Culture Media, and Yields

Numerous microorganisms have been considered as potential sources of vitamin B_{12}. Because of their rapid growth and their high productivity, two propionibacteria and one *Pseudomonas* cultivated on carbohydrates were selected rather than the *Streptomyces* and other genera for industrial purposes as shown in Table I. Other species have been recently examined mainly in relation to the need for organisms which can grow on cheaper nutrients than carbohydrates (food industries wastes, see Section II, B; hydrocarbons and alcohols, see Section II, C). In this respect, methanogenic bacteria grown on methanol might become serious challengers in the next decade.

The best strains are spontaneous or induced mutants, screened for their resistance to different agents: cobalt or manganese ions, antibiotics, etc. (Chinoin-Gyogyszer, 1965; Aries, 1974). Induced mutants are obtained with treatments by uv or X-rays or with such chemical agents as *N*-methyl-*N′*-nitro-*N*-nitrosoguanidine, nitrosoethylurea, ethylenimine,

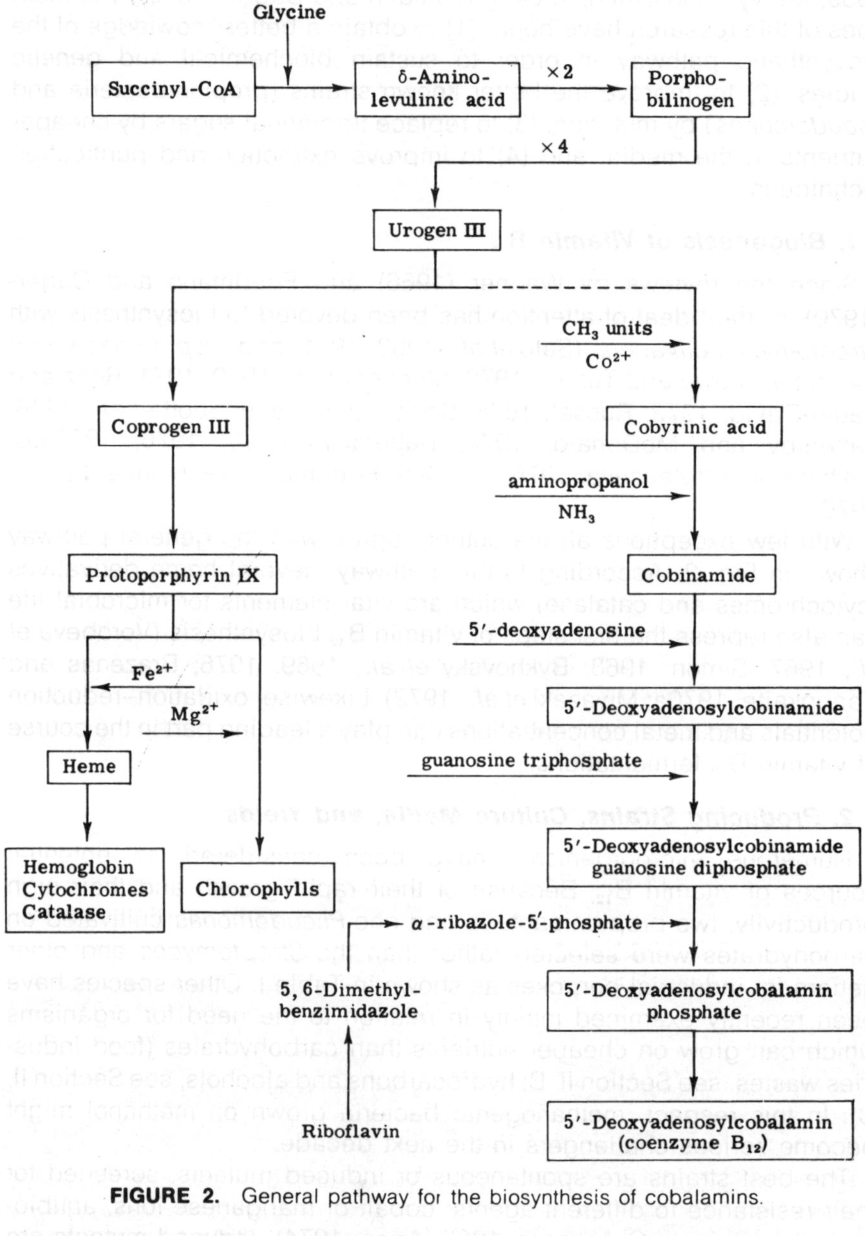

FIGURE 2. General pathway for the biosynthesis of cobalamins.

TABLE I. Vitamin B_{12} Production by Different Strains

Strain	Carbon source	Yield (mg/liter)	References
Micromonospora sp.	Glucose	11.5	Wagman *et al.* (1969)
Nocardia rugosa	Glucose–cane molasses	14	Farmaceutici Italia (1971a)
Propionibacterium freudenreichii	Glucose	25	Uclaf (1960)
Propionibacterium shermanii	Glucose	23	Speedie and Hull (1960)
Propionibacterium shermanii	Glucose	28	Chinoin-Gyogyszer (1965)
Propionibacterium shermanii	Glucose	39	Pierrel S.p.A. (1965b)
Propionibacterium vannielli	Glucose	25	Pliva T.F.K.P. (1964)
Pseudomonas denitrificans	Beet molasses	59	Merck and Co., Inc., (1971)
Streptomyces olivaceus	Glucose–lactose	8.5	Terada *et al.* (1959)
Mixed methanogenic bacteria	Methanol	35	Richter Gedeon V.G. (1975a,b)
Bacterium FM-02T	Methanol	2.6	Toraya *et al.* (1975)
Methanobacillus omelianskii	Methanol	8.8	Pantskhava and Bykhovsky (1966)
Protoaminobacter ruber	Methanol	2.5	Kojima *et al.* (1976)
Corynebacterium and *Rhodopseudomonas*	*n*-Paraffins	2.3	Nakao *et al.* (1974)
Nocardia gardneri	Hexadecane	4.5	Kyowa Hakko Kogyo Co., Ltd. (1970)

diethylsulfate, and mustard gas (Merck and Co., Inc., 1971; Gruzina *et al.*, 1974; Aries, 1974; Georgopapadakou *et al.*, 1976). These mutants exhibit new morphological or biochemical characteristics (size and color of colonies, antibiotic resistance, substrate utilization).

Whatever the strains and the culture conditions may be, it is necessary to add to the medium certain essential elements for vitamin biosynthesis: cobalt ions in all cases, and often 5,6-dimethylbenzimidazole. Supplementation with other potential precursors of corrinoids, such as glycine, threonine, δ-aminolevulinic acid, and aminopropanol, also can be beneficial.

3. Isolation of Vitamin B_{12}

During the last 20 years the extraction processes have been improved many times but usually they include the following main stages: (1) solubilization of cobalamins and conversion to cyanocobalamin and (2) isolation of a crude product, 80% pure (utilizable directly for animal feeding), followed by purification to a 95–98% level (for medical use).

In order to extract the vitamin B_{12}, the whole broth, or an aqueous suspension of harvested cells, is heated at 80°–120°C for 10–30 minutes at pH 6.5–8.5. The conversion to cyanocobalamin is obtained by treating the heated broth or cell suspension with cyanide or thiocyanate, often in the presence of sodium nitrite or chloramin B. If the cyanide treatment is postponed to a later stage in order to reduce tedious handling procedures, a conversion to the more stable sulfato- or sulfitocobalamin is advisable (Aschaffenburger Zellstoffwerke A.G., 1966; Aries, 1974).

The extraction operation units are classical (Robinson, 1955), and they are combined to achieve an efficient and inexpensive process. For instance, the following methods are used:

1. Adsorption on such different supports as Amberlite IRC 50, alumina, carbon, and elution by hydroalcoholic or hydrophenolic mixtures. Many new supports have been tested recently (Merck and Co., Inc., 1964; Olin Mathieson Chem. Corp., 1965; Aschaffenburger Zellstoffwerke A.G., 1966; Kamikubo, 1967; Rohm and Haas Co., 1968; Sato *et al.*, 1968, 1977; Zodrow *et al.*, 1968; Duric, 1969; Seisan Kaihatsu Kagaku Kenkyusho, 1970; Tortolani *et al.*, 1970a; Iashi, 1973; Vogelmann and Wagner, 1973; Lotti, 1975).
2. Extraction from aqueous solutions by phenol or cresol alone or in mixture with benzene, butanol, carbon tetrachloride, or chloroform (Stefaniak *et al.*, 1968; Seisan Kaihatsu Kagaku Kenkyusho, 1970; Merck and Co., Inc., 1971; Aries, 1974).
3. Precipitation or crystallization from various solutions by evapora-

tion, dilution with appropriate solvents or addition of such reagents as tannic acid or *p*-cresol (Olin Mathieson Chem. Corp., 1965; Pierrel S.p.A., 1965a). Some impurities can also be removed by common precipitating agents, such as calcium or zinc hydroxide.

An extraction process presently used is briefly outlined in Fig. 3.

B. Fermentation Processes from Carbohydrates

As previously indicated, propionibacteria and *Pseudomonas* are used for industrial vitamin B_{12} production.

1. Production by Propionibacteria

Several microaerophilic propionibacteria produce cobaltocorrinoids in conventional carbohydrate media correctly supplemented with cobalt, without aeration. However, the production of cobalamins, mainly adenosylcobalamin, depends on an internal or external supply of 5,6-dimethylbenzimidazole (DBI). For this process *Propionibacterium freudenreichii* ATCC 6207, *P. shermanii* ATCC 13673, and their substrains and mutants, which can synthesize their own DBI, are the most appropriate. Since aeration favors DBI formation, it is advisable to use a two-stage culture to obtain the highest yields. In the first stage a practically anaerobic culture is run to almost total depletion of sugar, in order to promote the growth of the bacteria and etiocobalamin biosynthesis. Then, in the second stage, an aeration shift leads to DBI biosynthesis and conversion of etiocobalamin to deoxyadenosylcobalamin. The two stages can be conducted batchwise in the same tank or continuously in two connected fermentors (Speedie and Hull, 1960). Anaerobically grown cells can also be collected by centrifugation and incubated, as a heavy suspension, with air and, if required, with DBI and cyanide. DBI supplementation is only required during the second stage, if selected propionibacteria are not able to synthesize their own, but, due to the inhibitory effect of DBI on the corrinoid biosynthesis, addition of DBI during the first stage must be precluded (Friedmann and Cagen, 1970).

Many improvements of this process have been attempted in different ways and several of them are described in detail by Noyes (1969). Little information is available on the strain preservation and culture maintenance conditions.

The fermentation media consist mainly of glucose or inverted molasses (10–100 gm/liter) with small amounts of ferrous, manganous, and magnesium salts in addition to cobalt salts (10–100 mg/liter), buffering or neutralizing agents, and nitrogenous compounds. Among these, yeast

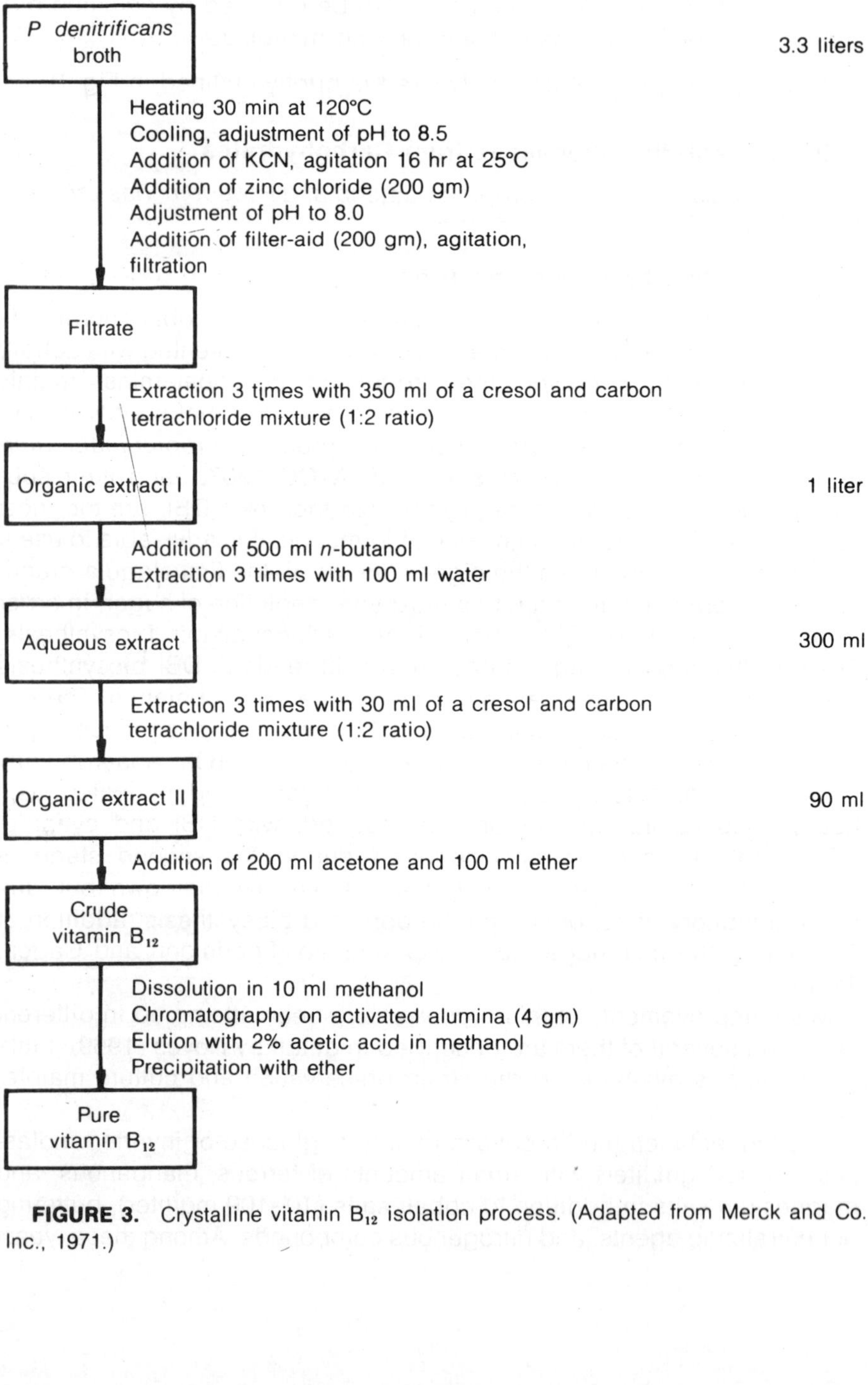

FIGURE 3. Crystalline vitamin B_{12} isolation process. (Adapted from Merck and Co., Inc., 1971.)

preparations, casein hydrolysates, and other traditional sources are currently used but corn steep liquor (30–70 gm/liter) is often preferred since it supplies some lactic and pantothenic acids, which enhance the growth of the bacteria. The latter acid, which stimulates corrinoids biosynthesis, may also be added as a supplement (Zodrow *et al.*, 1963). Generally the temperature of the culture is 30°C and the pH is controlled at around 6.5–7.0.

Recent studies have been conducted in order to increase yields by supplementing the above media with lactose (Danilova and Bankina, 1974), a large range of trace salts (Kotala and Luba, 1965), and some porphobilinogen precursors, such as glycine, and δ-aminolevulinic acid (Hercules Inc., 1968; Bykhovsky, 1969). Substituting the carbon and nitrogen sources by skim milk or lactoserum is also possible (Beskhov *et al.*, 1969, 1973; Popova and Ilieva, 1970; Popova and Bakchevanska, 1972; Giec *et al.*, 1976; Janicka *et al.*, 1976). These modifications may even be beneficial.

Reports indicate that propionibacteria produce from 25 to 40 mg of vitamin B_{12} per liter (see Table I). Even though the values presently reached at an industrial scale are higher than these, the 216 mg/liter claimed by Aries patent (1974) seems beyond any expectation.

Figure 4 gives an example of a current process using *P. shermanii;* the main culture is successively conducted with two modes of aeration.

2. *Production by* Pseudomonas

Several *Pseudomonas* species are claimed to be vitamin B_{12} producers but currently the most used is a *Pseudomonas denitrificans* strain (Miller and Rosenblum, 1960). In contrast with the propionibacteria fermentation, the growth of the *Pseudomonas* keeps pace with cobalamin biosynthesis under aerobic conditions throughout the culture. Accordingly, the fermentation is conducted with aeration and agitation in a single vat, batchwise or continuously. This strain only requires traditional components, such as sucrose, yeast extract and several metallic salts, in the growth medium but from 10 to 25 mg of DBI per liter and from 40 to 200 mg of cobaltous nitrate per liter must be supplemented at the starting of the culture in order to enhance vitamin production (McDaniel, 1961; Daniels, 1970). Likewise, betaine and, to some extent, choline have favorable effects in activating some biosynthetic steps or altering the membrane permeability (Demain *et al.*, 1968; Daniels, 1970; Demain and White, 1971; White and Demain, 1971; Kaplan and Birnbaum, 1973). Glutamic acid stimulates the growth of bacteria (Daniels, 1966; Koike and Hattori, 1975). Owing to their cheapness and high betaine and glutamic acid contents, beet molasses, a multivalent nutrient of choice,

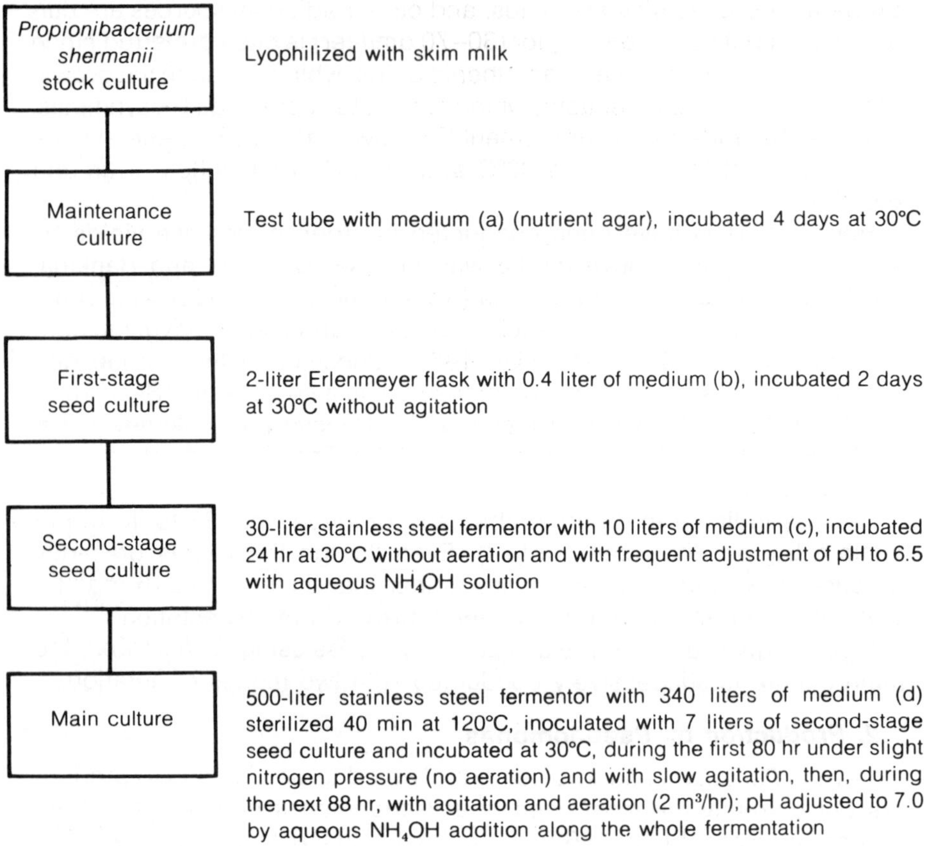

Medium (a): Tryptone, 10 gm; yeast extract, 10 gm; filtered tomato juice, 200 ml; agar, 15 gm; with tap water to 1 liter and pH adjusted to 7.2.
Medium (b): Medium (a) without agar.
Medium (c): Corn-steep liquor, 20 gm; dextrose, 90 gm; with tap water to 1 liter and pH adjusted to 6.5.
Medium (d): Corn-steep liquor, 40 gm; dextrose (sterilized separately), 100 gm; cobalt chloride, 20 mg; with tap water to 1 liter and pH adjusted to 7.0.

FIGURE 4. Vitamin B_{12} from *Propionibacterium shermanii*. Semipilot plant scale fermentation process. (Adapted from Speedie and Hull, 1960; and Pierrel S.p.A., 1965b.)

is preferentially used in industrial fermentations at a 60–120 gm/liter concentration in conjunction with from 2 to 5 gm of ammonium phosphate per liter and some oligoelements. Temperature optimum is 28°C, with a pH of about 7.0. By strain selection and mutation, the vitamin B_{12} production has jumped from 0.6 to 60 mg/liter within 12 years (Long, 1962; Lago and Demain, 1969; Merck and Co., Inc., 1971).

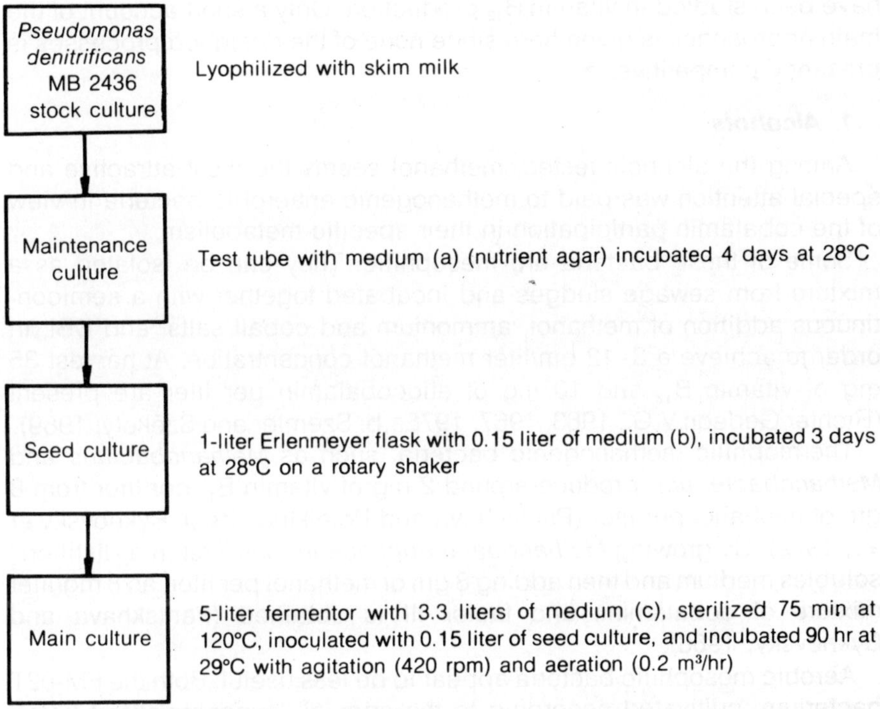

Medium (a): Beet molasses, 60 gm; brewers' yeast, 1 gm; N-Z-amine, 1 gm; diammonium phosphate, 2 gm; magnesium sulfate, 1 gm; manganese sulfate, 200 mg; zinc sulfate, 20 mg; molybdate sulfate, 5 mg; agar, 25 gm; with tap water to 1 liter and pH adjusted to 7.4 (White *et al.*, 1973).

Medium (b): Medium (a) without agar.

Medium (c): Beet molasses, 100 gm; yeast, 2 gm; diammonium phosphate, 5 gm; magnesium sulfate, 3 gm; manganese sulfate, 200 mg; cobalt nitrate, 188 mg; 5,6-dimethylbenzimidazole, 25 mg; zinc sulfate, 20 mg; sodium molybdate, 5 mg; with tap water to 1 liter and pH adjusted to 7.4.

FIGURE 5. Vitamin B_{12} from *Pseudomonas denitrificans*. Laboratory-scale fermentation process. (Adapted from Merck and Co., Inc., 1971.)

Such a process is outlined in Fig. 5. Using the extraction process outlined in Fig. 3, 98% pure cyanocobalamin is obtained with a 75% yield from a broth containing 59 mg/liter.

C. Fermentation Processes from Other Substrates

Continuing the general trend of replacing carbohydrates by cheaper nutrients in industrial fermentations, several alcohols and hydrocarbons have been studied in vitamin B_{12} production. Only a short account of the main approaches is given here since none of the designed processes is presently competitive.

1. Alcohols

Among the alcohols tested, methanol seems the most attractive and special attention was paid to methanogenic anaerobic bacteria in view of the cobalamin participation in their specific metabolism.

Some of these bacteria are mesophilic. They can be isolated as a mixture from sewage sludges and incubated together with a semicontinuous addition of methanol, ammonium and cobalt salts, and DBI, in order to achieve a 3–12 gm/liter methanol concentration. At harvest 35 mg of vitamin B_{12} and 10 mg of etiocobalamin per liter are present (Richter Gedeon V.G., 1963, 1967, 1975a,b; Szemler and Székely, 1969).

Thermophilic methanogenic bacteria, such as *Methanobacillus* and *Methanobacterium,* produce around 2 mg of vitamin B_{12} per liter from 8 gm of methanol per liter (Pantskhava and Pcjelkina, 1970; Bykhovsky *et al.*, 1972). By growing *Methanobacterium soehngenii* first in a distillers' solubles medium and then adding 8 gm of methanol per liter, an 8 mg/liter mixture of cobalamins and factor III is obtained (Pantskhava and Bykhovsky, 1966).

Aerobic mesophilic bacteria appear to be less useful. Both the FM-02T bacterium, cultivated according to the special "exponential-fed batch cultivation" technique, and *Protoaminobacter ruber* produce only 2.5 mg of vitamin B_{12} per liter from 12 gm of methanol per liter (Toraya *et al.*, 1975; Kojima *et al.*, 1976). With *Klebsiella* sp. 101, *Pseudomonas* AM1 and *Microcyclus eburneus,* yields do not exceed 0.2 mg/liter (Nishio *et al.*, 1975, 1976a; Kamikubo *et al.*, 1976).

However, the future of methanol utilization may be considered as promising in view of the mutational potentialities of the strains which have not been exploited.

Up to now, the interest of cultivating *Bacillus badius* or *Arthrobacter* species, or the methylotroph bacterium FM-02T, respectively, on ethanol, isopropanol, and propanediol has to be proved (Toraya *et al.*, 1975; Kojima *et al.*, 1976; Nishio *et al.*, 1976b).

2. Hydrocarbons

Despite the efforts of the past 10 years and the wide range of potentially useful strains, vitamin B_{12} production from hydrocarbons does not look attractive. Apart from yields of 4.5 mg/liter from 20 gm of hexadecane per liter by *Nocardia gardneri* (Kyowa Hakko Kogyo Co., Ltd., 1970) and 2.3 mg/liter from 100 gm of *n*-paraffins per liter by a mixed culture of *Corynebacterium* sp. and *Rhodopseudomonas spheroides* (Nakao *et al.*, 1974), the production from all other species does not exceed 0.5 mg/liter as, for example, with *Pseudomonas* (Morikawa and Kamikubo, 1969; Kanegafuchi Chem. Ind. Co., Ltd., 1972; Kamikubo *et al.*, 1976), mycobacteria (Fukui *et al.*, 1967; Mil'ko and Rabotnova, 1967), *Corynebacterium simplex* (Fujii *et al.*, 1966; Fujii and Fukui, 1970; Kyowa Hakko Kogyo Co., Ltd., 1973), *B. badius* (Kamikubo *et al.*, 1976; Nishio *et al.*, 1976b), and such phototrophic species as *R. spheroides* (Morikawa *et al.*, 1970) and *Rhodospirillum fulvum* (Uehisa, 1972).

III. PRODUCTION OF VITAMIN B_{12} DERIVATIVES AND ANTAGONISTS

Although formerly isolated from beef liver and microbial cells as their cyano derivative, cobalamins are most frequently encountered in nature as a mixture of cyano-, hydroxo-, deoxyadenosyl-, and methylcobalamins. These four compounds, for instance, are present in human serum with distributions of 0–10, 8–15, 22–39, and 39–62%, respectively, and they are bound more or less to specific transport agents, the transcobalamins (McDonald *et al.*, 1977). Each of these derivatives controls several enzymatic reactions which could play a role in man in the genesis of pernicious anemia, neuropathy, and growth deficiency (Barker, 1972; Kamikubo *et al.*, 1972; Fenton *et al.*, 1976). In order to clarify the situation and to obtain more powerful therapeutic agents, an intensive effort to produce these different cobalamins has been made.

Recent studies have also showed the existence, in some microbial cultures, of vitamin B_{12} antagonists which could well turn out to be useful tools in the study of the different aspects of vitamin B_{12} activity.

A. Cobalamin Derivatives

The various cobalamins are directly extracted from the broth used for cyanocobalamin production but without the introduction of cyanide. Since the isolation and separation steps in the mixtures are especially tedious, it is nowadays considered advisable to prepare the pure substances by biochemical or chemical transformation of cyanocobalamin.

1. Hydroxocobalamin

This compound was formerly isolated from cultures of antibiotic-producing *Streptomyces*, then from the cells of numerous bacteria, such as *P. shermanii* (Ilieva, 1971). During the extraction, successive transformations of native cobalamins in their sulfato, nitrato, and chloro derivatives are required before a final hydrolytic treatment by Amberlite IRA 400 (OH^-) to generate hydroxocobalamin (Kaczka *et al.*, 1956; Pierrel S.p.A., 1963a).

The chemical synthesis from cyanocobalamin is also carried out through sulfito and nitrito derivatives (Pierrel S.p.A., 1963b; Smith, 1965).

2. 5′-Deoxyadenosylcobalamin (Coenzyme B_{12})

Coenzyme B_{12} was recently reviewed by Abeles and Dolphin (1976). It is extracted directly from the broths of *P. shermanii* (Barker *et al.*, 1963; Chinoin-Gyogyszer, 1971; Ilieva and Popova, 1974), *P. freudenreichii* (Sifa, 1964), *Nocardia rugosa* (Farmaceutici Italia, 1971a), *N. gardneri* (Kyowa Hakko Kogyo Co., Ltd., 1970), *Klebsiella* sp. 101 (Nishio *et al.*, 1976a), *Methanobacillus kuzneceovii* (Pcjelkina *et al.*, 1970), and *Protoaminobacter ruber* (Shimizu and Sato, 1976), where it can represent up to 80% of the total native cobalamins.

Coenzyme B_{12} is also obtained by bioconversion of cyano- or hydroxocobalamin according to several processes using: (a) Whole broths of propionibacteria (Kaken Kagaku K.K., 1965; Sifa, 1965, 1967) (for instance, 42 mg of cyanocobalamin per liter are added to a 3-day-old anaerobic culture of *P. freudenreichii;* by incubating 2 days more with aeration, the coenzyme B_{12} content amounts to 52 mg/liter); (b) suspensions of *P. shermanii* cells in buffered glucose solutions (these are collected by centrifugation, washed with water and occasionally dried with acetone) (Fukui *et al.*, 1964; Pawelkiewicz, 1971); (c) cell-free extracts of *Clostridium tetanomorphum* (Weissbach *et al.*, 1961; Vitols *et al.*, 1966), *Bacillus megaterium,* and *P. shermanii* (Brady *et al.*, 1962; Brazenas and Kanopkaite, 1970b) in the presence of ATP or flavin cofactor.

When the concentration of added compound does not exceed 50 mg/liter, the transformation yield is near 100% from hydroxocobalamin and 50% from cyanocobalamin.

Coenzyme B_{12} is always endocellular; after harvest of the cells by centrifugation, extraction is performed in the cold by an acetone–water mixture, or at 80°–100°C during a short time by a 2% phenol aqueous solution or an ethanol–water mixture. Purification of coenzyme B_{12} is achieved by the methods described in Section II, A, 3. All operations must be conducted in the cold, under subdued light, and avoiding any drastic pH conditions and the presence of cyanide in order to reach a global yield of 80–85% in the extraction process (Kaken Kaguku K.K., 1965; Chinoin-Gyogyszer, 1971).

Presently coenzyme B_{12} is manufactured by condensing a 5′-ester of adenosine with vitamin B_{12r} or B_{12s}, which both result from cyano- or hydroxocobalamin reduction (Hoffmann-La Roche, 1963; Glaxo Group, Ltd., 1964).

3. Mecobalamin (Methylcobalamin)

The presence of mecobalamin was detected in whole broths of *E. coli, Streptomyces rimosus* (Lindstrand, 1964, 1965), propionibacteria (Skupin *et al.*, 1970), *M. kuzneceovii* (Pcjelkina *et al.*, 1970), and *Klebsiella* sp. 101 (Kamibuko *et al.*, 1975). In *S. olivaceus* and *P. ruber* cultures, mecobalamin present in 0.3 mg/liter amounts up to 50% of the total cobalamins during the growth period; this proportion is reduced to 5% when the culture is continued to reach the maximal yield of cobalamins (Sato *et al.*, 1971, 1977).

Cell-free extracts of *Methanosarcina barkeri* (Blaylock and Stadtman, 1966), *Clostridium thermoaceticum* (Kuratomi *et al.*, 1966; Ljungdahl and Wood, 1969), and *M. kuzneceovii* (Pantskhava *et al.*, 1973) easily convert cyanocobalamin or vitamin B_{12s} to mecobalamin.

Chemically mecobalamin is synthesized by reaction of methyl bromide with reduced forms of cyano- or hydroxocobalamin (Hoffmann-La Roche, 1963; Glaxo Group, Ltd., 1964). In aqueous solution mecobalamin is sensitive to light but does not react with cyanide (Barker, 1968; Lorenzi and Oddono, 1972).

B. Vitamin B_{12} Antagonists

Substances which inhibit, in special conditions, the growth of vitamin B_{12}-dependent microorganisms are produced by several bacteria. These antagonists belong to two groups. The first is constituted by corrinoids deprived of cobalt and found among photosynthetic bacteria such as *Chromatium* (Perlman and Toohey, 1966; Toohey, 1971, 1974). The second includes peptides and uncommon amino acids, such as *N*-5-hydroxy-L-arginine, 4-oxo-L-isoleucine, and the compound 102804, which are distributed among anthracomorph bacteria (Perlman *et al.*, 1974; Kageyama *et al.*, 1977). Cultures of all these bacteria do not require special conditions; at the moment they are still in an experimental stage.

Other vitamin B_{12} antagonists can also be formed by the chemical transformation of natural corrinoids (Ford, 1959; Koppenhagen and Pfiffner, 1970, 1971; Koppenhagen *et al.*, 1974).

IV. MARKETING PROSPECTS FOR COBALAMINS

The world market for cyanocobalamin is presently estimated to be 9000–10,000 kg/year, a major part of which is directed to the synthesis of other cobalamins. The pharmaceutical compositions take some 3500 kg of cyanocobalamin, 2000 kg of hydroxocobalamin, 1000 kg of coenzyme B_{12}, and a small amount of mecobalamin, the remaining part being used in animal feed. Apparently there is at present no prospect for any marked expansion of this market.

Some of the main pharmaceutical companies involved in the field are Chinoin (Hungary), Farmitalia (Italy), Glaxo (United Kingdom), Merck (United States), Rhône-Poulenc (France), Richter (Hungary), and Roussel (France).

The bulk selling price of cyanocobalamin dropped from $500 to $15 per gm between 1951 and 1963. Since 1974 the price has fluctuated between $6 and $1.80 per gm.

REFERENCES

Abeles, R. H., and Dolphin, D. (1976). *Acc. Chem. Res.* **9,** 114–120.
Aries, R. (1974). French Patent 2,209,842.
Aschaffenburger Zellstoffwerke A.G. (1966). French Patent 1,426,051.
Bacher, F. A., Boley, A. E., and Shonk, C. E. (1954). *Anal. Chem.* **26,** 1146–1149.
Barker, H. A. (1968). *In* "The Vitamins" (W. H. Sebrell, Jr. and R. S. Harris, eds.), 2nd ed., Vol. 2, pp. 184–212. Academic Press, New York.
Barker, H. A. (1972). *Annu. Rev. Biochem.* **41,** 55–90.
Barker, H. A., Smyth, R. D., and Hogenkamp, H. P. (1963). *Biochem. Prep.* **10,** 27–35.
Battersby, A. R., and McDonald, E. (1976). *Philos. Trans. R. Soc. London, Ser. B* **273,** 161–180.
Battersby, A. R., Hollenstein, R., McDonald, E., and Williams, D. C. (1976). *J. Chem. Soc., Chem. Commun.* **14,** 543–544.
Battersby, A. R., Ihara, M., McDonald, E., Redfern, J. R., and Golding, B. T. (1977a). *J. Chem. Soc., Perkin Trans. 1,* **2,** 158–166.
Battersby, A. R., McDonald, E., Hollenstein, R., Ihara, M., Satoh, F., and Williams, D. C. (1977b). *J. Chem. Soc., Perkin Trans. 1,* **2,** 166–178.
Beskhov, M. N., Karova, E. A., and Murgov, I. D. (1969). *Nauch. Tr., Vissh Inst. Khranit. Vkusova Prom-st., Plovdiv.* **16,** 337–351; *Chem. Abstr.* **77,** 59987 (1972).
Beskhov, M. N., Keskinova, D., Khristozova, T., and Ploshchakova, M. (1973). *Nauch. Tr., Vissh Inst. Khranit. Vkusova Prom-st., Plovdiv.* **20,** 211–216; *Chem. Abstr.* **83,** 129945 (1975).

Bilkus, D. I., and Mervyn, L. (1971). *In* "Cobalamins" (H. R. V. Arnstein, ed.), pp. 17–21. Churchill, London.
Blaylock, B. A., and Stadtman, T. C. (1966). *Arch. Biochem. Biophys.* **116,** 138–152.
Brady, R. O., Castanera, E. G., and Barker, H. A. (1962). *J. Biol. Chem.* **237,** 2325–2332.
Brazenas, G., and Kanopkaite, S. (1970a). *Liet. TSR Mokslu Akad. Darb. Ser. C.* **3,** 157–164; *Chem. Abstr.* **75,** 127033 (1971).
Brazenas, G., and Kanopkaite, S. (1970b). *Liet. TSR Mokslu Akad. Darb. Ser. C.* **3,** 165–173; *Chem. Abstr.* **75,** 127034 (1971).
Bykhovsky, V. Y. (1969). Russian Patent 227,528.
Bykhovsky, V. Y., and Zaitseva, N. I. (1976). *Prikl. Biokhim. Mikrobiol.* **12,** 365–371.
Bykhovsky, V. Y., Zaitseva, N. I., and Bukin, V. N. (1969). *Dokl. Akad. Nauk SSSR* **185,** 459–461.
Bykhovsky, V. Y., Zaitseva, N. I., and Yavorskaya, A. N. (1976). *Prikl. Biokhim. Mikrobiol.* **12,** 491–494.
Bykhovsky, V. Y., Pantskhava, E. S., Zaitseva, N. I., Pcjelkina, V. V., Fedorov, S. A., Golubovsky, S. A., and Yarovenko, M. L. (1972). Russian Patent 359,264.
Celletti, P., Moretti, G. P., and Petrangeli, B. (1976). *Farmaco, Ed. Prat.* **31,** 413–419.
Chinoin-Gyogyszer (1965). Hungarian Patent 152,435.
Chinoin-Gyogyszer (1971). French Patent 2,062,367.
Daniels, H. J. (1966). *Can. J. Microbiol.* **12,** 1095–1098.
Daniels, H. J. (1970). *Can. J. Microbiol.* **16,** 809–815.
Danilova, N. P., and Bankina, L. P. (1974). Russian Patent 454,250.
Demain, A. L., and White, R. F. (1971). *J. Bacteriol.* **107,** 456–460.
Demain, A. L., Daniels, H. J., Schnable, L., and White, R. F. (1968). *Nature (London)* **220,** 1324–1325.
Duric, N. (1969). *Process Biochem.* **4**(12), 35–50.
Farmaceutici Italia (1971a). French Patent 2,078,641.
Farmaceutici Italia (1971b). French Patent 2,083,992.
Fenton, W. A., Ambani, L. M., and Rosenberg, L. E. (1976). *J. Biol. Chem.* **251,** 6616–6623.
Ford, J. E. (1959). *J. Gen. Microbiol.* **21,** 693–701.
Ford, S. H., and Friedmann, H. C. (1976). *Arch. Biochem. Biophys.* **175,** 121–130.
Friedmann, H. C., and Cagen, L. M. (1970). *Annu. Rev. Microbiol.* **24,** 159–208.
Fujii, K., and Fukui, S. (1970). *Eur. J. Biochem.* **17,** 552–560.
Fujii, K., Shimizu, S., and Fukui, S. (1966). *Hakko Kogaku Zasshi.* **44,** 185–191; *Chem. Abstr.* **68,** 28441 (1968).
Fukui, S., Tamao, Y., Kato, T., Takahashi, T., and Shimizu, S. (1964). *J. Vitaminol.* **10,** 202–210.
Fukui, S., Shimizu, S., and Fujii, K. (1967). *Hakko Kogaku Zasshi.* **45,** 530–540; *Chem. Abstr.* **69,** 18021 (1968).
Georgopapadakou, N. H., Petrillo, J., Scott, A. I., and Low, B. (1976). *Genet. Res.* **28,** 93–100.
Giec, A., Weker, H., and Skupin, J. (1976). *Bull. Acad. Pol. Sci., Ser. Sci. Biol.* **24,** 497–503.
Glaxo Group, Ltd. (1964). British Patent 963,373.
Goldstein, S., and Duca, A. (1976). *J. Pharm. Sci.* **65,** 1831–1833.
Goldstein, S., Ciupitoiu, A., Vasilescu, V., and Duca, A. (1974). *Can. J. Pharm. Sci.* **9,** 96–98.
Gruzina, V. D., Erokhina, L. I., and Ponomareva, G. M. (1974). *Genetika* **10,** 121–127.
Hercules Inc. (1968). U.S. Patent 3,411,991.
Hermann, R., and Mueller, O. (1976). *Hoppe-Seyler's Z. Physiol. Chem.* **357,** 1695–1698.

Hodgkin, D. C., Pickworth, J., Robertson, J. H., Trueblood, K. N., and Prosen, R. J. (1955). *Nature (London)* **176,** 325–328.
Hoffmann-La Roche (1963). Belgian Patent 631,589.
Iashi (1973). German Patent 2,210,973.
Ilieva, M. (1971). *Nauch. Tr., Vissh Inst. Khranit Vkusova Prom-st., Plovdiv.* **18,** 371–377; *Chem. Abstr.* **79,** 40871 (1973).
Ilieva, M., and Popova, Y. (1974). *Nauch. Tr., Vissh Inst. Khranit Vkusova Prom-st., Plovdiv.* **21,** 105–112; *Chem. Abstr.* **85,** 141264 (1976).
Imfeld, M., Townsend, C. A., and Arigoni, D. (1976). *J. Chem. Soc., Chem. Commun.* **14,** 541–542.
Janicka, I., Maliszewska, M., and Pedziwilk, F. (1976). *Acta Microbiol. Pol.* **25,** 205–210.
Kaczka, E. A., Wolf, D. E., and Folkers, K. (1956). U.S. Patent 2,738,302 (to Merck and Co., Inc.).
Kageyama, M., Burg, K. A., and Perlman, D. (1977). *J. Antibiot.* **30,** 283–288.
Kaken Kagaku, K.K. (1965). Japanese Patent 65/4,440.
Kamikubo, T. (1967). Japanese Patent 67/14,802.
Kamikubo, T., Takeda, Y., Hayashi, M., and Friedrich, W. (1972). *Agric. Biol. Chem.* **36,** 164–165.
Kamikubo, T., Nishio, N., and Yano, T. (1975). Japanese Patent 75/132,186 (to Kyowa Hakko Kogyo Co., Ltd.).
Kamikubo, T., Hayashi, M., and Nishio, N. (1976). *Abstr., Int. Ferment. Symp., 5th, 1976,* No. 21-15, p. 401.
Kanegafuchi Chem. Ind. Co., Ltd. (1972). Japanese Patent 72/15,748.
Kaplan, L., and Birnbaum, J. (1973). *Abstr. Annu. Meet. Am. Soc. Microbiol.* p.8 (E 45).
Kato, T., and Shimizu, S. (1962). *Vitamins* **26,** 470–472.
Koike, I., and Hattori, A. (1975). *J. Gen. Microbiol.* **88,** 1–10.
Kojima, I., Sato, H., Maruhashi, K., and Fujiwara, Y. (1976). *Abstr., Int. Ferment. Symp., 5th, 1976,* No. 7-04, p. 134.
Koppenhagen, V. B., and Pfiffner, J. J. (1970). *J. Biol. Chem.* **245,** 5865–5873.
Koppenhagen, V. B., and Pfiffner, J. J. (1971). *Fed. Proc., Fed. Am. Soc. Exp. Biol.* **30,** 1088 (Abstr. 209).
Koppenhagen, V. B., Elsenhans, B., Wagner, F., and Pfiffner, J. J. (1974). *J. Biol. Chem.* **249,** 6532–6540.
Kotala, L., and Luba, J. (1965). *Acta Pol. Pharm.* **22,** 419–422.
Krieger, J. H. (1973). *Chem. & Eng. News* **51** (11), 16–19, 25, 27, and 29.
Kuratomi, K., Poston, J. M., and Stadtman, E. R. (1966). *Biochem. Biophys. Res. Commun.* **23,** 691–695.
Kyowa Hakko Kogyo Co., Ltd. (1970). Japanese Patent 70/36,159.
Kyowa Hakko Kogyo Co., Ltd. (1973). Japanese Patent 73/38,880.
Lago, B. D., and Demain, A. L. (1969). *J. Bacteriol.* **99,** 347–349.
Lindstrand, K. (1964). *Nature (London)* **204,** 188–189.
Lindstrand, K. (1965). *Acta Chem. Scand.* **19,** 1762–1763.
Ljungdahl, L. G., and Wood, H. G. (1969). *Annu. Rev. Microbiol.* **23,** 515–538.
Long, R. A. (1962). U.S. Patent 3,018,225 (to Merck and Co., Inc.).
Lorenzi, E., and Oddono, F. (1972). *Farmaco, Ed. Prat.* **27,** 280–297.
Lotti, B. (1975). *Boll. Chim. Farm.* **114,** 416–420.
Lowe, D. A., and Turner, J. M. (1970). *J. Gen. Microbiol.* **64,** 119–122.
McDaniel, L. E. (1961). U.S. Patent 3,000,793 (to Merck and Co., Inc.).
McDonald, C. M., Farquharson, J., Bessent, R. G., and Adams, J. F. (1977). *Clin. Sci. Mol. Med.* **52,** 215–218.

Matsumoto, J., Takuma, J., Takahashi, K., and Murakami, M. (1974). *Yamanouchi Seiyaku Kenkyu Hokoku.* **2,** 34–39; *Chem. Abstr.* **84,** 65304 (1976).
Maugh, T. H. (1973). *Science* **179,** 266–267.
Merck and Co., Inc. (1964). U.S. Patent 3,163,637.
Merck and Co., Inc. (1971). French Patent 2,038,828.
Mervyn, L., and Smith, E. L. (1964). *Prog. Ind. Microbiol.* **5,** 151–201.
Mil'ko, E. S., and Rabotnova, I. L. (1967). *Prikl. Biokhim. Mikrobiol.* **3,** 277–282.
Miller, I. M., and Rosenblum, C. (1960). U.S. Patent 2,939,822 (to Merck and Co., Inc.).
Miyazaki, A., Hayashi, M., and Kamikubo, T. (1972). *Vitamins* **45,** 131–135.
Morikawa, H., and Kamikubo, T. (1969). *Hakko Kogaku Zasshi.* **47,** 470–477; *Chem. Abstr.* **73,** 64967 (1970).
Morikawa, H., Hayashi, M., and Kamikubo, T. (1970). *Asahi Garasu Kogyo Gijutsu Shoreikai Kenkyu Hokoku.* **16,** 95–102; *Chem. Abstr.* **75,** 18560 (1971).
Mueller, G., Dieterle, W., and Siebke, G. (1970). *Z. Naturforsch., Teil B* **75,** 307–309.
Mueller, G., Gross, R., and Siebke, G. (1971). *Hoppe-Seyler's Z. Physiol. Chem.* **352,** 1720–1722.
Nakao, Y., Hisano, K., Kanemaru, T., and Yamano, T. (1974). Japanese Patent 74/15,796 (to Takeda Chem. Ind., Ltd.).
Nishio, N., Yano, T., and Kamikubo, T. (1975). *Agric. Biol. Chem.* **39,** 21–27.
Nishio, N., Yano, T., Hayashi, M., and Kamikubo, T. (1976a). *Agric. Biol. Chem.* **40,** 1035–1037.
Nishio, N., Ueda, M., Omae, Y., Hayashi, M., and Kamikubo, T. (1976b). *Agric. Biol. Chem.* **40,** 2037–2043.
Noyes, R. (1969). "Chemical Proçess Review," No. 40. Noyes Dev. Corp., Park Ridge, New Jersey.
Olin Mathieson Chem. Corp. (1965). U.S. Patent 3,164,582.
Pantskhava, E. S., and Bykhovsky, V. Y. (1966). *Prod. Mikrobnogo Sin., Akad. Nauk Latv. SSR, Inst. Mikrobiol.* 81–94; Chem. Abstr. **67,** 10335 (1967).
Pantskhava, E. S., and Pcjelkina, V. V. (1970). Russian Patent 265,043.
Pantskhava, E. S., Pcjelkina, V. V., and Bukin, V. N. (1973). *Biokhimiya* **38,** 507–514.
Pawelkiewicz, J. (1971). *In* "Methods in Enzymology" (S. P. Colowick and N. O. Kaplan, eds.), Vol. 18, Part C, pp. 99–101. Academic Press, New York.
Pcjelkina, V. V., Pantskhava, E. S., and Bukin, V. N. (1970). *Biokhimiya* **35,** 1007–1013.
Perlman, D. (1959). *Adv. Appl. Microbiol.* **1,** 87–122.
Perlman, D. (1967). *In* "Microbial Technology" (H. J. Peppler, ed.), pp. 283–287. Van Nostrand-Reinhold, Princeton, New Jersey.
Perlman, D., and Toohey, J. I. (1966). *Nature (London)* **212,** 300–301.
Perlman, D., Vlietinck, A. J., Matthews, H. W., and Lo, F. F. (1974). *J. Antibiot.* **27,** 826–832.
Pierrel S.p.A. (1963a). French Patent 1,325,304.
Pierrel S.p.A. (1963b). French Patent 1,325,308.
Pierrel S.p.A. (1965a). British Patent 1,007,971.
Pierrel S.p.A. (1965b). British Patent 1,007,972.
Pliva T.F.K.P. (1964). Belgian Patent 651,153.
Popova, Y., and Bakchevanska, S. (1972). *Nauch. Tr., Vissh Inst. Khranit. Vkusova Prom-st., Plovdiv.* **19,** 237–244; *Chem. Abstr.* **83,** 145681 (1975).
Popova, Y., and Ilieva, M. (1970). *Nauch. Tr., Vissh Inst. Khranit. Vkusova Prom-st., Plovdiv.* **17,** 151–159; *Chem. Abstr.* **79,** 30449 (1973).
Prescott, S. C., and Dunn, C. G. (1959). *In* "Industrial Microbiology," 3rd ed., pp. 482–496. McGraw-Hill, New York.
Renz, P., and Bauer-David, A. J. (1972). *Z. Naturforsch., Teil B* **27,** 539–544.

Richter Gedeon V.G. (1963). Hungarian Patent 151,065.
Richter Gedeon V.G. (1967). French Patent 1,503,402.
Richter Gedeon V.G. (1975a). Belgian Patent 821,535.
Richter Gedeon V.G. (1975b). Belgian Patent 821,536.
Rickes, E. L., Brink, N. G., Koniuszy, F. R., Wood, T. R., and Folkers, K. (1948). *Science* **107,** 396–397.
Robinson, F. M. (1955). *Encycl. Chem. Technol.* **14,** 813–828.
Rohm and Haas Co. (1968). Belgian Patent 710,151.
Rosenthal, H. L. (1968). *In* "The Vitamins" (W. H. Sebrell, Jr. and R. S. Harris, eds.), 2nd ed., Vol. 2, pp. 145–170. Academic Press, New York.
Russel, C. S. (1974). *J. Theor. Biol.* **47,** 145–151.
Sato, K., Shimizu, S., and Fukui, S. (1968). *Agric. Biol. Chem.* **32,** 1–11.
Sato, K., Ohmori, H., Shimizu, S., and Fukui, S. (1971). *Agric. Biol. Chem.* **35,** 333–350.
Sato, K., Ueda, S., and Shimizu, S. (1977). *Appl. Environ. Microbiol.* **33,** 515–521.
Scott, A. I. (1975). *Ann. N.Y. Acad. Sci.* **244,** 356–370.
Scott, A. I. (1976). *Philos. Trans. R. Soc. London, Ser. B* **273,** 303–318.
Scott, A. I., Kajiwara, M., Takahashi, T., Armitage, I. M., Demou, P., and Petrocine, D. (1976). *J. Chem. Soc., Chem. Commun.* **16,** 544–546.
Seisan Kaihatsu Kagaku Kenkyusho (1970). Japanese Patent 70/13,434.
Sheehan, T. L., and Hercules, D. M. (1977). *Anal. Chem.* **49,** 446–450.
Shimizu, S., and Sato, K. (1976). *Abstr., Int. Ferment. Symp., 5th, 1976,* No. 21-16, p. 402.
Sifa (1964). French Patent 1,368,892.
Sifa (1965). French Patent 1,422,040.
Sifa (1967). French Patent Addition 90,362 (addition to French Patent 1,422,040).
Simon, A. (1968). *Zentralbl. Bakteriol., Parasitenkd., Infektionskr. Hyg., Abt. 2* **122,** 143–154.
Skupin, J., Pędziwilk, F., and Jaszewski, B. (1970). *Bull. Acad. Pol. Sci., Ser. Sci. Biol.* **18,** 511–515.
Smith, E. L. (1948). *Nature (London)* **161,** 638–639.
Smith, E. L. (1965). U.S. Patent 3,167,539 (to Glaxo Lab., Ltd.).
Speedie, J. D., and Hull, G. W. (1960). U.S. Patent 2,951,017 (to Distillers Co., Ltd.).
Stefanjak, O., Lubienska, B., and Janicka, M. (1968). *Acta Microbiol. Pol.* **17,** 83–90.
Szemler, L. L., and Székely, A. D. (1969). *Process Biochem.* **4** (12), 25–27.
Terada, O., Ohishi, K., and Kinoshita, S. (1959). *Vitamins* **18,** 16–33.
Toohey, J. I. (1971). *In* "Methods in Enzymology" (S. P. Colowick and N. O. Kaplan, eds.), Vol. 18, Part C, pp. 71–75. Academic Press, New York.
Toohey, J. I. (1974). U.S. Patent 3,846,237.
Toraya, T., Yongsmith, B., Tanaka, A., and Fukui, S. (1975). *Appl. Microbiol.* **30,** 477–479.
Tortolani, G., and Ferri, P. G. (1974). *J. Chromatogr.* **88,** 430–433.
Tortolani, G., and Mantovani, V. (1974). *J. Chromatogr.* **92,** 201–206.
Tortolani, G., Bianchini, P., and Mantovani, V. (1970a). *J. Chromatogr.* **53,** 577–579.
Tortolani, G., Bianchini, P., and Mantovani, V. (1970b). *Farmaco, Ed. Prat.* **25,** 772–775.
Uclaf (1960). French Patent 1,264,016.
Uehisa, Y. (1972). Japanese Patent 72/50,397.
Vitols, E., Walker, G. A., and Huennekens, F. M. (1966). *J. Biol. Chem.* **241,** 1455–1461.
Vogelmann, H., and Wagner, F. (1973). *J. Chromatogr.* **76,** 359–379.
Vorobeva, L. I., Konovalova, L. V., and Lisenkova, L. L. (1967). *Prikl. Biokhim. Mikrobiol.* **3,** 264–269.
Wagman, G. H., Gannon, R. D., and Weinstein, M. J. (1969). *Appl. Microbiol.* **17,** 648–649.
Wagner, F. (1966). *Annu. Rev. Biochem.* **35,** 405–434.

Weissbach, H., Redfield, B., and Peterkofsky, A. (1961). *J. Biol. Chem.* **236,** PC40 (Prelim. Commun.).
White, R. F., and Demain, A. L. (1971). *Biochim. Biophys. Acta* **237,** 112–119.
White, R. F., Kaplan, L., and Birnbaum, J. (1973). *J. Bacteriol.* **113,** 218–223.
Whitlock, L. L., Melton, J. R., and Billings, T. J. (1976). *J. Assoc. Off. Anal. Chem.* **59,** 580–581.
Wipple, G. H., and Robscheit-Robbins, F. S. (1925). *Am. J. Physiol.* **72,** 395–407.
Zodrow, K., Stefaniak, O., Chelkowski, J., and Szczepska, K. (1963). *Acta Microbiol. Pol.* **12,** 263–266.
Zodrow, K., Stefaniak, O., Lubienska, B., and Janicka, M. (1968). *Acta Microbiol. Pol.* **17,** 91–94.

Chapter 16

Microbial Process for Riboflavin Production

D. PERLMAN

I. INTRODUCTION

Microbiologically produced riboflavin has long been available in yeast and related preparations in association with many other vitamins of the B-complex. It is a unique vitamin in that it can be produced *de novo* to a high concentration rather rapidly by certain microorganisms. Fermentation processes have been described in which the riboflavin content of the fermented medium amounted to more than 7 gm/liter. The organisms involved in these processes are ascomycetes, namely, *Eremothecium ashbyii* and *Ashbya gossypii*.

Although riboflavin is produced by many microorganisms, including bacteria, yeasts, and molds, other microorganisms require riboflavin for growth. The organisms which have been found to produce sufficient riboflavin to be of interest from a commercial standpoint are listed in Table I, where the yields are mentioned along with the sensitivity of the microorganism to iron.

Commercial fermentation processes for production of riboflavin or riboflavin concentrates are relatively recent, having been developed in the past 40 years. Aside from food yeasts, the first organisms employed

MICROBIAL TECHNOLOGY, 2nd ed., VOL. I
 ISBN 0-12-551501-4

TABLE I. Microorganisms Producing Considerable Amounts of Riboflavin and the Effects of Iron on Biosynthesis of the Vitamin

Microorganism	Riboflavin yield (mg/liter)	Optimum iron conc. (mg/liter)	References
Clostridium acetobutylicum	97	1–3	Meade *et al.* (1947)
Mycobacterium smegmatis	58	Not critical	Mayer and Rodbart (1946)
Mycocandida riboflavina	200	Not critical	McClary (1951)
Candida flareri	567	0.04–0.06	Levine *et al.* (1949)
Eremothecium ashbyii	2480	Not critical	Moss and Klein (1949)
Ashbya gossypii	6420	Not critical	Szczeniak *et al.* (1973)

primarily for riboflavin production was *Clostridium acetobutylicum* which, when grown in grain mashes or on whey low in iron, yielded residues containing 4–5 mg riboflavin per gram fermentation solids. These fermentations were succeeded in about 1940 by a process using *E. ashbyii* with yields of about 2 mg/ml. In 1946 processes using the ascomycete *A. gossypii* were started. In 1978 the only manufacturer using the microbiological process is Merck & Co., Inc. (United States). Companies manufacturing riboflavin by chemical synthesis include Hoffmann-La Roche Inc. (United States and Switzerland), Takeda Chemical Industries Ltd. (Japan), Pfizer, Inc (U.S.A.), and E. Merck (West Germany).

II. THE FERMENTATION PROCESS

Development of the current processes used for riboflavin production by *A. gossypii* NRRL Y-1056 focused on three aspects: (1) preparation of the culture medium; (2) selection of mutant cultures and optimization of inoculum preparation; and (3) optimization of the fermentation conditions, e.g., incubation temperature, aeration level, and fermentor design. The most important increases in fermentation productivity have resulted from the changes in the culture media and the selection of high riboflavin-producing mutant cultures.

Initial studies (Tanner *et al.*, 1949) showed that good growth of *A. gossypii* occurred in a medium containing glucose, corn steep liquor, and animal stick liquor, tankage, or meat scraps. It was also shown that a sterilization time of less than 30 minutes, a low concentration of "young inoculum," e.g., 2% v/v, and efficient aeration were important. Maximum yields of 700 mg/liter (10 days) were obtained when the medium was autoclaved for 15–30 minutes (as compared to 325 mg/liter when it was autoclaved for 90 minutes). Pilot-plant reproduction of these conditions resulted in yields of 500–850 mg/liter.

Further improvements resulted when enzymatically degraded collagen and lipids were used as energy sources together with accessory factors present in corn steep liquor. Distillers' solubles, and brewers' yeast were included in the medium with yields of the order of 4200 mg/liter. The lipid used as an energy source was proposed earlier for the *E. ashbyii* fermentation (Rudert, 1945; Phelps, 1949) and also worked well for the *A. gossypii* fermentation. Typical data on stimulation of riboflavin production as a result of including in the media various peptones are summarized in Table II, as is the effect of supplementation with glycine.

Following earlier studies, Pfeifer *et al.* (1950) suggested that some improvement in productivity was possible as a result of culture selection. Pridham and Raper (1952) reported on methods they used to obtain mutants yielding 200–300 mg of riboflavin per liter more than the parent. Since only 700 mutants were tested, the chance of obtaining an improved mutant was slim. In addition, they suggested several methods of handling the cultures, including: (1) incorporating a reducing agent,

TABLE II. Effect of Supplementing Growth Medium with Peptones and Glycine on Riboflavin Production by *Ashbya gossypii*

Nature of supplement	Riboflavin yield (mg/liter)
A. Peptones[a]	
Peptic hydrolysate of animal tissue	1520
Equal amounts of a peptic hydrolysate of animal tissue and pancreatic digest of casein	1280
Pancreatic digest of lactalbumin	1000
Pancreatic digests of casein	340
Pancreatic digest of gelatin	3620
Papaic digest of soybean meal	673
B. Glycine[b] (added in gm/liter)[c]	
0	3280
1	3640
2	3980
3	4200

[a] From Malzahn *et al.* (1959).
[b] From Malzahn *et al.* (1963).
[c] The basal medium contained (w/v): corn steep liquor solid, 2.25%; Wilson's peptone W-809, 3.5%; soybean oil, 4.5%.

e.g., sodium dithionate, into the stock culture media to decrease the production of low-yielding mutants; (2) reisolating superior substrains frequently; and (3) preserving proved strains by immediate lyophilization or storage under mineral oil. Malzahn *et al.* (1963) recommended a procedure in which the cultures were transferred at 4-day intervals on malt–yeast extract–glucose–agar slants. Cultures were frozen after 4 days' growth to maintain their flavinogenic capacity. They also noted that productivity was improved in mutants obtained by treatment of the cells with ultraviolet radiation, uranyl nitrate, nitrogen mustard, or ethyleneamine, with selection based on increased pigmentation of colonies when grown on the malt–yeast extract–agar medium. One isolate produced 3620 mg riboflavin per liter when soybean oil was the energy source, compared with 1420 mg/liter for the parent grown in the same medium.

The optimal fermentation conditions include aeration at about one-third volume of air per volume of liquid per minute, and agitation with three impellers at a rate of 1.0 hp/1000 liters of medium. When foaming was excessive, it was controlled by initial addition of emulsified silicone antifoam, and later by addition of soybean oil (which also acted as a nutrient). Sterilization of the medium was accomplished by heating at 121°C for 3 hours (which improved productivity), and the incubation temperature was maintained at 28°C for the 7-day incubation period (Malzahn *et al.*, 1963).

III. RECOVERY OF RIBOFLAVIN FROM FERMENTED MEDIA

Upon completion of the fermentation, the solids were dried to a crude product for animal-feed supplementation or processed to a United States Pharmacopoeia-grade product. In either case, the pH value of the fermented medium was adjusted to pH 4.5. For the feed-grade product, the broth was concentrated to about 30% solids and dried on double-drum driers.

When a crystalline product was required, the fermented broth was heated for 1 hour at 121°C to solubilize the riboflavin. Insoluble matter was removed by centrifugation, and riboflavin recovered by conversion to the less soluble form (Michaels *et al.*, 1936). Both chemical (Hines, 1945a) and microbiological (Hines, 1945b) methods of conversion have been used. The precipitated riboflavin was then dissolved in water or polar solvents (Dale, 1947) or an alkaline solution (Morehouse, 1958), oxidized by aeration, and recovered by recrystallization from the aqueous or polar solvent solution or by acidification of the alkaline solution.

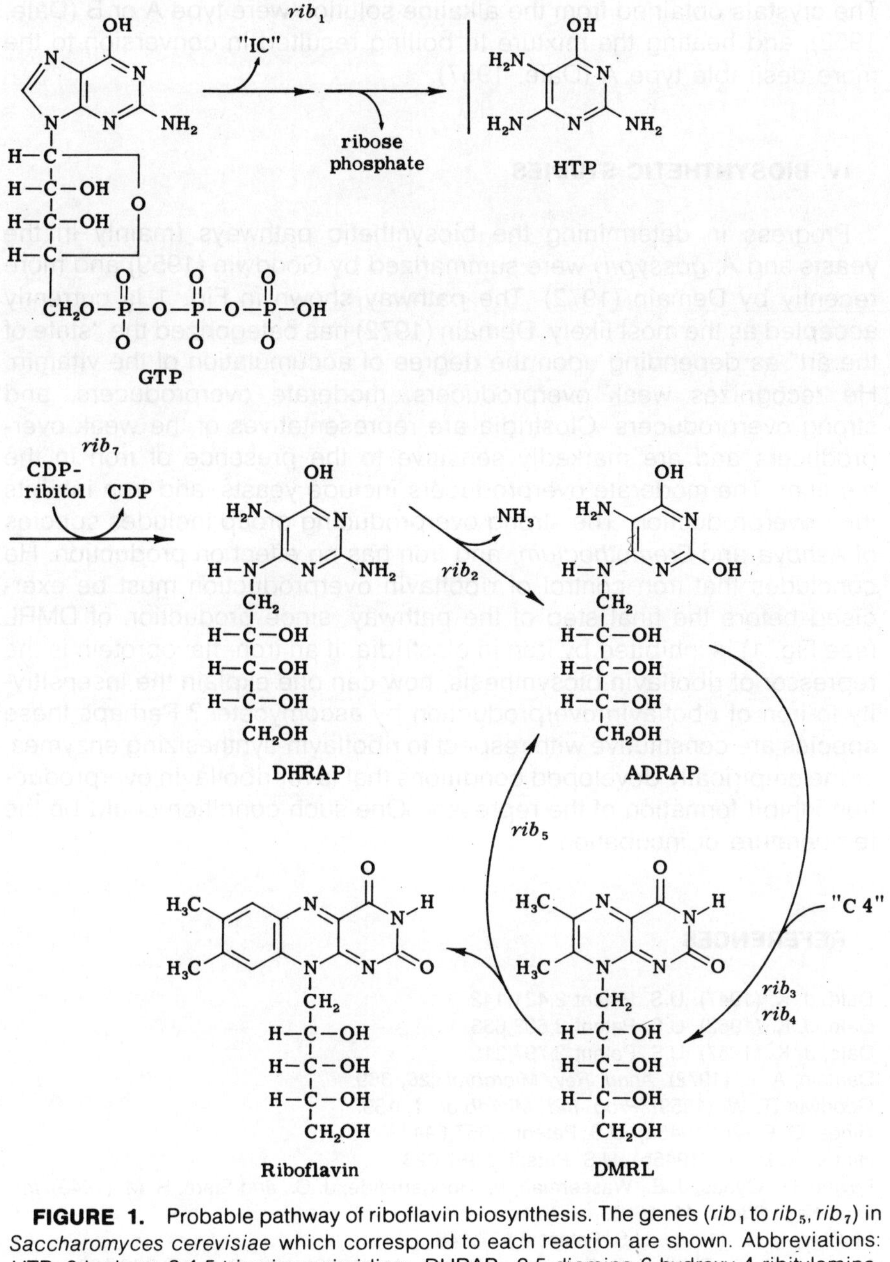

FIGURE 1. Probable pathway of riboflavin biosynthesis. The genes (*rib*$_1$ to *rib*$_5$, *rib*$_7$) in *Saccharomyces cerevisiae* which correspond to each reaction are shown. Abbreviations: HTP 6-hydroxy-2,4,5-triaminopyrimidine; DHRAP, 2,5-diamino-6-hydroxy-4-ribitylaminopyrimidine; ADRAP 5-amino-2,6-dihydroxy-4-ribitylaminopyrimidine; DMRL, 6,7-dimethyl-8-ribityllumizine.

The crystals obtained from the alkaline solution were type A or B (Dale, 1952), and heating the mixture to boiling resulted in conversion to the more desirable type A (Dale, 1957).

IV. BIOSYNTHETIC STUDIES

Progress in determining the biosynthetic pathways (mainly in the yeasts and *A. gossypii*) were summarized by Goodwin (1959) and more recently by Demain (1972). The pathway shown in Fig. 1 is currently accepted as the most likely. Demain (1972) has categorized the "state of the art" as depending upon the degree of accumulation of the vitamin. He recognizes weak overproducers, moderate overproducers, and strong overproducers. Clostridia are representatives of the weak overproducers and are markedly sensitive to the presence of iron in the medium. The moderate overproducers include yeasts, and iron inhibits their overproduction. The strong overproducing group includes species of *Ashbya* and *Eremothecium,* and iron has no effect on production. He concludes that iron control of riboflavin overproduction must be exercised before the final step of the pathway, since production of DMRL (see Fig. 1) is inhibited by iron in clostridia. If an iron–flavoprotein is the repressor of riboflavin biosynthesis, how can one explain the insensitivity to iron of riboflavin overproduction by ascomycetes? Perhaps these species are constitutive with respect to riboflavin-synthesizing enzymes, or the empirically developed conditions that favor riboflavin overproduction inhibit formation of the repressor. One such condition could be the temperature of incubation.

REFERENCES

Dale, J. K. (1947). U.S. Patent 2,421,142.
Dale, J. K. (1952). U.S. Patent 2,603,633.
Dale, J. K. (1957). U.S. Patent 2,797,215.
Demain, A. L. (1972). *Annu. Rev. Microbiol.* **26,** 369.
Goodwin, T. W. (1959). *Prog. Ind. Microbiol.* **1,** 139.
Hines, G. E., Jr. (1945a). U.S. Patent 2,367,644.
Hines, G. E., Jr. (1945b). U.S. Patent 2,387,023.
Levine, H., Oyaas, J. E., Wasserman, L., Hoogerheide, J. C., and Stern, R. M. (1949). *Ind. Eng. Chem.* **41,** 1665.
McClary, J. E. (1951). U.S. Patent 2,537,148.
Malzahn, R. C., Phillips, R. F., and Hanson, A. M. (1959). U.S. Patent 2,876,169.
Malzahn, R. C., Phillips, R. F., and Hanson, A. M. (1963). *Abstr. 63rd Annu. Meet. Am. Soc. Microbiol.* p. 21.

Mayer, R. L., and Rodbart, R. (1946). *Arch. Biochem.* **11,** 49.
Meade, R. E., Pollard, H. L., and Rodgers, N. E. (1947). U.S. Patent 2,433,680.
Michaels, L., Schubert, M., and Symthe, C. V. (1936). *J. Biol. Chem.* **116,** 587.
Morehouse, A. L. (1958). U.S. Patent 2,822,361.
Moss, A. R., and Klein, R. (1949). British Patent 615,847.
Pfeifer, V. F., Tanner, F. W., Jr., Vojnovich, C., and Traufler, D. H. (1950). *Ind. Eng. Chem.* **42,** 1776.
Phelps, A. S. (1949). U.S. Patent 2,473,818.
Pridham, T. G., and Raper, K. B. (1952). *Mycologia* **44,** 452.
Rudert, F. J. (1945). U.S. Patent 2,374,503.
Szczeniak, T., Karabin, L., and Wituch, K. (1973). Polish Patent 66,611.
Tanner, F. W., Jr., Vojnovich, C., and Van Lanen, J. M. (1949). *J. Bacteriol.* **58,** 737.

Chapter 17

Carotenoids

L. NINET
J. RENAUT

I. INTRODUCTION

Carotenoids form one of the most important families of natural pigments. They are liposoluble tetraterpenes, usually yellow to red in color, which are formed by the condensation of isoprenyl units. The first carotenoids were discovered around 1800 and now more than 400 different compounds are known (Isler, 1977), the majority of which consist of hydrocarbons (carotenes) or their oxygenated derivatives (xanthophylls), both containing 40 carbon atoms. The chemical structures of the main carotenoids related to fermentation processes are shown in Fig. 1.

Some of the carotenoids are normally present in food and these have essential biological functions, particularly as precursors of vitamin A. In view of these biological activities, they are used as food supplements to

MICROBIAL TECHNOLOGY, 2nd ed., VOL. I
 ISBN 0-12-551501-4

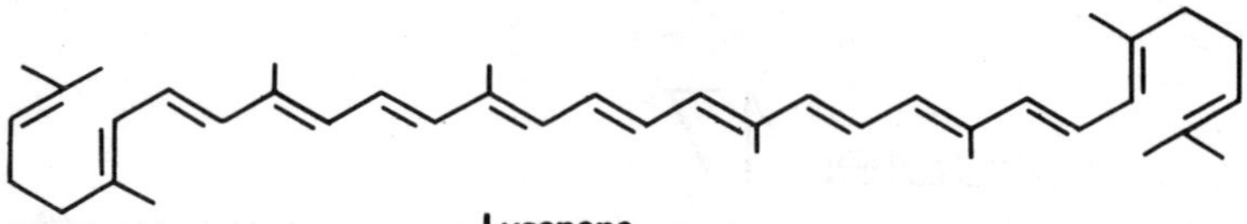

Lycopene

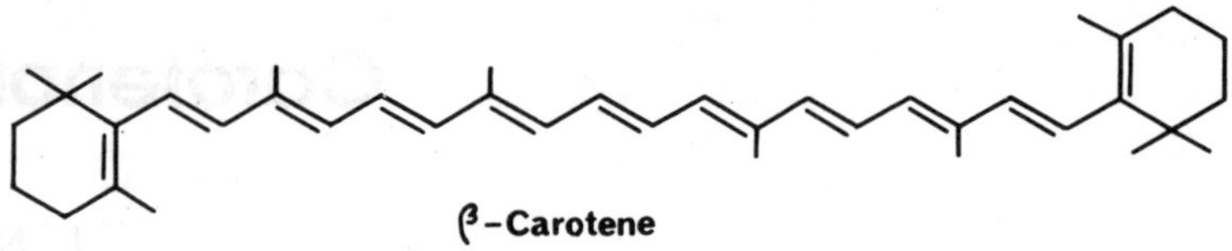

β-Carotene

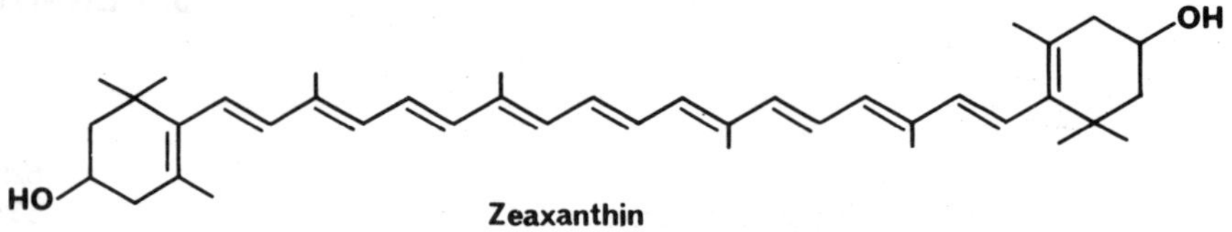

Zeaxanthin

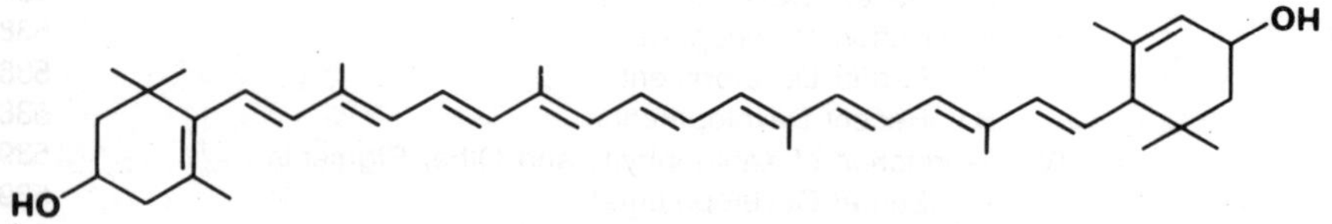

Lutein

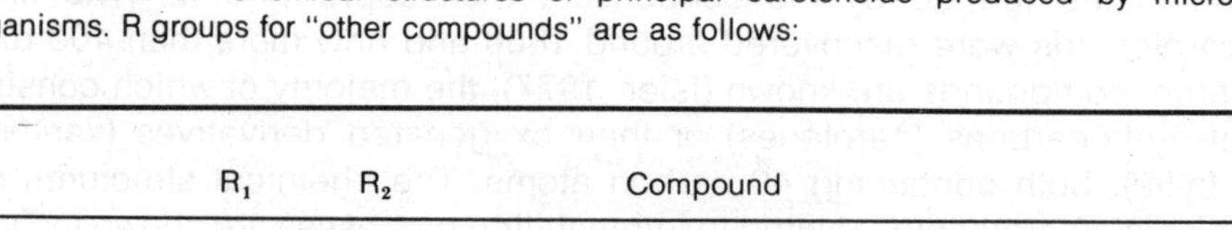

Other compounds

FIGURE 1. Chemical structures of principal carotenoids produced by microorganisms. R groups for "other compounds" are as follows:

R_1	R_2	Compound
H	H	Isorenieratene (leprotene)
OH	H	3-Hydroxyisorenieratene
OH	OH	3,3′-Dihydroxyisorenieratene

prevent or cure vitamin deficiencies. In addition other pigmented carotenoids are used both as food additives for intensifying or modifying the color in fat, oil, cheese, and beverages and also as animal feed supplement to enhance the color of such foods as egg yolks and chicken flesh. It is for these latter uses, as food and feed additives, that pure products derived from organic processes, rather than from plant sources, have played an ever-increasing role. However, since 1956, attempts have also been made to produce by fermentation either pure carotenoids as β-carotene, lycopene, and xanthophylls or crude colored biomass which may be added directly to the animal feed.

While the carotenoids are widely found in plants and animals, only microorganisms and plants have the necessary systems to synthesize a wide range of these substances. All of them arise from minor variations in the final steps of a common biosynthetic pathway. This pathway is summarized in Fig. 2 (Goodwin, 1972; Beytia and Porter, 1976). In practice β-carotene and related carotenes are mainly found in fungi and algae, while the xanthophylls occur almost exclusively in bacteria and algae (Ciegler, 1965a; Goodwin, 1972; Weedon, 1971). However the rates of synthesis vary considerably in different microorganisms so that the number of such organisms which can be used in industrial fermentation processes is limited (Ciegler, 1965a).

II. PRODUCTION OF β-CAROTENE

A. Earlier Developments

Many species of algae and fungi (e.g., *Neurospora crassa, Penicillium sclerotiorum, Phycomyces blakesleeanus*) and also yeasts (*Rhodotorula*) were considered for use in β-carotene production but were found to be unsuitable (Ciegler, 1965a).

Following the publication of the work by Barnett *et al.* (1956) on some particular fungi in the Mucorales group, the interest concentrated on the development of industrial fermentation with Choanephoraceae, mainly *Blakeslea trispora*. The use in the same culture of the two sexual forms of this species was found to lead to a remarkable increase in mycelial carotenoid content. The special conditions required for cultures in liquid media with mixed sexual forms of *B. trispora* have been shown by several authors, particularly from the Northern Regional Research Laboratory, Peoria, Illinois. First the medium should be viscous and rich in vegetable oils, kerosene, and surface-active agents. Second, α- or β-ionone (or another terpenoid) is added during the incubation period

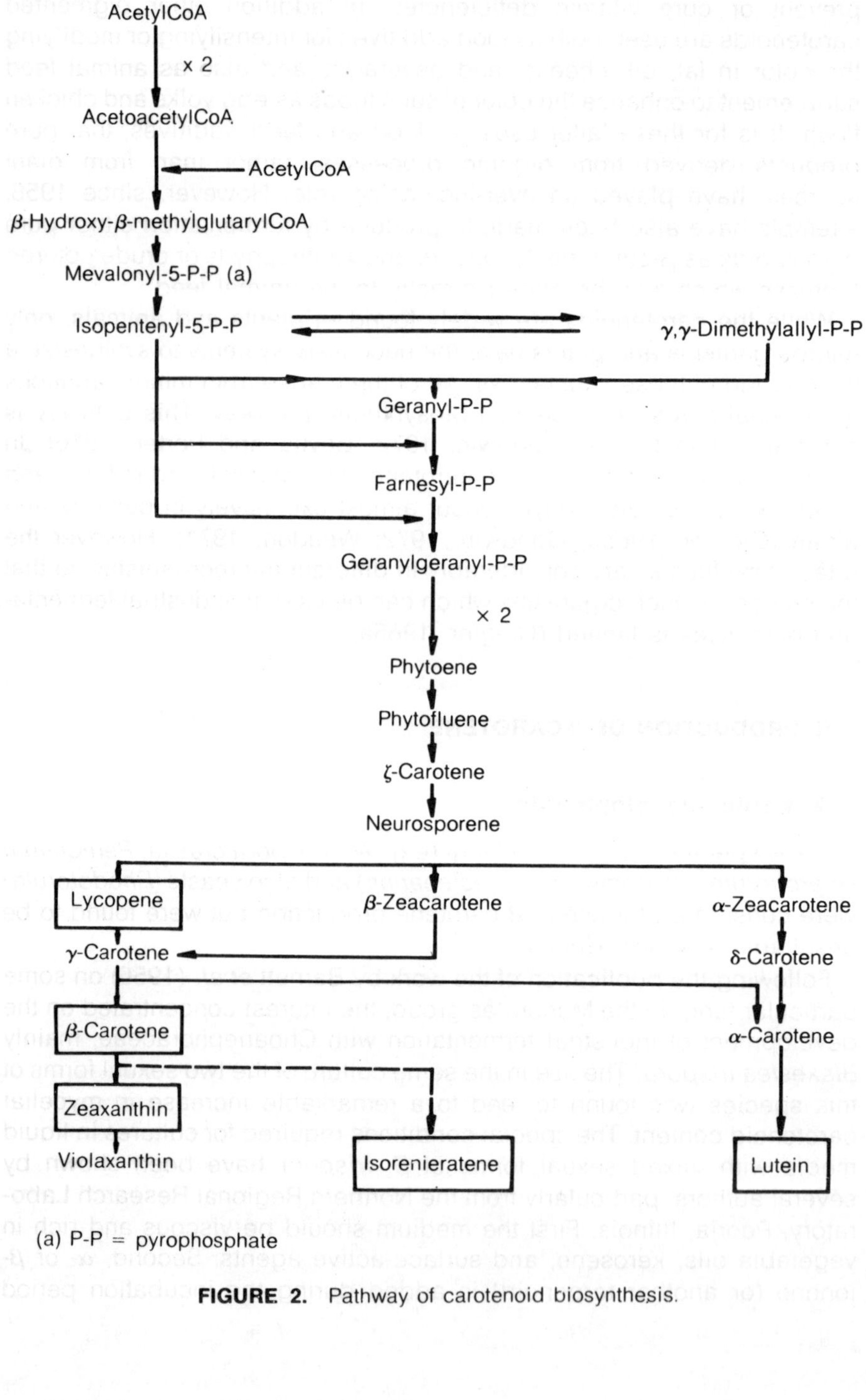

FIGURE 2. Pathway of carotenoid biosynthesis.

according to a previous study on *Phycomyces* (Mackinney *et al.*, 1952), together with an antioxidant to prevent oxidation of the β-carotene. The medium is inoculated with equal volumes of mature cultures of each sexual form of *B. trispora*. However there is no loss of yield if the mixed culture is seeded with an excess of *B. trispora* (−) form (Zajic, 1964) or if the ionones are replaced by the mycelia of various fungi or yeasts, whether unwashed or washed with hexane (Ciegler, 1965b). The yield of β-carotene with this procedure amounts to 1 gm/liter in the medium, i.e., 20 mg/gm of dried mycelium. The work in this area was reviewed by Ciegler (1965a) and Hanson (1967).

B. Recent Developments

1. *Fermentation*

Much work has been devoted to ways of simplifying and improving the fermentation process with *B. trispora* and, as shown below, the main approaches have been to modify the composition of the medium or to suppress the interdependence of the two sexual forms.

a. Modification of the Medium Composition. The medium normally contains from 1 to 8% of vegetable or animal oils, and these can be partially or totally replaced by cheaper oils, such as linoleic and linolenic acid-rich oils (autoclave oils and soapstock), which are by-products of oil refining (Pazola *et al.*, 1969; Bohinski, 1970). The yield of β-carotene with this substitution remains 1 gm/liter (Gutcho, 1973d).

b. Substitution of the β-Ionone. As β-ionone is not incorporated into β-carotene (Reyes *et al.*, 1964), chemically related compounds have been examined as replacement products. The Krebs cycle keto acids are only partially efficient as substitutes (Bjork and Neujahr, 1969), but bicyclic terpenes of turpentine (Cederberg and Neujahr, 1969) as well as other natural terpenes (Kolot *et al.*, 1973) are suitable for both partial and total replacement of β-ionone. Other substances tested include isoprene dimers and trimers (Nakamura *et al.*, 1974) and cyclohexane, cyclohexanone, and their trimethyl derivatives (Kolot *et al.*, 1973; Ninet *et al.*, 1963a, 1964, 1969). Among these, 2,6,6-trimethyl-1-acetylcyclohexene added at the level of 1 gm/liter is the most efficient at providing the same β-carotene yield as β-ionone itself. Cheap citrus fruit peels, rich in citric acid, were also added in the medium instead of ionone (Ciegler, 1965a). When this material was treated with alkali, the β-carotene content amounted to 1.7 gm/liter (Upjohn Co., 1965b).

c. New Activators of Carotene Production. The observation that carotene production was somewhat enhanced by the presence of dimethylformamide used as a solvent for the antioxidant prompted a systematic survey of numerous amides and nitrogenous compounds for similar but more effective activity. α-Pyrrolidone and succinimide, added as 3–4 gm/liter, showed some activity but the best results so far have been obtained with isoniazid, iproniazid, and 4-formylpyridine at concentrations of 1, 1, and 2 gm/liter, respectively. Interestingly, these activators, as well as being active in their own right, show an enhanced effect when used in combination with β-ionone or with 2,6,6-trimethyl-1-acetylcyclohexene giving a yield of β-carotene of 3 gm/liter, i.e., 30 mg/gm of dried mycelium (see Table I) (Ninet *et al.*, 1963b, 1965b, 1966, 1969). It is not yet known whether these substances are only permeation agents for *B. trispora* or act as inducers or activators of some enzymes operating in the carotenoid pathway. The detailed methods for utilization of these activators were described by Hanson (1967) and by Gutcho (1973c).

d. Suppression of the Sexual Interdependence. The (−) sexual form of *B. trispora* in the absence of the (+) form is capable of synthesizing significant quantities of β-carotene if the hormonal substances (trisporic acids), produced in cultures of the mixed forms, are added to the medium (Prieto *et al.*, 1964; Sebek and Jäger, 1964). These hormones can be extracted, with chloroform, from the filtered broth of a mated culture, the main component being trisporic acid C (Caglioti *et al.*, 1964) (see Fig. 3). Studies on the structure, biosynthesis, and mode of action of

TABLE I. Improvements in β-Carotene Production with New Activators[a]

Added substances	β-Carotene (mg/liter) with: None	Ionone (1 gm/liter)	TMACH[b] (1 gm/liter)
None	850	1550	1350
Succinimide	1350	2200	1850
Isoniazid	2200	3050	2950
Iproniazid	2200	3200	3200

[a] From Ninet *et al.* (1969).

[b] TMACH, 2,6,6-trimethyl-1-acetylcyclohexene. Strains : *Blakeslea trispora* NRRL 2456 (+) and NRRL 2457 (−). Basal culture medium: distillers' solubles, 60 gm; "Fox head" starch, 60 gm; soya bean oil, 35 gm; cotton seed oil, 35 gm; yeast extract, 1 gm; potassium phosphate, 0.5 gm; manganese sulfate, 0.2 gm; thiamin hydrochloride, 0.01 gm; kerosene, 20 gm; ethoxyquin, 0.5 gm; with tap water to 1 liter.

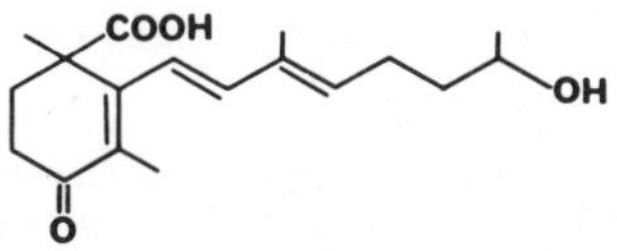

FIGURE 3. Structure of trisporic acid.

these hormones were summarized by Gooday (1974). They appear to arise from metabolism of β-carotene, or related substances, normally secreted by the fungus in mixed cultures (Feofilova *et al.*, 1976; Feofilova and Bekhtereva, 1976). At the same time, they behave similarly to β-ionone, acting competitively with it in derepressing the enzyme converting mevalonylpyrophosphate to γ, γ′-dimethylallylpyrophosphate (Rao and Modi, 1977) and play a role in the formation and the fusion of zygophores to zygospores. These findings were utilized in the development of a two-stage process for β-carotene production by Farmaceutici Italia (1965) and Jäger (1966). The trisporic acids are produced in a mixed culture of *B. trispora* (+) NRRL 2456 and (−) NRRL 9159 in a medium containing distillers' solubles, starch, cotton seed oil, and kerosene which gives a yield of 3.5 gm/liter. They are then extracted from the broth and added to a culture of (−) NRRL 9159 in the same medium, supplemented with β-ionone at a level of 0.5 gm/liter. This gives a β-carotene yield of 1.2 gm/liter (Farmaceutici Italia, 1965).

The highest yielding process so far reported is summarized in Fig. 4. However, despite the increased productivity so far achieved with the *B. trispora* cultures, this process cannot compete with chemical synthesis. This might be reversed by improving the yields through strain mutations but the multinucleate nature of the spores makes this difficult and only rare attempts have been made (Skryabin, 1967).

A very active ^{14}C-labeled β-carotene was prepared for metabolism and stability studies by adding [1,2 ^{14}C]sodium acetate to a 48-hour-old mixed culture of *B. trispora* (Purcell and Walter, 1971; Gutcho, 1973f).

2. *Extraction*

In contrast with the considerable literature on fermentation processes, there are relatively few publications on the extraction of carotenes from broth. The extraction technology of β-carotene is almost identical, whether it be from plant or from microbial sources, as it is always endocellular. For animal feedstuffs, the mycelium can be directly used after drying, but as the stability of the β-carotene in these biomasses is poor, antioxidants (e.g., 0.5 gm/liter of ethoxyquin) are added to the broth or to the derived biomass to stabilize the product.

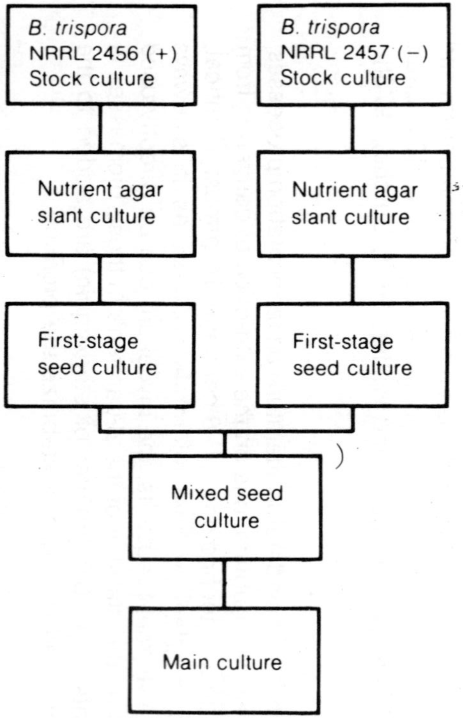

Spores mixed with sterile soil

Test tube with medium (a) incubated 7 days at 27°C

2-liter Erlenmeyer flask with 0.4 liter of medium (b) incubated 48 hours at 26°C on a rotary shaker

170-liter stainless steel fermentor with 120 liters of medium (b), inoculated with 0.4 liter of each first-stage seed culture and incubated 40 hours at 26°C with agitation (170 rpm) and aeration (8 m^3/hour)

800-liter stainless steel fermentor with 320 liters of medium (c) sterilized 55 minutes at 122°C, inoculated with 32 liters of the mixed seed culture, and incubated 185 hours at 26°C with agitation (210 rpm) and aeration (25 m^3/hour)

Medium (a): medium M_5 (from Hesseltine and Benjamin, 1959.)

Medium (b): Corn steep, 70 gm; corn starch, 50 gm; potassium dihydrogen phosphate, 0.5 gm; manganese sulfate, 0.1 gm; thiamin hydrochloride, 0.01 gm; with tap water to 1 liter.

Medium (c): Distillers' solubles, 70 gm; cornstarch, 60 gm; soya bean oil, 30 ml; cottonseed oil, 30 ml; ethoxyquin, 0.35 gm; manganese sulfate, 0.2 gm; thiamin hydrochloride, 0.5 mg; isoniazid, 0.6 gm; kerosene 20 ml; with tap water to 1 liter and pH adjusted to 6.3. The last two substances are each sterilized separately by filtration. After 48 hours of growth 1 gm/liter of β-ionone and 5 ml/liter of kerosene are added aseptically; then a total amount of 42 gm/liter of glucose, as an aqueous 53% solution, is continuously injected up to the end of the culture.

FIGURE 4. A high-yielding β-carotene process (Ninet *et al.*, 1965b).

More commonly the mycelium mass, after dehydration by vacuum drying or by treatment with low molecular weight aliphatic alcohols, is treated with hydrocarbon or chlorohydrocarbon solvents to extract the total carotenoids (of which 75–92% is β-carotene). This is followed by a series of precipitations and crystallizations with solvents mixtures to obtain crystalline β-carotene (Upjohn Co., 1965a). Such a process is illustrated in Fig. 5.

In some cases, the mycelium is extracted with refined vegetable oils to obtain an oily solution rich in carotenoids which is then treated by steam-stripping to remove odorous and volatile impurities (Pazola *et al.*, 1971).

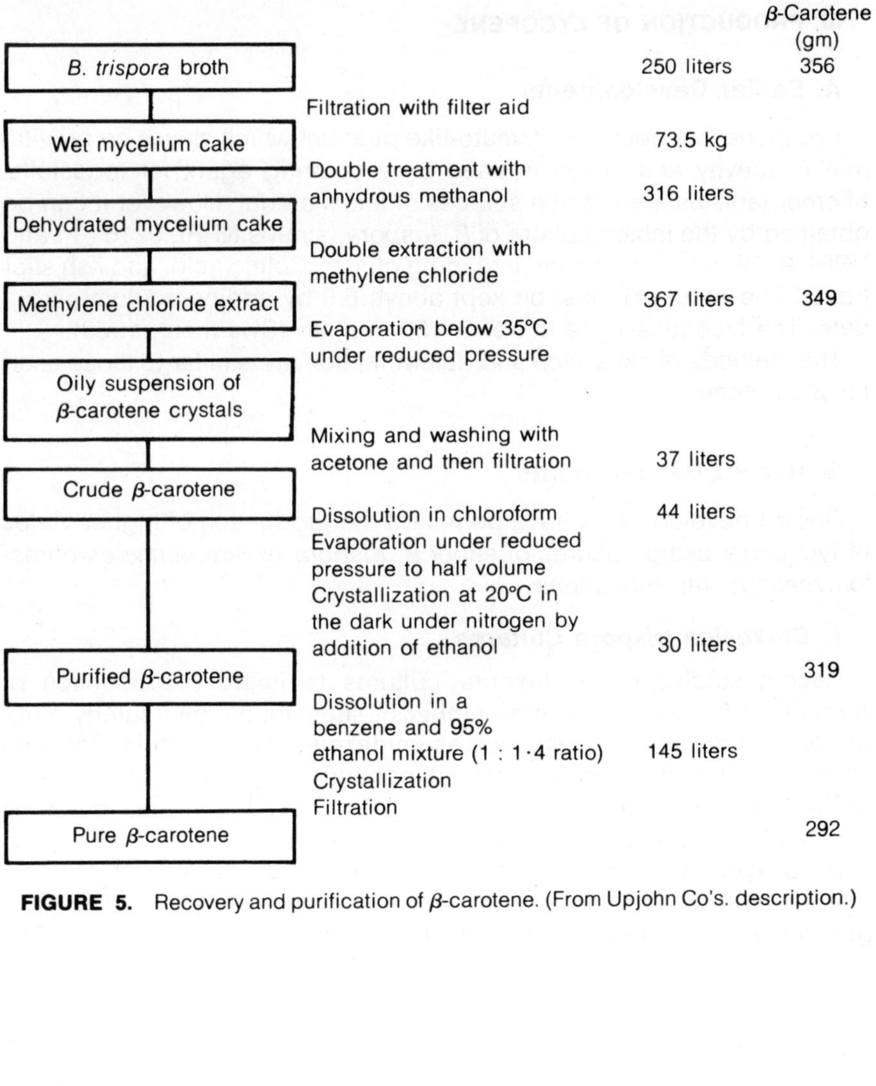

FIGURE 5. Recovery and purification of β-carotene. (From Upjohn Co's. description.)

3. β-Carotene Assay

The mixed cultures of *B. trispora* produce mainly the trans isomer of β-carotene. To determine its concentration in the broth the wet mycelium is harvested and either mechanically disintegrated or thoroughly dried. In both cases the carotenoids are extracted with an appropriate solvent and the extract is chromatographed on magnesium oxide and celite mixture or alumina columns, and the optical density of the eluates is measured at 450 nm. This final stage can also be used directly for assaying crude or pure preparation of β-carotene (Ciegler, 1965a; Liaaen-Jensen, 1971).

III. PRODUCTION OF LYCOPENE

A. Earlier Developments

Lycopene is a deep red, tomato-like pigment which shows no provitamin A activity and which is used as a coloring agent for foodstuffs. Microorganisms are not rich sources of this material. However it can be obtained by the mixed culture of *B. trispora,* strains NRRL 9216 (+) and NRRL 9159 (−), in a medium including glucose, oleic acid, and fish stick liquor. The broth pH must be kept above 6.6 by adding sodium carbonate. The lycopene yield is about 150 mg/liter (Swarthout, 1963).

The methods of extraction and determination are similar to those used for β-carotene.

B. Recent Developments

Recent developments have permitted the production of higher yields of lycopene, using cultures of either *B. trispora* or *Streptomyces chrestomyceticus* var. *rubescens.*

1. *Blakeslea trispora* Cultures

Recent studies on *B. trispora* cultures to improve production of β-carotene have shown that a number of substances, particularly nitrogenous heterocyclic bases (such as imidazole, pyridine, quinoline, and some of their substituted derivatives), can block carotenoid biosynthesis at the level of lycopene (Ninet *et al.*, 1965a, 1969). The position of the substitution is critical, some of the substituted substances having no effect on lycopene production or even enhancing β-carotene production. The effective agents, added to the broth at concentrations of 0.1–0.3 gm/liter, enable the production of lycopene to reach 1 gm/liter. The culture process is virtually identical with that for β-carotene production with the exception that β-ionone is absent, the latter having no effect on lycopene production. The details of the process are given by Gutcho (1973b).

Studies on the pigmentation of fruit and vegetables led to the discovery of other substances that can enhance the production of lycopene by fungi or photosynthetic bacteria (Coggins *et al.*, 1970; Hsu *et al.*, 1972, 1974; Yokoyama, 1974). These substances are essentially tertiary amines or their derivatives; for example, triethylamine, tributylamine, hordenine, 2-(4-chlorophenylthio)triethylamine, α-diethylaminopropiophenone, 4-[β-(diethylamino)ethoxy]benzophenone, and chalcone derivative of triethylamine. They act by derepressing the enzymes involved in carotene synthesis in *B. trispora* and by inhibiting the carotenoid cyclization enzymes (Elahi *et al.*, 1973). A 5-day mixed culture of *B. trispora* (+) NRRL 2895 and (−) NRRL 2896 in a medium based on cotton seed and corn flours and soya bean oil, supplemented at 48 hours with 1.5 gm/liter of triethylamine, will yield 0.4 gm lycopene per liter of broth, whereas in the presence of 1.25 gm/liter 4-(2-diethylaminoethyl)phenol (hordenine), mixed *B. trispora* (+) NRRL 9216 and (−) NRRL 9156 will provide 0.7 gm lycopene per liter (Ito *et al.*, 1973; Ito and Yamaguchi, 1973). Nicotine, which favors lycopene biosynthesis by several microorganisms, was apparently not tested with *Blakeslea* species (Goodwin, 1972; McDermott *et al.*, 1973; Elahi *et al.*, 1973).

2. *Streptomyces chrestomyceticus* subsp. *rubescens* Cultures

This strain was obtained after repeated mutagenic treatments of *Streptomyces chrestomyceticus.* After a 6-day growth in a medium including starch, soya bean flour, and ammonium sulfate, the yield was 0.5 gm lycopene per liter (Bianchi *et al.*, 1968; Gutcho, 1973b).

IV. PRODUCTION OF XANTHOPHYLLS AND OTHER PIGMENTS

A. Earlier Developments

Xanthophylls are mixtures of colored carotenoids which include lutein, canthaxanthin, cryptoxanthin, neoxanthin, violaxanthin, and zeaxanthin. These substances are coloring agents useful in poultry farming and are supplied by the ground maize kernels ordinarily included in the diet. However, it was also tempting to add to the feedstuffs the dried pigmented biomasses of one of the numerous microorganisms which produce xanthophylls.

The most extensively studied microorganisms are:

1. Basidiomycetes of the Dacrymycetaceae family; an example is *Dacrymyces deliquescens* which, when cultured over 5 days in liquid cultures containing glucose, glycerol, and corn steep liquor and under external illumination, provides 40 mg xanthophylls per liter (i.e., 4 mg/gm of dried mycelium) together with some β-carotene (Farrow and Tabenkin, 1961).

2. Green algae, particularly *Spongiococcum excentricum* and *Chlorella pyrenoidosa*. The latter, in a 7-day liquid culture with continuous addition of glucose, urea, and phosphate and under external illumination, yields 650 mg xanthophylls per liter (i.e., 2.1 mg/gm of dried biomass) (Kathrein, 1960, 1963, 1964; Farrow and Tabenkin, 1964a,b; Thériault, 1965).

Antioxidants are required, as with β-carotene feedstuffs, to prevent oxidation of the xanthophylls. For instance, ethoxyquin is added to the broth or the biomass at a 2.5% level.

However, none of these processes has as yet been used industrially.

B. Recent Developments

1. Zeaxanthin

Recent studies devoted to zeaxanthin production have concentrated on the use of marine flavobacteria. When those are grown in a medium based on glucose and corn steep, the zeaxanthin content amounts to 10–40 mg/liter and supplementation with palmitic esters, methionine, pyridoxine, and ferrous salts raises the zeaxanthin yields up to 190 mg/liter (i.e., 16.4 mg/gm of dried biomass) (Shocher and Wiss, 1972; Shepherd and Dasek, 1974; Shepherd *et al.*, 1974). When the temperature is reduced and the nutrients are added continuously, a content of 335 mg/liter can be obtained with a mutant strain of *Flavobacterium* (Dasek *et al.*, 1973). The methods of extraction and assay for zeaxanthin are similar to those for the carotenoids.

2. Other Xanthophylls

Studies relating to the production of carotenoid mixture of undetermined composition and which might be used as a feed additive in the form of the fermentation biomass are briefly discussed under this heading.

While the actinomycetes generally produce low yields of carotenoids (Krasil'nikov *et al.*, 1965; Garkavenko *et al.*, 1972), a *Streptomyces chrestomyceticus* var. *aurantioideus* mutant cultured over 5 days in a medium containing glucose and dried yeast gives 0.5 gm/liter of total carotenoids (Marnati *et al.*, 1967; Gutcho, 1973e), whereas *Streptomyces mediolani* cultured over 5 days in a medium containing dextrin, casein, and corn steep liquor yields 0.9 gm/liter with a large proportion of isorenieratene and its oxidation derivatives (Franceschi and Grein, 1969; Gutcho, 1973a).

A strain of *Mycobacterium phlei,* when grown over 7 days on beet molasses and urea and under illumination, yields 0.3 gm/liter of oxycarotenoids (Noury and Van Der Lande, 1967), whereas mycobacteria or corynebacteria grown on paraffins produce very low amounts (Kyowa Hakko Kogyo Co., Ltd., 1972; Yamada and Furukawa, 1973).

With a number of yeasts grown on various sugars, ethanol, or paraffins, the content is less than 5 mg/liter (i.e., 0.5 mg/gm of dried biomass) (Nikolaev, 1966; Vallet and Brabant, 1971; Bekers *et al.*, 1972; Protein Biosynthesis Institute, 1975).

Possibly a more extensive consideration of different activating agents and the development of improved mutant strains could lead to higher production in the future.

V. MARKETING PROSPECTS FOR CAROTENOIDS

The market for carotenoids is small at present, the worldwide sales of β-carotene amounting to some 27 tons/year, one-half of which is accounted for by the United States market. The carotenoids presently sold are those derived from synthetic and plant sources and consist mainly of β-carotene, canthaxanthin, citrinaxanthin, and two C 30 apocarotenoids. Most of the xanthophylls used are extracted from plant sources, such as maize kernel, alfalfa, xanthasouci, and algae. The fermentation derived products have not so far been exploited commercially, mainly due to the limited markets and relatively low prices of the competitive products. β-Carotene is mainly used as a coloring agent, first of all in the food industry and fairly less often in drugs and cosmetics manufacturing. As a provitamin A its market seems rather narrow. There are possibilities for its use in animal feedstuffs and as a source material in canthaxanthin synthesis. Lycopene is listed as a food coloring agent for some countries, but it is rarely used.

The other products are principally used in poultry farming as feedstuffs, the sales of citrinaxanthin for this use being 3–5 tons/year.

There is also some interest in the use of canthaxanthin as a "sun-tanning" agent taken orally.

As indicated above, fermentation production of carotenoids has not so far been of commercial value, and interest in studying new processes has been decreasing, except perhaps in Eastern Europe. However, any process producing a major improvement in yields or any factors markedly changing the price of the present raw materials could completely reverse this situation and make the fermentation derived carotenoids a viable proposition.

REFERENCES

Barnett, H. L., Lilly, V. G., and Krause, R. F. (1956). *Science* **123,** 141.
Bekers, M. J., Krause, I. J., Seile, M. K., Viesturs, U. E., Tschazkij, P. A., and Kluna, G. W. (1972). German Patent 2,059,387.
Beytia, E. D., and Porter, J. W. (1976). *Annu. Rev. Biochem.* **45,** 113–142.
Bianchi, M. L., Franceschi, G., Marnati, M. P., and Spalla, C. (1968). British Patent 1,128,440 (to Farmaceutici Italia).
Bjork, L., and Neujahr, H. Y. (1969). *Acta Chem. Scand.* **23,** 2908–2909.
Bohinski, R. C. (1970). U.S. Patent 3,492,202 (to Chas. Pfizer and Co., Inc.).
Caglioti, L., Cainelli, G., Camerino, B., Mondelli, R., Prieto, A., Quilico, A., Salvatori, T., and Selva, A. (1964). *Chim. Ind. (Milan)* **46,** 1–6.
Cederberg, E., and Neujahr, H. Y. (1969). *Acta Chem. Scand.* **23,** 957–961.
Ciegler, A. (1965a). *Adv. Appl. Microbiol.* **7,** 1–34.
Ciegler, A. (1965b). U.S. Patent 3,226,302 (to U.S.).
Coggins, C. W., Jr., Henning, G. L., and Yokoyama, H. (1970). *Science* **168,** 1589–1590.
Dasek, J., Shepherd, D., and Traelnes, K. R. (1973). Belgian Patent 790,289 (to Société des Produits Nestlé S.A.).
Elahi, M., Chichester, C. O., and Simpson, K. L. (1973). *Phytochemistry* **12,** 1627–1632.
Farmaceutici Italia (1965). French Patent 1,414,828.
Farrow, W. M., and Tabenkin, B. (1961). U.S. Patent 2,974,044 (to Hoffmann-La Roche and Co.).
Farrow, W. M., and Tabenkin, B. (1964a). French Patent 1,376,027 (to Hoffmann-La Roche and Co.).
Farrow, W. M., and Tabenkin, B. (1964b). Belgian Patent 639,531 (to Hoffmann-La Roche and Co.).
Feofilova, E. P., and Bekhtereva, M. N. (1976). *Mikrobiologiya* **45,** 557–558.
Feofilova, E. P., Fateeva, T. V., and Arbuzov, V. A. (1976). *Mikrobiologiya* **45,** 169–171.
Franceschi, G., and Grein, A. (1969). Belgian Patent 723,901 (to Farmaceutici Italia).
Garkavenko, A. I., Savchenko, L. F., and Razumovskii, P. N. (1972). Russian Patent 327,250.
Gooday, G. W. (1974). *Annu. Rev. Biochem.* **43,** 39–43.
Goodwin, T. W. (1972). *Prog. Ind. Microbiol.* **11,** 29–88.
Gutcho, S. J. (1973a). *Chem. Technol. Rev.* (Chemicals by Fermentation) **19,** 304.
Gutcho, S. J. (1973b). *Chem. Technol. Rev.* (Chemicals by Fermentation) **19,** 308–311.
Gutcho, S. J. (1973c). *Chem. Technol. Rev.* (Chemicals by Fermentation) **19,** 311–315.
Gutcho, S. J. (1973d). *Chem. Technol. Rev.* (Chemicals by Fermentation) **19,** 315–318.

Gutcho, S. J. (1973e). *Chem. Technol. Rev.* (Chemicals by Fermentation) **19,** 318–320.
Gutcho, S. J. (1973f). *Chem. Technol. Rev.* (Chemicals by Fermentation) **19,** 320–322.
Hanson, A. M. (1967). *In* "Microbial Technology" (H. J. Peppler, ed.), pp. 222–250. Van Nostrand-Reinhold, Princeton, New Jersey.
Hesseltine, C. W., and Benjamin, C. R. (1959). *Mycologia* **51,** 887–901.
Hsu, W. J., Yokoyama, H., and Coggins, C. W., Jr. (1972). *Phytochemistry* **11,** 2985–2990.
Hsu, W. J., Poling, S. M., and Yokoyama, H. (1974). *Phytochemistry* **13,** 415–419.
Isler, O. (1977). *Experientia* **33,** 555–573.
Ito, S., and Yamaguchi, Y. (1973). Japanese Patent 73/16,189 (to Takasago Perfumery Co., Ltd.).
Ito, S., Yamaguchi, Y., Yoshida, T., and Komatsu, A. (1973). Japanese Patent 73/16,190 (to Takasago Perfumery Co., Ltd.).
Jäger, H. K. (1966). French Patent 1,457,619 (to Upjohn Co.).
Kathrein, H. R. (1960). U.S. Patent 2,949,700 (to Grain Processing Corp.).
Kathrein, H. R. (1963). U.S. Patent 3,108,402 (to Grain Processing Corp.).
Kathrein, H. R. (1964). U.S. Patent 3,142,135 (to Grain Processing Corp.).
Kolot, F. B., Vakulova, L. A., and Samokhvalov, G. I. (1973). *In* "Vitaminy i Vitaminnye Preparaty" (V. A. Yakovlev, ed.), pp. 241–246. Izd. "Meditsina," Moscow.
Krasil'nikov, N. A., Korenyako, A. I., Gavrilova, O. A., Artamanova, O. I., Leonov, L. I., Nikitina, N. I., Dushenkova, L. I., Ulezlo, I. V., and Khokhlova, Y. M. (1965). Russian Patent 169,749.
Kyowa Hakko Kogyo Co., Ltd. (1972). Japanese Patent 72/04,503.
Liaaen-Jensen, S. (1971). *In* "Carotenoids" (O. Isler, ed.), pp. 61–73. Birkhäuser, Basel.
McDermott, J. C. B., Britton, G., and Goodwin, T. W. (1973). *J. Gen. Microbiol.* **77,** 161–171.
Mackinney, G., Nakayama, T., Buss, C. D., and Chichester, C. O. (1952). *J. Am. Chem. Soc.* **74,** 3456–3457.
Marnati, M. P., Prieto, A., and Spalla, C. (1967). U.S. Patent 3,330,737 (to Farmaceutici Italia).
Nakamura, T., Tsuji, M., Abe, H., Fujii, K., Makimoto, T., and Ootera, J. (1974). Japanese Patent 74/31,896 (to Kuraray Co., Ltd.).
Nikolaev, P. I. (1966). Russian Patent 178,449.
Ninet, L., Renaut, J., and Tissier, R. (1963a). French Patent 1,325,656 (to Rhône-Poulenc S. A.).
Ninet, L., Renaut, J., and Tissier, R. (1963b). French Patent 1,344, 264 (to Rhône-Poulenc S. A.).
Ninet, L., Renaut, J., and Tissier, R. (1964). French Patent 1,377,523 (to Rhône-Poulenc S. A.).
Ninet, L., Renaut, J., and Tissier, R. (1965a). French Patent 1,403,839 (to Rhône-Poulenc S. A.).
Ninet, L., Renaut, J., and Tissier, R. (1965b). French Patent 1,412,506 (to Rhône-Poulenc S. A.).
Ninet, L., Renaut, J., and Tissier, R. (1966). French Patent 1,449,879 (to Rhône-Poulenc S. A.).
Ninet, L., Renaut, J., and Tissier, R. (1969). *Biotechnol. Bioeng.* **11,** 1195–1210.
Noury and Van Der Lande N. V. Koninklijke Industrieele Maatschappij (1967). French Patent 1,467,317.
Pazola, Z., Switek, H., Janicki, J., and Michnikowska, W. (1969). U.S. Patent 3,476,646 (to Centralne Laboratorium Przemyslu Koncentratow Spozywczych-Poland).
Pazola, Z., Switek, H., and Janicki, J. (1971). East German Patent 76,661.
Prieto, A., Spalla, C., Bianchi, M., and Biffi, G. (1964). *Chem. Ind. (London)* **13,** 551.

Protein Biosynthesis Institute (1975). Russian Patent 448,219.
Purcell, A. E., and Walter, W. M., Jr. (1971). U.S. Patent 3,579,424 (to U.S.).
Rao, S., and Modi, V. V. (1977). *Experientia* **33,** 31–33.
Reyes, P., Chichester, C. O., and Nakayama, T. O. M. (1964). *Biochim. Biophys. Acta* **90,** 578–592.
Schocher, A. J., and Wiss, O. (1972). Belgian Patent 770,744 (to Hoffmann-La Roche and Co.).
Sebek, O. K., and Jäger, H. K. (1964). *Abstr. 148th Meet., Am. Chem. Soc.* pp. 9Q–10Q.
Shepherd, D., and Dasek, J. (1974). Belgian Patent 816,766 (to Société des Produits Nestlé S.A.).
Shepherd, D., Dasek, J., and Carels, M. (1974). Belgian Patent 816,767 (to Société des Produits Nestlé S.A.).
Skryabin, G. K. (1967). Russian Patent 185,316.
Swarthout, E. J. (1963). French Patent 1,333,942 (to Miles Laboratories, Inc.).
Thériault, R. J. (1965). *Appl. Microbiol.* **13,** 402–416.
Upjohn Co. (1965a). Dutch Patent 64/11,184.
Upjohn Co. (1965b). Dutch Patent 65/00,788.
Vallet, G., and Brabant, P. G. (1971). French Patent 2,057,482 (to Distillerie Brabant et Cie.).
Weedon, B. C. L. (1971). *In* "Carotenoids" (O. Isler, ed.), pp. 29–59. Birkhäuser, Basel.
Yamada, K., and Furukawa, T. (1973). Japanese Patent 73/21,515 (to Mitsui Petrochemical Industries, Ltd.).
Yokoyama, H. (1974). Cited in *Chem. Eng. News.* **52**(15), 19–21.
Zajic, J. E. (1964). U.S. Patent 3,128,236 (to Grain Processing Corp.).

Subject Index

A

D

E